RSMeans

Electrical Cost Data

27th Annual Edition

2004

Senior Editor
John H. Chiang, PE

Contributing Editors
Barbara Balboni
Robert A. Bastoni
Paul C. Crosscup
Robert J. Kuchta
Robert C. McNichols
Robert W. Mewis, CCC
Melville J. Mossman, PE
John J. Moylan
Jeannene D. Murphy
Stephen C. Plotner
Michael J. Regan
Eugene R. Spencer
Marshall J. Stetson
Phillip R. Waier, PE

Senior Engineering Operations Manager
John H. Ferguson, PE

President
Curtis B. Allen

Vice President, Product Management
Roger J. Grant

Sales Director
John M. Shea

Production Manager
Michael Kokernak

Production Coordinator
Marion E. Schofield

Technical Support
Thomas J. Dion
Jonathan Forgit
Mary Lou Geary
Gary L. Hoitt
Paula Reale-Camelio
Robin Richardson
Kathryn S. Rodriguez
Sheryl A. Rose

Book & Cover Design
Norman R. Forgit

First Printing

Foreword

RSMeans is a product line of Reed Construction Data, a leading provider of construction information, products, and services in North America and globally. Reed Construction Data's project information products include more than 100 regional editions, national construction data, sales leads, and local plan rooms in major business centers. Reed Construction Data's PlansDirect provides surveys, plans and specifications. The First Source suite of products consists of *First Source for Products*, SPEC-DATA™, MANU-SPEC™, CADBlocks, Manufacturer Catalogs and First Source Exchange (www.firstsourceexchange.com) for the selection of nationally available building products. Reed Construction Data also publishes ProFile, a database of more than 20,000 U.S. architectural firms. RSMeans provides construction cost data, training, and consulting services in print, CD-ROM and online. Reed Construction Data, headquartered in Atlanta, is owned by Reed Business Information (www.cahners.com), a leading provider of critical information and marketing solutions to business professionals in the media, manufacturing, electronics, construction and retail industries. Its market-leading properties include more than 135 business-to-business publications, over 125 Webzines and Web portals, as well as online services, custom publishing, directories, research and direct-marketing lists. Reed Business Information is a member of the Reed Elsevier plc group (NYSE: RUK and ENL)—a world-leading publisher and information provider operating in the science and medical, legal, education and business-to-business industry sectors.

Our Mission

Since 1942, RSMeans has been actively engaged in construction cost publishing and consulting throughout North America.

Today, over 50 years after the company began, our primary objective remains the same: to provide you, the construction and facilities professional, with the most current and comprehensive construction cost data possible.

Whether you are a contractor, an owner, an architect, an engineer, a facilities manager, or anyone else who needs a fast and reliable construction cost estimate, you'll find this publication to be a highly useful and necessary tool.

Today, with the constant flow of new construction methods and materials, it's difficult to find the time to look at and evaluate all the different construction cost possibilities. In addition, because labor and material costs keep changing, last year's cost information is not a reliable basis for today's estimate or budget.

That's why so many construction professionals turn to RSMeans. We keep track of the costs for you, along with a wide range of other key information, from city cost indexes . . . to productivity rates . . . to crew composition . . . to contractor's overhead and profit rates.

RSMeans performs these functions by collecting data from all facets of the industry, and organizing it in a format that is instantly accessible to you. From the preliminary budget to the detailed unit price estimate, you'll find the data in this book useful for all phases of construction cost determination.

The Staff, the Organization, and Our Services

When you purchase one of RSMeans' publications, you are in effect hiring the services of a full-time staff of construction and engineering professionals.

Our thoroughly experienced and highly qualified staff works daily at collecting, analyzing, and disseminating comprehensive cost information for your needs. These staff members have years of practical construction experience and engineering training prior to joining the firm. As a result, you can count on them not only for the cost figures, but also for additional background reference information that will help you create a realistic estimate.

The RSMeans organization is always prepared to help you solve construction problems through its five major divisions: Construction and Cost Data Publishing, Electronic Products and Services, Consulting Services, Insurance Services, and Educational Services.

Besides a full array of construction cost estimating books, Means also publishes a number of other reference works for the construction industry. Subjects include construction estimating and project and business management; special topics such as HVAC, roofing, plumbing, and hazardous waste remediation; and a library of facility management references.

In addition, you can access all of our construction cost data through your computer with RSMeans CostWorks 2004 CD-ROM, an electronic tool that offers over 50,000 lines of Means detailed construction cost data, along with assembly and whole building cost data. You can also access Means cost information from our Web site at www.rsmeans.com

What's more, you can increase your knowledge and improve your construction estimating and management performance with a Means Construction Seminar or In-House Training Program. These two-day seminar programs offer unparalleled opportunities for everyone in your organization to get updated on a wide variety of construction-related issues.

RSMeans also is a worldwide provider of construction cost management and analysis services for commercial and government owners and of claims and valuation services for insurers.

In short, RSMeans can provide you with the tools and expertise for constructing accurate and dependable construction estimates and budgets in a variety of ways.

Robert Snow Means Established a Tradition of Quality That Continues Today

Robert Snow Means spent years building his company, making certain he always delivered a quality product.

Today, at RSMeans, we do more than talk about the quality of our data and the usefulness of our books. We stand behind all of our data, from historical cost indexes... to construction materials and techniques... to current costs.

If you have any questions about our products or services, please call us toll-free at 1-800-334-3509. Our customer service representatives will be happy to assist you or visit our Web site at www.rsmeans.com

Table of Contents

UNIT PRICES

GENERAL REQUIREMENTS	1
SITE CONSTRUCTION	2
CONCRETE	3
METALS	5
WOOD & PLASTICS	6
THERMAL & MOISTURE PROTECTION	7
SPECIALTIES	10
EQUIPMENT	11
SPECIAL CONSTRUCTION	13
MECHANICAL	15
ELECTRICAL	16
SQUARE FOOT	17

ASSEMBLIES

SERVICES	D
BUILDING SITEWORK	G

REFERENCE INFORMATION

REFERENCE NUMBERS

CREWS

COST INDEXES

INDEX

Note: Changes to Section contents and line numbers have been made in Unit Cost and Assembly Sections of the 2004 publications to better comply with the industry standard organization systems, MasterFormat95 and UNIFORMAT II. While all the data you need is still here, it may have moved or been renumbered. Using the Index will help you find anything you have trouble locating. For a complete map of changes, visit our web site at **www.rsmeans.com/ links** and follow the instructions to obtain the mapping tables.

How the Book Is Built: An Overview

A Powerful Construction Tool

You have in your hands one of the most powerful construction tools available today. A successful project is built on the foundation of an accurate and dependable estimate. This book will enable you to construct just such an estimate.

For the casual user the book is designed to be:

- quickly and easily understood so you can get right to your estimate
- filled with valuable information so you can understand the necessary factors that go into the cost estimate

For the regular user, the book is designed to be:

- a handy desk reference that can be quickly referred to for key costs
- a comprehensive, fully reliable source of current construction costs and productivity rates, so you'll be prepared to estimate any project
- a source book for preliminary project cost, product selections, and alternate materials and methods

To meet all of these requirements we have organized the book into the following clearly defined sections.

Quick Start
This one-page section (see following page) can quickly get you started on your estimate.

How To Use the Book: The Details
This section contains an in-depth explanation of how the book is arranged . . . and how you can use it to determine a reliable construction cost estimate. It includes information about how we develop our cost figures and how to completely prepare your estimate.

Unit Price Section
All cost data has been divided into the 16 divisions according to the MasterFormat system of classification and numbering as developed by the Construction Specifications Institute (CSI) and Construction Specifications Canada (CSC). For a listing of these divisions and an outline of their subdivisions, see the Unit Price Section Table of Contents.

Estimating tips are included at the beginning of each division.

Division 17: Quick Project Estimates:
In addition to the 16 Unit Price Divisions there is a S.F. (Square Foot) and C.F. (Cubic Foot) Cost Division, Division 17. It contains costs for 59 different building types that allow you to quickly make a rough estimate for the overall cost of a project or its major components.

Assemblies Section
The cost data in this section has been organized in an "Assemblies" format. These assemblies are the functional elements of a building and are arranged according to the 7 divisions of the UNIFORMAT II classification system. For a complete explanation of a typical "Assemblies" page, see "How To Use the Assemblies Cost Tables."

Reference Section
This section includes information on Reference Numbers, Crew Listings, Historical Cost Indexes, City Cost Indexes, Location Factors and a listing of Abbreviations. It is visually identified by a vertical gray bar on the edge of pages.

Reference Numbers Following selected major classifications throughout the book are "reference numbers" shown in bold squares. These numbers refer you to related information in other sections.

In these other sections, you'll find related tables, explanations, and estimating information. Also included are alternate pricing methods and technical data, along with information on design and economy in construction. You'll also find helpful tips on estimating and construction.

It is recommended that you refer to the Reference Section if a "reference" number appears within the section you are estimating.

Crew Listings: This section lists all the crews referenced. For the purposes of this book, a crew is composed of more than one trade classification and/or the addition of equipment to any trade classification. Equipment is included in the cost of the crew. Costs are shown both with bare labor rates and with the installing contractor's overhead and profit added. For each, the total crew cost per eight-hour day and the composite cost per labor-hour are listed.

Historical Cost Indexes: These indexes provide you with data to adjust construction costs over time. If you know costs for a past project, you can use these indexes to estimate the cost to construct the same project today.

City Cost Indexes: Obviously, costs vary depending on the regional economy. You can adjust the "national average" costs in this book to 316 major cities throughout the U.S. and Canada by using the data in this section. How to use information is included.

Location Factors to quickly adjust the data to over 930 zip code areas, are included.

Abbreviations: A listing of the abbreviations and the terms they represent, is included.

Index
A comprehensive listing of all terms and subjects in this book to help you find what you need quickly.

The Scope of This Book
This book is designed to be comprehensive and easy to use. To that end we have made certain assumptions:

1. We have established material prices based on a "national average."
2. We have computed labor costs based on a 30-city "national average" of union wage rates.
3. We have targeted the data for projects of a certain size range.

For a more detailed explanation of how the cost data is developed, see "How To Use the Book: The Details."

Project Size
This book is aimed primarily at commercial and industrial projects costing $1,000,000 and up, or large multi-family housing projects. Costs are primarily for new construction or major renovation of buildings rather than repairs or minor alterations.

With reasonable exercise of judgment the figures can be used for any building work. *However, for civil engineering structures such as bridges, dams, highways, or the like, please refer to **Means Heavy Construction Cost Data.***

Quick Start

If you feel you are ready to use this book and don't think you need the detailed instructions that begin on the following page, this Quick Start section is for you.

These steps will allow you to get started estimating in a matter of minutes.

1 First, decide whether you require a Unit Price or Assemblies type estimate. Unit price estimating requires a breakdown of the work to individual items. Assemblies estimates combine individual items or components into building systems.

If you need to estimate each line item separately, follow the instructions for **Unit Prices.**

If you can use an estimate for the entire assembly or system, follow the instructions for **Assemblies.**

2 Find each cost data section you need in the Table of Contents (either for Unit Prices or Assemblies).

Unit Prices: The cost data for Unit Prices has been divided into 16 divisions according to the CSI MasterFormat.

Assemblies: The cost data for Assemblies has been divided into 7 divisions according to the UNIFORMAT II.

3 Turn to the indicated section and locate the line item or assemblies table you need for your estimate. Portions of a sample page layout from both the Unit Price Listings and the Assemblies Cost Tables appear below.

Unit Prices: If there is a reference number listed at the beginning of the section, it refers to additional information you may find useful. See the referenced section for additional information.

• Note the crew code designation. You'll find full descriptions of crews in the Crew Listings including labor-hour and equipment costs.

Assemblies: The Assemblies (*not* shown in full here) are generally separated into three parts: 1) an illustration of the system to be estimated; 2) the components and related costs of a typical system; and 3) the costs for similar systems with dimensional and/or size variations. The Assemblies Section also contains reference numbers for additional useful information.

4 Determine the total number of units your job will require.

Unit Prices: Note the unit of measure for the material you're using is listed under "Unit."

• **Bare Costs:** These figures show unit costs for materials and installation. Labor and equipment costs are calculated according to crew costs and average daily output. Bare costs do not contain allowances for overhead, profit or taxes.

• "Labor-hours" allows you to calculate the total labor-hours to complete that task. Just multiply the quantity of work by this figure for an estimate of activity duration.

Assemblies: Note the unit of measure for the assembly or system you're estimating is listed in the Assemblies table.

5 Then multiply the total units by. . .

Unit Prices: "Total Incl. O&P" which stands for the total cost including the installing contractor's overhead and profit. (See the "How To Use the Unit Price Pages" for a complete explanation.)

Assemblies: The "Total" in the right-hand column, which is the total cost including the installing contractor's overhead and profit. (See the "How To Use the Assemblies Cost Tables" section for a complete explanation.)

Material and equipment cost figures include a 10% markup. For labor markups, see the inside back cover of this book. If the work is to be subcontracted, add the general contractor's markup, typically 10%.

6 The price you calculate will be an estimate for either an individual item of work or a completed assembly or *system.*

7 Compile a list of all items or assemblies included in the total project. Summarize cost information, and add project overhead.

Localize costs by using The City Cost Indexes or Location Factors found in the Reference Section.

For a more complete explanation of the way costs are derived, please see the following sections.

Editors' Note: We urge you to spend time reading and understanding all of the supporting material and to take into consideration the reference material such as Crews Listing and the "reference numbers."

Unit Price Pages

Assemblies Pages

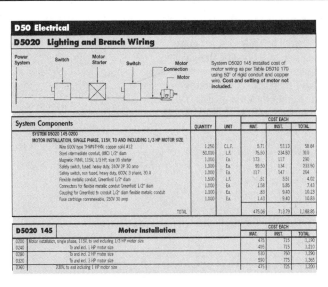

v

How to Use the Book: The Details

What's Behind the Numbers? The Development of Cost Data

The staff at RSMeans continuously monitors developments in the construction industry in order to ensure reliable, thorough and up-to-date cost information.

While *overall* construction costs may vary relative to general economic conditions, price fluctuations within the industry are dependent upon many factors. Individual price variations may, in fact, be opposite to overall economic trends. Therefore, costs are continually monitored and complete updates are published yearly. Also, new items are frequently added in response to changes in materials and methods.

Costs — $ (U.S.)

All costs represent U.S. national averages and are given in U.S. dollars. The Means City Cost Indexes can be used to adjust costs to a particular location. The City Cost Indexes for Canada can be used to adjust U.S. national averages to local costs in Canadian dollars.

Material Costs

The RSMeans staff contacts manufacturers, dealers, distributors, and contractors all across the U.S. and Canada to determine national average material costs. If you have access to current material costs for your specific location, you may wish to make adjustments to reflect differences from the national average. Included within material costs are fasteners for a normal installation. RSMeans engineers use manufacturers' recommendations, written specifications and/or standard construction practice for size and spacing of fasteners. Adjustments to material costs may be required for your specific application or location. Material costs do not include sales tax.

Labor Costs

Labor costs are based on the average of wage rates from 30 major U.S. cities. Rates are determined from labor union agreements or prevailing wages for construction trades for the current year. Rates along with overhead and profit markups are listed on the inside back cover of this book.

- If wage rates in your area vary from those used in this book, or if rate increases are expected within a given year, labor costs should be adjusted accordingly.

Labor costs reflect productivity based on actual working conditions. These figures include time spent during a normal workday on tasks other than actual installation, such as material receiving and handling, mobilization at site, site movement, breaks, and cleanup.

Productivity data is developed over an extended period so as not to be influenced by abnormal variations and reflects a typical average.

Equipment Costs

Equipment costs include not only rental, but also operating costs for equipment under normal use. The operating costs include parts and labor for routine servicing such as repair and replacement of pumps, filters and worn lines. Normal operating expendables such as fuel, lubricants, tires and electricity (where applicable) are also included. Extraordinary operating expendables with highly variable wear patterns such as diamond bits and blades are excluded. These costs are included under materials. Equipment rental rates are obtained from industry sources throughout North America—contractors, suppliers, dealers, manufacturers, and distributors.

Crew Equipment Cost/Day—The power equipment required for each crew is included in the crew cost. The daily cost for crew equipment is based on dividing the weekly bare rental rate by 5 (number of working days per week), and then adding the hourly operating cost times 8 (hours per day). This "Crew Equipment Cost/Day" is listed in Subdivision 01590.

Factors Affecting Costs

Costs can vary depending upon a number of variables. Here's how we have handled the main factors affecting costs.

Quality—The prices for materials and the workmanship upon which productivity is based represent sound construction work. They are also in line with U.S. government specifications.

Overtime—We have made no allowance for overtime. If you anticipate premium time or work beyond normal working hours, be sure to make an appropriate adjustment to your labor costs.

Productivity—The productivity, daily output, and labor-hour figures for each line item are based on working an eight-hour day in daylight hours in moderate temperatures. For work that extends beyond normal work hours or is performed under adverse conditions, productivity may decrease. (See the section in "How To Use the Unit Price Pages" for more on productivity.)

Size of Project—The size, scope of work, and type of construction project will have a significant impact on cost. Economies of scale can reduce costs for large projects. Unit costs can often run higher for small projects. Costs in this book are intended for the size and type of project as previously described in "How the Book Is Built: An Overview." Costs for projects of a significantly different size or type should be adjusted accordingly.

Location—Material prices in this book are for metropolitan areas. However, in dense urban areas, traffic and site storage limitations may increase costs. Beyond a 20-mile radius of large cities, extra trucking or transportation charges may also increase the material costs slightly. On the other hand, lower wage rates may be in effect. Be sure to consider both these factors when preparing an estimate, particularly if the job site is located in a central city or remote rural location.

In addition, highly specialized subcontract items may require travel and per diem expenses for mechanics.

Other factors—

- season of year
- contractor management
- weather conditions
- local union restrictions
- building code requirements
- availability of:
  - adequate energy
  - skilled labor
  - building materials
- owner's special requirements/restrictions
- safety requirements
- environmental considerations

Unpredictable Factors—General business conditions influence "in-place" costs of all items. Substitute materials and construction methods may have to be employed. These may affect the installed cost and/or life cycle costs. Such factors may be difficult to evaluate and cannot necessarily be predicted on the basis of the job's location in a particular section of the country. Thus, where these factors apply, you may find significant, but unavoidable cost variations for which you will have to apply a measure of judgment to your estimate.

Rounding of Costs

In general, all unit prices in excess of $5.00 have been rounded to make them easier to use and still maintain adequate precision of the results. The rounding rules we have chosen are in the following table.

Prices from . . .	Rounded to the nearest . . .
$.01 to $5.00	$.01
$5.01 to $20.00	$.05
$20.01 to $100.00	$.50
$100.01 to $300.00	$1.00
$300.01 to $1,000.00	$5.00
$1,000.01 to $10,000.00	$25.00
$10,000.01 to $50,000.00	$100.00
$50,000.01 and above	$500.00

Estimating Labor-Hours

The labor-hours expressed in this publication are based on Average Installation time, using an efficiency level of approximately 60-65% (see item 7 below) which has been found reasonable and acceptable by many contractors.

The book uses this National Efficiency Average to establish a consistent benchmark.

For bid situations, adjustments to this efficiency level should be the responsibility of the contractor bidding the project.

The unit labor-hour is divided in the following manner. A typical day for a crew might be:

1.	Study Plans	3%	14.4 min.
2.	Material Procurement	3%	14.4 min.
3.	Receiving and Storing	3%	14.4 min
4.	Mobilization	5%	24.0 min.
5.	Site Movement	5%	24.0 min.
6.	Layout and Marking	8%	38.4 min.
7.	Actual Installation	64%	307.2 min.
8.	Cleanup	3%	14.4 min.
9.	Breaks, Non-Productive	6%	28.8 min.
		100%	480.0 min.

If any of the percentages expressed in this breakdown do not apply to the particular work or project situation, then that percentage or a portion of it may be deducted from or added to labor hours.

Overhead, Profit & Contingencies

General—Prices given in this book are of two kinds: (1) BARE COSTS and (2) TOTAL INCLUDING INSTALLING CONTRACTOR'S OVERHEAD & PROFIT.

General Conditions—Cost data in this book is presented in two ways: Bare Costs and Total Cost including O&P (Overhead and Profit). General Conditions, when applicable, should also be added to the Total Cost including O&P. The costs for General Conditions are listed in Division 1 of the Unit Price Section and the Reference Section of this book. General Conditions for the *Installing Contractor* may range from 0% to 10% of the Total Cost including O&P. For the *General* or *Prime Contractor*, costs for General Conditions may range from 5% to 15% of the Total Cost including O&P, with a figure of 10% as the most typical allowance.

Overhead and Profit—Total Cost including O&P for the *Installing Contractor* is shown in the last column on both the Unit Price and the Assemblies pages of this book. This figure is the sum of the bare material cost plus 10% for profit, the base labor cost plus total overhead and profit, and the bare equipment cost plus 10% for profit. Details for the calculation of Overhead and Profit on labor are shown on the inside back cover and in the Reference Section of this book. (See the "How To Use the Unit Price Pages" for an example of this calculation.)

Subcontractors—Usually a considerable portion of all large jobs is subcontracted. In fact the percentage done by subs is constantly increasing and may run over 90%. Since the workers employed by these companies do nothing else but install their particular product, they soon become experts in that line. The result is, installation by these firms is accomplished so efficiently that the total in-place cost, even with the subcontractor's overhead and profit, is no more and often less than if the principal contractor had handled the installation himself. Also, the quality of the work may be higher.

Contingencies—The allowance for contingencies generally is to provide for indefinable construction difficulties. On alterations or repair jobs, 20% is none too much. If drawings are final and only field contingencies are being considered, 2% or 3% is probably sufficient and often nothing need be added. As far as the contract is concerned, future changes in plans will be covered by extras. The contractor should allow for inflationary price trends and possible material shortages during the course of the job. If drawings are not complete or approved, or a budget cost wanted, it is wise to add 5% to 10%. Contingencies, then, are a matter of judgment. Additional allowances are shown in the Unit Price Section 01250 for contingencies and job conditions and Reference Section R01250-010 for factors to convert prices for repair and remodeling jobs.

Final Checklist

Estimating can be a straightforward process provided you remember the basics. Here's a checklist of some of the items you should remember to do before completing your estimate.

Did you remember to . . .

- factor in the City Cost Index for your locale
- take into consideration which items have been marked up and by how much
- mark up the entire estimate sufficiently for your purposes
- read the background information on techniques and technical matters that could impact your project time span and cost
- include all components of your project in the final estimate
- double check your figures to be sure of your accuracy
- call RSMeans if you have any questions about your estimate or the data you've found in our publications.

Remember, RSMeans stands behind its publications. If you have any questions about your estimate . . .about the costs you've used from our books . . .or even about the technical aspects of the job that may affect your estimate, feel free to call the RSMeans editors at 1-800-334-3509.

Unit Price Section

Table of Contents

1

How to Use the Unit Price Pages

The following is a detailed explanation of a sample entry in the Unit Price Section. Next to each bold number below is the item being described with appropriate component of the sample entry following in parenthesis. Some prices are listed as bare costs, others as costs that include overhead and profit of the installing contractor. In most cases, if the work is to be subcontracted, the general contractor will need to add an additional markup (RSMeans suggests using 10%) to the figures in the column "Total Incl. O&P."

1 Division Number/Title (03300/Cast-In-Place Concrete)

Use the Unit Price Section Table of Contents to locate specific items. The sections are classified according to the CSI MasterFormat (1995 Edition).

2 Line Numbers (03310 240 3900)

Each unit price line item has been assigned a unique 12-digit code based on the CSI MasterFormat classification.

- Level One - CSI-MasterFormat Division
- Level Two - CSI

03300
03310-240-3900

- Means 12-digit Line Number
- Level Four - Means
- Level Three - CSI

3 Description (Concrete-In-Place, etc.)

Each line item is described in detail. Sub-items and additional sizes are indented beneath the appropriate line items. The first line or two after the main item (in boldface) may contain descriptive information that pertains to all line items beneath this boldface listing.

4 Reference Number Information

R03310 -010 You'll see reference numbers shown in bold rectangles at the beginning of some sections. These refer to related items in the Reference Section, visually identified by a vertical gray bar on the edge of pages.

The relation may be: (1) an estimating procedure that should be read before estimating, (2) an alternate pricing method, or (3) technical information.

The "R" designates the Reference Section. The numbers refer to the MasterFormat classification system.

It is strongly recommended that you review all reference numbers that appear within the section in which you are working.

Note: Not all reference numbers appear in all Means publications.

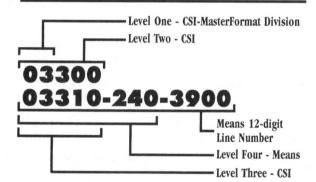

03300 | Cast-In-Place Concrete

| | | 03310 | Structural Concrete | CREW | DAILY OUTPUT | LABOR-HOURS | UNIT | 2004 BARE COSTS | | | | TOTAL INCL O&P | |
								MAT.	LABOR	EQUIP.	TOTAL		
240	0010	**CONCRETE IN PLACE** Including forms (4 uses), reinforcing											240
	0050	steel and finishing unless otherwise indicated	R03310 -010										
	0300	Beams, 5 kip per L.F., 10' span	R03310 -100	C-14A	15.	.04	C.Y.	233	425	8	706	975	
	0350	25' span		"	18.55	10.782		214	355	40	609		
	3520	For 7 to 20 stories high		C-8	21,200	.003	S.F.		.08	.03	.11		
	3800	Footings, spread under 1 C.Y.		C-14C	38.07	2.942	C.Y.	105	93	.68	198.68	263	
	3850	Over 5 C.Y.			81.04	1.382		96.50	43.50	.32	140.32	175	
	3900	Footings, strip, 18" x 9", unreinforced			40	2.800		91	88.50	.64	180.14	240	
	3920	18" x 9", reinforced			35	3.200		105	101	.74	206.74	275	
	3925	20" x 10", unreinforced			45	2.489		88	78.50	.57	167.07	221	
	3930	20" x 10", reinforced			40	2.800		99.50	88.50	.64	188.64	249	
	3935	24" x 12", unreinforced			55	2.036		87	64.50	.47	151.97	198	
	3940	24" x 12", reinforced			48	2.333		98.50	74	.54	173.04	226	
	3945	36" x 12", unreinforced			70	1.600		83.50	50.50	.37	134.37	172	
	3950	36" x 12", reinforced			60	1.867		94	59	.43	153.43	196	
	4000	Foundation mat, under 10 C.Y.			38.67	2.896		129	91.50	.67	221.17	287	

2

Crew (C-14C)

The "Crew" column designates the typical trade or crew used to install the item. If an installation can be accomplished by one trade and requires no power equipment, that trade and the number of workers are listed (for example, "2 Carpenters"). If an installation requires a composite crew, a crew code designation is listed (for example, "C-14C"). You'll find full details on all composite crews in the Crew Listings.

• For a complete list of all trades utilized in this book and their abbreviations, see the inside back cover.

Crews

Crew No.	Bare Costs		Incl. Subs O & P		Cost Per Labor-Hour	
Crew C-14C	Hr.	Daily	Hr.	Daily	Bare Costs	Incl. O&P
1 Carpenter Foreman (out)	$35.00	$280.00	$54.55	$436.40	$31.63	$49.56
6 Carpenters	33.00	1584.00	51.40	2467.20		
2 Rodmen (reinf.)	37.10	593.60	61.25	980.00		
4 Laborers	26.00	832.00	40.50	1296.00		
1 Cement Finisher	31.55	252.40	46.45	371.60		
1 Gas Engine Vibrator		26.00		28.60	.23	.26
112 L.H., Daily Totals		$3568.00		$5579.80	$31.86	$49.82

Productivity: Daily Output (40.0)/Labor-Hours (2.80)

The "Daily Output" represents the typical number of units the designated crew will install in a normal 8-hour day. To find out the number of days the given crew would require to complete the installation, divide your quantity by the daily output. For example:

Quantity	÷	Daily Output	=	Duration
100 C.Y.	÷	40.0/ Crew Day	=	2.50 Crew Days

The "Labor-Hours" figure represents the number of labor-hours required to install one unit of work. To find out the number of labor-hours required for your particular task, multiply the quantity of the item times the number of labor-hours shown. For example:

Quantity	x	Productivity Rate	=	Duration
100 C.Y.	x	2.80 Labor-Hours/ C.Y.	=	280 Labor-Hours

Unit (C.Y.)

The abbreviated designation indicates the unit of measure upon which the price, production, and crew are based (C.Y. = Cubic Yard). For a complete listing of abbreviations refer to the Abbreviations Listing in the Reference Section of this book.

Bare Costs:

Mat. (Bare Material Cost) (91)

The unit material cost is the "bare" material cost with no overhead and profit included. *Costs shown reflect national average material prices for January of the current year and include delivery to the job site. No sales taxes are included.*

Labor (88.50)

The unit labor cost is derived by multiplying bare labor-hour costs for Crew C-14C by labor-hour units. The bare labor-hour cost is found in the Crew Section under C-14C. (If a trade is listed, the hourly labor cost—the wage rate—is found on the inside back cover.)

Labor-Hour Cost Crew C-14C	x	Labor-Hour Units	=	Labor
$31.63	x	2.80	=	$88.50

Equip. (Equipment) (.64)

Equipment costs for each crew are listed in the description of each crew. Tools or equipment whose value justifies purchase or ownership by a contractor are considered overhead as shown on the inside back cover. The unit equipment cost is derived by multiplying the bare equipment hourly cost by the labor-hour units.

Equipment Cost Crew C-14C	x	Labor-Hour Units	=	Equip.
.23	x	2.80	=	$.64

Total (180.14)

The total of the bare costs is the arithmetic total of the three previous columns: mat., labor, and equip.

Material	+	Labor	+	Equip.	=	Total
$91	+	$88.50	+	$.64	=	$180.14

Total Costs Including O&P

This figure is the sum of the bare material cost plus 10% for profit; the bare labor cost plus total overhead and profit (per the inside back cover or, if a crew is listed, from the crew listings); and the bare equipment cost plus 10% for profit.

Material is Bare Material Cost + 10% = 91 + 9.10	=	$100.10
Labor for Crew C-14C = Labor-Hour Cost (49.56) x Labor-Hour Units (2.80)	=	$138.77
Equip. is Bare Equip. Cost + 10% = .64 + .06	=	$ .70
Total (Rounded)	=	$240

Division 1
General Requirements

Estimating Tips

The General Requirements of any contract are very important to both the bidder and the owner. These lay the ground rules under which the contract will be executed and have a significant influence on the cost of operations. Therefore, it is extremely important to thoroughly read and understand the General Requirements both before preparing an estimate and when the estimate is complete, to ascertain that nothing in the contract is overlooked. Caution should be exercised when applying items listed in Division 1 to an estimate. Many of the items are included in the unit prices listed in the other divisions such as mark-ups on labor and company overhead.

01200 Price & Payment Procedures

- When estimating historic preservation projects (depending on the condition of the existing structure and the owner's requirements), a 15-20% contingency or allowance is recommended, regardless of the stage of the drawings.

01300 Administrative Requirements

- Before determining a final cost estimate, it is a good practice to review all the items listed in subdivision 01300 to make final adjustments for items that may need customizing to specific job conditions.
- Historic preservation projects may require specialty labor and methods, as well as extra time to protect existing materials that must be preserved and/or restored. Some additional expenses may be incurred in architectural fees for facility surveys and other special inspections and analyses.

01330 Submittal Procedures

- Requirements for initial and periodic submittals can represent a significant cost to the General Requirements of a job. Thoroughly check the submittal specifications when estimating a project to determine any costs that should be included.

01400 Quality Requirements

- All projects will require some degree of Quality Control. This cost is not included in the unit cost of construction listed in each division. Depending upon the terms of the contract, the various costs of inspection and testing can be the responsibility of either the owner or the contractor. Be sure to include the required costs in your estimate.

01500 Temporary Facilities & Controls

- Barricades, access roads, safety nets, scaffolding, security and many more requirements for the execution of a safe project are elements of direct cost. These costs can easily be overlooked when preparing an estimate. When looking through the major classifications of this subdivision, determine which items apply to each division in your estimate.

01590 Equipment Rental

- This subdivision contains transportation, handling, storage, protection and product options and substitutions. Listed in this cost manual are average equipment rental rates for all types of equipment. This is useful information when estimating the time and materials requirement of any particular operation in order to establish a unit or total cost.
- A good rule of thumb is that weekly rental is 3 times daily rental and that monthly rental is 3 times weekly rental.
- The figures in the column for Crew Equipment Cost represent the rental rate used in determining the daily cost of equipment in a crew. It is calculated by dividing the weekly rate by 5 days and adding the hourly operating cost times 8 hours.

01770 Closeout Procedures

- When preparing an estimate, read the specifications to determine the requirements for Contract Closeout thoroughly. Final cleaning, record documentation, operation and maintenance data, warranties and bonds, and spare parts and maintenance materials can all be elements of cost for the completion of a contract. Do not overlook these in your estimate.

01830 Operations & Maintenance

- If maintenance and repair are included in your contract, they require special attention. To estimate the cost to remove and replace any unit usually requires a site visit to determine the accessibility and the specific difficulty at that location. Obstructions, dust control, safety, and often overtime hours must be considered when preparing your estimate.

Reference Numbers

Reference numbers are shown in bold squares at the beginning of some major classifications. These numbers refer to related items in the Reference Section. The reference information may be an estimating procedure, an alternate pricing method or technical information.

Note: Not all subdivisions listed here necessarily appear in this publication.

01100 | Summary

01107 | Professional Consultant

			CREW	DAILY OUTPUT	LABOR-HOURS	UNIT	2004 BARE COSTS				TOTAL INCL O&P	
							MAT.	LABOR	EQUIP.	TOTAL		
200	0010	**CONSTRUCTION MANAGEMENT FEES** $1,000,000 job, minimum				Project					4.50%	**200**
	0050	Maximum									7.50%	
	0060	For work to $10,000									10%	
	0300	$5,000,000 job, minimum									2.50%	
	0350	Maximum				▼					4%	
300	0010	**ENGINEERING FEES**										**300**
	0020	Educational planning consultant, minimum				Project					.50%	
	0100	Maximum				"					2.50%	
	0200	Electrical, minimum				Contrct					4.10%	
	0300	Maximum									10.10%	
	0400	Elevator & conveying systems, minimum									2.50%	
	0500	Maximum									5%	
	0600	Food service & kitchen equipment, minimum									8%	
	0700	Maximum									12%	
	1000	Mechanical (plumbing & HVAC), minimum									4.10%	
	1100	Maximum				▼					10.10%	

01200 | Price & Payment Procedures

01250 | Contract Modification Procedures

			CREW	DAILY OUTPUT	LABOR-HOURS	UNIT	2004 BARE COSTS				TOTAL INCL O&P	
							MAT.	LABOR	EQUIP.	TOTAL		
200	0010	**CONTINGENCIES** for estimate at conceptual stage				Project					20%	**200**
	0050	Schematic stage									15%	
	0100	Preliminary working drawing stage (Design Dev.)									10%	
	0150	Final working drawing stage				▼					3%	
300	0010	**CREWS** For building construction, see How To Use This Book										**300**
400	0010	**FACTORS** Cost adjustments										**400**
	0100	Add to construction costs for particular job requirements										
	0500	Cut & patch to match existing construction, add, minimum				Costs	2%	3%				
	0550	Maximum					5%	9%				
	0800	Dust protection, add, minimum					1%	2%				
	0850	Maximum					4%	11%				
	1100	Equipment usage curtailment, add, minimum					1%	1%				
	1150	Maximum					3%	10%				
	1400	Material handling & storage limitation, add, minimum					1%	1%				
	1450	Maximum					6%	7%				
	1700	Protection of existing work, add, minimum					2%	2%				
	1750	Maximum					5%	7%				
	2000	Shift work requirements, add, minimum						5%				
	2050	Maximum						30%				
	2300	Temporary shoring and bracing, add, minimum					2%	5%				
	2350	Maximum					5%	12%				
	2400	Work inside prisons and high security areas, add, minimum						30%				
	2450	Maximum				▼		50%				
500	0010	**JOB CONDITIONS** Modifications to total										**500**
	0020	project cost summaries										
	0100	Economic conditions, favorable, deduct				Project					2%	
	0200	Unfavorable, add				▼					5%	

Important: See the Reference Section for critical supporting data - Reference Nos., Crews, & City Cost Indexes

01200 | Price & Payment Procedures

01250 | Contract Modification Procedures

			CREW	DAILY OUTPUT	LABOR-HOURS	UNIT	2004 BARE COSTS MAT.	LABOR	EQUIP.	TOTAL	TOTAL INCL O&P	
500	0300	Hoisting conditions, favorable, deduct				Project					2%	500
	0400	Unfavorable, add									5%	
	0700	Labor availability, surplus, deduct									1%	
	0800	Shortage, add									10%	
	0900	Material storage area, available, deduct									1%	
	1000	Not available, add									2%	
	1100	Subcontractor availability, surplus, deduct									5%	
	1200	Shortage, add									12%	
	1300	Work space, available, deduct									2%	
	1400	Not available, add									5%	
600	0010	**OVERTIME** For early completion of projects or where	R01100 -110									600
	0020	labor shortages exist, add to usual labor, up to				Costs		100%				

01255 | Cost Indexes

			CREW	DAILY OUTPUT	LABOR-HOURS	UNIT	MAT.	LABOR	EQUIP.	TOTAL	TOTAL INCL O&P	
200	0010	**CONSTRUCTION COST INDEX** (Reference) over 930 zip code locations in										200
	0020	The U.S. and Canada, total bldg cost, min. (Clarksdale, MS)				%					65.50%	
	0050	Average									100%	
	0100	Maximum (New York, NY)									133.90%	
400	0010	**HISTORICAL COST INDEXES** (Reference) Back to 1954										400
500	0010	**LABOR INDEX** (Reference) For over 930 zip code locations in										500
	0020	the U.S. and Canada, minimum (Clarksdale, MS)				%		31.30%				
	0050	Average						100%				
	0100	Maximum (New York, NY)						163.20%				
600	0011	**MATERIAL INDEX** For over 930 zip code locations in										600
	0020	the U.S. and Canada, minimum (Elizabethtown, KY)				%	92.10%					
	0040	Average					100%					
	0060	Maximum (Ketchikan, AK)					145.20%					

01290 | Payment Procedures

			CREW	DAILY OUTPUT	LABOR-HOURS	UNIT	MAT.	LABOR	EQUIP.	TOTAL	TOTAL INCL O&P	
800	0010	**TAXES** Sales tax, State, average	R01100 -090			%	4.65%					800
	0050	Maximum					7%					
	0300	Unemployment, MA, combined Federal and State, minimum	R01100 -100					2.10%				
	0350	Average						6.20%				
	0400	Maximum						8%				

01300 | Administrative Requirements

01310 | Project Management/Coordination

			CREW	DAILY OUTPUT	LABOR-HOURS	UNIT	MAT.	LABOR	EQUIP.	TOTAL	TOTAL INCL O&P	
150	0010	**PERMITS** Rule of thumb, most cities, minimum				Job					.50%	150
	0100	Maximum				"					2%	
200	0010	**PERFORMANCE BOND** For buildings, minimum	R01100 -080			Job					.60%	200
	0100	Maximum				"					2.50%	
350	0010	**INSURANCE** Builders risk, standard, minimum	R01100 -040			Job					.22%	350
	0050	Maximum									.59%	
	0200	All-risk type, minimum	R01100 -060								.25%	
	0250	Maximum									.62%	
	0400	Contractor's equipment floater, minimum				Value					.50%	
	0450	Maximum				"					1.50%	

350		01310	Project Management/Coordination	CREW	DAILY OUTPUT	LABOR-HOURS	UNIT	2004 BARE COSTS MAT.	LABOR	EQUIP.	TOTAL	TOTAL INCL O&P	350
	0600		Public liability, average	R01100 -040			Job					1.55%	
	0800		Workers' compensation & employer's liability, average										
	0850		by trade, carpentry, general	R01100 -060			Payroll		18.51%				
	1000		Electrical						6.40%				
	1150		Insulation						15.24%				
	1450		Plumbing						7.78%				
	1550		Sheet metal work (HVAC)						11.09%				
500	0010	**MARK-UP** For General Contractors for change		R01100 -070									500
	0100		of scope of job as bid										
	0200		Extra work, by subcontractors, add				%					10%	
	0250		By General Contractor, add									15%	
	0400		Omitted work, by subcontractors, deduct all but									5%	
	0450		By General Contractor, deduct all but									7.50%	
	0600		Overtime work, by subcontractors, add									15%	
	0650		By General Contractor, add									10%	
	1000		Installing contractors, on his own labor, minimum						46.90%				
	1100		Maximum						87.30%				
600	0010	**OVERHEAD** As percent of direct costs, minimum		R01100 -070			%					5%	600
	0050		Average									13%	
	0100		Maximum									30%	
620	0010	**OVERHEAD & PROFIT** Allowance to add to items in this		R01100 -070									620
	0020		book that do not include Subs O&P, average				%					25%	
	0100		Allowance to add to items in this book that										
	0110		do include Subs O&P, minimum				%					5%	
	0150		Average									10%	
	0200		Maximum									15%	
	0300		Typical, by size of project, under $100,000									30%	
	0350		$500,000 project									25%	
	0400		$2,000,000 project									20%	
	0450		Over $10,000,000 project									15%	

01321 | Construction Photos

500	0010	**PHOTOGRAPHS** 8″ x 10″, 4 shots, 2 prints ea., std. mounting				Set	283			283	310	500
	0100	Hinged linen mounts					305			305	340	
	0200	8″ x 10″, 4 shots, 2 prints each, in color					330			330	360	
	0300	For I.D. slugs, add to all above					3.97			3.97	4.37	
	1500	Time lapse equipment, camera and projector, buy					3,575			3,575	3,925	
	1550	Rent per month					530			530	585	
	1700	Cameraman and film, including processing, B.&W.				Day	1,300			1,300	1,425	
	1720	Color				"	1,300			1,300	1,425	

050		01510	Temporary Utilities	CREW	DAILY OUTPUT	LABOR-HOURS	UNIT	2004 BARE COSTS MAT.	LABOR	EQUIP.	TOTAL	TOTAL INCL O&P	050
	0010	**TEMPORARY POWER EQUIP (PRO-RATED PER JOB)**											
	0020	Service, overhead feed, 3 use											
	0030	100 Amp		1 Elec	1.25	6.400	Ea.	550	252		802	980	
	0040	200 Amp		1	8		700	315		1,015	1,250	

Important: See the Reference Section for critical supporting data - Reference Nos., Crews, & City Cost Indexes

			DAILY	LABOR-				2004 BARE COSTS			TOTAL	
	01510	**Temporary Utilities**	**CREW**	**OUTPUT**	**HOURS**	**UNIT**	**MAT.**	**LABOR**	**EQUIP.**	**TOTAL**	**INCL O&P**	
050	0050	400 Amp	1 Elec	.75	10.667	Ea.	1,350	420		1,770	2,125	050
	0060	600 Amp	↓	.50	16	↓	1,700	630		2,330	2,800	
	0100	Underground feed, 3 use										
	0110	100 Amp	1 Elec	2	4	Ea.	525	158		683	815	
	0120	200 Amp		1.15	6.957		720	274		994	1,200	
	0130	400 Amp		1	8		1,350	315		1,665	1,950	
	0140	600 Amp		.75	10.667		1,550	420		1,970	2,325	
	0150	800 Amp		.50	16		2,600	630		3,230	3,800	
	0160	1000 Amp		.35	22.857		2,700	900		3,600	4,300	
	0170	1200 Amp		.25	32		2,950	1,250		4,200	5,125	
	0180	2000 Amp	↓	.20	40	↓	3,800	1,575		5,375	6,525	
	0200	Transformers, 3 use										
	0210	30 KVA	1 Elec	1	8	Ea.	565	315		880	1,100	
	0220	45 KVA		.75	10.667		675	420		1,095	1,375	
	0230	75 KVA		.50	16		1,025	630		1,655	2,075	
	0240	112.5 KVA	↓	.40	20	↓	1,350	790		2,140	2,650	
	0250	Feeder, PVC, CU wire										
	0260	60 Amp w/trench	1 Elec	96	.083	L.F.	3.78	3.28		7.06	9.05	
	0270	100 Amp w/trench		85	.094		5.05	3.71		8.76	11.10	
	0280	200 Amp w/trench		59	.136		10.65	5.35		16	19.65	
	0290	400 Amp w/trench	↓	42	.190	↓	22.50	7.50		30	35.50	
	0300	Feeder, PVC, aluminum wire										
	0310	60 Amp w/trench	1 Elec	96	.083	L.F.	3.53	3.28		6.81	8.75	
	0320	100 Amp w/trench		85	.094		4.08	3.71		7.79	10	
	0330	200 Amp w/trench		59	.136		8.60	5.35		13.95	17.40	
	0340	400 Amp w/trench		42	.190		14.75	7.50		22.25	27.50	
	0350	Feeder, EMT, CU wire										
	0360	60 Amp	1 Elec	90	.089	L.F.	3.63	3.50		7.13	9.20	
	0370	100 Amp		80	.100		6.80	3.94		10.74	13.35	
	0380	200 Amp		60	.133		13.10	5.25		18.35	22.50	
	0390	400 Amp	↓	35	.229	↓	17.15	9		26.15	32.50	
	0400	Feeder, EMT, Al wire										
	0410	60 Amp	1 Elec	90	.089	L.F.	2.93	3.50		6.43	8.40	
	0420	100 Amp		80	.100		5.90	3.94		9.84	12.35	
	0430	200 Amp		60	.133		11.05	5.25		16.30	19.95	
	0440	400 Amp	↓	35	.229	↓	15.05	9		24.05	30	
	0500	Equipment, 3 use										
	0510	Spider box 50 Amp	1 Elec	8	1	Ea.	168	39.50		207.50	244	
	0520	Lighting cord 100'		8	1		36.50	39.50		76	99	
	0530	Light stanchion	↓	8	1	↓	74	39.50		113.50	140	
	0540	Temporary cords, 100', 3 use										
	0550	Feeder cord, 50 Amp	1 Elec	16	.500	Ea.	156	19.70		175.70	202	
	0560	Feeder cord, 100 Amp		12	.667		450	26.50		476.50	535	
	0570	Tap cord, 50 Amp		12	.667		430	26.50		456.50	515	
	0580	Tap cord, 100 Amp	↓	6	1.333	↓	615	52.50		667.50	760	
	0590	Temporary cords, 50', 3 use										
	0600	Feeder cord, 50 Amp	1 Elec	16	.500	Ea.	91	19.70		110.70	130	
	0610	Feeder cord, 100 Amp		12	.667		224	26.50		250.50	286	
	0620	Tap cord, 50 Amp		12	.667		195	26.50		221.50	253	
	0630	Tap cord, 100 Amp	↓	6	1.333	↓	320	52.50		372.50	435	
	0700	Connections										
	0710	Compressor or pump										
	0720	30 Amp	1 Elec	7	1.143	Ea.	18.30	45		63.30	87	
	0730	60 Amp		5.30	1.509		42	59.50		101.50	135	
	0740	100 Amp	↓	4	2	↓	54	79		133	177	
	0750	Tower crane										

GENERAL REQUIREMENTS 1

		01510	Temporary Utilities	CREW	DAILY OUTPUT	LABOR-HOURS	UNIT	2004 BARE COSTS				TOTAL INCL O&P	
								MAT.	LABOR	EQUIP.	TOTAL		
050	0760		60 Amp	1 Elec	4.50	1.778	Ea.	42	70		112	150	050
	0770		100 Amp	"	3	2.667	"	54	105		159	216	
	0780		Manlift										
	0790		Single	1 Elec	3	2.667	Ea.	43	105		148	204	
	0800		Double	"	2	4	"	55.50	158		213.50	296	
	0810		Welder										
	0820		50 Amp w/disconnect	1 Elec	5	1.600	Ea.	270	63		333	390	
	0830		100 Amp w/disconnect		3.80	2.105		415	83		498	580	
	0840		200 Amp w/disconnect		2.50	3.200		660	126		786	915	
	0850		400 Amp w/disconnect	▼	1	8	▼	2,100	315		2,415	2,775	
	0860		Office trailer										
	0870		60 Amp	1 Elec	4.50	1.778	Ea.	84.50	70		154.50	197	
	0880		100 Amp		3	2.667		113	105		218	281	
	0890		200 Amp	▼	2	4	▼	500	158		658	785	
	0900		Lamping, add per floor				Total				600	660	
	0910		Maintenance, total temp. power cost				Job				5%		
800	0010		**TEMPORARY UTILITIES**										800
	0100		Heat, incl. fuel and operation, per week, 12 hrs. per day	1 Skwk	100	.080	CSF Flr	5.20	2.69		7.89	9.95	
	0200		24 hrs. per day	"	60	.133		7.85	4.49		12.34	15.60	
	0350		Lighting, incl. service lamps, wiring & outlets, minimum	1 Elec	34	.235		2.11	9.25		11.36	16.10	
	0360		Maximum	"	17	.471		4.60	18.55		23.15	32.50	
	0400		Power for temp lighting only, per month, min/month 6.6 KWH								.75	1.18	
	0450		Maximum/month 23.6 KWH								2.85	3.14	
	0600		Power for job duration incl. elevator, etc., minimum								47	51.70	
	0650		Maximum				▼				110	121	
	1000		Toilet, portable, see division 01590-400										

		01520	Construction Facilities										
500	0010		**OFFICE** Trailer, furnished, no hookups, 20' x 8', buy	2 Skwk	1	16	Ea.	6,125	540		6,665	7,600	500
	0250		Rent per month					153			153	169	
	0300		32' x 8', buy	2 Skwk	.70	22.857		9,225	770		9,995	11,300	
	0350		Rent per month					164			164	180	
	0400		50' x 10', buy	2 Skwk	.60	26.667		16,800	895		17,695	19,900	
	0450		Rent per month					260			260	286	
	0500		50' x 12', buy	2 Skwk	.50	32		20,000	1,075		21,075	23,700	
	0550		Rent per month					286			286	315	
	0700		For air conditioning, rent per month, add				▼	39.50			39.50	43.50	
	0800		For delivery, add per mile				Mile	1.53			1.53	1.68	
	1000		Portable buildings, prefab, on skids, economy, 8' x 8'	2 Carp	265	.060	S.F.	81.50	1.99		83.49	93	
	1100		Deluxe, 8' x 12'	"	150	.107	"	88.50	3.52		92.02	103	
	1200		Storage boxes, 20' x 8', buy	2 Skwk	1.80	8.889	Ea.	3,050	299		3,349	3,850	
	1250		Rent per month					74			74	81.50	
	1300		40' x 8', buy	2 Skwk	1.40	11.429		4,100	385		4,485	5,100	
	1350		Rent per month				▼	105			105	115	
550	0010		**FIELD OFFICE EXPENSE**										550
	0100		Field office expense, office equipment rental average				Month	139			139	153	
	0120		Office supplies, average				"	83.50			83.50	92	
	0125		Office trailer rental, see division 01520-500										
	0140		Telephone bill; avg. bill/month incl. long dist.				Month	204			204	224	
	0160		Field office lights & HVAC				"	94			94	103	

		01530	Temporary Construction										
700	0010		**PROTECTION** Stair tread, 2" x 12" planks, 1 use	1 Carp	75	.107	Tread	4.42	3.52		7.94	10.35	700
	0100		Exterior plywood, 1/2" thick, 1 use		65	.123		1.18	4.06		5.24	7.65	
	0200		3/4" thick, 1 use	▼	60	.133	▼	1.72	4.40		6.12	8.75	

Important: See the Reference Section for critical supporting data - Reference Nos., Crews, & City Cost Indexes

		01530	Temporary Construction	CREW	DAILY OUTPUT	LABOR-HOURS	UNIT	2004 BARE COSTS				TOTAL INCL O&P	
								MAT.	LABOR	EQUIP.	TOTAL		
900	0010		**WINTER PROTECTION** Reinforced plastic on wood										900
	0100		framing to close openings	2 Clab	750	.021	S.F.	.38	.55		.93	1.28	
	0200		Tarpaulins hung over scaffolding, 8 uses, not incl. scaffolding		1,500	.011		.18	.28		.46	.63	
	0300		Prefab fiberglass panels, steel frame, 8 uses		1,200	.013		.74	.35		1.09	1.35	

01540 | Construction Aids

				CREW	DAILY OUTPUT	LABOR-HOURS	UNIT	MAT.	LABOR	EQUIP.	TOTAL	TOTAL INCL O&P	
750	0010		**SCAFFOLDING** R01540 -100										750
	0015		Steel tubular, reg, rent/mo, no plank, incl erect or dismantle										
	0090		Building exterior, wall face, 1 to 5 stories, 6'-4" x 5' frames	3 Carp	24	1	C.S.F.	24.50	33		57.50	78	
	0200		6 to 12 stories	4 Carp	21.20	1.509		24.50	50		74.50	104	
	0310		13 to 20 stories	5 Carp	20	2		24.50	66		90.50	130	
	0460		Building interior, wall face area, up to 16' high	3 Carp	25	.960		24.50	31.50		56	76	
	0560		16' to 40' high		23	1.043		24.50	34.50		59	80	
	0800		Building interior floor area, up to 30' high		312	.077	C.C.F.	2.57	2.54		5.11	6.80	
	0900		Over 30' high	4 Carp	275	.116	"	2.57	3.84		6.41	8.85	
	0910		Steel tubular, heavy duty shoring, buy										
	0920		Frames 5' high 2' wide				Ea.	75			75	82.50	
	0925		5' high 4' wide					85			85	93.50	
	0930		6' high 2' wide					86			86	94.50	
	0935		6' high 4' wide					101			101	111	
	0940		Accessories										
	0945		Cross braces				Ea.	16			16	17.60	
	0950		U-head, 8" x 8"					17.50			17.50	19.25	
	0955		J-head, 4" x 8"					12.80			12.80	14.10	
	0960		Base plate, 8" x 8"					14.20			14.20	15.60	
	0965		Leveling jack					30.50			30.50	33.50	
	1000		Steel tubular, regular, buy										
	1100		Frames 3' high 5' wide				Ea.	58			58	64	
	1150		5' high 5' wide					67			67	73.50	
	1200		6'-4" high 5' wide					84			84	92.50	
	1350		7'-6" high 6' wide					145			145	160	
	1500		Accessories cross braces					15			15	16.50	
	1550		Guardrail post					15			15	16.50	
	1600		Guardrail 7' section					7.25			7.25	8	
	1650		Screw jacks & plates					24			24	26.50	
	1700		Sidearm brackets					28			28	31	
	1750		8" casters					33			33	36.50	
	1800		Plank 2" x 10" x 16'-0"					42.50			42.50	47	
	1900		Stairway section					245			245	270	
	1910		Stairway starter bar					29			29	32	
	1920		Stairway inside handrail					53			53	58.50	
	1930		Stairway outside handrail					73			73	80.50	
	1940		Walk-thru frame guardrail					37			37	40.50	
	2000		Steel tubular, regular, rent/mo.										
	2100		Frames 3' high 5' wide				Ea.	3.75			3.75	4.13	
	2150		5' high 5' wide					3.75			3.75	4.13	
	2200		6'-4" high 5' wide					3.75			3.75	4.13	
	2250		7'-6" high 6' wide					7			7	7.70	
	2500		Accessories, cross braces					.60			.60	.66	
	2550		Guardrail post					1			1	1.10	
	2600		Guardrail 7' section					.75			.75	.83	
	2650		Screw jacks & plates					1.50			1.50	1.65	
	2700		Sidearm brackets					1.50			1.50	1.65	
	2750		8" casters					6			6	6.60	
	2800		Outrigger for rolling tower					3			3	3.30	
	2850		Plank 2" x 10" x 16'-0"					5			5	5.50	

GENERAL REQUIREMENTS **1**

01540	Construction Aids	CREW	DAILY OUTPUT	LABOR-HOURS	UNIT	2004 BARE COSTS				TOTAL INCL O&P
						MAT.	LABOR	EQUIP.	TOTAL	
2900	Stairway section				Ea.	10			10	11
2910	Stairway starter bar					.10			.10	.11
2920	Stairway inside handrail					5			5	5.50
2930	Stairway outside handrail					5			5	5.50
2940	Walk-thru frame guardrail					2			2	2.20
3000	Steel tubular, heavy duty shoring, rent/mo.									
3250	5' high 2' & 4' wide				Ea.	5			5	5.50
3300	6' high 2' & 4' wide					5			5	5.50
3500	Accessories, cross braces					1			1	1.10
3600	U - head, 8" x 8"					1			1	1.10
3650	J - head, 4" x 8"					1			1	1.10
3700	Base plate, 8" x 8"					1			1	1.10
3750	Leveling jack					2			2	2.20
5700	Planks, 2x10x16'-0", labor only, erect or remove to 50' H	3 Carp	144	.167			5.50		5.50	8.55
5800	Over 50' high	4 Carp	160	.200			6.60		6.60	10.30

(R01540-100)

755										
0010	**SCAFFOLDING SPECIALTIES**									
1500	Sidewalk bridge using tubular steel									
1510	scaffold frames, including planking	3 Carp	45	.533	L.F.	4.72	17.60		22.32	32.50
1600	For 2 uses per month, deduct from all above					50%				
1700	For 1 use every 2 months, add to all above					100%				
1900	Catwalks, 20" wide, no guardrails, 7' span, buy				Ea.	125			125	138
2000	10' span, buy					160			160	176
3720	Putlog, standard, 8' span, with hangers, buy					61			61	67
3750	12' span, buy					92			92	101
3760	Trussed type, 16' span, buy					210			210	231
3790	22' span, buy					252			252	277
3795	Rent per month					30			30	33
3800	Rolling ladders with handrails, 30" wide, buy, 2 step					151			151	166
4000	7 step					510			510	560
4050	10 step					665			665	735
4100	Rolling towers, buy, 5' wide, 7' long, 10' high					1,150			1,150	1,250
4200	For 5' high added sections, to buy, add					190			190	209
4300	Complete incl. wheels, railings, outriggers,									
4350	21' high, to buy				Ea.	1,925			1,925	2,125
4400	Rent/month				"	145			145	159

780										
0010	**SWING STAGING**, 500 lb cap., 2' wide to 24' long, hand operated hoist									
0020	steel cable type, with 60' cables, buy				Ea.	4,250			4,250	4,675
0030	Rent per month				"	425			425	470
2200	Move swing staging (setup and remove)	E-4	2	16	Move		600	37.50	637.50	1,125

820										
0010	**SMALL TOOLS** As % of contractor's work, minimum				Total					.50%
0100	Maximum				"					2%

01550	Vehicular Access & Parking									
0010	**ROADS AND SIDEWALKS** Temporary									
0050	Roads, gravel fill, no surfacing, 4" gravel depth	B-14	715	.067	S.Y.	3.08	1.84	.29	5.21	6.55
0100	8" gravel depth	"	615	.078	"	6.15	2.14	.34	8.63	10.45
1000	Ramp, 3/4" plywood on 2" x 6" joists, 16" O.C.	2 Carp	300	.053	S.F.	1.10	1.76		2.86	3.94
1100	On 2" x 10" joists, 16" O.C.	"	275	.058		1.65	1.92		3.57	4.80
2200	Sidewalks, 2" x 12" planks, 2 uses	1 Carp	350	.023		.74	.75		1.49	1.99
2300	Exterior plywood, 2 uses, 1/2" thick		750	.011		.20	.35		.55	.77
2400	5/8" thick		650	.012		.23	.41		.64	.88
2500	3/4" thick		600	.013		.29	.44		.73	1.01

Important: See the Reference Section for critical supporting data - Reference Nos., Crews, & City Cost Indexes

01560	Barriers & Enclosures	CREW	DAILY OUTPUT	LABOR-HOURS	UNIT	2004 BARE COSTS				TOTAL INCL O&P	
						MAT.	LABOR	EQUIP.	TOTAL		
100	0010	**BARRICADES** 5' high, 3 rail @ 2" x 8", fixed	2 Carp	30	.533	L.F.	11	17.60		28.60	39.50	100
	0150	Movable		20	.800		11	26.50		37.50	53	
	1000	Guardrail, wooden, 3' high, 1" x 6", on 2" x 4" posts		200	.080		.78	2.64		3.42	4.97	
	1100	2" x 6", on 4" x 4" posts	↓	165	.097		1.74	3.20		4.94	6.90	
	1200	Portable metal with base pads, buy					15.50			15.50	17.05	
	1250	Typical installation, assume 10 reuses	2 Carp	600	.027	↓	1.60	.88		2.48	3.13	
	1300	Barricade tape, polyethelyne, 7 mil, 3" wide x 500' long roll				Ea.	25			25	27.50	
250	0010	**TEMPORARY FENCING** Chain link, 11 ga, 5' high	2 Clab	400	.040	L.F.	3.58	1.04		4.62	5.55	250
	0100	6' high		300	.053		3.45	1.39		4.84	5.95	
	0200	Rented chain link, 6' high, to 1000' (up to 12 mo.)		400	.040		2.26	1.04		3.30	4.11	
	0250	Over 1000' (up to 12 mo.)	↓	300	.053		1.64	1.39		3.03	3.96	
	0350	Plywood, painted, 2" x 4" frame, 4' high	A-4	135	.178		3.62	5.65		9.27	12.75	
	0400	4" x 4" frame, 8' high	"	110	.218		7.70	6.95		14.65	19.20	
	0500	Wire mesh on 4" x 4" posts, 4' high	2 Carp	100	.160		6.85	5.30		12.15	15.70	
	0550	8' high	"	80	.200	↓	10.25	6.60		16.85	21.50	
400	0010	**TEMPORARY CONSTRUCTION** See also division 01530										400

01580	Project Signs										
700	0010	**SIGNS** Hi-intensity reflectorized, no posts, buy				S.F.	16.35			16.35	18	700

GENERAL REQUIREMENTS 1

13

01590 | Equipment Rental

GENERAL REQUIREMENTS

		UNIT	HOURLY OPER. COST	RENT PER DAY	RENT PER WEEK	RENT PER MONTH	CREW EQUIPMENT COST/DAY	
100	0010	**CONCRETE EQUIPMENT RENTAL** R01590-100						100
	0100	without operators						
	0150	For batch plant, see div. 01590-500						
	0200	Bucket, concrete lightweight, 1/2 C.Y.	Ea.	.50	15.35	46	138	13.20
	0300	1 C.Y.		.55	18.65	56	168	15.60
	0400	1-1/2 C.Y.		.70	25.50	77	231	21
	0500	2 C.Y.		.80	30.50	92	276	24.80
	0580	8 C.Y.		4.25	202	605	1,825	155
	0600	Cart, concrete, self propelled, operator walking, 10 C.F.		1.85	56.50	170	510	48.80
	0700	Operator riding, 18 C.F.		2.85	86.50	260	780	74.80
	0800	Conveyer for concrete, portable, gas, 16" wide, 26' long		6.15	117	350	1,050	119.20
	0900	46' long		6.55	143	430	1,300	138.40
	1000	56' long		6.65	153	460	1,375	145.20
	1100	Core drill, electric, 2-1/2 H.P., 1" to 8" bit diameter		1.53	58.50	176	530	47.45
	1150	11 HP, 8" to 18" cores		6.73	87	260.80	780	106
	1200	Finisher, concrete floor, gas, riding trowel, 48" diameter		3.55	86.50	260	780	80.40
	1300	Gas, manual, 3 blade, 36" trowel		.80	25.50	76	228	21.60
	1400	4 blade, 48" trowel		1.15	30	90	270	27.20
	1500	Float, hand-operated (Bull float) 48" wide		.08	13.35	40	120	8.65
	1570	Curb builder, 14 H.P., gas, single screw		8.50	150	450	1,350	158
	1590	Double screw		9.25	190	570	1,700	188
	1600	Grinder, concrete and terrazzo, electric, floor		1.82	75	225	675	59.55
	1700	Wall grinder		.91	37.50	113	340	29.90
	1800	Mixer, powered, mortar and concrete, gas, 6 C.F., 18 H.P.		4.50	90	270	810	90
	1900	10 C.F., 25 H.P.		5.40	103	310	930	105.20
	2000	16 C.F.		5.70	127	380	1,150	121.60
	2100	Concrete, stationary, tilt drum, 2 C.Y.		5	202	605	1,825	161
	2120	Pump, concrete, truck mounted 4" line 80' boom		21.15	915	2,750	8,250	719.20
	2140	5" line, 110' boom		28.60	1,275	3,800	11,400	988.80
	2160	Mud jack, 50 C.F. per hr.		4.91	119	358	1,075	110.90
	2180	225 C.F. per hr.		7.70	159	476.80	1,425	156.95
	2190	Shotcrete pump rig, 12 CY/hr		10.80	235	705	2,125	227.40
	2600	Saw, concrete, manual, gas, 18 H.P.		2.85	33.50	100	300	42.80
	2650	Self-propelled, gas, 30 H.P.		5.40	81.50	245	735	92.20
	2700	Vibrators, concrete, electric, 60 cycle, 2 H.P.		.38	15	45	135	12.05
	2800	3 H.P.		.58	21.50	65	195	17.65
	2900	Gas engine, 5 H.P.		.75	24	72	216	20.40
	3000	8 H.P.		1.05	29.50	88	264	26
	3050	Vibrating screed, gas engine, 8HP		1.53	42.50	127	380	37.65
	3100	Concrete transit mixer, hydraulic drive						
	3120	6 x 4, 250 H.P., 8 C.Y., rear discharge		32.85	630	1,885	5,650	639.80
	3200	Front discharge		36.20	680	2,045	6,125	698.60
	3300	6 x 6, 285 H.P., 12 C.Y., rear discharge		35.10	640	1,915	5,750	663.80
	3400	Front discharge		36.90	690	2,070	6,200	709.20
200	0010	**EARTHWORK EQUIPMENT RENTAL** Without operators R01590-100						200
	0040	Aggregate spreader, push type 8' to 12' wide	Ea.	1.80	45	135	405	41.40
	0045	Tailgate type, 8' wide	"	1.75	31.50	95	285	33
	0050	Augers for truck or trailer mounting, vertical drilling						
	0055	Fence post auger, truck mounted	Ea.	8.30	485	1,455	4,375	357.40
	0060	4" to 36" diam., 54 H.P., gas, 10' spindle travel		28.35	645	1,940	5,825	614.80
	0070	14' spindle travel		32.25	805	2,410	7,225	740
	0075	Auger, truck mounted, vertical drilling, to 25' depth		138.75	2,950	8,835	26,500	2,877
	0080	Auger, horizontal boring machine, 12" to 36" diameter, 45 H.P.		15.45	193	580	1,750	239.60
	0090	12" to 48" diameter, 65 H.P.		21.70	345	1,040	3,125	381.60
	0100	Excavator, diesel hydraulic, crawler mounted, 1/2 C.Y. cap.		14.25	350	1,045	3,125	323
	0120	5/8 C.Y. capacity		17.50	465	1,390	4,175	418
	0140	3/4 C.Y. capacity		20.35	495	1,480	4,450	458.80
	0150	1 C.Y. capacity		22.30	515	1,545	4,625	487.40

Important: See the Reference Section for critical supporting data - Reference Nos., Crews, & City Cost Indexes

01590 | Equipment Rental

		UNIT	HOURLY OPER. COST	RENT PER DAY	RENT PER WEEK	RENT PER MONTH	CREW EQUIPMENT COST/DAY	
200								**200**
0200	1-1/2 C.Y. capacity	Ea.	29.30	750	2,245	6,725	683.40	
0300	2 C.Y. capacity		38.15	970	2,910	8,725	887.20	
0320	2-1/2 C.Y. capacity		49.75	1,300	3,915	11,700	1,181	
0340	3-1/2 C.Y. capacity		84.80	2,150	6,435	19,300	1,965	
0341	Attachments							
0342	Bucket thumbs		2.40	208	625	1,875	144.20	
0345	Grapples		1	220	658.80	1,975	139.75	
0350	Gradall type, truck mounted, 3 ton @ 15' radius, 5/8 C.Y.		35.55	900	2,695	8,075	823.40	
0370	1 C.Y. capacity		41.10	1,050	3,150	9,450	958.80	
0400	Backhoe-loader, 40 to 45 H.P., 5/8 C.Y. capacity		7.55	180	540	1,625	168.40	
0450	45 H.P. to 60 H.P., 3/4 C.Y. capacity		9.10	225	675	2,025	207.80	
0460	80 H.P., 1-1/4 C.Y. capacity		11.10	237	710	2,125	230.80	
0470	112 H.P., 1-1/2 C.Y. capacity		16.35	390	1,170	3,500	364.80	
0480	Attachments							
0482	Compactor, 20,000 lb		4.25	117	350	1,050	104	
0485	Hydraulic hammer, 750 ft-lbs		1.95	68.50	205	615	56.60	
0486	Hydraulic hammer, 1200 ft-lbs		4	133	400	1,200	112	
0500	Brush chipper, gas engine, 6" cutter head, 35 H.P.		5.90	93.50	280	840	103.20	
0550	12" cutter head, 130 H.P.		9.35	150	450	1,350	164.80	
0600	15" cutter head, 165 H.P.		13.40	158	475	1,425	202.20	
0750	Bucket, clamshell, general purpose, 3/8 C.Y.		.95	33.50	100	300	27.60	
0800	1/2 C.Y.		1.05	41.50	125	375	33.40	
0850	3/4 C.Y.		1.20	51.50	155	465	40.60	
0900	1 C.Y.		1.25	56.50	170	510	44	
0950	1-1/2 C.Y.		1.95	75	225	675	60.60	
1000	2 C.Y.		2.05	85	255	765	67.40	
1010	Bucket, dragline, medium duty, 1/2 C.Y.		.55	22.50	67	201	17.80	
1020	3/4 C.Y.		.55	23.50	71	213	18.60	
1030	1 C.Y.		.60	25.50	77	231	20.20	
1040	1-1/2 C.Y.		.90	38.50	115	345	30.20	
1050	2 C.Y.		1	43.50	130	390	34	
1070	3 C.Y.		1.45	60	180	540	47.60	
1200	Compactor, roller, 2 drum, 2000 lb., operator walking		6.10	137	410	1,225	130.80	
1250	Rammer compactor, gas, 1000 lb. blow		1.55	38.50	115	345	35.40	
1300	Vibratory plate, gas, 18" plate, 3000 lb. blow		1.45	28.50	85	255	28.60	
1350	21" plate, 5000 lb. blow		1.60	47.50	143	430	41.40	
1370	Curb builder/extruder, 14 H.P., gas, single screw		8.50	152	456	1,375	159.20	
1390	Double screw		9.25	191	572	1,725	188.40	
1500	Disc harrow attachment, for tractor		.35	59	177	530	38.20	
1750	Extractor, piling, see lines 2500 to 2750							
1810	Feller buncher, shearing & accumulating trees, 100 H.P.	Ea.	19.80	455	1,370	4,100	432.40	
1860	Grader, self-propelled, 25,000 lb.		16.25	410	1,235	3,700	377	
1910	30,000 lb.		18.25	475	1,430	4,300	432	
1920	40,000 lb.		28.20	735	2,205	6,625	666.60	
1930	55,000 lb.		37.70	1,025	3,060	9,175	913.60	
1950	Hammer, pavement demo., hyd., gas, self-prop., 1000 to 1250 lb.		20.15	390	1,170	3,500	395.20	
2000	Diesel 1300 to 1500 lb.		27.80	590	1,765	5,300	575.40	
2050	Pile driving hammer, steam or air, 4150 ft.-lb. @ 225 BPM		6.35	275	825	2,475	215.80	
2100	8750 ft.-lb. @ 145 BPM		8.25	450	1,350	4,050	336	
2150	15,000 ft.-lb. @ 60 BPM		8.60	485	1,455	4,375	359.80	
2200	24,450 ft.-lb. @ 111 BPM		11.35	535	1,605	4,825	411.80	
2250	Leads, 15,000 ft.-lb. hammers	L.F.	.03	2.10	6.29	18.85	1.50	
2300	24,450 ft.-lb. hammers and heavier	"	.05	3	9	27	2.20	
2350	Diesel type hammer, 22,400 ft.-lb.	Ea.	24	620	1,860	5,575	564	
2400	41,300 ft.-lb.		32.20	675	2,025	6,075	662.60	
2450	141,000 ft.-lb.		60.90	1,475	4,430	13,300	1,373	
2500	Vib. elec. hammer/extractor, 200 KW diesel generator, 34 H.P.		26.25	655	1,965	5,900	603	
2550	80 H.P.		44.80	960	2,885	8,650	935.40	

R01590 -100

GENERAL REQUIREMENTS 1

15

01590 | Equipment Rental

		UNIT	HOURLY OPER. COST	RENT PER DAY	RENT PER WEEK	RENT PER MONTH	CREW EQUIPMENT COST/DAY
200								**200**
2600	150 H.P.	Ea.	64	1,475	4,455	13,400	1,403
2700	Extractor, steam or air, 700 ft.-lb.		14.50	345	1,040	3,125	324
2750	1000 ft.-lb.		16.60	450	1,350	4,050	402.80
2800	Log chipper, up to 22" diam, 600 H.P.		41.93	1,450	4,360	13,100	1,207
2850	Logger, for skidding & stacking logs, 150 H.P.		33.80	790	2,375	7,125	745.40
2900	Rake, spring tooth, with tractor		7.04	214	643	1,925	184.90
3000	Roller, tandem, gas, 3 to 5 ton		4.90	115	345	1,025	108.20
3050	Diesel, 8 to 12 ton		6.95	213	640	1,925	183.60
3100	Towed type vibratory compactor, diesel, 50 HP, 72" smooth drum		31.15	600	1,805	5,425	610.20
3150	Sheepsfoot, double 60" x 60"		2.35	98.50	295	885	77.80
3170	Landfill compactor, 220 HP		43.05	1,175	3,505	10,500	1,045
3200	Pneumatic tire diesel roller, 12 ton		7.95	300	905	2,725	244.60
3250	21 to 25 ton		13	550	1,650	4,950	434
3300	Sheepsfoot roller, self-propelled, 4 wheel, 130 H.P.		31.95	845	2,535	7,600	762.60
3320	300 H.P.		43.50	1,200	3,590	10,800	1,066
3350	Vibratory steel drum & pneumatic tire, diesel, 18,000 lb.		16.45	320	960	2,875	323.60
3400	29,000 lb.		24.10	405	1,220	3,650	436.80
3410	Rotary mower, brush, 60", with tractor		10.40	232	695	2,075	222.20
3450	Scrapers, towed type, 9 to 12 C.Y. capacity		4.65	221	664	2,000	170
3500	12 to 17 C.Y. capacity		1.51	295	885.20	2,650	189.10
3550	Scrapers, self-propelled, 4 x 4 drive, 2 engine, 14 C.Y. capacity		76.50	1,425	4,280	12,800	1,468
3600	2 engine, 24 C.Y. capacity		116.80	2,275	6,815	20,400	2,297
3640	32 - 44 C.Y. capacity		139.15	2,650	7,935	23,800	2,700
3650	Self-loading, 11 C.Y. capacity		39.80	810	2,435	7,300	805.40
3700	22 C.Y. capacity		75.20	1,650	4,975	14,900	1,597
3710	Screening plant 110 hp. w / 5' x 10'screen		20.35	375	1,125	3,375	387.80
3720	5' x 16' screen	▼	22.40	475	1,425	4,275	464.20
3850	Shovels, see Cranes division 01590-600						
3860	Shovel/backhoe bucket, 1/2 C.Y.	Ea.	1.65	53.50	160	480	45.20
3870	3/4 C.Y.		1.75	60	180	540	50
3880	1 C.Y.		1.85	70	210	630	56.80
3890	1-1/2 C.Y.		1.95	83.50	250	750	65.60
3910	3 C.Y.		2.30	118	355	1,075	89.40
3950	Stump chipper, 18" deep, 30 H.P.		4.65	83.50	250	750	87.20
4110	Tractor, crawler, with bulldozer, torque converter, diesel 75 H.P.		14.40	310	925	2,775	300.20
4150	105 H.P.		19.25	465	1,400	4,200	434
4200	140 H.P.		23.90	585	1,755	5,275	542.20
4260	200 H.P.		35.55	965	2,895	8,675	863.40
4310	300 H.P.		46	1,225	3,655	11,000	1,099
4360	410 H.P.		64.80	1,600	4,775	14,300	1,473
4370	500 H.P.		83.45	2,075	6,205	18,600	1,909
4380	700 H.P.		128	3,375	10,115	30,300	3,047
4400	Loader, crawler, torque conv., diesel, 1-1/2 C.Y., 80 H.P.		13.45	320	965	2,900	300.60
4450	1-1/2 to 1-3/4 C.Y., 95 H.P.		15.55	385	1,150	3,450	354.40
4510	1-3/4 to 2-1/4 C.Y., 130 H.P.		21.25	615	1,840	5,525	538
4530	2-1/2 to 3-1/4 C.Y., 190 H.P.		32	840	2,525	7,575	761
4560	3-1/2 to 5 C.Y., 275 H.P.		42.85	1,200	3,590	10,800	1,061
4610	Tractor loader, wheel, torque conv., 4 x 4, 1 to 1-1/4 C.Y., 65 H.P.		9	200	600	1,800	192
4620	1-1/2 to 1-3/4 C.Y., 80 H.P.		11.15	245	735	2,200	236.20
4650	1-3/4 to 2 C.Y., 100 H.P.		12.35	278	835	2,500	265.80
4710	2-1/2 to 3-1/2 C.Y., 130 H.P.		13.45	315	940	2,825	295.60
4730	3 to 4-1/2 C.Y., 170 H.P.		19.05	475	1,430	4,300	438.40
4760	5-1/4 to 5-3/4 C.Y., 270 H.P.		32.10	695	2,090	6,275	674.80
4810	7 to 8 C.Y., 375 H.P.		57.75	1,275	3,830	11,500	1,228
4870	12-1/2 C.Y., 690 H.P.		82.45	2,025	6,085	18,300	1,877
4880	Wheeled, skid steer, 10 C.F., 30 H.P. gas		7.40	140	420	1,250	143.20
4890	1 C.Y., 78 H.P., diesel	▼	8.85	188	565	1,700	183.80
4891	Attachments for all skid steer loaders						

Reference note: R01590-100

Important: See the Reference Section for critical supporting data - Reference Nos., Crews, & City Cost Indexes

01590 | Equipment Rental

			UNIT	HOURLY OPER. COST	RENT PER DAY	RENT PER WEEK	RENT PER MONTH	CREW EQUIPMENT COST/DAY	
200	4892	Auger	Ea.	.39	65	195	585	42.10	**200**
	4893	Backhoe		.65	108	325	975	70.20	
	4894	Broom		.63	106	317	950	68.45	
	4895	Forks		.22	36.50	109	325	23.55	
	4896	Grapple		.50	82.50	248	745	53.60	
	4897	Concrete hammer		.96	161	482	1,450	104.10	
	4898	Tree spade		.93	155	464	1,400	100.25	
	4899	Trencher		.68	113	340	1,025	73.45	
	4900	Trencher, chain, boom type, gas, operator walking, 12 H.P.		2.30	68.50	205	615	59.40	
	4910	Operator riding, 40 H.P.		7.70	245	735	2,200	208.60	
	5000	Wheel type, diesel, 4' deep, 12" wide		41.30	730	2,185	6,550	767.40	
	5100	Diesel, 6' deep, 20" wide		59.45	1,425	4,250	12,800	1,326	
	5150	Ladder type, diesel, 5' deep, 8" wide		21.05	650	1,950	5,850	558.40	
	5200	Diesel, 8' deep, 16" wide		52.05	1,500	4,470	13,400	1,310	
	5210	Tree spade, self-propelled		9.60	267	800	2,400	236.80	
	5250	Truck, dump, tandem, 12 ton payload		20.25	272	815	2,450	325	
	5300	Three axle dump, 16 ton payload		27.95	420	1,265	3,800	476.60	
	5350	Dump trailer only, rear dump, 16-1/2 C.Y.		4.15	117	350	1,050	103.20	
	5400	20 C.Y.		4.55	133	400	1,200	116.40	
	5450	Flatbed, single axle, 1-1/2 ton rating		11	56.50	170	510	122	
	5500	3 ton rating		14.05	80	240	720	160.40	
	5550	Off highway rear dump, 25 ton capacity		40.55	995	2,980	8,950	920.40	
	5600	35 ton capacity		41.45	1,025	3,075	9,225	946.60	
	5610	50 ton capacity		53.40	1,325	3,995	12,000	1,226	
	5620	65 ton capacity		57.20	1,425	4,260	12,800	1,310	
	5630	100 ton capacity		73.40	1,800	5,380	16,100	1,663	
	6000	Vibratory plow, 25 H.P., walking		3.90	58.50	175	525	66.20	
400	0010	**GENERAL EQUIPMENT RENTAL** Without operators							**400**
	0150	Aerial lift, scissor type, to 15' high, 1000 lb. cap., electric	Ea.	2.20	43.50	130	390	43.60	
	0160	To 25' high, 2000 lb. capacity		2.60	63.50	190	570	58.80	
	0170	Telescoping boom to 40' high, 500 lb. capacity, gas		10.90	272	815	2,450	250.20	
	0180	To 45' high, 500 lb. capacity		11.70	315	940	2,825	281.60	
	0190	To 60' high, 600 lb. capacity		13.65	415	1,250	3,750	359.20	
	0195	Air compressor, portable, 6.5 CFM, electric		.31	17.65	53	159	13.10	
	0196	gasoline		.35	26.50	79	237	18.60	
	0200	Air compressor, portable, gas engine, 60 C.F.M.		4.10	32.50	98	294	52.40	
	0300	160 C.F.M.		7	43.50	130	390	82	
	0400	Diesel engine, rotary screw, 250 C.F.M.		7.95	107	320	960	127.60	
	0500	365 C.F.M.		9.95	130	390	1,175	157.60	
	0550	450 C.F.M.		13.40	158	475	1,425	202.20	
	0600	600 C.F.M.		19.50	217	650	1,950	286	
	0700	750 C.F.M.		21	228	685	2,050	305	
	0800	For silenced models, small sizes, add		3%	5%	5%	5%		
	0900	Large sizes, add		5%	7%	7%	7%		
	0920	Air tools and accessories							
	0930	Breaker, pavement, 60 lb.	Ea.	.35	14	42	126	11.20	
	0940	80 lb.		.35	17	51	153	13	
	0950	Drills, hand (jackhammer) 65 lb.		.45	14.35	43	129	12.20	
	0960	Track or wagon, swing boom, 4" drifter		33.90	605	1,815	5,450	634.20	
	0970	5" drifter		46.05	765	2,295	6,875	827.40	
	0975	Track mounted quarry drill, 6" diameter drill		48.65	825	2,480	7,450	885.20	
	0980	Dust control per drill		.79	12.35	37	111	13.70	
	0990	Hammer, chipping, 12 lb.		.40	21.50	64	192	16	
	1000	Hose, air with couplings, 50' long, 3/4" diameter		.03	5.35	16	48	3.45	
	1100	1" diameter		.03	5.35	16	48	3.45	
	1200	1-1/2" diameter		.05	7.65	23	69	5	
	1300	2" diameter		.10	16.65	50	150	10.80	

The R01590 -100 reference appears in the UNIT column area at rows 4892 and 0010.

01590 | Equipment Rental

		UNIT	HOURLY OPER. COST	RENT PER DAY	RENT PER WEEK	RENT PER MONTH	CREW EQUIPMENT COST/DAY	
400	1400	2-1/2" diameter	Ea.	.12	19.65	59	177	12.75	400
	1410	3" diameter		.16	27.50	82	246	17.70	
	1450	Drill, steel, 7/8" x 2'		.05	6.65	20	60	4.40	
	1460	7/8" x 6'		.06	8	24	72	5.30	
	1520	Moil points		.03	4.67	14	42	3.05	
	1525	Pneumatic nailer w/accessories		.41	27	81	243	19.50	
	1530	Sheeting driver for 60 lb. breaker		.10	6.95	20.80	62.50	4.95	
	1540	For 90 lb. breaker		.15	10	30	90	7.20	
	1550	Spade, 25 lb.		.35	6.65	20	60	6.80	
	1560	Tamper, single, 35 lb.		.49	32.50	98	294	23.50	
	1570	Triple, 140 lb.		.74	49	147	440	35.30	
	1580	Wrenches, impact, air powered, up to 3/4" bolt		.20	10	30	90	7.60	
	1590	Up to 1-1/4" bolt		.30	20	60	180	14.40	
	1600	Barricades, barrels, reflectorized, 1 to 50 barrels		.02	3.10	9.30	28	2	
	1610	100 to 200 barrels		.01	2.33	7	21	1.50	
	1620	Barrels with flashers, 1 to 50 barrels		.02	3.77	11.30	34	2.40	
	1630	100 to 200 barrels		.02	3	9	27	1.95	
	1640	Barrels with steady burn type C lights		.03	5	15	45	3.25	
	1650	Illuminated board, trailer mounted, with generator		.65	117	350	1,050	75.20	
	1670	Portable barricade, stock, with flashers, 1 to 6 units		.02	3.77	11.30	34	2.40	
	1680	25 to 50 units		.02	3.50	10.50	31.50	2.25	
	1690	Butt fusion machine, electric		21.70	435	1,300	3,900	433.60	
	1695	Electro fusion machine		8.60	173	520	1,550	172.80	
	1700	Carts, brick, hand powered, 1000 lb. capacity		.23	38.50	115	345	24.85	
	1800	Gas engine, 1500 lb., 7-1/2' lift		2.80	95	285	855	79.40	
	1822	Dehumidifier, medium, 6 Lb/Hr, 150 CFM		.68	41.50	124	370	30.25	
	1824	Large, 18 Lb/Hr, 600 CFM		1.36	82.50	248	745	60.50	
	1830	Distributor, asphalt, trailer mtd, 2000 gal., 38 H.P. diesel		7	253	760	2,275	208	
	1840	3000 gal., 38 H.P. diesel		8.30	290	870	2,600	240.40	
	1850	Drill, rotary hammer, electric, 1-1/2" diameter		.40	24.50	74	222	18	
	1860	Carbide bit for above		.03	5.35	16	48	3.45	
	1865	Rotary, crawler, 250 HP		81.10	1,675	5,055	15,200	1,660	
	1870	Emulsion sprayer, 65 gal., 5 H.P. gas engine		1.83	74	222	665	59.05	
	1880	200 gal., 5 H.P. engine		4.75	123	370	1,100	112	
	1900	Fencing, see division 01560-250 & 02820-000							
	1920	Floodlight, mercury vapor, or quartz, on tripod							
	1930	1000 watt	Ea.	.30	11.65	35	105	9.40	
	1940	2000 watt		.52	21.50	65	195	17.15	
	1950	Floodlights, trailer mounted with generator, 1 - 300 watt light		2.45	65	195	585	58.60	
	1960	2 - 1000 watt lights		3.30	108	325	975	91.40	
	2000	4 - 300 watt lights		2.75	76.50	230	690	68	
	2020	Forklift, wheeled, for brick, 18', 3000 lb., 2 wheel drive, gas		13.15	188	565	1,700	218.20	
	2040	28', 4000 lb., 4 wheel drive, diesel		11.35	248	745	2,225	239.80	
	2050	For rough terrain, 8000 lb., 16' lift, 68 HP		15	370	1,110	3,325	342	
	2060	For plant, 4 T. capacity, 80 H.P., 2 wheel drive, gas		7.50	95	285	855	117	
	2080	10 T. capacity, 120 H.P., 2 wheel drive, diesel		11.25	172	515	1,550	193	
	2100	Generator, electric, gas engine, 1.5 KW to 3 KW		1.65	21.50	64	192	26	
	2200	5 KW		2.25	31.50	95	285	37	
	2300	10 KW		3.70	68.50	205	615	70.60	
	2400	25 KW		7.10	92.50	278	835	112.40	
	2500	Diesel engine, 20 KW		5.95	63.50	190	570	85.60	
	2600	50 KW		11.15	70	210	630	131.20	
	2700	100 KW		16.45	80	240	720	179.60	
	2800	250 KW		46.45	143	430	1,300	457.60	
	2850	Hammer, hydraulic, for mounting on boom, to 500 ft.-lb.		1.75	63.50	190	570	52	
	2860	1000 ft.-lb.		3.10	102	305	915	85.80	
	2900	Heaters, space, oil or electric, 50 MBH		.92	12.35	37	111	14.75	
	3000	100 MBH		1.66	16.65	50	150	23.30	

Reference number box: R01590-100

Side tab: GENERAL REQUIREMENTS — 1

Important: See the Reference Section for critical supporting data - Reference Nos., Crews, & City Cost Indexes

01590 | Equipment Rental

		UNIT	HOURLY OPER. COST	RENT PER DAY	RENT PER WEEK	RENT PER MONTH	CREW EQUIPMENT COST/DAY	
400	3100	300 MBH	Ea.	5.33	35	105	315	63.65	**400**
	3150	500 MBH	R01590 -100	10.70	50	150	450	115.60	
	3200	Hose, water, suction with coupling, 20' long, 2" diameter		.02	5.65	17	51	3.55	
	3210	3" diameter		.03	8.65	26	78	5.45	
	3220	4" diameter		.04	11.65	35	105	7.30	
	3230	6" diameter		.09	23	69	207	14.50	
	3240	8" diameter		.26	43.50	130	390	28.10	
	3250	Discharge hose with coupling, 50' long, 2" diameter		.02	5	15	45	3.15	
	3260	3" diameter		.02	6	18	54	3.75	
	3270	4" diameter		.03	8.35	25	75	5.25	
	3280	6" diameter		.06	19	57	171	11.90	
	3290	8" diameter		.36	59.50	178	535	38.50	
	3295	Insulation blower		.11	7.35	22	66	5.30	
	3300	Ladders, extension type, 16' to 36' long		.16	26	78	234	16.90	
	3400	40' to 60' long		.19	31	93	279	20.10	
	3405	Lance for cutting concrete		2.84	109	327	980	88.10	
	3407	Lawn mower, rotary, 22", 5HP		.98	27.50	83	249	24.45	
	3408	48" self propelled		2.44	85.50	257	770	70.90	
	3410	Level, laser type, transit		1.24	82.50	248	745	59.50	
	3430	For pipe laying, manual leveling		.71	47.50	142	425	34.10	
	3440	Rotary beacon		.92	61	183	550	43.95	
	3460	Builders level with tripod and rod		.09	14.65	44	132	9.50	
	3500	Light towers, towable, with diesel generator, 2000 watt		2.75	76.50	230	690	68	
	3600	4000 watt		3.30	108	325	975	91.40	
	3700	Mixer, powered, plaster and mortar, 6 C.F., 7 H.P.		1.05	36.50	110	330	30.40	
	3800	10 C.F., 9 H.P.		1.30	53.50	160	480	42.40	
	3850	Nailer, pneumatic		.41	27	81	243	19.50	
	3900	Paint sprayers complete, 8 CFM		.69	45.50	137	410	32.90	
	4000	17 CFM		1.10	73.50	220	660	52.80	
	4020	Pavers, bituminous, rubber tires, 8' wide, 52 H.P., gas		25.75	770	2,315	6,950	669	
	4030	8' wide, 64 H.P., diesel		38.40	11.20	33.65	101	313.95	
	4050	Crawler, 10' wide, 78 H.P., gas		44.95	1,325	3,940	11,800	1,148	
	4060	10' wide, 87 H.P., diesel		58.05	1,650	4,965	14,900	1,457	
	4070	Concrete paver, 12' to 24' wide, 250 H.P.		51.75	1,225	3,710	11,100	1,156	
	4080	Placer-spreader-trimmer, 24' wide, 300 H.P.		64.05	1,775	5,340	16,000	1,580	
	4100	Pump, centrifugal gas pump, 1-1/2", 4 MGPH		2.50	35	105	315	41	
	4200	2", 8 MGPH		3.15	43.50	130	390	51.20	
	4300	3", 15 MGPH		3.35	45	135	405	53.80	
	4400	6", 90 MGPH		15.60	175	525	1,575	229.80	
	4500	Submersible electric pump, 1-1/4", 55 GPM		.35	20.50	62	186	15.20	
	4600	1-1/2", 83 GPM		.41	23.50	70	210	17.30	
	4700	2", 120 GPM		.57	29.50	88	264	22.15	
	4800	3", 300 GPM		.97	38.50	115	345	30.75	
	4900	4", 560 GPM		6.24	124	372	1,125	124.30	
	5000	6", 1590 GPM		9.05	185	555	1,675	183.40	
	5100	Diaphragm pump, gas, single, 1-1/2" diameter		.74	36	108	325	27.50	
	5200	2" diameter		2.40	45	135	405	46.20	
	5300	3" diameter		2.45	48.50	145	435	48.60	
	5400	Double, 4" diameter		4.30	88.50	265	795	87.40	
	5500	Trash pump, self-priming, gas, 2" diameter		2.90	35	105	315	44.20	
	5600	Diesel, 4" diameter		4.35	80	240	720	82.80	
	5650	Diesel, 6" diameter		9	128	385	1,150	149	
	5655	Grout Pump		3.90	38.50	116	350	54.40	
	5660	Rollers, see division 01590-200							
	5700	Salamanders, L.P. gas fired, 100,000 B.T.U.	Ea.	1.66	10.65	32	96	19.70	
	5705	50,000 BTU		1.25	7.65	23	69	14.60	
	5720	Sandblaster, portable, open top, 3 C.F. capacity		.40	20.50	62	186	15.60	
	5730	6 C.F. capacity		.65	30	90	270	23.20	

01590 | Equipment Rental

		UNIT	HOURLY OPER. COST	RENT PER DAY	RENT PER WEEK	RENT PER MONTH	CREW EQUIPMENT COST/DAY	
5740	Accessories for above	Ea.	.11	18	54	162	11.70	400
5750	Sander, floor		.70	16.35	49	147	15.40	
5760	Edger		.60	20	60	180	16.80	
5800	Saw, chain, gas engine, 18" long		1.15	16	48	144	18.80	
5900	36" long		.55	48.50	145	435	33.40	
5950	60" long		.55	50	150	450	34.40	
6000	Masonry, table mounted, 14" diameter, 5 H.P.		1.30	56	168	505	44	
6050	Portable cut-off, 8 H.P.		1.20	25.50	76	228	24.80	
6100	Circular, hand held, electric, 7-1/4" diameter	Ea.	.20	10	30	90	7.60	
6200	12" diameter		.27	14	42	126	10.55	
6250	Wall saw, w/hydraulic power, 10 H.P		2.08	97.50	292.40	875	75.10	
6275	Shot blaster, walk behind, 20" wide		1.04	445	1,330	4,000	274.30	
6300	Steam cleaner, 100 gallons per hour		2.20	63.50	190	570	55.60	
6310	200 gallons per hour		2.90	78.50	235	705	70.20	
6340	Tar Kettle/Pot, 400 gallon		2.69	51.50	155	465	52.50	
6350	Torch, cutting, acetylene-oxygen, 150' hose		1.50	13.35	40	120	20	
6360	Hourly operating cost includes tips and gas		8.10				64.80	
6410	Toilet, portable chemical		.10	17.35	52	156	11.20	
6420	Recycle flush type		.13	21.50	64	192	13.85	
6430	Toilet, fresh water flush, garden hose,		.14	24	72	216	15.50	
6440	Hoisted, non-flush, for high rise		.13	21	63	189	13.65	
6450	Toilet, trailers, minimum		.22	36	108	325	23.35	
6460	Maximum		.65	108	324	970	70	
6465	Tractor, farm with attachment		9.30	225	675	2,025	209.40	
6470	Trailer, office, see division 01520-500							
6500	Trailers, platform, flush deck, 2 axle, 25 ton capacity	Ea.	4.20	91.50	275	825	88.60	
6600	40 ton capacity		5.45	128	385	1,150	120.60	
6700	3 axle, 50 ton capacity		5.90	142	425	1,275	132.20	
6800	75 ton capacity		7.40	185	555	1,675	170.20	
6810	Trailer mounted cable reel for H.V. line work		4.38	209	626	1,875	160.25	
6820	Trailer mounted cable tensioning rig		8.61	410	1,230	3,700	314.90	
6830	Cable pulling rig		54.46	2,325	6,980	20,900	1,832	
6850	Trailer, storage, see division 01520-500							
6900	Water tank, engine driven discharge, 5000 gallons	Ea.	5.50	123	370	1,100	118	
6925	10,000 gallons		7.65	175	525	1,575	166.20	
6950	Water truck, off highway, 6000 gallons		49.05	715	2,150	6,450	822.40	
7010	Tram car for H.V. line work, powered, 2 conductor		5.68	113	340	1,025	113.45	
7020	Transit (builder's level) with tripod		.09	14.65	44	132	9.50	
7030	Trench box, 3000 lbs. 6'x8'		.42	70	210	630	45.35	
7040	7200 lbs. 6'x20'		.82	136	409	1,225	88.35	
7050	8000 lbs., 8' x 16'		.87	145	436	1,300	94.15	
7060	9500 lbs., 8'x20'		1.17	195	584	1,750	126.15	
7065	11,000 lbs., 8'x24'		1.31	218	654	1,950	141.30	
7070	12,000 lbs., 10' x 20'		1.63	272	817	2,450	176.45	
7100	Truck, pickup, 3/4 ton, 2 wheel drive		5.35	55	165	495	75.80	
7200	4 wheel drive		5.50	63.50	190	570	82	
7250	Crew carrier, 9 passenger		5.53	108	324.40	975	109.10	
7290	Tool van, 24,000 G.V.W.		9.09	94.50	283.20	850	129.35	
7300	Tractor, 4 x 2, 30 ton capacity, 195 H.P.		13.90	173	520	1,550	215.20	
7410	250 H.P.		18.90	272	815	2,450	314.20	
7500	6 x 2, 40 ton capacity, 240 H.P.		17.50	272	815	2,450	303	
7600	6 x 4, 45 ton capacity, 240 H.P.		21.45	267	800	2,400	331.60	
7620	Vacuum truck, hazardous material, 2500 gallon		6.72	315	943	2,825	242.35	
7625	5,000 gallon		9.09	420	1,256.80	3,775	324.10	
7640	Tractor, with A frame, boom and winch, 225 H.P.		14.30	233	700	2,100	254.40	
7650	Vacuum, H.E.P.A., 16 gal., wet/dry		.27	24	72	216	16.55	
7655	55 gal, wet/dry		.60	36	108	325	26.40	
7660	Water tank, portable		1	9.35	28	84	13.60	

Note for 5740: R01590-100

Important: See the Reference Section for critical supporting data - Reference Nos., Crews, & City Cost Indexes

01590 | Equipment Rental

		UNIT	HOURLY OPER. COST	RENT PER DAY	RENT PER WEEK	RENT PER MONTH	CREW EQUIPMENT COST/DAY	
400	7690	Large production vacuum loader, 3150 CFM R01590-100	Ea.	15.63	630	1,890	5,675	503.05	**400**
	7700	Welder, electric, 200 amp		3.74	57	171	515	64.10	
	7800	300 amp		5.22	60	180	540	77.75	
	7900	Gas engine, 200 amp		5.30	35.50	106	320	63.60	
	8000	300 amp		6.20	42.50	128	385	75.20	
	8100	Wheelbarrow, any size		.06	10.35	31	93	6.70	
	8200	Wrecking ball, 4000 lb.		1.85	68.50	205	615	55.80	
500	0010	**HIGHWAY EQUIPMENT RENTAL** R01590-100							**500**
	0050	Asphalt batch plant, portable drum mixer, 100 ton/hr.	Ea.	55.10	1,350	4,020	12,100	1,245	
	0060	200 ton/hr.		61.10	1,400	4,215	12,600	1,332	
	0070	300 ton/hr.		71.20	1,675	4,990	15,000	1,568	
	0100	Backhoe attachment, long stick, up to 185 HP, 10.5' long		.30	19.65	59	177	14.20	
	0140	Up to 250 HP, 12' long		.32	21.50	64	192	15.35	
	0180	Over 250 HP, 15' long		.42	27.50	83	249	19.95	
	0200	Special dipper arm, up to 100 HP, 32' long		.86	57.50	172	515	41.30	
	0240	Over 100 HP, 33' long		1.08	72	216	650	51.85	
	0300	Concrete batch plant, portable, electric, 200 CY/Hr		12.67	645	1,940	5,825	489.35	
	0500	Grader attachment, ripper/scarifier, rear mounted	Ea.						
	0520	Up to 135 HP		2.70	58.50	175	525	56.60	
	0540	Up to 180 HP		3.20	75	225	675	70.60	
	0580	Up to 250 HP		3.55	86.50	260	780	80.40	
	0700	Pvmt. removal bucket, for hyd. excavator, up to 90 HP		1.30	41.50	125	375	35.40	
	0740	Up to 200 HP		1.50	63.50	190	570	50	
	0780	Over 200 HP		1.65	76.50	230	690	59.20	
	0900	Aggregate spreader, self-propelled, 187 HP		37.15	790	2,370	7,100	771.20	
	1000	Chemical spreader, 3 C.Y.		2.20	80.50	242	725	66	
	1900	Hammermill, traveling, 250 HP		38.85	1,675	5,020	15,100	1,315	
	2000	Horizontal borer, 3" diam, 13 HP gas driven		3.85	53.50	160	480	62.80	
	2200	Hydromulchers, gas power, 3000 gal., for truck mounting		10.50	187	560	1,675	196	
	2400	Joint & crack cleaner, walk behind, 25 HP		2.05	46.50	140	420	44.40	
	2500	Filler, trailer mounted, 400 gal., 20 HP		6	182	545	1,625	157	
	3000	Paint striper, self propelled, double line, 30 HP		5.15	155	465	1,400	134.20	
	3200	Post drivers, 6" I-Beam frame, for truck mounting		8.15	415	1,250	3,750	315.20	
	3400	Road sweeper, self propelled, 8' wide, 90 HP		23.50	385	1,160	3,475	420	
	4000	Road mixer, self-propelled, 130 HP		28.35	590	1,770	5,300	580.80	
	4100	310 HP		53.85	1,975	5,895	17,700	1,610	
	4200	Cold mix paver, incl pug mill and bitumen tank,							
	4220	165 HP	Ea.	65.05	1,925	5,770	17,300	1,674	
	4250	Paver, asphalt, wheel or crawler, 130 H.P., diesel		58.05	1,650	4,965	14,900	1,457	
	4300	Paver, road widener, gas 1' to 6', 67 HP		28.70	635	1,900	5,700	609.60	
	4400	Diesel, 2' to 14', 88 HP		38.80	995	2,980	8,950	906.40	
	4600	Slipform pavers, curb and gutter, 2 track, 75 HP		22.95	650	1,950	5,850	573.60	
	4700	4 track, 165 HP		31.05	730	2,195	6,575	687.40	
	4800	Median barrier, 215 HP		31.50	755	2,260	6,775	704	
	4901	Trailer, low bed, 75 ton capacity		7.85	182	545	1,625	171.80	
	5000	Road planer, walk behind, 10" cutting width, 10 HP		2	26	78	234	31.60	
	5100	Self propelled, 12" cutting width, 64 HP		5.30	288	865	2,600	215.40	
	5200	Pavement profiler, 4' to 6' wide, 450 HP		141.40	2,800	8,380	25,100	2,807	
	5300	8' to 10' wide, 750 HP		225.75	4,250	12,725	38,200	4,351	
	5400	Roadway plate, steel, 1"x8'x20'		.06	9.65	29	87	6.30	
	5600	Stabilizer, self-propelled, 150 HP		25.25	560	1,685	5,050	539	
	5700	310 HP		42.50	1,200	3,605	10,800	1,061	
	5800	Striper, thermal, truck mounted 120 gal. paint, 150H.P.		32.40	485	1,450	4,350	549.20	
	6000	Tar kettle, 330 gal., trailer mounted		2.37	36.50	110	330	40.95	
	7000	Tunnel locomotive, diesel, 8 to 12 ton		20.45	560	1,675	5,025	498.60	
	7005	Electric, 10 ton		20.45	635	1,905	5,725	544.60	
	7010	Muck cars, 1/2 C.Y. capacity		1.50	20.50	62	186	24.40	

01590 | Equipment Rental

		UNIT	HOURLY OPER. COST	RENT PER DAY	RENT PER WEEK	RENT PER MONTH	CREW EQUIPMENT COST/DAY	
500								**500**
7020	1 C.Y. capacity	Ea. R01590 -100	1.70	29	87	261	31	
7030	2 C.Y. capacity		1.80	33.50	100	300	34.40	
7040	Side dump, 2 C.Y. capacity		2	41.50	125	375	41	
7050	3 C.Y. capacity		2.70	48.50	145	435	50.60	
7060	5 C.Y. capacity		3.80	61.50	185	555	67.40	
7100	Ventilating blower for tunnel, 7-1/2 H.P.		1.25	40	120	360	34	
7110	10 H.P.		1.44	41.50	125	375	36.50	
7120	20 H.P.		2.33	48.50	145	435	47.65	
7140	40 H.P.		4.07	73.50	220	660	76.55	
7160	60 H.P.		6.16	110	330	990	115.30	
7175	75 H.P.		7.89	148	445	1,325	152.10	
7180	200 H.P.		17.60	205	615	1,850	263.80	
7800	Windrow loader, elevating		32.70	905	2,720	8,150	805.60	
600								**600**
0010	**LIFTING AND HOISTING EQUIPMENT RENTAL**	R01590 -100						
0100	without operators							
0120	Aerial lift truck, 2 person, to 80'	Ea.	17.85	600	1,795	5,375	501.80	
0140	Boom work platform, 40' snorkel		9.25	205	615	1,850	197	
0150	Crane, flatbed mntd, 3 ton cap.		11.60	182	545	1,625	201.80	
0200	Crane, climbing, 106' jib, 6000 lb. capacity, 410 FPM		41.95	1,325	3,990	12,000	1,134	
0300	101' jib, 10,250 lb. capacity, 270 FPM		47.25	1,675	5,050	15,200	1,388	
0400	Tower, static, 130' high, 106' jib,							
0500	6200 lb. capacity at 400 FPM	Ea.	45.05	1,525	4,610	13,800	1,282	
0600	Crawler mounted, lattice boom, 1/2 C.Y., 15 tons at 12' radius		20.11	485	1,450	4,350	450.90	
0700	3/4 C.Y., 20 tons at 12' radius		26.81	650	1,950	5,850	604.50	
0800	1 C.Y., 25 tons at 12' radius		35.75	845	2,530	7,600	792	
0900	1-1/2 C.Y., 40 tons at 12' radius		39.90	995	2,980	8,950	915.20	
1000	2 C.Y., 50 tons at 12' radius		49.90	1,425	4,305	12,900	1,260	
1100	3 C.Y., 75 tons at 12' radius		47.75	1,400	4,175	12,500	1,217	
1200	100 ton capacity, 60' boom		61.05	1,850	5,570	16,700	1,602	
1300	165 ton capacity, 60' boom		90.10	2,400	7,235	21,700	2,168	
1400	200 ton capacity, 70' boom		95.85	2,625	7,855	23,600	2,338	
1500	350 ton capacity, 80' boom		138	3,750	11,215	33,600	3,347	
1600	Truck mounted, lattice boom, 6 x 4, 20 tons at 10' radius		20.93	800	2,400	7,200	647.45	
1700	25 tons at 10' radius		22.32	855	2,560	7,675	690.55	
1800	8 x 4, 30 tons at 10' radius		28.67	905	2,720	8,150	773.35	
1900	40 tons at 12' radius		24.56	960	2,880	8,650	772.50	
2000	8 x 4, 60 tons at 15' radius		30.91	1,100	3,290	9,875	905.30	
2050	82 tons at 15' radius		45.40	1,600	4,770	14,300	1,317	
2100	90 tons at 15' radius		46.24	1,750	5,230	15,700	1,416	
2200	115 tons at 15' radius		51.10	1,950	5,815	17,400	1,572	
2300	150 tons at 18' radius		47.45	2,050	6,150	18,500	1,610	
2350	165 tons at 18' radius		73.35	2,400	7,235	21,700	2,034	
2400	Truck mounted, hydraulic, 12 ton capacity		31	605	1,820	5,450	612	
2500	25 ton capacity		31.15	625	1,875	5,625	624.20	
2550	33 ton capacity		32	655	1,965	5,900	649	
2560	40 ton capacity		30.40	645	1,940	5,825	631.20	
2600	55 ton capacity		44.75	920	2,765	8,300	911	
2700	80 ton capacity		54	980	2,940	8,825	1,020	
2720	100 ton capacity		78.90	2,600	7,770	23,300	2,185	
2740	120 ton capacity		82.60	2,825	8,460	25,400	2,353	
2760	150 ton capacity		100.45	3,550	10,625	31,900	2,929	
2800	Self-propelled, 4 x 4, with telescoping boom, 5 ton		14.60	335	1,005	3,025	317.80	
2900	12-1/2 ton capacity		21.90	510	1,535	4,600	482.20	
3000	15 ton capacity		23.85	600	1,800	5,400	550.80	
3050	20 ton capacity		24.80	625	1,880	5,650	574.40	
3100	25 ton capacity		25.90	615	1,840	5,525	575.20	
3150	40 ton capacity		41.65	880	2,640	7,925	861.20	

Important: See the Reference Section for critical supporting data - Reference Nos., Crews, & City Cost Indexes

GENERAL REQUIREMENTS

1

01590	Equipment Rental	UNIT	HOURLY OPER. COST	RENT PER DAY	RENT PER WEEK	RENT PER MONTH	CREW EQUIPMENT COST/DAY
600	3200	Derricks, guy, 20 ton capacity, 60' boom, 75' mast	Ea.	12.33	325	976	2,925	293.85	**600**
	3300	100' boom, 115' mast		20.01	560	1,680	5,050	496.10
	3400	Stiffleg, 20 ton capacity, 70' boom, 37' mast		14.25	415	1,250	3,750	364
	3500	100' boom, 47' mast		22.53	680	2,040	6,125	588.25
	3550	Helicopter, small, lift to 1250 lbs. maximum, w/pilot		66.23	2,625	7,890	23,700	2,108
	3600	Hoists, chain type, overhead, manual, 3/4 ton		.10	3	9	27	2.60
	3900	10 ton		.55	16	48	144	14
	4000	Hoist and tower, 5000 lb. cap., portable electric, 40' high		4.14	188	563	1,700	145.70
	4100	For each added 10' section, add		.09	14.65	44	132	9.50
	4200	Hoist and single tubular tower, 5000 lb. electric, 100' high		5.58	262	785	2,350	201.65
	4300	For each added 6'-6" section, add		.15	24.50	74	222	16
	4400	Hoist and double tubular tower, 5000 lb., 100' high		5.98	288	865	2,600	220.85
	4500	For each added 6'-6" section, add		.17	27.50	83	249	17.95
	4550	Hoist and tower, mast type, 6000 lb., 100' high		6.47	299	897	2,700	231.15
	4570	For each added 10' section, add		.11	18	54	162	11.70
	4600	Hoist and tower, personnel, electric, 2000 lb., 100' @ 125 FPM		13.27	795	2,390	7,175	584.15
	4700	3000 lb., 100' @ 200 FPM		15.15	900	2,700	8,100	661.20
	4800	3000 lb., 150' @ 300 FPM		16.80	1,000	3,030	9,100	740.40
	4900	4000 lb., 100' @ 300 FPM		17.43	1,025	3,090	9,275	757.45
	5000	6000 lb., 100' @ 275 FPM		18.84	1,075	3,240	9,725	798.70
	5100	For added heights up to 500', add	L.F.	.01	1.67	5	15	1.10
	5200	Jacks, hydraulic, 20 ton	Ea.	.05	8	24	72	5.20
	5500	100 ton	"	.30	23	69	207	16.20
	6000	Jacks, hydraulic, climbing with 50' jackrods						
	6010	and control consoles, minimum 3 mo. rental						
	6100	30 ton capacity	Ea.	1.62	108	323	970	77.55
	6150	For each added 10' jackrod section, add		.05	3.33	10	30	2.40
	6300	50 ton capacity		2.60	173	520	1,550	124.80
	6350	For each added 10' jackrod section, add		.06	4	12	36	2.90
	6500	125 ton capacity		6.80	455	1,360	4,075	326.40
	6550	For each added 10' jackrod section, add		.47	31	93	279	22.35
	6600	Cable jack, 10 ton capacity with 200' cable		1.35	90	270	810	64.80
	6650	For each added 50' of cable, add		.15	9.65	29	87	7
700	0010	**WELLPOINT EQUIPMENT RENTAL** See also division 02240							**700**
	0020	Based on 2 months rental						
	0100	Combination jetting & wellpoint pump, 60 H.P. diesel	Ea.	8.91	267	801	2,400	231.50
	0200	High pressure gas jet pump, 200 H.P., 300 psi	"	15.79	228	684	2,050	263.10
	0300	Discharge pipe, 8" diameter	L.F.	.01	.43	1.29	3.87	.35
	0350	12" diameter		.01	.64	1.92	5.75	.45
	0400	Header pipe, flows up to 150 G.P.M., 4" diameter		.01	.39	1.18	3.54	.30
	0500	400 G.P.M., 6" diameter		.01	.46	1.39	4.17	.35
	0600	800 G.P.M., 8" diameter		.01	.64	1.92	5.75	.45
	0700	1500 G.P.M., 10" diameter		.01	.67	2.02	6.05	.50
	0800	2500 G.P.M., 12" diameter		.02	1.27	3.81	11.45	.90
	0900	4500 G.P.M., 16" diameter		.02	1.63	4.88	14.65	1.15
	0950	For quick coupling aluminum and plastic pipe, add		.03	1.68	5.05	15.15	1.25
	1100	Wellpoint, 25' long, with fittings & riser pipe, 1-1/2" or 2" diameter	Ea.	.05	3.36	10.08	30	2.40
	1200	Wellpoint pump, diesel powered, 4" diameter, 20 H.P.		4.33	154	462	1,375	127.05
	1300	6" diameter, 30 H.P.		5.66	191	573	1,725	159.90
	1400	8" suction, 40 H.P.		7.70	262	785	2,350	218.60
	1500	10" suction, 75 H.P.		10.55	305	918	2,750	268
	1600	12" suction, 100 H.P.		15.79	490	1,470	4,400	420.30
	1700	12" suction, 175 H.P.		20.97	540	1,620	4,850	491.75
800	0010	**MARINE EQUIPMENT RENTAL**							**800**
	0200	Barge, 400 Ton, 30' wide x 90' long	Ea.	16.15	240	720	2,150	273.20
	0240	800 Ton, 45' wide x 90' long		26.65	345	1,030	3,100	419.20
	2000	Tugboat, diesel, 100 HP		16.95	168	505	1,525	236.60

R01590-100

GENERAL REQUIREMENTS 1

		01590	Equipment Rental	UNIT	HOURLY OPER. COST	RENT PER DAY	RENT PER WEEK	RENT PER MONTH	CREW EQUIPMENT COST/DAY	
800	2040		250 HP	Ea.	31.45	315	945	2,825	440.60	800
	2080		380 HP	↓	73.10	925	2,780	8,350	1,141	

Important: See the Reference Section for critical supporting data - Reference Nos., Crews, & City Cost Indexes

01700 | Execution Requirements

01740 | Cleaning

		CREW	DAILY OUTPUT	LABOR-HOURS	UNIT	MAT.	LABOR	EQUIP.	TOTAL	TOTAL INCL O&P	
						2004 BARE COSTS					
500	0010	**CLEANING UP** After job completion, allow, minimum				Job					.30%	500
	0040	Maximum				"					1%	
	0050	Cleanup of floor area, continuous, per day, during const.	A-5	24	.750	M.S.F.	1.70	19.50	1.27	22.47	34	
	0100	Final by GC at end of job	"	11.50	1.565	"	2.65	40.50	2.65	45.80	69	

01800 | Facility Operation

01810 | Commissioning

		CREW	DAILY OUTPUT	LABOR-HOURS	UNIT	MAT.	LABOR	EQUIP.	TOTAL	TOTAL INCL O&P	
						2004 BARE COSTS					
100	0010	**COMMISSIONING** Including documentation of design intent										100
	0100	performance verification, O&M, training, min				Project					.50%	
	0150	Maximum				"					.75%	

01832 | Facilities Maintenance

		CREW	DAILY OUTPUT	LABOR-HOURS	UNIT	MAT.	LABOR	EQUIP.	TOTAL	TOTAL INCL O&P	
360	0010	**ELECTRICAL FACILITIES MAINTENANCE**										360
	0700	Cathodic protection systems										
	0720	Check and adjust reading on rectifier	1 Elec	20	.400	Ea.		15.75		15.75	23.50	
	0730	Check pipe to soil potential		20	.400			15.75		15.75	23.50	
	0740	Replace lead connection		4	2			79		79	117	
	0800	Control device, install		5.70	1.404			55.50		55.50	82.50	
	0810	Disassemble, clean and reinstall		7	1.143			45		45	67	
	0820	Replace		10.70	.748			29.50		29.50	44	
	0830	Trouble shoot		10	.800			31.50		31.50	47	
	0900	Demolition, for electrical demolition see Division 16055-300										
	1000	Distribution systems and equipment install or repair a breaker										
	1010	In power panels up to 200 amps	1 Elec	7	1.143	Ea.		45		45	67	
	1020	Over 200 amps		2	4			158		158	234	
	1030	Reset breaker or replace fuse		20	.400			15.75		15.75	23.50	
	1100	Megger test MCC (each stack)		4	2			79		79	117	
	1110	MCC vacuum and clean (each stack)		5.30	1.509			59.50		59.50	88.50	
	2500	Remove/replace or maint, road fixture & lamp		3	2.667		710	105		815	935	
	2510	Fluorescent fixture		7	1.143		57	45		102	130	
	2515	Relamp (fluor.) facility area each tube		60	.133		4.46	5.25		9.71	12.70	
	2518	Fluorescent fixture, clean (area)		44	.182		.08	7.15		7.23	10.75	
	2520	Incandescent fixture		11	.727		44.50	28.50		73	91.50	
	2530	Lamp (incadescent or fluorescent)		60	.133		2.57	5.25		7.82	10.65	
	2535	Replace cord in socket lamp		13	.615		1.92	24.50		26.42	38	
	2540	Ballast electronic type for two tubes		8	1		32.50	39.50		72	94.50	
	2541	Starter		30	.267		.91	10.50		11.41	16.65	
	2545	Replace other lighting parts		11	.727		11.65	28.50		40.15	55.50	
	2550	Switch		11	.727		4.76	28.50		33.26	48	
	2555	Receptacle		11	.727		4.71	28.50		33.21	47.50	
	2560	Floodlight		4	2		228	79		307	370	
	2570	Christmas lighting, indoor, per string		16	.500			19.70		19.70	29.50	
	2580	Outdoor		13	.615			24.50		24.50	36	
	2590	Test battery operated emergency lights		40	.200			7.90		7.90	11.70	
	2600	Repair/replace component in communication system		6	1.333		48.50	52.50		101	132	
	2700	Repair misc. appliances (incl. clocks, vent fan, blower, etc.)		6	1.333			52.50		52.50	78	
	2710	Reset clocks & timers		50	.160			6.30		6.30	9.40	
	2720	Adjust time delay relays		16	.500			19.70		19.70	29.50	

01832	Facilities Maintenance	CREW	DAILY OUTPUT	LABOR-HOURS	UNIT	2004 BARE COSTS				TOTAL INCL O&P	
						MAT.	LABOR	EQUIP.	TOTAL		
360	2730	Test specific gravity of lead-acid batteries	1 Elec	80	.100	Ea.		3.94		3.94	5.85	**360**
	3000	Motors and generators										
	3020	Disassemble, clean and reinstall motor, up to 1/4 HP	1 Elec	4	2	Ea.		79		79	117	
	3030	Up to 3/4 HP		3	2.667			105		105	156	
	3040	Up to 10 HP		2	4			158		158	234	
	3050	Replace part, up to 1/4 HP		6	1.333			52.50		52.50	78	
	3060	Up to 3/4 HP		4	2			79		79	117	
	3070	Up to 10 HP		3	2.667			105		105	156	
	3080	Megger test motor windings		5.33	1.501			59		59	88	
	3082	Motor vibration check		16	.500			19.70		19.70	29.50	
	3084	Oil motor bearings		25	.320			12.60		12.60	18.75	
	3086	Run test emergency generator for 30 minutes		11	.727			28.50		28.50	42.50	
	3090	Rewind motor, up to 1/4 HP		3	2.667			105		105	156	
	3100	Up to 3/4 HP		2	4			158		158	234	
	3110	Up to 10 HP		1.50	5.333			210		210	315	
	3150	Generator, repair or replace part		4	2			79		79	117	
	3160	Repair DC generator		2	4			158		158	234	
	4000	Stub pole, install or remove		3	2.667			105		105	156	
	4500	Transformer maintenance up to 15 kVA	▼	2.70	2.963	▼		117		117	174	

For information about Means Estimating Seminars, see yellow pages 12 and 13 in back of book

Important: See the Reference Section for critical supporting data - Reference Nos., Crews, & City Cost Indexes

Division 2
Site Construction

Estimating Tips

02200 Site Preparation

- If possible visit the site and take an inventory of the type, quantity and size of the trees. Certain trees may have a landscape resale value or firewood value. Stump disposal can be very expensive, particularly if they cannot be buried at the site. Consider using a bulldozer in lieu of hand cutting trees.
- Estimators should visit the site to determine the need for haul road, access, storage of materials, and security considerations. When estimating for access roads on unstable soil, consider using a geotextile stabilization fabric. It can greatly reduce the quantity of crushed stone or gravel. Sites of limited size and access can cause cost overruns due to lost productivity. Theft and damage is another consideration if the location is isolated. A temporary fence or security guards may be required. Investigate the site thoroughly.

02210 Subsurface Investigation

In preparing estimates on structures involving earthwork or foundations, all information concerning soil characteristics should be obtained. Look particularly for hazardous waste, evidence of prior dumping of debris, and previous stream beds.

02220 Selective Demolition

The costs shown for selective demolition do not include rubbish handling or disposal. These items should be estimated separately using Means data or other sources.

- Historic preservation often requires that the contractor remove materials from the existing structure, rehab them and replace them. The estimator must be aware of any related measures and precautions that must be taken when doing selective demolition, and cutting and patching. Requirements may include special handling and storage, as well as security.
- In addition to Section 02220, you can find selective demolition items in each division. Example: Roofing demolition is in division 7.

02300 Earthwork

- Estimating the actual cost of performing earthwork requires careful consideration of the variables involved. This includes items such as type of soil, whether or not water will be encountered, dewatering, whether or not banks need bracing, disposal of excavated earth, length of haul to fill or spoil sites, etc. If the project has large quantities of cut or fill, consider raising or lowering the site to reduce costs while paying close attention to the effect on site drainage and utilities if doing this.
- If the project has large quantities of fill, creating a borrow pit on the site can significantly lower the costs.
- It is very important to consider what time of year the project is scheduled for completion. Bad weather can create large cost overruns from dewatering, site repair and lost productivity from cold weather.

02500 Utility Services
02600 Drainage & Containment

- Never assume that the water, sewer and drainage lines will go in at the early stages of the project. Consider the site access needs before dividing the site in half with open trenches, loose pipe, and machinery obstructions. Always inspect the site to establish that the site drawings are complete. Check off all existing utilities on your drawing as you locate them. If you find any discrepancies, mark up the site plan for further research. Differing site conditions can be very costly if discovered later in the project.
- See also Section 02955 for restoration of pipe where removal/replacement may be undesirable.

02700 Bases, Ballasts, Pavements/Appurtenances

- When estimating paving, keep in mind the project schedule. If an asphaltic paving project is in a colder climate and runs through to the spring, consider placing the base course in the autumn, then topping it in the spring just prior to completion. This could save considerable costs in spring repair. Keep in mind that prices for asphalt and concrete are generally higher in the cold seasons.
- See also Sections 02960/02965.

02900 Planting

- The timing of planting and guarantee specifications often dictate the costs for establishing tree and shrub growth and a stand of grass or ground cover. Establish the work performance schedule to coincide with the local planting season. Maintenance and growth guarantees can add from 20% to 100% to the total landscaping cost. The cost to replace trees and shrubs can be as high as 5% of the total cost depending on the planting zone, soil conditions and time of year.

02960 & 02965 Flexible Pavement Surfacing Recovery

- Recycling of asphalt pavement is becoming very popular and is an alternative to removal and replacement of asphalt pavement. It can be a good value engineering proposal if removed pavement can be recycled either at the site or another site that is reasonably close to the project site.

Reference Numbers

Reference numbers are shown in bold squares at the beginning of some major classifications. These numbers refer to related items in the Reference Section. The reference information may be an estimating procedure, an alternate pricing method or technical information.

Note: Not all subdivisions listed here necessarily appear in this publication.

02080	Utility Materials	CREW	DAILY OUTPUT	LABOR-HOURS	UNIT	2004 BARE COSTS				TOTAL INCL O&P
						MAT.	LABOR	EQUIP.	TOTAL	
400	0010	**UTILITY BOXES** Precast concrete, 6" thick									**400**
	0050	5' x 10' x 6' high, I.D.	B-13	2	28	Ea.	1,450	785	310	2,545	3,150
	0100	6' x 10' x 6' high, I.D.		2	28		1,500	785	310	2,595	3,200
	0150	5' x 12' x 6' high, I.D.		2	28		1,600	785	310	2,695	3,300
	0200	6' x 12' x 6' high, I.D.		1.80	31.111		1,775	870	345	2,990	3,700
	0250	6' x 13' x 6' high, I.D.		1.50	37.333		2,350	1,050	415	3,815	4,625
	0300	8' x 14' x 7' high, I.D.	▼	1	56	▼	2,550	1,575	625	4,750	5,900
600	0010	**UTILITY ACCESSORIES**									**600**
	0400	Underground tape, detectable, reinforced, alum. foil core, 2"	1 Clab	150	.053	C.L.F.	1.43	1.39		2.82	3.73
	0500	6"	"	140	.057	"	3.58	1.49		5.07	6.25

02210	Subsurface Investigation	CREW	DAILY OUTPUT	LABOR-HOURS	UNIT	2004 BARE COSTS				TOTAL INCL O&P
						MAT.	LABOR	EQUIP.	TOTAL	
200	0010	**CORE DRILLING** Reinforced concrete slab, up to 6" thick slab									**200**
	0020	Including bit, layout and set up									
	0100	1" diameter core	B-89A	28	.571	Ea.	2.40	17.05	3.79	23.24	33.50
	0150	Each added inch thick, add		300	.053		.43	1.59	.35	2.37	3.34
	0300	3" diameter core		23	.696		5.35	21	4.61	30.96	43.50
	0350	Each added inch thick, add		186	.086		.96	2.57	.57	4.10	5.70
	0500	4" diameter core		19	.842		5.35	25	5.60	35.95	51
	0550	Each added inch thick, add		170	.094		1.21	2.81	.62	4.64	6.40
	0700	6" diameter core		14	1.143		8.80	34	7.60	50.40	71
	0750	Each added inch thick, add		140	.114		1.49	3.41	.76	5.66	7.75
	0900	8" diameter core		11	1.455		12	43.50	9.65	65.15	91.50
	0950	Each added inch thick, add		95	.168		2.02	5	1.12	8.14	11.25
	1100	10" diameter core		10	1.600		16	47.50	10.60	74.10	104
	1150	Each added inch thick, add		80	.200		2.65	5.95	1.33	9.93	13.70
	1300	12" diameter core		9	1.778		19.20	53	11.80	84	116
	1350	Each added inch thick, add		68	.235		3.18	7	1.56	11.74	16.10
	1500	14" diameter core		7	2.286		23.50	68	15.15	106.65	148
	1550	Each added inch thick, add		55	.291		4.04	8.70	1.93	14.67	20
	1700	18" diameter core		4	4		30	119	26.50	175.50	249
	1750	Each added inch thick, add	▼	28	.571		5.30	17.05	3.79	26.14	36.50
	1760	For horizontal holes, add to above				▼				30%	30%
	1770	Prestressed hollow core plank, 6" thick									
	1780	1" diameter core	B-89A	52	.308	Ea.	1.59	9.20	2.04	12.83	18.30
	1790	Each added inch thick, add		350	.046		.28	1.36	.30	1.94	2.76
	1800	3" diameter core		50	.320		3.51	9.55	2.12	15.18	21
	1810	Each added inch thick, add		240	.067		.58	1.99	.44	3.01	4.23
	1820	4" diameter core		48	.333		4.67	9.95	2.21	16.83	23
	1830	Each added inch thick, add		216	.074		.81	2.21	.49	3.51	4.87
	1840	6" diameter core		44	.364		5.80	10.85	2.41	19.06	26
	1850	Each added inch thick, add		175	.091		.96	2.73	.61	4.30	6
	1860	8" diameter core		32	.500		7.75	14.90	3.32	25.97	35
	1870	Each added inch thick, add		118	.136		1.34	4.04	.90	6.28	8.75
	1880	10" diameter core		28	.571		10.45	17.05	3.79	31.29	42
	1890	Each added inch thick, add	▼	99	.162	▼	1.44	4.82	1.07	7.33	10.25

2 SITE CONSTRUCTION

02210	Subsurface Investigation	CREW	DAILY OUTPUT	LABOR-HOURS	UNIT	2004 BARE COSTS				TOTAL INCL O&P	
						MAT.	LABOR	EQUIP.	TOTAL		
200	1900	12" diameter core	B-89A	22	.727	Ea.	12.75	21.50	4.82	39.07	53.50	**200**
	1910	Each added inch thick, add		85	.188	↓	2.12	5.60	1.25	8.97	12.45	
	1950	Minimum charge for above, 3" diameter core		7	2.286	Total		68	15.15	83.15	123	
	2000	4" diameter core		6.80	2.353			70	15.60	85.60	126	
	2050	6" diameter core		6	2.667			79.50	17.70	97.20	143	
	2100	8" diameter core		5.50	2.909			87	19.30	106.30	156	
	2150	10" diameter core		4.75	3.368			100	22.50	122.50	181	
	2200	12" diameter core		3.90	4.103			122	27	149	220	
	2250	14" diameter core		3.38	4.734			141	31.50	172.50	255	
	2300	18" diameter core	↓	3.15	5.079	↓		152	33.50	185.50	273	

02220	Site Demolition										
240	0010	**MINOR SITE DEMOLITION**										**240**
	0015	Minor site demolition, no hauling, abandon catch basin or manhole	B-6	7	3.429	Ea.		96	29.50	125.50	181	
	0020	Remove existing catch basin or manhole, masonry		4	6			168	52	220	315	
	0030	Catch basin or manhole frames and covers, stored		13	1.846			52	16	68	97	
	0040	Remove and reset		7	3.429	↓		96	29.50	125.50	181	
	4000	Sidewalk removal, bituminous, 2-1/2" thick		325	.074	S.Y.		2.07	.64	2.71	3.89	
	4050	Brick, set in mortar		185	.130			3.64	1.12	4.76	6.85	
	4100	Concrete, plain, 4"		160	.150			4.21	1.30	5.51	7.90	
	4200	Mesh reinforced	↓	150	.160	↓		4.49	1.39	5.88	8.40	

250	0010	**DEMOLISH, REMOVE PAVEMENT AND CURB**										**250**
	5010	Pavement removal, bituminous roads, 3" thick	B-38	690	.058	S.Y.		1.69	1.17	2.86	3.88	
	5050	4" to 6" thick		420	.095			2.78	1.92	4.70	6.40	
	5100	Bituminous driveways		640	.063			1.82	1.26	3.08	4.19	
	5200	Concrete to 6" thick, hydraulic hammer, mesh reinforced		255	.157			4.57	3.17	7.74	10.50	
	5300	Rod reinforced		200	.200	↓		5.85	4.04	9.89	13.40	
	5400	Concrete, 7" to 24" thick, plain		33	1.212	C.Y.		35.50	24.50	60	81	
	5500	Reinforced	↓	24	1.667	"		48.50	33.50	82	112	
	5600	With hand held air equipment, bituminous, to 6" thick	B-39	1,900	.025	S.F.		.69	.08	.77	1.16	
	5700	Concrete to 6" thick, no reinforcing		1,600	.030			.82	.10	.92	1.38	
	5800	Mesh reinforced		1,400	.034			.94	.11	1.05	1.58	
	5900	Rod reinforced	↓	765	.063	↓		1.72	.21	1.93	2.89	

310	0010	**SELECTIVE DEMOLITION, CUTOUT**										**310**
	0020	Concrete, elev. slab, light reinforcement, under 6 CF	B-9C	65	.615	C.F.		16.25	2.46	18.71	28	
	0050	Light reinforcing, over 6 C.F.	"	75	.533	"		14.10	2.13	16.23	24.50	
	6000	Walls, interior, not including re-framing,										
	6010	openings to 5 S.F.										
	6100	Drywall to 5/8" thick	1 Clab	24	.333	Ea.		8.65		8.65	13.50	
	6200	Paneling to 3/4" thick		20	.400			10.40		10.40	16.20	
	6300	Plaster, on gypsum lath		20	.400			10.40		10.40	16.20	
	6340	On wire lath	↓	14	.571	↓		14.85		14.85	23	

330	0010	**SELECTIVE DEMOLITION, DUMP CHARGES**										**330**
	0020	Dump charges, typical urban city, tipping fees only										
	0100	Building construction materials				Ton					70	
	0200	Trees, brush, lumber									50	
	0300	Rubbish only									60	
	0500	Reclamation station, usual charge				↓					85	

350	0010	**SELECTIVE DEMOLITION, RUBBISH HANDLING**										**350**
	0020	The following are to be added to the demolition prices										
	0400	Chute, circular, prefabricated steel, 18" diameter	B-1	40	.600	L.F.	28	16		44	56	
	0440	30" diameter	"	30	.800	"	37.50	21.50		59	74	
	0725	Dumpster, weekly rental, 1 dump/week, 20 C.Y. capacity (8 Tons)				Week					440	
	0800	30 C.Y. capacity (10 Tons)				↓					665	

2 SITE CONSTRUCTION

02220 | Site Demolition

		CREW	DAILY OUTPUT	LABOR-HOURS	UNIT	MAT.	2004 BARE COSTS LABOR	EQUIP.	TOTAL	TOTAL INCL O&P	
350	0840	40 C.Y. capacity (13 Tons)				Week					805	350
	1000	Dust partition, 6 mil polyethylene, 1" x 3" frame	2 Carp	2,000	.008	S.F.	.16	.26		.42	.58	
	1080	2" x 4" frame	"	2,000	.008	"	.26	.26		.52	.70	
	2000	Load, haul, and dump, 50' haul	2 Clab	24	.667	C.Y.		17.35		17.35	27	
	2040	100' haul		16.50	.970			25		25	39.50	
	2080	Over 100' haul, add per 100 L.F.		35.50	.451			11.70		11.70	18.25	
	2120	In elevators, per 10 floors, add	↓	140	.114			2.97		2.97	4.63	
	3000	Loading & trucking, including 2 mile haul, chute loaded	B-16	45	.711			18.90	10.60	29.50	41	
	3040	Hand loading truck, 50' haul	"	48	.667			17.75	9.95	27.70	38.50	
	3080	Machine loading truck	B-17	120	.267			7.35	4.44	11.79	16.20	
	5000	Haul, per mile, up to 8 C.Y. truck	B-34B	1,165	.007			.18	.41	.59	.73	
	5100	Over 8 C.Y. truck	"	1,550	.005	↓		.14	.31	.45	.55	
370	0010	**SELECTIVE DEMOLITION, TORCH CUTTING**										370
	0020	Steel, 1" thick plate	1 Clab	360	.022	L.F.	.18	.58		.76	1.10	
	0040	1" diameter bar	"	210	.038	Ea.		.99		.99	1.54	
	1000	Oxygen lance cutting, reinforced concrete walls										
	1040	12" to 16" thick walls	1 Clab	10	.800	L.F.		21		21	32.50	
	1080	24" thick walls	"	6	1.333	"		34.50		34.50	54	

02240 | Dewatering

		CREW	DAILY OUTPUT	LABOR-HOURS	UNIT	MAT.	2004 BARE COSTS LABOR	EQUIP.	TOTAL	TOTAL INCL O&P	
500	0010	**DEWATERING**										500
	0020	Excavate drainage trench, 2' wide, 2' deep	B-11C	90	.178	C.Y.		5.30	2.31	7.61	10.65	
	0100	2' wide, 3' deep, with backhoe loader	"	135	.119			3.54	1.54	5.08	7.10	
	0200	Excavate sump pits by hand, light soil	1 Clab	7.10	1.127			29.50		29.50	45.50	
	0300	Heavy soil	"	3.50	2.286	↓		59.50		59.50	92.50	
	0500	Pumping 8 hr., attended 2 hrs. per day, including 20 L.F.										
	0550	of suction hose & 100 L.F. discharge hose										
	0600	2" diaphragm pump used for 8 hours	B-10H	4	3	Day		93.50	14	107.50	157	
	0650	4" diaphragm pump used for 8 hours	B-10I	4	3			93.50	26.50	120	171	
	0800	8 hrs. attended, 2" diaphragm pump	B-10H	1	12			375	56	431	630	
	0900	3" centrifugal pump	B-10J	1	12			375	66.50	441.50	645	
	1000	4" diaphragm pump	B-10I	1	12			375	105	480	685	
	1100	6" centrifugal pump	B-10K	1	12	↓		375	268	643	865	

02305 | Equipment

		CREW	DAILY OUTPUT	LABOR-HOURS	UNIT	MAT.	2004 BARE COSTS LABOR	EQUIP.	TOTAL	TOTAL INCL O&P	
250	0010	**MOBILIZATION OR DEMOB.** (One or the other, unless noted) R01590 -100										250
	0015	Up to 25 mi haul dist (50 mi round trip for mob/demob crew)										
	0020	Dozer, loader, backhoe, excav., grader, paver, roller, 70 to 150 H.P.	B-34N	4	2	Ea.		53	111	164	204	
	0100	Above 150 HP	B-34K	3	2.667			70.50	168	238.50	292	
	0900	Shovel or dragline, 3/4 C.Y.	"	3.60	2.222			59	140	199	244	
	1100	Small equipment, placed in rear of, or towed by pickup truck	A-3A	8	1			25.50	10.25	35.75	50.50	
	1150	Equip up to 70 HP, on flatbed trailer behind pickup truck	A-3D	4	2			51.50	42.50	94	126	
	2000	Crane, truck-mounted, up to 75 ton (costs incl both mob & demob)	1 EQHV	3.60	2.222			77.50		77.50	116	
	2100	Crane, truck-mounted, over 75 ton	A-3E	2.50	6.400			196	33	229	335	
	2200	Crawler-mounted, up to 75 ton	A-3F	2	8			245	278	523	675	
	2300	Over 75 ton	A-3G	1.50	10.667	↓		325	390	715	925	
	2500	For each additional 5 miles haul distance, add	↓					10%	10%			

Important: See the Reference Section for critical supporting data - Reference Nos., Crews, & City Cost Indexes

02305 | Equipment

		CREW	DAILY OUTPUT	LABOR-HOURS	UNIT	MAT.	2004 BARE COSTS LABOR	EQUIP.	TOTAL	TOTAL INCL O&P	
250	3000	For large pieces of equipment, allow for assembly/knockdown R01590 0000										**250**
	3100	For mob/demob of micro-tunneling equip, see section 02441-40										

02315 | Excavation and Fill

			CREW	DAILY OUTPUT	LABOR-HOURS	UNIT	MAT.	LABOR	EQUIP.	TOTAL	TOTAL INCL O&P	
110	0010	**BACKFILL, GENERAL**										**110**
	0015	By hand, no compaction, light soil	1 Clab	14	.571	C.Y.		14.85		14.85	23	
	0100	Heavy soil		11	.727			18.90		18.90	29.50	
	0300	Compaction in 6" layers, hand tamp, add to above		20.60	.388			10.10		10.10	15.75	
	0400	Roller compaction operator walking, add	B-10A	100	.120			3.73	1.31	5.04	7.15	
	0500	Air tamp, add	B-9D	190	.211			5.55	.97	6.52	9.70	
	0600	Vibrating plate, add	A-1D	60	.133			3.47	.48	3.95	5.90	
	0800	Compaction in 12" layers, hand tamp, add to above	1 Clab	34	.235			6.10		6.10	9.55	
	0900	Roller compaction operator walking, add	B-10A	150	.080			2.49	.87	3.36	4.74	
	1000	Air tamp, add	B-9	285	.140			3.71	.56	4.27	6.35	
	1100	Vibrating plate, add	A-1E	90	.089			2.31	.46	2.77	4.11	
	1300	Dozer backfilling, bulk, up to 300' haul, no compaction	B-10B	1,200	.010			.31	.72	1.03	1.26	
	1400	Air tamped	B-11B	240	.067			1.94	.80	2.74	3.84	
	1600	Compacting backfill, 6" to 12" lifts, vibrating roller	B-10C	800	.015			.47	1.84	2.31	2.74	
	1700	Sheepsfoot roller	B-10D	750	.016			.50	1.25	1.75	2.14	
	1900	Dozer backfilling, trench, up to 300' haul, no compaction	B-10B	900	.013			.41	.96	1.37	1.69	
	2000	Air tamped	B-11B	235	.068			1.98	.81	2.79	3.92	
	2200	Compacting backfill, 6" to 12" lifts, vibrating roller	B-10C	700	.017			.53	2.10	2.63	3.13	
	2300	Sheepsfoot roller	B-10D	650	.018			.57	1.45	2.02	2.46	
462	0010	**EXCAVATION, STRUCTURAL**										**462**
	0015	Hand, pits to 6' deep, sandy soil	1 Clab	8	1	C.Y.		26		26	40.50	
	0100	Heavy soil or clay		4	2			52		52	81	
	0300	Pits 6' to 12' deep, sandy soil		5	1.600			41.50		41.50	65	
	0500	Heavy soil or clay		3	2.667			69.50		69.50	108	
	0700	Pits 12' to 18' deep, sandy soil		4	2			52		52	81	
	0900	Heavy soil or clay		2	4			104		104	162	
	6030	Common earth, hydraulic backhoe, 1/2 C.Y. bucket	B-12E	55	.291			9.30	5.85	15.15	20.50	
	6035	3/4 C.Y. bucket	B-12F	90	.178			5.70	5.10	10.80	14.15	
	6040	1 C.Y. bucket	B-12A	108	.148			4.74	4.51	9.25	12.10	
	6050	1-1/2 C.Y. bucket	B-12B	144	.111			3.56	4.75	8.31	10.55	
	6060	2 C.Y. bucket	B-12C	200	.080			2.56	4.44	7	8.75	
	6070	Sand and gravel, 3/4 C.Y. bucket	B-12F	100	.160			5.10	4.59	9.69	12.75	
	6080	1 C.Y. bucket	B-12A	120	.133			4.27	4.06	8.33	10.90	
	6090	1-1/2 C.Y. bucket	B-12B	160	.100			3.20	4.27	7.47	9.50	
	6100	2 C.Y. bucket	B-12C	220	.073			2.33	4.03	6.36	7.95	
	6110	Clay, till, or blasted rock, 3/4 C.Y. bucket	B-12F	80	.200			6.40	5.75	12.15	15.95	
	6120	1 C.Y. bucket	B-12A	95	.168			5.40	5.15	10.55	13.75	
	6130	1-1/2 C.Y. bucket	B-12B	130	.123			3.94	5.25	9.19	11.75	
	6140	2 C.Y. bucket	B-12C	175	.091			2.93	5.05	7.98	10	
490	0010	**HAULING**, excavated or borrow, loose cubic yards										**490**
	0012	no loading included, highway haulers										
	0020	6 C.Y. dump truck, 1/4 mile round trip, 5.0 loads/hr.	B-34A	195	.041	C.Y.		1.09	1.67	2.76	3.48	
	0030	1/2 mile round trip, 4.1 loads/hr.		160	.050			1.32	2.03	3.35	4.25	
	0040	1 mile round trip, 3.3 loads/hr.		130	.062			1.63	2.50	4.13	5.25	
	0100	2 mile round trip, 2.6 loads/hr.		100	.080			2.12	3.25	5.37	6.80	
	0150	3 mile round trip, 2.1 loads/hr.		80	.100			2.65	4.06	6.71	8.50	
	0200	4 mile round trip, 1.8 loads/hr.		70	.114			3.02	4.64	7.66	9.70	
	0310	12 C.Y. dump truck, 1/4 mile round trip 3.7 loads/hr.	B-34B	288	.028			.73	1.66	2.39	2.94	
	0400	2 mile round trip, 2.2 loads/hr.		180	.044			1.18	2.65	3.83	4.70	
	0450	3 mile round trip, 1.9 loads/hr.		170	.047			1.24	2.80	4.04	4.98	
	0500	4 mile round trip, 1.6 loads/hr.		125	.064			1.69	3.81	5.50	6.75	

			DAILY	LABOR-		2004 BARE COSTS				TOTAL	
02315	**Excavation and Fill**	CREW	OUTPUT	HOURS	UNIT	MAT.	LABOR	EQUIP.	TOTAL	INCL O&P	
490	1300	Hauling in medium traffic, add				C.Y.				20%	20%	**490**
	1400	Heavy traffic, add								30%	30%	
	1600	Grading at dump, or embankment if required, by dozer	B-10B	1,000	.012	↓		.37	.86	1.23	1.52	
	1800	Spotter at fill or cut, if required	1 Clab	8	1	Hr.		26		26	40.50	
620	0010	**EXCAVATING, UTILITY TRENCH** Common earth										**620**
	0050	Trenching with chain trencher, 12 H.P., operator walking										
	0100	4" wide trench, 12" deep	B-53	800	.010	L.F.		.32	.07	.39	.56	
	0150	18" deep		750	.011			.34	.08	.42	.61	
	0200	24" deep		700	.011			.37	.08	.45	.64	
	0300	6" wide trench, 12" deep		650	.012			.40	.09	.49	.70	
	0350	18" deep		600	.013			.43	.10	.53	.76	
	0400	24" deep		550	.015			.47	.11	.58	.82	
	0450	36" deep		450	.018			.57	.13	.70	1.01	
	0600	8" wide trench, 12" deep		475	.017			.54	.13	.67	.96	
	0650	18" deep		400	.020			.64	.15	.79	1.13	
	0700	24" deep		350	.023			.73	.17	.90	1.30	
	0750	36" deep	↓	300	.027	↓		.86	.20	1.06	1.51	
	1000	Backfill by hand including compaction, add										
	1050	4" wide trench, 12" deep	A-1G	800	.010	L.F.		.26	.04	.30	.46	
	1100	18" deep		530	.015			.39	.07	.46	.68	
	1150	24" deep		400	.020			.52	.09	.61	.91	
	1300	6" wide trench, 12" deep		540	.015			.39	.07	.46	.67	
	1350	18" deep		405	.020			.51	.09	.60	.90	
	1400	24" deep		270	.030			.77	.13	.90	1.34	
	1450	36" deep		180	.044			1.16	.20	1.36	2.02	
	1600	8" wide trench, 12" deep		400	.020			.52	.09	.61	.91	
	1650	18" deep		265	.030			.78	.13	.91	1.37	
	1700	24" deep		200	.040			1.04	.18	1.22	1.81	
	1750	36" deep	↓	135	.059	↓		1.54	.26	1.80	2.69	
	2000	Chain trencher, 40 H.P. operator riding										
	2050	6" wide trench and backfill, 12" deep	B-54	1,200	.007	L.F.		.21	.17	.38	.51	
	2100	18" deep		1,000	.008			.26	.21	.47	.62	
	2150	24" deep		975	.008			.26	.21	.47	.64	
	2200	36" deep		900	.009			.29	.23	.52	.69	
	2250	48" deep		750	.011			.34	.28	.62	.83	
	2300	60" deep		650	.012			.40	.32	.72	.95	
	2400	8" wide trench and backfill, 12" deep		1,000	.008			.26	.21	.47	.62	
	2450	18" deep		950	.008			.27	.22	.49	.65	
	2500	24" deep		900	.009			.29	.23	.52	.69	
	2550	36" deep		800	.010			.32	.26	.58	.77	
	2600	48" deep		650	.012			.40	.32	.72	.95	
	2700	12" wide trench and backfill, 12" deep		975	.008			.26	.21	.47	.64	
	2750	18" deep		860	.009			.30	.24	.54	.72	
	2800	24" deep		800	.010			.32	.26	.58	.77	
	2850	36" deep		725	.011			.35	.29	.64	.85	
	3000	16" wide trench and backfill, 12" deep		835	.010			.31	.25	.56	.73	
	3050	18" deep		750	.011			.34	.28	.62	.83	
	3100	24" deep	↓	700	.011	↓		.37	.30	.67	.88	
	3200	Compaction with vibratory plate, add								50%	50%	
	5100	Hand excavate and trim for pipe bells after trench excavation										
	5200	8" pipe	1 Clab	155	.052	L.F.		1.34		1.34	2.09	
	5300	18" pipe	"	130	.062	"		1.60		1.60	2.49	
640	0010	**UTILITY BEDDING** For pipe and conduit, not incl. compaction										**640**
	0050	Crushed or screened bank run gravel	B-6	150	.160	C.Y.	19.25	4.49	1.39	25.13	29.50	
	0100	Crushed stone 3/4" to 1/2"		150	.160		24.50	4.49	1.39	30.38	35.50	
	0200	Sand, dead or bank	↓	150	.160	↓	3.93	4.49	1.39	9.81	12.75	

Important: See the Reference Section for critical supporting data - Reference Nos., Crews, & City Cost Indexes

SITE CONSTRUCTION

2

02300 | Earthwork

		02315	Excavation and Fill	CREW	DAILY OUTPUT	LABOR-HOURS	UNIT	2004 BARE COSTS MAT.	LABOR	EQUIP.	TOTAL	TOTAL INCL O&P	
640	0500		Compacting bedding in trench	A-1D	90	.089	C.Y.		2.31	.32	2.63	3.95	640

02400 | Tunneling, Boring & Jacking

		02441	Microtunneling	CREW	DAILY OUTPUT	LABOR-HOURS	UNIT	2004 BARE COSTS MAT.	LABOR	EQUIP.	TOTAL	TOTAL INCL O&P	
400	0010		**MICROTUNNELING** Not including excavation, backfill, shoring,										400
	0020		or dewatering, average 50'/day, slurry method										
	0100		24" to 48" outside diameter, minimum				L.F.					640	
	0110		Adverse conditions, add				%					50%	
	1000		Rent microtunneling machine, average monthly lease				Month					85,500	
	1010		Operating technician				Day					640	
	1100		Mobilization and demobilization, minimum				Job					42,800	
	1110		Maximum				"					430,000	

02500 | Utility Services

		02580	Elec/Communication Structures	CREW	DAILY OUTPUT	LABOR-HOURS	UNIT	2004 BARE COSTS MAT.	LABOR	EQUIP.	TOTAL	TOTAL INCL O&P	
100	0010		**RADIO TOWERS**										100
	0020		Guyed, 50'h, 40 lb. sec., 70MPH basic wind spd.	2 Sswk	1	16	Ea.	1,650	595		2,245	2,875	
	0100		Wind load 90 MPH basic wind speed	"	1	16		1,650	595		2,245	2,875	
	0300		190' high, 40 lb. section, wind load 70 MPH basic wind speed	K-2	.33	72.727		4,450	2,475	485	7,410	9,700	
	0400		200' high, 70 lb. section, wind load 90 MPH basic wind speed		.33	72.727		8,950	2,475	485	11,910	14,700	
	0600		300' high, 70 lb. section, wind load 70 MPH basic wind speed		.20	120		12,700	4,075	800	17,575	21,900	
	0700		270' high, 90 lb. section, wind load 90 MPH basic wind speed		.20	120		14,700	4,075	800	19,575	24,000	
	0800		400' high, 100 lb. section, wind load 70 MPH basic wind speed		.14	171		21,400	5,825	1,150	28,375	34,800	
	0900		Self-supporting, 60' high, wind load 70 MPH basic wind speed		.80	30		3,425	1,025	200	4,650	5,750	
	0910		60' high, wind load 90MPH basic wind speed		.45	53.333		6,175	1,825	355	8,355	10,300	
	1000		120' high, wind load 70MPH basic wind speed		.40	60		8,425	2,050	400	10,875	13,200	
	1200		190' high, wind load 90 MPH basic wind speed	▼	.20	120		20,500	4,075	800	25,375	30,400	
	2000		For states west of Rocky Mountains, add for shipping				▼	10%					
300	0010		**UNDERGROUND STRUCTURES**, communication and power										300
	9960		Nylon polyethylene pull rope, 1/4"	2 Elec	2,000	.008	L.F.	.09	.32		.41	.57	
410	0010		**UNDERGROUND DUCTS AND MANHOLES**, In slab or duct bank										410
	0011		Not including excavation, backfill and cast in place concrete										
	1000		Direct burial										
	1010		PVC, schedule 40, w/coupling, 1/2" diameter	1 Elec	340	.024	L.F.	.29	.93		1.22	1.70	
	1020		3/4" diameter		290	.028		.38	1.09		1.47	2.04	
	1030		1" diameter		260	.031		.56	1.21		1.77	2.42	
	1040		1-1/2" diameter		210	.038		.92	1.50		2.42	3.24	
	1050		2" diameter	▼	180	.044		1.18	1.75		2.93	3.90	
	1060		3" diameter	2 Elec	240	.067		2.34	2.63		4.97	6.50	
	1070		4" diameter	▼	160	.100	▼	3.22	3.94		7.16	9.40	

SITE CONSTRUCTION 2

		CREW	DAILY OUTPUT	LABOR-HOURS	UNIT	2004 BARE COSTS				TOTAL INCL O&P
02580	Elec/Communication Structures					MAT.	LABOR	EQUIP.	TOTAL	
410 1080	5" diameter	2 Elec	120	.133	L.F.	4.72	5.25		9.97	13	410
1090	6" diameter	↓	90	.178	↓	6.30	7		13.30	17.30
1110	Elbows, 1/2" diameter	1 Elec	48	.167	Ea.	1.58	6.55		8.13	11.50
1120	3/4" diameter		38	.211		1.59	8.30		9.89	14.10
1130	1" diameter		32	.250		2.69	9.85		12.54	17.60
1140	1-1/2" diameter		21	.381		5.20	15		20.20	28
1150	2" diameter		16	.500		7.55	19.70		27.25	38
1160	3" diameter		12	.667		23	26.50		49.50	64.50
1170	4" diameter		9	.889		40	35		75	96
1180	5" diameter		8	1		70	39.50		109.50	136
1190	6" diameter		5	1.600		119	63		182	225
1210	Adapters, 1/2" diameter		52	.154		.51	6.05		6.56	9.55
1220	3/4" diameter		43	.186		.94	7.35		8.29	11.95
1230	1" diameter		39	.205		1.19	8.10		9.29	13.30
1240	1-1/2" diameter		35	.229		1.85	9		10.85	15.45
1250	2" diameter		26	.308		2.66	12.10		14.76	21
1260	3" diameter		20	.400		6.75	15.75		22.50	31
1270	4" diameter		14	.571		11.80	22.50		34.30	46.50
1280	5" diameter		12	.667		23	26.50		49.50	64.50
1290	6" diameter		9	.889		28	35		63	83
1340	Bell end & cap, 1-1/2" diameter		35	.229		7.75	9		16.75	22
1350	Bell end & plug, 2" diameter		26	.308		8.25	12.10		20.35	27
1360	3" diameter		20	.400		10.30	15.75		26.05	35
1370	4" diameter		14	.571		12.10	22.50		34.60	47
1380	5" diameter		12	.667		18.30	26.50		44.80	59
1390	6" diameter		9	.889		20.50	35		55.50	75
1450	Base spacer, 2" diameter		56	.143		1.41	5.65		7.06	9.90
1460	3" diameter		46	.174		1.56	6.85		8.41	11.90
1470	4" diameter		41	.195		1.70	7.70		9.40	13.30
1480	5" diameter		37	.216		1.86	8.50		10.36	14.70
1490	6" diameter		34	.235		3	9.25		12.25	17.10
1550	Intermediate spacer, 2" diameter		60	.133		1.37	5.25		6.62	9.30
1560	3" diameter		46	.174		1.49	6.85		8.34	11.85
1570	4" diameter		41	.195		1.64	7.70		9.34	13.25
1580	5" diameter		37	.216		1.77	8.50		10.27	14.60
1590	6" diameter		34	.235	↓	2.86	9.25		12.11	16.95
4010	PVC, schedule 80, w/coupling, 1/2" diameter		215	.037	L.F.	.46	1.47		1.93	2.69
4020	3/4" diameter		180	.044		.57	1.75		2.32	3.23
4030	1" diameter		145	.055		.88	2.17		3.05	4.20
4040	1-1/2" diameter		120	.067		1.68	2.63		4.31	5.75
4050	2" diameter	↓	100	.080		2.14	3.15		5.29	7.05
4060	3" diameter	2 Elec	130	.123		4.12	4.85		8.97	11.75
4070	4" diameter		90	.178		6	7		13	17
4080	5" diameter		70	.229		8.65	9		17.65	23
4090	6" diameter	↓	50	.320	↓	11.85	12.60		24.45	32
4110	Elbows, 1/2" diameter	1 Elec	29	.276	Ea.	2.42	10.85		13.27	18.80
4120	3/4" diameter		23	.348		3.39	13.70		17.09	24
4130	1" diameter		20	.400		5.25	15.75		21	29.50
4140	1-1/2" diameter		16	.500		10.45	19.70		30.15	41
4150	2" diameter		12	.667		13.90	26.50		40.40	54.50
4160	3" diameter		9	.889		48	35		83	105
4170	4" diameter		7	1.143		77	45		122	152
4180	5" diameter		6	1.333		114	52.50		166.50	203
4190	6" diameter		4	2		230	79		309	370
4210	Adapter, 1/2" diameter		39	.205		.51	8.10		8.61	12.55
4220	3/4" diameter	↓	33	.242	↓	.94	9.55		10.49	15.25

Important: See the Reference Section for critical supporting data - Reference Nos., Crews, & City Cost Indexes

02580 | Elec/Communication Structures

		CREW	DAILY OUTPUT	LABOR-HOURS	UNIT	2004 BARE COSTS				TOTAL INCL O&P
						MAT.	LABOR	EQUIP.	TOTAL	
410	4230	1" diameter	1 Elec	29	.276	Ea.	1.18	10.85		12.03	17.45
	4240	1-1/2" diameter		26	.308		1.84	12.10		13.94	20
	4250	2" diameter		23	.348		2.65	13.70		16.35	23.50
	4260	3" diameter		18	.444		6.75	17.50		24.25	33.50
	4270	4" diameter		13	.615		11.85	24.50		36.35	49
	4280	5" diameter		11	.727		23	28.50		51.50	68
	4290	6" diameter		8	1		28	39.50		67.50	89.50
	4310	Bell end & cap, 1-1/2" diameter		26	.308		7.75	12.10		19.85	26.50
	4320	Bell end & plug, 2" diameter		23	.348		8.25	13.70		21.95	29.50
	4330	3" diameter		18	.444		10.30	17.50		27.80	37.50
	4340	4" diameter		13	.615		12.10	24.50		36.60	49.50
	4350	5" diameter		11	.727		18.30	28.50		46.80	62.50
	4360	6" diameter		8	1		20.50	39.50		60	81.50
	4370	Base spacer, 2" diameter		42	.190		1.41	7.50		8.91	12.70
	4380	3" diameter		33	.242		1.56	9.55		11.11	15.90
	4390	4" diameter		29	.276		1.70	10.85		12.55	18
	4400	5" diameter		26	.308		1.86	12.10		13.96	20
	4410	6" diameter		25	.320		3	12.60		15.60	22
	4420	Intermediate spacer, 2" diameter		45	.178		1.37	7		8.37	11.90
	4430	3" diameter		34	.235		1.49	9.25		10.74	15.45
	4440	4" diameter		31	.258		1.64	10.15		11.79	16.90
	4450	5" diameter		28	.286		1.77	11.25		13.02	18.70
	4460	6" diameter	↓	25	.320	↓	2.86	12.60		15.46	22
420	0010	**ELECTRIC & TELEPHONE UNDERGROUND**, Not including excavation									
	0200	backfill and cast in place concrete									
	0250	For bedding see div. 02315									
	0380	Underground marking tape, 6" wide	R-19	2,530	.008	L.F.	.28	.31		.59	.77
	0400	Hand holes, precast concrete, with concrete cover									
	0600	2' x 2' x 3' deep	R-3	2.40	8.333	Ea.	250	320	66.50	636.50	830
	0800	3' x 3' x 3' deep		1.90	10.526		325	405	83.50	813.50	1,050
	1000	4' x 4' x 4' deep	↓	1.40	14.286	↓	650	555	114	1,319	1,675
	1200	Manholes, precast with iron racks & pulling irons, C.I. frame									
	1400	and cover, 4' x 6' x 7' deep	B-13	2	28	Ea.	1,250	785	310	2,345	2,925
	1600	6' x 8' x 7' deep		1.90	29.474		1,525	825	330	2,680	3,300
	1800	6' x 10' x 7' deep	↓	1.80	31.111	↓	1,725	870	345	2,940	3,600
	4200	Underground duct, banks ready for concrete fill, min. of 7.5"									
	4400	between conduits, ctr. to ctr.(for wire & cable see div. 16120)									
	4580	PVC, type EB, 1 @ 2" diameter	2 Elec	480	.033	L.F.	.70	1.31		2.01	2.72
	4600	2 @ 2" diameter		240	.067		1.40	2.63		4.03	5.45
	4800	4 @ 2" diameter		120	.133		2.80	5.25		8.05	10.90
	4900	1 @ 3" diameter		400	.040		.98	1.58		2.56	3.41
	5000	2 @ 3" diameter		200	.080		1.95	3.15		5.10	6.85
	5200	4 @ 3" diameter		100	.160		3.90	6.30		10.20	13.70
	5300	1 @ 4" diameter		320	.050		1.51	1.97		3.48	4.59
	5400	2 @ 4" diameter		160	.100		3.02	3.94		6.96	9.15
	5600	4 @ 4" diameter		80	.200		6.05	7.90		13.95	18.35
	5800	6 @ 4" diameter		54	.296		9.05	11.65		20.70	27.50
	5810	1 @ 5" diameter		260	.062		2.23	2.42		4.65	6.05
	5820	2 @ 5" diameter		130	.123		4.45	4.85		9.30	12.10
	5840	4 @ 5" diameter		70	.229		8.90	9		17.90	23
	5860	6 @ 5" diameter		50	.320		13.35	12.60		25.95	33.50
	5870	1 @ 6" diameter		200	.080		3.22	3.15		6.37	8.25
	5880	2 @ 6" diameter		100	.160		6.45	6.30		12.75	16.50
	5900	4 @ 6" diameter		50	.320		12.90	12.60		25.50	33
	5920	6 @ 6" diameter	↓	30	.533		19.30	21		40.30	53

R02580 -300 (at line 0380 reference box)

SITE CONSTRUCTION 2

35

420

02580	Elec/Communication Structures	CREW	DAILY OUTPUT	LABOR-HOURS	UNIT	MAT.	LABOR	EQUIP.	TOTAL	TOTAL INCL O&P	
						2004 BARE COSTS					
6200	Rigid galvanized steel, 2 @ 2" diameter R02580-300	2 Elec	180	.089	L.F.	10.20	3.50		13.70	16.40	420
6400	4 @ 2" diameter		90	.178		20.50	7		27.50	33	
6800	2 @ 3" diameter		100	.160		23.50	6.30		29.80	35.50	
7000	4 @ 3" diameter		50	.320		47	12.60		59.60	70.50	
7200	2 @ 4" diameter		70	.229		33	9		42	49.50	
7400	4 @ 4" diameter		34	.471		65.50	18.55		84.05	100	
7600	6 @ 4" diameter		22	.727		98.50	28.50		127	151	
7620	2 @ 5" diameter		60	.267		66.50	10.50		77	88.50	
7640	4 @ 5" diameter		30	.533		133	21		154	178	
7660	6 @ 5" diameter		18	.889		199	35		234	271	
7680	2 @ 6" diameter		40	.400		97	15.75		112.75	131	
7700	4 @ 6" diameter		20	.800		194	31.50		225.50	261	
7720	6 @ 6" diameter	▼	14	1.143	▼	291	45		336	385	
7800	For Cast-in-place Concrete - Add										
7810	Under 1 C.Y.	C-6	16	3	C.Y.	116	82	3.24	201.24	258	
7820	1 C.Y. - 5 C.Y.		19.20	2.500		105	68	2.70	175.70	224	
7830	Over 5 C.Y.	▼	24	2	▼	87	54.50	2.16	143.66	182	
7850	For Reinforcing Rods - Add										
7860	#4 to #7	2 Rodm	1.10	14.545	Ton	570	540		1,110	1,525	
7870	#8 to #14	"	1.50	10.667	"	555	395		950	1,275	
8000	Fittings, PVC type EB, elbow, 2" diameter	1 Elec	16	.500	Ea.	8.45	19.70		28.15	39	
8200	3" diameter		14	.571		14.75	22.50		37.25	50	
8400	4" diameter		12	.667		18.95	26.50		45.45	60	
8420	5" diameter		10	.800		40.50	31.50		72	91.50	
8440	6" diameter	▼	9	.889		77.50	35		112.50	137	
8500	Coupling, 2" diameter					1.60			1.60	1.76	
8600	3" diameter					2.43			2.43	2.67	
8700	4" diameter					3.87			3.87	4.26	
8720	5" diameter					7			7	7.70	
8740	6" diameter					12.10			12.10	13.30	
8800	Adapter, 2" diameter	1 Elec	26	.308		1.18	12.10		13.28	19.35	
9000	3" diameter		20	.400		5	15.75		20.75	29	
9200	4" diameter		16	.500		5.80	19.70		25.50	36	
9220	5" diameter		13	.615		22.50	24.50		47	61	
9240	6" diameter		10	.800		28	31.50		59.50	77.50	
9400	End bell, 2" diameter		16	.500		6.35	19.70		26.05	36.50	
9600	3" diameter		14	.571		7.45	22.50		29.95	41.50	
9800	4" diameter		12	.667		8.90	26.50		35.40	49	
9810	5" diameter		10	.800		13.95	31.50		45.45	62.50	
9820	6" diameter		8	1		15.30	39.50		54.80	75.50	
9830	5° angle coupling, 2" diameter		26	.308		9.60	12.10		21.70	28.50	
9840	3" diameter		20	.400		12.15	15.75		27.90	37	
9850	4" diameter		16	.500		14.35	19.70		34.05	45.50	
9860	5" diameter		13	.615		15.70	24.50		40.20	53.50	
9870	6" diameter		10	.800		16.10	31.50		47.60	64.50	
9880	Expansion joint, 2" diameter		16	.500		28	19.70		47.70	60.50	
9890	3" diameter		18	.444		49.50	17.50		67	80.50	
9900	4" diameter		12	.667		71.50	26.50		98	118	
9910	5" diameter		10	.800		111	31.50		142.50	169	
9920	6" diameter	▼	8	1		150	39.50		189.50	224	
9930	Heat bender, 2" diameter					400			400	440	
9940	6" diameter					1,150			1,150	1,275	
9950	Cement, quart				▼	12.75			12.75	14.05	

500

	UTILITY POLES										
0010											
6200	Poles, wood, preservative treatment, see also div. 16520, 20' high	R-3	3.10	6.452	Ea.	236	250	51.50	537.50	685	500

Important: See the Reference Section for critical supporting data - Reference Nos., Crews, & City Cost Indexes

02500 | Utility Services

			02580	Elec/Communication Structures	CREW	DAILY OUTPUT	LABOR-HOURS	UNIT	2004 BARE COSTS				TOTAL INCL O&P	
									MAT.	LABOR	EQUIP.	TOTAL		
500	6400			25' high	R-3	2.90	6.897	Ea.	250	267	55	572	735	500
	6600		R02580 -300	30' high		2.60	7.692		275	298	61	634	815	
	6800			35' high		2.40	8.333		360	320	66.50	746.50	950	
	7000			40' high		2.30	8.696		440	335	69	844	1,050	
	7200			45' high		1.70	11.765		535	455	93.50	1,083.50	1,375	
	7400			Cross arms with hardware & insulators										
	7600			4' long	1 Elec	2.50	3.200	Ea.	112	126		238	310	
	7800			5' long		2.40	3.333		130	131		261	340	
	8000			6' long		2.20	3.636		150	143		293	380	

02600 | Drainage & Containment

		02630	Storm Drainage	CREW	DAILY OUTPUT	LABOR-HOURS	UNIT	2004 BARE COSTS				TOTAL INCL O&P	
								MAT.	LABOR	EQUIP.	TOTAL		
110	0010		CATCH BASIN GRATES AND FRAMES not including footing, excavation										110
	1600		Frames & covers, C.I., 24" square, 500 lb.	B-6	7.80	3.077	Ea.	231	86.50	26.50	344	415	
	1700		26" D shape, 600 lb.		7	3.429		380	96	29.50	505.50	600	
	1800		Light traffic, 18" diameter, 100 lb.		10	2.400		125	67.50	21	213.50	265	
	1900		24" diameter, 300 lb.		8.70	2.759		170	77.50	24	271.50	335	
	2000		36" diameter, 900 lb.		5.80	4.138		425	116	36	577	685	
	2100		Heavy traffic, 24" diameter, 400 lb.		7.80	3.077		173	86.50	26.50	286	355	
	2200		36" diameter, 1150 lb.		3	8		510	224	69.50	803.50	980	
	2300		Mass. State standard, 26" diameter, 475 lb.		7	3.429		385	96	29.50	510.50	605	
	2400		30" diameter, 620 lb.		7	3.429		325	96	29.50	450.50	535	
	2500		Watertight, 24" diameter, 350 lb.		7.80	3.077		274	86.50	26.50	387	465	
	2600		26" diameter, 500 lb.		7	3.429		271	96	29.50	396.50	480	
	2700		32" diameter, 575 lb.		6	4		555	112	34.50	701.50	820	
	2800		3 piece cover & frame, 10" deep,										
	2900		1200 lbs., for heavy equipment	B-6	3	8	Ea.	860	224	69.50	1,153.50	1,375	
	3000		Raised for paving 1-1/4" to 2" high,										
	3100		4 piece expansion ring										
	3200		20" to 26" diameter	1 Clab	3	2.667	Ea.	114	69.50		183.50	233	
	3300		30" to 36" diameter	"	3	2.667	"	159	69.50		228.50	283	
	3320		Frames and covers, existing, raised for paving, 2", including										
	3340		row of brick, concrete collar, up to 12" wide frame	B-6	18	1.333	Ea.	35.50	37.50	11.55	84.55	109	
	3360		20" to 26" wide frame		11	2.182		46.50	61	18.90	126.40	166	
	3380		30" to 36" wide frame		9	2.667		58	75	23	156	204	
	3400		Inverts, single channel brick	D-1	3	5.333		65	160		225	315	
	3500		Concrete		5	3.200		50	96		146	201	
	3600		Triple channel, brick		2	8		98.50	239		337.50	475	
	3700		Concrete		3	5.333		87.50	160		247.50	340	
400	0010		STORM DRAINAGE MANHOLES, FRAMES & COVERS not including										400
	0020		footing, excavation, backfill (See line items for frame & cover)										
	0050		Brick, 4' inside diameter, 4' deep	D-1	1	16	Ea.	290	480		770	1,050	
	0100		6' deep		.70	22.857		405	685		1,090	1,500	
	0150		8' deep		.50	32		520	960		1,480	2,025	
	0200		For depths over 8', add		4	4	V.L.F.	116	120		236	310	
	0400		Concrete blocks (radial), 4' I.D., 4' deep		1.50	10.667	Ea.	243	320		563	750	
	0500		6' deep		1	16		320	480		800	1,075	

SITE CONSTRUCTION · **2**

		CREW	DAILY OUTPUT	LABOR-HOURS	UNIT	2004 BARE COSTS MAT.	LABOR	EQUIP.	TOTAL	TOTAL INCL O&P
02630	**Storm Drainage**									
0600	8' deep	D-1	.70	22.857	Ea.	400	685		1,085	1,500
0700	For depths over 8', add	↓	5.50	2.909	V.L.F.	40.50	87		127.50	178
0800	Concrete, cast in place, 4' x 4', 8" thick, 4' deep	C-14H	2	24	Ea.	355	785	12.95	1,152.95	1,625
0900	6' deep		1.50	32		510	1,050	17.30	1,577.30	2,200
1000	8' deep	↓	1	48	↓	735	1,575	26	2,336	3,300
1100	For depths over 8', add	↓	8	6	V.L.F.	83.50	196	3.24	282.74	400
1110	Precast, 4' I.D., 4' deep	B-22	4.10	7.317	Ea.	450	222	58	730	905
1120	6' deep		3	10		580	305	79.50	964.50	1,200
1130	8' deep		2	15	↓	675	455	119	1,249	1,575
1140	For depths over 8', add	↓	16	1.875	V.L.F.	95	57	14.90	166.90	209
1150	5' I.D., 4' deep	B-6	3	8	Ea.	480	224	69.50	773.50	945
1160	6' deep		2	12		645	335	104	1,084	1,350
1170	8' deep		1.50	16	↓	815	450	139	1,404	1,725
1180	For depths over 8', add		12	2	V.L.F.	106	56	17.30	179.30	223
1190	6' I.D., 4' deep		2	12	Ea.	785	335	104	1,224	1,500
1200	6' deep		1.50	16		1,025	450	139	1,614	1,975
1210	8' deep		1	24		1,250	675	208	2,133	2,625
1220	For depths over 8', add	↓	8	3	V.L.F.	164	84	26	274	340
1250	Slab tops, precast, 8" thick									
1300	4' diameter manhole	B-6	8	3	Ea.	161	84	26	271	335
1400	5' diameter manhole		7.50	3.200		320	90	27.50	437.50	520
1500	6' diameter manhole	↓	7	3.429		385	96	29.50	510.50	600
3800	Steps, heavyweight cast iron, 7" x 9"	1 Bric	40	.200		10	6.85		16.85	21.50
3900	8" x 9"		40	.200		15	6.85		21.85	27
3928	12" x 10-1/2"		40	.200		13.70	6.85		20.55	25.50
4000	Standard sizes, galvanized steel		40	.200		12.30	6.85		19.15	24
4100	Aluminum	↓	40	.200	↓	15.75	6.85		22.60	28

For information about Means Estimating Seminars, see yellow pages 12 and 13 in back of book

Important: See the Reference Section for critical supporting data - Reference Nos., Crews, & City Cost Indexes

Division 3
Concrete

Estimating Tips

General

- Carefully check all the plans and specifications. Concrete often appears on drawings other than structural drawings, including mechanical and electrical drawings for equipment pads. The cost of cutting and patching is often difficult to estimate. See Subdivision 02220 for demolition costs.
- Always obtain concrete prices from suppliers near the job site. A volume discount can often be negotiated depending upon competition in the area. Remember to add for waste, particularly for slabs and footings on grade.

03100 Concrete Forms & Accessories

- A primary cost for concrete construction is forming. Most jobs today are constructed with prefabricated forms. The selection of the forms best suited for the job and the total square feet of forms required for efficient concrete forming and placing are key elements in estimating concrete construction. Enough forms must be available for erection to make efficient use of the concrete placing equipment and crew.
- Concrete accessories for forming and placing depend upon the systems used. Study the plans and specifications to assure that all special accessory requirements have been included in the cost estimate such as anchor bolts, inserts and hangers.

03200 Concrete Reinforcement

- Ascertain that the reinforcing steel supplier has included all accessories, cutting, bending and an allowance for lapping, splicing and waste. A good rule of thumb is 10% for lapping, splicing and waste. Also, 10% waste should be allowed for welded wire fabric.

03300 Cast-in-Place Concrete

- When estimating structural concrete, pay particular attention to requirements for concrete additives, curing methods and surface treatments. Special consideration for climate, hot or cold, must be included in your estimate. Be sure to include requirements for concrete placing equipment and concrete finishing.

03400 Precast Concrete
03500 Cementitious Decks & Toppings

- The cost of hauling precast concrete structural members is often an important factor. For this reason, it is important to get a quote from the nearest supplier. It may become economically feasible to set up precasting beds on the site if the hauling costs are prohibitive.

Reference Numbers

Reference numbers are shown in bold squares at the beginning of some major classifications. These numbers refer to related items in the Reference Section. The reference information may be an estimating procedure, an alternate pricing method or technical information.

Note: Not all subdivisions listed here necessarily appear in this publication.

03100 | Concrete Forms & Accessories

03110 | Structural C.I.P. Forms

		CREW	DAILY OUTPUT	LABOR-HOURS	UNIT	2004 BARE COSTS MAT.	LABOR	EQUIP.	TOTAL	TOTAL INCL O&P	
425	0010	**FORMS IN PLACE, EQUIPMENT FOUNDATIONS** job built									425
	0020	1 use	C-2	160	.300	SFCA	2.52	9.65		12.17	17.80	
	0050	2 use		190	.253		1.39	8.15		9.54	14.15	
	0100	3 use		200	.240		1.01	7.70		8.71	13.15	
	0150	4 use	↓	205	.234	↓	.82	7.55		8.37	12.65	

03150 | Concrete Accessories

		CREW	DAILY OUTPUT	LABOR-HOURS	UNIT	2004 BARE COSTS MAT.	LABOR	EQUIP.	TOTAL	TOTAL INCL O&P	
620	0010	**ACCESSORIES, SLEEVES AND CHASES**									620
	0100	Plastic, 1 use, 9" long, 2" diameter	1 Carp	100	.080	Ea.	.53	2.64		3.17	4.69	
	0150	4" diameter		90	.089		1.56	2.93		4.49	6.30	
	0200	6" diameter		75	.107		2.75	3.52		6.27	8.55	
	0250	12" diameter		60	.133		18.05	4.40		22.45	26.50	
	5000	Sheet metal, 2" diameter		100	.080		.68	2.64		3.32	4.86	
	5100	4" diameter		90	.089		.85	2.93		3.78	5.50	
	5150	6" diameter		75	.107		1.23	3.52		4.75	6.85	
	5200	12" diameter		60	.133		2.46	4.40		6.86	9.55	
	6000	Steel pipe, 2" diameter		100	.080		2.44	2.64		5.08	6.80	
	6100	4" diameter		90	.089		8.30	2.93		11.23	13.70	
	6150	6" diameter		75	.107		15.45	3.52		18.97	22.50	
	6200	12" diameter	↓	60	.133	↓	38.50	4.40		42.90	49	

03300 | Cast-In-Place Concrete

03310 | Structural Concrete

		CREW	DAILY OUTPUT	LABOR-HOURS	UNIT	2004 BARE COSTS MAT.	LABOR	EQUIP.	TOTAL	TOTAL INCL O&P	
220	0010	**CONCRETE, READY MIX** Normal weight									220
	0020	2000 psi				C.Y.	69.50			69.50	76.50	
	0100	2500 psi					71			71	78	
	0150	3000 psi					72.50			72.50	80	
	0200	3500 psi				↓	74			74	81.50	
240	0010	**CONCRETE IN PLACE** Including forms (4 uses), reinforcing									240
	0050	steel and finishing unless otherwise indicated										
	3900	Footings, strip, 18" x 9", unreinforced	C-14C	40	2.800	C.Y.	91	88.50	.64	180.14	240	
	3920	18" x 9", reinforced		35	3.200		105	101	.74	206.74	275	
	3925	20" x 10", unreinforced		45	2.489		88	78.50	.57	167.07	221	
	3930	20" x 10", reinforced		40	2.800		99.50	88.50	.64	188.64	249	
	3935	24" x 12", unreinforced		55	2.036		87	64.50	.47	151.97	198	
	3940	24" x 12", reinforced		48	2.333		98.50	74	.54	173.04	226	
	3945	36" x 12", unreinforced		70	1.600		83.50	50.50	.37	134.37	172	
	3950	36" x 12", reinforced		60	1.867		94	59	.43	153.43	196	
	4000	Foundation mat, under 10 C.Y.		38.67	2.896		129	91.50	.67	221.17	287	
	4050	Over 20 C.Y.	↓	56.40	1.986		115	63	.46	178.46	226	
	4650	Slab on grade, not including finish, 4" thick	C-14E	60.75	1.449		88.50	47.50	.43	136.43	173	
	4700	6" thick	"	92	.957	↓	85	31	.29	116.29	143	
700	0010	**PLACING CONCRETE** and vibrating, including labor & equipment									700
	1900	Footings, continuous, shallow, direct chute	C-6	120	.400	C.Y.		10.90	.43	11.33	17.30	
	1950	Pumped	C-20	150	.427			11.90	5.15	17.05	24	
	2000	With crane and bucket	C-7	90	.800	↓		22.50	10.85	33.35	46.50	

3 CONCRETE

Important: See the Reference Section for critical supporting data - Reference Nos., Crews, & City Cost Indexes

03310 | Structural Concrete

		CREW	DAILY OUTPUT	LABOR-HOURS	UNIT	2004 BARE COSTS				TOTAL INCL O&P	
						MAT.	LABOR	EQUIP.	TOTAL		
700	2100	Footings, continuous, deep, direct chute	C-6	140	.343	C.Y.		9.35	.37	9.72	14.80	700
	2150	Pumped	C-20	160	.400			11.15	4.82	15.97	22.50	
	2200	With crane and bucket	C-7	110	.655			18.35	8.90	27.25	38	
	2900	Foundation mats, over 20 C.Y., direct chute	C-6	350	.137			3.74	.15	3.89	5.90	
	2950	Pumped	C-20	400	.160			4.46	1.93	6.39	8.95	
	3000	With crane and bucket	C-7	300	.240	↓		6.75	3.26	10.01	13.95	

03350 | Concrete Finishing

		CREW	DAILY OUTPUT	LABOR-HOURS	UNIT	2004 BARE COSTS				TOTAL INCL O&P	
						MAT.	LABOR	EQUIP.	TOTAL		
300	0010	**FINISHING FLOORS** Monolithic, screed finish	1 Cefi	900	.009	S.F.		.28		.28	.41	300
	0100	Screed and bull float (darby) finish		725	.011			.35		.35	.51	
	0150	Screed, float, and broom finish	↓	630	.013	↓		.40		.40	.59	
325	0010	**CONTROL JOINT**, concrete floor slab										325
	0100	Sawcut in green concrete										
	0120	1" depth	C-27	2,000	.008	L.F.		.25	.05	.30	.42	
	0140	1-1/2" depth		1,800	.009			.28	.05	.33	.47	
	0160	2" depth	↓	1,600	.010			.32	.06	.38	.52	
	0200	Clean out control joint of debris	C-28	6,000	.001	↓		.04		.04	.06	
	0300	Joint sealant										
	0320	Backer rod, polyethylene, 1/4" diameter	1 Cefi	460	.017	L.F.	.02	.55		.57	.83	
	0340	Sealant, polyurethane										
	0360	1/4" x 1/4" (308 LF/Gal)	1 Cefi	270	.030	L.F.	.15	.93		1.08	1.55	
	0380	1/4" x 1/2" (154 LF/Gal)	"	255	.031	"	.31	.99		1.30	1.80	

03920 | Concrete Resurfacing

		CREW	DAILY OUTPUT	LABOR-HOURS	UNIT	2004 BARE COSTS				TOTAL INCL O&P	
						MAT.	LABOR	EQUIP.	TOTAL		
600	0010	**PATCHING CONCRETE**										600
	0100	Floors, 1/4" thick, small areas, regular grout	1 Cefi	170	.047	S.F.	.07	1.48		1.55	2.27	
	0150	Epoxy grout	"	100	.080	"	4.05	2.52		6.57	8.20	
	0300	Slab on Grade, cut outs, up to 50 C.F.	2 Cefi	50	.320	C.F.	5.35	10.10		15.45	21	
	2000	Walls, including chipping, cleaning and epoxy grout										
	2100	Minimum	1 Cefi	65	.123	S.F.	3.29	3.88		7.17	9.30	
	2150	Average		50	.160		6.60	5.05		11.65	14.70	
	2200	Maximum	↓	40	.200		13.15	6.30		19.45	24	
	2510	Underlayment, P.C based self-leveling, 4100 psi, pumped, 1/4"	C-8	20,000	.003		1.39	.08	.04	1.51	1.69	
	2520	1/2"		19,000	.003		2.50	.09	.04	2.63	2.92	
	2530	3/4"		18,000	.003		3.89	.09	.04	4.02	4.46	
	2540	1"		17,000	.003		5.30	.10	.04	5.44	6	
	2550	1-1/2"	↓	15,000	.004		8.05	.11	.05	8.21	9.05	
	2560	Hand mix, 1/2"	C-18	4,000	.002		2.50	.06	.01	2.57	2.85	
	2610	Topping, P.C. based self-level/dry 6100 psi, pumped, 1/4"	C-8	20,000	.003		1.96	.08	.04	2.08	2.32	
	2620	1/2"		19,000	.003		3.53	.09	.04	3.66	4.05	
	2630	3/4"		18,000	.003		5.50	.09	.04	5.63	6.25	
	2660	1"		17,000	.003		7.45	.10	.04	7.59	8.40	
	2670	1-1/2"	↓	15,000	.004		11.35	.11	.05	11.51	12.70	
	2680	Hand mix, 1/2"	C-18	4,000	.002	↓	3.53	.06	.01	3.60	3.98	

For information about Means Estimating Seminars, see yellow pages 12 and 13 in back of book

For expanded coverage of these items see Means Concrete & Masonry Cost Data 2004

Division Notes

	CREW	DAILY OUTPUT	LABOR-HOURS	UNIT	2004 BARE COSTS				TOTAL INCL O&P
					MAT.	LABOR	EQUIP.	TOTAL	

Division 5
Metals

Estimating Tips

05050 Basic Metal Materials & Methods

- Nuts, bolts, washers, connection angles and plates can add a significant amount to both the tonnage of a structural steel job as well as the estimated cost. As a rule of thumb add 10% to the total weight to account for these accessories.
- Type 2 steel construction, commonly referred to as "simple construction," consists generally of field bolted connections with lateral bracing supplied by other elements of the building, such as masonry walls or x-bracing. The estimator should be aware, however, that shop connections may be accomplished by welding or bolting. The method may be particular to the fabrication shop and may have an impact on the estimated cost.

05200 Metal Joists

- In any given project the total weight of open web steel joists is determined by the loads to be supported and the design. However, economies can be realized in minimizing the amount of labor used to place the joists. This is done by maximizing the joist spacing and therefore minimizing the number of joists required to be installed on the job. Certain spacings and locations may be required by the design, but in other cases maximizing the spacing and keeping it as uniform as possible will keep the costs down.

05300 Metal Deck

- The takeoff and estimating of metal deck involves more than simply the area of the floor or roof and the type of deck specified or shown on the drawings. Many different sizes and types of openings may exist. Small openings for individual pipes or conduits may be drilled after the floor/roof is installed, but larger openings may require special deck lengths as well as reinforcing or structural support. The estimator should determine who will be supplying this reinforcing. Additionally, some deck terminations are part of the deck package, such as screed angles and pour stops, and others will be part of the steel contract, such as angles attached to structural members and cast-in-place angles and plates. The estimator must ensure that all pieces are accounted for in the complete estimate.

05500 Metal Fabrications

- The most economical steel stairs are those that use common materials, standard details and most importantly, a uniform and relatively simple method of field assembly. Commonly available A36 channels and plates are very good choices for the main stringers of the stairs, as are angles and tees for the carrier members. Risers and treads are usually made by specialty shops, and it is most economical to use a typical detail in as many places as possible. The stairs should be pre-assembled and shipped directly to the site. The field connections should be simple and straightforward to be accomplished efficiently and with a minimum of equipment and labor.

Reference Numbers

Reference numbers are shown in bold squares at the beginning of some major classifications. These numbers refer to related items in the Reference Section. The reference information may be an estimating procedure, an alternate pricing method or technical information.

Note: Not all subdivisions listed here necessarily appear in this publication.

05090	Metal Fastenings	CREW	DAILY OUTPUT	LABOR-HOURS	UNIT	MAT.	LABOR	EQUIP.	TOTAL	TOTAL INCL O&P	
150							2004 BARE COSTS				**150**
0010	**BOLTS & HEX NUTS** Steel, A307										
0100	1/4" diameter, 1/2" long				Ea.	.05			.05	.06	
0200	1" long					.06			.06	.07	
0300	2" long					.08			.08	.09	
0400	3" long					.12			.12	.13	
0500	4" long					.13			.13	.14	
0600	3/8" diameter, 1" long					.09			.09	.09	
0700	2" long					.11			.11	.12	
0800	3" long					.15			.15	.17	
0900	4" long					.19			.19	.21	
1000	5" long					.24			.24	.26	
1100	1/2" diameter, 1-1/2" long					.17			.17	.19	
1200	2" long					.19			.19	.21	
1300	4" long					.30			.30	.33	
1400	6" long					.41			.41	.45	
1500	8" long					.53			.53	.59	
1600	5/8" diameter, 1-1/2" long					.35			.35	.39	
1700	2" long					.38			.38	.42	
1800	4" long					.53			.53	.59	
1900	6" long					.67			.67	.74	
2000	8" long					.97			.97	1.07	
2100	10" long					1.21			1.21	1.33	
2200	3/4" diameter, 2" long					.55			.55	.61	
2300	4" long					.77			.77	.85	
2400	6" long					.98			.98	1.08	
2500	8" long					1.45			1.45	1.60	
2600	10" long					1.89			1.89	2.07	
2700	12" long					2.20			2.20	2.42	
2800	1" diameter, 3" long					1.44			1.44	1.58	
2900	6" long					2.21			2.21	2.43	
3000	12" long					4.16			4.16	4.58	
3100	For galvanized, add					75%					
3200	For stainless, add					350%					
300											**300**
0010	**CHEMICAL ANCHORS**, Includes layout & drilling										
1430	Chemical anchor, w/rod & epoxy cartridge, 3/4" diam. x 9-1/2" long	B-89A	27	.593	Ea.	10.25	17.70	3.93	31.88	43	
1435	1" diameter x 11-3/4" long		24	.667		19.65	19.90	4.42	43.97	57.50	
1440	1-1/4" diameter x 14" long		21	.762		37.50	22.50	5.05	65.05	82	
1445	Concrete anchor, w/rod & epoxy cartridge, 1-3/4" diameter x 15" long		20	.800		70.50	24	5.30	99.80	120	
1450	18" long		17	.941		84.50	28	6.25	118.75	143	
1455	2" diameter x 18" long		16	1		108	30	6.65	144.65	173	
1460	24" long		15	1.067		141	32	7.05	180.05	212	
340											**340**
0010	**DRILLING** For anchors, up to 4" deep, incl. bit and layout										
0050	in concrete or brick walls and floors, no anchor										
0100	Holes, 1/4" diameter	1 Carp	75	.107	Ea.	.08	3.52		3.60	5.60	
0150	For each additional inch of depth, add		430	.019		.02	.61		.63	.98	
0200	3/8" diameter		63	.127		.07	4.19		4.26	6.65	
0250	For each additional inch of depth, add		340	.024		.02	.78		.80	1.23	
0300	1/2" diameter		50	.160		.07	5.30		5.37	8.30	
0350	For each additional inch of depth, add		250	.032		.02	1.06		1.08	1.66	
0400	5/8" diameter		48	.167		.13	5.50		5.63	8.70	
0450	For each additional inch of depth, add		240	.033		.03	1.10		1.13	1.75	
0500	3/4" diameter		45	.178		.16	5.85		6.01	9.35	
0550	For each additional inch of depth, add		220	.036		.04	1.20		1.24	1.91	
0600	7/8" diameter		43	.186		.19	6.15		6.34	9.75	
0650	For each additional inch of depth, add		210	.038		.05	1.26		1.31	2.01	

METALS 5

		05090	Metal Fastenings	CREW	DAILY OUTPUT	LABOR-HOURS	UNIT	2004 BARE COSTS				TOTAL INCL O&P	
								MAT.	LABOR	EQUIP.	TOTAL		
340	0700		1" diameter	1 Carp	40	.200	Ea.	.22	6.60		6.82	10.55	**340**
	0750		For each additional inch of depth, add		190	.042		.06	1.39		1.45	2.22	
	0800		1-1/4" diameter		38	.211		.31	6.95		7.26	11.15	
	0850		For each additional inch of depth, add		180	.044		.08	1.47		1.55	2.37	
	0900		1-1/2" diameter		35	.229		.48	7.55		8.03	12.30	
	0950		For each additional inch of depth, add		165	.048		.12	1.60		1.72	2.62	
	1000		For ceiling installations, add						40%				
	1100		Drilling & layout for drywall/plaster walls, up to 1" deep, no anchor										
	1200		Holes, 1/4" diameter	1 Carp	150	.053	Ea.	.01	1.76		1.77	2.75	
	1300		3/8" diameter		140	.057		.01	1.89		1.90	2.95	
	1400		1/2" diameter		130	.062		.01	2.03		2.04	3.17	
	1500		3/4" diameter		120	.067		.02	2.20		2.22	3.45	
	1600		1" diameter		110	.073		.03	2.40		2.43	3.77	
	1700		1-1/4" diameter		100	.080		.04	2.64		2.68	4.15	
	1800		1-1/2" diameter		90	.089		.06	2.93		2.99	4.64	
	1900		For ceiling installations, add						40%				
	1910		Drilling & layout for steel, up to 1/4" deep, no anchor										
	1920		Holes, 1/4" diameter	1 Sswk	112	.071	Ea.	.10	2.65		2.75	4.86	
	1925		For each additional 1/4" depth, add		336	.024		.10	.88		.98	1.69	
	1930		3/8" diameter		104	.077		.12	2.86		2.98	5.25	
	1935		For each additional 1/4" depth, add		312	.026		.12	.95		1.07	1.84	
	1940		1/2" diameter		96	.083		.13	3.10		3.23	5.70	
	1945		For each additional 1/4" depth, add		288	.028		.13	1.03		1.16	1.99	
	1950		5/8" diameter		88	.091		.21	3.38		3.59	6.30	
	1955		For each additional 1/4" depth, add		264	.030		.21	1.13		1.34	2.25	
	1960		3/4" diameter		80	.100		.24	3.72		3.96	6.90	
	1965		For each additional 1/4" depth, add		240	.033		.24	1.24		1.48	2.48	
	1970		7/8" diameter		72	.111		.28	4.13		4.41	7.70	
	1975		For each additional 1/4" depth, add		216	.037		.28	1.38		1.66	2.78	
	1980		1" diameter		64	.125		.32	4.64		4.96	8.65	
	1985		For each additional 1/4" depth, add		192	.042		.32	1.55		1.87	3.12	
	1990		For drilling up, add						40%				
380	0010		**EXPANSION ANCHORS** & shields										**380**
	0100		Bolt anchors for concrete, brick or stone, no layout and drilling										
	0200		Expansion shields, zinc, 1/4" diameter, 1-5/16" long, single	1 Carp	90	.089	Ea.	.96	2.93		3.89	5.65	
	0300		1-3/8" long, double		85	.094		1.06	3.11		4.17	6	
	0400		3/8" diameter, 1-1/2" long, single		85	.094		1.58	3.11		4.69	6.60	
	0500		2" long, double		80	.100		1.95	3.30		5.25	7.30	
	0600		1/2" diameter, 2-1/16" long, single		80	.100		2.62	3.30		5.92	8.05	
	0700		2-1/2" long, double		75	.107		2.52	3.52		6.04	8.25	
	0800		5/8" diameter, 2-5/8" long, single		75	.107		3.74	3.52		7.26	9.60	
	0900		2-3/4" long, double		70	.114		3.74	3.77		7.51	9.95	
	1000		3/4" diameter, 2-3/4" long, single		70	.114		5.55	3.77		9.32	11.95	
	1100		3-15/16" long, double		65	.123		7.40	4.06		11.46	14.50	
	1500		Self drilling anchor, snap-off, for 1/4" diameter bolt		26	.308		.72	10.15		10.87	16.60	
	1600		3/8" diameter bolt		23	.348		1.04	11.50		12.54	19.05	
	1700		1/2" diameter bolt		20	.400		1.60	13.20		14.80	22.50	
	1800		5/8" diameter bolt		18	.444		2.68	14.65		17.33	26	
	1900		3/4" diameter bolt		16	.500		4.50	16.50		21	30.50	
	2100		Hollow wall anchors for gypsum wall board, plaster or tile										
	2300		1/8" diameter, short	1 Carp	160	.050	Ea.	.23	1.65		1.88	2.82	
	2400		Long		160	.050		.27	1.65		1.92	2.87	
	2500		3/16" diameter, short		150	.053		.48	1.76		2.24	3.27	
	2600		Long		150	.053		.52	1.76		2.28	3.31	
	2700		1/4" diameter, short		140	.057		.59	1.89		2.48	3.59	
	2800		Long		140	.057		.66	1.89		2.55	3.67	

METALS **5**

For expanded coverage of these items see *Means Building Construction Cost Data 2004*

05090 | Metal Fastenings

		CREW	DAILY OUTPUT	LABOR-HOURS	UNIT	2004 BARE COSTS				TOTAL INCL O&P	
						MAT.	LABOR	EQUIP.	TOTAL		
380	3000	Toggle bolts, bright steel, 1/8" diameter, 2" long	1 Carp	85	.094	Ea.	.21	3.11		3.32	5.05	**380**
	3100	4" long		80	.100		.32	3.30		3.62	5.50	
	3400	1/4" diameter, 3" long		75	.107		.41	3.52		3.93	5.95	
	3500	6" long		70	.114		.57	3.77		4.34	6.50	
	3600	3/8" diameter, 3" long		70	.114		.78	3.77		4.55	6.70	
	3700	6" long		60	.133		1.37	4.40		5.77	8.35	
	3800	1/2" diameter, 4" long		60	.133		2.03	4.40		6.43	9.10	
	3900	6" long		50	.160		3.35	5.30		8.65	11.90	
	4000	Nailing anchors										
	4100	Nylon nailing anchor, 1/4" diameter, 1" long	1 Carp	3.20	2.500	C	16.80	82.50		99.30	148	
	4200	1-1/2" long		2.80	2.857		21.50	94.50		116	171	
	4300	2" long		2.40	3.333		36	110		146	211	
	4400	Metal nailing anchor, 1/4" diameter, 1" long		3.20	2.500		25	82.50		107.50	157	
	4500	1-1/2" long		2.80	2.857		34	94.50		128.50	185	
	4600	2" long		2.40	3.333		43.50	110		153.50	219	
	5000	Screw anchors for concrete, masonry,										
	5100	stone & tile, no layout or drilling included										
	5200	Jute fiber, #6, #8, & #10, 1" long	1 Carp	240	.033	Ea.	.23	1.10		1.33	1.96	
	5400	#14, 2" long		160	.050		.52	1.65		2.17	3.14	
	5500	#16, 2" long		150	.053		.55	1.76		2.31	3.35	
	5600	#20, 2" long		140	.057		.87	1.89		2.76	3.90	
	5700	Lag screw shields, 1/4" diameter, short		90	.089		.38	2.93		3.31	4.99	
	5900	3/8" diameter, short		85	.094		.70	3.11		3.81	5.60	
	6100	1/2" diameter, short		80	.100		.97	3.30		4.27	6.20	
	6300	3/4" diameter, short		70	.114		2.71	3.77		6.48	8.85	
	6600	Lead, #6 & #8, 3/4" long		260	.031		.15	1.02		1.17	1.75	
	6700	#10 - #14, 1-1/2" long		200	.040		.22	1.32		1.54	2.30	
	6800	#16 & #18, 1-1/2" long		160	.050		.30	1.65		1.95	2.90	
	6900	Plastic, #6 & #8, 3/4" long		260	.031		.09	1.02		1.11	1.68	
	7100	#10 & #12, 1" long		220	.036		.12	1.20		1.32	2	
	8000	Wedge anchors, not including layout or drilling										
	8050	Carbon steel, 1/4" diameter, 1-3/4" long	1 Carp	150	.053	Ea.	.36	1.76		2.12	3.14	
	8200	5" long		145	.055		.96	1.82		2.78	3.90	
	8250	1/2" diameter, 2-3/4" long		140	.057		.84	1.89		2.73	3.86	
	8300	7" long		130	.062		1.43	2.03		3.46	4.73	
	8350	5/8" diameter, 3-1/2" long		130	.062		1.65	2.03		3.68	4.98	
	8400	8-1/2" long		115	.070		3.52	2.30		5.82	7.45	
	8450	3/4" diameter, 4-1/4" long		115	.070		1.99	2.30		4.29	5.75	
	8500	10" long		100	.080		4.53	2.64		7.17	9.10	
	8550	1" diameter, 6" long		100	.080		6.70	2.64		9.34	11.45	
	8575	9" long		80	.100		8.70	3.30		12	14.70	
	8600	12" long		80	.100		9.35	3.30		12.65	15.45	
460	0010	**LAG SCREWS**										**460**
	0020	Steel, 1/4" diameter, 2" long	1 Carp	200	.040	Ea.	.07	1.32		1.39	2.14	
	0100	3/8" diameter, 3" long		150	.053		.20	1.76		1.96	2.96	
	0200	1/2" diameter, 3" long		130	.062		.33	2.03		2.36	3.52	
	0300	5/8" diameter, 3" long		120	.067		.65	2.20		2.85	4.15	
500	0010	**MACHINE SCREWS**										**500**
	0020	Steel, round head, #8 x 1" long				C	1.98			1.98	2.18	
	0110	#8 x 2" long					4.31			4.31	4.74	
	0200	#10 x 1" long					2.83			2.83	3.11	
	0300	#10 x 2" long					5.30			5.30	5.80	
540	0010	**MACHINERY ANCHORS**, heavy duty, incl. sleeve, floating base nut,										**540**
	0020	lower stud & coupling nut, fiber plug, connecting stud, washer & nut.										

Important: See the Reference Section for critical supporting data - Reference Nos., Crews, & City Cost Indexes

05050 | Basic Metal Materials & Methods

05090 | Metal Fastenings

METALS 5

			CREW	DAILY OUTPUT	LABOR-HOURS	UNIT	MAT.	LABOR	EQUIP.	TOTAL	TOTAL INCL O&P	
540	0030	For flush mounted embedment in poured concrete heavy equip. pads.										540
	0200	Material only, 1/2" diameter stud & bolt				Ea.	46			46	50.50	
	0300	5/8" diameter					51			51	56	
	0500	3/4" diameter					58.50			58.50	64.50	
	0600	7/8" diameter					64			64	70.50	
	0800	1" diameter					67			67	74	
	0900	1-1/4" diameter				↓	89.50			89.50	98	
580	0010	POWDER ACTUATED Tools & fasteners										580
	0020	Stud driver, .22 caliber, buy, minimum				Ea.	280			280	310	
	0100	Maximum				"	450			450	495	
	0300	Powder charges for above, low velocity				C	15.80			15.80	17.40	
	0400	Standard velocity					22.50			22.50	24.50	
	0600	Drive pins & studs, 1/4" & 3/8" diam., to 3" long, minimum					10.30			10.30	11.35	
	0700	Maximum				↓	40.50			40.50	44.50	
	0800	Pneumatic stud driver for 1/8" diameter studs				Ea.	1,950			1,950	2,150	
	0900	Drive pins for above, 1/2" to 3/4" long				M	425			425	465	
600	0010	RIVETS										600
	0100	Aluminum rivet & mandrel, 1/2" grip length x 1/8" diameter				C	4.76			4.76	5.25	

05100 | Structural Metal Framing

05120 | Structural Steel

			CREW	DAILY OUTPUT	LABOR-HOURS	UNIT	MAT.	LABOR	EQUIP.	TOTAL	TOTAL INCL O&P	
520	0010	PIPE SUPPORT FRAMING										520
	0020	Under 10#/L.F.	E-4	3,900	.008	Lb.	.85	.31	.02	1.18	1.51	
	0200	10.1 to 15#/L.F.		4,300	.007		.84	.28	.02	1.14	1.45	
	0400	15.1 to 20#/L.F.		4,800	.007		.83	.25	.02	1.10	1.38	
	0600	Over 20#/L.F.	↓	5,400	.006	↓	.82	.22	.01	1.05	1.32	

05300 | Metal Deck

05310 | Steel Deck

			CREW	DAILY OUTPUT	LABOR-HOURS	UNIT	MAT.	LABOR	EQUIP.	TOTAL	TOTAL INCL O&P	
300	0010	METAL DECKING Steel decking										300
	0200	Cellular units, galvanized, 2" deep, 20-20 gauge, over 15 squares	E-4	1,460	.022	S.F.	3.34	.83	.05	4.22	5.20	
	0250	18-20 gauge		1,420	.023		3.80	.85	.05	4.70	5.75	
	0300	18-18 gauge		1,390	.023		3.91	.87	.05	4.83	5.90	
	0320	16-18 gauge		1,360	.024		4.65	.89	.06	5.60	6.75	
	0340	16-16 gauge		1,330	.024		5.20	.91	.06	6.17	7.40	
	0400	3" deep, galvanized, 20-20 gauge		1,375	.023		3.67	.88	.05	4.60	5.65	
	0500	18-20 gauge		1,350	.024		4.45	.89	.06	5.40	6.55	
	0600	18-18 gauge		1,290	.025		4.44	.93	.06	5.43	6.60	
	0700	16-18 gauge	↓	1,230	.026	↓	5	.98	.06	6.04	7.35	

For expanded coverage of these items see *Means Building Construction Cost Data 2004*

47

	05310	**Steel Deck**	CREW	DAILY OUTPUT	LABOR-HOURS	UNIT	2004 BARE COSTS				TOTAL INCL O&P	
							MAT.	LABOR	EQUIP.	TOTAL		
300	0800	16-16 gauge	E-4	1,150	.028	S.F.	5.45	1.05	.07	6.57	7.95	**300**
	1000	4-1/2″ deep, galvanized, 20-18 gauge		1,100	.029		5.15	1.10	.07	6.32	7.70	
	1100	18-18 gauge		1,040	.031		5.10	1.16	.07	6.33	7.75	
	1200	16-18 gauge		980	.033		5.75	1.23	.08	7.06	8.65	
	1300	16-16 gauge	▼	935	.034	▼	6.25	1.29	.08	7.62	9.30	
	1900	For multi-story or congested site, add						50%				

For information about Means Estimating Seminars, see yellow pages 12 and 13 in back of book

5

METALS

Important: See the Reference Section for critical supporting data - Reference Nos., Crews, & City Cost Indexes

Division 6
Wood & Plastics

Estimating Tips

06050 Basic Wood & Plastic Materials & Methods

- Common to any wood framed structure are the accessory connector items such as screws, nails, adhesives, hangers, connector plates, straps, angles and holdowns. For typical wood framed buildings, such as residential projects, the aggregate total for these items can be significant, especially in areas where seismic loading is a concern. For floor and wall framing, the material cost is based on 10 to 25 lbs. per MBF. Holdowns, hangers and other connectors should be taken off by the piece.

06100 Rough Carpentry

- Lumber is a traded commodity and therefore sensitive to supply and demand in the marketplace. Even in "budgetary" estimating of wood framed projects, it is advisable to call local suppliers for the latest market pricing.
- Common quantity units for wood framed projects are "thousand board feet" (MBF). A board foot is a volume of wood, 1″ x 1′ x 1′,

or 144 cubic inches. Board foot quantities are generally calculated using nominal material dimensions—dressed sizes are ignored. Board foot per lineal foot of any stick of lumber can be calculated by dividing the nominal cross sectional area by 12. As an example, 2,000 lineal feet of 2 x 12 equates to 4 MBF by dividing the nominal area, 2 x 12, by 12, which equals 2, and multiplying by 2,000 to give 4,000 board feet. This simple rule applies to all nominal dimensioned lumber.

- Waste is an issue of concern at the quantity takeoff for any area of construction. Framing lumber is sold in even foot lengths, i.e., 10′, 12′, 14′, 16′, and depending on spans, wall heights and the grade of lumber, waste is inevitable. A rule of thumb for lumber waste is 5% to 10% depending on material quality and the complexity of the framing.
- Wood in various forms and shapes is used in many projects, even where the main structural framing is steel, concrete or masonry. Plywood as a back-up partition material and 2x boards used as blocking and cant strips around roof edges are two

common examples. The estimator should ensure that the costs of all wood materials are included in the final estimate.

06200 Finish Carpentry

- It is necessary to consider the grade of workmanship when estimating labor costs for erecting millwork and interior finish. In practice, there are three grades: premium, custom and economy. The Means daily output for base and case moldings is in the range of 200 to 250 L.F. per carpenter per day. This is appropriate for most average custom grade projects. For premium projects an adjustment to productivity of 25% to 50% should be made depending on the complexity of the job.

Reference Numbers

Reference numbers are shown in bold squares at the beginning of some major classifications. These numbers refer to related items in the Reference Section. The reference information may be an estimating procedure, an alternate pricing method or technical information.

Note: Not all subdivisions listed here necessarily appear in this publication.

06052	Selective Demolition	CREW	DAILY OUTPUT	LABOR-HOURS	UNIT	2004 BARE COSTS				TOTAL INCL O&P
						MAT.	LABOR	EQUIP.	TOTAL	
600	0010	**NAILS** Prices of material only, based on 50# box purchase, copper, plain				Lb.	4.66			4.66	5.15	**600**
	0400	Stainless steel, plain					5.25			5.25	5.80
	0500	Box, 3d to 20d, bright					1.19			1.19	1.31
	0520	Galvanized					1.43			1.43	1.57
	0600	Common, 3d to 60d, plain					.78			.78	.86
	0700	Galvanized					.95			.95	1.05
	0800	Aluminum					3.38			3.38	3.72
	1000	Annular or spiral thread, 4d to 60d, plain					1.60			1.60	1.76
	1200	Galvanized					1.70			1.70	1.87
	1400	Drywall nails, plain					.70			.70	.77
	1600	Galvanized					1.49			1.49	1.64
	1800	Finish nails, 4d to 10d, plain					.81			.81	.89
	2000	Galvanized					1.12			1.12	1.23
	2100	Aluminum					4.10			4.10	4.51
	2300	Flooring nails, hardened steel, 2d to 10d, plain					1.37			1.37	1.51
	2400	Galvanized					2.21			2.21	2.43
	2500	Gypsum lath nails, 1-1/8", 13 ga. flathead, blued					1.36			1.36	1.50
	2600	Masonry nails, hardened steel, 3/4" to 3" long, plain					1.41			1.41	1.55
	2700	Galvanized					1.74			1.74	1.91
	5000	Add to prices above for cement coating					.07			.07	.08
	5200	Zinc or tin plating					.12			.12	.13
	5500	Vinyl coated sinkers, 8d to 16d					.56			.56	.62
650	0010	**NAILS** mat. only, for pneumatic tools, framing, per carton of 5000, 2"				Ea.	37			37	41	**650**
	0100	2-3/8"					42.50			42.50	46.50
	0200	Per carton of 4000, 3"					37.50			37.50	41.50
	0300	3-1/4"					40			40	44
	0400	Per carton of 5000, 2-3/8", galv.					57.50			57.50	63.50
	0500	Per carton of 4000, 3", galv.					65			65	71.50
	0600	3-1/4", galv.					80.50			80.50	88.50
	0700	Roofing, per carton of 7200, 1"					35			35	38.50
	0800	1-1/4"					32.50			32.50	36
	0900	1-1/2"					37.50			37.50	41.50
	1000	1-3/4"					45.50			45.50	50
700	0010	**SHEET METAL SCREWS** Steel, standard, #8 x 3/4", plain				C	2.81			2.81	3.09	**700**
	0100	Galvanized					3.44			3.44	3.78
	0300	#10 x 1", plain					3.76			3.76	4.14
	0400	Galvanized					4.35			4.35	4.79
	0600	With washers, #14 x 1", plain					10.20			10.20	11.20
	0700	Galvanized					10.20			10.20	11.20
	0900	#14 x 2", plain					10.20			10.20	11.20
	1000	Galvanized					18.15			18.15	19.95
	1500	Self-drilling, with washers, (pinch point) #8 x 3/4", plain					6.10			6.10	6.70
	1600	Galvanized					6.10			6.10	6.70
	1800	#10 x 3/4", plain					6.10			6.10	6.70
	1900	Galvanized					6.10			6.10	6.70
	3000	Stainless steel w/aluminum or neoprene washers, #14 x 1", plain					18.35			18.35	20
	3100	#14 x 2", plain					25			25	27.50
750	0010	**WOOD SCREWS** #8, 1" long, steel				C	3.36			3.36	3.70	**750**
	0100	Brass					11.20			11.20	12.35
	0200	#8, 2" long, steel					3.76			3.76	4.14
	0300	Brass					11.70			11.70	12.90
	0400	#10, 1" long, steel					4.30			4.30	4.73
	0500	Brass					23			23	25
	0600	#10, 2" long, steel					7.65			7.65	8.45
	0700	Brass					40.50			40.50	44.50

Important: See the Reference Section for critical supporting data - Reference Nos., Crews, & City Cost Indexes

06050 | Basic Wood / Plastic Materials / Methods

	06090	Wood & Plastic Fastenings	CREW	DAILY OUTPUT	LABOR-HOURS	UNIT	2004 BARE COSTS				TOTAL INCL O&P	
							MAT.	LABOR	EQUIP.	TOTAL		
750	0800	#10, 3" long, steel				C	11.95			11.95	13.15	**750**
	1000	#12, 2" long, steel					4.89			4.89	5.40	
	1100	Brass					16.30			16.30	17.95	
	1500	#12, 3" long, steel					16.20			16.20	17.85	
	2000	#12, 4" long, steel					29			29	32	

06100 | Rough Carpentry

	06160	Sheathing	CREW	DAILY OUTPUT	LABOR-HOURS	UNIT	2004 BARE COSTS				TOTAL INCL O&P	
							MAT.	LABOR	EQUIP.	TOTAL		
800	0010	**SHEATHING** Plywood on roof, CDX										**800**
	0030	5/16" thick	2 Carp	1,600	.010	S.F.	.42	.33		.75	.97	
	0035	Pneumatic nailed		1,952	.008		.42	.27		.69	.88	
	0050	3/8" thick		1,525	.010		.44	.35		.79	1.02	
	0055	Pneumatic nailed		1,860	.009		.44	.28		.72	.92	
	0100	1/2" thick		1,400	.011		.39	.38		.77	1.02	
	0105	Pneumatic nailed		1,708	.009		.39	.31		.70	.91	
	0200	5/8" thick		1,300	.012		.45	.41		.86	1.13	
	0205	Pneumatic nailed		1,586	.010		.45	.33		.78	1.02	
	0300	3/4" thick		1,200	.013		.57	.44		1.01	1.32	
	0305	Pneumatic nailed		1,464	.011		.57	.36		.93	1.19	
	0500	Plywood on walls with exterior CDX, 3/8" thick		1,200	.013		.42	.44		.86	1.15	
	0505	Pneumatic nailed		1,488	.011		.42	.35		.77	1.01	
	0600	1/2" thick		1,125	.014		.39	.47		.86	1.16	
	0605	Pneumatic nailed		1,395	.011		.39	.38		.77	1.02	
	0700	5/8" thick		1,050	.015		.45	.50		.95	1.28	
	0705	Pneumatic nailed		1,302	.012		.45	.41		.86	1.13	
	0800	3/4" thick		975	.016		.57	.54		1.11	1.47	
	0805	Pneumatic nailed		1,209	.013		.57	.44		1.01	1.31	

For information about Means Estimating Seminars, see yellow pages 12 and 13 in back of book

WOOD & PLASTICS 6

For expanded coverage of these items see *Means Interior Cost Data 2004*

Division Notes

	CREW	DAILY OUTPUT	LABOR-HOURS	UNIT	2004 BARE COSTS				TOTAL INCL O&P
					MAT.	LABOR	EQUIP.	TOTAL	

Division 7
Thermal & Moisture Protection

Estimating Tips

07100 Dampproofing & Waterproofing

- Be sure of the job specifications before pricing this subdivision. The difference in cost between waterproofing and dampproofing can be great. Waterproofing will hold back standing water. Dampproofing prevents the transmission of water vapor. Also included in this section are vapor retarding membranes.

07200 Thermal Protection

- Insulation and fireproofing products are measured by area, thickness, volume or R value. Specifications may only give what the specific R value should be in a certain situation. The estimator may need to choose the type of insulation to meet that R value.

07300 Shingles, Roof Tiles & Roof Coverings
07400 Roofing & Siding Panels

- Many roofing and siding products are bought and sold by the square. One square is equal to an area that measures 100 square feet.

This simple change in unit of measure could create a large error if the estimator is not observant. Accessories necessary for a complete installation must be figured into any calculations for both material and labor.

07500 Membrane Roofing
07600 Flashing & Sheet Metal
07700 Roof Specialties & Accessories

- The items in these subdivisions compose a roofing system. No one component completes the installation and all must be estimated. Built-up or single ply membrane roofing systems are made up of many products and installation trades. Wood blocking at roof perimeters or penetrations, parapet coverings, reglets, roof drains, gutters, downspouts, sheet metal flashing, skylights, smoke vents or roof hatches all need to be considered along with the roofing material. Several different installation trades will need to work together on the roofing system. Inherent difficulties in the scheduling and coordination of various trades must be accounted for when estimating labor costs.

07900 Joint Sealers

- To complete the weather-tight shell the sealants and caulkings must be estimated. Where different materials meet—at expansion joints, at flashing penetrations, and at hundreds of other locations throughout a construction project—they provide another line of defense against water penetration. Often, an entire system is based on the proper location and placement of caulking or sealants. The detail drawings that are included as part of a set of architectural plans, show typical locations for these materials. When caulking or sealants are shown at typical locations, this means the estimator must include them for all the locations where this detail is applicable. Be careful to keep different types of sealants separate, and remember to consider backer rods and primers if necessary.

Reference Numbers

Reference numbers are shown in bold squares at the beginning of some major classifications. These numbers refer to related items in the Reference Section. The reference information may be an estimating procedure, an alternate pricing method or technical information.

Note: Not all subdivisions listed here necessarily appear in this publication.

07840 | Firestopping

		CREW	DAILY OUTPUT	LABOR-HOURS	UNIT	MAT.	LABOR	EQUIP.	TOTAL	TOTAL INCL O&P
100	0010 **FIRESTOPPING** R07800-030									100
	0100 Metallic piping, non insulated									
	0110 Through walls, 2" diameter	1 Carp	16	.500	Ea.	9.75	16.50		26.25	36.50
	0120 4" diameter		14	.571		14.90	18.85		33.75	46
	0130 6" diameter		12	.667		20	22		42	56.50
	0140 12" diameter		10	.800		35.50	26.50		62	80
	0150 Through floors, 2" diameter		32	.250		5.95	8.25		14.20	19.35
	0160 4" diameter		28	.286		8.50	9.45		17.95	24
	0170 6" diameter		24	.333		11.20	11		22.20	29.50
	0180 12" diameter		20	.400		18.90	13.20		32.10	41.50
	0190 Metallic piping, insulated									
	0200 Through walls, 2" diameter	1 Carp	16	.500	Ea.	13.85	16.50		30.35	41
	0210 4" diameter		14	.571		19	18.85		37.85	50.50
	0220 6" diameter		12	.667		24	22		46	61
	0230 12" diameter		10	.800		39.50	26.50		66	84.50
	0240 Through floors, 2" diameter		32	.250		10	8.25		18.25	24
	0250 4" diameter		28	.286		12.60	9.45		22.05	28.50
	0260 6" diameter		24	.333		15.30	11		26.30	34
	0270 12" diameter		20	.400		18.90	13.20		32.10	41.50
	0280 Non metallic piping, non insulated									
	0290 Through walls, 2" diameter	1 Carp	12	.667	Ea.	40.50	22		62.50	79
	0300 4" diameter		10	.800		50.50	26.50		77	96.50
	0310 6" diameter		8	1		70.50	33		103.50	129
	0330 Through floors, 2" diameter		16	.500		31.50	16.50		48	60
	0340 4" diameter		6	1.333		39	44		83	112
	0350 6" diameter		6	1.333		47	44		91	120
	0370 Ductwork, insulated & non insulated, round									
	0380 Through walls, 6" diameter	1 Carp	12	.667	Ea.	20.50	22		42.50	57
	0390 12" diameter		10	.800		41	26.50		67.50	86
	0400 18" diameter		8	1		66.50	33		99.50	125
	0410 Through floors, 6" diameter		16	.500		11.20	16.50		27.70	38
	0420 12" diameter		14	.571		20.50	18.85		39.35	52
	0430 18" diameter		12	.667		35.50	22		57.50	74
	0440 Ductwork, insulated & non insulated, rectangular									
	0450 With stiffener/closure angle, through walls, 6" x 12"	1 Carp	8	1	Ea.	17	33		50	70
	0460 12" x 24"		6	1.333		22.50	44		66.50	93.50
	0470 24" x 48"		4	2		64.50	66		130.50	174
	0480 With stiffener/closure angle, through floors, 6" x 12"		10	.800		9.20	26.50		35.70	51
	0490 12" x 24"		8	1		16.55	33		49.55	69.50
	0500 24" x 48"		6	1.333		32.50	44		76.50	104
	0510 Multi trade openings									
	0520 Through walls, 6" x 12"	1 Carp	2	4	Ea.	35.50	132		167.50	246
	0530 12" x 24"	"	1	8		144	264		408	570
	0540 24" x 48"	2 Carp	1	16		575	530		1,105	1,450
	0550 48" x 96"	"	.75	21.333		2,300	705		3,005	3,650
	0560 Through floors, 6" x 12"	1 Carp	2	4		35.50	132		167.50	246
	0570 12" x 24"	"	1	8		144	264		408	570
	0580 24" x 48"	2 Carp	.75	21.333		575	705		1,280	1,725
	0590 48" x 96"	"	.50	32		2,300	1,050		3,350	4,200
	0600 Structural penetrations, through walls									
	0610 Steel beams, W8 x 10	1 Carp	8	1	Ea.	22.50	33		55.50	76
	0620 W12 x 14		6	1.333		35.50	44		79.50	108
	0630 W21 x 44		5	1.600		71.50	53		124.50	161
	0640 W36 x 135		3	2.667		173	88		261	330
	0650 Bar joists, 18" deep		6	1.333		32.50	44		76.50	105
	0660 24" deep		6	1.333		41	44		85	114

Important: See the Reference Section for critical supporting data - Reference Nos., Crews, & City Cost Indexes

07840	Firestopping	CREW	DAILY OUTPUT	LABOR-HOURS	UNIT	2004 BARE COSTS				TOTAL INCL O&P
						MAT.	LABOR	EQUIP.	TOTAL	
100 0670	36" deep	1 Carp	5	1.600	Ea.	61	53		114	150	100
0680	48" deep	↓	4	2	↓	71.50	66		137.50	182
0690	Construction joints, floor slab at exterior wall									
0700	Precast, brick, block or drywall exterior									
0710	2" wide joint	1 Carp	125	.064	L.F.	5.10	2.11		7.21	8.90
0720	4" wide joint	"	75	.107	"	10.20	3.52		13.72	16.70
0730	Metal panel, glass or curtain wall exterior									
0740	2" wide joint	1 Carp	40	.200	L.F.	12.05	6.60		18.65	23.50
0750	4" wide joint	"	25	.320	"	16.50	10.55		27.05	34.50
0760	Floor slab to drywall partition									
0770	Flat joint	1 Carp	100	.080	L.F.	5	2.64		7.64	9.60
0780	Fluted joint		50	.160		10.20	5.30		15.50	19.40
0790	Etched fluted joint	↓	75	.107	↓	6.65	3.52		10.17	12.80
0800	Floor slab to concrete/masonry partition									
0810	Flat joint	1 Carp	75	.107	L.F.	11.20	3.52		14.72	17.85
0820	Fluted joint	"	50	.160	"	13.25	5.30		18.55	23
0830	Concrete/CMU wall joints									
0840	1" wide	1 Carp	100	.080	L.F.	6.10	2.64		8.74	10.85
0850	2" wide		75	.107		11.20	3.52		14.72	17.85
0860	4" wide	↓	50	.160	↓	21.50	5.30		26.80	31.50
0870	Concrete/CMU floor joints									
0880	1" wide	1 Carp	200	.040	L.F.	3.06	1.32		4.38	5.45
0890	2" wide	↓	150	.053		5.60	1.76		7.36	8.90
0900	4" wide	↓	100	.080	↓	10.70	2.64		13.34	15.90

(R07800 -030)

07900 | Joint Sealers

07920	Joint Sealants	CREW	DAILY OUTPUT	LABOR-HOURS	UNIT	2004 BARE COSTS				TOTAL INCL O&P
						MAT.	LABOR	EQUIP.	TOTAL	
800 0010	CAULKING AND SEALANTS										800
3200	Polyurethane, 1 or 2 component				Gal.	47.50			47.50	52.50
3300	Cartridges				"	46.50			46.50	51.50
3500	Bulk, in place, 1/4" x 1/4"	1 Bric	150	.053	L.F.	.15	1.83		1.98	2.95
3600	1/2" x 1/4"		145	.055		.31	1.89		2.20	3.22
3800	3/4" x 3/8", 68 L.F./gal.		130	.062		.70	2.11		2.81	3.98
3900	1" x 1/2"	↓	110	.073	↓	1.24	2.49		3.73	5.15

For information about Means Estimating Seminars, see yellow pages 12 and 13 in back of book

THERMAL & MOISTURE PROTECTION **7**

Division Notes

	CREW	DAILY OUTPUT	LABOR-HOURS	UNIT	2004 BARE COSTS				TOTAL INCL O&P
					MAT.	LABOR	EQUIP.	TOTAL	
	CREW	DAILY OUTPUT	LABOR-HOURS	UNIT	2004 BARE COSTS				TOTAL INCL O&P
					MAT.	LABOR	EQUIP.	TOTAL	

Division 10 Specialties

Estimating Tips

General
- The items in this division are usually priced per square foot or each.
- Many items in Division 10 require some type of support system or special anchors that are not usually furnished with the item. The required anchors must be added to the estimate in the appropriate division.
- Some items in Division 10, such as lockers, may require assembly before installation. Verify the amount of assembly required. Assembly can often exceed installation time.

10150 Compartments & Cubicles
- Support angles and blocking are not included in the installation of toilet compartments, shower/dressing compartments or cubicles. Appropriate line items from Divisions 5 or 6 may need to be added to support the installations.
- Toilet partitions are priced by the stall. A stall consists of a side wall, pilaster and door with hardware. Toilet tissue holders and grab bars are extra.

10600 Partitions
- The required acoustical rating of a folding partition can have a significant impact on costs. Verify the sound transmission coefficient rating of the panel priced to the specification requirements.

10800 Toilet/Bath/Laundry Accessories
- Grab bar installation does not include supplemental blocking or backing to support the required load. When grab bars are installed at an existing facility provisions must be made to attach the grab bars to solid structure.

Reference Numbers
Reference numbers are shown in bold squares at the beginning of some major classifications. These numbers refer to related items in the Reference Section. The reference information may be an estimating procedure, an alternate pricing method or technical information.

Note: Not all subdivisions listed here necessarily appear in this publication.

				DAILY	LABOR-			2004 BARE COSTS				TOTAL
	10275	**Access Flooring**	CREW	OUTPUT	HOURS	UNIT	MAT.	LABOR	EQUIP.	TOTAL		INCL O&P
150	0010	**PEDESTAL ACCESS FLOORS** Computer room application, metal											**150**
	0020	Particle board or steel panels, no covering, under 6,000 S.F.	2 Carp	400	.040	S.F.	10.60	1.32		11.92		13.75
	0300	Metal covered, over 6,000 S.F.		450	.036		7.55	1.17		8.72		10.15
	0400	Aluminum, 24" panels		500	.032		25.50	1.06		26.56		30
	0600	For carpet covering, add					5.45			5.45		5.95
	0700	For vinyl floor covering, add					5.50			5.50		6.05
	0900	For high pressure laminate covering, add					3.67			3.67		4.04
	0910	For snap on stringer system, add	2 Carp	1,000	.016		1.25	.53		1.78		2.20
	0950	Office applications, to 8" high, steel panels,										
	0960	no covering, over 6,000 S.F.	2 Carp	500	.032	S.F.	8	1.06		9.06		10.45
	1000	Machine cutouts after initial installation	1 Carp	10	.800	Ea.	41	26.50		67.50		86
	1050	Pedestals, 6" to 12"	2 Carp	85	.188		7.70	6.20		13.90		18.20
	1100	Air conditioning grilles, 4" x 12"	1 Carp	17	.471		48.50	15.55		64.05		77
	1150	4" x 18"	"	14	.571		66.50	18.85		85.35		103
	1200	Approach ramps, minimum	2 Carp	85	.188	S.F.	24.50	6.20		30.70		36.50
	1300	Maximum	"	60	.267	"	29.50	8.80		38.30		45.50
	1500	Handrail, 2 rail, aluminum	1 Carp	15	.533	L.F.	88	17.60		105.60		125

				DAILY	LABOR-			2004 BARE COSTS				TOTAL
	10820	**Bath Accessories**	CREW	OUTPUT	HOURS	UNIT	MAT.	LABOR	EQUIP.	TOTAL		INCL O&P
400	0010	**MEDICINE CABINETS** With mirror, st. st. frame, 16" x 22", unlighted	1 Carp	14	.571	Ea.	69.50	18.85		88.35		106	**400**
	0100	Wood frame		14	.571		96.50	18.85		115.35		136
	0300	Sliding mirror doors, 20" x 16" x 4-3/4", unlighted		7	1.143		86.50	37.50		124		154
	0400	24" x 19" x 8-1/2", lighted		5	1.600		136	53		189		232
	0600	Triple door, 30" x 32", unlighted, plywood body		7	1.143		214	37.50		251.50		295
	0700	Steel body		7	1.143		282	37.50		319.50		370
	0900	Oak door, wood body, beveled mirror, single door		7	1.143		125	37.50		162.50		197
	1000	Double door		6	1.333		320	44		364		420
	1200	Hotel cabinets, stainless, with lower shelf, unlighted		10	.800		173	26.50		199.50		231
	1300	Lighted		5	1.600		256	53		309		365

For information about Means Estimating Seminars, see yellow pages 12 and 13 in back of book

10 SPECIALTIES

Important: See the Reference Section for critical supporting data - Reference Nos., Crews, & City Cost Indexes

Division 11
Equipment

Estimating Tips

General

- The items in this division are usually priced per square foot or each. Many of these items are purchased by the owner for installation by the contractor. Check the specifications for responsibilities, and include time for receiving, storage, installation and mechanical and electrical hook-ups in the appropriate divisions.

- Many items in Division 11 require some type of support system that is not usually furnished with the item. Examples of these systems include blocking for the attachment of casework and support angles for ceiling hung projection screens. The required blocking or supports must be added to the estimate in the appropriate division.

- Some items in Division 11 may require assembly or electrical hook-ups. Verify the amount of assembly required or the need for a hard electrical connection and add the appropriate costs.

Reference Numbers

Reference numbers are shown in bold squares at the beginning of some major classifications. These numbers refer to related items in the Reference Section. The reference information may be an estimating procedure, an alternate pricing method or technical information.

Note: Not all subdivisions listed here necessarily appear in this publication.

11060 | Theater & Stage Equipment

11063	Stage Equipment	CREW	DAILY OUTPUT	LABOR-HOURS	UNIT	2004 BARE COSTS				TOTAL INCL O&P
						MAT.	LABOR	EQUIP.	TOTAL	
600	0010 **STAGE EQUIPMENT**									600
	0050 Control boards with dimmers and breakers, minimum	1 Elec	1	8	Ea.	9,050	315		9,365	10,400
	0100 Average		.50	16		26,700	630		27,330	30,300
	0150 Maximum	↓	.20	40	↓	76,000	1,575		77,575	86,500
	2000 Lights, border, quartz, reflector, vented,									
	2100 colored or white	1 Elec	20	.400	L.F.	144	15.75		159.75	182
	2500 Spotlight, follow spot, with transformer, 2,100 watt	"	4	2	Ea.	1,375	79		1,454	1,625
	2600 For no transformer, deduct					495			495	545
	3000 Stationary spot, fresnel quartz, 6" lens	1 Elec	4	2		120	79		199	249
	3100 8" lens		4	2		188	79		267	325
	3500 Ellipsoidal quartz, 1,000W, 6" lens		4	2		249	79		328	390
	3600 12" lens		4	2		435	79		514	595
	4000 Strobe light, 1 to 15 flashes per second, quartz		3	2.667		560	105		665	775
	4500 Color wheel, portable, five hole, motorized	↓	4	2	↓	120	79		199	249

11110 | Commercial Laundry & Dry Cleaning Equipment

11119	Laundry Cleaning	CREW	DAILY OUTPUT	LABOR-HOURS	UNIT	2004 BARE COSTS				TOTAL INCL O&P
						MAT.	LABOR	EQUIP.	TOTAL	
450	0010 **LAUNDRY EQUIPMENT** Not incl. rough-in									450
	2000 Dry cleaners, electric, 20 lb. capacity	L-1	.20	80	Ea.	29,300	3,150		32,450	37,000
	2050 25 lb. capacity		.17	94.118		38,200	3,725		41,925	47,600
	2100 30 lb. capacity		.15	106		40,200	4,225		44,425	50,500
	2150 60 lb. capacity	↓	.09	177		63,000	7,025		70,025	80,000
	3500 Folders, blankets & sheets, minimum	1 Elec	.17	47.059		26,800	1,850		28,650	32,300
	3700 King size with automatic stacker		.10	80		46,400	3,150		49,550	55,500
	3800 For conveyor delivery, add		.45	17.778		5,525	700		6,225	7,125
	4500 Ironers, institutional, 110", single roll	↓	.20	40		25,400	1,575		26,975	30,400
	5000 Washers, residential, 4 cycle, average	1 Plum	3	2.667		630	106		736	850
	5300 Commercial, coin operated, average	"	3	2.667		1,000	106		1,106	1,250
	6000 Combination washer/extractor, 20 lb. capacity	L-6	1.50	8		3,625	315		3,940	4,450
	6100 30 lb. capacity		.80	15		7,000	595		7,595	8,575
	6200 50 lb. capacity		.68	17.647		8,700	700		9,400	10,600
	6300 75 lb. capacity		.30	40		18,000	1,575		19,575	22,100
	6350 125 lb. capacity	↓	.16	75	↓	21,800	2,975		24,775	28,400

11130 | Audio-Visual Equipment

11136	Projection Screens	CREW	DAILY OUTPUT	LABOR-HOURS	UNIT	2004 BARE COSTS				TOTAL INCL O&P
						MAT.	LABOR	EQUIP.	TOTAL	
600	0010 **MOVIE EQUIPMENT**									600
	0020 Changeover, minimum				Ea.	380			380	415
	0100 Maximum					735			735	810
	0800 Lamphouses, incl. rectifiers, xenon, 1,000 watt	1 Elec	2	4		5,175	158		5,333	5,900
	0900 1,600 watt		2	4		5,500	158		5,658	6,275
	1000 2,000 watt	↓	1.50	5.333	↓	5,925	210		6,135	6,825

11 EQUIPMENT

Important: See the Reference Section for critical supporting data - Reference Nos., Crews, & City Cost Indexes

11130 | Audio-Visual Equipment

11136 | Projection Screens

			CREW	DAILY OUTPUT	LABOR-HOURS	UNIT	2004 BARE COSTS				TOTAL INCL O&P	
							MAT.	LABOR	EQUIP.	TOTAL		
600	1100	4,000 watt	1 Elec	1.50	5.333	Ea.	7,325	210		7,535	8,375	600
	3700	Sound systems, incl. amplifier, mono, minimum		.90	8.889		2,700	350		3,050	3,475	
	3800	Dolby/Super Sound, maximum		.40	20		14,700	790		15,490	17,400	
	4100	Dual system, 2 channel, front surround, minimum		.70	11.429		3,775	450		4,225	4,825	
	4200	Dolby/Super Sound, 4 channel, maximum		.40	20		13,500	790		14,290	16,000	
	5300	Speakers, recessed behind screen, minimum		2	4		860	158		1,018	1,175	
	5400	Maximum		1	8		2,525	315		2,840	3,250	
	7000	For automation, varying sophistication, minimum	↓	1	8	System	1,925	315		2,240	2,600	
	7100	Maximum	2 Elec	.30	53.333	"	4,525	2,100		6,625	8,100	

11150 | Parking Control Equipment

11156 | Parking Equipment

			CREW	DAILY OUTPUT	LABOR-HOURS	UNIT	2004 BARE COSTS				TOTAL INCL O&P	
							MAT.	LABOR	EQUIP.	TOTAL		
600	0010	**PARKING EQUIPMENT**										600
	5000	Barrier gate with programmable controller	2 Elec	3	5.333	Ea.	3,050	210		3,260	3,675	
	5020	Industrial	"	3	5.333		4,150	210		4,360	4,900	
	5100	Card reader	1 Elec	2	4		1,675	158		1,833	2,050	
	5120	Proximity with customer display	2 Elec	1	16		5,050	630		5,680	6,500	
	5200	Cashier booth, average	B-22	1	30		8,800	910	239	9,949	11,300	
	5300	Collector station, pay on foot	2 Elec	.20	80		100,500	3,150		103,650	115,000	
	5320	Credit card only		.50	32		18,500	1,250		19,750	22,300	
	5500	Exit verifier	↓	1	16		16,000	630		16,630	18,500	
	5600	Fee computer	1 Elec	1.50	5.333		12,200	210		12,410	13,700	
	5700	Full sign, 4" letters	"	2	4		1,100	158		1,258	1,450	
	5800	Inductive loop	2 Elec	4	4		155	158		313	405	
	5900	Ticket spitter with time/date stamp, standard		2	8		5,800	315		6,115	6,850	
	5920	Mag stripe encoding	↓	2	8		17,000	315		17,315	19,200	
	5950	Vehicle detector, microprocessor based	1 Elec	3	2.667		355	105		460	545	
	6000	Parking control software, minimum		.50	16		20,100	630		20,730	23,000	
	6020	Maximum	↓	.20	40	↓	83,500	1,575		85,075	94,500	

11170 | Solid Waste Handling Equipment

11179 | Waste Handling Equipment

			CREW	DAILY OUTPUT	LABOR-HOURS	UNIT	2004 BARE COSTS				TOTAL INCL O&P	
							MAT.	LABOR	EQUIP.	TOTAL		
150	0010	**WASTE HANDLING**	L-4	1	24	Ea.						150
	0020	Compactors, 115 volt, 250#/hr., chute fed		1	24		8,600	735		9,335	10,600	
	1000	Heavy duty industrial compactor, 0.5 C.Y. capacity		1	24		5,525	735		6,260	7,225	
	1050	1.0 C.Y. capacity		1	24		8,500	735		9,235	10,500	
	1100	3 C.Y. capacity		.50	48		11,600	1,475		13,075	15,000	
	1150	5.0 C.Y. capacity		.50	48		14,300	1,475		15,775	18,100	
	1200	Combination shredder/compactor (5,000 lbs./hr.)	↓	.50	48		28,100	1,475		29,575	33,200	
	1400	For handling hazardous waste materials, 55 gallon drum packer, std.				↓	13,000			13,000	14,300	

11170 | Solid Waste Handling Equipment

11179 | Waste Handling Equipment

			CREW	DAILY OUTPUT	LABOR-HOURS	UNIT	2004 BARE COSTS MAT.	LABOR	EQUIP.	TOTAL	TOTAL INCL O&P	
150	1410	55 gallon drum packer w/HEPA filter				Ea.	16,300			16,300	17,900	150
	1420	55 gallon drum packer w/charcoal & HEPA filter					21,700			21,700	23,900	
	1430	All of the above made explosion proof, add					9,925			9,925	10,900	
	1500	Crematory, not including building, 1 place	Q-3	.20	160		46,300	6,050		52,350	60,000	
	1750	2 place	"	.10	320		66,000	12,100		78,100	90,500	
	5800	Shredder, industrial, minimum					16,800			16,800	18,500	
	5850	Maximum					90,500			90,500	99,500	
	5900	Baler, industrial, minimum					6,750			6,750	7,425	
	5950	Maximum				▼	394,000			394,000	433,500	

11400 | Food Service Equipment

11405 | Food Storage Equipment

			CREW	DAILY OUTPUT	LABOR-HOURS	UNIT	2004 BARE COSTS MAT.	LABOR	EQUIP.	TOTAL	TOTAL INCL O&P	
110	0010	**FOOD STORAGE EQUIPMENT**										110
	2350	Cooler, reach-in, beverage, 6' long	Q-1	6	2.667	Ea.	3,200	95		3,295	3,675	
	8310	With glass doors, 68 C.F.	"	4	4	"	6,925	143		7,068	7,850	

11410 | Food Preparation Equipment

			CREW	DAILY OUTPUT	LABOR-HOURS	UNIT	2004 BARE COSTS MAT.	LABOR	EQUIP.	TOTAL	TOTAL INCL O&P	
110	0010	**FOOD PREPARATION EQUIPMENT**										110
	1850	Coffee urn, twin 6 gallon urns	1 Plum	2	4	Ea.	6,625	158		6,783	7,550	
	3800	Food mixers, 20 quarts	L-7	7	4		2,750	128		2,878	3,200	
	3900	60 quarts	"	5	5.600	▼	10,000	179		10,179	11,300	

11420 | Food Cooking Equipment

			CREW	DAILY OUTPUT	LABOR-HOURS	UNIT	2004 BARE COSTS MAT.	LABOR	EQUIP.	TOTAL	TOTAL INCL O&P	
110	0010	**COOKING EQUIPMENT**										110
	0020	Bake oven, gas, one section	Q-1	8	2	Ea.	3,925	71.50		3,996.50	4,400	
	1300	Broiler, without oven, standard	"	8	2		3,775	71.50		3,846.50	4,250	
	1550	Infra-red	L-7	4	7		5,950	223		6,173	6,900	
	6350	20 Gal kettle w/steam jacket, tilting w/positive lock, S/S		7	4		6,075	128		6,203	6,875	
	6600	60 gallons		6	4.667		6,650	149		6,799	7,525	
	8850	Steamer, electric 27 KW		7	4		8,425	128		8,553	9,475	
	9100	Electric, 10 KW or gas 100,000 BTU	▼	5	5.600		4,000	179		4,179	4,675	
	9150	Toaster, conveyor type, 16-22 slices per minute				▼	945			945	1,050	
	9200	For deluxe models of above equipment, add					75%					
	9400	Rule of thumb: Equipment cost based										
	9410	on kitchen work area										
	9420	Office buildings, minimum	L-7	77	.364	S.F.	56.50	11.60		68.10	80	
	9450	Maximum		58	.483		95	15.40		110.40	129	
	9550	Public eating facilities, minimum		77	.364		74	11.60		85.60	99	
	9600	Maximum		46	.609		120	19.40		139.40	162	
	9750	Hospitals, minimum		58	.483		76	15.40		91.40	108	
	9800	Maximum	▼	39	.718	▼	127	23		150	176	
	3300	Food warmer, counter, 1.2 KW				Ea.	565			565	620	
	3550	1.6 KW				"	1,125			1,125	1,225	

11440 | Cleaning and Disposal Equipment

			CREW	DAILY OUTPUT	LABOR-HOURS	UNIT	2004 BARE COSTS MAT.	LABOR	EQUIP.	TOTAL	TOTAL INCL O&P	
110	0010	**CLEANING & DISPOSAL EQUIPMENT**										110
	2700	Dishwasher, commercial, rack type										

Important: See the Reference Section for critical supporting data - Reference Nos., Crews, & City Cost Indexes

11400 | Food Service Equipment

11440	Cleaning and Disposal Equipment	CREW	DAILY OUTPUT	LABOR-HOURS	UNIT	2004 BARE COSTS				TOTAL INCL O&P	
						MAT.	LABOR	EQUIP.	TOTAL		
110	2720	10 to 12 racks per hour	Q-1	3.20	5	Ea.	2,900	178		3,078	3,475	110
	2750	Semi-automatic 38 to 50 racks per hour	"	1.30	12.308	"	6,000	440		6,440	7,250	

11450 | Residential Equipment

11454	Residential Appliances	CREW	DAILY OUTPUT	LABOR-HOURS	UNIT	2004 BARE COSTS				TOTAL INCL O&P	
						MAT.	LABOR	EQUIP.	TOTAL		
500	0010	**RESIDENTIAL APPLIANCES**										500
	0020	Cooking range, 30" free standing, 1 oven, minimum	2 Clab	10	1.600	Ea.	241	41.50		282.50	330	
	0050	Maximum		4	4		1,450	104		1,554	1,725	
	0700	Free-standing, 1 oven, 21" wide range, minimum		10	1.600		247	41.50		288.50	335	
	0750	21" wide, maximum		4	4		273	104		377	460	
	0900	Counter top cook tops, 4 burner, standard, minimum	1 Elec	6	1.333		176	52.50		228.50	271	
	0950	Maximum		3	2.667		440	105		545	640	
	1050	As above, but with grille and griddle attachment, minimum		6	1.333		435	52.50		487.50	555	
	1100	Maximum		3	2.667		640	105		745	860	
	1200	Induction cooktop, 30" wide		3	2.667		520	105		625	725	
	1250	Microwave oven, minimum		4	2		75.50	79		154.50	200	
	1300	Maximum		2	4		390	158		548	665	
	2450	Dehumidifier, portable, automatic, 15 pint					149			149	164	
	2550	40 pint					167			167	183	
	2750	Dishwasher, built-in, 2 cycles, minimum	L-1	4	4		245	158		403	505	
	2800	Maximum		2	8		289	315		604	790	
	2950	4 or more cycles, minimum		4	4		262	158		420	525	
	2960	Average		4	4		350	158		508	620	
	3000	Maximum		2	8		535	315		850	1,050	
	3300	Garbage disposer, sink type, minimum		10	1.600		42.50	63		105.50	141	
	3350	Maximum		10	1.600		145	63		208	255	
	3550	Heater, electric, built-in, 1250 watt, ceiling type, minimum	1 Elec	4	2		70	79		149	194	
	3600	Maximum		3	2.667		114	105		219	282	
	3700	Wall type, minimum		4	2		99.50	79		178.50	226	
	3750	Maximum		3	2.667		132	105		237	300	
	3900	1500 watt wall type, with blower		4	2		123	79		202	252	
	3950	3000 watt		3	2.667		251	105		356	430	
	4150	Hood for range, 2 speed, vented, 30" wide, minimum	L-3	5	3.200		37.50	115		152.50	218	
	4200	Maximum		3	5.333		595	192		787	950	
	4300	42" wide, minimum		5	3.200		225	115		340	425	
	4330	Custom		5	3.200		625	115		740	860	
	4350	Maximum		3	5.333		760	192		952	1,125	
	4500	For ventless hood, 2 speed, add					15			15	16.50	
	4650	For vented 1 speed, deduct from maximum					39			39	43	
	4850	Humidifier, portable, 8 gallons per day					149			149	164	
	5000	15 gallons per day					179			179	197	
	6900	Water heater, electric, glass lined, 30 gallon, minimum	L-1	5	3.200		252	126		378	465	
	6950	Maximum		3	5.333		350	211		561	700	
	7100	80 gallon, minimum		2	8		485	315		800	1,000	
	7150	Maximum		1	16		670	630		1,300	1,675	

11460	Unit Kitchens	CREW	DAILY OUTPUT	LABOR-HOURS	UNIT	MAT.	LABOR	EQUIP.	TOTAL	TOTAL INCL O&P	
100	0010	**UNIT KITCHENS**										100
	1500	Combination range, refrigerator and sink, 30" wide, minimum	L-1	2	8	Ea.	730	315		1,045	1,275	

11450 | Residential Equipment

11460 | Unit Kitchens

			CREW	DAILY OUTPUT	LABOR-HOURS	UNIT	2004 BARE COSTS				TOTAL INCL O&P	
							MAT.	LABOR	EQUIP.	TOTAL		
100	1550	Maximum	L-1	1	16	Ea.	1,450	630		2,080	2,550	100
	1570	60" wide, average		1.40	11.429		2,400	450		2,850	3,325	
	1590	72" wide, average		1.20	13.333		2,725	525		3,250	3,775	
	1600	Office model, 48" wide		2	8		2,050	315		2,365	2,750	
	1620	Refrigerator and sink only	▼	2.40	6.667	▼	2,050	263		2,313	2,675	
	1640	Combination range, refrigerator, sink, microwave										
	1660	oven and ice maker	L-1	.80	20	Ea.	4,025	790		4,815	5,600	

11486 | Gymnasium Equipment

			CREW	DAILY OUTPUT	LABOR-HOURS	UNIT	2004 BARE COSTS				TOTAL INCL O&P	
							MAT.	LABOR	EQUIP.	TOTAL		
700	0010	**SCHOOL EQUIPMENT**										700
	0200	For exterior equipment, see Div. 02880										
	7000	Scoreboards, baseball, minimum	R-3	1.30	15.385	Ea.	2,625	595	122	3,342	3,925	
	7200	Maximum		.05	400		14,200	15,500	3,175	32,875	42,300	
	7300	Football, minimum		.86	23.256		3,600	900	185	4,685	5,525	
	7400	Maximum		.20	100		12,300	3,875	795	16,970	20,200	
	7500	Basketball (one side), minimum		2.07	9.662		2,000	375	77	2,452	2,875	
	7600	Maximum		.30	66.667		12,400	2,575	530	15,505	18,100	
	7700	Hockey-basketball (four sides), minimum		.25	80		5,650	3,100	635	9,385	11,600	
	7800	Maximum	▼	.15	133	▼	22,300	5,150	1,050	28,500	33,400	

11600 | Laboratory Equipment

11620 | Laboratory Equipment

			CREW	DAILY OUTPUT	LABOR-HOURS	UNIT	2004 BARE COSTS				TOTAL INCL O&P	
							MAT.	LABOR	EQUIP.	TOTAL		
110	0010	**LABORATORY EQUIPMENT**										110
	0700	Glassware washer, undercounter, minimum	L-1	1.80	8.889	Ea.	5,975	350		6,325	7,100	
	0710	Maximum	"	1	16		7,675	630		8,305	9,400	
	1400	Safety equipment, eye wash, hand held					294			294	325	
	1450	Deluge shower				▼	565			565	625	
	8000	Alternate pricing method: as percent of lab furniture										
	8050	Installation, not incl. plumbing & duct work				% Furn.					22%	
	8100	Plumbing, final connections, simple system									10%	
	8110	Moderately complex system									15%	
	8120	Complex system									20%	
	8150	Electrical, simple system									10%	
	8160	Moderately complex system									20%	
	8170	Complex system				▼					35%	

11700 | Medical Equipment

11710 | Medical Sterilizing Equipment

			CREW	DAILY OUTPUT	LABOR-HOURS	UNIT	2004 BARE COSTS				TOTAL INCL O&P	
							MAT.	LABOR	EQUIP.	TOTAL		
110	0010	**MEDICAL STERILIZING EQUIPMENT**										110
	6200	Steam generators, electric 10 KW to 180 KW, freestanding										

Important: See the Reference Section for critical supporting data - Reference Nos., Crews, & City Cost Indexes

11700 | Medical Equipment

		11710	Medical Sterilizing Equipment	CREW	DAILY OUTPUT	LABOR-HOURS	UNIT	2004 BARE COSTS				TOTAL INCL O&P	
								MAT.	LABOR	EQUIP.	TOTAL		
110	6250		Minimum	1 Elec	3	2.667	Ea.	6,575	105		6,680	7,400	110
	6300		Maximum	"	.70	11.429	"	23,700	450		24,150	26,800	

		11720	Examination & Treatment Equip.										
110	0010		EXAMINATION & TREATMENT EQUIPMENT										110
	6700		Surgical lights, doctor's office, single arm	2 Elec	2	8	Ea.	1,125	315		1,440	1,700	
	6750		Dual arm	"	1	16	"	4,550	630		5,180	5,950	

		11760	Operating Room Equipment										
110	0010		OPERATING ROOM EQUIPMENT										110
	6800		Surgical lights, major operating room, dual head, minimum	2 Elec	1	16	Ea.	14,400	630		15,030	16,700	
	6850		Maximum	"	1	16	"	23,900	630		24,530	27,200	

		11781	Mortuary Equipment										
110	0010		MORTUARY EQUIPMENT										110
	0015		Autopsy table, standard	1 Plum	1	8	Ea.	7,275	315		7,590	8,500	

For information about Means Estimating Seminars, see yellow pages 12 and 13 in back of book

EQUIPMENT 11

Division Notes

	CREW	DAILY OUTPUT	LABOR-HOURS	UNIT	2004 BARE COSTS				TOTAL INCL O&P
					MAT.	LABOR	EQUIP.	TOTAL	

Division 13
Special Construction

Estimating Tips

General

- The items and systems in this division are usually estimated, purchased, supplied and installed as a unit by one or more subcontractors. The estimator must ensure that all parties are operating from the same set of specifications and assumptions and that all necessary items are estimated and will be provided. Many times the complex items and systems are covered but the more common ones such as excavation or a crane are overlooked for the very reason that everyone assumes nobody could miss them. The estimator should be the central focus and be able to ensure that all systems are complete.

- Another area where problems can develop in this division is at the interface between systems. The estimator must ensure, for instance, that anchor bolts, nuts and washers are estimated and included for the air-supported structures and pre-engineered buildings to be bolted to their foundations.

Utility supply is a common area where essential items or pieces of equipment can be missed or overlooked due to the fact that each subcontractor may feel it is the others' responsibility. The estimator should also be aware of certain items which may be supplied as part of a package but installed by others, and ensure that the installing contractor's estimate includes the cost of installation. Conversely, the estimator must also ensure that items are not costed by two different subcontractors, resulting in an inflated overall estimate.

13120 Pre-Engineered Structures

- The foundations and floor slab, as well as rough mechanical and electrical, should be estimated, as this work is required for the assembly and erection of the structure. Generally, as noted in the book, the pre-engineered building comes as a shell and additional features must be included by the estimator. Here again, the estimator must have a clear understanding of the scope of each portion of the work and all the necessary interfaces.

13200 Storage Tanks

- The prices in this subdivision for above and below ground storage tanks do not include foundations or hold-down slabs. The estimator should refer to Divisions 2 and 3 for foundation system pricing. In addition to the foundations, required tank accessories such as tank gauges, leak detection devices, and additional manholes and piping must be added to the tank prices.

Reference Numbers

Reference numbers are shown in bold squares at the beginning of some major classifications. These numbers refer to related items in the Reference Section. The reference information may be an estimating procedure, an alternate pricing method or technical information.

Note: Not all subdivisions listed here necessarily appear in this publication.

13035	Special Purpose Rooms	CREW	DAILY OUTPUT	LABOR-HOURS	UNIT	2004 BARE COSTS				TOTAL INCL O&P
						MAT.	LABOR	EQUIP.	TOTAL	
200	**0010** **CLEAN ROOMS**									**200**
1100	Clean room, soft wall, 12' x 12', Class 100	1 Carp	.18	44.444	Ea.	12,400	1,475		13,875	15,900
1110	Class 1,000		.18	44.444		9,600	1,475		11,075	12,900
1120	Class 10,000		.21	38.095		8,175	1,250		9,425	11,000
1130	Class 100,000	↓	.21	38.095	↓	7,550	1,250		8,800	10,300
2800	Ceiling grid support, slotted channel struts 4'-0" O.C., ea. way				S.F.					6.50
3000	Ceiling panel, vinyl coated foil on mineral substrate									
3020	Sealed, non-perforated				S.F.					1.40
4000	Ceiling panel seal, silicone sealant, 150 L.F./gal.	1 Carp	150	.053	L.F.	.22	1.76		1.98	2.98
4100	Two sided adhesive tape	"	240	.033	"	.10	1.10		1.20	1.82
4200	Clips, one per panel				Ea.	.86			.86	.95
6000	HEPA filter,2'x4',99.97% eff.,3" dp beveled frame (silicone seal)					264			264	290
6040	6" deep skirted frame (channel seal)					305			305	335
6100	99.99% efficient, 3" deep beveled frame (silicone seal)					284			284	315
6140	6" deep skirted frame (channel seal)					325			325	355
6200	99.999% efficient, 3" deep beveled frame (silicone seal)					325			325	355
6240	6" deep skirted frame (channel seal)				↓	365			365	400
7000	Wall panel systems, including channel strut framing									
7020	Polyester coated aluminum, particle board				S.F.					20
7100	Porcelain coated aluminum, particle board									35
7400	Wall panel support, slotted channel struts, to 12' high				↓					18
800	**0010** **SAUNA** Prefabricated, incl. heater & controls, 7' high, 6' x 4', C/C	L-7	2.20	12.727	Ea.	3,450	405		3,855	4,425	**800**
0050	6' x 4', C/P		2	14		3,225	445		3,670	4,225
0400	6' x 5', C/C		2	14		3,875	445		4,320	4,950
0450	6' x 5', C/P		2	14		3,625	445		4,070	4,675
0600	6' x 6', C/C		1.80	15.556		4,125	495		4,620	5,300
0650	6' x 6', C/P		1.80	15.556		3,850	495		4,345	5,025
0800	6' x 9', C/C		1.60	17.500		5,175	560		5,735	6,550
0850	6' x 9', C/P		1.60	17.500		4,900	560		5,460	6,275
1000	8' x 12', C/C		1.10	25.455		8,000	810		8,810	10,100
1050	8' x 12', C/P		1.10	25.455		7,350	810		8,160	9,350
1200	8' x 8', C/C		1.40	20		6,100	640		6,740	7,675
1250	8' x 8', C/P		1.40	20		5,750	640		6,390	7,300
1400	8' x 10', C/C		1.20	23.333		6,775	745		7,520	8,600
1450	8' x 10', C/P		1.20	23.333		6,325	745		7,070	8,100
1600	10' x 12', C/C		1	28		8,475	895		9,370	10,700
1650	10' x 12', C/P	↓	1	28		7,675	895		8,570	9,825
2500	Heaters only (incl. above), wall mounted, to 200 C.F.					450			450	495
2750	To 300 C.F.					545			545	600
3000	Floor standing, to 720 C.F., 10,000 watts, w/controls	1 Elec	3	2.667		1,400	105		1,505	1,700
3250	To 1,000 C.F., 16,000 watts	"	3	2.667	↓	1,450	105		1,555	1,725
940	**0010** **STEAM BATH** Heater, timer & head, single, to 140 C.F.	1 Plum	1.20	6.667	Ea.	930	264		1,194	1,425	**940**
0500	To 300 C.F.		1.10	7.273		1,050	288		1,338	1,575
1000	Commercial size, with blow-down assembly, to 800 C.F.		.90	8.889		3,600	350		3,950	4,500
1500	To 2500 C.F.	↓	.80	10		5,700	395		6,095	6,875
2000	Multiple, motels, apts., 2 baths, w/ blow-down assm., 500 C.F.	Q-1	1.30	12.308		3,575	440		4,015	4,575
2500	4 baths	"	.70	22.857	↓	3,875	815		4,690	5,475

13 SPECIAL CONSTRUCTION

13091	X-Ray/Radio Freq Protection	CREW	DAILY OUTPUT	LABOR-HOURS	UNIT	2004 BARE COSTS				TOTAL INCL O&P
						MAT.	LABOR	EQUIP.	TOTAL	
600										**600**
0010	**SHIELDING LEAD**									
0100	Laminated lead in wood doors, 1/16" thick, no hardware				S.F.	23.50			23.50	25.50
0200	Lead lined door frame, not incl. hardware,									
0210	1/16" thick lead, butt prepared for hardware	1 Lath	2.40	3.333	Ea.	286	102		388	465
0300	Lead sheets, 1/16" thick	2 Lath	135	.119	S.F.	4.35	3.62		7.97	10.15
0400	1/8" thick		120	.133		8.90	4.07		12.97	15.85
0500	Lead shielding, 1/4" thick		135	.119		15.35	3.62		18.97	22.50
0550	1/2" thick	↓	120	.133	↓	30	4.07		34.07	39
0600	Lead glass, 1/4" thick, 2.0 mm LE, 12" x 16"	2 Glaz	13	1.231	Ea.	247	39.50		286.50	330
0700	24" x 36"		8	2		950	64		1,014	1,150
0800	36" x 60"		2	8		2,525	256		2,781	3,175
0850	Window frame with 1/16" lead and voice passage, 36" x 60"		2	8		435	256		691	870
0870	24" x 36" frame		8	2	↓	330	64		394	460
0900	Lead gypsum board, 5/8" thick with 1/16" lead		160	.100	S.F.	4.56	3.21		7.77	9.85
0910	1/8" lead	↓	140	.114		8.90	3.66		12.56	15.35
0930	1/32" lead	2 Lath	200	.080	↓	2.95	2.44		5.39	6.85
0950	Lead headed nails (average 1 lb. per sheet)				Lb.	5.25			5.25	5.80
1000	Butt joints in 1/8" lead or thicker, 2" batten strip x 7' long	2 Lath	240	.067	Ea.	12.65	2.04		14.69	16.90
1200	X-ray protection, average radiography or fluoroscopy									
1210	room, up to 300 S.F. floor, 1/16" lead, minimum	2 Lath	.25	64	Total	4,850	1,950		6,800	8,250
1500	Maximum, 7'-0" walls	"	.15	106	"	6,025	3,250		9,275	11,500
1600	Deep therapy X-ray room, 250 KV capacity,									
1800	up to 300 S.F. floor, 1/4" lead, minimum	2 Lath	.08	200	Total	16,400	6,100		22,500	27,100
1900	Maximum, 7'-0" walls	"	.06	266	"	21,100	8,150		29,250	35,300
2000	X-ray viewing panels, clear lead plastic									
2010	7 mm thick, 0.3 mm LE, 2.3 lbs/S.F.	H-3	139	.115	S.F.	123	3.26		126.26	141
2020	12 mm thick, 0.5 mm LE, 3.9 lbs/S.F.		82	.195		167	5.50		172.50	191
2030	18 mm thick, 0.8mm LE, 5.9 lbs/S.F.		54	.296		181	8.40		189.40	212
2040	22 mm thick, 1.0 mm LE, 7.2 lbs/S.F.		44	.364		185	10.30		195.30	219
2050	35 mm thick, 1.5 mm LE, 11.5 lbs/S.F.		28	.571		207	16.15		223.15	253
2060	46 mm thick, 2.0 mm LE, 15.0 lbs/S.F.	↓	21	.762	↓	273	21.50		294.50	335
2090	For panels 12 S.F. to 48 S.F., add crating charge				Ea.					50
4000	X-ray barriers, modular, panels mounted within framework for									
4002	attaching to floor, wall or ceiling, upper portion is clear lead									
4005	plastic window panels 48"H, lower portion is opaque leaded									
4008	steel panels 36"H, structural supports not incl.									
4010	1-section barrier, 36"W x 84"H overall									
4020	0.5 mm LE panels	H-3	6.40	2.500	Ea.	2,775	71		2,846	3,150
4030	0.8 mm LE panels		6.40	2.500		2,950	71		3,021	3,350
4040	1.0 mm LE panels		5.33	3.002		3,050	85		3,135	3,475
4050	1.5 mm LE panels	↓	5.33	3.002	↓	3,250	85		3,335	3,700
4060	2-section barrier, 72"W x 84"H overall									
4070	0.5 mm LE panels	H-3	4	4	Ea.	5,750	113		5,863	6,500
4080	0.8 mm LE panels		4	4		6,075	113		6,188	6,875
4090	1.0 mm LE panels		3.56	4.494		6,200	127		6,327	7,025
5000	1.5 mm LE panels	↓	3.20	5	↓	6,725	142		6,867	7,625
5010	3-section barrier, 108"W x 84"H overall									
5020	0.5 mm LE panels	H-3	3.20	5	Ea.	8,225	142		8,367	9,275
5030	0.8 mm LE panels		3.20	5		8,625	142		8,767	9,725
5040	1.0 mm LE panels		2.67	5.993		8,850	170		9,020	9,975
5050	1.5 mm LE panels	↓	2.46	6.504	↓	9,600	184		9,784	10,900
7000	X-ray barriers, mobile, mounted within framework w/casters on									
7005	bottom, clear lead plastic window panels on upper portion,									
7010	opaque on lower, 30"W x 75"H overall, incl. framework									
7020	24"H upper w/0.5 mm LE, 48"H lower w/0.8 mm LE	1 Carp	16	.500	Ea.	1,900	16.50		1,916.50	2,100
7030	48"W x 75"H overall, incl. framework									

13090 | Radiation Protection

13091	X-Ray/Radio Freq Protection	CREW	DAILY OUTPUT	LABOR-HOURS	UNIT	2004 BARE COSTS				TOTAL INCL O&P	
						MAT.	LABOR	EQUIP.	TOTAL		
600	7040	36"H upper w/0.5 mm LE, 36"H lower w/0.8 mm LE	1 Carp	16	.500	Ea.	3,325	16.50		3,341.50	3,700	**600**
	7050	36"H upper w/1.0 mm LE, 36"H lower w/1.5 mm LE	"	16	.500	"	4,025	16.50		4,041.50	4,475	
	7060	72"W x 75"H overall, incl. framework										
	7070	36"H upper w/0.5 mm LE, 36"H lower w/0.8 mm LE	1 Carp	16	.500	Ea.	3,950	16.50		3,966.50	4,375	
	7080	36"H upper w/1.0 mm LE, 36"H lower w/1.5 mm LE	"	16	.500	"	5,000	16.50		5,016.50	5,525	
700	0010	**SHIELDING, RADIO FREQUENCY**										**700**
	0020	Prefabricated or screen-type copper or steel, minimum	2 Carp	180	.089	SF Surf	24	2.93		26.93	31	
	0100	Average		155	.103		26	3.41		29.41	34	
	0150	Maximum		145	.110		31	3.64		34.64	39.50	

13100 | Lightning Protection

13101	Lightning Protection	CREW	DAILY OUTPUT	LABOR-HOURS	UNIT	2004 BARE COSTS				TOTAL INCL O&P	
						MAT.	LABOR	EQUIP.	TOTAL		
055	0010	**LIGHTNING PROTECTION**										**055**
	0200	Air terminals & base, copper										
	0400	3/8" diameter x 10" (to 75' high)	1 Elec	8	1	Ea.	23.50	39.50		63	84	
	0500	1/2" diameter x 12" (over 75' high)		8	1		36	39.50		75.50	98	
	1000	Aluminum, 1/2" diameter x 12" (to 75' high)		8	1		18	39.50		57.50	78.50	
	1100	5/8" diameter x 12" (over 75' high)		8	1		24	39.50		63.50	85	
	2000	Cable, copper, 220 lb. per thousand ft. (to 75' high)		320	.025	L.F.	1.07	.99		2.06	2.65	
	2100	375 lb. per thousand ft. (over 75' high)		230	.035		1.61	1.37		2.98	3.81	
	2500	Aluminum, 101 lb. per thousand ft. (to 75' high)		280	.029		.58	1.13		1.71	2.31	
	2600	199 lb. per thousand ft. (over 75' high)		240	.033		.94	1.31		2.25	2.98	
	3000	Arrester, 175 volt AC to ground		8	1	Ea.	36	39.50		75.50	98	
	3100	650 volt AC to ground		6.70	1.194	"	70	47		117	147	

13110 | Cathodic Protection

13111	Cathodic Protection	CREW	DAILY OUTPUT	LABOR-HOURS	UNIT	2004 BARE COSTS				TOTAL INCL O&P	
						MAT.	LABOR	EQUIP.	TOTAL		
050	0010	**CATHODIC PROTECTION**										**050**
	1000	Anodes, magnesium type, 9 #	R-15	18.50	2.595	Ea.	48	99.50	13.50	161	216	
	1010	17 #		13	3.692		88.50	141	19.25	248.75	330	
	1020	32 #		10	4.800		163	184	25	372	480	
	1030	48 #		7.20	6.667		245	255	34.50	534.50	690	
	1100	Graphite type w/ epoxy cap, 3" x 60" (32 #)	R-22	8.40	4.438		71.50	147		218.50	300	
	1110	4" x 80" (68 #)		6	6.213		149	206		355	475	
	1120	6" x 72" (80 #)		5.20	7.169		875	237		1,112	1,325	
	1130	6" x 36" (45 #)		9.60	3.883		445	129		574	680	
	2000	Rectifiers, silicon type, air cooled, 28 V/10 A	R-19	3.50	5.714		905	226		1,131	1,325	
	2010	20 V/20 A		3.50	5.714		925	226		1,151	1,350	
	2100	Oil immersed, 28 V/10 A		3	6.667		1,150	263		1,413	1,675	
	2110	20 V/20 A		3	6.667		1,225	263		1,488	1,750	
	3000	Anode backfill, coke breeze	R-22	3,850	.010	Lb.	.10	.32		.42	.59	

Important: See the Reference Section for critical supporting data - Reference Nos., Crews, & City Cost Indexes

13110 | Cathodic Protection

13111 | Cathodic Protection

			CREW	DAILY OUTPUT	LABOR-HOURS	UNIT	2004 BARE COSTS				TOTAL INCL O&P	
							MAT.	LABOR	EQUIP.	TOTAL		
050	4000	Cable, OR2, No. 8	R-22	2.40	15.533	M.L.F.	220	515		735	1,025	050
	4010	No. 6		2.40	15.533		310	515		825	1,125	
	4020	No. 4		2.40	15.533		420	515		935	1,225	
	4030	No. 2		2.40	15.533		690	515		1,205	1,525	
	4040	No. 1		2.20	16.945		860	560		1,420	1,800	
	4050	No. 1/0		2.20	16.945		1,100	560		1,660	2,075	
	4060	No. 2/0		2.20	16.945		1,300	560		1,860	2,300	
	4070	No. 4/0		2	18.640		2,250	615		2,865	3,400	
	5000	Test station, 7 terminal box, flush curb type w/lockable cover	R-19	12	1.667	Ea.	59	66		125	163	
	5010	Reference cell, 2″ dia PVC conduit, cplg, plug, set flush	"	4.80	4.167	"	32	165		197	280	

13150 | Swimming Pools

13151 | Swimming Pools

			CREW	DAILY OUTPUT	LABOR-HOURS	UNIT	2004 BARE COSTS				TOTAL INCL O&P	
							MAT.	LABOR	EQUIP.	TOTAL		
200	0010	**SWIMMING POOLS** Residential in-ground, vinyl lined, concrete sides										200
	0020	Sides including equipment, sand bottom	B-52	300	.187	SF Surf	10.60	5.70	1.27	17.57	22	
	0100	Metal or polystyrene sides	B-14	410	.117		8.85	3.20	.51	12.56	15.25	
	0200	Add for vermiculite bottom					.68			.68	.75	
	0500	Gunite bottom and sides, white plaster finish										
	0600	12′ x 30′ pool	B-52	145	.386	SF Surf	17.55	11.75	2.62	31.92	40.50	
	0720	16′ x 32′ pool		155	.361		15.85	11	2.45	29.30	37	
	0750	20′ x 40′ pool		250	.224		14.15	6.80	1.52	22.47	28	
	0810	Concrete bottom and sides, tile finish										
	0820	12′ x 30′ pool	B-52	80	.700	SF Surf	17.75	21.50	4.75	44	58	
	0830	16′ x 32′ pool		95	.589		14.65	17.95	4	36.60	48.50	
	0840	20′ x 40′ pool		130	.431		11.65	13.10	2.92	27.67	36.50	
	1100	Motel, gunite with plaster finish, incl. medium										
	1150	capacity filtration & chlorination	B-52	115	.487	SF Surf	21.50	14.80	3.31	39.61	50.50	
	1200	Municipal, gunite with plaster finish, incl. high										
	1250	capacity filtration & chlorination	B-52	100	.560	SF Surf	28	17.05	3.80	48.85	61.50	
	1350	Add for formed gutters				L.F.	48.50			48.50	53.50	
	1360	Add for stainless steel gutters				"	144			144	158	
	1700	Filtration and deck equipment only, as % of total				Total				20%	20%	
	1800	Deck equipment, rule of thumb, 20′ x 40′ pool				SF Pool					1.30	
	1900	5000 S.F. pool				"					1.90	
700	0010	**SWIMMING POOL EQUIPMENT** Diving stand, stainless steel, 3 meter	2 Carp	.40	40	Ea.	4,825	1,325		6,150	7,350	700
	2100	Lights, underwater, 12 volt, with transformer, 300 watt	1 Elec	.40	20		136	790		926	1,325	
	2200	110 volt, 500 watt, standard		.40	20		127	790		917	1,325	
	2400	Low water cutoff type		.40	20		130	790		920	1,325	
	2800	Heaters, see division 15510-880										

71

			DAILY	LABOR-		2004 BARE COSTS				TOTAL
13281	**Hazardous Material Remediation**	CREW	OUTPUT	HOURS	UNIT	MAT.	LABOR	EQUIP.	TOTAL	INCL O&P
155	0010	**WASTE PACKAGING, HANDLING, & DISPOSAL**									155
	0100	Collect and bag bulk material, 3 C.F. bags, by hand	A-9	400	.160	Ea.	1.15	5.75		6.90	10.35
	0200	Large production vacuum loader	A-12	880	.073		.80	2.62	.57	3.99	5.65
	1000	Double bag and decontaminate	A-9	960	.067		2.30	2.40		4.70	6.30
	2000	Containerize bagged material in drums, per 3 C.F. drum	"	800	.080		6.50	2.88		9.38	11.70
	3000	Cart bags 50' to dumpster	2 Asbe	400	.040			1.44		1.44	2.27
	5000	Disposal charges, not including haul, minimum				C.Y.					50
	5020	Maximum				"					175
	5100	Remove refrigerant from system	1 Plum	40	.200	Lb.		7.90		7.90	11.90
	9000	For type C (supplied air) respirator equipment, add				%					10%

13 **SPECIAL CONSTRUCTION**

			DAILY	LABOR-		2004 BARE COSTS				TOTAL
13720	**Detection & Alarm**	CREW	OUTPUT	HOURS	UNIT	MAT.	LABOR	EQUIP.	TOTAL	INCL O&P
065	0010	**DETECTION SYSTEMS**, not including wires & conduits									065
	0100	Burglar alarm, battery operated, mechanical trigger	1 Elec	4	2	Ea.	249	79		328	390
	0200	Electrical trigger		4	2		297	79		376	440
	0400	For outside key control, add		8	1		70.50	39.50		110	136
	0600	For remote signaling circuitry, add		8	1		112	39.50		151.50	182
	0800	Card reader, flush type, standard		2.70	2.963		835	117		952	1,100
	1000	Multi-code		2.70	2.963		1,075	117		1,192	1,350
	1010	Card reader, proximity type		2.70	2.963		460	117		577	680
	1200	Door switches, hinge switch		5.30	1.509		52.50	59.50		112	147
	1400	Magnetic switch		5.30	1.509		62	59.50		121.50	157
	1600	Exit control locks, horn alarm		4	2		310	79		389	455
	1800	Flashing light alarm		4	2		350	79		429	500
	2000	Indicating panels, 1 channel		2.70	2.963		330	117		447	540
	2200	10 channel	2 Elec	3.20	5		1,125	197		1,322	1,550
	2400	20 channel		2	8		2,200	315		2,515	2,900
	2600	40 channel		1.14	14.035		4,000	555		4,555	5,225
	2800	Ultrasonic motion detector, 12 volt	1 Elec	2.30	3.478		206	137		343	430
	3000	Infrared photoelectric detector		2.30	3.478		170	137		307	390
	3200	Passive infrared detector		2.30	3.478		254	137		391	485
	3400	Glass break alarm switch		8	1		42.50	39.50		82	106
	3420	Switchmats, 30" x 5'		5.30	1.509		76	59.50		135.50	172
	3440	30" x 25'		4	2		182	79		261	315
	3460	Police connect panel		4	2		219	79		298	360
	3480	Telephone dialer		5.30	1.509		345	59.50		404.50	470
	3500	Alarm bell		4	2		69.50	79		148.50	194
	3520	Siren		4	2		131	79		210	261
	3540	Microwave detector, 10' to 200'		2	4		600	158		758	895
	3560	10' to 350'		2	4		1,750	158		1,908	2,150
	3594	Fire, alarm control panel									
	3600	4 zone	2 Elec	2	8	Ea.	920	315		1,235	1,475
	3800	8 zone		1	16		1,400	630		2,030	2,500
	4000	12 zone		.67	23.988		1,825	945		2,770	3,400
	4020	Alarm device	1 Elec	8	1		122	39.50		161.50	193
	4050	Actuating device	"	8	1		292	39.50		331.50	380
	4160	Alarm command center multiplex, addressable w/o voice large	2 Elec	.33	48.048		31,500	1,900		33,400	37,500
	4170	addressable w/ voice large	"	.30	53.333		49,900	2,100		52,000	58,000

72   **Important: See the Reference Section for critical supporting data - Reference Nos., Crews, & City Cost Indexes**

13720 | Detection & Alarm

		CREW	DAILY OUTPUT	LABOR-HOURS	UNIT	2004 BARE COSTS				TOTAL INCL O&P	
						MAT.	LABOR	EQUIP.	TOTAL		
065	4175	Addressable interface device	1 Elec	7.25	1.103	Ea.	205	43.50		248.50	291	065
	4200	Battery and rack		4	2		690	79		769	875	
	4400	Automatic charger		8	1		445	39.50		484.50	550	
	4600	Signal bell		8	1		49.50	39.50		89	113	
	4800	Trouble buzzer or manual station		8	1		37	39.50		76.50	99	
	5000	Detector, rate of rise		8	1		33.50	39.50		73	95.50	
	5010	Detector, heat (addressable type)		7.25	1.103		150	43.50		193.50	230	
	5100	Fixed temperature		8	1		28	39.50		67.50	89.50	
	5200	Smoke detector, ceiling type		6.20	1.290		75	51		126	158	
	5240	Smoke detector (addressable type)		6	1.333		150	52.50		202.50	243	
	5400	Duct type		3.20	2.500		250	98.50		348.50	420	
	5420	Duct (addressable type)		3.20	2.500		285	98.50		383.50	460	
	5600	Strobe and horn		5.30	1.509		95	59.50		154.50	194	
	5610	Strobe and horn (ADA type)		5.30	1.509		95	59.50		154.50	194	
	5620	Visual alarm (ADA type)		6.70	1.194		47.50	47		94.50	123	
	5800	Fire alarm horn		6.70	1.194		36.50	47		83.50	110	
	6000	Door holder, electro-magnetic		4	2		77.50	79		156.50	203	
	6200	Combination holder and closer		3.20	2.500		430	98.50		528.50	620	
	6400	Code transmitter		4	2		690	79		769	875	
	6600	Drill switch		8	1		86.50	39.50		126	154	
	6800	Master box		2.70	2.963		3,100	117		3,217	3,575	
	7000	Break glass station		8	1		50	39.50		89.50	114	
	7010	Break glass station, addressable		7.25	1.103		68	43.50		111.50	140	
	7800	Remote annunciator, 8 zone lamp		1.80	4.444		175	175		350	455	
	8000	12 zone lamp	2 Elec	2.60	6.154		300	242		542	690	
	8200	16 zone lamp	"	2.20	7.273		300	287		587	755	

13832 | Direct Digital Controls

		CREW	DAILY OUTPUT	LABOR-HOURS	UNIT	2004 BARE COSTS				TOTAL INCL O&P	
						MAT.	LABOR	EQUIP.	TOTAL		
200	0010	**CONTROL COMPONENTS / DDC SYSTEMS** (Sub's quote incl. M & L)										200
	0100	Analog inputs										
	0110	Sensors (avg. 50' run in 1/2" EMT)										
	0120	Duct temperature				Ea.					353.73	
	0130	Space temperature									634.64	
	0140	Duct humidity, +/- 3%									665.85	
	0150	Space humidity, +/- 2%									1,019.59	
	0160	Duct static pressure									541	
	0170	C.F.M./Transducer									728.28	
	0172	Water temp. (see 15120 for well tap add)									624.24	
	0174	Water flow (see 15120 for circuit sensor add)									2,288.88	
	0176	Water pressure differential (see 15120 for tap add)									936.36	
	0177	Steam flow (see 15120 for circuit sensor add)									2,288.88	
	0178	Steam pressure (see 15120 for tap add)									977.97	
	0180	K.W./Transducer									1,300.50	
	0182	K.W.H. totalization (not incl. elec. meter pulse xmtr.)									597.72	
	0190	Space static pressure									1,020	
	1000	Analog outputs (avg. 50' run in 1/2" EMT)										
	1010	P/I Transducer				Ea.					603.43	
	1020	Analog output, matl. in MUX									291.31	

SPECIAL CONSTRUCTION 13

				DAILY	LABOR-		\multicolumn{4}{c}{2004 BARE COSTS}	TOTAL	
13832		**Direct Digital Controls**	CREW	OUTPUT	HOURS	UNIT	MAT.	LABOR	EQUIP.	TOTAL	INCL O&P	
200	1030	Pneumatic (not incl. control device)				Ea.					613.83	200
	1040	Electric (not incl control device)				↓					364.14	
	2000	Status (Alarms)										
	2100	Digital inputs (avg. 50' run in 1/2" EMT)										
	2110	Freeze				Ea.					416.16	
	2120	Fire									374.54	
	2130	Differential pressure, (air)									572.22	
	2140	Differential pressure, (water)									811.51	
	2150	Current sensor									416.16	
	2160	Duct high temperature thermostat									546.21	
	2170	Duct smoke detector				↓					676.26	
	2200	Digital output (avg. 50' run in 1/2" EMT)										
	2210	Start/stop				Ea.					326.40	
	2220	On/off (maintained contact)				"					561	
	3000	Controller M.U.X. panel, incl. function boards										
	3100	48 point				Ea.					5,045.94	
	3110	128 point				"					6,918.66	
	3200	D.D.C. controller (avg. 50' run in conduit)										
	3210	Mechanical room										
	3214	16 point controller (incl. 120v/1ph power supply)				Ea.					3,121.20	
	3229	32 point controller (incl. 120v/1ph power supply)				"					5,202	
	3230	Includes software programming and checkout										
	3260	Space										
	3266	V.A.V. terminal box (incl. space temp. sensor)				Ea.					806.31	
	3280	Host computer (avg. 50' run in conduit)										
	3281	Package complete with PC, keyboard,										
	3282	printer, color CRT, modem, basic software				Ea.					9,363.60	
	4000	Front end costs										
	4100	Computer (P.C.)/software program				Ea.					6,242.40	
	4200	Color graphics software									3,745.44	
	4300	Color graphics slides									468.18	
	4350	Additional dot matrix printer				↓					936.36	
	4400	Communications trunk cable				L.F.					3.64	
	4500	Engineering labor, (not incl. dftg.)				Point					79.07	
	4600	Calibration labor									79.07	
	4700	Start-up, checkout labor				↓					119.64	
	4800	Drafting labor, as req'd										
	5000	Communications bus (data transmission cable)										
	5010	#18 twisted shielded pair in 1/2" EMT conduit				C.L.F.					364.14	
	8000	Applications software										
	8050	Basic maintenance manager software (not incl. data base entry)				Ea.					1,872.72	
	8100	Time program				Point					6.55	
	8120	Duty cycle									13.05	
	8140	Optimum start/stop									39.53	
	8160	Demand limiting									19.56	
	8180	Enthalpy program				↓					39.53	
	8200	Boiler optimization				Ea.					1,170.45	
	8220	Chiller optimization				"					1,560.60	
	8240	Custom applications										
	8260	Cost varies with complexity										

13836		**Pneumatic Controls**										
200	0010	**CONTROL SYSTEMS, PNEUMATIC** (Sub's quote incl. mat. & labor)										200
	0011	and nominal 50 Ft. of tubing. Add control panelboard if req'd.										
	0100	Heating and Ventilating, split system										
	0200	Mixed air control, economizer cycle,panel readout,tubing										

Important: See the Reference Section for critical supporting data - Reference Nos., Crews, & City Cost Indexes

13836 | Pneumatic Controls

		CREW	DAILY OUTPUT	LABOR-HOURS	UNIT	2004 BARE COSTS MAT.	LABOR	EQUIP.	TOTAL	TOTAL INCL O&P
200	0220	Up to 10 tons	Q-19	.68	35.294	Ea.	2,900	1,300		4,200	5,125	200
	0240	For 10 to 20 tons		.63	37.915		3,100	1,400		4,500	5,500
	0260	For over 20 tons		.58	41.096		3,350	1,525		4,875	5,950
	0270	Enthalpy cycle, up to 10 tons		.50	48.387		3,200	1,800		5,000	6,175
	0280	For 10 to 20 tons		.46	52.174		3,450	1,925		5,375	6,675
	0290	For over 20 tons		.42	56.604		3,725	2,100		5,825	7,225
	0300	Heating coil, hot water, 3 way valve,									
	0320	Freezestat, limit control on discharge, readout	Q-5	.69	23.088	Ea.	2,150	825		2,975	3,600
	0500	Cooling coil, chilled water, room									
	0520	Thermostat, 3 way valve	Q-5	2	8	Ea.	955	286		1,241	1,475
	0600	Cooling tower, fan cycle, damper control,									
	0620	Control system including water readout in/out at panel	Q-19	.67	35.821	Ea.	3,800	1,325		5,125	6,150
	1000	Unit ventilator, day/night operation,									
	1100	freezestat, ASHRAE, cycle 2	Q-19	.91	26.374	Ea.	2,100	975		3,075	3,775
	2000	Compensated hot water from boiler, valve control,									
	2100	readout and reset at panel, up to 60 GPM	Q-19	.55	43.956	Ea.	3,925	1,625		5,550	6,750
	2120	For 120 GPM		.51	47.059		4,200	1,750		5,950	7,225
	2140	For 240 GPM		.49	49.180		4,400	1,825		6,225	7,575
	3000	Boiler room combustion air, damper to 5 SF, controls		1.36	17.582		1,900	650		2,550	3,050
	3500	Fan coil, heating and cooling valves, 4 pipe control system		3	8		855	296		1,151	1,375
	3600	Heat exchanger system controls		.86	27.907		1,850	1,025		2,875	3,575
	3900	Multizone control (one per zone), includes thermostat, damper									
	3910	motor and reset of discharge temperature	Q-5	.51	31.373	Ea.	1,900	1,125		3,025	3,775
	4000	Pneumatic thermostat, including controlling room radiator valve	"	2.43	6.593		570	236		806	985
	4040	Program energy saving optimizer	Q-19	1.21	19.786		4,750	730		5,480	6,325
	4060	Pump control system	"	3	8		875	296		1,171	1,400
	4080	Reheat coil control system, not incl coil	Q-5	2.43	6.593		745	236		981	1,175
	4500	Air supply for pneumatic control system									
	4600	Tank mounted duplex compressor, starter, alternator,									
	4620	piping, dryer, PRV station and filter									
	4630	1/2 HP	Q-19	.68	35.139	Ea.	7,050	1,300		8,350	9,700
	4640	3/4 HP		.64	37.383		7,400	1,375		8,775	10,200
	4650	1 HP		.61	39.539		8,075	1,450		9,525	11,100
	4660	1-1/2 HP		.57	41.739		8,600	1,550		10,150	11,800
	4680	3 HP		.55	43.956		11,700	1,625		13,325	15,300
	4690	5 HP		.42	57.143		20,500	2,125		22,625	25,700
	4800	Main air supply, includes 3/8" copper main and labor	Q-5	1.82	8.791	C.L.F.	237	315		552	730
	4810	If poly tubing used, deduct									30%
	7000	Static pressure control for air handling unit, includes pressure									
	7010	sensor, receiver controller, readout and damper motors	Q-19	.64	37.383	Ea.	5,625	1,375		7,000	8,250
	7020	If return air fan requires control, add									70%
	8600	VAV boxes, incl. thermostat, damper motor, reheat coil & tubing	Q-5	1.46	10.989		880	395		1,275	1,550
	8610	If no reheat coil, deduct									204

13838 | Pneumatic/Electric Controls

		CREW	DAILY OUTPUT	LABOR-HOURS	UNIT	2004 BARE COSTS MAT.	LABOR	EQUIP.	TOTAL	TOTAL INCL O&P
200	0010	**CONTROL COMPONENTS**										200
	0700	Controller, receiver									
	0850	Electric, single snap switch	1 Elec	4	2	Ea.	360	79		439	510
	0860	Dual snap switches	"	3	2.667	"	450	105		555	650
	3590	Sensor, electric operated									
	3620	Humidity	1 Elec	8	1	Ea.	27	39.50		66.50	88.50
	3650	Pressure		8	1		830	39.50		869.50	975
	3680	Temperature		10	.800		96	31.50		127.50	152
	5000	Thermostats									
	5200	24 hour, automatic, clock	1 Shee	8	1	Ea.	124	39		163	196

SPECIAL CONSTRUCTION 13

13838 | Pneumatic/Electric Controls

200			CREW	DAILY OUTPUT	LABOR-HOURS	UNIT	2004 BARE COSTS				TOTAL INCL O&P	200
							MAT.	LABOR	EQUIP.	TOTAL		
	5220	Electric, low voltage, 2 wire	1 Elec	13	.615	Ea.	18	24.50		42.50	56	
	5230	3 wire		10	.800		22	31.50		53.50	71.50	
	5420	Electric operated, humidity		8	1		50.50	39.50		90	114	
	5430	DPST		8	1		71	39.50		110.50	137	
	7090	Valves, motor controlled, including actuator										
	7100	Electric motor actuated										
	7200	Brass, two way, screwed										
	7210	1/2" pipe size	L-6	36	.333	Ea.	201	13.20		214.20	242	
	7220	3/4" pipe size		30	.400		294	15.80		309.80	350	
	7230	1" pipe size		28	.429		330	16.95		346.95	390	
	7240	1-1/2" pipe size		19	.632		370	25		395	450	
	7250	2" pipe size		16	.750		575	29.50		604.50	680	
	7350	Brass, three way, screwed										
	7360	1/2" pipe size	L-6	33	.364	Ea.	212	14.35		226.35	255	
	7370	3/4" pipe size		27	.444		238	17.55		255.55	288	
	7380	1" pipe size		25.50	.471		210	18.60		228.60	259	
	7384	1-1/4" pipe size		21	.571		266	22.50		288.50	325	
	7390	1-1/2" pipe size		17	.706		300	28		328	370	
	7400	2" pipe size		14	.857		465	34		499	565	
	7550	Iron body, two way, flanged										
	7560	2-1/2" pipe size	L-6	4	3	Ea.	355	119		474	570	
	7570	3" pipe size		3	4		470	158		628	750	
	7580	4" pipe size		2	6		655	237		892	1,075	
	7850	Iron body, three way, flanged										
	7860	2-1/2" pipe size	L-6	3	4	Ea.	770	158		928	1,075	
	7870	3" pipe size		2.50	4.800		930	190		1,120	1,300	
	7880	4" pipe size		2	6		1,200	237		1,437	1,675	

13851 | Detection & Alarm

050			CREW	DAILY OUTPUT	LABOR-HOURS	UNIT	2004 BARE COSTS				TOTAL INCL O&P	050
							MAT.	LABOR	EQUIP.	TOTAL		
	0010	**CLOCKS**										
	0080	12" diameter, single face	1 Elec	8	1	Ea.	75.50	39.50		115	142	
	0100	Double face	"	6.20	1.290	"	144	51		195	234	
055	0010	**CLOCK SYSTEMS**, not including wires & conduits										055
	0100	Time system components, master controller	1 Elec	.33	24.242	Ea.	1,625	955		2,580	3,225	
	0200	Program bell		8	1		53.50	39.50		93	118	
	0400	Combination clock & speaker		3.20	2.500		189	98.50		287.50	355	
	0600	Frequency generator		2	4		6,350	158		6,508	7,200	
	0800	Job time automatic stamp recorder, minimum		4	2		445	79		524	605	
	1000	Maximum		4	2		680	79		759	865	
	1200	Time stamp for correspondence, hand operated					340			340	375	
	1400	Fully automatic					495			495	545	
	1600	Master time clock system, clocks & bells, 20 room	4 Elec	.20	160		4,075	6,300		10,375	13,900	
	1800	50 room	"	.08	400		9,400	15,800		25,200	33,700	
	2000	Time clock, 100 cards in & out, 1 color	1 Elec	3.20	2.500		950	98.50		1,048.50	1,200	
	2200	2 colors		3.20	2.500		1,025	98.50		1,123.50	1,275	
	2400	With 3 circuit program device, minimum		2	4		375	158		533	650	

Important: See the Reference Section for critical supporting data - Reference Nos., Crews, & City Cost Indexes

	13851	Detection & Alarm	CREW	DAILY OUTPUT	LABOR-HOURS	UNIT	2004 BARE COSTS				TOTAL INCL O&P	
							MAT.	LABOR	EQUIP.	TOTAL		
055	2600	Maximum	1 Elec	2	4	Ea.	545	158		703	835	055
	2800	Metal rack for 25 cards		7	1.143		60	45		105	133	
	3000	Watchman's tour station		8	1		63	39.50		102.50	128	
	3200	Annunciator with zone indication		1	8		230	315		545	725	
	3400	Time clock with tape	▼	1	8	▼	610	315		925	1,150	

	13960	CO2 Fire Extinguishing	CREW	DAILY OUTPUT	LABOR-HOURS	UNIT	2004 BARE COSTS				TOTAL INCL O&P	
							MAT.	LABOR	EQUIP.	TOTAL		
200	0010	**AUTOMATIC FIRE SUPPRESSION SYSTEMS**										200
	0040	For detectors and control stations, see division 13850										
	0100	Control panel, single zone with batteries (2 zones det., 1 suppr.)	1 Elec	1	8	Ea.	1,275	315		1,590	1,875	
	0150	Multizone (4) with batteries (8 zones det., 4 suppr.)	"	.50	16		2,450	630		3,080	3,650	
	1000	Dispersion nozzle, CO2, 3" x 5"	1 Plum	18	.444		50	17.60		67.60	81.50	
	1100	FM200, 1-1/2"	"	14	.571		50	22.50		72.50	89	
	2000	Extinguisher, CO2 system, high pressure, 75 lb. cylinder	Q-1	6	2.667		950	95		1,045	1,200	
	2100	100 lb. cylinder	"	5	3.200	▼	1,025	114		1,139	1,300	
	2400	FM200 system, filled, with mounting bracket										
	2460	26 lb. container	Q-1	8	2	Ea.	1,800	71.50		1,871.50	2,075	
	2480	44 lb. container		7	2.286		2,400	81.50		2,481.50	2,775	
	2500	63 lb. container		6	2.667		2,800	95		2,895	3,225	
	2520	101 lb. container		5	3.200		3,750	114		3,864	4,300	
	2540	196 lb. container	▼	4	4		5,800	143		5,943	6,600	
	3000	Electro/mechanical release	L-1	4	4		125	158		283	375	
	3400	Manual pull station	1 Plum	6	1.333		45	53		98	129	
	4000	Pneumatic damper release	"	8	1	▼	175	39.50		214.50	253	
	6000	Average FM200 system, minimum				C.F.	1.25			1.25	1.38	
	6020	Maximum				"	2.50			2.50	2.75	

For information about Means Estimating Seminars, see yellow pages 12 and 13 in back of book

SPECIAL CONSTRUCTION **13**

Division Notes

	CREW	DAILY OUTPUT	LABOR-HOURS	UNIT	2004 BARE COSTS				TOTAL INCL O&P
					MAT.	LABOR	EQUIP.	TOTAL	

Division 15 Mechanical

Estimating Tips

15100 Building Services Piping

This subdivision is primarily basic pipe and related materials. The pipe may be used by any of the mechanical disciplines, i.e., plumbing, fire protection, heating, and air conditioning.

- The piping section lists the add to labor for elevated pipe installation. These adds apply to all elevated pipe, fittings, valves, insulation, etc., that are placed above 10' high. CAUTION: the correct percentage may vary for the same pipe. For example, the percentage add for the basic pipe installation should be based on the maximum height that the craftsman must install for that particular section. If the pipe is to be located 14' above the floor but it is suspended on threaded rod from beams, the bottom flange of which is 18' high (4' rods), then the height is actually 18' and the add is 20%. The pipe coverer, however, does not have to go above the 14' and so his add should be 10%.

- Most pipe is priced first as straight pipe with a joint (coupling, weld, etc.) every 10' and a hanger usually every 10'. There are exceptions with hanger spacing such as: for cast iron pipe (5') and plastic pipe (3 per 10'). Following each type of pipe there are several lines listing sizes and the amount to be subtracted to delete couplings and hangers. This is for pipe that is to be buried or supported together on trapeze hangers. The reason that the couplings are deleted is that these runs are usually long and frequently longer lengths of pipe are used. By deleting the couplings the estimator is expected to look up and add back the correct reduced number of couplings.

- When preparing an estimate it may be necessary to approximate the fittings. Fittings usually run between 25% and 50% of the cost of the pipe. The lower percentage is for simpler runs, and the higher number is for complex areas like mechanical rooms.

- For historic restoration projects, the systems must be as invisible as possible, and pathways must be sought for pipes, conduits, and ductwork. While installations in accessible spaces (such as basements and attics) are relatively straightforward to estimate, labor costs may be more difficult to determine when delivery systems must be concealed.

15400 Plumbing Fixtures & Equipment

- Plumbing fixture costs usually require two lines, the fixture itself and its "rough-in, supply and waste".

- In the Assemblies Section (Plumbing D2010) for the desired fixture, the System Components Group in the center of the page shows the fixture itself on the first line while the rest of the list (fittings, pipe, tubing, etc.) will total up to what we refer to in the Unit Price section as "Rough-in, supply, waste and vent". Note that for most fixtures we allow a nominal 5' of tubing to reach from the fixture to a main or riser.

- Remember that gas and oil fired units need venting.

15500 Heat Generation Equipment

- When estimating the cost of an HVAC system, check to see who is responsible for providing and installing the temperature control system. It is possible to overlook controls, assuming that they would be included in the electrical estimate.

- When looking up a boiler be careful on specified capacity. Some manufacturers rate their products on output while others use input.

- Include HVAC insulation for pipe, boiler and duct (wrap and liner).

- Be careful when looking up mechanical items to get the correct pressure rating and connection type (thread, weld, flange).

15700 Heating/Ventilation/ Air Conditioning Equipment

- Combination heating and cooling units are sized by the air conditioning requirements. (See Reference No. R15710-020 for preliminary sizing guide.)

- A ton of air conditioning is nominally 400 CFM.

- Rectangular duct is taken off by the linear foot for each size, but its cost is usually estimated by the pound. Remember that SMACNA standards now base duct on internal pressure.

- Prefabricated duct is estimated and purchased like pipe: straight sections and fittings.

- Note that cranes or other lifting equipment are not included on any lines in Division 15. For example, if a crane is required to lift a heavy piece of pipe into place high above a gym floor, or to put a rooftop unit on the roof of a four-story building, etc., it must be added. Due to the potential for extreme variation—from nothing additional required, to a major crane or helicopter—we feel that including a nominal amount for "lifting contingency" would be useless and detract from the accuracy of the estimate. When using equipment rental from Means do not forget to include the cost of the operator(s).

Reference Numbers

Reference numbers are shown in bold squares at the beginning of some major classifications. These numbers refer to related items in the Reference Section. The reference information may be an estimating procedure, an alternate pricing method or technical information.

Note: Not all subdivisions listed here necessarily appear in this publication.

15051	Mechanical General		CREW	DAILY OUTPUT	LABOR-HOURS	UNIT	2004 BARE COSTS				TOTAL INCL O&P	
							MAT.	LABOR	EQUIP.	TOTAL		
100	0010	**BOILERS, GENERAL** Prices do not include flue piping, elec. wiring,										100
	0020	gas or oil piping, boiler base, pad, or tankless unless noted										
	0100	Boiler H.P.: 10 KW = 34 lbs/steam/hr = 33,475 BTU/hr.										
	0120											
	0150	To convert SFR to BTU rating: Hot water, 150 x SFR;										
	0160	Forced hot water, 180 x SFR; steam, 240 x SFR										

15060	Hangers & Supports											
300	0010	**PIPE HANGERS AND SUPPORTS,** TYPE numbers per MSS-SP58										300
	0050	Brackets										
	0060	Beam side or wall, malleable iron, TYPE 34										
	0070	3/8" threaded rod size	1 Plum	48	.167	Ea.	2.88	6.60		9.48	13.05	
	0080	1/2" threaded rod size		48	.167		5.55	6.60		12.15	16	
	0090	5/8" threaded rod size		48	.167		7.40	6.60		14	18	
	0100	3/4" threaded rod size		48	.167		8.85	6.60		15.45	19.60	
	0110	7/8" threaded rod size		48	.167		10.60	6.60		17.20	21.50	
	0120	For concrete installation, add						30%				
	0150	Wall, welded steel, medium, TYPE 32										
	0160	0 size, 12" wide, 18" deep	1 Plum	34	.235	Ea.	216	9.30		225.30	252	
	0170	1 size, 18" wide 24" deep		34	.235		257	9.30		266.30	296	
	0180	2 size, 24" wide, 30" deep		34	.235		340	9.30		349.30	390	
	0300	Clamps										
	0310	C-clamp, for mounting on steel beam flange, w/locknut, TYPE 23										
	0320	3/8" threaded rod size	1 Plum	160	.050	Ea.	1.72	1.98		3.70	4.86	
	0330	1/2" threaded rod size		160	.050		1.57	1.98		3.55	4.70	
	0340	5/8" threaded rod size		160	.050		2.28	1.98		4.26	5.50	
	0350	3/4" threaded rod size		160	.050		3.29	1.98		5.27	6.60	
	0750	Riser or extension pipe, carbon steel, TYPE 8										
	0760	3/4" pipe size	1 Plum	48	.167	Ea.	2.82	6.60		9.42	13	
	0770	1" pipe size		47	.170		2.84	6.75		9.59	13.20	
	0780	1-1/4" pipe size		46	.174		3.46	6.90		10.36	14.15	
	0790	1-1/2" pipe size		45	.178		3.65	7.05		10.70	14.55	
	0800	2" pipe size		43	.186		3.83	7.35		11.18	15.25	
	0810	2-1/2" pipe size		41	.195		4.05	7.75		11.80	16.05	
	0820	3" pipe size		40	.200		4.41	7.90		12.31	16.75	
	0830	3-1/2" pipe size		39	.205		5.40	8.10		13.50	18.15	
	0840	4" pipe size		38	.211		5.55	8.35		13.90	18.60	
	0850	5" pipe size		37	.216		7.70	8.55		16.25	21.50	
	0860	6" pipe size		36	.222		9.25	8.80		18.05	23.50	
	1150	Insert, concrete										
	1160	Wedge type, carbon steel body, malleable iron nut										
	1170	1/4" threaded rod size	1 Plum	96	.083	Ea.	2.32	3.30		5.62	7.50	
	1180	3/8" threaded rod size		96	.083		2.50	3.30		5.80	7.70	
	1190	1/2" threaded rod size		96	.083		2.59	3.30		5.89	7.80	
	1200	5/8" threaded rod size		96	.083		2.67	3.30		5.97	7.90	
	1210	3/4" threaded rod size		96	.083		2.98	3.30		6.28	8.25	
	1220	7/8" threaded rod size		96	.083		3.93	3.30		7.23	9.25	
	1230	For galvanized, add					.90			.90	.99	
	2650	Rods, carbon steel										
	2660	Continuous thread										
	2670	1/4" thread size	1 Plum	144	.056	L.F.	.48	2.20		2.68	3.83	
	2680	3/8" thread size		144	.056		.53	2.20		2.73	3.88	
	2690	1/2" thread size		144	.056		.84	2.20		3.04	4.22	
	2700	5/8" thread size		144	.056		1.19	2.20		3.39	4.61	
	2710	3/4" thread size		144	.056		2.11	2.20		4.31	5.60	
	2720	7/8" thread size		144	.056		2.64	2.20		4.84	6.20	

15 MECHANICAL

Important: See the Reference Section for critical supporting data - Reference Nos., Crews, & City Cost Indexes

			DAILY	LABOR-		2004 BARE COSTS				TOTAL	
15060	**Hangers & Supports**	CREW	OUTPUT	HOURS	UNIT	MAT.	LABOR	EQUIP.	TOTAL	INCL O&P	
300	2725	1/4" thread size, bright finish	1 Plum	144	.056	L.F.	1	2.20		3.20	4.40	**300**
	2726	1/2" thread size, bright finish	↓	144	.056	↓	.88	2.20		3.08	4.27	
	2730	For galvanized, add					40%					
	2750	Both ends machine threaded 18" length										
	2760	3/8" thread size	1 Plum	240	.033	Ea.	3.94	1.32		5.26	6.30	
	2770	1/2" thread size		240	.033		6.30	1.32		7.62	8.95	
	2780	5/8" thread size		240	.033		9.05	1.32		10.37	12	
	2790	3/4" thread size		240	.033		13.90	1.32		15.22	17.30	
	2800	7/8" thread size		240	.033		15.55	1.32		16.87	19.10	
	2810	1" thread size	↓	240	.033	↓	24.50	1.32		25.82	28.50	
	4400	U-bolt, carbon steel										
	4410	Standard, with nuts, TYPE 42										
	4420	1/2" pipe size	1 Plum	160	.050	Ea.	1.12	1.98		3.10	4.20	
	4430	3/4" pipe size		158	.051		1.12	2		3.12	4.24	
	4450	1" pipe size		152	.053		1.22	2.08		3.30	4.47	
	4460	1-1/4" pipe size		148	.054		1.35	2.14		3.49	4.70	
	4470	1-1/2" pipe size		143	.056		1.40	2.22		3.62	4.87	
	4480	2" pipe size		139	.058		1.44	2.28		3.72	5	
	4490	2-1/2" pipe size		134	.060		2	2.36		4.36	5.75	
	4500	3" pipe size		128	.063		3.89	2.48		6.37	8	
	4510	3-1/2" pipe size		122	.066		4.01	2.60		6.61	8.30	
	4520	4" pipe size		117	.068		4.07	2.71		6.78	8.55	
	4530	5" pipe size		114	.070		4.14	2.78		6.92	8.70	
	4540	6" pipe size	↓	111	.072	↓	4.23	2.85		7.08	8.95	
	4580	For plastic coating on 1/2" thru 6" size, add					150%					

15400 | Plumbing Fixtures & Equipment

			DAILY	LABOR-		2004 BARE COSTS				TOTAL	
15411	**Commercial/Indust Fixtures**	CREW	OUTPUT	HOURS	UNIT	MAT.	LABOR	EQUIP.	TOTAL	INCL O&P	
400	0010	**HOT WATER DISPENSERS**										**400**
	0160	Commercial, 100 cup, 11.3 amp	1 Plum	14	.571	Ea.	291	22.50		313.50	355	
	3180	Household, 60 cup	"	14	.571	"	156	22.50		178.50	206	

15480	**Domestic Water Heaters**										
200	0010	**WATER HEATERS**										**200**
	1000	Residential, electric, glass lined tank, 5 yr, 10 gal., single element	1 Plum	2.30	3.478	Ea.	210	138		348	440	
	1040	20 gallon, single element		2.20	3.636		244	144		388	485	
	1060	30 gallon, double element		2.20	3.636		280	144		424	525	
	1080	40 gallon, double element		2	4		300	158		458	570	
	1100	52 gallon, double element		2	4		345	158		503	620	
	1120	66 gallon, double element		1.80	4.444		475	176		651	790	
	1140	80 gallon, double element		1.60	5		535	198		733	885	
	1180	120 gallon, double element	↓	1.40	5.714	↓	770	226		996	1,200	
	2000	Gas fired, foam lined tank, 10 yr, vent not incl.,										
	2040	30 gallon	1 Plum	2	4	Ea.	385	158		543	665	
	2060	40 gallon		1.90	4.211		405	167		572	695	
	2080	50 gallon		1.80	4.444		500	176		676	815	
	2100	75 gallon	↓	1.50	5.333	↓	745	211		956	1,125	

MECHANICAL 15

		CREW	DAILY OUTPUT	LABOR-HOURS	UNIT	2004 BARE COSTS				TOTAL INCL O&P	
15480	**Domestic Water Heaters**					MAT.	LABOR	EQUIP.	TOTAL		
200	2120	100 gallon	1 Plum	1.30	6.154	Ea.	1,325	244		1,569	1,850	200
	3000	Oil fired, glass lined tank, 5 yr, vent not included, 30 gallon		2	4		785	158		943	1,100	
	3040	50 gallon		1.80	4.444		1,375	176		1,551	1,775	
	3060	70 gallon		1.50	5.333		1,425	211		1,636	1,900	
	3080	85 gallon	▼	1.40	5.714	▼	4,100	226		4,326	4,850	
	4000	Commercial, 100° rise. NOTE: for each size tank, a range of										
	4010	heaters between the ones shown are available										
	4020	Electric										
	4100	5 gal., 3 kW, 12 GPH, 208V	1 Plum	2	4	Ea.	1,450	158		1,608	1,825	
	4120	10 gal., 6 kW, 25 GPH, 208V		2	4		1,600	158		1,758	2,000	
	4140	50 gal., 9 kW, 37 GPH, 208V		1.80	4.444		2,200	176		2,376	2,675	
	4160	50 gal., 36 kW, 148 GPH, 208V		1.80	4.444		3,350	176		3,526	3,950	
	4180	80 gal., 12 kW, 49 GPH, 208V		1.50	5.333		2,700	211		2,911	3,300	
	4200	80 gal., 36 kW, 148 GPH, 208V		1.50	5.333		3,750	211		3,961	4,425	
	4220	100 gal., 36 kW, 148 GPH, 208V		1.20	6.667		3,900	264		4,164	4,675	
	4240	120 gal., 36 kW, 148 GPH, 208V		1.20	6.667		4,050	264		4,314	4,875	
	4260	150 gal., 15 kW , 61 GPH, 480V		1	8		9,525	315		9,840	11,000	
	4280	150 gal., 120 kW, 490 GPH, 480V	▼	1	8		13,400	315		13,715	15,200	
	4300	200 gal., 15 kW, 61 GPH, 480V	Q-1	1.70	9.412		10,300	335		10,635	11,900	
	4320	200 gal., 120 kW , 490 GPH, 480V		1.70	9.412		14,100	335		14,435	16,000	
	4340	250 gal., 15 kW, 61 GPH, 480V		1.50	10.667		10,700	380		11,080	12,300	
	4360	250 gal., 150 kW, 615 GPH, 480V		1.50	10.667		15,500	380		15,880	17,700	
	4380	300 gal., 30 kW, 123 GPH, 480V		1.30	12.308		11,700	440		12,140	13,600	
	4400	300 gal., 180 kW, 738 GPH, 480V		1.30	12.308		17,000	440		17,440	19,400	
	4420	350 gal., 30 kW, 123 GPH, 480V		1.10	14.545		12,400	520		12,920	14,400	
	4440	350 gal, 180 kW, 738 GPH, 480V		1.10	14.545		17,500	520		18,020	20,000	
	4460	400 gal., 30 kW, 123 GPH, 480V		1	16		14,000	570		14,570	16,300	
	4480	400 gal., 210 kW, 860 GPH, 480V		1	16		20,200	570		20,770	23,100	
	4500	500 gal., 30 kW, 123 GPH, 480V		.80	20		16,400	715		17,115	19,100	
	4520	500 gal., 240 kW, 984 GPH, 480V	▼	.80	20		24,100	715		24,815	27,600	
	4540	600 gal., 30 kW, 123 GPH, 480V	Q-2	1.20	20		18,800	740		19,540	21,700	
	4560	600 gal., 300 kW, 1230 GPH, 480V		1.20	20		28,500	740		29,240	32,400	
	4580	700 gal., 30 kW, 123 GPH, 480V		1	24		19,700	885		20,585	23,000	
	4600	700 gal., 300 kW, 1230 GPH, 480V		1	24		29,600	885		30,485	33,900	
	4620	800 gal., 60 kW, 245 GPH, 480V		.90	26.667		21,300	985		22,285	25,000	
	4640	800 gal., 300 kW, 1230 GPH, 480V		.90	26.667		30,300	985		31,285	34,800	
	4660	1000 gal., 60 kW, 245 GPH, 480V		.70	34.286		23,400	1,275		24,675	27,600	
	4680	1000 gal., 480 kW, 1970 GPH, 480V		.70	34.286		31,500	1,275		32,775	36,600	
	4700	1250 gal., 60 kW, 245 GPH, 480V		.60	40		26,400	1,475		27,875	31,200	
	4720	1250 gal., 480 kW, 1970 GPH, 480V		.60	40		40,100	1,475		41,575	46,300	
	4740	1500 gal., 60 kW, 245 GPH, 480V		.50	48		34,800	1,775		36,575	41,000	
	4760	1500 gal., 480 kW, 1970 GPH, 480V	▼	.50	48		48,300	1,775		50,075	55,500	
	5400	Modulating step control, 2-5 steps	1 Elec	5.30	1.509		1,325	59.50		1,384.50	1,575	
	5440	6-10 steps		3.20	2.500		1,650	98.50		1,748.50	1,975	
	5460	11-15 steps		2.70	2.963		2,050	117		2,167	2,425	
	5480	16-20 steps	▼	1.60	5	▼	2,425	197		2,622	2,950	

15 MECHANICAL

Important: See the Reference Section for critical supporting data - Reference Nos., Crews, & City Cost Indexes

15510	Heating Boilers and Accessories	CREW	DAILY OUTPUT	LABOR-HOURS	UNIT	2004 BARE COSTS				TOTAL INCL O&P
						MAT.	LABOR	EQUIP.	TOTAL	
300	**0010 BOILERS, ELECTRIC, ASME** Standard controls and trim									**300**
1000	Steam, 6 KW, 20.5 MBH	Q-19	1.20	20	Ea.	2,625	740		3,365	3,975
1040	9 KW, 30.7 MBH		1.20	20		2,675	740		3,415	4,025
1060	18 KW, 61.4 MBH		1.20	20		2,750	740		3,490	4,125
1080	24 KW, 81.8 MBH		1.10	21.818		3,175	805		3,980	4,700
1120	36 KW, 123 MBH		1.10	21.818		3,550	805		4,355	5,100
1160	60 KW, 205 MBH		1	24		4,575	890		5,465	6,375
1220	112 KW, 382 MBH		.75	32		6,075	1,175		7,250	8,450
1240	148 KW, 505 MBH		.65	36.923		6,250	1,375		7,625	8,925
1260	168 KW, 573 MBH		.60	40		8,850	1,475		10,325	12,000
1280	222 KW, 758 MBH		.55	43.636		10,100	1,625		11,725	13,500
1300	296 KW, 1010 MBH		.45	53.333		10,200	1,975		12,175	14,200
1320	300 KW, 1023 MBH		.40	60		13,100	2,225		15,325	17,700
1340	370 KW, 1263 MBH		.35	68.571		14,100	2,525		16,625	19,300
1360	444 KW, 1515 MBH	▼	.30	80		15,100	2,950		18,050	21,000
1380	518 KW, 1768 MBH	Q-21	.36	88.889		17,100	3,350		20,450	23,800
1400	592 KW, 2020 MBH		.34	94.118		19,300	3,550		22,850	26,500
1420	666 KW, 2273 MBH		.32	100		20,300	3,775		24,075	28,100
1460	740 KW, 2526 MBH		.28	114		21,300	4,300		25,600	30,000
1480	814 KW, 2778 MBH		.25	128		23,900	4,825		28,725	33,500
1500	962 KW, 3283 MBH		.22	145		24,500	5,475		29,975	35,200
1520	1036 KW, 3536 MBH		.20	160		25,700	6,025		31,725	37,300
1540	1110 KW, 3788 MBH		.19	168		27,000	6,350		33,350	39,200
1560	2070 KW, 7063 MBH		.18	177		39,000	6,700		45,700	53,000
1580	2250 KW, 7677 MBH		.17	188		45,400	7,100		52,500	60,500
1600	2,340 KW, 7984 MBH	▼	.16	200		50,500	7,525		58,025	67,000
2000	Hot water, 6 KW, 20.5 MBH	Q-19	1.30	18.462		2,450	685		3,135	3,725
2020	9 KW, 30.7 MBH		1.30	18.462		2,475	685		3,160	3,750
2040	24 KW, 82 MBH		1.20	20		2,925	740		3,665	4,325
2060	30 KW, 103 MBH		1.20	20		3,200	740		3,940	4,600
2070	36 KW, 123 MBH		1.20	20		3,425	740		4,165	4,850
2080	48 KW, 164 MBH		1.10	21.818		4,775	805		5,580	6,450
2100	74 KW, 253 MBH		1.10	21.818		4,950	805		5,755	6,650
2120	93 KW, 317 MBH		1	24		5,975	890		6,865	7,900
2140	112 KW, 382 MBH		.90	26.667		6,050	985		7,035	8,125
2160	148 KW, 505 MBH		.75	32		7,100	1,175		8,275	9,600
2180	168 KW, 573 MBH		.65	36.923		7,300	1,375		8,675	10,100
2200	222 KW, 758 MBH		.60	40		8,075	1,475		9,550	11,100
2220	296 KW, 1010 MBH		.55	43.636		8,925	1,625		10,550	12,300
2280	370 KW, 1263 MBH		.40	60		9,900	2,225		12,125	14,200
2300	444 KW, 1515 MBH	▼	.35	68.571		12,000	2,525		14,525	17,000
2340	518 KW, 1768 MBH	Q-21	.44	72.727		13,200	2,750		15,950	18,700
2360	592 KW, 2020 MBH		.43	74.419		14,600	2,800		17,400	20,300
2400	666 KW, 2273 MBH		.40	80		15,700	3,025		18,725	21,700
2420	740 KW, 2526 MBH		.39	82.051		16,000	3,100		19,100	22,200
2440	814 KW, 2778 MBH		.38	84.211		16,500	3,175		19,675	23,000
2460	888 KW, 3031 MBH		.37	86.486		17,400	3,250		20,650	24,100
2480	962 KW, 3283 MBH		.36	88.889		19,500	3,350		22,850	26,500
2500	1,036 KW, 3536 MBH		.34	94.118		20,700	3,550		24,250	28,100
2520	1110 KW, 3788 MBH		.33	96.970		22,900	3,650		26,550	30,700
2540	1440 KW, 4915 MBH		.32	100		26,400	3,775		30,175	34,700
2560	1560 KW, 5323 MBH		.31	103		29,600	3,900		33,500	38,400
2580	1680 KW, 5733 MBH		.30	106		31,600	4,025		35,625	40,800
2600	1800 KW, 6143 MBH		.29	110		33,300	4,150		37,450	42,800
2620	1980 KW, 6757 MBH		.28	114		35,500	4,300		39,800	45,600
2640	2100 KW, 7167 MBH	▼	.27	118		39,700	4,475		44,175	50,500

MECHANICAL 15

15510 | Heating Boilers and Accessories

			CREW	DAILY OUTPUT	LABOR-HOURS	UNIT	MAT.	LABOR	EQUIP.	TOTAL	TOTAL INCL O&P	
300	2660	2220 KW, 7576 MBH	Q-21	.26	123	Ea.	41,000	4,650		45,650	52,000	300
	2680	2,400 KW, 8191 MBH		.25	128		43,200	4,825		48,025	55,000	
	2700	2610 KW, 8905 MBH		.24	133		46,500	5,025		51,525	58,500	
	2720	2790 KW, 9519 MBH		.23	139		48,600	5,250		53,850	61,500	
	2740	2970 KW, 10133 MBH		.21	152		49,700	5,750		55,450	63,000	
	2760	3150 KW, 10748 MBH		.19	168		54,500	6,350		60,850	69,500	
	2780	3240 KW, 11055 MBH		.18	177		55,500	6,700		62,200	71,000	
	2800	3420 KW, 11669 MBH		.17	188		59,000	7,100		66,100	75,000	
	2820	3,600 KW, 12,283 MBH		.16	200		61,000	7,525		68,525	78,500	
880	0010	**SWIMMING POOL HEATERS** Not including wiring, external										880
	0020	piping, base or pad,										
	2000	Electric, 12 KW, 4,800 gallon pool	Q-19	3	8	Ea.	1,550	296		1,846	2,175	
	2020	15 KW, 7,200 gallon pool		2.80	8.571		1,550	315		1,865	2,200	
	2040	24 KW, 9,600 gallon pool		2.40	10		2,100	370		2,470	2,875	
	2060	30 KW, 12,000 gallon pool		2	12		2,175	445		2,620	3,075	
	2080	35 KW, 14,400 gallon pool		1.60	15		2,175	555		2,730	3,225	
	2100	55 KW, 24,000 gallon pool		1.20	20		3,000	740		3,740	4,400	
	9000	To select pool heater: 12 BTUH x S.F. pool area										
	9010	X temperature differential =required output										
	9050	For electric, KW = gallons x 2.5 divided by 1000										
	9100	For family home type pool, double the										
	9110	Rated gallon capacity = 1/2°F rise per hour										

15530 | Furnaces

			CREW	DAILY OUTPUT	LABOR-HOURS	UNIT	MAT.	LABOR	EQUIP.	TOTAL	TOTAL INCL O&P	
400	0010	**FURNACES** Hot air heating, blowers, standard controls										400
	1000	Electric, UL listed										
	1020	10.2 MBH	Q-20	5	4	Ea.	315	143		458	565	
	1040	17.1 MBH		4.80	4.167		330	149		479	585	
	1060	27.3 MBH		4.60	4.348		395	156		551	670	
	1100	34.1 MBH		4.40	4.545		405	163		568	695	
	1120	51.6 MBH		4.20	4.762		505	171		676	815	
	1140	68.3 MBH		4	5		610	179		789	945	
	1160	85.3 MBH		3.80	5.263		670	189		859	1,025	

15700 | Heating/Ventilating/Air Conditioning Equipment

15760 | Terminal Heating & Cooling Units

			CREW	DAILY OUTPUT	LABOR-HOURS	UNIT	MAT.	LABOR	EQUIP.	TOTAL	TOTAL INCL O&P	
200	0010	**DUCT HEATERS** Electric, 480 V, 3 Ph										200
	0020	Finned tubular insert, 500°F										
	0100	8" wide x 6" high, 4.0 kW	Q-20	16	1.250	Ea.	540	45		585	665	
	0120	12" high, 8.0 kW		15	1.333		895	48		943	1,050	
	0140	18" high, 12.0 kW		14	1.429		1,250	51		1,301	1,450	
	0160	24" high, 16.0 kW		13	1.538		1,625	55		1,680	1,850	
	0180	30" high, 20.0 kW		12	1.667		1,975	59.50		2,034.50	2,275	
	0300	12" wide x 6" high, 6.7 kW		15	1.333		575	48		623	705	
	0320	12" high, 13.3 kW		14	1.429		925	51		976	1,100	
	0340	18" high, 20.0 kW		13	1.538		1,300	55		1,355	1,500	

Important: See the Reference Section for critical supporting data - Reference Nos., Crews, & City Cost Indexes

	15760	Terminal Heating & Cooling Units	CREW	DAILY OUTPUT	LABOR-HOURS	UNIT	2004 BARE COSTS				TOTAL INCL O&P	
							MAT.	LABOR	EQUIP.	TOTAL		
200	0360	24" high, 26.7 kW	Q-20	12	1.667	Ea.	1,675	59.50		1,734.50	1,925	200
	0380	30" high, 33.3 kW		11	1.818		2,050	65		2,115	2,350	
	0500	18" wide x 6" high, 13.3 kW		14	1.429		605	51		656	750	
	0520	12" high, 26.7 kW		13	1.538		1,050	55		1,105	1,250	
	0540	18" high, 40.0 kW		12	1.667		1,400	59.50		1,459.50	1,650	
	0560	24" high, 53.3 kW		11	1.818		1,875	65		1,940	2,175	
	0580	30" high, 66.7 kW		10	2		2,350	71.50		2,421.50	2,675	
	0700	24" wide x 6" high, 17.8 kW		13	1.538		670	55		725	820	
	0720	12" high, 35.6 kW		12	1.667		1,150	59.50		1,209.50	1,375	
	0740	18" high, 53.3 kW		11	1.818		1,600	65		1,665	1,850	
	0760	24" high, 71.1 kW		10	2		2,075	71.50		2,146.50	2,400	
	0780	30" high, 88.9 kW		9	2.222		2,550	79.50		2,629.50	2,925	
	0900	30" wide x 6" high, 22.2 kW		12	1.667		710	59.50		769.50	870	
	0920	12" high, 44.4 kW		11	1.818		1,225	65		1,290	1,425	
	0940	18" high, 66.7 kW		10	2		1,675	71.50		1,746.50	1,950	
	0960	24" high, 88.9 kW		9	2.222		2,175	79.50		2,254.50	2,525	
	0980	30" high, 111.0 kW	▼	8	2.500	▼	2,700	89.50		2,789.50	3,100	
	1400	Note decreased kW available for										
	1410	each duct size at same cost										
	1420	See line 5000 for modifications and accessories										
	2000	Finned tubular flange with insulated										
	2020	terminal box, 500°F										
	2100	12" wide x 36" high, 54 kW	Q-20	10	2	Ea.	2,650	71.50		2,721.50	3,025	
	2120	40" high, 60 kW		9	2.222		3,050	79.50		3,129.50	3,500	
	2200	24" wide x 36" high, 118.8 kW		9	2.222		2,925	79.50		3,004.50	3,350	
	2220	40" high, 132 kW		8	2.500		3,350	89.50		3,439.50	3,800	
	2400	36" wide x 8" high, 40 kW		11	1.818		1,350	65		1,415	1,575	
	2420	16" high, 80 kW		10	2		1,800	71.50		1,871.50	2,100	
	2440	24" high, 120 kW		9	2.222		2,375	79.50		2,454.50	2,725	
	2460	32" high, 160 kW		8	2.500		3,025	89.50		3,114.50	3,450	
	2480	36" high, 180 kW		7	2.857		3,725	102		3,827	4,250	
	2500	40" high, 200 kW		6	3.333		4,125	119		4,244	4,725	
	2600	40" wide x 8" high, 45 kW		11	1.818		1,400	65		1,465	1,650	
	2620	16" high, 90 kW		10	2		1,950	71.50		2,021.50	2,250	
	2640	24" high, 135 kW		9	2.222		2,475	79.50		2,554.50	2,850	
	2660	32" high, 180 kW		8	2.500		3,300	89.50		3,389.50	3,750	
	2680	36" high, 202.5 kW		7	2.857		3,800	102		3,902	4,325	
	2700	40" high, 225 kW		6	3.333		4,325	119		4,444	4,925	
	2800	48" wide x 8" high, 54.8 kW		10	2		1,475	71.50		1,546.50	1,725	
	2820	16" high, 109.8 kW		9	2.222		2,050	79.50		2,129.50	2,375	
	2840	24" high, 164.4 kW		8	2.500		2,625	89.50		2,714.50	3,000	
	2860	32" high, 219.2 kW		7	2.857		3,500	102		3,602	4,000	
	2880	36" high, 246.6 kW		6	3.333		4,025	119		4,144	4,600	
	2900	40" high, 274 kW		5	4		4,550	143		4,693	5,225	
	3000	56" wide x 8" high, 64 kW		9	2.222		1,700	79.50		1,779.50	1,975	
	3020	16" high, 128 kW		8	2.500		2,350	89.50		2,439.50	2,700	
	3040	24" high, 192 kW		7	2.857		2,825	102		2,927	3,275	
	3060	32" high, 256 kW		6	3.333		3,975	119		4,094	4,525	
	3080	36" high, 288kW		5	4		4,550	143		4,693	5,225	
	3100	40" high, 320kW		4	5		5,025	179		5,204	5,800	
	3200	64" wide x 8" high, 74kW		8	2.500		1,725	89.50		1,814.50	2,025	
	3220	16" high, 148kW		7	2.857		2,400	102		2,502	2,800	
	3240	24" high, 222kW		6	3.333		3,075	119		3,194	3,550	
	3260	32" high, 296kW		5	4		4,100	143		4,243	4,750	
	3280	36" high, 333kW		4	5		4,900	179		5,079	5,650	
	3300	40" high, 370kW	▼	3	6.667	▼	5,400	239		5,639	6,325	

MECHANICAL 15

15760	Terminal Heating & Cooling Units	CREW	DAILY OUTPUT	LABOR-HOURS	UNIT	2004 BARE COSTS				TOTAL INCL O&P	
						MAT.	LABOR	EQUIP.	TOTAL		
200	3800	Note decreased kW available for									**200**
	3820	each duct size at same cost									
	5000	Duct heater modifications and accessories									
	5120	T.C.O. limit auto or manual reset	Q-20	42	.476	Ea.	88.50	17.05		105.55	124
	5140	Thermostat		28	.714		380	25.50		405.50	455
	5160	Overheat thermocouple (removable)		7	2.857		535	102		637	745
	5180	Fan interlock relay		18	1.111		129	40		169	203
	5200	Air flow switch		20	1		112	36		148	178
	5220	Split terminal box cover	▼	100	.200	▼	37	7.15		44.15	51.50
	8000	To obtain BTU multiply kW by 3413									
250	0010	**ELECTRIC HEATING**, not incl. conduit or feed wiring									**250**
	1100	Rule of thumb: Baseboard units, including control	1 Elec	4.40	1.818	kW	77	71.50		148.50	192
	1300	Baseboard heaters, 2' long, 375 watt		8	1	Ea.	29	39.50		68.50	90.50
	1400	3' long, 500 watt		8	1		34.50	39.50		74	96.50
	1600	4' long, 750 watt		6.70	1.194		41	47		88	115
	1800	5' long, 935 watt		5.70	1.404		48.50	55.50		104	136
	2000	6' long, 1125 watt		5	1.600		54	63		117	154
	2200	7' long, 1310 watt		4.40	1.818		59	71.50		130.50	172
	2400	8' long, 1500 watt		4	2		68	79		147	192
	2600	9' long, 1680 watt		3.60	2.222		76	87.50		163.50	214
	2800	10' long, 1875 watt	▼	3.30	2.424	▼	84.50	95.50		180	235
	2950	Wall heaters with fan, 120 to 277 volt									
	3160	Recessed, residential, 750 watt	1 Elec	6	1.333	Ea.	84.50	52.50		137	171
	3170	1000 watt		6	1.333		87	52.50		139.50	174
	3180	1250 watt		5	1.600		95.50	63		158.50	199
	3190	1500 watt		4	2		95.50	79		174.50	222
	3210	2000 watt		4	2		100	79		179	227
	3230	2500 watt		3.50	2.286		180	90		270	330
	3240	3000 watt		3	2.667		300	105		405	485
	3250	4000 watt		2.70	2.963		330	117		447	535
	3260	Commercial, 750 watt		6	1.333		140	52.50		192.50	232
	3270	1000 watt		6	1.333		140	52.50		192.50	232
	3280	1250 watt		5	1.600		140	63		203	248
	3290	1500 watt		4	2		140	79		219	271
	3300	2000 watt		4	2		140	79		219	271
	3310	2500 watt		3.50	2.286		140	90		230	288
	3320	3000 watt		3	2.667		247	105		352	430
	3330	4000 watt		2.70	2.963		247	117		364	445
	3600	Thermostats, integral		16	.500		17.80	19.70		37.50	49
	3800	Line voltage, 1 pole		8	1		21	39.50		60.50	82
	3810	2 pole		8	1		27	39.50		66.50	88
	3820	Low voltage, 1 pole	▼	8	1	▼	23	39.50		62.50	84
	4000	Heat trace system, 400 degree									
	4020	115V, 2.5 watts per L.F.	1 Elec	530	.015	L.F.	6.70	.59		7.29	8.25
	4030	5 watts per L.F.		530	.015		6.70	.59		7.29	8.25
	4050	10 watts per L.F.		530	.015		6.70	.59		7.29	8.25
	4060	208V, 5 watts per L.F.		530	.015		6.70	.59		7.29	8.25
	4080	480V, 8 watts per L.F.	▼	530	.015	▼	6.70	.59		7.29	8.25
	4200	Heater raceway									
	4260	Heat transfer cement									
	4280	1 gallon				Ea.	52.50			52.50	57.50
	4300	5 gallon				"	214			214	236
	4320	Snap band, clamp									
	4340	3/4" pipe size	1 Elec	470	.017	Ea.		.67		.67	1
	4360	1" pipe size		444	.018			.71		.71	1.06
	4380	1-1/4" pipe size	▼	400	.020	▼		.79		.79	1.17

Boxes in table: R15770-200 (rows 4000/4020), R15770-201 (rows 4030/4050)

Important: See the Reference Section for critical supporting data - Reference Nos., Crews, & City Cost Indexes

		15760	Terminal Heating & Cooling Units	CREW	DAILY OUTPUT	LABOR-HOURS	UNIT	2004 BARE COSTS				TOTAL INCL O&P	
								MAT.	LABOR	EQUIP.	TOTAL		
250	4400		1-1/2" pipe size	1 Elec	355	.023	Ea.		.89		.89	1.32	**250**
	4420		2" pipe size		320	.025			.99		.99	1.47	
	4440		3" pipe size		160	.050			1.97		1.97	2.93	
	4460		4" pipe size		100	.080			3.15		3.15	4.69	
	4480		Thermostat NEMA 3R, 22 amp, 0-150 Deg, 10' cap.		8	1		179	39.50		218.50	255	
	4500		Thermostat NEMA 4X, 25 amp, 40 Deg, 5-1/2' cap.		7	1.143		179	45		224	263	
	4520		Thermostat NEMA 4X, 22 amp, 25-325 Deg, 10' cap.		7	1.143		480	45		525	595	
	4540		Thermostat NEMA 4X, 22 amp, 15-140 Deg,		6	1.333		405	52.50		457.50	525	
	4580		Thermostat NEMA 4,7,9, 22 amp, 25-325 Deg, 10' cap.		3.60	2.222		595	87.50		682.50	785	
	4600		Thermostat NEMA 4,7,9, 22 amp, 15-140 Deg,		3	2.667		590	105		695	800	
	4720		Fiberglass application tape, 36 yard roll		11	.727		54	28.50		82.50	102	
	5000		Radiant heating ceiling panels, 2' x 4', 500 watt		16	.500		195	19.70		214.70	244	
	5050		750 watt		16	.500		215	19.70		234.70	266	
	5200		For recessed plaster frame, add		32	.250		39	9.85		48.85	57.50	
	5300		Infrared quartz heaters, 120 volts, 1000 watts		6.70	1.194		119	47		166	201	
	5350		1500 watt		5	1.600		119	63		182	225	
	5400		240 volts, 1500 watt		5	1.600		119	63		182	225	
	5450		2000 watt		4	2		119	79		198	248	
	5500		3000 watt		3	2.667		139	105		244	310	
	5550		4000 watt		2.60	3.077		139	121		260	330	
	5570		Modulating control	▼	.80	10	▼	72.50	395		467.50	665	
	5600		Unit heaters, heavy duty, with fan & mounting bracket										
	5650		Single phase, 208-240-277 volt, 3 kW	1 Elec	3.20	2.500	Ea.	320	98.50		418.50	495	
	5750		5 kW		2.40	3.333		330	131		461	555	
	5800		7 kW		1.90	4.211		510	166		676	805	
	5850		10 kW		1.30	6.154		580	242		822	1,000	
	5950		15 kW		.90	8.889		940	350		1,290	1,550	
	6000		480 volt, 3 kW		3.30	2.424		400	95.50		495.50	580	
	6020		4 kW		3	2.667		360	105		465	550	
	6040		5 kW		2.60	3.077		365	121		486	585	
	6060		7 kW		2	4		545	158		703	835	
	6080		10 kW		1.40	5.714		595	225		820	990	
	6100		13 kW		1.10	7.273		940	287		1,227	1,450	
	6120		15 kW		1	8		940	315		1,255	1,500	
	6140		20 kW		.90	8.889		1,225	350		1,575	1,850	
	6300		3 phase, 208-240 volt, 5 kW		2.40	3.333		310	131		441	535	
	6320		7 kW		1.90	4.211		480	166		646	775	
	6340		10 kW		1.30	6.154		520	242		762	930	
	6360		15 kW		.90	8.889		870	350		1,220	1,475	
	6380		20 kW		.70	11.429		1,225	450		1,675	2,025	
	6400		25 kW		.50	16		1,450	630		2,080	2,550	
	6500		480 volt, 5 kW		2.60	3.077		435	121		556	660	
	6520		7 kW		2	4		550	158		708	840	
	6540		10 kW		1.40	5.714		580	225		805	975	
	6560		13 kW		1.10	7.273		940	287		1,227	1,450	
	6580		15 kW		1	8		940	315		1,255	1,500	
	6600		20 kW		.90	8.889		1,225	350		1,575	1,850	
	6620		25 kW		.60	13.333		1,450	525		1,975	2,375	
	6630		30 kW		.70	11.429		1,700	450		2,150	2,525	
	6640		40 kW		.60	13.333		2,150	525		2,675	3,150	
	6650		50 kW	▼	.50	16	▼	2,600	630		3,230	3,800	
	6800		Vertical discharge heaters, with fan										
	6820		Single phase, 208-240-277 volt, 10 kW	1 Elec	1.30	6.154	Ea.	535	242		777	950	
	6840		15 kW		.90	8.889		900	350		1,250	1,500	
	6900		3 phase, 208-240 volt, 10 kW		1.30	6.154		520	242		762	930	
	6920		15 kW		.90	8.889		870	350		1,220	1,475	

R15770-200

R15770-201

MECHANICAL 15

				DAILY	LABOR-			2004 BARE COSTS				TOTAL	
15760		**Terminal Heating & Cooling Units**	CREW	OUTPUT	HOURS	UNIT	MAT.	LABOR	EQUIP.	TOTAL		INCL O&P	
250	6940	20 kW		1 Elec	.70	11.429	Ea.	1,225	450		1,675	2,025	**250**
	6960	25 kW	R15770 -200		.50	16		1,450	630		2,080	2,550	
	6980	30 kW			.40	20		1,700	790		2,490	3,025	
	7000	40 kW	R15770 -201		.36	22.222		2,150	875		3,025	3,675	
	7020	50 kW			.32	25		2,600	985		3,585	4,325	
	7100	480 volt, 10 kW			1.40	5.714		580	225		805	975	
	7120	15 kW			1	8		940	315		1,255	1,500	
	7140	20 kW			.90	8.889		1,225	350		1,575	1,850	
	7160	25 kW			.60	13.333		1,450	525		1,975	2,375	
	7180	30 kW			.50	16		1,700	630		2,330	2,800	
	7200	40 kW			.40	20		2,150	790		2,940	3,550	
	7220	50 kW			.35	22.857		2,600	900		3,500	4,200	
	7410	Sill height convector heaters, 5" high x 2' long, 500 watt			6.70	1.194		222	47		269	315	
	7420	3' long, 750 watt			6.50	1.231		262	48.50		310.50	360	
	7430	4' long, 1000 watt			6.20	1.290		305	51		356	410	
	7440	5' long, 1250 watt			5.50	1.455		345	57.50		402.50	465	
	7450	6' long, 1500 watt			4.80	1.667		390	65.50		455.50	530	
	7460	8' long, 2000 watt			3.60	2.222		530	87.50		617.50	715	
	7470	10' long, 2500 watt		▼	3	2.667	▼	660	105		765	880	
	7900	Cabinet convector heaters, 240 volt											
	7920	3' long, 2000 watt		1 Elec	5.30	1.509	Ea.	1,375	59.50		1,434.50	1,625	
	7940	3000 watt			5.30	1.509		1,450	59.50		1,509.50	1,700	
	7960	4000 watt			5.30	1.509		1,500	59.50		1,559.50	1,750	
	7980	6000 watt			4.60	1.739		1,550	68.50		1,618.50	1,800	
	8000	8000 watt			4.60	1.739		1,625	68.50		1,693.50	1,875	
	8020	4' long, 4000 watt			4.60	1.739		1,500	68.50		1,568.50	1,775	
	8040	6000 watt			4	2		1,575	79		1,654	1,850	
	8060	8000 watt			4	2		1,650	79		1,729	1,925	
	8080	10,000 watt		▼	4	2	▼	1,650	79		1,729	1,950	
	8100	Available also in 208 or 277 volt											
	8200	Cabinet unit heaters, 120 to 277 volt, 1 pole,											
	8220	wall mounted, 2 kW		1 Elec	4.60	1.739	Ea.	1,375	68.50		1,443.50	1,625	
	8230	3 kW			4.60	1.739		1,450	68.50		1,518.50	1,700	
	8240	4 kW			4.40	1.818		1,500	71.50		1,571.50	1,750	
	8250	5 kW			4.40	1.818		1,550	71.50		1,621.50	1,800	
	8260	6 kW			4.20	1.905		1,550	75		1,625	1,800	
	8270	8 kW			4	2		1,625	79		1,704	1,900	
	8280	10 kW			3.80	2.105		1,650	83		1,733	1,950	
	8290	12 kW			3.50	2.286		1,700	90		1,790	1,975	
	8300	13.5 kW			2.90	2.759		1,725	109		1,834	2,050	
	8310	16 kW			2.70	2.963		1,750	117		1,867	2,100	
	8320	20 kW			2.30	3.478		2,650	137		2,787	3,125	
	8330	24 kW			1.90	4.211		2,700	166		2,866	3,200	
	8350	Recessed, 2 kW			4.40	1.818		1,375	71.50		1,446.50	1,625	
	8370	3 kW			4.40	1.818		1,450	71.50		1,521.50	1,700	
	8380	4 kW			4.20	1.905		1,500	75		1,575	1,750	
	8390	5 kW			4.20	1.905		1,550	75		1,625	1,800	
	8400	6 kW			4	2		1,575	79		1,654	1,850	
	8410	8 kW			3.80	2.105		1,650	83		1,733	1,925	
	8420	10 kW			3.50	2.286		2,025	90		2,115	2,350	
	8430	12 kW			2.90	2.759		2,050	109		2,159	2,400	
	8440	13.5 kW			2.70	2.963		2,075	117		2,192	2,450	
	8450	16 kW			2.30	3.478		2,075	137		2,212	2,475	
	8460	20 kW			1.90	4.211		2,500	166		2,666	3,000	
	8470	24 kW			1.60	5		2,575	197		2,772	3,125	
	8490	Ceiling mounted, 2 kW		▼	3.20	2.500	▼	1,375	98.50		1,473.50	1,675	

Important: See the Reference Section for critical supporting data - Reference Nos., Crews, & City Cost Indexes

	15760	Terminal Heating & Cooling Units	CREW	DAILY OUTPUT	LABOR-HOURS	UNIT	MAT.	LABOR	EQUIP.	TOTAL	TOTAL INCL O&P	
250	8510	3 kW R15770-200	1 Elec	3.20	2.500	Ea.	1,450	98.50		1,548.50	1,750	250
	8520	4 kW		3	2.667		1,500	105		1,605	1,800	
	8530	5 kW R15770-201		3	2.667		1,550	105		1,655	1,850	
	8540	6 kW		2.80	2.857		1,550	113		1,663	1,875	
	8550	8 kW		2.40	3.333		1,625	131		1,756	1,975	
	8560	10 kW		2.20	3.636		1,650	143		1,793	2,050	
	8570	12 kW		2	4		1,700	158		1,858	2,075	
	8580	13.5 kW		1.50	5.333		1,725	210		1,935	2,225	
	8590	16 kW		1.30	6.154		1,750	242		1,992	2,275	
	8600	20 kW		.90	8.889		2,500	350		2,850	3,275	
	8610	24 kW	▼	.60	13.333	▼	2,550	525		3,075	3,575	
	8630	208 to 480 V, 3 pole										
	8650	Wall mounted, 2 kW	1 Elec	4.60	1.739	Ea.	1,375	68.50		1,443.50	1,625	
	8670	3 kW		4.60	1.739		1,450	68.50		1,518.50	1,700	
	8680	4 kW		4.40	1.818		1,500	71.50		1,571.50	1,750	
	8690	5 kW		4.40	1.818		1,550	71.50		1,621.50	1,800	
	8700	6 kW		4.20	1.905		1,550	75		1,625	1,800	
	8710	8 kW		4	2		1,625	79		1,704	1,900	
	8720	10 kW		3.80	2.105		1,650	83		1,733	1,950	
	8730	12 kW		3.50	2.286		1,700	90		1,790	1,975	
	8740	13.5 kW		2.90	2.759		1,725	109		1,834	2,050	
	8750	16 kW		2.70	2.963		1,750	117		1,867	2,100	
	8760	20 kW		2.30	3.478		2,500	137		2,637	2,950	
	8770	24 kW		1.90	4.211		2,550	166		2,716	3,050	
	8790	Recessed, 2 kW		4.40	1.818		1,375	71.50		1,446.50	1,625	
	8810	3 kW		4.40	1.818		1,450	71.50		1,521.50	1,700	
	8820	4 kW		4.20	1.905		1,500	75		1,575	1,750	
	8830	5 kW		4.20	1.905		1,550	75		1,625	1,800	
	8840	6 kW		4	2		1,550	79		1,629	1,825	
	8850	8 kW		3.80	2.105		1,625	83		1,708	1,900	
	8860	10 kW		3.50	2.286		1,650	90		1,740	1,950	
	8870	12 kW		2.90	2.759		1,700	109		1,809	2,000	
	8880	13.5 kW		2.70	2.963		1,725	117		1,842	2,075	
	8890	16 kW		2.30	3.478		1,750	137		1,887	2,125	
	8900	20 kW		1.90	4.211		2,500	166		2,666	3,000	
	8920	24 kW		1.60	5		2,550	197		2,747	3,100	
	8940	Ceiling mount, 2 kW		3.20	2.500		1,375	98.50		1,473.50	1,675	
	8950	3 kW		3.20	2.500		1,450	98.50		1,548.50	1,750	
	8960	4 kW		3	2.667		1,500	105		1,605	1,800	
	8970	5 kW		3	2.667		1,550	105		1,655	1,850	
	8980	6 kW		2.80	2.857		1,550	113		1,663	1,875	
	8990	8 kW		2.40	3.333		1,625	131		1,756	1,975	
	9000	10 kW		2.20	3.636		1,650	143		1,793	2,050	
	9020	13.5 kW		1.50	5.333		1,725	210		1,935	2,225	
	9030	16 kW		1.30	6.154		1,750	242		1,992	2,275	
	9040	20 kW		.90	8.889		2,500	350		2,850	3,275	
	9060	24 kW	▼	.60	13.333	▼	2,550	525		3,075	3,575	
700	0010	**INFRARED UNIT**										700
	2000	Electric, single or three phase										
	2050	6 kW, 20,478 BTU	1 Elec	2.30	3.478	Ea.	370	137		507	615	
	2100	13.5 KW, 40,956 BTU		2.20	3.636		585	143		728	860	
	2150	24 KW, 81,912 BTU	▼	2	4	▼	1,250	158		1,408	1,600	

For expanded coverage of these items see *Means Mechanical or Plumbing Cost Data 2004*

15830	Fans	CREW	DAILY OUTPUT	LABOR-HOURS	UNIT	2004 BARE COSTS				TOTAL INCL O&P
						MAT.	LABOR	EQUIP.	TOTAL	
100	0010	**FANS**									100
	0020	Air conditioning and process air handling									
	0030	Axial flow, compact, low sound, 2.5" S.P.									
	0050	3,800 CFM, 5 HP	Q-20	3.40	5.882	Ea.	3,700	211		3,911	4,400
	0080	6,400 CFM, 5 HP		2.80	7.143		4,125	256		4,381	4,950
	0100	10,500 CFM, 7-1/2 HP		2.40	8.333		5,150	299		5,449	6,100
	0120	15,600 CFM, 10 HP		1.60	12.500		6,475	450		6,925	7,800
	0140	23,000 CFM, 15 HP		.70	28.571		9,925	1,025		10,950	12,500
	0160	28,000 CFM, 20 HP	▼	.40	50	▼	11,100	1,800		12,900	14,900
	0200	In-line centrifugal, supply/exhaust booster									
	0220	aluminum wheel/hub, disconnect switch, 1/4" S.P.									
	0240	500 CFM, 10" diameter connection	Q-20	3	6.667	Ea.	820	239		1,059	1,275
	0260	1,380 CFM, 12" diameter connection		2	10		870	360		1,230	1,500
	0280	1,520 CFM, 16" diameter connection		2	10		1,225	360		1,585	1,900
	0300	2,560 CFM, 18" diameter connection		1	20		1,025	715		1,740	2,225
	0320	3,480 CFM, 20" diameter connection		.80	25		1,225	895		2,120	2,725
	0326	5,080 CFM, 20" diameter connection		.75	26.667		1,325	955		2,280	2,925
	1500	Vaneaxial, low pressure, 2000 CFM, 1/2 HP		3.60	5.556		1,625	199		1,824	2,075
	1520	4,000 CFM, 1 HP		3.20	6.250		1,650	224		1,874	2,175
	1540	8,000 CFM, 2 HP		2.80	7.143		1,950	256		2,206	2,550
	1560	16,000 CFM, 5 HP	▼	2.40	8.333	▼	2,575	299		2,874	3,275
	2000	Blowers, direct drive with motor, complete									
	2020	1045 CFM @ .5" S.P., 1/5 HP	Q-20	18	1.111	Ea.	145	40		185	221
	2040	1385 CFM @ .5" S.P., 1/4 HP		18	1.111		147	40		187	223
	2060	1640 CFM @ .5" S.P., 1/3 HP		18	1.111		175	40		215	254
	2080	1760 CFM @ .5" S.P., 1/2 HP	▼	18	1.111	▼	190	40		230	270
	2090	4 speed									
	2100	1164 to 1739 CFM @ .5" S.P., 1/3 HP	Q-20	16	1.250	Ea.	251	45		296	345
	2120	1467 to 2218 CFM @ 1.0" S.P., 3/4 HP	"	14	1.429	"	325	51		376	440
	2500	Ceiling fan, right angle, extra quiet, 0.10" S.P.									
	2520	95 CFM	Q-20	20	1	Ea.	130	36		166	198
	2540	210 CFM		19	1.053		153	37.50		190.50	227
	2560	385 CFM		18	1.111		195	40		235	275
	2580	885 CFM		16	1.250		385	45		430	490
	2600	1,650 CFM		13	1.538		530	55		585	670
	2620	2,960 CFM	▼	11	1.818		710	65		775	880
	2680	For speed control switch, add	1 Elec	16	.500	▼	71	19.70		90.70	108
	3000	Paddle blade air circulator, 3 speed switch									
	3020	42", 5,000 CFM high, 3000 CFM low	1 Elec	2.40	3.333	Ea.	73.50	131		204.50	276
	3040	52", 6,500 CFM high, 4000 CFM low	"	2.20	3.636	"	74.50	143		217.50	295
	3100	For antique white motor, same cost									
	3200	For brass plated motor, same cost									
	3300	For light adaptor kit, add				Ea.	27.50			27.50	30
	3500	Centrifugal, airfoil, motor and drive, complete									
	3520	1000 CFM, 1/2 HP	Q-20	2.50	8	Ea.	905	287		1,192	1,425
	3540	2,000 CFM, 1 HP		2	10		970	360		1,330	1,625
	3560	4,000 CFM, 3 HP		1.80	11.111		1,250	400		1,650	1,975
	3580	8,000 CFM, 7-1/2 HP		1.40	14.286		1,825	510		2,335	2,800
	3600	12,000 CFM, 10 HP	▼	1	20	▼	2,700	715		3,415	4,075
	4500	Corrosive fume resistant, plastic									
	4600	roof ventilators, centrifugal, V belt drive, motor									
	4620	1/4" S.P., 250 CFM, 1/4 HP	Q-20	6	3.333	Ea.	2,425	119		2,544	2,825
	4640	895 CFM, 1/3 HP		5	4		2,625	143		2,768	3,100
	4660	1630 CFM, 1/2 HP		4	5		3,100	179		3,279	3,675
	4680	2240 CFM, 1 HP		3	6.667		3,225	239		3,464	3,925
	4700	3810 CFM, 2 HP	▼	2	10		3,600	360		3,960	4,500

Important: See the Reference Section for critical supporting data - Reference Nos., Crews, & City Cost Indexes

15830 | Fans

		CREW	DAILY OUTPUT	LABOR-HOURS	UNIT	2004 BARE COSTS				TOTAL INCL O&P
						MAT.	LABOR	EQUIP.	TOTAL	
100	4710	5000 CFM, 2 HP	Q-20	1.80	11.111	Ea.	7,550	400		7,950	8,900
	4715	8000 CFM, 5 HP		1.40	14.286		8,400	510		8,910	10,000
	4720	11760 CFM, 5 HP		1	20		8,525	715		9,240	10,500
	4740	18810 CFM, 10 HP	▼	.70	28.571	▼	8,850	1,025		9,875	11,300
	4800	For intermediate capacity, motors may be varied									
	4810	For explosion proof motor, add				Ea.	15%				
	5000	Utility set, centrifugal, V belt drive, motor									
	5020	1/4" S.P., 1200 CFM, 1/4 HP	Q-20	6	3.333	Ea.	2,550	119		2,669	2,975
	5040	1520 CFM, 1/3 HP		5	4		2,550	143		2,693	3,025
	5060	1850 CFM, 1/2 HP		4	5		2,550	179		2,729	3,100
	5080	2180 CFM, 3/4 HP		3	6.667		2,600	239		2,839	3,225
	5100	1/2" S.P., 3600 CFM, 1 HP		2	10		3,750	360		4,110	4,675
	5120	4250 CFM, 1-1/2 HP		1.60	12.500		3,800	450		4,250	4,850
	5140	4800 CFM, 2 HP		1.40	14.286		3,850	510		4,360	5,000
	5160	6920 CFM, 5 HP		1.30	15.385		3,975	550		4,525	5,225
	5180	7700 CFM, 7-1/2 HP	▼	1.20	16.667	▼	4,100	595		4,695	5,400
	5200	For explosion proof motor, add					15%				
	5500	Industrial exhauster, for air which may contain granular material									
	5520	1000 CFM, 1-1/2 HP	Q-20	2.50	8	Ea.	2,100	287		2,387	2,725
	5540	2000 CFM, 3 HP		2	10		2,525	360		2,885	3,325
	5560	4000 CFM, 7-1/2 HP		1.80	11.111		4,075	400		4,475	5,075
	5580	8000 CFM, 15 HP		1.40	14.286		6,375	510		6,885	7,775
	5600	12,000 CFM, 30 HP	▼	1	20	▼	8,850	715		9,565	10,800
	6000	Propeller exhaust, wall shutter, 1/4" S.P.									
	6020	Direct drive, two speed									
	6100	375 CFM, 1/10 HP	Q-20	10	2	Ea.	233	71.50		304.50	365
	6120	730 CFM, 1/7 HP		9	2.222		260	79.50		339.50	405
	6140	1000 CFM, 1/8 HP		8	2.500		360	89.50		449.50	530
	6160	1890 CFM, 1/4 HP		7	2.857		370	102		472	560
	6180	3275 CFM, 1/2 HP		6	3.333		375	119		494	590
	6200	4720 CFM, 1 HP	▼	5	4	▼	580	143		723	855
	6300	V-belt drive, 3 phase									
	6320	6175 CFM, 3/4 HP	Q-20	5	4	Ea.	485	143		628	755
	6340	7500 CFM, 3/4 HP		5	4		510	143		653	780
	6360	10,100 CFM, 1 HP		4.50	4.444		625	159		784	930
	6380	14,300 CFM, 1-1/2 HP		4	5		735	179		914	1,075
	6400	19,800 CFM, 2 HP		3	6.667		840	239		1,079	1,300
	6420	26,250 CFM, 3 HP		2.60	7.692		1,000	276		1,276	1,525
	6440	38,500 CFM, 5 HP		2.20	9.091		1,200	325		1,525	1,800
	6460	46,000 CFM, 7-1/2 HP		2	10		1,275	360		1,635	1,950
	6480	51,500 CFM, 10 HP	▼	1.80	11.111	▼	1,325	400		1,725	2,050
	6650	Residential, bath exhaust, grille, back draft damper									
	6660	50 CFM	Q-20	24	.833	Ea.	30	30		60	78.50
	6670	110 CFM		22	.909		50	32.50		82.50	105
	6680	Light combination, squirrel cage, 100 watt, 70 CFM	▼	24	.833	▼	61.50	30		91.50	113
	6700	Light/heater combination, ceiling mounted									
	6710	70 CFM, 1450 watt	Q-20	24	.833	Ea.	74.50	30		104.50	128
	6800	Heater combination, recessed, 70 CFM		24	.833		35.50	30		65.50	84.50
	6820	With 2 infrared bulbs		23	.870		53.50	31		84.50	107
	6900	Kitchen exhaust, grille, complete, 160 CFM		22	.909		62.50	32.50		95	118
	6910	180 CFM		20	1		53	36		89	113
	6920	270 CFM		18	1.111		95.50	40		135.50	166
	6930	350 CFM	▼	16	1.250	▼	135	45		180	216
	6940	Residential roof jacks and wall caps									
	6944	Wall cap with back draft damper									
	6946	3" & 4" dia. round duct	1 Shee	11	.727	Ea.	12.80	28		40.80	57.50

MECHANICAL 15

100

		CREW	DAILY OUTPUT	LABOR-HOURS	UNIT	2004 BARE COSTS				TOTAL INCL O&P
15830	**Fans**					MAT.	LABOR	EQUIP.	TOTAL	
6948	6" dia. round duct	1 Shee	11	.727	Ea.	31	28		59	77.50
6958	Roof jack with bird screen and back draft damper									
6960	3" & 4" dia. round duct	1 Shee	11	.727	Ea.	12.30	28		40.30	57
6962	3-1/4" x 10" rectangular duct	"	10	.800	"	22.50	31		53.50	72.50
6980	Transition									
6982	3-1/4" x 10" to 6" dia. round	1 Shee	20	.400	Ea.	13.85	15.50		29.35	39.50
7000	Roof exhauster, centrifugal, aluminum housing, 12" galvanized									
7020	curb, bird screen, back draft damper, 1/4" S.P.									
7100	Direct drive, 320 CFM, 11" sq. damper	Q-20	7	2.857	Ea.	320	102		422	505
7120	600 CFM, 11" sq. damper		6	3.333		325	119		444	535
7140	815 CFM, 13" sq. damper		5	4		325	143		468	575
7160	1450 CFM, 13" sq. damper		4.20	4.762		415	171		586	715
7180	2050 CFM, 16" sq. damper		4	5		415	179		594	730
7200	V-belt drive, 1650 CFM, 12" sq. damper		6	3.333		725	119		844	975
7220	2750 CFM, 21" sq. damper		5	4		840	143		983	1,150
7230	3500 CFM, 21" sq. damper		4.50	4.444		925	159		1,084	1,275
7240	4910 CFM, 23" sq. damper		4	5		1,150	179		1,329	1,550
7260	8525 CFM, 28" sq. damper		3	6.667		1,425	239		1,664	1,950
7280	13,760 CFM, 35" sq. damper		2	10		1,975	360		2,335	2,725
7300	20,558 CFM, 43" sq. damper		1	20		4,200	715		4,915	5,725
7320	For 2 speed winding, add					15%				
7340	For explosionproof motor, add				Ea.	330			330	360
7360	For belt driven, top discharge, add					15%				
7500	Utility set, steel construction, pedestal, 1/4" S.P.									
7520	Direct drive, 150 CFM, 1/8 HP	Q-20	6.40	3.125	Ea.	620	112		732	850
7540	485 CFM, 1/6 HP		5.80	3.448		780	124		904	1,050
7560	1950 CFM, 1/2 HP		4.80	4.167		915	149		1,064	1,225
7580	2410 CFM, 3/4 HP		4.40	4.545		1,700	163		1,863	2,100
7600	3328 CFM, 1-1/2 HP		3	6.667		1,875	239		2,114	2,450
7680	V-belt drive, drive cover, 3 phase									
7700	800 CFM, 1/4 HP	Q-20	6	3.333	Ea.	460	119		579	690
7720	1,300 CFM, 1/3 HP		5	4		485	143		628	755
7740	2,000 CFM, 1 HP		4.60	4.348		575	156		731	865
7760	2,900 CFM, 3/4 HP		4.20	4.762		775	171		946	1,100
7780	3,600 CFM, 3/4 HP		4	5		950	179		1,129	1,325
7800	4,800 CFM, 1 HP		3.50	5.714		1,125	205		1,330	1,525
7820	6,700 CFM, 1-1/2 HP		3	6.667		1,375	239		1,614	1,900
7830	7,500 CFM, 2 HP		2.50	8		1,875	287		2,162	2,500
7840	11,000 CFM, 3 HP		2	10		2,525	360		2,885	3,325
7860	13,000 CFM, 3 HP		1.60	12.500		2,550	450		3,000	3,475
7880	15,000 CFM, 5 HP		1	20		2,650	715		3,365	4,000
7900	17,000 CFM, 7-1/2 HP		.80	25		2,825	895		3,720	4,475
7920	20,000 CFM, 7-1/2 HP		.80	25		3,375	895		4,270	5,075
8000	Ventilation, residential									
8020	Attic, roof type									
8030	Aluminum dome, damper & curb									
8040	6" diameter, 300 CFM	1 Elec	16	.500	Ea.	238	19.70		257.70	291
8050	7" diameter, 450 CFM		15	.533		259	21		280	315
8060	9" diameter, 900 CFM		14	.571		415	22.50		437.50	495
8080	12" diameter, 1000 CFM (gravity)		10	.800		268	31.50		299.50	340
8090	16" diameter, 1500 CFM (gravity)		9	.889		325	35		360	405
8100	20" diameter, 2500 CFM (gravity)		8	1		395	39.50		434.50	495
8110	26" diameter, 4000 CFM (gravity)		7	1.143		480	45		525	595
8120	32" diameter, 6500 CFM (gravity)		6	1.333		660	52.50		712.50	805
8130	38" diameter, 8000 CFM (gravity)		5	1.600		975	63		1,038	1,175
8140	50" diameter, 13,000 CFM (gravity)		4	2		1,425	79		1,504	1,675

15 MECHANICAL

			DAILY	LABOR-		2004 BARE COSTS				TOTAL
15830	**Fans**	CREW	OUTPUT	HOURS	UNIT	MAT.	LABOR	EQUIP.	TOTAL	INCL O&P
8160	Plastic, ABS dome									
8180	1050 CFM	1 Elec	14	.571	Ea.	79	22.50		101.50	120
8200	1600 CFM	"	12	.667	"	118	26.50		144.50	169
8240	Attic, wall type, with shutter, one speed									
8250	12" diameter, 1000 CFM	1 Elec	14	.571	Ea.	187	22.50		209.50	240
8260	14" diameter, 1500 CFM		12	.667		203	26.50		229.50	262
8270	16" diameter, 2000 CFM	↓	9	.889	↓	229	35		264	305
8290	Whole house, wall type, with shutter, one speed									
8300	30" diameter, 4800 CFM	1 Elec	7	1.143	Ea.	490	45		535	605
8310	36" diameter, 7000 CFM		6	1.333		535	52.50		587.50	670
8320	42" diameter, 10,000 CFM		5	1.600		600	63		663	755
8330	48" diameter, 16,000 CFM	↓	4	2		745	79		824	935
8340	For two speed, add				↓	44.50			44.50	49
8350	Whole house, lay-down type, with shutter, one speed									
8360	30" diameter, 4500 CFM	1 Elec	8	1	Ea.	525	39.50		564.50	635
8370	36" diameter, 6500 CFM		7	1.143		560	45		605	685
8380	42" diameter, 9000 CFM		6	1.333		620	52.50		672.50	760
8390	48" diameter, 12,000 CFM	↓	5	1.600		700	63		763	865
8440	For two speed, add					33.50			33.50	36.50
8450	For 12 hour timer switch, add	1 Elec	32	.250	↓	33.50	9.85		43.35	51
8500	Wall exhausters, centrifugal, auto damper, 1/8" S.P.									
8520	Direct drive, 610 CFM, 1/20 HP	Q-20	14	1.429	Ea.	188	51		239	285
8540	796 CFM, 1/12 HP		13	1.538		195	55		250	298
8560	822 CFM, 1/6 HP		12	1.667		305	59.50		364.50	430
8580	1,320 CFM, 1/4 HP		12	1.667		310	59.50		369.50	430
8600	1756 CFM, 1/4 HP		11	1.818		360	65		425	500
8620	1983 CFM, 1/4 HP		10	2		435	71.50		506.50	585
8640	2900 CFM, 1/2 HP		9	2.222		525	79.50		604.50	695
8660	3307 CFM, 3/4 HP	↓	8	2.500	↓	575	89.50		664.50	770
9500	V-belt drive, 3 phase									
9520	2,800 CFM, 1/4 HP	Q-20	9	2.222	Ea.	855	79.50		934.50	1,050
9540	3,740 CFM, 1/2 HP		8	2.500		890	89.50		979.50	1,100
9560	4400 CFM, 3/4 HP		7	2.857		900	102		1,002	1,150
9580	5700 CFM, 1-1/2 HP	↓	6	3.333	↓	935	119		1,054	1,200

For information about Means Estimating Seminars, see yellow pages 12 and 13 in back of book

MECHANICAL 15

For expanded coverage of these items see *Means Mechanical or Plumbing Cost Data 2004*

Division Notes		CREW	DAILY OUTPUT	LABOR-HOURS	UNIT	2004 BARE COSTS				TOTAL INCL O&P
						MAT.	LABOR	EQUIP.	TOTAL	

Division 16
Electrical

Estimating Tips

16060 Grounding & Bonding
- When taking off grounding system, identify separately the type and size of wire and list each unique type of ground connection.

16100 Wiring Methods
- Conduit should be taken off in three main categories: power distribution, branch power, and branch lighting, so the estimator can concentrate on systems and components, therefore making it easier to ensure all items have been accounted for.
- For cost modifications for elevated conduit installation, add the percentages to labor according to the height of installation and only the quantities exceeding the different height levels, not to the total conduit quantities.
- Remember that aluminum wiring of equal ampacity is larger in diameter than copper and may require larger conduit.
- If more than three wires at a time are being pulled, deduct percentages from the labor hours of that grouping of wires.

- The estimator should take the weights of materials into consideration when completing a takeoff. Topics to consider include: How will the materials be supported? What methods of support are available? How high will the support structure have to reach? Will the final support structure be able to withstand the total burden? Is the support material included or separate from the fixture, equipment and material specified?

16200 Electrical Power
- Do not overlook the costs for equipment used in the installation. If scaffolding or highlifts are available in the field, contractors may use them in lieu of the proposed ladders and rolling staging.

16400 Low-Voltage Distribution
- Supports and concrete pads may be shown on drawings for the larger equipment, or the support system may be just a piece of plywood for the back of a panelboard. In either case, it must be included in the costs.

16500 Lighting
- Fixtures should be taken off room by room, using the fixture schedule, specifications, and the ceiling plan. For large concentrations of lighting fixtures in the same area deduct the percentages from labor hours.

16700 Communications
16800 Sound & Video
- When estimating material costs for special systems, it is always prudent to obtain manufacturers' quotations for equipment prices and special installation requirements which will affect the total costs.

Reference Numbers
Reference numbers are shown in bold squares at the beginning of some major classifications. These numbers refer to related items in the Reference Section. The reference information may be an estimating procedure, an alternate pricing method or technical information.

Note: Not all subdivisions listed here necessarily appear in this publication.

16055	Selective Demolition	CREW	DAILY OUTPUT	LABOR-HOURS	UNIT	2004 BARE COSTS				TOTAL INCL O&P
						MAT.	LABOR	EQUIP.	TOTAL	
300	**0010** ELECTRICAL DEMOLITION									**300**
0020	Conduit to 15' high, including fittings & hangers									
0100	Rigid galvanized steel, 1/2" to 1" diameter	1 Elec	242	.033	L.F.		1.30		1.30	1.94
0120	1-1/4" to 2"		200	.040			1.58		1.58	2.34
0140	2" to 4"		151	.053			2.09		2.09	3.10
0160	4" to 6"		80	.100			3.94		3.94	5.85
0200	Electric metallic tubing (EMT) 1/2" to 1"		394	.020			.80		.80	1.19
0220	1-1/4" to 1-1/2"		326	.025			.97		.97	1.44
0240	2" to 3"		236	.034			1.34		1.34	1.99
0260	3-1/2" to 4"		155	.052			2.03		2.03	3.02
0270	Armored cable, (BX) avg. 50' runs									
0280	#14, 2 wire	1 Elec	690	.012	L.F.		.46		.46	.68
0290	#14, 3 wire		571	.014			.55		.55	.82
0300	#12, 2 wire		605	.013			.52		.52	.77
0310	#12, 3 wire		514	.016			.61		.61	.91
0320	#10, 2 wire		514	.016			.61		.61	.91
0330	#10, 3 wire		425	.019			.74		.74	1.10
0340	#8, 3 wire		342	.023			.92		.92	1.37
0350	Non metallic sheathed cable (Romex)									
0360	#14, 2 wire	1 Elec	720	.011	L.F.		.44		.44	.65
0370	#14, 3 wire		657	.012			.48		.48	.71
0380	#12, 2 wire		629	.013			.50		.50	.75
0390	#10, 3 wire		450	.018			.70		.70	1.04
0400	Wiremold raceway, including fittings & hangers									
0420	No. 3000	1 Elec	250	.032	L.F.		1.26		1.26	1.88
0440	No. 4000		217	.037			1.45		1.45	2.16
0460	No. 6000		166	.048			1.90		1.90	2.82
0500	Channels, steel, including fittings & hangers									
0520	3/4" x 1-1/2"	1 Elec	308	.026	L.F.		1.02		1.02	1.52
0540	1-1/2" x 1-1/2"		269	.030			1.17		1.17	1.74
0560	1-1/2" x 1-7/8"		229	.035			1.38		1.38	2.05
0600	Copper bus duct, indoor, 3 phase									
0610	Including hangers & supports									
0620	225 amp	2 Elec	135	.119	L.F.		4.67		4.67	6.95
0640	400 amp		106	.151			5.95		5.95	8.85
0660	600 amp		86	.186			7.35		7.35	10.90
0680	1000 amp		60	.267			10.50		10.50	15.65
0700	1600 amp		40	.400			15.75		15.75	23.50
0720	3000 amp		10	1.600			63		63	94
0800	Plug-in switches, 600V 3 ph, incl. disconnecting									
0820	wire, conduit terminations, 30 amp	1 Elec	15.50	.516	Ea.		20.50		20.50	30.50
0840	60 amp		13.90	.576			22.50		22.50	33.50
0850	100 amp		10.40	.769			30.50		30.50	45
0860	200 amp		6.20	1.290			51		51	75.50
0880	400 amp	2 Elec	5.40	2.963			117		117	174
0900	600 amp		3.40	4.706			185		185	276
0920	800 amp		2.60	6.154			242		242	360
0940	1200 amp		2	8			315		315	470
0960	1600 amp		1.70	9.412			370		370	550
1010	Safety switches, 250 or 600V, incl. disconnection									
1050	of wire & conduit terminations									
1100	30 amp	1 Elec	12.30	.650	Ea.		25.50		25.50	38
1120	60 amp		8.80	.909			36		36	53.50
1140	100 amp		7.30	1.096			43		43	64
1160	200 amp		5	1.600			63		63	94
1180	400 amp	2 Elec	6.80	2.353			92.50		92.50	138

R16055 -300

16 ELECTRICAL

Important: See the Reference Section for critical supporting data - Reference Nos., Crews, & City Cost Indexes

16055	Selective Demolition		CREW	DAILY OUTPUT	LABOR-HOURS	UNIT	2004 BARE COSTS				TOTAL INCL O&P	
							MAT.	LABOR	EQUIP.	TOTAL		
300	1200	600 amp	R16055	2 Elec	4.60	3.478	Ea.		137		137	204	300
	1210	Panel boards, incl. removal of all breakers,	-300										
	1220	conduit terminations & wire connections											
	1230	3 wire, 120/240V, 100A, to 20 circuits		1 Elec	2.60	3.077	Ea.		121		121	180	
	1240	200 amps, to 42 circuits			1.30	6.154			242		242	360	
	1250	400 amps, to 42 circuits			1.10	7.273			287		287	425	
	1260	4 wire, 120/208V, 125A, to 20 circuits			2.40	3.333			131		131	195	
	1270	200 amps, to 42 circuits			1.20	6.667			263		263	390	
	1280	400 amps, to 42 circuits		▼	.96	8.333	▼		330		330	490	
	1300	Transformer, dry type, 1 ph, incl. removal of											
	1320	supports, wire & conduit terminations											
	1340	1 kVA		1 Elec	7.70	1.039	Ea.		41		41	61	
	1360	5 kVA			4.70	1.702			67		67	99.50	
	1380	10 kVA		▼	3.60	2.222			87.50		87.50	130	
	1400	37.5 kVA		2 Elec	3	5.333			210		210	315	
	1420	75 kVA		1 Elec	1.25	6.400	▼		252		252	375	
	1440	3 Phase to 600V, primary											
	1460	3 kVA		1 Elec	3.85	2.078	Ea.		82		82	122	
	1480	15 kVA		2 Elec	4.20	3.810			150		150	223	
	1500	30 kVA			3.50	4.571			180		180	268	
	1510	45 kVA		▼	3.10	5.161			203		203	300	
	1520	75 kVA		1 Elec	1.35	5.926			233		233	345	
	1530	112.5 kVA		R-3	2.90	6.897			267	55	322	460	
	1540	150 kVA			2.70	7.407			287	59	346	490	
	1550	300 kVA			1.80	11.111			430	88.50	518.50	735	
	1560	500 kVA			1.40	14.286			555	114	669	950	
	1570	750 kVA		▼	1.10	18.182	▼		705	145	850	1,200	
	1600	Pull boxes & cabinets, sheet metal, incl. removal											
	1620	of supports and conduit terminations											
	1640	6" x 6" x 4"		1 Elec	31.10	.257	Ea.		10.15		10.15	15.05	
	1660	12" x 12" x 4"			23.30	.343			13.55		13.55	20	
	1680	24" x 24" x 6"			12.30	.650			25.50		25.50	38	
	1700	36" x 36" x 8"			7.70	1.039			41		41	61	
	1720	Junction boxes, 4" sq. & oct.			80	.100			3.94		3.94	5.85	
	1740	Handy box			107	.075			2.95		2.95	4.38	
	1760	Switch box			107	.075			2.95		2.95	4.38	
	1780	Receptacle & switch plates			257	.031			1.23		1.23	1.82	
	1790	Receptacles & switches, 15 to 30 amp		▼	135	.059	▼		2.33		2.33	3.47	
	1800	Wire, THW-THWN-THHN, removed from											
	1810	in place conduit, to 15' high											
	1830	#14		1 Elec	65	.123	C.L.F.		4.85		4.85	7.20	
	1840	#12			55	.145			5.75		5.75	8.50	
	1850	#10			45.50	.176			6.95		6.95	10.30	
	1860	#8			40.40	.198			7.80		7.80	11.60	
	1870	#6		▼	32.60	.245			9.65		9.65	14.40	
	1880	#4		2 Elec	53	.302			11.90		11.90	17.70	
	1890	#3			50	.320			12.60		12.60	18.75	
	1900	#2			44.60	.359			14.15		14.15	21	
	1910	1/0			33.20	.482			19		19	28	
	1920	2/0			29.20	.548			21.50		21.50	32	
	1930	3/0			25	.640			25		25	37.50	
	1940	4/0			22	.727			28.50		28.50	42.50	
	1950	250 kcmil			20	.800			31.50		31.50	47	
	1960	300 kcmil			19	.842			33		33	49.50	
	1970	350 kcmil			18	.889			35		35	52	
	1980	400 kcmil		▼	17	.941	▼		37		37	55	

ELECTRICAL 16

16055 | Selective Demolition

			CREW	DAILY OUTPUT	LABOR-HOURS	UNIT	2004 BARE COSTS				TOTAL INCL O&P	
							MAT.	LABOR	EQUIP.	TOTAL		
300	1990	500 kcmil	2 Elec	16.20	.988	C.L.F.		39		39	58	300
	2000	Interior fluorescent fixtures, incl. supports (R16055-300)										
	2010	& whips, to 15' high										
	2100	Recessed drop-in 2' x 2', 2 lamp	2 Elec	35	.457	Ea.		18		18	27	
	2120	2' x 4', 2 lamp		33	.485			19.10		19.10	28.50	
	2140	2' x 4', 4 lamp		30	.533			21		21	31.50	
	2160	4' x 4', 4 lamp		20	.800			31.50		31.50	47	
	2180	Surface mount, acrylic lens & hinged frame										
	2200	1' x 4', 2 lamp	2 Elec	44	.364	Ea.		14.35		14.35	21.50	
	2220	2' x 2', 2 lamp		44	.364			14.35		14.35	21.50	
	2260	2' x 4', 4 lamp		33	.485			19.10		19.10	28.50	
	2280	4' x 4', 4 lamp		40	.400			15.75		15.75	23.50	
	2300	Strip fixtures, surface mount										
	2320	4' long, 1 lamp	2 Elec	53	.302	Ea.		11.90		11.90	17.70	
	2340	4' long, 2 lamp		50	.320			12.60		12.60	18.75	
	2360	8' long, 1 lamp		42	.381			15		15	22.50	
	2380	8' long, 2 lamp		40	.400			15.75		15.75	23.50	
	2400	Pendant mount, industrial, incl. removal										
	2410	of chain or rod hangers, to 15' high										
	2420	4' long, 2 lamp	2 Elec	35	.457	Ea.		18		18	27	
	2440	8' long, 2 lamp	"	27	.593	"		23.50		23.50	34.50	
	2460	Interior incandescent, surface, ceiling										
	2470	or wall mount, to 12' high										
	2480	Metal cylinder type, 75 Watt	2 Elec	62	.258	Ea.		10.15		10.15	15.10	
	2500	150 Watt	"	62	.258	"		10.15		10.15	15.10	
	2520	Metal halide, high bay										
	2540	400 Watt	2 Elec	15	1.067	Ea.		42		42	62.50	
	2560	1000 Watt		12	1.333			52.50		52.50	78	
	2580	150 Watt, low bay		20	.800			31.50		31.50	47	
	2600	Exterior fixtures, incandescent, wall mount										
	2620	100 Watt	2 Elec	50	.320	Ea.		12.60		12.60	18.75	
	2640	Quartz, 500 Watt		33	.485			19.10		19.10	28.50	
	2660	1500 Watt		27	.593			23.50		23.50	34.50	
	2680	Wall pack, mercury vapor										
	2700	175 Watt	2 Elec	25	.640	Ea.		25		25	37.50	
	2720	250 Watt	"	25	.640	"		25		25	37.50	
	2740	Minimum labor/equipment charge	1 Elec	4	2	Job		79		79	117	

16060 | Grounding & Bonding

			CREW	DAILY OUTPUT	LABOR-HOURS	UNIT	MAT.	LABOR	EQUIP.	TOTAL	TOTAL INCL O&P	
800	0010	**GROUNDING**										800
	0030	Rod, copper clad, 8' long, 1/2" diameter (R16060-800)	1 Elec	5.50	1.455	Ea.	13.10	57.50		70.60	99.50	
	0040	5/8" diameter		5.50	1.455		14.15	57.50		71.65	101	
	0050	3/4" diameter		5.30	1.509		27	59.50		86.50	118	
	0080	10' long, 1/2" diameter		4.80	1.667		16.90	65.50		82.40	116	
	0090	5/8" diameter		4.60	1.739		20.50	68.50		89	125	
	0100	3/4" diameter		4.40	1.818		30	71.50		101.50	140	
	0130	15' long, 3/4" diameter		4	2		79	79		158	204	
	0150	Coupling, bronze, 1/2" diameter					2.82			2.82	3.10	
	0160	5/8" diameter					3.90			3.90	4.29	
	0170	3/4" diameter					8.85			8.85	9.75	
	0190	Drive studs, 1/2" diameter					5.30			5.30	5.85	
	0210	5/8" diameter					6.45			6.45	7.10	
	0220	3/4" diameter					7.25			7.25	8	
	0230	Clamp, bronze, 1/2" diameter	1 Elec	32	.250		3.56	9.85		13.41	18.55	
	0240	5/8" diameter		32	.250		3.98	9.85		13.83	19.05	

Important: See the Reference Section for critical supporting data - Reference Nos., Crews, & City Cost Indexes

	16060	Grounding & Bonding	CREW	DAILY OUTPUT	LABOR-HOURS	UNIT	2004 BARE COSTS				TOTAL INCL O&P	
							MAT.	LABOR	EQUIP.	TOTAL		
800	0250	3/4" diameter	1 Elec	32	.250	Ea.	5.25	9.85		15.10	20.50	800
	0260	Wire ground bare armored, #8-1 conductor	R16060-800	2	4	C.L.F.	72	158		230	315	
	0270	#6-1 conductor		1.80	4.444		90.50	175		265.50	360	
	0280	#4-1 conductor		1.60	5		133	197		330	440	
	0320	Bare copper wire, #14 solid		14	.571		2.60	22.50		25.10	36.50	
	0330	#12		13	.615		3.60	24.50		28.10	40	
	0340	#10		12	.667		5.55	26.50		32.05	45	
	0350	#8		11	.727		9.15	28.50		37.65	52.50	
	0360	#6		10	.800		15.80	31.50		47.30	64.50	
	0370	#4		8	1		23	39.50		62.50	84	
	0380	#2		5	1.600		36.50	63		99.50	135	
	0390	Bare copper wire, #8 stranded		11	.727		9.80	28.50		38.30	53.50	
	0400	#6	▼	10	.800		16.60	31.50		48.10	65.50	
	0450	#4	2 Elec	16	1		26	39.50		65.50	87	
	0600	#2		10	1.600		38.50	63		101.50	137	
	0650	#1		9	1.778		44	70		114	153	
	0700	1/0		8	2		57.50	79		136.50	181	
	0750	2/0		7.20	2.222		74.50	87.50		162	212	
	0800	3/0		6.60	2.424		91.50	95.50		187	243	
	1000	4/0	▼	5.70	2.807		114	111		225	289	
	1200	250 kcmil	3 Elec	7.20	3.333		135	131		266	345	
	1210	300 kcmil		6.60	3.636		161	143		304	390	
	1220	350 kcmil		6	4		187	158		345	440	
	1230	400 kcmil		5.70	4.211		241	166		407	510	
	1240	500 kcmil		5.10	4.706		268	185		453	570	
	1250	600 kcmil		3.90	6.154		268	242		510	655	
	1260	750 kcmil		3.60	6.667		470	263		733	910	
	1270	1000 kcmil	▼	3	8		650	315		965	1,175	
	1350	Bare aluminum, #8 stranded	1 Elec	10	.800		7.45	31.50		38.95	55	
	1360	#6	"	9	.889		12.10	35		47.10	65.50	
	1370	#4	2 Elec	16	1		15.70	39.50		55.20	76	
	1380	#2		13	1.231		23.50	48.50		72	98	
	1390	#1		10.60	1.509		28	59.50		87.50	119	
	1400	1/0		9	1.778		32	70		102	140	
	1410	2/0		8	2		39.50	79		118.50	160	
	1420	3/0		7.20	2.222		50.50	87.50		138	186	
	1430	4/0	▼	6.60	2.424		63.50	95.50		159	212	
	1440	250 kcmil	3 Elec	9.30	2.581		91	102		193	251	
	1450	300 kcmil		8.70	2.759		113	109		222	286	
	1460	400 kcmil		7.50	3.200		135	126		261	335	
	1470	500 kcmil		6.90	3.478		173	137		310	395	
	1480	600 kcmil		6	4		199	158		357	455	
	1490	700 kcmil		5.70	4.211		223	166		389	490	
	1500	750 kcmil		5.10	4.706		237	185		422	535	
	1510	1000 kcmil	▼	4.80	5	▼	335	197		532	665	
	1800	Water pipe ground clamps, heavy duty										
	2000	Bronze, 1/2" to 1" diameter	1 Elec	8	1	Ea.	13.05	39.50		52.55	73	
	2100	1-1/4" to 2" diameter		8	1		17.20	39.50		56.70	77.50	
	2200	2-1/2" to 3" diameter		6	1.333		35.50	52.50		88	117	
	2730	Exothermic weld, 4/0 wire to 1" ground rod		7	1.143		8.55	45		53.55	76.50	
	2740	4/0 wire to building steel		7	1.143		6.80	45		51.80	74.50	
	2750	4/0 wire to motor frame		7	1.143		6.80	45		51.80	74.50	
	2760	4/0 wire to 4/0 wire		7	1.143		5.45	45		50.45	73	
	2770	4/0 wire to #4 wire		7	1.143		5.45	45		50.45	73	
	2780	4/0 wire to #8 wire	▼	7	1.143	▼	5.45	45		50.45	73	
	2790	Mold, reusable, for above					85			85	93.50	

ELECTRICAL 16

16 ELECTRICAL

16060	Grounding & Bonding	CREW	DAILY OUTPUT	LABOR-HOURS	UNIT	2004 BARE COSTS				TOTAL INCL O&P
						MAT.	LABOR	EQUIP.	TOTAL	
800 2800	Brazed connections, #6 wire R16060 -800	1 Elec	12	.667	Ea.	10.75	26.50		37.25	51	800
3000	#2 wire		10	.800		14.40	31.50		45.90	63
3100	3/0 wire		8	1		21.50	39.50		61	82.50
3200	4/0 wire		7	1.143		24.50	45		69.50	94
3400	250 kcmil wire		5	1.600		29	63		92	126
3600	500 kcmil wire		4	2		35.50	79		114.50	156
3700	Insulated ground wire, copper #14		13	.615	C.L.F.	3.70	24.50		28.20	40
3710	#12		11	.727		5.05	28.50		33.55	48
3720	#10		10	.800		7.60	31.50		39.10	55.50
3730	#8		8	1		12.65	39.50		52.15	72.50
3740	#6		6.50	1.231		20	48.50		68.50	94
3750	#4	2 Elec	10.60	1.509		30.50	59.50		90	122
3770	#2		9	1.778		47	70		117	156
3780	#1		8	2		59.50	79		138.50	183
3790	1/0		6.60	2.424		72	95.50		167.50	221
3800	2/0		5.80	2.759		88.50	109		197.50	259
3810	3/0		5	3.200		109	126		235	310
3820	4/0		4.40	3.636		138	143		281	365
3830	250 kcmil	3 Elec	6	4		168	158		326	420
3840	300 kcmil		5.70	4.211		199	166		365	465
3850	350 kcmil		5.40	4.444		230	175		405	515
3860	400 kcmil		5.10	4.706		273	185		458	575
3870	500 kcmil		4.80	5		320	197		517	645
3880	600 kcmil		3.90	6.154		390	242		632	790
3890	750 kcmil		3.30	7.273		520	287		807	995
3900	1000 kcmil		2.70	8.889		720	350		1,070	1,300
3950	Insulated ground wire, aluminum #8	1 Elec	9	.889		12.45	35		47.45	65.50
3960	#6	"	8	1		17.05	39.50		56.55	77.50
3970	#4	2 Elec	13	1.231		21	48.50		69.50	95.50
3980	#2		10.60	1.509		29	59.50		88.50	120
3990	#1		9	1.778		42	70		112	150
4000	1/0		8	2		50.50	79		129.50	173
4010	2/0		7.20	2.222		59.50	87.50		147	196
4020	3/0		6.60	2.424		74	95.50		169.50	224
4030	4/0		6.20	2.581		82.50	102		184.50	242
4040	250 kcmil	3 Elec	8.70	2.759		100	109		209	272
4050	300 kcmil		8.10	2.963		139	117		256	325
4060	350 kcmil		7.50	3.200		141	126		267	345
4070	400 kcmil		6.90	3.478		165	137		302	385
4080	500 kcmil		6	4		182	158		340	435
4090	600 kcmil		5.70	4.211		230	166		396	500
4100	700 kcmil		5.10	4.706		262	185		447	565
4110	750 kcmil		4.80	5		269	197		466	590
5000	Copper Electrolytic ground rod system									
5010	Includes augering hole, mixing bentonite clay,									
5020	Installing rod, and terminating ground wire									
5100	Straight Vertical type, 2" Dia.									
5120	8.5' long, Clamp Connection	1 Elec	2.67	2.996	Ea.	585	118		703	820
5130	With exothermic weld Connection		1.95	4.103		585	162		747	885
5140	10' long		2.35	3.404		650	134		784	915
5150	With exothermic weld Connection		1.78	4.494		650	177		827	980
5160	12' long		2.16	3.704		785	146		931	1,075
5170	With exothermic weld Connection		1.67	4.790		785	189		974	1,150
5180	20' long		1.74	4.598		1,200	181		1,381	1,600
5190	With exothermic weld Connection		1.40	5.714		1,225	225		1,450	1,675
5195	40' long with exothermic weld connection	2 Elec	2	8		2,350	315		2,665	3,050

Important: See the Reference Section for critical supporting data - Reference Nos., Crews, & City Cost Indexes

16060 | Grounding & Bonding

			CREW	DAILY OUTPUT	LABOR-HOURS	UNIT	2004 BARE COSTS				TOTAL INCL O&P	
							MAT.	LABOR	EQUIP.	TOTAL		
800	5200	L-Shaped, 2" Dia.										**800**
	5220	4' Vert. x 10' Horz., Clamp Connection R16060-800	1 Elec	5.33	1.501	Ea.	900	59		959	1,075	
	5230	With exothermic weld Connection	"	3.08	2.597	"	930	102		1,032	1,175	
	5300	Protective Box at grade level, with breather slots										
	5320	Round 12" long, Plastic	1 Elec	32	.250	Ea.	40	9.85		49.85	58.50	
	5330	Concrete	"	16	.500		50	19.70		69.70	84.50	
	5400	Bentonite Clay, 50# bag, 1 per 10' of rod					35			35	38.50	

16070 | Hangers & Supports

			CREW	DAILY OUTPUT	LABOR-HOURS	UNIT	MAT.	LABOR	EQUIP.	TOTAL	TOTAL INCL O&P	
310	0010	**FASTENERS** see division 05090 and 15060										**310**
320	0010	**HANGERS** Steel R16132-205										**320**
	0030	Conduit supports										
	0050	Strap w/2 holes, rigid steel conduit										
	0100	1/2" diameter	1 Elec	470	.017	Ea.	.18	.67		.85	1.20	
	0150	3/4" diameter		440	.018		.21	.72		.93	1.30	
	0200	1" diameter		400	.020		.35	.79		1.14	1.56	
	0300	1-1/4" diameter		355	.023		.55	.89		1.44	1.93	
	0350	1-1/2" diameter		320	.025		.65	.99		1.64	2.19	
	0400	2" diameter		266	.030		.76	1.19		1.95	2.60	
	0500	2-1/2" diameter		160	.050		1.56	1.97		3.53	4.65	
	0550	3" diameter		133	.060		1.84	2.37		4.21	5.55	
	0600	3-1/2" diameter		100	.080		2.54	3.15		5.69	7.50	
	0650	4" diameter		80	.100		2.74	3.94		6.68	8.85	
	0700	EMT, 1/2" diameter		470	.017		.11	.67		.78	1.12	
	0800	3/4" diameter		440	.018		.16	.72		.88	1.25	
	0850	1" diameter		400	.020		.23	.79		1.02	1.42	
	0900	1-1/4" diameter		355	.023		.35	.89		1.24	1.71	
	0950	1-1/2" diameter		320	.025		.41	.99		1.40	1.92	
	1000	2" diameter		266	.030		.63	1.19		1.82	2.45	
	1100	2-1/2" diameter		160	.050		.88	1.97		2.85	3.90	
	1150	3" diameter		133	.060		1.09	2.37		3.46	4.72	
	1200	3-1/2" diameter		100	.080		1.37	3.15		4.52	6.20	
	1250	4" diameter		80	.100		1.54	3.94		5.48	7.55	
	1400	Hanger, with bolt, 1/2" diameter		200	.040		.42	1.58		2	2.80	
	1450	3/4" diameter		190	.042		.43	1.66		2.09	2.94	
	1500	1" diameter		176	.045		.70	1.79		2.49	3.43	
	1550	1-1/4" diameter		160	.050		.95	1.97		2.92	3.98	
	1600	1-1/2" diameter		140	.057		1.20	2.25		3.45	4.67	
	1650	2" diameter		130	.062		1.32	2.42		3.74	5.05	
	1700	2-1/2" diameter		100	.080		1.74	3.15		4.89	6.60	
	1750	3" diameter		64	.125		2.03	4.93		6.96	9.60	
	1800	3-1/2" diameter		50	.160		3	6.30		9.30	12.70	
	1850	4" diameter		40	.200		6.85	7.90		14.75	19.25	
	1900	Riser clamps, conduit, 1/2" diameter		40	.200		6.20	7.90		14.10	18.50	
	1950	3/4" diameter		36	.222		6.20	8.75		14.95	19.80	
	2000	1" diameter		30	.267		6.25	10.50		16.75	22.50	
	2100	1-1/4" diameter		27	.296		8.30	11.65		19.95	26.50	
	2150	1-1/2" diameter		27	.296		8.80	11.65		20.45	27	
	2200	2" diameter		20	.400		9.20	15.75		24.95	33.50	
	2250	2-1/2" diameter		20	.400		9.60	15.75		25.35	34	
	2300	3" diameter		18	.444		10.40	17.50		27.90	37.50	
	2350	3-1/2" diameter		18	.444		13.20	17.50		30.70	40.50	
	2400	4" diameter		14	.571		13.70	22.50		36.20	48.50	
	2500	Threaded rod, painted, 1/4" diameter		260	.031	L.F.	1.31	1.21		2.52	3.24	

ELECTRICAL 16

	16070	Hangers & Supports		CREW	DAILY OUTPUT	LABOR-HOURS	UNIT	2004 BARE COSTS				TOTAL INCL O&P	
								MAT.	LABOR	EQUIP.	TOTAL		
320	2600	3/8" diameter	R16132 -205	1 Elec	200	.040	L.F.	1.66	1.58		3.24	4.17	320
	2700	1/2" diameter			140	.057		2.29	2.25		4.54	5.85	
	2800	5/8" diameter			100	.080		3.15	3.15		6.30	8.15	
	2900	3/4" diameter			60	.133		3.96	5.25		9.21	12.15	
	2940	Couplings painted, 1/4" diameter					C	130			130	143	
	2960	3/8" diameter						176			176	194	
	2970	1/2" diameter						268			268	295	
	2980	5/8" diameter						385			385	420	
	2990	3/4" diameter						565			565	625	
	3000	Nuts, galvanized, 1/4" diameter						8.80			8.80	9.70	
	3050	3/8" diameter						13.10			13.10	14.40	
	3100	1/2" diameter						29.50			29.50	32.50	
	3150	5/8" diameter						44.50			44.50	49	
	3200	3/4" diameter						66.50			66.50	73	
	3250	Washers, galvanized, 1/4" diameter						4.70			4.70	5.15	
	3300	3/8" diameter						9.80			9.80	10.80	
	3350	1/2" diameter						12.10			12.10	13.30	
	3400	5/8" diameter						36.50			36.50	40.50	
	3450	3/4" diameter						53			53	58.50	
	3500	Lock washers, galvanized, 1/4" diameter						3.90			3.90	4.29	
	3550	3/8" diameter						8.60			8.60	9.45	
	3600	1/2" diameter						11.80			11.80	13	
	3650	5/8" diameter						21.50			21.50	24	
	3700	3/4" diameter						34			34	37.50	
	3800	Channels, steel, 3/4" x 1-1/2"		1 Elec	80	.100	L.F.	3.23	3.94		7.17	9.40	
	3900	1-1/2" x 1-1/2"			70	.114		4.32	4.50		8.82	11.45	
	4000	1-7/8" x 1-1/2"			60	.133		9.30	5.25		14.55	18	
	4100	3" x 1-1/2"			50	.160		10.70	6.30		17	21	
	4200	Spring nuts, long, 1/4"			120	.067	Ea.	1.04	2.63		3.67	5.05	
	4250	3/8"			100	.080		1.85	3.15		5	6.75	
	4300	1/2"			80	.100		1.98	3.94		5.92	8.05	
	4350	Spring nuts, short, 1/4"			120	.067		1.19	2.63		3.82	5.20	
	4400	3/8"			100	.080		1.31	3.15		4.46	6.15	
	4450	1/2"			80	.100		1.45	3.94		5.39	7.45	
	4500	Closure strip			200	.040	L.F.	1.97	1.58		3.55	4.51	
	4550	End cap			60	.133	Ea.	.89	5.25		6.14	8.80	
	4600	End connector 3/4" conduit			40	.200		3.24	7.90		11.14	15.25	
	4650	Junction box, 1 channel			16	.500		20	19.70		39.70	51.50	
	4700	2 channel			14	.571		24.50	22.50		47	60	
	4750	3 channel			12	.667		28	26.50		54.50	69.50	
	4800	4 channel			10	.800		30.50	31.50		62	81	
	4850	Spliceplate			40	.200		7.60	7.90		15.50	20	
	4900	Continuous concrete insert, 1-1/2" deep, 1' long			16	.500		10.45	19.70		30.15	41	
	4950	2' long			14	.571		13.05	22.50		35.55	48	
	5000	3' long			12	.667		15.65	26.50		42.15	56	
	5050	4' long			10	.800		18.10	31.50		49.60	67	
	5100	6' long			8	1		27	39.50		66.50	88.50	
	5150	3/4" deep, 1' long			16	.500		9.45	19.70		29.15	40	
	5200	2' long			14	.571		11.70	22.50		34.20	46.50	
	5250	3' long			12	.667		14.05	26.50		40.55	54.50	
	5300	4' long			10	.800		16.20	31.50		47.70	65	
	5350	6' long			8	1		24	39.50		63.50	85	
	5400	90° angle fitting 2-1/8" x 2-1/8"			60	.133		1.79	5.25		7.04	9.75	
	5450	Supports, suspension rod type, small			60	.133		7.75	5.25		13	16.35	
	5500	Large			40	.200		8.55	7.90		16.45	21	
	5550	Beam clamp, small			60	.133		4.62	5.25		9.87	12.90	

16070 | Hangers & Supports

		CREW	DAILY OUTPUT	LABOR-HOURS	UNIT	2004 BARE COSTS				TOTAL INCL O&P
						MAT.	LABOR	EQUIP.	TOTAL	
320	5600	Large	1 Elec	40	.200	Ea.	4.90	7.90		12.80	17.10
	5650	U-support, small		60	.133		2.18	5.25		7.43	10.20
	5700	Large		40	.200		7.15	7.90		15.05	19.55
	5750	Concrete insert, cast, for up to 1/2" threaded rod		16	.500		2.40	19.70		22.10	32
	5800	Beam clamp, 1/4" clamp, 1/4" threaded drop rod		32	.250		1.97	9.85		11.82	16.80
	5900	3/8" clamp, 3/8" threaded drop rod		32	.250		4.70	9.85		14.55	19.80
	6000	Strap, rigid conduit, 1/2" diameter		540	.015		.91	.58		1.49	1.87
	6050	3/4" diameter		440	.018		1.02	.72		1.74	2.19
	6100	1" diameter		420	.019		1.09	.75		1.84	2.32
	6150	1-1/4" diameter		400	.020		1.26	.79		2.05	2.56
	6200	1-1/2" diameter		400	.020		1.43	.79		2.22	2.74
	6250	2" diameter		267	.030		1.62	1.18		2.80	3.54
	6300	2-1/2" diameter		267	.030		1.76	1.18		2.94	3.70
	6350	3" diameter		160	.050		1.94	1.97		3.91	5.05
	6400	3-1/2" diameter		133	.060		2.37	2.37		4.74	6.15
	6450	4" diameter		100	.080		2.64	3.15		5.79	7.60
	6500	5" diameter		80	.100		6.35	3.94		10.29	12.85
	6550	6" diameter		60	.133		7.20	5.25		12.45	15.75
	6600	EMT, 1/2" diameter		540	.015		.91	.58		1.49	1.87
	6650	3/4" diameter		440	.018		.97	.72		1.69	2.14
	6700	1" diameter		420	.019		1.09	.75		1.84	2.32
	6750	1-1/4" diameter		400	.020		1.22	.79		2.01	2.51
	6800	1-1/2" diameter		400	.020		1.48	.79		2.27	2.80
	6850	2" diameter		267	.030		1.59	1.18		2.77	3.51
	6900	2-1/2" diameter		267	.030		2.02	1.18		3.20	3.98
	6950	3" diameter		160	.050		2.12	1.97		4.09	5.25
	6970	3-1/2" diameter		133	.060		2.42	2.37		4.79	6.20
	6990	4" diameter		100	.080		2.78	3.15		5.93	7.75
	7000	Clip, 1 hole for rigid conduit, 1/2" diameter		500	.016		.55	.63		1.18	1.55
	7050	3/4" diameter		470	.017		.82	.67		1.49	1.90
	7100	1" diameter		440	.018		1.03	.72		1.75	2.20
	7150	1-1/4" diameter		400	.020		2.15	.79		2.94	3.54
	7200	1-1/2" diameter		355	.023		2.50	.89		3.39	4.07
	7250	2" diameter		320	.025		5	.99		5.99	6.95
	7300	2-1/2" diameter		266	.030		11.10	1.19		12.29	13.95
	7350	3" diameter		160	.050		15.90	1.97		17.87	20.50
	7400	3-1/2" diameter		133	.060		22.50	2.37		24.87	28.50
	7450	4" diameter		100	.080		50.50	3.15		53.65	60.50
	7500	5" diameter		80	.100		141	3.94		144.94	161
	7550	6" diameter		60	.133		152	5.25		157.25	175
	7820	Conduit hangers, with bolt & 12" rod, 1/2" diameter		150	.053		2.08	2.10		4.18	5.40
	7830	3/4" diameter		145	.055		2.09	2.17		4.26	5.55
	7840	1" diameter		135	.059		2.36	2.33		4.69	6.05
	7850	1-1/4" diameter		120	.067		2.61	2.63		5.24	6.80
	7860	1-1/2" diameter		110	.073		2.86	2.87		5.73	7.40
	7870	2" diameter		100	.080		3.61	3.15		6.76	8.65
	7880	2-1/2" diameter		80	.100		4.03	3.94		7.97	10.30
	7890	3" diameter		60	.133		4.32	5.25		9.57	12.55
	7900	3-1/2" diameter		45	.178		5.30	7		12.30	16.20
	7910	4" diameter		35	.229		9.15	9		18.15	23.50
	7920	5" diameter		30	.267		10	10.50		20.50	26.50
	7930	6" diameter		25	.320		19.05	12.60		31.65	40
	7950	Jay clamp, 1/2" diameter		32	.250		4.12	9.85		13.97	19.20
	7960	3/4" diameter		32	.250		4.12	9.85		13.97	19.20
	7970	1" diameter		32	.250		4.12	9.85		13.97	19.20
	7980	1-1/4" diameter		30	.267		5.90	10.50		16.40	22

R16132 -205

ELECTRICAL 16

	16070	Hangers & Supports		CREW	DAILY OUTPUT	LABOR-HOURS	UNIT	2004 BARE COSTS				TOTAL INCL O&P
								MAT.	LABOR	EQUIP.	TOTAL	
320	7990	1-1/2" diameter	R16132 -205	1 Elec	30	.267	Ea.	5.90	10.50		16.40	22	320
	8000	2" diameter			30	.267		7.40	10.50		17.90	24
	8010	2-1/2" diameter			28	.286		10.65	11.25		21.90	28.50
	8020	3" diameter			28	.286		14.50	11.25		25.75	32.50
	8030	3-1/2" diameter			25	.320		18.50	12.60		31.10	39.50
	8040	4" diameter			25	.320		32.50	12.60		45.10	55
	8050	5" diameter			20	.400		82	15.75		97.75	114
	8060	6" diameter			16	.500		155	19.70		174.70	200
	8070	Channels, 3/4" x 1-1/2" w/12" rods for 1/2" to 1" conduit			30	.267		6.50	10.50		17	23
	8080	1-1/2" x 1-1/2" w/12" rods for 1-1/4" to 2" conduit			28	.286		7.05	11.25		18.30	24.50
	8090	1-1/2" x 1-1/2" w/12" rods for 2-1/2" to 4" conduit			26	.308		8.55	12.10		20.65	27.50
	8100	1-1/2" x 1-7/8" w/12" rods for 5" to 6" conduit			24	.333		22.50	13.15		35.65	44.50
	8110	Beam clamp, conduit, plastic coated steel, 1/2" diam.			30	.267		11.95	10.50		22.45	29
	8120	3/4" diameter			30	.267		12.55	10.50		23.05	29.50
	8130	1" diameter			30	.267		12.70	10.50		23.20	29.50
	8140	1-1/4" diameter			28	.286		17.30	11.25		28.55	36
	8150	1-1/2" diameter			28	.286		21	11.25		32.25	40
	8160	2" diameter			28	.286		27.50	11.25		38.75	47.50
	8170	2-1/2" diameter			26	.308		30.50	12.10		42.60	51.50
	8180	3" diameter			26	.308		34	12.10		46.10	55.50
	8190	3-1/2" diameter			23	.348		35	13.70		48.70	59
	8200	4" diameter			23	.348		38.50	13.70		52.20	62.50
	8210	5" diameter		▼	18	.444	▼	113	17.50		130.50	150
	8220	Channels, plastic coated										
	8250	3/4" x 1-1/2", w/12" rods for 1/2" to 1" conduit		1 Elec	28	.286	Ea.	17.10	11.25		28.35	35.50
	8260	1-1/2" x 1-1/2", w/12" rods for 1-1/4" to 2" conduit			26	.308		19.20	12.10		31.30	39
	8270	1-1/2" x 1-1/2", w/12" rods for 2-1/2" to 3-1/2" conduit			24	.333		21	13.15		34.15	42.50
	8280	1-1/2" x 1-7/8", w/12" rods for 4" to 5" conduit			22	.364		35	14.35		49.35	60
	8290	1-1/2" x 1-7/8", w/12" rods for 6" conduit			20	.400		38	15.75		53.75	65.50
	8320	Conduit hangers, plastic coated steel, with bolt & 12" rod, 1/2" diam.			140	.057		9.80	2.25		12.05	14.15
	8330	3/4" diameter			135	.059		10.05	2.33		12.38	14.50
	8340	1" diameter			125	.064		10.35	2.52		12.87	15.15
	8350	1-1/4" diameter			110	.073		10.95	2.87		13.82	16.30
	8360	1-1/2" diameter			100	.080		12.35	3.15		15.50	18.30
	8370	2" diameter			90	.089		13.50	3.50		17	20
	8380	2-1/2" diameter			70	.114		16.15	4.50		20.65	24.50
	8390	3" diameter			50	.160		19.50	6.30		25.80	31
	8400	3-1/2" diameter			35	.229		20.50	9		29.50	36.50
	8410	4" diameter			25	.320		29	12.60		41.60	51
	8420	5" diameter			20	.400		32	15.75		47.75	58.50
	9000	Parallel type, conduit beam clamp, 1/2"			32	.250		3.65	9.85		13.50	18.65
	9010	3/4"			32	.250		4	9.85		13.85	19.05
	9020	1"			32	.250		4.25	9.85		14.10	19.35
	9030	1-1/4"			30	.267		5.90	10.50		16.40	22
	9040	1-1/2"			30	.267		6.70	10.50		17.20	23
	9050	2"			30	.267		8.65	10.50		19.15	25
	9060	2-1/2"			28	.286		11.50	11.25		22.75	29.50
	9070	3"			28	.286		14.65	11.25		25.90	33
	9090	4"			25	.320		18.75	12.60		31.35	39.50
	9110	Right angle, conduit beam clamp, 1/2"			32	.250		2.40	9.85		12.25	17.30
	9120	3/4"			32	.250		2.50	9.85		12.35	17.40
	9130	1"			32	.250		2.75	9.85		12.60	17.70
	9140	1-1/4"			30	.267		3.40	10.50		13.90	19.40
	9150	1-1/2"			30	.267		3.80	10.50		14.30	19.85
	9160	2"			30	.267		5.55	10.50		16.05	22
	9170	2-1/2"		▼	28	.286	▼	6.85	11.25		18.10	24.50

Important: See the Reference Section for critical supporting data - Reference Nos., Crews, & City Cost Indexes

16070 | Hangers & Supports

			CREW	DAILY OUTPUT	LABOR-HOURS	UNIT	2004 BARE COSTS MAT.	LABOR	EQUIP.	TOTAL	TOTAL INCL O&P	
320	9180	3"	1 Elec	28	.286	Ea.	7.90	11.25		19.15	25.50	320
	9190	3-1/2"	R16132-205	25	.320		9.95	12.60		22.55	29.50	
	9200	4"		25	.320		10.75	12.60		23.35	30.50	
	9230	Adjustable, conduit hanger, 1/2"		32	.250		2.60	9.85		12.45	17.50	
	9240	3/4"		32	.250		2.45	9.85		12.30	17.35	
	9250	1"		32	.250		3	9.85		12.85	17.95	
	9260	1-1/4"		30	.267		3.55	10.50		14.05	19.55	
	9270	1-1/2"		30	.267		4.10	10.50		14.60	20	
	9280	2"		30	.267		4.40	10.50		14.90	20.50	
	9290	2-1/2"		28	.286		5.50	11.25		16.75	23	
	9300	3"		28	.286		5.80	11.25		17.05	23	
	9310	3-1/2"		25	.320		7.55	12.60		20.15	27	
	9320	4"		25	.320		13.20	12.60		25.80	33.50	
	9330	5"		20	.400		17.20	15.75		32.95	42.50	
	9340	6"		16	.500		21.50	19.70		41.20	53	
	9350	Combination conduit hanger, 3/8"		32	.250		6	9.85		15.85	21.50	
	9360	Adjustable flange 3/8"		32	.250		7	9.85		16.85	22.50	

ELECTRICAL 16

16120 | Conductors & Cables

			CREW	DAILY OUTPUT	LABOR-HOURS	UNIT	2004 BARE COSTS MAT.	LABOR	EQUIP.	TOTAL	TOTAL INCL O&P	
120	0010	**ARMORED CABLE**										120
	0050	600 volt, copper (BX), #14, 2 conductor, solid	R16120-120	1 Elec	2.40	3.333	C.L.F.	52	131		183	253	
	0100	3 conductor, solid		2.20	3.636		69.50	143		212.50	290	
	0120	4 conductor, solid		2	4		100	158		258	345	
	0150	#12, 2 conductor, solid		2.30	3.478		53	137		190	263	
	0200	3 conductor, solid		2	4		83.50	158		241.50	325	
	0220	4 conductor, solid		1.80	4.444		120	175		295	390	
	0240	#12, 19 conductor, stranded		1.10	7.273		615	287		902	1,100	
	0250	#10, 2 conductor, solid		2	4		96.50	158		254.50	340	
	0300	3 conductor, solid		1.60	5		130	197		327	435	
	0320	4 conductor, solid		1.40	5.714		181	225		406	535	
	0350	#8, 3 conductor, solid		1.30	6.154		203	242		445	585	
	0370	4 conductor, stranded		1.10	7.273		290	287		577	745	
	0380	#6, 2 conductor, stranded		1.30	6.154		250	242		492	635	
	0400	3 conductor with PVC jacket, in cable tray, #6		3.10	2.581		279	102		381	455	
	0450	#4	2 Elec	5.40	2.963		360	117		477	570	
	0500	#2		4.60	3.478		480	137		617	730	
	0550	#1		4	4		650	158		808	950	
	0600	1/0		3.60	4.444		765	175		940	1,100	
	0650	2/0		3.40	4.706		925	185		1,110	1,300	
	0700	3/0		3.20	5		1,075	197		1,272	1,500	
	0750	4/0		3	5.333		1,250	210		1,460	1,700	
	0800	250 kcmil	3 Elec	3.60	6.667		1,400	263		1,663	1,925	
	0850	350 kcmil		3.30	7.273		1,850	287		2,137	2,475	
	0900	500 kcmil		3	8		2,425	315		2,740	3,125	
	0910	4 conductor with PVC jacket, in cable tray, #6	1 Elec	2.70	2.963		355	117		472	565	
	0920	#4	2 Elec	4.60	3.478		455	137		592	705	
	0930	#2		4	4		565	158		723	855	

			CREW	DAILY OUTPUT	LABOR-HOURS	UNIT	2004 BARE COSTS				TOTAL INCL O&P
	16120	**Conductors & Cables**					MAT.	LABOR	EQUIP.	TOTAL	
120	0940	#1	2 Elec	3.60	4.444	C.L.F.	825	175		1,000	1,175
	0950	1/0		3.40	4.706		940	185		1,125	1,300
	0960	2/0		3.20	5		1,075	197		1,272	1,475
	0970	3/0		3	5.333		1,275	210		1,485	1,725
	0980	4/0		2.40	6.667		1,625	263		1,888	2,200
	0990	250 kcmil	3 Elec	3.30	7.273		1,950	287		2,237	2,550
	1000	350 kcmil		3	8		2,300	315		2,615	3,000
	1010	500 kcmil		2.70	8.889		3,150	350		3,500	4,000
	1050	5 kV, copper, 3 conductor with PVC jacket,									
	1060	non-shielded, in cable tray, #4	2 Elec	380	.042	L.F.	5.50	1.66		7.16	8.50
	1100	#2		360	.044		7.15	1.75		8.90	10.45
	1200	#1		300	.053		9.10	2.10		11.20	13.15
	1400	1/0		290	.055		10.50	2.17		12.67	14.80
	1600	2/0		260	.062		12.15	2.42		14.57	16.95
	2000	4/0		240	.067		16.20	2.63		18.83	21.50
	2100	250 kcmil	3 Elec	330	.073		22	2.87		24.87	29
	2150	350 kcmil		315	.076		27.50	3		30.50	34.50
	2200	500 kcmil		270	.089		33.50	3.50		37	41.50
	2400	15 kV, copper, 3 conductor with PVC jacket galv steel armored									
	2500	grounded neutral, in cable tray, #2	2 Elec	300	.053	L.F.	11.55	2.10		13.65	15.85
	2600	#1		280	.057		12.30	2.25		14.55	16.90
	2800	1/0		260	.062		14.20	2.42		16.62	19.20
	2900	2/0		220	.073		18.70	2.87		21.57	25
	3000	4/0		190	.084		21	3.32		24.32	28.50
	3100	250 kcmil	3 Elec	270	.089		23.50	3.50		27	31
	3150	350 kcmil		240	.100		28	3.94		31.94	36.50
	3200	500 kcmil		210	.114		37.50	4.50		42	47.50
	3400	15 kV, copper, 3 conductor with PVC jacket,									
	3450	ungrounded neutral, in cable tray, #2	2 Elec	260	.062	L.F.	12.45	2.42		14.87	17.30
	3500	#1		230	.070		13.80	2.74		16.54	19.30
	3600	1/0		200	.080		15.80	3.15		18.95	22
	3700	2/0		190	.084		19.40	3.32		22.72	26.50
	3800	4/0		160	.100		23.50	3.94		27.44	31.50
	4000	250 kcmil	3 Elec	210	.114		27.50	4.50		32	36.50
	4050	350 kcmil		195	.123		36	4.85		40.85	46.50
	4100	500 kcmil		180	.133		43.50	5.25		48.75	56
	4200	600 volt, aluminum, 3 conductor in cable tray with PVC jacket									
	4300	#2	2 Elec	540	.030	L.F.	2.75	1.17		3.92	4.77
	4400	#1		460	.035		3.05	1.37		4.42	5.40
	4500	#1/0		400	.040		3.80	1.58		5.38	6.50
	4600	#2/0		360	.044		3.85	1.75		5.60	6.85
	4700	#3/0		340	.047		4.50	1.85		6.35	7.70
	4800	#4/0		320	.050		5.45	1.97		7.42	8.95
	4900	250 kcmil	3 Elec	450	.053		6.55	2.10		8.65	10.35
	5000	350 kcmil		360	.067		7.80	2.63		10.43	12.50
	5200	500 kcmil		330	.073		9.70	2.87		12.57	14.90
	5300	750 kcmil		285	.084		12.50	3.32		15.82	18.70
	5400	600 volt, aluminum, 4 conductor in cable tray with PVC jacket									
	5410	#2	2 Elec	520	.031	L.F.	3.15	1.21		4.36	5.25
	5430	#1		440	.036		3.85	1.43		5.28	6.35
	5450	1/0		380	.042		4.50	1.66		6.16	7.40
	5470	2/0		340	.047		4.60	1.85		6.45	7.80
	5480	3/0		320	.050		5.40	1.97		7.37	8.90
	5500	4/0		300	.053		6.35	2.10		8.45	10.15
	5520	250 kcmil	3 Elec	420	.057		6.80	2.25		9.05	10.85
	5540	350 kcmil		330	.073		8.80	2.87		11.67	13.95

R16120 -120

16 ELECTRICAL

Important: See the Reference Section for critical supporting data - Reference Nos., Crews, & City Cost Indexes

16120 | Conductors & Cables

			CREW	DAILY OUTPUT	LABOR-HOURS	UNIT	MAT.	LABOR	EQUIP.	TOTAL	TOTAL INCL O&P	
120	5560	500 kcmil	3 Elec	300	.080	L.F.	11	3.15		14.15	16.80	120
	5580	750 kcmil	↓	270	.089		15.15	3.50		18.65	22	
	5600	5 kV, aluminum, unshielded in cable tray, #2 with PVC jacket	2 Elec	380	.042		3.85	1.66		5.51	6.70	
	5700	#1 with PVC jacket		360	.044		4.30	1.75		6.05	7.35	
	5800	1/0 with PVC jacket		300	.053		4.40	2.10		6.50	7.95	
	6000	2/0 with PVC jacket		290	.055		4.50	2.17		6.67	8.20	
	6200	3/0 with PVC jacket		260	.062		5.40	2.42		7.82	9.55	
	6300	4/0 with PVC jacket	↓	240	.067		6.35	2.63		8.98	10.90	
	6400	250 kcmil with PVC jacket	3 Elec	330	.073		6.95	2.87		9.82	11.90	
	6500	350 kcmil with PVC jacket		315	.076		8.15	3		11.15	13.40	
	6600	500 kcmil with PVC jacket		300	.080		9.65	3.15		12.80	15.30	
	6800	750 kcmil with PVC jacket	↓	270	.089		11.90	3.50		15.40	18.30	
	6900	15 KV, aluminum, shielded-grounded, #2 with PVC jacket	2 Elec	320	.050		8.75	1.97		10.72	12.60	
	7000	#1 with PVC jacket		300	.053		9	2.10		11.10	13.05	
	7200	1/0 with PVC jacket		280	.057		9.70	2.25		11.95	14	
	7300	2/0 with PVC jacket		260	.062		9.85	2.42		12.27	14.45	
	7400	3/0 with PVC jacket		240	.067		11.05	2.63		13.68	16.05	
	7500	4/0 with PVC jacket	↓	220	.073		11.35	2.87		14.22	16.75	
	7600	250 kcmil with PVC jacket	3 Elec	300	.080		12.45	3.15		15.60	18.40	
	7700	350 kcmil with PVC jacket		270	.089		14.60	3.50		18.10	21.50	
	7800	500 kcmil with PVC jacket		240	.100		17.50	3.94		21.44	25	
	8000	750 kcmil with PVC jacket	↓	204	.118		21	4.64		25.64	30	
	8200	15 kV, aluminum, shielded-ungrounded, #1 with PVC jacket	2 Elec	250	.064		10.95	2.52		13.47	15.80	
	8300	1/0 with PVC jacket		230	.070		11.35	2.74		14.09	16.60	
	8400	2/0 with PVC jacket		210	.076		12.45	3		15.45	18.15	
	8500	3/0 with PVC jacket		200	.080		12.65	3.15		15.80	18.60	
	8600	4/0 with PVC jacket	↓	190	.084		13.75	3.32		17.07	20	
	8700	250 kcmil with PVC jacket	3 Elec	270	.089		14.75	3.50		18.25	21.50	
	8800	350 kcmil with PVC jacket		240	.100		16.85	3.94		20.79	24.50	
	8900	500 kcmil with PVC jacket		210	.114		20.50	4.50		25	29	
	8950	750 kcmil with PVC jacket	↓	174	.138	↓	25.50	5.45		30.95	36	
	9010	600 volt, copper (MC) steel clad, #14, 2 wire	1 Elec	2.40	3.333	C.L.F.	50.50	131		181.50	251	
	9020	3 wire		2.20	3.636		67	143		210	287	
	9030	4 wire		2	4		90.50	158		248.50	335	
	9040	#12, 2 wire		2.30	3.478		51	137		188	260	
	9050	3 wire		2	4		79	158		237	320	
	9060	4 wire		1.80	4.444		106	175		281	375	
	9070	#10, 2 wire		2	4		90	158		248	335	
	9080	3 wire		1.60	5		140	197		337	445	
	9090	4 wire	↓	1.40	5.714	↓	221	225		446	580	
	9091	For Health Care Facilities cable, add, minimum					5%					
	9092	Maximum					20%					
	9100	#8, 2 wire, stranded	1 Elec	1.80	4.444	C.L.F.	172	175		347	450	
	9110	3 wire, stranded		1.30	6.154		251	242		493	635	
	9120	4 wire, stranded		1.10	7.273		335	287		622	795	
	9130	#6, 2 wire, stranded		1.30	6.154		266	242		508	655	
	9200	600 volt, copper (MC) alum. clad, #14, 2 wire		2.65	3.019		50.50	119		169.50	233	
	9210	3 wire		2.45	3.265		67	129		196	265	
	9220	4 wire		2.20	3.636		90.50	143		233.50	315	
	9230	#12, 2 wire		2.55	3.137		51.50	124		175.50	241	
	9240	3 wire		2.20	3.636		79	143		222	300	
	9250	4 wire		2	4		106	158		264	350	
	9260	#10, 2 wire		2.20	3.636		90	143		233	310	
	9270	3 wire		1.80	4.444		140	175		315	415	
	9280	4 wire	↓	1.55	5.161	↓	221	203		424	545	

Table reference box: R16120 -120

ELECTRICAL 16

							2004 BARE COSTS				TOTAL
16120	**Conductors & Cables**	CREW	DAILY OUTPUT	LABOR-HOURS	UNIT	MAT.	LABOR	EQUIP.	TOTAL		INCL O&P
210	0010	**CABLE CONNECTORS**										210
	0100	600 volt, nonmetallic, #14-2 wire	1 Elec	160	.050	Ea.	.41	1.97		2.38	3.38
	0200	#14-3 wire to #12-2 wire		133	.060		.41	2.37		2.78	3.97
	0300	#12-3 wire to #10-2 wire		114	.070		.41	2.77		3.18	4.56
	0400	#10-3 wire to #14-4 and #12-4 wire		100	.080		.41	3.15		3.56	5.15
	0500	#8-3 wire to #10-4 wire		80	.100		1.14	3.94		5.08	7.10
	0600	#6-3 wire		40	.200		1.82	7.90		9.72	13.70
	0800	SER aluminum, 3 #8 insulated + 1 #8 ground		32	.250		1.87	9.85		11.72	16.70
	0900	3 #6 + 1 #6 ground		24	.333		1.87	13.15		15.02	21.50
	1000	3 #4 + 1 #6 ground		22	.364		2.92	14.35		17.27	24.50
	1100	3 #2 + 1 #4 ground		20	.400		5.30	15.75		21.05	29.50
	1200	3 1/0 + 1 #2 ground		18	.444		13.70	17.50		31.20	41
	1400	3 2/0 + 1 #1 ground		16	.500		13.70	19.70		33.40	44.50
	1600	3 4/0 + 1 #2/0 ground		14	.571		16.50	22.50		39	51.50
	1800	600 volt, armored, #14-2 wire		80	.100		.49	3.94		4.43	6.40
	2200	#14-4, #12-3 and #10-2 wire		40	.200		.49	7.90		8.39	12.25
	2400	#12-4, #10-3 and #8-2 wire		32	.250		1.17	9.85		11.02	15.95
	2600	#8-3 and #10-4 wire		26	.308		2.02	12.10		14.12	20.50
	2650	#8-4 wire		22	.364		3.60	14.35		17.95	25.50
	2700	PVC jacket connector, #6-3 wire, #6-4 wire		16	.500		9.70	19.70		29.40	40
	2800	#4-3 wire, #4-4 wire		16	.500		9.70	19.70		29.40	40
	2900	#2-3 wire		12	.667		9.70	26.50		36.20	49.50
	3000	#1-3 wire, #2-4 wire		12	.667		13.75	26.50		40.25	54
	3200	1/0-3 wire		11	.727		13.75	28.50		42.25	57.50
	3400	2/0-3 wire, 1/0-4 wire		10	.800		13.75	31.50		45.25	62
	3500	3/0-3 wire, 2/0-4 wire		9	.889		28.50	35		63.50	83
	3600	4/0-3 wire, 3/0-4 wire		7	1.143		28.50	45		73.50	98
	3800	250 kcmil-3 wire, 4/0-4 wire		6	1.333		39.50	52.50		92	121
	4000	350 kcmil-3 wire, 250 kcmil-4 wire		5	1.600		39.50	63		102.50	137
	4100	350 kcmil-4 wire		4	2		153	79		232	285
	4200	500 kcmil-3 wire		4	2		153	79		232	285
	4250	500 kcmil-4 wire, 750 kcmil-3 wire		3.50	2.286		200	90		290	355
	4300	750 kcmil-4 wire		3	2.667		200	105		305	375
	4400	5 kV, armored, #4		8	1		51	39.50		90.50	115
	4600	#2		8	1		51	39.50		90.50	115
	4800	#1		8	1		51	39.50		90.50	115
	5000	1/0		6.40	1.250		63	49.50		112.50	143
	5200	2/0		5.30	1.509		63	59.50		122.50	158
	5500	4/0		4	2		83	79		162	209
	5600	250 kcmil		3.60	2.222		102	87.50		189.50	242
	5650	350 kcmil		3.20	2.500		120	98.50		218.50	279
	5700	500 kcmil		2.50	3.200		120	126		246	320
	5720	750 kcmil		2.20	3.636		175	143		318	405
	5750	1000 kcmil		2	4		234	158		392	490
	5800	15 kV, armored, #1		4	2		80	79		159	205
	5900	1/0		4	2		80	79		159	205
	6000	3/0		3.60	2.222		103	87.50		190.50	243
	6100	4/0		3.40	2.353		103	92.50		195.50	251
	6200	250 kcmil		3.20	2.500		114	98.50		212.50	272
	6300	350 kcmil		2.70	2.963		132	117		249	320
	6400	500 kcmil	▼	2	4	▼	132	158		290	380
220	0010	**CABLE SPLICING** URD or similar, ideal conditions										220
	0100	#6 stranded to #1 stranded, 5 kV	1 Elec	4	2	Ea.	95	79		174	222
	0120	15 kV		3.60	2.222		95	87.50		182.50	235
	0140	25 kV	▼	3.20	2.500	▼	95	98.50		193.50	252

16 ELECTRICAL

			CREW	DAILY OUTPUT	LABOR-HOURS	UNIT	MAT.	LABOR	EQUIP.	TOTAL	TOTAL INCL O&P	
	16120	**Conductors & Cables**					2004 BARE COSTS					
220	0200	#1 stranded to 4/0 stranded, 5 kV	1 Elec	3.60	2.222	Ea.	99	87.50		186.50	239	**220**
	0210	15 kV		3.20	2.500		99	98.50		197.50	256	
	0220	25 kV		2.80	2.857		99	113		212	276	
	0300	4/0 stranded to 500 kcmil stranded, 5 kV		3.30	2.424		223	95.50		318.50	385	
	0310	15 kV		2.90	2.759		223	109		332	405	
	0320	25 kV		2.50	3.200		223	126		349	435	
	0400	500 kcmil, 5 kV		3.20	2.500		223	98.50		321.50	390	
	0410	15 kV		2.80	2.857		223	113		336	410	
	0420	25 kV		2.30	3.478		223	137		360	450	
	0500	600 kcmil, 5 kV		2.90	2.759		223	109		332	405	
	0510	15 kV		2.40	3.333		223	131		354	440	
	0520	25 kV		2	4		223	158		381	480	
	0600	750 kcmil, 5 kV		2.60	3.077		263	121		384	470	
	0610	15 kV		2.20	3.636		263	143		406	500	
	0620	25 kV		1.90	4.211		263	166		429	535	
	0700	1000 kcmil, 5 kV		2.30	3.478		284	137		421	515	
	0710	15 kV		1.90	4.211		284	166		450	555	
	0720	25 kV		1.60	5		284	197		481	605	
230	0010	**CABLE TERMINATIONS**										**230**
	0015	Wire connectors, screw type, #22 to #14	1 Elec	260	.031	Ea.	.07	1.21		1.28	1.88	
	0020	#18 to #12		240	.033		.08	1.31		1.39	2.04	
	0025	#18 to #10		240	.033		.12	1.31		1.43	2.08	
	0030	Screw-on connectors, insulated, #18 to #12		240	.033		.09	1.31		1.40	2.05	
	0035	#16 to #10		230	.035		.11	1.37		1.48	2.16	
	0040	#14 to #8		210	.038		.24	1.50		1.74	2.49	
	0045	#12 to #6		180	.044		.40	1.75		2.15	3.04	
	0050	Terminal lugs, solderless, #16 to #10		50	.160		.46	6.30		6.76	9.90	
	0100	#8 to #4		30	.267		.59	10.50		11.09	16.30	
	0150	#2 to #1		22	.364		.79	14.35		15.14	22.50	
	0200	1/0 to 2/0		16	.500		2.22	19.70		21.92	32	
	0250	3/0		12	.667		2.70	26.50		29.20	42	
	0300	4/0		11	.727		2.70	28.50		31.20	45.50	
	0350	250 kcmil		9	.889		2.70	35		37.70	55	
	0400	350 kcmil		7	1.143		3.50	45		48.50	71	
	0450	500 kcmil		6	1.333		6.80	52.50		59.30	85.50	
	0500	600 kcmil		5.80	1.379		7.30	54.50		61.80	89	
	0550	750 kcmil		5.20	1.538		8	60.50		68.50	99	
	0600	Split bolt connectors, tapped, #6		16	.500		2.95	19.70		22.65	33	
	0650	#4		14	.571		3.55	22.50		26.05	37.50	
	0700	#2		12	.667		5.40	26.50		31.90	45	
	0750	#1		11	.727		6.80	28.50		35.30	50	
	0800	1/0		10	.800		6.80	31.50		38.30	54.50	
	0850	2/0		9	.889		11.05	35		46.05	64	
	0900	3/0		7.20	1.111		15.40	44		59.40	82	
	1000	4/0		6.40	1.250		18.75	49.50		68.25	94	
	1100	250 kcmil		5.70	1.404		18.75	55.50		74.25	103	
	1200	300 kcmil		5.30	1.509		34.50	59.50		94	127	
	1400	350 kcmil		4.60	1.739		34.50	68.50		103	140	
	1500	500 kcmil		4	2		45	79		124	167	
	1600	Crimp 1 hole lugs, copper or aluminum, 600 volt										
	1620	#14	1 Elec	60	.133	Ea.	.30	5.25		5.55	8.15	
	1630	#12		50	.160		.42	6.30		6.72	9.85	
	1640	#10		45	.178		.42	7		7.42	10.85	
	1780	#8		36	.222		1.61	8.75		10.36	14.75	
	1800	#6		30	.267		2.60	10.50		13.10	18.50	
	2000	#4		27	.296		2.80	11.65		14.45	20.50	

ELECTRICAL 16

16120	Conductors & Cables	CREW	DAILY OUTPUT	LABOR-HOURS	UNIT	2004 BARE COSTS				TOTAL INCL O&P	
						MAT.	LABOR	EQUIP.	TOTAL		
230	2200	#2	1 Elec	24	.333	Ea.	3.80	13.15		16.95	23.50	230
	2400	#1		20	.400		4.60	15.75		20.35	28.50
	2500	1/0		17.50	.457		5.15	18		23.15	32.50
	2600	2/0		15	.533		6.20	21		27.20	38.50
	2800	3/0		12	.667		7.70	26.50		34.20	47.50
	3000	4/0		11	.727		8.35	28.50		36.85	51.50
	3200	250 kcmil		9	.889		9.90	35		44.90	63
	3400	300 kcmil		8	1		11.10	39.50		50.60	70.50
	3500	350 kcmil		7	1.143		11.50	45		56.50	79.50
	3600	400 kcmil		6.50	1.231		14.30	48.50		62.80	88
	3800	500 kcmil		6	1.333		17.40	52.50		69.90	97
	4000	600 kcmil		5.80	1.379		27.50	54.50		82	112
	4200	700 kcmil		5.50	1.455		30	57.50		87.50	118
	4400	750 kcmil	▼	5.20	1.538	▼	30	60.50		90.50	123
	4500	Crimp 2-way connectors, copper or alum., 600 volt,									
	4510	#14	1 Elec	60	.133	Ea.	.48	5.25		5.73	8.35
	4520	#12		50	.160		.76	6.30		7.06	10.25
	4530	#10		45	.178		.76	7		7.76	11.25
	4540	#8		27	.296		1.76	11.65		13.41	19.30
	4600	#6		25	.320		3.66	12.60		16.26	23
	4800	#4		23	.348		3.95	13.70		17.65	25
	5000	#2		20	.400		5.65	15.75		21.40	29.50
	5200	#1		16	.500		8.10	19.70		27.80	38.50
	5400	1/0		13	.615		8.95	24.50		33.45	46
	5420	2/0		12	.667		9.20	26.50		35.70	49
	5440	3/0		11	.727		11.05	28.50		39.55	54.50
	5460	4/0		10	.800		11.50	31.50		43	59.50
	5480	250 kcmil		9	.889		13.10	35		48.10	66.50
	5500	300 kcmil		8.50	.941		14.65	37		51.65	71
	5520	350 kcmil		8	1		15.20	39.50		54.70	75
	5540	400 kcmil		7.30	1.096		19.55	43		62.55	85.50
	5560	500 kcmil		6.20	1.290		24	51		75	102
	5580	600 kcmil		5.50	1.455		36	57.50		93.50	125
	5600	700 kcmil		4.50	1.778		37	70		107	145
	5620	750 kcmil		4	2		37	79		116	158
	7000	Compression equipment adapter, aluminum wire, #6		30	.267		7.15	10.50		17.65	23.50
	7020	#4		27	.296		7.50	11.65		19.15	25.50
	7040	#2		24	.333		7.95	13.15		21.10	28.50
	7060	#1		20	.400		9.15	15.75		24.90	33.50
	7080	1/0		18	.444		9.50	17.50		27	36.50
	7100	2/0		15	.533		18	21		39	51.50
	7140	4/0		11	.727		18	28.50		46.50	62.50
	7160	250 kcmil		9	.889		18	35		53	72
	7180	300 kcmil		8	1		21.50	39.50		61	82
	7200	350 kcmil		7	1.143		21.50	45		66.50	90.50
	7220	400 kcmil		6.50	1.231		27.50	48.50		76	103
	7240	500 kcmil		6	1.333		27.50	52.50		80	109
	7260	600 kcmil		5.80	1.379		38	54.50		92.50	123
	7280	750 kcmil	▼	5.20	1.538		38	60.50		98.50	132
	8000	Compression tool, hand					1,100			1,100	1,200
	8100	Hydraulic					2,525			2,525	2,775
	8500	Hydraulic dies				▼	206			206	227
240	0010	CABLE TERMINATIONS, 5 kV to 35 kV										240
	0100	Indoor, insulation diameter range .525" to 1.025"									
	0300	Padmount, 5 kV	1 Elec	8	1	Ea.	55	39.50		94.50	119
	0400	15 kV	▼	6.40	1.250	▼	99	49.50		148.50	183

Important: See the Reference Section for critical supporting data - Reference Nos., Crews, & City Cost Indexes

16 ELECTRICAL

		16120	Conductors & Cables	CREW	DAILY OUTPUT	LABOR-HOURS	UNIT	MAT.	2004 BARE COSTS LABOR	EQUIP.	TOTAL	TOTAL INCL O&P	
240	0500		25 kV	1 Elec	6	1.333	Ea.	115	52.50		167.50	205	**240**
	0600		35 kV	↓	5.60	1.429	↓	147	56.50		203.50	246	
	0700		insulation diameter range .975" to 1.570"										
	0800		Padmount, 5 kV	1 Elec	8	1	Ea.	67	39.50		106.50	132	
	0900		15 kV		6	1.333		132	52.50		184.50	223	
	1000		25 kV		5.60	1.429		152	56.50		208.50	251	
	1100		35 kV	↓	5.30	1.509	↓	182	59.50		241.50	289	
	1200		insulation diameter range 1.540" to 1.900"										
	1300		Padmount, 5 KV	1 Elec	7.40	1.081	Ea.	101	42.50		143.50	175	
	1400		15 kV		5.60	1.429		167	56.50		223.50	268	
	1500		25 kV		5.30	1.509		213	59.50		272.50	325	
	1600		35 kV	↓	5	1.600	↓	238	63		301	355	
	1700		Outdoor systems, #4 stranded to 1/0 stranded										
	1800		5 kV	1 Elec	7.40	1.081	Ea.	109	42.50		151.50	184	
	1900		15 kV		5.30	1.509		135	59.50		194.50	238	
	2000		25 kV		5	1.600		152	63		215	261	
	2100		35 kV		4.80	1.667		177	65.50		242.50	293	
	2200		#1 solid to 4/0 stranded, 5 kV		6.90	1.159		122	45.50		167.50	202	
	2300		15 kV		5	1.600		135	63		198	243	
	2400		25 kV		4.80	1.667		182	65.50		247.50	298	
	2500		35 kV		4.60	1.739		182	68.50		250.50	300	
	2600		3/0 solid to 350 kcmil stranded, 5 kV		6.40	1.250		152	49.50		201.50	241	
	2700		15 kV		4.80	1.667		152	65.50		217.50	265	
	2800		25 kV		4.60	1.739		213	68.50		281.50	335	
	2900		35 kV		4.40	1.818		213	71.50		284.50	340	
	3000		400 kcmil compact to 750 kcmil stranded, 5 kV		6	1.333		197	52.50		249.50	295	
	3100		15 kV		4.60	1.739		197	68.50		265.50	320	
	3200		25 kV		4.40	1.818		238	71.50		309.50	370	
	3300		35 kV		4.20	1.905		238	75		313	375	
	3400		1000 kcmil, 5 kV		5.60	1.429		213	56.50		269.50	320	
	3500		15 kV		4.40	1.818		228	71.50		299.50	360	
	3600		25 kV		4.20	1.905		238	75		313	375	
	3700		35 kV	↓	4	2	↓	238	79		317	380	
280	0010		**CONTROL CABLE**										**280**
	0020		600 volt, copper, #14 THWN wire with PVC jacket, 2 wires	1 Elec	9	.889	C.L.F.	12.80	35		47.80	66	
	0030		3 wires		8	1		18.20	39.50		57.70	78.50	
	0100		4 wires		7	1.143		22.50	45		67.50	92	
	0150		5 wires		6.50	1.231		27	48.50		75.50	102	
	0200		6 wires		6	1.333		37	52.50		89.50	119	
	0300		8 wires		5.30	1.509		45	59.50		104.50	138	
	0400		10 wires		4.80	1.667		54.50	65.50		120	158	
	0500		12 wires		4.30	1.860		61.50	73.50		135	177	
	0600		14 wires		3.80	2.105		75	83		158	206	
	0700		16 wires		3.50	2.286		88	90		178	231	
	0800		18 wires		3.30	2.424		92	95.50		187.50	243	
	0810		19 wires		3.10	2.581		100	102		202	261	
	0900		20 wires		3	2.667		113	105		218	280	
	1000		22 wires	↓	2.80	2.857	↓	115	113		228	294	
500	0010		**MINERAL INSULATED CABLE** 600 volt										**500**
	0100		1 conductor, #12	1 Elec	1.60	5	C.L.F.	208	197		405	520	
	0200		#10		1.60	5		271	197		468	590	
	0400		#8		1.50	5.333		320	210		530	665	
	0500		#6	↓	1.40	5.714		360	225		585	730	
	0600		#4	2 Elec	2.40	6.667	↓	495	263		758	935	

ELECTRICAL 16

16 ELECTRICAL

16120	Conductors & Cables	CREW	DAILY OUTPUT	LABOR-HOURS	UNIT	2004 BARE COSTS				TOTAL INCL O&P
						MAT.	LABOR	EQUIP.	TOTAL	
500										500
0800	#2	2 Elec	2.20	7.273	C.L.F.	660	287		947	1,150
0900	#1		2.10	7.619		755	300		1,055	1,275
1000	1/0		2	8		870	315		1,185	1,425
1100	2/0		1.90	8.421		1,025	330		1,355	1,650
1200	3/0		1.80	8.889		1,225	350		1,575	1,875
1400	4/0		1.60	10		1,400	395		1,795	2,125
1410	250 kcmil	3 Elec	2.40	10		1,600	395		1,995	2,325
1420	350 kcmil		1.95	12.308		1,950	485		2,435	2,875
1430	500 kcmil		1.95	12.308		2,450	485		2,935	3,400
1500	2 conductor, #12	1 Elec	1.40	5.714		430	225		655	810
1600	#10		1.20	6.667		520	263		783	960
1800	#8		1.10	7.273		650	287		937	1,150
2000	#6		1.05	7.619		825	300		1,125	1,350
2100	#4	2 Elec	2	8		1,075	315		1,390	1,675
2200	3 conductor, #12	1 Elec	1.20	6.667		540	263		803	985
2400	#10		1.10	7.273		625	287		912	1,100
2600	#8		1.05	7.619		755	300		1,055	1,275
2800	#6		1	8		980	315		1,295	1,550
3000	#4	2 Elec	1.80	8.889		1,300	350		1,650	1,950
3100	4 conductor, #12	1 Elec	1.20	6.667		575	263		838	1,025
3200	#10		1.10	7.273		680	287		967	1,175
3400	#8		1	8		890	315		1,205	1,450
3600	#6		.90	8.889		1,125	350		1,475	1,750
3620	7 conductor, #12		1.10	7.273		725	287		1,012	1,225
3640	#10		1	8		910	315		1,225	1,475
3800	M.I. terminations, 600 volt, 1 conductor, #12		8	1	Ea.	8.45	39.50		47.95	68
4000	#10		7.60	1.053		8.45	41.50		49.95	71
4100	#8		7.30	1.096		8.45	43		51.45	73.50
4200	#6		6.70	1.194		8.45	47		55.45	79.50
4400	#4		6.20	1.290		8.45	51		59.45	85
4600	#2		5.70	1.404		12.70	55.50		68.20	96.50
4800	#1		5.30	1.509		12.70	59.50		72.20	102
5000	1/0		5	1.600		12.70	63		75.70	108
5100	2/0		4.70	1.702		12.70	67		79.70	113
5200	3/0		4.30	1.860		12.70	73.50		86.20	123
5400	4/0		4	2		28	79		107	148
5410	250 kcmil		4	2		28	79		107	148
5420	350 kcmil		4	2		50.50	79		129.50	173
5430	500 kcmil		4	2		50.50	79		129.50	173
5500	2 conductor, #12		6.70	1.194		8.45	47		55.45	79.50
5600	#10		6.40	1.250		12.70	49.50		62.20	87.50
5800	#8		6.20	1.290		12.70	51		63.70	89.50
6000	#6		5.70	1.404		12.70	55.50		68.20	96.50
6200	#4		5.30	1.509		28	59.50		87.50	120
6400	3 conductor, #12		5.70	1.404		12.70	55.50		68.20	96.50
6500	#10		5.50	1.455		12.70	57.50		70.20	99
6600	#8		5.20	1.538		12.70	60.50		73.20	104
6800	#6		4.80	1.667		12.70	65.50		78.20	111
7200	#4		4.60	1.739		29	68.50		97.50	134
7400	4 conductor, #12		4.60	1.739		14.10	68.50		82.60	118
7500	#10		4.40	1.818		14.10	71.50		85.60	123
7600	#8		4.20	1.905		14.10	75		89.10	128
8400	#6		4	2		29.50	79		108.50	150
8500	7 conductor, #12		3.50	2.286		14.10	90		104.10	150
8600	#10		3	2.667		29.50	105		134.50	189
8800	Crimping tool, plier type					40			40	44

Important: See the Reference Section for critical supporting data - Reference Nos., Crews, & City Cost Indexes

			DAILY	LABOR-			2004 BARE COSTS				TOTAL	
16120	**Conductors & Cables**	CREW	OUTPUT	HOURS	UNIT	MAT.	LABOR	EQUIP.	TOTAL	INCL O&P	
500	9000	Stripping tool				Ea.	100			100	110	500
	9200	Hand vise				↓	45			45	49.50	
550	0010	**NON-METALLIC SHEATHED CABLE** 600 volt										550
	0100	Copper with ground wire, (Romex)										
	0150	#14, 2 conductor	1 Elec	2.70	2.963	C.L.F.	12.25	117		129.25	187	
	0200	3 conductor		2.40	3.333		21.50	131		152.50	219	
	0220	4 conductor		2.20	3.636		28	143		171	244	
	0250	#12, 2 conductor		2.50	3.200		17.25	126		143.25	207	
	0300	3 conductor		2.20	3.636		31	143		174	247	
	0320	4 conductor		2	4		42.50	158		200.50	281	
	0350	#10, 2 conductor		2.20	3.636		34	143		177	250	
	0400	3 conductor		1.80	4.444		47.50	175		222.50	310	
	0420	4 conductor		1.60	5		79.50	197		276.50	380	
	0450	#8, 3 conductor		1.50	5.333		91	210		301	415	
	0480	4 conductor		1.40	5.714		153	225		378	505	
	0500	#6, 3 conductor	↓	1.40	5.714		145	225		370	495	
	0520	#4, 3 conductor	2 Elec	2.40	6.667		218	263		481	630	
	0540	#2, 3 conductor	"	2.20	7.273	↓	325	287		612	780	
	0550	SE type SER aluminum cable, 3 RHW and										
	0600	1 bare neutral, 3 #8 & 1 #8	1 Elec	1.60	5	C.L.F.	92.50	197		289.50	395	
	0650	3 #6 & 1 #6	"	1.40	5.714		105	225		330	450	
	0700	3 #4 & 1 #6	2 Elec	2.40	6.667		117	263		380	520	
	0750	3 #2 & 1 #4		2.20	7.273		173	287		460	615	
	0800	3 #1/0 & 1 #2		2	8		261	315		576	755	
	0850	3 #2/0 & 1 #1		1.80	8.889		310	350		660	860	
	0900	3 #4/0 & 1 #2/0	↓	1.60	10		440	395		835	1,075	
	1450	UF underground feeder cable, copper with ground, #14, 2 conductor	1 Elec	4	2		15.90	79		94.90	135	
	1500	#12, 2 conductor		3.50	2.286		22	90		112	158	
	1550	#10, 2 conductor		3	2.667		41.50	105		146.50	202	
	1600	#14, 3 conductor		3.50	2.286		24	90		114	161	
	1650	#12, 3 conductor		3	2.667		34	105		139	193	
	1700	#10, 3 conductor		2.50	3.200		56	126		182	250	
	2400	SEU service entrance cable, copper 2 conductors, #8 + #8 neut.		1.50	5.333		78.50	210		288.50	400	
	2600	#6 + #8 neutral		1.30	6.154		116	242		358	490	
	2800	#6 + #6 neutral	↓	1.30	6.154		127	242		369	500	
	3000	#4 + #6 neutral	2 Elec	2.20	7.273		180	287		467	625	
	3200	#4 + #4 neutral		2.20	7.273		196	287		483	640	
	3400	#3 + #5 neutral		2.10	7.619		210	300		510	675	
	3600	#3 + #3 neutral		2.10	7.619		230	300		530	695	
	3800	#2 + #4 neutral		2	8		263	315		578	760	
	4000	#1 + #1 neutral		1.90	8.421		450	330		780	990	
	4200	1/0 + 1/0 neutral		1.80	8.889		530	350		880	1,100	
	4400	2/0 + 2/0 neutral		1.70	9.412		635	370		1,005	1,250	
	4600	3/0 + 3/0 neutral	↓	1.60	10		785	395		1,180	1,450	
	4800	Aluminum 2 conductors, #8 + #8 neutral	1 Elec	1.60	5		76	197		273	375	
	5000	#6 + #6 neutral	"	1.40	5.714		76.50	225		301.50	420	
	5100	#4 + #6 neutral	2 Elec	2.50	6.400		98.50	252		350.50	485	
	5200	#4 + #4 neutral		2.40	6.667		98.50	263		361.50	500	
	5300	#2 + #4 neutral		2.30	6.957		131	274		405	555	
	5400	#2 + #2 neutral		2.20	7.273		131	287		418	570	
	5450	1/0 + #2 neutral		2.10	7.619		200	300		500	665	
	5500	1/0 + 1/0 neutral		2	8		200	315		515	690	
	5550	2/0 + #1 neutral		1.90	8.421		229	330		559	745	
	5600	2/0 + 2/0 neutral		1.80	8.889		229	350		579	770	
	5800	3/0 + 1/0 neutral		1.70	9.412		293	370		663	870	
	6000	3/0 + 3/0 neutral	↓	1.70	9.412		293	370		663	870	

ELECTRICAL 16

			DAILY	LABOR-		2004 BARE COSTS				TOTAL	
16120	**Conductors & Cables**	CREW	OUTPUT	HOURS	UNIT	MAT.	LABOR	EQUIP.	TOTAL	INCL O&P	
550	6200	4/0 + 2/0 neutral	2 Elec	1.60	10	C.L.F.	320	395		715	940	550
	6400	4/0 + 4/0 neutral	↓	1.60	10	↓	320	395		715	940	
	6500	Service entrance cap for copper SEU										
	6600	100 amp	1 Elec	12	.667	Ea.	8.35	26.50		34.85	48	
	6700	150 amp		10	.800		12.10	31.50		43.60	60.50	
	6800	200 amp	↓	8	1	↓	18.30	39.50		57.80	78.50	
600	0010	**PORTABLE CORD** 600 volt										600
	0100	Type SO, #18, 2 conductor	1 Elec	980	.008	L.F.	.21	.32		.53	.71	
	0110	3 conductor		980	.008		.26	.32		.58	.77	
	0120	#16, 2 conductor		840	.010		.23	.38		.61	.81	
	0130	3 conductor		840	.010		.27	.38		.65	.86	
	0140	4 conductor		840	.010		.37	.38		.75	.97	
	0240	#14, 2 conductor		840	.010		.40	.38		.78	1	
	0250	3 conductor		840	.010		.44	.38		.82	1.04	
	0260	4 conductor		840	.010		.56	.38		.94	1.18	
	0280	#12, 2 conductor		840	.010		.52	.38		.90	1.13	
	0290	3 conductor		840	.010		.62	.38		1	1.24	
	0320	#10, 2 conductor		765	.010		.61	.41		1.02	1.28	
	0330	3 conductor		765	.010		.71	.41		1.12	1.39	
	0340	4 conductor		765	.010		.86	.41		1.27	1.56	
	0360	#8, 2 conductor		555	.014		1.28	.57		1.85	2.25	
	0370	3 conductor		540	.015		1.29	.58		1.87	2.29	
	0380	4 conductor		525	.015		1.56	.60		2.16	2.61	
	0400	#6, 2 conductor		525	.015		1.26	.60		1.86	2.28	
	0410	3 conductor		490	.016		1.60	.64		2.24	2.72	
	0420	4 conductor	↓	415	.019		2.15	.76		2.91	3.50	
	0440	#4, 2 conductor	2 Elec	830	.019		2	.76		2.76	3.33	
	0450	3 conductor		700	.023		3.05	.90		3.95	4.70	
	0460	4 conductor		660	.024		3.59	.96		4.55	5.35	
	0480	#2, 2 conductor		450	.036		2.95	1.40		4.35	5.35	
	0490	3 conductor		350	.046		3.62	1.80		5.42	6.65	
	0500	4 conductor	↓	280	.057	↓	4.46	2.25		6.71	8.25	
700	0010	**SHIELDED CABLE** Splicing & terminations not included										700
	0040	Copper, XLP shielding, 5 kV, #6	2 Elec	4.40	3.636	C.L.F.	84	143		227	305	
	0050	#4		4.40	3.636		109	143		252	335	
	0100	#2		4	4		129	158		287	375	
	0200	#1		4	4		148	158		306	395	
	0400	1/0		3.80	4.211		163	166		329	425	
	0600	2/0		3.60	4.444		205	175		380	485	
	0800	4/0	↓	3.20	5		267	197		464	585	
	1000	250 kcmil	3 Elec	4.50	5.333		330	210		540	680	
	1200	350 kcmil		3.90	6.154		425	242		667	830	
	1400	500 kcmil	↓	3.60	6.667		520	263		783	960	
	1600	15 kV, ungrounded neutral, #1	2 Elec	4	4		180	158		338	430	
	1800	1/0		3.80	4.211		215	166		381	485	
	2000	2/0		3.60	4.444		245	175		420	530	
	2200	4/0	↓	3.20	5		325	197		522	655	
	2400	250 kcmil	3 Elec	4.50	5.333		360	210		570	710	
	2600	350 kcmil		3.90	6.154		455	242		697	860	
	2800	500 kcmil	↓	3.60	6.667		560	263		823	1,000	
	3000	25 kV, grounded neutral, #1/0	2 Elec	3.60	4.444		297	175		472	585	
	3200	2/0		3.40	4.706		325	185		510	635	
	3400	4/0	↓	3	5.333		410	210		620	765	
	3600	250 kcmil	3 Elec	4.20	5.714		510	225		735	895	
	3800	350 kcmil		3.60	6.667		600	263		863	1,050	
	3900	500 kcmil	↓	3.30	7.273	↓	700	287		987	1,200	

16 ELECTRICAL

Important: See the Reference Section for critical supporting data - Reference Nos., Crews, & City Cost Indexes

16120 | Conductors & Cables

		CREW	DAILY OUTPUT	LABOR-HOURS	UNIT	MAT.	2004 BARE COSTS LABOR	EQUIP.	TOTAL	TOTAL INCL O&P
700 4000	35 kV, grounded neutral, #1/0	2 Elec	3.40	4.706	C.L.F.	315	185		500	620
4200	2/0		3.20	5		370	197		567	700
4400	4/0		2.80	5.714		465	225		690	845
4600	250 kcmil	3 Elec	3.90	6.154		545	242		787	960
4800	350 kcmil		3.30	7.273		655	287		942	1,150
5000	500 kcmil		3	8		770	315		1,085	1,325
5050	Aluminum, XLP shielding, 5 kV, #2	2 Elec	5	3.200		117	126		243	315
5070	#1		4.40	3.636		121	143		264	345
5090	1/0		4	4		140	158		298	390
5100	2/0		3.80	4.211		157	166		323	420
5150	4/0		3.60	4.444		186	175		361	465
5200	250 kcmil	3 Elec	4.80	5		225	197		422	540
5220	350 kcmil		4.50	5.333		265	210		475	605
5240	500 kcmil		3.90	6.154		340	242		582	735
5260	750 kcmil		3.60	6.667		450	263		713	885
5300	15 kV aluminum, XLP, #1	2 Elec	4.40	3.636		150	143		293	380
5320	1/0		4	4		155	158		313	405
5340	2/0		3.80	4.211		185	166		351	450
5360	4/0		3.60	4.444		205	175		380	485
5380	250 kcmil	3 Elec	4.80	5		245	197		442	565
5400	350 kcmil		4.50	5.333		275	210		485	620
5420	500 kcmil		3.90	6.154		380	242		622	780
5440	750 kcmil		3.60	6.667		505	263		768	945
750 0010	**SPECIAL WIRES & FITTINGS**									
0100	Fixture TFFN 600 volt 90°C stranded, #18	1 Elec	13	.615	C.L.F.	3.78	24.50		28.28	40
0150	#16		13	.615		4.91	24.50		29.41	41.50
0500	Thermostat, no jacket, twisted, #18-2 conductor		8	1		7.05	39.50		46.55	66.50
0550	#18-3 conductor		7	1.143		9.70	45		54.70	77.50
0600	#18-4 conductor		6.50	1.231		13.50	48.50		62	87
0650	#18-5 conductor		6	1.333		16.20	52.50		68.70	96
0700	#18-6 conductor		5.50	1.455		20	57.50		77.50	107
0750	#18-7 conductor		5	1.600		25.50	63		88.50	123
0800	#18-8 conductor		4.80	1.667		25.50	65.50		91	126
1500	Fire alarm FEP teflon 150 volt to 200°C									
1550	#22, 1 pair	1 Elec	10	.800	C.L.F.	52.50	31.50		84	105
1600	2 pair		8	1		87.50	39.50		127	155
1650	4 pair		7	1.143		134	45		179	214
1700	6 pair		6	1.333		174	52.50		226.50	269
1750	8 pair		5.50	1.455		219	57.50		276.50	325
1800	10 pair		5	1.600		264	63		327	385
1850	#18, 1 pair		8	1		60	39.50		99.50	125
1900	2 pair		6.50	1.231		115	48.50		163.50	199
1950	4 pair		4.80	1.667		174	65.50		239.50	289
2000	6 pair		4	2		229	79		308	370
2050	8 pair		3.50	2.286		299	90		389	465
2100	10 pair		3	2.667		345	105		450	530
2500	Tray cable, type TC, copper #14, 2 conductor		9	.889		12.80	35		47.80	66
2520	3 conductor		8	1		18.20	39.50		57.70	78.50
2540	4 conductor		7	1.143		22.50	45		67.50	92
2560	5 conductor		6.50	1.231		27	48.50		75.50	102
2580	6 conductor		6	1.333		37	52.50		89.50	119
2600	8 conductor		5.30	1.509		45	59.50		104.50	138
2620	10 conductor		4.80	1.667		54.50	65.50		120	158
2640	300 V, copper braided shield, PVC jacket									
2650	2 conductor #18 stranded	1 Elec	7	1.143	C.L.F.	39	45		84	110

16120	Conductors & Cables	CREW	DAILY OUTPUT	LABOR-HOURS	UNIT	2004 BARE COSTS				TOTAL INCL O&P	
						MAT.	LABOR	EQUIP.	TOTAL		
750 2660	3 conductor #18	1 Elec	6	1.333	C.L.F.	55	52.50		107.50	139	**750**
3000	Strain relief grip for cable										
3050	Cord, top, #12-3	1 Elec	40	.200	Ea.	13.15	7.90		21.05	26	
3060	#12-4		40	.200		13.15	7.90		21.05	26	
3070	#12-5		39	.205		15.30	8.10		23.40	29	
3100	#10-3		39	.205		15.30	8.10		23.40	29	
3110	#10-4		38	.211		15.30	8.30		23.60	29	
3120	#10-5		38	.211		16.65	8.30		24.95	30.50	
3200	Bottom, #12-3		40	.200		33.50	7.90		41.40	48.50	
3210	#12-4		40	.200		33.50	7.90		41.40	48.50	
3220	#12-5		39	.205		34	8.10		42.10	49.50	
3230	#10-3		39	.205		36.50	8.10		44.60	52	
3300	#10-4		38	.211		39	8.30		47.30	55.50	
3310	#10-5		38	.211		43	8.30		51.30	59.50	
3400	Cable ties, standard, 4" length		190	.042		.16	1.66		1.82	2.65	
3410	7" length		160	.050		.25	1.97		2.22	3.21	
3420	14.5" length		90	.089		.55	3.50		4.05	5.80	
3430	Heavy, 14.5" length	▼	80	.100	▼	.81	3.94		4.75	6.75	
755 0010	**MODULAR FLEXIBLE WIRING SYSTEM**										**755**
0050	Standard selector cable, 3 conductor, 15' long	1 Elec	40	.200	Ea.	38.50	7.90		46.40	54	
0100	Conversion Module		16	.500		14	19.70		33.70	45	
0150	Cable set, 3 conductor, 15' long		24	.333		15	13.15		28.15	36	
0200	Switching assembly, 3 conductor, 10' long	▼	32	.250	▼	29.50	9.85		39.35	47	
800 0010	**UNDERCARPET**										**800**
0020	Power System	R16120 -800									
0100	Cable flat, 3 conductor, #12, w/attached bottom shield	1 Elec	982	.008	L.F.	4.42	.32		4.74	5.35	
0200	Shield, top, steel		1,768	.005	"	4.80	.18		4.98	5.55	
0250	Splice, 3 conductor		48	.167	Ea.	14.65	6.55		21.20	26	
0300	Top shield		96	.083		1.30	3.28		4.58	6.30	
0350	Tap		40	.200		18.80	7.90		26.70	32	
0400	Insulating patch, splice, tap, & end		48	.167		46.50	6.55		53.05	61	
0450	Fold		230	.035			1.37		1.37	2.04	
0500	Top shield, tap & fold		96	.083		1.30	3.28		4.58	6.30	
0700	Transition, block assembly		77	.104		67.50	4.09		71.59	80.50	
0750	Receptacle frame & base		32	.250		37	9.85		46.85	55	
0800	Cover receptacle		120	.067		3.15	2.63		5.78	7.40	
0850	Cover blank		160	.050		3.70	1.97		5.67	7	
0860	Receptacle, direct connected, single		25	.320		79.50	12.60		92.10	106	
0870	Dual		16	.500		130	19.70		149.70	173	
0880	Combination Hi & Lo, tension		21	.381		96	15		111	128	
0900	Box, floor with cover		20	.400		80.50	15.75		96.25	112	
0920	Floor service w/barrier		4	2		227	79		306	365	
1000	Wall, surface, with cover		20	.400		53	15.75		68.75	81.50	
1100	Wall, flush, with cover		20	.400	▼	37	15.75		52.75	64	
1450	Cable flat, 5 conductor #12, w/attached bottom shield		800	.010	L.F.	7.25	.39		7.64	8.55	
1550	Shield, top, steel		1,768	.005	"	7.20	.18		7.38	8.15	
1600	Splice, 5 conductor		48	.167	Ea.	23.50	6.55		30.05	36	
1650	Top shield		96	.083		1.30	3.28		4.58	6.30	
1700	Tap		48	.167		31	6.55		37.55	44	
1750	Insulating patch, splice tap, & end		83	.096		45.50	3.80		49.30	55.50	
1800	Transition, block assembly		77	.104		48.50	4.09		52.59	59.50	
1850	Box, wall, flush with cover		20	.400	▼	49.50	15.75		65.25	77.50	
1900	Cable flat, 4 conductor, #12		933	.009	L.F.	5.95	.34		6.29	7.05	
1950	3 conductor #10		982	.008		5.15	.32		5.47	6.15	
1960	4 conductor #10	▼	933	.009	▼	6.75	.34		7.09	7.95	

Important: See the Reference Section for critical supporting data - Reference Nos., Crews, & City Cost Indexes

16 ELECTRICAL

16120	Conductors & Cables		CREW	DAILY OUTPUT	LABOR-HOURS	UNIT	2004 BARE COSTS				TOTAL INCL O&P
							MAT.	LABOR	EQUIP.	TOTAL	
800	1970	5 conductor #10		1 Elec	884	.009	L.F.	8.25	.36		8.61	9.65	**800**
	2500	Telephone System										
	2510	Transition fitting wall box, surface		1 Elec	24	.333	Ea.	37.50	13.15		50.65	60.50
	2520	Flush			24	.333		37.50	13.15		50.65	60.50
	2530	Flush, for PC board			24	.333		37.50	13.15		50.65	60.50
	2540	Floor service box		↓	4	2		206	79		285	345
	2550	Cover, surface						12.65			12.65	13.90
	2560	Flush						12.65			12.65	13.90
	2570	Flush for PC board						12.65			12.65	13.90
	2700	Floor fitting w/duplex jack & cover		1 Elec	21	.381		39	15		54	65.50
	2720	Low profile			53	.151		13.60	5.95		19.55	24
	2740	Miniature w/duplex jack			53	.151		21	5.95		26.95	32
	2760	25 pair kit			21	.381		41	15		56	67.50
	2780	Low profile			53	.151		13.90	5.95		19.85	24
	2800	Call director kit for 5 cable			19	.421		65.50	16.60		82.10	96.50
	2820	4 pair kit			19	.421		78	16.60		94.60	110
	2840	3 pair kit			19	.421		82.50	16.60		99.10	115
	2860	Comb. 25 pair & 3 cond power			21	.381		65.50	15		80.50	94.50
	2880	5 cond power			21	.381		74.50	15		89.50	105
	2900	PC board, 8-3 pair			161	.050		56.50	1.96		58.46	65.50
	2920	6-4 pair			161	.050		56.50	1.96		58.46	65.50
	2940	3 pair adapter			161	.050		51.50	1.96		53.46	59.50
	2950	Plug			77	.104		2.40	4.09		6.49	8.75
	2960	Couplers			321	.025	↓	6.75	.98		7.73	8.90
	3000	Bottom shield for 25 pr. cable			4,420	.002	L.F.	.67	.07		.74	.85
	3020	4 pair			4,420	.002		.32	.07		.39	.46
	3040	Top shield for 25 pr. cable			4,420	.002	↓	.67	.07		.74	.85
	3100	Cable assembly, double-end, 50', 25 pr.			11.80	.678	Ea.	199	26.50		225.50	259
	3110	3 pair			23.60	.339		58.50	13.35		71.85	84.50
	3120	4 pair			23.60	.339	↓	65.50	13.35		78.85	92
	3140	Bulk 3 pair			1,473	.005	L.F.	1	.21		1.21	1.42
	3160	4 pair		↓	1,473	.005	"	1.25	.21		1.46	1.70
	3500	Data System										
	3520	Cable 25 conductor w/conn. 40', 75 ohm		1 Elec	14.50	.552	Ea.	59	21.50		80.50	97.50
	3530	Single lead			22	.364		161	14.35		175.35	199
	3540	Dual lead		↓	22	.364	↓	200	14.35		214.35	242
	3560	Shields same for 25 cond. as 25 pair tele.										
	3570	Single & dual, none req'd.										
	3590	BNC coax connectors, Plug		1 Elec	40	.200	Ea.	8.55	7.90		16.45	21
	3600	TNC coax connectors, Plug		"	40	.200	"	11	7.90		18.90	24
	3700	Cable-bulk										
	3710	Single lead		1 Elec	1,473	.005	L.F.	2.15	.21		2.36	2.69
	3720	Dual lead		"	1,473	.005	"	3.10	.21		3.31	3.73
	3730	Hand tool crimp					Ea.	315			315	345
	3740	Hand tool notch					"	15.35			15.35	16.90
	3750	Boxes & floor fitting same as telephone										
	3790	Data cable notching, 90°		1 Elec	97	.082	Ea.		3.25		3.25	4.83
	3800	180°			60	.133			5.25		5.25	7.80
	8100	Drill floor			160	.050	↓	1.70	1.97		3.67	4.80
	8200	Marking floor			1,600	.005	L.F.		.20		.20	.29
	8300	Tape, hold down			6,400	.001	"	.13	.05		.18	.21
	8350	Tape primer, 500 ft. per can		↓	96	.083	Ea.	26	3.28		29.28	33.50
	8400	Tool, splicing		↓			"	207			207	228
900	0010	**WIRE**											**900**
	0020	600 volt type THW, copper, solid, #14		1 Elec	13	.615	C.L.F.	2.95	24.50		27.45	39.50

R16120-800

R16120-900

ELECTRICAL 16

16120 | Conductors & Cables

		CREW	DAILY OUTPUT	LABOR-HOURS	UNIT	2004 BARE COSTS				TOTAL INCL O&P
						MAT.	LABOR	EQUIP.	TOTAL	
900 0030	#12 — R16120-900	1 Elec	11	.727	C.L.F.	4.15	28.50		32.65	47
0040	#10		10	.800		6.50	31.50		38	54
0050	Stranded, #14 — R16120-910		13	.615		3.70	24.50		28.20	40
0100	#12		11	.727		5.05	28.50		33.55	48
0120	#10 — R16132-210		10	.800		7.60	31.50		39.10	55.50
0140	#8		8	1		12.65	39.50		52.15	72.50
0160	#6		6.50	1.231		20	48.50		68.50	94
0180	#4	2 Elec	10.60	1.509		30.50	59.50		90	122
0200	#3		10	1.600		37.50	63		100.50	136
0220	#2		9	1.778		47	70		117	156
0240	#1		8	2		59.50	79		138.50	183
0260	1/0		6.60	2.424		72	95.50		167.50	221
0280	2/0		5.80	2.759		88.50	109		197.50	259
0300	3/0		5	3.200		109	126		235	310
0350	4/0		4.40	3.636		138	143		281	365
0400	250 kcmil	3 Elec	6	4		168	158		326	420
0420	300 kcmil		5.70	4.211		199	166		365	465
0450	350 kcmil		5.40	4.444		230	175		405	515
0480	400 kcmil		5.10	4.706		273	185		458	575
0490	500 kcmil		4.80	5		320	197		517	645
0500	600 kcmil		3.90	6.154		390	242		632	790
0510	750 kcmil		3.30	7.273		520	287		807	995
0520	1000 kcmil		2.70	8.889		720	350		1,070	1,300
0530	Aluminum, stranded, #8	1 Elec	9	.889		12.45	35		47.45	65.50
0540	#6	"	8	1		17.05	39.50		56.55	77.50
0560	#4	2 Elec	13	1.231		21	48.50		69.50	95.50
0580	#2		10.60	1.509		29	59.50		88.50	120
0600	#1		9	1.778		42	70		112	150
0620	1/0		8	2		50.50	79		129.50	173
0640	2/0		7.20	2.222		59.50	87.50		147	196
0680	3/0		6.60	2.424		74	95.50		169.50	224
0700	4/0		6.20	2.581		82.50	102		184.50	242
0720	250 kcmil	3 Elec	8.70	2.759		100	109		209	272
0740	300 kcmil		8.10	2.963		139	117		256	325
0760	350 kcmil		7.50	3.200		141	126		267	345
0780	400 kcmil		6.90	3.478		165	137		302	385
0800	500 kcmil		6	4		182	158		340	435
0850	600 kcmil		5.70	4.211		230	166		396	500
0880	700 kcmil		5.10	4.706		262	185		447	565
0900	750 kcmil		4.80	5		269	197		466	590
0920	Type THWN-THHN, copper, solid, #14	1 Elec	13	.615		2.95	24.50		27.45	39.50
0940	#12		11	.727		4.15	28.50		32.65	47
0960	#10		10	.800		6.50	31.50		38	54
1000	Stranded, #14		13	.615		3.70	24.50		28.20	40
1200	#12		11	.727		5.05	28.50		33.55	48
1250	#10		10	.800		7.60	31.50		39.10	55.50
1300	#8		8	1		12.65	39.50		52.15	72.50
1350	#6		6.50	1.231		20	48.50		68.50	94
1400	#4	2 Elec	10.60	1.509		30.50	59.50		90	122
1450	#3		10	1.600		37.50	63		100.50	136
1500	#2		9	1.778		47	70		117	156
1550	#1		8	2		59.50	79		138.50	183
1600	1/0		6.60	2.424		72	95.50		167.50	221
1650	2/0		5.80	2.759		88.50	109		197.50	259
1700	3/0		5	3.200		109	126		235	310
2000	4/0		4.40	3.636		138	143		281	365

16 ELECTRICAL

		16120	Conductors & Cables		CREW	DAILY OUTPUT	LABOR-HOURS	UNIT	2004 BARE COSTS				TOTAL INCL O&P	
									MAT.	LABOR	EQUIP.	TOTAL		
900	2200	250 kcmil		R16120 -900	3 Elec	6	4	C.L.F.	168	158		326	420	**900**
	2400	300 kcmil				5.70	4.211		199	166		365	465	
	2600	350 kcmil		R16120 -910		5.40	4.444		230	175		405	515	
	2700	400 kcmil				5.10	4.706		273	185		458	575	
	2800	500 kcmil		R16132 -210		4.80	5		320	197		517	645	
	2900	600 volt, copper type XHHW, solid, #14			1 Elec	13	.615		5.80	24.50		30.30	42.50	
	2920	#12				11	.727		7	28.50		35.50	50	
	2940	#10				10	.800		9.60	31.50		41.10	57.50	
	3000	Stranded, #14				13	.615		6.40	24.50		30.90	43	
	3020	#12				11	.727		8.10	28.50		36.60	51.50	
	3040	#10				10	.800		11.50	31.50		43	59.50	
	3060	#8				8	1		18.90	39.50		58.40	79.50	
	3080	#6				6.50	1.231		26.50	48.50		75	102	
	3100	#4			2 Elec	10.60	1.509		40	59.50		99.50	133	
	3120	#2				9	1.778		59	70		129	169	
	3140	#1				8	2		75	79		154	200	
	3160	1/0				6.60	2.424		79	95.50		174.50	229	
	3180	2/0				5.80	2.759		98	109		207	270	
	3200	3/0				5	3.200		123	126		249	325	
	3220	4/0				4.40	3.636		152	143		295	380	
	3240	250 kcmil			3 Elec	6	4		187	158		345	440	
	3260	300 kcmil				5.70	4.211		215	166		381	485	
	3280	350 kcmil				5.40	4.444		251	175		426	535	
	3300	400 kcmil				5.10	4.706		289	185		474	595	
	3320	500 kcmil				4.80	5		355	197		552	685	
	3340	600 kcmil				3.90	6.154		475	242		717	885	
	3360	750 kcmil				3.30	7.273		585	287		872	1,075	
	3380	1000 kcmil				2.40	10		765	395		1,160	1,425	
	5000	600 volt, aluminum type XHHW, stranded, #8			1 Elec	9	.889		12.45	35		47.45	65.50	
	5020	#6			"	8	1		17.05	39.50		56.55	77.50	
	5040	#4			2 Elec	13	1.231		21	48.50		69.50	95.50	
	5060	#2				10.60	1.509		29	59.50		88.50	120	
	5080	#1				9	1.778		42	70		112	150	
	5100	1/0				8	2		50.50	79		129.50	173	
	5120	2/0				7.20	2.222		59.50	87.50		147	196	
	5140	3/0				6.60	2.424		74	95.50		169.50	224	
	5160	4/0				6.20	2.581		82	102		184	241	
	5180	250 kcmil			3 Elec	8.70	2.759		100	109		209	272	
	5200	300 kcmil				8.10	2.963		139	117		256	325	
	5220	350 kcmil				7.50	3.200		141	126		267	345	
	5240	400 kcmil				6.90	3.478		165	137		302	385	
	5260	500 kcmil				6	4		182	158		340	435	
	5280	600 kcmil				5.70	4.211		230	166		396	500	
	5300	700 kcmil				5.40	4.444		266	175		441	555	
	5320	750 kcmil				5.10	4.706		269	185		454	570	
	5340	1000 kcmil				3.60	6.667		400	263		663	830	
	5400	600 volt, copper type XLPE-USE(RHW), solid, #12			1 Elec	11	.727		8.25	28.50		36.75	51.50	
	5420	#10				10	.800		12	31.50		43.50	60	
	5440	Stranded, #14				13	.615		6.95	24.50		31.45	43.50	
	5460	#12				11	.727		9.50	28.50		38	53	
	5480	#10				10	.800		13	31.50		44.50	61.50	
	5500	#8				8	1		22	39.50		61.50	82.50	
	5520	#6				6.50	1.231		31.50	48.50		80	107	
	5540	#4			2 Elec	10.60	1.509		44.50	59.50		104	138	
	5560	#2				9	1.778		65	70		135	176	
	5580	#1				8	2		89	79		168	215	

ELECTRICAL 16

16120 | Conductors & Cables

			CREW	DAILY OUTPUT	LABOR-HOURS	UNIT	2004 BARE COSTS MAT.	LABOR	EQUIP.	TOTAL	TOTAL INCL O&P
900	5600	1/0	2 Elec	6.60	2.424	C.L.F.	91	95.50		186.50	242
	5620	2/0		5.80	2.759		111	109		220	284
	5640	3/0		5	3.200		137	126		263	340
	5660	4/0		4.40	3.636		169	143		312	400
	5680	250 kcmil	3 Elec	6	4		210	158		368	465
	5700	300 kcmil		5.70	4.211		248	166		414	520
	5720	350 kcmil		5.40	4.444		278	175		453	565
	5740	400 kcmil		5.10	4.706		330	185		515	635
	5760	500 kcmil		4.80	5		390	197		587	725
	5780	600 kcmil		3.90	6.154		530	242		772	945
	5800	750 kcmil		3.30	7.273		660	287		947	1,150
	5820	1000 kcmil		2.70	8.889		760	350		1,110	1,350
	5840	600 volt, type XLPE-USE (RHW) aluminum, stranded, #6	1 Elec	8	1		21	39.50		60.50	81.50
	5860	#4	2 Elec	13	1.231		25	48.50		73.50	99.50
	5880	#2		10.60	1.509		33	59.50		92.50	125
	5900	#1		9	1.778		48.50	70		118.50	157
	5920	1/0		8	2		57	79		136	180
	5940	2/0		7.20	2.222		67	87.50		154.50	204
	5960	3/0		6.60	2.424		82	95.50		177.50	232
	5980	4/0		6.20	2.581		92	102		194	252
	6000	250 kcmil	3 Elec	8.70	2.759		118	109		227	292
	6020	300 kcmil		8.10	2.963		162	117		279	350
	6040	350 kcmil		7.50	3.200		165	126		291	370
	6060	400 kcmil		6.90	3.478		192	137		329	415
	6080	500 kcmil		6	4		211	158		369	465
	6100	600 kcmil		5.70	4.211		270	166		436	545
	6110	700 kcmil		5.40	4.444		310	175		485	600
	6120	750 kcmil		5.10	4.706		315	185		500	620

Reference notes in column: R16120-900, R16120-910, R16132-210

16131 | Cable Trays

			CREW	DAILY OUTPUT	LABOR-HOURS	UNIT	2004 BARE COSTS MAT.	LABOR	EQUIP.	TOTAL	TOTAL INCL O&P
105	0010	**CABLE TRAY LADDER TYPE** w/ ftngs & supports, 4" dp., to 15' elev.									
	0100	For higher elevations, see 16131-130-9900									
	0160	Galvanized steel tray									
	0170	4" rung spacing, 6" wide	2 Elec	98	.163	L.F.	8.50	6.45		14.95	18.90
	0180	9" wide		92	.174		9.25	6.85		16.10	20.50
	0200	12" wide		86	.186		10.20	7.35		17.55	22
	0400	18" wide		82	.195		11.85	7.70		19.55	24.50
	0600	24" wide		78	.205		13.65	8.10		21.75	27
	0650	30" wide		68	.235		17.25	9.25		26.50	33
	0700	36" wide		60	.267		19.10	10.50		29.60	36.50
	0800	6" rung spacing, 6" wide		100	.160		7.70	6.30		14	17.90
	0850	9" wide		94	.170		8.50	6.70		15.20	19.30
	0860	12" wide		88	.182		8.90	7.15		16.05	20.50
	0870	18" wide		84	.190		10.15	7.50		17.65	22.50
	0880	24" wide		80	.200		11.35	7.90		19.25	24
	0890	30" wide		70	.229		13.50	9		22.50	28.50
	0900	36" wide		64	.250		14.75	9.85		24.60	31
	0910	9" rung spacing, 6" wide		102	.157		7.25	6.20		13.45	17.20
	0920	9" wide		98	.163		7.45	6.45		13.90	17.70
	0930	12" wide		94	.170		8.55	6.70		15.25	19.35
	0940	18" wide		90	.178		9.65	7		16.65	21
	0950	24" wide		86	.186		11	7.35		18.35	23
	0960	30" wide		80	.200		12	7.90		19.90	25
	0970	36" wide		74	.216		12.80	8.50		21.30	27
	0980	12" rung spacing, 6" wide		106	.151		7.15	5.95		13.10	16.70
	0990	9" wide		104	.154		7.30	6.05		13.35	17.05

Reference note in column: R16131-100

Important: See the Reference Section for critical supporting data - Reference Nos., Crews, & City Cost Indexes

16 ELECTRICAL

			CREW	DAILY OUTPUT	LABOR-HOURS	UNIT	2004 BARE COSTS				TOTAL INCL O&P	
							MAT.	LABOR	EQUIP.	TOTAL		
105		**16131	Cable Trays**											105
1000	12" wide	R16131-100	2 Elec	100	.160	L.F.	7.70	6.30		14	17.90
1010	18" wide			96	.167		8.30	6.55		14.85	18.85
1020	24" wide			94	.170		8.90	6.70		15.60	19.75
1030	30" wide			88	.182		10.10	7.15		17.25	22
1040	36" wide			84	.190		10.65	7.50		18.15	23
1041	18" rung spacing, 6" wide			108	.148		7.10	5.85		12.95	16.50
1042	9" wide			106	.151		7.20	5.95		13.15	16.75
1043	12" wide			102	.157		7.65	6.20		13.85	17.65
1044	18" wide			98	.163		8.05	6.45		14.50	18.40
1045	24" wide			96	.167		8.30	6.55		14.85	18.90
1046	30" wide			90	.178		8.75	7		15.75	20
1047	36" wide			86	.186	▼	9	7.35		16.35	21
1050	Elbows horiz. 9" rung spacing, 90°, 12" radius, 6" wide			9.60	1.667	Ea.	33	65.50		98.50	134
1060	9" wide			8.40	1.905		35	75		110	151
1070	12" wide			7.60	2.105		38	83		121	165
1080	18" wide			6.20	2.581		50	102		152	206
1090	24" wide			5.40	2.963		56	117		173	236
1100	30" wide			4.80	3.333		74	131		205	277
1110	36" wide			4.20	3.810		86	150		236	320
1120	90° 24" radius, 6" wide			9.20	1.739		68	68.50		136.50	177
1130	9" wide			8	2		70	79		149	194
1140	12" wide			7.20	2.222		72	87.50		159.50	209
1150	18" wide			5.80	2.759		78	109		187	248
1160	24" wide			5	3.200		84	126		210	281
1170	30" wide			4.40	3.636		91	143		234	315
1180	36" wide			3.80	4.211		108	166		274	365
1190	90° 36" radius, 6" wide			8.80	1.818		75	71.50		146.50	190
1200	9" wide			7.60	2.105		79	83		162	210
1210	12" wide			6.80	2.353		83	92.50		175.50	230
1220	18" wide			5.40	2.963		95	117		212	279
1230	24" wide			4.60	3.478		104	137		241	320
1240	30" wide			4	4		125	158		283	370
1250	36" wide			3.40	4.706		136	185		321	425
1260	45° 12" radius, 6" wide			13.20	1.212		25.50	48		73.50	99
1270	9" wide			11	1.455		26.50	57.50		84	114
1280	12" wide			9.60	1.667		27.50	65.50		93	128
1290	18" wide			7.60	2.105		29	83		112	155
1300	24" wide			6.20	2.581		37.50	102		139.50	192
1310	30" wide			5.40	2.963		48	117		165	227
1320	36" wide			4.60	3.478		50	137		187	259
1330	45° 24" radius, 6" wide			12.80	1.250		32.50	49.50		82	109
1340	9" wide			10.60	1.509		34.50	59.50		94	126
1350	12" wide			9.20	1.739		38.50	68.50		107	145
1360	18" wide			7.20	2.222		39.50	87.50		127	174
1370	24" wide			5.80	2.759		48	109		157	215
1380	30" wide			5	3.200		57	126		183	251
1390	36" wide			4.20	3.810		65.50	150		215.50	295
1400	45° 36" radius, 6" wide			12.40	1.290		46.50	51		97.50	127
1410	9" wide			10.20	1.569		49	62		111	146
1420	12" wide			8.80	1.818		50	71.50		121.50	162
1430	18" wide			6.80	2.353		54	92.50		146.50	198
1440	24" wide			5.40	2.963		57	117		174	237
1450	30" wide			4.60	3.478		74	137		211	286
1460	36" wide		▼	3.80	4.211	▼	82	166		248	335
1470	Elbows horizontal, 4" rung spacing, use 9" rung x 1.50										
1480	6" rung spacing use 9" rung x 1.20		▼								

ELECTRICAL 16

16131	Cable Trays		CREW	DAILY OUTPUT	LABOR-HOURS	UNIT	2004 BARE COSTS				TOTAL INCL O&P
							MAT.	LABOR	EQUIP.	TOTAL	
105	1490	12" rung spacing use 9" rung x .93										105
	1500	Elbows vert. 9" rung spacing, 90°, 12" radius, 6" wide	R16131-100	2 Elec	9.60	1.667	Ea.	45.50	65.50		111	148
	1510	9" wide			8.40	1.905		46.50	75		121.50	163
	1520	12" wide			7.60	2.105		48	83		131	176
	1530	18" wide			6.20	2.581		50	102		152	206
	1540	24" wide			5.40	2.963		52	117		169	231
	1550	30" wide			4.80	3.333		56	131		187	257
	1560	36" wide			4.20	3.810		58	150		208	287
	1570	24" radius, 6" wide			9.20	1.739		64	68.50		132.50	173
	1580	9" wide			8	2		66	79		145	190
	1590	12" wide			7.20	2.222		68	87.50		155.50	205
	1600	18" wide			5.80	2.759		72	109		181	241
	1610	24" wide			5	3.200		76	126		202	272
	1620	30" wide			4.40	3.636		79	143		222	300
	1630	36" wide			3.80	4.211		89	166		255	345
	1640	36" radius, 6" wide			8.80	1.818		86	71.50		157.50	202
	1650	9" wide			7.60	2.105		89	83		172	221
	1660	12" wide			6.80	2.353		90	92.50		182.50	237
	1670	18" wide			5.40	2.963		98	117		215	282
	1680	24" wide			4.60	3.478		99	137		236	315
	1690	30" wide			4	4		112	158		270	355
	1700	36" wide			3.40	4.706		120	185		305	410
	1710	Elbows vertical, 4" rung spacing, use 9" rung x 1.25										
	1720	6" rung spacing, use 9" rung x 1.15										
	1730	12" rung spacing, use 9" rung x .90										
	1740	Tee horizontal, 9" rung spacing, 12" radius, 6" wide		2 Elec	5	3.200	Ea.	70	126		196	265
	1750	9" wide			4.60	3.478		75	137		212	287
	1760	12" wide			4.40	3.636		79	143		222	300
	1770	18" wide			4	4		90	158		248	335
	1780	24" wide			3.60	4.444		106	175		281	375
	1790	30" wide			3.40	4.706		127	185		312	415
	1800	36" wide			3	5.333		144	210		354	475
	1810	24" radius, 6" wide			4.60	3.478		109	137		246	325
	1820	9" wide			4.20	3.810		114	150		264	350
	1830	12" wide			4	4		118	158		276	365
	1840	18" wide			3.60	4.444		128	175		303	400
	1850	24" wide			3.20	5		139	197		336	445
	1860	30" wide			3	5.333		151	210		361	480
	1870	36" wide			2.60	6.154		194	242		436	575
	1880	36" radius, 6" wide			4.20	3.810		163	150		313	400
	1890	9" wide			3.80	4.211		170	166		336	435
	1900	12" wide			3.60	4.444		181	175		356	460
	1910	18" wide			3.20	5		194	197		391	505
	1920	24" wide			2.80	5.714		219	225		444	575
	1930	30" wide			2.60	6.154		239	242		481	625
	1940	36" wide			2.20	7.273		257	287		544	710
	1980	Tee vertical, 9" rung spacing, 12" radius, 6" wide			5.40	2.963		119	117		236	305
	1990	9" wide			5.20	3.077		120	121		241	310
	2000	12" wide			5	3.200		121	126		247	320
	2010	18" wide			4.60	3.478		122	137		259	340
	2020	24" wide			4.40	3.636		125	143		268	350
	2030	30" wide			4	4		133	158		291	380
	2040	36" wide			3.60	4.444		138	175		313	410
	2050	24" radius, 6" wide			5	3.200		214	126		340	425
	2060	9" wide			4.80	3.333		217	131		348	435
	2070	12" wide			4.60	3.478		220	137		357	445

16 ELECTRICAL

Important: See the Reference Section for critical supporting data - Reference Nos., Crews, & City Cost Indexes

		16131	Cable Trays	CREW	DAILY OUTPUT	LABOR-HOURS	UNIT	2004 BARE COSTS				TOTAL INCL O&P	
								MAT.	LABOR	EQUIP.	TOTAL		
105	2080		18" wide	2 Elec	4.20	3.810	Ea.	227	150		377	475	105
	2090		24" wide		4	4		233	158		391	490	
	2100		30" wide		3.60	4.444		239	175		414	525	
	2110		36" wide		3.20	5		250	197		447	570	
	2120		36" radius, 6" wide		4.60	3.478		400	137		537	645	
	2130		9" wide		4.40	3.636		405	143		548	660	
	2140		12" wide		4.20	3.810		410	150		560	675	
	2150		18" wide		3.80	4.211		415	166		581	705	
	2160		24" wide		3.60	4.444		420	175		595	725	
	2170		30" wide		3.20	5		435	197		632	775	
	2180		36" wide		2.80	5.714		440	225		665	820	
	2190		Tee, 4" rung spacing, use 9" rung x 1.30										
	2200		6" rung spacing, use 9" rung x 1.20										
	2210		12" rung spacing, use 9" rung x .90										
	2220		Cross horizontal, 9" rung spacing, 12" radius, 6" wide	2 Elec	4	4	Ea.	86.50	158		244.50	330	
	2230		9" wide		3.80	4.211		90.50	166		256.50	345	
	2240		12" wide		3.60	4.444		99	175		274	370	
	2250		18" wide		3.40	4.706		108	185		293	395	
	2260		24" wide		3	5.333		120	210		330	445	
	2270		30" wide		2.80	5.714		145	225		370	495	
	2280		36" wide		2.60	6.154		161	242		403	535	
	2290		24" radius, 6" wide		3.60	4.444		168	175		343	445	
	2300		9" wide		3.40	4.706		173	185		358	465	
	2310		12" wide		3.20	5		178	197		375	490	
	2320		18" wide		3	5.333		190	210		400	525	
	2330		24" wide		2.60	6.154		202	242		444	580	
	2340		30" wide		2.40	6.667		215	263		478	625	
	2350		36" wide		2.20	7.273		250	287		537	700	
	2360		36" radius, 6" wide		3.20	5		234	197		431	550	
	2370		9" wide		3	5.333		244	210		454	585	
	2380		12" wide		2.80	5.714		255	225		480	615	
	2390		18" wide		2.60	6.154		281	242		523	670	
	2400		24" wide		2.20	7.273		310	287		597	770	
	2410		30" wide		2	8		340	315		655	840	
	2420		36" wide		1.80	8.889		365	350		715	920	
	2430		Cross horizontal, 4" rung spacing, use 9" rung x 1.30										
	2440		6" rung spacing, use 9" rung x 1.20										
	2450		12" rung spacing, use 9" rung x .90										
	2460		Reducer, 9" to 6" wide tray	2 Elec	13	1.231	Ea.	48	48.50		96.50	125	
	2470		12" to 9" wide tray		12	1.333		49	52.50		101.50	132	
	2480		18" to 12" wide tray		10.40	1.538		49	60.50		109.50	144	
	2490		24" to 18" wide tray		9	1.778		50	70		120	159	
	2500		30" to 24" wide tray		8	2		51	79		130	173	
	2510		36" to 30" wide tray		7	2.286		57	90		147	197	
	2511		Reducer, 18" to 6" wide tray		10.40	1.538		51	60.50		111.50	146	
	2512		24" to 12" wide tray		9	1.778		52	70		122	161	
	2513		30" to 18" wide tray		8	2		57	79		136	180	
	2514		30" to 12" wide tray		8	2		55	79		134	178	
	2515		36" to 24" wide tray		7	2.286		58	90		148	198	
	2516		36" to 18" wide tray		7	2.286		58	90		148	198	
	2517		36" to 12" wide tray		7	2.286		58	90		148	198	
	2520		Dropout or end plate, 6" wide		32	.500		5.35	19.70		25.05	35.50	
	2530		9" wide		28	.571		6.30	22.50		28.80	40.50	
	2540		12" wide		26	.615		6.90	24.50		31.40	43.50	
	2550		18" wide		22	.727		8.10	28.50		36.60	51.50	
	2560		24" wide		20	.800		9.80	31.50		41.30	58	

R16131 -100

ELECTRICAL **16**

16131 | Cable Trays

		CREW	DAILY OUTPUT	LABOR-HOURS	UNIT	2004 BARE COSTS				TOTAL INCL O&P
						MAT.	LABOR	EQUIP.	TOTAL	
2570	30" wide	2 Elec	18	.889	Ea.	10.70	35		45.70	64
2580	36" wide		16	1		12.30	39.50		51.80	72
2590	Tray connector		48	.333	▼	9.50	13.15		22.65	30
3200	Aluminum tray, 4" deep, 6" rung spacing, 6" wide		134	.119	L.F.	10.65	4.70		15.35	18.70
3210	9" wide		128	.125		11.30	4.93		16.23	19.75
3220	12" wide		124	.129		11.95	5.10		17.05	20.50
3230	18" wide		114	.140		13.30	5.55		18.85	23
3240	24" wide		106	.151		15.45	5.95		21.40	26
3250	30" wide		100	.160		17	6.30		23.30	28
3260	36" wide		94	.170		18.45	6.70		25.15	30.50
3270	9" rung spacing, 6" wide		140	.114		8.65	4.50		13.15	16.20
3280	9" wide		134	.119		9.05	4.70		13.75	16.95
3290	12" wide		130	.123		9.25	4.85		14.10	17.35
3300	18" wide		122	.131		10.80	5.15		15.95	19.55
3310	24" wide		116	.138		12.50	5.45		17.95	22
3320	30" wide		108	.148		13.85	5.85		19.70	24
3330	36" wide		100	.160		14.65	6.30		20.95	25.50
3340	12" rung spacing, 6" wide		146	.110		9.20	4.32		13.52	16.50
3350	9" wide		140	.114		9.30	4.50		13.80	16.95
3360	12" wide		134	.119		9.60	4.70		14.30	17.55
3370	18" wide		128	.125		10.15	4.93		15.08	18.50
3380	24" wide		124	.129		11.30	5.10		16.40	19.95
3390	30" wide		114	.140		12.10	5.55		17.65	21.50
3400	36" wide		106	.151		12.70	5.95		18.65	23
3401	18" rung spacing, 6" wide		150	.107		9.15	4.20		13.35	16.30
3402	9" wide tray		144	.111		9.25	4.38		13.63	16.65
3403	12" wide tray		140	.114		9.60	4.50		14.10	17.25
3404	18" wide tray		134	.119		10.10	4.70		14.80	18.10
3405	24" wide tray		130	.123		10.90	4.85		15.75	19.15
3406	30" wide tray		120	.133		11.25	5.25		16.50	20
3407	36" wide tray		110	.145	▼	11.75	5.75		17.50	21.50
3410	Elbows horiz. 9" rung spacing, 90° 12" radius, 6" wide		9.60	1.667	Ea.	41	65.50		106.50	143
3420	9" wide		8.40	1.905		42	75		117	159
3430	12" wide		7.60	2.105		50	83		133	178
3440	18" wide		6.20	2.581		56	102		158	213
3450	24" wide		5.40	2.963		69.50	117		186.50	251
3460	30" wide		4.80	3.333		74	131		205	276
3470	36" wide		4.20	3.810		89	150		239	320
3480	24" radius, 6" wide		9.20	1.739		73	68.50		141.50	183
3490	9" wide		8	2		76	79		155	201
3500	12" wide		7.20	2.222		79.50	87.50		167	218
3510	18" wide		5.80	2.759		86.50	109		195.50	257
3520	24" wide		5	3.200		95	126		221	293
3530	30" wide		4.40	3.636		103	143		246	325
3540	36" wide		3.80	4.211		117	166		283	375
3550	90°, 36" radius, 6" wide		8.80	1.818		80	71.50		151.50	195
3560	9" wide		7.60	2.105		82	83		165	213
3570	12" wide		6.80	2.353		92	92.50		184.50	239
3580	18" wide		5.40	2.963		103	117		220	287
3590	24" wide		4.60	3.478		116	137		253	330
3600	30" wide		4	4		128	158		286	375
3610	36" wide		3.40	4.706		147	185		332	440
3620	45° 12" radius, 6" wide		13.20	1.212		27	48		75	101
3630	9" wide		11	1.455		28	57.50		85.50	116
3640	12" wide		9.60	1.667		29	65.50		94.50	130
3650	18" wide		7.60	2.105	▼	34.50	83		117.50	161

105

R16131-100

Important: See the Reference Section for critical supporting data - Reference Nos., Crews, & City Cost Indexes

16 ELECTRICAL

16131	Cable Trays		CREW	DAILY OUTPUT	LABOR-HOURS	UNIT	2004 BARE COSTS				TOTAL INCL O&P
							MAT.	LABOR	EQUIP.	TOTAL	
105 3660	24" wide	R16131 -100	2 Elec	6.20	2.581	Ea.	42.50	102		144.50	198 **105**
3670	30" wide			5.40	2.963		49	117		166	228
3680	36" wide			4.60	3.478		51	137		188	260
3690	45° 24" radius, 6" wide			12.80	1.250		36	49.50		85.50	113
3700	9" wide			10.60	1.509		41	59.50		100.50	134
3710	12" wide			9.20	1.739		42	68.50		110.50	148
3720	18" wide			7.20	2.222		45	87.50		132.50	180
3730	24" wide			5.80	2.759		52	109		161	219
3740	30" wide			5	3.200		62	126		188	256
3750	36" wide			4.20	3.810		66	150		216	296
3760	45° 36" radius, 6" wide			12.40	1.290		49	51		100	130
3770	9" wide			10.20	1.569		50	62		112	147
3780	12" wide			8.80	1.818		51	71.50		122.50	163
3790	18" wide			6.80	2.353		56	92.50		148.50	200
3800	24" wide			5.40	2.963		69	117		186	250
3810	30" wide			4.60	3.478		73	137		210	285
3820	36" wide	↓		3.80	4.211	↓	80	166		246	335
3830	Elbows horizontal, 4" rung spacing, use 9" rung x 1.50										
3840	6" rung spacing, use 9" rung x 1.20										
3850	12" rung spacing, use 9" rung x .93										
3860	Elbows vertical 9" rung spacing, 90° 12" radius, 6" wide		2 Elec	9.60	1.667	Ea.	57	65.50		122.50	160
3870	9" wide			8.40	1.905		60	75		135	178
3880	12" wide			7.60	2.105		61	83		144	190
3890	18" wide			6.20	2.581		63	102		165	221
3900	24" wide			5.40	2.963		67	117		184	248
3910	30" wide			4.80	3.333		70	131		201	272
3920	36" wide			4.20	3.810		72	150		222	300
3930	24" radius, 6" wide			9.20	1.739		62	68.50		130.50	170
3940	9" wide			8	2		64	79		143	188
3950	12" wide			7.20	2.222		66	87.50		153.50	203
3960	18" wide			5.80	2.759		70	109		179	239
3970	24" wide			5	3.200		74	126		200	270
3980	30" wide			4.40	3.636		78	143		221	299
3990	36" wide			3.80	4.211		83	166		249	340
4000	36" radius, 6" wide			8.80	1.818		96	71.50		167.50	213
4010	9" wide			7.60	2.105		98	83		181	231
4020	12" wide			6.80	2.353		104	92.50		196.50	252
4030	18" wide			5.40	2.963		108	117		225	293
4040	24" wide			4.60	3.478		112	137		249	325
4050	30" wide			4	4		121	158		279	365
4060	36" wide	↓		3.40	4.706	↓	126	185		311	415
4070	Elbows vertical, 4" rung spacing, use 9" rung x 1.25										
4080	6" rung spacing, use 9" rung x 1.15										
4090	12" rung spacing, use 9" rung x .90										
4100	Tee horizontal 9" rung spacing, 12" radius, 6" wide		2 Elec	5	3.200	Ea.	63.50	126		189.50	258
4110	9" wide			4.60	3.478		68	137		205	279
4120	12" wide			4.40	3.636		69	143		212	289
4130	18" wide			4.20	3.810		82	150		232	315
4140	24" wide			4	4		90	158		248	335
4150	30" wide			3.60	4.444		105	175		280	375
4160	36" wide			3.40	4.706		126	185		311	415
4170	24" radius, 6" wide			4.60	3.478		108	137		245	325
4180	9" wide			4.20	3.810		114	150		264	350
4190	12" wide			4	4		117	158		275	365
4200	18" wide			3.80	4.211		128	166		294	390
4210	24" wide	↓		3.60	4.444	↓	140	175		315	415

16131 | Cable Trays

		CREW	DAILY OUTPUT	LABOR-HOURS	UNIT	2004 BARE COSTS				TOTAL INCL O&P
						MAT.	LABOR	EQUIP.	TOTAL	
4220	30" wide	2 Elec	3.20	5	Ea.	152	197		349	460
4230	36" wide		3	5.333		234	210		444	570
4240	36" radius, 6" wide		4.20	3.810		153	150		303	390
4250	9" wide		3.80	4.211		157	166		323	420
4260	12" wide		3.60	4.444		172	175		347	450
4270	18" wide		3.40	4.706		181	185		366	475
4280	24" wide		3.20	5		200	197		397	515
4290	30" wide		2.80	5.714		253	225		478	615
4300	36" wide		2.60	6.154		280	242		522	670
4310	Tee vertical 9" rung spacing, 12" radius, 6" wide		5.40	2.963		142	117		259	330
4320	9" wide		5.20	3.077		147	121		268	340
4330	12" wide		5	3.200		148	126		274	350
4340	18" wide		4.60	3.478		151	137		288	370
4350	24" wide		4.40	3.636		163	143		306	390
4360	30" wide		4.20	3.810		166	150		316	405
4370	36" wide		4	4		172	158		330	425
4380	24" radius, 6" wide		5	3.200		241	126		367	455
4390	9" wide		4.80	3.333		244	131		375	465
4400	12" wide		4.60	3.478		248	137		385	475
4410	18" wide		4.20	3.810		254	150		404	500
4420	24" wide		4	4		261	158		419	520
4430	30" wide		3.80	4.211		268	166		434	540
4440	36" wide		3.60	4.444		291	175		466	580
4450	36" radius, 6" wide		4.60	3.478		530	137		667	790
4460	9" wide		4.40	3.636		540	143		683	810
4470	12" wide		4.20	3.810		560	150		710	845
4480	18" wide		3.80	4.211		565	166		731	870
4490	24" wide		3.60	4.444		580	175		755	900
4500	30" wide		3.40	4.706		590	185		775	920
4510	36" wide		3.20	5		600	197		797	955
4520	Tees, 4" rung spacing, use 9" rung x 1.30									
4530	6" rung spacing, use 9" rung x 1.20									
4540	12" rung spacing, use 9" rung x .90									
4550	Cross horizontal 9" rung spacing, 12" radius, 6" wide	2 Elec	4.40	3.636	Ea.	88	143		231	310
4560	9" wide		4.20	3.810		98	150		248	330
4570	12" wide		4	4		99	158		257	345
4580	18" wide		3.60	4.444		120	175		295	390
4590	24" wide		3.40	4.706		124	185		309	410
4600	30" wide		3	5.333		146	210		356	475
4610	36" wide		2.80	5.714		182	225		407	535
4620	24" radius, 6" wide		4	4		169	158		327	420
4630	9" wide		3.80	4.211		175	166		341	440
4640	12" wide		3.60	4.444		180	175		355	460
4650	18" wide		3.20	5		189	197		386	500
4660	24" wide		3	5.333		206	210		416	540
4670	30" wide		2.60	6.154		220	242		462	600
4680	36" wide		2.40	6.667		249	263		512	665
4690	36" radius, 6" wide		3.60	4.444		200	175		375	480
4700	9" wide		3.40	4.706		208	185		393	505
4710	12" wide		3.20	5		220	197		417	535
4720	18" wide		2.80	5.714		235	225		460	595
4730	24" wide		2.60	6.154		291	242		533	680
4740	30" wide		2.20	7.273		320	287		607	780
4750	36" wide		2	8		360	315		675	865
4760	Cross horizontal, 4" rung spacing, use 9" rung x 1.30									
4770	6" rung spacing, use 9" rung x 1.20									

R16131 -100

Important: See the Reference Section for critical supporting data - Reference Nos., Crews, & City Cost Indexes

16 ELECTRICAL

16131 | Cable Trays

		CREW	DAILY OUTPUT	LABOR-HOURS	UNIT	2004 BARE COSTS				TOTAL INCL O&P
						MAT.	LABOR	EQUIP.	TOTAL	
105	4780	12" rung spacing, use 9" rung x .90 R16131-100									105
	4790	Reducer, 9" to 6" wide tray	2 Elec	16	1	Ea.	47	39.50		86.50	110
	4800	12" to 9" wide tray		14	1.143		49	45		94	121
	4810	18" to 12" wide tray		12.40	1.290		49	51		100	130
	4820	24" to 18" wide tray		10.60	1.509		50	59.50		109.50	144
	4830	30" to 24" wide tray		9.20	1.739		51	68.50		119.50	158
	4840	36" to 30" wide tray		8	2		54	79		133	177
	4841	Reducer, 18" to 6" wide tray		12.40	1.290		49	51		100	130
	4842	24" to 12" wide tray		10.60	1.509		50	59.50		109.50	144
	4843	30" to 18" wide tray		9.20	1.739		51	68.50		119.50	158
	4844	30" to 12" wide tray		9.20	1.739		51	68.50		119.50	158
	4845	36" to 24" wide tray		8	2		54	79		133	177
	4846	36" to 18" wide tray		8	2		54	79		133	177
	4847	36" to 12" wide tray		8	2		54	79		133	177
	4850	Dropout or end plate, 6" wide		32	.500		6.20	19.70		25.90	36.50
	4860	9" wide tray		28	.571		6.60	22.50		29.10	41
	4870	12" wide tray		26	.615		7.20	24.50		31.70	44
	4880	18" wide tray		22	.727		9.60	28.50		38.10	53
	4890	24" wide tray		20	.800		11.30	31.50		42.80	59.50
	4900	30" wide tray		18	.889		13.40	35		48.40	67
	4910	36" wide tray		16	1		14.90	39.50		54.40	75
	4920	Tray connector		48	.333		9	13.15		22.15	29.50
	8000	Elbow 36" radius horiz., 60°, 6" wide tray		10.60	1.509		68	59.50		127.50	164
	8010	9" wide tray		9	1.778		74	70		144	186
	8020	12" wide tray		7.80	2.051		76	81		157	204
	8030	18" wide tray		6.20	2.581		81	102		183	240
	8040	24" wide tray		5	3.200		95	126		221	293
	8050	30" wide tray		4.40	3.636		97	143		240	320
	8060	30°, 6" wide tray		14	1.143		52	45		97	124
	8070	9" wide tray		11.40	1.404		55	55.50		110.50	143
	8080	12" wide tray		9.80	1.633		56	64.50		120.50	157
	8090	18" wide tray		7.40	2.162		59	85		144	192
	8100	24" wide tray		5.80	2.759		72	109		181	241
	8110	30" wide tray		4.80	3.333		74	131		205	277
	8120	Adjustable, 6" wide tray		12.40	1.290		69	51		120	152
	8130	9" wide tray		10.20	1.569		73	62		135	173
	8140	12" wide tray		8.80	1.818		74	71.50		145.50	189
	8150	18" wide tray		6.80	2.353		81	92.50		173.50	227
	8160	24" wide tray		5.40	2.963		86	117		203	269
	8170	30" wide tray		4.60	3.478		96	137		233	310
	8180	Wye 36" radius horiz., 45°, 6" wide tray		4.60	3.478		75	137		212	287
	8190	9" wide tray		4.40	3.636		79	143		222	300
	8200	12" wide tray		4.20	3.810		82	150		232	315
	8210	18" wide tray		3.80	4.211		93	166		259	350
	8220	24" wide tray		3.60	4.444		104	175		279	375
	8230	30" wide tray		3.40	4.706		116	185		301	405
	8240	Elbow 36" radius vert. in/outside, 60°, 6" wide tray		10.60	1.509		75	59.50		134.50	171
	8250	9" wide tray		9	1.778		78	70		148	190
	8260	12" wide tray		7.80	2.051		79	81		160	207
	8270	18" wide tray		6.20	2.581		82	102		184	241
	8280	24" wide tray		5	3.200		87	126		213	284
	8290	30" wide tray		4.40	3.636		93	143		236	315
	8300	45°, 6" wide tray		12.40	1.290		67.50	51		118.50	150
	8310	9" wide tray		10.20	1.569		72	62		134	171
	8320	12" wide tray		8.80	1.818		73	71.50		144.50	187
	8330	18" wide tray		6.80	2.353		75	92.50		167.50	221

ELECTRICAL 16

	16131 | **Cable Trays**		CREW	DAILY OUTPUT	LABOR-HOURS	UNIT	2004 BARE COSTS				TOTAL INCL O&P	
							MAT.	LABOR	EQUIP.	TOTAL		
105	8340	24" wide tray	2 Elec	5.40	2.963	Ea.	79	117		196	261	**105**
	8350	30" wide tray		4.60	3.478		81	137		218	293	
	8360	30°, 6" wide tray		14	1.143		59	45		104	132	
	8370	9" wide tray		11.40	1.404		60.50	55.50		116	149	
	8380	12" wide tray		9.80	1.633		64.50	64.50		129	167	
	8390	18" wide tray		7.40	2.162		65.50	85		150.50	199	
	8400	24" wide tray		5.80	2.759		67	109		176	236	
	8410	30" wide tray		4.80	3.333		70.50	131		201.50	273	
	8660	Adjustable, 6" wide tray		12.40	1.290		67.50	51		118.50	150	
	8670	9" wide tray		10.20	1.569		72	62		134	171	
	8680	12" wide tray		8.80	1.818		73	71.50		144.50	188	
	8690	18" wide tray		6.80	2.353		81	92.50		173.50	227	
	8700	24" wide tray		5.40	2.963		86	117		203	269	
	8710	30" wide tray		4.60	3.478		95	137		232	310	
	8720	Cross, vertical, 24" radius, 6" wide tray		3.60	4.444		525	175		700	840	
	8730	9" wide tray		3.40	4.706		535	185		720	865	
	8740	12" wide tray		3.20	5		550	197		747	900	
	8750	18" wide tray		2.80	5.714		575	225		800	970	
	8760	24" wide tray		2.60	6.154		595	242		837	1,025	
	8770	30" wide tray		2.20	7.273		630	287		917	1,125	
	9200	Splice plate	1 Elec	48	.167	Pr.	9.40	6.55		15.95	20	
	9210	Expansion joint		48	.167		11.30	6.55		17.85	22	
	9220	Horizontal hinged		48	.167		9.40	6.55		15.95	20	
	9230	Vertical hinged		48	.167		11.30	6.55		17.85	22	
	9240	Ladder hanger, vertical		28	.286	Ea.	2.80	11.25		14.05	19.85	
	9250	Ladder to channel connector		24	.333		33	13.15		46.15	55.50	
	9260	Ladder to box connector 30" wide		19	.421		33	16.60		49.60	60.50	
	9270	24" wide		20	.400		33	15.75		48.75	59.50	
	9280	18" wide		21	.381		31.50	15		46.50	57.50	
	9290	12" wide		22	.364		29.50	14.35		43.85	54	
	9300	9" wide		23	.348		26.50	13.70		40.20	49.50	
	9310	6" wide		24	.333		26.50	13.15		39.65	48.50	
	9320	Ladder floor flange		24	.333		21.50	13.15		34.65	43	
	9330	Cable roller for tray 30" wide		10	.800		179	31.50		210.50	244	
	9340	24" wide		11	.727		152	28.50		180.50	210	
	9350	18" wide		12	.667		145	26.50		171.50	199	
	9360	12" wide		13	.615		114	24.50		138.50	161	
	9370	9" wide		14	.571		101	22.50		123.50	145	
	9380	6" wide		15	.533		82	21		103	122	
	9390	Pulley, single wheel		12	.667		187	26.50		213.50	245	
	9400	Triple wheel		10	.800		375	31.50		406.50	455	
	9440	Nylon cable tie, 14" long		80	.100		.38	3.94		4.32	6.25	
	9450	Ladder, hold down clamp		60	.133		6	5.25		11.25	14.40	
	9460	Cable clamp		60	.133		6.10	5.25		11.35	14.50	
	9470	Wall bracket, 30" wide tray		19	.421		33	16.60		49.60	61	
	9480	24" wide tray		20	.400		25.50	15.75		41.25	52	
	9490	18" wide tray		21	.381		23	15		38	48	
	9500	12" wide tray		22	.364		21.50	14.35		35.85	45	
	9510	9" wide tray		23	.348		18.90	13.70		32.60	41.50	
	9520	6" wide tray		24	.333		18.40	13.15		31.55	39.50	
110	0010	**CABLE TRAY SOLID BOTTOM** w/ ftngs & supports, 3" dp, to 15' high										**110**
	0200	For higher elevations, see 16131-130-9900										
	0220	Galvanized steel, tray, 6" wide	2 Elec	120	.133	L.F.	6.85	5.25		12.10	15.35	
	0240	12" wide		100	.160		8.80	6.30		15.10	19.05	
	0260	18" wide		70	.229		10.65	9		19.65	25	
	0280	24" wide		60	.267		12.50	10.50		23	29.50	

R16131 -100

Important: See the Reference Section for critical supporting data - Reference Nos., Crews, & City Cost Indexes

16 ELECTRICAL

			CREW	DAILY OUTPUT	LABOR-HOURS	UNIT	2004 BARE COSTS				TOTAL INCL O&P
16131	**Cable Trays**						MAT.	LABOR	EQUIP.	TOTAL	
110	0300	30" wide	2 Elec	50	.320	L.F.	14.90	12.60		27.50	35	110
	0320	36" wide		44	.364	↓	21.50	14.35		35.85	45
	0340	Elbow horizontal 90°, 12" radius, 6" wide		9.60	1.667	Ea.	52	65.50		117.50	155
	0360	12" wide		6.80	2.353		58	92.50		150.50	202
	0370	18" wide		5.40	2.963		66.50	117		183.50	248
	0380	24" wide		4.40	3.636		81	143		224	300
	0390	30" wide		3.80	4.211		97	166		263	355
	0400	36" wide		3.40	4.706		110	185		295	395
	0420	24" radius, 6" wide		9.20	1.739		74	68.50		142.50	184
	0440	12" wide		6.40	2.500		86	98.50		184.50	242
	0450	18" wide		5	3.200		100	126		226	298
	0460	24" wide		4	4		116	158		274	360
	0470	30" wide		3.40	4.706		136	185		321	425
	0480	36" wide		3	5.333		153	210		363	485
	0500	36" radius, 6" wide		8.80	1.818		108	71.50		179.50	226
	0520	12" wide		6	2.667		122	105		227	290
	0530	18" wide		4.60	3.478		149	137		286	370
	0540	24" wide		3.60	4.444		161	175		336	435
	0550	30" wide		3	5.333		195	210		405	530
	0560	36" wide		2.60	6.154		219	242		461	600
	0580	Elbow vertical 90°, 12" radius, 6" wide		9.60	1.667		63	65.50		128.50	167
	0600	12" wide		6.80	2.353		67	92.50		159.50	212
	0610	18" wide		5.40	2.963		68	117		185	249
	0620	24" wide		4.40	3.636		79	143		222	300
	0630	30" wide		3.80	4.211		84	166		250	340
	0640	36" wide		3.40	4.706		87	185		272	370
	0670	24" radius, 6" wide		9.20	1.739		89	68.50		157.50	200
	0690	12" wide		6.40	2.500		98	98.50		196.50	255
	0700	18" wide		5	3.200		106	126		232	305
	0710	24" wide		4	4		115	158		273	360
	0720	30" wide		3.40	4.706		122	185		307	410
	0730	36" wide		3	5.333		133	210		343	460
	0750	36" radius, 6" wide		8.80	1.818		122	71.50		193.50	241
	0770	12" wide		6.60	2.424		136	95.50		231.50	292
	0780	18" wide		4.60	3.478		149	137		286	370
	0790	24" wide		3.60	4.444		163	175		338	440
	0800	30" wide		3	5.333		178	210		388	510
	0810	36" wide		2.60	6.154		193	242		435	570
	0840	Tee horizontal, 12" radius, 6" wide		5	3.200		73	126		199	269
	0860	12" wide		4	4		79	158		237	320
	0870	18" wide		3.40	4.706		94	185		279	380
	0880	24" wide		2.80	5.714		105	225		330	450
	0890	30" wide		2.60	6.154		120	242		362	490
	0900	36" wide		2.20	7.273		137	287		424	575
	0940	24" radius, 6" wide		4.60	3.478		113	137		250	330
	0960	12" wide		3.60	4.444		132	175		307	405
	0970	18" wide		3	5.333		147	210		357	475
	0980	24" wide		2.40	6.667		199	263		462	610
	0990	30" wide		2.20	7.273		216	287		503	665
	1000	36" wide		1.80	8.889		234	350		584	775
	1020	36" radius, 6" wide		4.20	3.810		178	150		328	420
	1040	12" wide		3.20	5		199	197		396	510
	1050	18" wide		2.60	6.154		216	242		458	600
	1060	24" wide		2.20	7.273		278	287		565	730
	1070	30" wide		2	8		300	315		615	800
	1080	36" wide		1.60	10		310	395		705	930

R16131 -100

ELECTRICAL 16

		CREW	DAILY OUTPUT	LABOR-HOURS	UNIT	2004 BARE COSTS				TOTAL INCL O&P
16131	**Cable Trays**					MAT.	LABOR	EQUIP.	TOTAL	
110 1100	Tee vertical, 12" radius, 6" wide	2 Elec	5	3.200	Ea.	120	126		246	320 110
1120	12" wide [R16131 -100]		4	4		125	158		283	370
1130	18" wide		3.60	4.444		127	175		302	400
1140	24" wide		3.40	4.706		139	185		324	430
1150	30" wide		3	5.333		149	210		359	480
1160	36" wide		2.60	6.154		154	242		396	530
1180	24" radius, 6" wide		4.60	3.478		177	137		314	400
1200	12" wide		3.60	4.444		185	175		360	465
1210	18" wide		3.20	5		194	197		391	505
1220	24" wide		3	5.333		204	210		414	540
1230	30" wide		2.60	6.154		224	242		466	605
1240	36" wide		2.20	7.273		234	287		521	680
1260	36" radius, 6" wide		4.20	3.810		276	150		426	530
1280	12" wide		3.20	5		281	197		478	605
1290	18" wide		2.80	5.714		300	225		525	665
1300	24" wide		2.60	6.154		310	242		552	705
1310	30" wide		2.20	7.273		365	287		652	825
1320	36" wide		2	8		385	315		700	895
1340	Cross horizontal, 12" radius, 6" wide		4	4		87	158		245	330
1360	12" wide		3.40	4.706		95	185		280	380
1370	18" wide		2.80	5.714		111	225		336	455
1380	24" wide		2.40	6.667		125	263		388	530
1390	30" wide		2	8		139	315		454	625
1400	36" wide		1.80	8.889		153	350		503	690
1420	24" radius, 6" wide		3.60	4.444		159	175		334	435
1440	12" wide		3	5.333		178	210		388	510
1450	18" wide		2.40	6.667		199	263		462	610
1460	24" wide		2	8		255	315		570	750
1470	30" wide		1.80	8.889		276	350		626	825
1480	36" wide		1.60	10		291	395		686	905
1500	36" radius, 6" wide		3.20	5		260	197		457	580
1520	12" wide		2.60	6.154		286	242		528	675
1530	18" wide		2	8		310	315		625	815
1540	24" wide		1.80	8.889		370	350		720	925
1550	30" wide		1.60	10		400	395		795	1,025
1560	36" wide		1.40	11.429		425	450		875	1,150
1580	Drop out or end plate, 6" wide		32	.500		10.60	19.70		30.30	41
1600	12" wide		26	.615		13.20	24.50		37.70	50.50
1610	18" wide		22	.727		14.15	28.50		42.65	58
1620	24" wide		20	.800		16.70	31.50		48.20	65.50
1630	30" wide		18	.889		18.70	35		53.70	72.50
1640	36" wide		16	1		20	39.50		59.50	80.50
1660	Reducer, 12" to 6" wide		12	1.333		51	52.50		103.50	134
1680	18" to 12" wide		10.60	1.509		52	59.50		111.50	146
1700	18" to 6" wide		10.60	1.509		52	59.50		111.50	146
1720	24" to 18" wide		9.20	1.739		54	68.50		122.50	162
1740	24" to 12" wide		9.20	1.739		54	68.50		122.50	162
1760	30" to 24" wide		8	2		58	79		137	181
1780	30" to 18" wide		8	2		58	79		137	181
1800	30" to 12" wide		8	2		58	79		137	181
1820	36" to 30" wide		7.20	2.222		60	87.50		147.50	196
1840	36" to 24" wide		7.20	2.222		60	87.50		147.50	196
1860	36" to 18" wide		7.20	2.222		60	87.50		147.50	196
1880	36" to 12" wide		7.20	2.222		60	87.50		147.50	196
2000	Aluminum tray, 6" wide		150	.107	L.F.	8.30	4.20		12.50	15.40
2020	12" wide		130	.123		10.90	4.85		15.75	19.15

Important: See the Reference Section for critical supporting data - Reference Nos., Crews, & City Cost Indexes

16 ELECTRICAL

16131 | Cable Trays

		CREW	DAILY OUTPUT	LABOR-HOURS	UNIT	2004 BARE COSTS				TOTAL INCL O&P
						MAT.	LABOR	EQUIP.	TOTAL	
2030	18" wide	2 Elec	100	.160	L.F.	13.65	6.30		19.95	24.50
2040	24" wide		90	.178		17.20	7		24.20	29.50
2050	30" wide		70	.229		20	9		29	36
2060	36" wide		64	.250		25.50	9.85		35.35	42.50
2080	Elbow horizontal 90°, 12" radius, 6" wide		9.60	1.667	Ea.	66.50	65.50		132	171
2100	12" wide		7.60	2.105		75	83		158	206
2110	18" wide		6.80	2.353		92	92.50		184.50	239
2120	24" wide		5.80	2.759		105	109		214	278
2130	30" wide		5	3.200		128	126		254	330
2140	36" wide		4.40	3.636		144	143		287	370
2160	24" radius, 6" wide		9.20	1.739		95	68.50		163.50	207
2180	12" wide		7.20	2.222		112	87.50		199.50	253
2190	18" wide		6.40	2.500		128	98.50		226.50	288
2200	24" wide		5.40	2.963		158	117		275	350
2210	30" wide		4.60	3.478		181	137		318	405
2220	36" wide		4	4		208	158		366	465
2240	36" radius, 6" wide		8.80	1.818		140	71.50		211.50	261
2260	12" wide		6.80	2.353		168	92.50		260.50	325
2270	18" wide		6	2.667		200	105		305	375
2280	24" wide		5	3.200		214	126		340	425
2290	30" wide		4.20	3.810		242	150		392	490
2300	36" wide		3.60	4.444		284	175		459	570
2320	Elbow vertical 90°, 12" radius, 6" wide		9.60	1.667		79	65.50		144.50	185
2340	12" wide		7.60	2.105		82	83		165	213
2350	18" wide		6.80	2.353		93	92.50		185.50	240
2360	24" wide		5.80	2.759		98	109		207	270
2370	30" wide		5	3.200		104	126		230	300
2380	36" wide		4.40	3.636		107	143		250	330
2400	24" radius, 6" wide		9.20	1.739		112	68.50		180.50	225
2420	12" wide		7.20	2.222		122	87.50		209.50	264
2430	18" wide		6.40	2.500		131	98.50		229.50	291
2440	24" wide		5.40	2.963		142	117		259	330
2450	30" wide		4.60	3.478		151	137		288	370
2460	36" wide		4	4		166	158		324	415
2480	36" radius, 6" wide		8.80	1.818		148	71.50		219.50	270
2500	12" wide		6.80	2.353		166	92.50		258.50	320
2510	18" wide		6	2.667		181	105		286	355
2520	24" wide		5	3.200		188	126		314	395
2530	30" wide		4.20	3.810		208	150		358	450
2540	36" wide		3.60	4.444		216	175		391	500
2560	Tee horizontal, 12" radius, 6" wide		5	3.200		103	126		229	300
2580	12" wide		4.40	3.636		125	143		268	350
2590	18" wide		4	4		140	158		298	390
2600	24" wide		3.60	4.444		157	175		332	435
2610	30" wide		3	5.333		181	210		391	515
2620	36" wide		2.40	6.667		208	263		471	620
2640	24" radius, 6" wide		4.60	3.478		170	137		307	390
2660	12" wide		4	4		191	158		349	445
2670	18" wide		3.60	4.444		216	175		391	500
2680	24" wide		3	5.333		265	210		475	605
2690	30" wide		2.40	6.667		300	263		563	720
2700	36" wide		2.20	7.273		340	287		627	795
2720	36" radius, 6" wide		4.20	3.810		276	150		426	530
2740	12" wide		3.60	4.444		300	175		475	590
2750	18" wide		3.20	5		350	197		547	680
2760	24" wide		2.60	6.154		405	242		647	805

R16131-100

110

16 ELECTRICAL

			DAILY OUTPUT	LABOR-HOURS	UNIT	2004 BARE COSTS				TOTAL INCL O&P
	16131	**Cable Trays**				MAT.	LABOR	EQUIP.	TOTAL	
110	2770	30" wide	2	8	Ea.	445	315		760	960	110
	2780	36" wide	1.80	8.889		500	350		850	1,075
	2800	Tee vertical, 12" radius, 6" wide	5	3.200		147	126		273	350
	2820	12" wide	4.40	3.636		150	143		293	380
	2830	18" wide	4.20	3.810		156	150		306	395
	2840	24" wide	4	4		166	158		324	415
	2850	30" wide	3.60	4.444		181	175		356	460
	2860	36" wide	3	5.333		191	210		401	525
	2880	24" radius, 6" wide	4.60	3.478		216	137		353	440
	2900	12" wide	4	4		229	158		387	485
	2910	18" wide	3.80	4.211		244	166		410	515
	2920	24" wide	3.60	4.444		265	175		440	550
	2930	30" wide	3.20	5		291	197		488	615
	2940	36" wide	2.60	6.154		305	242		547	700
	2960	36" radius, 6" wide	4.20	3.810		340	150		490	595
	2980	12" wide	3.40	4.706		355	185		540	665
	2990	18" wide	3.40	4.706		375	185		560	685
	3000	24" wide	3.20	5		385	197		582	720
	3010	30" wide	2.80	5.714		415	225		640	795
	3020	36" wide	2.20	7.273		460	287		747	930
	3040	Cross horizontal, 12" radius, 6" wide	4.40	3.636		131	143		274	355
	3060	12" wide	4	4		148	158		306	395
	3070	18" wide	3.40	4.706		168	185		353	460
	3080	24" wide	2.80	5.714		191	225		416	545
	3090	30" wide	2.60	6.154		216	242		458	600
	3100	36" wide	2.20	7.273		239	287		526	690
	3120	24" radius, 6" wide	4	4		229	158		387	485
	3140	12" wide	3.60	4.444		265	175		440	550
	3150	18" wide	3	5.333		288	210		498	630
	3160	24" wide	2.40	6.667		350	263		613	775
	3170	30" wide	2.20	7.273		385	287		672	850
	3180	36" wide	1.80	8.889		400	350		750	960
	3200	36" radius, 6" wide	3.60	4.444		400	175		575	700
	3220	12" wide	3.20	5		420	197		617	760
	3230	18" wide	2.60	6.154		470	242		712	875
	3240	24" wide	2	8		510	315		825	1,025
	3250	30" wide	1.80	8.889		550	350		900	1,125
	3260	36" wide	1.60	10		645	395		1,040	1,300
	3280	Dropout, or end plate, 6" wide	32	.500		13.20	19.70		32.90	44
	3300	12" wide	26	.615		14.90	24.50		39.40	52.50
	3310	18" wide	22	.727		17.80	28.50		46.30	62
	3320	24" wide	20	.800		19.60	31.50		51.10	68.50
	3330	30" wide	18	.889		23	35		58	77
	3340	36" wide	16	1		26.50	39.50		66	87.50
	3380	Reducer, 12" to 6" wide	14	1.143		64.50	45		109.50	138
	3400	18" to 12" wide	12	1.333		66.50	52.50		119	152
	3420	18" to 6" wide	12	1.333		66.50	52.50		119	152
	3440	24" to 18" wide	10.60	1.509		70.50	59.50		130	167
	3460	24" to 12" wide	10.60	1.509		70.50	59.50		130	167
	3480	30" to 24" wide	9.20	1.739		77	68.50		145.50	187
	3500	30" to 18" wide	9.20	1.739		77	68.50		145.50	187
	3520	30" to 12" wide	9.20	1.739		77	68.50		145.50	187
	3540	36" to 30" wide	8	2		81	79		160	206
	3560	36" to 24" wide	8	2		82	79		161	207
	3580	36" to 18" wide	8	2		82	79		161	207
	3600	36" to 12" wide	8	2		84	79		163	210

Crew: 2 Elec — R16131-100

Important: See the Reference Section for critical supporting data - Reference Nos., Crews, & City Cost Indexes

16131 | Cable Trays

		CREW	DAILY OUTPUT	LABOR-HOURS	UNIT	2004 BARE COSTS				TOTAL INCL O&P
						MAT.	LABOR	EQUIP.	TOTAL	
0010	**CABLE TRAY TROUGH** vented, w/ ftngs & supports, 6" dp, to 15' high									
0020	For higher elevations, see 16131-130-9900									
0200	Galvanized steel, tray, 6" wide R16131-100	2 Elec	90	.178	L.F.	9	7		16	20.50
0240	12" wide		80	.200		11	7.90		18.90	24
0260	18" wide		70	.229		13.30	9		22.30	28
0280	24" wide		60	.267		15.20	10.50		25.70	32.50
0300	30" wide		50	.320		21.50	12.60		34.10	43
0320	36" wide		40	.400		24	15.75		39.75	50
0340	Elbow horizontal 90°, 12" radius, 6" wide		7.60	2.105	Ea.	55.50	83		138.50	184
0360	12" wide		5.60	2.857		66	113		179	240
0370	18" wide		4.40	3.636		76	143		219	297
0380	24" wide		3.60	4.444		94	175		269	365
0390	30" wide		3.20	5		106	197		303	410
0400	36" wide		2.80	5.714		123	225		348	470
0420	24" radius, 6" wide		7.20	2.222		82	87.50		169.50	220
0440	12" wide		5.20	3.077		95	121		216	285
0450	18" wide		4	4		114	158		272	360
0460	24" wide		3.20	5		129	197		326	435
0470	30" wide		2.80	5.714		149	225		374	500
0480	36" wide		2.40	6.667		173	263		436	580
0500	36" radius, 6" wide		6.80	2.353		123	92.50		215.50	273
0520	12" wide		4.80	3.333		140	131		271	350
0530	18" wide		3.60	4.444		168	175		343	445
0540	24" wide		2.80	5.714		178	225		403	530
0550	30" wide		2.40	6.667		200	263		463	610
0560	36" wide		2	8		223	315		538	715
0580	Elbow vertical 90°, 12" radius, 6" wide		7.60	2.105		72	83		155	202
0600	12" wide		5.60	2.857		80	113		193	255
0610	18" wide		4.40	3.636		81	143		224	300
0620	24" wide		3.60	4.444		93	175		268	360
0630	30" wide		3.20	5		95	197		292	400
0640	36" wide		2.80	5.714		100	225		325	445
0660	24" radius, 6" wide		7.20	2.222		103	87.50		190.50	243
0680	12" wide		5.20	3.077		109	121		230	300
0690	18" wide		4	4		121	158		279	365
0700	24" wide		3.20	5		125	197		322	430
0710	30" wide		2.80	5.714		140	225		365	490
0720	36" wide		2.40	6.667		147	263		410	550
0740	36" radius, 6" wide		6.80	2.353		139	92.50		231.50	291
0760	12" wide		4.80	3.333		148	131		279	360
0770	18" wide		3.60	4.444		164	175		339	440
0780	24" wide		2.80	5.714		176	225		401	530
0790	30" wide		2.40	6.667		180	263		443	590
0800	36" wide		2	8		200	315		515	690
0820	Tee horizontal, 12" radius, 6" wide		4	4		84	158		242	325
0840	12" wide		3.20	5		94	197		291	395
0850	18" wide		2.80	5.714		106	225		331	450
0860	24" wide		2.40	6.667		123	263		386	525
0870	30" wide		2.20	7.273		140	287		427	580
0880	36" wide		2	8		154	315		469	640
0900	24" radius, 6" wide		3.60	4.444		139	175		314	415
0920	12" wide		2.80	5.714		152	225		377	500
0930	18" wide		2.40	6.667		170	263		433	575
0940	24" wide		2	8		220	315		535	710
0950	30" wide		1.80	8.889		239	350		589	785
0960	36" wide		1.60	10		264	395		659	875

120

120

ELECTRICAL 16

			DAILY	LABOR-		2004 BARE COSTS				TOTAL
16131	**Cable Trays**	CREW	OUTPUT	HOURS	UNIT	MAT.	LABOR	EQUIP.	TOTAL	INCL O&P
120	0980	36" radius, 6" wide	2 Elec	3.20	5	Ea.	200	197		397	515	120
	1000	12" wide	R16131 -100	2.40	6.667		228	263		491	640
	1010	18" wide		2	8		249	315		564	745
	1020	24" wide		1.60	10		300	395		695	915
	1030	30" wide		1.40	11.429		320	450		770	1,025
	1040	36" wide		1.20	13.333		375	525		900	1,200
	1060	Tee vertical, 12" radius, 6" wide		4	4		140	158		298	390
	1080	12" wide		3.20	5		142	197		339	450
	1090	18" wide		3	5.333		146	210		356	475
	1100	24" wide		2.80	5.714		154	225		379	505
	1110	30" wide		2.60	6.154		163	242		405	540
	1120	36" wide		2.20	7.273		166	287		453	610
	1140	24" radius, 6" wide		3.60	4.444		188	175		363	465
	1160	12" wide		2.80	5.714		198	225		423	555
	1170	18" wide		2.60	6.154		207	242		449	590
	1180	24" wide		2.40	6.667		223	263		486	635
	1190	30" wide		2.20	7.273		234	287		521	680
	1200	36" wide		1.80	8.889		254	350		604	800
	1220	36" radius, 6" wide		3.20	5		300	197		497	625
	1240	12" wide		2.40	6.667		305	263		568	730
	1250	18" wide		2.20	7.273		320	287		607	775
	1260	24" wide		2	8		335	315		650	840
	1270	30" wide		1.80	8.889		390	350		740	950
	1280	36" wide		1.40	11.429		395	450		845	1,100
	1300	Cross horizontal, 12" radius, 6" wide		3.20	5		108	197		305	410
	1320	12" wide		2.80	5.714		109	225		334	455
	1330	18" wide		2.40	6.667		122	263		385	525
	1340	24" wide		2	8		128	315		443	610
	1350	30" wide		1.80	8.889		144	350		494	680
	1360	36" wide		1.60	10		161	395		556	760
	1380	24" radius, 6" wide		2.80	5.714		170	225		395	520
	1400	12" wide		2.40	6.667		178	263		441	585
	1410	18" wide		2	8		196	315		511	685
	1420	24" wide		1.60	10		249	395		644	860
	1430	30" wide		1.40	11.429		275	450		725	975
	1440	36" wide		1.20	13.333		291	525		816	1,100
	1460	36" radius, 6" wide		2.40	6.667		291	263		554	710
	1480	12" wide		2	8		300	315		615	800
	1490	18" wide		1.60	10		305	395		700	925
	1500	24" wide		1.20	13.333		370	525		895	1,175
	1510	30" wide		1	16		405	630		1,035	1,375
	1520	36" wide		.80	20		435	790		1,225	1,650
	1540	Dropout or end plate, 6" wide		26	.615		12.50	24.50		37	50
	1560	12" wide		22	.727		15.30	28.50		43.80	59.50
	1580	18" wide		20	.800		16.80	31.50		48.30	65.50
	1600	24" wide		18	.889		19.30	35		54.30	73
	1620	30" wide		16	1		20.50	39.50		60	81.50
	1640	36" wide		13.40	1.194		23	47		70	95
	1660	Reducer, 12" to 6" wide		9.40	1.702		52.50	67		119.50	158
	1680	18" to 12" wide		8.40	1.905		54.50	75		129.50	172
	1700	18" to 6" wide		8.40	1.905		54.50	75		129.50	172
	1720	24" to 18" wide		7.20	2.222		58	87.50		145.50	194
	1740	24" to 12" wide		7.20	2.222		58	87.50		145.50	194
	1760	30" to 24" wide		6.40	2.500		60.50	98.50		159	214
	1780	30" to 18" wide		6.40	2.500		60.50	98.50		159	214
	1800	30" to 12" wide		6.40	2.500		60.50	98.50		159	214

16 ELECTRICAL

16131	Cable Trays		CREW	DAILY OUTPUT	LABOR-HOURS	UNIT	2004 BARE COSTS				TOTAL INCL O&P
							MAT.	LABOR	EQUIP.	TOTAL	
120	1820	36" to 30" wide	R16131 -100	2 Elec	5.80	2.759	Ea.	64.50	109		173.50	233
	1840	36" to 24" wide			5.80	2.759		64.50	109		173.50	233
	1860	36" to 18" wide			5.80	2.759		65.50	109		174.50	234
	1880	36" to 12" wide			5.80	2.759		65.50	109		174.50	234
	2000	Aluminum, tray, vented, 6" wide			120	.133	L.F.	10.90	5.25		16.15	19.75
	2010	9" wide			110	.145		12.30	5.75		18.05	22
	2020	12" wide			100	.160		13.70	6.30		20	24.50
	2030	18" wide			90	.178		16.40	7		23.40	28.50
	2040	24" wide			80	.200		19.35	7.90		27.25	33
	2050	30" wide			70	.229		26	9		35	42
	2060	36" wide			60	.267		28.50	10.50		39	46.50
	2080	Elbow horiz. 90°, 12" radius, 6" wide			7.60	2.105	Ea.	72	83		155	202
	2090	9" wide			7	2.286		78	90		168	220
	2100	12" wide			6.20	2.581		83	102		185	243
	2110	18" wide			5.60	2.857		97	113		210	274
	2120	24" wide			4.60	3.478		112	137		249	325
	2130	30" wide			4	4		138	158		296	385
	2140	36" wide			3.60	4.444		151	175		326	425
	2160	24" radius, 6" wide			7.20	2.222		104	87.50		191.50	244
	2180	12" wide			5.80	2.759		122	109		231	296
	2190	18" wide			5.20	3.077		140	121		261	335
	2200	24" wide			4.20	3.810		160	150		310	400
	2210	30" wide			3.60	4.444		183	175		358	460
	2220	36" wide			3.20	5		198	197		395	510
	2240	36" radius, 6" wide			6.80	2.353		147	92.50		239.50	300
	2260	12" wide			5.40	2.963		167	117		284	360
	2270	18" wide			4.80	3.333		190	131		321	405
	2280	24" wide			3.80	4.211		208	166		374	475
	2290	30" wide			3.40	4.706		244	185		429	545
	2300	36" wide			2.80	5.714		265	225		490	625
	2320	Elbow vertical 90°, 12" radius, 6" wide			7.60	2.105		87	83		170	219
	2330	9" wide			7	2.286		93	90		183	236
	2340	12" wide			6.20	2.581		94	102		196	254
	2350	18" wide			5.60	2.857		97	113		210	274
	2360	24" wide			4.60	3.478		107	137		244	320
	2370	30" wide			4	4		112	158		270	355
	2380	36" wide			3.60	4.444		113	175		288	385
	2400	24" radius, 6" wide			7.20	2.222		118	87.50		205.50	260
	2420	12" wide			5.80	2.759		128	109		237	305
	2430	18" wide			5.20	3.077		139	121		260	335
	2440	24" wide			4.20	3.810		141	150		291	380
	2450	30" wide			3.60	4.444		149	175		324	425
	2460	36" wide			3.20	5		156	197		353	465
	2480	36" radius, 6" wide			6.80	2.353		146	92.50		238.50	299
	2500	12" wide			5.40	2.963		156	117		273	345
	2510	18" wide			4.80	3.333		175	131		306	390
	2520	24" wide			3.80	4.211		189	166		355	455
	2530	30" wide			3.40	4.706		208	185		393	505
	2540	36" wide			2.80	5.714		214	225		439	570
	2560	Tee horizontal, 12" radius, 6" wide			4	4		109	158		267	355
	2570	9" wide			3.80	4.211		112	166		278	370
	2580	12" wide			3.60	4.444		122	175		297	395
	2590	18" wide			3.20	5		141	197		338	450
	2600	24" wide			2.80	5.714		162	225		387	515
	2610	30" wide			2.40	6.667		174	263		437	580
	2620	36" wide			2.20	7.273		198	287		485	645

ELECTRICAL 16

			R16131 -100	CREW	DAILY OUTPUT	LABOR-HOURS	UNIT	2004 BARE COSTS				TOTAL INCL O&P	
	16131	**Cable Trays**						MAT.	LABOR	EQUIP.	TOTAL		
120	2640	24" radius, 6" wide		2 Elec	3.60	4.444	Ea.	166	175		341	445	120
	2660	12" wide			3.20	5		189	197		386	500	
	2670	18" wide			2.80	5.714		208	225		433	565	
	2680	24" wide			2.40	6.667		260	263		523	675	
	2690	30" wide			2	8		280	315		595	780	
	2700	36" wide			1.80	8.889		330	350		680	880	
	2720	36" radius, 6" wide			3.20	5		267	197		464	585	
	2740	12" wide			2.80	5.714		305	225		530	675	
	2750	18" wide			2.40	6.667		350	263		613	775	
	2760	24" wide			2	8		405	315		720	915	
	2770	30" wide			1.60	10		430	395		825	1,050	
	2780	36" wide			1.40	11.429		500	450		950	1,225	
	2800	Tee vertical, 12" radius, 6" wide			4	4		150	158		308	400	
	2810	9" wide			3.80	4.211		151	166		317	415	
	2820	12" wide			3.60	4.444		162	175		337	440	
	2830	18" wide			3.40	4.706		163	185		348	455	
	2840	24" wide			3.20	5		167	197		364	475	
	2850	30" wide			3	5.333		175	210		385	510	
	2860	36" wide			2.60	6.154		182	242		424	560	
	2880	24" radius, 6" wide			3.60	4.444		212	175		387	495	
	2900	12" wide			3.20	5		229	197		426	545	
	2910	18" wide			3	5.333		244	210		454	585	
	2920	24" wide			2.80	5.714		260	225		485	620	
	2930	30" wide			2.60	6.154		281	242		523	670	
	2940	36" wide			2.20	7.273		286	287		573	740	
	2960	36" radius, 6" wide			3.20	5		330	197		527	655	
	2980	12" wide			2.80	5.714		350	225		575	720	
	2990	18" wide			2.60	6.154		360	242		602	755	
	3000	24" wide			2.40	6.667		385	263		648	815	
	3010	30" wide			2.20	7.273		415	287		702	885	
	3020	36" wide			1.80	8.889		435	350		785	1,000	
	3040	Cross horizontal, 12" radius, 6" wide			3.60	4.444		136	175		311	410	
	3050	9" wide			3.40	4.706		146	185		331	435	
	3060	12" wide			3.20	5		150	197		347	460	
	3070	18" wide			2.80	5.714		157	225		382	510	
	3080	24" wide			2.40	6.667		175	263		438	585	
	3090	30" wide			2.20	7.273		212	287		499	660	
	3100	36" wide			1.80	8.889		256	350		606	800	
	3120	24" radius, 6" wide			3.20	5		244	197		441	560	
	3140	12" wide			2.80	5.714		267	225		492	630	
	3150	18" wide			2.40	6.667		288	263		551	705	
	3160	24" wide			2	8		335	315		650	835	
	3170	30" wide			1.80	8.889		360	350		710	915	
	3180	36" wide			1.40	11.429		405	450		855	1,125	
	3200	36" radius, 6" wide			2.80	5.714		405	225		630	780	
	3220	12" wide			2.40	6.667		425	263		688	860	
	3230	18" wide			2	8		465	315		780	980	
	3240	24" wide			1.60	10		535	395		930	1,175	
	3250	30" wide			1.40	11.429		580	450		1,030	1,300	
	3260	36" wide			1.20	13.333		675	525		1,200	1,525	
	3280	Dropout, or end plate, 6" wide			26	.615		13.60	24.50		38.10	51	
	3300	12" wide			22	.727		16.10	28.50		44.60	60	
	3310	18" wide			20	.800		19.40	31.50		50.90	68.50	
	3320	24" wide			18	.889		23.50	35		58.50	77.50	
	3330	30" wide			16	1		25	39.50		64.50	86	
	3340	36" wide			14	1.143		28.50	45		73.50	98.50	

16 ELECTRICAL

Important: See the Reference Section for critical supporting data - Reference Nos., Crews, & City Cost Indexes

16131 | Cable Trays

			DAILY	LABOR-		2004 BARE COSTS				TOTAL
		CREW	OUTPUT	HOURS	UNIT	MAT.	LABOR	EQUIP.	TOTAL	INCL O&P
120	3370	Reducer, 9" to 6" wide	R16131 -100	2 Elec	12	1.333	Ea.	66.50	52.50		119	152	120

		DAILY OUTPUT	LABOR-HOURS	MAT.	LABOR	TOTAL	TOTAL INCL O&P
3380	12" to 6" wide	11.40	1.404	70	55.50	125.50	160
3390	12" to 9" wide	11.40	1.404	70	55.50	125.50	160
3400	18" to 12" wide	9.60	1.667	75	65.50	140.50	180
3420	18" to 6" wide	9.60	1.667	75	65.50	140.50	180
3430	18" to 9" wide	9.60	1.667	75	65.50	140.50	180
3440	24" to 18" wide	8.40	1.905	80	75	155	200
3460	24" to 12" wide	8.40	1.905	80	75	155	200
3470	24" to 9" wide	8.40	1.905	81	75	156	201
3475	24" to 6" wide	8.40	1.905	81	75	156	201
3480	30" to 24" wide	7.20	2.222	82	87.50	169.50	220
3500	30" to 18" wide	7.20	2.222	82	87.50	169.50	220
3520	30" to 12" wide	7.20	2.222	84	87.50	171.50	223
3540	36" to 30" wide	6.40	2.500	85	98.50	183.50	241
3560	36" to 24" wide	6.40	2.500	85	98.50	183.50	241
3580	36" to 18" wide	6.40	2.500	85	98.50	183.50	241
3600	36" to 12" wide	6.40	2.500	85	98.50	183.50	241
3610	Elbow horizontal 60°, 12" radius, 6" wide	7.80	2.051	58	81	139	184
3620	9" wide	7.20	2.222	63.50	87.50	151	200
3630	12" wide	6.40	2.500	70	98.50	168.50	224
3640	18" wide	5.80	2.759	75	109	184	245
3650	24" wide	4.80	3.333	92	131	223	296
3680	Elbow horizontal 45°, 12" radius, 6" wide	8	2	50	79	129	172
3690	9" wide	7.40	2.162	53	85	138	186
3700	12" wide	6.60	2.424	56	95.50	151.50	204
3710	18" wide	6	2.667	63	105	168	226
3720	24" wide	5	3.200	73	126	199	269
3750	Elbow horizontal, 30° 12" radius, 6" wide	8.20	1.951	43.50	77	120.50	162
3760	9" wide	7.60	2.105	46	83	129	174
3770	12" wide	6.80	2.353	49	92.50	141.50	192
3780	18" wide	6.20	2.581	53	102	155	210
3790	24" wide	5.20	3.077	58	121	179	244
3820	Elbow vertical 60° in/outside, 12" radius, 6" wide	7.80	2.051	72	81	153	199
3830	9" wide	7.20	2.222	73	87.50	160.50	211
3840	12" wide	6.40	2.500	74	98.50	172.50	229
3850	18" wide	5.80	2.759	77	109	186	247
3860	24" wide	4.80	3.333	80	131	211	283
3890	Elbow vertical 45° in/outside, 12" radius, 6" wide	8	2	58	79	137	181
3900	9" wide	7.40	2.162	61.50	85	146.50	195
3910	12" wide	6.60	2.424	63.50	95.50	159	212
3920	18" wide	6	2.667	65.50	105	170.50	228
3930	24" wide	5	3.200	72	126	198	267
3960	Elbow vertical 30° in/outside, 12" radius, 6" wide	8.20	1.951	50	77	127	169
3970	9" wide	7.60	2.105	53	83	136	182
3980	12" wide	6.80	2.353	54	92.50	146.50	198
3990	18" wide	6.20	2.581	56	102	158	213
4000	24" wide	5.20	3.077	57	121	178	243
4250	Reducer, left or right hand, 24" to 18" wide	8.40	1.905	76	75	151	196
4260	24" to 12" wide	8.40	1.905	76	75	151	196
4270	24" to 9" wide	8.40	1.905	76	75	151	196
4280	24" to 6" wide	8.40	1.905	77	75	152	197
4290	18" to 12" wide	9.60	1.667	71	65.50	136.50	176
4300	18" to 9" wide	9.60	1.667	71	65.50	136.50	176
4310	18" to 6" wide	9.60	1.667	71	65.50	136.50	176
4320	12" to 9" wide	11.40	1.404	68	55.50	123.50	158
4330	12" to 6" wide	11.40	1.404	68	55.50	123.50	158

ELECTRICAL 16

			CREW	DAILY OUTPUT	LABOR-HOURS	UNIT	2004 BARE COSTS				TOTAL INCL O&P
		**16131	Cable Trays**						MAT.	LABOR	EQUIP.	TOTAL	
120	4340	9" to 6" wide	2 Elec	12	1.333	Ea.	66	52.50		118.50	151
	4350	Splice plate	1 Elec	48	.167		6.30	6.55		12.85	16.70
	4360	Splice plate, expansion joint		48	.167		6.45	6.55		13	16.85
	4370	Splice plate, hinged, horizontal		48	.167		5.20	6.55		11.75	15.45
	4380	Vertical		48	.167		7.40	6.55		13.95	17.90
	4390	Trough, hanger, vertical		28	.286		21	11.25		32.25	40
	4400	Box connector, 24" wide		20	.400		28	15.75		43.75	54.50
	4410	18" wide		21	.381		23.50	15		38.50	48
	4420	12" wide		22	.364		22.50	14.35		36.85	46
	4430	9" wide		23	.348		21.50	13.70		35.20	44
	4440	6" wide		24	.333		20.50	13.15		33.65	42
	4450	Floor flange		24	.333		21.50	13.15		34.65	43
	4460	Hold down clamp		60	.133		2.45	5.25		7.70	10.50
	4520	Wall bracket, 24" wide tray		20	.400		21.50	15.75		37.25	47
	4530	18" wide tray		21	.381		21	15		36	45.50
	4540	12" wide tray		22	.364		11.20	14.35		25.55	34
	4550	9" wide tray		23	.348		10	13.70		23.70	31.50
	4560	6" wide tray		24	.333		9.10	13.15		22.25	29.50
	5000	Cable channel aluminum, vented, 1-1/4" deep, 4" wide, straight		80	.100	L.F.	9	3.94		12.94	15.75
	5010	Elbow horizontal, 36" radius, 90°		5	1.600	Ea.	162	63		225	272
	5020	60°		5.50	1.455		124	57.50		181.50	221
	5030	45°		6	1.333		101	52.50		153.50	189
	5040	30°		6.50	1.231		87	48.50		135.50	168
	5050	Adjustable		6	1.333		83	52.50		135.50	170
	5060	Elbow vertical, 36" radius, 90°		5	1.600		174	63		237	285
	5070	60°		5.50	1.455		136	57.50		193.50	235
	5080	45°		6	1.333		112	52.50		164.50	201
	5090	30°		6.50	1.231		99	48.50		147.50	181
	5100	Adjustable		6	1.333		83	52.50		135.50	170
	5110	Splice plate, hinged, horizontal		48	.167		6.20	6.55		12.75	16.55
	5120	Splice plate, hinged, vertical		48	.167		8.80	6.55		15.35	19.45
	5130	Hanger, vertical		28	.286		9.70	11.25		20.95	27.50
	5140	Single		28	.286		14.75	11.25		26	33
	5150	Double		20	.400		15.10	15.75		30.85	40
	5160	Channel to box connector		24	.333		19.80	13.15		32.95	41.50
	5170	Hold down clip		80	.100		2.80	3.94		6.74	8.95
	5180	Wall bracket, single		28	.286		7.25	11.25		18.50	25
	5190	Double		20	.400		9.30	15.75		25.05	34
	5200	Cable roller		16	.500		114	19.70		133.70	155
	5210	Splice plate		48	.167		3.50	6.55		10.05	13.60
130	0010	**CABLE TRAY, COVERS AND DIVIDERS** To 15' high									**130**
	0011	For higher elevations, see 16131-130-9900									
	0100	Covers, ventilated galv. steel, straight, 6" wide tray size	2 Elec	520	.031	L.F.	3.30	1.21		4.51	5.45
	0200	9" wide tray size		460	.035		4.05	1.37		5.42	6.50
	0300	12" wide tray size		400	.040		4.85	1.58		6.43	7.70
	0400	18" wide tray size		300	.053		6.60	2.10		8.70	10.40
	0500	24" wide tray size		220	.073		8.10	2.87		10.97	13.15
	0600	30" wide tray size		180	.089		9.80	3.50		13.30	16
	0700	36" wide tray size		160	.100		11.20	3.94		15.14	18.15
	1000	Elbow horizontal 90°, 12" radius, 6" wide tray size		150	.107	Ea.	24.50	4.20		28.70	33.50
	1020	9" wide tray size		128	.125		27	4.93		31.93	37.50
	1040	12" wide tray size		108	.148		28.50	5.85		34.35	40
	1060	18" wide tray size		84	.190		39.50	7.50		47	54.50
	1080	24" wide tray size		66	.242		48	9.55		57.55	66.50
	1100	30" wide tray size		60	.267		59.50	10.50		70	80.50
	1120	36" wide tray size		50	.320		72	12.60		84.60	98

Reference notes: R16131-100 (line 4340/4350), R16131-100 (line 0010/0011)

16 ELECTRICAL

Important: See the Reference Section for critical supporting data - Reference Nos., Crews, & City Cost Indexes

16131 | Cable Trays

		CREW	DAILY OUTPUT	LABOR-HOURS	UNIT	MAT.	LABOR	EQUIP.	TOTAL	TOTAL INCL O&P	
130							2004 BARE COSTS					**139**
1160	24" radius, 6" wide tray size	2 Elec	136	.118	Ea.	40.50	4.64		45.14	51.50	
1180	9" wide tray size		116	.138		41.50	5.45		46.95	54	
1200	12" wide tray size		96	.167		46.50	6.55		53.05	61	
1220	18" wide tray size		76	.211		57	8.30		65.30	75.50	
1240	24" wide tray size		60	.267		69.50	10.50		80	92	
1260	30" wide tray size		52	.308		91.50	12.10		103.60	119	
1280	36" wide tray size		44	.364		106	14.35		120.35	139	
1320	36" radius, 6" wide tray size		120	.133		59.50	5.25		64.75	73	
1340	9" wide tray size		104	.154		65.50	6.05		71.55	81	
1360	12" wide tray size		84	.190		70	7.50		77.50	88	
1380	18" wide tray size		72	.222		89	8.75		97.75	111	
1400	24" wide tray size		52	.308		106	12.10		118.10	135	
1420	30" wide tray size		46	.348		126	13.70		139.70	160	
1440	36" wide tray size		40	.400		148	15.75		163.75	187	
1480	Elbow horizontal 45°, 12" radius, 6" wide tray size		150	.107		17.40	4.20		21.60	25.50	
1500	9" wide tray size		128	.125		21	4.93		25.93	30.50	
1520	12" wide tray size		108	.148		22.50	5.85		28.35	33.50	
1540	18" wide tray size		88	.182		27.50	7.15		34.65	41	
1560	24" wide tray size		76	.211		32	8.30		40.30	48	
1580	30" wide tray size		66	.242		38	9.55		47.55	56	
1600	36" wide tray size		60	.267		42.50	10.50		53	62.50	
1640	24" radius, 6" wide tray size		136	.118		25.50	4.64		30.14	35	
1660	9" wide tray size		116	.138		28.50	5.45		33.95	39.50	
1680	12" wide tray size		96	.167		32	6.55		38.55	45.50	
1700	18" wide tray size		80	.200		37	7.90		44.90	52	
1720	24" wide tray size		70	.229		42.50	9		51.50	60.50	
1740	30" wide tray size		60	.267		54	10.50		64.50	75	
1760	36" wide tray size		52	.308		59.50	12.10		71.60	83	
1800	36" radius, 6" wide tray size		120	.133		37	5.25		42.25	48.50	
1820	9" wide tray size		104	.154		41.50	6.05		47.55	55	
1840	12" wide tray size		84	.190		46	7.50		53.50	61.50	
1860	18" wide tray size		76	.211		52.50	8.30		60.80	70.50	
1880	24" wide tray size		62	.258		64	10.15		74.15	85.50	
1900	30" wide tray size		52	.308		69.50	12.10		81.60	94.50	
1920	36" wide tray size		48	.333		84	13.15		97.15	112	
1960	Elbow vertical 90°, 12" radius, 6" wide tray size		150	.107		20.50	4.20		24.70	29	
1980	9" wide tray size		128	.125		21	4.93		25.93	30.50	
2000	12" wide tray size		108	.148		22.50	5.85		28.35	33	
2020	18" wide tray size		88	.182		25.50	7.15		32.65	38.50	
2040	24" wide tray size		68	.235		26	9.25		35.25	42.50	
2060	30" wide tray size		60	.267		28.50	10.50		39	47	
2080	36" wide tray size		50	.320		37	12.60		49.60	59.50	
2120	24" radius, 6" wide tray size		136	.118		25	4.64		29.64	34.50	
2140	9" wide tray size		116	.138		27.50	5.45		32.95	38.50	
2160	12" wide tray size		96	.167		28.50	6.55		35.05	41.50	
2180	18" wide tray size		80	.200		38	7.90		45.90	53.50	
2200	24" wide tray size		62	.258		42.50	10.15		52.65	62	
2220	30" wide tray size		52	.308		48.50	12.10		60.60	71	
2240	36" wide tray size		44	.364		55	14.35		69.35	82	
2280	36" radius, 6" wide tray size		120	.133		29	5.25		34.25	40	
2300	9" wide tray size		104	.154		37	6.05		43.05	49.50	
2320	12" wide tray size		84	.190		41.50	7.50		49	57	
2340	18" wide tray size		76	.211		52.50	8.30		60.80	70.50	
2350	24" wide tray size		54	.296		58	11.65		69.65	81.50	
2360	30" wide tray size		46	.348		69.50	13.70		83.20	97	
2370	36" wide tray size		40	.400		82	15.75		97.75	114	

R16131 -100

			CREW	DAILY OUTPUT	LABOR-HOURS	UNIT	2004 BARE COSTS				TOTAL INCL O&P	
		16131 | Cable Trays					MAT.	LABOR	EQUIP.	TOTAL		
130	2400	Tee horizontal, 12" radius, 6" wide tray size	2 Elec	92	.174	Ea.	37	6.85		43.85	50.50	130
	2410	9" wide tray size		80	.200		38	7.90		45.90	53.50	
	2420	12" wide tray size		68	.235		43	9.25		52.25	61.50	
	2430	18" wide tray size		60	.267		52	10.50		62.50	72.50	
	2440	24" wide tray size		52	.308		65.50	12.10		77.60	90	
	2460	30" wide tray size		36	.444		77	17.50		94.50	111	
	2470	36" wide tray size		30	.533		93	21		114	134	
	2500	24" radius, 6" wide tray size		88	.182		58	7.15		65.15	74.50	
	2510	9" wide tray size		76	.211		66.50	8.30		74.80	86	
	2520	12" wide tray size		64	.250		68.50	9.85		78.35	90	
	2530	18" wide tray size		56	.286		88	11.25		99.25	114	
	2540	24" wide tray size		48	.333		135	13.15		148.15	169	
	2560	30" wide tray size		32	.500		154	19.70		173.70	199	
	2570	36" wide tray size		26	.615		170	24.50		194.50	223	
	2600	36" radius, 6" wide tray size		84	.190		105	7.50		112.50	127	
	2610	9" wide tray size		72	.222		107	8.75		115.75	131	
	2620	12" wide tray size		60	.267		119	10.50		129.50	147	
	2630	18" wide tray size		52	.308		139	12.10		151.10	171	
	2640	24" wide tray size		44	.364		183	14.35		197.35	223	
	2660	30" wide tray size		28	.571		199	22.50		221.50	253	
	2670	36" wide tray size		22	.727		226	28.50		254.50	292	
	2700	Cross horizontal, 12" radius, 6" wide tray size		68	.235		55	9.25		64.25	74.50	
	2710	9" wide tray size		64	.250		58	9.85		67.85	78.50	
	2720	12" wide tray size		60	.267		65	10.50		75.50	87	
	2730	18" wide tray size		52	.308		77	12.10		89.10	103	
	2740	24" wide tray size		36	.444		94	17.50		111.50	129	
	2760	30" wide tray size		30	.533		107	21		128	150	
	2770	36" wide tray size		28	.571		125	22.50		147.50	172	
	2800	24" radius, 6" wide tray size		64	.250		105	9.85		114.85	131	
	2810	9" wide tray size		60	.267		114	10.50		124.50	141	
	2820	12" wide tray size		56	.286		123	11.25		134.25	152	
	2830	18" wide tray size		48	.333		147	13.15		160.15	182	
	2840	24" wide tray size		32	.500		179	19.70		198.70	227	
	2860	30" wide tray size		26	.615		202	24.50		226.50	258	
	2870	36" wide tray size		24	.667		227	26.50		253.50	289	
	2900	36" radius, 6" wide tray size		60	.267		176	10.50		186.50	210	
	2910	9" wide tray size		56	.286		182	11.25		193.25	217	
	2920	12" wide tray size		52	.308		191	12.10		203.10	228	
	2930	18" wide tray size		44	.364		220	14.35		234.35	264	
	2940	24" wide tray size		28	.571		281	22.50		303.50	345	
	2960	30" wide tray size		22	.727		300	28.50		328.50	375	
	2970	36" wide tray size		20	.800		320	31.50		351.50	400	
	3000	Reducer, 9" to 6" wide tray size		128	.125		23.50	4.93		28.43	33	
	3010	12" to 6" wide tray size		108	.148		24.50	5.85		30.35	35.50	
	3020	12" to 9" wide tray size		108	.148		24.50	5.85		30.35	35.50	
	3030	18" to 12" wide tray size		88	.182		26	7.15		33.15	39	
	3050	18" to 6" wide tray size		88	.182		26	7.15		33.15	39	
	3060	24" to 18" wide tray size		80	.200		36.50	7.90		44.40	51.50	
	3070	24" to 12" wide tray size		80	.200		33.50	7.90		41.40	48	
	3090	30" to 24" wide tray size		70	.229		38.50	9		47.50	56	
	3100	30" to 18" wide tray size		70	.229		38.50	9		47.50	56	
	3110	30" to 12" wide tray size		70	.229		33.50	9		42.50	50	
	3140	36" to 30" wide tray size		64	.250		42	9.85		51.85	60.50	
	3150	36" to 24" wide tray size		64	.250		42	9.85		51.85	60.50	
	3160	36" to 18" wide tray size		64	.250		42	9.85		51.85	60.50	
	3170	36" to 12" wide tray size		64	.250		42	9.85		51.85	60.50	

R16131 -100

Important: See the Reference Section for critical supporting data - Reference Nos., Crews, & City Cost Indexes

16 ELECTRICAL

				DAILY	LABOR-			2004 BARE COSTS				TOTAL	
16131	**Cable Trays**		CREW	OUTPUT	HOURS	UNIT	MAT.	LABOR	EQUIP.	TOTAL	INCL O&P	
130	3250	Covers, aluminum, straight, 6" wide tray size	R16131 -100	2 Elec	520	.031	L.F.	3.20	1.21		4.41	5.30	130
	3270	9" wide tray size			460	.035		3.85	1.37		5.22	6.30	
	3290	12" wide tray size			400	.040		4.60	1.58		6.18	7.40	
	3310	18" wide tray size			320	.050		6.10	1.97		8.07	9.65	
	3330	24" wide tray size			260	.062		7.60	2.42		10.02	11.95	
	3350	30" wide tray size			200	.080		8.40	3.15		11.55	13.95	
	3370	36" wide tray size			180	.089	↓	8.90	3.50		12.40	15	
	3400	Elbow horizontal 90°, 12" radius, 6" wide tray size			150	.107	Ea.	24	4.20		28.20	33	
	3410	9" wide tray size			128	.125		25	4.93		29.93	35	
	3420	12" wide tray size			108	.148		27	5.85		32.85	38	
	3430	18" wide tray size			88	.182		35.50	7.15		42.65	49.50	
	3440	24" wide tray size			70	.229		44.50	9		53.50	62.50	
	3460	30" wide tray size			64	.250		53.50	9.85		63.35	73.50	
	3470	36" wide tray size			54	.296		65.50	11.65		77.15	89.50	
	3500	24" radius, 6" wide tray size			136	.118		33	4.64		37.64	43	
	3510	9" wide tray size			116	.138		40.50	5.45		45.95	52.50	
	3520	12" wide tray size			96	.167		44.50	6.55		51.05	59	
	3530	18" wide tray size			80	.200		53	7.90		60.90	70	
	3540	24" wide tray size			64	.250		65.50	9.85		75.35	86.50	
	3560	30" wide tray size			56	.286		80	11.25		91.25	105	
	3570	36" wide tray size			48	.333		99	13.15		112.15	129	
	3600	36" radius, 6" wide tray size			120	.133		56	5.25		61.25	70	
	3610	9" wide tray size			104	.154		61.50	6.05		67.55	76.50	
	3620	12" wide tray size			84	.190		69.50	7.50		77	87.50	
	3630	18" wide tray size			76	.211		83	8.30		91.30	104	
	3640	24" wide tray size			56	.286		101	11.25		112.25	128	
	3660	30" wide tray size			50	.320		116	12.60		128.60	147	
	3670	36" wide tray size			44	.364		136	14.35		150.35	172	
	3700	Elbow horizontal 45°, 12" radius, 6" wide tray size			150	.107		17.70	4.20		21.90	25.50	
	3710	9" wide tray size			128	.125		18.20	4.93		23.13	27.50	
	3720	12" wide tray size			108	.148		20.50	5.85		26.35	31	
	3730	18" wide tray size			88	.182		23.50	7.15		30.65	36	
	3740	24" wide tray size			80	.200		26	7.90		33.90	40	
	3760	30" wide tray size			70	.229		33	9		42	49.50	
	3770	36" wide tray size			64	.250		39.50	9.85		49.35	58	
	3800	24" radius, 6" wide tray size			136	.118		20.50	4.64		25.14	29.50	
	3810	9" wide tray size			116	.138		26	5.45		31.45	36.50	
	3820	12" wide tray size			96	.167		27	6.55		33.55	40	
	3830	18" wide tray size			80	.200		33	7.90		40.90	47.50	
	3840	24" wide tray size			72	.222		40.50	8.75		49.25	57.50	
	3860	30" wide tray size			64	.250		47	9.85		56.85	66	
	3870	36" wide tray size			56	.286		54	11.25		65.25	76.50	
	3900	36" radius, 6" wide tray size			120	.133		35.50	5.25		40.75	47	
	3910	9" wide tray size			104	.154		38.50	6.05		44.55	51.50	
	3920	12" wide tray size			84	.190		40.50	7.50		48	55.50	
	3930	18" wide tray size			76	.211		49	8.30		57.30	66.50	
	3940	24" wide tray size			64	.250		59.50	9.85		69.35	79.50	
	3960	30" wide tray size			56	.286		67.50	11.25		78.75	91.50	
	3970	36" wide tray size			50	.320		77	12.60		89.60	103	
	4000	Elbow vertical 90°, 12" radius, 6" wide tray size			150	.107		20.50	4.20		24.70	29	
	4010	9" wide tray size			128	.125		20.50	4.93		25.43	30	
	4020	12" wide tray size			108	.148		22.50	5.85		28.35	33	
	4030	18" wide tray size			88	.182		26	7.15		33.15	39	
	4040	24" wide tray size			70	.229		27	9		36	43.50	
	4060	30" wide tray size			64	.250		27.50	9.85		37.35	45	
	4070	36" wide tray size			54	.296		33.50	11.65		45.15	54	

ELECTRICAL 16

16131 | Cable Trays

		CREW	DAILY OUTPUT	LABOR-HOURS	UNIT	2004 BARE COSTS				TOTAL INCL O&P
						MAT.	LABOR	EQUIP.	TOTAL	
4100	24" radius, 6" wide tray size	2 Elec	136	.118	Ea.	23.50	4.64		28.14	32.50
4110	9" wide tray size		116	.138		25.50	5.45		30.95	36
4120	12" wide tray size		96	.167		27.50	6.55		34.05	40.50
4130	18" wide tray size		80	.200		33.50	7.90		41.40	48
4140	24" wide tray size		64	.250		36.50	9.85		46.35	54.50
4160	30" wide tray size		56	.286		47	11.25		58.25	68.50
4170	36" wide tray size		48	.333		51	13.15		64.15	75.50
4200	36" radius, 6" wide tray size		120	.133		27	5.25		32.25	38
4210	9" wide tray size		104	.154		33.50	6.05		39.55	45.50
4220	12" wide tray size		84	.190		38.50	7.50		46	53.50
4230	18" wide tray size		76	.211		47	8.30		55.30	64
4240	24" wide tray size		56	.286		55	11.25		66.25	77.50
4260	30" wide tray size		50	.320		69.50	12.60		82.10	95.50
4270	36" wide tray size		44	.364		75	14.35		89.35	104
4300	Tee horizontal, 12" radius, 6" wide tray size		108	.148		33.50	5.85		39.35	45
4310	9" wide tray size		88	.182		35.50	7.15		42.65	49.50
4320	12" wide tray size		80	.200		39.50	7.90		47.40	55
4330	18" wide tray size		68	.235		47	9.25		56.25	65.50
4340	24" wide tray size		56	.286		59.50	11.25		70.75	82
4360	30" wide tray size		44	.364		69.50	14.35		83.85	98
4370	36" wide tray size		36	.444		84	17.50		101.50	119
4400	24" radius, 6" wide tray size		96	.167		54	6.55		60.55	69.50
4410	9" wide tray size		80	.200		60.50	7.90		68.40	78
4420	12" wide tray size		72	.222		67.50	8.75		76.25	87.50
4430	18" wide tray size		60	.267		79	10.50		89.50	103
4440	24" wide tray size		48	.333		126	13.15		139.15	159
4460	30" wide tray size		40	.400		139	15.75		154.75	177
4470	36" wide tray size		32	.500		155	19.70		174.70	201
4500	36" radius, 6" wide tray size		88	.182		99	7.15		106.15	120
4510	9" wide tray size		72	.222		101	8.75		109.75	124
4520	12" wide tray size		64	.250		111	9.85		120.85	137
4530	18" wide tray size		56	.286		126	11.25		137.25	156
4540	24" wide tray size		44	.364		160	14.35		174.35	198
4560	30" wide tray size		36	.444		180	17.50		197.50	224
4570	36" wide tray size		28	.571		203	22.50		225.50	257
4600	Cross horizontal, 12" radius, 6" wide tray size		80	.200		51	7.90		58.90	67.50
4610	9" wide tray size		72	.222		54	8.75		62.75	72.50
4620	12" wide tray size		64	.250		60	9.85		69.85	80.50
4630	18" wide tray size		56	.286		73	11.25		84.25	97.50
4640	24" wide tray size		48	.333		84	13.15		97.15	112
4660	30" wide tray size		40	.400		100	15.75		115.75	134
4670	36" wide tray size		32	.500		115	19.70		134.70	157
4700	24" radius, 6" wide tray size		72	.222		100	8.75		108.75	123
4710	9" wide tray size		64	.250		107	9.85		116.85	133
4720	12" wide tray size		56	.286		115	11.25		126.25	144
4730	18" wide tray size		48	.333		134	13.15		147.15	167
4740	24" wide tray size		40	.400		160	15.75		175.75	200
4760	30" wide tray size		32	.500		187	19.70		206.70	236
4770	36" wide tray size		24	.667		203	26.50		229.50	262
4800	36" radius, 6" wide tray size		64	.250		160	9.85		169.85	191
4810	9" wide tray size		56	.286		167	11.25		178.25	201
4820	12" wide tray size		50	.320		180	12.60		192.60	217
4830	18" wide tray size		44	.364		198	14.35		212.35	240
4840	24" wide tray size		36	.444		244	17.50		261.50	294
4860	30" wide tray size		28	.571		281	22.50		303.50	345
4870	36" wide tray size		22	.727		305	28.50		333.50	385

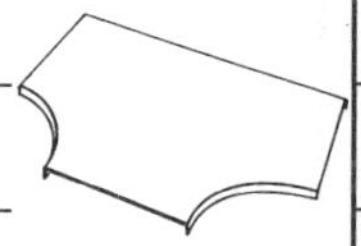

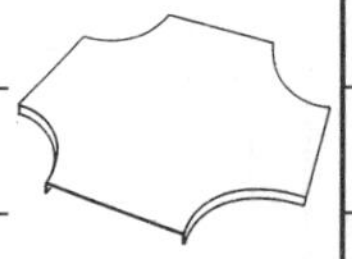

Reference box: R16131 -100

Margin markers: 130

16 ELECTRICAL

Important: See the Reference Section for critical supporting data - Reference Nos., Crews, & City Cost Indexes

				DAILY	LABOR-		2004 BARE COSTS				TOTAL
16131	**Cable Trays**		CREW	OUTPUT	HOURS	UNIT	MAT.	LABOR	EQUIP.	TOTAL	INCL O&P
130	4900	Reducer, 9" to 6" wide tray size	2 Elec	128	.125	Ea.	25.50	4.93		30.43	35.50	130
	4910	12" to 6" wide tray size		108	.148		26.50	5.85		32.35	37.50
	4920	12" to 9" wide tray size		108	.148		26.50	5.85		32.35	37.50
	4930	18" to 12" wide tray size		88	.182		29	7.15		36.15	42.50
	4950	18" to 6" wide tray size		88	.182		29	7.15		36.15	42.50
	4960	24" to 18" wide tray size		80	.200		37.50	7.90		45.40	52.50
	4970	24" to 12" wide tray size		80	.200		32	7.90		39.90	47
	4990	30" to 24" wide tray size		70	.229		39	9		48	56.50
	5000	30" to 18" wide tray size		70	.229		39	9		48	56.50
	5010	30" to 12" wide tray size		70	.229		39	9		48	56.50
	5040	36" to 30" wide tray size		64	.250		43	9.85		52.85	62
	5050	36" to 24" wide tray size		64	.250		43	9.85		52.85	62
	5060	36" to 18" wide tray size		64	.250		43	9.85		52.85	62
	5070	36" to 12" wide tray size		64	.250		43	9.85		52.85	62
	5710	Tray cover hold down clamp	1 Elec	60	.133		6.90	5.25		12.15	15.40
	8000	Divider strip, straight, galvanized, 3" deep		200	.040	L.F.	2.90	1.58		4.48	5.55
	8020	4" deep		180	.044		3.55	1.75		5.30	6.50
	8040	6" deep		160	.050		4.65	1.97		6.62	8.05
	8060	Aluminum, straight, 3" deep		210	.038		2.90	1.50		4.40	5.40
	8080	4" deep		190	.042		3.60	1.66		5.26	6.45
	8100	6" deep		170	.047		4.60	1.85		6.45	7.80
	8110	Divider strip vertical fitting 3" deep									
	8120	12" radius, galvanized, 30°	1 Elec	28	.286	Ea.	14.05	11.25		25.30	32
	8140	45°		27	.296		17.60	11.65		29.25	36.50
	8160	60°		26	.308		18.70	12.10		30.80	38.50
	8180	90°		25	.320		23.50	12.60		36.10	44.50
	8200	Aluminum, 30°		29	.276		11.90	10.85		22.75	29.50
	8220	45°		28	.286		14.05	11.25		25.30	32
	8240	60°		27	.296		16.40	11.65		28.05	35.50
	8260	90°		26	.308		21	12.10		33.10	41
	8280	24" radius, galvanized, 30°		25	.320		21.50	12.60		34.10	42.50
	8300	45°		24	.333		24	13.15		37.15	46
	8320	60°		23	.348		30	13.70		43.70	53.50
	8340	90°		22	.364		41	14.35		55.35	66.50
	8360	Aluminum, 30°		26	.308		19.80	12.10		31.90	40
	8380	45°		25	.320		23	12.60		35.60	44
	8400	60°		24	.333		28.50	13.15		41.65	51
	8420	90°		23	.348		39	13.70		52.70	63.50
	8440	36" radius, galvanized, 30°		22	.364		28.50	14.35		42.85	53
	8460	45°		21	.381		33.50	15		48.50	59
	8480	60°		20	.400		39	15.75		54.75	66.50
	8500	90°		19	.421		54	16.60		70.60	84
	8520	Aluminum, 30°		23	.348		30	13.70		43.70	53.50
	8540	45°		22	.364		39	14.35		53.35	64.50
	8560	60°		21	.381		50.50	15		65.50	78
	8570	90°		20	.400		66	15.75		81.75	96
	8590	Divider strip vertical fitting 4" deep									
	8600	12" radius, galvanized, 30°	1 Elec	27	.296	Ea.	17.90	11.65		29.55	37
	8610	45°		26	.308		21	12.10		33.10	41
	8620	60°		25	.320		23.50	12.60		36.10	44.50
	8630	90°		24	.333		28.50	13.15		41.65	51
	8640	Aluminum, 30°		28	.286		16.80	11.25		28.05	35.50
	8650	45°		27	.296		19.50	11.65		31.15	39
	8660	60°		26	.308		22.50	12.10		34.60	42.50
	8670	90°		25	.320		26.50	12.60		39.10	48
	8680	24" radius, galvanized, 30°		24	.333		28.50	13.15		41.65	51

R16131 -100

ELECTRICAL 16

				DAILY	LABOR-			2004 BARE COSTS				TOTAL
16131		**Cable Trays**	**CREW**	**OUTPUT**	**HOURS**	**UNIT**	**MAT.**	**LABOR**	**EQUIP.**	**TOTAL**	**INCL O&P**	
130	8690	45°	1 Elec	23	.348	Ea.	36	13.70		49.70	60	130
	8700	60°		22	.364		40.50	14.35		54.85	66	
	8710	90°		21	.381		54	15		69	82	
	8720	Aluminum, 30°		25	.320		26.50	12.60		39.10	48	
	8730	45°		24	.333		32	13.15		45.15	55	
	8740	60°		23	.348		38	13.70		51.70	62.50	
	8750	90°		22	.364		52	14.35		66.35	78.50	
	8760	36" radius, galvanized, 30°		23	.348		33.50	13.70		47.20	57	
	8770	45°		22	.364		38	14.35		52.35	63.50	
	8780	60°		21	.381		48	15		63	75	
	8790	90°		20	.400		65	15.75		80.75	95	
	8800	Aluminum, 30°		24	.333		40.50	13.15		53.65	64	
	8810	45°		23	.348		52	13.70		65.70	77.50	
	8820	60°		22	.364		63.50	14.35		77.85	91	
	8830	90°	▼	21	.381	▼	80	15		95	111	
	8840	Divider strip vertical fitting 6" deep										
	8850	12" radius, galvanized, 30°	1 Elec	24	.333	Ea.	19.50	13.15		32.65	41	
	8860	45°		23	.348		22	13.70		35.70	44.50	
	8870	60°		22	.364		25	14.35		39.35	49	
	8880	90°		21	.381		31	15		46	57	
	8890	Aluminum, 30°		25	.320		18.40	12.60		31	39	
	8900	45°		24	.333		22	13.15		35.15	43.50	
	8910	60°		23	.348		23	13.70		36.70	45.50	
	8920	90°		22	.364		27	14.35		41.35	51.50	
	8930	24" radius, galvanized, 30°		23	.348		28.50	13.70		42.20	52	
	8940	45°		22	.364		36	14.35		50.35	61	
	8950	60°		21	.381		41	15		56	67.50	
	8960	90°		20	.400		54	15.75		69.75	83	
	8970	Aluminum, 30°		24	.333		27	13.15		40.15	49.50	
	8980	45°		23	.348		37	13.70		50.70	61	
	8990	60°		22	.364		40.50	14.35		54.85	66	
	9000	90°		21	.381		55	15		70	83	
	9010	36" radius, galvanized, 30°		22	.364		33.50	14.35		47.85	58	
	9020	45°		21	.381		41	15		56	67.50	
	9030	60°		20	.400		54	15.75		69.75	83	
	9040	90°		19	.421		69.50	16.60		86.10	101	
	9050	Aluminum, 30°		23	.348		41	13.70		54.70	65.50	
	9060	45°		22	.364		55	14.35		69.35	82	
	9070	60°		21	.381		65	15		80	94	
	9080	90°		20	.400		78	15.75		93.75	110	
	9120	Divider strip, horizontal fitting, galvanized, 3" deep		33	.242		20	9.55		29.55	36	
	9130	4" deep		30	.267		22.50	10.50		33	40	
	9140	6" deep		27	.296		28.50	11.65		40.15	49	
	9150	Aluminum, 3" deep		35	.229		18.90	9		27.90	34.50	
	9160	4" deep		32	.250		21	9.85		30.85	37.50	
	9170	6" deep		29	.276	▼	27.50	10.85		38.35	46.50	
	9300	Divider strip protector	▼	300	.027	L.F.	2.05	1.05		3.10	3.82	
	9310	Fastener, ladder tray				Ea.	.37			.37	.41	
	9320	Trough or solid bottom tray				"	.25			.25	.28	
	9899											
	9900	Add to labor for higher elevated installation										
	9910	15' to 20' high add						10%				
	9920	20' to 25' high add						20%				
	9930	25' to 30' high add						25%				
	9940	30' to 35' high add						30%				
	9960	Over 40' high add						40%				

R16131-100

16132 | Conduit & Tubing

		CREW	DAILY OUTPUT	LABOR-HOURS	UNIT	2004 BARE COSTS				TOTAL INCL O&P
						MAT.	LABOR	EQUIP.	TOTAL	
205	0010 **CONDUIT** To 15' high, includes 2 terminations, 2 elbows and									205
	0020 11 beam clamps per 100 L.F. [R16132 -205]									
	0300 Aluminum, 1/2" diameter	1 Elec	100	.080	L.F.	1.28	3.15		4.43	6.10
	0500 3/4" diameter		90	.089		1.72	3.50		5.22	7.10
	0700 1" diameter		80	.100		2.30	3.94		6.24	8.40
	1000 1-1/4" diameter		70	.114		2.99	4.50		7.49	10
	1030 1-1/2" diameter		65	.123		3.69	4.85		8.54	11.25
	1050 2" diameter		60	.133		4.90	5.25		10.15	13.20
	1070 2-1/2" diameter		50	.160		7.70	6.30		14	17.85
	1100 3" diameter	2 Elec	90	.178		10.30	7		17.30	22
	1130 3-1/2" diameter		80	.200		12.90	7.90		20.80	26
	1140 4" diameter		70	.229		15.60	9		24.60	30.50
	1150 5" diameter		50	.320		28	12.60		40.60	50
	1160 6" diameter		40	.400		40	15.75		55.75	67.50
	1161 Field bends, 45° to 90°, 1/2" diameter	1 Elec	53	.151	Ea.		5.95		5.95	8.85
	1162 3/4" diameter		47	.170			6.70		6.70	9.95
	1163 1" diameter		44	.182			7.15		7.15	10.65
	1164 1-1/4" diameter		23	.348			13.70		13.70	20.50
	1165 1-1/2" diameter		21	.381			15		15	22.50
	1166 2" diameter		16	.500			19.70		19.70	29.50
	1170 Elbows, 1/2" diameter		40	.200		4.34	7.90		12.24	16.45
	1200 3/4" diameter		32	.250		5.90	9.85		15.75	21
	1230 1" diameter		28	.286		8.20	11.25		19.45	26
	1250 1-1/4" diameter		24	.333		13	13.15		26.15	34
	1270 1-1/2" diameter		20	.400		17.35	15.75		33.10	42.50
	1300 2" diameter		16	.500		25.50	19.70		45.20	57.50
	1330 2-1/2" diameter		12	.667		43	26.50		69.50	86.50
	1350 3" diameter		8	1		66.50	39.50		106	132
	1370 3-1/2" diameter		6	1.333		104	52.50		156.50	192
	1400 4" diameter		5	1.600		123	63		186	229
	1410 5" diameter		4	2		335	79		414	485
	1420 6" diameter		2.50	3.200		465	126		591	700
	1430 Couplings, 1/2" diameter					1.43			1.43	1.57
	1450 3/4" diameter					2.16			2.16	2.38
	1470 1" diameter					2.84			2.84	3.12
	1500 1-1/4" diameter					3.47			3.47	3.82
	1530 1-1/2" diameter					3.99			3.99	4.39
	1550 2" diameter					5.65			5.65	6.25
	1570 2-1/2" diameter					12.85			12.85	14.15
	1600 3" diameter					16.75			16.75	18.45
	1630 3-1/2" diameter					23			23	25.50
	1650 4" diameter					27.50			27.50	30.50
	1670 5" diameter					70			70	77
	1690 6" diameter					109			109	120
	1750 Rigid galvanized steel, 1/2" diameter	1 Elec	90	.089	L.F.	1.71	3.50		5.21	7.10
	1770 3/4" diameter		80	.100		1.99	3.94		5.93	8.05
	1800 1" diameter		65	.123		2.80	4.85		7.65	10.30
	1830 1-1/4" diameter		60	.133		3.81	5.25		9.06	12
	1850 1-1/2" diameter		55	.145		4.40	5.75		10.15	13.35
	1870 2" diameter		45	.178		5.90	7		12.90	16.85
	1900 2-1/2" diameter		35	.229		10.50	9		19.50	25
	1930 3" diameter	2 Elec	50	.320		12.90	12.60		25.50	33
	1950 3-1/2" diameter		44	.364		15.60	14.35		29.95	38.50
	1970 4" diameter		40	.400		18.45	15.75		34.20	44
	1980 5" diameter		30	.533		36	21		57	71
	1990 6" diameter		20	.800		54.50	31.50		86	107

ELECTRICAL 16

16132 | Conduit & Tubing

		CREW	DAILY OUTPUT	LABOR-HOURS	UNIT	2004 BARE COSTS				TOTAL INCL O&P
						MAT.	LABOR	EQUIP.	TOTAL	
1991	Field bends, 45° to 90°, 1/2" diameter R16132 -205	1 Elec	44	.182	Ea.		7.15		7.15	10.65
1992	3/4" diameter		40	.200			7.90		7.90	11.70
1993	1" diameter		36	.222			8.75		8.75	13
1994	1-1/4" diameter		19	.421			16.60		16.60	24.50
1995	1-1/2" diameter		18	.444			17.50		17.50	26
1996	2" diameter		13	.615			24.50		24.50	36
2000	Elbows, 1/2" diameter		32	.250		4.06	9.85		13.91	19.10
2030	3/4" diameter		28	.286		4.86	11.25		16.11	22
2050	1" diameter		24	.333		7.25	13.15		20.40	27.50
2070	1-1/4" diameter		18	.444		10.40	17.50		27.90	37.50
2100	1-1/2" diameter		16	.500		13.50	19.70		33.20	44.50
2130	2" diameter		12	.667		19.50	26.50		46	60.50
2150	2-1/2" diameter		8	1		35.50	39.50		75	98
2170	3" diameter		6	1.333		52	52.50		104.50	135
2200	3-1/2" diameter		4.20	1.905		81.50	75		156.50	202
2220	4" diameter		4	2		93	79		172	219
2230	5" diameter		3.50	2.286		220	90		310	375
2240	6" diameter		2	4		330	158		488	600
2250	Couplings, 1/2" diameter					1.21			1.21	1.33
2270	3/4" diameter					1.48			1.48	1.63
2300	1" diameter					2.19			2.19	2.41
2330	1-1/4" diameter					2.74			2.74	3.01
2350	1-1/2" diameter					3.46			3.46	3.81
2370	2" diameter					4.57			4.57	5.05
2400	2-1/2" diameter					10.70			10.70	11.75
2430	3" diameter					13.85			13.85	15.20
2450	3-1/2" diameter					18.55			18.55	20.50
2470	4" diameter					18.65			18.65	20.50
2480	5" diameter					43			43	47.50
2490	6" diameter					59.50			59.50	65.50
2500	Steel, intermediate conduit (IMC), 1/2" diameter	1 Elec	100	.080	L.F.	1.37	3.15		4.52	6.20
2530	3/4" diameter		90	.089		1.62	3.50		5.12	7
2550	1" diameter		70	.114		2.26	4.50		6.76	9.20
2570	1-1/4" diameter		65	.123		2.98	4.85		7.83	10.50
2600	1-1/2" diameter		60	.133		3.44	5.25		8.69	11.60
2630	2" diameter		50	.160		4.44	6.30		10.74	14.30
2650	2-1/2" diameter		40	.200		8.95	7.90		16.85	21.50
2670	3" diameter	2 Elec	60	.267		11.50	10.50		22	28.50
2700	3-1/2" diameter		54	.296		13.15	11.65		24.80	32
2730	4" diameter		50	.320		16.35	12.60		28.95	37
2731	Field bends, 45° to 90°, 1/2" diameter	1 Elec	44	.182	Ea.		7.15		7.15	10.65
2732	3/4" diameter		40	.200			7.90		7.90	11.70
2733	1" diameter		36	.222			8.75		8.75	13
2734	1-1/4" diameter		19	.421			16.60		16.60	24.50
2735	1-1/2" diameter		18	.444			17.50		17.50	26
2736	2" diameter		13	.615			24.50		24.50	36
2750	Elbows, 1/2" diameter		32	.250		4.07	9.85		13.92	19.15
2770	3/4" diameter		28	.286		4.34	11.25		15.59	21.50
2800	1" diameter		24	.333		7.15	13.15		20.30	27.50
2830	1-1/4" diameter		18	.444		11.60	17.50		29.10	39
2850	1-1/2" diameter		16	.500		13.05	19.70		32.75	44
2870	2" diameter		12	.667		18.80	26.50		45.30	59.50
2900	2-1/2" diameter		8	1		33.50	39.50		73	95
2930	3" diameter		6	1.333		51	52.50		103.50	134
2950	3-1/2" diameter		4.20	1.905		78	75		153	198
2970	4" diameter		4	2		90	79		169	217

Important: See the Reference Section for critical supporting data - Reference Nos., Crews, & City Cost Indexes

			DAILY	LABOR-		2004 BARE COSTS				TOTAL	
16132	**Conduit & Tubing**	CREW	OUTPUT	HOURS	UNIT	MAT.	LABOR	EQUIP.	TOTAL	INCL O&P	
205	3000	Couplings, 1/2" diameter	R16132			Ea.	1.21			1.21	1.33	205
	3030	3/4" diameter	-205				1.48			1.48	1.63	
	3050	1" diameter					2.19			2.19	2.41	
	3070	1-1/4" diameter					2.74			2.74	3.01	
	3100	1-1/2" diameter					3.46			3.46	3.81	
	3130	2" diameter					4.57			4.57	5.05	
	3150	2-1/2" diameter					10.70			10.70	11.75	
	3170	3" diameter					13.85			13.85	15.20	
	3200	3-1/2" diameter					18.55			18.55	20.50	
	3230	4" diameter					18.65			18.65	20.50	
	4100	Rigid steel, plastic coated, 40 mil. thick										
	4130	1/2" diameter	1 Elec	80	.100	L.F.	3.54	3.94		7.48	9.75	
	4150	3/4" diameter		70	.114		3.96	4.50		8.46	11.05	
	4170	1" diameter		55	.145		4.99	5.75		10.74	14	
	4200	1-1/4" diameter		50	.160		6.35	6.30		12.65	16.35	
	4230	1-1/2" diameter		45	.178		7.55	7		14.55	18.70	
	4250	2" diameter		35	.229		10	9		19	24.50	
	4270	2-1/2" diameter		25	.320		16.05	12.60		28.65	36.50	
	4300	3" diameter	2 Elec	44	.364		20	14.35		34.35	44	
	4330	3-1/2" diameter		40	.400		27.50	15.75		43.25	53.50	
	4350	4" diameter		36	.444		29.50	17.50		47	58.50	
	4370	5" diameter		30	.533		58.50	21		79.50	96	
	4400	Elbows, 1/2" diameter	1 Elec	28	.286	Ea.	10.40	11.25		21.65	28	
	4430	3/4" diameter		24	.333		10.80	13.15		23.95	31.50	
	4450	1" diameter		18	.444		12.40	17.50		29.90	39.50	
	4470	1-1/4" diameter		16	.500		15.20	19.70		34.90	46	
	4500	1-1/2" diameter		12	.667		18.70	26.50		45.20	59.50	
	4530	2" diameter		8	1		26	39.50		65.50	87	
	4550	2-1/2" diameter		6	1.333		49	52.50		101.50	132	
	4570	3" diameter		4.20	1.905		78.50	75		153.50	199	
	4600	3-1/2" diameter		4	2		103	79		182	230	
	4630	4" diameter		3.80	2.105		112	83		195	246	
	4650	5" diameter		3.50	2.286		271	90		361	430	
	4680	Couplings, 1/2" diameter					2.95			2.95	3.25	
	4700	3/4" diameter					3.10			3.10	3.41	
	4730	1" diameter					4.05			4.05	4.46	
	4750	1-1/4" diameter					4.70			4.70	5.15	
	4770	1-1/2" diameter					5.60			5.60	6.15	
	4800	2" diameter					8.20			8.20	9	
	4830	2-1/2" diameter					20.50			20.50	22.50	
	4850	3" diameter					24.50			24.50	27	
	4870	3-1/2" diameter					31.50			31.50	35	
	4900	4" diameter					37			37	40.50	
	4950	5" diameter					121			121	133	
	5000	Electric metallic tubing (EMT), 1/2" diameter	1 Elec	170	.047	L.F.	.40	1.85		2.25	3.19	
	5020	3/4" diameter		130	.062		.60	2.42		3.02	4.27	
	5040	1" diameter		115	.070		1.02	2.74		3.76	5.20	
	5060	1-1/4" diameter		100	.080		1.53	3.15		4.68	6.40	
	5080	1-1/2" diameter		90	.089		1.95	3.50		5.45	7.35	
	5100	2" diameter		80	.100		2.50	3.94		6.44	8.60	
	5120	2-1/2" diameter		60	.133		5.80	5.25		11.05	14.15	
	5140	3" diameter	2 Elec	100	.160		6.35	6.30		12.65	16.35	
	5160	3-1/2" diameter		90	.178		8.10	7		15.10	19.35	
	5180	4" diameter		80	.200		9	7.90		16.90	21.50	
	5200	Field bends, 45° to 90°, 1/2" diameter	1 Elec	89	.090	Ea.		3.54		3.54	5.25	
	5220	3/4" diameter		80	.100			3.94		3.94	5.85	

ELECTRICAL 16

147

		CREW	DAILY OUTPUT	LABOR-HOURS	UNIT	2004 BARE COSTS				TOTAL INCL O&P	
16132	**Conduit & Tubing**					MAT.	LABOR	EQUIP.	TOTAL		
205											**205**
5240	1" diameter	1 Elec	73	.110	Ea.		4.32		4.32	6.40	
5260	1-1/4" diameter		38	.211			8.30		8.30	12.35	
5280	1-1/2" diameter		36	.222			8.75		8.75	13	
5300	2" diameter		26	.308			12.10		12.10	18.05	
5320	Offsets, 1/2" diameter		65	.123			4.85		4.85	7.20	
5340	3/4" diameter		62	.129			5.10		5.10	7.55	
5360	1" diameter		53	.151			5.95		5.95	8.85	
5380	1-1/4" diameter		30	.267			10.50		10.50	15.65	
5400	1-1/2" diameter		28	.286			11.25		11.25	16.75	
5420	2" diameter		20	.400			15.75		15.75	23.50	
5700	Elbows, 1" diameter		40	.200		3.10	7.90		11	15.10	
5720	1-1/4" diameter		32	.250		4.50	9.85		14.35	19.60	
5740	1-1/2" diameter		24	.333		5.90	13.15		19.05	26	
5760	2" diameter		20	.400		9	15.75		24.75	33.50	
5780	2-1/2" diameter		12	.667		22	26.50		48.50	63	
5800	3" diameter		9	.889		32.50	35		67.50	88	
5820	3-1/2" diameter		7	1.143		44	45		89	116	
5840	4" diameter		6	1.333		51.50	52.50		104	135	
6200	Couplings, set screw, steel, 1/2" diameter					.68			.68	.75	
6220	3/4" diameter					1.02			1.02	1.12	
6240	1" diameter					1.70			1.70	1.87	
6260	1-1/4" diameter					3.57			3.57	3.93	
6280	1-1/2" diameter					4.96			4.96	5.45	
6300	2" diameter					6.65			6.65	7.30	
6320	2-1/2" diameter					24.50			24.50	26.50	
6340	3" diameter					21			21	23	
6360	3-1/2" diameter					23			23	25.50	
6380	4" diameter					25			25	27.50	
6500	Box connectors, set screw, steel, 1/2" diameter	1 Elec	120	.067		.53	2.63		3.16	4.49	
6520	3/4" diameter		110	.073		.88	2.87		3.75	5.25	
6540	1" diameter		90	.089		1.65	3.50		5.15	7	
6560	1-1/4" diameter		70	.114		3.22	4.50		7.72	10.25	
6580	1-1/2" diameter		60	.133		4.65	5.25		9.90	12.90	
6600	2" diameter		50	.160		6.40	6.30		12.70	16.45	
6620	2-1/2" diameter		36	.222		22.50	8.75		31.25	38	
6640	3" diameter		27	.296		27	11.65		38.65	47	
6680	3-1/2" diameter		21	.381		37.50	15		52.50	64	
6700	4" diameter		16	.500		42	19.70		61.70	75.50	
6740	Insulated box connectors, set screw, steel, 1/2" diameter		120	.067		.74	2.63		3.37	4.72	
6760	3/4" diameter		110	.073		1.16	2.87		4.03	5.55	
6780	1" diameter		90	.089		2.14	3.50		5.64	7.55	
6800	1-1/4" diameter		70	.114		3.89	4.50		8.39	11	
6820	1-1/2" diameter		60	.133		5.50	5.25		10.75	13.85	
6840	2" diameter		50	.160		8	6.30		14.30	18.20	
6860	2-1/2" diameter		36	.222		38	8.75		46.75	54.50	
6880	3" diameter		27	.296		45	11.65		56.65	66.50	
6900	3-1/2" diameter		21	.381		62.50	15		77.50	91	
6920	4" diameter		16	.500		68.50	19.70		88.20	105	
7000	EMT to conduit adapters, 1/2" diameter (compression)		70	.114		3.74	4.50		8.24	10.80	
7020	3/4" diameter		60	.133		5.30	5.25		10.55	13.65	
7040	1" diameter		50	.160		8.05	6.30		14.35	18.30	
7060	1-1/4" diameter		40	.200		14	7.90		21.90	27	
7080	1-1/2" diameter		30	.267		17.30	10.50		27.80	34.50	
7100	2" diameter		25	.320		25	12.60		37.60	46.50	
7200	EMT to Greenfield adapters, 1/2" to 3/8" diameter (compression)		90	.089		2.67	3.50		6.17	8.15	
7220	1/2" diameter		90	.089		5.05	3.50		8.55	10.75	

R16132
-205

Important: See the Reference Section for critical supporting data - Reference Nos., Crews, & City Cost Indexes

	16132	Conduit & Tubing	CREW	DAILY OUTPUT	LABOR-HOURS	UNIT	2004 BARE COSTS				TOTAL INCL O&P	
							MAT.	LABOR	EQUIP.	TOTAL		
205	7240	3/4" diameter	1 Elec	80	.100	Ea.	6.45	3.94		10.39	12.95	205
	7260	1" diameter		70	.114		17.50	4.50		22	26	
	7270	1-1/4" diameter		60	.133		21.50	5.25		26.75	31.50	
	7280	1-1/2" diameter		50	.160		24.50	6.30		30.80	36	
	7290	2" diameter		40	.200		36.50	7.90		44.40	51.50	
	7400	EMT,LB, LR or LL fittings with covers, 1/2" dia, set screw		24	.333		9.60	13.15		22.75	30	
	7420	3/4" diameter		20	.400		11.85	15.75		27.60	36.50	
	7440	1" diameter		16	.500		18.20	19.70		37.90	49.50	
	7450	1-1/4" diameter		13	.615		27	24.50		51.50	65.50	
	7460	1-1/2" diameter		11	.727		32	28.50		60.50	77.50	
	7470	2" diameter		9	.889		50.50	35		85.50	108	
	7600	EMT, "T" fittings with covers, 1/2" diameter, set screw		16	.500		12.40	19.70		32.10	43	
	7620	3/4" diameter		15	.533		15.75	21		36.75	49	
	7640	1" diameter		12	.667		23	26.50		49.50	64	
	7650	1-1/4" diameter		11	.727		32.50	28.50		61	78.50	
	7660	1-1/2" diameter		10	.800		40.50	31.50		72	91.50	
	7670	2" diameter		8	1		53.50	39.50		93	117	
	8000	EMT, expansion fittings, no jumper, 1/2" diameter		24	.333		37.50	13.15		50.65	61	
	8020	3/4" diameter		20	.400		44	15.75		59.75	72	
	8040	1" diameter		16	.500		51.50	19.70		71.20	86.50	
	8060	1-1/4" diameter		13	.615		70	24.50		94.50	113	
	8080	1-1/2" diameter		11	.727		95.50	28.50		124	148	
	8100	2" diameter		9	.889		142	35		177	208	
	8110	2-1/2" diameter		7	1.143		225	45		270	315	
	8120	3" diameter		6	1.333		277	52.50		329.50	385	
	8140	4" diameter		5	1.600		485	63		548	625	
	8200	Split adapter, 1/2" diameter		110	.073		2.25	2.87		5.12	6.75	
	8210	3/4" diameter		90	.089		2.40	3.50		5.90	7.85	
	8220	1" diameter		70	.114		3.50	4.50		8	10.55	
	8230	1-1/4" diameter		60	.133		5.30	5.25		10.55	13.65	
	8240	1-1/2" diameter		50	.160		8.15	6.30		14.45	18.35	
	8250	2" diameter		36	.222		24	8.75		32.75	39.50	
	8300	1 hole clips, 1/2" diameter		500	.016		.24	.63		.87	1.20	
	8320	3/4" diameter		470	.017		.32	.67		.99	1.35	
	8340	1" diameter		444	.018		.59	.71		1.30	1.71	
	8360	1-1/4" diameter		400	.020		.79	.79		1.58	2.04	
	8380	1-1/2" diameter		355	.023		1.20	.89		2.09	2.64	
	8400	2" diameter		320	.025		1.87	.99		2.86	3.53	
	8420	2-1/2" diameter		266	.030		3.49	1.19		4.68	5.60	
	8440	3" diameter		160	.050		4.39	1.97		6.36	7.75	
	8460	3-1/2" diameter		133	.060		6.90	2.37		9.27	11.10	
	8480	4" diameter		100	.080		8.95	3.15		12.10	14.55	
	8500	Clamp back spacers, 1/2" diameter		500	.016		.73	.63		1.36	1.74	
	8510	3/4" diameter		470	.017		.87	.67		1.54	1.96	
	8520	1" diameter		444	.018		1.53	.71		2.24	2.74	
	8530	1-1/4" diameter		400	.020		2.73	.79		3.52	4.17	
	8540	1-1/2" diameter		355	.023		3.06	.89		3.95	4.69	
	8550	2" diameter		320	.025		5.15	.99		6.14	7.10	
	8560	2-1/2" diameter		266	.030		9.90	1.19		11.09	12.65	
	8570	3" diameter		160	.050		14.20	1.97		16.17	18.55	
	8580	3-1/2" diameter		133	.060		19.10	2.37		21.47	24.50	
	8590	4" diameter		100	.080		41.50	3.15		44.65	50.50	
	8600	Offset connectors, 1/2" diameter		40	.200		2.63	7.90		10.53	14.60	
	8610	3/4" diameter		32	.250		3.80	9.85		13.65	18.85	
	8620	1" diameter		24	.333		7.30	13.15		20.45	27.50	
	8650	90° pulling elbows, female, 1/2" diameter, with gasket		24	.333		4.45	13.15		17.60	24.50	

R16132 -205

ELECTRICAL 16

		CREW	DAILY OUTPUT	LABOR-HOURS	UNIT	2004 BARE COSTS				TOTAL INCL O&P
16132	**Conduit & Tubing**					MAT.	LABOR	EQUIP.	TOTAL	
8660	3/4" diameter	1 Elec	20	.400	Ea.	6.95	15.75		22.70	31
8700	Couplings, compression, 1/2" diameter, steel					1.94			1.94	2.13
8710	3/4" diameter					2.70			2.70	2.97
8720	1" diameter					4.65			4.65	5.10
8730	1-1/4" diameter					8.80			8.80	9.70
8740	1-1/2" diameter					12.80			12.80	14.10
8750	2" diameter					17.40			17.40	19.15
8760	2-1/2" diameter					78			78	86
8770	3" diameter					97.50			97.50	107
8780	3-1/2" diameter					159			159	175
8790	4" diameter					162			162	178
8800	Box connectors, compression, 1/2" diam., steel	1 Elec	120	.067		2.28	2.63		4.91	6.40
8810	3/4" diameter		110	.073		3.30	2.87		6.17	7.90
8820	1" diameter		90	.089		5.65	3.50		9.15	11.40
8830	1-1/4" diameter		70	.114		11.25	4.50		15.75	19.10
8840	1-1/2" diameter		60	.133		16.40	5.25		21.65	26
8850	2" diameter		50	.160		23.50	6.30		29.80	35
8860	2-1/2" diameter		36	.222		73.50	8.75		82.25	93.50
8870	3" diameter		27	.296		101	11.65		112.65	128
8880	3-1/2" diameter		21	.381		152	15		167	190
8890	4" diameter		16	.500		155	19.70		174.70	201
8900	Box connectors, insulated compression, 1/2" diam., steel		120	.067		2.93	2.63		5.56	7.15
8910	3/4" diameter		110	.073		4	2.87		6.87	8.65
8920	1" diameter		90	.089		7	3.50		10.50	12.90
8930	1-1/4" diameter		70	.114		14.55	4.50		19.05	22.50
8940	1-1/2" diameter		60	.133		22	5.25		27.25	32
8950	2" diameter		50	.160		30.50	6.30		36.80	43
8960	2-1/2" diameter		36	.222		99.50	8.75		108.25	122
8970	3" diameter		27	.296		135	11.65		146.65	166
8980	3-1/2" diameter		21	.381		224	15		239	269
8990	4" diameter		16	.500	L.F.	201	19.70		220.70	251
9100	PVC, #40, 1/2" diameter		190	.042		.71	1.66		2.37	3.25
9110	3/4" diameter		145	.055		.83	2.17		3	4.14
9120	1" diameter		125	.064		1.21	2.52		3.73	5.10
9130	1-1/4" diameter		110	.073		1.59	2.87		4.46	6
9140	1-1/2" diameter		100	.080		1.88	3.15		5.03	6.75
9150	2" diameter		90	.089		2.45	3.50		5.95	7.90
9160	2-1/2" diameter		65	.123		3.86	4.85		8.71	11.45
9170	3" diameter	2 Elec	110	.145		4.43	5.75		10.18	13.40
9180	3-1/2" diameter		100	.160		5.75	6.30		12.05	15.75
9190	4" diameter		90	.178		6.55	7		13.55	17.60
9200	5" diameter		70	.229		9.30	9		18.30	23.50
9210	6" diameter		60	.267		12.55	10.50		23.05	29.50
9220	Elbows, 1/2" diameter	1 Elec	50	.160	Ea.	1.58	6.30		7.88	11.15
9230	3/4" diameter		42	.190		1.59	7.50		9.09	12.90
9240	1" diameter		35	.229		2.69	9		11.69	16.35
9250	1-1/4" diameter		28	.286		3.83	11.25		15.08	21
9260	1-1/2" diameter		20	.400		5.20	15.75		20.95	29
9270	2" diameter		16	.500		7.55	19.70		27.25	38
9280	2-1/2" diameter		11	.727		13.15	28.50		41.65	57
9290	3" diameter		9	.889		23	35		58	77.50
9300	3-1/2" diameter		7	1.143		32	45		77	102
9310	4" diameter		6	1.333		40	52.50		92.50	122
9320	5" diameter		4	2		70	79		149	194
9330	6" diameter		3	2.667		119	105		224	287
9340	Field bends, 45° & 90°, 1/2" diameter		45	.178			7		7	10.40

R16132-205

205

205

16 ELECTRICAL

Important: See the Reference Section for critical supporting data - Reference Nos., Crews, & City Cost Indexes

			CREW	DAILY OUTPUT	LABOR-HOURS	UNIT	2004 BARE COSTS				TOTAL INCL O&P	
	16132	**Conduit & Tubing**					MAT.	LABOR	EQUIP.	TOTAL		
205	9350	3/4" diameter	1 Elec	40	.200	Ea.		7.90		7.90	11.70	**205**
	9360	1" diameter		35	.229			9		9	13.40	
	9370	1-1/4" diameter		32	.250			9.85		9.85	14.65	
	9380	1-1/2" diameter		27	.296			11.65		11.65	17.35	
	9390	2" diameter		20	.400			15.75		15.75	23.50	
	9400	2-1/2" diameter		16	.500			19.70		19.70	29.50	
	9410	3" diameter		13	.615			24.50		24.50	36	
	9420	3-1/2" diameter		12	.667			26.50		26.50	39	
	9430	4" diameter		10	.800			31.50		31.50	47	
	9440	5" diameter		9	.889			35		35	52	
	9450	6" diameter		8	1			39.50		39.50	58.50	
	9460	PVC adapters, 1/2" diameter		50	.160		.51	6.30		6.81	9.95	
	9470	3/4" diameter		42	.190		.94	7.50		8.44	12.20	
	9480	1" diameter		38	.211		1.19	8.30		9.49	13.65	
	9490	1-1/4" diameter		35	.229		1.59	9		10.59	15.15	
	9500	1-1/2" diameter		32	.250		1.85	9.85		11.70	16.70	
	9510	2" diameter		27	.296		2.66	11.65		14.31	20.50	
	9520	2-1/2" diameter		23	.348		4.57	13.70		18.27	25.50	
	9530	3" diameter		18	.444		6.75	17.50		24.25	33.50	
	9540	3-1/2" diameter		13	.615		8.70	24.50		33.20	45.50	
	9550	4" diameter		11	.727		11.80	28.50		40.30	55.50	
	9560	5" diameter		8	1		23	39.50		62.50	84	
	9570	6" diameter		6	1.333		28	52.50		80.50	109	
	9580	PVC-LB, LR or LL fittings & covers										
	9590	1/2" diameter	1 Elec	20	.400	Ea.	3.70	15.75		19.45	27.50	
	9600	3/4" diameter		16	.500		5.30	19.70		25	35.50	
	9610	1" diameter		12	.667		5.80	26.50		32.30	45.50	
	9620	1-1/4" diameter		9	.889		8.30	35		43.30	61	
	9630	1-1/2" diameter		7	1.143		9.70	45		54.70	77.50	
	9640	2" diameter		6	1.333		17.15	52.50		69.65	97	
	9650	2-1/2" diameter		6	1.333		66	52.50		118.50	151	
	9660	3" diameter		5	1.600		68	63		131	169	
	9670	3-1/2" diameter		4	2		74.50	79		153.50	199	
	9680	4" diameter		3	2.667		75	105		180	238	
	9690	PVC-tee fitting & cover										
	9700	1/2"	1 Elec	14	.571	Ea.	5.05	22.50		27.55	39	
	9710	3/4"		13	.615		6.20	24.50		30.70	43	
	9720	1"		10	.800		6.35	31.50		37.85	54	
	9730	1-1/4"		9	.889		10.70	35		45.70	64	
	9740	1-1/2"		8	1		14.15	39.50		53.65	74	
	9750	2"		7	1.143		19.75	45		64.75	88.50	
	9760	PVC-reducers, 3/4" x 1/2" diameter					1.13			1.13	1.24	
	9770	1" x 1/2" diameter					2.50			2.50	2.75	
	9780	1" x 3/4" diameter					2.70			2.70	2.97	
	9790	1-1/4" x 3/4" diameter					3.50			3.50	3.85	
	9800	1-1/4" x 1" diameter					3.50			3.50	3.85	
	9810	1-1/2" x 1-1/4" diameter					3.65			3.65	4.02	
	9820	2" x 1-1/4" diameter					4.40			4.40	4.84	
	9830	2-1/2" x 2" diameter					14.30			14.30	15.75	
	9840	3" x 2" diameter					15.25			15.25	16.80	
	9850	4" x 3" diameter					19.05			19.05	21	
	9860	Cement, quart					12.75			12.75	14.05	
	9870	Gallon					54			54	59.50	
	9880	Heat bender, to 6" diameter					1,150			1,150	1,275	
	9900	Add to labor for higher elevated installation										
	9910	15' to 20' high, add						10%				

16132	Conduit & Tubing	CREW	DAILY OUTPUT	LABOR-HOURS	UNIT	2004 BARE COSTS				TOTAL INCL O&P	
						MAT.	LABOR	EQUIP.	TOTAL		
205	9920	20' to 25' high, add						20%				205

			CREW	DAILY OUTPUT	LABOR-HOURS	UNIT	MAT.	LABOR	EQUIP.	TOTAL	TOTAL INCL O&P	
205	9920	20' to 25' high, add	R16132 -205					20%				205
	9930	25' to 30' high, add						25%				
	9940	30' to 35' high, add						30%				
	9950	35' to 40' high, add						35%				
	9960	Over 40' high, add						40%				
	9980	Allow. for cond. ftngs., 5% min.-20% max.										
210	0010	**CONDUIT** To 15' high, includes couplings only										210
	0200	Electric metallic tubing, 1/2"diameter	1 Elec	435	.018	L.F.	.26	.72		.98	1.37	
	0220	3/4" diameter		253	.032		.45	1.25		1.70	2.35	
	0240	1" diameter		207	.039		.79	1.52		2.31	3.13	
	0260	1-1/4"diameter		173	.046		1.24	1.82		3.06	4.08	
	0280	1-1/2" diameter		153	.052		1.60	2.06		3.66	4.82	
	0300	2" diameter		130	.062		2.02	2.42		4.44	5.85	
	0320	2-1/2" diameter		92	.087		4.58	3.43		8.01	10.15	
	0340	3" diameter	2 Elec	148	.108		4.81	4.26		9.07	11.65	
	0360	3-1/2" diameter		134	.119		6.10	4.70		10.80	13.70	
	0380	4" diameter		114	.140		6.50	5.55		12.05	15.35	
	0500	Steel rigid galvanized, 1/2" diameter	1 Elec	146	.055		1.16	2.16		3.32	4.49	
	0520	3/4" diameter		125	.064		1.42	2.52		3.94	5.30	
	0540	1" diameter		93	.086		2.17	3.39		5.56	7.45	
	0560	1-1/4" diameter		88	.091		2.90	3.58		6.48	8.55	
	0580	1-1/2" diameter		80	.100		3.41	3.94		7.35	9.60	
	0600	2" diameter		65	.123		4.56	4.85		9.41	12.20	
	0620	2-1/2" diameter		48	.167		8.55	6.55		15.10	19.15	
	0640	3" diameter	2 Elec	64	.250		10.55	9.85		20.40	.26.50	
	0660	3-1/2" diameter		60	.267		12.70	10.50		23.20	29.50	
	0680	4" diameter		52	.308		14.40	12.10		26.50	34	
	0700	5" diameter		50	.320		28.50	12.60		41.10	50.50	
	0720	6" diameter		48	.333		41.50	13.15		54.65	65.50	
	1000	Steel intermediate conduit (IMC), 1/2 diameter	1 Elec	155	.052		.91	2.03		2.94	4.02	
	1010	3/4" diameter		130	.062		1.09	2.42		3.51	4.81	
	1020	1" diameter		100	.080		1.66	3.15		4.81	6.50	
	1030	1-1/4" diameter		93	.086		2.10	3.39		5.49	7.35	
	1040	1-1/2" diameter		85	.094		2.52	3.71		6.23	8.25	
	1050	2" diameter		70	.114		3.39	4.50		7.89	10.45	
	1060	2-1/2" diameter		53	.151		7.55	5.95		13.50	17.15	
	1070	3" diameter	2 Elec	80	.200		9.40	7.90		17.30	22	
	1080	3-1/2" diameter		70	.229		10.15	9		19.15	24.50	
	1090	4" diameter		60	.267		12.75	10.50		23.25	29.50	
220	0010	**CONDUIT NIPPLES** With locknuts and bushings										220
	0100	Aluminum, 1/2" diameter, close	1 Elec	36	.222	Ea.	3.02	8.75		11.77	16.30	
	0120	1-1/2" long		36	.222		3.06	8.75		11.81	16.35	
	0140	2" long		36	.222		3.17	8.75		11.92	16.50	
	0160	2-1/2" long		36	.222		3.47	8.75		12.22	16.80	
	0180	3" long		36	.222		3.57	8.75		12.32	16.95	
	0200	3-1/2" long		36	.222		3.73	8.75		12.48	17.10	
	0220	4" long		36	.222		3.89	8.75		12.64	17.30	
	0240	5" long		36	.222		4.22	8.75		12.97	17.65	
	0260	6" long		36	.222		4.36	8.75		13.11	17.80	
	0280	8" long		36	.222		5.30	8.75		14.05	18.85	
	0300	10" long		36	.222		6.10	8.75		14.85	19.70	
	0320	12" long		36	.222		6.80	8.75		15.55	20.50	
	0340	3/4" diameter, close		32	.250		4.33	9.85		14.18	19.40	
	0360	1-1/2" long		32	.250		4.41	9.85		14.26	19.50	
	0380	2" long		32	.250		4.50	9.85		14.35	19.60	

16 ELECTRICAL

Important: See the Reference Section for critical supporting data - Reference Nos., Crews, & City Cost Indexes

16132	Conduit & Tubing	CREW	DAILY OUTPUT	LABOR-HOURS	UNIT	2004 BARE COSTS				TOTAL INCL O&P
						MAT.	LABOR	EQUIP.	TOTAL	
0400	2-1/2" long	1 Elec	32	.250	Ea.	4.62	9.85		14.47	19.75
0420	3" long		32	.250		4.80	9.85		14.65	19.95
0440	3-1/2" long		32	.250		4.89	9.85		14.74	20
0460	4" long		32	.250		5.05	9.85		14.90	20
0480	5" long		32	.250		5.55	9.85		15.40	21
0500	6" long		32	.250		5.95	9.85		15.80	21
0520	8" long		32	.250		7.10	9.85		16.95	22.50
0540	10" long		32	.250		7.90	9.85		17.75	23.50
0560	12" long		32	.250		9.10	9.85		18.95	24.50
0580	1" diameter, close		27	.296		6.55	11.65		18.20	24.50
0600	2" long		27	.296		6.80	11.65		18.45	25
0620	2-1/2" long		27	.296		7.05	11.65		18.70	25
0640	3" long		27	.296		7.30	11.65		18.95	25.50
0660	3-1/2" long		27	.296		7.70	11.65		19.35	26
0680	4" long		27	.296		8	11.65		19.65	26
0700	5" long		27	.296		8.70	11.65		20.35	27
0720	6" long		27	.296		9.50	11.65		21.15	28
0740	8" long		27	.296		10.75	11.65		22.40	29
0760	10" long		27	.296		12.45	11.65		24.10	31
0780	12" long		27	.296		13.95	11.65		25.60	32.50
0800	1-1/4" diameter, close		23	.348		9.10	13.70		22.80	30.50
0820	2" long		23	.348		9.15	13.70		22.85	30.50
0840	2-1/2" long		23	.348		9.45	13.70		23.15	31
0860	3" long		23	.348		9.95	13.70		23.65	31.50
0880	3-1/2" long		23	.348		10.50	13.70		24.20	32
0900	4" long		23	.348		10.95	13.70		24.65	32.50
0920	5" long		23	.348		11.85	13.70		25.55	33.50
0940	6" long		23	.348		12.75	13.70		26.45	34.50
0960	8" long		23	.348		14.65	13.70		28.35	36.50
0980	10" long		23	.348		16.55	13.70		30.25	38.50
1000	12" long		23	.348		18.35	13.70		32.05	40.50
1020	1-1/2" diameter, close		20	.400		12.90	15.75		28.65	37.50
1040	2" long		20	.400		13	15.75		28.75	38
1060	2-1/2" long		20	.400		13.25	15.75		29	38
1080	3" long		20	.400		13.85	15.75		29.60	39
1100	3-1/2" long		20	.400		15.10	15.75		30.85	40
1120	4" long		20	.400		15.10	15.75		30.85	40
1140	5" long		20	.400		16	15.75		31.75	41
1160	6" long		20	.400		17.10	15.75		32.85	42.50
1180	8" long		20	.400		19.45	15.75		35.20	45
1200	10" long		20	.400		21.50	15.75		37.25	47.50
1220	12" long		20	.400		24	15.75		39.75	50
1240	2" diameter, close		18	.444		18.30	17.50		35.80	46
1260	2-1/2" long		18	.444		18.70	17.50		36.20	46.50
1280	3" long		18	.444		19.30	17.50		36.80	47.50
1300	3-1/2" long		18	.444		20	17.50		37.50	48.50
1320	4" long		18	.444		20.50	17.50		38	49
1340	5" long		18	.444		22	17.50		39.50	50.50
1360	6" long		18	.444		23.50	17.50		41	52
1380	8" long		18	.444		26.50	17.50		44	55
1400	10" long		18	.444		29	17.50		46.50	58
1420	12" long		18	.444		32	17.50		49.50	61
1440	2-1/2" diameter, close		15	.533		38.50	21		59.50	74
1460	3" long		15	.533		39	21		60	74.50
1480	3-1/2" long		15	.533		40	21		61	75.50
1500	4" long		15	.533		41	21		62	76.50

ELECTRICAL 16

220

		CREW	DAILY OUTPUT	LABOR-HOURS	UNIT	2004 BARE COSTS MAT.	LABOR	EQUIP.	TOTAL	TOTAL INCL O&P	
16132	**Conduit & Tubing**										
220 1520	5" long	1 Elec	15	.533	Ea.	42.50	21		63.50	78	**220**
1540	6" long		15	.533		44	21		65	79.50	
1560	8" long		15	.533		48	21		69	84.50	
1580	10" long		15	.533		52.50	21		73.50	89	
1600	12" long		15	.533		56	21		77	93.50	
1620	3" diameter, close		12	.667		46	26.50		72.50	89.50	
1640	3" long		12	.667		47	26.50		73.50	91	
1660	3-1/2" long		12	.667		49.50	26.50		76	93.50	
1680	4" long		12	.667		50	26.50		76.50	94	
1700	5" long		12	.667		52.50	26.50		79	96.50	
1720	6" long		12	.667		55	26.50		81.50	99.50	
1740	8" long		12	.667		60.50	26.50		87	106	
1760	10" long		12	.667		66	26.50		92.50	112	
1780	12" long		12	.667		71.50	26.50		98	118	
1800	3-1/2" diameter, close		11	.727		89.50	28.50		118	141	
1820	4" long		11	.727		94.50	28.50		123	147	
1840	5" long		11	.727		97	28.50		125.50	150	
1860	6" long		11	.727		100	28.50		128.50	153	
1880	8" long		11	.727		106	28.50		134.50	160	
1900	10" long		11	.727		113	28.50		141.50	168	
1920	12" long		11	.727		120	28.50		148.50	175	
1940	4" diameter, close		9	.889		107	35		142	169	
1960	4" long		9	.889		111	35		146	174	
1980	5" long		9	.889		114	35		149	178	
2000	6" long		9	.889		119	35		154	182	
2020	8" long		9	.889		126	35		161	191	
2040	10" long		9	.889		134	35		169	200	
2060	12" long		9	.889		142	35		177	209	
2080	5" diameter, close		7	1.143		236	45		281	325	
2100	5" long		7	1.143		242	45		287	335	
2120	6" long		7	1.143		245	45		290	335	
2140	8" long		7	1.143		257	45		302	350	
2160	10" long		7	1.143		269	45		314	365	
2180	12" long		7	1.143		280	45		325	375	
2200	6" diameter, close		6	1.333		395	52.50		447.50	515	
2220	5" long		6	1.333		410	52.50		462.50	530	
2240	6" long		6	1.333		410	52.50		462.50	535	
2260	8" long		6	1.333		430	52.50		482.50	550	
2280	10" long		6	1.333		445	52.50		497.50	570	
2300	12" long		6	1.333		455	52.50		507.50	580	
2320	Rigid galvanized steel, 1/2" diameter, close		32	.250		2.03	9.85		11.88	16.90	
2340	1-1/2" long		32	.250		2.17	9.85		12.02	17.05	
2360	2" long		32	.250		2.29	9.85		12.14	17.15	
2380	2-1/2" long		32	.250		2.37	9.85		12.22	17.25	
2400	3" long		32	.250		2.46	9.85		12.31	17.35	
2420	3-1/2" long		32	.250		2.57	9.85		12.42	17.50	
2440	4" long		32	.250		2.67	9.85		12.52	17.60	
2460	5" long		32	.250		2.84	9.85		12.69	17.75	
2480	6" long		32	.250		3.14	9.85		12.99	18.10	
2500	8" long		32	.250		4.50	9.85		14.35	19.60	
2520	10" long		32	.250		4.96	9.85		14.81	20	
2540	12" long		32	.250		5.50	9.85		15.35	20.50	
2560	3/4" diameter, close		27	.296		2.80	11.65		14.45	20.50	
2580	2" long		27	.296		2.98	11.65		14.63	20.50	
2600	2-1/2" long		27	.296		3.10	11.65		14.75	21	
2620	3" long		27	.296		3.21	11.65		14.86	21	

Important: See the Reference Section for critical supporting data - Reference Nos., Crews, & City Cost Indexes

16132	Conduit & Tubing	CREW	DAILY OUTPUT	LABOR-HOURS	UNIT	2004 BARE COSTS				TOTAL INCL O&P	
						MAT.	LABOR	EQUIP.	TOTAL		
220	2640	3-1/2" long	1 Elec	27	.296	Ea.	3.27	11.65		14.92	21	220
	2660	4" long		27	.296		3.46	11.65		15.11	21	
	2680	5" long		27	.296		3.70	11.65		15.35	21.50	
	2700	6" long		27	.296		4.01	11.65		15.66	22	
	2720	8" long		27	.296		5.40	11.65		17.05	23.50	
	2740	10" long		27	.296		6.10	11.65		17.75	24	
	2760	12" long		27	.296		6.65	11.65		18.30	24.50	
	2780	1" diameter, close		23	.348		4.29	13.70		17.99	25	
	2800	2" long		23	.348		4.44	13.70		18.14	25.50	
	2820	2-1/2" long		23	.348		4.57	13.70		18.27	25.50	
	2840	3" long		23	.348		4.77	13.70		18.47	26	
	2860	3-1/2" long		23	.348		5	13.70		18.70	26	
	2880	4" long		23	.348		5.20	13.70		18.90	26	
	2900	5" long		23	.348		5.50	13.70		19.20	26.50	
	2920	6" long		23	.348		5.70	13.70		19.40	27	
	2940	8" long		23	.348		7.35	13.70		21.05	28.50	
	2960	10" long		23	.348		8.65	13.70		22.35	30	
	2980	12" long		23	.348		9.45	13.70		23.15	31	
	3000	1-1/4" diameter, close		20	.400		5.75	15.75		21.50	30	
	3020	2" long		20	.400		5.90	15.75		21.65	30	
	3040	3" long		20	.400		6.25	15.75		22	30.50	
	3060	3-1/2" long		20	.400		6.60	15.75		22.35	31	
	3080	4" long		20	.400		6.75	15.75		22.50	31	
	3100	5" long		20	.400		7.20	15.75		22.95	31.50	
	3120	6" long		20	.400		7.60	15.75		23.35	32	
	3140	8" long		20	.400		9.90	15.75		25.65	34.50	
	3160	10" long		20	.400		11.50	15.75		27.25	36	
	3180	12" long		20	.400		12.70	15.75		28.45	37.50	
	3200	1-1/2" diameter, close		18	.444		8.20	17.50		25.70	35	
	3220	2" long		18	.444		8.35	17.50		25.85	35	
	3240	2-1/2" long		18	.444		8.65	17.50		26.15	35.50	
	3260	3" long		18	.444		8.85	17.50		26.35	35.50	
	3280	3-1/2" long		18	.444		9.25	17.50		26.75	36	
	3300	4" long		18	.444		9.55	17.50		27.05	36.50	
	3320	5" long		18	.444		10	17.50		27.50	37	
	3340	6" long		18	.444		10.90	17.50		28.40	38	
	3360	8" long		18	.444		13.50	17.50		31	41	
	3380	10" long		18	.444		15.05	17.50		32.55	42.50	
	3400	12" long		18	.444		15.80	17.50		33.30	43.50	
	3420	2" diameter, close		16	.500		11.80	19.70		31.50	42.50	
	3440	2-1/2" long		16	.500		12.25	19.70		31.95	43	
	3460	3" long		16	.500		12.75	19.70		32.45	43.50	
	3480	3-1/2" long		16	.500		13.30	19.70		33	44	
	3500	4" long		16	.500		13.70	19.70		33.40	44.50	
	3520	5" long		16	.500		14.50	19.70		34.20	45.50	
	3540	6" long		16	.500		15.25	19.70		34.95	46.50	
	3560	8" long		16	.500		18	19.70		37.70	49.50	
	3580	10" long		16	.500		19.80	19.70		39.50	51.50	
	3600	12" long		16	.500		21.50	19.70		41.20	53	
	3620	2-1/2" diameter, close		13	.615		28.50	24.50		53	67.50	
	3640	3" long		13	.615		28.50	24.50		53	67.50	
	3660	3-1/2" long		13	.615		30	24.50		54.50	69	
	3680	4" long		13	.615		30.50	24.50		55	69.50	
	3700	5" long		13	.615		32	24.50		56.50	71.50	
	3720	6" long		13	.615		33.50	24.50		58	73	
	3740	8" long		13	.615		37.50	24.50		62	77.50	

ELECTRICAL 16

			DAILY	LABOR-				2004 BARE COSTS				TOTAL	
16132	**Conduit & Tubing**												
		CREW	OUTPUT	HOURS	UNIT	MAT.	LABOR	EQUIP.		TOTAL		INCL O&P	
220	3760	10" long	1 Elec	13	.615	Ea.	40.50	24.50			65		80.50	220
	3780	12" long		13	.615		43.50	24.50			68		84	
	3800	3" diameter, close		12	.667		35	26.50			61.50		77.50	
	3820	3" long		12	.667		37	26.50			63.50		79.50	
	3900	3-1/2" long		12	.667		37	26.50			63.50		79.50	
	3920	4" long		12	.667		38	26.50			64.50		80.50	
	3940	5" long		12	.667		39.50	26.50			66		82.50	
	3960	6" long		12	.667		41.50	26.50			68		84.50	
	3980	8" long		12	.667		46	26.50			72.50		89.50	
	4000	10" long		12	.667		49.50	26.50			76		93.50	
	4020	12" long		12	.667		54.50	26.50			81		99	
	4040	3-1/2" diameter, close		10	.800		61	31.50			92.50		114	
	4060	4" long		10	.800		64	31.50			95.50		118	
	4080	5" long		10	.800		66	31.50			97.50		120	
	4100	6" long		10	.800		68.50	31.50			100		122	
	4120	8" long		10	.800		73	31.50			104.50		127	
	4140	10" long		10	.800		77.50	31.50			109		133	
	4160	12" long		10	.800		82	31.50			113.50		138	
	4180	4" diameter, close		8	1		74	39.50			113.50		140	
	4200	4" long		8	1		77	39.50			116.50		143	
	4220	5" long		8	1		79.50	39.50			119		146	
	4240	6" long		8	1		82	39.50			121.50		149	
	4260	8" long		8	1		87	39.50			126.50		154	
	4280	10" long		8	1		93	39.50			132.50		161	
	4300	12" long		8	1		99	39.50			138.50		168	
	4320	5" diameter, close		6	1.333		148	52.50			200.50		241	
	4340	5" long		6	1.333		159	52.50			211.50		253	
	4360	6" long		6	1.333		163	52.50			215.50		257	
	4380	8" long		6	1.333		168	52.50			220.50		263	
	4400	10" long		6	1.333		176	52.50			228.50		271	
	4420	12" long		6	1.333		186	52.50			238.50		283	
	4440	6" diameter, close		5	1.600		220	63			283		335	
	4460	5" long		5	1.600		232	63			295		350	
	4480	6" long		5	1.600		236	63			299		355	
	4500	8" long		5	1.600		244	63			307		360	
	4520	10" long		5	1.600		257	63			320		375	
	4540	12" long		5	1.600		264	63			327		385	
	4560	Plastic coated, 40 mil thick, 1/2" diameter, 2" long		32	.250		9.20	9.85			19.05		25	
	4580	2-1/2" long		32	.250		10.25	9.85			20.10		26	
	4600	3" long		32	.250		10.30	9.85			20.15		26	
	4680	3-1/2" long		32	.250		11.35	9.85			21.20		27	
	4700	4" long		32	.250		11.45	9.85			21.30		27	
	4720	5" long		32	.250		11.55	9.85			21.40		27.50	
	4740	6" long		32	.250		11.85	9.85			21.70		27.50	
	4760	8" long		32	.250		12.25	9.85			22.10		28	
	4780	10" long		32	.250		12.70	9.85			22.55		28.50	
	4800	12" long		32	.250		13.15	9.85			23		29	
	4820	3/4" diameter, 2" long		26	.308		9.95	12.10			22.05		29	
	4840	2-1/2" long		26	.308		11	12.10			23.10		30	
	4860	3" long		26	.308		11.10	12.10			23.20		30.50	
	4880	3-1/2" long		26	.308		12.15	12.10			24.25		31.50	
	4900	4" long		26	.308		12.25	12.10			24.35		31.50	
	4920	5" long		26	.308		12.40	12.10			24.50		31.50	
	4940	6" long		26	.308		12.70	12.10			24.80		32	
	4960	8" long		26	.308		13.10	12.10			25.20		32.50	
	4980	10" long		26	.308		13.55	12.10			25.65		33	

T6 ELECTRICAL

Important: See the Reference Section for critical supporting data - Reference Nos., Crews, & City Cost Indexes

16132	Conduit & Tubing	CREW	DAILY OUTPUT	LABOR-HOURS	UNIT	2004 BARE COSTS				TOTAL INCL O&P	
						MAT.	LABOR	EQUIP.	TOTAL		
220											**220**
5000	12" long	1 Elec	26	.308	Ea.	14.05	12.10		26.15	33.50	
5020	1" diameter, 2" long		22	.364		11.40	14.35		25.75	34	
5040	2-1/2" long		22	.364		12.50	14.35		26.85	35.50	
5060	3" long		22	.364		12.65	14.35		27	35.50	
5080	3-1/2" long		22	.364		13.75	14.35		28.10	36.50	
5100	4" long		22	.364		13.90	14.35		28.25	37	
5120	5" long		22	.364		14.05	14.35		28.40	37	
5140	6" long		22	.364		14.35	14.35		28.70	37.50	
5160	8" long		22	.364		15	14.35		29.35	38	
5180	10" long		22	.364		15.65	14.35		30	38.50	
5200	12" long		22	.364		17.25	14.35		31.60	40.50	
5220	1-1/4" diameter, 2" long		18	.444		14.40	17.50		31.90	42	
5240	2-1/2" long		18	.444		15.55	17.50		33.05	43	
5260	3" long		18	.444		15.70	17.50		33.20	43.50	
5280	3-1/2" long		18	.444		16.35	17.50		33.85	44	
5300	4" long		18	.444		16.95	17.50		34.45	44.50	
5320	5" long		18	.444		17.25	17.50		34.75	45	
5340	6" long		18	.444		17.80	17.50		35.30	45.50	
5360	8" long		18	.444		17.95	17.50		35.45	46	
5380	10" long		18	.444		19.90	17.50		37.40	48	
5400	12" long		18	.444		23	17.50		40.50	51.50	
5420	1-1/2" diameter, 2" long		16	.500		16.85	19.70		36.55	48	
5440	2-1/2" long		16	.500		18.05	19.70		37.75	49.50	
5460	3" long		16	.500		18.20	19.70		37.90	49.50	
5480	3-1/2" long		16	.500		18.90	19.70		38.60	50.50	
5500	4" long		16	.500		19.60	19.70		39.30	51	
5520	5" long		16	.500		19.90	19.70		39.60	51.50	
5540	6" long		16	.500		21.50	19.70		41.20	53	
5560	8" long		16	.500		22.50	19.70		42.20	54.50	
5580	10" long		16	.500		26.50	19.70		46.20	59	
5600	12" long		16	.500		30.50	19.70		50.20	63	
5620	2" diameter, 2-1/2" long		14	.571		23	22.50		45.50	58.50	
5640	3" long		14	.571		23	22.50		45.50	59	
5660	3-1/2" long		14	.571		24.50	22.50		47	60.50	
5680	4" long		14	.571		26.50	22.50		49	63	
5700	5" long		14	.571		26.50	22.50		49	63	
5720	6" long		14	.571		27.50	22.50		50	64	
5740	8" long		14	.571		30.50	22.50		53	67	
5760	10" long		14	.571		36	22.50		58.50	73	
5780	12" long		14	.571		40	22.50		62.50	77.50	
5800	2-1/2" diameter, 3-1/2" long		12	.667		45	26.50		71.50	88.50	
5820	4" long		12	.667		46	26.50		72.50	89.50	
5840	5" long		12	.667		52	26.50		78.50	96.50	
5860	6" long		12	.667		53	26.50		79.50	97.50	
5880	8" long		12	.667		59.50	26.50		86	104	
5900	10" long		12	.667		64	26.50		90.50	110	
5920	12" long		12	.667		71.50	26.50		98	118	
5940	3" diameter, 3-1/2" long		11	.727		54	28.50		82.50	102	
5960	4" long		11	.727		56	28.50		84.50	105	
5980	5" long		11	.727		59.50	28.50		88	108	
6000	6" long		11	.727		64	28.50		92.50	113	
6020	8" long		11	.727		70.50	28.50		99	120	
6040	10" long		11	.727		79.50	28.50		108	130	
6060	12" long		11	.727		90.50	28.50		119	142	
6080	3-1/2" diameter, 4" long		9	.889		87.50	35		122.50	148	
6100	5" long		9	.889		88	35		123	149	

ELECTRICAL 16

16132 | Conduit & Tubing

		CREW	DAILY OUTPUT	LABOR-HOURS	UNIT	2004 BARE COSTS				TOTAL INCL O&P
						MAT.	LABOR	EQUIP.	TOTAL	
220											**220**
6120	6" long	1 Elec	9	.889	Ea.	92.50	35		127.50	154
6140	8" long		9	.889		103	35		138	165
6160	10" long		9	.889		114	35		149	177
6180	12" long		9	.889		120	35		155	184
6200	4" diameter, 4" long		7.50	1.067		101	42		143	174
6220	5" long		7.50	1.067		103	42		145	177
6240	6" long		7.50	1.067		110	42		152	184
6260	8" long		7.50	1.067		121	42		163	196
6280	10" long		7.50	1.067		139	42		181	216
6300	12" long		7.50	1.067		145	42		187	222
6320	5" diameter, 5" long		5.50	1.455		191	57.50		248.50	295
6340	6" long		5.50	1.455		201	57.50		258.50	305
6360	8" long		5.50	1.455		208	57.50		265.50	315
6380	10" long		5.50	1.455		222	57.50		279.50	330
6400	12" long		5.50	1.455		225	57.50		282.50	330
6420	6" diameter, 5" long		4.50	1.778		274	70		344	405
6440	6" long		4.50	1.778		284	70		354	420
6460	8" long		4.50	1.778		294	70		364	430
6480	10" long		4.50	1.778		310	70		380	445
6500	12" long		4.50	1.778		325	70		395	460
230	0010	**CONDUIT IN CONCRETE SLAB** Including terminations,									**230**
	0020	fittings and supports	R16132 -230							
3230	PVC, schedule 40, 1/2" diameter	1 Elec	270	.030	L.F.	.43	1.17		1.60	2.22
3250	3/4" diameter		230	.035		.52	1.37		1.89	2.61
3270	1" diameter		200	.040		.69	1.58		2.27	3.10
3300	1-1/4" diameter		170	.047		.97	1.85		2.82	3.83
3330	1-1/2" diameter		140	.057		1.20	2.25		3.45	4.66
3350	2" diameter		120	.067		1.51	2.63		4.14	5.55
3370	2-1/2" diameter		90	.089		2.49	3.50		5.99	7.95
3400	3" diameter	2 Elec	160	.100		3.26	3.94		7.20	9.45
3430	3-1/2" diameter		120	.133		4.32	5.25		9.57	12.55
3440	4" diameter		100	.160		4.91	6.30		11.21	14.80
3450	5" diameter		80	.200		7.35	7.90		15.25	19.75
3460	6" diameter		60	.267		10.90	10.50		21.40	27.50
3530	Sweeps, 1" diameter, 30" radius	1 Elec	32	.250	Ea.	9.65	9.85		19.50	25.50
3550	1-1/4" diameter		24	.333		12.80	13.15		25.95	33.50
3570	1-1/2" diameter		21	.381		16	15		31	40
3600	2" diameter		18	.444		20	17.50		37.50	48
3630	2-1/2" diameter		14	.571		25	22.50		47.50	61
3650	3" diameter		10	.800		31	31.50		62.50	81.50
3670	3-1/2" diameter		8	1		39	39.50		78.50	102
3700	4" diameter		7	1.143		48.50	45		93.50	121
3710	5" diameter		6	1.333		75.50	52.50		128	161
3730	Couplings, 1/2" diameter					.41			.41	.45
3750	3/4" diameter					.50			.50	.55
3770	1" diameter					.77			.77	.85
3800	1-1/4" diameter					1.02			1.02	1.12
3830	1-1/2" diameter					1.40			1.40	1.54
3850	2" diameter					1.86			1.86	2.05
3870	2-1/2" diameter					3.30			3.30	3.63
3900	3" diameter					5.40			5.40	5.95
3930	3-1/2" diameter					6			6	6.60
3950	4" diameter					8.35			8.35	9.20
3960	5" diameter					21			21	23.50
3970	6" diameter					27			27	30
4030	End bells 1" diameter, PVC	1 Elec	60	.133		3.45	5.25		8.70	11.60

Important: See the Reference Section for critical supporting data - Reference Nos., Crews, & City Cost Indexes

				DAILY	LABOR-		2004 BARE COSTS				TOTAL	
16132	**Conduit & Tubing**		CREW	OUTPUT	HOURS	UNIT	MAT.	LABOR	EQUIP.	TOTAL	INCL O&P	
230	4050	1-1/4" diameter	R16132 -230	1 Elec	53	.151	Ea.	4.25	5.95		10.20	13.55	**230**
	4100	1-1/2" diameter			48	.167		4.25	6.55		10.80	14.45	
	4150	2" diameter			34	.235		6.35	9.25		15.60	21	
	4170	2-1/2" diameter			27	.296		7.10	11.65		18.75	25	
	4200	3" diameter			20	.400		7.45	15.75		23.20	31.50	
	4250	3-1/2" diameter			16	.500		8.20	19.70		27.90	38.50	
	4300	4" diameter			14	.571		8.90	22.50		31.40	43.50	
	4310	5" diameter			12	.667		13.95	26.50		40.45	54.50	
	4320	6" diameter			9	.889		15.30	35		50.30	69	
	4350	Rigid galvanized steel, 1/2" diameter			200	.040	L.F.	1.43	1.58		3.01	3.92	
	4400	3/4" diameter			170	.047		1.72	1.85		3.57	4.65	
	4450	1" diameter			130	.062		2.53	2.42		4.95	6.40	
	4500	1-1/4" diameter			110	.073		3.33	2.87		6.20	7.95	
	4600	1-1/2" diameter			100	.080		3.93	3.15		7.08	9	
	4800	2" diameter			90	.089		5.25	3.50		8.75	10.95	
240	0010	**CONDUIT IN TRENCH** Includes terminations and fittings	R16132 -240										**240**
	0020	Does not include excavation or backfill, see div. 02315											
	0200	Rigid galvanized steel, 2" diameter		1 Elec	150	.053	L.F.	5.05	2.10		7.15	8.70	
	0400	2-1/2" diameter		"	100	.080		9.55	3.15		12.70	15.20	
	0600	3" diameter		2 Elec	160	.100		11.95	3.94		15.89	19	
	0800	3-1/2" diameter			140	.114		14.90	4.50		19.40	23	
	1000	4" diameter			100	.160		16.95	6.30		23.25	28	
	1200	5" diameter			80	.200		34.50	7.90		42.40	49.50	
	1400	6" diameter			60	.267		51	10.50		61.50	71.50	
250	0010	**CONDUIT FITTINGS FOR RIGID GALVANIZED STEEL**											**250**
	0050	Standard, locknuts, 1/2" diameter					Ea.	.19			.19	.21	
	0100	3/4" diameter						.33			.33	.36	
	0300	1" diameter						.53			.53	.58	
	0500	1-1/4" diameter						.72			.72	.79	
	0700	1-1/2" diameter						1.19			1.19	1.31	
	1000	2" diameter						1.73			1.73	1.90	
	1030	2-1/2" diameter						4.84			4.84	5.30	
	1050	3" diameter						6.20			6.20	6.80	
	1070	3-1/2" diameter						10.75			10.75	11.85	
	1100	4" diameter						13.05			13.05	14.40	
	1110	5" diameter						28			28	31	
	1120	6" diameter						48			48	52.50	
	1130	Bushings, plastic, 1/2" diameter		1 Elec	40	.200		.19	7.90		8.09	11.90	
	1150	3/4" diameter			32	.250		.31	9.85		10.16	15	
	1170	1" diameter			28	.286		.51	11.25		11.76	17.30	
	1200	1-1/4" diameter			24	.333		.77	13.15		13.92	20.50	
	1230	1-1/2" diameter			18	.444		1.05	17.50		18.55	27	
	1250	2" diameter			15	.533		1.95	21		22.95	33.50	
	1270	2-1/2" diameter			13	.615		4.53	24.50		29.03	41	
	1300	3" diameter			12	.667		4.68	26.50		31.18	44	
	1330	3-1/2" diameter			11	.727		6.15	28.50		34.65	49.50	
	1350	4" diameter			9	.889		7.60	35		42.60	60.50	
	1360	5" diameter			7	1.143		16.50	45		61.50	85	
	1370	6" diameter			5	1.600		31.50	63		94.50	129	
	1390	Steel, 1/2" diameter			40	.200		.46	7.90		8.36	12.20	
	1400	3/4" diameter			32	.250		.60	9.85		10.45	15.30	
	1430	1" diameter			28	.286		.91	11.25		12.16	17.75	
	1450	Steel insulated, 1-1/4" diameter			24	.333		5.45	13.15		18.60	25.50	
	1470	1-1/2" diameter			18	.444		6.90	17.50		24.40	33.50	

ELECTRICAL 16

ELECTRICAL

16

		CREW	DAILY OUTPUT	LABOR-HOURS	UNIT	2004 BARE COSTS				TOTAL INCL O&P
16132	**Conduit & Tubing**					MAT.	LABOR	EQUIP.	TOTAL	
1500	2" diameter	1 Elec	15	.533	Ea.	10	21		31	42.50
1530	2-1/2" diameter		13	.615		18.35	24.50		42.85	56
1550	3" diameter		12	.667		25	26.50		51.50	66.50
1570	3-1/2" diameter		11	.727		31.50	28.50		60	77.50
1600	4" diameter		9	.889		40	35		75	96
1610	5" diameter		7	1.143		69.50	45		114.50	144
1620	6" diameter		5	1.600		107	63		170	212
1630	Sealing locknuts, 1/2" diameter		40	.200		1.34	7.90		9.24	13.15
1650	3/4" diameter		32	.250		1.57	9.85		11.42	16.40
1670	1" diameter		28	.286		2.29	11.25		13.54	19.25
1700	1-1/4" diameter		24	.333		3.92	13.15		17.07	24
1730	1-1/2" diameter		18	.444		4.94	17.50		22.44	31.50
1750	2" diameter		15	.533		6.25	21		27.25	38.50
1760	Grounding bushing, insulated, 1/2" diameter		32	.250		4.25	9.85		14.10	19.35
1770	3/4" diameter		28	.286		5.35	11.25		16.60	22.50
1780	1" diameter		20	.400		6.05	15.75		21.80	30
1800	1-1/4" diameter		18	.444		7.60	17.50		25.10	34.50
1830	1-1/2" diameter		16	.500		8.20	19.70		27.90	38.50
1850	2" diameter		13	.615		10	24.50		34.50	47
1870	2-1/2" diameter		12	.667		18.10	26.50		44.60	59
1900	3" diameter		11	.727		23	28.50		51.50	68
1930	3-1/2" diameter		9	.889		30.50	35		65.50	85.50
1950	4" diameter		8	1		32.50	39.50		72	94.50
1960	5" diameter		6	1.333		80.50	52.50		133	167
1970	6" diameter		4	2		123	79		202	252
1990	Coupling, with set screw, 1/2" diameter		50	.160		4.20	6.30		10.50	14
2000	3/4" diameter		40	.200		5.30	7.90		13.20	17.55
2030	1" diameter		35	.229		8.75	9		17.75	23
2050	1-1/4" diameter		28	.286		14.90	11.25		26.15	33
2070	1-1/2" diameter		23	.348		19.20	13.70		32.90	41.50
2090	2" diameter		20	.400		42.50	15.75		58.25	70.50
2100	2-1/2" diameter		18	.444		92	17.50		109.50	127
2110	3" diameter		15	.533		109	21		130	152
2120	3-1/2" diameter		12	.667		156	26.50		182.50	210
2130	4" diameter		10	.800		203	31.50		234.50	270
2140	5" diameter		9	.889		263	35		298	340
2150	6" diameter		8	1		340	39.50		379.50	435
2160	Box connector with set screw, plain, 1/2" diameter		70	.114		2.75	4.50		7.25	9.75
2170	3/4" diameter		60	.133		3.90	5.25		9.15	12.10
2180	1" diameter		50	.160		6.20	6.30		12.50	16.20
2190	Insulated, 1-1/4" diameter		40	.200		12.75	7.90		20.65	26
2200	1-1/2" diameter		30	.267		18.30	10.50		28.80	35.50
2210	2" diameter		20	.400		36.50	15.75		52.25	63.50
2220	2-1/2" diameter		18	.444		88.50	17.50		106	123
2230	3" diameter		15	.533		117	21		138	161
2240	3-1/2" diameter		12	.667		165	26.50		191.50	220
2250	4" diameter		10	.800		217	31.50		248.50	285
2260	5" diameter		9	.889		261	35		296	340
2270	6" diameter		8	1		335	39.50		374.50	430
2280	LB, LR or LL fittings & covers, 1/2" diameter		16	.500		8.70	19.70		28.40	39
2290	3/4" diameter		13	.615		10.50	24.50		35	47.50
2300	1" diameter		11	.727		15.50	28.50		44	59.50
2330	1-1/4" diameter		8	1		23	39.50		62.50	84
2350	1-1/2" diameter		6	1.333		29	52.50		81.50	110
2370	2" diameter		5	1.600		48	63		111	147
2380	2-1/2" diameter		4	2		97	79		176	223

250

Important: See the Reference Section for critical supporting data - Reference Nos., Crews, & City Cost Indexes

16132	Conduit & Tubing	CREW	DAILY OUTPUT	LABOR-HOURS	UNIT	2004 BARE COSTS				TOTAL INCL O&P
						MAT.	LABOR	EQUIP.	TOTAL	
250	2390	3" diameter	1 Elec	3.50	2.286	Ea.	125	90		215	272
	2400	3-1/2" diameter		3	2.667		201	105		306	375
	2410	4" diameter		2.50	3.200		233	126		359	445
	2420	T fittings, with cover, 1/2" diameter		12	.667		10.50	26.50		37	50.50
	2430	3/4" diameter		11	.727		12.55	28.50		41.05	56.50
	2440	1" diameter		9	.889		18.60	35		53.60	72.50
	2450	1-1/4" diameter		6	1.333		26	52.50		78.50	107
	2470	1-1/2" diameter		5	1.600		33.50	63		96.50	131
	2500	2" diameter		4	2		53	79		132	175
	2510	2-1/2" diameter		3.50	2.286		101	90		191	245
	2520	3" diameter		3	2.667		142	105		247	310
	2530	3-1/2" diameter		2.50	3.200		238	126		364	450
	2540	4" diameter		2	4		251	158		409	510
	2550	Nipples chase, plain, 1/2" diameter		40	.200		.82	7.90		8.72	12.60
	2560	3/4" diameter		32	.250		1.03	9.85		10.88	15.80
	2570	1" diameter		28	.286		2.08	11.25		13.33	19.05
	2600	Insulated, 1-1/4" diameter		24	.333		7.80	13.15		20.95	28
	2630	1-1/2" diameter		18	.444		10.30	17.50		27.80	37.50
	2650	2" diameter		15	.533		15.50	21		36.50	48.50
	2660	2-1/2" diameter		12	.667		37	26.50		63.50	79.50
	2670	3" diameter		10	.800		40	31.50		71.50	91
	2680	3-1/2" diameter		9	.889		56	35		91	114
	2690	4" diameter		8	1		86	39.50		125.50	153
	2700	5" diameter		7	1.143		201	45		246	289
	2710	6" diameter		6	1.333		310	52.50		362.50	425
	2720	Nipples offset, plain, 1/2" diameter		40	.200		4.05	7.90		11.95	16.15
	2730	3/4" diameter		32	.250		4.23	9.85		14.08	19.30
	2740	1" diameter		24	.333		5.45	13.15		18.60	25.50
	2750	Insulated, 1-1/4" diameter		20	.400		29.50	15.75		45.25	55.50
	2760	1-1/2" diameter		18	.444		36	17.50		53.50	65.50
	2770	2" diameter		16	.500		57	19.70		76.70	92
	2780	3" diameter		14	.571		115	22.50		137.50	161
	2850	Coupling, expansion, 1/2" diameter		12	.667		37.50	26.50		64	80.50
	2880	3/4" diameter		10	.800		44	31.50		75.50	95.50
	2900	1" diameter		8	1		51.50	39.50		91	116
	2920	1-1/4" diameter		6.40	1.250		70	49.50		119.50	151
	2940	1-1/2" diameter		5.30	1.509		95.50	59.50		155	194
	2960	2" diameter		4.60	1.739		142	68.50		210.50	258
	2980	2-1/2" diameter		3.60	2.222		225	87.50		312.50	380
	3000	3" diameter		3	2.667		277	105		382	460
	3020	3-1/2" diameter		2.80	2.857		405	113		518	615
	3040	4" diameter		2.40	3.333		485	131		616	725
	3060	5" diameter		2	4		815	158		973	1,125
	3080	6" diameter		1.80	4.444		1,625	175		1,800	2,050
	3100	Expansion deflection, 1/2" diameter		12	.667		134	26.50		160.50	186
	3120	3/4" diameter		12	.667		144	26.50		170.50	197
	3140	1" diameter		10	.800		182	31.50		213.50	247
	3160	1-1/4" diameter		6.40	1.250		226	49.50		275.50	325
	3180	1-1/2" diameter		5.30	1.509		239	59.50		298.50	350
	3200	2" diameter		4.60	1.739		310	68.50		378.50	440
	3220	2-1/2" diameter		3.60	2.222		435	87.50		522.50	610
	3240	3" diameter		3	2.667		550	105		655	760
	3260	3-1/2" diameter		2.80	2.857		655	113		768	885
	3280	4" diameter		2.40	3.333		730	131		861	995
	3300	5" diameter		2	4		1,200	158		1,358	1,550
	3320	6" diameter		1.80	4.444		2,000	175		2,175	2,450

ELECTRICAL 16

250

		CREW	DAILY OUTPUT	LABOR-HOURS	UNIT	2004 BARE COSTS				TOTAL INCL O&P
16132	**Conduit & Tubing**					MAT.	LABOR	EQUIP.	TOTAL	
3340	Ericson, 1/2" diameter	1 Elec	16	.500	Ea.	3.08	19.70		22.78	33
3360	3/4" diameter		14	.571		3.96	22.50		26.46	38
3380	1" diameter		11	.727		7.30	28.50		35.80	50.50
3400	1-1/4" diameter		8	1		15.55	39.50		55.05	75.50
3420	1-1/2" diameter		7	1.143		18.65	45		63.65	87.50
3440	2" diameter		5	1.600		37.50	63		100.50	135
3460	2-1/2" diameter		4	2		91	79		170	217
3480	3" diameter		3.50	2.286		138	90		228	286
3500	3-1/2" diameter		3	2.667		230	105		335	410
3520	4" diameter		2.70	2.963		270	117		387	470
3540	5" diameter		2.50	3.200		500	126		626	740
3560	6" diameter		2.30	3.478		665	137		802	935
3580	Split, 1/2" diameter		32	.250		3.75	9.85		13.60	18.80
3600	3/4" diameter		27	.296		4.65	11.65		16.30	22.50
3620	1" diameter		20	.400		9.10	15.75		24.85	33.50
3640	1-1/4" diameter		16	.500		13.10	19.70		32.80	44
3660	1-1/2" diameter		14	.571		16.40	22.50		38.90	51.50
3680	2" diameter		12	.667		33	26.50		59.50	75
3700	2-1/2" diameter		10	.800		76	31.50		107.50	131
3720	3" diameter		9	.889		114	35		149	177
3740	3-1/2" diameter		8	1		184	39.50		223.50	261
3760	4" diameter		7	1.143		217	45		262	305
3780	5" diameter		6	1.333		300	52.50		352.50	410
3800	6" diameter		5	1.600		405	63		468	540
4600	Reducing bushings, 3/4" to 1/2" diameter		54	.148		.91	5.85		6.76	9.70
4620	1" to 3/4" diameter		46	.174		1.39	6.85		8.24	11.75
4640	1-1/4" to 1" diameter		40	.200		3.24	7.90		11.14	15.25
4660	1-1/2" to 1-1/4" diameter		36	.222		4.66	8.75		13.41	18.15
4680	2" to 1-1/2" diameter		32	.250		10.55	9.85		20.40	26.50
4740	2-1/2" to 2" diameter		30	.267		15.20	10.50		25.70	32.50
4760	3" to 2-1/2" diameter		28	.286		19.75	11.25		31	38.50
4800	Through-wall seal, 1/2" diameter		8	1		127	39.50		166.50	199
4820	3/4" diameter		7.50	1.067		127	42		169	203
4840	1" diameter		6.50	1.231		127	48.50		175.50	212
4860	1-1/4" diameter		5.50	1.455		176	57.50		233.50	279
4880	1-1/2" diameter		5	1.600		176	63		239	288
4900	2" diameter		4.20	1.905		191	75		266	320
4920	2-1/2" diameter		3.50	2.286		204	90		294	360
4940	3" diameter		3	2.667		204	105		309	380
4960	3-1/2" diameter		2.50	3.200		320	126		446	540
4980	4" diameter		2	4		365	158		523	640
5000	5" diameter		1.50	5.333		465	210		675	830
5020	6" diameter		1	8		465	315		780	985
5100	Cable supports, 2 or more wires									
5120	1-1/2" diameter	1 Elec	8	1	Ea.	56	39.50		95.50	120
5140	2" diameter		6	1.333		82.50	52.50		135	169
5160	2-1/2" diameter		4	2		92.50	79		171.50	219
5180	3" diameter		3.50	2.286		122	90		212	268
5200	3-1/2" diameter		2.60	3.077		165	121		286	360
5220	4" diameter		2	4		206	158		364	460
5240	5" diameter		1.50	5.333		385	210		595	740
5260	6" diameter		1	8		645	315		960	1,175
5280	Service entrance cap, 1/2" diameter		16	.500		6.80	19.70		26.50	37
5300	3/4" diameter		13	.615		7.85	24.50		32.35	44.50
5320	1" diameter		10	.800		7.90	31.50		39.40	55.50
5340	1-1/4" diameter		8	1		13.70	39.50		53.20	73.50

250

	16132	Conduit & Tubing	CREW	DAILY OUTPUT	LABOR-HOURS	UNIT	2004 BARE COSTS				TOTAL INCL O&P	
							MAT.	LABOR	EQUIP.	TOTAL		
250	5360	1-1/2" diameter	1 Elec	6.50	1.231	Ea.	19.20	48.50		67.70	93	**250**
	5380	2" diameter		5.50	1.455		37.50	57.50		95	127	
	5400	2-1/2" diameter		4	2		117	79		196	246	
	5420	3" diameter		3.40	2.353		178	92.50		270.50	335	
	5440	3-1/2" diameter		3	2.667		218	105		323	395	
	5460	4" diameter		2.70	2.963		330	117		447	540	
	5600	Fire stop fittings, to 3/4" diameter		24	.333		84	13.15		97.15	112	
	5610	1" diameter		22	.364		103	14.35		117.35	135	
	5620	1-1/2" diameter		20	.400		130	15.75		145.75	167	
	5640	2" diameter		16	.500		201	19.70		220.70	251	
	5660	3" diameter		12	.667		252	26.50		278.50	315	
	5680	4" diameter		10	.800		335	31.50		366.50	410	
	5700	6" diameter		8	1		575	39.50		614.50	695	
	5750	90° pull elbows steel, female, 1/2" diameter		16	.500		7.85	19.70		27.55	38	
	5760	3/4" diameter		13	.615		9.15	24.50		33.65	46	
	5780	1" diameter		11	.727		15.50	28.50		44	59.50	
	5800	1-1/4" diameter		8	1		23	39.50		62.50	84	
	5820	1-1/2" diameter		6	1.333		32.50	52.50		85	114	
	5840	2" diameter	▼	5	1.600	▼	52.50	63		115.50	152	
	6000	Explosion proof, flexible coupling										
	6010	1/2" diameter, 4" long	1 Elec	12	.667	Ea.	87.50	26.50		114	135	
	6020	6" long		12	.667		97	26.50		123.50	146	
	6050	12" long		12	.667		121	26.50		147.50	172	
	6070	18" long		12	.667		152	26.50		178.50	206	
	6090	24" long		12	.667		179	26.50		205.50	236	
	6110	30" long		12	.667		207	26.50		233.50	267	
	6130	36" long		12	.667		235	26.50		261.50	298	
	6140	3/4" diameter, 4" long		10	.800		105	31.50		136.50	163	
	6150	6" long		10	.800		119	31.50		150.50	178	
	6180	12" long		10	.800		156	31.50		187.50	219	
	6200	18" long		10	.800		195	31.50		226.50	262	
	6220	24" long		10	.800		235	31.50		266.50	305	
	6240	30" long		10	.800		272	31.50		303.50	345	
	6260	36" long		10	.800		310	31.50		341.50	390	
	6270	1" diameter, 6" long		8	1		212	39.50		251.50	292	
	6300	12" long		8	1		268	39.50		307.50	355	
	6320	18" long		8	1		320	39.50		359.50	415	
	6340	24" long		8	1		375	39.50		414.50	475	
	6360	30" long		8	1		515	39.50		554.50	625	
	6380	36" long		8	1		580	39.50		619.50	700	
	6390	1-1/4" diameter, 12" long		6.40	1.250		405	49.50		454.50	525	
	6410	18" long		6.40	1.250		555	49.50		604.50	685	
	6430	24" long		6.40	1.250		660	49.50		709.50	805	
	6450	30" long		6.40	1.250		750	49.50		799.50	900	
	6470	36" long		6.40	1.250		940	49.50		989.50	1,100	
	6480	1-1/2" diameter, 12" long		5.30	1.509		540	59.50		599.50	685	
	6500	18" long		5.30	1.509		635	59.50		694.50	785	
	6520	24" long		5.30	1.509		720	59.50		779.50	880	
	6540	30" long		5.30	1.509		965	59.50		1,024.50	1,150	
	6560	36" long		5.30	1.509		1,100	59.50		1,159.50	1,325	
	6570	2" diameter, 12" long		4.60	1.739		830	68.50		898.50	1,025	
	6590	18" long		4.60	1.739		1,025	68.50		1,093.50	1,225	
	6610	24" long		4.60	1.739		1,125	68.50		1,193.50	1,325	
	6630	30" long		4.60	1.739		1,225	68.50		1,293.50	1,450	
	6650	36" long		4.60	1.739		1,500	68.50		1,568.50	1,750	
	7000	Close up plug, 1/2" diameter, explosion proof	▼	40	.200	▼	1.80	7.90		9.70	13.70	

ELECTRICAL 16

16132 | Conduit & Tubing

		CREW	DAILY OUTPUT	LABOR-HOURS	UNIT	2004 BARE COSTS				TOTAL INCL O&P
						MAT.	LABOR	EQUIP.	TOTAL	
250	7010	3/4" diameter	1 Elec	32	.250	Ea.	2	9.85		11.85	16.85	**250**
	7020	1" diameter		28	.286		2.50	11.25		13.75	19.50
	7030	1-1/4" diameter		24	.333		2.65	13.15		15.80	22.50
	7040	1-1/2" diameter		18	.444		3.80	17.50		21.30	30
	7050	2" diameter		15	.533		6.45	21		27.45	38.50
	7060	2-1/2" diameter		13	.615		10.50	24.50		35	47.50
	7070	3" diameter		12	.667		15.35	26.50		41.85	56
	7080	3-1/2" diameter		11	.727		19.20	28.50		47.70	63.50
	7090	4" diameter		9	.889		25.50	35		60.50	80.50
	7091	Elbow female, 45°, 1/2"		16	.500		7.55	19.70		27.25	38
	7092	3/4"		13	.615		8.15	24.50		32.65	45
	7093	1"		11	.727		11.50	28.50		40	55
	7094	1-1/4"		8	1		16.20	39.50		55.70	76.50
	7095	1-1/2"		6	1.333		16.75	52.50		69.25	96.50
	7096	2"		5	1.600		20.50	63		83.50	117
	7097	2-1/2"		4.50	1.778		56	70		126	166
	7098	3"		4.20	1.905		60.50	75		135.50	179
	7099	3-1/2"		4	2		92	79		171	218
	7100	4"		3.80	2.105		112	83		195	246
	7101	90°, 1/2"		16	.500		7.15	19.70		26.85	37.50
	7102	3/4"		13	.615		7.80	24.50		32.30	44.50
	7103	1"		11	.727		10.55	28.50		39.05	54
	7104	1-1/4"		8	1		17.25	39.50		56.75	77.50
	7105	1-1/2"		6	1.333		25	52.50		77.50	106
	7106	2"		5	1.600		39.50	63		102.50	138
	7107	2-1/2"		4.50	1.778		72	70		142	183
	7110	Elbows 90°, long male & female, 1/2" diameter, explosion proof		16	.500		10.55	19.70		30.25	41
	7120	3/4" diameter		13	.615		11.85	24.50		36.35	49
	7130	1" diameter		11	.727		17.15	28.50		45.65	61.50
	7140	1-1/4" diameter		8	1		19.70	39.50		59.20	80
	7150	1-1/2" diameter		6	1.333		29.50	52.50		82	111
	7160	2" diameter		5	1.600		44.50	63		107.50	143
	7170	Capped elbow, 1/2" diameter, explosion proof		11	.727		10.30	28.50		38.80	54
	7180	3/4" diameter		8	1		11.50	39.50		51	71
	7190	1" diameter		6	1.333		14.25	52.50		66.75	93.50
	7200	1-1/4" diameter		5	1.600		33.50	63		96.50	131
	7210	Pulling elbow, 1/2" diameter, explosion proof		11	.727		49.50	28.50		78	97
	7220	3/4" diameter		8	1		49.50	39.50		89	113
	7230	1" diameter		6	1.333		130	52.50		182.50	221
	7240	1-1/4" diameter		5	1.600		140	63		203	248
	7250	1-1/2" diameter		5	1.600		198	63		261	310
	7260	2" diameter		4	2		206	79		285	345
	7270	2-1/2" diameter		3.50	2.286		470	90		560	655
	7280	3" diameter		3	2.667		455	105		560	655
	7290	3-1/2" diameter		2.50	3.200		850	126		976	1,125
	7300	4" diameter		2.20	3.636		830	143		973	1,125
	7310	LB conduit body, 1/2" diameter		11	.727		30.50	28.50		59	76
	7320	3/4" diameter		8	1		35	39.50		74.50	97
	7330	T conduit body, 1/2" diameter		9	.889		32	35		67	87
	7340	3/4" diameter		6	1.333		37.50	52.50		90	120
	7350	Explosion proof, round box w/cover, 3 threaded hubs, 1/2" diameter		8	1		33	39.50		72.50	94.50
	7351	3/4" diameter		8	1		36	39.50		75.50	98
	7352	1" diameter		7.50	1.067		44.50	42		86.50	111
	7353	1-1/4" diameter		7	1.143		81	45		126	156
	7354	1-1/2" diameter		7	1.143		148	45		193	230
	7355	2" diameter		6	1.333		153	52.50		205.50	246

16132	Conduit & Tubing	CREW	DAILY OUTPUT	LABOR-HOURS	UNIT	2004 BARE COSTS				TOTAL INCL O&P	
						MAT.	LABOR	EQUIP.	TOTAL		
7356	Round box w/cover & mtng flange, 3 threaded hubs, 1/2" diameter	1 Elec	8	1	Ea.	48	39.50		87.50	112	
7357	3/4" diameter		8	1		50	39.50		89.50	114	
7358	4 threaded hubs, 1" diameter		7	1.143		48.50	45		93.50	121	
7400	Unions, 1/2" diameter		20	.400		7.85	15.75		23.60	32	
7410	3/4" - 1/2" diameter		16	.500		11.20	19.70		30.90	42	
7420	3/4" diameter		16	.500		10.75	19.70		30.45	41.50	
7430	1" diameter		14	.571		19.45	22.50		41.95	55	
7440	1-1/4" diameter		12	.667		29.50	26.50		56	71	
7450	1-1/2" diameter		10	.800		37.50	31.50		69	88.50	
7460	2" diameter		8.50	.941		48.50	37		85.50	108	
7480	2-1/2" diameter		8	1		72.50	39.50		112	139	
7490	3" diameter		7	1.143		100	45		145	177	
7500	3-1/2" diameter		6	1.333		166	52.50		218.50	261	
7510	4" diameter		5	1.600		184	63		247	296	
7680	Reducer, 3/4" to 1/2"		54	.148		1.80	5.85		7.65	10.70	
7690	1" to 1/2"		46	.174		2.55	6.85		9.40	13	
7700	1" to 3/4"		46	.174		2.55	6.85		9.40	13	
7710	1-1/4" to 3/4"		40	.200		3.70	7.90		11.60	15.75	
7720	1-1/4" to 1"		40	.200		3.70	7.90		11.60	15.75	
7730	1-1/2" to 1"		36	.222		6	8.75		14.75	19.60	
7740	1-1/2" to 1-1/4"		36	.222		6	8.75		14.75	19.60	
7750	2" to 3/4"		32	.250		11.70	9.85		21.55	27.50	
7760	2" to 1-1/4"		32	.250		8.20	9.85		18.05	23.50	
7770	2" to 1-1/2"		32	.250		8.20	9.85		18.05	23.50	
7780	2-1/2" to 1-1/2"		30	.267		13.95	10.50		24.45	31	
7790	3" to 2"		30	.267		16.40	10.50		26.90	33.50	
7800	3-1/2" to 2-1/2"		28	.286		32	11.25		43.25	52.50	
7810	4" to 3"		28	.286		37.50	11.25		48.75	58.50	
7820	Sealing fitting, vertical/horizontal, 1/2" diameter		14.50	.552		12.20	21.50		33.70	46	
7830	3/4" diameter		13.30	.601		14.35	23.50		37.85	51.50	
7840	1" diameter		11.40	.702		18.50	27.50		46	61.50	
7850	1-1/4" diameter		10	.800		22.50	31.50		54	71.50	
7860	1-1/2" diameter		8.80	.909		34	36		70	91	
7870	2" diameter		8	1		44	39.50		83.50	107	
7880	2-1/2" diameter		6.70	1.194		61.50	47		108.50	138	
7890	3" diameter		5.70	1.404		75.50	55.50		131	166	
7900	3-1/2" diameter		4.70	1.702		201	67		268	320	
7910	4" diameter		4	2		315	79		394	465	
7920	Sealing hubs, 1" by 1-1/2"		12	.667		20.50	26.50		47	61.50	
7930	1-1/4" by 2"		10	.800		23	31.50		54.50	72.50	
7940	1-1/2" by 2"		9	.889		44	35		79	101	
7950	2" by 2-1/2"		8	1		58.50	39.50		98	123	
7960	3" by 4"		7	1.143		100	45		145	177	
7970	4" by 5"		6	1.333		209	52.50		261.50	310	
7980	Drain, 1/2"		32	.250		37	9.85		46.85	55.50	
7990	Breather, 1/2"		32	.250		37	9.85		46.85	55.50	
8000	Plastic coated 40 mil thick										
8010	LB, LR or LL conduit body w/cover, 1/2" diameter	1 Elec	13	.615	Ea.	27	24.50		51.50	65.50	
8020	3/4" diameter		11	.727		32	28.50		60.50	78	
8030	1" diameter		8	1		43	39.50		82.50	106	
8040	1-1/4" diameter		6	1.333		62	52.50		114.50	146	
8050	1-1/2" diameter		5	1.600		76	63		139	178	
8060	2" diameter		4.50	1.778		112	70		182	227	
8070	2-1/2" diameter		4	2		206	79		285	345	
8080	3" diameter		3.50	2.286		259	90		349	420	
8090	3-1/2" diameter		3	2.667		375	105		480	570	

250

16132 | Conduit & Tubing

		CREW	DAILY OUTPUT	LABOR-HOURS	UNIT	2004 BARE COSTS				TOTAL INCL O&P
						MAT.	LABOR	EQUIP.	TOTAL	
250 8100	4" diameter	1 Elec	2.50	3.200	Ea.	420	126		546	655	**250**
8150	T conduit body with cover, 1/2" diameter		11	.727		33.50	28.50		62	79
8160	3/4" diameter		9	.889		42	35		77	98
8170	1" diameter		6	1.333		50	52.50		102.50	133
8180	1-1/4" diameter		5	1.600		70	63		133	171
8190	1-1/2" diameter		4.50	1.778		88	70		158	201
8200	2" diameter		4	2		128	79		207	258
8210	2-1/2" diameter		3.50	2.286		216	90		306	370
8220	3" diameter		3	2.667		288	105		393	470
8230	3-1/2" diameter		2.50	3.200		385	126		511	610
8240	4" diameter		2	4		415	158		573	690
8300	FS conduit body, 1 gang, 3/4" diameter		11	.727		33	28.50		61.50	79
8310	1" diameter		10	.800		36	31.50		67.50	86.50
8350	2 gang, 3/4" diameter		9	.889		57	35		92	115
8360	1" diameter		8	1		68	39.50		107.50	134
8400	Duplex receptacle cover		64	.125		24	4.93		28.93	34
8410	Switch cover		64	.125		33	4.93		37.93	44
8420	Switch, vaportight cover		53	.151		76	5.95		81.95	92.50
8430	Blank, cover		64	.125		22	4.93		26.93	31.50
8520	FSC conduit body, 1 gang, 3/4" diameter		10	.800		39	31.50		70.50	90
8530	1" diameter		9	.889		44	35		79	101
8550	2 gang, 3/4" diameter		8	1		66	39.50		105.50	131
8560	1" diameter		7	1.143		73	45		118	148
8590	Conduit hubs, 1/2" diameter		18	.444		22	17.50		39.50	50
8600	3/4" diameter		16	.500		25	19.70		44.70	57
8610	1" diameter		14	.571		31	22.50		53.50	67.50
8620	1-1/4" diameter		12	.667		35.50	26.50		62	78
8630	1-1/2" diameter		10	.800		41	31.50		72.50	92
8640	2" diameter		8.80	.909		58.50	36		94.50	118
8650	2-1/2" diameter		8.50	.941		92	37		129	156
8660	3" diameter		8	1		126	39.50		165.50	198
8670	3-1/2" diameter		7.50	1.067		163	42		205	242
8680	4" diameter		7	1.143		200	45		245	287
8690	5" diameter		6	1.333		240	52.50		292.50	340
8700	Plastic coated 40 mil thick									
8710	Pipe strap, stamped 1 hole, 1/2" diameter	1 Elec	470	.017	Ea.	5.50	.67		6.17	7.05
8720	3/4" diameter		440	.018		5.55	.72		6.27	7.15
8730	1" diameter		400	.020		7.05	.79		7.84	8.90
8740	1-1/4" diameter		355	.023		8.70	.89		9.59	10.85
8750	1-1/2" diameter		320	.025		10.70	.99		11.69	13.20
8760	2" diameter		266	.030		11.05	1.19		12.24	13.90
8770	2-1/2" diameter		200	.040		13.30	1.58		14.88	17
8780	3" diameter		133	.060		14	2.37		16.37	18.90
8790	3-1/2" diameter		110	.073		15	2.87		17.87	21
8800	4" diameter		90	.089		19.35	3.50		22.85	26.50
8810	5" diameter		70	.114		91	4.50		95.50	107
8840	Clamp back spacers, 3/4" diameter		440	.018		7.80	.72		8.52	9.65
8850	1" diameter		400	.020		9.30	.79		10.09	11.40
8860	1-1/4" diameter		355	.023		12.35	.89		13.24	14.85
8870	1-1/2" diameter		320	.025		16.90	.99		17.89	20
8880	2" diameter		266	.030		26.50	1.19		27.69	31.50
8900	3" diameter		133	.060		60	2.37		62.37	69.50
8920	4" diameter		90	.089		87.50	3.50		91	101
8950	Touch-up plastic coating, spray, 12 oz.					25			25	27
8960	Sealing fittings, 1/2" diameter	1 Elec	11	.727		36	28.50		64.50	82
8970	3/4" diameter		9	.889		37	35		72	92.50

Important: See the Reference Section for critical supporting data - Reference Nos., Crews, & City Cost Indexes

16 ELECTRICAL

16132 | Conduit & Tubing

			CREW	DAILY OUTPUT	LABOR-HOURS	UNIT	MAT.	LABOR	EQUIP.	TOTAL	TOTAL INCL O&P	
250	8980	1" diameter	1 Elec	7.50	1.067	Ea.	44	42		86	111	250
	8990	1-1/4" diameter		6.50	1.231		53	48.50		101.50	131	
	9000	1-1/2" diameter		5.50	1.455		73	57.50		130.50	166	
	9010	2" diameter		4.80	1.667		88.50	65.50		154	195	
	9020	2-1/2" diameter		4	2		128	79		207	258	
	9030	3" diameter		3.50	2.286		161	90		251	310	
	9040	3-1/2" diameter		3	2.667		405	105		510	600	
	9050	4" diameter		2.50	3.200		655	126		781	910	
	9060	5" diameter		1.70	4.706		855	185		1,040	1,225	
	9070	Unions, 1/2" diameter		18	.444		30	17.50		47.50	59	
	9080	3/4" diameter		15	.533		30.50	21		51.50	65	
	9090	1" diameter		13	.615		41	24.50		65.50	81	
	9100	1-1/4" diameter		11	.727		66	28.50		94.50	115	
	9110	1-1/2" diameter		9.50	.842		79.50	33		112.50	137	
	9120	2" diameter		8	1		106	39.50		145.50	176	
	9130	2-1/2" diameter		7.50	1.067		160	42		202	239	
	9140	3" diameter		6.80	1.176		221	46.50		267.50	310	
	9150	3-1/2" diameter		5.80	1.379		274	54.50		328.50	380	
	9160	4" diameter		4.80	1.667		350	65.50		415.50	485	
	9170	5" diameter		4	2		565	79		644	740	
260	0010	**CUTTING AND DRILLING**										260
	0100	Hole drilling to 10' high, concrete wall										
	0110	8" thick, 1/2" pipe size	R-31	12	.667	Ea.	3.26	26.50	3.95	33.71	47	
	0120	3/4" pipe size		12	.667		3.26	26.50	3.95	33.71	47	
	0130	1" pipe size		9.50	.842		6.70	33	4.99	44.69	62.50	
	0140	1-1/4" pipe size		9.50	.842		6.70	33	4.99	44.69	62.50	
	0150	1-1/2" pipe size		9.50	.842		6.70	33	4.99	44.69	62.50	
	0160	2" pipe size		4.40	1.818		7.25	71.50	10.80	89.55	127	
	0170	2-1/2" pipe size		4.40	1.818		7.25	71.50	10.80	89.55	127	
	0180	3" pipe size		4.40	1.818		7.25	71.50	10.80	89.55	127	
	0190	3-1/2" pipe size		3.30	2.424		7.75	95.50	14.40	117.65	166	
	0200	4" pipe size		3.30	2.424		7.75	95.50	14.40	117.65	166	
	0500	12" thick, 1/2" pipe size		9.40	.851		4.98	33.50	5.05	43.53	61	
	0520	3/4" pipe size		9.40	.851		4.98	33.50	5.05	43.53	61	
	0540	1" pipe size		7.30	1.096		9.45	43	6.50	58.95	81.50	
	0560	1-1/4" pipe size		7.30	1.096		9.45	43	6.50	58.95	81.50	
	0570	1-1/2" pipe size		7.30	1.096		9.45	43	6.50	58.95	81.50	
	0580	2" pipe size		3.60	2.222		11.10	87.50	13.20	111.80	157	
	0590	2-1/2" pipe size		3.60	2.222		11.10	87.50	13.20	111.80	157	
	0600	3" pipe size		3.60	2.222		11.10	87.50	13.20	111.80	157	
	0610	3-1/2" pipe size		2.80	2.857		12.60	113	16.95	142.55	200	
	0630	4" pipe size		2.50	3.200		12.60	126	19	157.60	223	
	0650	16" thick, 1/2" pipe size		7.60	1.053		6.70	41.50	6.25	54.45	75.50	
	0670	3/4" pipe size		7	1.143		6.70	45	6.80	58.50	82	
	0690	1" pipe size		6	1.333		12.25	52.50	7.90	72.65	100	
	0710	1-1/4" pipe size		5.50	1.455		12.25	57.50	8.65	78.40	108	
	0730	1-1/2" pipe size		5.50	1.455		12.25	57.50	8.65	78.40	108	
	0750	2" pipe size		3	2.667		14.95	105	15.80	135.75	190	
	0770	2-1/2" pipe size		2.70	2.963		14.95	117	17.55	149.50	210	
	0790	3" pipe size		2.50	3.200		14.95	126	19	159.95	225	
	0810	3-1/2" pipe size		2.30	3.478		17.45	137	20.50	174.95	246	
	0830	4" pipe size		2	4		17.45	158	23.50	198.95	279	
	0850	20" thick, 1/2" pipe size		6.40	1.250		8.40	49.50	7.40	65.30	91	
	0870	3/4" pipe size		6	1.333		8.40	52.50	7.90	68.80	96	
	0890	1" pipe size		5	1.600		15	63	9.50	87.50	121	
	0910	1-1/4" pipe size		4.80	1.667		15	65.50	9.90	90.40	125	

ELECTRICAL 16

			DAILY	LABOR-		2004 BARE COSTS				TOTAL	
16132	**Conduit & Tubing**	CREW	OUTPUT	HOURS	UNIT	MAT.	LABOR	EQUIP.	TOTAL	INCL O&P	
260	0930	1-1/2" pipe size	R-31	4.60	1.739	Ea.	15	68.50	10.30	93.80	130	260
	0950	2" pipe size		2.70	2.963		18.75	117	17.55	153.30	214	
	0970	2-1/2" pipe size		2.40	3.333		18.75	131	19.75	169.50	237	
	0990	3" pipe size		2.20	3.636		18.75	143	21.50	183.25	257	
	1010	3-1/2" pipe size		2	4		22.50	158	23.50	204	285	
	1030	4" pipe size		1.70	4.706		22.50	185	28	235.50	330	
	1050	24" thick, 1/2" pipe size		5.50	1.455		10.15	57.50	8.65	76.30	106	
	1070	3/4" pipe size		5.10	1.569		10.15	62	9.30	81.45	113	
	1090	1" pipe size		4.30	1.860		17.75	73.50	11.05	102.30	141	
	1110	1-1/4" pipe size		4	2		17.75	79	11.85	108.60	150	
	1130	1-1/2" pipe size		4	2		17.75	79	11.85	108.60	150	
	1150	2" pipe size		2.40	3.333		22.50	131	19.75	173.25	242	
	1170	2-1/2" pipe size		2.20	3.636		22.50	143	21.50	187	262	
	1190	3" pipe size		2	4		22.50	158	23.50	204	285	
	1210	3-1/2" pipe size		1.80	4.444		27	175	26.50	228.50	320	
	1230	4" pipe size		1.50	5.333		27	210	31.50	268.50	380	
	1500	Brick wall, 8" thick, 1/2" pipe size		18	.444		3.26	17.50	2.64	23.40	32.50	
	1520	3/4" pipe size		18	.444		3.26	17.50	2.64	23.40	32.50	
	1540	1" pipe size		13.30	.601		6.70	23.50	3.57	33.77	47	
	1560	1-1/4" pipe size		13.30	.601		6.70	23.50	3.57	33.77	47	
	1580	1-1/2" pipe size		13.30	.601		6.70	23.50	3.57	33.77	47	
	1600	2" pipe size		5.70	1.404		7.25	55.50	8.30	71.05	99.50	
	1620	2-1/2" pipe size		5.70	1.404		7.25	55.50	8.30	71.05	99.50	
	1640	3" pipe size		5.70	1.404		7.25	55.50	8.30	71.05	99.50	
	1660	3-1/2" pipe size		4.40	1.818		7.75	71.50	10.80	90.05	127	
	1680	4" pipe size		4	2		7.75	79	11.85	98.60	139	
	1700	12" thick, 1/2" pipe size		14.50	.552		4.98	21.50	3.27	29.75	41.50	
	1720	3/4" pipe size		14.50	.552		4.98	21.50	3.27	29.75	41.50	
	1740	1" pipe size		11	.727		9.45	28.50	4.31	42.26	57.50	
	1760	1-1/4" pipe size		11	.727		9.45	28.50	4.31	42.26	57.50	
	1780	1-1/2" pipe size		11	.727		9.45	28.50	4.31	42.26	57.50	
	1800	2" pipe size		5	1.600		11.10	63	9.50	83.60	117	
	1820	2-1/2" pipe size		5	1.600		11.10	63	9.50	83.60	117	
	1840	3" pipe size		5	1.600		11.10	63	9.50	83.60	117	
	1860	3-1/2" pipe size		3.80	2.105		12.60	83	12.50	108.10	151	
	1880	4" pipe size		3.30	2.424		12.60	95.50	14.40	122.50	172	
	1900	16" thick, 1/2" pipe size		12.30	.650		6.70	25.50	3.86	36.06	49.50	
	1920	3/4" pipe size		12.30	.650		6.70	25.50	3.86	36.06	49.50	
	1940	1" pipe size		9.30	.860		12.25	34	5.10	51.35	69.50	
	1960	1-1/4" pipe size		9.30	.860		12.25	34	5.10	51.35	69.50	
	1980	1-1/2" pipe size		9.30	.860		12.25	34	5.10	51.35	69.50	
	2000	2" pipe size		4.40	1.818		14.95	71.50	10.80	97.25	135	
	2010	2-1/2" pipe size		4.40	1.818		14.95	71.50	10.80	97.25	135	
	2030	3" pipe size		4.40	1.818		14.95	71.50	10.80	97.25	135	
	2050	3-1/2" pipe size		3.30	2.424		17.45	95.50	14.40	127.35	177	
	2070	4" pipe size		3	2.667		17.45	105	15.80	138.25	193	
	2090	20" thick, 1/2" pipe size		10.70	.748		8.40	29.50	4.43	42.33	58	
	2110	3/4" pipe size		10.70	.748		8.40	29.50	4.43	42.33	58	
	2130	1" pipe size		8	1		15	39.50	5.95	60.45	81.50	
	2150	1-1/4" pipe size		8	1		15	39.50	5.95	60.45	81.50	
	2170	1-1/2" pipe size		8	1		15	39.50	5.95	60.45	81.50	
	2190	2" pipe size		4	2		18.75	79	11.85	109.60	151	
	2210	2-1/2" pipe size		4	2		18.75	79	11.85	109.60	151	
	2230	3" pipe size		4	2		18.75	79	11.85	109.60	151	
	2250	3-1/2" pipe size		3	2.667		22.50	105	15.80	143.30	198	
	2270	4" wipe size		2.70	2.963		22.50	117	17.55	157.05	218	

Important: See the Reference Section for critical supporting data - Reference Nos., Crews, & City Cost Indexes

16 ELECTRICAL

			DAILY	LABOR-		2004 BARE COSTS				TOTAL	
	16132	**Conduit & Tubing**	CREW	OUTPUT	HOURS	UNIT	MAT.	LABOR	EQUIP.	TOTAL	INCL O&P
260	2290	24" thick, 1/2" pipe size	R-31	9.40	.851	Ea.	10.15	33.50	5.05	48.70	66.50	**260**
	2310	3/4" pipe size		9.40	.851		10.15	33.50	5.05	48.70	66.50	
	2330	1" pipe size		7.10	1.127		17.75	44.50	6.70	68.95	93	
	2350	1-1/4" pipe size		7.10	1.127		17.75	44.50	6.70	68.95	93	
	2370	1-1/2" pipe size		7.10	1.127		17.75	44.50	6.70	68.95	93	
	2390	2" pipe size		3.60	2.222		22.50	87.50	13.20	123.20	170	
	2410	2-1/2" pipe size		3.60	2.222		22.50	87.50	13.20	123.20	170	
	2430	3" pipe size		3.60	2.222		22.50	87.50	13.20	123.20	170	
	2450	3-1/2" pipe size		2.80	2.857		27	113	16.95	156.95	216	
	2470	4" pipe size	▼	2.50	3.200	▼	27	126	19	172	239	
	3000	Knockouts to 8' high, metal boxes & enclosures										
	3020	With hole saw, 1/2" pipe size	1 Elec	53	.151	Ea.		5.95		5.95	8.85	
	3040	3/4" pipe size		47	.170			6.70		6.70	9.95	
	3050	1" pipe size		40	.200			7.90		7.90	11.70	
	3060	1-1/4" pipe size		36	.222			8.75		8.75	13	
	3070	1-1/2" pipe size		32	.250			9.85		9.85	14.65	
	3080	2" pipe size		27	.296			11.65		11.65	17.35	
	3090	2-1/2" pipe size		20	.400			15.75		15.75	23.50	
	4010	3" pipe size		16	.500			19.70		19.70	29.50	
	4030	3-1/2" pipe size		13	.615			24.50		24.50	36	
	4050	4" pipe size		11	.727			28.50		28.50	42.50	
	4070	With hand punch set, 1/2" pipe size		40	.200			7.90		7.90	11.70	
	4090	3/4" pipe size		32	.250			9.85		9.85	14.65	
	4110	1" pipe size		30	.267			10.50		10.50	15.65	
	4130	1-1/4" pipe size		28	.286			11.25		11.25	16.75	
	4150	1-1/2" pipe size		26	.308			12.10		12.10	18.05	
	4170	2" pipe size		20	.400			15.75		15.75	23.50	
	4190	2-1/2" pipe size		17	.471			18.55		18.55	27.50	
	4200	3" pipe size		15	.533			21		21	31.50	
	4220	3-1/2" pipe size		12	.667			26.50		26.50	39	
	4240	4" pipe size		10	.800			31.50		31.50	47	
	4260	With hydraulic punch, 1/2" pipe size		44	.182			7.15		7.15	10.65	
	4280	3/4" pipe size		38	.211			8.30		8.30	12.35	
	4300	1" pipe size		38	.211			8.30		8.30	12.35	
	4320	1-1/4" pipe size		38	.211			8.30		8.30	12.35	
	4340	1-1/2" pipe size		38	.211			8.30		8.30	12.35	
	4360	2" pipe size		32	.250			9.85		9.85	14.65	
	4380	2-1/2" pipe size		27	.296			11.65		11.65	17.35	
	4400	3" pipe size		23	.348			13.70		13.70	20.50	
	4420	3-1/2" pipe size		20	.400			15.75		15.75	23.50	
	4440	4" pipe size	▼	18	.444	▼		17.50		17.50	26	
300	0010	**ELECTRICAL NONMETALLIC TUBING (ENT)**										**300**
	0050	Flexible, 1/2" diameter	1 Elec	270	.030	L.F.	.40	1.17		1.57	2.18	
	0100	3/4" diameter		230	.035		.55	1.37		1.92	2.65	
	0200	1" diameter		145	.055		.89	2.17		3.06	4.21	
	0210	1-1/4" diameter		125	.064		1.23	2.52		3.75	5.10	
	0220	1-1/2" diameter		100	.080		1.67	3.15		4.82	6.55	
	0230	2" diameter		75	.107	▼	1.95	4.20		6.15	8.40	
	0300	Connectors, to outlet box, 1/2" diameter		230	.035	Ea.	.63	1.37		2	2.73	
	0310	3/4" diameter		210	.038		1.31	1.50		2.81	3.67	
	0320	1" diameter		200	.040		1.88	1.58		3.46	4.41	
	0400	Couplings, to conduit, 1/2" diameter		145	.055		.78	2.17		2.95	4.09	
	0410	3/4" diameter		130	.062		1.02	2.42		3.44	4.73	
	0420	1" diameter	▼	125	.064	▼	1.80	2.52		4.32	5.75	

ELECTRICAL 16

			CREW	DAILY OUTPUT	LABOR-HOURS	UNIT	2004 BARE COSTS				TOTAL INCL O&P
16132	**Conduit & Tubing**						MAT.	LABOR	EQUIP.	TOTAL	
320	0010	**FLEXIBLE METALLIC CONDUIT**										320
	0050	Steel, 3/8″ diameter	1 Elec	200	.040	L.F.	.25	1.58		1.83	2.62
	0100	1/2″ diameter		200	.040		.31	1.58		1.89	2.68
	0200	3/4″ diameter		160	.050		.43	1.97		2.40	3.40
	0250	1″ diameter		100	.080		.84	3.15		3.99	5.60
	0300	1-1/4″ diameter		70	.114		.92	4.50		5.42	7.70
	0350	1-1/2″ diameter		50	.160		1.59	6.30		7.89	11.15
	0370	2″ diameter		40	.200		1.94	7.90		9.84	13.85
	0380	2-1/2″ diameter		30	.267		2.33	10.50		12.83	18.20
	0390	3″ diameter	2 Elec	50	.320		3.54	12.60		16.14	22.50
	0400	3-1/2″ diameter		40	.400		4.46	15.75		20.21	28.50
	0410	4″ diameter		30	.533		6	21		27	38
	0420	Connectors, plain, 3/8″ diameter	1 Elec	100	.080	Ea.	.95	3.15		4.10	5.75
	0430	1/2″ diameter		80	.100		1.44	3.94		5.38	7.45
	0440	3/4″ diameter		70	.114		1.60	4.50		6.10	8.45
	0450	1″ diameter		50	.160		3.93	6.30		10.23	13.70
	0452	1-1/4″ diameter		45	.178		6.65	7		13.65	17.70
	0454	1-1/2″ diameter		40	.200		9.35	7.90		17.25	22
	0456	2″ diameter		28	.286		13.30	11.25		24.55	31.50
	0458	2-1/2″ diameter		25	.320		27	12.60		39.60	48.50
	0460	3″ diameter		20	.400		37.50	15.75		53.25	65
	0462	3-1/2″ diameter		16	.500		109	19.70		128.70	150
	0464	4″ diameter		13	.615		138	24.50		162.50	188
	0490	Insulated, 1″ diameter		40	.200		6.45	7.90		14.35	18.80
	0500	1-1/4″ diameter		40	.200		11.95	7.90		19.85	25
	0550	1-1/2″ diameter		32	.250		17.70	9.85		27.55	34
	0600	2″ diameter		23	.348		25.50	13.70		39.20	48.50
	0610	2-1/2″ diameter		20	.400		53.50	15.75		69.25	82.50
	0620	3″ diameter		17	.471		70.50	18.55		89.05	105
	0630	3-1/2″ diameter		13	.615		177	24.50		201.50	231
	0640	4″ diameter		10	.800		225	31.50		256.50	295
	0650	Connectors 90°, plain, 3/8″ diameter		80	.100		1.65	3.94		5.59	7.65
	0660	1/2″ diameter		60	.133		2.75	5.25		8	10.85
	0700	3/4″ diameter		50	.160		4.95	6.30		11.25	14.85
	0750	1″ diameter		40	.200		8.70	7.90		16.60	21.50
	0790	Insulated, 1″ diameter		40	.200		13.45	7.90		21.35	26.50
	0800	1-1/4″ diameter		30	.267		28.50	10.50		39	47
	0850	1-1/2″ diameter		23	.348		50	13.70		63.70	75.50
	0900	2″ diameter		18	.444		66.50	17.50		84	99
	0910	2-1/2″ diameter		16	.500		131	19.70		150.70	174
	0920	3″ diameter		14	.571		163	22.50		185.50	214
	0930	3-1/2″ diameter		11	.727		475	28.50		503.50	570
	0940	4″ diameter		8	1		715	39.50		754.50	845
	0960	Couplings, to flexible conduit, 1/2″ diameter		50	.160		.75	6.30		7.05	10.25
	0970	3/4″ diameter		40	.200		1.31	7.90		9.21	13.15
	0980	1″ diameter		35	.229		2.94	9		11.94	16.65
	0990	1-1/4″ diameter		28	.286		4.35	11.25		15.60	21.50
	1000	1-1/2″ diameter		23	.348		5.80	13.70		19.50	27
	1010	2″ diameter		20	.400		11	15.75		26.75	35.50
	1020	2-1/2″ diameter		18	.444		15.15	17.50		32.65	42.50
	1030	3″ diameter		15	.533		38	21		59	73.50
	1070	Sealtite, 3/8″ diameter		140	.057	L.F.	1.84	2.25		4.09	5.35
	1080	1/2″ diameter		140	.057		1.96	2.25		4.21	5.50
	1090	3/4″ diameter		100	.080		2.86	3.15		6.01	7.85
	1100	1″ diameter		70	.114		4.16	4.50		8.66	11.30
	1200	1-1/4″ diameter		50	.160		5.65	6.30		11.95	15.60

Important: See the Reference Section for critical supporting data - Reference Nos., Crews, & City Cost Indexes

16 ELECTRICAL

16132 | Conduit & Tubing

		CREW	DAILY OUTPUT	LABOR-HOURS	UNIT	MAT.	LABOR	EQUIP.	TOTAL	TOTAL INCL O&P	
320							2004 BARE COSTS				**320**
1300	1-1/2" diameter	1 Elec	40	.200	L.F.	6.80	7.90		14.70	19.15	
1400	2" diameter		30	.267		8.35	10.50		18.85	25	
1410	2-1/2" diameter	↓	27	.296		16.65	11.65		28.30	35.50	
1420	3" diameter	2 Elec	50	.320		23	12.60		35.60	44	
1440	4" diameter	"	30	.533	↓	32	21		53	67	
1490	Connectors, plain, 3/8" diameter	1 Elec	70	.114	Ea.	2.47	4.50		6.97	9.40	
1500	1/2" diameter		70	.114		2.43	4.50		6.93	9.35	
1700	3/4" diameter		50	.160		3.56	6.30		9.86	13.30	
1900	1" diameter		40	.200		6.10	7.90		14	18.40	
1910	Insulated, 1" diameter		40	.200		7.90	7.90		15.80	20.50	
2000	1-1/4" diameter		32	.250		12.35	9.85		22.20	28.50	
2100	1-1/2" diameter		27	.296		16.10	11.65		27.75	35	
2200	2" diameter		20	.400		29	15.75		44.75	55	
2210	2-1/2" diameter		15	.533		170	21		191	219	
2220	3" diameter		12	.667		190	26.50		216.50	248	
2240	4" diameter		8	1		236	39.50		275.50	320	
2290	Connectors, 90°, 3/8" diameter		70	.114		4.98	4.50		9.48	12.20	
2300	1/2" diameter		70	.114		4.63	4.50		9.13	11.80	
2400	3/4" diameter		50	.160		7.15	6.30		13.45	17.25	
2600	1" diameter		40	.200		15.30	7.90		23.20	28.50	
2790	Insulated, 1" diameter		40	.200		20.50	7.90		28.40	34	
2800	1-1/4" diameter		32	.250		33.50	9.85		43.35	51.50	
3000	1-1/2" diameter		27	.296		28	11.65		39.65	48.50	
3100	2" diameter		20	.400		42	15.75		57.75	70	
3110	2-1/2" diameter		14	.571		197	22.50		219.50	251	
3120	3" diameter		11	.727		228	28.50		256.50	294	
3140	4" diameter		7	1.143		295	45		340	390	
4300	Coupling sealtite to rigid, 1/2" diameter		20	.400		4.30	15.75		20.05	28	
4500	3/4" diameter		18	.444		6.10	17.50		23.60	32.50	
4800	1" diameter		14	.571		8.40	22.50		30.90	43	
4900	1-1/4" diameter		12	.667		14.20	26.50		40.70	54.50	
5000	1-1/2" diameter		11	.727		25.50	28.50		54	70.50	
5100	2" diameter		10	.800		44	31.50		75.50	95	
5110	2-1/2" diameter		9.50	.842		168	33		201	235	
5120	3" diameter		9	.889		181	35		216	251	
5130	3-1/2" diameter		9	.889		222	35		257	296	
5140	4" diameter	↓	8.50	.941	↓	222	37		259	299	

16133 | Multi-outlet Assemblies

		CREW	DAILY OUTPUT	LABOR-HOURS	UNIT	MAT.	LABOR	EQUIP.	TOTAL	TOTAL INCL O&P	
540											**540**
0010	**TRENCH DUCT** Steel with cover										
0020	Standard adjustable, depths to 4"										
0100	Straight, single compartment, 9" wide	2 Elec	40	.400	L.F.	72	15.75		87.75	103	
0200	12" wide		32	.500		82	19.70		101.70	120	
0400	18" wide		26	.615		109	24.50		133.50	156	
0600	24" wide		22	.727		140	28.50		168.50	197	
0700	27" wide		21	.762		151	30		181	211	
0800	30" wide		20	.800		169	31.50		200.50	233	
1000	36" wide		16	1		203	39.50		242.50	282	
1020	Two compartment, 9" wide		38	.421		82	16.60		98.60	115	
1030	12" wide		30	.533		93	21		114	134	
1040	18" wide		24	.667		115	26.50		141.50	166	
1050	24" wide		20	.800		150	31.50		181.50	212	
1060	30" wide		18	.889		182	35		217	252	
1070	36" wide		14	1.143		212	45		257	300	
1090	Three compartment, 9" wide	↓	36	.444	↓	93	17.50		110.50	128	

ELECTRICAL 16

16133 | Multi-outlet Assemblies

		CREW	DAILY OUTPUT	LABOR-HOURS	UNIT	2004 BARE COSTS MAT.	LABOR	EQUIP.	TOTAL	TOTAL INCL O&P	
540	**1100**	12" wide	2 Elec	28	.571	L.F.	103	22.50		125.50	147	**540**
	1110	18" wide		22	.727		129	28.50		157.50	185	
	1120	24" wide		18	.889		162	35		197	230	
	1130	30" wide		16	1		197	39.50		236.50	276	
	1140	36" wide		12	1.333		230	52.50		282.50	330	
	1200	Horizontal elbow, 9" wide		5.40	2.963	Ea.	267	117		384	470	
	1400	12" wide		4.60	3.478		310	137		447	545	
	1600	18" wide		4	4		395	158		553	670	
	1800	24" wide		3.20	5		555	197		752	905	
	1900	27" wide		3	5.333		630	210		840	1,000	
	2000	30" wide		2.60	6.154		740	242		982	1,175	
	2200	36" wide		2.40	6.667		970	263		1,233	1,475	
	2220	Two compartment, 9" wide		3.80	4.211		430	166		596	720	
	2230	12" wide		3	5.333		475	210		685	840	
	2240	18" wide		2.40	6.667		575	263		838	1,025	
	2250	24" wide		2	8		725	315		1,040	1,275	
	2260	30" wide		1.80	8.889		975	350		1,325	1,600	
	2270	36" wide		1.60	10		1,175	395		1,570	1,875	
	2290	Three compartment, 9" wide		3.60	4.444		450	175		625	755	
	2300	12" wide		2.80	5.714		505	225		730	890	
	2310	18" wide		2.20	7.273		615	287		902	1,100	
	2320	24" wide		1.80	8.889		780	350		1,130	1,375	
	2330	30" wide		1.60	10		1,000	395		1,395	1,700	
	2350	36" wide		1.40	11.429		1,225	450		1,675	2,025	
	2400	Vertical elbow, 9" wide		5.40	2.963		93	117		210	276	
	2600	12" wide		4.60	3.478		101	137		238	315	
	2800	18" wide		4	4		116	158		274	360	
	3000	24" wide		3.20	5		144	197		341	450	
	3100	27" wide		3	5.333		151	210		361	480	
	3200	30" wide		2.60	6.154		160	242		402	535	
	3400	36" wide		2.40	6.667		175	263		438	585	
	3600	Cross, 9" wide		4	4		440	158		598	715	
	3800	12" wide		3.20	5		465	197		662	805	
	4000	18" wide		2.60	6.154		555	242		797	970	
	4200	24" wide		2.20	7.273		705	287		992	1,200	
	4300	27" wide		2.20	7.273		825	287		1,112	1,325	
	4400	30" wide		2	8		915	315		1,230	1,475	
	4600	36" wide		1.80	8.889		1,150	350		1,500	1,775	
	4620	Two compartment, 9" wide		3.80	4.211		455	166		621	750	
	4630	12" wide		3	5.333		480	210		690	845	
	4640	18" wide		2.40	6.667		580	263		843	1,025	
	4650	24" wide		2	8		740	315		1,055	1,275	
	4660	30" wide		1.80	8.889		975	350		1,325	1,600	
	4670	36" wide		1.60	10		1,175	395		1,570	1,875	
	4690	Three compartment, 9" wide		3.60	4.444		465	175		640	770	
	4700	12" wide		2.80	5.714		530	225		755	920	
	4710	18" wide		2.20	7.273		630	287		917	1,125	
	4720	24" wide		1.80	8.889		785	350		1,135	1,375	
	4730	30" wide		1.60	10		1,025	395		1,420	1,700	
	4740	36" wide		1.40	11.429		1,275	450		1,725	2,075	
	4800	End closure, 9" wide		14.40	1.111		27.50	44		71.50	95	
	5000	12" wide		12	1.333		31.50	52.50		84	113	
	5200	18" wide		10	1.600		48	63		111	147	
	5400	24" wide		8	2		63	79		142	187	
	5500	27" wide		7	2.286		73	90		163	215	
	5600	30" wide		6.60	2.424		79	95.50		174.50	229	

Important: See the Reference Section for critical supporting data - Reference Nos., Crews, & City Cost Indexes

16133	Multi-outlet Assemblies	CREW	DAILY OUTPUT	LABOR-HOURS	UNIT	2004 BARE COSTS				TOTAL INCL O&P	
						MAT.	LABOR	EQUIP.	TOTAL		
540	5800	36" wide	2 Elec	5.80	2.759	Ea.	94	109		203	265	**540**
	6000	Tees, 9" wide		4	4		266	158		424	525	
	6200	12" wide		3.60	4.444		310	175		485	600	
	6400	18" wide		3.20	5		395	197		592	730	
	6600	24" wide		3	5.333		570	210		780	940	
	6700	27" wide		2.80	5.714		630	225		855	1,025	
	6800	30" wide		2.60	6.154		740	242		982	1,175	
	7000	36" wide		2	8		970	315		1,285	1,550	
	7020	Two compartment, 9" wide		3.80	4.211		310	166		476	585	
	7030	12" wide		3.40	4.706		330	185		515	640	
	7040	18" wide		3	5.333		445	210		655	805	
	7050	24" wide		2.80	5.714		600	225		825	995	
	7060	30" wide		2.40	6.667		815	263		1,078	1,300	
	7070	36" wide		1.90	8.421		1,025	330		1,355	1,625	
	7090	Three compartment, 9" wide		3.60	4.444		355	175		530	650	
	7100	12" wide		3.20	5		370	197		567	700	
	7110	18" wide		2.80	5.714		470	225		695	850	
	7120	24" wide		2.60	6.154		645	242		887	1,075	
	7130	30" wide		2.20	7.273		845	287		1,132	1,350	
	7140	36" wide		1.80	8.889		1,075	350		1,425	1,700	
	7200	Riser, and cabinet connector, 9" wide		5.40	2.963		116	117		233	300	
	7400	12" wide		4.60	3.478		135	137		272	355	
	7600	18" wide		4	4		166	158		324	415	
	7800	24" wide		3.20	5		201	197		398	515	
	7900	27" wide		3	5.333		208	210		418	545	
	8000	30" wide		2.60	6.154		232	242		474	615	
	8200	36" wide	▼	2	8		270	315		585	765	
	8400	Insert assembly, cell to conduit adapter, 1-1/4"	1 Elec	16	.500	▼	46	19.70		65.70	80	
	8500	Adjustable partition	"	320	.025	L.F.	16.30	.99		17.29	19.40	
	8600	Depth of duct over 4", per 1", add					7.15			7.15	7.85	
	8700	Support post	1 Elec	240	.033		16.30	1.31		17.61	19.90	
	8800	Cover double tile trim, 2 sides					25			25	27.50	
	8900	4 sides					75			75	82.50	
	9160	Trench duct 3-1/2" x 4-1/2", add					6.75			6.75	7.45	
	9170	Trench duct 4" x 5", add					6.75			6.75	7.45	
	9200	For carpet trim, add					23			23	25.50	
	9210	For double carpet trim, add				▼	70			70	77	
560	0010	**UNDERFLOOR DUCT**	R16133 -560									**560**
	0020											
	0100	Duct, 1-3/8" x 3-1/8" blank, standard	2 Elec	160	.100	L.F.	8.80	3.94		12.74	15.55	
	0200	1-3/8" x 7-1/4" blank, super duct		120	.133		17.60	5.25		22.85	27	
	0400	7/8" or 1-3/8" insert type, 24" O.C., 1-3/8" x 3-1/8", std.		140	.114		11.80	4.50		16.30	19.70	
	0600	1-3/8" x 7-1/4", super duct	▼	100	.160	▼	20.50	6.30		26.80	32	
	0800	Junction box, single duct, 1 level, 3-1/8"	1 Elec	4	2	Ea.	265	79		344	410	
	0820	3-1/8" x 7-1/4"		4	2		310	79		389	455	
	0840	2 level, 3-1/8" upper & lower		3.20	2.500		310	98.50		408.50	485	
	0860	3-1/8" upper, 7-1/4" lower		2.70	2.963		310	117		427	515	
	0880	Carpet pan for above		80	.100		219	3.94		222.94	247	
	0900	Terrazzo pan for above		67	.119		515	4.70		519.70	570	
	1000	Junction box, single duct, 1 level, 7-1/4"		2.70	2.963		310	117		427	515	
	1020	2 level, 7-1/4" upper & lower		2.70	2.963		355	117		472	565	
	1040	2 duct, two 3-1/8" upper & lower		3.20	2.500		470	98.50		568.50	660	
	1200	1 level, 2 duct, 3-1/8"		3.20	2.500		355	98.50		453.50	535	
	1220	Carpet pan for above boxes		80	.100		219	3.94		222.94	247	
	1240	Terrazzo pan for above boxes	▼	67	.119	▼	515	4.70		519.70	570	

ELECTRICAL 16

16133 | Multi-outlet Assemblies

			CREW	DAILY OUTPUT	LABOR-HOURS	UNIT	2004 BARE COSTS				TOTAL INCL O&P	
							MAT.	LABOR	EQUIP.	TOTAL		
560	1260	Junction box, 1 level, two 3-1/8″ x one 3-1/8″ + one 7-1/4″	1 Elec	2.30	3.478	Ea.	600	137		737	865	560
	1280	2 level, two 3-1/8″ upper, one 3-1/8″ + one 7-1/4″ lower		2	4		660	158		818	960	
	1300	Carpet pan for above boxes		80	.100		219	3.94		222.94	247	
	1320	Terrazzo pan for above boxes		67	.119		515	4.70		519.70	570	
	1400	Junction box, 1 level, 2 duct, 7-1/4″		2.30	3.478		910	137		1,047	1,200	
	1420	Two 3-1/8″ + one 7-1/4″		2	4		910	158		1,068	1,225	
	1440	Carpet pan for above		80	.100		219	3.94		222.94	247	
	1460	Terrazzo pan for above		67	.119		515	4.70		519.70	570	
	1580	Junction box, 1 level, one 3-1/8″ + one 7-1/4″ x same		2.30	3.478		600	137		737	865	
	1600	Triple duct, 3-1/8″		2.30	3.478		600	137		737	865	
	1700	Junction box, 1 level, one 3-1/8″ + two 7-1/4″		2	4		1,000	158		1,158	1,350	
	1720	Carpet pan for above		80	.100		219	3.94		222.94	247	
	1740	Terrazzo pan for above		67	.119		515	4.70		519.70	570	
	1800	Insert to conduit adapter, 3/4″ & 1″		32	.250		21.50	9.85		31.35	38.50	
	2000	Support, single cell		27	.296		32.50	11.65		44.15	53.50	
	2200	Super duct		16	.500		32.50	19.70		52.20	65.50	
	2400	Double cell		16	.500		32.50	19.70		52.20	65.50	
	2600	Triple cell		11	.727		32.50	28.50		61	78.50	
	2800	Vertical elbow, standard duct		10	.800		58.50	31.50		90	111	
	3000	Super duct		8	1		58.50	39.50		98	123	
	3200	Cabinet connector, standard duct		32	.250		44	9.85		53.85	63	
	3400	Super duct		27	.296		44	11.65		55.65	66	
	3600	Conduit adapter, 1″ to 1-1/4″		32	.250		44	9.85		53.85	63	
	3800	2″ to 1-1/4″		27	.296		52.50	11.65		64.15	75.50	
	4000	Outlet, low tension (tele, computer, etc.)		8	1		61.50	39.50		101	127	
	4200	High tension, receptacle (120 volt)		8	1		61.50	39.50		101	127	
	4300	End closure, standard duct		160	.050		2.40	1.97		4.37	5.55	
	4310	Super duct		160	.050		4.65	1.97		6.62	8.05	
	4350	Elbow, horiz., standard duct		26	.308		147	12.10		159.10	180	
	4360	Super duct		26	.308		147	12.10		159.10	180	
	4380	Elbow, offset, standard duct		26	.308		58.50	12.10		70.60	82	
	4390	Super duct		26	.308		58.50	12.10		70.60	82	
	4400	Marker screw assembly for inserts		50	.160		10.30	6.30		16.60	21	
	4410	Y take off, standard duct		26	.308		88	12.10		100.10	115	
	4420	Super duct		26	.308		88	12.10		100.10	115	
	4430	Box opening plug, standard duct		160	.050		8.20	1.97		10.17	11.95	
	4440	Super duct		160	.050		8.20	1.97		10.17	11.95	
	4450	Sleeve coupling, standard duct		160	.050		22	1.97		23.97	27	
	4460	Super duct		160	.050		27	1.97		28.97	33	
	4470	Conduit adapter, standard duct, 3/4″		32	.250		44	9.85		53.85	63	
	4480	1″ or 1-1/4″		32	.250		44	9.85		53.85	63	
	4500	1-1/2″		32	.250		44	9.85		53.85	63	
580	0010	**WIRING DUCT** Plastic										580
	1250	PVC, snap-in slots, adhesive backed										
	1270	1-1/2″W x 2″H	2 Elec	120	.133	L.F.	4.80	5.25		10.05	13.10	
	1280	1-1/2″W x 3″H		120	.133		5.90	5.25		11.15	14.30	
	1290	1-1/2″W x 4″H		120	.133		7.20	5.25		12.45	15.70	
	1300	2″W x 1″H		120	.133		4.70	5.25		9.95	12.95	
	1310	2″W x 1-1/2″H		120	.133		4.90	5.25		10.15	13.20	
	1320	2″W x 2″H		120	.133		5.05	5.25		10.30	13.35	
	1330	2″W x 2-1/2″H		120	.133		6.10	5.25		11.35	14.50	
	1340	2″W x 3″H		120	.133		6.30	5.25		11.55	14.75	
	1350	2″W x 4″H		120	.133		7.40	5.25		12.65	15.95	
	1360	2-1/2″W x 3″H		120	.133		6.70	5.25		11.95	15.15	
	1370	3″W x 1″H		110	.145		5.20	5.75		10.95	14.20	
	1380	3″W x 1-1/4″H		110	.145		6.05	5.75		11.80	15.15	

R16133-560

Important: See the Reference Section for critical supporting data - Reference Nos., Crews, & City Cost Indexes

16133	Multi-outlet Assemblies	CREW	DAILY OUTPUT	LABOR-HOURS	UNIT	2004 BARE COSTS				TOTAL INCL O&P
						MAT.	LABOR	EQUIP.	TOTAL	
580	1390	3"W x 2"H	2 Elec	110	.145	L.F.	5.95	5.75		11.70	15.05	580
	1400	3"W x 3"H		110	.145		7.10	5.75		12.85	16.30
	1410	3"W x 4"H		110	.145		8.60	5.75		14.35	17.95
	1420	3"W x 5"H		110	.145		11.40	5.75		17.15	21
	1430	4"W x 1-1/2"H		100	.160		6.15	6.30		12.45	16.15
	1440	4"W x 2"H		100	.160		6.95	6.30		13.25	17.05
	1450	4"W x 3"H		100	.160		8.05	6.30		14.35	18.25
	1460	4"W x 4"H		100	.160		9.65	6.30		15.95	20
	1470	4"W x 5"H		100	.160		12.95	6.30		19.25	23.50
	1550	Cover, 1-1/2"W		200	.080		.90	3.15		4.05	5.70
	1560	2"W		200	.080		1.19	3.15		4.34	6
	1570	2-1/2"W		200	.080		1.52	3.15		4.67	6.35
	1580	3"W		200	.080		1.78	3.15		4.93	6.65
	1590	4"W		200	.080		2.20	3.15		5.35	7.10
800	0010	**SURFACE RACEWAY**									800
	0090	Metal, straight section									
	0100	No. 500	1 Elec	100	.080	L.F.	.73	3.15		3.88	5.50
	0110	No. 700		100	.080		.82	3.15		3.97	5.60
	0400	No. 1500, small pancake		90	.089		1.50	3.50		5	6.85
	0600	No. 2000, base & cover, blank		90	.089		1.47	3.50		4.97	6.80
	0610	Receptacle, 6" O.C.		40	.200		8.20	7.90		16.10	21
	0620	12" O.C.		44	.182		5.35	7.15		12.50	16.55
	0630	18" O.C.		46	.174		5.65	6.85		12.50	16.40
	0650	30" O.C.		50	.160		3.90	6.30		10.20	13.70
	0670	No. 2200, base & cover, blank		80	.100		2.15	3.94		6.09	8.20
	0700	Receptacle, 18" O.C.		36	.222		5.30	8.75		14.05	18.85
	0720	30" O.C.		40	.200		5	7.90		12.90	17.20
	0800	No. 3000, base & cover, blank		75	.107		2.95	4.20		7.15	9.50
	0810	Receptacle, 6" O.C.		45	.178		25.50	7		32.50	38.50
	0820	12" O.C.		62	.129		14.30	5.10		19.40	23.50
	0830	18" O.C.		64	.125		11.50	4.93		16.43	20
	0840	24" O.C.		66	.121		8.65	4.78		13.43	16.60
	0850	30" O.C.		68	.118		7.80	4.64		12.44	15.50
	0860	60" O.C.		70	.114		5.70	4.50		10.20	12.95
	1000	No. 4000, base & cover, blank		65	.123		4.80	4.85		9.65	12.50
	1010	Receptacle, 6" O.C.		41	.195		35	7.70		42.70	50
	1020	12" O.C.		52	.154		21	6.05		27.05	32
	1030	18" O.C.		54	.148		17.45	5.85		23.30	28
	1040	24" O.C.		56	.143		14.30	5.65		19.95	24
	1050	30" O.C.		58	.138		12.70	5.45		18.15	22
	1060	60" O.C.		60	.133		9.95	5.25		15.20	18.75
	1200	No. 6000, base & cover, blank		50	.160		8.05	6.30		14.35	18.25
	1210	Receptacle, 6" O.C.		30	.267		43	10.50		53.50	63
	1220	12" O.C.		37	.216		27.50	8.50		36	43
	1230	18" O.C.		39	.205		23.50	8.10		31.60	38
	1240	24" O.C.		41	.195		19.20	7.70		26.90	32.50
	1250	30" O.C.		43	.186		18.40	7.35		25.75	31
	1260	60" O.C.		45	.178		14.20	7		21.20	26
	2400	Fittings, elbows, No. 500		40	.200	Ea.	1.33	7.90		9.23	13.15
	2800	Elbow cover, No. 2000		40	.200		2.55	7.90		10.45	14.50
	2880	Tee, No. 500		42	.190		2.56	7.50		10.06	13.95
	2900	No. 2000		27	.296		8.05	11.65		19.70	26
	3000	Switch box, No. 500		16	.500		8.40	19.70		28.10	39
	3400	Telephone outlet, No. 1500		16	.500		9.65	19.70		29.35	40
	3600	Junction box, No. 1500		16	.500		6.70	19.70		26.40	37
	3800	Plugmold wired sections, No. 2000									

ELECTRICAL 16

16133	Multi-outlet Assemblies	CREW	DAILY OUTPUT	LABOR-HOURS	UNIT	2004 BARE COSTS				TOTAL INCL O&P
						MAT.	LABOR	EQUIP.	TOTAL	
800											800
4000	1 circuit, 6 outlets, 3 ft. long	1 Elec	8	1	Ea.	24.50	39.50		64	85.50
4100	2 circuits, 8 outlets, 6 ft. long		5.30	1.509		41	59.50		100.50	134
4110	Tele-power pole, alum, w/ 2 recept, 10'		4	2		129	79		208	259
4120	12'		3.85	2.078		155	82		237	293
4130	15'		3.70	2.162		193	85		278	340
4140	Steel, w/ 2 recept, 10'		4	2		88	79		167	214
4150	One phone fitting, 10'		4	2		120	79		199	249
4160	Alum, 4 outlets, 10'		3.70	2.162		161	85		246	305
4300	Overhead distribution systems, 125 volt									
4800	No. 2000, entrance end fitting	1 Elec	20	.400	Ea.	3.40	15.75		19.15	27
5000	Blank end fitting		40	.200		1.50	7.90		9.40	13.35
5200	Supporting clip		40	.200		.82	7.90		8.72	12.60
5800	No. 3000, entrance end fitting		20	.400		6.20	15.75		21.95	30.50
6000	Blank end fitting		40	.200		1.84	7.90		9.74	13.70
6020	Internal elbow		20	.400		8.40	15.75		24.15	33
6030	External elbow		20	.400		11.50	15.75		27.25	36
6040	Device bracket		53	.151		2.85	5.95		8.80	12
6400	Hanger clamp		32	.250		3.93	9.85		13.78	18.95
7000	Base		90	.089	L.F.	3.05	3.50		6.55	8.55
7200	Divider		100	.080	"	.59	3.15		3.74	5.35
7400	Entrance end fitting		16	.500	Ea.	14.80	19.70		34.50	46
7600	Blank end fitting		40	.200		4.35	7.90		12.25	16.50
7610	Recp. & tele. cover		53	.151		7.35	5.95		13.30	16.95
7620	External elbow		16	.500		23	19.70		42.70	55
7630	Coupling		53	.151		3.50	5.95		9.45	12.70
7640	Divider clip & coupling		80	.100		.69	3.94		4.63	6.60
7650	Panel connector		16	.500		14.65	19.70		34.35	45.50
7800	Take off connector		16	.500		43.50	19.70		63.20	77.50
8000	No.6000, take off connector		16	.500		52.50	19.70		72.20	87.50
8100	Take off fitting		16	.500		40.50	19.70		60.20	74
8200	Hanger clamp		32	.250		9.20	9.85		19.05	25
8230	Coupling					4.38			4.38	4.82
8240	One gang device plate	1 Elec	53	.151		5.80	5.95		11.75	15.25
8250	Two gang device plate		40	.200		6.80	7.90		14.70	19.20
8260	Blank end fitting		40	.200		6.15	7.90		14.05	18.45
8270	Combination elbow		14	.571		24	22.50		46.50	60
8300	Panel connector		16	.500		10.55	19.70		30.25	41
8500	Chan-L-Wire system installed in 1-5/8" x 1-5/8" strut. Strut									
8600	not incl., 30 amp, 4 wire, 3 phase	1 Elec	200	.040	L.F.	3.30	1.58		4.88	5.95
8700	Junction box		8	1	Ea.	22	39.50		61.50	83
8800	Insulating end cap		40	.200		5.90	7.90		13.80	18.20
8900	Strut splice plate		40	.200		7.75	7.90		15.65	20.50
9000	Tap		40	.200		15.40	7.90		23.30	28.50
9100	Fixture hanger		60	.133		6.65	5.25		11.90	15.10
9200	Pulling tool					59.50			59.50	65.50
9300	Non-metallic, straight section									
9310	7/16" x 7/8", base & cover, blank	1 Elec	160	.050	L.F.	1.14	1.97		3.11	4.18
9320	Base & cover w/ adhesive		160	.050		1.33	1.97		3.30	4.39
9340	7/16" x 1-5/16", base & cover, blank		145	.055		1.33	2.17		3.50	4.69
9350	Base & cover w/ adhesive		145	.055		1.55	2.17		3.72	4.94
9370	11/16" x 2-1/4", base & cover, blank		130	.062		1.87	2.42		4.29	5.65
9380	Base & cover w/ adhesive		130	.062		2.15	2.42		4.57	6
9400	Fittings, elbows, 7/16" x 7/8"		50	.160	Ea.	1.33	6.30		7.63	10.85
9410	7/16" x 1-5/16"		45	.178		1.38	7		8.38	11.90
9420	11/16" x 2-1/4"		40	.200		1.50	7.90		9.40	13.35
9430	Tees, 7/16" x 7/8"		35	.229		1.75	9		10.75	15.35

16 ELECTRICAL

Important: See the Reference Section for critical supporting data - Reference Nos., Crews, & City Cost Indexes

			CREW	DAILY OUTPUT	LABOR-HOURS	UNIT	2004 BARE COSTS				TOTAL INCL O&P	
							MAT.	LABOR	EQUIP.	TOTAL		
800	**16133**	**Multi-outlet Assemblies**										**800**
	9440	7/16" x 1-5/16"	1 Elec	32	.250	Ea.	1.80	9.85		11.65	16.65	
	9450	11/16" x 2-1/4"		30	.267		1.82	10.50		12.32	17.65	
	9460	Cover clip, 7/16" x 7/8"		80	.100		.35	3.94		4.29	6.25	
	9470	7/16" x 1-5/16"		72	.111		.32	4.38		4.70	6.85	
	9480	11/16" x 2-1/4"		64	.125		.53	4.93		5.46	7.95	
	9490	Blank end, 7/16" x 7/8"		50	.160		.50	6.30		6.80	9.95	
	9510	11/16" x 2-1/4"		40	.200		.85	7.90		8.75	12.65	
	9520	Round fixture box 5.5" dia x 1"		25	.320		8.40	12.60		21	28	
	9530	Device box, 1 gang		30	.267		3.71	10.50		14.21	19.75	
	9540	2 gang	▼	25	.320	▼	5.50	12.60		18.10	25	
150	**16134**	**Wireway & Aux Gutters**										**150**
	0010	**WIREWAY** to 15' high R16134 -150										
	0020	For higher elevations, see 16131-130-9900										
	0100	Screw cover, NEMA 1 w/ fittings and supports, 2-1/2" x 2-1/2"	1 Elec	45	.178	L.F.	9.90	7		16.90	21.50	
	0200	4" x 4"	"	40	.200		10.50	7.90		18.40	23.50	
	0400	6" x 6"	2 Elec	60	.267		17.85	10.50		28.35	35.50	
	0600	8" x 8"		40	.400		30	15.75		45.75	56.50	
	0620	10" x 10"		30	.533		38	21		59	73.50	
	0640	12" x 12"	▼	20	.800	▼	53.50	31.50		85	106	
	0800	Elbows, 90°, 2-1/2"	1 Elec	24	.333	Ea.	30.50	13.15		43.65	53	
	1000	4"		20	.400		35	15.75		50.75	62	
	1200	6"		18	.444		39.50	17.50		57	69.50	
	1400	8"		16	.500		63.50	19.70		83.20	99.50	
	1420	10"		12	.667		71	26.50		97.50	117	
	1440	12"		10	.800		115	31.50		146.50	174	
	1500	Elbows, 45°, 2-1/2"		24	.333		29.50	13.15		42.65	51.50	
	1510	4"		20	.400		38	15.75		53.75	65	
	1520	6"		18	.444		39.50	17.50		57	69.50	
	1530	8"		16	.500		63.50	19.70		83.20	99.50	
	1540	10"		12	.667		68	26.50		94.50	114	
	1550	12"		10	.800		161	31.50		192.50	224	
	1600	"T" box, 2-1/2"		18	.444		35.50	17.50		53	65	
	1800	4"		16	.500		42.50	19.70		62.20	76.50	
	2000	6"		14	.571		48	22.50		70.50	86.50	
	2200	8"		12	.667		89	26.50		115.50	137	
	2220	10"		10	.800		105	31.50		136.50	163	
	2240	12"		8	1		170	39.50		209.50	246	
	2300	Cross, 2-1/2"		16	.500		39.50	19.70		59.20	73	
	2310	4"		14	.571		48	22.50		70.50	86.50	
	2320	6"		12	.667		59.50	26.50		86	105	
	2400	Panel adapter, 2-1/2"		24	.333		13.15	13.15		26.30	34	
	2600	4"		20	.400		14.80	15.75		30.55	40	
	2800	6"		18	.444		19.30	17.50		36.80	47	
	3000	8"		16	.500		27.50	19.70		47.20	60	
	3020	10"		14	.571		30.50	22.50		53	67.50	
	3040	12"		12	.667		44.50	26.50		71	88	
	3200	Reducer, 4" to 2-1/2"		24	.333		20.50	13.15		33.65	42.50	
	3400	6" to 4"		20	.400		39.50	15.75		55.25	67	
	3600	8" to 6"		18	.444		45.50	17.50		63	76	
	3620	10" to 8"		16	.500		53	19.70		72.70	87.50	
	3640	12" to 10"		14	.571		66	22.50		88.50	106	
	3780	End cap, 2-1/2"		24	.333		4.40	13.15		17.55	24.50	
	3800	4"		20	.400		5.40	15.75		21.15	29.50	
	4000	6"		18	.444		6.50	17.50		24	33	
	4200	8"	▼	16	.500	▼	8.90	19.70		28.60	39.50	

ELECTRICAL 16

16134 | Wireway & Aux Gutters

		CREW	DAILY OUTPUT	LABOR-HOURS	UNIT	2004 BARE COSTS				TOTAL INCL O&P
						MAT.	LABOR	EQUIP.	TOTAL	
150										**150**
4220	10"	1 Elec	14	.571	Ea.	10.65	22.50		33.15	45
4240	12"		12	.667		19.90	26.50		46.40	61
4300	U-connector, 2-1/2"		200	.040		4.40	1.58		5.98	7.20
4320	4"		200	.040		5.40	1.58		6.98	8.30
4340	6"		180	.044		6.50	1.75		8.25	9.75
4360	8"		170	.047		11	1.85		12.85	14.85
4380	10"		150	.053		11	2.10		13.10	15.25
4400	12"		130	.062		26.50	2.42		28.92	32.50
4420	Hanger, 2-1/2"		100	.080		7.70	3.15		10.85	13.15
4430	4"		100	.080		9.80	3.15		12.95	15.50
4440	6"		80	.100		17.60	3.94		21.54	25
4450	8"		65	.123		23	4.85		27.85	32.50
4460	10"		50	.160		29.50	6.30		35.80	42
4470	12"		40	.200	▼	62	7.90		69.90	79.50
4475	Screw cover, NEMA 3R w/ fittings and supports, 4" x 4"		36	.222	L.F.	22.50	8.75		31.25	37.50
4480	6" x 6"	2 Elec	55	.291		29.50	11.45		40.95	49.50
4485	8" x 8"		36	.444		47.50	17.50		65	78.50
4490	12" x 12"	▼	18	.889		66	35		101	125
4500	Hinged cover, with fittings and supports, 2-1/2" x 2-1/2"	1 Elec	60	.133		9.90	5.25		15.15	18.70
4520	4" x 4"	"	45	.178		10.50	7		17.50	22
4540	6" x 6"	2 Elec	80	.200		17.85	7.90		25.75	31.50
4560	8" x 8"		60	.267		30	10.50		40.50	48.50
4580	10" x 10"		50	.320		38	12.60		50.60	61
4600	12" x 12"	▼	24	.667	▼	53.50	26.50		80	98
4700	Elbows 90°, hinged cover, 2-1/2" x 2-1/2"	1 Elec	32	.250	Ea.	30.50	9.85		40.35	48
4720	4"		27	.296		35	11.65		46.65	56
4730	6"		23	.348		39.50	13.70		53.20	64
4740	8"		18	.444		63.50	17.50		81	96
4750	10"		14	.571		71	22.50		93.50	112
4760	12"		12	.667		115	26.50		141.50	166
4800	Tee box, hinged cover, 2-1/2" x 2-1/2"		23	.348		35.50	13.70		49.20	59.50
4810	4"		20	.400		42.50	15.75		58.25	70.50
4820	6"		18	.444		48	17.50		65.50	79
4830	8"		16	.500		89	19.70		108.70	128
4840	10"		12	.667		105	26.50		131.50	155
4860	12"		10	.800		170	31.50		201.50	234
4880	Cross box, hinged cover, 2-1/2" x 2-1/2"		18	.444		39.50	17.50		57	69.50
4900	4"		16	.500		48	19.70		67.70	82.50
4920	6"		13	.615		59.50	24.50		84	102
4940	8"		11	.727		88	28.50		116.50	140
4960	10"		10	.800		151	31.50		182.50	213
4980	12"		9	.889	▼	166	35		201	235
5000	Flanged, oil tite w/screw cover, 2-1/2" x 2-1/2"		40	.200	L.F.	21.50	7.90		29.40	35
5020	4" x 4"	▼	35	.229		25	9		34	41
5040	6" x 6"	2 Elec	60	.267		36	10.50		46.50	55
5060	8" x 8"	"	50	.320	▼	49.50	12.60		62.10	73.50
5120	Elbows 90°, flanged, 2-1/2" x 2-1/2"	1 Elec	23	.348	Ea.	45.50	13.70		59.20	71
5140	4"		20	.400		58	15.75		73.75	87.50
5160	6"		18	.444		72	17.50		89.50	105
5180	8"		15	.533		106	21		127	149
5240	Tee box, flanged, 2-1/2" x 2-1/2"		18	.444		63	17.50		80.50	95
5260	4"		16	.500		72	19.70		91.70	109
5280	6"		15	.533		97	21		118	139
5300	8"		13	.615		140	24.50		164.50	190
5360	Cross box, flanged, 2-1/2" x 2-1/2"		15	.533		77.50	21		98.50	117
5380	4"	▼	13	.615		101	24.50		125.50	148

R16134
-150

Important: See the Reference Section for critical supporting data - Reference Nos., Crews, & City Cost Indexes

16 ELECTRICAL

16100 | Wiring Methods

16134 | Wireway & Aux Gutters

		CREW	DAILY OUTPUT	LABOR-HOURS	UNIT	2004 BARE COSTS				TOTAL INCL O&P	
						MAT.	LABOR	EQUIP.	TOTAL		
150	5400	6"	1 Elec	12	.667	Ea.	131	26.50		157.50	184	**150**
	5420	8"		10	.800		154	31.50		185.50	216	
	5480	Flange gasket, 2-1/2"		160	.050		2.27	1.97		4.24	5.45	
	5500	4"		80	.100		3.08	3.94		7.02	9.25	
	5520	6"		53	.151		4.13	5.95		10.08	13.40	
	5530	8"		40	.200		5.50	7.90		13.40	17.75	

R16134-150

16136 | Boxes

		CREW	DAILY OUTPUT	LABOR-HOURS	UNIT	MAT.	LABOR	EQUIP.	TOTAL	TOTAL INCL O&P	
600	0010	**OUTLET BOXES**									**600**
	0020	Pressed steel, octagon, 4"	1 Elec	20	.400	Ea.	1.48	15.75		17.23	25
	0040	For Romex or BX		20	.400		2.23	15.75		17.98	26
	0050	For Romex or BX, with bracket		20	.400		2.93	15.75		18.68	26.50
	0060	Covers, blank		64	.125		.62	4.93		5.55	8.05
	0100	Extension rings		40	.200		2.39	7.90		10.29	14.35
	0150	Square, 4"		20	.400		2.10	15.75		17.85	26
	0160	For Romex or BX		20	.400		3.75	15.75		19.50	27.50
	0170	For Romex or BX, with bracket		20	.400		4.04	15.75		19.79	28
	0200	Extension rings		40	.200		2.48	7.90		10.38	14.45
	0220	2-1/8" deep, 1" KO		20	.400		3.31	15.75		19.06	27
	0250	Covers, blank		64	.125		.70	4.93		5.63	8.10
	0260	Raised device		64	.125		1.53	4.93		6.46	9.05
	0300	Plaster rings		64	.125		1.14	4.93		6.07	8.60
	0350	Square, 4-11/16"		20	.400		3.60	15.75		19.35	27.50
	0370	2-1/8" deep, 3/4" to 1-1/4" KO		20	.400		4.88	15.75		20.63	29
	0400	Extension rings		40	.200		5.45	7.90		13.35	17.70
	0450	Covers, blank		53	.151		1.29	5.95		7.24	10.25
	0460	Raised device		53	.151		3.64	5.95		9.59	12.85
	0500	Plaster rings		53	.151		3.61	5.95		9.56	12.80
	0550	Handy box		27	.296		1.81	11.65		13.46	19.35
	0560	Covers, device		64	.125		.57	4.93		5.50	8
	0600	Extension rings		54	.148		2.31	5.85		8.16	11.25
	0650	Switchbox		27	.296		2.31	11.65		13.96	19.90
	0660	Romex or BX		27	.296		2.55	11.65		14.20	20
	0670	with bracket		27	.296		3.67	11.65		15.32	21.50
	0680	Partition, metal		27	.296		1.44	11.65		13.09	18.95
	0700	Masonry, 1 gang, 2-1/2" deep		27	.296		5.30	11.65		16.95	23
	0710	3-1/2" deep		27	.296		4.65	11.65		16.30	22.50
	0750	2 gang, 2-1/2" deep		20	.400		8.25	15.75		24	32.50
	0760	3-1/2" deep		20	.400		7.75	15.75		23.50	32
	0800	3 gang, 2-1/2" deep		13	.615		9.50	24.50		34	46.50
	0850	4 gang, 2-1/2" deep		10	.800		11.40	31.50		42.90	59.50
	0860	5 gang, 2-1/2" deep		9	.889		12.65	35		47.65	66
	0870	6 gang, 2-1/2" deep		8	1		18.45	39.50		57.95	79
	0880	Masonry thru-the-wall, 1 gang, 4" block		16	.500		11.30	19.70		31	42
	0890	6" block		16	.500		12.85	19.70		32.55	43.50
	0900	8" block		16	.500		14.40	19.70		34.10	45.50
	0920	2 gang, 6" block		16	.500		12.20	19.70		31.90	43
	0940	Bar hanger with 3/8" stud, for wood and masonry boxes		53	.151		2.77	5.95		8.72	11.90
	0950	Concrete, set flush, 4" deep		20	.400		8.90	15.75		24.65	33.50
	1000	Plate with 3/8" stud		80	.100		3.20	3.94		7.14	9.35
	1100	Concrete, floor, 1 gang		5.30	1.509		61.50	59.50		121	157
	1150	2 gang		4	2		97	79		176	224
	1200	3 gang		2.70	2.963		138	117		255	325
	1250	For duplex receptacle, pedestal mounted, add		24	.333		58	13.15		71.15	83.50
	1270	Flush mounted, add		27	.296		15.85	11.65		27.50	35
	1300	For telephone, pedestal mounted, add		30	.267		62	10.50		72.50	83.50

R16136-600

		16136	Boxes	CREW	DAILY OUTPUT	LABOR-HOURS	UNIT	2004 BARE COSTS MAT.	LABOR	EQUIP.	TOTAL	TOTAL INCL O&P	
600	1350	Carpet flange, 1 gang		1 Elec	53	.151	Ea.	32.50	5.95		38.45	45	600
	1400	Cast, 1 gang, FS (2" deep), 1/2" hub	R16136 -600		12	.667		12.70	26.50		39.20	53	
	1410	3/4" hub			12	.667		13.40	26.50		39.90	54	
	1420	FD (2-11/16" deep), 1/2" hub			12	.667		15.25	26.50		41.75	56	
	1430	3/4" hub			12	.667		16.60	26.50		43.10	57.50	
	1450	2 gang, FS, 1/2" hub			10	.800		23	31.50		54.50	72	
	1460	3/4" hub			10	.800		23.50	31.50		55	73	
	1470	FD, 1/2" hub			10	.800		27.50	31.50		59	77.50	
	1480	3/4" hub			10	.800		27.50	31.50		59	77	
	1500	3 gang, FS, 3/4" hub			9	.889		35.50	35		70.50	91	
	1510	Switch cover, 1 gang, FS			64	.125		3.40	4.93		8.33	11.10	
	1520	2 gang			53	.151		5.50	5.95		11.45	14.90	
	1530	Duplex receptacle cover, 1 gang, FS			64	.125		3.40	4.93		8.33	11.10	
	1540	2 gang, FS			53	.151		5.55	5.95		11.50	14.95	
	1550	Weatherproof switch cover			64	.125		4.30	4.93		9.23	12.10	
	1600	Weatherproof receptacle cover			64	.125		4.30	4.93		9.23	12.10	
	1750	FSC, 1 gang, 1/2" hub			11	.727		14.10	28.50		42.60	58	
	1760	3/4" hub			11	.727		15.40	28.50		43.90	59.50	
	1770	2 gang, 1/2" hub			9	.889		25	35		60	79.50	
	1780	3/4" hub			9	.889		26.50	35		61.50	81.50	
	1790	FDC, 1 gang, 1/2" hub			11	.727		16.80	28.50		45.30	61	
	1800	3/4" hub			11	.727		18.10	28.50		46.60	62.50	
	1810	2 gang, 1/2" hub			9	.889		31	35		66	86	
	1820	3/4" hub			9	.889		29	35		64	83.50	
	2000	Poke-thru fitting, fire rated, for 3-3/4" floor			6.80	1.176		92.50	46.50		139	171	
	2040	For 7" floor			6.80	1.176		92.50	46.50		139	171	
	2100	Pedestal, 15 amp, duplex receptacle & blank plate			5.25	1.524		96	60		156	195	
	2120	Duplex receptacle and telephone plate			5.25	1.524		96.50	60		156.50	196	
	2140	Pedestal, 20 amp, duplex recept. & phone plate			5	1.600		97	63		160	200	
	2160	Telephone plate, both sides			5.25	1.524		91.50	60		151.50	191	
	2200	Abandonment plate			32	.250		28	9.85		37.85	45.50	
620	0010	**OUTLET BOXES, PLASTIC**											620
	0050	4" diameter, round with 2 mounting nails		1 Elec	25	.320	Ea.	1.86	12.60		14.46	21	
	0100	Bar hanger mounted			25	.320		3.28	12.60		15.88	22.50	
	0200	4", square with 2 mounting nails			25	.320		3	12.60		15.60	22	
	0300	Plaster ring			64	.125		1.03	4.93		5.96	8.50	
	0400	Switch box with 2 mounting nails, 1 gang			30	.267		1.36	10.50		11.86	17.15	
	0500	2 gang			25	.320		2.28	12.60		14.88	21.50	
	0600	3 gang			20	.400		3.60	15.75		19.35	27.50	
	0700	Old work box			30	.267		2.40	10.50		12.90	18.30	
	1400	PVC, FSS, 1 gang, 1/2" hub			14	.571		9.30	22.50		31.80	44	
	1410	3/4" hub			14	.571		9.90	22.50		32.40	44.50	
	1420	FD, 1 gang for variable terminations			14	.571		9	22.50		31.50	43.50	
	1450	FS, 2 gang for variable terminations			12	.667		8.45	26.50		34.95	48.50	
	1480	Blank cover, FS, 1 gang			64	.125		3.20	4.93		8.13	10.85	
	1500	2 gang			53	.151		3.65	5.95		9.60	12.85	
	1510	Switch cover, FS, 1 gang			64	.125		7.95	4.93		12.88	16.10	
	1520	2 gang			53	.151		12.60	5.95		18.55	22.50	
	1530	Duplex receptacle cover, FS, 1 gang			64	.125		11.30	4.93		16.23	19.80	
	1540	2 gang			53	.151		12.35	5.95		18.30	22.50	
	1750	FSC, 1 gang, 1/2" hub			13	.615		9	24.50		33.50	46	
	1760	3/4" hub			13	.615		9.90	24.50		34.40	47	
	1770	FSC, 2 gang, 1/2" hub			11	.727		10.65	28.50		39.15	54	
	1780	3/4" hub			11	.727		10.85	28.50		39.35	54.50	
	1790	FDC, 1 gang, 1/2" hub			13	.615		11.25	24.50		35.75	48.50	

16 ELECTRICAL

Important: See the Reference Section for critical supporting data - Reference Nos., Crews, & City Cost Indexes

		16136 | Boxes	CREW	DAILY OUTPUT	LABOR-HOURS	UNIT	2004 BARE COSTS				TOTAL INCL O&P	
							MAT.	LABOR	EQUIP.	TOTAL		
620	1800	3/4" hub	1 Elec	13	.615	Ea.	11.45	24.50		35.95	48.50	620
	1810	Weatherproof, T box w/ 3 holes		14	.571		9.05	22.50		31.55	43.50	
	1820	4" diameter round w/ 5 holes	▼	14	.571	▼	13.25	22.50		35.75	48	
700	0010	**PULL BOXES & CABINETS**										700
	0100	Sheet metal, pull box, NEMA 1, type SC, 6" W x 6" H x 4" D R16136-700	1 Elec	8	1	Ea.	9.65	39.50		49.15	69	
	0180	6" W x 8" H x 4" D		8	1		11.30	39.50		50.80	71	
	0200	8" W x 8" H x 4" D		8	1		13.20	39.50		52.70	73	
	0210	10" W x 10" H x 4" D		7	1.143		17.40	45		62.40	86	
	0220	12" W x 12" H x 4" D		6.50	1.231		22.50	48.50		71	96.50	
	0230	15" W x 15" H x 4" D		5.20	1.538		30	60.50		90.50	123	
	0240	18" W x 18" H x 4" D		4.40	1.818		40	71.50		111.50	151	
	0250	6" W x 6" H x 6" D		8	1		11.65	39.50		51.15	71.50	
	0260	8" W x 8" H x 6" D		7.50	1.067		15.90	42		57.90	80	
	0270	10" W x 10" H x 6" D		5.50	1.455		20.50	57.50		78	108	
	0300	10" W x 12" H x 6" D		5.30	1.509		23.50	59.50		83	114	
	0310	12" W x 12" H x 6" D		5.20	1.538		26	60.50		86.50	119	
	0320	15" W x 15" H x 6" D		4.60	1.739		37.50	68.50		106	143	
	0330	18" W x 18" H x 6" D		4.20	1.905		46.50	75		121.50	164	
	0340	24" W x 24" H x 6" D		3.20	2.500		94.50	98.50		193	251	
	0350	12" W x 12" H x 8" D		5	1.600		29	63		92	126	
	0360	15" W x 15" H x 8" D		4.50	1.778		60	70		130	170	
	0370	18" W x 18" H x 8" D		4	2		92.50	79		171.50	219	
	0380	24" W x 18" H x 6" D		3.70	2.162		73.50	85		158.50	208	
	0400	16" W x 20" H x 8" D		4	2		88	79		167	214	
	0500	20" W x 24" H x 8" D		3.20	2.500		103	98.50		201.50	260	
	0510	24" W x 24" H x 8" D		3	2.667		105	105		210	272	
	0600	24" W x 36" H x 8" D		2.70	2.963		145	117		262	335	
	0610	30" W x 30" H x 8" D		2.70	2.963		194	117		311	385	
	0620	36" W x 36" H x 8" D		2	4		237	158		395	495	
	0630	24" W x 24" H x 10" D		2.50	3.200		121	126		247	320	
	0650	Hinged cabinets, NEMA 1, 6" W x 6" H x 4" D		8	1		9.80	39.50		49.30	69.50	
	0660	8" W x 8" H x 4" D		8	1		13.45	39.50		52.95	73.50	
	0670	10" W x 10" H x 4" D		7	1.143		17.75	45		62.75	86.50	
	0680	12" W x 12" H x 4" D		6	1.333		23	52.50		75.50	103	
	0690	15" W x 15" H x 4" D		5.20	1.538		35.50	60.50		96	129	
	0700	18" W x 18" H x 4" D		4.40	1.818		40.50	71.50		112	152	
	0710	6" W x 6" H x 6" D		8	1		11.90	39.50		51.40	71.50	
	0720	8" W x 8" H x 6" D		7.50	1.067		16.15	42		58.15	80.50	
	0730	10" W x 10" H x 6" D		5.50	1.455		21	57.50		78.50	108	
	0740	12" W x 12" H x 6" D		5.20	1.538		26.50	60.50		87	120	
	0800	12" W x 16" H x 6" D		4.70	1.702		32	67		99	135	
	0810	15" W x 15" H x 6" D		4.60	1.739		37.50	68.50		106	144	
	0820	18" W x 18" H x 6" D		4.20	1.905		47.50	75		122.50	165	
	1000	20" W x 20" H x 6" D		3.60	2.222		65	87.50		152.50	202	
	1010	24" W x 24" H x 6" D		3.20	2.500		96.50	98.50		195	253	
	1020	12" W x 12" H x 8" D		5	1.600		55.50	63		118.50	155	
	1030	15" W x 15" H x 8" D		4.50	1.778		78.50	70		148.50	191	
	1040	18" W x 18" H x 8" D		4	2		111	79		190	239	
	1200	20" W x 20" H x 8" D		3.20	2.500		130	98.50		228.50	290	
	1210	24" W x 24" H x 8" D		3	2.667		148	105		253	320	
	1220	30" W x 30" H x 8" D		2.70	2.963		211	117		328	405	
	1400	24" W x 36" H x 8" D		2.70	2.963		208	117		325	405	
	1600	24" W x 42" H x 8" D		2	4		315	158		473	580	
	1610	36" W x 36" H x 8" D	▼	2	4	▼	290	158		448	555	
	2100	Pull box, NEMA 3R, type SC, raintight & weatherproof										

ELECTRICAL 16

181

			DAILY	LABOR-		2004 BARE COSTS				TOTAL
16136	**Boxes**	CREW	OUTPUT	HOURS	UNIT	MAT.	LABOR	EQUIP.	TOTAL	INCL O&P
700 2150	6" L x 6" W x 6" D R16136 -700	1 Elec	10	.800	Ea.	16.50	31.50		48	65	**700**
2200	8" L x 6" W x 6" D		8	1		20.50	39.50		60	81
2250	10" L x 6" W x 6" D		7	1.143		27.50	45		72.50	97
2300	12" L x 12" W x 6" D		5	1.600		38.50	63		101.50	137
2350	16" L x 16" W x 6" D		4.50	1.778		77	70		147	189
2400	20" L x 20" W x 6" D		4	2		106	79		185	234
2450	24" L x 18" W x 8" D		3	2.667		116	105		221	284
2500	24" L x 24" W x 10" D		2.50	3.200		155	126		281	360
2550	30" L x 24" W x 12" D		2	4		283	158		441	545
2600	36" L x 36" W x 12" D		1.50	5.333		370	210		580	725
2800	Cast iron, pull boxes for surface mounting									
3000	NEMA 4, watertight & dust tight									
3050	6" L x 6" W x 6" D	1 Elec	4	2	Ea.	155	79		234	288
3100	8" L x 6" W x 6" D		3.20	2.500		210	98.50		308.50	380
3150	10" L x 6" W x 6" D		2.50	3.200		261	126		387	475
3200	12" L x 12" W x 6" D		2.30	3.478		440	137		577	690
3250	16" L x 16" W x 6" D		1.30	6.154		905	242		1,147	1,350
3300	20" L x 20" W x 6" D		.80	10		1,675	395		2,070	2,425
3350	24" L x 18" W x 8" D		.70	11.429		1,775	450		2,225	2,625
3400	24" L x 24" W x 10" D		.50	16		2,700	630		3,330	3,925
3450	30" L x 24" W x 12" D		.40	20		4,525	790		5,315	6,175
3500	36" L x 36" W x 12" D		.20	40		5,600	1,575		7,175	8,525
3510	NEMA 4 clamp cover, 6" L x 6" W x 4" D		4	2		174	79		253	310
3520	8" L x 6" W x 4" D		4	2		205	79		284	345
4000	NEMA 7, explosionproof									
4050	6" L x 6" W x 6" D	1 Elec	2	4	Ea.	490	158		648	775
4100	8" L x 6" W x 6" D		1.80	4.444		560	175		735	880
4150	10" L x 6" W x 6" D		1.60	5		690	197		887	1,050
4200	12" L x 12" W x 6" D		1	8		1,125	315		1,440	1,725
4250	16" L x 14" W x 6" D		.60	13.333		1,575	525		2,100	2,500
4300	18" L x 18" W x 8" D		.50	16		2,775	630		3,405	4,000
4350	24" L x 18" W x 8" D		.40	20		3,700	790		4,490	5,250
4400	24" L x 24" W x 10" D		.30	26.667		4,800	1,050		5,850	6,850
4450	30" L x 24" W x 12" D		.20	40		7,325	1,575		8,900	10,400
5000	NEMA 9, dust tight 6" L x 6" W x 6" D		3.20	2.500		140	98.50		238.50	300
5050	8" L x 6" W x 6" D		2.70	2.963		168	117		285	360
5100	10" L x 6" W x 6" D		2	4		225	158		383	480
5150	12" L x 12" W x 6" D		1.60	5		445	197		642	785
5200	16" L x 16" W x 6" D		1	8		765	315		1,080	1,325
5250	18" L x 18" W x 8" D		.70	11.429		1,200	450		1,650	2,000
5300	24" L x 18" W x 8" D		.60	13.333		1,700	525		2,225	2,625
5350	24" L x 24" W x 10" D		.40	20		2,175	790		2,965	3,575
5400	30" L x 24" W x 12" D		.30	26.667		3,450	1,050		4,500	5,350
6000	J.I.C. wiring boxes, NEMA 12, dust tight & drip tight									
6050	6" L x 8" W x 4" D	1 Elec	10	.800	Ea.	38.50	31.50		70	89.50
6100	8" L x 10" W x 4" D		8	1		48	39.50		87.50	112
6150	12" L x 14" W x 6" D		5.30	1.509		78.50	59.50		138	175
6200	14" L x 16" W x 6" D		4.70	1.702		93.50	67		160.50	203
6250	16" L x 20" W x 6" D		4.40	1.818		199	71.50		270.50	325
6300	24" L x 30" W x 6" D		3.20	2.500		293	98.50		391.50	465
6350	24" L x 30" W x 8" D		2.90	2.759		315	109		424	505
6400	24" L x 36" W x 8" D		2.70	2.963		345	117		462	555
6450	24" L x 42" W x 8" D		2.30	3.478		380	137		517	620
6500	24" L x 48" W x 8" D		2	4		415	158		573	690
7000	Cabinets, current transformer									
7050	Single door, 24" H x 24" W x 10" D	1 Elec	1.60	5	Ea.	121	197		318	425

Important: See the Reference Section for critical supporting data - Reference Nos., Crews, & City Cost Indexes

16136	Boxes		CREW	DAILY OUTPUT	LABOR-HOURS	UNIT	2004 BARE COSTS				TOTAL INCL O&P
							MAT.	LABOR	EQUIP.	TOTAL	
700	**7100**	30" H x 24" W x 10" D	1 Elec	1.30	6.154	Ea.	127	242		369	500
	7150	36" H x 24" W x 10" D		1.10	7.273		142	287		429	580
	7200	30" H x 30" W x 10" D		1	8		148	315		463	635
	7250	36" H x 30" W x 10" D		.90	8.889		200	350		550	740
	7300	36" H x 36" W x 10" D		.80	10		218	395		613	825
	7500	Double door, 48" H x 36" W x 10" D		.60	13.333		410	525		935	1,225
	7550	24" H x 24" W x 12" D	▼	1	8	▼	205	315		520	695
	7600	Telephone with wood backboard									
	7620	Single door, 12" H x 12" W x 4" D	1 Elec	5.30	1.509	Ea.	93	59.50		152.50	191
	7650	18" H x 12" W x 4" D		4.70	1.702		128	67		195	241
	7700	24" H x 12" W x 4" D		4.20	1.905		145	75		220	272
	7720	18" H x 18" W x 4" D		4.20	1.905		157	75		232	285
	7750	24" H x 18" W x 4" D		4	2		214	79		293	350
	7780	36" H x 36" W x 4" D		3.60	2.222		295	87.50		382.50	455
	7800	24" H x 24" W x 6" D		3.60	2.222		282	87.50		369.50	440
	7820	30" H x 24" W x 6" D		3.20	2.500		365	98.50		463.50	545
	7850	30" H x 30" W x 6" D		2.70	2.963		420	117		537	640
	7880	36" H x 30" W x 6" D		2.50	3.200		450	126		576	685
	7900	48" H x 36" W x 6" D		2.20	3.636		790	143		933	1,075
	7920	Double door, 48" H x 36" W x 6" D	▼	2	4	▼	820	158		978	1,150
	8000	NEMA 12, double door, floor mounted									
	8020	54" H x 42" W x 8" D	2 Elec	6	2.667	Ea.	810	105		915	1,050
	8040	60" H x 48" W x 8" D		5.40	2.963		1,100	117		1,217	1,400
	8060	60" H x 48" W x 10" D		5.40	2.963		1,100	117		1,217	1,400
	8080	60" H x 60" W x 10" D		5	3.200		1,275	126		1,401	1,600
	8100	72" H x 60" W x 10" D		4	4		1,450	158		1,608	1,825
	8120	72" H x 72" W x 10" D		3.40	4.706		1,625	185		1,810	2,050
	8140	60" H x 48" W x 12" D		3.40	4.706		1,150	185		1,335	1,525
	8160	60" H x 60" W x 12" D		3.20	5		1,275	197		1,472	1,700
	8180	72" H x 60" W x 12" D		3	5.333		1,300	210		1,510	1,750
	8200	72" H x 72" W x 12" D		3	5.333		1,450	210		1,660	1,925
	8220	60" H x 48" W x 16" D		3.20	5		1,200	197		1,397	1,625
	8240	72" H x 72" W x 16" D		2.60	6.154		1,725	242		1,967	2,250
	8260	60" H x 48" W x 20" D		3	5.333		1,325	210		1,535	1,775
	8280	72" H x 72" W x 20" D		2.20	7.273		1,875	287		2,162	2,475
	8300	60" H x 48" W x 24" D		2.60	6.154		1,375	242		1,617	1,875
	8320	72" H x 72" W x 24" D	▼	2	8	▼	1,975	315		2,290	2,625
	8340	Pushbutton enclosure, oiltight									
	8360	3-1/2" H x 3-1/4" W x 2-3/4" D, for 1 P.B.	1 Elec	12	.667	Ea.	37	26.50		63.50	79.50
	8380	5-3/4" H x 3-1/4" W x 2-3/4" D, for 2 P.B.		11	.727		41.50	28.50		70	88
	8400	8" H x 3-1/4" W x 2-3/4" D, for 3 P.B.		10.50	.762		50.50	30		80.50	100
	8420	10-1/4" H x 3-1/4" W x 2-3/4" D, for 4 P.B.		10.50	.762		52	30		82	102
	8440	7-1/4" H x 6-1/4" W x 3" D, for 4 P.B.		10	.800		55	31.50		86.50	108
	8460	12-1/2" H x 3-1/4" W x 3" D, for 5 P.B.		9	.889		58	35		93	116
	8480	9-1/2" H x 6-1/4" W x 3" D, for 6 P.B.		8.50	.941		66.50	37		103.50	128
	8500	9-1/2" H x 8-1/2" W x 3" D, for 9 P.B.		8	1		71.50	39.50		111	137
	8510	11-3/4" H x 8-1/2" W x 3" D, for 12 P.B.		7	1.143		79	45		124	154
	8520	11-3/4" H x 10-3/4" W x 3" D, for 16 P.B.		6.50	1.231		103	48.50		151.50	185
	8540	14" H x 10-3/4" W x 3" D, for 20 P.B.		5	1.600		116	63		179	222
	8560	14" H x 13" W x 3" D, for 25 P.B.	▼	4.50	1.778	▼	132	70		202	249
	8580	Sloping front pushbutton enclosures									
	8600	3-1/2" H x 7-3/4" W x 4-7/8" D, for 3 P.B	1 Elec	10	.800	Ea.	61	31.50		92.50	114
	8620	7-1/4" H x 8-1/2" W x 6-3/4" D, for 6 P.B.		8	1		89	39.50		128.50	157
	8640	9-1/2" H x 8-1/2" W x 7-7/8" D, for 9 P.B.		7	1.143		99	45		144	176
	8660	11-1/4" H x 8-1/2" W x 9" D, for 12 P.B.		5	1.600		116	63		179	222
	8680	11-3/4" H x 10" W x 9" D, for 16 P.B.	▼	5	1.600		130	63		193	237

R16136 -700

ELECTRICAL 16

16136 | Boxes

		CREW	DAILY OUTPUT	LABOR-HOURS	UNIT	2004 BARE COSTS MAT.	LABOR	EQUIP.	TOTAL	TOTAL INCL O&P	
700	8700	11-3/4" H x 13" W x 9" D, for 20 P.B.	1 Elec	5	1.600	Ea.	151	63		214	260	**700**
	8720	14" H x 13" W x 10-1/8" D, for 25 P.B.	↓	4.50	1.778	↓	170	70		240	291	
	8740	Pedestals, not including P.B. enclosure or base										
	8760	Straight column 4" x 4"	1 Elec	4.50	1.778	Ea.	154	70		224	273	
	8780	6" x 6"		4	2		232	79		311	370	
	8800	Angled column 4" x 4"		4.50	1.778		181	70		251	305	
	8820	6" x 6"		4	2		274	79		353	415	
	8840	Pedestal, base 18" x 18"		10	.800		94	31.50		125.50	150	
	8860	24" x 24"	↓	9	.889	↓	212	35		247	285	
	8900	Electronic rack enclosures										
	8920	72" H x 25" W x 24" D	1 Elec	1.50	5.333	Ea.	1,600	210		1,810	2,075	
	8940	72" H x 30" W x 24" D		1.50	5.333		1,700	210		1,910	2,175	
	8960	72" H x 25" W x 30" D		1.30	6.154		1,725	242		1,967	2,225	
	8980	72" H x 25" W x 36" D		1.20	6.667		1,825	263		2,088	2,425	
	9000	72" H x 30" W x 36" D	↓	1.20	6.667	↓	1,950	263		2,213	2,550	
	9020	NEMA 12 & 4 enclosure panels										
	9040	12" x 24"	1 Elec	20	.400	Ea.	22	15.75		37.75	47.50	
	9060	16" x 12"		20	.400		15.60	15.75		31.35	40.50	
	9080	20" x 16"		20	.400		23	15.75		38.75	48.50	
	9100	20" x 20"		19	.421		27.50	16.60		44.10	55	
	9120	24" x 20"		18	.444		35.50	17.50		53	65.50	
	9140	24" x 24"		17	.471		41	18.55		59.55	72.50	
	9160	30" x 20"		16	.500		43	19.70		62.70	76.50	
	9180	30" x 24"		16	.500		49.50	19.70		69.20	84	
	9200	36" x 24"		15	.533		59.50	21		80.50	97	
	9220	36" x 30"		15	.533		79.50	21		100.50	119	
	9240	42" x 24"		15	.533		68.50	21		89.50	107	
	9260	42" x 30"		14	.571		91	22.50		113.50	134	
	9280	42" x 36"		14	.571		107	22.50		129.50	152	
	9300	48" x 24"		14	.571		77.50	22.50		100	119	
	9320	48" x 30"		14	.571		102	22.50		124.50	146	
	9340	48" x 36"		13	.615		121	24.50		145.50	169	
	9360	60" x 36"	↓	12	.667	↓	148	26.50		174.50	202	
	9400	Wiring trough steel JIC, clamp cover										
	9490	4" x 4", 12" long	1 Elec	12	.667	Ea.	53.50	26.50		80	98	
	9510	24" long		10	.800		68.50	31.50		100	123	
	9530	36" long		8	1		83.50	39.50		123	151	
	9540	48" long		7	1.143		98.50	45		143.50	175	
	9550	60" long		6	1.333		113	52.50		165.50	202	
	9560	6" x 6", 12" long		11	.727		70.50	28.50		99	120	
	9580	24" long		9	.889		92	35		127	153	
	9600	36" long		7	1.143		116	45		161	194	
	9610	48" long		6	1.333		139	52.50		191.50	231	
	9620	60" long	↓	5	1.600		163	63		226	273	
720	0010	**PULL BOXES & CABINETS** Nonmetallic										**720**
	0080	Enclosures fiberglass NEMA 4X										
	0100	Wall mount, quick release latch door, 20"H x 16"W x 6"D	1 Elec	4.80	1.667	Ea.	385	65.50		450.50	520	
	0110	20"H x 20"W x 6"D		4.50	1.778		455	70		525	605	
	0120	24"H x 20"W x 6"D		4.20	1.905		490	75		565	645	
	0130	20"H x 16"W x 8"D		4.50	1.778		435	70		505	580	
	0140	20"H x 20"W x 8"D		4.20	1.905		480	75		555	640	
	0150	24"H x 24"W x 8"D		3.80	2.105		540	83		623	720	
	0160	30"H x 24"W x 8"D		3.20	2.500		600	98.50		698.50	805	
	0170	36"H x 30"W x 8"D		3	2.667		840	105		945	1,075	
	0180	20"H x 16"W x 10"D		3.50	2.286		495	90		585	680	
	0190	20"H x 20"W x 10"D	↓	3.20	2.500	↓	535	98.50		633.50	735	

R16136 -700

16 ELECTRICAL

Important: See the Reference Section for critical supporting data - Reference Nos., Crews, & City Cost Indexes

16136	Boxes	CREW	DAILY OUTPUT	LABOR-HOURS	UNIT	2004 BARE COSTS				TOTAL INCL O&P
						MAT.	LABOR	EQUIP.	TOTAL	
720	0200	24"H x 20"W x 10"D	1 Elec	3	2.667	Ea.	575	105		680	785	720
0210	30"H x 24"W x 10"D		2.80	2.857		650	113		763	880
0220	20"H x 16"W x 12"D		3	2.667		545	105		650	755
0230	20"H x 20"W x 12"D		2.80	2.857		575	113		688	800
0240	24"H x 24"W x 12"D		2.60	3.077		630	121		751	870
0250	30"H x 24"W x 12"D		2.40	3.333		695	131		826	960
0260	36"H x 30"W x 12"D		2.20	3.636		960	143		1,103	1,275
0270	36"H x 36"W x 12"D		2.10	3.810		1,050	150		1,200	1,400
0280	48"H x 36"W x 12"D		2	4		1,250	158		1,408	1,600
0290	60"H x 36"W x 12"D		1.80	4.444		1,375	175		1,550	1,775
0300	30"H x 24"W x 16"D		1.40	5.714		825	225		1,050	1,250
0310	48"H x 36"W x 16"D		1.20	6.667		1,375	263		1,638	1,925
0320	60"H x 36"W x 16"D		1	8		1,550	315		1,865	2,200
0480	Freestanding, one door, 72"H x 25"W x 25"D		.80	10		2,675	395		3,070	3,525
0490	Two doors with two panels, 72"H x 49"W x 24"D		.50	16		6,350	630		6,980	7,925
0500	Floor stand kits, for NEMA 4 & 12, 20"W or more		24	.333		78.50	13.15		91.65	106
0510	6"H x 10"D		24	.333		86	13.15		99.15	114
0520	6"H x 12"D		24	.333		98	13.15		111.15	128
0530	6"H x 18"D		24	.333		116	13.15		129.15	148
0540	12"H x 8"D		22	.364		101	14.35		115.35	133
0550	12"H x 10"D		22	.364		109	14.35		123.35	142
0560	12"H x 12"D		22	.364		117	14.35		131.35	150
0570	12"H x 16"D		22	.364		132	14.35		146.35	167
0580	12"H x 18"D		22	.364		139	14.35		153.35	175
0590	12"H x 20"D		22	.364		147	14.35		161.35	184
0600	18"H x 8"D		20	.400		124	15.75		139.75	160
0610	18"H x 10"D		20	.400		132	15.75		147.75	169
0620	18"H x 12"D		20	.400		139	15.75		154.75	177
0630	18"H x 16"D		20	.400		155	15.75		170.75	195
0640	24"H x 8"D		16	.500		147	19.70		166.70	192
0650	24"H x 10"D		16	.500		155	19.70		174.70	201
0660	24"H x 12"D		16	.500		162	19.70		181.70	208
0670	24"H x 16"D		16	.500		178	19.70		197.70	226
0680	Small, screw cover, 5-1/2"H x 4"W x 4-15/16"D		12	.667		47.50	26.50		74	91
0690	7-1/2"H x 4"W x 4-15/16"D		12	.667		49.50	26.50		76	93.50
0700	7-1/2"H x 6"W x 5-3/16"D		10	.800		57	31.50		88.50	110
0710	9-1/2"H x 6"W x 5-11/16"D		10	.800		59.50	31.50		91	112
0720	11-1/2"H x 8"W x 6-11/16"D		8	1		87	39.50		126.50	154
0730	13-1/2"H x 10"W x 7-3/16"D		7	1.143		107	45		152	185
0740	15-1/2"H x 12"W x 8-3/16"D		6	1.333		142	52.50		194.50	234
0750	17-1/2"H x 14"W x 8-11/16"D		5	1.600		162	63		225	272
0760	Screw cover with window, 6"H x 4"W x 5"D		12	.667		76.50	26.50		103	124
0770	8"H x 4"W x 5"D		11	.727		82	28.50		110.50	133
0780	8"H x 6"W x 5"D		11	.727		116	28.50		144.50	171
0790	10"H x 6"W x 6"D		10	.800		122	31.50		153.50	181
0800	12"H x 8"W x 7"D		8	1		157	39.50		196.50	232
0810	14"H x 10"W x 7"D		7	1.143		187	45		232	273
0820	16"H x 12"W x 8"D		6	1.333		226	52.50		278.50	325
0830	18"H x 14"W x 9"D		5	1.600		269	63		332	390
0840	Quick-release latch cover, 5-1/2"H x 4"W x 5"D		12	.667		61.50	26.50		88	107
0850	7-1/2"H x 4"W x 5"D		12	.667		63.50	26.50		90	109
0860	7-1/2"H x 6"W x 5-1/4"D		10	.800		73.50	31.50		105	128
0870	9-1/2"H x 6"W x 5-3/4"D		10	.800		76.50	31.50		108	131
0880	11-1/2"H x 8"W x 6-3/4"D		8	1		112	39.50		151.50	182
0890	13-1/2"H x 10"W x 7-1/4"D		7	1.143		138	45		183	219
0900	15-1/2"H x 12"W x 8-1/4"D		6	1.333		177	52.50		229.50	273

ELECTRICAL 16

16136	Boxes	CREW	DAILY OUTPUT	LABOR-HOURS	UNIT	2004 BARE COSTS				TOTAL INCL O&P	
						MAT.	LABOR	EQUIP.	TOTAL		
720	0910	17-1/2"H x 14"W x 8-3/4"D	1 Elec	5	1.600	Ea.	210	63		273	325	**720**
	0920	Pushbutton, 1 hole 5-1/2"H x 4"W x 4-15/16"D		12	.667		45	26.50		71.50	88.50
	0930	2 hole 7-1/2"H x 4"W x 4-15/16"D		11	.727		51	28.50		79.50	98.50
	0940	4 hole 7-1/2"H x 6"W x 5-3/16"D		10.50	.762		60.50	30		90.50	112
	0950	6 hole 9-1/2"H x 6"W x 5-11/16"D		9	.889		74	35		109	134
	0960	8 hole 11-1/2"H x 8"W x 6-11/16"D		8.50	.941		93.50	37		130.50	158
	0970	12 hole 13-1/2"H x 10"W x 7-3/16"D		8	1		122	39.50		161.50	193
	0980	20 hole 15-1/2"H x 12"W x 8-3/16"D		5	1.600		166	63		229	277
	0990	30 hole 17-1/2"H x 14"W x 8-11/16"D	▼	4.50	1.778	▼	184	70		254	305
	1450	Enclosures polyester NEMA 4X									
	1460	Small, screw cover,									
	1500	3-15/16"H x 3-15/16"W x 3-1/16"D	1 Elec	12	.667	Ea.	38	26.50		64.50	80.50
	1510	5-3/16"H x 3-5/16"W x 3-1/16"D		12	.667		36.50	26.50		63	79.50
	1520	5-7/8"H x 3-7/8"W x 4-3/16"D		12	.667		39	26.50		65.50	82
	1530	5-7/8"H x 5-7/8"W x 4-3/16"D		12	.667		44.50	26.50		71	88
	1540	7-5/8"H x 3-5/16"W x 3-1/16"D		12	.667		40.50	26.50		67	83.50
	1550	10-3/16"H x 3-5/16"W x 3-1/16"D		10	.800		48.50	31.50		80	101
	1560	Clear cover, 3-15/16"H x 3-15/16"W x 2-7/8"D		12	.667		42.50	26.50		69	86
	1570	5-3/16"H x 3-5/16"W x 2-7/8"D		12	.667		45.50	26.50		72	89.50
	1580	5-7/8"H x 3-7/8"W x 4"D		12	.667		57	26.50		83.50	102
	1590	5-7/8"H x 5-7/8"W x 4"D		12	.667		70	26.50		96.50	116
	1600	7-5/8"H x 3-5/16"W x 2-7/8"D		12	.667		51	26.50		77.50	95
	1610	10-3/16"H x 3-5/16"W x 2-7/8"D		10	.800		64	31.50		95.50	118
	1620	Pushbutton, 1 hole, 5-5/16"H x 3-5/16"W x 3-1/16"D		12	.667		33.50	26.50		60	76
	1630	2 hole, 7-5/8"H x 3-5/16"W x 3-1/8"D		11	.727		37.50	28.50		66	83.50
	1640	3 hole, 10-3/16"H x 3-5/16"W x 3-1/16"D		10.50	.762		44.50	30		74.50	93.50
	8000	Wireway fiberglass, straight sect. screwcover, 12" L, 4" W x 4" D		40	.200		114	7.90		121.90	137
	8010	6" W x 6" D		30	.267		153	10.50		163.50	184
	8020	24" L, 4" W x 4" D		20	.400		145	15.75		160.75	184
	8030	6" W x 6" D		15	.533		224	21		245	278
	8040	36" L, 4" W x 4" D		13.30	.601		177	23.50		200.50	231
	8050	6" W x 6" D		10	.800		281	31.50		312.50	355
	8060	48" L, 4" W x 4" D		10	.800		208	31.50		239.50	276
	8070	6" W x 6" D		7.50	1.067		340	42		382	440
	8080	60" L, 4" W x 4" D		8	1		240	39.50		279.50	325
	8090	6" W x 6" D		6	1.333		395	52.50		447.50	515
	8100	Elbow, 90°, 4" W x 4" D		20	.400		113	15.75		128.75	148
	8110	6" W x 6" D		18	.444		227	17.50		244.50	276
	8120	Elbow, 45°, 4" W x 4" D		20	.400		113	15.75		128.75	148
	8130	6" W x 6" D		18	.444		223	17.50		240.50	271
	8140	Tee, 4" W x 4" D		16	.500		145	19.70		164.70	190
	8150	6" W x 6" D		14	.571		262	22.50		284.50	320
	8160	Cross, 4" W x 4" D		14	.571		204	22.50		226.50	258
	8170	6" W x 6" D		12	.667		380	26.50		406.50	460
	8180	Cut-off fitting, w/flange & adhesive, 4" W x 4" D		18	.444		88	17.50		105.50	123
	8190	6" W x 6" D		16	.500		177	19.70		196.70	225
	8200	Flexible ftng., hvy neoprene coated nylon, 4" W x 4" D		20	.400		145	15.75		160.75	184
	8210	6" W x 6" D		18	.444		217	17.50		234.50	265
	8220	Closure plate, fiberglass, 4" W x 4" D		20	.400		28	15.75		43.75	54.50
	8230	6" W x 6" D		18	.444		31.50	17.50		49	61
	8240	Box connector, stainless steel type 304, 4" W x 4" D		20	.400		49	15.75		64.75	77.50
	8250	6" W x 6" D		18	.444		54	17.50		71.50	85
	8260	Hanger, 4" W x 4" D		100	.080		12.20	3.15		15.35	18.10
	8270	6" W x 6" D		80	.100		14.60	3.94		18.54	22
	8280	Straight tube section fiberglass, 4"W x 4"D, 12" long		40	.200		99	7.90		106.90	121
	8290	24" long	▼	20	.400	▼	115	15.75		130.75	151

Important: See the Reference Section for critical supporting data - Reference Nos., Crews, & City Cost Indexes

16 ELECTRICAL

			CREW	DAILY OUTPUT	LABOR-HOURS	UNIT	2004 BARE COSTS				TOTAL INCL O&P	
		16136 Boxes					MAT.	LABOR	EQUIP.	TOTAL		
720	8300	36" long	1 Elec	13.30	.601	Ea.	143	23.50		166.50	193	**720**
	8310	48" long		10	.800		158	31.50		189.50	221	
	8320	60" long		8	1		179	39.50		218.50	256	
	8330	120" long		4	2		274	79		353	415	

16139 | Residential Wiring

			CREW	DAILY OUTPUT	LABOR-HOURS	UNIT	MAT.	LABOR	EQUIP.	TOTAL	TOTAL INCL O&P	
700	0010	**RESIDENTIAL WIRING**										**700**
	0020	20' avg. runs and #14/2 wiring incl. unless otherwise noted										
	1000	Service & panel, includes 24' SE-AL cable, service eye, meter,										
	1010	Socket, panel board, main bkr., ground rod, 15 or 20 amp										
	1020	1-pole circuit breakers, and misc. hardware										
	1100	100 amp, with 10 branch breakers	1 Elec	1.19	6.723	Ea.	410	265		675	850	
	1110	With PVC conduit and wire		.92	8.696		445	345		790	1,000	
	1120	With RGS conduit and wire		.73	10.959		575	430		1,005	1,275	
	1150	150 amp, with 14 branch breakers		1.03	7.767		645	305		950	1,175	
	1170	With PVC conduit and wire		.82	9.756		715	385		1,100	1,350	
	1180	With RGS conduit and wire		.67	11.940		955	470		1,425	1,750	
	1200	200 amp, with 18 branch breakers	2 Elec	1.80	8.889		840	350		1,190	1,450	
	1220	With PVC conduit and wire		1.46	10.959		915	430		1,345	1,650	
	1230	With RGS conduit and wire		1.24	12.903		1,225	510		1,735	2,100	
	1800	Lightning surge suppressor for above services, add	1 Elec	32	.250		40	9.85		49.85	58.50	
	2000	Switch devices										
	2100	Single pole, 15 amp, Ivory, with a 1-gang box, cover plate,										
	2110	Type NM (Romex) cable	1 Elec	17.10	.468	Ea.	6.75	18.45		25.20	35	
	2120	Type MC (BX) cable		14.30	.559		18.05	22		40.05	53	
	2130	EMT & wire		5.71	1.401		17.85	55		72.85	102	
	2150	3-way, #14/3, type NM cable		14.55	.550		10.10	21.50		31.60	43	
	2170	Type MC cable		12.31	.650		23	25.50		48.50	63.50	
	2180	EMT & wire		5	1.600		19.95	63		82.95	116	
	2200	4-way, #14/3, type NM cable		14.55	.550		23	21.50		44.50	57.50	
	2220	Type MC cable		12.31	.650		36	25.50		61.50	77.50	
	2230	EMT & wire		5	1.600		33	63		96	130	
	2250	S.P., 20 amp, #12/2, type NM cable		13.33	.600		11.90	23.50		35.40	48	
	2270	Type MC cable		11.43	.700		22.50	27.50		50	65.50	
	2280	EMT & wire		4.85	1.649		23.50	65		88.50	122	
	2290	S.P. rotary dimmer, 600W, no wiring		17	.471		16	18.55		34.55	45	
	2300	S.P. rotary dimmer, 600W, type NM cable		14.55	.550		18.45	21.50		39.95	52.50	
	2320	Type MC cable		12.31	.650		29.50	25.50		55	70.50	
	2330	EMT & wire		5	1.600		30	63		93	127	
	2350	3-way rotary dimmer, type NM cable		13.33	.600		16.15	23.50		39.65	53	
	2370	Type MC cable		11.43	.700		27.50	27.50		55	71	
	2380	EMT & wire		4.85	1.649		28	65		93	127	
	2400	Interval timer wall switch, 20 amp, 1-30 min., #12/2										
	2410	Type NM cable	1 Elec	14.55	.550	Ea.	29	21.50		50.50	64	
	2420	Type MC cable		12.31	.650		37.50	25.50		63	79	
	2430	EMT & wire		5	1.600		40.50	63		103.50	139	
	2500	Decorator style										
	2510	S.P., 15 amp, type NM cable	1 Elec	17.10	.468	Ea.	10.30	18.45		28.75	39	
	2520	Type MC cable		14.30	.559		21.50	22		43.50	57	
	2530	EMT & wire		5.71	1.401		21.50	55		76.50	106	
	2550	3-way, #14/3, type NM cable		14.55	.550		13.65	21.50		35.15	47	
	2570	Type MC cable		12.31	.650		26.50	25.50		52	67	
	2580	EMT & wire		5	1.600		23.50	63		86.50	120	
	2600	4-way, #14/3, type NM cable		14.55	.550		26.50	21.50		48	61	
	2620	Type MC cable		12.31	.650		39.50	25.50		65	81.50	
	2630	EMT & wire		5	1.600		36.50	63		99.50	134	

ELECTRICAL 16

16139	Residential Wiring	CREW	DAILY OUTPUT	LABOR-HOURS	UNIT	MAT.	LABOR	EQUIP.	TOTAL	TOTAL INCL O&P	
700								2004 BARE COSTS				**700**
2650	S.P., 20 amp, #12/2, type NM cable	1 Elec	13.33	.600	Ea.	15.45	23.50		38.95	52	
2670	Type MC cable		11.43	.700		26	27.50		53.50	69.50	
2680	EMT & wire		4.85	1.649		27	65		92	126	
2700	S.P., slide dimmer, type NM cable		17.10	.468		25.50	18.45		43.95	55.50	
2720	Type MC cable		14.30	.559		37	22		59	73.50	
2730	EMT & wire		5.71	1.401		37	55		92	123	
2750	S.P., touch dimmer, type NM cable		17.10	.468		22	18.45		40.45	51.50	
2770	Type MC cable		14.30	.559		33	22		55	69.50	
2780	EMT & wire		5.71	1.401		33.50	55		88.50	119	
2800	3-way touch dimmer, type NM cable		13.33	.600		40	23.50		63.50	79	
2820	Type MC cable		11.43	.700		51	27.50		78.50	97.50	
2830	EMT & wire	▼	4.85	1.649	▼	51.50	65		116.50	153	
3000	Combination devices										
3100	S.P. switch/15 amp recpt., Ivory, 1-gang box, plate										
3110	Type NM cable	1 Elec	11.43	.700	Ea.	14.80	27.50		42.30	57.50	
3120	Type MC cable		10	.800		26	31.50		57.50	75.50	
3130	EMT & wire		4.40	1.818		26.50	71.50		98	136	
3150	S.P. switch/pilot light, type NM cable		11.43	.700		15.45	27.50		42.95	58	
3170	Type MC cable		10	.800		26.50	31.50		58	76.50	
3180	EMT & wire		4.43	1.806		27	71		98	136	
3190	2-S.P. switches, 2-#14/2, no wiring		14	.571		5.70	22.50		28.20	40	
3200	2-S.P. switches, 2-#14/2, type NM cables		10	.800		16.55	31.50		48.05	65	
3220	Type MC cable		8.89	.900		36	35.50		71.50	92	
3230	EMT & wire		4.10	1.951		27.50	77		104.50	145	
3250	3-way switch/15 amp recpt., #14/3, type NM cable		10	.800		21	31.50		52.50	70	
3270	Type MC cable		8.89	.900		34	35.50		69.50	90	
3280	EMT & wire		4.10	1.951		31	77		108	148	
3300	2-3 way switches, 2-#14/3, type NM cables		8.89	.900		27.50	35.50		63	82.50	
3320	Type MC cable		8	1		50	39.50		89.50	114	
3330	EMT & wire		4	2		34.50	79		113.50	155	
3350	S.P. switch/20 amp recpt., #12/2, type NM cable		10	.800		25	31.50		56.50	74.50	
3370	Type MC cable		8.89	.900		33.50	35.50		69	89	
3380	EMT & wire	▼	4.10	1.951	▼	36.50	77		113.50	155	
3400	Decorator style										
3410	S.P. switch/15 amp recpt., type NM cable	1 Elec	11.43	.700	Ea.	18.35	27.50		45.85	61	
3420	Type MC cable		10	.800		29.50	31.50		61	79.50	
3430	EMT & wire		4.40	1.818		30	71.50		101.50	140	
3450	S.P. switch/pilot light, type NM cable		11.43	.700		19	27.50		46.50	62	
3470	Type MC cable		10	.800		30.50	31.50		62	80.50	
3480	EMT & wire		4.40	1.818		30.50	71.50		102	141	
3500	2-S.P. switches, 2-#14/2, type NM cables		10	.800		20	31.50		51.50	69	
3520	Type MC cable		8.89	.900		39.50	35.50		75	96	
3530	EMT & wire		4.10	1.951		31	77		108	148	
3550	3-way/15 amp recpt., #14/3, type NM cable		10	.800		24.50	31.50		56	74	
3570	Type MC cable		8.89	.900		37.50	35.50		73	93.50	
3580	EMT & wire		4.10	1.951		34.50	77		111.50	152	
3650	2-3 way switches, 2-#14/3, type NM cables		8.89	.900		31	35.50		66.50	86.50	
3670	Type MC cable		8	1		53.50	39.50		93	118	
3680	EMT & wire		4	2		38.50	79		117.50	159	
3700	S.P. switch/20 amp recpt., #12/2, type NM cable		10	.800		29	31.50		60.50	78.50	
3720	Type MC cable		8.89	.900		37	35.50		72.50	93	
3730	EMT & wire	▼	4.10	1.951	▼	40	77		117	158	
4000	Receptacle devices										
4010	Duplex outlet, 15 amp recpt., Ivory, 1-gang box, plate										
4015	Type NM cable	1 Elec	14.55	.550	Ea.	5.25	21.50		26.75	38	
4020	Type MC cable	▼	12.31	.650	▼	16.55	25.50		42.05	56	

Important: See the Reference Section for critical supporting data - Reference Nos., Crews, & City Cost Indexes

			DAILY	LABOR-		2004 BARE COSTS				TOTAL	
16139	**Residential Wiring**	CREW	OUTPUT	HOURS	UNIT	MAT.	LABOR	EQUIP.	TOTAL	INCL O&P	
700	4030	EMT & wire	1 Elec	5.33	1.501	Ea.	16.35	59		75.35	106	700
	4050	With #12/2, type NM cable		12.31	.650		6.25	25.50		31.75	45	
	4070	Type MC cable		10.67	.750		16.75	29.50		46.25	62.50	
	4080	EMT & wire		4.71	1.699		17.65	67		84.65	119	
	4100	20 amp recpt., #12/2, type NM cable		12.31	.650		12.15	25.50		37.65	51.50	
	4120	Type MC cable		10.67	.750		22.50	29.50		52	69	
	4130	EMT & wire	▼	4.71	1.699	▼	23.50	67		90.50	126	
	4140	For GFI see line 4300 below										
	4150	Decorator style, 15 amp recpt., type NM cable	1 Elec	14.55	.550	Ea.	8.80	21.50		30.30	41.50	
	4170	Type MC cable		12.31	.650		20	25.50		45.50	60	
	4180	EMT & wire		5.33	1.501		19.90	59		78.90	110	
	4200	With #12/2, type NM cable		12.31	.650		9.80	25.50		35.30	49	
	4220	Type MC cable		10.67	.750		20.50	29.50		50	66.50	
	4230	EMT & wire		4.71	1.699		21	67		88	123	
	4250	20 amp recpt. #12/2, type NM cable		12.31	.650		15.70	25.50		41.20	55.50	
	4270	Type MC cable		10.67	.750		26	29.50		55.50	73	
	4280	EMT & wire		4.71	1.699		27	67		94	130	
	4300	GFI, 15 amp recpt., type NM cable		12.31	.650		33	25.50		58.50	74	
	4320	Type MC cable		10.67	.750		44	29.50		73.50	92.50	
	4330	EMT & wire		4.71	1.699		44	67		111	148	
	4350	GFI with #12/2, type NM cable		10.67	.750		34	29.50		63.50	81.50	
	4370	Type MC cable		9.20	.870		44.50	34.50		79	100	
	4380	EMT & wire		4.21	1.900		45.50	75		120.50	161	
	4400	20 amp recpt., #12/2 type NM cable		10.67	.750		35.50	29.50		65	83	
	4420	Type MC cable		9.20	.870		46	34.50		80.50	102	
	4430	EMT & wire		4.21	1.900		47	75		122	163	
	4500	Weather-proof cover for above receptacles, add	▼	32	.250	▼	4.40	9.85		14.25	19.50	
	4550	Air conditioner outlet, 20 amp-240 volt recpt.										
	4560	30' of #12/2, 2 pole circuit breaker										
	4570	Type NM cable	1 Elec	10	.800	Ea.	40.50	31.50		72	91.50	
	4580	Type MC cable		9	.889		54.50	35		89.50	112	
	4590	EMT & wire		4	2		51.50	79		130.50	174	
	4600	Decorator style, type NM cable		10	.800		44.50	31.50		76	96	
	4620	Type MC cable		9	.889		58.50	35		93.50	117	
	4630	EMT & wire	▼	4	2	▼	55.50	79		134.50	178	
	4650	Dryer outlet, 30 amp-240 volt recpt., 20' of #10/3										
	4660	2 pole circuit breaker										
	4670	Type NM cable	1 Elec	6.41	1.248	Ea.	47	49		96	125	
	4680	Type MC cable		5.71	1.401		60.50	55		115.50	149	
	4690	EMT & wire	▼	3.48	2.299	▼	52.50	90.50		143	193	
	4700	Range outlet, 50 amp-240 volt recpt., 30' of #8/3										
	4710	Type NM cable	1 Elec	4.21	1.900	Ea.	68.50	75		143.50	187	
	4720	Type MC cable		4	2		106	79		185	234	
	4730	EMT & wire		2.96	2.703		67	106		173	232	
	4750	Central vacuum outlet, Type NM cable		6.40	1.250		42	49.50		91.50	120	
	4770	Type MC cable		5.71	1.401		64	55		119	153	
	4780	EMT & wire	▼	3.48	2.299	▼	53.50	90.50		144	194	
	4800	30 amp-110 volt locking recpt., #10/2 circ. bkr.										
	4810	Type NM cable	1 Elec	6.20	1.290	Ea.	50	51		101	131	
	4820	Type MC cable		5.40	1.481		76	58.50		134.50	171	
	4830	EMT & wire	▼	3.20	2.500	▼	61	98.50		159.50	214	
	4900	Low voltage outlets										
	4910	Telephone recpt., 20' of 4/C phone wire	1 Elec	26	.308	Ea.	6.70	12.10		18.80	25.50	
	4920	TV recpt., 20' of RG59U coax wire, F type connector	"	16	.500	"	11.15	19.70		30.85	42	
	4950	Door bell chime, transformer, 2 buttons, 60' of bellwire										
	4970	Economy model	1 Elec	11.50	.696	Ea.	51.50	27.50		79	98	

ELECTRICAL 16

16139 | Residential Wiring

		CREW	DAILY OUTPUT	LABOR-HOURS	UNIT	2004 BARE COSTS				TOTAL INCL O&P
						MAT.	LABOR	EQUIP.	TOTAL	
700	4980	Custom model	1 Elec	11.50	.696	Ea.	82.50	27.50		110	132
	4990	Luxury model, 3 buttons	↓	9.50	.842	↓	224	33		257	296
	6000	Lighting outlets									
	6050	Wire only (for fixture), type NM cable	1 Elec	32	.250	Ea.	3.31	9.85		13.16	18.30
	6070	Type MC cable		24	.333		12.65	13.15		25.80	33.50
	6080	EMT & wire		10	.800		11.45	31.50		42.95	59.50
	6100	Box (4"), and wire (for fixture), type NM cable		25	.320		7.35	12.60		19.95	27
	6120	Type MC cable		20	.400		16.70	15.75		32.45	42
	6130	EMT & wire	↓	11	.727	↓	15.45	28.50		43.95	59.50
	6200	Fixtures (use with lines 6050 or 6100 above)									
	6210	Canopy style, economy grade	1 Elec	40	.200	Ea.	25.50	7.90		33.40	39.50
	6220	Custom grade		40	.200		46	7.90		53.90	62
	6250	Dining room chandelier, economy grade		19	.421		76	16.60		92.60	108
	6260	Custom grade		19	.421		225	16.60		241.60	273
	6270	Luxury grade		15	.533		495	21		516	575
	6310	Kitchen fixture (fluorescent), economy grade		30	.267		51.50	10.50		62	72
	6320	Custom grade		25	.320		161	12.60		173.60	196
	6350	Outdoor, wall mounted, economy grade		30	.267		27	10.50		37.50	45
	6360	Custom grade		30	.267		101	10.50		111.50	127
	6370	Luxury grade		25	.320		227	12.60		239.60	269
	6410	Outdoor PAR floodlights, 1 lamp, 150 watt		20	.400		21	15.75		36.75	46.50
	6420	2 lamp, 150 watt each		20	.400		35	15.75		50.75	62
	6430	For infrared security sensor, add		32	.250		87	9.85		96.85	111
	6450	Outdoor, quartz-halogen, 300 watt flood		20	.400		38	15.75		53.75	65.50
	6600	Recessed downlight, round, pre-wired, 50 or 75 watt trim		30	.267		35	10.50		45.50	54
	6610	With shower light trim		30	.267		43	10.50		53.50	63
	6620	With wall washer trim		28	.286		52.50	11.25		63.75	75
	6630	With eye-ball trim	↓	28	.286		52.50	11.25		63.75	75
	6640	For direct contact with insulation, add					1.60			1.60	1.76
	6700	Porcelain lamp holder	1 Elec	40	.200		3.50	7.90		11.40	15.55
	6710	With pull switch		40	.200		3.84	7.90		11.74	15.90
	6750	Fluorescent strip, 1-20 watt tube, wrap around diffuser, 24"		24	.333		51.50	13.15		64.65	76
	6760	1-40 watt tube, 48"		24	.333		65	13.15		78.15	91
	6770	2-40 watt tubes, 48"		20	.400		79	15.75		94.75	111
	6780	With residential ballast		20	.400		89.50	15.75		105.25	122
	6800	Bathroom heat lamp, 1-250 watt		28	.286		34.50	11.25		45.75	55
	6810	2-250 watt lamps	↓	28	.286	↓	55	11.25		66.25	77.50
	6820	For timer switch, see line 2400									
	6900	Outdoor post lamp, incl. post, fixture, 35' of #14/2									
	6910	Type NMC cable	1 Elec	3.50	2.286	Ea.	181	90		271	335
	6920	Photo-eye, add		27	.296		29	11.65		40.65	49.50
	6950	Clock dial time switch, 24 hr., w/enclosure, type NM cable		11.43	.700		52.50	27.50		80	98.50
	6970	Type MC cable		11	.727		63.50	28.50		92	113
	6980	EMT & wire	↓	4.85	1.649	↓	63.50	65		128.50	166
	7000	Alarm systems									
	7050	Smoke detectors, box, #14/3, type NM cable	1 Elec	14.55	.550	Ea.	28	21.50		49.50	62.50
	7070	Type MC cable		12.31	.650		38.50	25.50		64	80
	7080	EMT & wire	↓	5	1.600		35.50	63		98.50	133
	7090	For relay output to security system, add				↓	11.75			11.75	12.95
	8000	Residential equipment									
	8050	Disposal hook-up, incl. switch, outlet box, 3' of flex									
	8060	20 amp-1 pole circ. bkr., and 25' of #12/2									
	8070	Type NM cable	1 Elec	10	.800	Ea.	19.60	31.50		51.10	68.50
	8080	Type MC cable		8	1		32	39.50		71.50	93.50
	8090	EMT & wire	↓	5	1.600	↓	32.50	63		95.50	130
	8100	Trash compactor or dishwasher hook-up, incl. outlet box,									

Important: See the Reference Section for critical supporting data - Reference Nos., Crews, & City Cost Indexes

			DAILY	LABOR-		2004 BARE COSTS				TOTAL	
16139	**Residential Wiring**	CREW	OUTPUT	HOURS	UNIT	MAT.	LABOR	EQUIP.	TOTAL	INCL O&P	
700	8110	3' of flex, 15 amp-1 pole circ. bkr., and 25' of #14/2										700
	8120	Type NM cable	1 Elec	10	.800	Ea.	13.95	31.50		45.45	62.50	
	8130	Type MC cable		8	1		27.50	39.50		67	88.50	
	8140	EMT & wire	↓	5	1.600	↓	27	63		90	124	
	8150	Hot water sink dispensor hook-up, use line 8100										
	8200	Vent/exhaust fan hook-up, type NM cable	1 Elec	32	.250	Ea.	3.31	9.85		13.16	18.30	
	8220	Type MC cable		24	.333		12.65	13.15		25.80	33.50	
	8230	EMT & wire	↓	10	.800	↓	11.45	31.50		42.95	59.50	
	8250	Bathroom vent fan, 50 CFM (use with above hook-up)										
	8260	Economy model	1 Elec	15	.533	Ea.	21.50	21		42.50	55	
	8270	Low noise model		15	.533		30	21		51	64.50	
	8280	Custom model	↓	12	.667	↓	111	26.50		137.50	161	
	8300	Bathroom or kitchen vent fan, 110 CFM										
	8310	Economy model	1 Elec	15	.533	Ea.	56.50	21		77.50	93.50	
	8320	Low noise model	"	15	.533	"	75	21		96	114	
	8350	Paddle fan, variable speed (w/o lights)										
	8360	Economy model (AC motor)	1 Elec	10	.800	Ea.	100	31.50		131.50	157	
	8370	Custom model (AC motor)		10	.800		173	31.50		204.50	237	
	8380	Luxury model (DC motor)		8	1		340	39.50		379.50	435	
	8390	Remote speed switch for above, add	↓	12	.667	↓	24.50	26.50		51	66	
	8500	Whole house exhaust fan, ceiling mount, 36", variable speed										
	8510	Remote switch, incl. shutters, 20 amp-1 pole circ. bkr.										
	8520	30' of #12/2, type NM cable	1 Elec	4	2	Ea.	610	79		689	785	
	8530	Type MC cable		3.50	2.286		625	90		715	820	
	8540	EMT & wire	↓	3	2.667	↓	625	105		730	845	
	8600	Whirlpool tub hook-up, incl. timer switch, outlet box										
	8610	3' of flex, 20 amp-1 pole GFI circ. bkr.										
	8620	30' of #12/2, type NM cable	1 Elec	5	1.600	Ea.	79.50	63		142.50	182	
	8630	Type MC cable		4.20	1.905		88.50	75		163.50	209	
	8640	EMT & wire	↓	3.40	2.353	↓	89.50	92.50		182	236	
	8650	Hot water heater hook-up, incl. 1-2 pole circ. bkr., box;										
	8660	3' of flex, 20' of #10/2, type NM cable	1 Elec	5	1.600	Ea.	20.50	63		83.50	117	
	8670	Type MC cable		4.20	1.905		36.50	75		111.50	152	
	8680	EMT & wire	↓	3.40	2.353	↓	28.50	92.50		121	170	
	9000	Heating/air conditioning										
	9050	Furnace/boiler hook-up, incl. firestat, local on-off switch										
	9060	Emergency switch, and 40' of type NM cable	1 Elec	4	2	Ea.	42.50	79		121.50	164	
	9070	Type MC cable		3.50	2.286		62	90		152	202	
	9080	EMT & wire	↓	1.50	5.333	↓	61	210		271	385	
	9100	Air conditioner hook-up, incl. local 60 amp disc. switch										
	9110	3' sealtite, 40 amp, 2 pole circuit breaker										
	9130	40' of #8/2, type NM cable	1 Elec	3.50	2.286	Ea.	142	90		232	290	
	9140	Type MC cable		3	2.667		199	105		304	375	
	9150	EMT & wire	↓	1.30	6.154	↓	152	242		394	525	
	9200	Heat pump hook-up, incl-1-40 & 1-100 amp 2 pole circ. bkr.										
	9210	Local disconnect switch, 3' sealtite										
	9220	40' of #8/2 & 30' of #3/2										
	9230	Type NM cable	1 Elec	1.30	6.154	Ea.	325	242		567	715	
	9240	Type MC cable		1.08	7.407		500	292		792	985	
	9250	EMT & wire	↓	.94	8.511	↓	335	335		670	870	
	9500	Thermostat hook-up, using low voltage wire										
	9520	Heating only	1 Elec	24	.333	Ea.	4.74	13.15		17.89	25	
	9530	Heating/cooling	"	20	.400	"	5.70	15.75		21.45	30	

ELECTRICAL 16

			DAILY	LABOR-		2004 BARE COSTS				TOTAL	
16140	**Wiring Devices**	CREW	OUTPUT	HOURS	UNIT	MAT.	LABOR	EQUIP.	TOTAL	INCL O&P	
500	**0010**	**LOW VOLTAGE SWITCHING**										**500**
	3600	Relays, 120 V or 277 V standard	1 Elec	12	.667	Ea.	26	26.50		52.50	67.50
	3800	Flush switch, standard		40	.200		9.05	7.90		16.95	21.50
	4000	Interchangeable		40	.200		11.80	7.90		19.70	24.50
	4100	Surface switch, standard		40	.200		6.60	7.90		14.50	19
	4200	Transformer 115 V to 25 V		12	.667		93	26.50		119.50	141
	4400	Master control, 12 circuit, manual		4	2		94	79		173	220
	4500	25 circuit, motorized		4	2		102	79		181	229
	4600	Rectifier, silicon		12	.667		30.50	26.50		57	72.50
	4800	Switchplates, 1 gang, 1, 2 or 3 switch, plastic		80	.100		3	3.94		6.94	9.15
	5000	Stainless steel		80	.100		8.10	3.94		12.04	14.75
	5400	2 gang, 3 switch, stainless steel		53	.151		15.65	5.95		21.60	26
	5500	4 switch, plastic		53	.151		6.70	5.95		12.65	16.20
	5600	2 gang, 4 switch, stainless steel		53	.151		16.35	5.95		22.30	27
	5700	6 switch, stainless steel		53	.151		36	5.95		41.95	48.50
	5800	3 gang, 9 switch, stainless steel		32	.250		50	9.85		59.85	69.50
	5900	Receptacle, triple, 1 return, 1 feed		26	.308		32	12.10		44.10	53
	6000	2 feed		20	.400		32	15.75		47.75	58.50
	6100	Relay gang boxes, flush or surface, 6 gang		5.30	1.509		55	59.50		114.50	149
	6200	12 gang		4.70	1.702		58	67		125	164
	6400	18 gang		4	2		72	79		151	196
	6500	Frame, to hold up to 6 relays		12	.667		48	26.50		74.50	92
	7200	Control wire, 2 conductor		6.30	1.270	C.L.F.	12.45	50		62.45	88
	7400	3 conductor		5	1.600		19	63		82	115
	7600	19 conductor		2.50	3.200		150	126		276	355
	7800	26 conductor		2	4		230	158		388	485
	8000	Weatherproof, 3 conductor		5	1.600		44	63		107	143
910	**0010**	**WIRING DEVICES** R16140-910										**910**
	0200	Toggle switch, quiet type, single pole, 15 amp	1 Elec	40	.200	Ea.	4.69	7.90		12.59	16.85
	0500	20 amp		27	.296		6.80	11.65		18.45	25
	0510	30 amp		23	.348		19	13.70		32.70	41.50
	0530	Lock handle, 20 amp		27	.296		21.50	11.65		33.15	41
	0540	Security key, 20 amp		26	.308		53.50	12.10		65.60	77
	0550	Rocker, 15 amp		40	.200		5.05	7.90		12.95	17.25
	0560	20 amp		27	.296		11.70	11.65		23.35	30
	0600	3 way, 15 amp		23	.348		6.70	13.70		20.40	28
	0800	20 amp		18	.444		7.75	17.50		25.25	34.50
	0810	30 amp		9	.889		27	35		62	82
	0830	Lock handle, 20 amp		18	.444		24	17.50		41.50	52
	0840	Security key, 20 amp		17	.471		58	18.55		76.55	91.50
	0850	Rocker, 15 amp		23	.348		7.10	13.70		20.80	28.50
	0860	20 amp		18	.444		17	17.50		34.50	44.50
	0900	4 way, 15 amp		15	.533		20	21		41	53.50
	1000	20 amp		11	.727		33.50	28.50		62	79.50
	1020	Lock handle, 20 amp		11	.727		53	28.50		81.50	101
	1030	Rocker, 15 amp		15	.533		27.50	21		48.50	61.50
	1040	20 amp		11	.727		42	28.50		70.50	88.50
	1100	Toggle switch, quiet type, double pole, 15 amp		15	.533		13.80	21		34.80	46.50
	1200	20 amp		11	.727		13.45	28.50		41.95	57.50
	1210	30 amp		9	.889		27.50	35		62.50	82.50
	1230	Lock handle, 20 amp		11	.727		24.50	28.50		53	69
	1250	Security key, 20 amp		10	.800		58	31.50		89.50	111
	1420	Toggle switch quiet type, 1 pole, 2 throw center off, 15 amp		23	.348		46	13.70		59.70	71
	1440	20 amp		18	.444		50.50	17.50		68	81.50
	1460	Lock handle, 20 amp		18	.444		51	17.50		68.50	82

Important: See the Reference Section for critical supporting data - Reference Nos., Crews, & City Cost Indexes

16140 | Wiring Devices

		CREW	DAILY OUTPUT	LABOR-HOURS	UNIT	2004 BARE COSTS				TOTAL INCL O&P	
						MAT.	LABOR	EQUIP.	TOTAL		
910	1480	Momentary contact, 15 amp	1 Elec	23	.348	Ea.	19.95	13.70		33.65	42.50	910
	1500	20 amp		18	.444		27	17.50		44.50	56	
	1520	Momentary contact, lock handle, 20 amp		18	.444		35	17.50		52.50	64.50	
	1650	Dimmer switch, 120 volt, incandescent, 600 watt, 1 pole		16	.500		10.80	19.70		30.50	41.50	
	1700	600 watt, 3 way		12	.667		8.50	26.50		35	48.50	
	1750	1000 watt, 1 pole		16	.500		44.50	19.70		64.20	78.50	
	1800	1000 watt, 3 way		12	.667		64	26.50		90.50	110	
	2000	1500 watt, 1 pole		11	.727		87	28.50		115.50	138	
	2100	2000 watt, 1 pole		8	1		130	39.50		169.50	202	
	2110	Fluorescent, 600 watt		15	.533		67	21		88	105	
	2120	1000 watt		15	.533		83.50	21		104.50	124	
	2130	1500 watt		10	.800		157	31.50		188.50	220	
	2160	Explosionproof, toggle switch, wall, single pole 20 amp		5.30	1.509		133	59.50		192.50	236	
	2180	Receptacle, single outlet, 20 amp		5.30	1.509		211	59.50		270.50	320	
	2190	30 amp		4	2		415	79		494	570	
	2290	60 amp		2.50	3.200		465	126		591	700	
	2360	Plug, 20 amp		16	.500		96	19.70		115.70	135	
	2370	30 amp		12	.667		170	26.50		196.50	226	
	2380	60 amp		8	1		228	39.50		267.50	310	
	2410	Furnace, thermal cutoff switch with plate		26	.308		12.40	12.10		24.50	31.50	
	2460	Receptacle, duplex, 120 volt, grounded, 15 amp		40	.200		1.14	7.90		9.04	12.95	
	2470	20 amp		27	.296		7.05	11.65		18.70	25	
	2480	Ground fault interrupting, 15 amp		27	.296		29	11.65		40.65	49	
	2482	20 amp		27	.296		30.50	11.65		42.15	51	
	2486	Clock receptacle, 15 amp		40	.200		18.50	7.90		26.40	32	
	2490	Dryer, 30 amp		15	.533		10.35	21		31.35	43	
	2500	Range, 50 amp		11	.727		10.75	28.50		39.25	54.50	
	2530	Surge suppresser receptacle, duplex, 20 amp		27	.296		41	11.65		52.65	63	
	2540	Isolated ground receptacle, duplex, 20 amp		27	.296		20	11.65		31.65	39.50	
	2550	Simplex, 20 amp		27	.296		22.50	11.65		34.15	42.50	
	2560	Simplex, 30 amp		15	.533		26.50	21		47.50	60.50	
	2600	Wall plates, stainless steel, 1 gang		80	.100		1.80	3.94		5.74	7.85	
	2800	2 gang		53	.151		4.10	5.95		10.05	13.35	
	3000	3 gang		32	.250		5.50	9.85		15.35	20.50	
	3100	4 gang		27	.296		8.45	11.65		20.10	26.50	
	3110	Brown plastic, 1 gang		80	.100		.30	3.94		4.24	6.20	
	3120	2 gang		53	.151		.77	5.95		6.72	9.70	
	3130	3 gang		32	.250		1.14	9.85		10.99	15.90	
	3140	4 gang		27	.296		1.78	11.65		13.43	19.30	
	3150	Brushed brass, 1 gang		80	.100		3.85	3.94		7.79	10.10	
	3160	Anodized aluminum, 1 gang		80	.100		2.35	3.94		6.29	8.45	
	3170	Switch cover, weatherproof, 1 gang		60	.133		7.25	5.25		12.50	15.80	
	3180	Vandal proof lock, 1 gang		60	.133		6.55	5.25		11.80	15	
	3200	Lampholder, keyless		26	.308		9.20	12.10		21.30	28	
	3400	Pullchain with receptacle		22	.364		8.90	14.35		23.25	31.50	
	3500	Pilot light, neon with jewel		27	.296		10.90	11.65		22.55	29.50	
	3600	Receptacle, 20 amp, 250 volt, NEMA 6		27	.296		15	11.65		26.65	34	
	3620	277 volt NEMA 7		27	.296		17.10	11.65		28.75	36	
	3640	125/250 volt NEMA 10		27	.296		17.95	11.65		29.60	37	
	3680	125/250 volt NEMA 14		25	.320		22.50	12.60		35.10	44	
	3700	3 pole, 250 volt NEMA 15		25	.320		23.50	12.60		36.10	45	
	3720	120/208 volt NEMA 18		25	.320		25	12.60		37.60	46	
	3740	30 amp, 125 volt NEMA 5		15	.533		15.95	21		36.95	49	
	3760	250 volt NEMA 6		15	.533		18.25	21		39.25	51.50	
	3780	277 volt NEMA 7		15	.533		23.50	21		44.50	57	
	3820	125/250 volt NEMA 14		14	.571		50.50	22.50		73	89	

R16140 -910

ELECTRICAL 16

	16140	Wiring Devices	CREW	DAILY OUTPUT	LABOR-HOURS	UNIT	2004 BARE COSTS				TOTAL INCL O&P	
							MAT.	LABOR	EQUIP.	TOTAL		
910	3840	3 pole, 250 volt NEMA 15	1 Elec	14	.571	Ea.	47	22.50		69.50	85	910
	3880	50 amp, 125 volt NEMA 5		11	.727		21	28.50		49.50	65.50	
	3900	250 volt NEMA 6		11	.727		21	28.50		49.50	65.50	
	3920	277 volt NEMA 7		11	.727		27	28.50		55.50	72.50	
	3960	125/250 volt NEMA 14		10	.800		65	31.50		96.50	119	
	3980	3 pole 250 volt NEMA 15		10	.800		64	31.50		95.50	118	
	4020	60 amp, 125/250 volt, NEMA 14		8	1		72	39.50		111.50	138	
	4040	3 pole, 250 volt NEMA 15		8	1		75.50	39.50		115	142	
	4060	120/208 volt NEMA 18		8	1		66	39.50		105.50	131	
	4100	Receptacle locking, 20 amp, 125 volt NEMA L5		27	.296		16.10	11.65		27.75	35	
	4120	250 volt NEMA L6		27	.296		16.10	11.65		27.75	35	
	4140	277 volt NEMA L7		27	.296		16.10	11.65		27.75	35	
	4150	3 pole, 250 volt, NEMA L11		27	.296		25.50	11.65		37.15	45.50	
	4160	20 amp, 480 volt NEMA L8		27	.296		20.50	11.65		32.15	40	
	4180	600 volt NEMA L9		27	.296		30.50	11.65		42.15	51.50	
	4200	125/250 volt NEMA L10		27	.296		25.50	11.65		37.15	45.50	
	4230	125/250 volt NEMA L14		25	.320		22.50	12.60		35.10	44	
	4280	250 volt NEMA L15		25	.320		22.50	12.60		35.10	44	
	4300	480 volt NEMA L16		25	.320		27	12.60		39.60	48.50	
	4320	3 phase, 120/208 volt NEMA L18		25	.320		34.50	12.60		47.10	57	
	4340	277/480 volt NEMA L19		25	.320		35	12.60		47.60	57.50	
	4360	347/600 volt NEMA L20		25	.320		35	12.60		47.60	57.50	
	4380	120/208 volt NEMA L21		23	.348		27.50	13.70		41.20	50.50	
	4400	277/480 volt NEMA L22		23	.348		33	13.70		46.70	57	
	4420	347/600 volt NEMA L23		23	.348		33	13.70		46.70	57	
	4440	30 amp, 125 volt NEMA L5		15	.533		23	21		44	56.50	
	4460	250 volt NEMA L6		15	.533		24.50	21		45.50	58.50	
	4480	277 volt NEMA L7		15	.533		24.50	21		45.50	58.50	
	4500	480 volt NEMA L8		15	.533		29.50	21		50.50	64	
	4520	600 volt NEMA L9		15	.533		29.50	21		50.50	64	
	4540	125/250 volt NEMA L10		15	.533		31.50	21		52.50	66	
	4560	3 phase, 250 volt NEMA L11		15	.533		31.50	21		52.50	66	
	4620	125/250 volt NEMA L14		14	.571		35.50	22.50		58	72.50	
	4640	250 volt NEMA L15		14	.571		35.50	22.50		58	72.50	
	4660	480 volt NEMA L16		14	.571		42	22.50		64.50	80	
	4680	600 volt NEMA L17		14	.571		41	22.50		63.50	78.50	
	4700	120/208 volt NEMA L18		14	.571		51	22.50		73.50	89.50	
	4720	277/480 volt NEMA L19		14	.571		51	22.50		73.50	89.50	
	4740	347/600 volt NEMA L20		14	.571		51.50	22.50		74	90	
	4760	120/208 volt NEMA L21		13	.615		38	24.50		62.50	78	
	4780	277/480 volt NEMA L22		13	.615		45.50	24.50		70	86	
	4800	347/600 volt NEMA L23		13	.615		56.50	24.50		81	98.50	
	4840	Receptacle, corrosion resistant, 15 or 20 amp, 125 volt NEMA L5		27	.296		23	11.65		34.65	43	
	4860	250 volt NEMA L6		27	.296		23	11.65		34.65	43	
	4900	Receptacle, cover plate, phenolic plastic, NEMA 5 & 6		80	.100		.46	3.94		4.40	6.35	
	4910	NEMA 7-23		80	.100		.53	3.94		4.47	6.45	
	4920	Stainless steel, NEMA 5 & 6		80	.100		1.76	3.94		5.70	7.80	
	4930	NEMA 7-23		80	.100		2.75	3.94		6.69	8.90	
	4940	Brushed brass NEMA 5 & 6		80	.100		4.35	3.94		8.29	10.65	
	4950	NEMA 7-23		80	.100		4.70	3.94		8.64	11	
	4960	Anodized aluminum, NEMA 5 & 6		80	.100		2.40	3.94		6.34	8.50	
	4970	NEMA 7-23		80	.100		3.90	3.94		7.84	10.15	
	4980	Weatherproof NEMA 7-23		60	.133		26.50	5.25		31.75	37	
	5100	Plug, 20 amp, 250 volt, NEMA 6		30	.267		13.65	10.50		24.15	30.50	
	5110	277 volt NEMA 7		30	.267		14.95	10.50		25.45	32	
	5120	3 pole, 120/250 volt, NEMA 10		26	.308		19.10	12.10		31.20	39	

R16140 -910

Important: See the Reference Section for critical supporting data - Reference Nos., Crews, & City Cost Indexes

			CREW	DAILY OUTPUT	LABOR-HOURS	UNIT	2004 BARE COSTS				TOTAL INCL O&P	
16140	**Wiring Devices**						MAT.	LABOR	EQUIP.	TOTAL		
910	5130	125/250 volt NEMA 14	R16140 -910	1 Elec	26	.308	Ea.	38	12.10		50.10	59.50	910
	5140	250 volt NEMA 15			26	.308		40	12.10		52.10	62	
	5150	120/208 volt NEMA 8			26	.308		41	12.10		53.10	63	
	5160	30 amp, 125 volt NEMA 5			13	.615		39	24.50		63.50	78.50	
	5170	250 volt NEMA 6			13	.615		40	24.50		64.50	80	
	5180	277 volt NEMA 7			13	.615		49.50	24.50		74	90.50	
	5190	125/250 volt NEMA 14			13	.615		44.50	24.50		69	85	
	5200	3 pole, 250 volt NEMA 15			12	.667		46.50	26.50		73	90.50	
	5210	50 amp, 125 volt NEMA 5			9	.889		48.50	35		83.50	105	
	5220	250 volt NEMA 6			9	.889		50.50	35		85.50	108	
	5230	277 volt NEMA 7			9	.889		54	35		89	112	
	5240	125/250 volt NEMA 14			9	.889		54	35		89	112	
	5250	3 pole, 250 volt NEMA 15			8	1		56.50	39.50		96	121	
	5260	60 amp, 125/250 volt NEMA 14			7	1.143		62.50	45		107.50	136	
	5270	3 pole, 250 volt NEMA 15			7	1.143		66.50	45		111.50	141	
	5280	120/208 volt NEMA 18			7	1.143		65	45		110	139	
	5300	Plug angle, 20 amp, 250 volt NEMA 6			30	.267		18.80	10.50		29.30	36	
	5310	30 amp, 125 volt NEMA 5			13	.615		39	24.50		63.50	78.50	
	5320	250 volt NEMA 6			13	.615		40	24.50		64.50	80	
	5330	277 volt NEMA 7			13	.615		49.50	24.50		74	90.50	
	5340	125/250 volt NEMA 14			13	.615		49.50	24.50		74	90	
	5350	3 pole, 250 volt NEMA 15			12	.667		52	26.50		78.50	96	
	5360	50 amp, 125 volt NEMA 5			9	.889		42.50	35		77.50	99	
	5370	250 volt NEMA 6			9	.889		42	35		77	98	
	5380	277 volt NEMA 7			9	.889		54.50	35		89.50	112	
	5390	125/250 volt NEMA 14			9	.889		59.50	35		94.50	118	
	5400	3 pole, 250 volt NEMA 15			8	1		62.50	39.50		102	128	
	5410	60 amp, 125/250 volt NEMA 14			7	1.143		70	45		115	144	
	5420	3 pole, 250 volt NEMA 15			7	1.143		72.50	45		117.50	147	
	5430	120/208 volt NEMA 18			7	1.143		72.50	45		117.50	147	
	5500	Plug, locking, 20 amp, 125 volt NEMA L5			30	.267		12.60	10.50		23.10	29.50	
	5510	250 volt NEMA L6			30	.267		12.60	10.50		23.10	29.50	
	5520	277 volt NEMA L7			30	.267		12.60	10.50		23.10	29.50	
	5530	480 volt NEMA L8			30	.267		15.80	10.50		26.30	33	
	5540	600 volt NEMA L9			30	.267		18.75	10.50		29.25	36	
	5550	3 pole, 125/250 volt NEMA L10			26	.308		17.20	12.10		29.30	37	
	5560	250 volt NEMA L11			26	.308		17.20	12.10		29.30	37	
	5570	480 volt NEMA L12			26	.308		20	12.10		32.10	40	
	5580	125/250 volt NEMA L14			26	.308		19.15	12.10		31.25	39	
	5590	250 volt NEMA L15			26	.308		20.50	12.10		32.60	40.50	
	5600	480 volt NEMA L16			26	.308		24	12.10		36.10	44.50	
	5610	4 pole, 120/208 volt NEMA L18			24	.333		30.50	13.15		43.65	53.50	
	5620	277/480 volt NEMA L19			24	.333		30.50	13.15		43.65	53.50	
	5630	347/600 volt NEMA L20			24	.333		30.50	13.15		43.65	53.50	
	5640	120/208 volt NEMA L21			24	.333		24.50	13.15		37.65	46.50	
	5650	277/480 volt NEMA L22			24	.333		29.50	13.15		42.65	52	
	5660	347/600 volt NEMA L23			24	.333		29.50	13.15		42.65	52	
	5670	30 amp, 125 volt NEMA L5			13	.615		19.60	24.50		44.10	57.50	
	5680	250 volt NEMA L6			13	.615		19.75	24.50		44.25	57.50	
	5690	277 volt NEMA L7			13	.615		19.25	24.50		43.75	57	
	5700	480 volt NEMA L8			13	.615		23.50	24.50		48	61.50	
	5710	600 volt NEMA L9			13	.615		22.50	24.50		47	61	
	5720	3 pole, 125/250 volt NEMA L10			11	.727		18.60	28.50		47.10	63	
	5730	250 volt NEMA L11			11	.727		18.60	28.50		47.10	63	
	5760	125/250 volt NEMA L14			11	.727		26	28.50		54.50	71.50	
	5770	250 volt NEMA L15			11	.727		26.50	28.50		55	71.50	

			CREW	DAILY OUTPUT	LABOR-HOURS	UNIT	2004 BARE COSTS				TOTAL INCL O&P	
16140	**Wiring Devices**						MAT.	LABOR	EQUIP.	TOTAL		
910	5780	480 volt NEMA L16	1 Elec	11	.727	Ea.	31	28.50		59.50	76.50	910
	5790	600 volt NEMA L17		11	.727		31.50	28.50		60	77	
	5800	4 pole, 120/208 volt NEMA L18		10	.800		38.50	31.50		70	89.50	
	5810	120/208 volt NEMA L19		10	.800		38.50	31.50		70	89.50	
	5820	347/600 volt NEMA L20		10	.800		38.50	31.50		70	89.50	
	5830	120/208 volt NEMA L21		10	.800		29	31.50		60.50	79	
	5840	277/480 volt NEMA L22		10	.800		35	31.50		66.50	85.50	
	5850	347/600 volt NEMA L23		10	.800		43.50	31.50		75	94.50	
	6000	Connector, 20 amp, 250 volt NEMA 6		30	.267		21.50	10.50		32	39.50	
	6010	277 volt NEMA 7		30	.267		24.50	10.50		35	42.50	
	6020	3 pole, 120/250 volt NEMA 10		26	.308		27.50	12.10		39.60	48.50	
	6030	125/250 volt NEMA 14		26	.308		27.50	12.10		39.60	48.50	
	6040	250 volt NEMA 15		26	.308		27.50	12.10		39.60	48.50	
	6050	120/208 volt NEMA 18		26	.308		30	12.10		42.10	51.50	
	6060	30 amp, 125 volt NEMA 5		13	.615		48	24.50		72.50	88.50	
	6070	250 volt NEMA 6		13	.615		48	24.50		72.50	88.50	
	6080	277 volt NEMA 7		13	.615		48	24.50		72.50	88.50	
	6110	50 amp, 125 volt NEMA 5		9	.889		66	35		101	125	
	6120	250 volt NEMA 6		9	.889		66	35		101	125	
	6130	277 volt NEMA 7		9	.889		66	35		101	125	
	6200	Connector, locking, 20 amp, 125 volt NEMA L5		30	.267		19.60	10.50		30.10	37	
	6210	250 volt NEMA L6		30	.267		19.60	10.50		30.10	37	
	6220	277 volt NEMA L7		30	.267		19.60	10.50		30.10	37	
	6230	480 volt NEMA L8		30	.267		29.50	10.50		40	47.50	
	6240	600 volt NEMA L9		30	.267		29.50	10.50		40	47.50	
	6250	3 pole, 125/250 volt NEMA L10		26	.308		27	12.10		39.10	47.50	
	6260	250 volt NEMA L11		26	.308		27	12.10		39.10	47.50	
	6280	125/250 volt NEMA L14		26	.308		27	12.10		39.10	47.50	
	6290	250 volt NEMA L15		26	.308		27	12.10		39.10	47.50	
	6300	480 volt NEMA L16		26	.308		32.50	12.10		44.60	53.50	
	6310	4 pole, 120/208 volt NEMA L18		24	.333		42.50	13.15		55.65	66.50	
	6320	277/480 volt NEMA L19		24	.333		42.50	13.15		55.65	66.50	
	6330	347/600 volt NEMA L20		24	.333		43	13.15		56.15	66.50	
	6340	120/208 volt NEMA L21		24	.333		42.50	13.15		55.65	66	
	6350	277/480 volt NEMA L22		24	.333		51	13.15		64.15	75.50	
	6360	347/600 volt NEMA L23		24	.333		51	13.15		64.15	75.50	
	6370	30 amp, 125 volt NEMA L5		13	.615		39	24.50		63.50	79	
	6380	250 volt NEMA L6		13	.615		39.50	24.50		64	79	
	6390	277 volt NEMA L7		13	.615		39.50	24.50		64	79	
	6400	480 volt NEMA L8		13	.615		47	24.50		71.50	87.50	
	6410	600 volt NEMA L9		13	.615		47	24.50		71.50	87.50	
	6420	3 pole, 125/250 volt NEMA L10		11	.727		52	28.50		80.50	99.50	
	6430	250 volt NEMA L11		11	.727		52	28.50		80.50	99.50	
	6460	125/250 volt NEMA L14		11	.727		54	28.50		82.50	102	
	6470	250 volt NEMA L15		11	.727		54.50	28.50		83	102	
	6480	480 volt NEMA L16		11	.727		65.50	28.50		94	115	
	6490	600 volt NEMA L17		11	.727		65.50	28.50		94	115	
	6500	4 pole, 120/208 volt NEMA L18		10	.800		81.50	31.50		113	137	
	6510	120/208 volt NEMA L19		10	.800		81.50	31.50		113	137	
	6520	347/600 volt NEMA L20		10	.800		81.50	31.50		113	137	
	6530	120/208 volt NEMA L21		10	.800		53.50	31.50		85	106	
	6540	277/480 volt NEMA L22		10	.800		64.50	31.50		96	118	
	6550	347/600 volt NEMA L23		10	.800		80	31.50		111.50	135	
	7000	Receptacle computer, 250 volt, 15 amp, 3 pole 4 wire		8	1		62	39.50		101.50	127	
	7010	20 amp, 2 pole 3 wire		8	1		62	39.50		101.50	127	
	7020	30 amp, 2 pole 3 wire		6.50	1.231		97	48.50		145.50	179	

R16140 -910

Important: See the Reference Section for critical supporting data - Reference Nos., Crews, & City Cost Indexes

16 ELECTRICAL

16140 | Wiring Devices

			CREW	DAILY OUTPUT	LABOR-HOURS	UNIT	MAT.	LABOR	EQUIP.	TOTAL	TOTAL INCL O&P	
910	7030	30 amp, 3 pole 4 wire	R16140 -910	1 Elec	6.50	1.231	Ea.	104	48.50		152.50	186	910
	7040	60 amp, 3 pole 4 wire			4.50	1.778		174	70		244	295	
	7050	100 amp, 3 pole 4 wire			3	2.667		220	105		325	400	
	7100	Connector computer, 250 volt, 15 amp, 3 pole 4 wire			27	.296		86	11.65		97.65	112	
	7110	20 amp, 2 pole 3 wire			27	.296		86	11.65		97.65	112	
	7120	30 amp, 2 pole 3 wire			15	.533		136	21		157	181	
	7130	30 amp, 3 pole 4 wire			15	.533		128	21		149	173	
	7140	60 amp, 3 pole 4 wire			8	1		219	39.50		258.50	300	
	7150	100 amp, 3 pole 4 wire			4	2		305	79		384	450	
	7200	Plug, computer, 250 volt, 15 amp, 3 pole 4 wire			27	.296		77.50	11.65		89.15	102	
	7210	20 amp, 2 pole, 3 wire			27	.296		77.50	11.65		89.15	102	
	7220	30 amp, 2 pole, 3 wire			15	.533		113	21		134	157	
	7230	30 amp, 3 pole, 4 wire			15	.533		122	21		143	167	
	7240	60 amp, 3 pole, 4 wire			8	1		184	39.50		223.50	261	
	7250	100 amp, 3 pole, 4 wire			4	2		239	79		318	380	
	7300	Connector adapter to flexible conduit, 1/2"			60	.133		3.28	5.25		8.53	11.40	
	7310	3/4"			50	.160		4.80	6.30		11.10	14.70	
	7320	1-1/4"			30	.267		12.25	10.50		22.75	29	
	7330	1-1/2"			23	.348		17.55	13.70		31.25	40	

16150 | Wiring Connections

			CREW	DAILY OUTPUT	LABOR-HOURS	UNIT	MAT.	LABOR	EQUIP.	TOTAL	TOTAL INCL O&P	
275	0010	**MOTOR CONNECTIONS**											275
	0020	Flexible conduit and fittings, 115 volt, 1 phase, up to 1 HP motor		1 Elec	8	1	Ea.	4.34	39.50		43.84	63.50	
	0050	2 HP motor	R16150 -275		6.50	1.231		4.41	48.50		52.91	77	
	0100	3 HP motor			5.50	1.455		7.45	57.50		64.95	93	
	0120	230 volt, 10 HP motor, 3 phase			4.20	1.905		8	75		83	121	
	0150	15 HP motor			3.30	2.424		15.45	95.50		110.95	159	
	0200	25 HP motor			2.70	2.963		17.35	117		134.35	193	
	0400	50 HP motor			2.20	3.636		47.50	143		190.50	265	
	0600	100 HP motor			1.50	5.333		122	210		332	450	
	1500	460 volt, 5 HP motor, 3 phase			8	1		4.83	39.50		44.33	64	
	1520	10 HP motor			8	1		4.83	39.50		44.33	64	
	1530	25 HP motor			6	1.333		9.25	52.50		61.75	88	
	1540	30 HP motor			6	1.333		9.25	52.50		61.75	88	
	1550	40 HP motor			5	1.600		14.50	63		77.50	110	
	1560	50 HP motor			5	1.600		16.15	63		79.15	112	
	1570	60 HP motor			3.80	2.105		21.50	83		104.50	147	
	1580	75 HP motor			3.50	2.286		28.50	90		118.50	166	
	1590	100 HP motor			2.50	3.200		40.50	126		166.50	233	
	1600	125 HP motor			2	4		53	158		211	292	
	1610	150 HP motor			1.80	4.444		55	175		230	320	
	1620	200 HP motor			1.50	5.333		95.50	210		305.50	420	
	2005	460 Volt, 5 HP motor, 3 Phase, w/sealtite			8	1		9.30	39.50		48.80	68.50	
	2010	10 HP motor			8	1		9.30	39.50		48.80	68.50	
	2015	25 HP motor			6	1.333		16.80	52.50		69.30	96.50	
	2020	30 HP motor			6	1.333		16.80	52.50		69.30	96.50	
	2025	40 HP motor			5	1.600		30.50	63		93.50	128	
	2030	50 HP motor			5	1.600		29	63		92	126	
	2035	60 HP motor			3.80	2.105		44	83		127	171	
	2040	75 HP motor			3.50	2.286		47	90		137	186	
	2045	100 HP motor			2.50	3.200		62	126		188	256	
	2055	150 HP motor			1.80	4.444		67	175		242	335	
	2060	200 HP motor			1.50	5.333		405	210		615	760	

		CREW	DAILY OUTPUT	LABOR-HOURS	UNIT	2004 BARE COSTS				TOTAL INCL O&P
						MAT.	LABOR	EQUIP.	TOTAL	
600	**0010**	**METER CENTERS AND SOCKETS**									**600**
0100	Sockets, single position, 4 terminal, 100 amp	1 Elec	3.20	2.500	Ea.	31.50	98.50		130	182
0200	150 amp		2.30	3.478		35.50	137		172.50	243
0300	200 amp		1.90	4.211		47.50	166		213.50	299
0400	20 amp		3.20	2.500		75.50	98.50		174	230
0500	Double position, 4 terminal, 100 amp		2.80	2.857		124	113		237	305
0600	150 amp		2.10	3.810		141	150		291	380
0700	200 amp		1.70	4.706		300	185		485	610
0800	Trans-socket, 13 terminal, 3 CT mounts, 400 amp	▼	1	8	▼	755	315		1,070	1,300
0900	800 amp	2 Elec	1.20	13.333		795	525		1,320	1,650
2000	Meter center, main fusible switch, 1P 3W 120/240 volt									
2030	400 amp	2 Elec	1.60	10	Ea.	1,150	395		1,545	1,825
2040	600 amp		1.10	14.545		2,000	575		2,575	3,050
2050	800 amp		.90	17.778		3,125	700		3,825	4,475
2060	Rainproof 1P 3W 120/240 volt, 400 amp		1.60	10		1,150	395		1,545	1,825
2070	600 amp		1.10	14.545		2,000	575		2,575	3,050
2080	800 amp		.90	17.778		3,125	700		3,825	4,475
2100	3P 4W 120/208V, 400 amp		1.60	10		1,300	395		1,695	2,025
2110	600 amp		1.10	14.545		2,450	575		3,025	3,525
2120	800 amp		.90	17.778		4,525	700		5,225	6,050
2130	Rainproof 3P 4W 120/208V, 400 amp		1.60	10		1,300	395		1,695	2,025
2140	600 amp		1.10	14.545		2,450	575		3,025	3,525
2150	800 amp	▼	.90	17.778	▼	4,525	700		5,225	6,050
2170	Main circuit breaker, 1P 3W 120/240V									
2180	400 amp	2 Elec	1.60	10	Ea.	2,050	395		2,445	2,825
2190	600 amp		1.10	14.545		2,750	575		3,325	3,875
2200	800 amp		.90	17.778		3,200	700		3,900	4,575
2210	1000 amp		.80	20		4,425	790		5,215	6,050
2220	1200 amp		.76	21.053		5,975	830		6,805	7,800
2230	1600 amp		.68	23.529		9,800	925		10,725	12,200
2240	Rainproof 1P 3W 120/240V, 400 amp		1.60	10		2,050	395		2,445	2,825
2250	600 amp		1.10	14.545		2,750	575		3,325	3,875
2260	800 amp		.90	17.778		3,200	700		3,900	4,575
2270	1000 amp		.80	20		4,425	790		5,215	6,050
2280	1200 amp		.76	21.053		5,975	830		6,805	7,800
2300	3P 4W 120/208V, 400 amp		1.60	10		2,325	395		2,720	3,125
2310	600 amp		1.10	14.545		3,275	575		3,850	4,450
2320	800 amp		.90	17.778		3,900	700		4,600	5,350
2330	1000 amp		.80	20		5,125	790		5,915	6,800
2340	1200 amp		.76	21.053		6,550	830		7,380	8,425
2350	1600 amp		.68	23.529		9,800	925		10,725	12,200
2360	Rainproof 3P 4W 120/208V, 400 amp		1.60	10		2,325	395		2,720	3,125
2370	600 amp		1.10	14.545		3,275	575		3,850	4,450
2380	800 amp		.90	17.778		3,900	700		4,600	5,350
2390	1000 amp		.76	21.053		5,125	830		5,955	6,850
2400	1200 amp	▼	.68	23.529	▼	6,550	925		7,475	8,575
2420	Main lugs terminal box, 1P 3W 120/240V									
2430	800 amp	2 Elec	.94	17.021	Ea.	385	670		1,055	1,425
2440	1200 amp		.72	22.222		770	875		1,645	2,150
2450	Rainproof 1P 3W 120/240V, 225 amp		2.40	6.667		261	263		524	680
2460	800 amp		.94	17.021		385	670		1,055	1,425
2470	1200 amp		.72	22.222		770	875		1,645	2,150
2500	3P 4W 120/208V, 800 amp		.94	17.021		425	670		1,095	1,475
2510	1200 amp		.72	22.222		860	875		1,735	2,250
2520	Rainproof 3P 4W 120/208V, 225 amp		2.40	6.667		261	263		524	680
2530	800 amp	▼	.94	17.021		425	670		1,095	1,475

16210	Electrical Utility Services	CREW	DAILY OUTPUT	LABOR-HOURS	UNIT	2004 BARE COSTS				TOTAL INCL O&P	
						MAT.	LABOR	EQUIP.	TOTAL		
600											**600**
2540	1200 amp	2 Elec	.72	22.222	Ea.	860	875		1,735	2,250	
2590	Basic meter device										
2600	1P 3W 120/240V 4 jaw 125A sockets, 3 meter	2 Elec	1	16	Ea.	440	630		1,070	1,425	
2610	4 meter		.90	17.778		525	700		1,225	1,625	
2620	5 meter		.80	20		660	790		1,450	1,900	
2630	6 meter		.60	26.667		760	1,050		1,810	2,400	
2640	7 meter		.56	28.571		965	1,125		2,090	2,750	
2650	8 meter		.52	30.769		1,050	1,200		2,250	2,950	
2660	10 meter	▼	.48	33.333	▼	1,325	1,325		2,650	3,400	
2680	Rainproof 1P 3W 120/240V 4 jaw 125A sockets										
2690	3 meter	2 Elec	1	16	Ea.	440	630		1,070	1,425	
2700	4 meter		.90	17.778		525	700		1,225	1,625	
2710	6 meter		.60	26.667		760	1,050		1,810	2,400	
2720	7 meter		.56	28.571		965	1,125		2,090	2,750	
2730	8 meter	▼	.52	30.769	▼	1,050	1,200		2,250	2,950	
2750	1P 3W 120/240V 4 jaw sockets										
2760	with 125A circuit breaker, 3 meter	2 Elec	1	16	Ea.	825	630		1,455	1,850	
2770	4 meter		.90	17.778		1,050	700		1,750	2,200	
2780	5 meter		.80	20		1,300	790		2,090	2,600	
2790	6 meter		.60	26.667		1,525	1,050		2,575	3,250	
2800	7 meter		.56	28.571		1,850	1,125		2,975	3,725	
2810	8 meter		.52	30.769		2,075	1,200		3,275	4,075	
2820	10 meter	▼	.48	33.333	▼	2,600	1,325		3,925	4,800	
2830	Rainproof 1P 3W 120/240V 4 jaw sockets										
2840	with 125A circuit breaker, 3 meter	2 Elec	1	16	Ea.	825	630		1,455	1,850	
2850	4 meter		.90	17.778		1,050	700		1,750	2,200	
2870	6 meter		.60	26.667		1,525	1,050		2,575	3,250	
2880	7 meter		.56	28.571		1,850	1,125		2,975	3,725	
2890	8 meter	▼	.52	30.769		2,075	1,200		3,275	4,075	
2920	1P 3W on 3P 4W 120/208V system 5 jaw										
2930	125A sockets, 3 meter	2 Elec	1	16	Ea.	440	630		1,070	1,425	
2940	4 meter		.90	17.778		525	700		1,225	1,625	
2950	5 meter		.80	20		660	790		1,450	1,900	
2960	6 meter		.60	26.667		760	1,050		1,810	2,400	
2970	7 meter		.56	28.571		965	1,125		2,090	2,750	
2980	8 meter		.52	30.769		1,050	1,200		2,250	2,950	
2990	10 meter	▼	.48	33.333	▼	1,325	1,325		2,650	3,400	
3000	Rainproof 1P 3W on 3P 4W 120/208V system										
3020	5 jaw 125A sockets, 3 meter	2 Elec	1	16	Ea.	440	630		1,070	1,425	
3030	4 meter		.90	17.778		525	700		1,225	1,625	
3050	6 meter		.60	26.667		760	1,050		1,810	2,400	
3060	7 meter		.56	28.571		965	1,125		2,090	2,750	
3070	8 meter	▼	.52	30.769	▼	1,050	1,200		2,250	2,950	
3090	1P 3W on 3P 4W 120/208V system 5 jaw sockets										
3100	With 125A circuit breaker, 3 meter	2 Elec	1	16	Ea.	825	630		1,455	1,850	
3110	4 meter		.90	17.778		1,050	700		1,750	2,200	
3120	5 meter		.80	20		1,300	790		2,090	2,600	
3130	6 meter		.60	26.667		1,525	1,050		2,575	3,250	
3140	7 meter		.56	28.571		1,850	1,125		2,975	3,725	
3150	8 meter		.52	30.769		2,075	1,200		3,275	4,075	
3160	10 meter	▼	.48	33.333	▼	2,600	1,325		3,925	4,800	
3170	Rainproof 1P 3W on 3P 4W 120/208V system										
3180	5 jaw sockets w/125A circuit breaker, 3 meter	2 Elec	1	16	Ea.	825	630		1,455	1,850	
3190	4 meter		.90	17.778		1,050	700		1,750	2,200	
3210	6 meter		.60	26.667		1,525	1,050		2,575	3,250	
3220	7 meter		.56	28.571		1,850	1,125		2,975	3,725	

ELECTRICAL 16

16 ELECTRICAL

	16210	Electrical Utility Services	CREW	DAILY OUTPUT	LABOR-HOURS	UNIT	2004 BARE COSTS MAT.	LABOR	EQUIP.	TOTAL	TOTAL INCL O&P	
600	3230	8 meter	2 Elec	.52	30.769	Ea.	2,075	1,200		3,275	4,075	**600**
	3250	1P 3W 120/240V 4 jaw sockets										
	3260	with 200A circuit breaker, 3 meter	2 Elec	1	16	Ea.	1,225	630		1,855	2,300	
	3270	4 meter		.90	17.778		1,675	700		2,375	2,875	
	3290	6 meter		.60	26.667		2,475	1,050		3,525	4,300	
	3300	7 meter		.56	28.571		2,900	1,125		4,025	4,875	
	3310	8 meter		.56	28.571		3,325	1,125		4,450	5,350	
	3330	Rainproof 1P 3W 120/240V 4 jaw sockets										
	3350	with 200A circuit breaker, 3 meter	2 Elec	1	16	Ea.	1,225	630		1,855	2,300	
	3360	4 meter		.90	17.778		1,675	700		2,375	2,875	
	3380	6 meter		.60	26.667		2,475	1,050		3,525	4,300	
	3390	7 meter		.56	28.571		2,900	1,125		4,025	4,875	
	3400	8 meter		.52	30.769		3,325	1,200		4,525	5,475	
	3420	1P 3W on 3P 4W 120/208V 5 jaw sockets										
	3430	with 200A circuit breaker, 3 meter	2 Elec	1	16	Ea.	1,225	630		1,855	2,300	
	3440	4 meter		.90	17.778		1,675	700		2,375	2,875	
	3460	6 meter		.60	26.667		2,475	1,050		3,525	4,300	
	3470	7 meter		.56	28.571		2,900	1,125		4,025	4,875	
	3480	8 meter		.52	30.769		3,325	1,200		4,525	5,475	
	3500	Rainproof 1P 3W on 3P 4W 120/208V 5 jaw socket										
	3510	with 200A circuit breaker, 3 meter	2 Elec	1	16	Ea.	1,225	630		1,855	2,300	
	3520	4 meter		.90	17.778		1,675	700		2,375	2,875	
	3540	6 meter		.60	26.667		2,475	1,050		3,525	4,300	
	3550	7 meter		.56	28.571		2,900	1,125		4,025	4,875	
	3560	8 meter		.52	30.769		3,325	1,200		4,525	5,475	
	3600	Automatic circuit closing, add					53			53	58.50	
	3610	Manual circuit closing, add					60.50			60.50	66.50	
	3650	Branch meter device										
	3660	3P 4W 208/120 or 240/120 V 7 jaw sockets										
	3670	with 200A circuit breaker, 2 meter	2 Elec	.90	17.778	Ea.	2,050	700		2,750	3,325	
	3680	3 meter		.80	20		3,100	790		3,890	4,575	
	3690	4 meter		.70	22.857		4,125	900		5,025	5,875	
	3700	Main circuit breaker 42,000 rms, 400 amp		1.60	10		1,800	395		2,195	2,550	
	3710	600 amp		1.10	14.545		2,900	575		3,475	4,050	
	3720	800 amp		.90	17.778		3,900	700		4,600	5,350	
	3730	Rainproof main circ. breaker 42,000 rms, 400 amp		1.60	10		1,800	395		2,195	2,550	
	3740	600 amp		1.10	14.545		2,900	575		3,475	4,050	
	3750	800 amp		.90	17.778		3,900	700		4,600	5,350	
	3760	Main circuit breaker 65,000 rms, 400 amp		1.60	10		2,450	395		2,845	3,275	
	3770	600 amp		1.10	14.545		3,450	575		4,025	4,625	
	3780	800 amp		.90	17.778		3,900	700		4,600	5,350	
	3790	1000 amp		.80	20		5,125	790		5,915	6,800	
	3800	1200 amp		.76	21.053		6,550	830		7,380	8,425	
	3810	1600 amp		.68	23.529		9,800	925		10,725	12,200	
	3820	Rainproof main circ. breaker 65,000 rms, 400 amp		1.60	10		2,450	395		2,845	3,275	
	3830	600 amp		1.10	14.545		3,450	575		4,025	4,625	
	3840	800 amp		.90	17.778		3,900	700		4,600	5,325	
	3850	1000 amp		.80	20		5,125	790		5,915	6,800	
	3860	1200 amp		.76	21.053		6,550	830		7,380	8,425	
	3880	Main circuit breaker 100,000 rms, 400 amp		1.60	10		2,450	395		2,845	3,275	
	3890	600 amp		1.10	14.545		3,450	575		4,025	4,625	
	3900	800 amp		.90	17.778		4,050	700		4,750	5,500	
	3910	Rainproof main circ. breaker 100,000 rms, 400 amp		1.60	10		2,450	395		2,845	3,275	
	3920	600 amp		1.10	14.545		3,450	575		4,025	4,625	
	3930	800 amp		.90	17.778		4,050	700		4,750	5,500	
	3940	Main lugs terminal box, 800 amp		.94	17.021		425	670		1,095	1,475	

Important: See the Reference Section for critical supporting data - Reference Nos., Crews, & City Cost Indexes

		CREW	DAILY OUTPUT	LABOR-HOURS	UNIT	2004 BARE COSTS				TOTAL INCL O&P	
	16210	**Electrical Utility Services**				MAT.	LABOR	EQUIP.	TOTAL		
600	3950	1600 amp	2 Elec	.72	22.222	Ea.	1,525	875		2,400	2,975	**600**
	3960	Rainproof, 800 amp		.94	17.021		425	670		1,095	1,475	
	3970	1600 amp	↓	.72	22.222	↓	1,525	875		2,400	2,975	

	16220	**Motors & Generators**									
600	0010	**HANDLING** Add to normal labor cost for restricted areas									**600**
	5000	Motors									
	5100	1/2 HP, 23 pounds	1 Elec	4	2	Ea.		79		79	117	
	5110	3/4 HP, 28 pounds		4	2			79		79	117	
	5120	1 HP, 33 pounds		4	2			79		79	117	
	5130	1-1/2 HP, 44 pounds		3.20	2.500			98.50		98.50	147	
	5140	2 HP, 56 pounds		3	2.667			105		105	156	
	5150	3 HP, 71 pounds		2.30	3.478			137		137	204	
	5160	5 HP, 82 pounds		1.90	4.211			166		166	247	
	5170	7-1/2 HP, 124 pounds		1.50	5.333			210		210	315	
	5180	10 HP, 144 pounds		1.20	6.667			263		263	390	
	5190	15 HP, 185 pounds	↓	1	8			315		315	470	
	5200	20 HP, 214 pounds	2 Elec	1.50	10.667			420		420	625	
	5210	25 HP, 266 pounds		1.40	11.429			450		450	670	
	5220	30 HP, 310 pounds		1.20	13.333			525		525	780	
	5230	40 HP, 400 pounds		1	16			630		630	940	
	5240	50 HP, 450 pounds		.90	17.778			700		700	1,050	
	5250	75 HP, 680 pounds	↓	.80	20			790		790	1,175	
	5260	100 HP, 870 pounds	3 Elec	1	24			945		945	1,400	
	5270	125 HP, 940 pounds		.80	30			1,175		1,175	1,750	
	5280	150 HP, 1200 pounds		.70	34.286			1,350		1,350	2,000	
	5290	175 HP, 1300 pounds		.60	40			1,575		1,575	2,350	
	5300	200 HP, 1400 pounds	↓	.50	48	↓		1,900		1,900	2,825	

	610										
610	0010	**MOTORS** 230/460 volts, 60 HZ									**610**
	0050	Dripproof, premium efficiency, 1.15 service factor									
	0060	1800 RPM, 1/4 HP	1 Elec	5.33	1.501	Ea.	128	59		187	229	
	0070	1/3 HP		5.33	1.501		140	59		199	242	
	0080	1/2 HP		5.33	1.501		165	59		224	270	
	0090	3/4 HP		5.33	1.501		184	59		243	290	
	0100	1 HP		4.50	1.778		194	70		264	315	
	0150	2 HP		4.50	1.778		228	70		298	355	
	0200	3 HP		4.50	1.778		229	70		299	355	
	0250	5 HP		4.50	1.778		295	70		365	430	
	0300	7.5 HP		4.20	1.905		430	75		505	580	
	0350	10 HP		4	2		530	79		609	700	
	0400	15 HP	↓	3.20	2.500		670	98.50		768.50	885	
	0450	20 HP	2 Elec	5.20	3.077		840	121		961	1,100	
	0500	25 HP		5	3.200		975	126		1,101	1,275	
	0550	30 HP		4.80	3.333		1,125	131		1,256	1,450	
	0600	40 HP		4	4		1,700	158		1,858	2,100	
	0650	50 HP		3.20	5		1,925	197		2,122	2,425	
	0700	60 HP		2.80	5.714		2,525	225		2,750	3,100	
	0750	75 HP	↓	2.40	6.667		2,875	263		3,138	3,550	
	0800	100 HP	3 Elec	2.70	8.889		3,775	350		4,125	4,675	
	0850	125 HP		2.10	11.429		4,375	450		4,825	5,500	
	0900	150 HP		1.80	13.333		5,600	525		6,125	6,925	
	0950	200 HP	↓	1.50	16		7,225	630		7,855	8,900	
	1000	1200 RPM, 1 HP	1 Elec	4.50	1.778		225	70		295	350	
	1050	2 HP		4.50	1.778	↓	320	70		390	455	

ELECTRICAL 16

			CREW	DAILY OUTPUT	LABOR-HOURS	UNIT	2004 BARE COSTS				TOTAL INCL O&P	
	16220	**Motors & Generators**					MAT.	LABOR	EQUIP.	TOTAL		
610	1100	3 HP	1 Elec	4.50	1.778	Ea.	425	70		495	575	610
	1150	5 HP		4.50	1.778		510	70		580	670	
	1200	3600 RPM, 2 HP		4.50	1.778		231	70		301	360	
	1250	3 HP		4.50	1.778		264	70		334	395	
	1300	5 HP	▼	4.50	1.778	▼	295	70		365	430	
	1350	Totally enclosed, premium efficiency 1.15 service factor										
	1360	1800 RPM, 1/4 HP	1 Elec	5.33	1.501	Ea.	130	59		189	231	
	1370	1/3 HP		5.33	1.501		145	59		204	248	
	1380	1/2 HP		5.33	1.501		171	59		230	276	
	1390	3/4 HP		5.33	1.501		184	59		243	290	
	1400	1 HP		4.50	1.778		233	70		303	360	
	1450	2 HP		4.50	1.778		275	70		345	410	
	1500	3 HP		4.50	1.778		310	70		380	445	
	1550	5 HP		4.50	1.778		355	70		425	495	
	1600	7.5 HP		4.20	1.905		510	75		585	670	
	1650	10 HP		4	2		615	79		694	790	
	1700	15 HP	▼	3.20	2.500		820	98.50		918.50	1,050	
	1750	20 HP	2 Elec	5.20	3.077		1,000	121		1,121	1,275	
	1800	25 HP		5	3.200		1,225	126		1,351	1,550	
	1850	30 HP		4.80	3.333		1,425	131		1,556	1,775	
	1900	40 HP		4	4		1,850	158		2,008	2,250	
	1950	50 HP		3.20	5		2,275	197		2,472	2,800	
	2000	60 HP		2.80	5.714		3,400	225		3,625	4,075	
	2050	75 HP	▼	2.40	6.667		4,350	263		4,613	5,175	
	2100	100 HP	3 Elec	2.70	8.889		5,175	350		5,525	6,200	
	2150	125 HP		2.10	11.429		7,500	450		7,950	8,925	
	2200	150 HP		1.80	13.333		8,500	525		9,025	10,100	
	2250	200 HP	▼	1.50	16		10,200	630		10,830	12,100	
	2300	1200 RPM, 1 HP	1 Elec	4.50	1.778		272	70		342	405	
	2350	2 HP		4.50	1.778		330	70		400	470	
	2400	3 HP		4.50	1.778		440	70		510	590	
	2450	5 HP		4.50	1.778		650	70		720	820	
	2500	3600 RPM, 2 HP		4.50	1.778		269	70		339	400	
	2550	3 HP		4.50	1.778		315	70		385	450	
	2600	5 HP	▼	4.50	1.778	▼	390	70		460	535	
900	0010	**VARIABLE FREQUENCY DRIVES/ADJUSTABLE FREQUENCY DRIVES**										900
	0100	Enclosed (NEMA 1), 460 volt, for 3 HP motor size	1 Elec	.80	10	Ea.	1,475	395		1,870	2,175	
	0110	5 HP motor size		.80	10		1,600	395		1,995	2,325	
	0120	7.5 HP motor size		.67	11.940		1,875	470		2,345	2,775	
	0130	10 HP motor size	▼	.67	11.940		1,875	470		2,345	2,775	
	0140	15 HP motor size	2 Elec	.89	17.978		2,175	710		2,885	3,450	
	0150	20 HP motor size		.89	17.978		3,225	710		3,935	4,600	
	0160	25 HP motor size		.67	23.881		3,750	940		4,690	5,525	
	0170	30 HP motor size		.67	23.881		4,575	940		5,515	6,425	
	0180	40 HP motor size		.67	23.881		6,700	940		7,640	8,775	
	0190	50 HP motor size	▼	.53	30.189		7,300	1,200		8,500	9,800	
	0200	60 HP motor size	R-3	.56	35.714		8,000	1,375	284	9,659	11,200	
	0210	75 HP motor size		.56	35.714		11,000	1,375	284	12,659	14,500	
	0220	100 HP motor size		.50	40		11,200	1,550	320	13,070	15,000	
	0230	125 HP motor size		.50	40		12,400	1,550	320	14,270	16,300	
	0240	150 HP motor size		.50	40		15,800	1,550	320	17,670	20,100	
	0250	200 HP motor size	▼	.42	47.619		19,600	1,850	380	21,830	24,700	
	1100	Custom-engineered, 460 volt, for 3 HP motor size	1 Elec	.56	14.286		2,425	565		2,990	3,475	
	1110	5 HP motor size		.56	14.286		2,550	565		3,115	3,625	
	1120	7.5 HP motor size	▼	.47	17.021		3,050	670		3,720	4,350	

			DAILY	LABOR-		2004 BARE COSTS				TOTAL	
	16220	**Motors & Generators**									
			CREW	OUTPUT	HOURS	UNIT	MAT.	LABOR	EQUIP.	TOTAL	INCL O&P
900	1130	10 HP motor size	1 Elec	.47	17.021	Ea.	3,050	670		3,720	4,350	**900**
	1140	15 HP motor size	2 Elec	.62	25.806		3,350	1,025		4,375	5,175
	1150	20 HP motor size		.62	25.806		4,575	1,025		5,600	6,525
	1160	25 HP motor size		.47	34.043		4,950	1,350		6,300	7,450
	1170	30 HP motor size		.47	34.043		6,750	1,350		8,100	9,425
	1180	40 HP motor size		.47	34.043		7,550	1,350		8,900	10,300
	1190	50 HP motor size		.37	43.243		7,550	1,700		9,250	10,800
	1200	60 HP motor size	R-3	.39	51.282		11,300	1,975	410	13,685	15,800
	1210	75 HP motor size		.39	51.282		13,000	1,975	410	15,385	17,700
	1220	100 HP motor size		.35	57.143		13,600	2,200	455	16,255	18,700
	1230	125 HP motor size		.35	57.143		14,300	2,200	455	16,955	19,500
	1240	150 HP motor size		.35	57.143		16,100	2,200	455	18,755	21,500
	1250	200 HP motor size		.29	68.966		22,200	2,675	550	25,425	29,000
	2000	For complex & special design systems to meet specific									
	2010	requirements, obtain quote from vendor.									

			DAILY	LABOR-		2004 BARE COSTS				TOTAL	
	16230	**Generator Assemblies**									
450	0010	**GENERATOR SET** R16230 -450									**450**
	0020	Gas or gasoline operated, includes battery,									
	0050	charger, muffler & transfer switch									
	0200	3 phase 4 wire, 277/480 volt, 7.5 kW	R-3	.83	24.096	Ea.	6,000	930	192	7,122	8,200
	0300	11.5 kW		.71	28.169		8,500	1,100	224	9,824	11,200
	0400	20 kW		.63	31.746		10,000	1,225	252	11,477	13,100
	0500	35 kW		.55	36.364		12,000	1,400	289	13,689	15,600
	0520	60 kW		.50	40		16,500	1,550	320	18,370	20,900
	0600	80 kW		.40	50		20,500	1,925	400	22,825	25,900
	0700	100 kW		.33	60.606		22,500	2,350	480	25,330	28,800
	0800	125 kW		.28	71.429		46,000	2,775	570	49,345	55,500
	0900	185 kW		.25	80		61,000	3,100	635	64,735	72,500
	2000	Diesel engine, including battery, charger,									
	2010	muffler, automatic transfer switch & day tank, 30 kW	R-3	.55	36.364	Ea.	16,000	1,400	289	17,689	20,000
	2100	50 kW		.42	47.619		19,700	1,850	380	21,930	24,900
	2110	60 kW		.39	51.282		21,300	1,975	410	23,685	26,800
	2200	75 kW		.35	57.143		25,700	2,200	455	28,355	32,100
	2300	100 kW		.31	64.516		28,500	2,500	515	31,515	35,700
	2400	125 kW		.29	68.966		30,000	2,675	550	33,225	37,600
	2500	150 kW		.26	76.923		34,400	2,975	610	37,985	42,900
	2600	175 kW		.25	80		37,500	3,100	635	41,235	46,600
	2700	200 kW		.24	83.333		38,700	3,225	665	42,590	48,100
	2800	250 kW		.23	86.957		45,600	3,375	690	49,665	56,000
	2850	275 kW		.22	90.909		45,600	3,525	725	49,850	56,000
	2900	300 kW		.22	90.909		49,300	3,525	725	53,550	60,000
	3000	350 kW		.20	100		55,500	3,875	795	60,170	68,000
	3100	400 kW		.19	105		69,000	4,075	835	73,910	82,500
	3200	500 kW		.18	111		86,500	4,300	885	91,685	102,500
	3220	600 kW		.17	117		113,500	4,550	935	118,985	133,000
	3230	650 kW	R-13	.38	110		135,000	4,125	425	139,550	155,000
	3240	750 kW		.38	110		142,500	4,125	425	147,050	163,500
	3250	800 kW		.36	116		148,000	4,350	450	152,800	170,000
	3260	900 kW		.31	135		171,000	5,050	525	176,575	196,000
	3270	1000 kW		.27	155		176,000	5,800	600	182,400	203,500

	16260	**Static Power Converters**									
800	0010	**UNINTERRUPTIBLE POWER SUPPLY/CONDITIONER TRANSFORMERS**									**800**
	0100	Volt. regulating, isolating trans., w/invert. & 10 min. battery pack									

ELECTRICAL 16

16260 | Static Power Converters

		CREW	DAILY OUTPUT	LABOR-HOURS	UNIT	2004 BARE COSTS				TOTAL INCL O&P	
						MAT.	LABOR	EQUIP.	TOTAL		
800	0110	Single-phase, 120 V, 0.35 kVA	1 Elec	2.29	3.493	Ea.	900	138		1,038	1,200	800
	0120	0.5 kVA		2	4		945	158		1,103	1,275	
	0130	For additional 55 min. battery, add to .35 kVA		2.29	3.493		530	138		668	790	
	0140	Add to 0.5 kVA		1.14	7.018		560	276		836	1,025	
	0150	Single-phase, 120 V, 0.75 kVA		.80	10		1,200	395		1,595	1,900	
	0160	1.0 kVA		.80	10		1,700	395		2,095	2,450	
	0170	1.5 kVA	2 Elec	1.14	14.035		2,950	555		3,505	4,050	
	0180	2 kVA	"	.89	17.978		3,200	710		3,910	4,550	
	0190	3 kVA	R-3	.63	31.746		3,875	1,225	252	5,352	6,350	
	0200	5 kVA		.42	47.619		5,600	1,850	380	7,830	9,350	
	0210	7.5 kVA		.33	60.606		7,150	2,350	480	9,980	11,900	
	0220	10 kVA		.28	71.429		7,600	2,775	570	10,945	13,100	
	0230	15 kVA		.22	90.909		10,500	3,525	725	14,750	17,600	
	0500	For options & accessories add to above, minimum									10%	
	0520	Maximum									35%	
	0600	For complex & special design systems to meet specific										
	0610	requirements, obtain quote from vendor										

16270 | Transformers

		CREW	DAILY OUTPUT	LABOR-HOURS	UNIT	MAT.	LABOR	EQUIP.	TOTAL	TOTAL INCL O&P	
100	0010	**BUCK-BOOST TRANSFORMER**										100
	0100	Single phase, 120/240 V primary, 12/24 V secondary										
	0200	0.10 kVA	1 Elec	8	1	Ea.	54	39.50		93.50	118	
	0400	0.25 kVA		5.70	1.404		79.50	55.50		135	170	
	0600	0.50 kVA		4	2		109	79		188	237	
	0800	0.75 kVA		3.10	2.581		140	102		242	305	
	1000	1.0 kVA		2	4		174	158		332	425	
	1200	1.5 kVA		1.80	4.444		213	175		388	495	
	1400	2.0 kVA		1.60	5		258	197		455	575	
	1600	3.0 kVA		1.40	5.714		345	225		570	715	
	1800	5.0 kVA		1.20	6.667		455	263		718	890	
	2000	3 phase, 240 V primary, 208/120 V secondary, 15 kVA	2 Elec	2.40	6.667		1,375	263		1,638	1,900	
	2200	30 kVA		1.60	10		1,525	395		1,920	2,250	
	2400	45 kVA		1.40	11.429		1,850	450		2,300	2,700	
	2600	75 kVA		1.20	13.333		2,225	525		2,750	3,225	
	2800	112.5 kVA	R-3	1.40	14.286		2,775	555	114	3,444	4,000	
	3000	150 kVA		1.10	18.182		3,675	705	145	4,525	5,250	
	3200	225 kVA		1	20		4,800	775	159	5,734	6,600	
	3400	300 kVA		.90	22.222		6,525	860	177	7,562	8,650	
200	0010	**DRY TYPE TRANSFORMER**										200
	0050	Single phase, 240/480 volt primary, 120/240 volt secondary										
	0100	1 kVA	1 Elec	2	4	Ea.	163	158		321	415	
	0300	2 kVA		1.60	5		244	197		441	560	
	0500	3 kVA		1.40	5.714		305	225		530	670	
	0700	5 kVA		1.20	6.667		415	263		678	850	
	0900	7.5 kVA	2 Elec	2.20	7.273		580	287		867	1,050	
	1100	10 kVA		1.60	10		740	395		1,135	1,400	
	1300	15 kVA		1.20	13.333		1,000	525		1,525	1,875	
	1500	25 kVA		1	16		1,300	630		1,930	2,375	
	1700	37.5 kVA		.80	20		1,675	790		2,465	3,025	
	1900	50 kVA		.70	22.857		2,000	900		2,900	3,550	
	2100	75 kVA		.65	24.615		2,650	970		3,620	4,375	
	2110	100 kVA	R-3	.90	22.222		3,450	860	177	4,487	5,275	
	2120	167 kVA	"	.80	25		5,725	965	199	6,889	7,975	
	2190	480V primary 120/240V secondary, nonvent., 15 kVA	2 Elec	1.20	13.333		985	525		1,510	1,850	

R16260-800

R16270-600

Important: See the Reference Section for critical supporting data - Reference Nos., Crews, & City Cost Indexes

16270	Transformers	CREW	DAILY OUTPUT	LABOR-HOURS	UNIT	2004 BARE COSTS				TOTAL INCL O&P
						MAT.	LABOR	EQUIP.	TOTAL	
200 2200	25 kVA	2 Elec	.90	17.778	Ea.	1,450	700		2,150	2,650	200
2210	37 kVA		.75	21.333		1,725	840		2,565	3,150
2220	50 kVA		.65	24.615		2,050	970		3,020	3,700
2230	75 kVA		.60	26.667		2,700	1,050		3,750	4,550
2240	100 kVA		.50	32		3,525	1,250		4,775	5,750
2250	Low operating temperature(80°C), 25 kVA		1	16		2,625	630		3,255	3,850
2260	37 kVA		.80	20		2,850	790		3,640	4,300
2270	50 kVA		.70	22.857		3,700	900		4,600	5,425
2280	75 kVA		.65	24.615		5,900	970		6,870	7,950
2290	100 kVA		.55	29.091		6,150	1,150		7,300	8,450
2300	3 phase, 480 volt primary 120/208 volt secondary									
2310	Ventilated, 3 kVA	1 Elec	1	8	Ea.	525	315		840	1,050
2700	6 kVA		.80	10		720	395		1,115	1,375
2900	9 kVA		.70	11.429		820	450		1,270	1,575
3100	15 kVA	2 Elec	1.10	14.545		1,100	575		1,675	2,050
3300	30 kVA		.90	17.778		1,275	700		1,975	2,450
3500	45 kVA		.80	20		1,525	790		2,315	2,875
3700	75 kVA		.70	22.857		2,300	900		3,200	3,900
3900	112.5 kVA	R-3	.90	22.222		3,075	860	177	4,112	4,850
4100	150 kVA		.85	23.529		4,000	910	187	5,097	5,950
4300	225 kVA		.65	30.769		5,425	1,200	245	6,870	8,025
4500	300 kVA		.55	36.364		6,850	1,400	289	8,539	9,975
4700	500 kVA		.45	44.444		11,400	1,725	355	13,480	15,500
4800	750 kVA		.35	57.143		19,900	2,200	455	22,555	25,700
4820	1000 kVA		.32	62.500		22,800	2,425	495	25,720	29,200
4850	K-4 rated, 15 kVA	2 Elec	1.10	14.545		1,225	575		1,800	2,200
4855	30 kVA		.90	17.778		1,875	700		2,575	3,100
4860	45 kVA		.80	20		2,175	790		2,965	3,550
4865	75 kVA		.70	22.857		2,950	900		3,850	4,600
4870	112.5 kVA	R-3	.90	22.222		4,925	860	177	5,962	6,900
4875	150 kVA		.85	23.529		6,600	910	187	7,697	8,800
4880	225 kVA		.65	30.769		10,000	1,200	245	11,445	13,000
4885	300 kVA		.55	36.364		13,900	1,400	289	15,589	17,700
4890	500 kVA		.45	44.444		19,200	1,725	355	21,280	24,100
4900	K-13 rated, 15 kVA	2 Elec	1.10	14.545		1,600	575		2,175	2,625
4905	30 kVA		.90	17.778		2,200	700		2,900	3,475
4910	45 kVA		.80	20		2,600	790		3,390	4,025
4915	75 kVA		.70	22.857		3,675	900		4,575	5,375
4920	112.5 kVA	R-3	.90	22.222		7,550	860	177	8,587	9,800
4925	150 kVA		.85	23.529		9,175	910	187	10,272	11,700
4930	225 kVA		.65	30.769		13,400	1,200	245	14,845	16,700
4935	300 kVA		.55	36.364		18,800	1,400	289	20,489	23,100
4940	500 kVA		.45	44.444		28,700	1,725	355	30,780	34,600
5020	480 volt primary 120/208 volt secondary									
5030	Nonventilated, 15 kVA	2 Elec	1.10	14.545	Ea.	2,250	575		2,825	3,325
5040	30 kVA		.80	20		2,525	790		3,315	3,975
5050	45 kVA		.70	22.857		3,450	900		4,350	5,150
5060	75 kVA		.65	24.615		5,025	970		5,995	6,975
5070	112.5 kVA	R-3	.85	23.529		6,850	910	187	7,947	9,075
5081	150 kVA		.85	23.529		8,275	910	187	9,372	10,700
5090	225 kVA		.60	33.333		10,100	1,300	265	11,665	13,300
5100	300 kVA		.50	40		12,100	1,550	320	13,970	16,000
5200	Low operating temperature (80°C), 30 kVA	2 Elec	.90	17.778		2,075	700		2,775	3,325
5210	45 kVA		.80	20		3,100	790		3,890	4,600
5220	75 kVA		.70	22.857		4,125	900		5,025	5,900
5230	112.5 kVA	R-3	.90	22.222		7,300	860	177	8,337	9,525

R16270-600

16270	Transformers		CREW	DAILY OUTPUT	LABOR-HOURS	UNIT	2004 BARE COSTS				TOTAL INCL O&P	
							MAT.	LABOR	EQUIP.	TOTAL		
200	5240	150 kVA	R-3	.85	23.529	Ea.	9,250	910	187	10,347	11,800	200
	5250	225 kVA		.65	30.769		12,800	1,200	245	14,245	16,000	
	5260	300 kVA		.55	36.364		15,300	1,400	289	16,989	19,200	
	5270	500 kVA		.45	44.444		23,000	1,725	355	25,080	28,300	
	5380	3 phase, 5 kV primary 277/480 volt secondary										
	5400	High voltage, 112.5 kVA	R-3	.85	23.529	Ea.	8,925	910	187	10,022	11,400	
	5410	150 kVA		.65	30.769		9,700	1,200	245	11,145	12,700	
	5420	225 kVA		.55	36.364		11,500	1,400	289	13,189	15,000	
	5430	300 kVA		.45	44.444		14,700	1,725	355	16,780	19,100	
	5440	500 kVA		.35	57.143		18,900	2,200	455	21,555	24,600	
	5450	750 kVA		.32	62.500		25,900	2,425	495	28,820	32,600	
	5460	1000 kVA		.30	66.667		30,400	2,575	530	33,505	37,800	
	5470	1500 kVA		.27	74.074		35,400	2,875	590	38,865	43,800	
	5480	2000 kVA		.25	80		41,500	3,100	635	45,235	51,000	
	5490	2500 kVA		.20	100		46,700	3,875	795	51,370	58,000	
	5500	3000 kVA		.18	111		61,500	4,300	885	66,685	75,000	
	5590	15 kV primary 277/480 volt secondary										
	5600	High voltage, 112.5 kVA	R-3	.85	23.529	Ea.	13,300	910	187	14,397	16,300	
	5610	150 kVA		.65	30.769		15,600	1,200	245	17,045	19,100	
	5620	225 kVA		.55	36.364		17,600	1,400	289	19,289	21,700	
	5630	300 kVA		.45	44.444		20,700	1,725	355	22,780	25,700	
	5640	500 kVA		.35	57.143		26,200	2,200	455	28,855	32,600	
	5650	750 kVA		.32	62.500		34,200	2,425	495	37,120	41,700	
	5660	1000 kVA		.30	66.667		39,000	2,575	530	42,105	47,300	
	5670	1500 kVA		.27	74.074		44,900	2,875	590	48,365	54,500	
	5680	2000 kVA		.25	80		49,900	3,100	635	53,635	60,500	
	5690	2500 kVA		.20	100		58,000	3,875	795	62,670	70,000	
	5700	3000 kVA		.18	111		69,000	4,300	885	74,185	83,000	
	6000	2400V primary, 480V secondary, 300 KVA		.45	44.444		14,700	1,725	355	16,780	19,100	
	6010	500 KVA		.35	57.143		18,900	2,200	455	21,555	24,600	
	6020	750 KVA		.32	62.500		24,000	2,425	495	26,920	30,500	
300	0010	**ISOLATING PANELS** used with isolating transformers										300
	0020	For hospital applications										
	0100	Critical care area, 8 circuit, 3 kVA	1 Elec	.58	13.793	Ea.	5,800	545		6,345	7,175	
	0200	5 kVA		.54	14.815		5,950	585		6,535	7,425	
	0400	7.5 kVA		.52	15.385		6,075	605		6,680	7,575	
	0600	10 kVA		.44	18.182		6,275	715		6,990	7,975	
	0800	Operating room power & lighting, 8 circuit, 3 kVA		.58	13.793		4,850	545		5,395	6,125	
	1000	5 kVA		.54	14.815		5,100	585		5,685	6,500	
	1200	7.5 kVA		.52	15.385		5,575	605		6,180	7,050	
	1400	10 kVA		.44	18.182		5,950	715		6,665	7,625	
	1600	X-ray systems, 15 kVA, 90 amp		.44	18.182		12,400	715		13,115	14,700	
	1800	25 kVA, 125 amp		.36	22.222		12,700	875		13,575	15,300	
310	0010	**ISOLATING TRANSFORMER**										310
	0100	Single phase, 120/240 volt primary, 120/240 volt secondary										
	0200	0.50 kVA	1 Elec	4	2	Ea.	227	79		306	365	
	0400	1 kVA		2	4		283	158		441	545	
	0600	2 kVA		1.60	5		425	197		622	760	
	0800	3 kVA		1.40	5.714		485	225		710	870	
	1000	5 kVA		1.20	6.667		640	263		903	1,100	
	1200	7.5 kVA		1.10	7.273		810	287		1,097	1,325	
	1400	10 kVA		.80	10		1,025	395		1,420	1,700	
	1600	15 kVA		.60	13.333		1,325	525		1,850	2,225	
	1800	25 kVA		.50	16		1,900	630		2,530	3,050	
	1810	37.5 kVA	2 Elec	.80	20		3,150	790		3,940	4,625	

R16270 -600

Important: See the Reference Section for critical supporting data - Reference Nos., Crews, & City Cost Indexes

		16270	Transformers	CREW	DAILY OUTPUT	LABOR-HOURS	UNIT	2004 BARE COSTS				TOTAL INCL O&P	
								MAT.	LABOR	EQUIP.	TOTAL		
310	1820		75 kVA	2 Elec	.65	24.615	Ea.	4,725	970		5,695	6,650	310
	1830		3 phase, 120/240 primary, 120/208V secondary, 112.5 kVA [R16270-600]	R-3	.90	22.222		5,950	860	177	6,987	8,025	
	1840		150 kVA		.85	23.529		7,575	910	187	8,672	9,875	
	1850		225 kVA		.65	30.769		10,500	1,200	245	11,945	13,600	
	1860		300 kVA		.55	36.364		14,000	1,400	289	15,689	17,800	
	1870		500 kVA		.45	44.444		23,400	1,725	355	25,480	28,700	
	1880		750 kVA		.35	57.143		23,600	2,200	455	26,255	29,800	
600	0010		**OIL FILLED TRANSFORMER** Pad mounted, primary delta or Y,										600
	0050		5 kV or 15 kV, with taps, 277/480 V secondary, 3 phase [R16270-600]										
	0100		150 kVA	R-3	.65	30.769	Ea.	6,300	1,200	245	7,745	8,975	
	0110		225 kVA		.55	36.364		6,875	1,400	289	8,564	10,000	
	0200		300 kVA		.45	44.444		8,600	1,725	355	10,680	12,400	
	0300		500 kVA		.40	50		12,600	1,925	400	14,925	17,100	
	0400		750 kVA		.38	52.632		15,500	2,025	420	17,945	20,500	
	0500		1000 kVA		.26	76.923		18,300	2,975	610	21,885	25,300	
	0600		1500 kVA		.23	86.957		21,800	3,375	690	25,865	29,800	
	0700		2000 kVA		.20	100		27,500	3,875	795	32,170	37,000	
	0710		2500 kVA		.19	105		33,300	4,075	835	38,210	43,600	
	0720		3000 kVA		.17	117		40,100	4,550	935	45,585	52,000	
	0800		3750 kVA		.16	125		51,500	4,825	995	57,320	65,000	
610	0010		**TRANSFORMER, LIQUID-FILLED** Pad mounted										610
	0020		5 kV or 15 kV primary, 277/480 volt secondary, 3 phase										
	0050		225 kVA	R-3	.55	36.364	Ea.	9,175	1,400	289	10,864	12,500	
	0100		300 kVA		.45	44.444		10,900	1,725	355	12,980	15,000	
	0200		500 kVA		.40	50		13,800	1,925	400	16,125	18,400	
	0250		750 kVA		.38	52.632		17,800	2,025	420	20,245	23,000	
	0300		1000 kVA		.26	76.923		20,600	2,975	610	24,185	27,800	
	0350		1500 kVA		.23	86.957		24,100	3,375	690	28,165	32,300	
	0400		2000 kVA		.20	100		29,800	3,875	795	34,470	39,500	
	0450		2500 kVA		.19	105		35,500	4,075	835	40,410	46,100	
620	0010		**TRANSFORMER HANDLING** Add to normal labor cost in restricted areas										620
	5000		Transformers										
	5150		15 kVA, approximately 200 pounds	2 Elec	2.70	5.926	Ea.		233		233	345	
	5160		25 kVA, approximately 300 pounds		2.50	6.400			252		252	375	
	5170		37.5 kVA, approximately 400 pounds		2.30	6.957			274		274	410	
	5180		50 kVA, approximately 500 pounds		2	8			315		315	470	
	5190		75 kVA, approximately 600 pounds		1.80	8.889			350		350	520	
	5200		100 kVA, approximately 700 pounds		1.60	10			395		395	585	
	5210		112.5 kVA, approximately 800 pounds	3 Elec	2.20	10.909			430		430	640	
	5220		125 kVA, approximately 900 pounds		2	12			475		475	705	
	5230		150 kVA, approximately 1000 pounds		1.80	13.333			525		525	780	
	5240		167 kVA, approximately 1200 pounds		1.60	15			590		590	880	
	5250		200 kVA, approximately 1400 pounds		1.40	17.143			675		675	1,000	
	5260		225 kVA, approximately 1600 pounds		1.30	18.462			725		725	1,075	
	5270		250 kVA, approximately 1800 pounds		1.10	21.818			860		860	1,275	
	5280		300 kVA, approximately 2000 pounds		1	24			945		945	1,400	
	5290		500 kVA, approximately 3000 pounds		.75	32			1,250		1,250	1,875	
	5300		600 kVA, approximately 3500 pounds		.67	35.821			1,400		1,400	2,100	
	5310		750 kVA, approximately 4000 pounds		.60	40			1,575		1,575	2,350	
	5320		1000 kVA, approximately 5000 pounds		.50	48			1,900		1,900	2,825	

16280 | Power Filters & Conditioners

100	0010	**AUTOMATIC VOLTAGE REGULATORS**											100
	0100	Computer grade, solid state, variable trans. volt. regulator											

ELECTRICAL 16

			DAILY	LABOR-		2004 BARE COSTS				TOTAL	
16280	**Power Filters & Conditioners**	CREW	OUTPUT	HOURS	UNIT	MAT.	LABOR	EQUIP.	TOTAL	INCL O&P	
100	0110	Single-phase, 120 V, 8.6 kVA	2 Elec	1.33	12.030	Ea.	4,400	475		4,875	5,550	**100**
	0120	17.3 kVA		1.14	14.035		5,200	555		5,755	6,550	
	0130	208/240 V, 7.5/8.6 kVA		1.33	12.030		4,400	475		4,875	5,550	
	0140	13.5/15.6 kVA		1.33	12.030		5,200	475		5,675	6,425	
	0150	27.0/31.2 kVA		1.14	14.035		6,550	555		7,105	8,050	
	0210	Two-phase, single control, 208/240 V, 15.0/17.3 kVA		1.14	14.035		5,200	555		5,755	6,550	
	0220	Individual phase control, 15.0/17.3 kVA	▼	1.14	14.035		5,200	555		5,755	6,550	
	0230	30.0/34.6 kVA	3 Elec	1.33	18.045		6,550	710		7,260	8,275	
	0310	Three-phase single control, 208/240 V, 26/30 kVA	2 Elec	1	16		5,200	630		5,830	6,675	
	0320	380/480 V, 24/30 kVA	"	1	16		5,200	630		5,830	6,675	
	0330	43/54 kVA	3 Elec	1.33	18.045		9,450	710		10,160	11,500	
	0340	Individual phase control, 208 V, 26 kVA	"	1.33	18.045		5,200	710		5,910	6,775	
	0350	52 kVA	R-3	.91	21.978		6,550	850	175	7,575	8,700	
	0360	340/480 V, 24/30 kVA	2 Elec	1	16		5,200	630		5,830	6,675	
	0370	43/54 kVA	"	1	16		6,550	630		7,180	8,175	
	0380	48/60 kVA	3 Elec	1.33	18.045		9,500	710		10,210	11,600	
	0390	86/108 kVA	R-3	.91	21.978	▼	10,600	850	175	11,625	13,100	
	0500	Standard grade, solid state, variable transformer volt. regulator										
	0510	Single-phase, 115 V, 2.3 kVA	1 Elec	2.29	3.493	Ea.	1,625	138		1,763	2,000	
	0520	4.2 kVA		2	4		2,600	158		2,758	3,075	
	0530	6.6 kVA		1.14	7.018		3,200	276		3,476	3,900	
	0540	13.0 kVA	▼	1.14	7.018		4,125	276		4,401	4,950	
	0550	16.6 kVA	2 Elec	1.23	13.008		4,875	515		5,390	6,125	
	0610	230 V, 8.3 kVA		1.33	12.030		4,125	475		4,600	5,250	
	0620	21.4 kVA		1.23	13.008		4,875	515		5,390	6,125	
	0630	29.9 kVA		1.23	13.008		4,875	515		5,390	6,125	
	0710	460 V, 9.2 kVA		1.33	12.030		4,125	475		4,600	5,250	
	0720	20.7 kVA	▼	1.23	13.008		4,875	515		5,390	6,125	
	0810	Three-phase, 230 V, 13.1 kVA	3 Elec	1.41	17.021		4,125	670		4,795	5,550	
	0820	19.1 kVA		1.41	17.021		4,875	670		5,545	6,375	
	0830	25.1 kVA	▼	1.60	15		4,875	590		5,465	6,250	
	0840	57.8 kVA	R-3	.95	21.053		8,850	815	167	9,832	11,100	
	0850	74.9 kVA	"	.91	21.978		8,850	850	175	9,875	11,200	
	0910	460 V, 14.3 kVA	3 Elec	1.41	17.021		4,125	670		4,795	5,550	
	0920	19.1 kVA		1.41	17.021		4,875	670		5,545	6,375	
	0930	27.9 kVA	▼	1.50	16		4,875	630		5,505	6,325	
	0940	59.8 kVA	R-3	1	20		8,850	775	159	9,784	11,100	
	0950	79.7 kVA		.95	21.053		9,900	815	167	10,882	12,300	
	0960	118 kVA	▼	.95	21.053	▼	10,400	815	167	11,382	12,900	
	1000	Laboratory grade, precision, electronic voltage regulator										
	1110	Single-phase, 115 V, 0.5 kVA	1 Elec	2.29	3.493	Ea.	1,350	138		1,488	1,675	
	1120	1.0 kVA		2	4		1,450	158		1,608	1,800	
	1130	3.0 kVA	▼	.80	10		2,000	395		2,395	2,775	
	1140	6.0 kVA	2 Elec	1.46	10.959		3,725	430		4,155	4,750	
	1150	10.0 kVA	3 Elec	1	24		4,825	945		5,770	6,725	
	1160	15.0 kVA	"	1.50	16		5,525	630		6,155	7,025	
	1210	230 V, 3.0 kVA	1 Elec	.80	10		2,275	395		2,670	3,075	
	1220	6.0 kVA	2 Elec	1.46	10.959		3,800	430		4,230	4,825	
	1230	10.0 kVA	3 Elec	1.71	14.035		5,050	555		5,605	6,375	
	1240	15.0 kVA	"	1.60	15	▼	5,700	590		6,290	7,150	
300	0010	**CAPACITORS** Indoor										**300**
	0020	240 volts, single & 3 phase, 0.5 kVAR	1 Elec	2.70	2.963	Ea.	278	117		395	480	
	0100	1.0 kVAR		2.70	2.963		335	117		452	545	
	0150	2.5 kVAR		2	4		400	158		558	675	
	0200	5.0 kVAR		1.80	4.444		465	175		640	770	
	0250	7.5 kVAR	▼	1.60	5	▼	530	197		727	875	

16280	Power Filters & Conditioners	CREW	DAILY OUTPUT	LABOR-HOURS	UNIT	2004 BARE COSTS				TOTAL INCL O&P	
						MAT.	LABOR	EQUIP.	TOTAL		
300	0300	10 kVAR	1 Elec	1.50	5.333	Ea.	635	210		845	1,025	**300**
	0350	15 kVAR		1.30	6.154		875	242		1,117	1,325	
	0400	20 kVAR		1.10	7.273		1,100	287		1,387	1,625	
	0450	25 kVAR		1	8		1,300	315		1,615	1,900	
	1000	480 volts, single & 3 phase, 1 kVAR		2.70	2.963		253	117		370	450	
	1050	2 kVAR		2.70	2.963		291	117		408	495	
	1100	5 kVAR		2	4		370	158		528	640	
	1150	7.5 kVAR		2	4		405	158		563	680	
	1200	10 kVAR		2	4		445	158		603	725	
	1250	15 kVAR		2	4		550	158		708	840	
	1300	20 kVAR		1.60	5		610	197		807	970	
	1350	30 kVAR		1.50	5.333		765	210		975	1,150	
	1400	40 kVAR		1.20	6.667		975	263		1,238	1,475	
	1450	50 kVAR		1.10	7.273		1,100	287		1,387	1,650	
	2000	600 volts, single & 3 phase, 1 kVAR		2.70	2.963		261	117		378	460	
	2050	2 kVAR		2.70	2.963		300	117		417	505	
	2100	5 kVAR		2	4		370	158		528	640	
	2150	7.5 kVAR		2	4		405	158		563	680	
	2200	10 kVAR		2	4		445	158		603	725	
	2250	15 kVAR		1.60	5		550	197		747	900	
	2300	20 kVAR		1.60	5		610	197		807	970	
	2350	25 kVAR		1.50	5.333		685	210		895	1,075	
	2400	35 kVAR		1.40	5.714		870	225		1,095	1,300	
	2450	50 kVAR		1.30	6.154		1,100	242		1,342	1,575	
340	0010	**COMPUTER ISOLATION TRANSFORMER**										**340**
	0100	Computer grade										
	0110	Single-phase, 120/240 V, 0.5 kVA	1 Elec	4	2	Ea.	345	79		424	495	
	0120	1.0 kVA		2.67	2.996		490	118		608	715	
	0130	2.5 kVA		2	4		745	158		903	1,050	
	0140	5 kVA		1.14	7.018		790	276		1,066	1,275	
360	0010	**COMPUTER REGULATOR TRANSFORMER**										**360**
	0100	Ferro-resonant, constant voltage, variable transformer										
	0110	Single-phase, 240 V, 0.5 kVA	1 Elec	2.67	2.996	Ea.	440	118		558	660	
	0120	1.0 kVA		2	4		605	158		763	900	
	0130	2.0 kVA		1	8		1,025	315		1,340	1,600	
	0210	Plug-in unit 120 V, 0.14 kVA		8	1		254	39.50		293.50	340	
	0220	0.25 kVA		8	1		297	39.50		336.50	385	
	0230	0.5 kVA		8	1		440	39.50		479.50	545	
	0240	1.0 kVA		5.33	1.501		605	59		664	755	
	0250	2.0 kVA		4	2		1,025	79		1,104	1,250	
600	0010	**POWER CONDITIONER TRANSFORMER**										**600**
	0100	Electronic solid state, buck-boost, transformer, w/tap switch										
	0110	Single-phase, 115 V, 3.0 kVA, + or - 3% accuracy	2 Elec	1.60	10	Ea.	2,375	395		2,770	3,200	
	0120	208, 220, 230, or 240 V, 5.0 kVA, + or - 1.5% accuracy	3 Elec	1.60	15		3,075	590		3,665	4,250	
	0130	5.0 kVA, + or - 6% accuracy	2 Elec	1.14	14.035		2,750	555		3,305	3,850	
	0140	7.5 kVA, + or - 1.5% accuracy	3 Elec	1.50	16		3,900	630		4,530	5,250	
	0150	7.5 kVA, + or - 6% accuracy		1.60	15		3,250	590		3,840	4,450	
	0160	10.0 kVA, + or - 1.5% accuracy		1.33	18.045		5,200	710		5,910	6,775	
	0170	10.0 kVA, + or - 6% accuracy		1.41	17.021		4,425	670		5,095	5,875	
820	0010	**TRANSIENT SUPPRESSOR/VOLTAGE REGULATOR** (without isolation)										**820**
	0110	Single-phase, 115 V, 1.0 kVA	1 Elec	2.67	2.996	Ea.	880	118		998	1,150	
	0120	2.0 kVA		2.29	3.493		1,200	138		1,338	1,525	
	0130	4.0 kVA		2.13	3.756		1,475	148		1,623	1,850	
	0140	220 V, 1.0 kVA		2.67	2.996		880	118		998	1,150	
	0150	2.0 kVA		2.29	3.493		1,200	138		1,338	1,525	

ELECTRICAL 16

209

16280 | Power Filters & Conditioners

			CREW	DAILY OUTPUT	LABOR-HOURS	UNIT	2004 BARE COSTS				TOTAL INCL O&P	
							MAT.	LABOR	EQUIP.	TOTAL		
820	0160	4.0 kVA	1 Elec	2.13	3.756	Ea.	1,550	148		1,698	1,950	820
	0210	Plug-in unit, 120 V, 1.0 kVA		8	1		840	39.50		879.50	985	
	0220	2.0 kVA	↓	8	1	↓	1,175	39.50		1,214.50	1,350	
840	0010	**TRANSIENT VOLTAGE SUPPRESSOR TRANSFORMER**										840
	0110	Single-phase, 120 V, 1.8 kVA	1 Elec	4	2	Ea.	490	79		569	650	
	0120	3.6 kVA		4	2		865	79		944	1,075	
	0130	7.2 kVA		3.20	2.500		1,125	98.50		1,223.50	1,375	
	0150	240 V, 3.6 kVA		4	2		865	79		944	1,075	
	0160	7.2 kVA		4	2		1,125	79		1,204	1,350	
	0170	14.4 kVA		3.20	2.500		1,425	98.50		1,523.50	1,725	
	0210	Plug-in unit, 120 V, 1.8 kVA	↓	8	1	↓	470	39.50		509.50	575	

16290 | Power Measurement & Control

			CREW	DAILY OUTPUT	LABOR-HOURS	UNIT	MAT.	LABOR	EQUIP.	TOTAL	TOTAL INCL O&P	
800	0010	**SWITCHBOARD INSTRUMENTS** 3 phase, 4 wire										800
	0100	AC indicating, ammeter & switch	1 Elec	8	1	Ea.	1,275	39.50		1,314.50	1,450	
	0200	Voltmeter & switch		8	1		1,275	39.50		1,314.50	1,450	
	0300	Wattmeter		8	1		2,500	39.50		2,539.50	2,800	
	0400	AC recording, ammeter		4	2		4,450	79		4,529	5,025	
	0500	Voltmeter		4	2		4,450	79		4,529	5,025	
	0600	Ground fault protection, zero sequence		2.70	2.963		3,925	117		4,042	4,500	
	0700	Ground return path		2.70	2.963		3,925	117		4,042	4,500	
	0800	3 current transformers, 5 to 800 amp		2	4		1,825	158		1,983	2,250	
	0900	1000 to 1500 amp		1.30	6.154		2,625	242		2,867	3,250	
	1200	2000 to 4000 amp		1	8		3,100	315		3,415	3,900	
	1300	Fused potential transformer, maximum 600 volt	↓	8	1	↓	690	39.50		729.50	820	
860	0010	**VOLTAGE MONITOR SYSTEMS** (test equipment)										860
	0100	AC voltage monitor system, 120/240 V, one-channel				Ea.	3,000			3,000	3,300	
	0110	Modem adapter					375			375	415	
	0120	Add-on detector only					1,575			1,575	1,725	
	0150	AC voltage remote monitor sys., 3 channel, 120, 230, or 480 V					5,450			5,450	6,000	
	0160	With internal modem					5,750			5,750	6,325	
	0170	Combination temperature and humidity probe					845			845	930	
	0180	Add-on detector only					3,950			3,950	4,350	
	0190	With internal modem				↓	4,300			4,300	4,725	

16300 | Transmission & Distribution

16310 | Transmission & Dist Accessories

			CREW	DAILY OUTPUT	LABOR-HOURS	UNIT	2004 BARE COSTS				TOTAL INCL O&P	
							MAT.	LABOR	EQUIP.	TOTAL		
600	0010	**LINE POLES & FIXTURES**										600
	0100	Digging holes in earth, average [R16310 -600]	R-5	25.14	3.500	Ea.		119	57.50	176.50	243	
	0105	In rock, average	"	4.51	19.512			665	320	985	1,350	
	0200	Wood poles, material handling and spotting	R-7	6.49	7.396			201	16.80	217.80	330	
	0220	Erect wood poles & backfill holes, in earth	R-5	6.77	12.999		1,025	445	214	1,684	2,025	
	0250	In rock	"	5.87	14.991	↓	1,025	510	247	1,782	2,175	
	0260	Disposal of surplus material	R-7	20.87	2.300	Mile		62.50	5.20	67.70	102	
	0300	Crossarms for wood pole structure										
	0310	Material handling and spotting	R-7	14.55	3.299	Ea.		89.50	7.50	97	145	
	0320	Install crossarms	R-5	11	8	"	340	272	132	744	930	

Important: See the Reference Section for critical supporting data - Reference Nos., Crews, & City Cost Indexes

16310			Transmission & Dist Accessories	CREW	DAILY OUTPUT	LABOR-HOURS	UNIT	2004 BARE COSTS				TOTAL INCL O&P	
								MAT.	LABOR	EQUIP.	TOTAL		
600	0330		Disposal of surplus material	R-7	40	1.200	Mile		32.50	2.72	35.22	53	**600**
	0400		Formed plate pole structure [R16310 -600]										
	0410		Material handling and spotting	R-7	2.40	20	Ea.		540	45.50	585.50	885	
	0420		Erect steel plate pole	R-5	1.95	45.128		6,400	1,525	745	8,670	10,100	
	0500		Guys, anchors and hardware for pole, in earth	↓	7.04	12.500		385	425	206	1,016	1,300	
	0510		In rock	↓	17.96	4.900	↓	465	167	80.50	712.50	850	
	0900		Foundations for line poles										
	0920		Excavation, in earth	R-5	135.38	.650	C.Y.		22	10.70	32.70	45.50	
	0940		In rock	"	20	4.400	"		150	72.50	222.50	305	
	0950		See also Division 02400										
	0960		Concrete foundations	R-5	11	8	C.Y.	90	272	132	494	655	
	0970		See also Division 03300	↓									
610	0010		**LINE TOWERS & FIXTURES**										**610**
	0100		Excavation and backfill, earth	R-5	135.38	.650	C.Y.		22	10.70	32.70	45.50	
	0105		Rock		21.46	4.101	"		140	67.50	207.50	285	
	0200		Steel footings (grillage) in earth		3.91	22.506	Ton	1,275	765	370	2,410	2,950	
	0205		In rock	↓	3.20	27.500	"	1,275	935	455	2,665	3,300	
	0290		See also Division 02400										
	0300		Rock anchors	R-5	5.87	14.991	Ea.	340	510	247	1,097	1,425	
	0400		Concrete foundations	"	12.85	6.848	C.Y.	87	233	113	433	570	
	0490		See also Division 03300										
	0500		Towers-material handling and spotting	R-7	22.56	2.128	Ton		57.50	4.83	62.33	94	
	0540		Steel tower erection	R-5	7.65	11.503		1,225	390	189	1,804	2,150	
	0550		Lace and box		7.10	12.394	↓	1,225	420	204	1,849	2,200	
	0560		Painting total structure	↓	1.47	59.864	Ea.	273	2,050	985	3,308	4,450	
	0570		Disposal of surplus material	R-7	20.87	2.300	Mile		62.50	5.20	67.70	102	
	0600		Special towers-material handling and spotting	"	12.31	3.899	Ton		106	8.85	114.85	172	
	0640		Special steel structure erection	R-6	6.52	13.497		1,500	460	460	2,420	2,850	
	0650		Special steel lace and box	"	6.29	13.990	↓	1,500	475	475	2,450	2,900	
	0670		Disposal of surplus material	R-7	7.87	6.099	Mile		165	13.85	178.85	269	
700	0010		**OVERHEAD LINE CONDUCTORS & DEVICES** [R16310 -600]										**700**
	0100		Conductors, primary circuits										
	0110		Material handling and spotting	R-5	9.78	8.998	W.Mile		305	148	453	625	
	0120		For river crossing, add		11	8			272	132	404	555	
	0150		Conductors, per wire, 210 to 636 kcmil		1.96	44.898		6,925	1,525	740	9,190	10,700	
	0160		795 to 954 kcmil		1.87	47.059		10,200	1,600	775	12,575	14,500	
	0170		1000 to 1600 kcmil		1.47	59.864		16,700	2,050	985	19,735	22,600	
	0180		Over 1600 kcmil		1.35	65.185		22,200	2,225	1,075	25,500	28,900	
	0200		For river crossing, add, 210 to 636 kcmil		1.24	70.968			2,425	1,175	3,600	4,900	
	0220		795 to 954 kcmil		1.09	80.734			2,750	1,325	4,075	5,575	
	0230		1000 to 1600 kcmil		.97	90.722			3,100	1,500	4,600	6,300	
	0240		Over 1600 kcmil	↓	.87	101	↓		3,450	1,675	5,125	7,000	
	0300		Joints and dead ends	R-8	6	8	Ea.	1,000	276	31	1,307	1,550	
	0400		Sagging	R-5	7.33	12.001	W.Mile		410	198	608	830	
	0500		Clipping, per structure, 69 kV	R-10	9.60	5	Ea.		185	47	232	330	
	0510		161 kV		5.33	9.006			335	84.50	419.50	595	
	0520		345 to 500 kV	↓	2.53	18.972			700	178	878	1,250	
	0600		Make and install jumpers, per structure, 69 kV	R-8	3.20	15		271	520	58	849	1,150	
	0620		161 kV		1.20	40		540	1,375	154	2,069	2,850	
	0640		345 to 500 kV	↓	.32	150		910	5,175	580	6,665	9,400	
	0700		Spacers	R-10	68.57	.700		54	26	6.55	86.55	105	
	0720		For river crossings, add	"	60	.800	↓		29.50	7.50	37	53	
	0800		Installing pulling line (500 kV only)	R-9	1.45	44.138	W.Mile	505	1,425	128	2,058	2,850	
	0810		Disposal of surplus material	R-7	6.96	6.897	Mile		187	15.65	202.65	305	
	0820		With trailer mounted reel stands	"	13.71	3.501	"		95	7.95	102.95	155	
	0900		Insulators and hardware, primary circuits	↓									

ELECTRICAL 16

16310	Transmission & Dist Accessories	CREW	DAILY OUTPUT	LABOR-HOURS	UNIT	2004 BARE COSTS				TOTAL INCL O&P	
						MAT.	LABOR	EQUIP.	TOTAL		
700	0920	Material handling and spotting, 69 kV R16310-600	R-7	480	.100	Ea.		2.71	.23	2.94	4.41	700
	0930	161 kV		685.71	.070			1.90	.16	2.06	3.09	
	0950	345 to 500 kV		960	.050			1.36	.11	1.47	2.21	
	1000	Disk insulators, 69 kV	R-5	880	.100		55	3.41	1.65	60.06	67.50	
	1020	161 kV		977.78	.090		63	3.06	1.48	67.54	75.50	
	1040	345 to 500 kV		1,100	.080		63	2.72	1.32	67.04	75	
	1060	See Div. 16360-800-7400 for pin or pedestal insulator										
	1100	Install disk insulator at river crossing, add										
	1110	69 kV	R-5	586.67	.150	Ea.		5.10	2.47	7.57	10.40	
	1120	161 kV		880	.100			3.41	1.65	5.06	6.90	
	1140	345 to 500 kV		880	.100			3.41	1.65	5.06	6.90	
	1150	Disposal of surplus material	R-7	41.74	1.150	Mile		31	2.61	33.61	51	
	1300	Overhead ground wire installation										
	1320	Material handling and spotting	R-7	5.65	8.496	W.Mile		230	19.30	249.30	375	
	1340	Overhead ground wire	R-5	1.76	50		3,025	1,700	825	5,550	6,775	
	1350	At river crossing, add		1.17	75.214			2,550	1,250	3,800	5,200	
	1360	Disposal of surplus material		41.74	2.108	Mile		72	34.50	106.50	146	
	1400	Installing conductors, underbuilt circuits										
	1420	Material handling and spotting	R-7	5.65	8.496	W.Mile		230	19.30	249.30	375	
	1440	Conductors, per wire, 210 to 636 kcmil	R-5	1.96	44.898		6,925	1,525	740	9,190	10,700	
	1450	795 to 954 kcmil		1.87	47.059		10,200	1,600	775	12,575	14,500	
	1460	1000 to 1600 kcmil		1.47	59.864		16,700	2,050	985	19,735	22,600	
	1470	Over 1600 kcmil		1.35	65.185		22,200	2,225	1,075	25,500	28,900	
	1500	Joints and dead ends	R-8	6	8	Ea.	1,000	276	31	1,307	1,550	
	1550	Sagging	R-5	8.80	10	W.Mile		340	165	505	690	
	1600	Clipping, per structure, 69 kV	R-10	9.60	5	Ea.		185	47	232	330	
	1620	161 kV		5.33	9.006			335	84.50	419.50	595	
	1640	345 to 500 kV		2.53	18.972			700	178	878	1,250	
	1700	Making and installing jumpers, per structure, 69 kV	R-8	5.87	8.177		271	282	31.50	584.50	760	
	1720	161 kV		.96	50		540	1,725	193	2,458	3,400	
	1740	345 to 500 kV		.32	150		910	5,175	580	6,665	9,400	
	1800	Spacers	R-10	96	.500		54	18.50	4.68	77.18	92	
	1810	Disposal of surplus material	R-7	6.96	6.897	Mile		187	15.65	202.65	305	
	2000	Insulators and hardware for underbuilt circuits										
	2100	Material handling and spotting	R-7	1,200	.040	Ea.		1.08	.09	1.17	1.77	
	2150	Disk insulators, 69 kV	R-8	600	.080		55	2.76	.31	58.07	65	
	2160	161 kV		686	.070		63	2.42	.27	65.69	73.50	
	2170	345 to 500 kV		800	.060		63	2.07	.23	65.30	73	
	2180	Disposal of surplus material	R-7	41.74	1.150	Mile		31	2.61	33.61	51	
	2300	Sectionalizing switches, 69 kV	R-5	1.26	69.841	Ea.	14,200	2,375	1,150	17,725	20,500	
	2310	161 kV		.80	110		16,100	3,750	1,800	21,650	25,300	
	2500	Protective devices		5.50	16		4,650	545	263	5,458	6,200	
	2600	Clearance poles, 8 poles per mile										
	2650	In earth, 69 kV	R-5	1.16	75.862	Mile	3,825	2,575	1,250	7,650	9,450	
	2660	161 kV	"	.64	137		6,275	4,675	2,275	13,225	16,500	
	2670	345 to 500 kV	R-6	.48	183		7,550	6,250	6,250	20,050	24,600	
	2800	In rock, 69 kV	R-5	.69	127		3,825	4,350	2,100	10,275	13,000	
	2820	161 kV	"	.35	251		6,275	8,550	4,150	18,975	24,400	
	2840	345 to 500 kV	R-6	.24	366		7,550	12,500	12,500	32,550	40,900	
850	0010	**TRANSMISSION LINE RIGHT OF WAY**										850
	0100	Clearing right of way	B-87	6.67	5.997	Acre		193	380	573	710	
	0200	Restoration & seeding	B-10D	4	3	"	890	93.50	235	1,218.50	1,375	

Important: See the Reference Section for critical supporting data - Reference Nos., Crews, & City Cost Indexes

16 ELECTRICAL

16330	Med-Voltage Sw & Prot	CREW	DAILY OUTPUT	LABOR-HOURS	UNIT	2004 BARE COSTS				TOTAL INCL O&P
						MAT.	LABOR	EQUIP.	TOTAL	
760										**760**
0010	**SWITCHGEAR,** Incorporate switch with cable connections, transformer,									
0100	& Low Voltage section									
0200	Load interrupter switch, 600 amp, 2 position									
0300	NEMA 1, 4.8 KV, 300 kVA & below w/CLF fuses	R-3	.40	50	Ea.	14,000	1,925	400	16,325	18,700
0400	400 kVA & above w/CLF fuses		.38	52.632		15,600	2,025	420	18,045	20,600
0500	Non fusible		.41	48.780		11,300	1,875	390	13,565	15,800
0600	13.8 kV, 300 kVA & below		.38	52.632		17,500	2,025	420	19,945	22,800
0700	400 kVA & above		.36	55.556		17,500	2,150	440	20,090	23,000
0800	Non fusible		.40	50		13,300	1,925	400	15,625	17,900
0900	Cable lugs for 2 feeders 4.8 kV or 13.8 kV	1 Elec	8	1		425	39.50		464.50	530
1000	Pothead, one 3 conductor or three 1 conductor		4	2		2,050	79		2,129	2,375
1100	Two 3 conductor or six 1 conductor		2	4		4,025	158		4,183	4,650
1200	Key interlocks		8	1		470	39.50		509.50	580
1300	Lightning arresters, Distribution class (no charge)									
1400	Intermediate class or line type 4.8 kV	1 Elec	2.70	2.963	Ea.	2,300	117		2,417	2,700
1500	13.8 kV		2	4		3,050	158		3,208	3,575
1600	Station class, 4.8 kV		2.70	2.963		3,925	117		4,042	4,500
1700	13.8 kV		2	4		6,750	158		6,908	7,650
1800	Transformers, 4800 volts to 480/277 volts, 75 kVA	R-3	.68	29.412		12,000	1,150	234	13,384	15,100
1900	112.5 kVA		.65	30.769		14,600	1,200	245	16,045	18,000
2000	150 kVA		.57	35.088		16,600	1,350	279	18,229	20,600
2100	225 kVA		.48	41.667		19,100	1,600	330	21,030	23,800
2200	300 kVA		.41	48.780		21,300	1,875	390	23,565	26,800
2300	500 kVA		.36	55.556		28,100	2,150	440	30,690	34,600
2400	750 kVA		.29	68.966		31,800	2,675	550	35,025	39,600
2500	13,800 volts to 480/277 volts, 75 kVA		.61	32.787		16,800	1,275	261	18,336	20,700
2600	112.5 kVA		.55	36.364		22,300	1,400	289	23,989	27,000
2700	150 kVA		.49	40.816		22,600	1,575	325	24,500	27,500
2800	225 kVA		.41	48.780		26,000	1,875	390	28,265	31,900
2900	300 kVA		.37	54.054		26,600	2,100	430	29,130	32,800
3000	500 kVA		.31	64.516		29,400	2,500	515	32,415	36,600
3100	750 kVA		.26	76.923		32,400	2,975	610	35,985	40,700
3200	Forced air cooling & temperature alarm	1 Elec	1	8		2,600	315		2,915	3,325
3300	Low voltage components									
3400	Maximum panel height 49-1/2", single or twin row									
3500	Breaker heights, type FA or FH, 6"									
3600	type KA or KH, 8"									
3700	type LA, 11"									
3800	type MA, 14"									
3900	Breakers, 2 pole, 15 to 60 amp, type FA	1 Elec	5.60	1.429	Ea.	291	56.50		347.50	405
4000	70 to 100 amp, type FA		4.20	1.905		355	75		430	500
4100	15 to 60 amp, type FH		5.60	1.429		455	56.50		511.50	585
4200	70 to 100 amp, type FH		4.20	1.905		540	75		615	705
4300	125 to 225 amp, type KA		3.40	2.353		820	92.50		912.50	1,050
4400	125 to 225 amp, type KH		3.40	2.353		1,900	92.50		1,992.50	2,225
4500	125 to 400 amp, type LA		2.50	3.200		1,500	126		1,626	1,875
4600	125 to 600 amp, type MA		1.80	4.444		2,475	175		2,650	2,975
4700	700 & 800 amp, type MA		1.50	5.333		3,125	210		3,335	3,750
4800	3 pole, 15 to 60 amp, type FA		5.30	1.509		360	59.50		419.50	485
4900	70 to 100 amp, type FA		4	2		435	79		514	590
5000	15 to 60 amp, type FH		5.30	1.509		540	59.50		599.50	680
5100	70 to 100 amp, type FH		4	2		610	79		689	785
5200	125 to 225 amp, type KA		3.20	2.500		1,025	98.50		1,123.50	1,275
5300	125 to 225 amp, type KH		3.20	2.500		2,300	98.50		2,398.50	2,675
5400	125 to 400 amp, type LA		2.30	3.478		1,850	137		1,987	2,225
5500	125 to 600 amp, type MA		1.60	5		3,050	197		3,247	3,650

ELECTRICAL 16

				DAILY	LABOR-			2004 BARE COSTS				TOTAL	
	16330	**Med-Voltage Sw & Prot**	CREW	OUTPUT	HOURS	UNIT	MAT.	LABOR	EQUIP.	TOTAL		INCL O&P	
760	5600	700 & 800 amp, type MA	1 Elec	1.30	6.154	Ea.	3,950	242		4,192		4,700	760

	16360	**Unit Substations**											
800	0010	**SUBSTATION EQUIPMENT**											800
	1000	Main conversion equipment											
	1050	Power transformers, 13 to 26 kV	R-11	1.72	32.558	MVA	16,100	1,200	420	17,720		20,000	
	1060	46 kV		3.50	16		15,100	590	206	15,896		17,700	
	1070	69 kV		3.11	18.006		12,900	665	232	13,797		15,500	
	1080	110 kV		3.29	17.021		12,300	630	219	13,149		14,700	
	1090	161 kV		4.31	12.993		11,400	480	167	12,047		13,400	
	1100	500 kV		7	8		11,300	295	103	11,698		13,100	
	1200	Grounding transformers		3.11	18.006	Ea.	71,500	665	232	72,397		80,000	
	1300	Station capacitors											
	1350	Synchronous, 13 to 26 kV	R-11	3.11	18.006	MVAR	4,500	665	232	5,397		6,200	
	1360	46 kV		3.33	16.817		5,725	620	217	6,562		7,475	
	1370	69 kV		3.81	14.698		5,650	540	189	6,379		7,225	
	1380	161 kV		6.51	8.602		5,275	315	111	5,701		6,400	
	1390	500 kV		10.37	5.400		4,600	199	69.50	4,868.50		5,425	
	1450	Static, 13 to 26 kV		3.11	18.006		3,825	665	232	4,722		5,450	
	1460	46 kV		3.01	18.605		4,825	685	240	5,750		6,600	
	1470	69 kV		3.81	14.698		4,675	540	189	5,404		6,175	
	1480	161 kV		6.51	8.602		4,350	315	111	4,776		5,375	
	1490	500 kV		10.37	5.400		3,975	199	69.50	4,243.50		4,750	
	1600	Voltage regulators, 13 to 26 kV		.75	74.667	Ea.	170,500	2,750	960	174,210		193,000	
	2000	Power circuit breakers											
	2050	Oil circuit breakers, 13 to 26 kV	R-11	1.12	50	Ea.	38,100	1,850	645	40,595		45,400	
	2060	46 kV		.75	74.667		55,500	2,750	960	59,210		66,000	
	2070	69 kV		.45	124		129,000	4,600	1,600	135,200		150,500	
	2080	161 kV		.16	350		197,500	12,900	4,500	214,900		242,000	
	2090	500 kV		.06	933		739,500	34,400	12,000	785,900		878,000	
	2100	Air circuit breakers, 13 to 26 kV		.56	100		39,800	3,700	1,300	44,800		51,000	
	2110	161 kV		.24	233		171,000	8,600	3,000	182,600		204,500	
	2150	Gas circuit breakers, 13 to 26 kV		.56	100		149,500	3,700	1,300	154,500		171,500	
	2160	161 kV		.08	700		196,500	25,800	9,025	231,325		264,500	
	2170	500 kV		.04	1,400		692,500	51,500	18,000	762,000		859,000	
	2200	Vacuum circuit breakers, 13 to 26 kV		.56	100		32,600	3,700	1,300	37,600		42,900	
	3000	Disconnecting switches											
	3050	Gang operated switches											
	3060	Manual operation, 13 to 26 kV	R-11	1.65	33.939	Ea.	9,175	1,250	435	10,860		12,500	
	3070	46 kV		1.12	50		15,300	1,850	645	17,795		20,300	
	3080	69 kV		.80	70		17,100	2,575	900	20,575		23,700	
	3090	161 kV		.56	100		20,600	3,700	1,300	25,600		29,700	
	3100	500 kV		.14	400		56,000	14,800	5,150	75,950		89,500	
	3110	Motor operation, 161 kV		.51	109		31,000	4,050	1,425	36,475		41,700	
	3120	500 kV		.28	200		85,500	7,375	2,575	95,450		108,500	
	3250	Circuit switches, 161 kV		.41	136		62,000	5,050	1,750	68,800		77,500	
	3300	Single pole switches											
	3350	Disconnecting switches, 13 to 26 kV	R-11	28	2	Ea.	8,075	74	26	8,175		9,025	
	3360	46 kV		8	7		14,200	258	90	14,548		16,100	
	3370	69 kV		5.60	10		15,800	370	129	16,299		18,100	
	3380	161 kV		2.80	20		60,000	740	258	60,998		67,500	
	3390	500 kV		.22	254		171,000	9,400	3,275	183,675		205,500	
	3450	Grounding switches, 46 kV		5.60	10		21,700	370	129	22,199		24,500	
	3460	69 kV		3.73	15.013		22,500	555	193	23,248		25,800	
	3470	161 kV		2.24	25		23,800	925	320	25,045		27,900	

16 ELECTRICAL

Important: See the Reference Section for critical supporting data - Reference Nos., Crews, & City Cost Indexes

16360	Unit Substations	CREW	DAILY OUTPUT	LABOR-HOURS	UNIT	2004 BARE COSTS				TOTAL INCL O&P
						MAT.	LABOR	EQUIP.	TOTAL	
800	3480	500 kV	R-11	.62	90.323	Ea.	29,600	3,325	1,175	34,100	38,900	800
4000	Instrument transformers									
4050	Current transformers, 13 to 26 kV	R-11	14	4	Ea.	2,100	148	51.50	2,299.50	2,575
4060	46 kV		9.33	6.002		6,125	221	77.50	6,423.50	7,175
4070	69 kV		7	8		6,350	295	103	6,748	7,550
4080	161 kV		1.87	29.947		20,600	1,100	385	22,085	24,800
4100	Potential transformers, 13 to 26 kV		11.20	5		3,000	185	64.50	3,249.50	3,650
4110	46 kV		8	7		6,175	258	90	6,523	7,275
4120	69 kV		6.22	9.003		6,550	330	116	6,996	7,825
4130	161 kV		2.24	25		14,100	925	320	15,345	17,200
4140	500 kV	▼	1.40	40	▼	42,200	1,475	515	44,190	49,200
7000	Conduit, conductors, and insulators									
7100	Conduit, metallic	R-11	560	.100	Lb.	1.89	3.69	1.29	6.87	9
7110	Non-metallic	"	800	.070	"	1.75	2.58	.90	5.23	6.80
7190	See also Division 16131									
7200	Wire and cable	R-11	700	.080	Lb.	1.70	2.95	1.03	5.68	7.40
7290	See also Division 16100									
7300	Bus	R-11	590	.095	Lb.	2.45	3.50	1.22	7.17	9.30
7390	See also Division 16450									
7400	Insulators, pedestal type	R-11	112	.500	Ea.		18.45	6.45	24.90	34.50
7490	See also Line 16310-700-1000									
7500	Grounding systems	R-11	280	.200	Lb.	7.10	7.40	2.58	17.08	21.50
7590	See also Division 16060									
7600	Manholes	R-11	4.15	13.494	Ea.	2,825	500	174	3,499	4,025
7690	See also Division 02080									
7700	Cable tray	R-11	40	1.400	L.F.	11.55	51.50	18.05	81.10	110
7790	See also Division 16131									
8000	Protective equipment									
8050	Lightning arresters, 13 to 26 kV	R-11	18.67	2.999	Ea.	1,025	111	38.50	1,174.50	1,325
8060	46 kV		14	4		2,800	148	51.50	2,999.50	3,350
8070	69 kV		11.20	5		3,575	185	64.50	3,824.50	4,275
8080	161 kV		5.60	10		4,950	370	129	5,449	6,125
8090	500 kV		1.40	40		16,700	1,475	515	18,690	21,200
8150	Reactors and resistors, 13 to 26 kV		28	2		2,325	74	26	2,425	2,725
8160	46 kV		4.31	12.993		7,050	480	167	7,697	8,650
8170	69 kV		2.80	20		11,500	740	258	12,498	14,000
8180	161 kV		2.24	25		13,100	925	320	14,345	16,100
8190	500 kV		.08	700		52,500	25,800	9,025	87,325	106,500
8250	Fuses, 13 to 26 kV		18.67	2.999		1,375	111	38.50	1,524.50	1,725
8260	46 kV		11.20	5		1,575	185	64.50	1,824.50	2,075
8270	69 kV		8	7		1,650	258	90	1,998	2,275
8280	161 kV	▼	4.67	11.991	▼	2,075	440	154	2,669	3,100
9000	Station service equipment									
9100	Conversion equipment									
9110	Station service transformers	R-11	5.60	10	Ea.	62,000	370	129	62,499	68,500
9120	Battery chargers		11.20	5	"	2,550	185	64.50	2,799.50	3,175
9200	Control batteries	▼	14	4	K.A.H.	58.50	148	51.50	258	340
9210										

16410	Encl Switches & Circuit Breakers	CREW	DAILY OUTPUT	LABOR-HOURS	UNIT	2004 BARE COSTS				TOTAL INCL O&P
						MAT.	LABOR	EQUIP.	TOTAL	
200	**0010**	**CIRCUIT BREAKERS** (in enclosure)									**200**
0100	Enclosed (NEMA 1), 600 volt, 3 pole, 30 amp	1 Elec	3.20	2.500	Ea.	410	98.50		508.50	600
0200	60 amp		2.80	2.857		410	113		523	620
0400	100 amp		2.30	3.478		470	137		607	725
0600	225 amp		1.50	5.333		1,100	210		1,310	1,525
0700	400 amp	2 Elec	1.60	10		1,850	395		2,245	2,625
0800	600 amp		1.20	13.333		2,700	525		3,225	3,725
1000	800 amp		.94	17.021		3,500	670		4,170	4,850
1200	1000 amp		.84	19.048		4,425	750		5,175	6,000
1220	1200 amp		.80	20		5,675	790		6,465	7,400
1240	1600 amp		.72	22.222		11,200	875		12,075	13,600
1260	2000 amp		.64	25		12,100	985		13,085	14,800
1400	1200 amp with ground fault		.80	20		11,500	790		12,290	13,900
1600	1600 amp with ground fault		.72	22.222		12,700	875		13,575	15,200
1800	2000 amp with ground fault		.64	25		13,600	985		14,585	16,400
2000	Disconnect, 240 volt 3 pole, 5 HP motor	1 Elec	3.20	2.500		310	98.50		408.50	490
2020	10 HP motor		3.20	2.500		310	98.50		408.50	490
2040	15 HP motor		2.80	2.857		310	113		423	510
2060	20 HP motor		2.30	3.478		380	137		517	620
2080	25 HP motor		2.30	3.478		380	137		517	620
2100	30 HP motor		2.30	3.478		380	137		517	620
2120	40 HP motor		2	4		645	158		803	945
2140	50 HP motor		1.50	5.333		645	210		855	1,025
2160	60 HP motor		1.50	5.333		1,500	210		1,710	1,975
2180	75 HP motor	2 Elec	2	8		1,500	315		1,815	2,125
2200	100 HP motor		1.60	10		1,500	395		1,895	2,225
2220	125 HP motor		1.60	10		1,500	395		1,895	2,225
2240	150 HP motor		1.20	13.333		2,900	525		3,425	3,975
2260	200 HP motor		1.20	13.333		3,800	525		4,325	4,950
2300	Enclosed (NEMA 7), explosion proof, 600 volt 3 pole, 50 amp	1 Elec	2.30	3.478		915	137		1,052	1,200
2350	100 amp		1.50	5.333		950	210		1,160	1,375
2400	150 amp		1	8		2,275	315		2,590	3,000
2450	250 amp	2 Elec	1.60	10		2,850	395		3,245	3,725
2500	400 amp	"	1.20	13.333		3,175	525		3,700	4,275
800	**0010**	**SAFETY SWITCHES**	R16410-800								**800**
0100	General duty 240 volt, 3 pole NEMA 1, fusible, 30 amp	1 Elec	3.20	2.500	Ea.	74.50	98.50		173	229
0200	60 amp		2.30	3.478		126	137		263	345
0300	100 amp		1.90	4.211		217	166		383	485
0400	200 amp		1.30	6.154		465	242		707	875
0500	400 amp	2 Elec	1.80	8.889		1,175	350		1,525	1,825
0600	600 amp	"	1.20	13.333		2,200	525		2,725	3,175
0610	Nonfusible, 30 amp	1 Elec	3.20	2.500		60.50	98.50		159	214
0650	60 amp		2.30	3.478		80	137		217	292
0700	100 amp		1.90	4.211		185	166		351	450
0750	200 amp		1.30	6.154		345	242		587	735
0800	400 amp	2 Elec	1.80	8.889		825	350		1,175	1,425
0850	600 amp	"	1.20	13.333		1,575	525		2,100	2,500
1100	Heavy duty, 600 volt, 3 pole NEMA 1 nonfused									
1110	30 amp	1 Elec	3.20	2.500	Ea.	106	98.50		204.50	264
1500	60 amp		2.30	3.478		192	137		329	415
1700	100 amp		1.90	4.211		300	166		466	575
1900	200 amp		1.30	6.154		460	242		702	865
2100	400 amp	2 Elec	1.80	8.889		1,025	350		1,375	1,650
2300	600 amp		1.20	13.333		1,825	525		2,350	2,800
2500	800 amp		.94	17.021		3,725	670		4,395	5,100
2700	1200 amp		.80	20		5,000	790		5,790	6,675

16 ELECTRICAL

Important: See the Reference Section for critical supporting data - Reference Nos., Crews, & City Cost Indexes

16410	Encl Switches & Circuit Breakers	CREW	DAILY OUTPUT	LABOR-HOURS	UNIT	2004 BARE COSTS				TOTAL INCL O&P
						MAT.	LABOR	EQUIP.	TOTAL	
2900	Heavy duty, 240 volt, 3 pole NEMA 1 fusible R16410-800									
2910	30 amp	1 Elec	3.20	2.500	Ea.	120	98.50		218.50	279
3000	60 amp		2.30	3.478		204	137		341	430
3300	100 amp		1.90	4.211		320	166		486	600
3500	200 amp		1.30	6.154		550	242		792	965
3700	400 amp	2 Elec	1.80	8.889		1,425	350		1,775	2,075
3900	600 amp		1.20	13.333		2,425	525		2,950	3,450
4100	800 amp		.94	17.021		5,725	670		6,395	7,300
4300	1200 amp		.80	20		7,500	790		8,290	9,425
4340	2 pole fused, 30 amp	1 Elec	3.50	2.286		90.50	90		180.50	234
4350	600 volt, 3 pole, fusible, 30 amp		3.20	2.500		204	98.50		302.50	370
4380	60 amp		2.30	3.478		247	137		384	475
4400	100 amp		1.90	4.211		450	166		616	740
4420	200 amp		1.30	6.154		655	242		897	1,075
4440	400 amp	2 Elec	1.80	8.889		1,700	350		2,050	2,400
4450	600 amp		1.20	13.333		2,850	525		3,375	3,925
4460	800 amp		.94	17.021		5,725	670		6,395	7,300
4480	1200 amp		.80	20		7,500	790		8,290	9,425
4500	240 volt 3 pole NEMA 3R (no hubs), fusible									
4510	30 amp	1 Elec	3.10	2.581	Ea.	215	102		317	390
4700	60 amp		2.20	3.636		340	143		483	590
4900	100 amp		1.80	4.444		485	175		660	795
5100	200 amp		1.20	6.667		670	263		933	1,125
5300	400 amp	2 Elec	1.60	10		1,475	395		1,870	2,200
5500	600 amp	"	1	16		3,100	630		3,730	4,350
5510	Heavy duty, 600 volt, 3 pole 3ph. NEMA 3R fusible, 30 amp	1 Elec	3.10	2.581		345	102		447	525
5520	60 amp		2.20	3.636		400	143		543	655
5530	100 amp		1.80	4.444		625	175		800	950
5540	200 amp		1.20	6.667		865	263		1,128	1,350
5550	400 amp	2 Elec	1.60	10		2,025	395		2,420	2,800
5700	600 volt, 3 pole NEMA 3R nonfused									
5710	30 amp	1 Elec	3.10	2.581	Ea.	191	102		293	360
5900	60 amp		2.20	3.636		335	143		478	580
6100	100 amp		1.80	4.444		465	175		640	775
6300	200 amp		1.20	6.667		565	263		828	1,025
6500	400 amp	2 Elec	1.60	10		1,400	395		1,795	2,125
6700	600 amp	"	1	16		2,825	630		3,455	4,050
6900	600 volt, 6 pole NEMA 3R nonfused, 30 amp	1 Elec	2.70	2.963		1,200	117		1,317	1,500
7100	60 amp		2	4		1,400	158		1,558	1,775
7300	100 amp		1.50	5.333		1,725	210		1,935	2,225
7500	200 amp		1.20	6.667		3,875	263		4,138	4,675
7600	600 volt, 3 pole NEMA 7 explosion proof nonfused									
7610	30 amp	1 Elec	2.20	3.636	Ea.	950	143		1,093	1,275
7620	60 amp		1.80	4.444		1,150	175		1,325	1,500
7630	100 amp		1.20	6.667		1,375	263		1,638	1,925
7640	200 amp		.80	10		2,875	395		3,270	3,750
7710	600 volt 6 pole, NEMA 3R fusible, 30 amp		2.70	2.963		1,375	117		1,492	1,675
7900	60 amp		2	4		1,625	158		1,783	2,025
8100	100 amp		1.50	5.333		2,000	210		2,210	2,500
8110	240 volt 3 pole, NEMA 12 fusible, 30 amp		3.10	2.581		263	102		365	440
8120	60 amp		2.20	3.636		355	143		498	605
8130	100 amp		1.80	4.444		520	175		695	835
8140	200 amp		1.20	6.667		745	263		1,008	1,200
8150	400 amp	2 Elec	1.60	10		1,725	395		2,120	2,450
8160	600 amp	"	1	16		2,775	630		3,405	4,000
8180	600 volt 3 pole, NEMA 12 fused, 30 amp	1 Elec	3.10	2.581		345	102		447	530

ELECTRICAL 16

16 ELECTRICAL

		Encl Switches & Circuit Breakers	CREW	DAILY OUTPUT	LABOR-HOURS	UNIT	MAT.	LABOR	EQUIP.	TOTAL	TOTAL INCL O&P	
16410								**2004 BARE COSTS**				
800	8190	60 amp R16410-800	1 Elec	2.20	3.636	Ea.	355	143		498	605	800
	8200	100 amp		1.80	4.444		570	175		745	890	
	8210	200 amp		1.20	6.667		890	263		1,153	1,375	
	8220	400 amp	2 Elec	1.60	10		1,950	395		2,345	2,725	
	8230	600 amp	"	1	16		3,300	630		3,930	4,575	
	8240	600 volt 3 pole, NEMA 12 nonfused, 30 amp	1 Elec	3.10	2.581		244	102		346	420	
	8250	60 amp		2.20	3.636		315	143		458	560	
	8260	100 amp		1.80	4.444		450	175		625	755	
	8270	200 amp		1.20	6.667		595	263		858	1,050	
	8280	400 amp	2 Elec	1.60	10		1,475	395		1,870	2,200	
	8290	600 amp	"	1	16		2,400	630		3,030	3,600	
	8310	600 volt, 3 pole NEMA 4 fusible, 30 amp	1 Elec	3	2.667		945	105		1,050	1,200	
	8320	60 amp		2.20	3.636		1,050	143		1,193	1,375	
	8330	100 amp		1.80	4.444		2,325	175		2,500	2,825	
	8340	200 amp		1.20	6.667		2,975	263		3,238	3,675	
	8350	400 amp	2 Elec	1.60	10		5,850	395		6,245	7,000	
	8360	600 volt 3 pole NEMA 4 nonfused, 30 amp	1 Elec	3	2.667		785	105		890	1,025	
	8370	60 amp		2.20	3.636		935	143		1,078	1,250	
	8380	100 amp		1.80	4.444		1,900	175		2,075	2,350	
	8390	200 amp		1.20	6.667		2,600	263		2,863	3,250	
	8400	400 amp	2 Elec	1.60	10		5,275	395		5,670	6,375	
	8490	Motor starters, manual, single phase, NEMA 1	1 Elec	6.40	1.250		47	49.50		96.50	126	
	8500	NEMA 4		4	2		139	79		218	270	
	8700	NEMA 7		4	2		152	79		231	285	
	8900	NEMA 1 with pilot		6.40	1.250		64	49.50		113.50	144	
	8920	3 pole, NEMA 1, 230/460 volt, 5 HP, size 0		3.50	2.286		147	90		237	296	
	8940	10 HP, size 1		2	4		175	158		333	425	
	9010	Disc. switch, 600V 3 pole fusible, 30 amp, to 10 HP motor		3.20	2.500		217	98.50		315.50	385	
	9050	60 amp, to 30 HP motor		2.30	3.478		500	137		637	755	
	9070	100 amp, to 60 HP motor		1.90	4.211		500	166		666	795	
	9100	200 amp, to 125 HP motor		1.30	6.154		750	242		992	1,175	
	9110	400 amp, to 200 HP motor	2 Elec	1.80	8.889		1,900	350		2,250	2,600	
840	0010	**TIME SWITCHES**										840
	0100	Single pole, single throw, 24 hour dial	1 Elec	4	2	Ea.	82	79		161	207	
	0200	24 hour dial with reserve power		3.60	2.222		350	87.50		437.50	515	
	0300	Astronomic dial		3.60	2.222		141	87.50		228.50	285	
	0400	Astronomic dial with reserve power		3.30	2.424		455	95.50		550.50	640	
	0500	7 day calendar dial		3.30	2.424		126	95.50		221.50	281	
	0600	7 day calendar dial with reserve power		3.20	2.500		385	98.50		483.50	565	
	0700	Photo cell 2000 watt		8	1		15.15	39.50		54.65	75	
	1080	Load management device, 4 loads		2	4		800	158		958	1,125	
	1100	Load management device, 8 loads		1	8		1,300	315		1,615	1,900	
16415		**Transfer Switches**										
100	0010	**NON-AUTOMATIC TRANSFER SWITCHES** enclosed										100
	0100	Manual operated, 480 volt 3 pole, 30 amp	1 Elec	2.30	3.478	Ea.	1,275	137		1,412	1,600	
	0150	60 amp		1.90	4.211		1,275	166		1,441	1,650	
	0200	100 amp		1.30	6.154		1,275	242		1,517	1,750	
	0250	200 amp	2 Elec	2	8		2,550	315		2,865	3,300	
	0300	400 amp		1.60	10		3,725	395		4,120	4,675	
	0350	600 amp		1	16		5,650	630		6,280	7,150	
	1000	250 volt 3 pole, 30 amp	1 Elec	2.30	3.478		1,275	137		1,412	1,600	
	1100	60 amp		1.90	4.211		1,275	166		1,441	1,650	
	1150	100 amp		1.30	6.154		1,275	242		1,517	1,750	
	1200	200 amp	2 Elec	2	8		2,550	315		2,865	3,300	
	1300	600 amp	"	1	16		5,650	630		6,280	7,150	

Important: See the Reference Section for critical supporting data - Reference Nos., Crews, & City Cost Indexes

16415	Transfer Switches	CREW	DAILY OUTPUT	LABOR-HOURS	UNIT	2004 BARE COSTS				TOTAL INCL O&P	
						MAT.	LABOR	EQUIP.	TOTAL		
100	1500	Electrically operated, 480 volt 3 pole, 60 amp	1 Elec	1.90	4.211	Ea.	2,450	166		2,616	2,950	100
	1600	100 amp	"	1.30	6.154		2,450	242		2,692	3,050	
	1650	200 amp	2 Elec	2	8		4,050	315		4,365	4,925	
	1700	400 amp		1.60	10		5,675	395		6,070	6,800	
	1750	600 amp	↓	1	16		8,150	630		8,780	9,925	
	2000	250 volt 3 pole, 30 amp	1 Elec	2.30	3.478		2,450	137		2,587	2,900	
	2050	60 amp	"	1.90	4.211		2,450	166		2,616	2,950	
	2150	200 amp	2 Elec	2	8		4,050	315		4,365	4,925	
	2200	400 amp		1.60	10		5,675	395		6,070	6,800	
	2250	600 amp	↓	1	16		8,150	630		8,780	9,925	
	2500	NEMA 3R, 480 volt 3 pole, 60 amp	1 Elec	1.80	4.444		2,900	175		3,075	3,425	
	2550	100 amp	"	1.20	6.667		2,900	263		3,163	3,575	
	2600	200 amp	2 Elec	1.80	8.889		4,575	350		4,925	5,575	
	2650	400 amp	"	1.40	11.429		6,200	450		6,650	7,500	
	2800	NEMA 3R, 250 volt 3 pole solid state, 100 amp	1 Elec	1.20	6.667		2,900	263		3,163	3,575	
	2850	150 amp	2 Elec	1.80	8.889		3,850	350		4,200	4,750	
	2900	250 volt 2 pole solid state, 100 amp	1 Elec	1.30	6.154		2,825	242		3,067	3,450	
	2950	150 amp	2 Elec	2	8	↓	3,750	315		4,065	4,600	
600	0010	**AUTOMATIC TRANSFER SWITCHES**										600
	0015	Switches, enclosed 120/240 volt, 2 pole, 30 amp [R16415-600]	1 Elec	2.40	3.333	Ea.	2,600	131		2,731	3,075	
	0020	70 amp		2	4		2,600	158		2,758	3,100	
	0030	100 amp	↓	1.35	5.926		2,600	233		2,833	3,225	
	0040	225 amp	2 Elec	2.10	7.619		3,850	300		4,150	4,675	
	0050	400 amp		1.70	9.412		5,475	370		5,845	6,575	
	0060	600 amp		1.06	15.094		8,050	595		8,645	9,725	
	0070	800 amp	↓	.84	19.048		9,450	750		10,200	11,500	
	0100	Switches, enclosed 480 volt, 3 pole, 30 amp	1 Elec	2.30	3.478		2,900	137		3,037	3,400	
	0200	60 amp		1.90	4.211		2,900	166		3,066	3,450	
	0300	100 amp	↓	1.30	6.154		2,900	242		3,142	3,550	
	0400	150 amp	2 Elec	2.40	6.667		3,575	263		3,838	4,325	
	0500	225 amp		2	8		4,575	315		4,890	5,500	
	0600	260 amp		2	8		5,300	315		5,615	6,300	
	0700	400 amp		1.60	10		6,175	395		6,570	7,375	
	0800	600 amp		1	16		8,950	630		9,580	10,800	
	0900	800 amp		.80	20		10,500	790		11,290	12,800	
	1000	1000 amp		.76	21.053		13,900	830		14,730	16,500	
	1100	1200 amp		.70	22.857		19,000	900		19,900	22,300	
	1200	1600 amp		.60	26.667		21,600	1,050		22,650	25,400	
	1300	2000 amp	↓	.50	32		24,200	1,250		25,450	28,500	
	1600	Accessories, time delay on engine starting					226			226	249	
	1700	Adjustable time delay on retransfer					226			226	249	
	1800	Shunt trips for customer connections					405			405	445	
	1900	Maintenance select switch					92			92	101	
	2000	Auxiliary contact when normal fails					106			106	117	
	2100	Pilot light-emergency					92			92	101	
	2200	Pilot light-normal					92			92	101	
	2300	Auxiliary contact-closed on normal					106			106	117	
	2400	Auxiliary contact-closed on emergency					106			106	117	
	2500	Emergency source sensing, frequency relay	↓			↓	470			470	515	

16420	Enclosed Controllers										
200	0010	**CONTACTORS, AC** Enclosed (NEMA 1)										200
	0050	Lighting, 600 volt 3 pole, electrically held										
	0100	20 amp	1 Elec	4	2	Ea.	219	79		298	360	
	0200	30 amp	↓	3.60	2.222	↓	232	87.50		319.50	385	

16420	Enclosed Controllers	CREW	DAILY OUTPUT	LABOR-HOURS	UNIT	2004 BARE COSTS				TOTAL INCL O&P
						MAT.	LABOR	EQUIP.	TOTAL	
200 0300	60 amp	1 Elec	3	2.667	Ea.	465	105		570	665	**200**
0400	100 amp		2.50	3.200		770	126		896	1,025
0500	200 amp	▼	1.40	5.714		1,800	225		2,025	2,300
0600	300 amp	2 Elec	1.60	10		3,850	395		4,245	4,800
0800	600 volt 3 pole, mechanically held, 30 amp	1 Elec	3.60	2.222		350	87.50		437.50	515
0900	60 amp		3	2.667		695	105		800	915
1000	75 amp		2.80	2.857		770	113		883	1,000
1100	100 amp		2.50	3.200		985	126		1,111	1,275
1200	150 amp		2	4		2,700	158		2,858	3,200
1300	200 amp		1.40	5.714		2,700	225		2,925	3,300
1500	Magnetic with auxiliary contact, size 00, 9 amp		4	2		160	79		239	293
1600	Size 0, 18 amp		4	2		191	79		270	325
1700	Size 1, 27 amp		3.60	2.222		217	87.50		304.50	370
1800	Size 2, 45 amp		3	2.667		400	105		505	600
1900	Size 3, 90 amp		2.50	3.200		650	126		776	905
2000	Size 4, 135 amp	▼	2.30	3.478		1,475	137		1,612	1,825
2100	Size 5, 270 amp	2 Elec	1.80	8.889		3,125	350		3,475	3,975
2200	Size 6, 540 amp		1.20	13.333		9,150	525		9,675	10,900
2300	Size 7, 810 amp		1	16		12,300	630		12,930	14,400
2310	Size 8, 1215 amp	▼	.80	20		19,100	790		19,890	22,200
2500	Magnetic, 240 volt, 1-2 pole, .75 HP motor	1 Elec	4	2		129	79		208	259
2520	2 HP motor		3.60	2.222		144	87.50		231.50	289
2540	5 HP motor		2.50	3.200		350	126		476	575
2560	10 HP motor		1.40	5.714		580	225		805	970
2600	240 volt or less, 3 pole, .75 HP motor		4	2		129	79		208	259
2620	5 HP motor		3.60	2.222		160	87.50		247.50	305
2640	10 HP motor		3.60	2.222		186	87.50		273.50	335
2660	15 HP motor		2.50	3.200		370	126		496	600
2700	25 HP motor	▼	2.50	3.200		370	126		496	600
2720	30 HP motor	2 Elec	2.80	5.714		620	225		845	1,025
2740	40 HP motor		2.80	5.714		620	225		845	1,025
2760	50 HP motor		1.60	10		620	395		1,015	1,275
2800	75 HP motor		1.60	10		1,450	395		1,845	2,175
2820	100 HP motor		1	16		1,450	630		2,080	2,550
2860	150 HP motor		1	16		3,100	630		3,730	4,350
2880	200 HP motor	▼	1	16		3,100	630		3,730	4,350
3000	600 volt, 3 pole, 5 HP motor	1 Elec	4	2		160	79		239	293
3020	10 HP motor		3.60	2.222		186	87.50		273.50	335
3040	25 HP motor		3	2.667		370	105		475	565
3100	50 HP motor	▼	2.50	3.200		620	126		746	870
3160	100 HP motor	2 Elec	2.80	5.714		1,450	225		1,675	1,925
3220	200 HP motor	"	1.60	10	▼	3,100	395		3,495	3,975
220 0010	**CONTROL STATIONS**										**220**
0050	NEMA 1, heavy duty, stop/start	1 Elec	8	1	Ea.	119	39.50		158.50	190
0100	Stop/start, pilot light		6.20	1.290		162	51		213	254
0200	Hand/off/automatic		6.20	1.290		87.50	51		138.50	172
0400	Stop/start/reverse		5.30	1.509		160	59.50		219.50	265
0500	NEMA 7, heavy duty, stop/start		6	1.333		188	52.50		240.50	285
0600	Stop/start, pilot light		4	2		230	79		309	370
0700	NEMA 7 or 9, 1 element		6	1.333		167	52.50		219.50	262
0800	2 element		6	1.333		199	52.50		251.50	297
0900	3 element		4	2		445	79		524	605
0910	Selector switch, 2 position		6	1.333		167	52.50		219.50	262
0920	3 position		4	2		172	79		251	305
0930	Oiltight, 1 element		8	1		82.50	39.50		122	150
0940	2 element	▼	6.20	1.290	▼	119	51		170	207

Important: See the Reference Section for critical supporting data - Reference Nos., Crews, & City Cost Indexes

		**16420	Enclosed Controllers**	CREW	DAILY OUTPUT	LABOR-HOURS	UNIT	MAT.	LABOR	EQUIP.	TOTAL	TOTAL INCL O&P	
							2004 BARE COSTS					
220	0950	3 element	1 Elec	5.30	1.509	Ea.	160	59.50		219.50	265	**220**
	0960	Selector switch, 2 position		6.20	1.290		87.50	51		138.50	172	
	0970	3 position	↓	5.30	1.509	↓	87.50	59.50		147	185	
240	0010	**CONTROL SWITCHES** Field installed										**240**
	6000	Push button 600V 10A, momentary contact										
	6150	Standard operator with colored button	1 Elec	34	.235	Ea.	14.60	9.25		23.85	30	
	6160	With single block 1NO 1NC		18	.444		31	17.50		48.50	60	
	6170	With double block 2NO 2NC		15	.533		47	21		68	83	
	6180	Stnd operator w/mushroom button 1-9/16" diam.	↓	34	.235	↓	31	9.25		40.25	48	
	6190	Stnd operator w/mushroom button 2-1/4" diam.										
	6200	With single block 1NO 1NC	1 Elec	18	.444	Ea.	47	17.50		64.50	77.50	
	6210	With double block 2NO 2NC		15	.533		63	21		84	101	
	6500	Maintained contact, selector operator		34	.235		38.50	9.25		47.75	56.50	
	6510	With single block 1NO 1NC		18	.444		52	17.50		69.50	83	
	6520	With double block 2NO 2NC		15	.533		65.50	21		86.50	104	
	6560	Spring-return selector operator		34	.235		38.50	9.25		47.75	56.50	
	6570	With single block 1NO 1NC		18	.444		52	17.50		69.50	83	
	6580	With double block 2NO 2NC	↓	15	.533	↓	65.50	21		86.50	104	
	6620	Transformer operator w/illuminated										
	6630	button 6V #12 lamp	1 Elec	32	.250	Ea.	65.50	9.85		75.35	86.50	
	6640	With single block 1NO 1NC w/guard		16	.500		95.50	19.70		115.20	135	
	6650	With double block 2NO 2NC w/guard		13	.615		112	24.50		136.50	159	
	6690	Combination operator		34	.235		38.50	9.25		47.75	56.50	
	6700	With single block 1NO 1NC		18	.444		52	17.50		69.50	83	
	6710	With double block 2NO 2NC	↓	15	.533	↓	65.50	21		86.50	104	
	9000	Indicating light unit, full voltage										
	9010	110-125V front mount	1 Elec	32	.250	Ea.	52	9.85		61.85	71.50	
	9020	130V resistor type		32	.250		47	9.85		56.85	66	
	9030	6V transformer type	↓	32	.250	↓	58	9.85		67.85	78	
800	0010	**RELAYS** Enclosed (NEMA 1)										**800**
	0050	600 volt AC, 1 pole, 12 amp	1 Elec	5.30	1.509	Ea.	81	59.50		140.50	178	
	0100	2 pole, 12 amp		5	1.600		81	63		144	183	
	0200	4 pole, 10 amp		4.50	1.778		108	70		178	223	
	0500	250 volt DC, 1 pole, 15 amp		5.30	1.509		106	59.50		165.50	206	
	0600	2 pole, 10 amp		5	1.600		102	63		165	206	
	0700	4 pole, 4 amp	↓	4.50	1.778	↓	134	70		204	252	

		**16440	Swbds, Panels & Control Centers**										
500	0010	**LOAD CENTERS** (residential type)										**500**
	0100	3 wire, 120/240V, 1 phase, including 1 pole plug-in breakers R16440 -500										
	0200	100 amp main lugs, indoor, 8 circuits	1 Elec	1.40	5.714	Ea.	118	225		343	465	
	0300	12 circuits		1.20	6.667		165	263		428	570	
	0400	Rainproof, 8 circuits		1.40	5.714		142	225		367	490	
	0500	12 circuits	↓	1.20	6.667		200	263		463	610	
	0600	200 amp main lugs, indoor, 16 circuits	R-1A	1.80	8.889		272	284		556	730	
	0700	20 circuits		1.50	10.667		340	340		680	890	
	0800	24 circuits		1.30	12.308		450	395		845	1,075	
	0900	30 circuits		1.20	13.333		515	425		940	1,200	
	1000	42 circuits		.80	20		715	640		1,355	1,750	
	1200	Rainproof, 16 circuits		1.80	8.889		325	284		609	785	
	1300	20 circuits		1.50	10.667		390	340		730	940	
	1400	24 circuits		1.30	12.308		580	395		975	1,225	
	1500	30 circuits		1.20	13.333		650	425		1,075	1,350	
	1600	42 circuits	↓	.80	20	↓	850	640		1,490	1,900	

ELECTRICAL 16

		16440 | Swbds, Panels & Control Centers		CREW	DAILY OUTPUT	LABOR-HOURS	UNIT	2004 BARE COSTS				TOTAL INCL O&P	
								MAT.	LABOR	EQUIP.	TOTAL		
500	1800	400 amp main lugs, indoor, 42 circuits	R16440 -500	R-1A	.72	22.222	Ea.	1,000	710		1,710	2,175	500
	1900	Rainproof, 42 circuit		↓	.72	22.222	↓	1,175	710		1,885	2,375	
	2200	Plug in breakers, 20 amp, 1 pole, 4 wire, 120/208 volts											
	2210	125 amp main lugs, indoor, 12 circuits		1 Elec	1.20	6.667	Ea.	256	263		519	670	
	2300	18 circuits			.80	10		365	395		760	985	
	2400	Rainproof, 12 circuits			1.20	6.667		294	263		557	715	
	2500	18 circuits		↓	.80	10		395	395		790	1,025	
	2600	200 amp main lugs, indoor, 24 circuits		R-1A	1.30	12.308		490	395		885	1,125	
	2700	30 circuits			1.20	13.333		555	425		980	1,250	
	2800	36 circuits			1	16		675	510		1,185	1,525	
	2900	42 circuits			.80	20		740	640		1,380	1,775	
	3000	Rainproof, 24 circuits			1.30	12.308		540	395		935	1,200	
	3100	30 circuits			1.20	13.333		610	425		1,035	1,325	
	3200	36 circuits			1	16		845	510		1,355	1,700	
	3300	42 circuits			.80	20		915	640		1,555	1,975	
	3500	400 amp main lugs, indoor, 42 circuits			.72	22.222		1,125	710		1,835	2,300	
	3600	Rainproof, 42 circuits		↓	.72	22.222	↓	1,325	710		2,035	2,550	
	3700	Plug-in breakers, 20 amp, 1 pole, 3 wire, 120/240 volts											
	3800	100 amp main breaker, indoor, 12 circuits		1 Elec	1.20	6.667	Ea.	274	263		537	690	
	3900	18 circuits		"	.80	10		370	395		765	990	
	4000	200 amp main breaker, indoor, 20 circuits		R-1A	1.50	10.667		495	340		835	1,050	
	4200	24 circuits			1.30	12.308		600	395		995	1,250	
	4300	30 circuits			1.20	13.333		670	425		1,095	1,375	
	4400	40 circuits			.90	17.778		835	570		1,405	1,775	
	4500	Rainproof, 20 circuits			1.50	10.667		540	340		880	1,100	
	4600	24 circuits			1.30	12.308		675	395		1,070	1,325	
	4700	30 circuits			1.20	13.333		740	425		1,165	1,450	
	4800	40 circuits			.90	17.778		910	570		1,480	1,850	
	5000	400 amp main breaker, indoor, 42 circuits			.72	22.222		2,400	710		3,110	3,700	
	5100	Rainproof, 42 circuits		↓	.72	22.222	↓	2,675	710		3,385	4,025	
	5300	Plug in breakers, 20 amp, 1 pole, 4 wire, 120/208 volts											
	5400	200 amp main breaker, indoor, 30 circuits		R-1A	1.20	13.333	Ea.	1,025	425		1,450	1,775	
	5500	42 circuits			.80	20		1,250	640		1,890	2,350	
	5600	Rainproof, 30 circuits			1.20	13.333		1,075	425		1,500	1,850	
	5700	42 circuits		↓	.80	20	↓	1,325	640		1,965	2,425	
620	0010	**MOTOR CONTROL CENTER COMPONENTS**	R16440 -640										620
	0100	Starter, size 1, FVNR, NEMA 1, type A, fusible		1 Elec	2.70	2.963	Ea.	945	117		1,062	1,200	
	0120	Circuit breaker			2.70	2.963		1,025	117		1,142	1,300	
	0140	Type B, fusible			2.70	2.963		1,025	117		1,142	1,325	
	0160	Circuit breaker			2.70	2.963		1,125	117		1,242	1,425	
	0180	NEMA 12, type A, fusible			2.60	3.077		960	121		1,081	1,225	
	0200	Circuit breaker			2.60	3.077		1,050	121		1,171	1,325	
	0220	Type B, fusible			2.60	3.077		1,050	121		1,171	1,325	
	0240	Circuit breaker			2.60	3.077		1,150	121		1,271	1,425	
	0300	Starter, size 1, FVR, NEMA 1, type A, fusible			2	4		1,350	158		1,508	1,725	
	0320	Circuit breaker			2	4		1,350	158		1,508	1,725	
	0340	Type B, fusible			2	4		1,500	158		1,658	1,850	
	0360	Circuit breaker			2	4		1,500	158		1,658	1,850	
	0380	NEMA 12, type A, fusible			1.90	4.211		1,375	166		1,541	1,750	
	0400	Circuit breaker			1.90	4.211		1,375	166		1,541	1,750	
	0420	Type B, fusible			1.90	4.211		1,500	166		1,666	1,900	
	0440	Circuit breaker		↓	1.90	4.211	↓	1,500	166		1,666	1,900	
	0490	Starter size 1, 2 speed, separate winding											
	0500	NEMA 1, type A, fusible		1 Elec	2.60	3.077	Ea.	1,775	121		1,896	2,125	
	0520	Circuit breaker		↓	2.60	3.077	↓	1,775	121		1,896	2,125	

Important: See the Reference Section for critical supporting data - Reference Nos., Crews, & City Cost Indexes

16 ELECTRICAL

16440	Swbds, Panels & Control Centers	CREW	DAILY OUTPUT	LABOR-HOURS	UNIT	2004 BARE COSTS				TOTAL INCL O&P
						MAT.	LABOR	EQUIP.	TOTAL	
620	0540	Type B, fusible	R16440 -640	1 Elec	2.60	3.077	Ea.	1,950	121		2,071	2,325	**620**
0560	Circuit breaker		2.60	3.077		1,950	121		2,071	2,325
0580	NEMA 12, type A, fusible		2.50	3.200		1,825	126		1,951	2,200
0600	Circuit breaker		2.50	3.200		1,825	126		1,951	2,200
0620	Type B, fusible		2.50	3.200		2,000	126		2,126	2,400
0640	Circuit breaker		2.50	3.200		2,000	126		2,126	2,400
0650	Starter size 1, 2 speed, consequent pole									
0660	NEMA 1, type A, fusible	1 Elec	2.60	3.077	Ea.	1,775	121		1,896	2,125
0680	Circuit breaker		2.60	3.077		1,775	121		1,896	2,125
0700	Type B, fusible		2.60	3.077		1,950	121		2,071	2,325
0720	Circuit breaker		2.60	3.077		1,950	121		2,071	2,325
0740	NEMA 12, type A, fusible		2.50	3.200		1,825	126		1,951	2,200
0760	Circuit breaker		2.50	3.200		1,825	126		1,951	2,200
0780	Type B, fusible		2.50	3.200		1,975	126		2,101	2,375
0800	Circuit breaker		2.50	3.200		2,000	126		2,126	2,400
0810	Starter size 1, 2 speed, space only									
0820	NEMA 1, type A, fusible	1 Elec	16	.500	Ea.	400	19.70		419.70	470
0840	Circuit breaker		16	.500		400	19.70		419.70	470
0860	Type B, fusible		16	.500		400	19.70		419.70	470
0880	Circuit breaker		16	.500		400	19.70		419.70	470
0900	NEMA 12, type A, fusible		15	.533		415	21		436	490
0920	Circuit breaker		15	.533		415	21		436	490
0940	Type B, fusible		15	.533		415	21		436	490
0960	Circuit breaker		15	.533		415	21		436	490
1100	Starter size 2, FVNR, NEMA 1, type A, fusible	2 Elec	4	4		1,075	158		1,233	1,400
1120	Circuit breaker		4	4		1,175	158		1,333	1,500
1140	Type B, fusible		4	4		1,175	158		1,333	1,525
1160	Circuit breaker		4	4		1,275	158		1,433	1,625
1180	NEMA 12, type A, fusible		3.80	4.211		1,075	166		1,241	1,450
1200	Circuit breaker		3.80	4.211		1,175	166		1,341	1,550
1220	Type B, fusible		3.80	4.211		1,175	166		1,341	1,550
1240	Circuit breaker		3.80	4.211		1,300	166		1,466	1,675
1300	FVR, NEMA 1, type A, fusible		3.20	5		1,800	197		1,997	2,275
1320	Circuit breaker		3.20	5		1,800	197		1,997	2,275
1340	Type B, fusible		3.20	5		1,975	197		2,172	2,475
1360	Circuit breaker		3.20	5		1,975	197		2,172	2,475
1380	NEMA type 12, type A, fusible		3	5.333		1,825	210		2,035	2,350
1400	Circuit breaker		3	5.333		1,825	210		2,035	2,350
1420	Type B, fusible		3	5.333		2,025	210		2,235	2,550
1440	Circuit breaker		3	5.333		2,025	210		2,235	2,550
1490	Starter size 2, 2 speed, separate winding									
1500	NEMA 1, type A, fusible	2 Elec	3.80	4.211	Ea.	2,025	166		2,191	2,475
1520	Circuit breaker		3.80	4.211		2,025	166		2,191	2,475
1540	Type B, fusible		3.80	4.211		2,225	166		2,391	2,675
1560	Circuit breaker		3.80	4.211		2,225	166		2,391	2,675
1570	NEMA 12, type A, fusible		3.60	4.444		2,050	175		2,225	2,500
1580	Circuit breaker		3.60	4.444		2,050	175		2,225	2,500
1600	Type B, fusible		3.60	4.444		2,250	175		2,425	2,725
1620	Circuit breaker		3.60	4.444		2,250	175		2,425	2,725
1630	Starter size 2, 2 speed, consequent pole									
1640	NEMA 1, type A, fusible	2 Elec	3.80	4.211	Ea.	2,275	166		2,441	2,775
1660	Circuit breaker		3.80	4.211		2,275	166		2,441	2,775
1680	Type B, fusible		3.80	4.211		2,425	166		2,591	2,900
1700	Circuit breaker		3.80	4.211		2,425	166		2,591	2,900
1720	NEMA 12, type A, fusible		3.80	4.211		2,300	166		2,466	2,775
1740	Circuit breaker		3.60	4.444		2,325	175		2,500	2,800

ELECTRICAL 16

			CREW	DAILY OUTPUT	LABOR-HOURS	UNIT	2004 BARE COSTS				TOTAL INCL O&P	
		**16440	Swbds, Panels & Control Centers**					MAT.	LABOR	EQUIP.	TOTAL		
620	1760	Type B, fusible	2 Elec	3.60	4.444	Ea.	2,450	175		2,625	2,950	**620**
	1780	Circuit breaker	R16440 -640 ↓	3.60	4.444	↓	2,450	175		2,625	2,950	
	1830	Starter size 2, autotransformer										
	1840	NEMA 1, type A, fusible	2 Elec	3.40	4.706	Ea.	3,575	185		3,760	4,200	
	1860	Circuit breaker		3.40	4.706		3,650	185		3,835	4,300	
	1880	Type B, fusible		3.40	4.706		3,900	185		4,085	4,550	
	1900	Circuit breaker		3.40	4.706		3,900	185		4,085	4,550	
	1920	NEMA 12, type A, fusible		3.20	5		3,625	197		3,822	4,300	
	1940	Circuit breaker		3.20	5		3,625	197		3,822	4,300	
	1960	Type B, fusible		3.20	5		3,950	197		4,147	4,650	
	1980	Circuit breaker	▼	3.20	5	▼	3,950	197		4,147	4,650	
	2030	Starter size 2, space only										
	2040	NEMA 1, type A, fusible	1 Elec	16	.500	Ea.	400	19.70		419.70	470	
	2060	Circuit breaker		16	.500		400	19.70		419.70	470	
	2080	Type B, fusible		16	.500		400	19.70		419.70	470	
	2100	Circuit breaker		16	.500		400	19.70		419.70	470	
	2120	NEMA 12, type A, fusible		15	.533		415	21		436	490	
	2140	Circuit breaker		15	.533		415	21		436	490	
	2160	Type B, fusible		15	.533		415	21		436	490	
	2180	Circuit breaker	▼	15	.533		415	21		436	490	
	2300	Starter size 3, FVNR, NEMA 1, type A, fusible	2 Elec	2	8		2,025	315		2,340	2,700	
	2320	Circuit breaker		2	8		1,800	315		2,115	2,450	
	2340	Type B, fusible		2	8		2,225	315		2,540	2,925	
	2360	Circuit breaker		2	8		2,000	315		2,315	2,650	
	2380	NEMA 12, type A, fusible		1.90	8.421		2,075	330		2,405	2,775	
	2400	Circuit breaker		1.90	8.421		1,825	330		2,155	2,525	
	2420	Type B, fusible		1.90	8.421		2,300	330		2,630	3,050	
	2440	Circuit breaker		1.90	8.421		2,025	330		2,355	2,725	
	2500	Starter size 3, FVR, NEMA 1, type A, fusible		1.60	10		2,775	395		3,170	3,625	
	2520	Circuit breaker		1.60	10		2,650	395		3,045	3,475	
	2540	Type B, fusible		1.60	10		3,025	395		3,420	3,900	
	2560	Circuit breaker		1.60	10		2,900	395		3,295	3,750	
	2580	NEMA 12, type A, fusible		1.50	10.667		2,825	420		3,245	3,725	
	2600	Circuit breaker		1.50	10.667		2,700	420		3,120	3,600	
	2620	Type B, fusible		1.50	10.667		3,075	420		3,495	4,000	
	2640	Circuit breaker	▼	1.50	10.667	▼	3,175	420		3,595	4,125	
	2690	Starter size 3, 2 speed, separate winding										
	2700	NEMA 1, type A, fusible	2 Elec	2	8	Ea.	3,650	315		3,965	4,500	
	2720	Circuit breaker		2	8		3,225	315		3,540	4,025	
	2740	Type B, fusible		2	8		4,000	315		4,315	4,875	
	2760	Circuit breaker		2	8		3,525	315		3,840	4,350	
	2780	NEMA 12, type A, fusible		1.90	8.421		3,725	330		4,055	4,600	
	2800	Circuit breaker		1.90	8.421		3,300	330		3,630	4,125	
	2820	Type B, fusible		1.90	8.421		4,075	330		4,405	4,975	
	2840	Circuit breaker	▼	1.90	8.421	▼	3,600	330		3,930	4,450	
	2850	Starter size 3, 2 speed, consequent pole										
	2860	NEMA 1, type A, fusible	2 Elec	2	8	Ea.	4,075	315		4,390	4,950	
	2880	Circuit breaker		2	8		3,650	315		3,965	4,475	
	2900	Type B, fusible		2	8		4,425	315		4,740	5,325	
	2920	Circuit breaker		2	8		3,950	315		4,265	4,825	
	2940	NEMA 12, type A, fusible		1.90	8.421		4,150	330		4,480	5,075	
	2960	Circuit breaker		1.90	8.421		3,700	330		4,030	4,575	
	2980	Type B, fusible		1.90	8.421		4,500	330		4,830	5,450	
	3000	Circuit breaker		1.90	8.421		4,025	330		4,355	4,925	
	3100	Starter size 3, autotransformer, NEMA 1, type A, fusible		1.60	10		5,075	395		5,470	6,150	
	3120	Circuit breaker	▼	1.60	10	▼	5,075	395		5,470	6,150	

Important: See the Reference Section for critical supporting data - Reference Nos., Crews, & City Cost Indexes

16440	Swbds, Panels & Control Centers		CREW	DAILY OUTPUT	LABOR-HOURS	UNIT	2004 BARE COSTS				TOTAL INCL O&P
							MAT.	LABOR	EQUIP.	TOTAL	
620	3140	Type B, fusible	R16440 -640	2 Elec	1.60	10	Ea.	5,525	395		5,920	6,675	620
3160	Circuit breaker			1.60	10		5,525	395		5,920	6,675
3180	NEMA 12, type A, fusible			1.50	10.667		5,150	420		5,570	6,300
3200	Circuit breaker			1.50	10.667		5,150	420		5,570	6,300
3220	Type B, fusible			1.50	10.667		5,625	420		6,045	6,825
3240	Circuit breaker			1.50	10.667		5,625	420		6,045	6,825
3260	Starter size 3, space only, NEMA 1, type A, fusible		1 Elec	15	.533		785	21		806	895
3280	Circuit breaker			15	.533		610	21		631	700
3300	Type B, fusible			15	.533		785	21		806	895
3320	Circuit breaker			15	.533		610	21		631	700
3340	NEMA 12, type A, fusible			14	.571		835	22.50		857.50	955
3360	Circuit breaker			14	.571		650	22.50		672.50	750
3380	Type B, fusible			14	.571		835	22.50		857.50	955
3400	Circuit breaker			14	.571		650	22.50		672.50	750
3500	Starter size 4, FVNR, NEMA 1, type A, fusible		2 Elec	1.60	10		3,025	395		3,420	3,900
3520	Circuit breaker			1.60	10		2,750	395		3,145	3,600
3540	Type B, fusible			1.60	10		3,325	395		3,720	4,225
3560	Circuit breaker			1.60	10		3,025	395		3,420	3,900
3580	NEMA 12, type A, fusible			1.50	10.667		3,100	420		3,520	4,050
3600	Circuit breaker			1.50	10.667		2,800	420		3,220	3,700
3620	Type B, fusible			1.50	10.667		3,400	420		3,820	4,350
3640	Circuit breaker			1.50	10.667		3,075	420		3,495	4,000
3700	Starter size 4, FVR, NEMA 1, type A, fusible			1.20	13.333		4,150	525		4,675	5,325
3720	Circuit breaker			1.20	13.333		3,725	525		4,250	4,875
3740	Type B, fusible			1.20	13.333		4,525	525		5,050	5,750
3760	Circuit breaker			1.20	13.333		4,125	525		4,650	5,300
3780	NEMA 12, type A, fusible			1.16	13.793		4,250	545		4,795	5,475
3800	Circuit breaker			1.16	13.793		3,800	545		4,345	4,975
3820	Type B, fusible			1.16	13.793		3,825	545		4,370	5,025
3840	Circuit breaker			1.16	13.793		4,175	545		4,720	5,375
3890	Starter size 4, 2 speed, separate windings										
3900	NEMA 1, type A, fusible		2 Elec	1.60	10	Ea.	5,225	395		5,620	6,325
3920	Circuit breaker			1.60	10		3,925	395		4,320	4,900
3940	Type B, fusible			1.60	10		5,750	395		6,145	6,900
3960	Circuit breaker			1.60	10		4,300	395		4,695	5,300
3980	NEMA 12, type A, fusible			1.50	10.667		5,325	420		5,745	6,475
4000	Circuit breaker			1.50	10.667		4,000	420		4,420	5,025
4020	Type B, fusible			1.50	10.667		5,850	420		6,270	7,050
4040	Circuit breaker			1.50	10.667		4,375	420		4,795	5,425
4050	Starter size 4, 2 speed, consequent pole										
4060	NEMA 1, type A, fusible		2 Elec	1.60	10	Ea.	6,100	395		6,495	7,300
4080	Circuit breaker			1.60	10		4,550	395		4,945	5,575
4100	Type B, fusible			1.60	10		6,725	395		7,120	7,975
4120	Circuit breaker			1.60	10		4,975	395		5,370	6,050
4140	NEMA 12, type A, fusible			1.50	10.667		6,200	420		6,620	7,450
4160	Circuit breaker			1.50	10.667		4,600	420		5,020	5,700
4180	Type B, fusible			1.50	10.667		6,825	420		7,245	8,125
4200	Circuit breaker			1.50	10.667		5,050	420		5,470	6,175
4300	Starter size 4, autotransformer, NEMA 1, type A, fusible			1.30	12.308		6,025	485		6,510	7,350
4320	Circuit breaker			1.30	12.308		6,075	485		6,560	7,400
4340	Type B, fusible			1.30	12.308		6,625	485		7,110	8,025
4360	Circuit breaker			1.30	12.308		6,625	485		7,110	8,025
4380	NEMA 12, type A, fusible			1.24	12.903		6,125	510		6,635	7,475
4400	Circuit breaker			1.24	12.903		6,150	510		6,660	7,525
4420	Type B, fusible			1.24	12.903		6,725	510		7,235	8,150
4440	Circuit breaker			1.24	12.903		6,700	510		7,210	8,125

ELECTRICAL 16

	16440	Swbds, Panels & Control Centers	CREW	DAILY OUTPUT	LABOR-HOURS	UNIT	2004 BARE COSTS				TOTAL INCL O&P	
							MAT.	LABOR	EQUIP.	TOTAL		
620	4500	Starter size 4, space only, NEMA 1, type A, fusible	1 Elec	14	.571	Ea.	1,100	22.50		1,122.50	1,225	**620**
	4520	Circuit breaker	R16440 -640	14	.571		785	22.50		807.50	900	
	4540	Type B, fusible		14	.571		1,100	22.50		1,122.50	1,225	
	4560	Circuit breaker		14	.571		785	22.50		807.50	900	
	4580	NEMA 12, type A, fusible		13	.615		1,150	24.50		1,174.50	1,300	
	4600	Circuit breaker		13	.615		835	24.50		859.50	955	
	4620	Type B, fusible		13	.615		1,150	24.50		1,174.50	1,300	
	4640	Circuit breaker		13	.615		835	24.50		859.50	955	
	4800	Starter size 5, FVNR, NEMA 1, type A, fusible	2 Elec	1	16		6,250	630		6,880	7,825	
	4820	Circuit breaker		1	16		4,450	630		5,080	5,825	
	4840	Type B, fusible		1	16		6,875	630		7,505	8,500	
	4860	Circuit breaker		1	16		4,900	630		5,530	6,325	
	4880	NEMA 12, type A, fusible		.96	16.667		6,350	655		7,005	7,950	
	4900	Circuit breaker		.96	16.667		4,525	655		5,180	5,950	
	4920	Type B, fusible		.96	16.667		6,950	655		7,605	8,625	
	4940	Circuit breaker		.96	16.667		4,975	655		5,630	6,450	
	5000	Starter size 5, FVR, NEMA 1, type A, fusible		.80	20		9,225	790		10,015	11,300	
	5020	Circuit breaker		.80	20		7,250	790		8,040	9,125	
	5040	Type B, fusible		.80	20		10,100	790		10,890	12,300	
	5060	Circuit breaker		.80	20		7,950	790		8,740	9,900	
	5080	NEMA 12, type A, fusible		.76	21.053		9,400	830		10,230	11,500	
	5100	Circuit breaker		.76	21.053		7,350	830		8,180	9,300	
	5120	Type B, fusible		.76	21.053		10,300	830		11,130	12,500	
	5140	Circuit breaker		.76	21.053		8,050	830		8,880	10,100	
	5190	Starter size 5, 2 speed, separate windings										
	5200	NEMA 1, type A, fusible	2 Elec	1	16	Ea.	12,700	630		13,330	14,900	
	5220	Circuit breaker		1	16		9,475	630		10,105	11,300	
	5240	Type B, fusible		1	16		14,000	630		14,630	16,300	
	5260	Circuit breaker		1	16		10,400	630		11,030	12,400	
	5280	NEMA 12, type A, fusible		.96	16.667		12,900	655		13,555	15,200	
	5300	Circuit breaker		.96	16.667		9,600	655		10,255	11,600	
	5320	Type B, fusible		.96	16.667		14,100	655		14,755	16,600	
	5340	Circuit breaker		.96	16.667		10,500	655		11,155	12,600	
	5400	Starter size 5, autotransformer, NEMA 1, type A, fusible		.70	22.857		10,400	900		11,300	12,800	
	5420	Circuit breaker		.70	22.857		8,650	900		9,550	10,900	
	5440	Type B, fusible		.70	22.857		11,400	900		12,300	14,000	
	5460	Circuit breaker		.70	22.857		9,500	900		10,400	11,800	
	5480	NEMA 12, type A, fusible		.68	23.529		10,500	925		11,425	13,000	
	5500	Circuit breaker		.68	23.529		8,725	925		9,650	11,000	
	5520	Type B, fusible		.68	23.529		11,500	925		12,425	14,100	
	5540	Circuit breakers		.68	23.529		9,575	925		10,500	11,900	
	5600	Starter size 5, space only, NEMA 1, type A, fusible	1 Elec	12	.667		1,400	26.50		1,426.50	1,600	
	5620	Circuit breaker		12	.667		1,400	26.50		1,426.50	1,600	
	5640	Type B, fusible		12	.667		1,400	26.50		1,426.50	1,600	
	5660	Circuit breaker		12	.667		915	26.50		941.50	1,050	
	5680	NEMA 12, type A, fusible		11	.727		1,475	28.50		1,503.50	1,675	
	5700	Circuit breaker		11	.727		975	28.50		1,003.50	1,125	
	5720	Type B, fusible		11	.727		1,475	28.50		1,503.50	1,675	
	5740	Circuit breaker		11	.727		975	28.50		1,003.50	1,125	
	5800	Fuse, light contactor NEMA 1, type A, 30 amp		2.70	2.963		1,075	117		1,192	1,375	
	5820	60 amp		2	4		1,225	158		1,383	1,575	
	5840	100 amp		1	8		2,325	315		2,640	3,025	
	5860	200 amp		.80	10		4,950	395		5,345	6,025	
	5880	Type B, 30 amp		2.70	2.963		1,175	117		1,292	1,475	
	5900	60 amp		2	4		1,325	158		1,483	1,675	
	5920	100 amp		1	8		2,550	315		2,865	3,300	

Important: See the Reference Section for critical supporting data - Reference Nos., Crews, & City Cost Indexes

		CREW	DAILY OUTPUT	LABOR-HOURS	UNIT	2004 BARE COSTS				TOTAL INCL O&P	
						MAT.	LABOR	EQUIP.	TOTAL		
16440	**Swbds, Panels & Control Centers**										
620	5940	200 amp	2 Elec	1.60	10	Ea.	5,425	395		5,820	6,525	620
	5960	NEMA 12, type A, 30 amp	1 Elec	2.60	3.077		1,100	121		1,221	1,375	
	5980	60 amp		1.90	4.211		1,250	166		1,416	1,625	
	6000	100 amp	↓	.95	8.421		2,375	330		2,705	3,100	
	6020	200 amp	2 Elec	1.50	10.667		5,025	420		5,445	6,150	
	6040	Type B, 30 amp	1 Elec	2.60	3.077		1,200	121		1,321	1,500	
	6060	60 amp		1.90	4.211		1,325	166		1,491	1,725	
	6080	100 amp	↓	.95	8.421		2,600	330		2,930	3,375	
	6100	200 amp	2 Elec	1.50	10.667		5,500	420		5,920	6,675	
	6200	Circuit breaker, light contactor NEMA 1, type A, 30 amp	1 Elec	2.70	2.963		1,175	117		1,292	1,475	
	6220	60 amp		2	4		1,350	158		1,508	1,700	
	6240	100 amp	↓	1	8		2,075	315		2,390	2,750	
	6260	200 amp	2 Elec	1.60	10		4,100	395		4,495	5,075	
	6280	Type B, 30 amp	1 Elec	2.70	2.963		1,300	117		1,417	1,600	
	6300	60 amp		2	4		1,475	158		1,633	1,850	
	6320	100 amp	↓	1	8		2,250	315		2,565	2,950	
	6340	200 amp	2 Elec	1.60	10		4,475	395		4,870	5,500	
	6360	NEMA 12, type A, 30 amp	1 Elec	2.60	3.077		1,200	121		1,321	1,500	
	6380	60 amp		1.90	4.211		1,350	166		1,516	1,750	
	6400	100 amp	↓	.95	8.421		2,100	330		2,430	2,825	
	6420	200 amp	2 Elec	1.50	10.667		4,125	420		4,545	5,150	
	6440	Type B, 30 amp	1 Elec	2.60	3.077		1,400	121		1,521	1,725	
	6460	60 amp		1.90	4.211		1,500	166		1,666	1,900	
	6480	100 amp	↓	.95	8.421		2,300	330		2,630	3,025	
	6500	200 amp	2 Elec	1.50	10.667		4,525	420		4,945	5,600	
	6600	Fusible switch, NEMA 1, type A, 30 amp	1 Elec	5.30	1.509		715	59.50		774.50	880	
	6620	60 amp		5	1.600		765	63		828	935	
	6640	100 amp		4	2		845	79		924	1,050	
	6660	200 amp	↓	3.20	2.500		1,450	98.50		1,548.50	1,750	
	6680	400 amp	2 Elec	4.60	3.478		3,425	137		3,562	3,975	
	6700	600 amp		3.20	5		3,725	197		3,922	4,375	
	6720	800 amp	↓	2.60	6.154		9,325	242		9,567	10,700	
	6740	NEMA 12, type A, 30 amp	1 Elec	5.20	1.538		735	60.50		795.50	900	
	6760	60 amp		4.90	1.633		785	64.50		849.50	955	
	6780	100 amp		3.90	2.051		865	81		946	1,075	
	6800	200 amp	↓	3.10	2.581		1,500	102		1,602	1,800	
	6820	400 amp	2 Elec	4.40	3.636		3,500	143		3,643	4,075	
	6840	600 amp		3	5.333		3,825	210		4,035	4,525	
	6860	800 amp	↓	2.40	6.667		9,425	263		9,688	10,800	
	6900	Circuit breaker, NEMA 1, type A, 30 amp	1 Elec	5.30	1.509		660	59.50		719.50	815	
	6920	60 amp		5	1.600		660	63		723	820	
	6940	100 amp		4	2		660	79		739	840	
	6960	225 amp	↓	3.20	2.500		1,175	98.50		1,273.50	1,450	
	6980	400 amp	2 Elec	4.60	3.478		2,225	137		2,362	2,650	
	7000	600 amp		3.20	5		2,600	197		2,797	3,175	
	7020	800 amp	↓	2.60	6.154		5,450	242		5,692	6,350	
	7040	NEMA 12, type A, 30 amp	1 Elec	5.20	1.538		675	60.50		735.50	835	
	7060	60 amp		4.90	1.633		675	64.50		739.50	840	
	7080	100 amp		3.90	2.051		675	81		756	865	
	7100	225 amp	↓	3.10	2.581		1,200	102		1,302	1,475	
	7120	400 amp	2 Elec	4.40	3.636		2,275	143		2,418	2,725	
	7140	600 amp		3	5.333		2,650	210		2,860	3,225	
	7160	800 amp		2.40	6.667		5,550	263		5,813	6,525	
	7300	Incoming line, main lug only, 600 amp, alum, NEMA 1		1.60	10		975	395		1,370	1,650	
	7320	NEMA 12		1.50	10.667		995	420		1,415	1,725	
	7340	Copper, NEMA 1	↓ ↓	1.60	10	↓	1,025	395		1,420	1,700	

Note cell at row 5940: R16440-640

ELECTRICAL 16

16440 | Swbds, Panels & Control Centers

		CREW	DAILY OUTPUT	LABOR-HOURS	UNIT	2004 BARE COSTS				TOTAL INCL O&P
						MAT.	LABOR	EQUIP.	TOTAL	
620	7360	800 amp, alum., NEMA 1 [R16440-640]	2 Elec	1.50	10.667	Ea.	2,450	420		2,870	3,325	**620**
	7380	NEMA 12		1.40	11.429		2,475	450		2,925	3,400
	7400	Copper, NEMA 1		1.50	10.667		2,550	420		2,970	3,425
	7420	1200 amp, copper, NEMA 1		1.40	11.429		2,625	450		3,075	3,575
	7440	Incoming line, fusible switch, 400 amp, alum., NEMA 1		1.20	13.333		2,875	525		3,400	3,925
	7460	NEMA 12		1.10	14.545		2,925	575		3,500	4,075
	7480	Copper, NEMA 1		1.20	13.333		2,925	525		3,450	3,975
	7500	600 amp, alum., NEMA 1		1.10	14.545		3,500	575		4,075	4,700
	7520	NEMA 12		1	16		3,575	630		4,205	4,875
	7540	Copper, NEMA 1		1.10	14.545		3,550	575		4,125	4,750
	7560	Incoming line, circuit breaker, 225 amp, alum., NEMA 1		1.20	13.333		1,750	525		2,275	2,700
	7580	NEMA 12		1.10	14.545		1,775	575		2,350	2,800
	7600	Copper, NEMA 1		1.20	13.333		1,800	525		2,325	2,750
	7620	400 amp, alum., NEMA 1		1.20	13.333		2,625	525		3,150	3,650
	7640	NEMA 12		1.10	14.545		2,650	575		3,225	3,775
	7660	Copper, NEMA 1		1.20	13.333		2,675	525		3,200	3,700
	7680	600 amp, alum., NEMA 1		1.10	14.545		3,050	575		3,625	4,200
	7700	NEMA 12		1	16		3,100	630		3,730	4,350
	7720	Copper, NEMA 1		1.10	14.545		3,100	575		3,675	4,275
	7740	800 amp, copper, NEMA 1		.90	17.778		5,450	700		6,150	7,050
	7760	Incoming line, for copper bus, add					96.50			96.50	106
	7780	For 65000 amp bus bracing, add					143			143	157
	7800	For NEMA 3R enclosure, add					3,850			3,850	4,225
	7820	For NEMA 12 enclosure, add					106			106	116
	7840	For 1/4" x 1" ground bus, add	1 Elec	16	.500		80	19.70		99.70	117
	7860	For 1/4" x 2" ground bus, add	"	12	.667		80	26.50		106.50	127
	7880	Main rating basic section, alum., NEMA 1, 600 amp	2 Elec	1.60	10			395		395	585
	7900	800 amp		1.40	11.429		223	450		673	915
	7920	1200 amp		1.20	13.333		440	525		965	1,250
	7940	For copper bus, add					330			330	360
	7960	For 65000 amp bus bracing, add					260			260	286
	7980	For NEMA 3R enclosure, add					3,850			3,850	4,225
	8000	For NEMA 12, enclosure, add					124			124	136
	8020	For 1/4" x 1" ground bus, add	1 Elec	16	.500		80	19.70		99.70	117
	8040	For 1/4" x 2" ground bus, add		12	.667		80	26.50		106.50	127
	8060	Unit devices, pilot light, standard		16	.500		69	19.70		88.70	105
	8080	Pilot light, push to test		16	.500		96.50	19.70		116.20	136
	8100	Pilot light, standard, and push button		12	.667		165	26.50		191.50	221
	8120	Pilot light, push to test, and push button		12	.667		193	26.50		219.50	251
	8140	Pilot light, standard, and select switch		12	.667		165	26.50		191.50	221
	8160	Pilot light, push to test, and select switch		12	.667		193	26.50		219.50	251
640	0010	**MOTOR CONTROL CENTER** Consists of starters & structures [R16440-640]										**640**
	0050	Starters, class 1, type B, comb. MCP, FVNR, with									
	0100	control transformer, 10 HP, size 1, 12" high	1 Elec	2.70	2.963	Ea.	1,100	117		1,217	1,375
	0200	25 HP, size 2, 18" high	2 Elec	4	4		1,250	158		1,408	1,600
	0300	50 HP, size 3, 24" high		2	8		1,925	315		2,240	2,600
	0350	75 HP, size 4, 24" high		1.60	10		2,575	395		2,970	3,400
	0400	100 HP, size 4, 30" high		1.40	11.429		3,400	450		3,850	4,425
	0500	200 HP, size 5, 48" high		1	16		5,075	630		5,705	6,525
	0600	400 HP, size 6, 72" high		.80	20		10,400	790		11,190	12,700
	0800	Structures, 600 amp, 22,000 rms, takes any									
	0900	combination of starters up to 72" high	2 Elec	1.60	10	Ea.	1,800	395		2,195	2,550
	1000	Back to back, 72" front & 66" back	"	1.20	13.333		2,425	525		2,950	3,450
	1100	For copper bus add per structure					172			172	189
	1200	For NEMA 12, add per structure					99			99	109

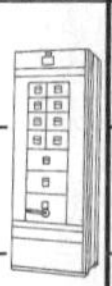

(Left margin vertical tab: 16 ELECTRICAL)

	16440	Swbds, Panels & Control Centers		CREW	DAILY OUTPUT	LABOR- HOURS	UNIT	2004 BARE COSTS				TOTAL INCL O&P	
								MAT.	LABOR	EQUIP.	TOTAL		
640	1300	For 42,000 rms, add per structure	R16440 -640				Ea.	166			166	183	640
	1400	For 100,000 rms, size 1 & 2, add						440			440	485	
	1500	Size 3, add						710			710	780	
	1600	Size 4, add						570			570	630	
	1700	For pilot lights, add per starter		1 Elec	16	.500		77	19.70		96.70	114	
	1800	For push button, add per starter			16	.500		77	19.70		96.70	114	
	1900	For auxilliary contacts, add per starter			16	.500		201	19.70		220.70	251	
660	0010	**MOTOR STARTERS & CONTROLS**	R16440 -660										660
	0050	Magnetic, FVNR, with enclosure and heaters, 480 volt											
	0080	2 HP, size 00		1 Elec	3.50	2.286	Ea.	173	90		263	325	
	0100	5 HP, size 0			2.30	3.478		209	137		346	435	
	0200	10 HP, size 1			1.60	5		235	197		432	550	
	0300	25 HP, size 2		2 Elec	2.20	7.273		440	287		727	910	
	0400	50 HP, size 3			1.80	8.889		720	350		1,070	1,300	
	0500	100 HP, size 4			1.20	13.333		1,600	525		2,125	2,525	
	0600	200 HP, size 5			.90	17.778		3,725	700		4,425	5,150	
	0610	400 HP, size 6			.80	20		10,500	790		11,290	12,700	
	0620	NEMA 7, 5 HP, size 0		1 Elec	1.60	5		865	197		1,062	1,250	
	0630	10 HP, size 1		"	1.10	7.273		905	287		1,192	1,425	
	0640	25 HP, size 2		2 Elec	1.80	8.889		1,450	350		1,800	2,125	
	0650	50 HP, size 3			1.20	13.333		2,150	525		2,675	3,150	
	0660	100 HP, size 4			.90	17.778		3,525	700		4,225	4,925	
	0670	200 HP, size 5			.50	32		8,250	1,250		9,500	11,000	
	0700	Combination, with motor circuit protectors, 5 HP, size 0		1 Elec	1.80	4.444		680	175		855	1,000	
	0800	10 HP, size 1		"	1.30	6.154		705	242		947	1,125	
	0900	25 HP, size 2		2 Elec	2	8		990	315		1,305	1,550	
	1000	50 HP, size 3			1.32	12.121		1,425	480		1,905	2,275	
	1200	100 HP, size 4			.80	20		3,100	790		3,890	4,600	
	1220	NEMA 7, 5 HP, size 0		1 Elec	1.30	6.154		1,525	242		1,767	2,025	
	1230	10 HP, size 1		"	1	8		1,575	315		1,890	2,200	
	1240	25 HP, size 2		2 Elec	1.32	12.121		2,075	480		2,555	3,000	
	1250	50 HP, size 3			.80	20		3,425	790		4,215	4,950	
	1260	100 HP, size 4			.60	26.667		5,325	1,050		6,375	7,425	
	1270	200 HP, size 5			.40	40		11,500	1,575		13,075	15,100	
	1400	Combination, with fused switch, 5 HP, size 0		1 Elec	1.80	4.444		520	175		695	830	
	1600	10 HP, size 1		"	1.30	6.154		555	242		797	970	
	1800	25 HP, size 2		2 Elec	2	8		900	315		1,215	1,450	
	2000	50 HP, size 3			1.32	12.121		1,525	480		2,005	2,375	
	2200	100 HP, size 4			.80	20		2,650	790		3,440	4,100	
	2610	NEMA 4, with start-stop pushbutton size 1		1 Elec	1.30	6.154		1,400	242		1,642	1,875	
	2620	Size 2		2 Elec	2	8		1,875	315		2,190	2,525	
	2630	Size 3			1.32	12.121		2,950	480		3,430	3,950	
	2640	Size 4			.80	20		4,550	790		5,340	6,175	
	2650	NEMA 4, FVNR, including control transformer											
	2660	Size 1		2 Elec	2.60	6.154	Ea.	1,275	242		1,517	1,750	
	2670	Size 2			2	8		1,825	315		2,140	2,475	
	2680	Size 3			1.32	12.121		2,925	480		3,405	3,900	
	2690	Size 4			.80	20		4,500	790		5,290	6,125	
	2710	Magnetic, FVR, control circuit transformer, NEMA 1, size 1		1 Elec	1.30	6.154		755	242		997	1,200	
	2720	Size 2		2 Elec	2	8		1,200	315		1,515	1,800	
	2730	Size 3			1.32	12.121		1,825	480		2,305	2,725	
	2740	Size 4			.80	20		4,000	790		4,790	5,575	
	2760	NEMA 4, size 1		1 Elec	1.10	7.273		1,075	287		1,362	1,625	
	2770	Size 2		2 Elec	1.60	10		1,750	395		2,145	2,500	
	2780	Size 3			1.20	13.333		2,625	525		3,150	3,675	

ELECTRICAL 16

	16440	Swbds, Panels & Control Centers	CREW	DAILY OUTPUT	LABOR-HOURS	UNIT	2004 BARE COSTS				TOTAL INCL O&P	
							MAT.	LABOR	EQUIP.	TOTAL		
660	2790	Size 4	2 Elec	.70	22.857	Ea.	5,400	900		6,300	7,300	**660**
	2820	NEMA 12, size 1	1 Elec	1.10	7.273		890	287		1,177	1,400	
	2830	Size 2	2 Elec	1.60	10		1,400	395		1,795	2,125	
	2840	Size 3		1.20	13.333		2,225	525		2,750	3,225	
	2850	Size 4		.70	22.857		4,575	900		5,475	6,375	
	2870	Combination FVR, fused, w/control XFMR & PB, NEMA 1, size 1	1 Elec	1	8		1,300	315		1,615	1,900	
	2880	Size 2	2 Elec	1.50	10.667		1,850	420		2,270	2,675	
	2890	Size 3		1.10	14.545		2,750	575		3,325	3,875	
	2900	Size 4		.70	22.857		5,600	900		6,500	7,500	
	2910	NEMA 4, size 1	1 Elec	.90	8.889		1,925	350		2,275	2,650	
	2920	Size 2	2 Elec	1.40	11.429		2,825	450		3,275	3,775	
	2930	Size 3		1	16		4,450	630		5,080	5,825	
	2940	Size 4		.60	26.667		7,375	1,050		8,425	9,700	
	2950	NEMA 12, size 1	1 Elec	1	8		1,475	315		1,790	2,100	
	2960	Size 2	2 Elec	1.40	11.429		2,075	450		2,525	2,975	
	2970	Size 3		1	16		3,050	630		3,680	4,325	
	2980	Size 4		.60	26.667		6,075	1,050		7,125	8,250	
	3010	Manual, single phase, w/pilot, 1 pole 120V NEMA 1	1 Elec	6.40	1.250		54	49.50		103.50	133	
	3020	NEMA 4		4	2		305	79		384	450	
	3030	2 pole, 120/240 V, NEMA 1		6.40	1.250		59.50	49.50		109	139	
	3040	NEMA 4		4	2		310	79		389	455	
	3041	3 phase, 3 pole 600V, NEMA 1		5.50	1.455		212	57.50		269.50	320	
	3042	NEMA 4		3.50	2.286		415	90		505	590	
	3043	NEMA 12		3.50	2.286		248	90		338	405	
	3070	Auxiliary contact, normally open					62			62	68	
	3500	Magnetic FVNR with NEMA 12, enclosure & heaters, 480 volt										
	3600	5 HP, size 0	1 Elec	2.20	3.636	Ea.	197	143		340	430	
	3700	10 HP, size 1	"	1.50	5.333		297	210		507	640	
	3800	25 HP, size 2	2 Elec	2	8		555	315		870	1,075	
	3900	50 HP, size 3		1.60	10		855	395		1,250	1,525	
	4000	100 HP, size 4		1	16		2,050	630		2,680	3,200	
	4100	200 HP, size 5		.80	20		4,900	790		5,690	6,575	
	4200	Combination, with motor circuit protectors, 5 HP, size 0	1 Elec	1.70	4.706		655	185		840	995	
	4300	10 HP, size 1	"	1.20	6.667		680	263		943	1,125	
	4400	25 HP, size 2	2 Elec	1.80	8.889		1,025	350		1,375	1,650	
	4500	50 HP, size 3		1.20	13.333		1,650	525		2,175	2,600	
	4600	100 HP, size 4		.74	21.622		3,725	850		4,575	5,375	
	4700	Combination, with fused switch, 5 HP, size 0	1 Elec	1.70	4.706		630	185		815	970	
	4800	10 HP, size 1	"	1.20	6.667		660	263		923	1,125	
	4900	25 HP, size 2	2 Elec	1.80	8.889		1,000	350		1,350	1,625	
	5000	50 HP, size 3		1.20	13.333		1,600	525		2,125	2,550	
	5100	100 HP, size 4		.74	21.622		3,250	850		4,100	4,850	
	5200	Factory installed controls, adders to size 0 thru 5										
	5300	Start-stop push button	1 Elec	32	.250	Ea.	41.50	9.85		51.35	60	
	5400	Hand-off-auto-selector switch		32	.250		41.50	9.85		51.35	60	
	5500	Pilot light		32	.250		77.50	9.85		87.35	99.50	
	5600	Start-stop-pilot		32	.250		119	9.85		128.85	146	
	5700	Auxiliary contact, NO or NC		32	.250		57	9.85		66.85	77	
	5800	NO-NC		32	.250		114	9.85		123.85	140	
	5810	Magnetic FVR, NEMA 7, w/heaters, size 1		.66	12.121		1,575	480		2,055	2,450	
	5830	Size 2	2 Elec	1.10	14.545		2,625	575		3,200	3,750	
	5840	Size 3		.70	22.857		4,200	900		5,100	5,975	
	5850	Size 4		.60	26.667		6,400	1,050		7,450	8,625	
	5860	Combination w/circuit breakers, heaters, control XFMR PB, size 1	1 Elec	.60	13.333		1,450	525		1,975	2,375	
	5870	Size 2	2 Elec	.80	20		1,850	790		2,640	3,225	
	5880	Size 3		.50	32		2,475	1,250		3,725	4,600	

R16440 -660

Important: See the Reference Section for critical supporting data - Reference Nos., Crews, & City Cost Indexes

			CREW	DAILY OUTPUT	LABOR-HOURS	UNIT	2004 BARE COSTS				TOTAL INCL O&P	
	16440	**Swbds, Panels & Control Centers**					MAT.	LABOR	EQUIP.	TOTAL		
660	5890	Size 4	2 Elec	.40	40	Ea.	4,900	1,575		6,475	7,750	**660**
	5900	Manual, 240 volt, .75 HP motor	1 Elec	4	2		44	79		123	166	
	5910	2 HP motor		4	2		119	79		198	248	
	6000	Magnetic, 240 volt, 1 or 2 pole, .75 HP motor		4	2		173	79		252	305	
	6020	2 HP motor		4	2		191	79		270	325	
	6040	5 HP motor		3	2.667		273	105		378	455	
	6060	10 HP motor		2.30	3.478		680	137		817	950	
	6100	3 pole, .75 HP motor		3	2.667		173	105		278	345	
	6120	5 HP motor		2.30	3.478		235	137		372	460	
	6140	10 HP motor		1.60	5		440	197		637	780	
	6160	15 HP motor		1.60	5		440	197		637	780	
	6180	20 HP motor		1.10	7.273		720	287		1,007	1,225	
	6200	25 HP motor	2 Elec	2.20	7.273		720	287		1,007	1,225	
	6210	30 HP motor		1.80	8.889		720	350		1,070	1,300	
	6220	40 HP motor		1.80	8.889		1,600	350		1,950	2,275	
	6230	50 HP motor		1.80	8.889		1,600	350		1,950	2,275	
	6240	60 HP motor		1.20	13.333		3,725	525		4,250	4,875	
	6250	75 HP motor		1.20	13.333		3,725	525		4,250	4,875	
	6260	100 HP motor		1.20	13.333		3,725	525		4,250	4,875	
	6270	125 HP motor		.90	17.778		10,500	700		11,200	12,600	
	6280	150 HP motor		.90	17.778		10,500	700		11,200	12,600	
	6290	200 HP motor		.90	17.778		10,500	700		11,200	12,600	
	6400	Starter & nonfused disconnect, 240 volt, 1-2 pole, .75 HP motor	1 Elec	2	4		233	158		391	490	
	6410	2 HP motor		2	4		251	158		409	510	
	6420	5 HP motor		1.80	4.444		335	175		510	625	
	6430	10 HP motor		1.40	5.714		760	225		985	1,175	
	6440	3 pole, .75 HP motor		1.60	5		233	197		430	550	
	6450	5 HP motor		1.40	5.714		295	225		520	660	
	6460	10 HP motor		1.10	7.273		520	287		807	1,000	
	6470	15 HP motor		1	8		520	315		835	1,050	
	6480	20 HP motor	2 Elec	1.50	10.667		905	420		1,325	1,625	
	6490	25 HP motor		1.50	10.667		905	420		1,325	1,625	
	6500	30 HP motor		1.30	12.308		905	485		1,390	1,725	
	6510	40 HP motor		1.24	12.903		1,950	510		2,460	2,875	
	6520	50 HP motor		1.12	14.286		1,950	565		2,515	2,950	
	6530	60 HP motor		.90	17.778		4,075	700		4,775	5,525	
	6540	75 HP motor		.76	21.053		4,075	830		4,905	5,700	
	6550	100 HP motor		.70	22.857		4,075	900		4,975	5,825	
	6560	125 HP motor		.60	26.667		11,300	1,050		12,350	14,000	
	6570	150 HP motor		.52	30.769		11,300	1,200		12,500	14,200	
	6580	200 HP motor		.50	32		11,300	1,250		12,550	14,300	
	6600	Starter & fused disconnect, 240 volt, 1-2 pole, .75 HP motor	1 Elec	2	4		247	158		405	505	
	6610	2 HP motor		2	4		266	158		424	525	
	6620	5 HP motor		1.80	4.444		350	175		525	645	
	6630	10 HP motor		1.40	5.714		805	225		1,030	1,225	
	6640	3 pole, .75 HP motor		1.60	5		247	197		444	565	
	6650	5 HP motor		1.40	5.714		310	225		535	675	
	6660	10 HP motor		1.10	7.273		565	287		852	1,050	
	6690	15 HP motor		1	8		565	315		880	1,100	
	6700	20 HP motor	2 Elec	1.60	10		935	395		1,330	1,600	
	6710	25 HP motor		1.60	10		935	395		1,330	1,600	
	6720	30 HP motor		1.40	11.429		935	450		1,385	1,700	
	6730	40 HP motor		1.20	13.333		2,075	525		2,600	3,050	
	6740	50 HP motor		1.20	13.333		2,075	525		2,600	3,050	
	6750	60 HP motor		.90	17.778		4,200	700		4,900	5,675	
	6760	75 HP motor		.90	17.778		4,200	700		4,900	5,675	

R16440 -660

ELECTRICAL **16**

		CREW	DAILY OUTPUT	LABOR-HOURS	UNIT	2004 BARE COSTS				TOTAL INCL O&P	
						MAT.	LABOR	EQUIP.	TOTAL		
660	6770	100 HP motor	R16440 -660	2 Elec	.70	22.857	Ea.	4,200	900		5,100	5,975	660
	6780	125 HP motor		↓	.54	29.630	↓	11,700	1,175		12,875	14,500	
	6790	Combination starter & nonfusible disconnect										
	6800	240 volt, 1-2 pole, .75 HP motor		1 Elec	2	4	Ea.	495	158		653	775
	6810	2 HP motor			2	4		495	158		653	775
	6820	5 HP motor			1.50	5.333		520	210		730	885
	6830	10 HP motor			1.20	6.667		810	263		1,073	1,275
	6840	3 pole, .75 HP motor			1.80	4.444		495	175		670	800
	6850	5 HP motor			1.30	6.154		520	242		762	930
	6860	10 HP motor			1	8		810	315		1,125	1,350
	6870	15 HP motor		▼	1	8		810	315		1,125	1,350
	6880	20 HP motor		2 Elec	1.32	12.121		1,325	480		1,805	2,150
	6890	25 HP motor			1.32	12.121		1,325	480		1,805	2,150
	6900	30 HP motor			1.32	12.121		1,325	480		1,805	2,150
	6910	40 HP motor			.80	20		2,550	790		3,340	3,975
	6920	50 HP motor			.80	20		2,550	790		3,340	3,975
	6930	60 HP motor			.70	22.857		5,675	900		6,575	7,600
	6940	75 HP motor			.70	22.857		5,675	900		6,575	7,600
	6950	100 HP motor			.70	22.857		5,675	900		6,575	7,600
	6960	125 HP motor			.60	26.667		14,900	1,050		15,950	18,000
	6970	150 HP motor			.60	26.667		14,900	1,050		15,950	18,000
	6980	200 HP motor		▼	.60	26.667	↓	14,900	1,050		15,950	18,000
	6990	Combination starter and fused disconnect										
	7000	240 volt, 1-2 pole, .75 HP motor		1 Elec	2	4	Ea.	510	158		668	795
	7010	2 HP motor			2	4		510	158		668	795
	7020	5 HP motor			1.50	5.333		535	210		745	900
	7030	10 HP motor			1.20	6.667		830	263		1,093	1,300
	7040	3 pole, .75 HP motor			1.80	4.444		510	175		685	820
	7050	5 HP motor			1.30	6.154		535	242		777	945
	7060	10 HP motor			1	8		830	315		1,145	1,375
	7070	15 HP motor		▼	1	8		830	315		1,145	1,375
	7080	20 HP motor		2 Elec	1.32	12.121		1,375	480		1,855	2,225
	7090	25 HP motor			1.32	12.121		1,375	480		1,855	2,225
	7100	30 HP motor			1.32	12.121		1,375	480		1,855	2,225
	7110	40 HP motor			.80	20		2,625	790		3,415	4,075
	7120	50 HP motor			.80	20		2,625	790		3,415	4,075
	7130	60 HP motor			.80	20		5,850	790		6,640	7,625
	7140	75 HP motor			.70	22.857		5,850	900		6,750	7,800
	7150	100 HP motor			.70	22.857		5,850	900		6,750	7,800
	7160	125 HP motor			.70	22.857		15,500	900		16,400	18,400
	7170	150 HP motor			.60	26.667		15,500	1,050		16,550	18,600
	7180	200 HP motor		▼	.60	26.667	↓	15,500	1,050		16,550	18,600
	7190	Combination starter & circuit breaker disconnect										
	7200	240 volt, 1-2 pole, .75 HP motor		1 Elec	2	4	Ea.	515	158		673	800
	7210	2 HP motor			2	4		515	158		673	800
	7220	5 HP motor			1.50	5.333		540	210		750	910
	7230	10 HP motor			1.20	6.667		825	263		1,088	1,300
	7240	3 pole, .75 HP motor			1.80	4.444		530	175		705	840
	7250	5 HP motor			1.30	6.154		555	242		797	970
	7260	10 HP motor			1	8		840	315		1,155	1,400
	7270	15 HP motor		▼	1	8		840	315		1,155	1,400
	7280	20 HP motor		2 Elec	1.32	12.121		1,425	480		1,905	2,275
	7290	25 HP motor			1.32	12.121		1,425	480		1,905	2,275
	7300	30 HP motor			1.32	12.121		1,425	480		1,905	2,275
	7310	40 HP motor			.80	20		3,100	790		3,890	4,600
	7320	50 HP motor		▼	.80	20	↓	3,100	790		3,890	4,600

Important: See the Reference Section for critical supporting data - Reference Nos., Crews, & City Cost Indexes

16440	Swbds, Panels & Control Centers		CREW	DAILY OUTPUT	LABOR-HOURS	UNIT	2004 BARE COSTS				TOTAL INCL O&P
							MAT.	LABOR	EQUIP.	TOTAL	
660 7330	60 HP motor	R16440 -660	2 Elec	.80	20	Ea.	7,175	790		7,965	9,050	**660**
7340	75 HP motor			.70	22.857		7,175	900		8,075	9,225	
7350	100 HP motor			.70	22.857		7,175	900		8,075	9,225	
7360	125 HP motor			.70	22.857		15,500	900		16,400	18,500	
7370	150 HP motor			.60	26.667		15,500	1,050		16,550	18,700	
7380	200 HP motor			.60	26.667		15,500	1,050		16,550	18,700	
7400	Magnetic FVNR with enclosure & heaters, 2 pole,											
7410	230 volt, 1 HP size 00		1 Elec	4	2	Ea.	157	79		236	290	
7420	2 HP, size 0			4	2		175	79		254	310	
7430	3 HP, size 1			3	2.667		201	105		306	375	
7440	5 HP, size 1p			3	2.667		258	105		363	440	
7450	115 volt, 1/3 HP, size 00			4	2		157	79		236	290	
7460	1 HP, size 0			4	2		175	79		254	310	
7470	2 HP, size 1			3	2.667		201	105		306	375	
7480	3 HP, size 1P			3	2.667		250	105		355	430	
7500	3 pole, 480 volt, 600 HP, size 7		2 Elec	.70	22.857		13,900	900		14,800	16,700	
7590	Magnetic FVNR with heater, NEMA 1											
7600	600 volt, 3 pole, 5 HP motor		1 Elec	2.30	3.478	Ea.	209	137		346	435	
7610	10 HP motor		"	1.60	5		235	197		432	550	
7620	25 HP motor		2 Elec	2.20	7.273		440	287		727	910	
7630	30 HP motor			1.80	8.889		720	350		1,070	1,300	
7640	40 HP motor			1.80	8.889		720	350		1,070	1,300	
7650	50 HP motor			1.80	8.889		720	350		1,070	1,300	
7660	60 HP motor			1.20	13.333		1,600	525		2,125	2,525	
7670	75 HP motor			1.20	13.333		1,600	525		2,125	2,525	
7680	100 HP motor			1.20	13.333		1,600	525		2,125	2,525	
7690	125 HP motor			.90	17.778		3,725	700		4,425	5,150	
7700	150 HP motor			.90	17.778		3,725	700		4,425	5,150	
7710	200 HP motor			.90	17.778		3,725	700		4,425	5,150	
7750	Starter & nonfused disconnect, 600 volt, 3 pole, 5 HP motor		1 Elec	1.40	5.714		315	225		540	680	
7760	10 HP motor		"	1.10	7.273		340	287		627	800	
7770	25 HP motor		2 Elec	1.50	10.667		545	420		965	1,225	
7780	30 HP motor			1.30	12.308		910	485		1,395	1,725	
7790	40 HP motor			1.30	12.308		910	485		1,395	1,725	
7800	50 HP motor			1.30	12.308		910	485		1,395	1,725	
7810	60 HP motor			.92	17.391		1,900	685		2,585	3,125	
7820	75 HP motor			.92	17.391		1,900	685		2,585	3,125	
7830	100 HP motor			.84	19.048		1,900	750		2,650	3,225	
7840	125 HP motor			.70	22.857		4,200	900		5,100	5,975	
7850	150 HP motor			.70	22.857		4,200	900		5,100	5,975	
7860	200 HP motor			.60	26.667		4,200	1,050		5,250	6,200	
7870	Starter & fused disconnect, 600 volt, 3 pole, 5 HP motor		1 Elec	1.40	5.714		415	225		640	790	
7880	10 HP motor		"	1.10	7.273		440	287		727	905	
7890	25 HP motor		2 Elec	1.50	10.667		645	420		1,065	1,325	
7900	30 HP motor			1.30	12.308		965	485		1,450	1,800	
7910	40 HP motor			1.30	12.308		965	485		1,450	1,800	
7920	50 HP motor			1.30	12.308		965	485		1,450	1,800	
7930	60 HP motor			.92	17.391		2,050	685		2,735	3,275	
7940	75 HP motor			.92	17.391		2,050	685		2,735	3,275	
7950	100 HP motor			.84	19.048		2,050	750		2,800	3,375	
7960	125 HP motor			.70	22.857		4,400	900		5,300	6,175	
7970	150 HP motor			.70	22.857		4,400	900		5,300	6,175	
7980	200 HP motor			.60	26.667		4,400	1,050		5,450	6,400	
7990	Combination starter and nonfusible disconnect											
8000	600 volt, 3 pole, 5 HP motor		1 Elec	1.80	4.444	Ea.	495	175		670	800	
8010	10 HP motor		"	1.30	6.154		520	242		762	930	

ELECTRICAL **16**

			DAILY	LABOR-			2004 BARE COSTS				TOTAL	
	16440	**Swbds, Panels & Control Centers**	CREW	OUTPUT	HOURS	UNIT	MAT.	LABOR	EQUIP.	TOTAL	INCL O&P	
660	8020	25 HP motor	2 Elec	2	8	Ea.	810	315		1,125	1,350	660
	8030	30 HP motor		1.32	12.121		1,375	480		1,855	2,225	
	8040	40 HP motor		1.32	12.121		1,375	480		1,855	2,225	
	8050	50 HP motor		1.32	12.121		1,325	480		1,805	2,150	
	8060	60 HP motor		.80	20		2,625	790		3,415	4,075	
	8070	75 HP motor		.80	20		2,625	790		3,415	4,075	
	8080	100 HP motor		.80	20		2,550	790		3,340	3,975	
	8090	125 HP motor		.70	22.857		5,850	900		6,750	7,800	
	8100	150 HP motor		.70	22.857		5,850	900		6,750	7,800	
	8110	200 HP motor		.70	22.857		5,850	900		6,750	7,800	
	8140	Combination starter and fused disconnect										
	8150	600 volt, 3 pole, 5 HP motor	1 Elec	1.80	4.444	Ea.	510	175		685	820	
	8160	10 HP motor	"	1.30	6.154		555	242		797	970	
	8170	25 HP motor	2 Elec	2	8		900	315		1,215	1,450	
	8180	30 HP motor		1.32	12.121		1,400	480		1,880	2,250	
	8190	40 HP motor		1.32	12.121		1,525	480		2,005	2,375	
	8200	50 HP motor		1.32	12.121		1,525	480		2,005	2,375	
	8210	60 HP motor		.80	20		2,650	790		3,440	4,100	
	8220	75 HP motor		.80	20		2,650	790		3,440	4,100	
	8230	100 HP motor		.80	20		2,650	790		3,440	4,100	
	8240	125 HP motor		.70	22.857		5,850	900		6,750	7,800	
	8250	150 HP motor		.70	22.857		5,850	900		6,750	7,800	
	8260	200 HP motor		.70	22.857		5,850	900		6,750	7,800	
	8290	Combination starter & circuit breaker disconnect										
	8300	600 volt, 3 pole, 5 HP motor	1 Elec	1.80	4.444	Ea.	680	175		855	1,000	
	8310	10 HP motor	"	1.30	6.154		705	242		947	1,125	
	8320	25 HP motor	2 Elec	2	8		990	315		1,305	1,550	
	8330	30 HP motor		1.32	12.121		1,425	480		1,905	2,275	
	8340	40 HP motor		1.32	12.121		1,425	480		1,905	2,275	
	8350	50 HP motor		1.32	12.121		1,425	480		1,905	2,275	
	8360	60 HP motor		.80	20		3,100	790		3,890	4,600	
	8370	75 HP motor		.80	20		3,100	790		3,890	4,600	
	8380	100 HP motor		.80	20		3,100	790		3,890	4,600	
	8390	125 HP motor		.70	22.857		7,175	900		8,075	9,225	
	8400	150 HP motor		.70	22.857		7,175	900		8,075	9,225	
	8410	200 HP motor		.70	22.857		7,175	900		8,075	9,225	
	8430	Starter & circuit breaker disconnect										
	8440	600 volt, 3 pole, 5 HP motor	1 Elec	1.40	5.714	Ea.	620	225		845	1,025	
	8450	10 HP motor	"	1.10	7.273		645	287		932	1,125	
	8460	25 HP motor	2 Elec	1.50	10.667		855	420		1,275	1,575	
	8470	30 HP motor		1.30	12.308		1,125	485		1,610	1,975	
	8480	40 HP motor		1.30	12.308		1,125	485		1,610	1,975	
	8490	50 HP motor		1.30	12.308		1,125	485		1,610	1,975	
	8500	60 HP motor		.92	17.391		2,675	685		3,360	3,975	
	8510	75 HP motor		.92	17.391		2,675	685		3,360	3,975	
	8520	100 HP motor		.84	19.048		2,675	750		3,425	4,075	
	8530	125 HP motor		.70	22.857		4,825	900		5,725	6,650	
	8540	150 HP motor		.70	22.857		4,825	900		5,725	6,650	
	8550	200 HP motor		.60	26.667		5,600	1,050		6,650	7,725	
	8900	240 volt, 1-2 pole, .75 HP motor	1 Elec	2	4		585	158		743	880	
	8910	2 HP motor		2	4		605	158		763	900	
	8920	5 HP motor		1.80	4.444		685	175		860	1,025	
	8930	10 HP motor		1.40	5.714		1,100	225		1,325	1,525	
	8950	3 pole, .75 HP motor		1.60	5		585	197		782	940	
	8970	5 HP motor		1.40	5.714		645	225		870	1,050	
	8980	10 HP motor		1.10	7.273		855	287		1,142	1,375	

R16440 -660

16 ELECTRICAL

				DAILY	LABOR-		2004 BARE COSTS				TOTAL	
16440	**Swbds, Panels & Control Centers**		CREW	OUTPUT	HOURS	UNIT	MAT.	LABOR	EQUIP.	TOTAL	INCL O&P	
660	8990	15 HP motor	R16440 -660	1 Elec	1	8	Ea.	855	315		1,170	1,400	**660**
	9100	20 HP motor		2 Elec	1.50	10.667		1,200	420		1,620	1,925	
	9110	25 HP motor			1.50	10.667		1,200	420		1,620	1,925	
	9120	30 HP motor			1.30	12.308		1,200	485		1,685	2,025	
	9130	40 HP motor			1.24	12.903		2,675	510		3,185	3,700	
	9140	50 HP motor			1.12	14.286		2,675	565		3,240	3,775	
	9150	60 HP motor			.90	17.778		5,600	700		6,300	7,200	
	9160	75 HP motor			.76	21.053		5,600	830		6,430	7,375	
	9170	100 HP motor			.70	22.857		5,600	900		6,500	7,500	
	9180	125 HP motor			.60	26.667		13,200	1,050		14,250	16,100	
	9190	150 HP motor			.52	30.769		13,200	1,200		14,400	16,300	
	9200	200 HP motor			.50	32		13,200	1,250		14,450	16,400	
700	0010	**PANELBOARD & LOAD CENTER CIRCUIT BREAKERS**	R16440 -820										**700**
	0050	Bolt-on, 10,000 amp IC, 120 volt, 1 pole											
	0100	15 to 50 amp		1 Elec	10	.800	Ea.	10.60	31.50		42.10	58.50	
	0200	60 amp			8	1		10.60	39.50		50.10	70	
	0300	70 amp			8	1		20	39.50		59.50	80.50	
	0350	240 volt, 2 pole											
	0400	15 to 50 amp		1 Elec	8	1	Ea.	23.50	39.50		63	84	
	0500	60 amp			7.50	1.067		23.50	42		65.50	88	
	0600	80 to 100 amp			5	1.600		60	63		123	160	
	0700	3 pole, 15 to 60 amp			6.20	1.290		73.50	51		124.50	157	
	0800	70 amp			5	1.600		93	63		156	196	
	0900	80 to 100 amp			3.60	2.222		105	87.50		192.50	246	
	1000	22,000 amp I.C., 240 volt, 2 pole, 70 - 225 amp			2.70	2.963		455	117		572	675	
	1100	3 pole, 70 - 225 amp			2.30	3.478		505	137		642	760	
	1200	14,000 amp I.C., 277 volts, 1 pole, 15 - 30 amp			8	1		28	39.50		67.50	89.50	
	1300	22,000 amp I.C., 480 volts, 2 pole, 70 - 225 amp			2.70	2.963		455	117		572	675	
	1400	3 pole, 70 - 225 amp			2.30	3.478		560	137		697	820	
	2000	Plug-in panel or load center, 120/240 volt, to 60 amp, 1 pole			12	.667		8.20	26.50		34.70	48	
	2010	2 pole			9	.889		18.55	35		53.55	72.50	
	2020	3 pole			7.50	1.067		64	42		106	133	
	2030	100 amp, 2 pole			6	1.333		86	52.50		138.50	173	
	2040	3 pole			4.50	1.778		95.50	70		165.50	209	
	2050	150 amp, 2 pole			3	2.667		157	105		262	330	
	2060	Plug-in tandem, 120/240 V, 2-15 A, 1 pole			11	.727		21	28.50		49.50	65.50	
	2070	1-15 A & 1-20 A			11	.727		21	28.50		49.50	65.50	
	2080	2-20 A			11	.727		21	28.50		49.50	65.50	
	2100	High interrupting capacity, 120/240 volt, plug-in, 30 amp, 1 pole			12	.667		16.15	26.50		42.65	57	
	2110	60 amp, 2 pole			9	.889		37.50	35		72.50	93.50	
	2120	3 pole			7.50	1.067		97.50	42		139.50	170	
	2130	100 amp, 2 pole			6	1.333		108	52.50		160.50	197	
	2140	3 pole			4.50	1.778		140	70		210	258	
	2150	125 amp, 2 pole			3	2.667		288	105		393	470	
	2200	Bolt-on, 30 amp, 1 pole			10	.800		20.50	31.50		52	69.50	
	2210	60 amp, 2 pole			7.50	1.067		43.50	42		85.50	111	
	2220	3 pole			6.20	1.290		112	51		163	200	
	2230	100 amp, 2 pole			5	1.600		125	63		188	231	
	2240	3 pole			3.60	2.222		163	87.50		250.50	310	
	2300	Ground fault, 240 volt, 30 amp, 1 pole			7	1.143		63.50	45		108.50	137	
	2310	2 pole			6	1.333		114	52.50		166.50	203	
	2350	Key operated, 240 volt, 1 pole, 30 amp			7	1.143		62.50	45		107.50	136	
	2360	Switched neutral, 240 volt, 30 amp, 2 pole			6	1.333		26.50	52.50		79	107	
	2370	3 pole			5.50	1.455		40	57.50		97.50	129	
	2400	Shunt trip, for 240 volt breaker, 60 amp, 1 pole			4	2		52	79		131	174	
	2410	2 pole			3.50	2.286		52	90		142	191	

16440 | Swbds, Panels & Control Centers

			CREW	DAILY OUTPUT	LABOR-HOURS	UNIT	2004 BARE COSTS				TOTAL INCL O&P
							MAT.	LABOR	EQUIP.	TOTAL	
700	2420	3 pole	1 Elec	3	2.667	Ea.	52	105		157	213	700
	2430	100 amp, 2 pole		3	2.667		52	105		157	213
	2440	3 pole		2.50	3.200		52	126		178	245
	2450	150 amp, 2 pole		2	4		123	158		281	370
	2500	Auxiliary switch, for 240 volt breaker, 60 amp, 1 pole		4	2		47	79		126	169
	2510	2 pole		3.50	2.286		47	90		137	186
	2520	3 pole		3	2.667		47	105		152	208
	2530	100 amp, 2 pole		3	2.667		47	105		152	208
	2540	3 pole		2.50	3.200		47	126		173	240
	2550	150 amp, 2 pole		2	4		78.50	158		236.50	320
	2600	Panel or load center, 277/480 volt, plug-in, 30 amp, 1 pole		12	.667		38	26.50		64.50	81
	2610	60 amp, 2 pole		9	.889		114	35		149	178
	2620	3 pole		7.50	1.067		167	42		209	247
	2650	Bolt-on, 60 amp, 2 pole		7.50	1.067		114	42		156	189
	2660	3 pole		6.20	1.290		167	51		218	260
	2700	I-line, 277/480 volt, 30 amp, 1 pole		8	1		43.50	39.50		83	107
	2710	60 amp, 2 pole		7.50	1.067		154	42		196	232
	2720	3 pole		6.20	1.290		200	51		251	296
	2730	100 amp, 1 pole		7.50	1.067		84.50	42		126.50	155
	2740	2 pole		5	1.600		200	63		263	315
	2750	3 pole		3.50	2.286		236	90		326	395
	2800	High interrupting capacity, 277/480 volt, plug-in, 30 amp, 1 pole		12	.667		220	26.50		246.50	282
	2810	60 amp, 2 pole		9	.889		340	35		375	425
	2820	3 pole		7	1.143		395	45		440	500
	2830	Bolt-on, 30 amp, 1 pole		8	1		220	39.50		259.50	300
	2840	60 amp, 2 pole		7.50	1.067		340	42		382	440
	2850	3 pole		6.20	1.290		395	51		446	510
	2900	I-line, 30 amp, 1 pole		8	1		220	39.50		259.50	300
	2910	60 amp, 2 pole		7.50	1.067		315	42		357	410
	2920	3 pole		6.20	1.290		350	51		401	460
	2930	100 amp, 1 pole		7.50	1.067		355	42		397	455
	2940	2 pole		5	1.600		355	63		418	485
	2950	3 pole		3.60	2.222		395	87.50		482.50	565
	2960	Shunt trip, 277/480V breaker, remote oper., 30 amp, 1 pole		4	2		236	79		315	375
	2970	60 amp, 2 pole		3.50	2.286		345	90		435	515
	2980	3 pole		3	2.667		390	105		495	585
	2990	100 amp, 1 pole		3.50	2.286		277	90		367	440
	3000	2 pole		3	2.667		390	105		495	585
	3010	3 pole		2.50	3.200		430	126		556	660
	3050	Under voltage trip, 277/480 volt breaker, 30 amp, 1 pole		4	2		236	79		315	375
	3060	60 amp, 2 pole		3.50	2.286		345	90		435	515
	3070	3 pole		3	2.667		390	105		495	585
	3080	100 amp, 1 pole		3.50	2.286		277	90		367	440
	3090	2 pole		3	2.667		390	105		495	585
	3100	3 pole		2.50	3.200		430	126		556	660
	3150	Motor operated, 277/480 volt breaker, 30 amp, 1 pole		4	2		420	79		499	575
	3160	60 amp, 2 pole		3.50	2.286		530	90		620	720
	3170	3 pole		3	2.667		575	105		680	790
	3180	100 amp, 1 pole		3.50	2.286		460	90		550	640
	3190	2 pole		3	2.667		575	105		680	790
	3200	3 pole		2.50	3.200		610	126		736	865
	3250	Panelboard spacers, per pole		40	.200		3.18	7.90		11.08	15.20
720	0010	**PANELBOARDS** (Commercial use)										720
	0050	NQOD, w/20 amp 1 pole bolt-on circuit breakers									
	0100	3 wire, 120/240 volts, 100 amp main lugs									
	0150	10 circuits	1 Elec	1	8	Ea.	400	315		715	910

Note on rows 2420, 2700, 2800, 2960, 3250, 0050 — R16440-820, R16440-500 reference numbers appear in the description area.

Important: See the Reference Section for critical supporting data - Reference Nos., Crews, & City Cost Indexes

16440	Swbds, Panels & Control Centers	CREW	DAILY OUTPUT	LABOR-HOURS	UNIT	2004 BARE COSTS				TOTAL INCL O&P
						MAT.	LABOR	EQUIP.	TOTAL	
720	0200	14 circuits	R16440 -500	1 Elec	.88	9.091	Ea.	470	360		830	1,050	720

		CREW	DAILY OUTPUT	LABOR-HOURS	UNIT	MAT.	LABOR	EQUIP.	TOTAL	TOTAL INCL O&P
0200	14 circuits (R16440-500)	1 Elec	.88	9.091	Ea.	470	360		830	1,050
0250	18 circuits		.75	10.667		510	420		930	1,200
0300	20 circuits		.65	12.308		575	485		1,060	1,350
0350	225 amp main lugs, 24 circuits	2 Elec	1.20	13.333		655	525		1,180	1,500
0400	30 circuits		.90	17.778		760	700		1,460	1,900
0450	36 circuits		.80	20		870	790		1,660	2,125
0500	38 circuits		.72	22.222		935	875		1,810	2,325
0550	42 circuits		.66	24.242		980	955		1,935	2,500
0600	4 wire, 120/208 volts, 100 amp main lugs, 12 circuits	1 Elec	1	8		450	315		765	965
0650	16 circuits		.75	10.667		520	420		940	1,200
0700	20 circuits		.65	12.308		605	485		1,090	1,400
0750	24 circuits		.60	13.333		660	525		1,185	1,500
0800	30 circuits		.53	15.094		760	595		1,355	1,725
0850	225 amp main lugs, 32 circuits	2 Elec	.90	17.778		855	700		1,555	2,000
0900	34 circuits		.84	19.048		875	750		1,625	2,100
0950	36 circuits		.80	20		895	790		1,685	2,150
1000	42 circuits		.68	23.529		1,000	925		1,925	2,475
1040	225 amp main lugs, NEMA 7, 12 circuits		1	16		2,400	630		3,030	3,600
1100	24 circuits		.40	40		3,000	1,575		4,575	5,650
1200	NEHB,w/20 amp, 1 pole bolt-on circuit breakers									
1250	4 wire, 277/480 volts, 100 amp main lugs, 12 circuits	1 Elec	.88	9.091	Ea.	865	360		1,225	1,475
1300	20 circuits	"	.60	13.333		1,275	525		1,800	2,175
1350	225 amp main lugs, 24 circuits	2 Elec	.90	17.778		1,475	700		2,175	2,675
1400	30 circuits		.80	20		1,775	790		2,565	3,125
1450	36 circuits		.72	22.222		2,075	875		2,950	3,575
1500	42 circuits		.60	26.667		2,350	1,050		3,400	4,175
1510	225 amp main lugs, NEMA 7, 12 circuits		.90	17.778		3,475	700		4,175	4,850
1590	24 circuits		.30	53.333		4,200	2,100		6,300	7,725
1600	NQOD panel, w/20 amp, 1 pole, circuit breakers									
1650	3 wire, 120/240 volt with main circuit breaker									
1700	100 amp main, 12 circuits	1 Elec	.80	10	Ea.	555	395		950	1,200
1750	20 circuits	"	.60	13.333		710	525		1,235	1,550
1800	225 amp main, 30 circuits	2 Elec	.68	23.529		1,350	925		2,275	2,875
1850	42 circuits		.52	30.769		1,575	1,200		2,775	3,525
1900	400 amp main, 30 circuits		.54	29.630		1,875	1,175		3,050	3,800
1950	42 circuits		.50	32		2,100	1,250		3,350	4,175
2000	4 wire, 120/208 volts with main circuit breaker									
2050	100 amp main, 24 circuits	1 Elec	.47	17.021	Ea.	830	670		1,500	1,900
2100	30 circuits	"	.40	20		935	790		1,725	2,200
2200	225 amp main, 32 circuits	2 Elec	.72	22.222		1,575	875		2,450	3,050
2250	42 circuits		.56	28.571		1,725	1,125		2,850	3,575
2300	400 amp main, 42 circuits		.48	33.333		2,350	1,325		3,675	4,525
2350	600 amp main, 42 circuits		.40	40		3,475	1,575		5,050	6,175
2400	NEHB, with 20 amp, 1 pole circuit breaker									
2450	4 wire, 277/480 volts with main circuit breaker									
2500	100 amp main, 24 circuits	1 Elec	.42	19.048	Ea.	1,700	750		2,450	3,000
2550	30 circuits	"	.38	21.053		2,000	830		2,830	3,425
2600	225 amp main, 30 circuits	2 Elec	.72	22.222		2,525	875		3,400	4,075
2650	42 circuits		.56	28.571		3,100	1,125		4,225	5,100
2700	400 amp main, 42 circuits		.46	34.783		3,725	1,375		5,100	6,150
2750	600 amp main, 42 circuits		.38	42.105		5,125	1,650		6,775	8,125
2900	Note: the following line items don't include branch circuit breakers									
2910	For branch circuit breakers information, refer to									
2920	Division 16440-700									
3010	Main lug, no main breaker, 240 volt, 1 pole, 3 wire, 100 amp	1 Elec	2.30	3.478	Ea.	360	137		497	600
3020	225 amp	2 Elec	2.40	6.667		440	263		703	870

ELECTRICAL 16

16440 | Swbds, Panels & Control Centers

			CREW	DAILY OUTPUT	LABOR-HOURS	UNIT	2004 BARE COSTS				TOTAL INCL O&P	
							MAT.	LABOR	EQUIP.	TOTAL		
720	3030	400 amp	2 Elec	1.80	8.889	Ea.	655	350		1,005	1,250	720
	3060	3 pole, 3 wire, 100 amp	1 Elec	2.30	3.478		390	137		527	635	
	3070	225 amp	2 Elec	2.40	6.667		465	263		728	900	
	3080	400 amp		1.80	8.889		715	350		1,065	1,300	
	3090	600 amp	↓	1.60	10		830	395		1,225	1,500	
	3110	3 pole, 4 wire, 100 amp	1 Elec	2.30	3.478		435	137		572	685	
	3120	225 amp	2 Elec	2.40	6.667		550	263		813	995	
	3130	400 amp		1.80	8.889		715	350		1,065	1,300	
	3140	600 amp	↓	1.60	10		830	395		1,225	1,500	
	3160	480 volt, 3 pole, 3 wire, 100 amp	1 Elec	2.30	3.478		525	137		662	780	
	3170	225 amp	2 Elec	2.40	6.667		640	263		903	1,100	
	3180	400 amp		1.80	8.889		905	350		1,255	1,525	
	3190	600 amp	↓	1.60	10		1,000	395		1,395	1,675	
	3210	277/480 volt, 3 pole, 4 wire, 100 amp	1 Elec	2.30	3.478		510	137		647	765	
	3220	225 amp	2 Elec	2.40	6.667		625	263		888	1,075	
	3230	400 amp		1.80	8.889		880	350		1,230	1,500	
	3240	600 amp	↓	1.60	10		985	395		1,380	1,650	
	3260	Main circuit breaker, 240 volt, 1 pole, 3 wire, 100 amp	1 Elec	2	4		495	158		653	780	
	3270	225 amp	2 Elec	2	8		1,025	315		1,340	1,625	
	3280	400 amp	"	1.60	10		1,650	395		2,045	2,375	
	3310	3 pole, 3 wire, 100 amp	1 Elec	2	4		570	158		728	860	
	3320	225 amp	2 Elec	2	8		1,200	315		1,515	1,775	
	3330	400 amp		1.60	10		1,875	395		2,270	2,650	
	3360	120/208 volt, 3 pole, 4 wire, 100 amp		4	4		570	158		728	860	
	3370	225 amp		2	8		1,200	315		1,515	1,775	
	3380	400 amp	↓	1.60	10		1,875	395		2,270	2,650	
	3410	480 volt, 3 pole, 3 wire, 100 amp	1 Elec	2	4		805	158		963	1,125	
	3420	225 amp	2 Elec	2	8		1,400	315		1,715	2,025	
	3430	400 amp		1.60	10		2,125	395		2,520	2,900	
	3460	277/480 volt, 3 pole, 4 wire, 100 amp		4	4		790	158		948	1,100	
	3470	225 amp		2	8		1,375	315		1,690	1,975	
	3480	400 amp	↓	1.60	10		2,125	395		2,520	2,925	
	3510	Main circuit breaker, HIC, 240 volt, 1 pole, 3 wire, 100 amp	1 Elec	2	4		765	158		923	1,075	
	3520	225 amp	2 Elec	2	8		2,050	315		2,365	2,750	
	3530	400 amp	"	1.60	10		2,775	395		3,170	3,625	
	3560	3 pole, 3 wire, 100 amp	1 Elec	2	4		860	158		1,018	1,175	
	3570	225 amp	2 Elec	2	8		2,325	315		2,640	3,025	
	3580	400 amp	"	1.60	10		3,075	395		3,470	3,975	
	3610	120/208 volt, 3 pole, 4 wire, 100 amp	1 Elec	2	4		860	158		1,018	1,175	
	3620	225 amp	2 Elec	2	8		2,325	315		2,640	3,025	
	3630	400 amp	"	1.60	10		3,075	395		3,470	3,975	
	3660	480 volt, 3 pole, 3 wire, 100 amp	1 Elec	2	4		1,300	158		1,458	1,650	
	3670	225 amp	2 Elec	2	8		2,575	315		2,890	3,325	
	3680	400 amp	"	1.60	10		3,300	395		3,695	4,200	
	3710	277/480 volt, 3 pole, 4 wire, 100 amp	1 Elec	2	4		1,225	158		1,383	1,575	
	3720	225 amp	2 Elec	2	8		2,450	315		2,765	3,175	
	3730	400 amp	"	1.60	10		3,275	395		3,670	4,175	
	3760	Main circuit breaker, shunt trip, 100 amp	1 Elec	1.20	6.667		640	263		903	1,100	
	3770	225 amp	2 Elec	1.60	10		1,525	395		1,920	2,250	
	3780	400 amp	"	1.40	11.429		2,200	450		2,650	3,100	
800	0010	**SWITCHBOARDS** distribution section										800
	0100	Aluminum bus bars, not including breakers										
	0160	Subfeed lug-rated at 60 amp	2 Elec	1.30	12.308	Ea.	1,375	485		1,860	2,225	
	0170	100 amp		1.26	12.698		1,375	500		1,875	2,250	
	0180	200 amp		1.20	13.333		1,375	525		1,900	2,275	
	0190	400 amp	↓	1.10	14.545		1,375	575		1,950	2,350	

Reference marks in table: R16440-500 (at row 3060), R16440-800 (at row 0010)

16 ELECTRICAL

Important: See the Reference Section for critical supporting data - Reference Nos., Crews, & City Cost Indexes

16440 | Swbds, Panels & Control Centers

			CREW	DAILY OUTPUT	LABOR-HOURS	UNIT	MAT.	LABOR	EQUIP.	TOTAL	TOTAL INCL O&P	
800	0200	120/208 or 277/480 volt, 4 wire, 600 amp R16440-800	2 Elec	1	16	Ea.	1,375	630		2,005	2,475	800
	0300	800 amp		.88	18.182		1,750	715		2,465	3,000	
	0400	1000 amp		.80	20		1,775	790		2,565	3,125	
	0500	1200 amp		.72	22.222		2,500	875		3,375	4,050	
	0600	1600 amp		.66	24.242		2,925	955		3,880	4,650	
	0700	2000 amp		.62	25.806		3,425	1,025		4,450	5,250	
	0800	2500 amp		.60	26.667		3,875	1,050		4,925	5,825	
	0900	3000 amp		.56	28.571		4,675	1,125		5,800	6,825	
	0950	4000 amp		.52	30.769		6,850	1,200		8,050	9,325	
820	0010	**SWITCHBOARDS** feeder section group mounted devices R16440-820										820
	0030	Circuit breakers										
	0160	FA frame, 15 to 60 amp, 240 volt, 1 pole	1 Elec	8	1	Ea.	75.50	39.50		115	142	
	0170	2 pole		7	1.143		151	45		196	233	
	0180	3 pole		5.30	1.509		218	59.50		277.50	330	
	0210	480 volt, 1 pole		8	1		95.50	39.50		135	164	
	0220	2 pole		7	1.143		248	45		293	340	
	0230	3 pole		5.30	1.509		315	59.50		374.50	440	
	0260	600 volt, 2 pole		7	1.143		297	45		342	390	
	0270	3 pole		5.30	1.509		370	59.50		429.50	495	
	0280	FA frame, 70 to 100 amp, 240 volt, 1 pole		7	1.143		99.50	45		144.50	177	
	0310	2 pole		5	1.600		235	63		298	350	
	0320	3 pole		4	2		268	79		347	410	
	0330	480 volt, 1 pole		7	1.143		119	45		164	198	
	0360	2 pole		5	1.600		315	63		378	445	
	0370	3 pole		4	2		380	79		459	530	
	0380	600 volt, 2 pole		5	1.600		360	63		423	490	
	0410	3 pole		4	2		445	79		524	600	
	0420	KA frame, 70 to 225 amp		3.20	2.500		835	98.50		933.50	1,075	
	0430	LA frame, 125 to 400 amp		2.30	3.478		1,875	137		2,012	2,275	
	0460	MA frame, 450 to 600 amp		1.60	5		3,100	197		3,297	3,725	
	0470	700 to 800 amp		1.30	6.154		4,050	242		4,292	4,800	
	0480	MAL frame, 1000 amp		1	8		4,200	315		4,515	5,100	
	0490	PA frame, 1200 amp		.80	10		8,525	395		8,920	9,950	
	0500	Branch circuit, fusible switch, 600 volt, double 30/30 amp		4	2		810	79		889	1,000	
	0550	60/60 amp		3.20	2.500		820	98.50		918.50	1,050	
	0600	100/100 amp		2.70	2.963		1,150	117		1,267	1,425	
	0650	Single, 30 amp		5.30	1.509		505	59.50		564.50	645	
	0700	60 amp		4.70	1.702		505	67		572	655	
	0750	100 amp		4	2		775	79		854	965	
	0800	200 amp		2.70	2.963		1,125	117		1,242	1,425	
	0850	400 amp		2.30	3.478		2,150	137		2,287	2,550	
	0900	600 amp		1.80	4.444		2,550	175		2,725	3,050	
	0950	800 amp		1.30	6.154		4,150	242		4,392	4,925	
	1000	1200 amp		.80	10		4,975	395		5,370	6,025	
	1080	Branch circuit, circuit breakers, high interrupting capacity										
	1100	60 amp, 240, 480 or 600 volt, 1 pole	1 Elec	8	1	Ea.	193	39.50		232.50	272	
	1120	2 pole		7	1.143		460	45		505	575	
	1140	3 pole		5.30	1.509		550	59.50		609.50	695	
	1150	100 amp, 240, 480 or 600 volt, 1 pole		7	1.143		214	45		259	305	
	1160	2 pole		5	1.600		555	63		618	705	
	1180	3 pole		4	2		620	79		699	800	
	1200	225 amp, 240, 480 or 600 volt, 2 pole		3.50	2.286		1,925	90		2,015	2,250	
	1220	3 pole		3.20	2.500		2,350	98.50		2,448.50	2,725	
	1240	400 amp, 240, 480 or 600 volt, 2 pole		2.50	3.200		2,225	126		2,351	2,650	
	1260	3 pole		2.30	3.478		3,100	137		3,237	3,625	

ELECTRICAL 16

		16440	Swbds, Panels & Control Centers	CREW	DAILY OUTPUT	LABOR-HOURS	UNIT	MAT.	LABOR	EQUIP.	TOTAL	TOTAL INCL O&P	
820	1280		600 amp, 240, 480 or 600 volt, 2 pole _R16440 -820_	1 Elec	1.80	4.444	Ea.	3,150	175		3,325	3,700	820
	1300		3 pole		1.60	5		3,775	197		3,972	4,450	
	1320		800 amp, 240, 480 or 600 volt, 2 pole		1.50	5.333		3,550	210		3,760	4,225	
	1340		3 pole		1.30	6.154		4,800	242		5,042	5,650	
	1360		1000 amp, 240, 480 or 600 volt, 2 pole		1.10	7.273		6,975	287		7,262	8,100	
	1380		3 pole		1	8		7,625	315		7,940	8,850	
	1400		1200 amp, 240, 480 or 600 volt, 2 pole		.90	8.889		4,750	350		5,100	5,750	
	1420		3 pole		.80	10		8,175	395		8,570	9,575	
	1700		Fusible switch, 240 V, 60 amp, 2 pole		3.20	2.500		287	98.50		385.50	460	
	1720		3 pole		3	2.667		395	105		500	590	
	1740		100 amp, 2 pole		2.70	2.963		291	117		408	495	
	1760		3 pole		2.50	3.200		440	126		566	675	
	1780		200 amp, 2 pole		2	4		610	158		768	905	
	1800		3 pole		1.90	4.211		710	166		876	1,025	
	1820		400 amp, 2 pole		1.50	5.333		1,150	210		1,360	1,575	
	1840		3 pole		1.30	6.154		1,425	242		1,667	1,925	
	1860		600 amp, 2 pole		1	8		1,625	315		1,940	2,275	
	1880		3 pole		.90	8.889		2,000	350		2,350	2,725	
	1900		240-600 V, 800 amp, 2 pole		.70	11.429		3,625	450		4,075	4,650	
	1920		3 pole		.60	13.333		4,425	525		4,950	5,650	
	2000		600 V, 60 amp, 2 pole		3.20	2.500		435	98.50		533.50	620	
	2040		100 amp, 2 pole		2.70	2.963		445	117		562	665	
	2080		200 amp, 2 pole		2	4		775	158		933	1,100	
	2120		400 amp, 2 pole		1.50	5.333		1,550	210		1,760	2,025	
	2160		600 amp, 2 pole		1	8		1,875	315		2,190	2,525	
	2500		Branch circuit, circuit breakers, 60 amp, 600 volt, 3 pole		5.30	1.509		415	59.50		474.50	545	
	2520		240, 480 or 600 volt, 1 pole		8	1		67	39.50		106.50	133	
	2540		240 volt, 2 pole		7	1.143		115	45		160	194	
	2560		480 or 600 volt, 2 pole		7	1.143		244	45		289	335	
	2580		240 volt, 3 pole		5.30	1.509		187	59.50		246.50	295	
	2600		480 volt, 3 pole		5.30	1.509		325	59.50		384.50	450	
	2620		100 amp, 600 volt, 2 pole		5	1.600		310	63		373	440	
	2640		3 pole		4	2		395	79		474	550	
	2660		480 volt, 2 pole		5	1.600		269	63		332	390	
	2680		240 volt, 2 pole		5	1.600		154	63		217	263	
	2700		3 pole		4	2		247	79		326	390	
	2720		480 volt, 3 pole		4	2		360	79		439	515	
	2740		225 amp, 240, 480 or 600 volt, 2 pole		3.50	2.286		410	90		500	585	
	2760		3 pole		3.20	2.500		470	98.50		568.50	660	
	2780		400 amp, 240, 480 or 600 volt, 2 pole		2.50	3.200		945	126		1,071	1,250	
	2800		3 pole		2.30	3.478		1,100	137		1,237	1,400	
	2820		600 amp, 240 or 480 volt, 2 pole		1.80	4.444		1,525	175		1,700	1,950	
	2840		3 pole		1.60	5		1,875	197		2,072	2,375	
	2860		800 amp, 240, 480 volt or 600 volt, 2 pole		1.50	5.333		2,300	210		2,510	2,875	
	2880		3 pole		1.30	6.154		2,700	242		2,942	3,325	
	2900		1000 amp, 240, 480 or 600 volt, 2 pole		1.10	7.273		2,825	287		3,112	3,550	
	2920		480 volt, 600 volt, 3 pole		1	8		3,250	315		3,565	4,050	
	2940		1200 amp, 240, 480 or 600 volt, 2 pole		.90	8.889		4,000	350		4,350	4,925	
	2960		3 pole		.80	10		4,350	395		4,745	5,375	
	2980		600 volt, 3 pole		.80	10		4,350	395		4,745	5,375	
840	0010		**SWITCHBOARDS** Incoming main service section _R16440 -840_										840
	0100		Aluminum bus bars, not including CT's or PT's										
	0200		No main disconnect, includes CT compartment										
	0300		120/208 volt, 4 wire, 600 amp	2 Elec	1	16	Ea.	2,750	630		3,380	3,975	
	0400		800 amp		.88	18.182		2,750	715		3,465	4,100	
	0500		1000 amp		.80	20		3,325	790		4,115	4,825	

Important: See the Reference Section for critical supporting data - Reference Nos., Crews, & City Cost Indexes

		16440	Swbds, Panels & Control Centers	CREW	DAILY OUTPUT	LABOR-HOURS	UNIT	2004 BARE COSTS				TOTAL INCL O&P	
								MAT.	LABOR	EQUIP.	TOTAL		
840	0600		1200 amp	2 Elec	.72	22.222	Ea.	3,325	875		4,200	4,950	**840**
	0700		1600 amp		.66	24.242		3,325	955		4,280	5,075	
	0800		2000 amp		.62	25.806		3,575	1,025		4,600	5,425	
	1000		3000 amp		.56	28.571		4,700	1,125		5,825	6,850	
	1200	277/480 volt, 4 wire, 600 amp			1	16		2,750	630		3,380	3,975	
	1300		800 amp		.88	18.182		2,750	715		3,465	4,100	
	1400		1000 amp		.80	20		3,500	790		4,290	5,025	
	1500		1200 amp		.72	22.222		3,500	875		4,375	5,150	
	1600		1600 amp		.66	24.242		3,500	955		4,455	5,275	
	1700		2000 amp		.62	25.806		3,575	1,025		4,600	5,425	
	1800		3000 amp		.56	28.571		4,700	1,125		5,825	6,850	
	1900		4000 amp		.52	30.769		5,825	1,200		7,025	8,225	
	2000	Fused switch & CT compartment											
	2100		120/208 volt, 4 wire, 400 amp	2 Elec	1.12	14.286	Ea.	3,550	565		4,115	4,725	
	2200		600 amp		.94	17.021		4,200	670		4,870	5,625	
	2300		800 amp		.84	19.048		7,450	750		8,200	9,325	
	2400		1200 amp		.68	23.529		9,700	925		10,625	12,100	
	2500		277/480 volt, 4 wire, 400 amp		1.14	14.035		3,725	555		4,280	4,925	
	2600		600 amp		.94	17.021		4,250	670		4,920	5,675	
	2700		800 amp		.84	19.048		7,450	750		8,200	9,325	
	2800		1200 amp		.68	23.529		9,700	925		10,625	12,100	
	2900	Pressure switch & CT compartment											
	3000		120/208 volt, 4 wire, 800 amp	2 Elec	.80	20	Ea.	6,700	790		7,490	8,550	
	3100		1200 amp		.66	24.242		13,000	955		13,955	15,700	
	3200		1600 amp		.62	25.806		13,800	1,025		14,825	16,700	
	3300		2000 amp		.56	28.571		14,700	1,125		15,825	17,800	
	3310		2500 amp		.50	32		18,000	1,250		19,250	21,700	
	3320		3000 amp		.44	36.364		24,300	1,425		25,725	28,800	
	3330		4000 amp		.40	40		31,200	1,575		32,775	36,700	
	3400		277/480 volt, 4 wire, 800 amp, with ground fault		.80	20		11,100	790		11,890	13,400	
	3600		1200 amp, with ground fault		.66	24.242		14,300	955		15,255	17,100	
	4000		1600 amp, with ground fault		.62	25.806		15,500	1,025		16,525	18,600	
	4200		2000 amp, with ground fault		.56	28.571		16,800	1,125		17,925	20,100	
	4400	Circuit breaker, molded case & CT compartment											
	4600		3 pole, 4 wire, 600 amp	2 Elec	.94	17.021	Ea.	5,775	670		6,445	7,350	
	4800		800 amp		.84	19.048		6,900	750		7,650	8,725	
	5000		1200 amp		.68	23.529		9,425	925		10,350	11,800	
	5100	Copper bus bars, not incl. CT's or PT's, add, minimum						15%					
860	0010	**SWITCHBOARDS** (in plant distribution)											**860**
	0100	Main lugs only, to 600 volt, 3 pole, 3 wire, 200 amp		2 Elec	1.20	13.333	Ea.	1,525	525		2,050	2,450	
	0110		400 amp		1.20	13.333		1,525	525		2,050	2,450	
	0120		600 amp		1.20	13.333		1,550	525		2,075	2,500	
	0130		800 amp		1.08	14.815		1,700	585		2,285	2,725	
	0140		1200 amp		.92	17.391		2,050	685		2,735	3,275	
	0150		1600 amp		.86	18.605		2,700	735		3,435	4,050	
	0160		2000 amp		.82	19.512		2,975	770		3,745	4,425	
	0250		To 480 volt, 3 pole, 4 wire, 200 amp		1.20	13.333		1,325	525		1,850	2,225	
	0260		400 amp		1.20	13.333		1,525	525		2,050	2,450	
	0270		600 amp		1.20	13.333		1,650	525		2,175	2,575	
	0280		800 amp		1.08	14.815		1,800	585		2,385	2,850	
	0290		1200 amp		.92	17.391		2,225	685		2,910	3,475	
	0300		1600 amp		.86	18.605		2,525	735		3,260	3,900	
	0310		2000 amp		.82	19.512		2,950	770		3,720	4,400	
	0400	Main circuit breaker, to 600 volt, 3 pole, 3 wire, 200 amp		1.20	13.333		2,100	525		2,625	3,100		
	0410		400 amp		1.14	14.035		2,100	555		2,655	3,150	
	0420		600 amp		1.10	14.545		2,675	575		3,250	3,800	

R16440
-840

ELECTRICAL 16

	16440	Swbds, Panels & Control Centers	CREW	DAILY OUTPUT	LABOR-HOURS	UNIT	2004 BARE COSTS				TOTAL INCL O&P	
							MAT.	LABOR	EQUIP.	TOTAL		
860	0430	800 amp	2 Elec	1.04	15.385	Ea.	4,475	605		5,080	5,825	860
	0440	1200 amp		.88	18.182		5,850	715		6,565	7,500	
	0450	1600 amp		.84	19.048		9,375	750		10,125	11,400	
	0460	2000 amp		.80	20		10,000	790		10,790	12,200	
	0550	277/480 volt, 3 pole, 4 wire, 200 amp		1.20	13.333		2,225	525		2,750	3,225	
	0560	400 amp		1.14	14.035		2,225	555		2,780	3,275	
	0570	600 amp		1.10	14.545		2,775	575		3,350	3,900	
	0580	800 amp		1.04	15.385		4,675	605		5,280	6,050	
	0590	1200 amp		.88	18.182		6,050	715		6,765	7,725	
	0600	1600 amp		.84	19.048		9,375	750		10,125	11,400	
	0610	2000 amp		.80	20		10,000	790		10,790	12,200	
	0700	Main fusible switch w/fuse, 208/240 volt, 3 pole, 3 wire, 200 amp		1.20	13.333		2,325	525		2,850	3,325	
	0710	400 amp		1.14	14.035		2,325	555		2,880	3,375	
	0720	600 amp		1.10	14.545		2,825	575		3,400	3,950	
	0730	800 amp		1.04	15.385		5,800	605		6,405	7,275	
	0740	1200 amp		.88	18.182		6,725	715		7,440	8,475	
	0800	120/208, 120/240 volt, 3 pole, 4 wire, 200 amp		1.20	13.333		2,075	525		2,600	3,050	
	0810	400 amp		1.14	14.035		2,075	555		2,630	3,100	
	0820	600 amp		1.10	14.545		2,700	575		3,275	3,825	
	0830	800 amp		1.04	15.385		4,300	605		4,905	5,625	
	0840	1200 amp		.88	18.182		4,975	715		5,690	6,550	
	0900	480 or 600 volt, 3 pole, 3 wire, 200 amp		1.20	13.333		2,225	525		2,750	3,225	
	0910	400 amp		1.14	14.035		2,225	555		2,780	3,275	
	0920	600 amp		1.10	14.545		2,675	575		3,250	3,800	
	0930	800 amp		1.04	15.385		4,125	605		4,730	5,450	
	0940	1200 amp		.88	18.182		4,800	715		5,515	6,350	
	1000	277 or 480 volt, 3 pole, 4 wire, 200 amp		1.20	13.333		2,325	525		2,850	3,325	
	1010	400 amp		1.14	14.035		2,325	555		2,880	3,375	
	1020	600 amp		1.10	14.545		2,800	575		3,375	3,925	
	1030	800 amp		1.04	15.385		4,300	605		4,905	5,625	
	1040	1200 amp		.88	18.182		4,975	715		5,690	6,550	
	1120	1600 amp		.76	21.053		9,100	830		9,930	11,200	
	1130	2000 amp		.68	23.529		12,000	925		12,925	14,600	
	1150	Pressure switch, bolted, 3 pole, 208/240 volt, 3 wire, 800 amp		.96	16.667		6,750	655		7,405	8,400	
	1160	1200 amp		.80	20		8,600	790		9,390	10,700	
	1170	1600 amp		.76	21.053		9,825	830		10,655	12,000	
	1180	2000 amp		.68	23.529		11,200	925		12,125	13,700	
	1200	120/208 or 120/240 volt, 3 pole, 4 wire, 800 amp		.96	16.667		5,325	655		5,980	6,825	
	1210	1200 amp		.80	20		6,225	790		7,015	8,025	
	1220	1600 amp		.76	21.053		9,825	830		10,655	12,000	
	1230	2000 amp		.68	23.529		11,200	925		12,125	13,700	
	1300	480 or 600 volt, 3 wire, 800 amp		.96	16.667		6,750	655		7,405	8,400	
	1310	1200 amp		.80	20		9,450	790		10,240	11,600	
	1320	1600 amp		.76	21.053		10,600	830		11,430	12,900	
	1330	2000 amp		.68	23.529		12,000	925		12,925	14,600	
	1400	277-480 volt, 4 wire, 800 amp		.96	16.667		6,750	655		7,405	8,400	
	1410	1200 amp		.80	20		9,450	790		10,240	11,600	
	1420	1600 amp		.76	21.053		10,600	830		11,430	12,900	
	1430	2000 amp		.68	23.529		12,000	925		12,925	14,600	
	1500	Main ground fault protector, 1200-2000 amp		5.40	2.963		4,325	117		4,442	4,925	
	1600	Busway connection, 200 amp		5.40	2.963		325	117		442	530	
	1610	400 amp		4.60	3.478		325	137		462	560	
	1620	600 amp		4	4		325	158		483	590	
	1630	800 amp		3.20	5		325	197		522	650	
	1640	1200 amp		2.60	6.154		325	242		567	715	
	1650	1600 amp		2.40	6.667		670	263		933	1,125	

16440 | Swbds, Panels & Control Centers

		CREW	DAILY OUTPUT	LABOR-HOURS	UNIT	2004 BARE COSTS				TOTAL INCL O&P
						MAT.	LABOR	EQUIP.	TOTAL	
1660	2000 amp	2 Elec	2	8	Ea.	670	315		985	1,200
1700	Shunt trip for remote operation 200 amp		8	2		860	79		939	1,050
1710	400 amp		8	2		1,375	79		1,454	1,650
1720	600 amp		8	2		1,625	79		1,704	1,900
1730	800 amp		8	2		2,100	79		2,179	2,425
1740	1200-2000 amp		8	2		4,425	79		4,504	5,000
1800	Motor operated main breaker 200 amp		8	2		2,225	79		2,304	2,575
1810	400 amp		8	2		2,225	79		2,304	2,575
1820	600 amp		8	2		2,275	79		2,354	2,625
1830	800 amp		8	2		2,675	79		2,754	3,075
1840	1200-2000 amp		8	2		4,500	79		4,579	5,075
1900	Current/potential transformer metering compartment 200-800 amp		5.40	2.963		1,450	117		1,567	1,775
1940	1200 amp		5.40	2.963		2,000	117		2,117	2,375
1950	1600-2000 amp		5.40	2.963		2,000	117		2,117	2,375
2000	With watt meter 200-800 amp		4	4		5,700	158		5,858	6,500
2040	1200 amp		4	4		7,050	158		7,208	7,975
2050	1600-2000 amp	▼	4	4		7,050	158		7,208	7,975
2100	Split bus 60-200 amp	1 Elec	5.30	1.509		260	59.50		319.50	375
2130	400 amp	2 Elec	4.60	3.478		450	137		587	700
2140	600 amp		3.60	4.444		550	175		725	865
2150	800 amp		2.60	6.154		700	242		942	1,125
2170	1200 amp	▼	2	8		800	315		1,115	1,350
2250	Contactor control 60 amp	1 Elec	2	4		1,700	158		1,858	2,100
2260	100 amp		1.50	5.333		1,925	210		2,135	2,425
2270	200 amp	▼	1	8		2,875	315		3,190	3,650
2280	400 amp	2 Elec	1	16		8,775	630		9,405	10,600
2290	600 amp		.84	19.048		9,825	750		10,575	11,900
2300	800 amp		.72	22.222		11,600	875		12,475	14,100
2500	Modifier, two distribution sections, add		.80	20		2,500	790		3,290	3,925
2520	Three distribution sections, add		.40	40		5,625	1,575		7,200	8,525
2560	Auxiliary pull section, 20", add		2	8		1,175	315		1,490	1,750
2580	24", add		1.80	8.889		1,175	350		1,525	1,800
2600	30", add		1.60	10		1,175	395		1,570	1,850
2620	36", add		1.40	11.429		1,400	450		1,850	2,200
2640	Dog house, 12", add		2.40	6.667		300	263		563	720
2660	18", add	▼	2	8	▼	600	315		915	1,125
3000	Transition section between switchboard and transformer									
3050	or motor control center, 4 wire alum. bus, 600 amp	2 Elec	1.14	14.035	Ea.	1,975	555		2,530	3,000
3100	800 amp		1	16		2,225	630		2,855	3,400
3150	1000 amp		.88	18.182		2,500	715		3,215	3,825
3200	1200 amp		.80	20		2,750	790		3,540	4,200
3250	1600 amp		.72	22.222		3,250	875		4,125	4,875
3300	2000 amp		.66	24.242		3,725	955		4,680	5,525
3350	2500 amp		.62	25.806		4,350	1,025		5,375	6,275
3400	3000 amp		.56	28.571		5,000	1,125		6,125	7,175
4000	Weatherproof construction, per vertical section	▼	1.76	9.091	▼	2,725	360		3,085	3,525

16450 | Enclosed Bus Assemblies

		CREW	DAILY OUTPUT	LABOR-HOURS	UNIT	2004 BARE COSTS				TOTAL INCL O&P
						MAT.	LABOR	EQUIP.	TOTAL	
0010	**ALUMINUM BUS DUCT** 10 ft. long									
0050	3 pole 4 wire, plug-in/indoor, straight section, 225 amp	2 Elec	44	.364	L.F.	124	14.35		138.35	158
0100	400 amp		36	.444		124	17.50		141.50	162
0150	600 amp		32	.500		124	19.70		143.70	166
0200	800 amp		26	.615		143	24.50		167.50	193
0250	1000 amp		24	.667		162	26.50		188.50	217
0300	1350 amp		22	.727		247	28.50		275.50	315
0310	1600 amp	▼	18	.889	▼	285	35		320	365

R16450 -100

860

100

	16450	Enclosed Bus Assemblies	CREW	DAILY OUTPUT	LABOR-HOURS	UNIT	2004 BARE COSTS				TOTAL INCL O&P	
							MAT.	LABOR	EQUIP.	TOTAL		
100	0320	2000 amp	2 Elec	16	1	L.F.	335	39.50		374.50	425	100
	0330	2500 amp	R16450 -100	14	1.143		430	45		475	535	
	0340	3000 amp		12	1.333		485	52.50		537.50	615	
	0350	Feeder, 600 amp		34	.471		114	18.55		132.55	154	
	0400	800 amp		28	.571		133	22.50		155.50	181	
	0450	1000 amp		26	.615		152	24.50		176.50	203	
	0455	1200 amp		25	.640		152	25		177	205	
	0500	1350 amp		24	.667		238	26.50		264.50	300	
	0550	1600 amp		20	.800		276	31.50		307.50	350	
	0600	2000 amp		18	.889		320	35		355	400	
	0620	2500 amp		14	1.143		420	45		465	525	
	0630	3000 amp		12	1.333		475	52.50		527.50	605	
	0640	4000 amp		10	1.600	▼	685	63		748	850	
	0650	Elbow, 225 amp		4.40	3.636	Ea.	675	143		818	955	
	0700	400 amp		3.80	4.211		675	166		841	985	
	0750	600 amp		3.40	4.706		675	185		860	1,025	
	0800	800 amp		3	5.333		700	210		910	1,100	
	0850	1000 amp		2.80	5.714		910	225		1,135	1,325	
	0870	1200 amp		2.70	5.926		1,125	233		1,358	1,575	
	0900	1350 amp		2.60	6.154		1,200	242		1,442	1,675	
	0950	1600 amp		2.40	6.667		1,425	263		1,688	1,975	
	1000	2000 amp		2	8		1,575	315		1,890	2,200	
	1020	2500 amp		1.80	8.889		1,875	350		2,225	2,575	
	1030	3000 amp		1.60	10		2,150	395		2,545	2,950	
	1040	4000 amp		1.40	11.429		3,475	450		3,925	4,500	
	1100	Cable tap box end, 225 amp		3.60	4.444		945	175		1,120	1,300	
	1150	400 amp		3.20	5		945	197		1,142	1,350	
	1200	600 amp		2.60	6.154		945	242		1,187	1,400	
	1250	800 amp		2.20	7.273		1,025	287		1,312	1,550	
	1300	1000 amp		2	8		1,125	315		1,440	1,700	
	1320	1200 amp		2	8		1,325	315		1,640	1,950	
	1350	1350 amp		1.60	10		1,450	395		1,845	2,175	
	1400	1600 amp		1.40	11.429		1,650	450		2,100	2,475	
	1450	2000 amp		1.20	13.333		1,875	525		2,400	2,850	
	1460	2500 amp		1	16		2,300	630		2,930	3,475	
	1470	3000 amp		.80	20		2,525	790		3,315	3,950	
	1480	4000 amp		.60	26.667		3,375	1,050		4,425	5,275	
	1500	Switchboard stub, 225 amp		5.80	2.759		830	109		939	1,075	
	1550	400 amp		5.40	2.963		830	117		947	1,100	
	1600	600 amp		4.60	3.478		830	137		967	1,125	
	1650	800 amp		4	4		940	158		1,098	1,250	
	1700	1000 amp		3.20	5		1,050	197		1,247	1,450	
	1720	1200 amp		3.10	5.161		1,175	203		1,378	1,575	
	1750	1350 amp		3	5.333		1,300	210		1,510	1,750	
	1800	1600 amp		2.60	6.154		1,450	242		1,692	1,950	
	1850	2000 amp		2.40	6.667		1,650	263		1,913	2,225	
	1860	2500 amp		2.20	7.273		1,975	287		2,262	2,600	
	1870	3000 amp		2	8		2,300	315		2,615	3,000	
	1880	4000 amp		1.80	8.889		2,875	350		3,225	3,700	
	1890	Tee fittings, 225 amp		3.20	5		900	197		1,097	1,275	
	1900	400 amp		2.80	5.714		900	225		1,125	1,325	
	1950	600 amp		2.60	6.154		900	242		1,142	1,350	
	2000	800 amp		2.40	6.667		960	263		1,223	1,450	
	2050	1000 amp		2.20	7.273		1,025	287		1,312	1,550	
	2070	1200 amp		2.10	7.619		1,200	300		1,500	1,750	
	2100	1350 amp		2	8		1,650	315		1,965	2,275	

Important: See the Reference Section for critical supporting data - Reference Nos., Crews, & City Cost Indexes

			CREW	DAILY OUTPUT	LABOR-HOURS	UNIT	2004 BARE COSTS				TOTAL INCL O&P	
16450	**Enclosed Bus Assemblies**						MAT.	LABOR	EQUIP.	TOTAL		
100	2150	1600 amp	2 Elec	1.60	10	Ea.	1,975	395		2,370	2,750	100
	2200	2000 amp	R16450-100	1.20	13.333		2,200	525		2,725	3,200	
	2220	2500 amp		1	16		2,625	630		3,255	3,825	
	2230	3000 amp		.80	20		3,000	790		3,790	4,475	
	2240	4000 amp		.60	26.667		4,975	1,050		6,025	7,050	
	2300	Wall flange, 600 amp		20	.800		112	31.50		143.50	170	
	2310	800 amp		16	1		112	39.50		151.50	182	
	2320	1000 amp		13	1.231		112	48.50		160.50	195	
	2325	1200 amp		12	1.333		112	52.50		164.50	201	
	2330	1350 amp		10.80	1.481		112	58.50		170.50	210	
	2340	1600 amp		9	1.778		112	70		182	227	
	2350	2000 amp		8	2		112	79		191	240	
	2360	2500 amp		6.60	2.424		112	95.50		207.50	265	
	2370	3000 amp		5.40	2.963		112	117		229	297	
	2380	4000 amp		4	4		112	158		270	355	
	2390	5000 amp		3	5.333		112	210		322	440	
	2400	Vapor barrier		8	2		290	79		369	435	
	2420	Roof flange kit		4	4		550	158		708	840	
	2600	Expansion fitting, 225 amp		10	1.600		1,100	63		1,163	1,325	
	2610	400 amp		8	2		1,100	79		1,179	1,350	
	2620	600 amp		6	2.667		1,100	105		1,205	1,375	
	2630	800 amp		4.60	3.478		1,275	137		1,412	1,625	
	2640	1000 amp		4	4		1,425	158		1,583	1,800	
	2650	1350 amp		3.60	4.444		2,100	175		2,275	2,550	
	2660	1600 amp		3.20	5		2,525	197		2,722	3,075	
	2670	2000 amp		2.80	5.714		2,800	225		3,025	3,400	
	2680	2500 amp		2.40	6.667		3,375	263		3,638	4,125	
	2690	3000 amp		2	8		3,900	315		4,215	4,750	
	2700	4000 amp		1.60	10		5,125	395		5,520	6,225	
	2800	Reducer nonfused, 400 amp		8	2		885	79		964	1,100	
	2810	600 amp		6	2.667		885	105		990	1,125	
	2820	800 amp		4.60	3.478		1,075	137		1,212	1,375	
	2830	1000 amp		4	4		1,225	158		1,383	1,575	
	2840	1350 amp		3.60	4.444		1,625	175		1,800	2,050	
	2850	1600 amp		3.20	5		2,200	197		2,397	2,725	
	2860	2000 amp		2.80	5.714		2,525	225		2,750	3,100	
	2870	2500 amp		2.40	6.667		3,175	263		3,438	3,900	
	2880	3000 amp		2	8		3,675	315		3,990	4,525	
	2890	4000 amp		1.60	10		4,900	395		5,295	5,975	
	2950	Reducer fuse included, 225 amp		4.40	3.636		2,600	143		2,743	3,075	
	2960	400 amp		4.20	3.810		2,650	150		2,800	3,125	
	2970	600 amp		3.60	4.444		3,125	175		3,300	3,675	
	2980	800 amp		3.20	5		4,950	197		5,147	5,750	
	2990	1000 amp		3	5.333		5,675	210		5,885	6,550	
	3000	1200 amp		2.80	5.714		5,675	225		5,900	6,550	
	3010	1600 amp		2.20	7.273		12,900	287		13,187	14,600	
	3020	2000 amp		1.80	8.889		14,300	350		14,650	16,300	
	3100	Reducer circuit breaker, 225 amp		4.40	3.636		2,550	143		2,693	3,025	
	3110	400 amp		4.20	3.810		3,100	150		3,250	3,650	
	3120	600 amp		3.60	4.444		4,425	175		4,600	5,100	
	3130	800 amp		3.20	5		5,175	197		5,372	6,000	
	3140	1000 amp		3	5.333		5,875	210		6,085	6,775	
	3150	1200 amp		2.80	5.714		7,075	225		7,300	8,100	
	3160	1600 amp		2.20	7.273		10,400	287		10,687	11,900	
	3170	2000 amp		1.80	8.889		11,400	350		11,750	13,100	
	3250	Reducer circuit breaker, 75,000 AIC, 225 amp		4.40	3.636		4,000	143		4,143	4,625	

ELECTRICAL 16

16450	Enclosed Bus Assemblies		CREW	DAILY OUTPUT	LABOR-HOURS	UNIT	2004 BARE COSTS				TOTAL INCL O&P
							MAT.	LABOR	EQUIP.	TOTAL	
3260	400 amp		2 Elec	4.20	3.810	Ea.	4,000	150		4,150	4,625
3270	600 amp	R16450 -100		3.60	4.444		5,350	175		5,525	6,125
3280	800 amp			3.20	5		5,875	197		6,072	6,750
3290	1000 amp			3	5.333		9,450	210		9,660	10,700
3300	1200 amp			2.80	5.714		9,450	225		9,675	10,700
3310	1600 amp			2.20	7.273		10,400	287		10,687	11,900
3320	2000 amp			1.80	8.889		11,400	350		11,750	13,100
3400	Reducer circuit breaker CLF 225 amp			4.40	3.636		4,100	143		4,243	4,750
3410	400 amp			4.20	3.810		4,875	150		5,025	5,600
3420	600 amp			3.60	4.444		7,300	175		7,475	8,275
3430	800 amp			3.20	5		7,625	197		7,822	8,675
3440	1000 amp			3	5.333		7,950	210		8,160	9,075
3450	1200 amp			2.80	5.714		10,400	225		10,625	11,700
3460	1600 amp			2.20	7.273		10,400	287		10,687	11,900
3470	2000 amp		▼	1.80	8.889	▼	11,400	350		11,750	13,100
3550	Ground bus added to bus duct, 225 amp			320	.050	L.F.	33.50	1.97		35.47	40
3560	400 amp			320	.050		33.50	1.97		35.47	40
3570	600 amp			280	.057		33.50	2.25		35.75	40.50
3580	800 amp			240	.067		33.50	2.63		36.13	41
3590	1000 amp			200	.080		33.50	3.15		36.65	41.50
3600	1350 amp			180	.089		33.50	3.50		37	42
3610	1600 amp			160	.100		33.50	3.94		37.44	43
3620	2000 amp			160	.100		33.50	3.94		37.44	43
3630	2500 amp			140	.114		33.50	4.50		38	43.50
3640	3000 amp			120	.133		33.50	5.25		38.75	45
3650	4000 amp			100	.160		33.50	6.30		39.80	46.50
3810	High short circuit, 400 amp			36	.444		114	17.50		131.50	152
3820	600 amp			32	.500		114	19.70		133.70	156
3830	800 amp			26	.615		133	24.50		157.50	183
3840	1000 amp			24	.667		143	26.50		169.50	196
3850	1350 amp			22	.727		190	28.50		218.50	252
3860	1600 amp			18	.889		219	35		254	293
3870	2000 amp			16	1		257	39.50		296.50	340
3880	2500 amp			14	1.143		430	45		475	535
3890	3000 amp		▼	12	1.333	▼	485	52.50		537.50	615
3920	Cross, 225 amp			5.60	2.857	Ea.	1,350	113		1,463	1,675
3930	400 amp			4.60	3.478		1,350	137		1,487	1,700
3940	600 amp			4	4		1,350	158		1,508	1,725
3950	800 amp			3.40	4.706		1,425	185		1,610	1,850
3960	1000 amp			3	5.333		1,500	210		1,710	1,975
3970	1350 amp			2.80	5.714		2,425	225		2,650	3,000
3980	1600 amp			2.20	7.273		2,875	287		3,162	3,600
3990	2000 amp			1.80	8.889		3,150	350		3,500	4,000
4000	2500 amp			1.60	10		3,750	395		4,145	4,700
4010	3000 amp			1.20	13.333		4,300	525		4,825	5,500
4020	4000 amp			1	16		6,575	630		7,205	8,175
4040	Cable tap box center, 225 amp			3.60	4.444		945	175		1,120	1,300
4050	400 amp			3.20	5		945	197		1,142	1,350
4060	600 amp			2.60	6.154		945	242		1,187	1,400
4070	800 amp			2.20	7.273		1,025	287		1,312	1,550
4080	1000 amp			2	8		1,125	315		1,440	1,700
4090	1350 amp			1.60	10		1,450	395		1,845	2,175
4100	1600 amp			1.40	11.429		1,650	450		2,100	2,475
4110	2000 amp			1.20	13.333		1,875	525		2,400	2,850
4120	2500 amp			1	16		2,300	630		2,930	3,475
4130	3000 amp		▼	.80	20	▼	2,525	790		3,315	3,950

Important: See the Reference Section for critical supporting data - Reference Nos., Crews, & City Cost Indexes

16 ELECTRICAL

	16450	Enclosed Bus Assemblies		CREW	DAILY OUTPUT	LABOR-HOURS	UNIT	2004 BARE COSTS				TOTAL INCL O&P	
								MAT.	LABOR	EQUIP.	TOTAL		
100	4140	4000 amp	R16450 -100	2 Elec	.60	26.667	Ea.	3,375	1,050		4,425	5,275	100
	4500	Weatherproof, feeder, 600 amp			30	.533	L.F.	137	21		158	182	
	4520	800 amp			24	.667		160	26.50		186.50	215	
	4540	1000 amp			22	.727		183	28.50		211.50	244	
	4550	1200 amp			21	.762		253	30		283	325	
	4560	1350 amp			20	.800		285	31.50		316.50	360	
	4580	1600 amp			17	.941		330	37		367	420	
	4600	2000 amp			16	1		390	39.50		429.50	485	
	4620	2500 amp			12	1.333		505	52.50		557.50	635	
	4640	3000 amp			10	1.600		570	63		633	725	
	4660	4000 amp			8	2		820	79		899	1,025	
	5000	3 pole, 3 wire, feeder, 600 amp			40	.400		105	15.75		120.75	139	
	5010	800 amp			32	.500		124	19.70		143.70	166	
	5020	1000 amp			30	.533		133	21		154	179	
	5025	1200 amp			29	.552		133	21.50		154.50	180	
	5030	1350 amp			28	.571		181	22.50		203.50	233	
	5040	1600 amp			24	.667		209	26.50		235.50	269	
	5050	2000 amp			20	.800		247	31.50		278.50	320	
	5060	2500 amp			16	1		345	39.50		384.50	435	
	5070	3000 amp			14	1.143		410	45		455	515	
	5080	4000 amp			12	1.333		495	52.50		547.50	625	
	5200	Plug-in type, 225 amp			50	.320		114	12.60		126.60	145	
	5210	400 amp			42	.381		114	15		129	149	
	5220	600 amp			36	.444		114	17.50		131.50	152	
	5230	800 amp			30	.533		133	21		154	179	
	5240	1000 amp			28	.571		143	22.50		165.50	191	
	5245	1200 amp			27	.593		162	23.50		185.50	213	
	5250	1350 amp			26	.615		190	24.50		214.50	245	
	5260	1600 amp			20	.800		219	31.50		250.50	288	
	5270	2000 amp			18	.889		257	35		292	335	
	5280	2500 amp			16	1		350	39.50		389.50	445	
	5290	3000 amp			14	1.143		420	45		465	525	
	5300	4000 amp			12	1.333		505	52.50		557.50	635	
	5330	High short circuit, 400 amp			42	.381		114	15		129	149	
	5340	600 amp			36	.444		114	17.50		131.50	152	
	5350	800 amp			30	.533		133	21		154	179	
	5360	1000 amp			28	.571		143	22.50		165.50	191	
	5370	1350 amp			26	.615		190	24.50		214.50	245	
	5380	1600 amp			20	.800		219	31.50		250.50	288	
	5390	2000 amp			18	.889		257	35		292	335	
	5400	2500 amp			16	1		405	39.50		444.50	505	
	5410	3000 amp			14	1.143		420	45		465	525	
	5440	Elbow, 225 amp			5	3.200	Ea.	580	126		706	830	
	5450	400 amp			4.40	3.636		580	143		723	855	
	5460	600 amp			4	4		580	158		738	875	
	5470	800 amp			3.40	4.706		620	185		805	955	
	5480	1000 amp			3.20	5		635	197		832	995	
	5485	1200 amp			3.10	5.161		675	203		878	1,050	
	5490	1350 amp			3	5.333		735	210		945	1,125	
	5500	1600 amp			2.80	5.714		1,125	225		1,350	1,550	
	5510	2000 amp			2.40	6.667		1,225	263		1,488	1,750	
	5520	2500 amp			2	8		1,525	315		1,840	2,150	
	5530	3000 amp			1.80	8.889		1,800	350		2,150	2,525	
	5540	4000 amp			1.60	10		2,425	395		2,820	3,225	
	5560	Tee fittings, 225 amp			3.60	4.444		810	175		985	1,150	
	5570	400 amp			3.20	5		810	197		1,007	1,175	

ELECTRICAL 16

16450 | Enclosed Bus Assemblies

		CREW	DAILY OUTPUT	LABOR-HOURS	UNIT	2004 BARE COSTS				TOTAL INCL O&P
						MAT.	LABOR	EQUIP.	TOTAL	
100 5580	600 amp	2 Elec	3	5.333	Ea.	810	210		1,020	1,200 100
5590	800 amp		2.80	5.714		865	225		1,090	1,275
5600	1000 amp		2.60	6.154		895	242		1,137	1,350
5605	1200 amp		2.50	6.400		955	252		1,207	1,425
5610	1350 amp		2.40	6.667		1,325	263		1,588	1,850
5620	1600 amp		1.80	8.889		1,550	350		1,900	2,250
5630	2000 amp		1.40	11.429		1,750	450		2,200	2,600
5640	2500 amp		1.20	13.333		2,175	525		2,700	3,150
5650	3000 amp		1	16		2,550	630		3,180	3,750
5660	4000 amp		.70	22.857		3,725	900		4,625	5,450
5680	Cross, 225 amp		6.40	2.500		1,300	98.50		1,398.50	1,575
5690	400 amp		5.40	2.963		1,300	117		1,417	1,600
5700	600 amp		4.60	3.478		1,300	137		1,437	1,625
5710	800 amp		4	4		1,375	158		1,533	1,750
5720	1000 amp		3.60	4.444		1,425	175		1,600	1,825
5730	1350 amp		3.20	5		2,125	197		2,322	2,625
5740	1600 amp		2.60	6.154		2,475	242		2,717	3,075
5750	2000 amp		2.20	7.273		2,700	287		2,987	3,400
5760	2500 amp		1.80	8.889		3,275	350		3,625	4,125
5770	3000 amp		1.40	11.429		3,900	450		4,350	4,975
5780	4000 amp		1.20	13.333		5,425	525		5,950	6,750
5800	Expansion fitting, 225 amp		11.60	1.379		950	54.50		1,004.50	1,125
5810	400 amp		9.20	1.739		950	68.50		1,018.50	1,150
5820	600 amp		7	2.286		950	90		1,040	1,175
5830	800 amp		5.20	3.077		1,125	121		1,246	1,400
5840	1000 amp		4.60	3.478		1,225	137		1,362	1,550
5850	1350 amp		4.20	3.810		1,500	150		1,650	1,875
5860	1600 amp		3.60	4.444		1,850	175		2,025	2,275
5870	2000 amp		3.20	5		2,200	197		2,397	2,700
5880	2500 amp		2.80	5.714		2,475	225		2,700	3,050
5890	3000 amp		2.40	6.667		2,900	263		3,163	3,575
5900	4000 amp		1.80	8.889		3,975	350		4,325	4,900
5940	Reducer, nonfused, 400 amp		9.20	1.739		760	68.50		828.50	935
5950	600 amp		7	2.286		760	90		850	970
5960	800 amp		5.20	3.077		800	121		921	1,050
5970	1000 amp		4.60	3.478		980	137		1,117	1,275
5980	1350 amp		4.20	3.810		1,450	150		1,600	1,825
5990	1600 amp		3.60	4.444		1,650	175		1,825	2,075
6000	2000 amp		3.20	5		1,925	197		2,122	2,425
6010	2500 amp		2.80	5.714		2,475	225		2,700	3,050
6020	3000 amp		2.20	7.273		2,900	287		3,187	3,600
6030	4000 amp		1.80	8.889		3,975	350		4,325	4,900
6050	Reducer, fuse included, 225 amp		5	3.200		2,000	126		2,126	2,400
6060	400 amp		4.80	3.333		2,650	131		2,781	3,100
6070	600 amp		4.20	3.810		3,375	150		3,525	3,950
6080	800 amp		3.60	4.444		5,250	175		5,425	6,000
6090	1000 amp		3.40	4.706		5,700	185		5,885	6,550
6100	1350 amp		3.20	5		10,400	197		10,597	11,800
6110	1600 amp		2.60	6.154		12,400	242		12,642	14,000
6120	2000 amp		2	8		14,300	315		14,615	16,300
6160	Reducer, circuit breaker, 225 amp		5	3.200		2,450	126		2,576	2,900
6170	400 amp		4.80	3.333		3,000	131		3,131	3,500
6180	600 amp		4.20	3.810		4,300	150		4,450	4,950
6190	800 amp		3.60	4.444		5,025	175		5,200	5,775
6200	1000 amp		3.40	4.706		5,700	185		5,885	6,550
6210	1350 amp		3.20	5		6,900	197		7,097	7,900

R16450-100

Important: See the Reference Section for critical supporting data - Reference Nos., Crews, & City Cost Indexes

16 ELECTRICAL

			CREW	DAILY OUTPUT	LABOR-HOURS	UNIT	2004 BARE COSTS				TOTAL INCL O&P	
							MAT.	LABOR	EQUIP.	TOTAL		
16450 | Enclosed Bus Assemblies												
100	6220	1600 amp	R16450 -100	2 Elec	2.60	6.154	Ea.	10,300	242		10,542	11,700	100
	6230	2000 amp			2	8		11,200	315		11,515	12,900	
	6270	Cable tap box center, 225 amp			4.20	3.810		910	150		1,060	1,225	
	6280	400 amp			3.60	4.444		910	175		1,085	1,250	
	6290	600 amp			3	5.333		910	210		1,120	1,325	
	6300	800 amp			2.60	6.154		995	242		1,237	1,450	
	6310	1000 amp			2.40	6.667		1,050	263		1,313	1,550	
	6320	1350 amp			1.80	8.889		1,275	350		1,625	1,925	
	6330	1600 amp			1.60	10		1,425	395		1,820	2,150	
	6340	2000 amp			1.40	11.429		1,625	450		2,075	2,475	
	6350	2500 amp			1.20	13.333		2,025	525		2,550	3,025	
	6360	3000 amp			1	16		2,300	630		2,930	3,500	
	6370	4000 amp			.70	22.857		2,725	900		3,625	4,350	
	6390	Cable tap box end, 225 amp			4.20	3.810		555	150		705	835	
	6400	400 amp			3.60	4.444		555	175		730	870	
	6410	600 amp			3	5.333		555	210		765	925	
	6420	800 amp			2.60	6.154		610	242		852	1,025	
	6430	1000 amp			2.40	6.667		655	263		918	1,125	
	6435	1200 amp			2.10	7.619		715	300		1,015	1,225	
	6440	1350 amp			1.80	8.889		790	350		1,140	1,400	
	6450	1600 amp			1.60	10		895	395		1,290	1,575	
	6460	2000 amp			1.40	11.429		1,025	450		1,475	1,800	
	6470	2500 amp			1.20	13.333		1,175	525		1,700	2,075	
	6480	3000 amp			1	16		1,400	630		2,030	2,475	
	6490	4000 amp			.70	22.857		1,675	900		2,575	3,175	
	7000	Weatherproof, feeder, 600 amp			34	.471	L.F.	135	18.55		153.55	176	
	7020	800 amp			28	.571		149	22.50		171.50	198	
	7040	1000 amp			26	.615		160	24.50		184.50	212	
	7050	1200 amp			25	.640		187	25		212	244	
	7060	1350 amp			24	.667		217	26.50		243.50	278	
	7080	1600 amp			20	.800		251	31.50		282.50	325	
	7100	2000 amp			18	.889		297	35		332	375	
	7120	2500 amp			14	1.143		410	45		455	515	
	7140	3000 amp			12	1.333		490	52.50		542.50	620	
	7160	4000 amp			10	1.600		595	63		658	750	
200	0010	**BUS DUCT** 100 amp and less, aluminum or copper, plug-in											200
	0080	Bus duct, 3 pole 3 wire, 100 amp		1 Elec	42	.190	L.F.	21	7.50		28.50	34	
	0110	Elbow			4	2	Ea.	53.50	79		132.50	176	
	0120	Tee			2	4		78	158		236	320	
	0130	Wall flange			8	1		6.90	39.50		46.40	66	
	0140	Ground kit			16	.500		14.70	19.70		34.40	45.50	
	0180	3 pole 4 wire, 100 amp			40	.200	L.F.	24	7.90		31.90	37.50	
	0200	Cable tap box			3.10	2.581	Ea.	65	102		167	223	
	0300	End closure			16	.500		8.65	19.70		28.35	39	
	0400	Elbow			4	2		53.50	79		132.50	176	
	0500	Tee			2	4		78	158		236	320	
	0600	Hangers			10	.800		5.20	31.50		36.70	52.50	
	0700	Circuit breakers, 15 to 50 amp, 1 pole			8	1		138	39.50		177.50	211	
	0800	15 to 60 amp, 2 pole			6.70	1.194		540	47		587	665	
	0900	3 pole			5.30	1.509		540	59.50		599.50	685	
	1000	60 to 100 amp, 1 pole			6.70	1.194		540	47		587	665	
	1100	70 to 100 amp, 2 pole			5.30	1.509		540	59.50		599.50	685	
	1200	3 pole			4.50	1.778		540	70		610	700	
	1220	Switch, nonfused, 3 pole, 4 wire			8	1		73.50	39.50		113	140	
	1240	Fused, 3 fuses, 4 wire, 30 amp			8	1		266	39.50		305.50	350	

ELECTRICAL 16

			DAILY	LABOR-		2004 BARE COSTS				TOTAL	
16450	**Enclosed Bus Assemblies**	CREW	OUTPUT	HOURS	UNIT	MAT.	LABOR	EQUIP.	TOTAL	INCL O&P	
200	1260	60 amp	1 Elec	5.30	1.509	Ea.	279	59.50		338.50	395	**200**
	1280	100 amp		4.50	1.778		465	70		535	615	
	1300	Plug, fusible, 3 pole 250 volt, 30 amp		5.30	1.509		266	59.50		325.50	380	
	1310	60 amp		5.30	1.509		315	59.50		374.50	435	
	1320	100 amp		4.50	1.778		465	70		535	615	
	1330	3 pole 480 volt, 30 amp		5.30	1.509		305	59.50		364.50	425	
	1340	60 amp		5.30	1.509		330	59.50		389.50	450	
	1350	100 amp		4.50	1.778		475	70		545	630	
	1360	Circuit breaker, 3 pole 250 volt, 60 amp		5.30	1.509		540	59.50		599.50	685	
	1370	3 pole 480 volt, 100 amp		4.50	1.778		540	70		610	700	
	2000	Bus duct, 2 wire, 250 volt, 30 amp		60	.133	L.F.	2.40	5.25		7.65	10.45	
	2100	60 amp		50	.160		2.40	6.30		8.70	12.05	
	2200	300 volt, 30 amp		60	.133		2.40	5.25		7.65	10.45	
	2300	60 amp		50	.160		2.40	6.30		8.70	12.05	
	2400	3 wire, 250 volt, 30 amp		60	.133		2.40	5.25		7.65	10.45	
	2500	60 amp		50	.160		2.40	6.30		8.70	12.05	
	2600	480/277 volt, 30 amp		60	.133		2.40	5.25		7.65	10.45	
	2700	60 amp		50	.160		2.40	6.30		8.70	12.05	
	2750	End feed, 300 volt 2 wire max. 30 amp		6	1.333	Ea.	45	52.50		97.50	128	
	2800	60 amp		5.50	1.455		45	57.50		102.50	135	
	2850	30 amp miniature		6	1.333		45	52.50		97.50	128	
	2900	3 wire, 30 amp		6	1.333		56.50	52.50		109	141	
	2950	60 amp		5.50	1.455		56.50	57.50		114	148	
	3000	30 amp miniature		6	1.333		56.50	52.50		109	141	
	3050	Center feed, 300 volt 2 wire, 30 amp		6	1.333		62.50	52.50		115	147	
	3100	60 amp		5.50	1.455		62.50	57.50		120	154	
	3150	3 wire, 30 amp		6	1.333		71	52.50		123.50	156	
	3200	60 amp		5.50	1.455		71	57.50		128.50	163	
	3220	Elbow, 30 amp		6	1.333		43.50	52.50		96	126	
	3240	60 amp		5.50	1.455		43.50	57.50		101	133	
	3260	End cap		40	.200		8.65	7.90		16.55	21	
	3280	Strength beam, 10 ft.		15	.533		24	21		45	58	
	3300	Hanger		24	.333		5.20	13.15		18.35	25.50	
	3320	Tap box, nonfusible		6.30	1.270		73.50	50		123.50	156	
	3340	Fusible switch 30 amp, 1 fuse		6	1.333		320	52.50		372.50	430	
	3360	2 fuse		6	1.333		325	52.50		377.50	435	
	3380	3 fuse		6	1.333		330	52.50		382.50	440	
	3400	Circuit breaker handle on cover, 1 pole		6	1.333		52	52.50		104.50	135	
	3420	2 pole		6	1.333		56	52.50		108.50	140	
	3440	3 pole		6	1.333		73.50	52.50		126	159	
	3460	Circuit breaker external operhandle, 1 pole		6	1.333		52	52.50		104.50	135	
	3480	2 pole		6	1.333		56	52.50		108.50	140	
	3500	3 pole		6	1.333		73.50	52.50		126	159	
	3520	Terminal plug only		16	.500		9.50	19.70		29.20	40	
	3540	Terminal with receptacle		16	.500		12.35	19.70		32.05	43	
	3560	Fixture plug		16	.500		8.50	19.70		28.20	39	
	4000	Copper bus duct, lighting, 2 wire 300 volt, 20 amp		70	.114	L.F.	2.38	4.50		6.88	9.30	
	4020	35 amp		60	.133		2.38	5.25		7.63	10.40	
	4040	50 amp		55	.145		2.38	5.75		8.13	11.10	
	4060	60 amp		50	.160		2.38	6.30		8.68	12	
	4080	3 wire 300 volt, 20 amp		70	.114		2.60	4.50		7.10	9.55	
	4100	35 amp		60	.133		2.60	5.25		7.85	10.65	
	4120	50 amp		55	.145		2.60	5.75		8.35	11.35	
	4140	60 amp		50	.160		2.60	6.30		8.90	12.25	
	4160	Feeder in box, end, 1 circuit		6	1.333	Ea.	52	52.50		104.50	135	
	4180	2 circuit		5.50	1.455		53.50	57.50		111	144	

16 ELECTRICAL

Important: See the Reference Section for critical supporting data - Reference Nos., Crews, & City Cost Indexes

	16450	Enclosed Bus Assemblies	CREW	DAILY OUTPUT	LABOR-HOURS	UNIT	2004 BARE COSTS				TOTAL INCL O&P	
							MAT.	LABOR	EQUIP.	TOTAL		
200	4200	Center, 1 circuit	1 Elec	6	1.333	Ea.	71	52.50		123.50	156	**200**
	4220	2 circuit		5.50	1.455		73.50	57.50		131	166	
	4240	End cap		40	.200		8.65	7.90		16.55	21	
	4260	Hanger, surface mount		24	.333		5.20	13.15		18.35	25.50	
	4280	Coupling	▼	40	.200	▼	6.55	7.90		14.45	18.95	
300	0010	**COPPER BUS DUCT**										**300**
	0100	3 pole 4 wire, weatherproof, feeder duct, 600 amp	2 Elec	24	.667	L.F.	145	26.50		171.50	199	
	0110	800 amp		18	.889		176	35		211	246	
	0120	1000 amp		17	.941		197	37		234	272	
	0125	1200 amp		16.50	.970		240	38		278	320	
	0130	1350 amp		16	1		291	39.50		330.50	380	
	0140	1600 amp		12	1.333		330	52.50		382.50	445	
	0150	2000 amp		10	1.600		425	63		488	565	
	0160	2500 amp		7	2.286		530	90		620	715	
	0170	3000 amp		5	3.200		645	126		771	900	
	0180	4000 amp		3.60	4.444		850	175		1,025	1,200	
	0200	Plug-in/indoor, bus duct high short circuit, 400 amp		32	.500		138	19.70		157.70	182	
	0210	600 amp		26	.615		138	24.50		162.50	188	
	0220	800 amp		20	.800		164	31.50		195.50	228	
	0230	1000 amp		18	.889		182	35		217	252	
	0240	1350 amp		16	1		260	39.50		299.50	345	
	0250	1600 amp		12	1.333		294	52.50		346.50	405	
	0260	2000 amp		10	1.600		370	63		433	505	
	0270	2500 amp		8	2		460	79		539	620	
	0280	3000 amp		6	2.667	▼	555	105		660	765	
	0310	Cross, 225 amp		3	5.333	Ea.	1,650	210		1,860	2,125	
	0320	400 amp		2.80	5.714		1,650	225		1,875	2,125	
	0330	600 amp		2.60	6.154		1,650	242		1,892	2,150	
	0340	800 amp		2.20	7.273		1,775	287		2,062	2,375	
	0350	1000 amp		2	8		1,975	315		2,290	2,650	
	0360	1350 amp		1.80	8.889		2,250	350		2,600	3,000	
	0370	1600 amp		1.70	9.412		2,500	370		2,870	3,300	
	0380	2000 amp		1.60	10		4,100	395		4,495	5,075	
	0390	2500 amp		1.40	11.429		4,950	450		5,400	6,125	
	0400	3000 amp		1.20	13.333		5,725	525		6,250	7,075	
	0410	4000 amp		1	16		7,475	630		8,105	9,150	
	0430	Expansion fitting, 225 amp		5.40	2.963		895	117		1,012	1,150	
	0440	400 amp		4.60	3.478		1,025	137		1,162	1,325	
	0450	600 amp		4	4		1,250	158		1,408	1,600	
	0460	800 amp		3.40	4.706		1,475	185		1,660	1,900	
	0470	1000 amp		3	5.333		1,700	210		1,910	2,200	
	0480	1350 amp		2.80	5.714		2,125	225		2,350	2,675	
	0490	1600 amp		2.60	6.154		2,975	242		3,217	3,625	
	0500	2000 amp		2.20	7.273		3,425	287		3,712	4,175	
	0510	2500 amp		1.80	8.889		4,150	350		4,500	5,100	
	0520	3000 amp		1.60	10		4,925	395		5,320	5,975	
	0530	4000 amp		1.20	13.333		6,325	525		6,850	7,725	
	0550	Reducer nonfused, 225 amp		5.40	2.963		985	117		1,102	1,250	
	0560	400 amp		4.60	3.478		985	137		1,122	1,275	
	0570	600 amp		4	4		985	158		1,143	1,300	
	0580	800 amp		3.40	4.706		1,200	185		1,385	1,575	
	0590	1000 amp		3	5.333		1,425	210		1,635	1,900	
	0600	1350 amp		2.80	5.714		2,175	225		2,400	2,700	
	0610	1600 amp		2.60	6.154		2,500	242		2,742	3,100	
	0620	2000 amp	▼	2.20	7.273	▼	3,000	287		3,287	3,725	

ELECTRICAL 16

	16450	Enclosed Bus Assemblies	CREW	DAILY OUTPUT	LABOR-HOURS	UNIT	2004 BARE COSTS				TOTAL INCL O&P	
							MAT.	LABOR	EQUIP.	TOTAL		
300	0630	2500 amp	2 Elec	1.80	8.889	Ea.	3,775	350		4,125	4,675	300
	0640	3000 amp		1.60	10		4,500	395		4,895	5,525	
	0650	4000 amp		1.20	13.333		5,850	525		6,375	7,225	
	0670	Reducer fuse included, 225 amp		4.40	3.636		2,100	143		2,243	2,550	
	0680	400 amp		4.20	3.810		2,625	150		2,775	3,125	
	0690	600 amp		3.60	4.444		3,250	175		3,425	3,825	
	0700	800 amp		3.20	5		4,600	197		4,797	5,350	
	0710	1000 amp		3	5.333		5,750	210		5,960	6,650	
	0720	1350 amp		2.80	5.714		5,825	225		6,050	6,750	
	0730	1600 amp		2.20	7.273		12,700	287		12,987	14,400	
	0740	2000 amp		1.80	8.889		14,300	350		14,650	16,300	
	0790	Reducer, circuit breaker, 225 amp		4.40	3.636		2,675	143		2,818	3,150	
	0800	400 amp		4.20	3.810		3,100	150		3,250	3,650	
	0810	600 amp		3.60	4.444		4,425	175		4,600	5,100	
	0820	800 amp		3.20	5		5,175	197		5,372	6,000	
	0830	1000 amp		3	5.333		5,875	210		6,085	6,775	
	0840	1350 amp		2.80	5.714		7,075	225		7,300	8,100	
	0850	1600 amp		2.20	7.273		10,400	287		10,687	11,900	
	0860	2000 amp		1.80	8.889		11,400	350		11,750	13,100	
	0910	Cable tap box, center, 225 amp		3.20	5		990	197		1,187	1,400	
	0920	400 amp		2.60	6.154		990	242		1,232	1,450	
	0930	600 amp		2.20	7.273		990	287		1,277	1,525	
	0940	800 amp		2	8		1,100	315		1,415	1,675	
	0950	1000 amp		1.60	10		1,175	395		1,570	1,875	
	0960	1350 amp		1.40	11.429		1,500	450		1,950	2,325	
	0970	1600 amp		1.20	13.333		1,675	525		2,200	2,625	
	0980	2000 amp		1	16		2,025	630		2,655	3,175	
	1040	2500 amp		.80	20		2,400	790		3,190	3,800	
	1060	3000 amp		.60	26.667		2,750	1,050		3,800	4,600	
	1080	4000 amp		.40	40		3,475	1,575		5,050	6,175	
	1800	3 pole 3 wire, feeder duct, weatherproof, 600 amp		28	.571	L.F.	145	22.50		167.50	194	
	1820	800 amp		22	.727		176	28.50		204.50	237	
	1840	1000 amp		20	.800		197	31.50		228.50	264	
	1850	1200 amp		19	.842		204	33		237	275	
	1860	1350 amp		18	.889		291	35		326	370	
	1880	1600 amp		14	1.143		330	45		375	430	
	1900	2000 amp		12	1.333		425	52.50		477.50	550	
	1920	2500 amp		8	2		530	79		609	695	
	1940	3000 amp		6	2.667		645	105		750	865	
	1960	4000 amp		4	4		850	158		1,008	1,175	
	2000	Feeder duct/indoor, 600 amp		32	.500		121	19.70		140.70	163	
	2010	800 amp		26	.615		147	24.50		171.50	198	
	2020	1000 amp		24	.667		164	26.50		190.50	220	
	2025	1200 amp		22	.727		216	28.50		244.50	281	
	2030	1350 amp		20	.800		242	31.50		273.50	315	
	2040	1600 amp		16	1		277	39.50		316.50	365	
	2050	2000 amp		14	1.143		355	45		400	455	
	2060	2500 amp		10	1.600		440	63		503	580	
	2070	3000 amp		8	2		535	79		614	705	
	2080	4000 amp		6	2.667		710	105		815	935	
	2090	5000 amp		5	3.200		865	126		991	1,150	
	2200	Bus duct plug-in/indoor, 225 amp		46	.348		138	13.70		151.70	173	
	2210	400 amp		36	.444		138	17.50		155.50	178	
	2220	600 amp		30	.533		138	21		159	184	
	2230	800 amp		24	.667		164	26.50		190.50	220	
	2240	1000 amp		20	.800		182	31.50		213.50	247	

Important: See the Reference Section for critical supporting data - Reference Nos., Crews, & City Cost Indexes

16 ELECTRICAL

		CREW	DAILY OUTPUT	LABOR-HOURS	UNIT	MAT.	LABOR	EQUIP.	TOTAL	TOTAL INCL O&P	
16450	**Enclosed Bus Assemblies**					2004 BARE COSTS					
300	2250	1350 amp	2 Elec	18	.889	L.F.	260	35		295	335	300
2260	1600 amp		14	1.143		294	45		339	390	
2270	2000 amp		12	1.333		370	52.50		422.50	490	
2280	2500 amp		10	1.600		460	63		523	600	
2290	3000 amp		8	2		555	79		634	725	
2330	High short circuit, 400 amp		36	.444		138	17.50		155.50	178	
2340	600 amp		30	.533		138	21		159	184	
2350	800 amp		24	.667		164	26.50		190.50	220	
2360	1000 amp		20	.800		182	31.50		213.50	247	
2370	1350 amp		18	.889		260	35		295	335	
2380	1600 amp		14	1.143		294	45		339	390	
2390	2000 amp		12	1.333		370	52.50		422.50	490	
2400	2500 amp		10	1.600		460	63		523	600	
2410	3000 amp		8	2	▼	555	79		634	725	
2440	Elbows, 225 amp		4.60	3.478	Ea.	690	137		827	965	
2450	400 amp		4.20	3.810		690	150		840	985	
2460	600 amp		3.60	4.444		690	175		865	1,025	
2470	800 amp		3.20	5		745	197		942	1,125	
2480	1000 amp		3	5.333		780	210		990	1,175	
2485	1200 amp		2.90	5.517		840	217		1,057	1,250	
2490	1350 amp		2.80	5.714		935	225		1,160	1,350	
2500	1600 amp		2.60	6.154		1,000	242		1,242	1,450	
2510	2000 amp		2	8		1,225	315		1,540	1,825	
2520	2500 amp		1.80	8.889		1,850	350		2,200	2,550	
2530	3000 amp		1.60	10		2,125	395		2,520	2,925	
2540	4000 amp		1.40	11.429		2,725	450		3,175	3,675	
2560	Tee fittings, 225 amp		2.80	5.714		815	225		1,040	1,225	
2570	400 amp		2.40	6.667		815	263		1,078	1,300	
2580	600 amp		2	8		815	315		1,130	1,375	
2590	800 amp		1.80	8.889		890	350		1,240	1,500	
2600	1000 amp		1.60	10		995	395		1,390	1,675	
2605	1200 amp		1.50	10.667		1,050	420		1,470	1,775	
2610	1350 amp		1.40	11.429		1,175	450		1,625	1,950	
2620	1600 amp		1.20	13.333		1,350	525		1,875	2,275	
2630	2000 amp		1	16		2,250	630		2,880	3,425	
2640	2500 amp		.70	22.857		2,650	900		3,550	4,250	
2650	3000 amp		.60	26.667		3,075	1,050		4,125	4,950	
2660	4000 amp		.50	32		3,975	1,250		5,225	6,225	
2680	Cross, 225 amp		3.60	4.444		1,250	175		1,425	1,625	
2690	400 amp		3.20	5		1,250	197		1,447	1,675	
2700	600 amp		3	5.333		1,250	210		1,460	1,700	
2710	800 amp		2.60	6.154		1,500	242		1,742	1,975	
2720	1000 amp		2.40	6.667		1,550	263		1,813	2,125	
2730	1350 amp		2.20	7.273		1,875	287		2,162	2,475	
2740	1600 amp		2	8		2,000	315		2,315	2,675	
2750	2000 amp		1.80	8.889		3,175	350		3,525	4,025	
2760	2500 amp		1.60	10		3,700	395		4,095	4,650	
2770	3000 amp		1.40	11.429		4,275	450		4,725	5,375	
2780	4000 amp		1	16		5,450	630		6,080	6,950	
2800	Expansion fitting, 225 amp		6.40	2.500		980	98.50		1,078.50	1,225	
2810	400 amp		5.40	2.963		995	117		1,112	1,275	
2820	600 amp		4.60	3.478		995	137		1,132	1,300	
2830	800 amp		4	4		1,200	158		1,358	1,525	
2840	1000 amp		3.60	4.444		1,300	175		1,475	1,675	
2850	1350 amp		3.20	5		1,625	197		1,822	2,100	
2860	1600 amp	▼	3	5.333	▼	1,775	210		1,985	2,300	

16450	Enclosed Bus Assemblies	CREW	DAILY OUTPUT	LABOR-HOURS	UNIT	2004 BARE COSTS				TOTAL INCL O&P	
						MAT.	LABOR	EQUIP.	TOTAL		
300											300
2870	2000 amp	2 Elec	2.60	6.154	Ea.	2,125	242		2,367	2,675	
2880	2500 amp		2.20	7.273		3,025	287		3,312	3,750	
2890	3000 amp		1.80	8.889		3,550	350		3,900	4,425	
2900	4000 amp		1.40	11.429		4,525	450		4,975	5,650	
2920	Reducer nonfused, 225 amp		6.40	2.500		810	98.50		908.50	1,025	
2930	400 amp		5.40	2.963		810	117		927	1,075	
2940	600 amp		4.60	3.478		810	137		947	1,100	
2950	800 amp		4	4		940	158		1,098	1,250	
2960	1000 amp		3.60	4.444		1,050	175		1,225	1,425	
2970	1350 amp		3.20	5		1,400	197		1,597	1,850	
2980	1600 amp		3	5.333		1,575	210		1,785	2,075	
2990	2000 amp		2.60	6.154		1,900	242		2,142	2,450	
3000	2500 amp		2.20	7.273		2,800	287		3,087	3,500	
3010	3000 amp		1.80	8.889		3,300	350		3,650	4,150	
3020	4000 amp		1.40	11.429		4,275	450		4,725	5,400	
3040	Reducer fuse included, 225 amp		5	3.200		1,925	126		2,051	2,300	
3050	400 amp		4.80	3.333		2,550	131		2,681	3,000	
3060	600 amp		4.20	3.810		3,025	150		3,175	3,550	
3070	800 amp		3.60	4.444		4,325	175		4,500	5,000	
3080	1000 amp		3.40	4.706		5,125	185		5,310	5,900	
3090	1350 amp		3.20	5		5,225	197		5,422	6,025	
3100	1600 amp		2.60	6.154		12,100	242		12,342	13,800	
3110	2000 amp		2	8		13,800	315		14,115	15,600	
3160	Reducer circuit breaker, 225 amp		5	3.200		2,450	126		2,576	2,900	
3170	400 amp		4.80	3.333		3,000	131		3,131	3,500	
3180	600 amp		4.20	3.810		4,300	150		4,450	4,950	
3190	800 amp		3.60	4.444		5,025	175		5,200	5,775	
3200	1000 amp		3.40	4.706		5,700	185		5,885	6,550	
3210	1350 amp		3.20	5		6,900	197		7,097	7,900	
3220	1600 amp		2.60	6.154		10,300	242		10,542	11,700	
3230	2000 amp		2	8		11,200	315		11,515	12,900	
3280	3 pole, 3 wire, cable tap box center, 225 amp		3.60	4.444		1,125	175		1,300	1,500	
3290	400 amp		3	5.333		1,125	210		1,335	1,575	
3300	600 amp		2.60	6.154		1,125	242		1,367	1,600	
3310	800 amp		2.40	6.667		1,275	263		1,538	1,800	
3320	1000 amp		1.80	8.889		1,375	350		1,725	2,025	
3330	1350 amp		1.60	10		1,750	395		2,145	2,525	
3340	1600 amp		1.40	11.429		1,975	450		2,425	2,850	
3350	2000 amp		1.20	13.333		2,400	525		2,925	3,400	
3360	2500 amp		1	16		2,850	630		3,480	4,075	
3370	3000 amp		.70	22.857		3,300	900		4,200	5,000	
3380	4000 amp		.50	32		4,200	1,250		5,450	6,500	
3400	Cable tap box end, 225 amp		3.60	4.444		610	175		785	935	
3410	400 amp		3	5.333		610	210		820	990	
3420	600 amp		2.60	6.154		685	242		927	1,125	
3430	800 amp		2.40	6.667		685	263		948	1,150	
3440	1000 amp		1.80	8.889		755	350		1,105	1,350	
3445	1200 amp		1.70	9.412		850	370		1,220	1,475	
3450	1350 amp		1.60	10		945	395		1,340	1,625	
3460	1600 amp		1.40	11.429		1,075	450		1,525	1,875	
3470	2000 amp		1.20	13.333		1,250	525		1,775	2,125	
3480	2500 amp		1	16		1,475	630		2,105	2,575	
3490	3000 amp		.70	22.857		1,700	900		2,600	3,225	
3500	4000 amp		.50	32		2,175	1,250		3,425	4,275	
4600	Plug-in, fusible switch w/3 fuses, 3 pole, 250 volt, 30 amp	1 Elec	4	2		240	79		319	380	
4610	60 amp		3.60	2.222		310	87.50		397.50	475	

Important: See the Reference Section for critical supporting data - Reference Nos., Crews, & City Cost Indexes

16450	**Enclosed Bus Assemblies**	CREW	DAILY OUTPUT	LABOR-HOURS	UNIT	2004 BARE COSTS				TOTAL INCL O&P	
						MAT.	LABOR	EQUIP.	TOTAL		
4620	100 amp	1 Elec	2.70	2.963	Ea.	450	117		567	670	300
4630	200 amp	2 Elec	3.20	5		750	197		947	1,125	
4640	400 amp		1.40	11.429		1,950	450		2,400	2,825	
4650	600 amp	↓	.90	17.778		2,725	700		3,425	4,025	
4700	4 pole, 120/208 volt, 30 amp	1 Elec	3.90	2.051		325	81		406	480	
4710	60 amp		3.50	2.286		355	90		445	525	
4720	100 amp	↓	2.60	3.077		490	121		611	720	
4730	200 amp	2 Elec	3	5.333		825	210		1,035	1,225	
4740	400 amp		1.30	12.308		1,925	485		2,410	2,850	
4750	600 amp	↓	.80	20		2,725	790		3,515	4,150	
4800	3 pole, 480 volt, 30 amp	1 Elec	4	2		246	79		325	385	
4810	60 amp		3.60	2.222		260	87.50		347.50	415	
4820	100 amp	↓	2.70	2.963		440	117		557	655	
4830	200 amp	2 Elec	3.20	5		750	197		947	1,125	
4840	400 amp		1.40	11.429		1,750	450		2,200	2,600	
4850	600 amp		.90	17.778		2,500	700		3,200	3,800	
4860	800 amp		.66	24.242		6,450	955		7,405	8,525	
4870	1000 amp		.60	26.667		7,175	1,050		8,225	9,475	
4880	1200 amp		.50	32		7,200	1,250		8,450	9,800	
4890	1600 amp	↓	.44	36.364		8,325	1,425		9,750	11,300	
4900	4 pole, 277/480 volt, 30 amp	1 Elec	3.90	2.051		355	81		436	510	
4910	60 amp		3.50	2.286		380	90		470	550	
4920	100 amp	↓	2.60	3.077		550	121		671	785	
4930	200 amp	2 Elec	3	5.333		1,100	210		1,310	1,525	
4940	400 amp		1.30	12.308		2,075	485		2,560	3,000	
4950	600 amp		.80	20		2,825	790		3,615	4,275	
5050	800 amp		.60	26.667		9,325	1,050		10,375	11,900	
5060	1000 amp		.56	28.571		10,700	1,125		11,825	13,500	
5070	1200 amp		.48	33.333		10,800	1,325		12,125	13,900	
5080	1600 amp	↓	.42	38.095		12,600	1,500		14,100	16,000	
5150	Fusible with starter, 3 pole 250 volt, 30 amp	1 Elec	3.50	2.286		1,275	90		1,365	1,525	
5160	60 amp		3.20	2.500		1,350	98.50		1,448.50	1,625	
5170	100 amp	↓	2.50	3.200		1,525	126		1,651	1,875	
5180	200 amp	2 Elec	2.80	5.714		2,500	225		2,725	3,075	
5200	3 pole 480 volt, 30 amp	1 Elec	3.50	2.286		1,275	90		1,365	1,525	
5210	60 amp		3.20	2.500		1,350	98.50		1,448.50	1,625	
5220	100 amp	↓	2.50	3.200		1,525	126		1,651	1,875	
5230	200 amp	2 Elec	2.80	5.714		2,500	225		2,725	3,075	
5300	Fusible with contactor, 3 pole 250 volt, 30 amp	1 Elec	3.50	2.286		1,225	90		1,315	1,475	
5310	60 amp		3.20	2.500		1,575	98.50		1,673.50	1,875	
5320	100 amp	↓	2.50	3.200		2,200	126		2,326	2,625	
5330	200 amp	2 Elec	2.80	5.714		2,500	225		2,725	3,100	
5400	3 pole 480 volt, 30 amp	1 Elec	3.50	2.286		1,325	90		1,415	1,575	
5410	60 amp		3.20	2.500		1,875	98.50		1,973.50	2,200	
5420	100 amp	↓	2.50	3.200		2,575	126		2,701	3,025	
5430	200 amp	2 Elec	2.80	5.714		2,625	225		2,850	3,225	
5450	Fusible with capacitor, 3 pole 250 volt, 30 amp	1 Elec	3	2.667		3,200	105		3,305	3,675	
5460	60 amp		2	4		3,725	158		3,883	4,325	
5500	3 pole 480 volt, 30 amp		3	2.667		2,675	105		2,780	3,100	
5510	60 amp		2	4		3,350	158		3,508	3,925	
5600	Circuit breaker, 3 pole, 250 volt, 60 amp		4.50	1.778		335	70		405	475	
5610	100 amp		3.20	2.500		415	98.50		513.50	600	
5650	4 pole, 120/208 volt, 60 amp		4.40	1.818		380	71.50		451.50	525	
5660	100 amp		3.10	2.581		450	102		552	645	
5700	3 pole, 4 wire 277/480 volt, 60 amp		4.30	1.860		510	73.50		583.50	675	
5710	100 amp	↓	3	2.667	↓	570	105		675	780	

16450	Enclosed Bus Assemblies	CREW	DAILY OUTPUT	LABOR-HOURS	UNIT	2004 BARE COSTS				TOTAL INCL O&P	
						MAT.	LABOR	EQUIP.	TOTAL		
300	5720	225 amp	2 Elec	3.20	5	Ea.	1,250	197		1,447	1,700	**300**
5730	400 amp		1.20	13.333		2,625	525		3,150	3,650	
5740	600 amp		.96	16.667		3,525	655		4,180	4,850	
5750	700 amp		.60	26.667		4,450	1,050		5,500	6,475	
5760	800 amp		.60	26.667		4,450	1,050		5,500	6,475	
5770	900 amp		.54	29.630		5,875	1,175		7,050	8,175	
5780	1000 amp		.54	29.630		5,875	1,175		7,050	8,175	
5790	1200 amp	▼	.42	38.095		7,075	1,500		8,575	10,000	
5810	Circuit breaker w/HIC fuses, 3 pole 480 volt, 60 amp	1 Elec	4.40	1.818		645	71.50		716.50	815	
5820	100 amp	"	3.10	2.581		705	102		807	925	
5830	225 amp	2 Elec	3.40	4.706		2,250	185		2,435	2,750	
5840	400 amp		1.40	11.429		3,575	450		4,025	4,600	
5850	600 amp		1	16		3,625	630		4,255	4,925	
5860	700 amp		.64	25		4,800	985		5,785	6,750	
5870	800 amp		.64	25		4,800	985		5,785	6,750	
5880	900 amp		.56	28.571		10,400	1,125		11,525	13,100	
5890	1000 amp	▼	.56	28.571		10,400	1,125		11,525	13,100	
5950	3 pole 4 wire 277/480 volt, 60 amp	1 Elec	4.30	1.860		645	73.50		718.50	820	
5960	100 amp	"	3	2.667		705	105		810	930	
5970	225 amp	2 Elec	3	5.333		2,250	210		2,460	2,800	
5980	400 amp		1.10	14.545		3,575	575		4,150	4,775	
5990	600 amp		.94	17.021		3,625	670		4,295	4,975	
6000	700 amp		.58	27.586		4,800	1,075		5,875	6,900	
6010	800 amp		.58	27.586		4,800	1,075		5,875	6,900	
6020	900 amp		.52	30.769		10,400	1,200		11,600	13,200	
6030	1000 amp		.52	30.769		10,400	1,200		11,600	13,200	
6040	1200 amp	▼	.40	40		10,400	1,575		11,975	13,800	
6100	Circuit breaker with starter, 3 pole 250 volt, 60 amp	1 Elec	3.20	2.500		985	98.50		1,083.50	1,225	
6110	100 amp	"	2.50	3.200		1,350	126		1,476	1,700	
6120	225 amp	2 Elec	3	5.333		1,750	210		1,960	2,250	
6130	3 pole 480 volt, 60 amp	1 Elec	3.20	2.500		980	98.50		1,078.50	1,225	
6140	100 amp	"	2.50	3.200		1,350	126		1,476	1,700	
6150	225 amp	2 Elec	3	5.333		1,750	210		1,960	2,250	
6200	Circuit breaker with contactor, 3 pole 250 volt, 60 amp	1 Elec	3.20	2.500		930	98.50		1,028.50	1,175	
6210	100 amp	"	2.50	3.200		1,250	126		1,376	1,575	
6220	225 amp	2 Elec	3	5.333		1,625	210		1,835	2,125	
6250	3 pole 480 volt, 60 amp	1 Elec	3.20	2.500		930	98.50		1,028.50	1,175	
6260	100 amp	"	2.50	3.200		1,250	126		1,376	1,575	
6270	225 amp	2 Elec	3	5.333		1,625	210		1,835	2,125	
6300	Circuit breaker with capacitor, 3 pole 250 volt, 60 amp	1 Elec	2	4		3,825	158		3,983	4,425	
6310	3 pole 480 volt, 60 amp		2	4		3,550	158		3,708	4,125	
6400	Add control transformer with pilot light to above starter		16	.500		277	19.70		296.70	335	
6410	Switch, fusible, mechanically held contactor optional		16	.500		730	19.70		749.70	835	
6430	Circuit breaker, mechanically held contactor optional		16	.500		730	19.70		749.70	835	
6450	Ground neutralizer, 3 pole	▼	16	.500	▼	33.50	19.70		53.20	66.50	
320	0010	**COPPER BUS DUCT** 10 ft long										**320**
0050	3 pole 4 wire, plug-in/indoor, straight section, 225 amp	2 Elec	40	.400	L.F.	138	15.75		153.75	176	
1000	400 amp		32	.500		138	19.70		157.70	182	
1500	600 amp		26	.615		138	24.50		162.50	188	
2400	800 amp		20	.800		164	31.50		195.50	228	
2450	1000 amp		18	.889		182	35		217	252	
2470	1200 amp		17	.941		234	37		271	310	
2500	1350 amp		16	1		260	39.50		299.50	345	
2510	1600 amp		12	1.333		294	52.50		346.50	405	
2520	2000 amp	▼	10	1.600	▼	370	63		433	505	

Important: See the Reference Section for critical supporting data - Reference Nos., Crews, & City Cost Indexes

16450	Enclosed Bus Assemblies	CREW	DAILY OUTPUT	LABOR-HOURS	UNIT	2004 BARE COSTS				TOTAL INCL O&P
						MAT.	LABOR	EQUIP.	TOTAL	
320	2530	2500 amp	2 Elec	8	2	L.F.	460	79		539	620	320
2540	3000 amp		6	2.667		555	105		660	765
2550	Feeder, 600 amp		28	.571		121	22.50		143.50	167
2600	800 amp		22	.727		147	28.50		175.50	205
2700	1000 amp		20	.800		164	31.50		195.50	228
2750	1200 amp		19	.842		216	33		249	288
2800	1350 amp		18	.889		242	35		277	320
2900	1600 amp		14	1.143		277	45		322	370
3000	2000 amp		12	1.333		355	52.50		407.50	470
3010	2500 amp		8	2		440	79		519	600
3020	3000 amp		6	2.667		535	105		640	745
3030	4000 amp		4	4		710	158		868	1,025
3040	5000 amp		2	8	▼	865	315		1,180	1,425
3100	Elbows, 225 amp		4	4	Ea.	820	158		978	1,125
3200	400 amp		3.60	4.444		820	175		995	1,150
3300	600 amp		3.20	5		820	197		1,017	1,200
3400	800 amp		2.80	5.714		890	225		1,115	1,325
3500	1000 amp		2.60	6.154		995	242		1,237	1,450
3550	1200 amp		2.50	6.400		1,100	252		1,352	1,575
3600	1350 amp		2.40	6.667		1,175	263		1,438	1,675
3700	1600 amp		2.20	7.273		1,275	287		1,562	1,825
3800	2000 amp		1.80	8.889		1,575	350		1,925	2,250
3810	2500 amp		1.60	10		2,475	395		2,870	3,300
3820	3000 amp		1.40	11.429		2,875	450		3,325	3,825
3830	4000 amp		1.20	13.333		3,700	525		4,225	4,850
3840	5000 amp		1	16		6,000	630		6,630	7,550
4000	End box, 225 amp		34	.471		111	18.55		129.55	150
4100	400 amp		32	.500		111	19.70		130.70	152
4200	600 amp		28	.571		111	22.50		133.50	156
4300	800 amp		26	.615		111	24.50		135.50	158
4400	1000 amp		24	.667		111	26.50		137.50	161
4410	1200 amp		23	.696		111	27.50		138.50	163
4500	1350 amp		22	.727		111	28.50		139.50	165
4600	1600 amp		20	.800		111	31.50		142.50	169
4700	2000 amp		18	.889		136	35		171	201
4710	2500 amp		16	1		136	39.50		175.50	208
4720	3000 amp		14	1.143		136	45		181	216
4730	4000 amp		12	1.333		164	52.50		216.50	259
4740	5000 amp		10	1.600		164	63		227	275
4800	Cable tap box end, 225 amp		3.20	5		820	197		1,017	1,200
5000	400 amp		2.60	6.154		820	242		1,062	1,250
5100	600 amp		2.20	7.273		820	287		1,107	1,325
5200	800 amp		2	8		920	315		1,235	1,475
5300	1000 amp		1.60	10		1,025	395		1,420	1,700
5350	1200 amp		1.50	10.667		1,150	420		1,570	1,875
5400	1350 amp		1.40	11.429		1,225	450		1,675	2,000
5500	1600 amp		1.20	13.333		1,375	525		1,900	2,275
5600	2000 amp		1	16		1,550	630		2,180	2,650
5610	2500 amp		.80	20		1,850	790		2,640	3,225
5620	3000 amp		.60	26.667		2,100	1,050		3,150	3,875
5630	4000 amp		.40	40		2,675	1,575		4,250	5,275
5640	5000 amp		.20	80		3,125	3,150		6,275	8,125
5700	Switchboard stub, 225 amp		5.40	2.963		830	117		947	1,100
5800	400 amp		4.60	3.478		830	137		967	1,125
5900	600 amp		4	4		830	158		988	1,150
6000	800 amp		3.20	5	▼	1,000	197		1,197	1,400

ELECTRICAL 16

16450	Enclosed Bus Assemblies	CREW	DAILY OUTPUT	LABOR-HOURS	UNIT	2004 BARE COSTS				TOTAL INCL O&P
						MAT.	LABOR	EQUIP.	TOTAL	
320 6100	1000 amp	2 Elec	3	5.333	Ea.	1,175	210		1,385	1,600 **320**
6150	1200 amp		2.80	5.714		1,350	225		1,575	1,825
6200	1350 amp		2.60	6.154		1,500	242		1,742	2,000
6300	1600 amp		2.40	6.667		1,700	263		1,963	2,250
6400	2000 amp		2	8		2,050	315		2,365	2,725
6410	2500 amp		1.80	8.889		2,500	350		2,850	3,250
6420	3000 amp		1.60	10		2,925	395		3,320	3,775
6430	4000 amp		1.40	11.429		3,800	450		4,250	4,850
6440	5000 amp		1.20	13.333		4,650	525		5,175	5,900
6490	Tee fittings, 225 amp		2.40	6.667		1,125	263		1,388	1,650
6500	400 amp		2	8		1,125	315		1,440	1,725
6600	600 amp		1.80	8.889		1,125	350		1,475	1,775
6700	800 amp		1.60	10		1,300	395		1,695	2,000
6750	1000 amp		1.40	11.429		1,500	450		1,950	2,325
6770	1200 amp		1.30	12.308		1,675	485		2,160	2,575
6800	1350 amp		1.20	13.333		1,875	525		2,400	2,825
7000	1600 amp		1	16		2,125	630		2,755	3,300
7100	2000 amp		.80	20		2,525	790		3,315	3,950
7110	2500 amp		.60	26.667		3,100	1,050		4,150	5,000
7120	3000 amp		.50	32		3,625	1,250		4,875	5,850
7130	4000 amp		.40	40		4,675	1,575		6,250	7,500
7140	5000 amp		.20	80		5,525	3,150		8,675	10,800
7200	Plug-in fusible switches w/3 fuses, 600 volt, 3 pole, 30 amp	1 Elec	4	2		495	79		574	655
7300	60 amp		3.60	2.222		545	87.50		632.50	730
7400	100 amp		2.70	2.963		845	117		962	1,100
7500	200 amp	2 Elec	3.20	5		1,500	197		1,697	1,950
7600	400 amp		1.40	11.429		4,425	450		4,875	5,550
7700	600 amp		.90	17.778		5,025	700		5,725	6,575
7800	800 amp		.66	24.242		7,050	955		8,005	9,175
7900	1200 amp		.50	32		13,200	1,250		14,450	16,400
7910	1600 amp		.44	36.364		13,200	1,425		14,625	16,600
8000	Plug-in circuit breakers, molded case, 15 to 50 amp	1 Elec	4.40	1.818		465	71.50		536.50	620
8100	70 to 100 amp	"	3.10	2.581		520	102		622	720
8200	150 to 225 amp	2 Elec	3.40	4.706		1,375	185		1,560	1,800
8300	250 to 400 amp		1.40	11.429		2,475	450		2,925	3,375
8400	500 to 600 amp		1	16		3,325	630		3,955	4,625
8500	700 to 800 amp		.64	25		4,100	985		5,085	6,000
8600	900 to 1000 amp		.56	28.571		5,875	1,125		7,000	8,125
8700	1200 amp		.44	36.364		7,075	1,425		8,500	9,900
8720	1400 amp		.40	40		10,400	1,575		11,975	13,900
8730	1600 amp		.40	40		11,400	1,575		12,975	15,000
8750	Circuit breakers, with current limiting fuse, 15 to 50 amp	1 Elec	4.40	1.818		935	71.50		1,006.50	1,125
8760	70 to 100 amp	"	3.10	2.581		1,100	102		1,202	1,375
8770	150 to 225 amp	2 Elec	3.40	4.706		2,375	185		2,560	2,900
8780	250 to 400 amp		1.40	11.429		3,675	450		4,125	4,725
8790	500 to 600 amp		1	16		4,250	630		4,880	5,600
8800	700 to 800 amp		.64	25		6,975	985		7,960	9,125
8810	900 to 1000 amp		.56	28.571		7,950	1,125		9,075	10,400
8850	Combination starter FVNR, fusible switch, NEMA size 0, 30 amp	1 Elec	2	4		1,275	158		1,433	1,625
8860	NEMA size 1, 60 amp		1.80	4.444		1,350	175		1,525	1,725
8870	NEMA size 2, 100 amp		1.30	6.154		1,675	242		1,917	2,200
8880	NEMA size 3, 200 amp	2 Elec	2	8		2,725	315		3,040	3,475
8900	Circuit breaker, NEMA size 0, 30 amp	1 Elec	2	4		1,300	158		1,458	1,650
8910	NEMA size 1, 60 amp		1.80	4.444		1,350	175		1,525	1,725
8920	NEMA size 2, 100 amp		1.30	6.154		1,925	242		2,167	2,475
8930	NEMA size 3, 200 amp	2 Elec	2	8		2,450	315		2,765	3,175

16
ELECTRICAL

16450 | Enclosed Bus Assemblies

			CREW	DAILY OUTPUT	LABOR-HOURS	UNIT	2004 BARE COSTS				TOTAL INCL O&P	
							MAT.	LABOR	EQUIP.	TOTAL		
320	8950	Combination contactor, fusible switch, NEMA size 0, 30 amp	1 Elec	2	4	Ea.	850	158		1,008	1,175	320
	8960	NEMA size 1, 60 amp		1.80	4.444		875	175		1,050	1,225	
	8970	NEMA size 2, 100 amp	▼	1.30	6.154		1,300	242		1,542	1,775	
	8980	NEMA size 3, 200 amp	2 Elec	2	8		1,500	315		1,815	2,125	
	9000	Circuit breaker, NEMA size 0, 30 amp	1 Elec	2	4		975	158		1,133	1,300	
	9010	NEMA size 1, 60 amp		1.80	4.444		1,000	175		1,175	1,350	
	9020	NEMA size 2, 100 amp	▼	1.30	6.154		1,550	242		1,792	2,050	
	9030	NEMA size 3, 200 amp	2 Elec	2	8		1,875	315		2,190	2,550	
	9050	Control transformer for above, NEMA size 0, 30 amp	1 Elec	8	1		143	39.50		182.50	216	
	9060	NEMA size 1, 60 amp		8	1		143	39.50		182.50	216	
	9070	NEMA size 2, 100 amp	▼	7	1.143		199	45		244	286	
	9080	NEMA size 3, 200 amp	2 Elec	14	1.143		277	45		322	370	
	9100	Comb. fusible switch & lighting control, electrically held, 30 amp	1 Elec	2	4		585	158		743	875	
	9110	60 amp		1.80	4.444		840	175		1,015	1,175	
	9120	100 amp	▼	1.30	6.154		1,075	242		1,317	1,550	
	9130	200 amp	2 Elec	2	8		2,650	315		2,965	3,400	
	9150	Mechanically held, 30 amp	1 Elec	2	4		730	158		888	1,050	
	9160	60 amp		1.80	4.444		1,100	175		1,275	1,450	
	9170	100 amp	▼	1.30	6.154		1,400	242		1,642	1,900	
	9180	200 amp	2 Elec	2	8	▼	2,850	315		3,165	3,625	
	9200	Ground bus added to bus duct, 225 amp		320	.050	L.F.	26	1.97		27.97	31.50	
	9210	400 amp		240	.067		26	2.63		28.63	32.50	
	9220	600 amp		240	.067		26	2.63		28.63	32.50	
	9230	800 amp		160	.100		30.50	3.94		34.44	39.50	
	9240	1000 amp		160	.100		34.50	3.94		38.44	44	
	9250	1350 amp		140	.114		52	4.50		56.50	63.50	
	9260	1600 amp		120	.133		56	5.25		61.25	70	
	9270	2000 amp		110	.145		73.50	5.75		79.25	89.50	
	9280	2500 amp		100	.160		91	6.30		97.30	109	
	9290	3000 amp		90	.178		108	7		115	129	
	9300	4000 amp		80	.200		143	7.90		150.90	169	
	9310	5000 amp	▼	70	.229		173	9		182	203	
	9320	High short circuit bracing, add				▼	13.50			13.50	14.85	
360	0010	**COPPER OR ALUMINUM BUS DUCT FITTINGS**										360
	0100	Flange, wall, with vapor barrier, 225 amp	2 Elec	6.20	2.581	Ea.	520	102		622	720	
	0110	400 amp		6	2.667		520	105		625	725	
	0120	600 amp		5.80	2.759		520	109		629	730	
	0130	800 amp		5.40	2.963		520	117		637	745	
	0140	1000 amp		5	3.200		520	126		646	760	
	0145	1200 amp		4.80	3.333		520	131		651	765	
	0150	1350 amp		4.60	3.478		520	137		657	775	
	0160	1600 amp		4.20	3.810		520	150		670	795	
	0170	2000 amp		4	4		520	158		678	805	
	0180	2500 amp		3.60	4.444		520	175		695	830	
	0190	3000 amp		3.20	5		520	197		717	865	
	0200	4000 amp		2.60	6.154		605	242		847	1,025	
	0300	Roof, 225 amp		6.20	2.581		550	102		652	755	
	0310	400 amp		6	2.667		550	105		655	760	
	0320	600 amp		5.80	2.759		550	109		659	765	
	0330	800 amp		5.40	2.963		550	117		667	780	
	0340	1000 amp		5	3.200		550	126		676	795	
	0345	1200 amp		4.80	3.333		550	131		681	800	
	0350	1350 amp		4.60	3.478		550	137		687	810	
	0360	1600 amp		4.20	3.810		550	150		700	830	
	0370	2000 amp	▼	4	4	▼	550	158		708	840	

ELECTRICAL 16

	16450	Enclosed Bus Assemblies	CREW	DAILY OUTPUT	LABOR-HOURS	UNIT	2004 BARE COSTS				TOTAL INCL O&P
							MAT.	LABOR	EQUIP.	TOTAL	
360	0380	2500 amp	2 Elec	3.60	4.444	Ea.	550	175		725	865
	0390	3000 amp		3.20	5		550	197		747	900
	0400	4000 amp		2.60	6.154		550	242		792	965
	0420	Support, floor mounted, 225 amp		20	.800		112	31.50		143.50	170
	0430	400 amp		20	.800		112	31.50		143.50	170
	0440	600 amp		18	.889		112	35		147	175
	0450	800 amp		16	1		112	39.50		151.50	182
	0460	1000 amp		13	1.231		112	48.50		160.50	195
	0465	1200 amp		11.80	1.356		112	53.50		165.50	203
	0470	1350 amp		10.60	1.509		112	59.50		171.50	212
	0480	1600 amp		9.20	1.739		112	68.50		180.50	225
	0490	2000 amp		8	2		112	79		191	240
	0500	2500 amp		6.40	2.500		112	98.50		210.50	270
	0510	3000 amp		5.40	2.963		112	117		229	297
	0520	4000 amp		4	4		112	158		270	355
	0540	Weather stop, 225 amp		12	1.333		335	52.50		387.50	450
	0550	400 amp		10	1.600		335	63		398	465
	0560	600 amp		9	1.778		335	70		405	475
	0570	800 amp		8	2		335	79		414	485
	0580	1000 amp		6.40	2.500		335	98.50		433.50	515
	0585	1200 amp		5.90	2.712		335	107		442	530
	0590	1350 amp		5.40	2.963		335	117		452	545
	0600	1600 amp		4.60	3.478		335	137		472	575
	0610	2000 amp		4	4		335	158		493	605
	0620	2500 amp		3.20	5		335	197		532	665
	0630	3000 amp		2.60	6.154		335	242		577	730
	0640	4000 amp		2	8		335	315		650	840
	0660	End closure, 225 amp		34	.471		111	18.55		129.55	150
	0670	400 amp		32	.500		111	19.70		130.70	152
	0680	600 amp		28	.571		111	22.50		133.50	156
	0690	800 amp		26	.615		111	24.50		135.50	158
	0700	1000 amp		24	.667		111	26.50		137.50	161
	0705	1200 amp		23	.696		111	27.50		138.50	163
	0710	1350 amp		22	.727		111	28.50		139.50	165
	0720	1600 amp		20	.800		111	31.50		142.50	169
	0730	2000 amp		18	.889		136	35		171	201
	0740	2500 amp		16	1		136	39.50		175.50	208
	0750	3000 amp		14	1.143		136	45		181	216
	0760	4000 amp		12	1.333		164	52.50		216.50	259
	0780	Switchboard stub, 3 pole 3 wire, 225 amp		6	2.667		665	105		770	890
	0790	400 amp		5.20	3.077		665	121		786	915
	0800	600 amp		4.60	3.478		665	137		802	940
	0810	800 amp		3.60	4.444		805	175		980	1,150
	0820	1000 amp		3.40	4.706		940	185		1,125	1,300
	0825	1200 amp		3.20	5		940	197		1,137	1,325
	0830	1350 amp		3	5.333		1,125	210		1,335	1,575
	0840	1600 amp		2.80	5.714		1,325	225		1,550	1,775
	0850	2000 amp		2.40	6.667		1,550	263		1,813	2,100
	0860	2500 amp		2	8		1,900	315		2,215	2,550
	0870	3000 amp		1.80	8.889		2,150	350		2,500	2,900
	0880	4000 amp		1.60	10		2,800	395		3,195	3,650
	0890	5000 amp		1.40	11.429		3,425	450		3,875	4,450
	0900	3 pole 4 wire, 225 amp		5.40	2.963		830	117		947	1,100
	0910	400 amp		4.60	3.478		830	137		967	1,125
	0920	600 amp		4	4		830	158		988	1,150
	0930	800 amp		3.20	5		1,000	197		1,197	1,400

Important: See the Reference Section for critical supporting data - Reference Nos., Crews, & City Cost Indexes

		CREW	DAILY OUTPUT	LABOR-HOURS	UNIT	2004 BARE COSTS				TOTAL INCL O&P
16450	**Enclosed Bus Assemblies**					MAT.	LABOR	EQUIP.	TOTAL	
0940	1000 amp	2 Elec	3	5.333	Ea.	1,175	210		1,385	1,600
0950	1350 amp		2.60	6.154		1,500	242		1,742	2,000
0960	1600 amp		2.40	6.667		1,700	263		1,963	2,250
0970	2000 amp		2	8		2,050	315		2,365	2,725
0980	2500 amp		1.80	8.889		2,500	350		2,850	3,275
0990	3000 amp		1.60	10		2,925	395		3,320	3,775
1000	4000 amp		1.40	11.429		3,800	450		4,250	4,850
1050	Service head, weatherproof, 3 pole 3 wire, 225 amp		3	5.333		1,125	210		1,335	1,575
1060	400 amp		2.80	5.714		1,125	225		1,350	1,575
1070	600 amp		2.60	6.154		1,125	242		1,367	1,600
1080	800 amp		2.40	6.667		1,275	263		1,538	1,800
1090	1000 amp		2	8		1,375	315		1,690	1,975
1100	1350 amp		1.80	8.889		1,775	350		2,125	2,475
1110	1600 amp		1.60	10		2,000	395		2,395	2,775
1120	2000 amp		1.40	11.429		2,425	450		2,875	3,350
1130	2500 amp		1.20	13.333		2,900	525		3,425	3,950
1140	3000 amp		.90	17.778		3,375	700		4,075	4,775
1150	4000 amp		.70	22.857		4,300	900		5,200	6,075
1200	3 pole 4 wire, 225 amp		2.60	6.154		1,275	242		1,517	1,750
1210	400 amp		2.40	6.667		1,275	263		1,538	1,800
1220	600 amp		2.20	7.273		1,275	287		1,562	1,825
1230	800 amp		2	8		1,475	315		1,790	2,100
1240	1000 amp		1.70	9.412		1,700	370		2,070	2,400
1250	1350 amp		1.50	10.667		2,000	420		2,420	2,850
1260	1600 amp		1.40	11.429		2,200	450		2,650	3,075
1270	2000 amp		1.20	13.333		2,925	525		3,450	4,000
1280	2500 amp		1	16		3,600	630		4,230	4,925
1290	3000 amp		.80	20		4,225	790		5,015	5,825
1300	4000 amp		.60	26.667		5,500	1,050		6,550	7,600
1350	Flanged end, 3 pole 3 wire, 225 amp		6	2.667		615	105		720	835
1360	400 amp		5.20	3.077		615	121		736	860
1370	600 amp		4.60	3.478		615	137		752	885
1380	800 amp		3.60	4.444		680	175		855	1,000
1390	1000 amp		3.40	4.706		770	185		955	1,125
1395	1200 amp		3.20	5		855	197		1,052	1,250
1400	1350 amp		3	5.333		920	210		1,130	1,325
1410	1600 amp		2.80	5.714		1,050	225		1,275	1,475
1420	2000 amp		2.40	6.667		1,225	263		1,488	1,750
1430	2500 amp		2	8		1,425	315		1,740	2,050
1440	3000 amp		1.80	8.889		1,650	350		2,000	2,325
1450	4000 amp		1.60	10		2,025	395		2,420	2,825
1500	3 pole 4 wire, 225 amp		5.40	2.963		705	117		822	950
1510	400 amp		4.60	3.478		705	137		842	980
1520	600 amp		4	4		705	158		863	1,000
1530	800 amp		3.20	5		835	197		1,032	1,200
1540	1000 amp		3	5.333		930	210		1,140	1,350
1545	1200 amp		2.80	5.714		1,050	225		1,275	1,475
1550	1350 amp		2.60	6.154		1,125	242		1,367	1,575
1560	1600 amp		2.40	6.667		1,300	263		1,563	1,825
1570	2000 amp		2	8		1,525	315		1,840	2,150
1580	2500 amp		1.80	8.889		1,825	350		2,175	2,525
1590	3000 amp		1.60	10		2,075	395		2,470	2,875
1600	4000 amp		1.40	11.429		2,700	450		3,150	3,650
1650	Hanger, standard, 225 amp		64	.250		13	9.85		22.85	29
1660	400 amp		48	.333		13	13.15		26.15	34
1670	600 amp		40	.400		13	15.75		28.75	38

360

ELECTRICAL 16

261

16450 | Enclosed Bus Assemblies

		CREW	DAILY OUTPUT	LABOR-HOURS	UNIT	2004 BARE COSTS				TOTAL INCL O&P
						MAT.	LABOR	EQUIP.	TOTAL	
360 1680	800 amp	2 Elec	32	.500	Ea.	13	19.70		32.70	44	**360**
1690	1000 amp		24	.667		13	26.50		39.50	53.50
1695	1200 amp		22	.727		13	28.50		41.50	57
1700	1350 amp		20	.800		13	31.50		44.50	61.50
1710	1600 amp		20	.800		13	31.50		44.50	61.50
1720	2000 amp		18	.889		13	35		48	66.50
1730	2500 amp		16	1		13	39.50		52.50	73
1740	3000 amp		16	1		13	39.50		52.50	73
1750	4000 amp		16	1		13	39.50		52.50	73
1800	Spring type, 225 amp		16	1		58	39.50		97.50	123
1810	400 amp		14	1.143		58	45		103	131
1820	600 amp		14	1.143		58	45		103	131
1830	800 amp		14	1.143		58	45		103	131
1840	1000 amp		14	1.143		58	45		103	131
1845	1200 amp		14	1.143		58	45		103	131
1850	1350 amp		14	1.143		58	45		103	131
1860	1600 amp		12	1.333		58	52.50		110.50	142
1870	2000 amp		12	1.333		58	52.50		110.50	142
1880	2500 amp		12	1.333		58	52.50		110.50	142
1890	3000 amp		10	1.600		58	63		121	158
1900	4000 amp	▼	10	1.600	▼	58	63		121	158
500 0010	**FEEDRAIL**, 12 foot mounting										**500**
0050	Trolley busway, 3 pole									
0100	300 volt 60 amp, plain, 10 ft. lengths	1 Elec	50	.160	L.F.	20	6.30		26.30	31.50
0300	Door track		50	.160		22	6.30		28.30	33.50
0500	Curved track	▼	30	.267	▼	236	10.50		246.50	276
0700	Coupling				Ea.	10.60			10.60	11.65
0900	Center feed	1 Elec	5.30	1.509		28.50	59.50		88	120
1100	End feed		5.30	1.509		34	59.50		93.50	126
1300	Hanger set		24	.333	▼	4.07	13.15		17.22	24
3000	600 volt 100 amp, plain, 10 ft. lengths		35	.229	L.F.	40	9		49	57.50
3300	Door track	▼	35	.229	"	46.50	9		55.50	64.50
3700	Coupling				Ea.	33.50			33.50	36.50
4000	End cap	1 Elec	40	.200		28.50	7.90		36.40	43
4200	End feed		4	2		174	79		253	310
4500	Trolley, 600 volt, 20 amp		5.30	1.509		225	59.50		284.50	335
4700	30 amp		5.30	1.509		225	59.50		284.50	335
4900	Duplex, 40 amp		4	2		475	79		554	635
5000	60 amp		4	2		475	79		554	635
5300	Fusible, 20 amp		4	2		575	79		654	745
5500	30 amp		4	2		575	79		654	745
5900	300 volt, 20 amp		5.30	1.509		175	59.50		234.50	282
6000	30 amp		5.30	1.509		195	59.50		254.50	305
6300	Fusible, 20 amp		4.70	1.702		290	67		357	420
6500	30 amp		4.70	1.702	▼	415	67		482	555
7300	Busway, 250 volt 50 amp, 2 wire	▼	70	.114	L.F.	11.70	4.50		16.20	19.60
7330	Coupling				Ea.	23			23	25
7340	Center feed	1 Elec	6	1.333		261	52.50		313.50	365
7350	End feed		6	1.333		47	52.50		99.50	130
7360	End cap		40	.200		17.10	7.90		25	30.50
7370	Hanger set		24	.333	▼	4.07	13.15		17.22	24
7400	125/250 volt 50 amp, 3 wire		60	.133	L.F.	12.30	5.25		17.55	21.50
7430	Coupling		6	1.333	Ea.	27.50	52.50		80	109
7440	Center feed		6	1.333		278	52.50		330.50	385
7450	End feed	▼	6	1.333	▼	50.50	52.50		103	134

Important: See the Reference Section for critical supporting data - Reference Nos., Crews, & City Cost Indexes

16 ELECTRICAL

		16450	Enclosed Bus Assemblies	CREW	DAILY OUTPUT	LABOR-HOURS	UNIT	2004 BARE COSTS MAT.	LABOR	EQUIP.	TOTAL	TOTAL INCL O&P	
500	7460		End cap	1 Elec	40	.200	Ea.	18.70	7.90		26.60	32	500
	7470		Hanger set		24	.333		4.07	13.15		17.22	24	
	7480		Trolley, 250 volt, 2 pole, 20 amp		6	1.333		28.50	52.50		81	110	
	7490		30 amp		6	1.333		28.50	52.50		81	110	
	7500		125/250 volt, 3 pole, 20 amp		6	1.333		30	52.50		82.50	111	
	7510		30 amp		6	1.333		30	52.50		82.50	111	
	8000		Cleaning tools, 300 volt, dust remover					53			53	58	
	8100		Bus bar cleaner					93			93	102	
	8300		600 volt, dust remover, 60 amp					196			196	216	
	8400		100 amp					269			269	296	
	8600		Bus bar cleaner, 60 amp					239			239	262	
	8700		100 amp					330			330	365	

		16490	Low V Dist Components/Accessories										
300	0010	**FUSES**											300
	0020		Cartridge, nonrenewable										
	0050		250 volt, 30 amp	1 Elec	50	.160	Ea.	1.30	6.30		7.60	10.85	
	0100		60 amp		50	.160		2.18	6.30		8.48	11.80	
	0150		100 amp		40	.200		9.20	7.90		17.10	22	
	0200		200 amp		36	.222		22.50	8.75		31.25	37.50	
	0250		400 amp		30	.267		51.50	10.50		62	72.50	
	0300		600 amp		24	.333		85.50	13.15		98.65	114	
	0400		600 volt, 30 amp		40	.200		7.50	7.90		15.40	19.95	
	0450		60 amp		40	.200		10.55	7.90		18.45	23.50	
	0500		100 amp		36	.222		23	8.75		31.75	38	
	0550		200 amp		30	.267		55.50	10.50		66	76.50	
	0600		400 amp		24	.333		111	13.15		124.15	142	
	0650		600 amp		20	.400		166	15.75		181.75	207	
	0800		Dual element, time delay, 250 volt, 30 amp		50	.160		3.26	6.30		9.56	13	
	0850		60 amp		50	.160		6	6.30		12.30	16	
	0900		100 amp		40	.200		13.40	7.90		21.30	26.50	
	0950		200 amp		36	.222		29.50	8.75		38.25	45.50	
	1000		400 amp		30	.267		53	10.50		63.50	74	
	1050		600 amp		24	.333		81.50	13.15		94.65	109	
	1300		600 volt, 15 to 30 amp		40	.200		7.20	7.90		15.10	19.60	
	1350		35 to 60 amp		40	.200		12.40	7.90		20.30	25.50	
	1400		70 to 100 amp		36	.222		25.50	8.75		34.25	41	
	1450		110 to 200 amp		30	.267		51.50	10.50		62	72	
	1500		225 to 400 amp		24	.333		103	13.15		116.15	133	
	1550		600 amp		20	.400		148	15.75		163.75	187	
	1800		Class RK1, high capacity, 250 volt, 30 amp		50	.160		4.43	6.30		10.73	14.25	
	1850		60 amp		50	.160		8.10	6.30		14.40	18.30	
	1900		100 amp		40	.200		18.20	7.90		26.10	31.50	
	1950		200 amp		36	.222		40	8.75		48.75	57	
	2000		400 amp		30	.267		72.50	10.50		83	95	
	2050		600 amp		24	.333		111	13.15		124.15	142	
	2200		600 volt, 30 amp		40	.200		10.15	7.90		18.05	23	
	2250		60 amp		40	.200		17.35	7.90		25.25	31	
	2300		100 amp		36	.222		36	8.75		44.75	52.50	
	2350		200 amp		30	.267		72	10.50		82.50	94.50	
	2400		400 amp		24	.333		144	13.15		157.15	178	
	2450		600 amp		20	.400		207	15.75		222.75	252	
	2700		Class J, CLF, 250 or 600 volt, 30 amp		40	.200		12.70	7.90		20.60	25.50	
	2750		60 amp		40	.200		21	7.90		28.90	34.50	
	2800		100 amp		36	.222		31.50	8.75		40.25	47.50	
	2850		200 amp		30	.267		59	10.50		69.50	80.50	

ELECTRICAL 16

16490	Low V Dist Components/Accessories	CREW	DAILY OUTPUT	LABOR-HOURS	UNIT	2004 BARE COSTS				TOTAL INCL O&P	
						MAT.	LABOR	EQUIP.	TOTAL		
300	2900	400 amp	1 Elec	24	.333	Ea.	130	13.15		143.15	163	300
	2950	600 amp		20	.400		184	15.75		199.75	226	
	3100	Class L, 250 or 600 volt, 601 to 1200 amp		16	.500		340	19.70		359.70	405	
	3150	1500-1600 amp		13	.615		435	24.50		459.50	515	
	3200	1800-2000 amp		10	.800		585	31.50		616.50	685	
	3250	2500 amp		10	.800		775	31.50		806.50	895	
	3300	3000 amp		8	1		895	39.50		934.50	1,050	
	3350	3500-4000 amp		8	1		1,225	39.50		1,264.50	1,400	
	3400	4500-5000 amp		6.70	1.194		1,925	47		1,972	2,200	
	3450	6000 amp		5.70	1.404		3,175	55.50		3,230.50	3,575	
	3600	Plug, 120 volt, 1 to 10 amp		50	.160		1.01	6.30		7.31	10.50	
	3650	15 to 30 amp		50	.160		1.35	6.30		7.65	10.90	
	3700	Dual element 0.3 to 14 amp		50	.160		6.20	6.30		12.50	16.20	
	3750	15 to 30 amp		50	.160		1.69	6.30		7.99	11.25	
	3800	Fustat, 120 volt, 15 to 30 amp		50	.160		2.39	6.30		8.69	12.05	
	3850	0.3 to 14 amp		50	.160		6.25	6.30		12.55	16.25	
	3900	Adapters 0.3 to 10 amp		50	.160		3.04	6.30		9.34	12.75	
	3950	15 to 30 amp		50	.160		2.63	6.30		8.93	12.30	

16510	Interior Luminaires	CREW	DAILY OUTPUT	LABOR-HOURS	UNIT	2004 BARE COSTS				TOTAL INCL O&P	
						MAT.	LABOR	EQUIP.	TOTAL		
300	0010	**FIXTURE HANGERS**										300
	0220	Box hub cover	1 Elec	32	.250	Ea.	2.95	9.85		12.80	17.90	
	0240	Canopy		12	.667		5.50	26.50		32	45	
	0260	Connecting block		40	.200		2.75	7.90		10.65	14.75	
	0280	Cushion hanger		16	.500		17	19.70		36.70	48	
	0300	Box hanger, with mounting strap		8	1		4.55	39.50		44.05	63.50	
	0320	Connecting block		40	.200		1.05	7.90		8.95	12.85	
	0340	Flexible, 1/2" diameter, 4" long		12	.667		8.35	26.50		34.85	48	
	0360	6" long		12	.667		9	26.50		35.50	49	
	0380	8" long		12	.667		10.05	26.50		36.55	50	
	0400	10" long		12	.667		10.65	26.50		37.15	50.50	
	0420	12" long		12	.667		11.55	26.50		38.05	51.50	
	0440	15" long		12	.667		12.20	26.50		38.70	52.50	
	0460	18" long		12	.667		13.90	26.50		40.40	54.50	
	0480	3/4" diameter, 4" long		10	.800		10.10	31.50		41.60	58	
	0500	6" long		10	.800		11.35	31.50		42.85	59.50	
	0520	8" long		10	.800		12.40	31.50		43.90	60.50	
	0540	10" long		10	.800		13.15	31.50		44.65	61.50	
	0560	12" long		10	.800		14.35	31.50		45.85	63	
	0580	15" long		10	.800		15.80	31.50		47.30	64.50	
	0600	18" long		10	.800		17.70	31.50		49.20	66.50	
430	0010	**INTERIOR H.I.D. FIXTURES** Incl. lamps, and mounting hardware										430
	0700	High pressure sodium, recessed, round, 70 watt	1 Elec	3.50	2.286	Ea.	360	90		450	530	
	0720	100 watt		3.50	2.286		380	90		470	550	
	0740	150 watt		3.20	2.500		425	98.50		523.50	615	
	0760	Square, 70 watt		3.60	2.222		360	87.50		447.50	525	
	0780	100 watt		3.60	2.222		380	87.50		467.50	550	

16 ELECTRICAL

16510	Interior Luminaires	CREW	DAILY OUTPUT	LABOR-HOURS	UNIT	2004 BARE COSTS				TOTAL INCL O&P
						MAT.	LABOR	EQUIP.	TOTAL	
0820	250 watt	1 Elec	3	2.667	Ea.	535	105		640	740
0840	1000 watt	2 Elec	4.80	3.333		1,025	131		1,156	1,325
0860	Surface, round, 70 watt	1 Elec	3	2.667		510	105		615	715
0880	100 watt		3	2.667		520	105		625	730
0900	150 watt		2.70	2.963		545	117		662	775
0920	Square, 70 watt		3	2.667		495	105		600	700
0940	100 watt		3	2.667		520	105		625	730
0980	250 watt		2.50	3.200		525	126		651	770
1040	Pendent, round, 70 watt		3	2.667		540	105		645	750
1060	100 watt		3	2.667		520	105		625	725
1080	150 watt		2.70	2.963		530	117		647	760
1100	Square, 70 watt		3	2.667		555	105		660	765
1120	100 watt		3	2.667		560	105		665	775
1140	150 watt		2.70	2.963		580	117		697	810
1160	250 watt		2.50	3.200		780	126		906	1,050
1180	400 watt		2.40	3.333		820	131		951	1,100
1220	Wall, round, 70 watt		3	2.667		465	105		570	670
1240	100 watt		3	2.667		475	105		580	680
1260	150 watt		2.70	2.963		490	117		607	710
1300	Square, 70 watt		3	2.667		495	105		600	700
1320	100 watt		3	2.667		520	105		625	725
1340	150 watt		2.40	3.333		550	131		681	800
1360	250 watt		2.50	3.200		560	126		686	805
1380	400 watt	2 Elec	4.80	3.333		690	131		821	955
1400	1000 watt	"	3.60	4.444		975	175		1,150	1,325
1500	Metal halide, recessed, round, 175 watt	1 Elec	3.40	2.353		360	92.50		452.50	535
1520	250 watt	"	3.20	2.500		400	98.50		498.50	585
1540	400 watt	2 Elec	5.80	2.759		565	109		674	785
1580	Square, 175 watt	1 Elec	3.40	2.353		330	92.50		422.50	500
1640	Surface, round, 175 watt		2.90	2.759		440	109		549	645
1660	250 watt		2.70	2.963		665	117		782	905
1680	400 watt	2 Elec	4.80	3.333		780	131		911	1,050
1720	Square, 175 watt	1 Elec	2.90	2.759		480	109		589	690
1800	Pendent, round, 175 watt		2.90	2.759		520	109		629	735
1820	250 watt		2.70	2.963		720	117		837	965
1840	400 watt	2 Elec	4.80	3.333		800	131		931	1,075
1880	Square, 175 watt	1 Elec	2.90	2.759		505	109		614	720
1900	250 watt	"	2.70	2.963		510	117		627	740
1920	400 watt	2 Elec	4.80	3.333		825	131		956	1,100
1980	Wall, round, 175 watt	1 Elec	2.90	2.759		440	109		549	645
2000	250 watt	"	2.70	2.963		660	117		777	905
2020	400 watt	2 Elec	4.80	3.333		680	131		811	945
2060	Square, 175 watt	1 Elec	2.90	2.759		465	109		574	670
2080	250 watt	"	2.70	2.963		480	117		597	705
2100	400 watt	2 Elec	4.80	3.333		675	131		806	940
2800	High pressure sodium, recessed, 70 watt	1 Elec	3.50	2.286		440	90		530	620
2820	100 watt		3.50	2.286		450	90		540	630
2840	150 watt		3.20	2.500		460	98.50		558.50	655
2900	Surface, 70 watt		3	2.667		495	105		600	700
2920	100 watt		3	2.667		515	105		620	725
2940	150 watt		2.70	2.963		540	117		657	770
3000	Pendent, 70 watt		3	2.667		490	105		595	695
3020	100 watt		3	2.667		505	105		610	715
3040	150 watt		2.70	2.963		535	117		652	760
3100	Wall, 70 watt		3	2.667		525	105		630	735
3120	100 watt		3	2.667		550	105		655	760

430

ELECTRICAL 16

16510	Interior Luminaires	CREW	DAILY OUTPUT	LABOR-HOURS	UNIT	2004 BARE COSTS				TOTAL INCL O&P	
						MAT.	LABOR	EQUIP.	TOTAL		
430	3140	150 watt	1 Elec	2.70	2.963	Ea.	570	117		687	805	**430**
	3200	Metal halide, recessed, 175 watt		3.40	2.353		425	92.50		517.50	610	
	3220	250 watt	▼	3.20	2.500		455	98.50		553.50	645	
	3240	400 watt	2 Elec	5.80	2.759		565	109		674	785	
	3260	1000 watt	"	4.80	3.333		1,025	131		1,156	1,350	
	3280	Surface, 175 watt	1 Elec	2.90	2.759		430	109		539	635	
	3300	250 watt	"	2.70	2.963		665	117		782	905	
	3320	400 watt	2 Elec	4.80	3.333		795	131		926	1,075	
	3340	1000 watt	"	3.60	4.444		1,175	175		1,350	1,525	
	3360	Pendent, 175 watt	1 Elec	2.90	2.759		430	109		539	635	
	3380	250 watt	"	2.70	2.963		665	117		782	905	
	3400	400 watt	2 Elec	4.80	3.333		790	131		921	1,075	
	3420	1000 watt	"	3.60	4.444		1,275	175		1,450	1,650	
	3440	Wall, 175 watt	1 Elec	2.90	2.759		470	109		579	675	
	3460	250 watt	"	2.70	2.963		700	117		817	945	
	3480	400 watt	2 Elec	4.80	3.333		830	131		961	1,100	
	3500	1000 watt	"	3.60	4.444		1,325	175		1,500	1,725	
440	0010	**INTERIOR LIGHTING FIXTURES** Including lamps, mounting										**440**
	0030	hardware and connections	R16510 -440									
	0100	Fluorescent, C.W. lamps, troffer, recess mounted in grid, RS										
	0200	Acrylic lens, 1'W x 4'L, two 40 watt	1 Elec	5.70	1.404	Ea.	45	55.50		100.50	132	
	0210	1'W x 4'L, three 40 watt		5.40	1.481		52.50	58.50		111	145	
	0300	2'W x 2'L, two U40 watt		5.70	1.404		48	55.50		103.50	136	
	0400	2'W x 4'L, two 40 watt		5.30	1.509		48	59.50		107.50	142	
	0500	2'W x 4'L, three 40 watt		5	1.600		51.50	63		114.50	151	
	0600	2'W x 4'L, four 40 watt	▼	4.70	1.702		54.50	67		121.50	160	
	0700	4'W x 4'L, four 40 watt	2 Elec	6.40	2.500		288	98.50		386.50	460	
	0800	4'W x 4'L, six 40 watt		6.20	2.581		298	102		400	480	
	0900	4'W x 4'L, eight 40 watt	▼	5.80	2.759		310	109		419	500	
	0910	Acrylic lens, 1'W x 4'L, two 32 watt	1 Elec	5.70	1.404		53.50	55.50		109	142	
	0930	2'W x 2'L, two U32 watt		5.70	1.404		76	55.50		131.50	166	
	0940	2'W x 4'L, two 32 watt		5.30	1.509		68	59.50		127.50	164	
	0950	2'W x 4'L, three 32 watt		5	1.600		72	63		135	173	
	0960	2'W x 4'L, four 32 watt	▼	4.70	1.702	▼	74	67		141	181	
	1000	Surface mounted, RS										
	1030	Acrylic lens with hinged & latched door frame										
	1100	1'W x 4'L, two 40 watt	1 Elec	7	1.143	Ea.	70	45		115	144	
	1110	1'W x 4'L, three 40 watt		6.70	1.194		72	47		119	149	
	1200	2'W x 2'L, two U40 watt		7	1.143		75	45		120	150	
	1300	2'W x 4'L, two 40 watt		6.20	1.290		86	51		137	170	
	1400	2'W x 4'L, three 40 watt		5.70	1.404		87	55.50		142.50	178	
	1500	2'W x 4'L, four 40 watt	▼	5.30	1.509		89	59.50		148.50	187	
	1600	4'W x 4'L, four 40 watt	2 Elec	7.20	2.222		400	87.50		487.50	570	
	1700	4'W x 4'L, six 40 watt		6.60	2.424		435	95.50		530.50	615	
	1800	4'W x 4'L, eight 40 watt		6.20	2.581		450	102		552	645	
	1900	2'W x 8'L, four 40 watt		6.40	2.500	▼	155	98.50		253.50	315	
	2000	2'W x 8'L, eight 40 watt	▼	6.20	2.581	▼	178	102		280	345	
	2010	Acrylic wrap around lens										
	2020	6"W x 4'L, one 40 watt	1 Elec	8	1	Ea.	64	39.50		103.50	129	
	2030	6"W x 8'L, two 40 watt	2 Elec	8	2		70	79		149	194	
	2040	11"W x 4'L, two 40 watt	1 Elec	7	1.143		42	45		87	113	
	2050	11"W x 8'L, four 40 watt	2 Elec	6.60	2.424		70	95.50		165.50	219	
	2060	16"W x 4'L, four 40 watt	1 Elec	5.30	1.509		70	59.50		129.50	166	
	2070	16"W x 8'L, eight 40 watt	2 Elec	6.40	2.500		152	98.50		250.50	315	
	2080	2'W x 2'L, two U40 watt	1 Elec	7	1.143	▼	86	45		131	162	

Important: See the Reference Section for critical supporting data - Reference Nos., Crews, & City Cost Indexes

				DAILY	LABOR-			2004 BARE COSTS				TOTAL
16510		**Interior Luminaires**	CREW	OUTPUT	HOURS	UNIT	MAT.	LABOR	EQUIP.	TOTAL		INCL O&P
440	2100	Strip fixture											440
	2130	Surface mounted [R16510 -440]										
	2200	4' long, one 40 watt RS	1 Elec	8.50	.941	Ea.	26.50	37		63.50		84
	2300	4' long, two 40 watt RS		8	1		28.50	39.50		68		89.50
	2400	4' long, one 40 watt, SL		8	1		38.50	39.50		78		101
	2500	4' long, two 40 watt, SL		7	1.143		52.50	45		97.50		125
	2600	8' long, one 75 watt, SL	2 Elec	13.40	1.194		39.50	47		86.50		114
	2700	8' long, two 75 watt, SL	"	12.40	1.290		47.50	51		98.50		128
	2800	4' long, two 60 watt, HO	1 Elec	6.70	1.194		77	47		124		155
	2900	8' long, two 110 watt, HO	2 Elec	10.60	1.509		81	59.50		140.50		178
	2910	4' long, two 115 watt, VHO	1 Elec	6.50	1.231		107	48.50		155.50		190
	2920	8' long, two 215 watt, VHO	2 Elec	10.40	1.538		101	60.50		161.50		201
	3000	Pendent mounted, industrial, white porcelain enamel										
	3100	4' long, two 40 watt, RS	1 Elec	5.70	1.404	Ea.	43.50	55.50		99		131
	3200	4' long, two 60 watt, HO	"	5	1.600		69	63		132		170
	3300	8' long, two 75 watt, SL	2 Elec	8.80	1.818		82	71.50		153.50		197
	3400	8' long, two 110 watt, HO	"	8	2		105	79		184		233
	3410	Acrylic finish, 4' long, two 40 watt, RS	1 Elec	5.70	1.404		72	55.50		127.50		162
	3420	4' long, two 60 watt, HO		5	1.600		133	63		196		240
	3430	4' long, two 115 watt, VHO		4.80	1.667		172	65.50		237.50		287
	3440	8' long, two 75 watt, SL	2 Elec	8.80	1.818		141	71.50		212.50		262
	3450	8' long, two 110 watt, HO		8	2		157	79		236		290
	3460	8' long, two 215 watt, VHO		7.60	2.105		222	83		305		365
	3470	Troffer, air handling, 2'W x 4'L with four 40 watt, RS	1 Elec	4	2		77	79		156		202
	3480	2'W x 2'L with two U40 watt RS		5.50	1.455		67	57.50		124.50		159
	3490	Air connector insulated, 5" diameter		20	.400		52.50	15.75		68.25		81.50
	3500	6" diameter		20	.400		53.50	15.75		69.25		82.50
	3510	Troffer parabolic lay-in, 1'W x 4'L with one F40		5.70	1.404		66	55.50		121.50		155
	3520	1'W x 4'L with two F40		5.30	1.509		70	59.50		129.50		166
	3530	2'W x 4'L with three F40		5	1.600		98	63		161		202
	3535	Downlight, recess mounted		8	1		85	39.50		124.50		152
	3540	Wall washer, recess mounted		8	1		85	39.50		124.50		152
	3550	Direct/indirect, 4' long, stl., pendent mtd.		5	1.600		115	63		178		221
	3560	4' long, alum., pendent mtd.		5	1.600		285	63		348		410
	3565	Prefabricated cove, 4' long, stl. continuous row		5	1.600		173	63		236		284
	3570	4' long, alum. continuous row		5	1.600		295	63		358		420
	4220	Metal halide, integral ballast, ceiling, recess mounted										
	4230	prismatic glass lens, floating door										
	4240	2'W x 2'L, 250 watt	1 Elec	3.20	2.500	Ea.	267	98.50		365.50		440
	4250	2'W x 2'L, 400 watt	2 Elec	5.80	2.759		305	109		414		495
	4260	Surface mounted, 2'W x 2'L, 250 watt	1 Elec	2.70	2.963		267	117		384		470
	4270	400 watt	2 Elec	4.80	3.333		315	131		446		540
	4280	High bay, aluminum reflector,										
	4290	Single unit, 400 watt	2 Elec	4.60	3.478	Ea.	320	137		457		555
	4300	Single unit, 1000 watt		4	4		460	158		618		740
	4310	Twin unit, 400 watt		3.20	5		640	197		837		1,000
	4320	Low bay, aluminum reflector, 250W DX lamp	1 Elec	3.20	2.500		310	98.50		408.50		485
	4330	400 watt lamp	2 Elec	5	3.200		450	126		576		685
	4340	High pressure sodium integral ballast ceiling, recess mounted										
	4350	prismatic glass lens, floating door										
	4360	2'W x 2'L, 150 watt lamp	1 Elec	3.20	2.500	Ea.	325	98.50		423.50		505
	4370	2'W x 2'L, 400 watt lamp	2 Elec	5.80	2.759		390	109		499		590
	4380	Surface mounted, 2'W x 2'L, 150 watt lamp	1 Elec	2.70	2.963		365	117		482		580
	4390	400 watt lamp	2 Elec	4.80	3.333		410	131		541		645
	4400	High bay, aluminum reflector,										
	4410	Single unit, 400 watt lamp	2 Elec	4.60	3.478	Ea.	295	137		432		530

ELECTRICAL 16

267

16510	Interior Luminaires		CREW	DAILY OUTPUT	LABOR-HOURS	UNIT	2004 BARE COSTS				TOTAL INCL O&P
							MAT.	LABOR	EQUIP.	TOTAL	
4430	Single unit, 1000 watt lamp	R16510 -440	2 Elec	4	4	Ea.	425	158		583	705
4440	Low bay, aluminum reflector, 150 watt lamp		1 Elec	3.20	2.500		255	98.50		353.50	430
4445	High bay H.I.D. quartz restrike		"	16	.500	↓	127	19.70		146.70	170
4450	Incandescent, high hat can, round alzak reflector, prewired										
4470	100 watt		1 Elec	8	1	Ea.	56.50	39.50		96	121
4480	150 watt			8	1		81	39.50		120.50	148
4500	300 watt			6.70	1.194		184	47		231	272
4520	Round with reflector and baffles, 150 watt			8	1		38	39.50		77.50	101
4540	Round with concentric louver, 150 watt PAR		↓	8	1	↓	61	39.50		100.50	126
4600	Square glass lens with metal trim, prewired										
4630	100 watt		1 Elec	6.70	1.194	Ea.	45	47		92	120
4700	200 watt			6.70	1.194		74	47		121	152
4800	300 watt			5.70	1.404		115	55.50		170.50	210
4810	500 watt			5	1.600		225	63		288	340
4900	Ceiling/wall, surface mounted, metal cylinder, 75 watt			10	.800		46	31.50		77.50	97.50
4920	150 watt			10	.800		66	31.50		97.50	120
4930	300 watt			8	1		128	39.50		167.50	200
5000	500 watt			6.70	1.194		305	47		352	405
5010	Square, 100 watt			8	1		96	39.50		135.50	165
5020	150 watt			8	1		105	39.50		144.50	175
5030	300 watt			7	1.143		284	45		329	375
5040	500 watt		↓	6	1.333	↓	284	52.50		336.50	390
5200	Ceiling, surface mounted, opal glass drum										
5300	8", one 60 watt lamp		1 Elec	10	.800	Ea.	34	31.50		65.50	84.50
5400	10", two 60 watt lamps			8	1		38	39.50		77.50	101
5500	12", four 60 watt lamps			6.70	1.194		55	47		102	131
5510	Pendent, round, 100 watt			8	1		96	39.50		135.50	165
5520	150 watt			8	1		97	39.50		136.50	166
5530	300 watt			6.70	1.194		135	47		182	219
5540	500 watt			5.50	1.455		257	57.50		314.50	370
5550	Square, 100 watt			6.70	1.194		120	47		167	202
5560	150 watt			6.70	1.194		125	47		172	208
5570	300 watt			5.70	1.404		185	55.50		240.50	287
5580	500 watt			5	1.600		250	63		313	370
5600	Wall, round, 100 watt			8	1		56	39.50		95.50	120
5620	300 watt			8	1		90	39.50		129.50	158
5630	500 watt			6.70	1.194		320	47		367	425
5640	Square, 100 watt			8	1		91	39.50		130.50	159
5650	150 watt			8	1		93.50	39.50		133	162
5660	300 watt			7	1.143		131	45		176	211
5670	500 watt			6	1.333		232	52.50		284.50	335
6010	Vapor tight, incandescent, ceiling mounted, 200 watt			6.20	1.290		48	51		99	129
6020	Recessed, 200 watt			6.70	1.194		85	47		132	164
6030	Pendent, 200 watt			6.70	1.194		48	47		95	123
6040	Wall, 200 watt			8	1		58	39.50		97.50	123
6100	Fluorescent, surface mounted, 2 lamps, 4'L, RS, 40 watt			3.20	2.500		90	98.50		188.50	246
6110	Industrial, 2 lamps 4' long in tandem, 430 MA			2.20	3.636		165	143		308	395
6130	2 lamps 4' long, 800 MA			1.90	4.211		141	166		307	400
6160	Pendent, indust, 2 lamps 4'L in tandem, 430 MA			1.90	4.211		193	166		359	460
6170	2 lamps 4' long, 430 MA			2.30	3.478		129	137		266	345
6180	2 lamps 4' long, 800 MA		↓	1.70	4.706	↓	163	185		348	455
6300	Explosionproof										
6310	Metal halide with ballast, ceiling, surface mounted, 175 W		1 Elec	2.90	2.759	Ea.	655	109		764	880
6320	250 watt		"	2.70	2.963	↓	785	117		902	1,050
6330	400 watt		2 Elec	4.80	3.333		840	131		971	1,125
6340	Ceiling, pendent mounted, 175 watt		1 Elec	2.60	3.077		620	121		741	865

440

16 ELECTRICAL

Important: See the Reference Section for critical supporting data - Reference Nos., Crews, & City Cost Indexes

	16510	Interior Luminaires		CREW	DAILY OUTPUT	LABOR-HOURS	UNIT	2004 BARE COSTS				TOTAL INCL O&P	
								MAT.	LABOR	EQUIP.	TOTAL		
440	6350	250 watt	R16510 -440	1 Elec	2.40	3.333	Ea.	755	131		886	1,025	440
	6360	400 watt		2 Elec	4.20	3.810		810	150		960	1,125	
	6370	Wall, surface mounted, 175 watt		1 Elec	2.90	2.759		700	109		809	930	
	6380	250 watt		"	2.70	2.963		835	117		952	1,100	
	6390	400 watt		2 Elec	4.80	3.333		890	131		1,021	1,175	
	6400	High pressure sodium, ceiling surface mounted, 70 watt		1 Elec	3	2.667		705	105		810	930	
	6410	100 watt			3	2.667		730	105		835	960	
	6420	150 watt			2.70	2.963		755	117		872	1,000	
	6430	Pendent mounted, 70 watt			2.70	2.963		650	117		767	890	
	6440	100 watt			2.70	2.963		700	117		817	945	
	6450	150 watt			2.40	3.333		725	131		856	990	
	6460	Wall mounted, 70 watt			3	2.667		755	105		860	985	
	6470	100 watt			3	2.667		780	105		885	1,025	
	6480	150 watt			2.70	2.963		805	117		922	1,050	
	6510	Incandescent, ceiling mounted, 200 watt			4	2		630	79		709	810	
	6520	Pendent mounted, 200 watt			3.50	2.286		540	90		630	730	
	6530	Wall mounted, 200 watt			4	2		625	79		704	805	
	6600	Fluorescent, RS, 4' long, ceiling mounted, two 40 watt			2.70	2.963		1,600	117		1,717	1,925	
	6610	Three 40 watt			2.20	3.636		2,300	143		2,443	2,750	
	6620	Four 40 watt			1.90	4.211		2,950	166		3,116	3,500	
	6630	Pendent mounted, two 40 watt			2.30	3.478		1,875	137		2,012	2,250	
	6640	Three 40 watt			1.90	4.211		2,650	166		2,816	3,175	
	6650	Four 40 watt			1.70	4.706		3,500	185		3,685	4,125	
	6850	Vandalproof, surface mounted, fluorescent, two 40 watt			3.20	2.500		190	98.50		288.50	355	
	6860	Incandescent, one 150 watt			8	1		53	39.50		92.50	117	
	6900	Mirror light, fluorescent, RS, acrylic enclosure, two 40 watt			8	1		84.50	39.50		124	152	
	6910	One 40 watt			8	1		66	39.50		105.50	131	
	6920	One 20 watt			12	.667		52	26.50		78.50	96.50	
	7000	Low bay, aluminum reflector. 70 watt, high pressure sodium			4	2		231	79		310	370	
	7010	250 watt			3.20	2.500		310	98.50		408.50	485	
	7020	400 watt		2 Elec	5	3.200		315	126		441	535	
	7500	Ballast replacement, by weight of ballast, to 15' high											
	7520	Indoor fluorescent, less than 2 lbs.		1 Elec	10	.800	Ea.		31.50		31.50	47	
	7540	Two 40W, watt reducer, 2 to 5 lbs.			9.40	.851		22	33.50		55.50	74	
	7560	Two F96 slimline, over 5 lbs.			8	1		32.50	39.50		72	94.50	
	7580	Vaportite ballast, less than 2 lbs.			9.40	.851			33.50		33.50	50	
	7600	2 lbs. to 5 lbs.			8.90	.899			35.50		35.50	52.50	
	7620	Over 5 lbs.			7.60	1.053			41.50		41.50	61.50	
	7630	Electronic ballast for two tubes			8	1		32.50	39.50		72	94.50	
	7640	Dimmable ballast one lamp			8	1		45	39.50		84.50	108	
	7650	Dimmable ballast two-lamp			7.60	1.053		73	41.50		114.50	142	
	7690	Emergency ballast (factory installed in fixture)						121			121	133	
	7990	Decorator											
	8000	Pendent RLM in colors, shallow dome, 12" diam. 100 W		1 Elec	8	1	Ea.	63	39.50		102.50	128	
	8010	Regular dome, 12" diam., 100 watt			8	1		65	39.50		104.50	130	
	8020	16" diam., 200 watt			7	1.143		66	45		111	140	
	8030	18" diam., 500 watt			6	1.333		80	52.50		132.50	166	
	8100	Picture framing light, minimum			16	.500		71.50	19.70		91.20	108	
	8110	Maximum			16	.500		94.50	19.70		114.20	134	
	8150	Miniature low voltage, recessed, pinhole			8	1		126	39.50		165.50	198	
	8160	Star			8	1		104	39.50		143.50	173	
	8170	Adjustable cone			8	1		141	39.50		180.50	214	
	8180	Eyeball			8	1		115	39.50		154.50	186	
	8190	Cone			8	1		101	39.50		140.50	170	
	8200	Coilex baffle			8	1		99	39.50		138.50	168	
	8210	Surface mounted, adjustable cylinder			8	1		102	39.50		141.50	171	

ELECTRICAL 16

			DAILY	LABOR-		2004 BARE COSTS				TOTAL
	16510	**Interior Luminaires**	CREW	OUTPUT	HOURS	UNIT	MAT.	LABOR	EQUIP.	TOTAL	INCL O&P

			CREW	OUTPUT	HOURS	UNIT	MAT.	LABOR	EQUIP.	TOTAL	INCL O&P	
440	8250	Chandeliers, incandescent R16510 -440										**440**
	8260	24" diam. x 42" high, 6 light candle	1 Elec	6	1.333	Ea.	298	52.50		350.50	410	
	8270	24" diam. x 42" high, 6 light candle w/glass shade		6	1.333		230	52.50		282.50	330	
	8280	17" diam. x 12" high, 8 light w/glass panels		8	1		245	39.50		284.50	330	
	8290	32" diam. x 48"H, 10 light bohemian lead crystal		4	2		505	79		584	670	
	8300	27" diam. x 29"H, 10 light bohemian lead crystal		4	2		470	79		549	630	
	8310	21" diam. x 9" high 6 light sculptured ice crystal	▼	8	1	▼	420	39.50		459.50	520	
	8500	Accent lights, on floor or edge, 0.5W low volt incandescent										
	8520	incl. transformer & fastenings, based on 100' lengths										
	8550	Lights in clear tubing, 12" on center	1 Elec	230	.035	L.F.	6.40	1.37		7.77	9.10	
	8560	6" on center		160	.050		8.35	1.97		10.32	12.15	
	8570	4" on center		130	.062		12.75	2.42		15.17	17.65	
	8580	3" on center		125	.064		14.15	2.52		16.67	19.30	
	8590	2" on center		100	.080		20.50	3.15		23.65	27	
	8600	Carpet, lights both sides 6" OC, in alum. extrusion		270	.030		19.30	1.17		20.47	22.50	
	8610	In bronze extrusion		270	.030		22	1.17		23.17	25.50	
	8620	Carpet-bare floor, lights 18" OC, in alum. extrusion		270	.030		15.40	1.17		16.57	18.70	
	8630	In bronze extrusion		270	.030		18	1.17		19.17	21.50	
	8640	Carpet edge-wall, lights 6" OC in alum. extrusion		270	.030		19.30	1.17		20.47	22.50	
	8650	In bronze extrusion		270	.030		22	1.17		23.17	25.50	
	8660	Bare floor, lights 18" OC, in aluminum extrusion		300	.027		15.40	1.05		16.45	18.50	
	8670	In bronze extrusion		300	.027		18	1.05		19.05	21.50	
	8680	Bare floor conduit, aluminum extrusion		300	.027		5.10	1.05		6.15	7.15	
	8690	In bronze extrusion		300	.027	▼	10.20	1.05		11.25	12.75	
	8700	Step edge to 36", lights 6" OC, in alum. extrusion		100	.080	Ea.	51.50	3.15		54.65	61	
	8710	In bronze extrusion		100	.080		53.50	3.15		56.65	63.50	
	8720	Step edge to 54", lights 6" OC, in alum. extrusion		100	.080		77	3.15		80.15	89	
	8730	In bronze extrusion		100	.080		81.50	3.15		84.65	94	
	8740	Step edge to 72", lights 6" OC, in alum. extrusion		100	.080		103	3.15		106.15	118	
	8750	In bronze extrusion		100	.080		112	3.15		115.15	129	
	8760	Connector, male		32	.250		1.90	9.85		11.75	16.75	
	8770	Female with pigtail		32	.250		4	9.85		13.85	19.05	
	8780	Clamps		400	.020		.38	.79		1.17	1.59	
	8790	Transformers, 50 watt		8	1		54	39.50		93.50	118	
	8800	250 watt		4	2		185	79		264	320	
	8810	1000 watt	▼	2.70	2.963	▼	340	117		457	550	
800	0010	**RESIDENTIAL FIXTURES**										**800**
	0400	Fluorescent, interior, surface, circline, 32 watt & 40 watt	1 Elec	20	.400	Ea.	76.50	15.75		92.25	108	
	0500	2' x 2', two U 40 watt		8	1		93	39.50		132.50	161	
	0700	Shallow under cabinet, two 20 watt		16	.500		40.50	19.70		60.20	74.50	
	0900	Wall mounted, 4'L, one 40 watt, with baffle		10	.800		99	31.50		130.50	156	
	2000	Incandescent, exterior lantern, wall mounted, 60 watt		16	.500		31	19.70		50.70	63.50	
	2100	Post light, 150W, with 7' post		4	2		110	79		189	238	
	2500	Lamp holder, weatherproof with 150W PAR		16	.500		16.90	19.70		36.60	48	
	2550	With reflector and guard		12	.667		52	26.50		78.50	96	
	2600	Interior pendent, globe with shade, 150 watt	▼	20	.400	▼	117	15.75		132.75	153	

	16520	**Exterior Luminaires**										
300	0010	**EXTERIOR FIXTURES** With lamps										**300**
	0200	Wall mounted, incandescent, 100 watt	1 Elec	8	1	Ea.	28	39.50		67.50	89.50	
	0400	Quartz, 500 watt		5.30	1.509		53.50	59.50		113	148	
	0420	1500 watt		4.20	1.905		97	75		172	219	
	1100	Wall pack, low pressure sodium, 35 watt		4	2		214	79		293	350	
	1150	55 watt		4	2		255	79		334	400	
	1160	High pressure sodium, 70 watt		4	2		185	79		264	320	
	1170	150 watt	▼	4	2		228	79		307	370	

Important: See the Reference Section for critical supporting data - Reference Nos., Crews, & City Cost Indexes

16520	Exterior Luminaires	CREW	DAILY OUTPUT	LABOR-HOURS	UNIT	2004 BARE COSTS				TOTAL INCL O&P
						MAT.	LABOR	EQUIP.	TOTAL	
300 1180	Metal Halide, 175 watt	1 Elec	4	2	Ea.	255	79		334	400 300
1190	250 watt	↓	4	2	↓	260	79		339	405
1200	Floodlights with ballast and lamp,									
1400	pole mounted, pole not included									
1950	Metal halide, 175 watt	1 Elec	2.70	2.963	Ea.	300	117		417	505
2000	400 watt	2 Elec	4.40	3.636		350	143		493	600
2200	1000 watt		4	4		505	158		663	790
2210	1500 watt	↓	3.70	4.324		530	170		700	840
2250	Low pressure sodium, 55 watt	1 Elec	2.70	2.963		485	117		602	710
2270	90 watt		2	4		535	158		693	825
2290	180 watt		2	4		680	158		838	985
2340	High pressure sodium, 70 watt		2.70	2.963		193	117		310	385
2360	100 watt		2.70	2.963		198	117		315	390
2380	150 watt	↓	2.70	2.963		220	117		337	415
2400	400 watt	2 Elec	4.40	3.636		340	143		483	585
2600	1000 watt	"	4	4		500	158		658	785
2610	Incandescent, 300 watt	1 Elec	4	2		76.50	79		155.50	201
2620	500 watt	"	4	2		122	79		201	251
2630	1000 watt	2 Elec	6	2.667		131	105		236	300
2640	1500 watt	"	6	2.667		144	105		249	315
2650	Roadway area luminaire, low pressure sodium, 135 watt	1 Elec	2	4		535	158		693	825
2700	180 watt	"	2	4		565	158		723	855
2750	Metal halide, 400 watt	2 Elec	4.40	3.636		445	143		588	700
2760	1000 watt		4	4		500	158		658	785
2780	High pressure sodium, 400 watt		4.40	3.636		460	143		603	720
2790	1000 watt	↓	4	4	↓	525	158		683	815
2800	Light poles, anchor base									
2820	not including concrete bases									
2840	Aluminum pole, 8' high	1 Elec	4	2	Ea.	470	79		549	635
2850	10' high		4	2		495	79		574	655
2860	12' high		3.80	2.105		515	83		598	690
2870	14' high		3.40	2.353		535	92.50		627.50	730
2880	16' high	↓	3	2.667		590	105		695	800
3000	20' high	R-3	2.90	6.897		615	267	55	937	1,125
3200	30' high		2.60	7.692		1,200	298	61	1,559	1,825
3400	35' high		2.30	8.696		1,300	335	69	1,704	2,000
3600	40' high	↓	2	10		1,475	385	79.50	1,939.50	2,300
3800	Bracket arms, 1 arm	1 Elec	8	1		80.50	39.50		120	147
4000	2 arms		8	1		162	39.50		201.50	237
4200	3 arms		5.30	1.509		243	59.50		302.50	355
4400	4 arms		5.30	1.509		325	59.50		384.50	445
4500	Steel pole, galvanized, 8' high		3.80	2.105		420	83		503	585
4510	10' high		3.70	2.162		440	85		525	610
4520	12' high		3.40	2.353		475	92.50		567.50	665
4530	14' high		3.10	2.581		505	102		607	705
4540	16' high		2.90	2.759		535	109		644	750
4550	18' high	↓	2.70	2.963		565	117		682	795
4600	20' high	R-3	2.60	7.692		745	298	61	1,104	1,325
4800	30' high		2.30	8.696		875	335	69	1,279	1,550
5000	35' high		2.20	9.091		960	350	72.50	1,382.50	1,650
5200	40' high	↓	1.70	11.765		1,175	455	93.50	1,723.50	2,075
5400	Bracket arms, 1 arm	1 Elec	8	1		122	39.50		161.50	193
5600	2 arms		8	1		189	39.50		228.50	267
5800	3 arms		5.30	1.509		205	59.50		264.50	315
6000	4 arms	↓	5.30	1.509		285	59.50		344.50	405
6100	Fiberglass pole, 1 or 2 fixtures, 20' high	R-3	4	5	↓	350	193	40	583	715

	16520	Exterior Luminaires	CREW	DAILY OUTPUT	LABOR-HOURS	UNIT	2004 BARE COSTS				TOTAL INCL O&P	
							MAT.	LABOR	EQUIP.	TOTAL		
300	6200	30' high	R-3	3.60	5.556	Ea.	550	215	44	809	975	300
	6300	35' high		3.20	6.250		690	242	49.50	981.50	1,175	
	6400	40' high	↓	2.80	7.143		840	276	57	1,173	1,400	
	6420	Wood pole, 4-1/2" x 5-1/8", 8' high	1 Elec	6	1.333		230	52.50		282.50	330	
	6430	10' high		6	1.333		260	52.50		312.50	365	
	6440	12' high		5.70	1.404		330	55.50		385.50	450	
	6450	15' high		5	1.600		385	63		448	520	
	6460	20' high	↓	4	2	↓	465	79		544	625	
	6500	Bollard light, lamp & ballast, 42" high with polycarbonate lens										
	6800	Metal halide, 175 watt	1 Elec	3	2.667	Ea.	595	105		700	810	
	6900	High pressure sodium, 70 watt		3	2.667		610	105		715	825	
	7000	100 watt		3	2.667		610	105		715	825	
	7100	150 watt		3	2.667		595	105		700	810	
	7200	Incandescent, 150 watt	↓	3	2.667	↓	430	105		535	630	
	7300	Transformer bases, not including concrete bases										
	7320	Maximum pole size, steel, 40' high	1 Elec	2	4	Ea.	1,025	158		1,183	1,350	
	7340	Cast aluminum, 30' high		3	2.667		555	105		660	770	
	7350	40' high	↓	2.50	3.200	↓	840	126		966	1,125	
	7380	Landscape recessed uplight, incl. housing, ballast, transformer										
	7390	& reflector										
	7420	Incandescent, 250 watt	1 Elec	5	1.600	Ea.	420	63		483	555	
	7440	Quartz, 250 watt		5	1.600		400	63		463	535	
	7460	500 watt	↓	4	2	↓	410	79		489	565	
	7500	Replacement (H.I.D.) ballasts,										
	7510	Multi-tap 120/208/240/277 volt										
	7550	High pressure sodium, 70 watt	1 Elec	10	.800	Ea.	118	31.50		149.50	176	
	7560	100 watt		9.40	.851		123	33.50		156.50	185	
	7570	150 watt		9	.889		131	35		166	196	
	7580	250 watt		8.50	.941		198	37		235	273	
	7590	400 watt		7	1.143		222	45		267	310	
	7600	1000 watt		6	1.333		320	52.50		372.50	430	
	7610	Metal halide, 175 watt		8	1		71	39.50		110.50	137	
	7620	250 watt		8	1		92	39.50		131.50	160	
	7630	400 watt		7	1.143		115	45		160	194	
	7640	1000 watt		6	1.333		197	52.50		249.50	295	
	7650	1500 watt		5	1.600		250	63		313	370	
	7810	Walkway luminaire, square 16", metal halide 250 watt		2.70	2.963		480	117		597	705	
	7820	High pressure sodium, 70 watt		3	2.667		550	105		655	760	
	7830	100 watt		3	2.667		560	105		665	770	
	7840	150 watt		3	2.667		560	105		665	770	
	7850	200 watt		3	2.667		565	105		670	775	
	7910	Round 19", metal halide, 250 watt		2.70	2.963		700	117		817	945	
	7920	High pressure sodium, 70 watt		3	2.667		770	105		875	1,000	
	7930	100 watt		3	2.667		770	105		875	1,000	
	7940	150 watt		3	2.667		775	105		880	1,000	
	7950	250 watt		2.70	2.963		810	117		927	1,075	
	8000	Sphere 14" opal, incandescent, 200 watt		4	2		226	79		305	365	
	8020	Sphere 18" opal, incandescent, 300 watt		3.50	2.286		273	90		363	435	
	8040	Sphere 16" clear, high pressure sodium, 70 watt		3	2.667		475	105		580	675	
	8050	100 watt		3	2.667		505	105		610	710	
	8100	Cube 16" opal, incandescent, 300 watt		3.50	2.286		300	90		390	465	
	8120	High pressure sodium, 70 watt		3	2.667		440	105		545	640	
	8130	100 watt		3	2.667		450	105		555	650	
	8230	Lantern, high pressure sodium, 70 watt		3	2.667		390	105		495	585	
	8240	100 watt		3	2.667		420	105		525	615	
	8250	150 watt	↓	3	2.667	↓	395	105		500	590	

Important: See the Reference Section for critical supporting data - Reference Nos., Crews, & City Cost Indexes

16520 | Exterior Luminaires

			CREW	DAILY OUTPUT	LABOR-HOURS	UNIT	2004 BARE COSTS MAT.	LABOR	EQUIP.	TOTAL	TOTAL INCL O&P	
300	8260	250 watt	1 Elec	2.70	2.963	Ea.	550	117		667	780	300
	8270	Incandescent, 300 watt		3.50	2.286		290	90		380	455	
	8330	Reflector 22" w/globe, high pressure sodium, 70 watt		3	2.667		370	105		475	560	
	8340	100 watt		3	2.667		375	105		480	565	
	8350	150 watt		3	2.667		380	105		485	570	
	8360	250 watt		2.70	2.963		485	117		602	705	

16525 | Aviation Lighting

			CREW	DAILY OUTPUT	LABOR-HOURS	UNIT	2004 BARE COSTS MAT.	LABOR	EQUIP.	TOTAL	TOTAL INCL O&P	
100	0010	**AIRPORT LIGHTING**										100
	0100	Runway centerline, bidir., semi-flush, 200 W, w/shallow insert base	R-22	12.40	3.006	Ea.	1,075	99.50		1,174.50	1,350	
	0120	Flush, 200 W, w/shallow insert base		12.40	3.006		1,075	99.50		1,174.50	1,350	
	0130	for mounting in base housing		18.64	2		570	66		636	730	
	0150	Touchdown zone light, unidirectional, 200 W, w/shallow insert base		12.40	3.006		890	99.50		989.50	1,125	
	0160	115 W		12.40	3.006		945	99.50		1,044.50	1,200	
	0180	Unidirectional, 200 W, for mounting in base housing		18.60	2.004		435	66.50		501.50	580	
	0190	115 W		18.60	2.004		470	66.50		536.50	615	
	0210	Runway edge & threshold light, bidir., 200 W, for base housing		9.36	3.983		615	132		747	875	
	0240	Threshold & approach light, unidir., 200 W, for base housing		9.36	3.983		360	132		492	595	
	0260	Runway edge, bi-directional, 2-115 W, for base housing		12.40	3.006		950	99.50		1,049.50	1,200	
	0280	Runway threshold & end, bidir., 2-115 W, for base housing		12.40	3.006		1,050	99.50		1,149.50	1,300	
	0370	45 W, flush, for mounting in base housing		18.64	2		515	66		581	665	
	0380	115 W		18.64	2		525	66		591	680	
	1200	Wind cone, 12' lighted assembly, rigid, w/obstruction light	R-21	1.36	24.118		6,025	950	42	7,017	8,100	
	1210	Without obstruction light		1.52	21.579		5,050	850	38	5,938	6,875	
	1220	Unlighted assembly, w/obstruction light		1.68	19.524		5,525	770	34	6,329	7,275	
	1230	Without obstruction light		1.84	17.826		5,050	700	31	5,781	6,625	
	1240	Wind cone slip fitter, 2-1/2" pipe		21.84	1.502		66.50	59	2.63	128.13	164	
	1250	Wind cone sock, 12' x 3', cotton		6.56	5		365	197	8.75	570.75	710	
	1260	Nylon		6.56	5		370	197	8.75	575.75	715	

16530 | Emergency Lighting

			CREW	DAILY OUTPUT	LABOR-HOURS	UNIT	2004 BARE COSTS MAT.	LABOR	EQUIP.	TOTAL	TOTAL INCL O&P	
320	0010	**EXIT AND EMERGENCY LIGHTING**										320
	0080	Exit light ceiling or wall mount, incandescent, single face	1 Elec	8	1	Ea.	38.50	39.50		78	101	
	0100	Double face		6.70	1.194		44	47		91	119	
	0120	Explosion proof		3.80	2.105		395	83		478	560	
	0150	Fluorescent, single face		8	1		62	39.50		101.50	127	
	0160	Double face		6.70	1.194		65	47		112	142	
	0200	L.E.D. standard, single face		8	1		62	39.50		101.50	127	
	0220	Double face		6.70	1.194		65	47		112	142	
	0240	L.E.D. w/battery unit, single face		4.40	1.818		105	71.50		176.50	223	
	0260	Double face		4	2		107	79		186	235	
	0300	Emergency light units, battery operated										
	0350	Twin sealed beam light, 25 watt, 6 volt each										
	0500	Lead battery operated	1 Elec	4	2	Ea.	110	79		189	238	
	0700	Nickel cadmium battery operated		4	2		505	79		584	670	
	0780	Additional remote mount, sealed beam, 25W 6V		26.70	.300		23	11.80		34.80	43	
	0790	Twin sealed beam light, 25W 6V each		26.70	.300		40	11.80		51.80	61.50	
	0900	Self-contained fluorescent lamp pack		10	.800		121	31.50		152.50	180	

16550 | Special Purpose Lighting

			CREW	DAILY OUTPUT	LABOR-HOURS	UNIT	2004 BARE COSTS MAT.	LABOR	EQUIP.	TOTAL	TOTAL INCL O&P	
820	0010	**TRACK LIGHTING**										820
	0080	Track, 1 circuit, 4' section	1 Elec	6.70	1.194	Ea.	36.50	47		83.50	111	
	0100	8' section		5.30	1.509		62	59.50		121.50	157	
	0200	12' section		4.40	1.818		95.50	71.50		167	212	

16550 | Special Purpose Lighting

		CREW	DAILY OUTPUT	LABOR-HOURS	UNIT	2004 BARE COSTS				TOTAL INCL O&P	
						MAT.	LABOR	EQUIP.	TOTAL		
820	0300	3 circuits, 4' section	1 Elec	6.70	1.194	Ea.	48	47		95	123	820
	0400	8' section		5.30	1.509		74	59.50		133.50	170	
	0500	12' section		4.40	1.818		148	71.50		219.50	270	
	1000	Feed kit, surface mounting		16	.500		9.10	19.70		28.80	39.50	
	1100	End cover		24	.333		3.20	13.15		16.35	23	
	1200	Feed kit, stem mounting, 1 circuit		16	.500		25	19.70		44.70	57	
	1300	3 circuit		16	.500		25	19.70		44.70	57	
	2000	Electrical joiner, for continuous runs, 1 circuit		32	.250		11.90	9.85		21.75	28	
	2100	3 circuit		32	.250		28	9.85		37.85	45	
	2200	Fixtures, spotlight, 75W PAR halogen		16	.500		87	19.70		106.70	125	
	2210	50W MR16 halogen		16	.500		106	19.70		125.70	147	
	3000	Wall washer, 250 watt tungsten halogen		16	.500		101	19.70		120.70	141	
	3100	Low voltage, 25/50 watt, 1 circuit		16	.500		102	19.70		121.70	142	
	3120	3 circuit	▼	16	.500	▼	105	19.70		124.70	146	

16580 | Lighting Accessories

		CREW	DAILY OUTPUT	LABOR-HOURS	UNIT	MAT.	LABOR	EQUIP.	TOTAL	TOTAL INCL O&P	
200	0010	**ENERGY SAVING LIGHTING DEVICES**										200
	0100	Occupancy sensors infrared, ceiling mounted	1 Elec	7	1.143	Ea.	88	45		133	164	
	0150	Automatic wall switches		24	.333		52	13.15		65.15	77	
	0200	Remote power pack		10	.800		24.50	31.50		56	73.50	
	0250	Photoelectric control, S.P.S.T. 120 V		8	1		11.95	39.50		51.45	71.50	
	0300	S.P.S.T. 208 V/277 V		8	1		15.15	39.50		54.65	75	
	0350	D.P.S.T. 120 V		6	1.333		121	52.50		173.50	211	
	0400	D.P.S.T. 208 V/277 V		6	1.333		125	52.50		177.50	216	
	0450	S.P.D.T. 208 V/277 V	▼	6	1.333	▼	149	52.50		201.50	242	
300	0010	**FIXTURE WHIPS**										300
	0080	3/8" Greenfield, 2 connectors, 6' long										
	0100	TFFN wire, three #18	1 Elec	32	.250	Ea.	7.35	9.85		17.20	23	
	0150	Four #18		28	.286		7.65	11.25		18.90	25	
	0200	Three #16		32	.250		7.35	9.85		17.20	23	
	0250	Four #16		28	.286		8	11.25		19.25	25.50	
	0300	THHN wire, three #14		32	.250		10	9.85		19.85	25.50	
	0350	Four #14	▼	28	.286	▼	10.80	11.25		22.05	28.50	

16585 | Lamps

		CREW	DAILY OUTPUT	LABOR-HOURS	UNIT	MAT.	LABOR	EQUIP.	TOTAL	TOTAL INCL O&P	
600	0010	**LAMPS**										600
	0080	Fluorescent, rapid start, cool white, 2' long, 20 watt	1 Elec	1	8	C	300	315		615	800	
	0100	4' long, 40 watt		.90	8.889		273	350		623	820	
	0120	3' long, 30 watt		.90	8.889		340	350		690	895	
	0125	3' long, 25 watt energy saver		.90	8.889		465	350		815	1,025	
	0150	U-40 watt		.80	10		845	395		1,240	1,500	
	0155	U-34 watt energy saver		.80	10		845	395		1,240	1,500	
	0170	4' long, 34 watt energy saver		.90	8.889		273	350		623	820	
	0176	2' long, T8, 17 W engergy saver		1	8		380	315		695	890	
	0178	3' long, T8, 25 W energy saver		.90	8.889		380	350		730	940	
	0180	4' long, T8, 32 watt energy saver		.90	8.889		238	350		588	780	
	0200	Slimline, 4' long, 40 watt		.90	8.889		650	350		1,000	1,225	
	0210	4' long, 30 watt energy saver		.90	8.889		650	350		1,000	1,225	
	0300	8' long, 75 watt		.80	10		715	395		1,110	1,375	
	0350	8' long, 60 watt energy saver		.80	10		455	395		850	1,075	
	0400	High output, 4' long, 60 watt		.90	8.889		865	350		1,215	1,475	
	0410	8' long, 95 watt energy saver		.80	10		790	395		1,185	1,450	
	0500	8' long, 110 watt		.80	10		790	395		1,185	1,450	
	0512	2' long, T5, 14 watt energy saver		1	8		1,025	315		1,340	1,600	
	0514	3' long, T5, 21 watt energy saver	▼	.90	8.889	▼	1,025	350		1,375	1,650	

Important: See the Reference Section for critical supporting data - Reference Nos., Crews, & City Cost Indexes

16 ELECTRICAL

16585	Lamps	CREW	DAILY OUTPUT	LABOR- HOURS	UNIT	2004 BARE COSTS				TOTAL INCL O&P
						MAT.	LABOR	EQUIP.	TOTAL	
0516	4' long, T5, 28 watt energy saver	1 Elec	.90	8.889	C	875	350		1,225	1,475
0520	Very high output, 4' long, 110 watt		.90	8.889		2,075	350		2,425	2,800
0525	8' long, 195 watt energy saver		.70	11.429		2,175	450		2,625	3,075
0550	8' long, 215 watt		.70	11.429		2,125	450		2,575	3,000
0554	Full spectrum, 4' long, 60 watt		.90	8.889		1,525	350		1,875	2,200
0556	6' long, 85 watt		.90	8.889		1,675	350		2,025	2,350
0558	8' long, 110 watt		.80	10		1,450	395		1,845	2,150
0560	Twin tube compact lamp		.90	8.889		495	350		845	1,075
0570	Double twin tube compact lamp		.80	10		1,200	395		1,595	1,900
0600	Mercury vapor, mogul base, deluxe white, 100 watt		.30	26.667		3,200	1,050		4,250	5,100
0650	175 watt		.30	26.667		2,375	1,050		3,425	4,175
0700	250 watt		.30	26.667		4,225	1,050		5,275	6,225
0800	400 watt		.30	26.667		3,400	1,050		4,450	5,300
0900	1000 watt		.20	40		7,925	1,575		9,500	11,100
1000	Metal halide, mogul base, 175 watt		.30	26.667		4,000	1,050		5,050	5,950
1100	250 watt		.30	26.667		4,500	1,050		5,550	6,525
1200	400 watt		.30	26.667		4,275	1,050		5,325	6,275
1300	1000 watt		.20	40		11,600	1,575		13,175	15,100
1320	1000 watt, 125,000 initial lumens		.20	40		16,100	1,575		17,675	20,100
1330	1500 watt		.20	40		15,100	1,575		16,675	19,000
1350	High pressure sodium, 70 watt		.30	26.667		4,625	1,050		5,675	6,675
1360	100 watt		.30	26.667		5,450	1,050		6,500	7,550
1370	150 watt		.30	26.667		4,975	1,050		6,025	7,050
1380	250 watt		.30	26.667		5,275	1,050		6,325	7,400
1400	400 watt		.30	26.667		5,425	1,050		6,475	7,550
1450	1000 watt		.20	40		16,800	1,575		18,375	20,900
1500	Low pressure sodium, 35 watt		.30	26.667		7,250	1,050		8,300	9,550
1550	55 watt		.30	26.667		7,975	1,050		9,025	10,400
1600	90 watt		.30	26.667		8,925	1,050		9,975	11,400
1650	135 watt		.20	40		11,400	1,575		12,975	15,000
1700	180 watt		.20	40		12,500	1,575		14,075	16,200
1750	Quartz line, clear, 500 watt		1.10	7.273		975	287		1,262	1,500
1760	1500 watt		.20	40		4,275	1,575		5,850	7,050
1762	Spot, MR 16, 50 watt		1.30	6.154		925	242		1,167	1,375
1770	Tungsten halogen, T4, 400 watt		1.10	7.273		3,775	287		4,062	4,575
1775	T3, 1200 watt		.30	26.667		4,575	1,050		5,625	6,625
1778	PAR 30, 50 watt		1.30	6.154		1,050	242		1,292	1,525
1780	PAR 38, 90 watt		1.30	6.154		945	242		1,187	1,400
1800	Incandescent, interior, A21, 100 watt		1.60	5		183	197		380	495
1900	A21, 150 watt		1.60	5		188	197		385	500
2000	A23, 200 watt		1.60	5		263	197		460	580
2200	PS 30, 300 watt		1.60	5		630	197		827	990
2210	PS 35, 500 watt		1.60	5		1,075	197		1,272	1,475
2230	PS 52, 1000 watt		1.30	6.154		2,350	242		2,592	2,950
2240	PS 52, 1500 watt		1.30	6.154		6,350	242		6,592	7,350
2300	R30, 75 watt		1.30	6.154		760	242		1,002	1,200
2400	R40, 100 watt		1.30	6.154		760	242		1,002	1,200
2500	Exterior, PAR 38, 75 watt		1.30	6.154		1,350	242		1,592	1,825
2600	PAR 38, 150 watt		1.30	6.154		1,500	242		1,742	2,000
2700	PAR 46, 200 watt		1.10	7.273		2,950	287		3,237	3,675
2800	PAR 56, 300 watt		1.10	7.273		3,675	287		3,962	4,475
3000	Guards, fluorescent lamp, 4' long		1	8		700	315		1,015	1,250
3200	8' long		.90	8.889		1,400	350		1,750	2,050

600

ELECTRICAL 16

16710	Communication Circuits	CREW	DAILY OUTPUT	LABOR-HOURS	UNIT	2004 BARE COSTS				TOTAL INCL O&P
						MAT.	LABOR	EQUIP.	TOTAL	
400	**0010**	**FIBER OPTICS**									**400**
0020	Fiber optics cable only. Added costs depend on the type of fiber									
0030	special connectors, optical modems, and networking parts.									
0040	Specialized tools & techniques cause installation costs to vary.									
0070	Cable, minimum, bulk simplex R16120-400	1 Elec	8	1	C.L.F.	24	39.50		63.50	85
0080	Cable, maximum, bulk plenum quad	"	2.29	3.493	"	117	138		255	335
0150	Fiber optic jumper				Ea.	55.50			55.50	61
0200	Fiber optic pigtail					30			30	33
0300	Fiber optic connector	1 Elec	24	.333		16.90	13.15		30.05	38
0350	Fiber optic finger splice		32	.250		32	9.85		41.85	49.50
0400	Transceiver (low cost bi-directional)		8	1		425	39.50		464.50	530
0450	Multi-channel rack enclosure (10 modules)		2	4		475	158		633	760
0500	Fiber optic patch panel (12 ports)		6	1.333		178	52.50		230.50	274
750	**0010**	**COMMUNICATION CABLES & FITTINGS**									**750**
0100	Communication outlets, voice/data devises not included									
0120	Voice/Data outlets, single opening	1 Elec	48	.167	Ea.	6.10	6.55		12.65	16.45
0140	Two jack openings		48	.167		2.70	6.55		9.25	12.70
0160	One jack & one 3/4" round opening		48	.167		6.10	6.55		12.65	16.45
0180	One jack & one twinaxial opening		48	.167		6.10	6.55		12.65	16.45
0200	One jack & one connector cabling opening		48	.167		6.10	6.55		12.65	16.45
0220	Two 3/8" coaxial openings		48	.167		6.10	6.55		12.65	16.45
0300	Data outlets, single opening		48	.167		6.10	6.55		12.65	16.45
0320	One 25-pin subminiature opening		48	.167		6.10	6.55		12.65	16.45
2200	Telephone twisted, PVC insulation, #22-2 conductor		10	.800	C.L.F.	6	31.50		37.50	53.50
2250	#22-3 conductor		9	.889		7.90	35		42.90	60.50
2300	#22-4 conductor		8	1		9.40	39.50		48.90	69
2350	#18-2 conductor		9	.889		7.30	35		42.30	60
2370	Telephone jack, eight pins		32	.250	Ea.	6.25	9.85		16.10	21.50
5000	High performance unshielded twisted pair (UTP)									
5100	Category 3, #24, 2 pair solid, PVC jacket R16120-750	1 Elec	10	.800	C.L.F.	3	31.50		34.50	50.50
5200	4 pair solid		7	1.143		4.90	45		49.90	72.50
5300	25 pair solid		3	2.667		26.50	105		131.50	185
5400	2 pair solid, plenum		10	.800		5.40	31.50		36.90	53
5500	4 pair solid		7	1.143		7.30	45		52.30	75
5600	25 pair solid		3	2.667		44.50	105		149.50	205
5700	4 pair stranded, PVC jacket		7	1.143		14	45		59	82.50
7000	Category 5, #24, 4 pair solid, PVC jacket		7	1.143		7.80	45		52.80	75.50
7100	4 pair solid, plenum		7	1.143		24	45		69	93.50
7200	4 pair stranded, PVC jacket		7	1.143		11.70	45		56.70	80
7210	Category 5e, #24, 4 pair solid, PVC jacket		7	1.143		6.90	45		51.90	74.50
7212	4 pair solid, plenum		7	1.143		19.90	45		64.90	89
7214	4 pair stranded, PVC jacket		7	1.143		13.10	45		58.10	81.50
7240	Category 6, #24, 4 pair solid, PVC jacket		7	1.143		16.40	45		61.40	85
7242	4 pair solid, plenum		7	1.143		44.50	45		89.50	116
7244	4 pair stranded, PVC jacket		7	1.143		17.80	45		62.80	86.50
7300	Category 5, connector, UTP RJ-45		80	.100	Ea.	.90	3.94		4.84	6.85
7302	shielded RJ-45		72	.111		2.55	4.38		6.93	9.30
7310	Category 3, jack, UTP RJ-45		72	.111		2.65	4.38		7.03	9.40
7312	Category 5		65	.123		3.95	4.85		8.80	11.55
7314	Category 5e		65	.123		3.95	4.85		8.80	11.55
7316	Category 6		65	.123		3.95	4.85		8.80	11.55
7322	Category 5, jack, shielded RJ-45		60	.133		4.95	5.25		10.20	13.25
7324	Category 5e		60	.133		4.95	5.25		10.20	13.25
7326	Category 6		60	.133		4.95	5.25		10.20	13.25

16 ELECTRICAL

Important: See the Reference Section for critical supporting data - Reference Nos., Crews, & City Cost Indexes

16720	Tel and Intercomm Equipment	CREW	DAILY OUTPUT	LABOR-HOURS	UNIT	2004 BARE COSTS				TOTAL INCL O&P	
						MAT.	LABOR	EQUIP.	TOTAL		
600	0010	**NURSE CALL SYSTEMS**										600
0100	Single bedside call station	1 Elec	8	1	Ea.	178	39.50		217.50	255
0200	Ceiling speaker station		8	1		52.50	39.50		92	117
0400	Emergency call station		8	1		89	39.50		128.50	156
0600	Pillow speaker		8	1		164	39.50		203.50	239
0800	Double bedside call station		4	2		162	79		241	295
1000	Duty station		4	2		131	79		210	261
1200	Standard call button		8	1		66	39.50		105.50	131
1400	Lights, corridor, dome or zone indicator	▼	8	1	▼	37	39.50		76.50	99
1600	Master control station for 20 stations	2 Elec	.65	24.615	Total	3,175	970		4,145	4,950

ELECTRICAL 16

16810	Sound and Video Circuits	CREW	DAILY OUTPUT	LABOR-HOURS	UNIT	2004 BARE COSTS				TOTAL INCL O&P	
						MAT.	LABOR	EQUIP.	TOTAL		
750	0010	**SOUND AND VIDEO CABLES & FITTINGS**										750
0900	TV antenna lead-in, 300 ohm, #20-2 conductor	1 Elec	7	1.143	C.L.F.	12	45		57	80
0950	Coaxial, feeder outlet		7	1.143		14.40	45		59.40	83
1000	Coaxial, main riser		6	1.333		21	52.50		73.50	101
1100	Sound, shielded with drain, #22-2 conductor		8	1		14.40	39.50		53.90	74.50
1150	#22-3 conductor		7.50	1.067		19.65	42		61.65	84
1200	#22-4 conductor		6.50	1.231		23.50	48.50		72	98
1250	Nonshielded, #22-2 conductor		10	.800		8.60	31.50		40.10	56.50
1300	#22-3 conductor		9	.889		11.70	35		46.70	65
1350	#22-4 conductor		8	1		14.80	39.50		54.30	75
1400	Microphone cable	▼	8	1	▼	51	39.50		90.50	115
3500	Coaxial connectors, 50 ohm impedance quick disconnect									
3540	BNC plug, for RG A/U #58 cable	1 Elec	42	.190	Ea.	3.05	7.50		10.55	14.50
3550	RG A/U #59 cable		42	.190		3.05	7.50		10.55	14.50
3560	RG A/U #62 cable		42	.190		3.05	7.50		10.55	14.50
3600	BNC jack, for RG A/U #58 cable		42	.190		3.15	7.50		10.65	14.60
3610	RG A/U #59 cable		42	.190		3.15	7.50		10.65	14.60
3620	RG A/U #62 cable		42	.190		3.15	7.50		10.65	14.60
3660	BNC panel jack, for RG A/U #58 cable		40	.200		4.90	7.90		12.80	17.10
3670	RG A/U #59 cable		40	.200		4.90	7.90		12.80	17.10
3680	RG A/U #62 cable		40	.200		4.90	7.90		12.80	17.10
3720	BNC bulkhead jack, for RG A/U #58 cable		40	.200		5.70	7.90		13.60	17.95
3730	RG A/U #59 cable		40	.200		5.70	7.90		13.60	17.95
3740	RG A/U #62 cable		40	.200	▼	5.70	7.90		13.60	17.95
3850	Coaxial cable, RG A/U 58, 50 ohm		8	1	C.L.F.	21	39.50		60.50	81.50
3860	RG A/U 59, 75 ohm		8	1		19.50	39.50		59	80
3870	RG A/U 62, 93 ohm		8	1		21	39.50		60.50	81.50
3875	RG 6/U, 75 ohm		8	1		17.80	39.50		57.30	78
3950	RG A/U 58, 50 ohm fire rated		8	1		45.50	39.50		85	109
3960	RG A/U 59, 75 ohm fire rated		8	1		79.50	39.50		119	146
3970	RG A/U 62, 93 ohm fire rated	▼	8	1	▼	64.50	39.50		104	130

16820	Sound Reinforcement	CREW	DAILY OUTPUT	LABOR-HOURS	UNIT	2004 BARE COSTS				TOTAL INCL O&P	
						MAT.	LABOR	EQUIP.	TOTAL		
300	0010	**DOORBELL SYSTEM** Incl. transformer, button & signal										300
0100	6" bell	1 Elec	4	2	Ea.	83	79		162	209

			CREW	DAILY OUTPUT	LABOR-HOURS	UNIT	2004 BARE COSTS				TOTAL INCL O&P	
16820		**Sound Reinforcement**					MAT.	LABOR	EQUIP.	TOTAL		
300	0200	Buzzer	1 Elec	4	2	Ea.	66	79		145	190	**300**
	1000	Door chimes, 2 notes, minimum		16	.500		22	19.70		41.70	53.50	
	1020	Maximum		12	.667		115	26.50		141.50	166	
	1100	Tube type, 3 tube system		12	.667		163	26.50		189.50	218	
	1180	4 tube system		10	.800		261	31.50		292.50	335	
	1900	For transformer & button, minimum add		5	1.600		12.40	63		75.40	108	
	1960	Maximum, add		4.50	1.778		37	70		107	145	
	3000	For push button only, minimum		24	.333		2.42	13.15		15.57	22	
	3100	Maximum		20	.400		19.45	15.75		35.20	45	
	3200	Bell transformer		16	.500		17.20	19.70		36.90	48.50	
800	0010	**PUBLIC ADDRESS SYSTEM**										**800**
	0100	Conventional, office	1 Elec	5.33	1.501	Speaker	97.50	59		156.50	195	
	0200	Industrial	"	2.70	2.963	"	188	117		305	380	
	0400	Explosionproof system is 3 times cost of central control										
	0600	Installation costs run about 120% of material cost										
840	0010	**SOUND SYSTEM** not including rough-in wires, cables & conduits										**840**
	0100	Components, outlet, projector	1 Elec	8	1	Ea.	45	39.50		84.50	108	
	0200	Microphone		4	2		50	79		129	172	
	0400	Speakers, ceiling or wall		8	1		85	39.50		124.50	152	
	0600	Trumpets		4	2		158	79		237	291	
	0800	Privacy switch		8	1		63	39.50		102.50	128	
	1000	Monitor panel		4	2		281	79		360	425	
	1200	Antenna, AM/FM		4	2		157	79		236	290	
	1400	Volume control		8	1		63	39.50		102.50	128	
	1600	Amplifier, 250 watts		1	8		1,250	315		1,565	1,850	
	1800	Cabinets		1	8		610	315		925	1,150	
	2000	Intercom, 25 station capacity, master station	2 Elec	2	8		1,475	315		1,790	2,100	
	2020	11 station capacity	"	4	4		690	158		848	995	
	2200	Remote station	1 Elec	8	1		118	39.50		157.50	189	
	2400	Intercom outlets		8	1		69.50	39.50		109	135	
	2600	Handset		4	2		230	79		309	370	
	2800	Emergency call system, 12 zones, annunciator		1.30	6.154		690	242		932	1,125	
	3000	Bell		5.30	1.509		71.50	59.50		131	167	
	3200	Light or relay		8	1		35.50	39.50		75	98	
	3400	Transformer		4	2		157	79		236	290	
	3600	House telephone, talking station		1.60	5		335	197		532	665	
	3800	Press to talk, release to listen		5.30	1.509		78.50	59.50		138	175	
	4000	System-on button					47			47	51.50	
	4200	Door release	1 Elec	4	2		84	79		163	210	
	4400	Combination speaker and microphone		8	1		143	39.50		182.50	216	
	4600	Termination box		3.20	2.500		45	98.50		143.50	197	
	4800	Amplifier or power supply		5.30	1.509		515	59.50		574.50	660	
	5000	Vestibule door unit		16	.500	Name	95	19.70		114.70	135	
	5200	Strip cabinet		27	.296	Ea.	180	11.65		191.65	214	
	5400	Directory		16	.500		84.50	19.70		104.20	123	
	6000	Master door, button buzzer type, 100 unit	2 Elec	.54	29.630		860	1,175		2,035	2,675	
	6020	200 unit		.30	53.333		1,625	2,100		3,725	4,900	
	6040	300 unit		.20	80		2,475	3,150		5,625	7,425	
	6060	Transformer	1 Elec	8	1		22	39.50		61.50	82.50	
	6080	Door opener		5.30	1.509		31.50	59.50		91	123	
	6100	Buzzer with door release and plate		4	2		31.50	79		110.50	152	
	6200	Intercom type, 100 unit	2 Elec	.54	29.630		1,075	1,175		2,250	2,900	
	6220	200 unit		.30	53.333		2,100	2,100		4,200	5,425	
	6240	300 unit		.20	80		3,150	3,150		6,300	8,175	
	6260	Amplifier	1 Elec	2	4		157	158		315	405	

Important: See the Reference Section for critical supporting data - Reference Nos., Crews, & City Cost Indexes

		16820 | **Sound Reinforcement**	CREW	DAILY OUTPUT	LABOR-HOURS	UNIT	2004 BARE COSTS				TOTAL INCL O&P	
							MAT.	LABOR	EQUIP.	TOTAL		
840	6280	Speaker with door release	1 Elec	4	2	Ea.	47	79		126	169	**840**

		16850 | **Television Equipment**										
600	0010	**T.V. SYSTEMS** not including rough-in wires, cables & conduits										**600**
	0100	Master TV antenna system										
	0200	VHF reception & distribution, 12 outlets	1 Elec	6	1.333	Outlet	155	52.50		207.50	249	
	0400	30 outlets		10	.800		102	31.50		133.50	159	
	0600	100 outlets		13	.615		104	24.50		128.50	150	
	0800	VHF & UHF reception & distribution, 12 outlets		6	1.333		154	52.50		206.50	247	
	1000	30 outlets		10	.800		102	31.50		133.50	159	
	1200	100 outlets		13	.615		104	24.50		128.50	150	
	1400	School and deluxe systems, 12 outlets		2.40	3.333		203	131		334	420	
	1600	30 outlets		4	2		178	79		257	315	
	1800	80 outlets		5.30	1.509	↓	171	59.50		230.50	277	
	1900	Amplifier		4	2	Ea.	510	79		589	675	
	1910	Antenna	↓	2	4	"	245	158		403	505	
	2000	Closed circuit, surveillance, one station (camera & monitor)	2 Elec	2.60	6.154	Total	985	242		1,227	1,425	
	2200	For additional camera stations, add	1 Elec	2.70	2.963	Ea.	550	117		667	780	
	2400	Industrial quality, one station (camera & monitor)	2 Elec	2.60	6.154	Total	2,050	242		2,292	2,600	
	2600	For additional camera stations, add	1 Elec	2.70	2.963	Ea.	1,250	117		1,367	1,550	
	2610	For low light, add		2.70	2.963		1,000	117		1,117	1,275	
	2620	For very low light, add		2.70	2.963		7,400	117		7,517	8,300	
	2800	For weatherproof camera station, add		1.30	6.154		770	242		1,012	1,200	
	3000	For pan and tilt, add		1.30	6.154		2,000	242		2,242	2,550	
	3200	For zoom lens - remote control, add, minimum		2	4		1,825	158		1,983	2,250	
	3400	Maximum		2	4		6,700	158		6,858	7,600	
	3410	For automatic iris for low light, add	↓	2	4	↓	1,600	158		1,758	2,000	
	3600	Educational T.V. studio, basic 3 camera system, black & white,										
	3800	electrical & electronic equip. only, minimum	4 Elec	.80	40	Total	9,525	1,575		11,100	12,900	
	4000	Maximum (full console)		.28	114		40,600	4,500		45,100	51,500	
	4100	As above, but color system, minimum		.28	114		54,000	4,500		58,500	65,500	
	4120	Maximum	↓	.12	266	↓	233,500	10,500		244,000	272,500	
	4200	For film chain, black & white, add	1 Elec	1	8	Ea.	10,900	315		11,215	12,500	
	4250	Color, add		.25	32		13,300	1,250		14,550	16,500	
	4400	For video tape recorders, add, minimum	↓	1	8		2,300	315		2,615	3,000	
	4600	Maximum	4 Elec	.40	80	↓	19,100	3,150		22,250	25,700	

ELECTRICAL 16

For information about Means Estimating Seminars, see yellow pages 12 and 13 in back of book

Division Notes

	CREW	DAILY OUTPUT	LABOR-HOURS	UNIT	2004 BARE COSTS				TOTAL INCL O&P
					MAT.	LABOR	EQUIP.	TOTAL	

Division 17
Square Foot & Cubic Foot Costs

Estimating Tips

- The cost figures in Division 17 were derived from approximately 11,200 projects contained in the Means database of completed construction projects, and include the contractor's overhead and profit, but do not generally include architectural fees or land costs. The figures have been adjusted to January of the current year. New projects are added to our files each year, and outdated projects are discarded. For this reason, certain costs may not show a uniform annual progression. In no case are all subdivisions of a project listed.

- These projects were located throughout the U.S. and reflect a tremendous variation in square foot (S.F.) and cubic foot (C.F.) costs. This is due to differences, not only in labor and material costs, but also in individual owners' requirements. For instance, a bank in a large city would have different features than one in a rural area. This is true of all the different types of buildings analyzed. Therefore, caution should be exercised when using Division 17 costs. For example, for court houses, costs in the database are local court house costs and will not apply to the larger, more elaborate federal court houses. As a general rule, the projects in the 1/4 column do not include any site work or equipment, while the projects in the 3/4 column may include both equipment and site work. The median figures do not generally include site work.

- None of the figures "go with" any others. All individual cost items were computed and tabulated separately. Thus the sum of the median figures for Plumbing, HVAC and Electrical will not normally total up to the total Mechanical and Electrical costs arrived at by separate analysis and tabulation of the projects.

- Each building was analyzed as to total and component costs and percentages. The figures were arranged in ascending order with the results tabulated as shown. The 1/4 column shows that 25% of the projects had lower costs, 75% higher. The 3/4 column shows that 75% of the projects had lower costs, 25% had higher. The median column shows that 50% of the projects had lower costs, 50% had higher.

- There are two times when square foot costs are useful. The first is in the conceptual stage when no details are available. Then square foot costs make a useful starting point. The second is after the bids are in and the costs can be worked back into their appropriate units for information purposes. As soon as details become available in the project design, the square foot approach should be discontinued and the project priced as to its particular components. When more precision is required or for estimating the replacement cost of specific buildings, the current edition of *Means Square Foot Costs* should be used.

- In using the figures in Division 17, it is recommended that the median column be used for preliminary figures if no additional information is available. The median figures, when multiplied by the total city construction cost index figures (see City Cost Indexes) and then multiplied by the project size modifier in Reference Number R17100-100, should present a fairly accurate base figure, which would then have to be adjusted in view of the estimator's experience, local economic conditions, code requirements and the owner's particular requirements. There is no need to factor the percentage figures, as these should remain constant from city to city. All tabulations mentioning air conditioning had at least partial air conditioning.

- The editors of this book would greatly appreciate receiving cost figures on one or more of your recent projects which would then be included in the averages for next year. All cost figures received will be kept confidential except that they will be averaged with other similar projects to arrive at S.F. and C.F. cost figures for next year's book. See the last page of the book for details and the discount available for submitting one or more of your projects.

17100 | S.F. & C.F. Costs		UNIT	UNIT COSTS			% OF TOTAL		
			1/4	MEDIAN	3/4	1/4	MEDIAN	3/4
010	**0010**	**APARTMENTS** Low Rise (1 to 3 story) R17100-100	S.F.	48	60.50	81				**010**
	0020	Total project cost	C.F.	4.30	5.75	7.05			
	0100	Site work	S.F.	4.30	6	9.50	5.05%	10.20%	13.85%
	0500	Masonry		.94	2.33	3.82	1.54%	3.35%	6.05%
	1500	Finishes		5.05	6.90	8.65	9.05%	10.60%	12.70%
	1800	Equipment		1.56	2.31	3.49	2.73%	4.07%	6%
	2720	Plumbing		3.69	4.88	6.10	6.60%	8.90%	10.20%
	2770	Heating, ventilating, air conditioning		2.38	2.93	4.31	4.20%	5.60%	7.60%
	2900	Electrical		2.77	3.62	4.95	5.20%	6.65%	8.40%
	3100	Total: Mechanical & Electrical	↓	9.60	12.20	15.20	15.90%	17.75%	22%
	9000	Per apartment unit, total cost	Apt.	44,200	67,000	100,000			
	9500	Total: Mechanical & Electrical	"	8,450	13,300	17,400			
020	**0010**	**APARTMENTS** Mid Rise (4 to 7 story) R17100-100	S.F.	65	77	93.50				**020**
	0020	Total project costs	C.F.	5.10	7	9.40			
	0100	Site work	S.F.	2.54	5.05	10.15	5.50%	6.70%	10.10%
	0500	Masonry		4.23	5.95	8.30	5.90%	7.60%	10.60%
	1500	Finishes		8	11.15	13.15	10.75%	13.50%	17.85%
	1800	Equipment		2.09	2.98	3.93	2.80%	3.48%	4.70%
	2500	Conveying equipment		1.54	1.85	2.17	2.11%	2.29%	2.83%
	2720	Plumbing		3.73	6	6.65	5.65%	7.20%	8.20%
	2900	Electrical		4.39	6	7	6.70%	7.50%	8.95%
	3100	Total: Mechanical & Electrical	↓	12.65	16.80	21	18.50%	21%	25%
	9000	Per apartment unit, total cost	Apt.	72,500	85,000	142,000			
	9500	Total: Mechanical & Electrical	"	13,600	15,700	19,900			
030	**0010**	**APARTMENTS** High Rise (8 to 24 story) R17100-100	S.F.	72	87	106				**030**
	0020	Total project costs	C.F.	6.15	8.55	10.40			
	0100	Site work	S.F.	2.21	4.23	6.15	2.58%	4.84%	6.15%
	0500	Masonry		4.17	7.60	9.30	5.30%	9.70%	11.10%
	1500	Finishes		8	10	11.80	10.50%	11.80%	13.70%
	1800	Equipment		2.32	2.85	3.78	2.78%	3.58%	4.35%
	2500	Conveying equipment		1.64	2.49	3.55	2.20%	2.80%	3.40%
	2720	Plumbing		5.40	6.25	8.90	7%	9.10%	10.60%
	2900	Electrical		4.95	6.25	8.45	6.45%	7.70%	8.90%
	3100	Total: Mechanical & Electrical	↓	14.60	18.95	22.50	17.80%	22%	24%
	9000	Per apartment unit, total cost	Apt.	75,500	84,000	114,500			
	9500	Total: Mechanical & Electrical	"	16,600	18,500	19,600			
040	**0010**	**AUDITORIUMS** R17100-100	S.F.	74	101	143				**040**
	0020	Total project costs	C.F.	4.69	6.55	9.85			
	2720	Plumbing	S.F.	4.75	6.55	8.35	6.30%	7.90%	9.10%
	2900	Electrical		6.05	8.40	11.25	7.65%	11.40%	14.40%
	3100	Total: Mechanical & Electrical	↓	11.65	15.55	27	14.40%	18.50%	23.60%
050	**0010**	**AUTOMOTIVE SALES** R17100-100	S.F.	55.50	75.50	92.50				**050**
	0020	Total project costs	C.F.	3.79	4.48	5.70			
	2720	Plumbing	S.F.	2.70	4.39	4.79	4.70%	6.50%	7%
	2770	Heating, ventilating, air conditioning		3.90	6.05	6.45	4.61%	10%	10.35%
	2900	Electrical		4.31	7	7.70	7.40%	9.95%	12.40%
	3100	Total: Mechanical & Electrical	↓	12.45	17.50	22	19.15%	20.50%	26%
060	**0010**	**BANKS** R17100-100	S.F.	109	135	170				**060**
	0020	Total project costs	C.F.	7.75	10.55	13.95			
	0100	Site work	S.F.	12.10	20	29.50	7.50%	13.90%	17.60%
	0500	Masonry		5.65	10.85	20.50	3.46%	8.05%	11.40%
	1500	Finishes		10.10	13.15	16.80	5.85%	8.55%	11.25%
	1800	Equipment		4.35	9	20.50	3.30%	7.70%	12.50%
	2720	Plumbing		3.41	4.87	7.10	2.80%	3.90%	4.93%
	2770	Heating, ventilating, air conditioning		6.50	8.65	11.55	4.80%	7.20%	8.50%
	2900	Electrical		10.30	13.75	17.90	8.20%	10.10%	12%
	3100	Total: Mechanical & Electrical	↓	24.50	33	40	17.45%	19.45%	24%

See R17100-100 and City Cost Indexes in the Reference Section

17100		S.F. & C.F. Costs		UNIT	UNIT COSTS			% OF TOTAL			
					1/4	MEDIAN	3/4	1/4	MEDIAN	3/4	
060	3500	See also division 11020 & 11030									060
130	0010	**CHURCHES**	R17100 -100	S.F.	73	92.50	120				130
	0020	Total project costs		C.F.	4.54	5.75	7.55				
	1800	Equipment		S.F.	.94	2.10	4.60	1.03%	2.33%	4.50%	
	2720	Plumbing			2.85	3.99	5.90	3.60%	5%	6.30%	
	2770	Heating, ventilating, air conditioning			6.65	8.70	12.80	7.50%	10%	12.20%	
	2900	Electrical			6.05	8.45	11.30	7.35%	8.80%	10.85%	
	3100	Total: Mechanical & Electrical		↓	15.95	24	34	18.30%	22%	25.40%	
	3500	See also division 11040									
150	0010	**CLUBS, COUNTRY**	R17100 -100	S.F.	78.50	95	119				150
	0020	Total project costs		C.F.	6.35	7.70	10.65				
	2720	Plumbing		S.F.	5.05	7.05	16.05	6%	9.60%	10%	
	2900	Electrical			6.25	8.85	11.65	7.70%	9.65%	11.40%	
	3100	Total: Mechanical & Electrical		↓	18.85	33	41	19%	26.50%	29.50%	
170	0010	**CLUBS, SOCIAL** Fraternal	R17100 -100	S.F.	62.50	90	117				170
	0020	Total project costs		C.F.	3.92	5.95	7.10				
	2720	Plumbing		S.F.	3.94	4.91	5.95	5.20%	6.80%	8.30%	
	2770	Heating, ventilating, air conditioning			5.70	6.90	8.85	8.70%	11%	14.40%	
	2900	Electrical			4.99	7.75	9.40	7.30%	9.50%	11.45%	
	3100	Total: Mechanical & Electrical		↓	13.90	24.50	26.50	19.80%	21%	23%	
180	0010	**CLUBS, Y.M.C.A.**	R17100 -100	S.F.	78.50	106	130				180
	0020	Total project costs		C.F.	3.63	6.05	9.05				
	2720	Plumbing		S.F.	5.25	9.90	11.10	6.70%	9.70%	16%	
	2900	Electrical			5.95	8.15	11.60	6.45%	9.25%	10.80%	
	3100	Total: Mechanical & Electrical		↓	17.65	22.50	30	17.10%	18.40%	28.50%	
190	0010	**COLLEGES** Classrooms & Administration	R17100 -100	S.F.	90	118	157				190
	0020	Total project costs		C.F.	6.55	9.15	14.75				
	0500	Masonry		S.F.	5.85	11.50	12.95	5.65%	11.90%	12.10%	
	2720	Plumbing			4.41	8.65	16	4.80%	8.90%	11.80%	
	2900	Electrical			7.35	11.30	13.60	8.40%	10.30%	12.15%	
	3100	Total: Mechanical & Electrical		↓	15.60	30.50	44	14.20%	24.70%	32.50%	
210	0010	**COLLEGES** Science, Engineering, Laboratories	R17100 -100	S.F.	148	173	214				210
	0020	Total project costs		C.F.	8.50	12.40	14.05				
	1800	Equipment		S.F.	8.25	18.65	20.50	5.30%	9.70%	15%	
	2900	Electrical			12.20	17.30	26.50	7.80%	9.50%	12.10%	
	3100	Total: Mechanical & Electrical		↓	48.50	53.50	83	28.50%	31.50%	41%	
	3500	See also division 11600									
230	0010	**COLLEGES** Student Unions	R17100 -100	S.F.	94.50	132	155				230
	0020	Total project costs		C.F.	5.25	6.90	8.90				
	3100	Total: Mechanical & Electrical		S.F.	25	35.50	38.50	17.40%	23.90%	29%	
250	0010	**COMMUNITY CENTERS**	R17100 -100	"	71	96	129				250
	0020	Total project costs		C.F.	5.10	7.25	9.40				
	1800	Equipment		S.F.	1.96	3.33	5.40	1.69%	3.12%	6%	
	2720	Plumbing			3.78	6.45	9.40	4.85%	7.10%	9.55%	
	2770	Heating, ventilating, air conditioning			6.25	8.95	12.30	6.80%	10.35%	12.90%	
	2900	Electrical			6.20	8.35	12.95	7.10%	8.90%	11.15%	
	3100	Total: Mechanical & Electrical		↓	23	27.50	39.50	21%	26%	32.50%	
280	0010	**COURT HOUSES**	R17100 -100	S.F.	111	129	150				280
	0020	Total project costs		C.F.	8.60	10.30	14.30				
	2720	Plumbing		S.F.	5.35	7.45	10.75	7.45%	8.20%	9.55%	
	2900	Electrical			10.45	12.55	15.70	8.90%	9.80%	11.25%	
	3100	Total: Mechanical & Electrical		↓	27	30	41.50	22.50%	26%	30%	

SQUARE FOOT 17

17100 | S.F. & C.F. Costs

			UNIT	UNIT COSTS			% OF TOTAL		
				1/4	MEDIAN	3/4	1/4	MEDIAN	3/4
300	0010	**DEPARTMENT STORES** R17100-100	S.F.	41.50	56	71				**300**
	0020	Total project costs	C.F.	2.24	3.02	4.25			
	2720	Plumbing	S.F.	1.35	1.63	2.48	1.82%	4.21%	5.90%
	2770	Heating, ventilating, air conditioning		3.78	5.85	8.80	8.30%	12.35%	14.80%
	2900	Electrical		4.77	6.55	7.70	10.45%	12.15%	14.95%
	3100	Total: Mechanical & Electrical	↓	8.40	10.75	18.80	13.20%	21.50%	50%
310	0010	**DORMITORIES** Low Rise (1 to 3 story) R17100-100	S.F.	68	100	121				**310**
	0020	Total project costs	C.F.	4.71	7.20	10.80			
	2720	Plumbing	S.F.	4.74	6.35	8	8.05%	9%	9.65%
	2770	Heating, ventilating, air conditioning		5	6	8	4.61%	8.05%	10%
	2900	Electrical		5.05	7.65	9.65	6.70%	9.10%	9.55%
	3100	Total: Mechanical & Electrical	↓	26.50	27.50	29	22%	26%	29%
	9000	Per bed, total cost	Bed	33,400	37,100	79,500			
320	0010	**DORMITORIES** Mid Rise (4 to 8 story) R17100-100	S.F.	96.50	126	153				**320**
	0020	Total project costs	C.F.	10.65	11.70	14			
	2900	Electrical	S.F.	7.15	11.25	11.65	7.35%	9.45%	10.20%
	3100	Total: Mechanical & Electrical	"	21.50	28.50	37.50	19.50%	25.50%	30.50%
	9000	Per bed, total cost	Bed	13,800	32,000	65,000			
340	0010	**FACTORIES** R17100-100	S.F.	37	54.50	84				**340**
	0020	Total project costs	C.F.	2.39	3.50	5.80			
	0100	Site work	S.F.	4.18	7.60	12.45	7.30%	11.45%	17.95%
	2720	Plumbing		2.13	3.72	6.25	3.73%	5.80%	8.30%
	2770	Heating, ventilating, air conditioning		3.84	5.50	7.45	6.55%	8.45%	11.35%
	2900	Electrical		4.55	7.30	11.20	8.55%	10.50%	14.20%
	3100	Total: Mechanical & Electrical	↓	10.80	17.35	26.50	22%	29%	35.50%
360	0010	**FIRE STATIONS** R17100-100	S.F.	70	96	130				**360**
	0020	Total project costs	C.F.	4.07	5.90	7.75			
	0500	Masonry	S.F.	10.65	19.15	25.50	8.60%	12.40%	16.95%
	1140	Roofing		2.37	6.40	7.30	1.90%	4.90%	5%
	1580	Painting		1.94	2.75	2.81	.99%	2.24%	2.42%
	1800	Equipment		1.56	2.17	5.25	1.10%	2.50%	4.80%
	2720	Plumbing		4.60	6.80	9.95	6.20%	7.70%	9.70%
	2770	Heating, ventilating, air conditioning		3.96	6.50	9.85	5.15%	7.50%	9.25%
	2900	Electrical		5.15	8.90	11.70	6.80%	8.50%	10.95%
	3100	Total: Mechanical & Electrical	↓	25.50	30	35.50	19.60%	23%	27%
370	0010	**FRATERNITY HOUSES** and Sorority Houses R17100-100	S.F.	72.50	93.50	132				**370**
	0020	Total project costs	C.F.	6.95	7.55	9.65			
	2720	Plumbing	S.F.	5.45	6.30	11.50	7.55%	8%	10.90%
	2900	Electrical		4.79	10.35	12.70	6.60%	9.90%	10.70%
	3100	Total: Mechanical & Electrical	↓	13.30	18.80	22.50	14.60%	19.90%	20.70%
380	0010	**FUNERAL HOMES** R17100-100	S.F.	76.50	105	190				**380**
	0020	Total project costs	C.F.	7.80	8.70	16.70			
	2900	Electrical	S.F.	3.39	6.20	7.30	4.44%	5.95%	11.05%
	3100	Total: Mechanical & Electrical	"	12.30	18.20	24	12.90%	17.80%	18.80%
390	0010	**GARAGES, COMMERCIAL** (Service) R17100-100	S.F.	43	68	92.50				**390**
	0020	Total project costs	C.F.	2.92	4.22	6.15			
	1800	Equipment	S.F.	2.45	5.50	8.55	3%	6.80%	8.30%
	2720	Plumbing		3	4.63	8.45	4.70%	7.40%	10.45%
	2730	Heating & ventilating		3.95	5.40	7.45	5.25%	6.85%	9.55%
	2900	Electrical		4.18	6.40	9.15	7.05%	9.40%	10.65%
	3100	Total: Mechanical & Electrical	↓	9.35	17.35	26	13.60%	17.40%	27%
400	0010	**GARAGES, MUNICIPAL** (Repair) R17100-100	S.F.	64	86	120				**400**
	0020	Total project costs	C.F.	3.98	5.05	8.25			

17 SQUARE FOOT

See R17100-100 and City Cost Indexes in the Reference Section

17100 | S.F. & C.F. Costs

			UNIT	UNIT COSTS			% OF TOTAL			
				1/4	MEDIAN	3/4	1/4	MEDIAN	3/4	
400	0500	Masonry	S.F.	6	11.70	18.15	4.03%	8.60%	11.80%	400
	2720	Plumbing		2.86	5.50	10.35	3.59%	6.70%	7.95%	
	2730	Heating & ventilating		4.89	7.10	13.65	6.15%	7.45%	13.50%	
	2900	Electrical		4.71	7.40	10.75	6.90%	9.25%	11.15%	
	3100	Total: Mechanical & Electrical	▼	15.15	24	44	21.50%	25.50%	28.50%	
410	0010	**GARAGES, PARKING** R17100-100	S.F.	24.50	36	62				410
	0020	Total project costs	C.F.	2.33	3.17	4.61				
	2720	Plumbing	S.F.	.64	1.09	1.68	1.72%	3.05%	3.85%	
	2900	Electrical		1.30	1.61	2.57	4.27%	4.98%	6.10%	
	3100	Total: Mechanical & Electrical	▼	2.17	3.97	5.10	6.90%	8.80%	11.05%	
	3200									
	9000	Per car, total cost	Car	10,700	13,200	16,800				
430	0010	**GYMNASIUMS** R17100-100	S.F.	69	91	112				430
	0020	Total project costs	C.F.	3.44	4.55	5.75				
	1800	Equipment	S.F.	1.45	2.96	5.90	2.03%	3.35%	5%	
	2720	Plumbing		4.36	5.40	6.60	4.63%	6.75%	7.85%	
	2770	Heating, ventilating, air conditioning		4.70	7.15	14.35	9%	11.10%	22.60%	
	2900	Electrical		4.74	7	8.75	6.15%	8.30%	10.30%	
	3100	Total: Mechanical & Electrical	▼	18.75	26.50	31	19.60%	26%	29.50%	
	3500	See also division 11480								
460	0010	**HOSPITALS** R17100-100	S.F.	139	160	248				460
	0020	Total project costs	C.F.	10.05	12.50	17.90				
	1800	Equipment	S.F.	3.36	6.45	11.15	.62%	2.63%	4.80%	
	2720	Plumbing		11.85	15.95	20.50	7.05%	8%	10.75%	
	2770	Heating, ventilating, air conditioning		16.75	22	30	7.65%	10.05%	16.50%	
	2900	Electrical		14.05	18.30	29.50	9.20%	11.45%	13.15%	
	3100	Total: Mechanical & Electrical	▼	40.50	54	88	26.50%	32.50%	36.50%	
	9000	Per bed or person, total cost	Bed	51,000	108,500	173,500				
	9900	See also division 11700								
480	0010	**HOUSING** For the Elderly R17100-100	S.F.	65	82.50	101				480
	0020	Total project costs	C.F.	4.65	6.40	8.30				
	0100	Site work	S.F.	4.83	7.35	10.30	6.10%	8.05%	12.10%	
	0500	Masonry		1.99	7.40	10.85	1.75%	4.16%	10.75%	
	1800	Equipment		1.57	2.17	3.43	1.88%	3.26%	4.47%	
	2510	Conveying systems		1.59	2.13	2.91	1.78%	2.20%	2.84%	
	2720	Plumbing		4.84	6.15	8.05	8.20%	9.70%	10.40%	
	2730	Heating, ventilating, air conditioning		2.48	3.52	5.25	3.25%	5.60%	7.25%	
	2900	Electrical		4.86	6.60	8.30	7.20%	8.40%	10.20%	
	3100	Total: Mechanical & Electrical	▼	16.70	19.70	26.50	18.10%	22.50%	29%	
	9000	Per rental unit, total cost	Unit	60,500	71,000	79,000				
	9500	Total: Mechanical & Electrical	"	13,300	15,500	18,100				
500	0010	**HOUSING** Public (Low Rise) R17100-100	S.F.	54.50	76	99				500
	0020	Total project costs	C.F.	4.93	6.10	7.70				
	0100	Site work	S.F.	6.95	10.05	16.25	7.10%	11%	15.55%	
	1800	Equipment		1.49	2.43	3.87	2.03%	2.90%	3.99%	
	2720	Plumbing		3.95	5.35	6.60	7.15%	9.05%	11.45%	
	2730	Heating, ventilating, air conditioning		1.98	3.85	4.22	4.26%	6.05%	6.45%	
	2900	Electrical		3.27	4.93	6.85	5.50%	6.75%	8.50%	
	3100	Total: Mechanical & Electrical	▼	15.70	20	23.50	14.70%	19.20%	26.50%	
	9000	Per apartment, total cost	Apt.	60,000	68,500	86,000				
	9500	Total: Mechanical & Electrical	"	12,800	15,800	17,500				
510	0010	**ICE SKATING RINKS** R17100-100	S.F.	48.50	109	120				510
	0020	Total project costs	C.F.	3.44	3.52	4.06				
	2720	Plumbing	S.F.	1.75	3.28	3.35	3.23%	5.65%	6.75%	
	2900	Electrical	▼	5	7.70	8.15	6.80%	10.15%	15.05%	

SQUARE FOOT 17

For expanded coverage of these items see *Means Square Foot Costs 2004*

17100 | S.F. & C.F. Costs

			UNIT	UNIT COSTS			% OF TOTAL			
				1/4	MEDIAN	3/4	1/4	MEDIAN	3/4	
510	3100	Total: Mechanical & Electrical	S.F.	8.50	12	15	18.95%	18.95%	18.95%	**510**
520	0010	**JAILS** R17100 -100	S.F.	144	184	237				**520**
	0020	Total project costs	C.F.	13.45	17.90	22				
	1800	Equipment	S.F.	5.95	16.45	28	3.11%	8.95%	13.40%	
	2720	Plumbing		14.50	18.40	24.50	7.05%	8.90%	9.60%	
	2770	Heating, ventilating, air conditioning		12.85	17.15	33	8%	9.45%	12.20%	
	2900	Electrical		15.65	20	24.50	9.80%	11.70%	14.95%	
	3100	Total: Mechanical & Electrical	↓	40	71	84	27.50%	30%	31%	
530	0010	**LIBRARIES** R17100 -100	S.F.	88.50	114	147				**530**
	0020	Total project costs	C.F.	6.15	7.45	9.85				
	0500	Masonry	S.F.	6.10	12.45	21	5.60%	8.25%	12.60%	
	1800	Equipment		1.23	3.31	5.15	.43%	1.50%	4.16%	
	2720	Plumbing		3.32	4.84	6.50	3.38%	4.26%	5.50%	
	2770	Heating, ventilating, air conditioning		7.35	12.45	16.25	7.80%	10.95%	12.80%	
	2900	Electrical		9.05	11.65	14.80	8.40%	10.25%	11.85%	
	3100	Total: Mechanical & Electrical	↓	27	34	42.50	19.65%	23%	26.50%	
540	0010	**LIVING, ASSISTED** R17100 -100	S.F.	86.50	99	117				**540**
	0020	Total project costs	C.F.	7.20	8.20	9.65				
	0500	Masonry	S.F.	2.45	2.92	3.43	2.37%	3.16%	3.86%	
	1800	Equipment		1.85	2.20	2.61	2.06%	2.44%	2.60%	
	2720	Plumbing		7	9.35	9.90	6.05%	8.15%	10.60%	
	2770	Heating, ventilating, air conditioning		8.30	8.65	9.50	7.95%	9.35%	9.70%	
	2900	Electrical		8.20	9.30	10.60	9.05%	10.30%	10.80%	
	3100	Total: Mechanical & Electrical	↓	23.50	28.50	31	25%	30%	32%	
550	0010	**MEDICAL CLINICS** R17100 -100	S.F.	83	106	135				**550**
	0020	Total project costs	C.F.	6.20	8.70	11.05				
	1800	Equipment	S.F.	2.29	4.81	7.50	1.10%	3.50%	6.90%	
	2720	Plumbing		5.65	7.85	10.95	6%	8.35%	10.35%	
	2770	Heating, ventilating, air conditioning		6.70	8.80	12.95	6.70%	8.85%	11.80%	
	2900	Electrical		7.10	10.35	13.65	8.15%	10%	12.40%	
	3100	Total: Mechanical & Electrical	↓	23	32	44	22.50%	27.50%	33.50%	
	3500	See also division 11700									
570	0010	**MEDICAL OFFICES** R17100 -100	S.F.	78	97.50	120				**570**
	0020	Total project costs	C.F.	5.80	8.05	10.90				
	1800	Equipment	S.F.	2.77	5.20	7.40	1.50%	4.97%	6.50%	
	2720	Plumbing		4.41	6.80	9.15	5.60%	6.80%	8.50%	
	2770	Heating, ventilating, air conditioning		5.30	7.70	10.15	6.20%	8.25%	9.70%	
	2900	Electrical		6.35	9.25	12.65	7.35%	9.75%	11.10%	
	3100	Total: Mechanical & Electrical	↓	17.10	24	37.50	19.65%	23.50%	30.50%	
590	0010	**MOTELS** R17100 -100	S.F.	50.50	74	96				**590**
	0020	Total project costs	C.F.	4.48	6	10.10				
	2720	Plumbing	S.F.	5.10	6.50	7.75	9.45%	10.60%	12.55%	
	2770	Heating, ventilating, air conditioning		3.11	4.63	8.30	5.60%	5.60%	10%	
	2900	Electrical		4.79	6.10	7.90	7.45%	9.20%	10.80%	
	3100	Total: Mechanical & Electrical	↓	16.15	20	34.50	18.50%	21%	25.50%	
	5000										
	9000	Per rental unit, total cost	Unit	25,600	48,900	52,500				
	9500	Total: Mechanical & Electrical	"	5,000	7,550	8,775				
600	0010	**NURSING HOMES** R17100 -100	S.F.	78.50	103	122				**600**
	0020	Total project costs	C.F.	6	7.95	10.80				
	1800	Equipment	S.F.	2.54	3.39	5.50	2.02%	3.63%	5.75%	
	2720	Plumbing		7.15	10.25	12.55	9.40%	10.25%	12.20%	
	2770	Heating, ventilating, air conditioning		7.10	10.80	14.35	10.65%	11.80%	12.70%	
	2900	Electrical	↓	7.80	10.05	13.55	9.15%	10.55%	12.55%	

17 SQUARE FOOT

17100 | S.F. & C.F. Costs

			UNIT	UNIT COSTS 1/4	UNIT COSTS MEDIAN	UNIT COSTS 3/4	% OF TOTAL 1/4	% OF TOTAL MEDIAN	% OF TOTAL 3/4	
600	3100	Total: Mechanical & Electrical	S.F.	19.20	26	43.50	26%	29.50%	30.50%	600
	9000	Per bed or person, total cost	Bed	35,900	45,000	58,500				
610	0010	**OFFICES** Low Rise (1 to 4 story) R17100-100	S.F.	66.50	86	111				610
	0020	Total project costs	C.F.	4.62	6.70	8.80				
	0100	Site work	S.F.	5.15	8.85	13.05	6.20%	9.70%	13.55%	
	0500	Masonry	"	2.21	5.15	9.80	2.55%	5.05%	7.90%	
	1800	Equipment	S.F.	.80	1.44	4.06	1.10%	1.53%	3.93%	
	2720	Plumbing		2.43	3.69	5.65	3.66%	4.50%	6.10%	
	2770	Heating, ventilating, air conditioning		5.30	7.40	10.80	7.25%	10.40%	11.80%	
	2900	Electrical		5.50	7.80	11	7.50%	9.60%	11.40%	
	3100	Total: Mechanical & Electrical	↓	14	19.75	29	18%	22.50%	26.50%	
620	0010	**OFFICES** Mid Rise (5 to 10 story) R17100-100	S.F.	70	85	116				620
	0020	Total project costs	C.F.	4.98	6.25	9				
	2720	Plumbing	S.F.	2.11	3.43	4.85	2.76%	4.34%	4.50%	
	2770	Heating, ventilating, air conditioning		5.35	7.65	12.20	7.65%	9.40%	11%	
	2900	Electrical		5.20	6.70	8.85	6.20%	7.80%	10%	
	3100	Total: Mechanical & Electrical	↓	13.15	16.85	34	19.15%	21%	24.50%	
630	0010	**OFFICES** High Rise (11 to 20 story) R17100-100	S.F.	86.50	109	134				630
	0020	Total project costs	C.F.	6.05	7.55	10.85				
	2900	Electrical	S.F.	5.40	6.40	9.50	6.20%	7.85%	10.50%	
	3100	Total: Mechanical & Electrical	"	16.90	20.50	38	16.90%	18.45%	26%	
640	0010	**POLICE STATIONS** R17100-100	S.F.	103	136	172				640
	0020	Total project costs	C.F.	8.25	10.10	14				
	0500	Masonry	S.F.	12.10	17.90	23	7.80%	10.65%	12.75%	
	1800	Equipment		1.65	7.25	11.45	1.43%	4.07%	6.70%	
	2720	Plumbing		6	11.55	14.40	6.60%	7%	11.95%	
	2770	Heating, ventilating, air conditioning		9.05	12.05	16.35	5.85%	10.55%	11.70%	
	2900	Electrical		11.30	16.90	21.50	9.75%	11.25%	13.95%	
	3100	Total: Mechanical & Electrical	↓	37	44.50	60.50	28.50%	32%	32.50%	
650	0010	**POST OFFICES** R17100-100	S.F.	79.50	101	126				650
	0020	Total project costs	C.F.	4.92	6.25	7.55				
	2720	Plumbing	S.F.	3.68	4.73	5.75	4.01%	5%	5.60%	
	2770	Heating, ventilating, air conditioning		5.75	7.10	7.90	6.65%	7.15%	9.35%	
	2900	Electrical		6.75	9.80	11.30	7.25%	9.05%	10.60%	
	3100	Total: Mechanical & Electrical	↓	19.50	26	29.50	16.25%	19%	22.50%	
660	0010	**POWER PLANTS** R17100-100	S.F.	580	750	1,375				660
	0020	Total project costs	C.F.	15.65	34	73				
	2900	Electrical	S.F.	40	85	127	9.30%	16.95%	25.50%	
	8100	Total: Mechanical & Electrical	"	100	320	715	32.50%	32.50%	52.50%	
670	0010	**RELIGIOUS EDUCATION** R17100-100	S.F.	64.50	84	106				670
	0020	Total project costs	C.F.	3.76	5.25	6.80				
	2720	Plumbing	S.F.	2.76	3.91	5.50	4.31%	4.92%	7%	
	2770	Heating, ventilating, air conditioning		6.45	7.90	11.15	9.75%	11.45%	12.35%	
	2900	Electrical		5.25	6.90	9.70	7.60%	9.10%	10.35%	
	3100	Total: Mechanical & Electrical	↓	17.95	28	30	22%	24.50%	27.50%	
690	0010	**RESEARCH** Laboratories and facilities R17100-100	S.F.	98.50	142	206				690
	0020	Total project costs	C.F.	7.70	14.85	17.40				
	1800	Equipment	S.F.	4.38	8.65	21	.94%	4.96%	8.80%	
	2720	Plumbing		9.90	12.65	20.50	4.46%	8%	9.80%	
	2770	Heating, ventilating, air conditioning		8.90	30	35.50	6.85%	10.80%	17%	
	2900	Electrical		11.55	19.05	32.50	9.55%	11.10%	14%	
	3100	Total: Mechanical & Electrical	↓	32	63.50	95	29%	35.50%	42%	
700	0010	**RESTAURANTS** R17100-100	S.F.	97	122	156				700
	0020	Total project costs	C.F.	8.05	10.40	13.50				
	1800	Equipment	S.F.	6.20	15.20	22.50	6.80%	13%	15.65%	
	2720	Plumbing	↓	7.80	9.20	13.10	6.10%	8.55%	9.20%	

SQUARE FOOT 17

For expanded coverage of these items see *Means Square Foot Costs 2004*

			UNIT	UNIT COSTS			% OF TOTAL		
				1/4	MEDIAN	3/4	1/4	MEDIAN	3/4
700	2770	Heating, ventilating, air conditioning	S.F.	10	13.30	17.35	8.65%	12%	12.40%	**700**
	2900	Electrical		10.30	12.35	16.25	8.60%	10.80%	11.55%	
	3100	Total: Mechanical & Electrical	↓	31	33	42.50	19.25%	24%	29.50%	
	9000	Per seat unit, total cost	Seat	3,400	4,675	5,800				
	9500	Total: Mechanical & Electrical	"	880	1,175	1,375				
720	0010	**RETAIL STORES** R17100-100	S.F.	44.50	60	79				**720**
	0020	Total project costs	C.F.	3.11	4.32	6.05				
	2720	Plumbing	S.F.	1.62	2.70	4.62	3.09%	4.60%	6.15%	
	2770	Heating, ventilating, air conditioning		3.50	4.79	7.20	6.85%	8.75%	10%	
	2900	Electrical		4.01	5.65	7.90	7.30%	10%	11.55%	
	3100	Total: Mechanical & Electrical	↓	10.70	13.75	18.45	17.10%	21.50%	23.50%	
740	0010	**SCHOOLS** Elementary R17100-100	S.F.	70.50	88.50	108				**740**
	0020	Total project costs	C.F.	4.77	6.15	7.90				
	0500	Masonry	S.F.	6.55	10.30	15.35	4.89%	9.95%	14.10%	
	1800	Equipment		2.20	3.76	6.70	1.76%	3.20%	5.15%	
	2720	Plumbing		4.08	5.90	7.90	5.65%	7%	9.30%	
	2730	Heating, ventilating, air conditioning		6.30	10	13.95	8%	10.80%	13.55%	
	2900	Electrical		6.70	8.75	11.25	8.40%	10.05%	11.80%	
	3100	Total: Mechanical & Electrical	↓	24	29	35.50	23.50%	27.50%	30%	
	9000	Per pupil, total cost	Ea.	7,050	12,500	37,300				
	9500	Total: Mechanical & Electrical	"	2,375	3,000	11,900				
760	0010	**SCHOOLS** Junior High & Middle R17100-100	S.F.	73	91.50	107				**760**
	0020	Total project costs	C.F.	4.65	6.25	7.05				
	0500	Masonry	S.F.	8.05	11.05	13.80	8.85%	11.10%	14%	
	1800	Equipment		2.40	3.87	5.95	1.81%	3.33%	5.15%	
	2720	Plumbing		4.77	5.40	7	5.50%	6.95%	7.25%	
	2770	Heating, ventilating, air conditioning		5.50	10.60	14.80	9.45%	12.75%	17.45%	
	2900	Electrical		6.95	8.70	10.90	7.90%	9.25%	10.60%	
	3100	Total: Mechanical & Electrical	↓	21	29.50	33.50	22.50%	25.50%	29%	
	9000	Per pupil, total cost	Ea.	9,550	12,900	16,800				
780	0010	**SCHOOLS** Senior High R17100-100	S.F.	76.50	98	124				**780**
	0020	Total project costs	C.F.	5	6.95	11.55				
	1800	Equipment	S.F.	2.06	4.84	7.15	1.86%	3.29%	4.80%	
	2720	Plumbing		4	5.30	12.05	5%	6.05%	8.20%	
	2770	Heating, ventilating, air conditioning		8.95	10.25	19.55	8.95%	13.80%	15%	
	2900	Electrical		7.70	10	16.55	8.65%	10.05%	11.95%	
	3100	Total: Mechanical & Electrical	↓	26.50	30.50	51	22%	26.50%	28.50%	
	9000	Per pupil, total cost	Ea.	7,375	12,500	18,800				
800	0010	**SCHOOLS** Vocational R17100-100	S.F.	63	89.50	112				**800**
	0020	Total project costs	C.F.	3.93	5.55	7.65				
	0500	Masonry	S.F.	3.72	9.20	14.05	3.20%	4.04%	10.95%	
	1800	Equipment	"	1.98	2.68	6.85	1.12%	3.25%	5.85%	
	2720	Plumbing	S.F.	4.12	6.20	9.35	5.40%	7.05%	8.55%	
	2770	Heating, ventilating, air conditioning		5.65	10.55	17.65	8.60%	12.05%	14.65%	
	2900	Electrical		6.45	9	12.70	8.50%	10.40%	13.05%	
	3100	Total: Mechanical & Electrical	↓	23	25.50	43.50	23.50%	29.50%	31%	
	9000	Per pupil, total cost	Ea.	8,825	23,600	35,200				
830	0010	**SPORTS ARENAS** R17100-100	S.F.	54	74	116				**830**
	0020	Total project costs	C.F.	3	5.50	6.90				
	2720	Plumbing	S.F.	3.21	4.87	10.30	5.80%	7.50%	8.50%	
	2770	Heating, ventilating, air conditioning		6.90	8.20	11.35	5.85%	10.45%	13.55%	
	2900	Electrical		6.10	7.85	10.10	8.75%	10.40%	12.25%	
	3100	Total: Mechanical & Electrical	↓	14.35	25.50	33	21.50%	25%	27.50%	
850	0010	**SUPERMARKETS** R17100-100	S.F.	50.50	60	74				**850**
	0020	Total project costs	C.F.	2.85	3.44	5.20				

17 SQUARE FOOT

17100 | S.F. & C.F. Costs			UNIT	UNIT COSTS			% OF TOTAL		
				1/4	MEDIAN	3/4	1/4	MEDIAN	3/4
850	2720	Plumbing	S.F.	2.85	3.72	4.19	5.75%	6%	7.60%	**850**
	2770	Heating, ventilating, air conditioning		4.20	5.60	6.80	8.60%	8.60%	10.45%	
	2900	Electrical		6.40	7.35	8.90	10.40%	12.95%	13.60%	
	3100	Total: Mechanical & Electrical	↓	16.40	17.85	25	20.50%	26.50%	31%	
860	0010	**SWIMMING POOLS** R17100 -100	S.F.	90	139	273				**860**
	0020	Total project costs	C.F.	6.65	8.25	9				
	2720	Plumbing	S.F.	7.65	8.75	12.20	4.80%	9.70%	12.30%	
	2900	Electrical		6.20	10.10	14.65	5.75%	6.60%	8%	
	3100	Total: Mechanical & Electrical	↓	15.15	38.50	52.50	11.15%	14.10%	23.50%	
870	0010	**TELEPHONE EXCHANGES** R17100 -100	S.F.	110	163	204				**870**
	0020	Total project costs	C.F.	6.85	10.45	15.10				
	2720	Plumbing	S.F.	4.67	7.50	10.50	5.40%	5.95%	6.90%	
	2770	Heating, ventilating, air conditioning		10.75	21.50	27	11.80%	16.05%	18.40%	
	2900	Electrical		11.20	17.10	31.50	10.30%	13.75%	17.85%	
	3100	Total: Mechanical & Electrical	↓	33	62.50	88.50	29.50%	33.50%	44.50%	
910	0010	**THEATERS** R17100 -100	S.F.	69	88.50	131				**910**
	0020	Total project costs	C.F.	3.24	4.72	6.95				
	2720	Plumbing	S.F.	2.30	2.50	10.20	2.92%	4.60%	5.70%	
	2770	Heating, ventilating, air conditioning		6.70	8.15	10.05	8%	12.25%	13.40%	
	2900	Electrical		6.05	8.15	16.60	8.20%	11%	12.45%	
	3100	Total: Mechanical & Electrical	↓	15.55	23	47.50	23%	26.50%	27.50%	
940	0010	**TOWN HALLS** City Halls & Municipal Buildings R17100 -100	S.F.	82.50	104	139				**940**
	0020	Total project costs	C.F.	7.05	8.75	11.85				
	2720	Plumbing	S.F.	3.22	6.05	11.10	4.22%	6.10%	8%	
	2770	Heating, ventilating, air conditioning		5.80	11.55	16.90	7.05%	9.05%	13.45%	
	2900	Electrical		7.30	10.80	14.40	8%	9.50%	12.15%	
	3100	Total: Mechanical & Electrical	↓	24.50	29	38	22%	26.50%	31%	
970	0010	**WAREHOUSES** And Storage Buildings R17100 -100	S.F.	30.50	44.50	62				**970**
	0020	Total project costs	C.F.	1.66	2.36	3.85				
	0100	Site work	S.F.	2.97	5.90	8.90	6%	12.55%	19.85%	
	0500	Masonry		1.80	4.09	8.85	3.50%	7.40%	12.30%	
	1800	Equipment		.47	1	5.60	1.16%	2.40%	6.55%	
	2720	Plumbing		.96	1.72	3.29	2.81%	4.80%	6.40%	
	2730	Heating, ventilating, air conditioning		1.15	3.09	4.15	2.41%	5%	8.40%	
	2900	Electrical		1.85	3.30	5.30	5.45%	7.20%	10.05%	
	3100	Total: Mechanical & Electrical	↓	4.75	7.30	15.95	10.70%	18.35%	26%	
990	0010	**WAREHOUSE & OFFICES** Combination R17100 -100	S.F.	35.50	47	66				**990**
	0020	Total project costs	C.F.	1.85	2.63	3.89				
	1800	Equipment	S.F.	.63	1.19	1.84	.54%	1.38%	2.40%	
	2720	Plumbing		1.40	2.43	3.72	3.82%	4.80%	6.45%	
	2770	Heating, ventilating, air conditioning		2.16	3.38	4.73	5%	5.65%	9.60%	
	2900	Electrical		2.37	3.48	5.60	5.90%	8%	10.25%	
	3100	Total: Mechanical & Electrical	↓	6.55	9.35	15.45	14.40%	19.95%	24%	

SQUARE FOOT 17

For information about Means Estimating Seminars, see yellow pages 12 and 13 in back of book

For expanded coverage of these items see _Means Square Foot Costs 2004_

Division Notes

		CREW	DAILY OUTPUT	LABOR-HOURS	UNIT	2004 BARE COSTS				TOTAL INCL O&P
						MAT.	LABOR	EQUIP.	TOTAL	

Assemblies Section

Table of Contents

How to Use the Assemblies Cost Tables

The following is a detailed explanation of a sample Assemblies Cost Table. Most Assembly Tables are separated into three parts: 1) an illustration of the system to be estimated; 2) the components and related costs of a typical system; and 3) the costs for similar systems with dimensional and/or size variations. For costs of the components that comprise these systems or "assemblies," refer to the Unit Price Section. Next to each bold number below is the item being described with the appropriate component of the sample entry following in parenthesis. In most cases, if the work is to be subcontracted, the general contractor will need to add an additional markup (R.S. Means suggests using 10%) to the "Total" figures.

1 System/Line Numbers (D5020 145 0200)

Each Assemblies Cost Line has been assigned a unique identification number based on the UNIFORMAT II classification system.

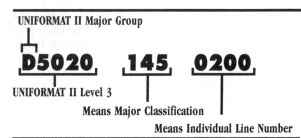

UNIFORMAT II Major Group

D5020   145   0200

UNIFORMAT II Level 3 — Means Major Classification — Means Individual Line Number

---

D50 Electrical
D5020 Lighting and Branch Wiring

Power System — Switch — Motor Starter — Switch — Motor Connection — Motor

System D5020 145 installed cost of motor wiring as per Table D5010 170 using 50' of rigid conduit and copper wire. **Cost and setting of motor not included.**

System Components	QUANTITY	UNIT	COST EACH MAT.	COST EACH INST.	COST EACH TOTAL
SYSTEM D5020 145 0200					
MOTOR INSTALLATION, SINGLE PHASE, 115V, TO AND INCLUDING 1/3 HP MOTOR SIZE					
Wire 600V type THWN-THHN, copper solid #12	1.250	C.L.F.	5.71	53.13	58.84
Steel intermediate conduit, (IMC) 1/2" diam	50.000	L.F.	75.50	234.50	310
Magnetic FVNR, 115V, 1/3 HP, size 00 starter	1.000	Ea.	173	117	290
Safety switch, fused, heavy duty, 240V 2P 30 amp	1.000	Ea.	99.50	134	233.50
Safety switch, non fused, heavy duty, 600V, 3 phase, 30 A	1.000	Ea.	117	147	264
Flexible metallic conduit, Greenfield 1/2" diam	1.500	L.F.	.51	3.51	4.02
Connectors for flexible metallic conduit Greenfield 1/2" diam	1.000	Ea.			
Coupling for Greenfield to conduit 1/2" diam flexible metalic conduit	1.000	Ea.	.83		
Fuse cartridge nonrenewable, 250V 30 amp	1.000	Ea.	1.43	9.40	
TOTAL			475.06	713.79	1,188.85

D5020 145	Motor Installation	MAT.	INST.	TOTAL
0200	Motor installation, single phase, 115V, to and including 1/3 HP motor size	475	715	1,190
0240	To and incl. 1 HP motor size	495	715	1,210
0280	To and incl. 2 HP motor size	530	760	1,290
0320	To and incl. 3 HP motor size	590	775	1,365
0360	230V, to and including 1 HP motor size	475	725	1,200
0400	To and incl. 2 HP motor size	500	725	1,225
0440	To and incl. 3 HP motor size	560	780	1,340
0520	Three phase, 200V, to and including 1-1/2 HP motor size	560	795	1,355
0560	To and incl. 3 HP motor size	600	865	1,465
0600	To and incl. 5 HP motor size	640	965	1,605
0640	To and incl. 7-1/2 HP motor size	650	985	1,635

Illustration

2 At the top of most assembly pages is an illustration, a brief description, and the design criteria used to develop the cost.

System Components

3 The components of a typical system are listed separately to show what has been included in the development of the total system price. The table below contains prices for other similar systems with dimensional and/or size variations.

Quantity

4 This is the number of line item units required for one system unit. For example, we assume that it will take 50 linear feet of Steel Intermediate (IMC) Conduit for each motor to be connected to the other items listed here.

Unit of Measure for Each Item

5 The abbreviated designation indicates the unit of measure, as defined by industry standards, upon which the price of the component is based. For example, wire is priced by C.L.F. (100 linear feet) while conduit is priced by L.F. (linear feet) and the starter and switches are priced by each unit. For a complete listing of abbreviations, see the Reference Section.

Unit of Measure for Each System (Cost Each)

6 Costs shown in the three right hand columns have been adjusted by the component quantity and unit of measure for the entire system. In this example, "Cost Each" is the unit of measure for this system or "assembly."

Materials (475)

7 This column contains the Materials Cost of each component. These cost figures are bare costs plus 10% for profit.

Installation (715)

8 Installation includes labor and equipment plus the installing contractor's overhead and profit. Equipment costs are the bare rental costs plus 10% for profit. The labor overhead and profit is defined on the inside back cover of this book.

Total (1,190)

9 The figure in this column is the sum of the material and installation costs.

Material Cost	+	Installation Cost	=	Total
$475.00	+	$715.00	=	$1,190.00

D SERVICES

D2020 Domestic Water Distribution

Installation includes piping and fittings within 10' of heater. Electric water heaters do not require venting.

1 Kilowatt hour will raise:			
Gallons of Water	Degrees F	Gallons of Water	Degrees F
4.1	100°	6.8	60°
4.5	90°	8.2	50°
5.1	80°	10.0	40°
5.9	70°		

System Components	QUANTITY	UNIT	COST EACH		
			MAT.	INST.	TOTAL
SYSTEM D2020 210 1780					
ELECTRIC WATER HEATER, RESIDENTIAL, 100° F RISE					
10 GALLON TANK, 7 GPH					
Water heater, residential electric, glass lined tank, 10 Gal	1.000	Ea.	231	207	438
Copper tubing, type L, solder joint, hanger 10' OC 1/2" diam	30.000	L.F.	33.30	175.50	208.80
Wrought copper 90° elbow for solder joints 1/2" diam.	4.000	Ea.	1.32	96	97.32
Wrought copper Tee for solder joints, 1/2" diam	2.000	Ea.	1.12	73	74.12
Wrought copper union for soldered joints, 1/2" diam	2.000	Ea.	6.94	50	56.94
Valve, gate, bronze, 125 lb, NRS, soldered 1/2" diam	2.000	Ea.	35.20	39.60	74.80
Relief valve, bronze, press & temp, self-close, 3/4" IPS	1.000	Ea.	81	17	98
Wrought copper adapter, CTS to MPT 3/4" IPS	1.000	Ea.	1.14	28	29.14
Copper tubing, type L, solder joints, 3/4" diam	1.000	L.F.	1.56	6.25	7.81
Wrought copper 90° elbow for solder joints 3/4" diam	1.000	Ea.	.74	25	25.74
TOTAL			393.32	717.35	1,110.67

D2020 210	Electric Water Heaters - Residential Systems		COST EACH		
			MAT.	INST.	TOTAL
1760	Electric water heater, residential, 100° F rise				
1780	10 gallon tank, 7 GPH		395	715	1,110
1820	20 gallon tank, 7 GPH	R15100	455	765	1,220
1860	30 gallon tank, 7 GPH	-120	545	805	1,350
1900	40 gallon tank, 8 GPH		610	885	1,495
1940	52 gallon tank, 10 GPH		665	890	1,555
1980	66 gallon tank, 13 GPH		995	1,025	2,020
2020	80 gallon tank, 16 GPH		1,075	1,075	2,150
2060	120 gallon tank, 23 GPH		1,625	1,250	2,875

SERVICES

D

Important: See the Reference Section for critical supporting data - Reference Numbers and City Cost Indexes

D2020 Domestic Water Distribution

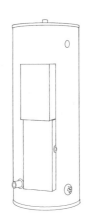

Systems below include piping and fittings within 10' of heater. Electric water heaters do not require venting.

System Components	QUANTITY	UNIT	COST EACH		
			MAT.	INST.	TOTAL
SYSTEM D2020 240 1820					
ELECTRIC WATER HEATER, COMMERCIAL, 100° F RISE					
50 GALLON TANK, 9 KW, 37 GPH					
Water heater, commercial, electric, 50 Gal, 9 KW, 37 GPH	1.000	Ea.	2,400	264	2,664
Copper tubing, type L, solder joint, hanger 10' OC, 3/4" diam	34.000	L.F.	53.04	212.50	265.54
Wrought copper 90° elbow for solder joints 3/4" diam	5.000	Ea.	3.70	125	128.70
Wrought copper Tee for solder joints, 3/4" diam	2.000	Ea.	2.72	79	81.72
Wrought copper union for soldered joints, 3/4" diam	2.000	Ea.	8.70	53	61.70
Valve, gate, bronze, 125 lb, NRS, soldered 3/4" diam	2.000	Ea.	40	48	88
Relief valve, bronze, press & temp, self-close, 3/4" IPS	1.000	Ea.	81	17	98
Wrought copper adapter, copper tubing to male, 3/4" IPS	1.000	Ea.	1.14	28	29.14
TOTAL			2,590.30	826.50	3,416.80

D2020 240	Electric Water Heaters - Commercial Systems		COST EACH		
			MAT.	INST.	TOTAL
1800	Electric water heater, commercial, 100° F rise				
1820	50 gallon tank, 9 KW 37 GPH		2,600	825	3,425
1860	80 gal, 12 KW 49 GPH	R15100 -110	3,275	1,025	4,300
1900	36 KW 147 GPH		4,450	1,100	5,550
1940	120 gal, 36 KW 147 GPH	R15100 -120	4,825	1,200	6,025
1980	150 gal, 120 KW 490 GPH		15,100	1,275	16,375
2020	200 gal, 120 KW 490 GPH		15,900	1,300	17,200
2060	250 gal, 150 KW 615 GPH		17,600	1,525	19,125
2100	300 gal, 180 KW 738 GPH		19,200	1,600	20,800
2140	350 gal, 30 KW 123 GPH		14,100	1,750	15,850
2180	180 KW 738 GPH		19,700	1,750	21,450
2220	500 gal, 30 KW 123 GPH		18,500	2,050	20,550
2260	240 KW 984 GPH		27,000	2,050	29,050
2300	700 gal, 30 KW 123 GPH		22,400	2,325	24,725
2340	300 KW 1230 GPH		33,300	2,325	35,625
2380	1000 gal, 60 KW 245 GPH		27,000	3,250	30,250
2420	480 KW 1970 GPH		36,000	3,250	39,250
2460	1500 gal, 60 KW 245 GPH		39,600	4,025	43,625
2500	480 KW 1970 GPH		54,500	4,025	58,525

SERVICES

D

D3020 Heat Generating Systems

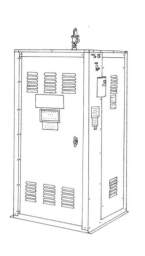

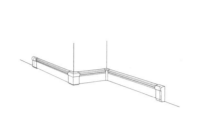

Boiler   **Baseboard Radiation**

**Small Electric Boiler
System Considerations:**
1. Terminal units are fin tube baseboard radiation rated at 720 BTU/hr with 200° water temperature or 820 BTU/hr steam.
2. Primary use being for residential or smaller supplementary areas, the floor levels are based on 7-1/2' ceiling heights.
3. All distribution piping is copper for boilers through 205 MBH. All piping for larger systems is steel pipe.

System Components	QUANTITY	UNIT	COST EACH		
			MAT.	INST.	TOTAL
SYSTEM D3020 102 1120					
SMALL HEATING SYSTEM, HYDRONIC, ELECTRIC BOILER					
1,480 S.F., 61 MBH, STEAM, 1 FLOOR					
Boiler, electric steam, std cntrls, trim, ftngs and valves, 18 KW, 61.4 MBH	1.000	Ea.	3,327.50	1,210	4,537.50
Copper tubing type L, solder joint, hanger 10'OC, 1-1/4" diam	160.000	L.F.	430.40	1,312	1,742.40
Radiation, 3/4" copper tube w/alum fin baseboard pkg 7" high	60.000	L.F.	375	888	1,263
Rough in baseboard panel or fin tube with valves & traps	10.000	Set	985	4,500	5,485
Pipe covering, calcium silicate w/cover 1" wall 1-1/4" diam	160.000	L.F.	368	792	1,160
Low water cut-off, quick hookup, in gage glass tappings	1.000	Ea.	174	30	204
TOTAL			5,659.90	8,732	14,391.90
COST PER S.F.			3.82	5.90	9.72

D3020 102	Small Heating Systems, Hydronic, Electric Boilers	COST PER S.F.		
		MAT.	INST.	TOTAL
1100	Small heating systems, hydronic, electric boilers			
1120	Steam, 1 floor, 1480 S.F., 61 M.B.H.	3.82	5.90	9.72
1160	3,000 S.F., 123 M.B.H.	2.82	5.15	7.97
1200	5,000 S.F., 205 M.B.H.	2.40	4.77	7.17
1240	2 floors, 12,400 S.F., 512 M.B.H.	1.94	4.73	6.67
1280	3 floors, 24,800 S.F., 1023 M.B.H.	2.05	4.68	6.73
1320	34,750 S.F., 1,433 M.B.H.	1.84	4.54	6.38
1360	Hot water, 1 floor, 1,000 S.F., 41 M.B.H.	6.40	3.31	9.71
1400	2,500 S.F., 103 M.B.H.	4.37	5.90	10.27
1440	2 floors, 4,850 S.F., 205 M.B.H.	4.22	7.05	11.27
1480	3 floors, 9,700 S.F., 410 M.B.H.	3.87	7.30	11.17

SERVICES D

D3020 Heat Generating Systems

Boiler

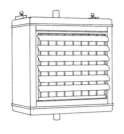

Unit Heater

Large Electric Boiler System Considerations:

1. Terminal units are all unit heaters of the same size. Quantities are varied to accommodate total requirements.
2. All air is circulated through the heaters a minimum of three times per hour.
3. As the capacities are adequate for commercial use, floor levels are based on 10' ceiling heights.
4. All distribution piping is black steel pipe.

System Components	QUANTITY	UNIT	COST EACH		
			MAT.	INST.	TOTAL
SYSTEM D3020 104 1240					
LARGE HEATING SYSTEM, HYDRONIC, ELECTRIC BOILER					
9,280 S.F., 150 KW, 510 MBH, 1 FLOOR					
Boiler, electric hot water, std ctrls, trim, ftngs, valves, 150 KW, 510 MBH	1.000	Ea.	11,737.50	2,662.50	14,400
Expansion tank, painted steel, 60 Gal capacity ASME	1.000	Ea.	2,475	143	2,618
Circulating pump, CI, close cpld, 50 GPM, 2 HP, 2" pipe conn	1.000	Ea.	1,250	285	1,535
Unit heater, 1 speed propeller, horizontal, 200° EWT, 72.7 MBH	7.000	Ea.	4,480	1,092	5,572
Unit heater piping hookup with controls	7.000	Set	2,520	7,175	9,695
Pipe, steel, black, schedule 40, welded, 2-1/2" diam	380.000	L.F.	1,774.60	7,607.60	9,382.20
Pipe covering, calcium silicate w/cover, 1" wall, 2-1/2" diam	380.000	L.F.	1,083	1,938	3,021
TOTAL			25,320.10	20,903.10	46,223.20
COST PER S.F.			2.73	2.25	4.98

D3020 104	Large Heating Systems, Hydronic, Electric Boilers	COST PER S.F.		
		MAT.	INST.	TOTAL
1230	Large heating systems, hydronic, electric boilers			
1240	9,280 S.F., 150 K.W., 510 M.B.H., 1 floor	2.73	2.25	4.98
1280	14,900 S.F., 240 K.W., 820 M.B.H., 2 floors	2.82	3.68	6.50
1320	18,600 S.F., 296 K.W., 1,010 M.B.H., 3 floors	2.74	4.01	6.75
1360	26,100 S.F., 420 K.W., 1,432 M.B.H., 4 floors	2.71	3.92	6.63
1400	39,100 S.F., 666 K.W., 2,273 M.B.H., 4 floors	2.35	3.28	5.63
1440	57,700 S.F., 900 K.W., 3,071 M.B.H., 5 floors	2.18	3.24	5.42
1480	111,700 S.F., 1,800 K.W., 6,148 M.B.H., 6 floors	2	2.77	4.77
1520	149,000 S.F., 2,400 K.W., 8,191 M.B.H., 8 floors	1.99	2.77	4.76
1560	223,300 S.F., 3,600 K.W., 12,283 M.B.H., 14 floors	2.06	3.16	5.22

SERVICES

D

D4090 Other Fire Protection Systems

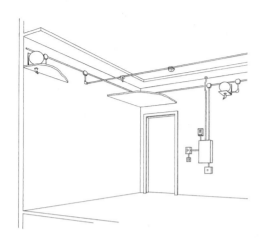

General: Automatic fire protection (suppression) systems other than water sprinklers may be desired for special environments, high risk areas, isolated locations or unusual hazards. Some typical applications would include:

Paint dip tanks
Securities vaults
Electronic data processing
Tape and data storage
Transformer rooms
Spray booths
Petroleum storage
High rack storage

Piping and wiring costs are dependent on the individual application and must be added to the component costs shown below.

All areas are assumed to be open.

D4090 910	Unit Components	COST EACH		
		MAT.	INST.	TOTAL
0020	Detectors with brackets			
0040	Fixed temperature heat detector	31	58.50	89.50
0060	Rate of temperature rise detector	37	58.50	95.50
0080	Ion detector (smoke) detector	82.50	75.50	158
0200	Extinguisher agent			
0240	200 lb FM200, container	6,375	214	6,589
0280	75 lb carbon dioxide cylinder	1,050	143	1,193
0320	Dispersion nozzle			
0340	FM200 1-1/2" dispersion nozzle	55	34	89
0380	Carbon dioxide 3" x 5" dispersion nozzle	55	26.50	81.50
0420	Control station			
0440	Single zone control station with batteries	1,400	470	1,870
0470	Multizone (4) control station with batteries	2,700	940	3,640
0490				
0500	Electric mechanical release	138	236	374
0520				
0550	Manual pull station	49.50	79.50	129
0570				
0640	Battery standby power 10" x 10" x 17"	760	117	877
0700				
0740	Bell signalling device	54.50	58.50	113

D4090 920	FM200 Systems	COST PER C.F.		
		MAT.	INST.	TOTAL
0820	Average FM200 system, minimum			1.38
0840	Maximum			2.75

Figure D5010-111   Typical Overhead Service Entrance

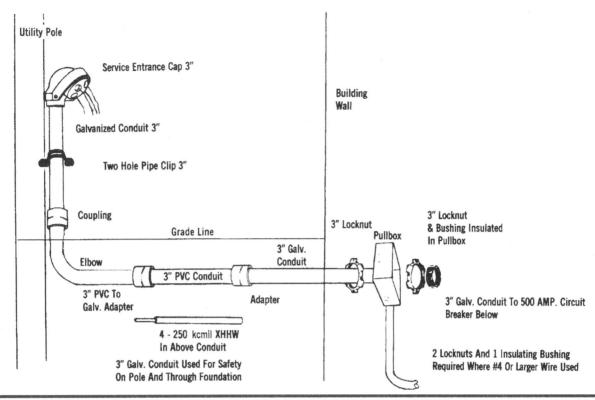

- Utility Pole
- Service Entrance Cap 3"
- Galvanized Conduit 3"
- Two Hole Pipe Clip 3"
- Coupling
- Building Wall
- Grade Line
- 3" Locknut
- Pullbox
- 3" Locknut & Bushing Insulated In Pullbox
- 3" Galv. Conduit
- Elbow
- 3" PVC Conduit
- Adapter
- 3" PVC To Galv. Adapter
- 3" Galv. Conduit To 500 AMP. Circuit Breaker Below
- 4 - 250 kcmil XHHW In Above Conduit
- 3" Galv. Conduit Used For Safety On Pole And Through Foundation
- 2 Locknuts And 1 Insulating Bushing Required Where #4 Or Larger Wire Used

Figure D5010-112   Typical Commercial Electric System

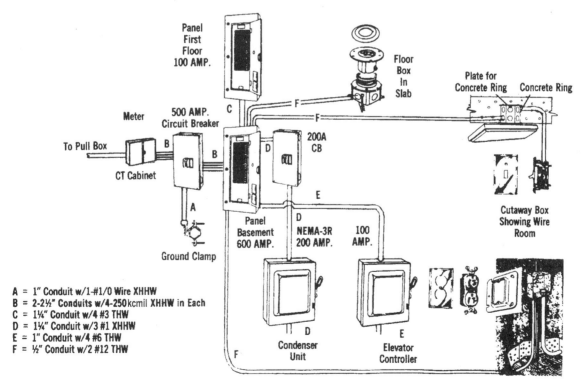

- Panel First Floor 100 AMP.
- Floor Box In Slab
- Plate for Concrete Ring
- Concrete Ring
- Meter
- 500 AMP. Circuit Breaker
- 200A CB
- To Pull Box
- CT Cabinet
- Cutaway Box Showing Wire Room
- Ground Clamp
- Panel Basement 600 AMP.
- NEMA-3R 200 AMP.
- 100 AMP.
- Condenser Unit
- Elevator Controller

A = 1" Conduit w/1-#1/0 Wire XHHW
B = 2-2½" Conduits w/4-250 kcmil XHHW in Each
C = 1¼" Conduit w/4 #3 THW
D = 1¼" Conduit w/3 #1 XHHW
E = 1" Conduit w/4 #6 THW
F = ½" Conduit w/2 #12 THW

Figure D5010-113 Preliminary Procedure

1. Determine building size and use
2. Develop total load in watts
   a. Lighting
   b. Receptacle
   c. Air Conditioning
   d. Elevator
   e. Other power requirements
3. Determine best voltage available from utility company.
4. Determine cost from tables for loads (a) thru (e) above.
5. Determine size of service from formulas (D5010-116).
6. Determine costs for service, panels, and feeders from tables.

Figure D5010-114 Office Building 90′ x 210′, 3 story, w/garage

Garage Area = 18,900 S.F.
Office Area = 56,700 S.F.
Elevator = 2 @ 125 FPM

Tables	Power Required	Watts
D5010-1151	Garage Lighting .5 Watt/S.F.	9,450
	Office Lighting 3 Watts/S.F.	170,100
D5010-1151	Office Receptacles 2 Watts/S.F.	113,400
D5010-1151, R16050-060	Low Rise Office A.C. 4.3 Watts/S.F.	243,810
D5010-1152, 1153	Elevators - 2 @ 20 HP = 2 @ 17,404 Watts/Ea.	34,808
D5010-1151	Misc. Motors + Power 1.2 Watts/S.F.	68,040
	Total	639,608 Watts

Voltage Available

277/480V, 3 Phase, 4 Wire

Formula

D5010-116
$$\text{Amperes} = \frac{\text{Watts}}{\text{Volts x Power Factor x 1.73}} = \frac{639,608}{480\text{V x .8 x 1.73}} = 963 \text{ Amps}$$

Use 1200 Amp Service

System	Description	Unit Cost	Unit	Total
D5020 210 0200	Garage Lighting (Interpolated)	$.98	S.F.	$ 18,522
D5020 210 0280	Office Lighting	5.79	S.F.	328,293
D5020 115 0880	Receptacle-Undercarpet	3.09	S.F.	175,203
D5020 140 0280	Air Conditioning	.41	S.F.	23,247
D5020 135 0320	Misc. Pwr.	.23	S.F.	13,041
D5020 145 2120	Elevators - 2 @ 20HP	2,145.00	Ea.	4,290
D5010 120 0480	Service-1200 Amp (add 25% for 277/480V)	14,175.00	Ea.	17,719
D5010 230 0480	Feeder - Assume 200 Ft.	181.00	Ft.	36,200
D5010 240 0320	Panels - 1200 Amp (add 20% for 277/480V)	23,200.00	Ea.	27,840
D5030 910 0400	Fire Detection	24,025.00	Ea.	24,025
	Total			$668,380
	or			$668,400

SERVICES D

Table D5010-1151 Nominal Watts Per S.F. for Electric Systems for Various Building Types

Type Construction	1. Lighting	2. Devices	3. HVAC	4. Misc.	5. Elevator	Total Watts
Apartment, luxury high rise	2	2.2	3	1		
Apartment, low rise	2	2	3	1		
Auditorium	2.5	1	3.3	.8		
Bank, branch office	3	2.1	5.7	1.4		
Bank, main office	2.5	1.5	5.7	1.4		
Church	1.8	.8	3.3	.8		
College, science building	3	3	5.3	1.3		
College, library	2.5	.8	5.7	1.4		
College, physical education center	2	1	4.5	1.1		
Department store	2.5	.9	4	1		
Dormitory, college	1.5	1.2	4	1		
Drive-in donut shop	3	4	6.8	1.7		
Garage, commercial	.5	.5	0	.5		
Hospital, general	2	4.5	5	1.3		
Hospital, pediatric	3	3.8	5	1.3		
Hotel, airport	2	1	5	1.3		
Housing for the elderly	2	1.2	4	1		
Manufacturing, food processing	3	1	4.5	1.1		
Manufacturing, apparel	2	1	4.5	1.1		
Manufacturing, tools	4	1	4.5	1.1		
Medical clinic	2.5	1.5	3.2	1		
Nursing home	2	1.6	4	1		
Office building, hi rise	3	2	4.7	1.2		
Office building, low rise	3	2	4.3	1.2		
Radio-TV studio	3.8	2.2	7.6	1.9		
Restaurant	2.5	2	6.8	1.7		
Retail store	2.5	.9	5.5	1.4		
School, elementary	3	1.9	5.3	1.3		
School, junior high	3	1.5	5.3	1.3		
School, senior high	2.3	1.7	5.3	1.3		
Supermarket	3	1	4	1		
Telephone exchange	1	.6	4.5	1.1		
Theater	2.5	1	3.3	.8		
Town Hall	2	1.9	5.3	1.3		
U.S. Post Office	3	2	5	1.3		
Warehouse, grocery	1	.6	0	.5		

Rule of Thumb: 1 KVA = 1 HP (Single Phase)

Three Phase:

Watts = 1.73 x Volts x Current x Power Factor x Efficiency

$$\text{Horsepower} = \frac{\text{Volts x Current x 1.73 x Power Factor}}{746 \text{ Watts}}$$

Table D5010-1152 Horsepower Requirements for Elevators with 3 Phase Motors

Type	Maximum Travel Height in Ft.	Travel Speeds in FPM	Capacity of Cars in Lbs.		
			1200	1500	1800
Hydraulic	70	70	10	15	15
		85	15	15	15
		100	15	15	20
		110	20	20	20
		125	20	20	20
		150	25	25	25
		175	25	30	30
		200	30	30	40
Geared Traction	300	200			
		350			
			2000	2500	3000
Hydraulic	70	70	15	20	20
		85	20	20	25
		100	20	25	30
		110	20	25	30
		125	25	30	40
		150	30	40	50
		175	40	50	50
		200	40	50	60
Geared Traction	300	200	10	10	15
		350	15	15	23
			3500	4000	4500
Hydraulic	70	70	20	25	30
		85	25	30	30
		100	30	40	40
		110	40	40	50
		125	40	50	50
		150	50	50	60
		175	60		
		200	60		
Geared Traction	300	200	15		23
		350	23		35

The power factor of electric motors varies from 80% to 90% in larger size motors. The efficiency likewise varies from 80% on a small motor to 90% on a large motor.

Table D5010-1153 Watts per Motor

90% Power Factor & Efficiency @ 200 or 460V			
HP	Watts	HP	Watts
10	9024	30	25784
15	13537	40	33519
20	17404	50	41899
25	21916	60	49634

SERVICES

D

Table D5010-116 Electrical Formulas

Ohm's Law

Ohm's Law is a method of explaining the relation existing between voltage, current, and resistance in an electrical circuit. It is practically the basis of all electrical calculations. The term "electromotive force" is often used to designate pressure in volts. This formula can be expressed in various forms.

To find the current in amperes:

$$\text{Current} = \frac{\text{Voltage}}{\text{Resistance}} \quad \text{or} \quad \text{Amperes} = \frac{\text{Volts}}{\text{Ohms}} \quad \text{or} \quad I = \frac{E}{R}$$

The flow of current in amperes through any circuit is equal to the voltage or electromotive force divided by the resistance of that circuit.

To find the pressure or voltage:

$$\text{Voltage} = \text{Current x Resistance} \quad \text{or} \quad \text{Volts} = \text{Amperes x Ohms}$$
$$\text{or} \quad E = I \times R$$

The voltage required to force a current through a circuit is equal to the resistance of the circuit multiplied by the current.

To find the resistance:

$$\text{Resistance} = \frac{\text{Voltage}}{\text{Current}} \quad \text{or} \quad \text{Ohms} = \frac{\text{Volts}}{\text{Amperes}} \quad \text{or} \quad R = \frac{E}{I}$$

The resistance of a circuit is equal to the voltage divided by the current flowing through that circuit.

Power Formulas

$$\text{One horsepower} = 746 \text{ watts} \qquad \text{One kilowatt} = 1000 \text{ watts}$$

The power factor of electric motors varies from 80% to 90% in the larger size motors.

Single-Phase Alternating Current Circuits

Power in Watts = Volts x Amperes x Power Factor

To find current in amperes:

$$\text{Current} = \frac{\text{Watts}}{\text{Volts x Power Factor}} \quad \text{or}$$

$$\text{Amperes} = \frac{\text{Watts}}{\text{Volts x Power Factor}} \quad \text{or} \quad I = \frac{W}{E \times PF}$$

To find current of a motor, single phase:

$$\text{Current} = \frac{\text{Horsepower x 746}}{\text{Volts x Power Factor x Efficiency}} \quad \text{or}$$

$$I = \frac{HP \times 746}{E \times PF \times Eff.}$$

To find horsepower of a motor, single phase:

$$\text{Horsepower} = \frac{\text{Volts x Current x Power Factor x Efficiency}}{746 \text{ Watts}}$$

$$HP = \frac{E \times I \times PF \times Eff.}{746}$$

To find power in watts of a motor, single phase:

Watts = Volts x Current x Power Factor x Efficiency   or
Watts = E x I x PF x Eff.

To find single phase kVA:

$$1 \text{ Phase kVA} = \frac{\text{Volts x Amps}}{1000}$$

Three-Phase Alternating Current Circuits

Power in Watts = Volts x Amperes x Power Factor x 1.73

To find current in amperes in each wire:

$$\text{Current} = \frac{\text{Watts}}{\text{Voltage x Power Factor x 1.73}} \quad \text{or}$$

$$\text{Amperes} = \frac{\text{Watts}}{\text{Volts x Power Factor x 1.73}} \quad \text{or} \quad I = \frac{W}{E \times PF \times 1.73}$$

To find current of a motor, 3 phase:

$$\text{Current} = \frac{\text{Horsepower x 746}}{\text{Volts x Power Factor x Efficiency x 1.73}} \quad \text{or}$$

$$I = \frac{HP \times 746}{E \times PF \times Eff. \times 1.73}$$

To find horsepower of a motor, 3 phase:

$$\text{Horsepower} = \frac{\text{Volts x Current x 1.73 x Power Factor}}{746 \text{ Watts}}$$

$$HP = \frac{E \times I \times 1.73 \times PF}{746}$$

To find power in watts of a motor, 3 phase:

Watts = Volts x Current x 1.73 x Power Factor x Efficiency   or
Watts = E x I x 1.73 x PF x Eff.

To find 3 phase kVA:

$$3 \text{ phase kVA} = \frac{\text{Volts x Amps x 1.73}}{1000} \quad \text{or}$$

$$kVA = \frac{V \times A \times 1.73}{1000}$$

Power Factor (PF) is the percentage ratio of the measured watts (effective power) to the volt-amperes (apparent watts)

$$\text{Power Factor} = \frac{\text{Watts}}{\text{Volts x Amperes}} \times 100\%$$

A Conceptual Estimate of the costs for a building, when final drawings are not available, can be quickly figured by using **Table D5010-117 Cost Per S.F. for Electrical Systems for Various Building Types.** The following definitions apply to this table.

1. **Service And Distribution:** This system includes the incoming primary feeder from the power company, the main building transformer, metering arrangement, switchboards, distribution panel boards, stepdown transformers, power and lighting panels. Items marked (*) include the cost of the primary feeder and transformer. In all other projects the cost of the primary feeder and transformer is paid for by the local power company.

2. **Lighting:** Includes all interior fixtures for decor, illumination, exit and emergency lighting. Fixtures for exterior building lighting are included but parking area lighting is not included unless mentioned. See also Section D5020 for detailed analysis of lighting requirements and costs.

3. **Devices:** Includes all outlet boxes, receptacles, switches for lighting control, dimmers and cover plates.

4. **Equipment Connections:** Includes all materials and equipment for making connections for Heating, Ventilating and Air Conditioning, Food Service and other motorized items requiring connections.

5. **Basic Materials:** This category includes all disconnect power switches not part of service equipment, raceways for wires, pull boxes, junction boxes, supports, fittings, grounding materials, wireways, busways, wire and cable systems.

6. **Special Systems:** Includes installed equipment only for the particular system such as fire detection and alarm, sound, emergency generator and others as listed in the table.

Figure D5010-117   Cost per S.F. for Electric Systems for Various Building Types

Type Construction	1. Service & Distrib.	2. Lighting	3. Devices	4. Equipment Connections	5. Basic Materials	6. Special Systems Fire Alarm & Detection	6. Special Systems Lightning Protection	6. Special Systems Master TV Antenna
Apartment, luxury high rise	$1.55	$1.01	$ .76	$ .92	$2.42	$ .48		$.33
Apartment, low rise	.90	.84	.69	.78	1.41	.40		
Auditorium	1.97	5.09	.63	1.32	2.90	.61		
Bank, branch office	2.37	5.65	1.02	1.35	2.75	1.76		
Bank, main office	1.78	3.07	.33	.57	2.97	.91		
Church	1.23	3.12	.39	.34	1.43	.91		
* College, science building	2.31	4.13	1.34	1.08	3.32	.80		
* College library	1.67	2.29	.30	.62	1.78	.91		
* College, physical education center	2.62	3.20	.41	.51	1.39	.51		
Department store	.87	2.21	.30	.90	2.39	.40		
* Dormitory, college	1.18	2.84	.31	.57	2.34	.64		.40
Drive-in donut shop	3.29	8.49	1.47	1.36	3.83	—		
Garage, commercial	.44	1.05	.22	.42	.82	—		
* Hospital, general	6.37	4.37	1.67	1.06	4.74	.56	$.18	
* Hospital, pediatric	5.58	6.51	1.45	3.91	8.80	.63		.49
* Hotel, airport	2.50	3.58	.31	.53	3.48	.52	.29	.45
Housing for the elderly	.71	.85	.40	1.03	2.97	.63		.39
Manufacturing, food processing	1.59	4.47	.28	1.95	3.24	.39		
Manufacturing, apparel	1.05	2.34	.34	.77	1.74	.35		
Manufacturing, tools	2.42	5.46	.32	.91	2.90	.41		
Medical clinic	.97	1.88	.51	1.40	2.28	.62		
Nursing home	1.65	3.53	.51	.41	2.90	.86		.33
Office Building	2.21	4.78	.28	.77	3.04	.46	.26	
Radio-TV studio	1.59	4.93	.76	1.42	3.58	.59		
Restaurant	5.92	4.68	.92	2.16	4.31	.35		
Retail Store	1.25	2.49	.31	.53	1.35	—		
School, elementary	2.13	4.41	.62	.53	3.62	.55		.25
School, junior high	1.30	3.69	.31	.96	2.87	.64		
* School, senior high	1.40	2.89	.53	1.25	3.20	.57		
Supermarket	1.42	2.50	.37	2.08	2.77	.28		
* Telephone Exchange	3.49	1.05	.23	.90	1.89	1.05		
Theater	2.73	3.41	.64	1.81	2.81	.78		
Town Hall	1.68	2.70	.64	.65	3.72	.51		
* U.S. Post Office	4.90	3.47	.65	.99	2.66	.51		
Warehouse, grocery	.92	1.51	.23	.53	1.98	.33		

*Includes cost of primary feeder and transformer. Cont'd. on next page.

SERVICES

D

COST ASSUMPTIONS:

Each of the projects analyzed in Table D5010-117 were bid within the last 10 years in the Northeastern part of the United States. Bid prices have been adjusted to Jan. 1 levels. The list of projects is by no means all-inclusive, yet by carefully examining the various systems for a particular building type, certain cost relationships will emerge. The use of Section K1010 with the S.F. and C.F. electrical costs should produce a budget S.F. cost for the electrical portion of a job that is consistent with the amount of design information normally available at the conceptual estimate stage.

Figure D5010-117 Cost per S.F. for Electric Systems for Various Building Types (cont.)

Type Construction	6. Special Systems, (cont.)						
	Intercom Systems	Sound Systems	Closed Circuit TV	Snow Melting	Emergency Generator	Security	Master Clock Sys.
Apartment, luxury high rise	$ .62						
Apartment, low rise	.45						
Auditorium		$1.65	$ .76		$1.20		
Bank, branch office	.86		1.75			$1.48	
Bank, main office	.49		.38		.97	.81	$.35
Church	.60						
* College, science building	.63				1.23		.40
* College, library					.63		
* College, physical education center		.85					
Department store					.29		
* Dormitory, college	.84						
Drive-in donut shop							.18
Garage, commercial							.15
Hospital, general	.64		.28		1.66		
* Hospital, pediatric	4.35	.45	.48		1.04		
* Hotel, airport	.63				.62		
Housing for the elderly	.76						
Manufacturing, food processing		.30			2.15		
Manufacturing apparel		.39					
Manufacturing, tools		.49		$.33			
Medical clinic							
Nursing home	1.45				.55		
Office Building		.26			.55	.28	.15
Radio-TV studio	.85				1.36		.61
Restaurant		.39					
Retail Store							
School, elementary		.28					.28
School, junior high		.71			.46		.48
* School, senior high	.58		.39		.63	.35	.36
Supermarket		.31			.58	.39	
* Telephone exchange					5.47	.22	
Theater		.57					
Town Hall							.28
* U.S. Post Office	.56			.14	.61		
Warehouse, grocery	.37						

*Includes cost of primary feeder and transformer. Cont'd. on next page.

D5010   Electrical Service/Distribution

General: Variations in the following square foot costs are due to the type of structural systems of the buildings, geographical location, local electrical codes, designer's preference for specific materials and equipment, and the owner's particular requirements.

Figure D5010-117 Cost per S.F. for Total Electric Systems for Various Building Types (cont.)

Type Construction	Basic Description	Total Floor Area in Square Feet	Total Cost per Square Foot for Total Electric Systems
Apartment building, luxury high rise	All electric, 18 floors, 86 1 B.R., 34 2 B.R.	115,000	$ 8.09
Apartment building, low rise	All electric, 2 floors, 44 units, 1 & 2 B.R.	40,200	5.47
Auditorium	All electric, 1200 person capacity	28,000	16.13
Bank, branch office	All electric, 1 floor	2,700	18.99
Bank, main office	All electric, 8 floors	54,900	12.63
Church	All electric, incl. Sunday school	17,700	8.02
*College, science building	All electric, 3-1/2 floors, 47 rooms	27,500	15.24
*College, library	All electric	33,500	8.20
*College, physical education center	All electric	22,000	9.49
Department store	Gas heat, 1 floor	85,800	7.36
*Dormitory, college	All electric, 125 rooms	63,000	9.12
Drive-in donut shop	Gas heat, incl. parking area lighting	1,500	18.62
Garage, commercial	All electric	52,300	3.10
*Hospital, general	Steam heat, 4 story garage, 300 beds	540,000	21.53
*Hospital, pediatric	Steam heat, 6 stories	278,000	33.69
Hotel, airport	All electric, 625 guest rooms	536,000	12.91
Housing for the elderly	All electric, 7 floors, 100 1 B.R. units	67,000	7.74
Manufacturing, food processing	Electric heat, 1 floor	9,600	14.37
Manufacturing, apparel	Electric heat, 1 floor	28,000	6.98
Manufacturing, tools	Electric heat, 2 floors	42,000	13.24
Medical clinic	Electric heat, 2 floors	22,700	7.66
Nursing home	Gas heat, 3 floors, 60 beds	21,000	12.19
Office building	All electric, 15 floors	311,200	13.04
Radio-TV studio	Electric heat, 3 floors	54,000	15.69
Restaurant	All electric	2,900	18.73
Retail store	All electric	3,000	5.93
School, elementary	All electric, 1 floor	39,500	12.67
School, junior high	All electric, 1 floor	49,500	11.42
*School, senior high	All electric, 1 floor	158,300	12.15
Supermarket	Gas heat	30,600	10.70
*Telephone exchange	Gas heat, 300 kW emergency generator	24,800	14.30
Theater	Electric heat, twin cinema	14,000	12.75
Town Hall	All electric	20,000	10.18
*U.S. Post Office	All electric	495,000	14.49
Warehouse, grocery	All electric	96,400	5.87

*Includes cost of primary feeder and transformer.

SERVICES

D

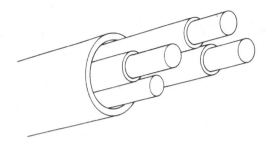

System Components	QUANTITY	UNIT	COST PER L.F.		
			MAT.	INST.	TOTAL
SYSTEM D5010 110 0200					
HIGH VOLTAGE CABLE, NEUTRAL AND CONDUIT INCLUDED, COPPER #2, 5 kV					
Cable, copper, XLP shield, 5 kV, #2, no connections	.030	C.L.F.	4.26	7.02	11.28
Wire 600 volt, type THW, copper, stranded, #4	.010	C.L.F.	.34	.89	1.23
Rigid galv steel conduit to 15'H, 2"diam, w/term, ftng, suppt & scaffold	1.000	L.F.	6.45	10.40	16.85
TOTAL			11.05	18.31	29.36

D5010 110	High Voltage Shielded Conductors	COST PER L.F.		
		MAT.	INST.	TOTAL
0200	High voltage cable, neutral & conduit included, copper #2, 5 kV	11.05	18.30	29.35
0240	Copper #1, 5 kV	11.85	18.45	30.30
0280	15 kV	20.50	27	47.50
0320	Copper 1/0, 5 kV	12.35	18.85	31.20
0360	15 kV	22	27	49
0400	25 kV	24.50	27.50	52
0440	35 kV	28	31	59
0480	Copper 2/0, 5 kV	19	22.50	41.50
0520	15 kV	23	27.50	50.50
0560	25 kV	28.50	31	59.50
0600	35 kV	33.50	33.50	67
0640	Copper 4/0, 5 kV	24	29	53
0680	15 kV	29	32	61
0720	25 kV	31.50	32.50	64
0760	35 kV	37	35	72
0800	Copper 250 kcmil, 5 kV	26.50	30	56.50
0840	15 kV	30	33	63
0880	25 kV	38.50	35.50	74
0920	35 kV	58.50	44	102.50
0960	Copper 350 kcmil, 5 kV	30	32	62
1000	15 kV	37.50	36.50	74
1040	25 kV	42	37.50	79.50
1080	35 kV	63	46.50	109.50
1120	Copper 500 kcmil, 5 kV	37	36	73
1160	15 kV	60.50	46	106.50
1200	25 kV	65	47	112
1240	35 kV	67.50	48	115.50

Important: See the Reference Section for critical supporting data - Reference Numbers and City Cost Indexes

D5010 Electrical Service/Distribution

Service Entrance Cap

Conduit

Circuit Breaker or Safety Switch

Meter Socket

Ground Rod

System Components			COST EACH		
	QUANTITY	UNIT	MAT.	INST.	TOTAL
SYSTEM D5010 120 0220					
SERVICE INSTALLATION, INCLUDES BREAKERS, METERING, 20' CONDUIT & WIRE					
3 PHASE, 4 WIRE, 60 A					
Circuit breaker, enclosed (NEMA 1), 600 volt, 3 pole, 60 A	1.000	Ea.	455	167	622
Meter socket, single position, 4 terminal, 100 A	1.000	Ea.	35	147	182
Rigid galvanized steel conduit, 3/4", including fittings	20.000	L.F.	43.80	117	160.80
Wire, 600V type XHHW, copper stranded #6	.900	C.L.F.	26.55	64.80	91.35
Service entrance cap 3/4" diameter	1.000	Ea.	8.65	36	44.65
Conduit LB fitting with cover, 3/4" diameter	1.000	Ea.	11.55	36	47.55
Ground rod, copper clad, 8' long, 3/4" diameter	1.000	Ea.	29.50	88.50	118
Ground rod clamp, bronze, 3/4" diameter	1.000	Ea.	5.80	14.65	20.45
Ground wire, bare armored, #6-1 conductor	.200	C.L.F.	20	52	72
TOTAL			635.85	722.95	1,358.80

D5010 120	Electric Service, 3 Phase - 4 Wire	COST EACH		
		MAT.	INST.	TOTAL
0200	Service installation, includes breakers, metering, 20' conduit & wire			
0220	3 phase, 4 wire, 120/208 volts, 60 A	635	725	1,360
0240	100 A	780	870	1,650
0280	200 A	1,100	1,350	2,450
0320	400 A	2,600	2,450	5,050
0360	600 A	4,925	3,325	8,250
0400	800 A	6,150	4,000	10,150
0440	1000 A	7,850	4,625	12,475
0480	1200 A	9,450	4,725	14,175
0520	1600 A	18,500	6,750	25,250
0560	2000 A	20,700	7,725	28,425
0570	Add 25% for 277/480 volt			
0610	1 phase, 3 wire, 120/240 volts, 100 A	405	785	1,190
0620	200 A	840	1,150	1,990

SERVICES

D

D50 Electrical

D5010 Electrical Service/Distribution

System Components			COST PER L.F.		
	QUANTITY	UNIT	MAT.	INST.	TOTAL
SYSTEM D5010 230 0200					
FEEDERS, INCLUDING STEEL CONDUIT & WIRE, 60 A					
Rigid galvanized steel conduit, 3/4", including fittings	1.000	L.F.	2.19	5.85	8.04
Wire 600 volt, type XHHW copper stranded #6	.040	C.L.F.	1.18	2.88	4.06
TOTAL			3.37	8.73	12.10

D5010 230	Feeder Installation	COST PER L.F.		
		MAT.	INST.	TOTAL
0200	Feeder installation 600 V, including RGS conduit and XHHW wire, 60 A	3.37	8.75	12.12
0240	100 A	5.85	11.55	17.40
0280	200 A	11.25	17.90	29.15
0320	400 A	22.50	36	58.50
0360	600 A	48.50	58.50	107
0400	800 A	63	70	133
0440	1000 A	78.50	89.50	168
0480	1200 A	89.50	91.50	181
0520	1600 A	126	140	266
0560	2000 A	157	179	336
1200	Branch installation 600 V, including EMT conduit and THW wire, 15 A	.65	4.46	5.11
1240	20 A	.65	4.46	5.11
1280	30 A	.99	5.50	6.49
1320	50 A	1.62	6.30	7.92
1360	65 A	1.86	6.70	8.56
1400	85 A	2.84	7.95	10.79
1440	100 A	3.38	8.40	11.78
1480	130 A	4.33	9.45	13.78
1520	150 A	5.35	10.85	16.20
1560	200 A	6.55	12.20	18.75

D5010 Electrical Service/Distribution

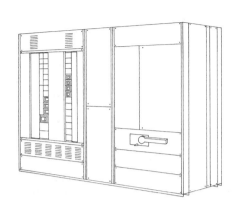

System Components	QUANTITY	UNIT	COST EACH		
			MAT.	INST.	TOTAL
SYSTEM D5010 240 0240					
SWITCHGEAR INSTALLATION, INCL SWBD, PANELS & CIRC BREAKERS, 600 A					
Panelboard, NQOD 225A 4W 120/208V main CB, w/20A bkrs 42 circ	1.000	Ea.	1,900	1,675	3,575
Switchboard, alum. bus bars, 120/208V, 4 wire, 600V	1.000	Ea.	3,025	940	3,965
Distribution sect., alum. bus bar, 120/208 or 277/480 V, 4 wire, 600A	1.000	Ea.	1,525	940	2,465
Feeder section circuit breakers, KA frame, 70 to 225 A	3.000	Ea.	2,760	441	3,201
TOTAL			9,210	3,996	13,206

D5010 240	Switchgear	COST EACH		
		MAT.	INST.	TOTAL
0200	Switchgear inst., incl. swbd., panels & circ bkr, 400 A, 120/208volt	3,475	2,950	6,425
0240	600 A	9,200	4,000	13,200
0280	800 A	11,600	5,700	17,300
0320	1200 A	14,500	8,700	23,200
0360	1600 A	19,800	12,200	32,000
0400	2000 A	25,000	15,500	40,500
0410	Add 20% for 277/480 volt			

D5020 Lighting and Branch Wiring

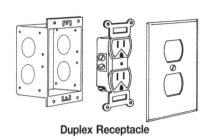

Duplex Receptacle

System Components	QUANTITY	UNIT	COST PER S.F.		
			MAT.	INST.	TOTAL
SYSTEM D5020 110 0200					
RECEPTACLES INCL. PLATE, BOX, CONDUIT, WIRE & TRANS. WHEN REQUIRED					
2.5 PER 1000 S.F., .3 WATTS PER S.F.					
Steel intermediate conduit, (IMC) 1/2" diam	167.000	L.F.	.25	.78	1.03
Wire 600V type THWN-THHN, copper solid #12	3.382	C.L.F.	.02	.14	.16
Wiring device, receptacle, duplex, 120V grounded, 15 amp	2.500	Ea.		.03	.03
Wall plate, 1 gang, brown plastic	2.500	Ea.		.01	.01
Steel outlet box 4" square	2.500	Ea.	.01	.06	.07
Steel outlet box 4" plaster rings	2.500	Ea.		.02	.02
TOTAL			.28	1.04	1.32

D5020 110	Receptacle (by Wattage)	COST PER S.F.		
		MAT.	INST.	TOTAL
0190	Receptacles include plate, box, conduit, wire & transformer when required			
0200	2.5 per 1000 S.F., .3 watts per S.F.	.28	1.04	1.32
0240	With transformer	.32	1.09	1.41
0280	4 per 1000 S.F., .5 watts per S.F.	.33	1.22	1.55
0320	With transformer	.39	1.30	1.69
0360	5 per 1000 S.F., .6 watts per S.F.	.38	1.44	1.82
0400	With transformer	.46	1.54	2
0440	8 per 1000 S.F., .9 watts per S.F.	.39	1.60	1.99
0480	With transformer	.50	1.74	2.24
0520	10 per 1000 S.F., 1.2 watts per S.F.	.40	1.74	2.14
0560	With transformer	.58	1.97	2.55
0600	16.5 per 1000 S.F., 2.0 watts per S.F.	.48	2.16	2.64
0640	With transformer	.79	2.56	3.35
0680	20 per 1000 S.F., 2.4 watts per S.F.	.52	2.36	2.88
0720	With transformer	.88	2.83	3.71

SERVICES D

D5020   Lighting and Branch Wiring

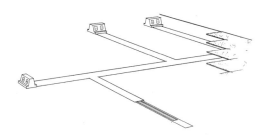

Underfloor Receptacle System

Description: Table D5020 115 includes installed costs of raceways and copper wire from panel to and including receptacle.

National Electrical Code prohibits use of undercarpet system in residential, school or hospital buildings. Can only be used with carpet squares.

Low density = (1) Outlet per 259 S.F. of floor area.

High density = (1) Outlet per 127 S.F. of floor area.

System Components			COST PER S.F.		
	QUANTITY	UNIT	MAT.	INST.	TOTAL
SYSTEM D5020 115 0200					
RECEPTACLE SYSTEMS, UNDERFLOOR DUCT, 5′ ON CENTER, LOW DENSITY					
Underfloor duct 3-1/8″ x 7/8″ w/insert 24″ on center	.190	L.F.	2.47	1.27	3.74
Underfloor duct junction box, single duct, 3-1/8″	.003	Ea.	.88	.35	1.23
Underfloor junction box carpet pan	.003	Ea.	.72	.02	.74
Underfloor duct outlet, high tension receptacle	.004	Ea.	.27	.23	.50
Wire 600V type THWN-THHN copper solid #12	.010	C.L.F.	.05	.43	.48
Vertical elbow for underfloor duct, 3-1/8″, included					
Underfloor duct conduit adapter, 2″ x 1-1/4″, included					
TOTAL			4.39	2.30	6.69

D5020 115	Receptacles	COST PER S.F.		
		MAT.	INST.	TOTAL
0200	Receptacle systems, underfloor duct, 5′ on center, low density	4.39	2.30	6.69
0240	High density	4.70	2.96	7.66
0280	7′ on center, low density	3.46	1.97	5.43
0320	High density	3.77	2.63	6.40
0400	Poke thru fittings, low density	.89	1.08	1.97
0440	High density	1.78	2.09	3.87
0520	Telepoles, using Romex, low density	.73	.70	1.43
0560	High density	1.46	1.40	2.86
0600	Using EMT, low density	.76	.93	1.69
0640	High density	1.54	1.86	3.40
0720	Conduit system with floor boxes, low density	.76	.79	1.55
0760	High density	1.52	1.59	3.11
0840	Undercarpet power system, 3 conductor with 5 conductor feeder, low density	1.30	.28	1.58
0880	High density	2.53	.56	3.09

SERVICES

D

D5020 Lighting and Branch Wiring

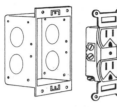
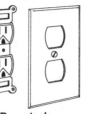

Duplex Receptacle

Wall Switch

System Components	QUANTITY	UNIT	COST PER S.F.		
			MAT.	INST.	TOTAL
SYSTEM D5020 120 0520					
RECEPTACLES AND WALL SWITCHES					
4 RECEPTACLES PER 400 S.F.					
Steel intermediate conduit, (IMC), 1/2" diam	.220	L.F.	.33	1.03	1.36
Wire, 600 volt, type THWN-THHN, copper, solid #12	.005	C.L.F.	.02	.21	.23
Steel outlet box 4" square	.010	Ea.	.02	.24	.26
Steel outlet box, 4" square, plaster rings	.010	Ea.	.01	.07	.08
Receptacle, duplex, 120 volt grounded, 15 amp	.010	Ea.	.01	.12	.13
Wall plate, 1 gang, brown plastic	.010	Ea.		.06	.06
TOTAL			.39	1.73	2.12

D5020 120	Receptacles & Wall Switches	COST PER S.F.		
		MAT.	INST.	TOTAL
0520	Receptacles and wall switches, 400 S.F., 4 receptacles	.39	1.73	2.12
0560	6 receptacles	.44	2.02	2.46
0600	8 receptacles	.50	2.35	2.85
0640	1 switch	.09	.39	.48
0680	600 S.F., 6 receptacles	.39	1.73	2.12
0720	8 receptacles	.43	1.97	2.40
0760	10 receptacles	.47	2.20	2.67
0800	2 switches	.13	.52	.65
0840	1000 S.F., 10 receptacles	.39	1.73	2.12
0880	12 receptacles	.40	1.82	2.22
0920	14 receptacles	.43	1.96	2.39
0960	2 switches	.06	.27	.33
1000	1600 S.F., 12 receptacles	.36	1.54	1.90
1040	14 receptacles	.37	1.68	2.05
1080	16 receptacles	.39	1.73	2.12
1120	4 switches	.07	.32	.39
1160	2000 S.F., 14 receptacles	.37	1.57	1.94
1200	16 receptacles	.37	1.59	1.96
1240	18 receptacles	.37	1.68	2.05
1280	4 switches	.06	.27	.33
1320	3000 S.F., 12 receptacles	.31	1.20	1.51
1360	18 receptacles	.36	1.49	1.85
1400	24 receptacles	.37	1.59	1.96
1440	6 switches	.06	.27	.33
1480	3600 S.F., 20 receptacles	.35	1.44	1.79
1520	24 receptacles	.36	1.48	1.84
1560	28 receptacles	.37	1.53	1.90
1600	8 switches	.07	.32	.39
1640	4000 S.F., 16 receptacles	.31	1.20	1.51
1680	24 receptacles	.36	1.49	1.85

D50 Electrical

D5020 Lighting and Branch Wiring

D5020 120	Receptacles & Wall Switches	COST PER S.F.		
		MAT.	INST.	TOTAL
1720	30 receptacles	.36	1.54	1.90
1760	8 switches	.06	.27	.33
1800	5000 S.F., 20 receptacles	.31	1.20	1.51
1840	26 receptacles	.36	1.45	1.81
1880	30 receptacles	.36	1.49	1.85
1920	10 switches	.06	.27	.33

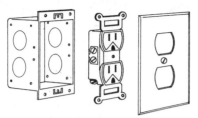

Duplex Receptacle

Wall Switch

System Components	QUANTITY	UNIT	COST PER EACH		
			MAT.	INST.	TOTAL
SYSTEM D5020 125 0520					
RECEPTACLES AND WALL SWITCHES					
Electric metallic tubing conduit, (EMT), 3/4" diam	22.000	L.F.	14.52	79.42	93.94
Wire, 600 volt, type THWN-THHN, copper, solid #12	.630	C.L.F.	2.88	26.78	29.66
Steel outlet box 4" square	1.000	Ea.	2.31	23.50	25.81
Steel outlet box, 4" square, plaster rings	1.000	Ea.	1.25	7.35	8.60
Receptacle, duplex, 120 volt grounded, 15 amp	1.000	Ea.	1.25	11.70	12.95
Wall plate, 1 gang, brown plastic	1.000	Ea.	.33	5.85	6.18
Receptacle duplex 120 V grounded, 15 A					
TOTAL			22.54	154.60	177.14

D5020 125	Receptacles & Switches by Each	COST PER EACH		
		MAT.	INST.	TOTAL
0460	Receptacles & Switches, with box, plate, 3/4" EMT conduit & wire			
0520	Receptacle duplex 120 V grounded, 15 A	22.50	155	177.50
0560	20 A	29	160	189
0600	Receptacle duplex ground fault interrupting, 15 A	53	160	213
0640	20 A	55	160	215
0680	Toggle switch single, 15 A	26.50	155	181.50
0720	20 A	29	160	189
0760	3 way switch, 15 A	28.50	163	191.50
0800	20 A	30	169	199
0840	4 way switch, 15 A	43.50	174	217.50
0880	20 A	58.50	185	243.50

Important: See the Reference Section for critical supporting data - Reference Numbers and City Cost Indexes

D5020 Lighting and Branch Wiring

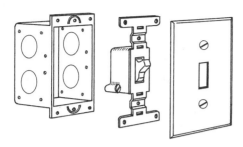

Description: Table D5020 130 includes the cost for switch, plate, box, conduit in slab or EMT exposed and copper wire. Add 20% for exposed conduit.

No power required for switches.

Federal energy guidelines recommend the maximum lighting area controlled per switch shall not exceed 1000 S.F. and that areas over 500 S.F. shall be so controlled that total illumination can be reduced by at least 50%.

System Components	QUANTITY	UNIT	COST PER S.F. MAT.	COST PER S.F. INST.	COST PER S.F. TOTAL
SYSTEM D5020 130 0360					
WALL SWITCHES, 5.0 PER 1000 S.F.					
Steel, intermediate conduit (IMC), 1/2" diameter	88.000	L.F.	.13	.41	.54
Wire, 600V type THWN-THHN, copper solid #12	1.710	C.L.F.	.01	.07	.08
Toggle switch, single pole, 15 amp	5.000	Ea.	.03	.06	.09
Wall plate, 1 gang, brown plastic	5.000	Ea.		.03	.03
Steel outlet box 4" plaster rings	5.000	Ea.	.01	.12	.13
Plaster rings	5.000	Ea.	.01	.04	.05
TOTAL		S.F.	.19	.73	.92

D5020 130	Wall Switch by Sq. Ft.	COST PER S.F. MAT.	COST PER S.F. INST.	COST PER S.F. TOTAL
0200	Wall switches, 1.0 per 1000 S.F.	.04	.17	.21
0240	1.2 per 1000 S.F.	.04	.19	.23
0280	2.0 per 1000 S.F.	.06	.27	.33
0320	2.5 per 1000 S.F.	.08	.33	.41
0360	5.0 per 1000 S.F.	.19	.73	.92
0400	10.0 per 1000 S.F.	.37	1.47	1.84

SERVICES

D

D5020   Lighting and Branch Wiring

System D5020 135 includes all wiring and connections.

System Components	QUANTITY	UNIT	COST PER S.F.		
			MAT.	INST.	TOTAL
SYSTEM D5020 135 0200					
MISCELLANEOUS POWER, TO .5 WATTS					
Steel intermediate conduit, (IMC) 1/2" diam	15.000	L.F.	.02	.07	.09
Wire 600V type THWN-THHN, copper solid #12	.325	C.L.F.		.01	.01
TOTAL			.02	.08	.10

D5020 135	Miscellaneous Power	COST PER S.F.		
		MAT.	INST.	TOTAL
0200	Miscellaneous power, to .5 watts	.02	.08	.10
0240	.8 watts	.03	.12	.15
0280	1 watt	.04	.16	.20
0320	1.2 watts	.05	.18	.23
0360	1.5 watts	.06	.22	.28
0400	1.8 watts	.07	.26	.33
0440	2 watts	.08	.31	.39
0480	2.5 watts	.11	.37	.48
0520	3 watts	.13	.44	.57

D5020 Lighting and Branch Wiring

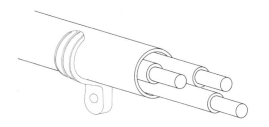

System D5020 140 includes all wiring and connections for central air conditioning units.

System Components			COST PER S.F.		
	QUANTITY	UNIT	MAT.	INST.	TOTAL
SYSTEM D5020 140 0200					
CENTRAL AIR CONDITIONING POWER, 1 WATT					
Steel intermediate conduit, 1/2″ diam.	.030	L.F.	.05	.14	.19
Wire 600V type THWN-THHN, copper solid #12	.001	C.L.F.		.04	.04
TOTAL			.05	.18	.23

D5020 140	Central A. C. Power (by Wattage)	COST PER S.F.		
		MAT.	INST.	TOTAL
0200	Central air conditioning power, 1 watt	.05	.18	.23
0220	2 watts	.05	.21	.26
0240	3 watts	.07	.24	.31
0280	4 watts	.10	.31	.41
0320	6 watts	.19	.44	.63
0360	8 watts	.22	.46	.68
0400	10 watts	.28	.53	.81

SERVICES

D

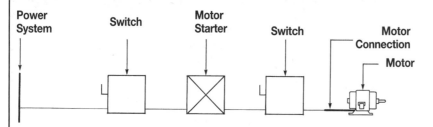

System D5020 145 installed cost of motor wiring as per Table D5010 170 using 50' of rigid conduit and copper wire. **Cost and setting of motor not included.**

System Components	QUANTITY	UNIT	COST EACH		
			MAT.	INST.	TOTAL
SYSTEM D5020 145 0200					
MOTOR INSTALLATION, SINGLE PHASE, 115V, TO AND INCLUDING 1/3 HP MOTOR SIZE					
Wire 600V type THWN-THHN, copper solid #12	1.250	C.L.F.	5.71	53.13	58.84
Steel intermediate conduit, (IMC) 1/2" diam	50.000	L.F.	75.50	234.50	310
Magnetic FVNR, 115V, 1/3 HP, size 00 starter	1.000	Ea.	173	117	290
Safety switch, fused, heavy duty, 240V 2P 30 amp	1.000	Ea.	99.50	134	233.50
Safety switch, non fused, heavy duty, 600V, 3 phase, 30 A	1.000	Ea.	117	147	264
Flexible metallic conduit, Greenfield 1/2" diam	1.500	L.F.	.51	3.51	4.02
Connectors for flexible metallic conduit Greenfield 1/2" diam	1.000	Ea.	1.58	5.85	7.43
Coupling for Greenfield to conduit 1/2" diam flexible metalic conduit	1.000	Ea.	.83	9.40	10.23
Fuse cartridge nonrenewable, 250V 30 amp	1.000	Ea.	1.43	9.40	10.83
TOTAL			475.06	713.79	1,188.85

D5020 145	Motor Installation	COST EACH		
		MAT.	INST.	TOTAL
0200	Motor installation, single phase, 115V, to and including 1/3 HP motor size	475	715	1,190
0240	To and incl. 1 HP motor size	495	715	1,210
0280	To and incl. 2 HP motor size	530	760	1,290
0320	To and incl. 3 HP motor size	590	775	1,365
0360	230V, to and including 1 HP motor size	475	725	1,200
0400	To and incl. 2 HP motor size	500	725	1,225
0440	To and incl. 3 HP motor size	560	780	1,340
0520	Three phase, 200V, to and including 1-1/2 HP motor size	560	795	1,355
0560	To and incl. 3 HP motor size	600	865	1,465
0600	To and incl. 5 HP motor size	640	965	1,605
0640	To and incl. 7-1/2 HP motor size	650	985	1,635
0680	To and incl. 10 HP motor size	1,050	1,225	2,275
0720	To and incl. 15 HP motor size	1,375	1,375	2,750
0760	To and incl. 20 HP motor size	1,725	1,575	3,300
0800	To and incl. 25 HP motor size	1,725	1,575	3,300
0840	To and incl. 30 HP motor size	2,725	1,875	4,600
0880	To and incl. 40 HP motor size	3,300	2,200	5,500
0920	To and incl. 50 HP motor size	5,750	2,575	8,325
0960	To and incl. 60 HP motor size	5,900	2,725	8,625
1000	To and incl. 75 HP motor size	7,675	3,125	10,800
1040	To and incl. 100 HP motor size	15,700	3,650	19,350
1080	To and incl. 125 HP motor size	16,000	4,025	20,025
1120	To and incl. 150 HP motor size	19,300	4,725	24,025
1160	To and incl. 200 HP motor size	23,200	5,775	28,975
1240	230V, to and including 1-1/2 HP motor size	530	785	1,315
1280	To and incl. 3 HP motor size	570	855	1,425
1320	To and incl. 5 HP motor size	610	955	1,565
1360	To and incl. 7-1/2 HP motor size	610	955	1,565
1400	To and incl. 10 HP motor size	950	1,175	2,125
1440	To and incl. 15 HP motor size	1,075	1,275	2,350
1480	To and incl. 20 HP motor size	1,600	1,525	3,125
1520	To and incl. 25 HP motor size	1,725	1,575	3,300

SERVICES D

D5020 Lighting and Branch Wiring

D5020 145	Motor Installation	COST EACH		
		MAT.	INST.	TOTAL
1560	To and incl. 30 HP motor size	1,725	1,575	3,300
1600	To and incl. 40 HP motor size	3,250	2,150	5,400
1640	To and incl. 50 HP motor size	3,350	2,275	5,625
1680	To and incl. 60 HP motor size	5,775	2,575	8,350
1720	To and incl. 75 HP motor size	7,000	2,925	9,925
1760	To and incl. 100 HP motor size	7,975	3,275	11,250
1800	To and incl. 125 HP motor size	15,900	3,775	19,675
1840	To and incl. 150 HP motor size	17,200	4,300	21,500
1880	To and incl. 200 HP motor size	18,600	4,750	23,350
1960	460V, to and including 2 HP motor size	640	795	1,435
2000	To and incl. 5 HP motor size	680	865	1,545
2040	To and incl. 10 HP motor size	710	950	1,660
2080	To and incl. 15 HP motor size	940	1,100	2,040
2120	To and incl. 20 HP motor size	970	1,175	2,145
2160	To and incl. 25 HP motor size	1,075	1,225	2,300
2200	To and incl. 30 HP motor size	1,375	1,325	2,700
2240	To and incl. 40 HP motor size	1,725	1,425	3,150
2280	To and incl. 50 HP motor size	1,900	1,575	3,475
2320	To and incl. 60 HP motor size	2,875	1,850	4,725
2360	To and incl. 75 HP motor size	3,275	2,050	5,325
2400	To and incl. 100 HP motor size	3,525	2,275	5,800
2440	To and incl. 125 HP motor size	5,950	2,600	8,550
2480	To and incl. 150 HP motor size	7,375	2,900	10,275
2520	To and incl. 200 HP motor size	8,475	3,275	11,750
2600	575V, to and including 2 HP motor size	640	795	1,435
2640	To and incl. 5 HP motor size	680	865	1,545
2680	To and incl. 10 HP motor size	710	950	1,660
2720	To and incl. 20 HP motor size	940	1,100	2,040
2760	To and incl. 25 HP motor size	970	1,175	2,145
2800	To and incl. 30 HP motor size	1,375	1,325	2,700
2840	To and incl. 50 HP motor size	1,450	1,375	2,825
2880	To and incl. 60 HP motor size	2,850	1,850	4,700
2920	To and incl. 75 HP motor size	2,875	1,850	4,725
2960	To and incl. 100 HP motor size	3,275	2,050	5,325
3000	To and incl. 125 HP motor size	5,875	2,550	8,425
3040	To and incl. 150 HP motor size	5,950	2,600	8,550
3080	To and incl. 200 HP motor size	7,475	2,950	10,425

SERVICES

D

D5020   Lighting and Branch Wiring

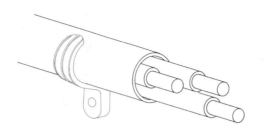

System Components	QUANTITY	UNIT	COST PER L.F.		
			MAT.	INST.	TOTAL
SYSTEM D5020 155 0200					
MOTOR FEEDER SYSTEMS, SINGLE PHASE, UP TO 115V, 1HP OR 230V, 2HP					
Steel intermediate conduit, (IMC) 1/2" diam	1.000	L.F.	1.51	4.69	6.20
Wire 600V type THWN-THHN, copper solid #12	.020	C.L.F.	.09	.85	.94
TOTAL			1.60	5.54	7.14

D5020 155	Motor Feeder	COST PER L.F.		
		MAT.	INST.	TOTAL
0200	Motor feeder systems, single phase, feed up to 115V 1HP or 230V 2 HP	1.60	5.55	7.15
0240	115V 2HP, 230V 3HP	1.65	5.65	7.30
0280	115V 3HP	1.79	5.85	7.64
0360	Three phase, feed to 200V 3HP, 230V 5HP, 460V 10HP, 575V 10HP	1.65	5.95	7.60
0440	200V 5HP, 230V 7.5HP, 460V 15HP, 575V 20HP	1.72	6.10	7.82
0520	200V 10HP, 230V 10HP, 460V 30HP, 575V 30HP	1.93	6.45	8.38
0600	200V 15HP, 230V 15HP, 460V 40HP, 575V 50HP	2.44	7.35	9.79
0680	200V 20HP, 230V 25HP, 460V 50HP, 575V 60HP	3.50	9.35	12.85
0760	200V 25HP, 230V 30HP, 460V 60HP, 575V 75HP	3.74	9.50	13.24
0840	200V 30HP	4.44	9.80	14.24
0920	230V 40HP, 460V 75HP, 575V 100HP	5.25	10.70	15.95
1000	200V 40HP	5.90	11.45	17.35
1080	230V 50HP, 460V 100HP, 575V 125HP	6.70	12.65	19.35
1160	200V 50HP, 230V 60HP, 460V 125HP, 575V 150HP	7.85	13.45	21.30
1240	200V 60HP, 460V 150HP	9.45	15.80	25.25
1320	230V 75HP, 575V 200HP	11.05	16.40	27.45
1400	200V 75HP	12	16.80	28.80
1480	230V 100HP, 460V 200HP	17.40	19.50	36.90
1560	200V 100HP	23	24.50	47.50
1640	230V 125HP	25.50	26.50	52
1720	200V 125HP, 230V 150HP	27	30.50	57.50
1800	200V 150HP	29.50	33	62.50
1880	200V 200HP	46.50	49	95.50
1960	230V 200HP	39	36.50	75.50

D5020 Lighting and Branch Wiring

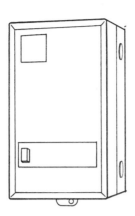

Starters are full voltage, type NEMA 1 for general purpose indoor application with motor overload protection and include mounting and wire connections.

System Components	QUANTITY	UNIT	COST EACH		
			MAT.	INST.	TOTAL
SYSTEM D5020 160 0200					
MAGNETIC STARTER,SIZE 00 TO 1/3 HP, 1 PHASE 115V OR 1 HP 230V	1.000	Ea.	173	117	290
TOTAL			173	117	290

D5020 160	Magnetic Starter	COST EACH		
		MAT.	INST.	TOTAL
0200	Magnetic starter, size 00, to 1/3 HP, 1 phase, 115V or 1 HP 230V	173	117	290
0280	Size 00, to 1-1/2 HP, 3 phase, 200-230V or 2 HP 460-575V	190	134	324
0360	Size 0, to 1 HP, 1 phase, 115V or 2 HP 230V	193	117	310
0440	Size 0, to 3 HP, 3 phase, 200-230V or 5 HP 460-575V	230	204	434
0520	Size 1, to 2 HP, 1 phase, 115V or 3 HP 230V	221	156	377
0600	Size 1, to 7-1/2 HP, 3 phase, 200-230V or 10 HP 460-575V	258	293	551
0680	Size 2, to 10 HP, 3 phase, 200V, 15 HP-230V or 25 HP 460-575V	485	425	910
0760	Size 3, to 25 HP, 3 phase, 200V, 30 HP-230V or 50 HP 460-575V	790	520	1,310
0840	Size 4, to 40 HP, 3 phase, 200V, 50 HP-230V or 100 HP 460-575V	1,750	780	2,530
0920	Size 5, to 75 HP, 3 phase, 200V, 100 HP-230V or 200 HP 460-575V	4,100	1,050	5,150
1000	Size 6, to 150 HP, 3 phase, 200V, 200 HP-230V or 400 HP 460-575V	11,500	1,175	12,675

SERVICES

D

D5020 Lighting and Branch Wiring

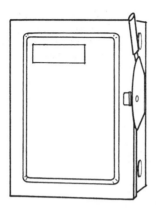

Safety switches are type NEMA 1 for general purpose indoor application, and include time delay fuses, insulation and wire terminations.

System Components	QUANTITY	UNIT	COST EACH		
			MAT.	INST.	TOTAL
SYSTEM D5020 165 0200					
SAFETY SWITCH, 30A FUSED, 1 PHASE, 115V OR 230V.					
Safety switch fused, hvy duty, 240V 2p 30 amp	1.000	Ea.	99.50	134	233.50
Fuse, dual element time delay 250V, 30 amp	2.000	Ea.	7.18	18.80	25.98
TOTAL			106.68	152.80	259.48

D5020 165	Safety Switches	COST EACH		
		MAT.	INST.	TOTAL
0200	Safety switch, 30A fused, 1 phase, 2HP 115V or 3HP, 230V.	107	153	260
0280	3 phase, 5HP, 200V or 7 1/2HP, 230V	143	175	318
0360	15HP, 460V or 20HP, 575V	248	182	430
0440	60A fused, 3 phase, 15HP 200V or 15HP 230V	244	232	476
0520	30HP 460V or 40HP 575V	315	239	554
0600	100A fused, 3 phase, 20HP 200V or 25HP 230V	400	282	682
0680	50HP 460V or 60HP 575V	580	286	866
0760	200A fused, 3 phase, 50HP 200V or 60HP 230V	705	400	1,105
0840	125HP 460V or 150HP 575V	890	405	1,295
0920	400A fused, 3 phase, 100HP 200V or 125HP 230V	1,725	565	2,290
1000	250HP 460V or 350HP 575V	2,225	580	2,805

Important: See the Reference Section for critical supporting data - Reference Numbers and City Cost Indexes

D5020 Lighting and Branch Wiring

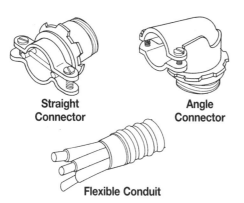

Straight Connector

Angle Connector

Flexible Conduit

Table below includes costs for the flexible conduit. Not included are wire terminations and testing motor for correct rotation.

System Components	QUANTITY	UNIT	COST EACH		
			MAT.	INST.	TOTAL
SYSTEM D5020 170 0200					
MOTOR CONNECTIONS, SINGLE PHASE, 115V/230V UP TO 1 HP					
Motor connection, flexible conduit & fittings, 1 HP motor 115V	1.000	Ea.	4.53	55.58	60.11
TOTAL			4.53	55.58	60.11

D5020 170	Motor Connections	COST EACH		
		MAT.	INST.	TOTAL
0200	Motor connections, single phase, 115/230V, up to 1 HP	4.53	55.50	60.03
0240	Up to 3 HP	4.37	65	69.37
0280	Three phase, 200/230/460/575V, up to 3 HP	4.85	72	76.85
0320	Up to 5 HP	4.85	72	76.85
0360	Up to 7-1/2 HP	8.20	85	93.20
0400	Up to 10 HP	8.80	112	120.80
0440	Up to 15 HP	17	142	159
0480	Up to 25 HP	19.10	174	193.10
0520	Up to 50 HP	52	213	265
0560	Up to 100 HP	134	315	449

D5020 Lighting and Branch Wiring

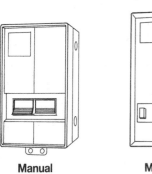

Manual Starter

Magnetic Starter

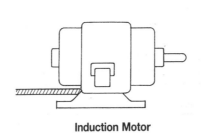

Induction Motor

For 230/460 Volt A.C., 3 phase, 60 cycle ball bearing squirrel cage induction motors, NEMA Class B standard line. Installation included.

No conduit, wire, or terminations included.

System Components	QUANTITY	UNIT	COST EACH		
			MAT.	INST.	TOTAL
SYSTEM D5020 175 0220					
MOTOR, DRIPPROOF CLASS B INSULATION, 1.15 SERVICE FACTOR, WITH STARTER					
1 H.P., 1200 RPM WITH MANUAL STARTER					
Motor, dripproof, class B insul, 1.15 serv fact, 1200 RPM, 1 HP	1.000	Ea.	248	104	352
Motor starter, manual, 3 phase, 1 HP motor	1.000	Ea.	162	134	296
TOTAL			410	238	648

D5020 175	Motor & Starter	COST EACH		
		MAT.	INST.	TOTAL
0190	Motor, dripproof, premium efficient, 1.15 service factor			
0200	1 HP, 1200 RPM, motor only	248	104	352
0220	With manual starter	410	238	648
0240	With magnetic starter	440	238	678
0260	1800 RPM, motor only	213	104	317
0280	With manual starter	375	238	613
0300	With magnetic starter	405	238	643
0320	2 HP, 1200 RPM, motor only	350	104	454
0340	With manual starter	510	238	748
0360	With magnetic starter	580	310	890
0380	1800 RPM, motor only	251	104	355
0400	With manual starter	415	238	653
0420	With magnetic starter	480	310	790
0440	3600 RPM, motor only	254	104	358
0460	With manual starter	415	238	653
0480	With magnetic starter	485	310	795
0500	3 HP, 1200 RPM, motor only	470	104	574
0520	With manual starter	630	238	868
0540	With magnetic starter	700	310	1,010
0560	1800 RPM, motor only	252	104	356
0580	With manual starter	415	238	653
0600	With magnetic starter	480	310	790
0620	3600 RPM, motor only	290	104	394
0640	With manual starter	450	238	688
0660	With magnetic starter	520	310	830
0680	5 HP, 1200 RPM, motor only	565	104	669
0700	With manual starter	760	340	1,100
0720	With magnetic starter	825	395	1,220
0740	1800 RPM, motor only	325	104	429
0760	With manual starter	520	340	860
0780	With magnetic starter	585	395	980
0800	3600 RPM, motor only	325	104	429

Important: See the Reference Section for critical supporting data - Reference Numbers and City Cost Indexes

D5020 Lighting and Branch Wiring

D5020 175	Motor & Starter	COST EACH		
		MAT.	INST.	TOTAL
0820	With manual starter	520	340	860
0840	With magnetic starter	585	395	980
0860	7.5 HP, 1800 RPM, motor only	470	112	582
0880	With manual starter	665	345	1,010
0900	With magnetic starter	955	535	1,490
0920	10 HP, 1800 RPM, motor only	585	117	702
0940	With manual starter	780	350	1,130
0960	With magnetic starter	1,075	540	1,615
0980	15 HP, 1800 RPM, motor only	740	147	887
1000	With magnetic starter	1,225	570	1,795
1040	20 HP, 1800 RPM, motor only	925	180	1,105
1060	With magnetic starter	1,725	700	2,425
1100	25 HP, 1800 RPM, motor only	1,075	188	1,263
1120	With magnetic starter	1,875	710	2,585
1160	30 HP, 1800 RPM, motor only	1,250	195	1,445
1180	With magnetic starter	2,050	715	2,765
1220	40 HP, 1800 RPM, motor only	1,875	234	2,109
1240	With magnetic starter	3,625	1,025	4,650
1280	50 HP, 1800 RPM, motor only	2,125	293	2,418
1300	With magnetic starter	3,875	1,075	4,950
1340	60 HP, 1800 RPM, motor only	2,775	335	3,110
1360	With magnetic starter	6,875	1,375	8,250
1400	75 HP, 1800 RPM, motor only	3,150	390	3,540
1420	With magnetic starter	7,250	1,450	8,700
1460	100 HP, 1800 RPM, motor only	4,150	520	4,670
1480	With magnetic starter	8,250	1,575	9,825
1520	125 HP, 1800 RPM, motor only	4,825	670	5,495
1540	With magnetic starter	16,300	1,850	18,150
1580	150 HP, 1800 RPM, motor only	6,150	780	6,930
1600	With magnetic starter	17,700	1,950	19,650
1640	200 HP, 1800 RPM, motor only	7,950	940	8,890
1660	With magnetic starter	19,500	2,125	21,625
1680	Totally encl, premium efficient, 1.0 ser. fac., 1HP, 1200RPM, motor only	299	104	403
1700	With manual starter	460	238	698
1720	With magnetic starter	490	238	728
1740	1800 RPM, motor only	256	104	360
1760	With manual starter	420	238	658
1780	With magnetic starter	445	238	683
1800	2 HP, 1200 RPM, motor only	365	104	469
1820	With manual starter	525	238	763
1840	With magnetic starter	595	310	905
1860	1800 RPM, motor only	305	104	409
1880	With manual starter	465	238	703
1900	With magnetic starter	535	310	845
1920	3600 RPM, motor only	296	104	400
1940	With manual starter	460	238	698
1960	With magnetic starter	525	310	835
1980	3 HP, 1200 RPM, motor only	485	104	589
2000	With manual starter	645	238	883
2020	With magnetic starter	715	310	1,025
2040	1800 RPM, motor only	340	104	444
2060	With manual starter	500	238	738
2080	With magnetic starter	570	310	880
2100	3600 RPM, motor only	345	104	449
2120	With manual starter	505	238	743
2140	With magnetic starter	575	310	885
2160	5 HP, 1200 RPM, motor only	715	104	819
2180	With manual starter	910	340	1,250

SERVICES

D

D5020 Lighting and Branch Wiring

D5020 175	Motor & Starter	COST EACH		
		MAT.	INST.	TOTAL
2200	With magnetic starter	975	395	1,370
2220	1800 RPM, motor only	390	104	494
2240	With manual starter	585	340	925
2260	With magnetic starter	650	395	1,045
2280	3600 RPM, motor only	430	104	534
2300	With manual starter	625	340	965
2320	With magnetic starter	690	395	1,085
2340	7.5 HP, 1800 RPM, motor only	560	112	672
2360	With manual starter	755	345	1,100
2380	With magnetic starter	1,050	535	1,585
2400	10 HP, 1800 RPM, motor only	675	117	792
2420	With manual starter	870	350	1,220
2440	With magnetic starter	1,150	540	1,690
2460	15 HP, 1800 RPM, motor only	905	147	1,052
2480	With magnetic starter	1,400	570	1,970
2500	20 HP, 1800 RPM, motor only	1,100	180	1,280
2520	With magnetic starter	1,900	700	2,600
2540	25 HP, 1800 RPM, motor only	1,350	188	1,538
2560	With magnetic starter	2,150	710	2,860
2580	30 HP, 1800 RPM, motor only	1,575	195	1,770
2600	With magnetic starter	2,375	715	3,090
2620	40 HP, 1800 RPM, motor only	2,025	234	2,259
2640	With magnetic starter	3,775	1,025	4,800
2660	50 HP, 1800 RPM, motor only	2,500	293	2,793
2680	With magnetic starter	4,250	1,075	5,325
2700	60 HP, 1800 RPM, motor only	3,750	335	4,085
2720	With magnetic starter	7,850	1,375	9,225
2740	75 HP, 1800 RPM, motor only	4,775	390	5,165
2760	With magnetic starter	8,875	1,450	10,325
2780	100 HP, 1800 RPM, motor only	5,675	520	6,195
2800	With magnetic starter	9,775	1,575	11,350
2820	125 HP, 1800 RPM, motor only	8,250	670	8,920
2840	With magnetic starter	19,800	1,850	21,650
2860	150 HP, 1800 RPM, motor only	9,350	780	10,130
2880	With magnetic starter	20,900	1,950	22,850
2900	200 HP, 1800 RPM, motor only	11,200	940	12,140
2920	With magnetic starter	22,700	2,125	24,825

Important: See the Reference Section for critical supporting data - Reference Numbers and City Cost Indexes

General: The cost of the lighting portion of the electrical costs is dependent upon:
1. The footcandle requirement of the proposed building.
2. The type of fixtures required.
3. The ceiling heights of the building.
4. Reflectance value of ceilings, walls and floors.
5. Fixture efficiencies and spacing vs. mounting height ratios.

Footcandle Requirements: See Table D5020-204 for Footcandle and Watts per S.F. determination.

Table D5020-201   IESNA* Recommended Illumination Levels in Footcandles

Commercial Buildings			Industrial Buildings		
Type	Description	Footcandles	Type	Description	Footcandles
Bank	Lobby	50	Assembly Areas	Rough bench & machine work	50
	Customer Areas	70		Medium bench & machine work	100
	Teller Stations	150		Fine bench & machine work	500
	Accounting Areas	150	Inspection Areas	Ordinary	50
Offices	Routine Work	100		Difficult	100
	Accounting	150		Highly Difficult	200
	Drafting	200	Material Handling	Loading	20
	Corridors, Halls, Washrooms	30		Stock Picking	30
Schools	Reading or Writing	70		Packing, Wrapping	50
	Drafting, Labs, Shops	100	Stairways	Service Areas	20
	Libraries	70	Washrooms	Service Areas	20
	Auditoriums, Assembly	15	Storage Areas	Inactive	5
	Auditoriums, Exhibition	30		Active, Rough, Bulky	10
Stores	Circulation Areas	30		Active, Medium	20
	Stock Rooms	30		Active, Fine	50
	Merchandise Areas, Service	100	Garages	Active Traffic Areas	20
	Self-Service Areas	200		Service & Repair	100

*IESNA - Illuminating Engineering Society of North America

SERVICES

D

Table D5020-202 General Lighting Loads by Occupancies

Type of Occupancy	Unit Load per S.F. (Watts)
Armories and Auditoriums	1
Banks	5
Barber Shops and Beauty Parlors	3
Churches	1
Clubs	2
Court Rooms	2
*Dwelling Units	3
Garages — Commercial (storage)	½
Hospitals	2
*Hotels and Motels, including apartment houses without provisions for cooking by tenants	2
Industrial Commercial (Loft) Buildings	2
Lodge Rooms	1½
Office Buildings	5
Restaurants	2
Schools	3
Stores	3
Warehouses (storage)	¼
*In any of the above occupancies except one-family dwellings and individual dwelling units of multi-family dwellings:	
Assembly Halls and Auditoriums	1
Halls, Corridors, Closets	½
Storage Spaces	¼

Table D5020-203 Lighting Limit (Connected Load) for Listed Occupancies: New Building Proposed Energy Conservation Guideline

Type of Use	Maximum Watts per S.F.
Interior	3.00
Category A: Classrooms, office areas, automotive mechanical areas, museums, conference rooms, drafting rooms, clerical areas, laboratories, merchandising areas, kitchens, examining rooms, book stacks, athletic facilities.	
Category B: Auditoriums, waiting areas, spectator areas, restrooms, dining areas, transportation terminals, working corridors in prisons and hospitals, book storage areas, active inventory storage, hospital bedrooms, hotel and motel bedrooms, enclosed shopping mall concourse areas, stairways.	1.00
Category C: Corridors, lobbies, elevators, inactive storage areas.	0.50
Category D: Indoor parking.	0.25
Exterior	
Category E: Building perimeter: wall-wash, facade, canopy.	5.00 (per linear foot)
Category F: Outdoor parking.	0.10

Table D5020-204   Procedure for Calculating Footcandles and Watts Per Square Foot

1. Initial footcandles = No. of fixtures × lamps per fixture × lumens per lamp × coefficient of utilization ÷ square feet
2. Maintained footcandles = initial footcandles × maintenance factor
3. Watts per square foot = No. of fixtures × lamps × (lamp watts + ballast watts) ÷ square feet

Example: To find footcandles and watts per S.F. for an office 20′ x 20′ with 11 fluorescent fixtures each having 4–40 watt C.W. lamps.

Based on good reflectance and clean conditions:

Lumens per lamp = 40 watt cool white at 3150 lumens per lamp RD5020-250, Table RD5020-251

Coefficient of utilization = .42 (varies from .62 for light colored areas to .27 for dark)

Maintenance factor = .75 (varies from .80 for clean areas with good maintenance to .50 for poor)

Ballast loss = 8 watts per lamp. (Varies with manufacturer. See manufacturers' catalog.)

1. Initial footcandles:

$$\frac{11 \times 4 \times 3150 \times .42}{400} = \frac{58,212}{400} = 145 \text{ footcandles}$$

2. Maintained footcandles:

$$145 \times .75 = 109 \text{ footcandles}$$

3. Watts per S.F.

$$\frac{11 \times 4 \,(40 + 8)}{400} = \frac{2,112}{400} = 5.3 \text{ watts per S.F.}$$

Table D5020-205   Approximate Watts Per Square Foot for Popular Fixture Types

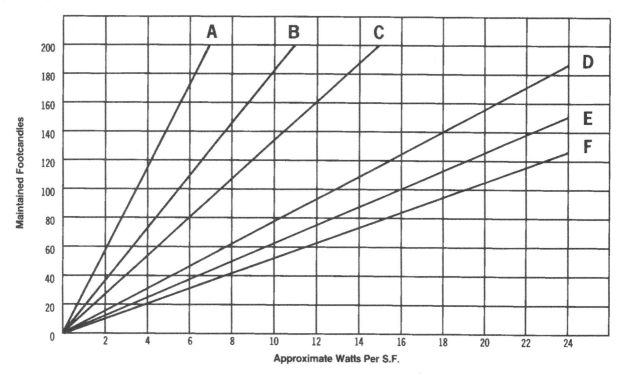

Due to the many variables involved, use for preliminary estimating only:
a. Fluorescent – industrial System D5020 208
b. Fluorescent – lens unit System D5020 208 Fixture types B & C
c. Fluorescent – louvered unit
d. Incandescent – open reflector System D5020 214, Type D
e. Incandescent – lens unit System D5020 214, Type A
f. Incandescent – down light System D5020 214, Type B

D5020 Lighting and Branch Wiring

A. Strip Fixture

B. Surface Mounted

C. Recessed

D. Pendent Mounted

Design Assumptions:

1. A 100 footcandle average maintained level of illumination.
2. Ceiling heights range from 9' to 11'.
3. Average reflectance values are assumed for ceilings, walls and floors.
4. Cool white (CW) fluorescent lamps with 3150 lumens for 40 watt lamps and 6300 lumens for 8' slimline lamps.
5. Four 40 watt lamps per 4' fixture and two 8' lamps per 8' fixture.
6. Average fixture efficiency values and spacing to mounting height ratios.
7. Installation labor is average U.S. rate as of January 1.

System Components	QUANTITY	UNIT	COST PER S.F.		
			MAT.	INST.	TOTAL
SYSTEM D5020 208 0520					
FLUORESCENT FIXTURES MOUNTED 9'-11" ABOVE FLOOR, 100 FC					
TYPE A, 8 FIXTURES PER 400 S.F.					
Steel intermediate conduit, (IMC) 1/2" diam	.404	L.F.	.61	1.89	2.50
Wire, 600V, type THWN-THHN, copper, solid, #12	.008	C.L.F.	.04	.34	.38
Fluorescent strip fixture 8' long, surface mounted, two 75W SL	.020	Ea.	1.05	1.51	2.56
Steel outlet box 4" concrete	.020	Ea.	.20	.47	.67
Steel outlet box plate with stud, 4" concrete	.020	Ea.	.07	.12	.19
TOTAL			1.97	4.33	6.30

D5020 208	Fluorescent Fixtures (by Type)	COST PER S.F.		
		MAT.	INST.	TOTAL
0520	Fluorescent fixtures, type A, 8 fixtures per 400 S.F.	1.97	4.33	6.30
0560	11 fixtures per 600 S.F.	1.86	4.20	6.06
0600	17 fixtures per 1000 S.F.	1.79	4.13	5.92
0640	23 fixtures per 1600 S.F.	1.63	3.91	5.54
0680	28 fixtures per 2000 S.F.	1.63	3.91	5.54
0720	41 fixtures per 3000 S.F.	1.59	3.91	5.50
0800	53 fixtures per 4000 S.F.	1.56	3.81	5.37
0840	64 fixtures per 5000 S.F.	1.56	3.81	5.37
0880	Type B, 11 fixtures per 400 S.F.	3.99	6.35	10.34
0920	15 fixtures per 600 S.F.	3.70	6.10	9.80
0960	24 fixtures per 1000 S.F.	3.60	6.05	9.65
1000	35 fixtures per 1600 S.F.	3.37	5.80	9.17
1040	42 fixtures per 2000 S.F.	3.30	5.80	9.10
1080	61 fixtures per 3000 S.F.	3.31	5.60	8.91
1160	80 fixtures per 4000 S.F.	3.20	5.70	8.90
1200	98 fixtures per 5000 S.F.	3.19	5.70	8.89
1240	Type C, 11 fixtures per 400 S.F.	2.94	6.70	9.64
1280	14 fixtures per 600 S.F.	2.63	6.25	8.88
1320	23 fixtures per 1000 S.F.	2.62	6.20	8.82
1360	34 fixtures per 1600 S.F.	2.53	6.10	8.63
1400	43 fixtures per 2000 S.F.	2.55	6.05	8.60
1440	63 fixtures per 3000 S.F.	2.48	6	8.48
1520	81 fixtures per 4000 S.F.	2.43	5.90	8.33
1560	101 fixtures per 5000 S.F.	2.43	5.90	8.33
1600	Type D, 8 fixtures per 400 S.F.	2.78	5.20	7.98
1640	12 fixtures per 600 S.F.	2.77	5.15	7.92
1680	19 fixtures per 1000 S.F.	2.67	5.05	7.72
1720	27 fixtures per 1600 S.F.	2.51	4.90	7.41
1760	34 fixtures per 2000 S.F.	2.49	4.86	7.35
1800	48 fixtures per 3000 S.F.	2.39	4.72	7.11
1880	64 fixtures per 4000 S.F.	2.39	4.72	7.11
1920	79 fixtures per 5000 S.F.	2.39	4.72	7.11

SERVICES D

D5020 Lighting and Branch Wiring

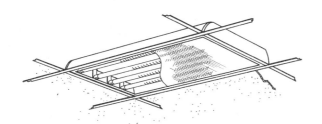

Type C. Recessed, mounted on grid ceiling suspension system, 2' x 4', four 40 watt lamps, acrylic prismatic diffusers.

5.3 watts per S.F. for 100 footcandles.

3 watts per S.F. for 57 footcandles.

System Components			COST PER S.F.		
	QUANTITY	UNIT	MAT.	INST.	TOTAL
SYSTEM D5020 210 0200					
FLUORESCENT FIXTURES RECESS MOUNTED IN CEILING					
1 WATT PER S.F., 20 FC, 5 FIXTURES PER 1000 S.F.					
Steel intermediate conduit, (IMC) 1/2" diam	.128	L.F.	.19	.60	.79
Wire, 600 volt, type THW, copper, solid, #12	.003	C.L.F.	.01	.13	.14
Fluorescent fixture, recessed, 2'x4', four 40W, w/lens, for grid ceiling	.005	Ea.	.30	.50	.80
Steel outlet box 4" square	.005	Ea.	.05	.12	.17
Fixture whip, Greenfield w/#12 THHN wire	.005	Ea.	.02	.03	.05
TOTAL			.57	1.38	1.95

D5020 210	Fluorescent Fixtures (by Wattage)	COST PER S.F.		
		MAT.	INST.	TOTAL
0190	Fluorescent fixtures recess mounted in ceiling			
0200	1 watt per S.F., 20 FC, 5 fixtures per 1000 S.F.	.57	1.38	1.95
0240	2 watts per S.F., 40 FC, 10 fixtures per 1000 S.F.	1.15	2.71	3.86
0280	3 watts per S.F., 60 FC, 15 fixtures per 1000 S.F	1.72	4.07	5.79
0320	4 watts per S.F., 80 FC, 20 fixtures per 1000 S.F.	2.29	5.40	7.69
0400	5 watts per S.F., 100 FC, 25 fixtures per 1000 S.F.	2.87	6.80	9.67

SERVICES

D

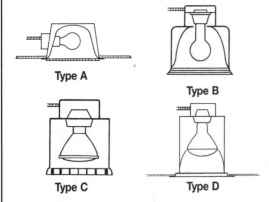

Type A. Recessed wide distribution reflector with flat glass lens 150 W.

Maximum spacing = 1.2 x mounting height.

13 watts per S.F. for 100 footcandles.

Type B. Recessed reflector down light with baffles 150 W.

Maximum spacing = 0.8 x mounting height.

18 watts per S.F. for 100 footcandles.

Type A

Type B

Type C. Recessed PAR–38 flood lamp with concentric louver 150 W.

Maximum spacing = 0.5 x mounting height.

19 watts per S.F. for 100 footcandles.

Type D. Recessed R–40 flood lamp with reflector skirt.

Maximum spacing = 0.7 x mounting height.

15 watts per S.F. for 100 footcandles.

Type C

Type D

System Components	QUANTITY	UNIT	COST PER S.F.		
			MAT.	INST.	TOTAL
SYSTEM D5020 214 0400					
INCANDESCENT FIXTURE RECESS MOUNTED, 100 FC					
TYPE A, 34 FIXTURES PER 400 S.F.					
Steel intermediate conduit, (IMC) 1/2" diam	1.060	L.F.	1.60	4.97	6.57
Wire, 600V, type THWN-THHN, copper, solid, #12	.033	C.L.F.	.15	1.40	1.55
Steel outlet box 4" square	.085	Ea.	.83	2	2.83
Fixture whip, Greenfield w/#12 THHN wire	.085	Ea.	.30	.50	.80
Incandescent fixture, recessed, w/lens, prewired, square trim, 200W	.085	Ea.	6.93	5.95	12.88
TOTAL			9.81	14.82	24.63

D5020 214	Incandescent Fixture (by Type)	COST PER S.F.		
		MAT.	INST.	TOTAL
0380	Incandescent fixture recess mounted, 100 FC			
0400	Type A, 34 fixtures per 400 S.F.	9.80	14.80	24.60
0440	49 fixtures per 600 S.F.	9.55	14.60	24.15
0480	63 fixtures per 800 S.F.	9.30	14.40	23.70
0520	90 fixtures per 1200 S.F.	8.95	14.10	23.05
0560	116 fixtures per 1600 S.F.	8.80	14	22.80
0600	143 fixtures per 2000 S.F.	8.70	13.95	22.65
0640	Type B, 47 fixtures per 400 S.F.	8.85	18.70	27.55
0680	66 fixtures per 600 S.F.	8.45	18.25	26.70
0720	88 fixtures per 800 S.F.	8.45	18.25	26.70
0760	127 fixtures per 1200 S.F.	8.30	18.05	26.35
0800	160 fixtures per 1600 S.F.	8.15	17.95	26.10
0840	206 fixtures per 2000 S.F.	8.15	17.85	26
0880	Type C, 51 fixtures per 400 S.F.	12.55	19.50	32.05
0920	74 fixtures per 600 S.F.	12.20	19.20	31.40
0960	97 fixtures per 800 S.F.	12.05	19.05	31.10
1000	142 fixtures per 1200 S.F.	11.85	18.90	30.75
1040	186 fixtures per 1600 S.F.	11.70	18.80	30.50
1080	230 fixtures per 2000 S.F.	11.65	18.75	30.40
1120	Type D, 39 fixtures per 400 S.F.	11.90	15.40	27.30
1160	57 fixtures per 600 S.F.	11.60	15.20	26.80
1200	75 fixtures per 800 S.F.	11.50	15.20	26.70
1240	109 fixtures per 1200 S.F.	11.25	15	26.25
1280	143 fixtures per 1600 S.F.	11.05	14.85	25.90
1320	176 fixtures per 2000 S.F.	10.95	14.80	25.75

D5020 Lighting and Branch Wiring

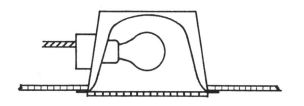

Type A. Recessed, wide distribution reflector with flat glass lens.

150 watt inside frost—2500 lumens per lamp.

PS–25 extended service lamp.

Maximum spacing = 1.2 x mounting height.

13 watts per S.F. for 100 footcandles.

System Components	QUANTITY	UNIT	COST PER S.F.		
			MAT.	INST.	TOTAL
SYSTEM D5020 216 0200					
INCANDESCENT FIXTURE RECESS MOUNTED, TYPE A					
1 WATT PER S.F., 8 FC, 6 FIXT PER 1000 S.F.					
Steel intermediate conduit, (IMC) 1/2" diam	.091	L.F.	.14	.43	.57
Wire, 600V, type THWN-THHN, copper, solid, #12	.002	C.L.F.	.01	.09	.10
Incandescent fixture, recessed, w/lens, prewired, square trim, 200W	.006	Ea.	.49	.42	.91
Steel outlet box 4" square	.006	Ea.	.06	.14	.20
Fixture whip, Greenfield w/#12 THHN wire	.006	Ea.	.02	.04	.06
TOTAL			.72	1.12	1.84

D5020 216	Incandescent Fixture (by Wattage)	COST PER S.F.		
		MAT.	INST.	TOTAL
0190	Incandescent fixture recess mounted, type A			
0200	1 watt per S.F., 8 FC, 6 fixtures per 1000 S.F.	.72	1.12	1.84
0240	2 watt per S.F., 16 FC, 12 fixtures per 1000 S.F.	1.43	2.21	3.64
0280	3 watt per S.F., 24 FC, 18 fixtures, per 1000 S.F.	2.14	3.28	5.42
0320	4 watt per S.F., 32 FC, 24 fixtures per 1000 S.F.	2.86	4.39	7.25
0400	5 watt per S.F., 40 FC, 30 fixtures per 1000 S.F.	3.58	5.50	9.08

SERVICES

D

D5020 Lighting and Branch Wiring

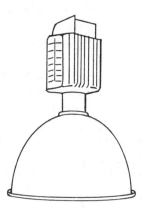

HIGH BAY FIXTURES

B. Metal halide 400 watt

C. High Pressure sodium 400 watt

E. Metal halide 1000 watt

F. High pressure sodium 1000 watt

G. Metal halide 1000 watt
   125,000 lumen lamp

System Components	QUANTITY	UNIT	COST PER S.F.		
			MAT.	INST.	TOTAL
SYSTEM D5020 220 0880					
HIGH INTENSITY DISCHARGE FIXTURE, 8'-10' ABOVE WORK PLANE, 100 FC					
TYPE B, 12 FIXTURES PER 900 S.F.					
Steel intermediate conduit, (IMC) 1/2" diam	.460	L.F.	.69	2.16	2.85
Wire, 600V, type THWN-THHN, copper, solid, #10	.009	C.L.F.	.06	.42	.48
Steel outlet box 4" concrete	.009	Ea.	.09	.21	.30
Steel outlet box plate with stud 4" concrete	.009	Ea.	.03	.05	.08
Metal halide, hi bay, aluminum reflector, 400 W lamp	.009	Ea.	3.15	1.84	4.99
TOTAL			4.02	4.68	8.70

D5020 220	H.I.D. Fixture, High Bay, 8'-10' (by Type)	COST PER S.F.		
		MAT.	INST.	TOTAL
0500	High intensity discharge fixture, 8'-10' above work plane, 100 FC			
0880	Type B, 8 fixtures per 900 S.F.	4.02	4.68	8.70
0920	15 fixtures per 1800 S.F.	3.67	4.50	8.17
0960	24 fixtures per 3000 S.F.	3.67	4.50	8.17
1000	31 fixtures per 4000 S.F.	3.68	4.49	8.17
1040	38 fixtures per 5000 S.F.	3.68	4.49	8.17
1080	60 fixtures per 8000 S.F.	3.68	4.49	8.17
1120	72 fixtures per 10000 S.F.	3.28	4.16	7.44
1160	115 fixtures per 16000 S.F.	3.28	4.16	7.44
1200	230 fixtures per 32000 S.F.	3.28	4.16	7.44
1240	Type C, 4 fixtures per 900 S.F.	1.89	2.85	4.74
1280	8 fixtures per 1800 S.F.	1.89	2.85	4.74
1320	13 fixtures per 3000 S.F.	1.89	2.85	4.74
1360	17 fixtures per 4000 S.F.	1.89	2.85	4.74
1400	21 fixtures per 5000 S.F.	1.84	2.62	4.46
1440	33 fixtures per 8000 S.F.	1.84	2.58	4.42
1480	40 fixtures per 10000 S.F.	1.80	2.44	4.24
1520	63 fixtures per 16000 S.F.	1.80	2.44	4.24
1560	126 fixtures per 32000 S.F.	1.80	2.44	4.24

Important: See the Reference Section for critical supporting data - Reference Numbers and City Cost Indexes

SERVICES D

D5020 Lighting and Branch Wiring

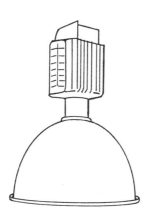

HIGH BAY FIXTURES

B. Metal halide 400 watt

C. High Pressure sodium 400 watt

E. Metal halide 1000 watt

F. High pressure sodium 1000 watt

G. Metal halide 1000 watt
125,000 lumen lamp

System Components			COST PER S.F.		
	QUANTITY	UNIT	MAT.	INST.	TOTAL
SYSTEM D5020 222 0240					
HIGH INTENSITY DISCHARGE FIXTURE, 8'-10' ABOVE WORK PLANE					
1 WATT/S.F., TYPE B, 29 FC, 2 FIXTURES/1000 S.F.					
Steel intermediate conduit, (IMC) 1/2" diam	.100	L.F.	.15	.47	.62
Wire, 600V, type THWN-THHN, copper, solid, #10	.002	C.L.F.	.01	.09	.10
Steel outlet box 4" concrete	.002	Ea.	.02	.05	.07
Steel outlet box plate with stud, 4" concrete	.002	Ea.	.01	.01	.02
Metal halide, hi bay, aluminum reflector, 400 W lamp	.002	Ea.	.70	.41	1.11
TOTAL			.89	1.03	1.92

D5020 222	H.I.D. Fixture, High Bay, 8'-10' (by Wattage)		COST PER S.F.		
			MAT.	INST.	TOTAL
0190	High intensity discharge fixture, 8'-10' above work plane				
0240	1 watt/S.F., type B, 29 FC, 2 fixtures/1000 S.F.		.89	1.03	1.92
0280	Type C, 54 FC, 2 fixtures/1000 S.F.		.78	.79	1.57
0400	2 watt/S.F., type B, 59 FC, 4 fixtures/1000 S.F.		1.76	1.99	3.75
0440	Type C, 108 FC, 4 fixtures/1000 S.F.	R16510 -105	1.52	1.51	3.03
0560	3 watt/S.F., type B, 103 FC, 7 fixtures/1000 S.F.		2.99	3.18	6.17
0600	Type C, 189 FC, 6 fixtures/1000 S.F.		2.45	1.96	4.41
0720	4 watt/S.F., type B, 133 FC, 9 fixtures/1000 S.F.		3.88	4.17	8.05
0760	Type C, 243 FC, 9 fixtures/1000 S.F.		3.43	3.39	6.82
0880	5 watt/S.F., type B, 162 FC, 11 fixtures/1000 S.F.		4.76	5.15	9.91
0920	Type C, 297 FC, 11 fixtures/1000 S.F.		4.20	4.19	8.39

SERVICES

D

D5020 Lighting and Branch Wiring

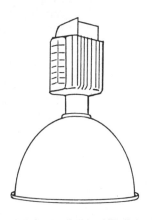

HIGH BAY FIXTURES

B. Metal halide 400 watt

C. High Pressure sodium 400 watt

E. Metal halide 1000 watt

F. High pressure sodium 1000 watt

G. Metal halide 1000 watt
125,000 lumen lamp

System Components	QUANTITY	UNIT	COST PER S.F.		
			MAT.	INST.	TOTAL
SYSTEM D5020 224 1240					
HIGH INTENSITY DISCHARGE FIXTURE, 16' ABOVE WORK PLANE, 100 FC					
TYPE C, 5 FIXTURES PER 900 S.F.					
Steel intermediate conduit, (IMC) 1/2" diam	.260	L.F.	.39	1.22	1.61
Wire, 600V, type THWN-THHN, copper, solid, #10	.007	C.L.F.	.05	.33	.38
Steel outlet box 4" concrete	.006	Ea.	.06	.14	.20
Steel outlet box plate with stud, 4" concrete	.006	Ea.	.02	.04	.06
High pressure sodium, hi bay, aluminum reflector, 400 W lamp	.006	Ea.	1.95	1.22	3.17
TOTAL			2.47	2.95	5.42

D5020 224	H.I.D. Fixture, High Bay, 16' (by Type)	COST PER S.F.		
		MAT.	INST.	TOTAL
0510	High intensity discharge fixture, 16' above work plane, 100 FC			
1240	Type C, 5 fixtures per 900 S.F.	2.47	2.95	5.42
1280	9 fixtures per 1800 S.F.	2.26	3.10	5.36
1320	15 fixtures per 3000 S.F.	2.26	3.10	5.36
1360	18 fixtures per 4000 S.F.	2.15	2.77	4.92
1400	22 fixtures per 5000 S.F.	2.15	2.77	4.92
1440	36 fixtures per 8000 S.F.	2.15	2.77	4.92
1480	42 fixtures per 10,000 S.F.	1.91	2.86	4.77
1520	65 fixtures per 16,000 S.F.	1.91	2.86	4.77
1600	Type G, 4 fixtures per 900 S.F.	3.12	4.67	7.79
1640	6 fixtures per 1800 S.F.	2.58	4.35	6.93
1720	9 fixtures per 4000 S.F.	2.12	4.24	6.36
1760	11 fixtures per 5000 S.F.	2.12	4.24	6.36
1840	21 fixtures per 10,000 S.F.	1.98	3.82	5.80
1880	33 fixtures per 16,000 S.F.	1.98	3.82	5.80

Important: See the Reference Section for critical supporting data - Reference Numbers and City Cost Indexes

D5020 Lighting and Branch Wiring

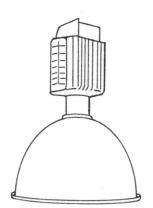

HIGH BAY FIXTURES

B. Metal halide 400 watt

C. High Pressure sodium 400 watt

E. Metal halide 1000 watt

F. High pressure sodium 1000 watt

G. Metal halide 1000 watt
125,000 lumen lamp

System Components	QUANTITY	UNIT	COST PER S.F.		
			MAT.	INST.	TOTAL
SYSTEM D5020 226 0240					
HIGH INTENSITY DISCHARGE FIXTURE, 16' ABOVE WORK PLANE					
1 WATT/S.F., TYPE E, 42 FC, 1 FIXTURE/1000 S.F.					
Steel intermediate conduit, (IMC) 1/2" diam	.160	L.F.	.24	.75	.99
Wire, 600V, type THWN-THHN, copper, solid, #10	.003	C.L.F.	.02	.14	.16
Steel outlet box 4" concrete	.001	Ea.	.01	.02	.03
Steel outlet box plate with stud, 4" concrete	.001	Ea.		.01	.01
Metal halide, hi bay, aluminum reflector, 1000 W lamp	.001	Ea.	.51	.23	.74
TOTAL			.78	1.15	1.93

D5020 226	H.I.D. Fixture, High Bay, 16' (by Wattage)		COST PER S.F.		
			MAT.	INST.	TOTAL
0190	High intensity discharge fixture, 16' above work plane				
0240	1 watt/S.F., type E, 42 FC, 1 fixture/1000 S.F.		.78	1.15	1.93
0280	Type G, 52 FC, 1 fixture/1000 S.F.	R16510	.78	1.15	1.93
0320	Type C, 54 FC, 2 fixture/1000 S.F.	-105	.92	1.29	2.21
0440	2 watt/S.F., type E, 84 FC, 2 fixture/1000 S.F.		1.57	2.36	3.93
0480	Type G, 105 FC, 2 fixture/1000 S.F.		1.57	2.36	3.93
0520	Type C, 108 FC, 4 fixture/1000 S.F.		1.83	2.58	4.41
0640	3 watt/S.F., type E, 126 FC, 3 fixture/1000 S.F.		2.35	3.51	5.86
0680	Type G, 157 FC, 3 fixture/1000 S.F.		2.35	3.51	5.86
0720	Type C, 162 FC, 6 fixture/1000 S.F.		2.75	3.87	6.62
0840	4 watt/S.F., type E, 168 FC, 4 fixture/1000 S.F.		3.14	4.71	7.85
0880	Type G, 210 FC, 4 fixture/1000 S.F.		3.14	4.71	7.85
0920	Type C, 243 FC, 9 fixture/1000 S.F.		4.02	5.40	9.42
1040	5 watt/S.F., type E, 210 FC, 5 fixture/1000 S.F.		3.93	5.85	9.78
1080	Type G, 262 FC, 5 fixture/1000 S.F.		3.93	5.85	9.78
1120	Type C, 297 FC, 11 fixture/1000 S.F.		4.94	6.70	11.64

SERVICES

D

D5020 Lighting and Branch Wiring

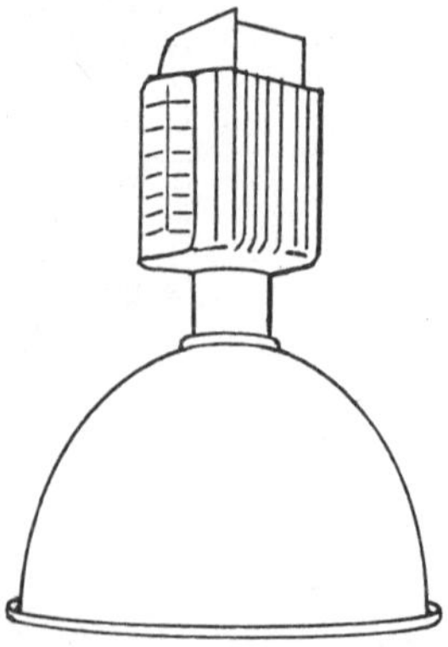

HIGH BAY FIXTURES

B. Metal halide 400 watt
C. High Pressure sodium 400 watt
E. Metal halide 1000 watt
F. High pressure sodium 1000 watt
G. Metal halide 1000 watt
125,000 lumen lamp

System Components	QUANTITY	UNIT	COST PER S.F. MAT.	COST PER S.F. INST.	COST PER S.F. TOTAL
SYSTEM D5020 228 1240					
HIGH INTENSITY DISCHARGE FIXTURE, 20′ ABOVE WORK PLANE, 100 FC					
TYPE C, 6 FIXTURES PER 900 S.F.					
Steel intermediate conduit, (IMC) 1/2″ diam	.350	L.F.	.53	1.64	2.17
Wire, 600V, type THWN-THHN, copper, solid, #10.	.011	C.L.F.	.08	.52	.60
Steel outlet box 4″ concrete	.007	Ea.	.07	.16	.23
Steel outlet box plate with stud, 4″ concrete	.007	Ea.	.02	.04	.06
High pressure sodium, hi bay, aluminum reflector, 400 W lamp	.007	Ea.	2.28	1.43	3.71
TOTAL			2.98	3.79	6.77

D5020 228	H.I.D. Fixture, High Bay, 20′ (by Type)	COST PER S.F. MAT.	COST PER S.F. INST.	COST PER S.F. TOTAL
0510	High intensity discharge fixture 20′ above work plane, 100 FC			
1240	Type C, 6 fixtures per 900 S.F.	2.98	3.79	6.77
1280	10 fixtures per 1800 S.F. R16510 -105	2.64	3.56	6.20
1320	16 fixtures per 3000 S.F.	2.33	3.42	5.75
1360	20 fixtures per 4000 S.F.	2.33	3.42	5.75
1400	24 fixtures per 5000 S.F.	2.33	3.42	5.75
1440	38 fixtures per 8000 S.F.	2.29	3.28	5.57
1520	68 fixtures per 16000 S.F.	2.12	3.61	5.73
1560	132 fixtures per 32000 S.F.	2.12	3.61	5.73
1600	Type G, 4 fixtures per 900 S.F.	3.07	4.57	7.64
1640	6 fixtures per 1800 S.F.	2.57	4.36	6.93
1680	7 fixtures per 3000 S.F.	2.52	4.22	6.74
1720	10 fixtures per 4000 S.F.	2.51	4.17	6.68
1760	11 fixtures per 5000 S.F.	2.17	4.48	6.65
1800	18 fixtures per 8000 S.F.	2.17	4.48	6.65
1840	22 fixtures per 10000 S.F.	2.17	4.48	6.65
1880	34 fixtures per 16000 S.F.	2.12	4.33	6.45
1920	66 fixtures per 32000 S.F.	2	3.96	5.96

Important: See the Reference Section for critical supporting data - Reference Numbers and City Cost Indexes

D5020 Lighting and Branch Wiring

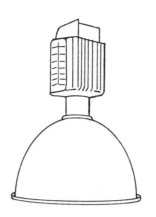

HIGH BAY FIXTURES

B. Metal halide 400 watt

C. High Pressure sodium 400 watt

E. Metal halide 1000 watt

F. High pressure sodium 1000 watt

G. Metal halide 1000 watt
125,000 lumen lamp

System Components	QUANTITY	UNIT	COST PER S.F.		
			MAT.	INST.	TOTAL
SYSTEM D5020 230 0240					
HIGH INTENSITY DISCHARGE FIXTURE, 20' ABOVE WORK PLANE					
1 WATT/S.F., TYPE E, 40 FC, 1 FIXTURE 1000 S.F.					
Steel intermediate conduit, (IMC) 1/2" diam	.160	L.F.	.24	.75	.99
Wire, 600V, type THWN-THHN, copper, solid, #10	.005	C.L.F.	.04	.24	.28
Steel outlet box 4" concrete	.001	Ea.	.01	.02	.03
Steel outlet box plate with stud, 4" concrete	.001	Ea.		.01	.01
Metal halide, hi bay, aluminum reflector, 1000 W lamp	.001	Ea.	.51	.23	.74
TOTAL			.80	1.25	2.05

D5020 230	H.I.D. Fixture, High Bay, 20' (by Wattage)		COST PER S.F.		
			MAT.	INST.	TOTAL
0190	High intensity discharge fixture, 20' above work plane				
0240	1 watt/S.F., type E, 40 FC, 1 fixture/1000 S.F.		.80	1.25	2.05
0280	Type G, 50 FC, 1 fixture/1000 S.F.		.80	1.25	2.05
0320	Type C, 52 FC, 2 fixtures/1000 S.F.	R16510 -105	.96	1.45	2.41
0440	2 watt/S.F., type E, 81 FC, 2 fixtures/1000 S.F.		1.59	2.50	4.09
0480	Type G, 101 FC, 2 fixtures/1000 S.F.		1.59	2.50	4.09
0520	Type C, 104 FC, 4 fixtures/1000 S.F.		1.89	2.83	4.72
0640	3 watt/S.F., type E, 121 FC, 3 fixtures/1000 S.F.		2.39	3.75	6.14
0680	Type G, 151 FC, 3 fixtures/1000 S.F.		2.39	3.75	6.14
0720	Type C, 155 FC, 6 fixtures/1000 S.F.		2.85	4.28	7.13
0840	4 watt/S.F., type E, 161 FC, 4 fixtures/1000 S.F.		3.18	4.99	8.17
0880	Type G, 202 FC, 4 fixtures/1000 S.F.		3.18	4.99	8.17
0920	Type C, 233 FC, 9 fixtures/1000 S.F.		4.13	5.90	10.03
1040	5 watt/S.F., type E, 202 FC, 5 fixtures/1000 S.F.		3.99	6.25	10.24
1080	Type G, 252 FC, 5 fixtures/1000 S.F.		3.99	6.25	10.24
1120	Type C, 285 FC, 11 fixtures/1000 S.F.		5.10	7.35	12.45

D50 Electrical

D5020 Lighting and Branch Wiring

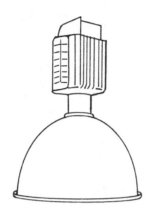

HIGH BAY FIXTURES

B. Metal halide 400 watt

C. High Pressure sodium 400 watt

E. Metal halide 1000 watt

F. High pressure sodium 1000 watt

G. Metal halide 1000 watt
   125,000 lumen lamp

System Components	QUANTITY	UNIT	COST PER S.F.		
			MAT.	INST.	TOTAL
SYSTEM D5020 232 1240					
HIGH INTENSITY DISCHARGE FIXTURE, 30' ABOVE WORK PLANE, 100 FC					
TYPE F, 4 FIXTURES PER 900 S.F.					
Steel intermediate conduit, (IMC) 1/2" diam	.580	L.F.	.88	2.72	3.60
Wire, 600V, type THWN-THHN, copper, solid, #10	.018	C.L.F.	.13	.85	.98
Steel outlet box 4" concrete	.004	Ea.	.04	.09	.13
Steel outlet box plate with stud, 4" concrete	.004	Ea.	.01	.02	.03
High pressure sodium, hi bay, aluminum, 1000 W lamp	.004	Ea.	1.88	.94	2.82
TOTAL			2.94	4.62	7.56

D5020 232	H.I.D. Fixture, High Bay, 30' (by Type)	COST PER S.F.		
		MAT.	INST.	TOTAL
0510	High intensity discharge fixture, 30' above work plane, 100 FC			
1240	Type F, 4 fixtures per 900 S.F.	2.94	4.62	7.56
1280	6 fixtures per 1800 S.F.	2.47	4.41	6.88
1320	8 fixtures per 3000 S.F.	2.47	4.41	6.88
1360	9 fixtures per 4000 S.F.	2.01	4.19	6.20
1400	10 fixtures per 5000 S.F.	2.01	4.19	6.20
1440	17 fixtures per 8000 S.F.	2.01	4.19	6.20
1480	18 fixtures per 10,000 S.F.	1.90	3.87	5.77
1520	27 fixtures per 16,000 S.F.	1.90	3.87	5.77
1560	52 fixtures per 32000 S.F.	1.89	3.82	5.71
1600	Type G, 4 fixtures per 900 S.F.	3.17	4.90	8.07
1640	6 fixtures per 1800 S.F.	2.60	4.45	7.05
1680	9 fixtures per 3000 S.F.	2.56	4.31	6.87
1720	11 fixtures per 4000 S.F.	2.51	4.17	6.68
1760	13 fixtures per 5000 S.F.	2.51	4.17	6.68
1800	21 fixtures per 8000 S.F.	2.51	4.17	6.68
1840	23 fixtures per 10,000 S.F.	2.07	4.43	6.50
1880	36 fixtures per 16,000 S.F.	2.07	4.43	6.50
1920	70 fixtures per 32000 S.F.	2.07	4.43	6.50

R16510 -105

Important: See the Reference Section for critical supporting data - Reference Numbers and City Cost Indexes

D5020 Lighting and Branch Wiring

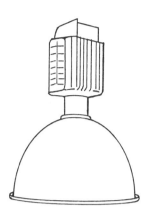

HIGH BAY FIXTURES

B. Metal halide 400 watt

C. High Pressure sodium 400 watt

E. Metal halide 1000 watt

F. High pressure sodium 1000 watt

G. Metal halide 1000 watt
   125,000 lumen lamp

System Components			COST PER S.F.		
	QUANTITY	UNIT	MAT.	INST.	TOTAL
SYSTEM D5020 234 0240					
HIGH INTENSITY DISCHARGE FIXTURE, 30′ ABOVE WORK PLANE					
1 WATT/S.F., TYPE E, 37 FC, 1 FIXTURE/1000 S.F.					
Steel intermediate conduit, (IMC) 1/2″ diam	.196	L.F.	.30	.92	1.22
Wire, 600V type THWN-THHN, copper, solid, #10	.006	C.L.F.	.04	.28	.32
Steel outlet box 4″ concrete	.001	Ea.	.01	.02	.03
Steel outlet box plate with stud, 4″ concrete	.001	Ea.		.01	.01
Metal halide, hi bay, aluminum reflector, 1000 W lamp	.001	Ea.	.51	.23	.74
TOTAL			.86	1.46	2.32

D5020 234	H.I.D. Fixture, High Bay, 30′ (by Wattage)		COST PER S.F.		
			MAT.	INST.	TOTAL
0190	High intensity discharge fixture, 30′ above work plane				
0240	1 watt/S.F., type E, 37 FC, 1 fixture/1000 S.F.		.86	1.46	2.32
0280	Type G, 45 FC., 1 fixture/1000 S.F.	R16510 -105	.86	1.46	2.32
0320	Type F, 50 FC, 1 fixture/1000 S.F.		.73	1.14	1.87
0440	2 watt/S.F., type E, 74 FC, 2 fixtures/1000 S.F.		1.72	2.93	4.65
0480	Type G, 92 FC, 2 fixtures/1000 S.F.		1.72	2.93	4.65
0520	Type F, 100 FC, 2 fixtures/1000 S.F.		1.48	2.34	3.82
0640	3 watt/S.F., type E, 110 FC, 3 fixtures/1000 S.F.		2.59	4.44	7.03
0680	Type G, 138FC, 3 fixtures/1000 S.F.		2.59	4.44	7.03
0720	Type F, 150 FC, 3 fixtures/1000 S.F.		2.21	3.48	5.69
0840	4 watt/S.F., type E, 148 FC, 4 fixtures/1000 S.F.		3.43	5.90	9.33
0880	Type G, 185 FC, 4 fixtures/1000 S.F.		3.43	5.90	9.33
0920	Type F, 200 FC, 4 fixtures/1000 S.F.		2.95	4.68	7.63
1040	5 watt/S.F., type E, 185 FC, 5 fixtures/1000 S.F.		4.30	7.40	11.70
1080	Type G, 230 FC, 5 fixtures/1000 S.F.		4.30	7.40	11.70
1120	Type F, 250 FC, 5 fixtures/1000 S.F.		3.70	5.80	9.50

SERVICES

D

LOW BAY FIXTURES

J. Metal halide 250 watt

K. High pressure sodium 150 watt

System Components	QUANTITY	UNIT	COST PER S.F.		
			MAT.	INST.	TOTAL
SYSTEM D5020 236 0920					
HIGH INTENSITY DISCHARGE FIXTURE, 8'-10' ABOVE WORK PLANE, 50 FC					
TYPE J, 13 FIXTURES PER 1800 S.F.					
Steel intermediate conduit, (IMC) 1/2" diam	.550	L.F.	.83	2.58	3.41
Wire, 600V, type THWN-THHN, copper, solid, #10	.012	C.L.F.	.09	.56	.65
Steel outlet box 4" concrete	.007	Ea.	.07	.16	.23
Steel outlet box plate with stud, 4" concrete	.007	Ea.	.02	.04	.06
Metal halide, lo bay, aluminum reflector, 250 W DX lamp	.007	Ea.	2.38	1.03	3.41
TOTAL			3.39	4.37	7.76

D5020 236	H.I.D. Fixture, Low Bay, 8'-10' (by Type)	COST PER S.F.		
		MAT.	INST.	TOTAL
0510	High intensity discharge fixture, 8'-10' above work plane, 50 FC			
0880	Type J, 7 fixtures per 900 S.F.	3.71	4.43	8.14
0920	13 fixtures per 1800 S.F.	3.39	4.37	7.76
0960	21 fixtures per 3000 S.F.	3.39	4.37	7.76
1000	28 fixtures per 4000 S.F.	3.39	4.37	7.76
1040	35 fixtures per 5000 S.F.	3.39	4.37	7.76
1120	62 fixtures per 10,000 S.F.	3.04	4.20	7.24
1160	99 fixtures per 16,000 S.F.	3.04	4.20	7.24
1200	199 fixtures per 32,000 S.F.	3.04	4.20	7.24
1240	Type K, 9 fixtures per 900 S.F.	3.54	3.79	7.33
1280	16 fixtures per 1800 S.F.	3.25	3.65	6.90
1320	26 fixtures per 3000 S.F.	3.24	3.60	6.84
1360	31 fixtures per 4000 S.F.	2.98	3.54	6.52
1400	39 fixtures per 5000 S.F.	2.98	3.54	6.52
1440	62 fixtures per 8000 S.F.	2.98	3.54	6.52
1480	78 fixtures per 10,000 S.F.	2.98	3.54	6.52
1520	124 fixtures per 16,000 S.F.	2.95	3.44	6.39
1560	248 fixtures per 32,000 S.F.	2.81	3.81	6.62

(note at row 0920/0960: R16510 -105)

Important: See the Reference Section for critical supporting data - Reference Numbers and City Cost Indexes

D5020 Lighting and Branch Wiring

LOW BAY FIXTURES

J. Metal halide 250 watt

K. High pressure sodium 150 watt

System Components	QUANTITY	UNIT	COST PER S.F.		
			MAT.	INST.	TOTAL
SYSTEM D5020 238 0240					
HIGH INTENSITY DISCHARGE FIXTURE, 8'-10' ABOVE WORK PLANE					
1 WATT/S.F., TYPE J, 30 FC, 4 FIXTURES/1000 S.F.					
Steel intermediate conduit, (IMC) 1/2″ diam	.280	L.F.	.42	1.31	1.73
Wire, 600V, type THWN-THHN, copper, solid, #10	.008	C.L.F.	.06	.38	.44
Steel outlet box 4″ concrete	.004	Ea.	.04	.09	.13
Steel outlet box plate with stud, 4″ concrete	.004	Ea.	.01	.02	.03
Metal halide, lo bay, aluminum reflector, 250 W DX lamp	.004	Ea.	1.36	.59	1.95
TOTAL			1.89	2.39	4.28

D5020 238	H.I.D. Fixture, Low Bay, 8'-10' (by Wattage)		COST PER S.F.		
			MAT.	INST.	TOTAL
0190	High intensity discharge fixture, 8'-10' above work plane				
0240	1 watt/S.F., type J, 30 FC, 4 fixtures/1000 S.F.		1.89	2.39	4.28
0280	Type K, 29 FC, 5 fixtures/1000 S.F.	R16510	1.84	2.15	3.99
0400	2 watt/S.F., type J, 52 FC, 7 fixtures/1000 S.F.	-105	3.35	4.38	7.73
0440	Type K, 63 FC, 11 fixtures/1000 S.F.		3.96	4.52	8.48
0560	3 watt/S.F., type J, 81 FC, 11 fixtures/1000 S.F.		5.20	6.60	11.80
0600	Type K, 92 FC, 16 fixtures/1000 S.F.		5.80	6.65	12.45
0720	4 watt/S.F., type J, 103 FC, 14 fixtures/1000 S.F.		6.70	8.75	15.45
0760	Type K, 127 FC, 22 fixtures/1000 S.F.		7.95	9.05	17
0880	5 watt/S.F., type J, 133 FC, 18 fixtures/1000 S.F.		8.55	11	19.55
0920	Type K, 155 FC, 27 fixtures/1000 S.F.		9.75	11.20	20.95

D SERVICES

D5020 Lighting and Branch Wiring

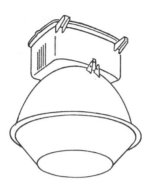

LOW BAY FIXTURES

J. Metal halide 250 watt

K. High pressure sodium 150 watt

System Components	QUANTITY	UNIT	COST PER S.F.		
			MAT.	INST.	TOTAL
SYSTEM D5020 240 0880					
HIGH INTENSITY DISCHARGE FIXTURE, 16′ ABOVE WORK PLANE, 50 FC					
TYPE J, 9 FIXTURES PER 900 S.F.					
Steel intermediate conduit, (IMC) 1/2″ diam	.630	L.F.	.95	2.95	3.90
Wire, 600V type, THWN-THHN, copper, solid, #10	.012	C.L.F.	.09	.56	.65
Steel outlet box 4″ concrete	.010	Ea.	.10	.24	.34
Steel outlet box plate with stud, 4″ concrete	.010	Ea.	.04	.06	.10
Metal halide, lo bay, aluminum reflector, 250 W DX lamp	.010	Ea.	3.40	1.47	4.87
TOTAL			4.58	5.28	9.86

D5020 240	H.I.D. Fixture, Low Bay, 16′ (by Type)	COST PER S.F.		
		MAT.	INST.	TOTAL
0510	High intensity discharge fixture, 16′ above work plane, 50 FC			
0880	Type J, 9 fixtures per 900 S.F.	4.58	5.30	9.88
0920	14 fixtures per 1800 S.F.	3.87	4.98	8.85
0960	24 fixtures per 3000 S.F.	3.90	5.10	9
1000	32 fixtures per 4000 S.F.	3.90	5.10	9
1040	35 fixtures per 5000 S.F.	3.56	4.94	8.50
1080	56 fixtures per 8000 S.F.	3.56	4.94	8.50
1120	70 fixtures per 10,000 S.F.	3.55	4.94	8.49
1160	111 fixtures per 16,000 S.F.	3.55	4.94	8.49
1200	222 fixtures per 32,000 S.F.	3.55	4.94	8.49
1240	Type K, 11 fixtures per 900 S.F.	4.36	4.93	9.29
1280	20 fixtures per 1800 S.F.	4.01	4.56	8.57
1320	29 fixtures per 3000 S.F.	3.74	4.49	8.23
1360	39 fixtures per 4000 S.F.	3.73	4.44	8.17
1400	44 fixtures per 5000 S.F.	3.45	4.35	7.80
1440	62 fixtures per 8000 S.F.	3.45	4.35	7.80
1480	87 fixtures per 10,000 S.F.	3.45	4.35	7.80
1520	138 fixtures per 16,000 S.F.	3.45	4.35	7.80

R16510 -105 (note at rows 0920/0960)

D5020 Lighting and Branch Wiring

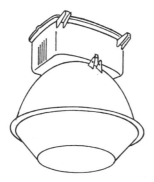

LOW BAY FIXTURES
J. Metal halide 250 watt
K. High pressure sodium 150 watt

System Components	QUANTITY	UNIT	COST PER S.F. MAT.	COST PER S.F. INST.	COST PER S.F. TOTAL
SYSTEM D5020 242 0240					
HIGH INTENSITY DISCHARGE FIXTURE, 16' ABOVE WORK PLANE					
1 WATT/S.F., TYPE J, 28 FC, 4 FIXTURES/1000 S.F.					
Steel intermediate conduit, (IMC) 1/2" diam	.328	L.F.	.50	1.54	2.04
Wire, 600V, type THWN-THHN, copper, solid, #10	.010	C.L.F.	.07	.47	.54
Steel outlet box 4" concrete	.004	Ea.	.04	.09	.13
Steel outlet box plate with stud, 4" concrete	.004	Ea.	.01	.02	.03
Metal halide, lo bay, aluminum reflector, 250 W DX lamp	.004	Ea.	1.36	.59	1.95
TOTAL			1.98	2.71	4.69

D5020 242	H.I.D. Fixture, Low Bay, 16' (by Wattage)		COST PER S.F. MAT.	COST PER S.F. INST.	COST PER S.F. TOTAL
0190	High intensity discharge fixture, mounted 16' above work plane				
0240	1 watt/S.F., type J, 28 FC, 4 fixt./1000 S.F.		1.98	2.71	4.69
0280	Type K, 27 FC, 5 fixt./1000 S.F.	R16510 -105	2.09	3.05	5.14
0400	2 watt/S.F., type J, 48 FC, 7 fixt/1000 S.F.		3.60	5.25	8.85
0440	Type K, 58 FC, 11 fixt/1000 S.F.		4.45	6.20	10.65
0560	3 watt/S.F., type J, 75 FC, 11 fixt/1000 S.F.		5.60	7.95	13.55
0600	Type K, 85 FC, 16 fixt/1000 S.F.		6.55	9.25	15.80
0720	4 watt/S.F., type J, 95 FC, 14 fixt/1000 S.F.		7.20	10.50	17.70
0760	Type K, 117 FC, 22 fixt/1000 S.F.		8.90	12.40	21.30
0880	5 watt/S.F., type J, 122 FC, 18 fixt/1000 S.F.		9.20	13.20	22.40
0920	Type K, 143 FC, 27 fixt/1000 S.F.		10.95	15.45	26.40

SERVICES

D

D5030 Communications and Security

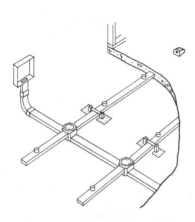

Description: System below includes telephone fitting installed. Does not include cable.

When poke thru fittings and telepoles are used for power, they can also be used for telephones at a negligible additional cost.

System Components	QUANTITY	UNIT	COST PER S.F.		
			MAT.	INST.	TOTAL
SYSTEM D5030 310 0200					
TELEPHONE SYSTEMS, UNDERFLOOR DUCT, 5' ON CENTER, LOW DENSITY					
Underfloor duct 7-1/4" w/insert 2' O.C. 1-3/8" x 7-1/4" super duct	.190	L.F.	4.28	1.79	6.07
Vertical elbow for underfloor superduct, 7-1/4", included					
Underfloor duct conduit adapter, 2" x 1-1/4", included					
Underfloor duct junction box, single duct, 7-1/4" x 3 1/8"	.003	Ea.	1.02	.35	1.37
Underfloor junction box carpet pan	.003	Ea.	.72	.02	.74
Underfloor duct outlet, low tension	.004	Ea.	.27	.23	.50
TOTAL			6.29	2.39	8.68

D5030 310	Telephone Systems	COST PER S.F.		
		MAT.	INST.	TOTAL
0200	Telephone systems, underfloor duct, 5' on center, low density	6.30	2.39	8.69
0240	5' on center, high density	6.55	2.63	9.18
0280	7' on center, low density	5.05	1.97	7.02
0320	7' on center, high density	5.30	2.21	7.51
0400	Poke thru fittings, low density	.85	.79	1.64
0440	High density	1.71	1.57	3.28
0520	Telepoles, low density	.71	.51	1.22
0560	High density	1.42	1.02	2.44
0640	Conduit system with floor boxes, low density	.84	.83	1.67
0680	High density	1.69	1.65	3.34

SERVICES D

D5030 Communications and Security

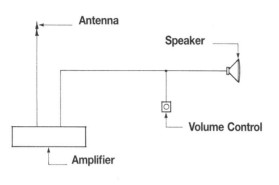

Sound System Includes AM–FM antenna, outlets, rigid conduit, and copper wire.
Fire Detection System Includes pull stations, signals, smoke and heat detectors, rigid conduit, and copper wire.
Intercom System Includes master and remote stations, rigid conduit, and copper wire.
Master Clock System Includes clocks, bells, rigid conduit, and copper wire.
Master TV Antenna Includes antenna, VHF–UHF reception and distribution, rigid conduit, and copper wire.

System Components	QUANTITY	UNIT	COST EACH		
			MAT.	INST.	TOTAL
SYSTEM D5030 910 0220					
SOUND SYSTEM, INCLUDES OUTLETS, BOXES, CONDUIT & WIRE					
Steel intermediate conduit, (IMC) 1/2" diam	1200.000	L.F.	1,812	5,628	7,440
Wire sound shielded w/drain, #22-2 conductor	15.500	C.L.F.	245.68	906.75	1,152.43
Sound system speakers ceiling or wall	12.000	Ea.	1,122	702	1,824
Sound system volume control	12.000	Ea.	834	702	1,536
Sound system amplifier, 250 Watts	1.000	Ea.	1,375	470	1,845
Sound system antenna, AM FM	1.000	Ea.	173	117	290
Sound system monitor panel	1.000	Ea.	310	117	427
Sound system cabinet	1.000	Ea.	675	470	1,145
Steel outlet box 4" square	12.000	Ea.	27.72	282	309.72
Steel outlet box 4" plaster rings	12.000	Ea.	15	88.20	103.20
TOTAL			6,589.40	9,482.95	16,072.35

D5030 910	Communication & Alarm Systems	COST EACH		
		MAT.	INST.	TOTAL
0200	Communication & alarm systems, includes outlets, boxes, conduit & wire			
0210	Sound system, 6 outlets	4,725	5,925	10,650
0220	12 outlets	6,600	9,475	16,075
0240	30 outlets	11,400	17,900	29,300
0280	100 outlets	34,300	60,000	94,300
0320	Fire detection systems, 12 detectors	2,300	4,925	7,225
0360	25 detectors	4,025	8,300	12,325
0400	50 detectors	7,725	16,300	24,025
0440	100 detectors	14,000	29,400	43,400
0480	Intercom systems, 6 stations	2,600	4,050	6,650
0520	12 stations	5,300	8,075	13,375
0560	25 stations	9,025	15,500	24,525
0600	50 stations	17,500	29,200	46,700
0640	100 stations	34,500	57,000	91,500
0680	Master clock systems, 6 rooms	3,875	6,625	10,500
0720	12 rooms	5,800	11,300	17,100
0760	20 rooms	7,700	16,000	23,700
0800	30 rooms	12,500	29,400	41,900
0840	50 rooms	20,000	49,600	69,600
0880	100 rooms	38,300	98,500	136,800
0920	Master TV antenna systems, 6 outlets	2,275	4,175	6,450
0960	12 outlets	4,225	7,800	12,025
1000	30 outlets	8,450	18,100	26,550
1040	100 outlets	28,000	59,000	87,000

SERVICES

D

D5090 Other Electrical Systems

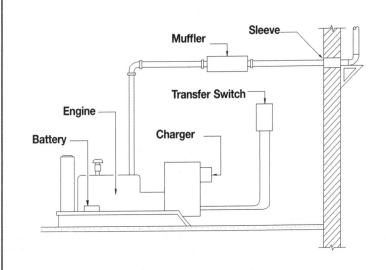

Muffler

Sleeve

Transfer Switch

Engine

Charger

Battery

Description: System below tabulates the installed cost for generators by kW. Included in costs are battery, charger, muffler, and transfer switch.

No conduit, wire, or terminations included.

System Components	QUANTITY	UNIT	COST PER kW		
			MAT.	INST.	TOTAL
SYSTEM D5090 210 0200					
GENERATOR SET, INCL. BATTERY, CHARGER, MUFFLER & TRANSFER SWITCH					
GAS/GASOLINE OPER., 3 PHASE, 4 WIRE, 277/480V, 7.5 kW					
Generator set, gas, 3 phase, 4 wire, 277/480V, 7.5 kW	.133	Ea.	880	214.80	1,094.80
TOTAL			880	214.80	1,094.80

D5090 210	Generators (by kW)	COST PER kW		
		MAT.	INST.	TOTAL
0190	Generator sets, include battery, charger, muffler & transfer switch			
0200	Gas/gasoline operated, 3 phase, 4 wire, 277/480 volt, 7.5 kW	880	215	1,095
0240	11.5 kW	815	163	978
0280	20 kW	550	105	655
0320	35 kW	375	69	444
0360	80 kW	283	41.50	324.50
0400	100 kW	248	40.50	288.50
0440	125 kW	405	38	443
0480	185 kW	360	29	389
0560	Diesel engine with fuel tank, 30 kW	585	80.50	665.50
0600	50 kW	435	63.50	498.50
0640	75 kW	375	50.50	425.50
0680	100 kW	315	43	358
0720	125 kW	264	37	301
0760	150 kW	252	34	286
0800	175 kW	236	30.50	266.50
0840	200 kW	213	27.50	240.50
0880	250 kW	200	23	223
0920	300 kW	180	20	200
0960	350 kW	176	19	195
1000	400 kW	189	17.50	206.50
1040	500 kW	190	14.75	204.75

Important: See the Reference Section for critical supporting data - Reference Numbers and City Cost Indexes

D5090 Other Electrical Systems

Low Density and Medium Density Baseboard Radiation

The costs shown in Table below are based on the following system considerations:

1. The heat loss per square foot is based on approximately 34 BTU/hr. per S.F. of floor or 10 watts per S.F. of floor.
2. Baseboard radiation is based on the low watt density type rated 187 watts per L.F. and the medium density type rated 250 watts per L.F.
3. Thermostat is not included.
4. Wiring costs include branch circuit wiring.

System Components	QUANTITY	UNIT	COST PER S.F.		
			MAT.	INST.	TOTAL
SYSTEM D5090 510 1000					
ELECTRIC BASEBOARD RADIATION, LOW DENSITY, 900 S.F., 31 MBH, 9 kW					
Electric baseboard radiator, 5' long 935 watt	.011	Ea.	.59	.91	1.50
Steel intermediate conduit, (IMC) 1/2" diam	.170	L.F.	.26	.80	1.06
Wire 600 volt, type THW, copper, solid, #12	.005	C.L.F.	.02	.21	.23
TOTAL			.87	1.92	2.79

D5090 510	Electric Baseboard Radiation (Low Density)	COST PER S.F.		
		MAT.	INST.	TOTAL
1000	Electric baseboard radiation, low density, 900 S.F., 31 MBH, 9 kW	.87	1.92	2.79
1200	1500 S.F., 51 MBH, 15 kW	.85	1.87	2.72
1400	2100 S.F., 72 MBH, 21 kW	.78	1.68	2.46
1600	3000 S.F., 102 MBH, 30 kW	.70	1.52	2.22
2000	Medium density, 900 S.F., 31 MBH, 9 kW	.76	1.75	2.51
2200	1500 S.F., 51 MBH, 15 kW	.74	1.70	2.44
2400	2100 S.F., 72 MBH, 21 kW	.72	1.59	2.31
2600	3000 S.F., 102 MBH, 30 kW	.65	1.44	2.09

SERVICES

D

351

D5090   Other Electrical Systems

Commercial Duty Baseboard Radiation
The costs shown in Table below are based on the following system considerations:

1. The heat loss per square foot is based on approximately 41 BTU/hr. per S.F. of floor or 12 watts per S.F.
2. The baseboard radiation is of the commercial duty type rated 250 watts per L.F. served by 277 volt, single phase power.
3. Thermostat is not included.
4. Wiring costs include branch circuit wiring.

System Components	QUANTITY	UNIT	COST PER S.F.		
			MAT.	INST.	TOTAL
SYSTEM D5090 520 1000					
ELECTRIC BASEBOARD RADIATION, MEDIUM DENSITY, 1230 S.F., 51 MBH, 15 kW					
Electric baseboard radiator, 5' long	.013	Ea.	.70	1.07	1.77
Steel intermediate conduit, (IMC) 1/2" diam	.154	L.F.	.23	.72	.95
Wire 600 volt, type THW, copper, solid, #12	.004	C.L.F.	.02	.17	.19
TOTAL			.95	1.96	2.91

D5090 520	Electric Baseboard Radiation (Medium Density)	COST PER S.F.		
		MAT.	INST.	TOTAL
1000	Electric baseboard radiation, medium density, 1230 SF, 51 MBH, 15 kW	.95	1.96	2.91
1200	2500 S.F. floor area, 106 MBH, 31 kW	.88	1.84	2.72
1400	3700 S.F. floor area, 157 MBH, 46 kW	.86	1.79	2.65
1600	4800 S.F. floor area, 201 MBH, 59 kW	.80	1.66	2.46
1800	11,300 S.F. floor area, 464 MBH, 136 kW	.76	1.52	2.28
2000	30,000 S.F. floor area, 1229 MBH, 360 kW	.74	1.49	2.23

For information about Means Estimating Seminars, see yellow pages 12 and 13 in back of book

Important: See the Reference Section for critical supporting data - Reference Numbers and City Cost Indexes

G BUILDING SITEWORK

G1030 Site Earthwork

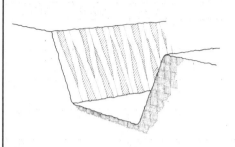

Trenching Systems are shown on a cost per linear foot basis. The systems include: excavation; backfill and removal of spoil; and compaction for various depths and trench bottom widths. The backfill has been reduced to accommodate a pipe of suitable diameter and bedding.

The slope for trench sides varies from 0:1 to 2:1.

The Expanded System Listing shows Trenching Systems that range from 2' to 12' in width. Depths range from 2' to 25'.

System Components	QUANTITY	UNIT	COST PER L.F. EQUIP.	COST PER L.F. LABOR	COST PER L.F. TOTAL
SYSTEM G1030 805 1310					
TRENCHING, BACKHOE, 0 TO 1 SLOPE, 2' WIDE, 2' DP, 3/8 C.Y. BUCKET					
Excavation, trench, hyd. backhoe, track mtd., 3/8 C.Y. bucket	.174	C.Y.	.26	.85	1.11
Backfill and load spoil, from stockpile	.174	C.Y.	.09	.25	.34
Compaction by rammer tamper, 8" lifts, 4 passes	.014	C.Y.		.03	.03
Remove excess spoil, 6 C.Y. dump truck, 2 mile roundtrip	.160	C.Y.	.57	.52	1.09
TOTAL			.92	1.65	2.57

G1030 805	Trenching	COST PER L.F. EQUIP.	COST PER L.F. LABOR	COST PER L.F. TOTAL
1310	Trenching, backhoe, 0 to 1 slope, 2' wide, 2' deep, 3/8 C.Y. bucket	.92	1.65	2.57
1320	3' deep, 3/8 C.Y. bucket	1.14	2.41	3.55
1330	4' deep, 3/8 C.Y. bucket	1.34	3.17	4.51
1340	6' deep, 3/8 C.Y. bucket	1.62	4.07	5.69
1350	8' deep, 1/2 C.Y. bucket	1.89	5.15	7.04
1360	10' deep, 1 C.Y. bucket	2.94	6.25	9.19
1400	4' wide, 2' deep, 3/8 C.Y. bucket	1.96	3.30	5.26
1410	3' deep, 3/8 C.Y. bucket	2.77	4.91	7.68
1420	4' deep, 1/2 C.Y. bucket	2.70	4.94	7.64
1430	6' deep, 1/2 C.Y. bucket	3.41	7.55	10.96
1440	8' deep, 1/2 C.Y. bucket	5.50	9.90	15.40
1450	10' deep, 1 C.Y. bucket	6.15	12.25	18.40
1460	12' deep, 1 C.Y. bucket	7.80	15.75	23.55
1470	15' deep, 1-1/2 C.Y. bucket	6.60	13.45	20.05
1480	18' deep, 2-1/2 C.Y. bucket	8.55	19.25	27.80
1520	6' wide, 6' deep, 5/8 C.Y. bucket	7.15	10.55	17.70
1530	8' deep, 3/4 C.Y. bucket	8.65	12.80	21.45
1540	10' deep, 1 C.Y. bucket	9.30	15.40	24.70
1550	12' deep, 1-1/4 C.Y. bucket	8.15	12.10	20.25
1560	16' deep, 2 C.Y. bucket	10.80	13.75	24.55
1570	20' deep, 3-1/2 C.Y. bucket	13.65	24.50	38.15
1580	24' deep, 3-1/2 C.Y. bucket	22	41.50	63.50
1640	8' wide, 12' deep, 1-1/4 C.Y. bucket	12.05	16.50	28.55
1650	15' deep, 1-1/2 C.Y. bucket	14.20	20.50	34.70
1660	18' deep, 2-1/2 C.Y. bucket	16.75	20.50	37.25
1680	24' deep, 3-1/2 C.Y. bucket	30	55.50	85.50
1730	10' wide, 20' deep, 3-1/2 C.Y. bucket	24.50	41.50	66
1740	24' deep, 3-1/2 C.Y. bucket	38.50	69.50	108
1780	12' wide, 20' deep, 3-1/2 C.Y. bucket	30	50	80
1790	25' deep, bucket	51	92.50	143.50
1800	1/2 to 1 slope, 2' wide, 2' deep, 3/8 C.Y. bucket	1.25	2.11	3.36
1810	3' deep, 3/8 C.Y. bucket	1.73	3.48	5.21
1820	4' deep, 3/8 C.Y. bucket	2.18	5.10	7.28
1840	6' deep, 3/8 C.Y. bucket	3.07	7.90	10.97

G1030 Site Earthwork

G1030 805	Trenching	COST PER L.F.		
		EQUIP.	LABOR	TOTAL
1860	8' deep, 1/2 C.Y. bucket	4.24	12	16.24
1880	10' deep, 1 C.Y. bucket	8.15	16.85	25
2300	4' wide, 2' deep, 3/8 C.Y. bucket	2.08	3.52	5.60
2310	3' deep, 3/8 C.Y. bucket	3.43	5.80	9.23
2320	4' deep, 1/2 C.Y. bucket	3.54	6.05	9.59
2340	6' deep, 1/2 C.Y. bucket	4.87	10.55	15.42
2360	8' deep, 1/2 C.Y. bucket	8.45	14.60	23.05
2380	10' deep, 1 C.Y. bucket	11.05	21.50	32.55
2400	12' deep, 1 C.Y. bucket	11.05	23	34.05
2430	15' deep, 1-1/2 C.Y. bucket	14.70	28.50	43.20
2460	18' deep, 2-1/2 C.Y. bucket	20.50	34.50	55
2840	6' wide, 6' deep, 5/8 C.Y. bucket	8.50	11.65	20.15
2860	8' deep, 3/4 C.Y. bucket	12.05	17.60	29.65
2880	10' deep, 1 C.Y. bucket	14	23	37
2900	12' deep, 1-1/4 C.Y. bucket	13.05	19.25	32.30
2940	16' deep, 2 C.Y. bucket	20.50	31	51.50
2980	20' deep, 3-1/2 C.Y. bucket	29.50	53	82.50
3020	24' deep, 3-1/2 C.Y. bucket	53	99	152
3100	8' wide, 12' deep, 1-1/4 C.Y. bucket	17.60	23.50	41.10
3120	15' deep, 1-1/2 C.Y. bucket	22.50	31.50	54
3140	18' deep, 2-1/2 C.Y. bucket	29	34.50	63.50
3180	24' deep, 3-1/2 C.Y. bucket	61.50	111	172.50
3270	10' wide, 20' deep, 3-1/2 C.Y. bucket	38.50	66	104.50
3280	24' deep, 3-1/2 C.Y. bucket	70.50	125	195.50
3370	12' wide, 20' deep, 3-1/2 C.Y. bucket	43.50	73.50	117
3380	25' deep, 3-1/2 C.Y. bucket	82	145	227
3500	1 to 1 slope, 2' wide, 2' deep, 3/8 C.Y. bucket	1.46	2.47	3.93
3520	3' deep, 3/8 C.Y. bucket	2.24	4.31	6.55
3540	4' deep, 3/8 C.Y. bucket	2.88	6.50	9.38
3560	6' deep, 3/8 C.Y. bucket	4.22	10.40	14.62
3580	8' deep, 1/2 C.Y. bucket	6	16.35	22.35
3600	10' deep, 1 C.Y. bucket	13.65	27	40.65
3800	4' wide, 2' deep, 3/8 C.Y. bucket	2.49	4.21	6.70
3820	3' deep, 3/8 C.Y. bucket	4.38	7.40	11.78
3840	4' deep, 1/2 C.Y. bucket	4.31	6.80	11.11
3860	6' deep, 1/2 C.Y. bucket	6.10	12.65	18.75
3880	8' deep, 1/2 C.Y. bucket	11.40	19.20	30.60
3900	10' deep, 1 C.Y. bucket	14.85	27.50	42.35
3920	12' deep, 1 C.Y. bucket	21.50	40.50	62
3940	15' deep, 1-1/2 C.Y. bucket	21	37.50	58.50
3960	18' deep, 2-1/2 C.Y. bucket	30	45	75
4030	6' wide, 6' deep, 5/8 C.Y. bucket	10.30	14.20	24.50
4040	8' deep, 3/4 C.Y. bucket	15.05	23	38.05
4050	10' deep, 1 C.Y. bucket	18.15	31.50	49.65
4060	12' deep, 1-1/4 C.Y. bucket	17.50	28.50	46
4070	16' deep, 2 C.Y. bucket	28.50	40	68.50
4080	20' deep, 3-1/2 C.Y. bucket	42.50	84	126.50
4090	24' deep, 3-1/2 C.Y. bucket	80	163	243
4500	8' wide, 12' deep, 1-1/4 C.Y. bucket	22.50	32	54.50
4550	15' deep, 1-1/2 C.Y. bucket	29.50	46	75.50
4600	18' deep, 2-1/2 C.Y. bucket	39.50	53	92.50
4650	24' deep, 3-1/2 C.Y. bucket	88	174	262
4800	10' wide, 20' deep, 3-1/2 C.Y. bucket	54	97.50	151.50
4850	24' deep, 3-1/2 C.Y. bucket	95.50	185	280.50
4950	12' wide, 20' deep, 3-1/2 C.Y. bucket	59	104	163
4980	25' deep, 3-1/2 C.Y. bucket	116	225	341
5000	1-1/2 to 1 slope, 2' wide, 2' deep, 3/8 C.Y. bucket	1.69	2.86	4.55
5020	3' deep, 3/8 C.Y. bucket	2.74	5.10	7.84

BUILDING SITEWORK

G

355

G1030 805	Trenching	COST PER L.F.		
		EQUIP.	LABOR	TOTAL
5040	4' deep, 3/8 C.Y. bucket	3.52	7.65	11.17
5060	6' deep, 3/8 C.Y. bucket	5.20	12.20	17.40
5080	8' deep, 1/2 C.Y. bucket	7.45	19.30	26.75
5100	10' deep, 1 C.Y. bucket	15.20	27.50	42.70
5300	4' wide, 2' deep, 3/8 C.Y. bucket	2.35	3.96	6.31
5320	3' deep, 3/8 C.Y. bucket	4.27	7.20	11.47
5340	4' deep, 1/2 C.Y. bucket	5.15	7.95	13.10
5360	6' deep, 1/2 C.Y. bucket	7.25	14.30	21.55
5380	8' deep, 1/2 C.Y. bucket	13.75	22	35.75
5400	10' deep, 1 C.Y. bucket	18	31.50	49.50
5420	12' deep, 1 C.Y. bucket	26.50	46.50	73
5450	15' deep, 1-1/2 C.Y. bucket	25.50	41.50	67
5480	18' deep, 2-1/2 C.Y. bucket	37	49	86
5660	6' wide, 6' deep, 5/8 C.Y. bucket	11.95	15.90	27.85
5680	8' deep, 3/4 C.Y. bucket	17.60	25.50	43.10
5700	10' deep, 1 C.Y. bucket	21.50	35.50	57
5720	12' deep, 1-1/4 C.Y. bucket	20.50	30	50.50
5760	16' deep, 2 C.Y. bucket	34	43	77
5800	20' deep, 3-1/2 C.Y. bucket	52	95.50	147.50
5840	24' deep, 3-1/2 C.Y. bucket	99	186	285
6020	8' wide, 12' deep, 1-1/4 C.Y. bucket	26.50	35.50	62
6050	15' deep, 1-1/2 C.Y. bucket	35.50	50	85.50
6080	18' deep, 2-1/2 C.Y. bucket	47	57	104
6140	24' deep, 3-1/2 C.Y. bucket	108	196	304
6300	10' wide, 20' deep, 3-1/2 C.Y. bucket	64	108	172
6350	24' deep, 3-1/2 C.Y. bucket	115	206	321
6450	12' wide, 20' deep, 3-1/2 C.Y. bucket	70.50	115	185.50
6480	25' deep, 3-1/2 C.Y. bucket	139	248	387
6600	2 to 1 slope, 2' wide, 2' deep, 3/8 C.Y. bucket	2.68	2.64	5.32
6620	3' deep, 3/8 C.Y. bucket	3.15	5.65	8.80
6640	4' deep, 3/8 C.Y. bucket	4.03	8.50	12.53
6660	6' deep, 3/8 C.Y. bucket	5.55	12.90	18.45
6680	8' deep, 1/2 C.Y. bucket	8.50	21.50	30
6700	10' deep, 1 C.Y. bucket	17.25	31.50	48.75
6900	4' wide, 2' deep, 3/8 C.Y. bucket	2.35	3.95	6.30
6920	3' deep, 3/8 C.Y. bucket	4.38	7.40	11.78
6940	4' deep, 1/2 C.Y. bucket	5.75	8.50	14.25
6960	6' deep, 1/2 C.Y. bucket	8.10	15.50	23.60
6980	8' deep, 1/2 C.Y. bucket	15.25	24	39.25
7000	10' deep, 1 C.Y. bucket	20	35	55
7020	12' deep, 1 C.Y. bucket	29.50	52	81.50
7050	15' deep, 1-1/2 C.Y. bucket	29	46.50	75.50
7080	18' deep, 2-1/2 C.Y. bucket	42	55	97
7260	6' wide, 6' deep, 5/8 C.Y. bucket	13.20	16.95	30.15
7280	8' deep, 3/4 C.Y. bucket	19.35	27.50	46.85
7300	10' deep, 3/4 C.Y. bucket	25.50	39.50	65
7320	12' deep, 1-1/4 C.Y. bucket	23	34	57
7360	16' deep, 2 C.Y. bucket	38	48	86
7400	20' deep, 3-1/2 C.Y. bucket	58.50	107	165.50
7440	24' deep, 3-1/2 C.Y. bucket	112	209	321
7620	8' wide, 12' deep, 1-1/4 C.Y. bucket	29.50	38.50	68
7650	15' deep, 1-1/2 C.Y. bucket	39	55	94
7680	18' deep, 2-1/2 C.Y. bucket	52.50	63	115.50
7740	24' deep, 3-1/2 C.Y. bucket	120	218	338
7920	10' wide, 20' deep, 3-1/2 C.Y. bucket	71	119	190
7940	24' deep, 3-1/2 C.Y. bucket	127	227	354

Important: See the Reference Section for critical supporting data - Reference Numbers and City Cost Indexes

G40 Site Electrical Utilities

G4020 Site Lighting

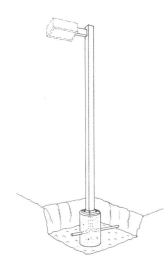

Table G4020 210 Procedure for Calculating Floodlights Required for Various Footcandles

Poles should not be spaced more than 4 times the fixture mounting height for good light distribution. To maintain 1 footcandle over a large area use these watts per square foot:

Incandescent	0.15
Metal Halide	0.032
Mercury Vapor	0.05
High Pressure Sodium	0.024

Estimating Chart

Select Lamp type.

Determine total square feet.

Chart will show quantity of fixtures to provide 1 footcandle initial, at intersection of lines. Multiply fixture quantity by desired footcandle level.

Chart based on use of wide beam luminaires in an area whose dimensions are large compared to mounting height and is approximate only.

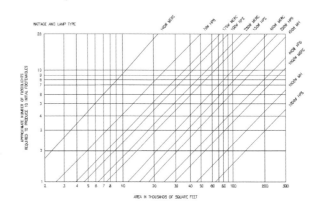

System Components			COST EACH		
	QUANTITY	UNIT	MAT.	INST.	TOTAL
SYSTEM G4020 210 0200					
LIGHT POLES, ALUMINUM, 20' HIGH, 1 ARM BRACKET					
Aluminum light pole, 20', no concrete base	1.000	Ea.	675	460.50	1,135.50
Bracket arm for Aluminum light pole	1.000	Ea.	88.50	58.50	147
Excavation by hand, pits to 6' deep, heavy soil or clay	2.368	C.Y.		191.81	191.81
Footing, concrete incl forms, reinforcing, spread, under 1 C.Y.	.465	C.Y.	52.08	67.47	119.55
Backfill by hand	1.903	C.Y.		56.14	56.14
Compaction vibrating plate	1.903	C.Y.		7.82	7.82
TOTAL			815.58	842.24	1,657.82

G4020 210	Light Pole (Installed)	COST EACH		
		MAT.	INST.	TOTAL
0200	Light pole, aluminum, 20' high, 1 arm bracket	815	840	1,655
0240	2 arm brackets	905	840	1,745
0280	3 arm brackets	995	870	1,865
0320	4 arm brackets	1,075	870	1,945
0360	30' high, 1 arm bracket	1,475	1,050	2,525
0400	2 arm brackets	1,575	1,050	2,625
0440	3 arm brackets	1,650	1,100	2,750
0480	4 arm brackets	1,750	1,100	2,850
0680	40' high, 1 arm bracket	1,800	1,425	3,225
0720	2 arm brackets	1,875	1,425	3,300
0760	3 arm brackets	1,975	1,450	3,425
0800	4 arm brackets	2,050	1,450	3,500
0840	Steel, 20' high, 1 arm bracket	1,000	895	1,895
0880	2 arm brackets	1,075	895	1,970
0920	3 arm brackets	1,100	925	2,025
0960	4 arm brackets	1,175	925	2,100
1000	30' high, 1 arm bracket	1,175	1,125	2,300
1040	2 arm brackets	1,250	1,125	2,375
1080	3 arm brackets	1,250	1,150	2,400
1120	4 arm brackets	1,350	1,150	2,500
1320	40' high, 1 arm bracket	1,500	1,525	3,025
1360	2 arm brackets	1,575	1,525	3,100

BUILDING SITEWORK

G

357

G4020 Site Lighting

G4020 210	Light Pole (Installed)	COST EACH		
		MAT.	INST.	TOTAL
1400	3 arm brackets	1,600	1,550	3,150
1440	4 arm brackets	1,700	1,550	3,250

For information about Means Estimating Seminars, see yellow pages 12 and 13 in back of book

Important: See the Reference Section for critical supporting data - Reference Numbers and City Cost Indexes

BUILDING SITEWORK
G

Reference Section

All the reference information is in one section making it easy to find what you need to know . . . and easy to use the book on a daily basis. This section is visually identified by a vertical gray bar on the edge of pages.

In the reference number information that follows, you'll see the background that relates to the "reference numbers" that appeared in the Unit Price Section. You'll find reference tables, explanations and estimating information that support how we arrived at the unit price data. Also included are alternate pricing methods, technical data and estimating procedures along with information on design and economy in construction.

Also in this Reference Section, we've included Crew Listings, a full listing of all the crews, equipment and their costs, Historical Cost Indexes for cost comparisons over time, City Cost Indexes and Location Factors for adjusting costs to the region you are in, and an explanation of all abbreviations used in the book.

Table of Contents

R01100-005 Tips for Accurate Estimating

1. Use pre-printed or columnar forms for orderly sequence of dimensions and locations and for recording telephone quotations.

2. Use only the front side of each paper or form except for certain pre-printed summary forms.

3. Be consistent in listing dimensions: For example, length x width x height. This helps in rechecking to ensure that, the total length of partitions is appropriate for the building area.

4. Use printed (rather than measured) dimensions where given.

5. Add up multiple printed dimensions for a single entry where possible.

6. Measure all other dimensions carefully.

7. Use each set of dimensions to calculate multiple related quantities.

8. Convert foot and inch measurements to decimal feet when listing. Memorize decimal equivalents to .01 parts of a foot (1/8″ equals approximately .01′).

9. Do not "round off" quantities until the final summary.

10. Mark drawings with different colors as items are taken off.

11. Keep similar items together, different items separate.

12. Identify location and drawing numbers to aid in future checking for completeness.

13. Measure or list everything on the drawings or mentioned in the specifications.

14. It may be necessary to list items not called for to make the job complete.

15. Be alert for: Notes on plans such as N.T.S. (not to scale); changes in scale throughout the drawings; reduced size drawings; discrepancies between the specifications and the drawings.

16. Develop a consistent pattern of performing an estimate. For example:
    a. Start the quantity takeoff at the lower floor and move to the next higher floor.
    b. Proceed from the main section of the building to the wings.
    c. Proceed from south to north or vice versa, clockwise or counterclockwise.
    d. Take off floor plan quantities first, elevations next, then detail drawings.

17. List all gross dimensions that can be either used again for different quantities, or used as a rough check of other quantities for verification (exterior perimeter, gross floor area, individual floor areas, etc.).

18. Utilize design symmetry or repetition (repetitive floors, repetitive wings, symmetrical design around a center line, similar room layouts, etc.). Note: Extreme caution is needed here so as not to omit or duplicate an area.

19. Do not convert units until the final total is obtained. For instance, when estimating concrete work, keep all units to the nearest cubic foot, then summarize and convert to cubic yards.

20. When figuring alternatives, it is best to total all items involved in the basic system, then total all items involved in the alternates. Therefore you work with positive numbers in all cases. When adds and deducts are used, it is often confusing whether to add or subtract a portion of an item; especially on a complicated or involved alternate.

R01100-040 Builder's Risk Insurance

Builder's Risk Insurance is insurance on a building during construction. Premiums are paid by the owner or the contractor. Blasting, collapse and underground insurance would raise total insurance costs above those listed. Floater policy for materials delivered to the job runs $.75 to $1.25 per $100 value. Contractor equipment insurance runs $.50 to $1.50 per $100 value. Insurance for miscellaneous tools to $1,500 value runs from $3.00 to $7.50 per $100 value.

Tabulated below are New England Builder's Risk insurance rates in dollars per $100 value for $1,000 deductible. For $25,000 deductible, rates can be reduced 13% to 34%. On contracts over $1,000,000, rates may be lower than those tabulated. Policies are written annually for the total completed value in place. For "all risk" insurance (excluding flood, earthquake and certain other perils) add $.025 to total rates below.

Coverage	Frame Construction (Class 1)			Brick Construction (Class 4)			Fire Resistive (Class 6)		
	Range		Average	Range		Average	Range		Average
Fire Insurance	$.350 to	$ .850	$.600	$.158 to	$.189	$.174	$.052 to	$.080	$.070
Extended Coverage	.115 to	.200	.158	.080 to	.105	.101	.081 to	.105	.100
Vandalism	.012 to	.016	.014	.008 to	.011	.011	.008 to	.011	.010
Total Annual Rate	$.477 to	$1.066	$.772	$.246 to	$.305	$.286	$.141 to	$.196	$.180

R01100-060　Workers' Compensation Insurance Rates by Trade

The table below tabulates the national averages for Workers' Compensation insurance rates by trade and type of building. The average "Insurance Rate" is multiplied by the "% of Building Cost" for each trade. This produces the "Workers' Compensation Cost" by % of total labor cost, to be added for each trade by building type to determine the weighted average Workers' Compensation rate for the building types analyzed.

Trade	Insurance Rate (% Labor Cost)			% of Building Cost			Workers' Compensation		
	Range		Average	Office Bldgs.	Schools & Apts.	Mfg.	Office Bldgs.	Schools & Apts.	Mfg.
Excavation, Grading, etc.	3.5 % to	18.5%	10.3%	4.8%	4.9%	4.5%	.49%	.50%	.46%
Piles & Foundations	7.3 to	76.7	22.9	7.1	5.2	8.7	1.63	1.19	1.99
Concrete	5.7 to	35.2	15.8	5.0	14.8	3.7	.79	2.34	.58
Masonry	5.1 to	31.0	15.0	6.9	7.5	1.9	1.04	1.13	.29
Structural Steel	7.2 to	112.0	38.9	10.7	3.9	17.6	4.16	1.52	6.85
Miscellaneous & Ornamental Metals	5.2 to	25.4	12.6	2.8	4.0	3.6	.35	.50	.45
Carpentry & Millwork	6.7 to	53.2	18.5	3.7	4.0	0.5	.68	.74	.09
Metal or Composition Siding	5.3 to	35.5	16.1	2.3	0.3	4.3	.37	.05	.69
Roofing	7.3 to	77.1	31.8	2.3	2.6	3.1	.73	.83	.99
Doors & Hardware	4.5 to	25.3	10.9	0.9	1.4	0.4	.10	.15	.04
Sash & Glazing	4.9 to	38.0	13.8	3.5	4.0	1.0	.48	.55	.14
Lath & Plaster	4.0 to	45.5	14.6	3.3	6.9	0.8	.48	1.01	.12
Tile, Marble & Floors	3.7 to	23.3	9.6	2.6	3.0	0.5	.25	.29	.05
Acoustical Ceilings	2.5 to	24.5	10.7	2.4	0.2	0.3	.26	.02	.03
Painting	4.3 to	29.6	12.9	1.5	1.6	1.6	.19	.21	.21
Interior Partitions	6.7 to	53.2	18.5	3.9	4.3	4.4	.72	.80	.81
Miscellaneous Items	2.5 to	110.1	17.3	5.2	3.7	9.7	.90	.64	1.68
Elevators	2.5 to	15.3	7.2	2.1	1.1	2.2	.15	.08	.16
Sprinklers	2.8 to	23.1	9.0	0.5	—	2.0	.05	—	.18
Plumbing	3.0 to	12.5	7.8	4.9	7.2	5.2	.38	.56	.41
Heat., Vent., Air Conditioning	4.0 to	28.1	11.1	13.5	11.0	12.9	1.50	1.22	1.43
Electrical	2.7 to	12.5	6.4	10.1	8.4	11.1	.65	.54	.71
Total	2.5 % to	110.1%	—	100.0%	100.0%	100.0%	16.35%	14.87%	18.36%
						Overall Weighted Average	16.53%		

Workers' Compensation Insurance Rates by States

The table below lists the weighted average Workers' Compensation base rate for each state with a factor comparing this with the national average of 16.2%.

State	Weighted Average	Factor	State	Weighted Average	Factor	State	Weighted Average	Factor
Alabama	28.0%	173	Kentucky	16.6%	102	North Dakota	12.9%	80
Alaska	17.9	110	Louisiana	28.2	174	Ohio	12.7	78
Arizona	7.1	44	Maine	20.8	128	Oklahoma	22.2	137
Arkansas	14.7	91	Maryland	12.0	74	Oregon	14.3	88
California	18.8	116	Massachusetts	14.6	90	Pennsylvania	16.0	99
Colorado	14.0	86	Michigan	19.0	117	Rhode Island	21.2	131
Connecticut	24.9	154	Minnesota	27.7	171	South Carolina	13.9	86
Delaware	11.9	73	Mississippi	17.0	105	South Dakota	14.5	90
District of Columbia	19.1	118	Missouri	20.7	128	Tennessee	16.4	101
Florida	31.4	194	Montana	20.6	127	Texas	14.6	90
Georgia	23.0	142	Nebraska	20.1	124	Utah	12.3	76
Hawaii	16.9	104	Nevada	16.2	100	Vermont	20.0	123
Idaho	10.4	64	New Hampshire	22.6	140	Virginia	12.5	77
Illinois	18.2	112	New Jersey	10.6	65	Washington	10.6	65
Indiana	6.2	38	New Mexico	15.2	94	West Virginia	12.8	79
Iowa	12.0	74	New York	13.5	83	Wisconsin	15.9	98
Kansas	8.8	54	North Carolina	13.7	85	Wyoming	8.0	49
			Weighted Average for U.S. is	16.5% of payroll = 100%				

Rates in the following table are the base or manual costs per $100 of payroll for Workers' Compensation in each state. Rates are usually applied to straight time wages only and not to premium time wages and bonuses.

The weighted average skilled worker rate for 35 trades is 16.2%. For bidding purposes, apply the full value of Workers' Compensation directly to total labor costs, or if labor is 38%, materials 42% and overhead and profit 20% of total cost, carry 38/80 x 16.2% =7.7% of cost (before overhead and profit) into overhead. Rates vary not only from state to state but also with the experience rating of the contractor.

Rates are the most current available at the time of publication.

R01100-060 Workers' Compensation Insurance Rates by Trade and State (cont.)

State	Carpentry – 3 stories or less	Carpentry – interior cab. work	Carpentry – general	Concrete Work – NOC	Concrete Work – flat (flr., sdwk.)	Electrical Wiring – inside	Excavation – earth NOC	Excavation – rock	Glaziers	Insulation Work	Lathing	Masonry	Painting & Decorating	Pile Driving	Plastering	Plumbing	Roofing	Sheet Metal Work (HVAC)	Steel Erection – door & sash	Steel Erection – inter., ornam.	Steel Erection – structure	Steel Erection – NOC	Tile Work – (interior ceramic)	Waterproofing	Wrecking
	5651	5437	5403	5213	5221	5190	6217	6217	5462	5479	5443	5022	5474	6003	5480	5183	5551	5538	5102	5102	5040	5057	5348	9014	5701
AL	23.03	13.25	33.49	13.23	10.32	9.17	15.14	15.14	38.02	19.34	13.00	30.97	28.83	29.60	45.45	11.88	68.79	28.14	25.40	25.40	44.99	44.16	15.68	7.34	44.99
AK	12.28	13.50	11.97	11.89	9.38	9.65	18.53	18.53	16.81	21.09	8.51	16.75	13.92	51.57	13.48	8.75	36.97	9.15	11.46	11.46	29.37	18.17	8.49	8.15	29.37
AZ	6.40	4.59	13.14	6.28	3.94	3.56	5.17	5.17	6.37	9.71	4.46	6.18	4.25	8.37	5.21	3.64	11.20	4.75	7.99	7.99	12.75	6.86	3.76	2.71	50.31
AR	12.83	10.62	14.66	14.60	6.75	6.24	9.07	9.07	15.62	26.86	10.49	10.75	9.85	13.56	13.96	6.36	22.13	10.19	8.32	8.32	37.69	30.03	7.13	4.00	37.69
CA	28.28	9.07	28.28	13.18	13.18	9.93	7.88	7.88	16.37	23.34	10.90	14.55	19.45	21.18	18.12	11.59	39.79	15.33	13.55	13.55	23.93	21.91	7.74	19.45	21.91
CO	17.74	9.16	12.15	13.17	8.11	5.71	11.90	11.90	10.18	15.27	5.87	14.10	11.24	19.75	11.07	8.33	25.02	13.10	8.13	8.13	34.92	15.33	8.02	6.34	34.92
CT	21.93	17.69	29.90	28.54	17.17	9.55	12.88	12.88	18.03	32.51	15.16	28.12	18.82	29.36	25.85	11.47	52.53	14.89	16.46	16.46	70.71	21.51	13.29	7.07	50.86
DE	11.48	11.48	9.88	9.99	7.61	5.06	7.82	7.82	9.69	9.88	10.52	9.25	12.70	16.84	10.52	6.05	21.69	8.41	10.11	10.11	23.77	10.11	8.23	9.25	23.37
DC	12.49	11.98	16.79	21.64	21.97	7.25	8.71	8.71	31.70	12.33	8.71	19.06	9.35	23.54	15.51	11.65	25.17	8.78	17.99	17.99	54.08	22.86	23.27	3.99	54.08
FL	35.52	25.27	36.16	35.18	17.49	12.49	16.03	16.03	27.03	25.63	15.21	27.00	24.86	68.52	42.18	12.45	53.69	21.04	17.19	17.19	67.47	46.35	12.40	10.76	67.47
GA	32.80	16.96	24.66	16.82	12.15	9.13	14.76	14.76	18.12	19.52	19.99	21.28	18.47	30.93	20.75	11.27	40.25	15.09	14.49	14.49	38.29	55.91	10.34	9.92	38.29
HI	17.38	11.62	28.20	13.74	12.27	7.10	7.73	7.73	20.55	21.19	10.94	17.87	11.33	20.15	15.61	6.06	33.24	8.02	11.11	11.11	31.71	21.70	9.78	11.54	31.71
ID	9.86	5.74	13.99	9.13	6.98	4.97	5.36	5.36	9.28	7.65	4.97	8.01	7.10	11.84	8.68	3.92	27.36	8.02	9.87	9.87	26.44	12.60	4.78	4.98	26.44
IL	15.59	13.15	18.76	27.06	10.81	8.23	9.43	9.43	15.78	13.70	9.70	16.83	10.09	25.35	13.11	9.87	29.38	14.82	15.55	15.55	50.68	23.78	15.66	4.69	50.68
IN	5.29	4.52	6.96	5.65	3.48	2.67	3.50	3.50	4.94	3.74	2.67	5.12	4.84	7.38	4.04	2.97	11.69	4.41	5.18	5.18	21.38	10.19	3.69	2.46	21.38
IA	10.40	5.30	11.50	12.21	6.38	3.96	6.30	6.30	13.38	8.84	5.29	7.81	8.26	9.58	8.44	4.85	17.87	6.26	10.24	10.24	48.08	29.00	6.20	4.52	29.58
KS	10.94	7.55	10.15	7.52	5.93	3.67	5.22	5.22	6.91	8.78	4.57	7.23	6.46	8.35	9.08	4.91	18.79	6.05	7.00	7.00	20.07	12.71	4.25	3.02	20.07
KY	17.33	12.77	19.21	16.77	6.05	6.85	17.98	17.98	17.33	19.31	11.00	10.81	10.91	19.93	17.05	5.48	17.33	13.46	13.29	13.29	36.48	24.51	12.79	3.95	36.48
LA	23.14	24.93	53.17	26.43	15.61	9.88	17.49	17.49	20.14	21.43	24.51	28.33	29.58	31.25	22.37	8.64	77.12	21.20	18.98	18.98	51.76	24.21	13.77	13.78	66.41
ME	13.43	10.42	43.59	24.29	11.12	5.03	12.19	12.19	13.35	15.39	14.03	17.26	16.07	31.44	17.95	7.32	32.07	8.82	15.08	15.08	37.84	61.81	11.66	6.06	37.84
MD	10.55	5.95	10.55	11.35	5.05	5.15	9.25	9.25	13.20	13.05	6.45	11.35	6.75	27.45	6.65	5.55	22.80	7.00	9.15	9.15	26.80	18.50	6.35	3.50	26.80
MA	10.62	6.96	16.60	17.84	9.34	3.69	6.41	6.41	8.63	14.02	6.80	14.44	8.26	14.80	5.69	5.09	33.29	8.79	14.31	14.31	40.78	35.81	11.47	3.86	38.14
MI	21.68	13.59	19.80	22.12	9.35	5.71	11.69	11.69	13.87	12.38	13.59	19.33	14.93	39.06	15.59	8.24	41.33	10.41	11.36	11.36	39.06	29.15	11.31	6.40	39.06
MN	20.92	21.58	36.74	16.39	13.64	6.79	15.72	15.72	18.94	22.75	23.83	25.13	18.89	29.83	23.83	10.56	65.54	10.92	16.66	16.66	112.03	36.34	17.83	6.43	132.92
MS	15.88	14.44	19.71	11.45	9.13	7.19	10.97	10.97	11.49	12.48	7.98	13.65	12.76	23.19	17.21	7.92	31.71	19.53	10.85	10.85	42.19	33.21	9.10	7.01	42.19
MO	18.67	12.07	17.22	17.39	12.13	7.88	11.53	11.53	13.12	21.70	14.67	18.39	15.05	25.80	16.18	9.66	36.85	13.49	14.36	14.36	69.65	43.83	6.53	8.15	69.65
MT	22.03	10.06	20.00	12.80	10.83	5.54	17.20	17.20	11.51	16.19	19.02	14.58	12.03	76.71	13.82	8.93	59.20	9.44	9.73	9.73	41.23	17.93	6.54	5.33	41.23
NE	21.22	10.97	20.95	28.60	9.65	8.32	12.20	12.20	15.50	24.47	11.30	18.50	12.32	23.35	14.75	11.17	36.72	13.05	13.15	13.15	52.27	38.35	9.97	6.15	50.40
NV	17.85	9.06	13.89	10.87	10.44	7.67	11.47	11.47	14.46	17.06	7.20	12.16	11.24	13.14	12.28	10.71	23.68	20.06	16.04	16.04	36.06	33.10	9.57	7.25	44.29
NH	16.93	11.48	21.49	32.83	11.47	6.81	17.55	17.55	12.44	28.41	8.80	21.86	15.11	17.52	21.89	11.37	61.69	12.49	14.85	14.85	59.93	32.72	14.21	6.54	59.93
NJ	11.19	7.77	11.19	9.15	6.78	3.85	6.97	6.97	7.19	10.59	9.00	12.02	8.97	13.42	9.00	5.75	28.36	6.67	9.38	9.38	17.22	10.28	4.46	4.54	23.36
NM	22.18	6.84	14.66	12.96	8.24	6.41	8.27	8.27	12.43	11.53	6.24	12.92	10.22	16.44	9.38	7.70	28.79	10.65	19.01	19.01	43.96	21.54	5.89	6.30	43.96
NY	13.46	6.24	14.87	17.22	11.81	5.86	8.25	8.25	10.46	8.73	15.59	17.70	12.38	15.54	8.98	7.46	28.94	14.33	8.80	8.80	14.52	20.59	8.83	6.21	29.99
NC	13.68	10.74	18.04	13.14	6.63	7.71	8.44	8.44	10.19	11.90	7.68	9.44	9.58	15.60	15.16	8.08	25.69	10.77	8.00	8.00	41.99	18.79	7.12	4.14	41.99
ND	9.60	9.60	19.60	5.81	5.81	4.00	5.39	5.39	9.60	9.60	9.04	8.48	6.30	20.12	9.04	5.35	21.53	5.35	20.12	20.12	20.12	20.12	9.60	21.53	14.48
OH	6.07	12.00	9.68	12.78	9.99	5.55	8.44	8.44	9.99	12.41	2.48	11.45	15.04	25.09	4.63	6.40	21.67	8.11	10.30	10.30	26.21	16.39	8.70	15.16	26.21
OK	26.16	11.79	19.33	15.51	11.50	7.41	18.45	18.45	13.55	18.00	12.05	16.80	17.10	36.26	18.29	8.72	42.26	12.45	11.17	11.17	69.45	46.76	11.42	8.20	69.45
OR	17.06	9.31	16.98	13.61	8.81	5.00	10.55	10.55	16.83	8.87	7.87	15.13	14.15	16.07	11.45	6.07	22.99	11.17	10.25	10.25	35.39	16.67	10.80	5.00	35.39
PA	8.69	8.69	12.27	13.48	8.94	6.46	7.91	7.91	15.69	12.63	18.94	11.65	13.04	16.76	10.78	7.18	25.87	12.58	13.96	13.96	59.05	13.96	8.22	18.96	81.34
RI	19.53	11.65	18.07	18.23	16.24	4.43	10.38	10.38	12.85	22.78	11.97	25.11	24.13	37.66	17.25	8.52	33.92	10.29	14.07	14.07	59.49	37.50	14.36	7.70	78.79
SC	19.10	13.03	19.24	12.88	6.19	7.02	8.32	8.32	12.22	10.69	7.27	9.55	11.19	16.64	18.89	6.88	28.98	11.42	9.49	9.49	21.80	23.66	6.23	4.65	21.80
SD	15.40	7.40	15.04	18.64	5.48	5.64	11.42	11.42	9.88	13.70	7.72	8.28	13.08	25.78	11.65	8.56	20.66	8.33	11.93	11.93	45.92	17.38	6.33	4.17	45.92
TN	26.71	12.31	20.38	16.92	8.67	7.33	12.26	12.26	10.17	12.94	10.97	14.30	12.72	15.20	15.23	9.52	33.46	12.19	9.28	9.28	39.46	21.96	7.22	5.73	39.46
TX	16.89	11.33	13.25	12.10	8.85	7.36	10.49	10.49	11.61	15.68	9.53	13.68	9.89	13.95	20.99	8.06	22.80	14.58	11.23	11.23	33.39	14.78	7.29	8.09	17.86
UT	9.45	6.23	10.52	7.68	7.99	7.05	6.38	6.38	9.55	10.95	12.32	13.75	15.29	17.24	10.60	6.41	26.27	6.30	8.99	8.99	23.51	23.51	6.06	5.83	26.53
VT	19.90	11.34	19.46	34.13	12.01	6.10	10.61	10.61	21.25	25.54	10.85	19.53	10.41	23.91	18.23	10.10	28.73	12.62	13.94	13.94	47.05	37.23	8.52	9.59	47.05
VA	11.10	8.05	11.35	12.89	6.35	4.15	7.42	7.42	7.80	7.18	12.99	8.00	10.15	16.73	7.63	6.04	22.40	7.96	13.93	13.93	35.15	20.77	9.28	3.45	35.15
WA	7.57	7.57	8.66	6.22	7.51	2.85	8.31	8.31	13.01	7.88	8.66	12.82	9.07	18.93	10.38	4.68	19.63	4.02	12.96	12.96	9.25	9.25	9.89	10.23	9.25
WV	12.00	12.00	12.00	23.80	23.80	5.13	7.74	7.74	6.08	6.08	16.88	11.57	12.20	12.89	12.20	4.77	13.44	6.08	14.98	14.98	9.84	13.15	16.02	3.88	13.27
WI	11.29	9.54	18.46	11.96	10.19	4.78	7.16	7.16	14.43	14.33	10.61	17.44	11.44	19.25	12.41	6.40	40.73	7.27	16.38	16.38	39.18	19.60	13.98	5.38	39.18
WY	7.29	7.29	7.29	7.29	7.29	7.29	7.29	7.29	7.29	7.29	7.29	7.29	7.29	7.29	7.29	7.29	7.29	7.29	7.29	7.29	7.29	7.29	7.29	7.29	7.29
AVG.	16.06	10.91	18.51	15.79	9.94	6.40	10.34	10.34	13.82	15.24	10.71	14.97	12.89	22.94	14.62	7.78	31.75	11.09	12.61	12.61	38.86	24.78	9.63	7.27	40.51

R01100-060 Workers' Compensation (cont.) (Canada in Canadian dollars)

Province		Alberta	British Columbia	Manitoba	Ontario	New Brunswick	Newfndld. & Labrador	Northwest Territories	Nova Scotia	Prince Edward Island	Quebec	Saskat-chewan	Yukon
Carpentry—3 stories or less	Rate	2.72	4.63	3.80	5.00	4.40	7.20	4.62	8.06	7.25	12.69	5.01	3.25
	Code	25401	721028	40102	723	422	403	4-41	4226	401	80110	B12-02	202
Carpentry—interior cab. work	Rate	1.17	4.85	3.80	5.00	3.48	7.20	4.62	5.69	3.49	12.69	3.25	3.25
	Code	42133	721021	40102	723	427	403	4-41	4274	402	80110	B11-27	202
CARPENTRY—general	Rate	2.72	4.63	3.80	5.00	4.40	7.20	4.62	8.06	7.25	12.69	5.01	3.25
	Code	25401	721028	40102	723	422	403	4-41	4226	401	80110	B12-02	202
CONCRETE WORK—NOC	Rate	7.65	6.48	5.76	17.18	4.40	7.20	4.62	4.97	7.25	13.89	6.50	3.25
	Code	42104	721010	40110	748	422	403	4-41	4224	401	80100	B13-14	203
CONCRETE WORK—flat (flr. sidewalk)	Rate	7.65	6.48	5.76	17.18	4.40	7.20	4.62	4.97	7.25	13.89	6.50	3.25
	Code	42104	721010	40110	748	422	403	4-41	4224	401	80100	B13-14	203
ELECTRICAL Wiring—inside	Rate	2.98	2.90	2.03	3.03	1.74	4.83	3.46	2.61	3.49	5.69	3.25	2.35
	Code	42124	721019	40203	704	426	400	4-46	4261	402	80170	B11-05	206
EXCAVATION—earth NOC	Rate	2.57	3.40	3.80	4.21	2.93	7.20	3.46	4.25	3.61	7.71	3.70	3.25
	Code	40604	721031	40706	711	421	403	4-43	4214	404	80030	R11-06	207
EXCAVATION—rock	Rate	2.57	3.40	3.80	4.21	2.93	7.20	3.46	4.25	3.61	7.71	3.70	3.25
	Code	40604	721031	40706	711	421	403	4-43	4214	404	80030	R11-06	207
GLAZIERS	Rate	2.79	3.21	3.77	8.42	6.01	5.12	4.62	8.06	3.49	14.40	6.50	2.35
	Code	42121	715020	40109	751	423	402	4-41	4233	402	80150	B13-04	212
INSULATION WORK	Rate	2.89	5.80	3.80	8.42	6.01	5.12	4.62	8.06	7.25	14.15	5.01	3.25
	Code	42184	721029	40102	751	423	402	4-41	4234	401	80120	B12-07	202
LATHING	Rate	8.79	8.29	3.80	5.00	3.48	5.12	4.62	5.69	3.49	14.15	6.50	3.25
	Code	42135	721033	40102	723	427	402	4-41	4271	402	80120	B13-16	202
MASONRY	Rate	7.65	8.29	3.80	12.36	6.01	7.20	4.62	8.06	7.25	13.89	6.50	3.25
	Code	42102	721037	40102	741	423	403	4-41	4231	401	80100	B13-18	202
PAINTING & DECORATING	Rate	4.07	5.80	4.20	7.09	3.28	5.12	4.62	5.69	3.49	12.69	5.01	3.25
	Code	42111	721041	40105	719	427	402	4-41	4275	402	80110	B12-01	202
PILE DRIVING	Rate	7.65	13.55	3.80	5.84	4.40	10.89	3.46	4.97	7.25	7.71	6.50	3.25
	Code	42159	722004	40706	732	422	404	4-43	4221	401	80030	B13-10	202
PLASTERING	Rate	8.79	8.29	4.46	7.09	3.28	5.12	4.62	5.69	3.49	12.69	5.01	3.25
	Code	42135	721042	40108	719	427	402	4-41	4271	402	80110	B12-21	202
PLUMBING	Rate	1.99	4.30	2.37	3.96	2.30	4.60	3.46	2.94	3.49	7.13	3.25	1.40
	Code	42122	721043	40204	707	424	401	4-46	4241	402	80160	B11-01	214
ROOFING	Rate	9.70	8.74	5.42	12.36	8.05	7.20	4.62	8.06	7.25	22.17	6.50	3.25
	Code	42118	721036	40403	728	430	403	4-41	4236	401	80130	B13-20	202
SHEET METAL WORK (HVAC)	Rate	1.99	4.30	5.14	3.96	2.30	4.60	3.46	2.94	3.49	7.13	3.25	2.35
	Code	42117	721043	40402	707	424	401	4-46	4244	402	80160	B11-07	208
STEEL ERECTION—door & sash	Rate	3.84	13.55	6.56	17.18	4.40	10.89	4.62	8.06	7.25	32.47	6.50	3.25
	Code	42106	722005	40502	748	422	404	4-41	4227	401	80080	B13-22	202
STEEL ERECTION—inter., ornam.	Rate	3.84	13.55	6.56	17.18	4.40	7.20	4.62	8.06	7.25	32.47	6.50	3.25
	Code	42106	722005	40502	748	422	403	4-41	4227	401	80080	B13-22	202
STEEL ERECTION—structure	Rate	3.84	13.55	6.56	17.18	4.40	10.89	4.62	8.06	7.25	32.47	6.50	3.25
	Code	42106	722005	40502	748	422	404	4-41	4227	401	80080	B13-22	202
STEEL ERECTION—NOC	Rate	3.84	13.55	6.56	17.18	4.40	10.89	4.62	8.06	7.25	32.47	6.50	3.25
	Code	42106	722005	40502	748	422	404	4-41	4227	401	80080	B13-22	202
TILE WORK—inter. (ceramic)	Rate	2.94	3.27	1.88	7.09	3.28	5.12	4.62	5.69	3.49	12.69	6.50	3.25
	Code	42113	721054	40103	719	427	402	4-41	4276	402	80110	B13-01	202
WATERPROOFING	Rate	4.07	5.80	3.80	5.00	6.01	7.20	4.62	8.06	3.49	22.17	5.01	3.25
	Code	42139	721016	40102	723	423	403	4-41	4239	402	80130	B12-17	202
WRECKING	Rate	2.57	6.08	5.15	17.18	2.93	7.20	3.46	4.25	7.25	12.69	6.50	3.25
	Code	40604	721005	40106	748	421	403	4-43	4211	401	80110	B13-09	202

R01100-070 Contractor's Overhead & Profit

Below are the **average** installing contractor's percentage mark-ups applied to base labor rates to arrive at typical billing rates.

Column A: Labor rates are based on union wages averaged for 30 major U.S. cities. Base rates including fringe benefits are listed hourly and daily. These figures are the sum of the wage rate and employer-paid fringe benefits such as vacation pay, employer-paid health and welfare costs, pension costs, plus appropriate training and industry advancement funds costs.

Column B: Workers' Compensation rates are the national average of state rates established for each trade.

Column C: Column C lists average fixed overhead figures for all trades. Included are Federal and State Unemployment costs set at 6.2%; Social Security Taxes (FICA) set at 7.65%; Builder's Risk Insurance costs set at 0.44%; and Public Liability costs set at 2.02%. All the percentages except those for Social Security Taxes vary from state to state as well as from company to company.

Columns D and E: Percentages in Columns D and E are based on the presumption that the installing contractor has annual billing of $4,000,000 and up. Overhead percentages may increase with smaller annual billing. The overhead percentages for any given contractor may vary greatly and depend on a number of factors, such as the contractor's annual volume, engineering and logistical support costs, and staff requirements. The figures for overhead and profit will also vary depending on the type of job, the job location, and the prevailing economic conditions. All factors should be examined very carefully for each job.

Column F: Column F lists the total of Columns B, C, D, and E.

Column G: Column G is Column A (hourly base labor rate) multiplied by the percentage in Column F (O&P percentage).

Column H: Column H is the total of Column A (hourly base labor rate) plus Column G (Total O&P).

Column I: Column I is Column H multiplied by eight hours.

		A		B	C	D	E	F	G	H	I
		Base Rate Incl. Fringes		Work-ers' Comp. Ins.	Average Fixed Over-head	Over-head	Profit	Total Overhead & Profit		Rate with O & P	
Abbr.	Trade	Hourly	Daily					%	Amount	Hourly	Daily
Skwk	Skilled Workers Average (35 trades)	$33.65	$269.20	16.2%	16.3%	13.0%	10.0%	55.5%	$18.70	$52.35	$418.80
	Helpers Average (5 trades)	24.55	196.40	17.9		11.0		55.2	13.55	38.10	304.80
	Foreman Average, Inside ($.50 over trade)	34.15	273.20	16.2		13.0		55.5	18.95	53.10	424.80
	Foreman Average, Outside ($2.00 over trade)	35.65	285.20	16.2		13.0		55.5	19.80	55.45	443.60
Clab	Common Building Laborers	26.00	208.00	18.5		11.0		55.8	14.50	40.50	324.00
Asbe	Asbestos/Insulation Workers/Pipe Coverers	36.00	288.00	15.2		16.0		57.5	20.70	56.70	453.60
Boil	Boilermakers	41.25	330.00	13.1		16.0		55.4	22.85	64.10	512.80
Bric	Bricklayers	34.25	274.00	15.0		11.0		52.3	17.90	52.15	417.20
Brhe	Bricklayer Helpers	25.60	204.80	15.0		11.0		52.3	13.40	39.00	312.00
Carp	Carpenters	33.00	264.00	18.5		11.0		55.8	18.40	51.40	411.20
Cefi	Cement Finishers	31.55	252.40	9.9		11.0		47.2	14.90	46.45	371.60
Elec	Electricians	39.40	315.20	6.4		16.0		48.7	19.20	58.60	468.80
Elev	Elevator Constructors	41.60	332.80	7.2		16.0		49.5	20.60	62.20	497.60
Eqhv	Equipment Operators, Crane or Shovel	34.80	278.40	10.3		14.0		50.6	17.60	52.40	419.20
Eqmd	Equipment Operators, Medium Equipment	33.65	269.20	10.3		14.0		50.6	17.05	50.70	405.60
Eqlt	Equipment Operators, Light Equipment	32.15	257.20	10.3		14.0		50.6	16.25	48.40	387.20
Eqol	Equipment Operators, Oilers	29.20	233.60	10.3		14.0		50.6	14.80	44.00	352.00
Eqmm	Equipment Operators, Master Mechanics	35.20	281.60	10.3		14.0		50.6	17.80	53.00	424.00
Glaz	Glaziers	32.05	256.40	13.8		11.0		51.1	16.40	48.45	387.60
Lath	Lathers	30.55	244.40	10.7		11.0		48.0	14.65	45.20	361.60
Marb	Marble Setters	31.95	255.60	15.0		11.0		52.3	16.70	48.65	389.20
Mill	Millwrights	34.35	274.80	10.3		11.0		47.6	16.35	50.70	405.60
Mstz	Mosaic and Terrazzo Workers	31.60	252.80	9.6		11.0		46.9	14.80	46.40	371.20
Pord	Painters, Ordinary	29.60	236.80	12.9		11.0		50.2	14.85	44.45	355.60
Psst	Painters, Structural Steel	30.05	240.40	50.0		11.0		87.3	26.25	56.30	450.40
Pape	Paper Hangers	29.35	234.80	12.9		11.0		50.2	14.75	44.10	352.80
Pile	Pile Drivers	32.05	256.40	22.9		16.0		65.2	20.90	52.95	423.60
Plas	Plasterers	29.95	239.60	14.6		11.0		51.9	15.55	45.50	364.00
Plah	Plasterer Helpers	25.80	206.40	14.6		11.0		51.9	13.40	39.20	313.60
Plum	Plumbers	39.60	316.80	7.8		16.0		50.1	19.85	59.45	475.60
Rodm	Rodmen (Reinforcing)	37.10	296.80	24.8		14.0		65.1	24.15	61.25	490.00
Rofc	Roofers, Composition	28.45	227.60	31.8		11.0		69.1	19.65	48.10	384.80
Rots	Roofers, Tile and Slate	28.65	229.20	31.8		11.0		69.1	19.80	48.45	387.60
Rohe	Roofer Helpers (Composition)	20.95	167.60	31.8		11.0		69.1	14.50	35.45	283.60
Shee	Sheet Metal Workers	38.80	310.40	11.1		16.0		53.4	20.70	59.50	476.00
Spri	Sprinkler Installers	39.25	314.00	9.0		16.0		51.3	20.15	59.40	475.20
Stpi	Steamfitters or Pipefitters	39.75	318.00	7.8		16.0		50.1	19.90	59.65	477.20
Ston	Stone Masons	33.85	270.80	15.0		11.0		52.3	17.70	51.55	412.40
Sswk	Structural Steel Workers	37.15	297.20	38.9		14.0		79.2	29.40	66.55	532.40
Tilf	Tile Layers	31.50	252.00	9.6		11.0		46.9	14.75	46.25	370.00
Tilh	Tile Layer Helpers	24.45	195.60	9.6		11.0		46.9	11.45	35.90	287.20
Trlt	Truck Drivers, Light	25.70	205.60	15.1		11.0		52.4	13.45	39.15	313.20
Trhv	Truck Drivers, Heavy	26.45	211.60	15.1		11.0		52.4	13.85	40.30	322.40
Sswl	Welders, Structural Steel	37.15	297.20	38.9	▼	14.0	▼	79.2	29.40	66.55	532.40
Wrck	*Wrecking	26.00	208.00	40.5		11.0		77.8	20.25	46.25	370.00

*Not included in Averages.

GENERAL REQUIREMENTS

REFERENCE NOS.

R01100-080 Performance Bond

This table shows the cost of a Performance Bond for a construction job scheduled to be completed in 12 months. Add 1% of the premium cost per month for jobs requiring more than 12 months to complete. The rates are "standard" rates offered to contractors that the bonding company considers financially sound and capable of doing the work. Preferred rates are offered by some bonding companies based upon financial strength of the contractor. Actual rates vary from contractor to contractor and from bonding company to bonding company. Contractors should prequalify through a bonding agency before submitting a bid on a contract that requires a bond.

Contract Amount		Building Construction Class B Projects			Highways & Bridges						
					Class A New Construction			Class A-1 Highway Resurfacing			
First $ 100,000 bid		$25.00 per M			$15.00 per M			$9.40 per M			
Next 400,000 bid		$ 2,500	plus	$15.00 per M	$ 1,500	plus	$10.00 per M	$ 940	plus	$7.20 per M	
Next 2,000,000 bid		8,500	plus	10.00 per M	5,500	plus	7.00 per M	3,820	plus	5.00 per M	
Next 2,500,000 bid		28,500	plus	7.50 per M	19,500	plus	5.50 per M	15,820	plus	4.50 per M	
Next 2,500,000 bid		47,250	plus	7.00 per M	33,250	plus	5.00 per M	28,320	plus	4.50 per M	
Over 7,500,000 bid		64,750	plus	6.00 per M	45,750	plus	4.50 per M	39,570	plus	4.00 per M	

R01100-090 Sales Tax by State

State sales tax on materials is tabulated below (5 states have no sales tax). Many states allow local jurisdictions, such as a county or city, to levy additional sales tax.

Some projects may be sales tax exempt, particularly those constructed with public funds.

State	Tax (%)	State	Tax (%)	State	Tax (%)	State	Tax (%)
Alabama	4	Illinois	6.25	Montana	0	Rhode Island	7
Alaska	0	Indiana	5	Nebraska	5	South Carolina	5
Arizona	5	Iowa	5	Nevada	6.5	South Dakota	4
Arkansas	4.625	Kansas	4.9	New Hampshire	0	Tennessee	6
California	6	Kentucky	6	New Jersey	6	Texas	6.25
Colorado	3	Louisiana	4	New Mexico	5	Utah	4.75
Connecticut	6	Maine	5	New York	4	Vermont	5
Delaware	0	Maryland	5	North Carolina	4	Virginia	3.5
District of Columbia	5.75	Massachusetts	5	North Dakota	5	Washington	6.5
Florida	6	Michigan	6	Ohio	5	West Virginia	6
Georgia	4	Minnesota	6.5	Oklahoma	4.5	Wisconsin	5
Hawaii	4	Mississippi	7	Oregon	0	Wyoming	4
Idaho	5	Missouri	4.225	Pennsylvania	6	Average	4.65 %

Sales Tax by Province (Canada)

GST - a value-added tax, which the government imposes on most goods and services provided in or imported into Canada.
PST - a retail sales tax, which five of the provinces impose on the price of most goods and some services.

QST - a value-added tax, similar to the federal GST, which Quebec imposes.
HST - Three provinces have combined their retail sales tax with the federal GST into one harmonized tax.

Province	PST (%)	QST (%)	GST(%)	HST(%)
Alberta	0	0	7	0
British Columbia	7.5	0	7	0
Manitoba	7	0	7	0
New Brunswick	0	0	0	15
Newfoundland	0	0	0	15
Northwest Territories	0	0	7	0
Nova Scotia	0	0	0	15
Ontario	8	0	7	0
Prince Edward Island	10	0	7	0
Quebec	0	7.5	7	0
Saskatchewan	6	0	7	0
Yukon	0	0	7	0

R01100-100 Unemployment Taxes and Social Security Taxes

Mass. State Unemployment tax ranges from 1.325% to 7.225% plus an experience rating assessment the following year, on the first $10,800 of wages. Federal Unemployment tax is 6.2% of the first $7,000 of wages. This is reduced by a credit for payment to the state. The minimum Federal Unemployment tax is 0.8% after all credits.

Combined rates in Mass. thus vary from 2.125% to 8.025% of the first $10,800 of wages. Combined average U.S. rate is about 6.2% of the first $7,000. Contractors with permanent workers will pay less since the average annual wages for skilled workers is $33.65 x 2,000 hours or about $67,300 per year. The average combined rate for U.S. would thus be 6.2% x $7,000 ÷ $67,300 = 0.6% of total wages for permanent employees.

Rates vary not only from state to state but also with the experience rating of the contractor.

Social Security (FICA) for 2004 is estimated at time of publication to be 7.65% of wages up to $87,000.

R01100-110   Overtime

One way to improve the completion date of a project or eliminate negative float from a schedule is to compress activity duration times. This can be achieved by increasing the crew size or working overtime with the proposed crew.

To determine the costs of working overtime to compress activity duration times, consider the following examples. Below is an overtime efficiency and cost chart based on a five, six, or seven day week with an eight through twelve hour day. Payroll percentage increases for time and one half and double time are shown for the various working days.

Days per Week	Hours per Day	Production Efficiency					Payroll Cost Factors	
		1 Week	2 Weeks	3 Weeks	4 Weeks	Average 4 Weeks	@ 1-1/2 Times	@ 2 Times
	8	100%	100%	100%	100%	100 %	100 %	100 %
	9	100	100	95	90	96.25	105.6	111.1
5	10	100	95	90	85	91.25	110.0	120.0
	11	95	90	75	65	81.25	113.6	127.3
	12	90	85	70	60	76.25	116.7	133.3
	8	100	100	95	90	96.25	108.3	116.7
	9	100	95	90	85	92.50	113.0	125.9
6	10	95	90	85	80	87.50	116.7	133.3
	11	95	85	70	65	78.75	119.7	139.4
	12	90	80	65	60	73.75	122.2	144.4
	8	100	95	85	75	88.75	114.3	128.6
	9	95	90	80	70	83.75	118.3	136.5
7	10	90	85	75	65	78.75	121.4	142.9
	11	85	80	65	60	72.50	124.0	148.1
	12	85	75	60	55	68.75	126.2	152.4

R01100-710   Unit Gross Area Requirements

The figures in the table below indicate typical ranges in square feet as a function of the "occupant" unit. This table is best used in the preliminary design stages to help determine the probable size requirement for the total project. See R17100-100 for the typical total size ranges for various types of buildings.

Building Type	Unit	Gross Area in S.F.		
		1/4	Median	3/4
Apartments	Unit	660	860	1,100
Auditorium & Play Theaters	Seat	18	25	38
Bowling Alleys	Lane		940	
Churches & Synagogues	Seat	20	28	39
Dormitories	Bed	200	230	275
Fraternity & Sorority Houses	Bed	220	315	370
Garages, Parking	Car	325	355	385
Hospitals	Bed	685	850	1,075
Hotels	Rental Unit	475	600	710
Housing for the elderly	Unit	515	635	755
Housing, Public	Unit	700	875	1,030
Ice Skating Rinks	Total	27,000	30,000	36,000
Motels	Rental Unit	360	465	620
Nursing Homes	Bed	290	350	450
Restaurants	Seat	23	29	39
Schools, Elementary	Pupil	65	77	90
Junior High & Middle		85	110	129
Senior High		102	130	145
Vocational	↓	110	135	195
Shooting Ranges	Point		450	
Theaters & Movies	Seat		15	

R01107-030 Engineering Fees

Typical **Structural Engineering Fees** based on type of construction and total project size. These fees are included in Architectural Fees.

Type of Construction	Total Project Size (in thousands of dollars)			
	$500	$500-$1,000	$1,000-$5,000	Over $5000
Industrial buildings, factories & warehouses	Technical payroll times 2.0 to 2.5	1.60%	1.25%	1.00%
Hotels, apartments, offices, dormitories, hospitals, public buildings, food stores		2.00%	1.70%	1.20%
Museums, banks, churches and cathedrals		2.00%	1.75%	1.25%
Thin shells, prestressed concrete, earthquake resistive		2.00%	1.75%	1.50%
Parking ramps, auditoriums, stadiums, convention halls, hangars & boiler houses		2.50%	2.00%	1.75%
Special buildings, major alterations, underpinning & future expansion		Add to above 0.5%	Add to above 0.5%	Add to above 0.5%

For complex reinforced concrete or unusually complicated structures, add 20% to 50%.

Typical **Mechanical and Electrical Engineering Fees** are based on the size of the subcontract. The fee structure for Mechanical Engineering is shown below; Electrical Engineering fees range from 2 to 5%. These fees are included in Architectural Fees.

Type of Construction	Subcontract Size							
	$25,000	$50,000	$100,000	$225,000	$350,000	$500,000	$750,000	$1,000,000
Simple structures	6.4%	5.7%	4.8%	4.5%	4.4%	4.3%	4.2%	4.1%
Intermediate structures	8.0	7.3	6.5	5.6	5.1	5.0	4.9	4.8
Complex structures	12.0	9.0	9.0	8.0	7.5	7.5	7.0	7.0

For renovations, add 15% to 25% to applicable fee.

R01250-010   Repair and Remodeling

Cost figures are based on new construction utilizing the most cost-effective combination of labor, equipment and material with the work scheduled in proper sequence to allow the various trades to accomplish their work in an efficient manner.

The costs for repair and remodeling work must be modified due to the following factors that may be present in any given repair and remodeling project.

1. Equipment usage curtailment due to the physical limitations of the project, with only hand-operated equipment being used.

2. Increased requirement for shoring and bracing to hold up the building while structural changes are being made and to allow for temporary storage of construction materials on above-grade floors.

3. Material handling becomes more costly due to having to move within the confines of an enclosed building. For multi-story construction, low capacity elevators and stairwells may be the only access to the upper floors.

4. Large amount of cutting and patching and attempting to match the existing construction is required. It is often more economical to remove entire walls rather than create many new door and window openings. This sort of trade-off has to be carefully analyzed.

5. Cost of protection of completed work is increased since the usual sequence of construction usually cannot be accomplished.

6. Economies of scale usually associated with new construction may not be present. If small quantities of components must be custom fabricated due to job requirements, unit costs will naturally increase. Also, if only small work areas are available at a given time, job scheduling between trades becomes difficult and subcontractor quotations may reflect the excessive start-up and shut-down phases of the job.

7. Work may have to be done on other than normal shifts and may have to be done around an existing production facility which has to stay in production during the course of the repair and remodeling.

8. Dust and noise protection of adjoining non-construction areas can involve substantial special protection and alter usual construction methods.

9. Job may be delayed due to unexpected conditions discovered during demolition or removal. These delays ultimately increase construction costs.

10. Piping and ductwork runs may not be as simple as for new construction. Wiring may have to be snaked through walls and floors.

11. Matching "existing construction" may be impossible because materials may no longer be manufactured. Substitutions may be expensive.

12. Weather protection of existing structure requires additional temporary structures to protect building at openings.

13. On small projects, because of local conditions, it may be necessary to pay a tradesman for a minimum of four hours for a task that is completed in one hour.

All of the above areas can contribute to increased costs for a repair and remodeling project. Each of the above factors should be considered in the planning, bidding and construction stage in order to minimize the increased costs associated with repair and remodeling jobs.

R01540-100   Steel Tubular Scaffolding

On new construction, tubular scaffolding is efficient up to 60′ high or five stories. Above this it is usually better to use a hung scaffolding if construction permits. Swing scaffolding operations may interfere with tenants. In this case, the tubular is more practical at all heights.

In repairing or cleaning the front of an existing building the cost of tubular scaffolding per S.F. of building front increases as the height increases above the first tier. The first tier cost is relatively high due to leveling and alignment.

The minimum efficient crew for erection is three workers. For heights over 50′, a crew of four is more efficient. Use two or more on top and two at the bottom for handing up or hoisting. Four workers can erect and dismantle about nine frames per hour up to five stories. From five to eight stories they will average six frames per hour. With 7′ horizontal spacing this will run about 400 S.F. and 265 S.F. of wall surface, respectively. Time for placing planks must be added to the above. On heights above 50′, five planks can be placed per labor-hour.

The table below shows the number of pieces required to erect tubular steel scaffolding for 1000 S.F. of building frontage. This area is made up of a scaffolding system that is 12 frames (11 bays) long by 2 frames high.

For jobs under twenty-five frames, add 50% to rental cost. Rental rates will be lower for jobs over three months duration. Large quantities for long periods can reduce rental rates by 20%.

Description of Component	CSI Line Item	Number of Pieces for 1000 S.F. of Building Front	Unit
5′ Wide Standard Frame, 6′-4″ High	01540-750-2200	24	Ea.
Leveling Jack & Plate	01540-750-2650	24	
Cross Brace	01540-750-2500	44	
Side Arm Bracket, 21″	01540-750-2700	12	
Guardrail Post	01540-750-2550	12	
Guardrail, 7′ section	01540-750-2600	22	
Stairway Section	01540-750-2900	2	
Stairway Starter Bar	01540-750-2910	1	
Stairway Inside Handrail	01540-750-2920	2	
Stairway Outside Handrail	01540-750-2930	2	
Walk-Thru Frame Guardrail	01540-750-2940	2	

Scaffolding is often used as falsework over 15′ high during construction of cast-in-place concrete beams and slabs. Two foot wide scaffolding is generally used for heavy beam construction. The span between frames depends upon the load to be carried with a maximum span of 5′.

Heavy duty shoring frames with a capacity of 10,000#/leg can be spaced up to 10′ O.C. depending upon form support design and loading.

Scaffolding used as horizontal shoring requires less than half the material required with conventional shoring.

On new construction, erection is done by carpenters.

Rolling towers supporting horizontal shores can reduce labor and speed the job. For maintenance work, catwalks with spans up to 70′ can be supported by the rolling towers.

R01590-100 Contractor Equipment

Rental Rates shown in Division 01590 pertain to late model high quality machines in excellent working condition, rented from equipment dealers. Rental rates from contractors may be substantially lower than the rental rates from equipment dealers depending upon economic conditions; for older, less productive machines, reduce rates by a maximum of 15%. Any overtime must be added to the base rates. For shift work, rates are lower. Usual rule of thumb is 150% of one shift rate for two shifts; 200% for three shifts.

For periods of less than one week, operated equipment is usually more economical to rent than renting bare equipment and hiring an operator.

Costs to move equipment to a job site (mobilization) or from a job site (demobilization) are not included in rental rates in Division 01590, nor in any Equipment costs on any Unit Price line items or crew listings. These costs can be found in section 02305-250. If a piece of equipment is already at a job site, it is not appropriate to utilize mob/demob costs in an estimate again.

Rental rates vary throughout the country with larger cities generally having lower rates. Lease plans for new equipment are available for periods in excess of six months with a percentage of payments applying toward purchase.

Monthly rental rates vary from 2% to 5% of the cost of the equipment depending on the anticipated life of the equipment and its wearing parts. Weekly rates are about 1/3 the monthly rates and daily rental rates about 1/3 the weekly rate.

The hourly operating costs for each piece of equipment include costs to the user such as fuel, oil, lubrication, normal expendables for the equipment, and a percentage of mechanic's wages chargeable to maintenance. The hourly operating costs listed do not include the operator's wages.

The daily cost for equipment used in the standard crews is figured by dividing the weekly rate by five, then adding eight times the hourly operating cost to give the total daily equipment cost, not including the operator. This figure is in the right hand column of Division 01590 under Crew Equipment Cost/Day.

Pile Driving rates shown for pile hammer and extractor do not include leads, crane, boiler or compressor. Vibratory pile driving requires an added field specialist during set-up and pile driving operation for the electric model. The hydraulic model requires a field specialist for set-up only. Up to 125 reuses of sheet piling are possible using vibratory drivers. For normal conditions, crane capacity for hammer type and size are as follows.

Crane Capacity	Hammer Type and Size		
	Air or Steam	Diesel	Vibratory
25 ton	to 8,750 ft.-lb.		70 H.P.
40 ton	15,000 ft.-lb.	to 32,000 ft.-lb.	170 H.P.
60 ton	25,000 ft.-lb.		300 H.P.
100 ton		112,000 ft.-lb.	

Cranes should be specified for the job by size, building and site characteristics, availability, performance characteristics, and duration of time required.

Backhoes & Shovels rent for about the same as equivalent size cranes but maintenance and operating expense is higher. Crane operators rate must be adjusted for high boom heights. Average adjustments: for 150' boom add 2% per hour; over 185', add 4% per hour; over 210', add 6% per hour; over 250', add 8% per hour and over 295', add 12% per hour.

Tower Cranes of the climbing or static type have jibs from 50' to 200' and capacities at maximum reach range from 4,000 to 14,000 pounds. Lifting capacities increase up to maximum load as the hook radius decreases.

Typical rental rates, based on purchase price are about 2% to 3% per month.

Erection and dismantling runs between 500 and 2000 labor hours. Climbing operation takes 10 labor hours per 20' climb. Crane dead time is about 5 hours per 40' climb. If crane is bolted to side of the building add cost of ties and extra mast sections. Climbing cranes have from 80' to 180' of mast while static cranes have 80' to 800' of mast.

Truck Cranes can be converted to tower cranes by using tower attachments. Mast heights over 400' have been used. See Division 01590-600 for rental rates of high boom cranes.

A single 100' high material **Hoist and Tower** can be erected and dismantled in about 400 labor hours; a double 100' high hoist and tower in about

600 labor hours. Erection times for additional heights are 3 and 4 labor hours per vertical foot respectively up to 150', and 4 to 5 labor hours per vertical foot over 150'. A 40' high portable Buck hoist takes about 160 labor hours to erect and dismantle. Additional heights take 2 labor hours per vertical foot to 80' and 3 labor hours per vertical foot for the next 100'. Most material hoists do not meet local code requirements for carrying personnel.

A 150' high **Personnel Hoist** requires about 500 to 800 labor hours to erect and dismantle. Budget erection time at 5 labor hours per vertical foot for all trades. Local code requirements or labor scarcity requiring overtime can add up to 50% to any of the above erection costs.

Earthmoving Equipment: The selection of earthmoving equipment depends upon the type and quantity of material, moisture content, haul distance, haul road, time available, and equipment available. Short haul cut and fill operations may require dozers only, while another operation may require excavators, a fleet of trucks, and spreading and compaction equipment. Stockpiled material and granular material are easily excavated with front end loaders. Scrapers are most economically used with hauls between 300' and 1-1/2 miles if adequate haul roads can be maintained. Shovels are often used for blasted rock and any material where a vertical face of 8' or more can be excavated. Special conditions may dictate the use of draglines, clamshells, or backhoes. Spreading and compaction equipment must be matched to the soil characteristics, the compaction required and the rate the fill is being supplied.

GENERAL REQUIREMENTS

REFERENCE NOS.

R02580-300 Concrete for Conduit Encasement

Table below lists C.Y. of concrete for 100 L.F. of trench. Conduits separation center to center should meet 7.5″ (N.E.C.).

Number of Conduits	1	2	3	4	6	8	9	Number of Conduits
Trench Dimension	11.5″ x 11.5″	11.5″ x 19″	11.5″ x 27″	19″ x 19″	19″ x 27″	19″ x 38″	27″ x 27″	Trench Dimension
Conduit Diameter 2.0″	3.29	5.39	7.64	8.83	12.51	17.66	17.72	Conduit Diameter 2.0″
2.5″	3.23	5.29	7.49	8.62	12.19	17.23	17.25	2.5″
3.0″	3.15	5.13	7.24	8.29	11.71	16.59	16.52	3.0″
3.5″	3.08	4.97	7.02	7.99	11.26	15.98	15.84	3.5″
4.0″	2.99	4.80	6.76	7.65	10.74	15.30	15.07	4.0″
5.0″	2.78	4.37	6.11	6.78	9.44	13.57	13.12	5.0″
6.0″	2.52	3.84	5.33	5.74	7.87	11.48	10.77	6.0″

R07800-030 Firestopping

Firestopping is the sealing of structural, mechanical, electrical and other penetrations through fire-rated assemblies. The basic components of firestop systems are safing insulation and firestop sealant on both sides of wall penetrations and the top side of floor penetrations.

Pipe penetrations are assumed to be through concrete, grout, or joint compound and can be sleeved or unsleeved. Costs for the penetrations and sleeves are not included. An annular space of 1″ is assumed. Escutcheons are not included.

Metallic pipe is assumed to be copper, aluminum, cast iron or similar metallic material. Insulated metallic pipe is assumed to be covered with a thermal insulating jacket of varying thickness and materials.

Non-metallic pipe is assumed to be PVC, CPVC, FR Polypropylene or similar plastic piping material. Intumescent firestop sealant or wrap strips are included. Collars on both sides of wall penetrations and a sheet metal plate on the underside of floor penetrations are included.

Ductwork is assumed to be sheet metal, stainless steel or similar metallic material. Duct penetrations are assumed to be through concrete, grout or joint compound. Costs for penetrations and sleeves are not included. An annular space of 1/2″ is assumed.

Multi-trade openings include costs for sheet metal forms, firestop mortar, wrap strips, collars and sealants as necessary.

Structural penetrations joints are assumed to be 1/2″ or less. CMU walls are assumed to be within 1-1/2″ of metal deck. Drywall walls are assumed to be tight to the underside of metal decking.

Metal panel, glass or curtain wall systems include a spandrel area of 5′ filled with mineral wool foil-faced insulation. Fasteners and stiffeners are included.

7

THERMAL & MOISTURE PROTECTION

REFERENCE NOS.

R15100-110 Hot Water Consumption Rates

Type of Building	Size Factor	Maximum Hourly Demand	Average Day Demand
Apartment Dwellings	No. of Apartments: Up to 20 21 to 50 51 to 75 76 to 100 101 to 200 201 up	 12.0 Gal. per apt. 10.0 Gal. per apt. 8.5 Gal. per apt. 7.0 Gal. per apt. 6.0 Gal. per apt. 5.0 Gal. per apt.	 42.0 Gal. per apt. 40.0 Gal. per apt. 38.0 Gal. per apt. 37.0 Gal. per apt. 36.0 Gal. per apt. 35.0 Gal. per apt.
Dormitories	Men Women	3.8 Gal. per man 5.0 Gal. per woman	13.1 Gal. per man 12.3 Gal. per woman
Hospitals	Per bed	23.0 Gal. per patient	90.0 Gal. per patient
Hotels	Single room with bath Double room with bath	17.0 Gal. per unit 27.0 Gal. per unit	50.0 Gal. per unit 80.0 Gal. per unit
Motels	No. of units: Up to 20 21 to 100 101 Up	 6.0 Gal. per unit 5.0 Gal. per unit 4.0 Gal. per unit	 20.0 Gal. per unit 14.0 Gal. per unit 10.0 Gal. per unit
Nursing Homes		4.5 Gal. per bed	18.4 Gal. per bed
Office buildings		0.4 Gal. per person	1.0 Gal. per person
Restaurants	Full meal type Drive-in snack type	1.5 Gal./max. meals/hr. 0.7 Gal./max. meals/hr.	2.4 Gal. per meal 0.7 Gal. per meal
Schools	Elementary Secondary & High	0.6 Gal. per student 1.0 Gal. per student	0.6 Gal. per student 1.8 Gal. per student

For evaluation purposes, recovery rate and storage capacity are inversely proportional. Water heaters should be sized so that the maximum hourly demand anticipated can be met in addition to allowance for the heat loss from the pipes and storage tank.

R15100-120 Fixture Demands in Gallons Per Fixture Per Hour

Table below is based on 140°F final temperature except for dishwashers in public places (*) where 180°F water is mandatory.

Fixture	Apartment House	Club	Gym	Hospital	Hotel	Indust. Plant	Office	Private Home	School
Bathtubs	20	20	30	20	20			20	
Dishwashers, automatic	15	50-150*		50-150*	50-200*	20-100*		15	20-100*
Kitchen sink	10	20		20	30	20	20	10	20
Laundry, stationary tubs	20	28		28	28			20	
Laundry, automatic wash	75	75		100	150			75	
Private lavatory	2	2	2	2	2	2	2	2	2
Public lavatory	4	6	8	6	8	12	6		15
Showers	30	150	225	75	75	225	30	30	225
Service sink	20	20		20	30	20	20	15	20
Demand factor	0.30	0.30	0.40	0.25	0.25	0.40	0.30	0.30	0.40
Storage capacity factor	1.25	0.90	1.00	0.60	0.80	1.00	2.00	0.70	1.00

To obtain the probable maximum demand multiply the total demands for the fixtures (gal./fixture/hour) by the demand factor. The heater should have a heating capacity in gallons per hour equal to this maximum. The storage tank should have a capacity in gallons equal to the probable maximum demand multiplied by the storage capacity factor.

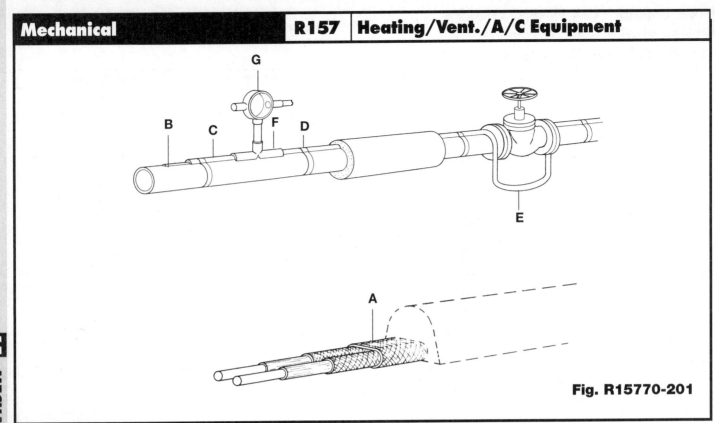

Fig. R15770-201

R15770-200   Heat Trace Systems

Before you can determine the cost of a HEAT TRACE installation
the method of attachment must be established. There are (4) common
methods:

1. Cable is simply attached to the pipe with polyester tape
   every 12′.
2. Cable is attached with a continuous cover of 2″ wide
   aluminum tape.
3. Cable is attached with factory extruded heat transfer cement
   and covered with metallic raceway with clips every 10′.
4. Cable is attached between layers of pipe insulation using either clips or
   polyester tape.

Example: Components for method 3 must include:

A. Heat trace cable by voltage and watts per linear foot.
B. Heat transfer cement, 1 gallon per 60 linear feet of cover.
C. Metallic raceway by size and type.
D. Raceway clips by size of pipe.

When taking off linear foot lengths of cable add the following for each valve
in the system. (E)

In all of the above methods each component of the system must be priced
individually.

SCREWED OR WELDED VALVE:			FLANGED VALVE:			BUTTERFLY VALVES:		
1/2″	=	6″	1/2″	=	1′ -0″	1/2″	=	0′
3/4″	=	9″	3/4″	=	1′ -6″	3/4″	=	0′
1″	=	1′ -0″	1″	=	2′ -0″	1″	=	1′ -0″
1-1/2″	=	1′ -6″	1-1/2″	=	2′ -6″	1-1/2″	=	1′ -6″
2″	=	2′	2″	=	2′ -6″	2″	=	2′ -0″
2-1/2″	=	2′ -6″	2-1/2″	=	3′ -0″	2-1/2″	=	2′ -6″
3″	=	2′ -6″	3″	=	3′ -6″	3″	=	2′ -6″
4″	=	4′ -0″	4″	=	4′ -0″	4″	=	3′ -0″
6″	=	7′ -0″	6″	=	8′ -0″	6″	=	3′ -6″
8″	=	9′ -6″	8″	=	11′ -0″	8″	=	4′ -0″
10″	=	12′ -6″	10″	=	14′ -0″	10″	=	4′ -0″
12″	=	15′ -0″	12″	=	16′ -6″	12″	=	5′ -0″
14″	=	18′ -0″	14″	=	19′ -6″	14″	=	5′ -6″
16″	=	21′ -6″	16″	=	23′ -0″	16″	=	6′ -0″
18″	=	25′ -6″	18″	=	27′ -0″	18″	=	6′ -6″
20″	=	28′ -6″	20″	=	30′ -0″	20″	=	7′ -0″
24″	=	34′ -0″	24″	=	36′ -0″	24″	=	8′ -0″
30″	=	40′ -0″	30″	=	42′ -0″	30″	=	10′ -0″

R15770-200 Heat Trace Systems (cont.)

Add the following quantities of heat transfer cement to linear foot totals for each valve:

Nominal Valve Size	Gallons of Cement per Valve
1/2"	0.14
3/4"	0.21
1"	0.29
1-1/2"	0.36
2"	0.43
2-1/2"	0.70
3"	0.71
4"	1.00
6"	1.43
8"	1.48
10"	1.50
12"	1.60
14"	1.75
16"	2.00
18"	2.25
20"	2.50
24"	3.00
30"	3.75

The following must be added to the list of components to accurately price HEAT TRACE systems:

1. Expediter fitting and clamp fasteners (F)
2. Junction box and nipple connected to expediter fitting (G)
3. Field installed terminal blocks within junction box
4. Ground lugs
5. Piping from power source to expediter fitting
6. Controls
7. Thermostats
8. Branch wiring
9. Cable splices
10. End of cable terminations
11. Branch piping fittings and boxes

Deduct the following percentages from labor if cable lengths in the same area exceed:

150' to 250'	10%	351' to 500'	20%
251' to 350'	15%	Over 500'	25%

Add the following percentages to labor for elevated installations:

15' to 20' high	10%	31' to 35' high	40%
21' to 25' high	20%	36' to 40' high	50%
26' to 30' high	30%	Over 40' high	60%

R15770-201 Spiral-Wrapped Heat Trace Cable (Pitch Table)

In order to increase the amount of heat, occasionally heat trace cable is wrapped in a spiral fashion around a pipe; increasing the number of feet of heater cable per linear foot of pipe.

Engineers first determine the heat loss per foot of pipe (based on the insulating material, its thickness, and the temperature differential across it). A ratio is then calculated by the formula:

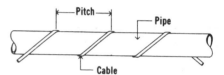

$$\text{Feet of Heat Trace per Foot of Pipe} = \frac{\text{Watts/Foot of Heat Loss}}{\text{Watts/Foot of the Cable}}$$

The linear distance between wraps (pitch) is then taken from a chart or table. Generally, the pitch is listed on a drawing leaving the estimator to calculate the total length of heat tape required. An approximation may be taken from this table.

Feet of Heat Trace Per Foot of Pipe																	
	Nominal Pipe Size in Inches																
Pitch In Inches	1	1¼	1½	2	2½	3	4	6	8	10	12	14	16	18	20	24
3.5	1.80															
4	1.65															
5	1.46	1.60	1.80													
6	1.34	1.45	1.55	1.75												
7	1.25	1.35	1.43	1.57	1.75											
8	1.20	1.28	1.34	1.45	1.60	1.80										
9	1.16	1.23	1.28	1.37	1.51	1.68										
10	1.13	1.19	1.24	1.32	1.44	1.57	1.82									
15	1.06	1.08	1.10	1.15	1.21	1.29	1.42	1.78								
20	1.04	1.05	1.06	1.08	1.13	1.17	1.25	1.49	1.73							
25		1.04	1.04	1.06	1.08	1.11	1.17	1.33	1.51	1.72						
30				1.04	1.05	1.07	1.12	1.24	1.37	1.54	1.70					
35					1.06	1.06	1.09	1.17	1.28	1.42	1.54	1.80				
40						1.05	1.07	1.14	1.22	1.33	1.44	1.64	1.78			
50							1.05	1.09	1.15	1.22	1.29	1.52	1.64	1.75		
60								1.06	1.11	1.16	1.21	1.35	1.44	1.53	1.64	1.83
70								1.05	1.08	1.12	1.17	1.25	1.31	1.39	1.46	1.62
80									1.06	1.09	1.13	1.19	1.24	1.30	1.35	1.47
90									1.04	1.06	1.10	1.15	1.19	1.24	1.28	1.38
100										1.05	1.08	1.10	1.16	1.19	1.23	1.32
												1.13	1.15	1.19	1.23

Note: Common practice would normally limit the lower end of the table to 5% of additional heat and above 80% an engineer would likely opt for two (2) parallel cables.

MECHANICAL 15

REFERENCE NOS.

R16050-010   Electric Circuit Voltages

General: The following method provides the user with a simple non-technical means of obtaining comparative costs of wiring circuits. The circuits considered serve the electrical loads of motors, electric heating, lighting and transformers, for example, that require low voltage 60 Hertz alternating current.

The method used here is suitable only for obtaining estimated costs. It is **not** intended to be used as a substitute for electrical engineering design applications.

Conduit and wire circuits can represent from twenty to thirty percent of the total building electrical cost. By following the described steps and using the tables the user can translate the various types of electric circuits into estimated costs.

Wire Size: Wire size is a function of the electric load which is usually listed in one of the following units:

1. Amperes (A)
2. Watts (W)
3. Kilowatts (kW)
4. Volt amperes (VA)
5. Kilovolt amperes (kVA)
6. Horsepower (HP)

These units of electric load must be converted to amperes in order to obtain the size of wire necessary to carry the load. To convert electric load units to amperes one must have an understanding of the voltage classification of the power source and the voltage characteristics of the electrical equipment or load to be energized. The seven A.C. circuits commonly used are illustrated in Figures R16050-011 thru R16050-017 showing the tranformer load voltage and the point of use voltage at the point on the circuit where the load is connected. The difference between the source and point of use voltages is attributed to the circuit voltage drop and is considered to be approximately 4%.

Motor Voltages: Motor voltages are listed by their point of use voltage and not the power source voltage.

For example: 460 volts instead of 480 volts
200 instead of 208 volts
115 volts instead of 120 volts

Lighting and Heating Voltages: Lighting and heating equipment voltages are listed by the power source voltage and not the point of wire voltage.

For example: 480, 277, 120 volt lighting
480 volt heating or air conditioning unit
208 volt heating unit

Transformer Voltages: Transformer primary (input) and secondary (output) voltages are listed by the power source voltage.

For example: Single phase 10 kVA
Primary 240/480 volts
Secondary 120/240 volts

In this case, the primary voltage may be 240 volts with a 120 volts secondary or may be 480 volts with either a 120V or a 240V secondary.

For example: Three phase 10 kVA
Primary 480 volts
Secondary 208Y/120 volts

In this case the transformer is suitable for connection to a circuit with a 3 phase 3 wire or 3 phase 4 wire circuit with a 480 voltage. This application will provide a secondary circuit of 3 phase 4 wire with 208 volts between phase wires and 120 volts between any phase wire and the neutral (white) wire.

R16050-011

3 Wire, 1 Phase, 120/240 Volt System

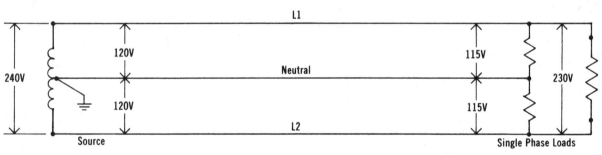

R16050-012

4 wire, 3 Phase, 208Y/120 Volt System

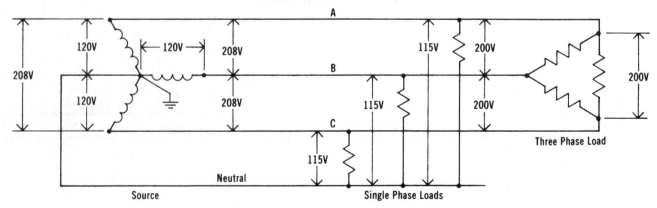

R16050-013

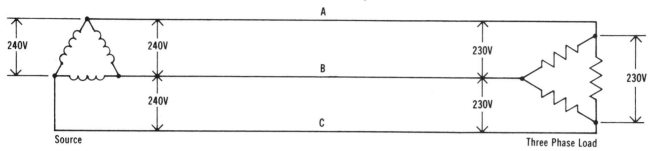

3 Wire, 3 Phase 240 Volt System

R16050-014

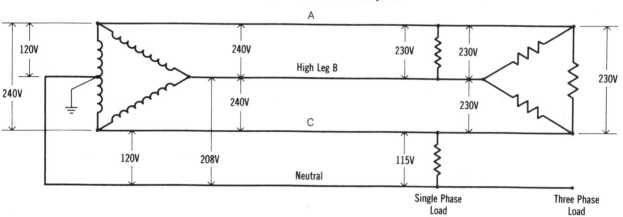

4 Wire, 3 Phase, 240/120 Volt System

R16050-015

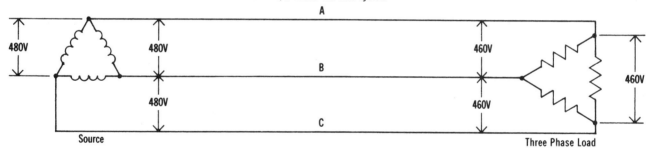

3 Wire, 3 Phase 480 Volt System

R16050-016

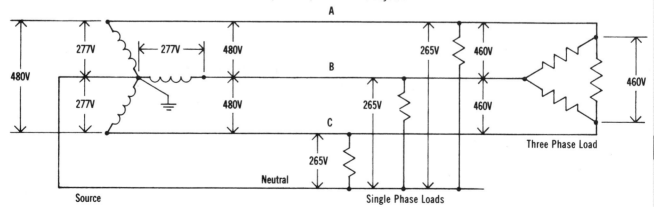

4 Wire, 3 Phase, 480Y/277 Volt System

R16050-017   Electric Circuit Voltages (cont.)

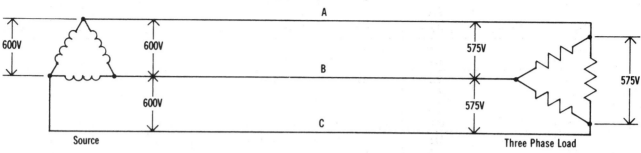

3 Wire, 3 Phase, 600 Volt System

R16050-020   kW Value/Cost Determination

General: Lighting and electric heating loads are expressed in watts and kilowatts.

Cost Determination:

The proper ampere values can be obtained as follows:

1. Convert watts to kilowatts
   (watts 1000 ÷ kilowatts)
2. Determine voltage rating of equipment.
3. Determine whether equipment is single phase or three phase.
4. Refer to Table R16050-021 to find ampere value from kW, Ton and Btu/hr. values.
5. Determine type of wire insulation – TW, THW, THWN.
6. Determine if wire is copper or aluminum.
7. Refer to Table R16120-910 to obtain copper or aluminum wire size from ampere values.
8. Next refer to Table R16132-210 for the proper conduit size to accommodate the number and size of wires in each particular case.
9. Next refer to unit cost data for the per linear foot cost of the conduit.
10. Next refer to unit cost data for the per linear foot cost of the wire. Multiply cost of wire per L.F. x number of wires in the circuits to obtain total wire cost per L.F.

11. Add values obtained in Step 9 and 10 for total cost per linear foot for conduit and wire x length of circuit = Total Cost.

Notes:

1. 1 Phase refers to single phase, 2 wire circuits.
2. 3 Phase refers to three phase, 3 wire circuits.
3. For circuits which operate continuously for 3 hours or more, multiply the ampere values by 1.25 for a given kw requirement.
4. For kW ratings not listed, add ampere values.

   For example: Find the ampere value of
   9 kW at 208 volt, single phase.

$$\begin{array}{r} 4 \text{ kW} = 19.2A \\ 5 \text{ kW} = 24.0A \\ \hline 9 \text{ kW} = 43.2A \end{array}$$

5. "Length of Circuit" refers to the one way distance of the run, not to the total sum of wire lengths.

R16050-021 Ampere Values as Determined by kW Requirements, BTU/HR or Ton, Voltage and Phase Values

			Ampere Values						
			120V	208V		240V		277V	480V
kW	Ton	BTU/HR	1 Phase	1 Phase	3 Phase	1 Phase	3 Phase	1 Phase	3 Phase
0.5	.1422	1,707	4.2A	2.4A	1.4A	2.1A	1.2A	1.8A	0.6A
0.75	.2133	2,560	6.2	3.6	2.1	3.1	1.9	2.7	.9
1.0	.2844	3,413	8.3	4.9	2.8	4.2	2.4	3.6	1.2
1.25	.3555	4,266	10.4	6.0	3.5	5.2	3.0	4.5	1.5
1.5	.4266	5,120	12.5	7.2	4.2	6.3	3.1	5.4	1.8
2.0	.5688	6,826	16.6	9.7	5.6	8.3	4.8	7.2	2.4
2.5	.7110	8,533	20.8	12.0	7.0	10.4	6.1	9.1	3.1
3.0	.8532	10,239	25.0	14.4	8.4	12.5	7.2	10.8	3.6
4.0	1.1376	13,652	33.4	19.2	11.1	16.7	9.6	14.4	4.8
5.0	1.4220	17,065	41.6	24.0	13.9	20.8	12.1	18.1	6.1
7.5	2.1331	25,598	62.4	36.0	20.8	31.2	18.8	27.0	9.0
10.0	2.8441	34,130	83.2	48.0	27.7	41.6	24.0	36.5	12.0
12.5	3.5552	42,663	104.2	60.1	35.0	52.1	30.0	45.1	15.0
15.0	4.2662	51,195	124.8	72.0	41.6	62.4	37.6	54.0	18.0
20.0	5.6883	68,260	166.4	96.0	55.4	83.2	48.0	73.0	24.0
25.0	7.1104	85,325	208.4	120.2	70.0	104.2	60.0	90.2	30.0
30.0	8.5325	102,390		144.0	83.2	124.8	75.2	108.0	36.0
35.0	9.9545	119,455		168.0	97.1	145.6	87.3	126.0	42.1
40.0	11.3766	136,520		192.0	110.8	166.4	96.0	146.0	48.0
45.0	12.7987	153,585			124.8	187.5	112.8	162.0	54.0
50.0	14.2208	170,650			140.0	208.4	120.0	180.4	60.0
60.0	17.0650	204,780			166.4		150.4	216.0	72.0
70.0	19.9091	238,910			194.2		174.6		84.2
80.0	22.7533	273,040			221.6		192.0		96.0
90.0	25.5975	307,170					225.6		108.0
100.0	28.4416	341,300							120.0

R16050-025 kVA Value/Cost Determination

General: Control transformers are listed in VA. Step-down and power transformers are listed in kVA.

Cost Determination:

1. Convert VA to kVA. Volt amperes (VA) ÷ 1000 = Kilovolt amperes (kVA).
2. Determine voltage rating of equipment.
3. Determine whether equipment is single phase or three phase.
4. Refer to Table R16050-026 to find ampere value from kVA value.
5. Determine type of wire insulation – TW, THW, THWN.
6. Determine if wire is copper or aluminum.
7. Refer to Table R16120-910 to obtain copper or aluminum wire size from ampere values.

8. Next refer to Table R16132-210 for the proper conduit size to accommodate the number and size of wires in each particular case.
9. Next refer to unit price data for the per linear foot cost of the conduit.
10. Next refer to unit price data for the per linear foot cost of the wire. Multiply cost of wire per L.F. x number of wires in the circuits to obtain total wire cost.
11. Add values obtained in Step 9 and 10 for total cost per linear foot for conduit and wire x length of circuit = Total Cost.

Example: A transformer rated 10 kVA 480 volts primary, 240 volts secondary, 3 phase has the capacity to furnish the following:

1. Primary amperes = 10 kVA x 1.20 = 12 amperes (from Table R16050-026)
2. Secondary amperes = 10 kVA x 2.40 = 24 amperes (from Table R16050-026)

Note: Transformers can deliver generally 125% of their rated kVA. For instance, a 10 kVA rated transformer can safely deliver 12.5 kVA.

R16050-026 Multiplier Values for kVA to Amperes Determined by Voltage and Phase Values

Volts	Multiplier for Circuits	
	2 Wire, 1 Phase	3 Wire, 3 Phase
115	8.70	
120	8.30	
230	4.30	2.51
240	4.16	2.40
200	5.00	2.89
208	4.80	2.77
265	3.77	2.18
277	3.60	2.08
460	2.17	1.26
480	2.08	1.20
575	1.74	1.00
600	1.66	0.96

ELECTRICAL

REFERENCE NOS.

R16050-030 HP Value/Cost Determination

General: Motors can be powered by any of the seven systems shown in Figure R16050-011 thru Figure R16050-017 provided the motor voltage characteristics are compatible with the power system characteristics.

Cost Determination:

Motor Amperes for the various size H.P. and voltage are listed in Table R16050-031. To find the amperes, locate the required H.P. rating and locate the amperes under the appropriate circuit characteristics.

For example:

    A. 100 H.P., 3 phase, 460 volt motor = 124 amperes (Table R16050-031)

    B. 10 H.P., 3 phase, 200 volt motor = 32.2 amperes (Table R16050-031)

Motor Wire Size: After the amperes are found in Table R16050-031 the amperes must be increased 25% to compensate for power losses. Next refer to Table R16120-910. Find the appropriate insulation column for copper or aluminum wire to determine the proper wire size.

For example:

    A. 100 H.P., 3 phase, 460 volt motor has an ampere value of 124 amperes from Table R16050-031

    B. 124A x 1.25 = 155 amperes

    C. Refer to Table R16120-910 for THW or THWN wire insulations to find the proper wire size. For a 155 ampere load using copper wire a size 2/0 wire is needed.

    D. For the 3 phase motor three wires of 2/0 size are required.

Conduit Size: To obtain the proper conduit size for the wires and type of insulation used, refer to Table R16132-210.

For example: For the 100 H.P., 460V, 3 phase motor, it was determined that three 2/0 wires are required. Assuming THWN insulated copper wire, use Table R16132-210 to determine that three 2/0 wires require 1-1/2″ conduit.

Material Cost of the conduit and wire system depends on:
1. Wire size required
2. Copper or aluminum wire
3. Wire insulation type selected
4. Steel or plastic conduit
5. Type of conduit raceway selected.

Labor Cost of the conduit and wire system depends on:
1. Type and size of conduit
2. Type and size of wires installed
3. Location and height of installation in building or depth of trench
4. Support system for conduit.

R16050-031 Ampere Values Determined by Horsepower, Voltage and Phase Values

H.P.	Amperes							
	Single Phase			Three Phase				
	115V	208V	230V	200V	208V	230V	460V	575V
1/6	4.4A	2.4A	2.2A					
1/4	5.8	3.2	2.9					
1/3	7.2	4.0	3.6					
1/2	9.8	5.4	4.9	2.5A	2.4A	2.2A	1.1A	0.9A
3/4	13.8	7.6	6.9	3.7	3.5	3.2	1.6	1.3
1	16	8.8	8	4.8	4.6	4.2	2.1	1.7
1-1/2	20	11	10	6.9	6.6	6.0	3.0	2.4
2	24	13.2	12	7.8	7.5	6.8	3.4	2.7
3	34	18.7	17	11.0	10.6	9.6	4.8	3.9
5	56	30.8	28	17.5	16.7	15.2	7.6	6.1
7-1/2	80	44	40	25.3	24.2	22	11	9
10	100	55	50	32.2	30.8	28	14	11
15				48.3	46.2	42	21	17
20				62.1	59.4	54	27	22
25				78.2	74.8	68	34	27
30				92	88	80	40	32
40				120	114	104	52	41
50				150	143	130	65	52
60				177	169	154	77	62
75				221	211	192	96	77
100				285	273	248	124	99
125				359	343	312	156	125
150				414	396	360	180	144
200				552	528	480	240	192
250							302	242
300							361	289
350							414	336
400							477	382

ELECTRICAL 16

REFERENCE NOS.

R16050-030 Cost Determination (cont.)

Magnetic starters, switches, and motor connection:

To complete the cost picture from H.P. to Costs additional items must be added to the cost of the conduit and wire system to arrive at a total cost.

1. Table D5020-160 Magnetic Starters Installed Cost lists the various size starters for single phase and three phase motors.
2. Table D5020-165 Heavy Duty Safety Switches Installed Cost lists safety switches required at the beginning of a motor circuit and also one required in the vicinity of the motor location.
3. Table D5020-170 Motor Connection lists the various costs for single and three phase motors.

Worksheet to obtain total motor wiring costs:

It is assumed that the motors or motor driven equipment are furnished and installed under other sections for this estimate and the following work is done under this section:

1. Conduit
2. Wire (add 10% for additional wire beyond conduit ends for connections to switches, boxes, starters, etc.)
3. Starters
4. Safety switches
5. Motor connections

Figure R16050-032

				Cost	
Item	Type	Size	Quantity	Unit	Total
Wire					
Conduit					
Switch					
Starter					
Switch					
Motor Connection					
Other					
Total Cost					

Worksheet for Motor Circuits

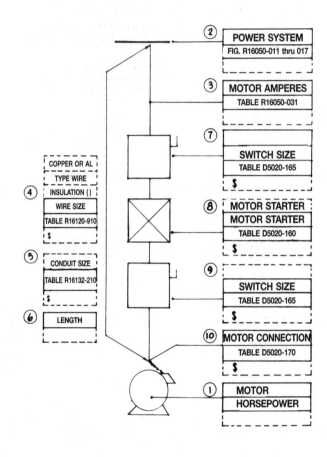

Maximum Horsepower for Starter Size by Voltage

Starter	Maximum HP (3φ)			
Size	208V	240V	480V	600V
00	1½	1½	2	2
0	3	3	5	5
1	7½	7½	10	10
2	10	15	25	25
3	25	30	50	50
4	40	50	100	100
5		100	200	200
6		200	300	300
7		300	600	600
8		450	900	900
8L		700	1500	1500

R16050-040 Standard Electrical Enclosure Types

NEMA Enclosures

Electrical enclosures serve two basic purposes; they protect people from accidental contact with enclosed electrical devices and connections, and they protect the enclosed devices and connections from specified external conditions. The National Electrical Manufacturers Association (NEMA) has established the following standards. Because these descriptions are not intended to be complete representations of NEMA listings, consultation of NEMA literature is advised for detailed information.

The following definitions and descriptions pertain to NONHAZARDOUS locations.

NEMA Type 1: General purpose enclosures intended for use indoors, primarily to prevent accidental contact of personnel with the enclosed equipment in areas that do not involve unusual conditions.

NEMA Type 2: Dripproof indoor enclosures intended to protect the enclosed equipment against dripping noncorrosive liquids and falling dirt.

NEMA Type 3: Dustproof, raintight and sleet-resistant (ice-resistant) enclosures intended for use outdoors to protect the enclosed equipment against wind-blown dust, rain, sleet, and external ice formation.

NEMA Type 3R: Rainproof and sleet-resistant (ice-resistant) enclosures which are intended for use outdoors to protect the enclosed equipment against rain. These enclosures are constructed so that the accumulation and melting of sleet (ice) will not damage the enclosure and its internal mechanisms.

NEMA Type 3S: Enclosures intended for outdoor use to provide limited protection against wind-blown dust, rain, and sleet (ice) and to allow operation of external mechanisms when ice-laden.

NEMA Type 4: Watertight and dust-tight enclosures intended for use indoors and out – to protect the enclosed equipment against splashing water, see page of water, falling or hose-directed water, and severe external condensation.

NEMA Type 4X: Watertight, dust-tight, and corrosion-resistant indoor and outdoor enclosures featuring the same provisions as Type 4 enclosures, plus corrosion resistance.

NEMA Type 5: Indoor enclosures intended primarily to provide limited protection against dust and falling dirt.

NEMA Type 6: Enclosures intended for indoor and outdoor use – primarily to provide limited protection against the entry of water during occasional temporary submersion at a limited depth.

NEMA Type 6R: Enclosures intended for indoor and outdoor use – primarily to provide limited protection against the entry of water during prolonged submersion at a limited depth.

NEMA Type 11: Enclosures intended for indoor use – primarily to provide, by means of oil immersion, limited protection to enclosed equipment against the corrosive effects of liquids and gases.

NEMA Type 12: Dust-tight and driptight indoor enclosures intended for use indoors in industrial locations to protect the enclosed equipment against fibers, flyings, lint, dust, and dirt, as well as light splashing, see page, dripping, and external condensation of noncorrosive liquids.

NEMA Type 13: Oil-tight and dust-tight indoor enclosures intended primarily to house pilot devices, such as limit switches, foot switches, push buttons, selector switches, and pilot lights, and to protect these devices against lint and dust, see page, external condensation, and sprayed water, oil, and noncorrosive coolant.

The following definitions and descriptions pertain to HAZARDOUS, or CLASSIFIED, locations:

NEMA Type 7: Enclosures intended to use in indoor locations classified as Class 1, Groups A, B, C, or D, as defined in the National Electrical Code.

NEMA Type 9: Enclosures intended for use in indoor locations classified as Class 2, Groups E, F, or G, as defined in the National Electrical Code.

R16050-060 Central Air Conditioning Watts per S.F., BTU's per Hour per S.F. of Floor Area and S.F. per Ton of Air Conditioning

Type Building	Watts per S.F.	BTUH per S.F.	S.F. per Ton	Type Building	Watts per S.F.	BTUH per S.F.	S.F. per Ton	Type Building	Watts per S.F.	BTUH per S.F.	S.F. per Ton
Apartments, Individual	3	26	450	Dormitory, Rooms	4.5	40	300	Libraries	5.7	50	240
Corridors	2.5	22	550	Corridors	3.4	30	400	Low Rise Office, Ext.	4.3	38	320
Auditoriums & Theaters	3.3	40	300/18*	Dress Shops	4.9	43	280	Interior	3.8	33	360
Banks	5.7	50	240	Drug Stores	9	80	150	Medical Centers	3.2	28	425
Barber Shops	5.5	48	250	Factories	4.5	40	300	Motels	3.2	28	425
Bars & Taverns	15	133	90	High Rise Off.-Ext. Rms.	5.2	46	263	Office (small suite)	4.9	43	280
Beauty Parlors	7.6	66	180	Interior Rooms	4.2	37	325	Post Office, Int. Office	4.9	42	285
Bowling Alleys	7.8	68	175	Hospitals, Core	4.9	43	280	Central Area	5.3	46	260
Churches	3.3	36	330/20*	Perimeter	5.3	46	260	Residences	2.3	20	600
Cocktail Lounges	7.8	68	175	Hotels, Guest Rooms	5	44	275	Restaurants	6.8	60	200
Computer Rooms	16	141	85	Public Spaces	6.2	55	220	Schools & Colleges	5.3	46	260
Dental Offices	6	52	230	Corridors	3.4	30	400	Shoe Stores	6.2	55	220
Dept. Stores, Basement	4	34	350	Industrial Plants, Offices	4.3	38	320	Shop'g. Ctrs., Sup. Mkts.	4	34	350
Main Floor	4.5	40	300	General Offices	4	34	350	Retail Stores	5.5	48	250
Upper Floor	3.4	30	400	Plant Areas	4.5	40	300	Specialty Shops	6.8	60	200

*Persons per ton

12,000 BTUH = 1 ton of air conditioning

R16055-300 Electrical Demolition (Removal for Replacement)

The purpose of this reference number is to provide a guide to users for electrical "removal for replacement" by applying the rule of thumb: 1/3 of new installation time (typical range from 20% to 50%) for removal. Remember to use reasonable judgment when applying the suggested percentage factor. For example:

Contractors have been requested to remove an existing fluorescent lighting fixture and replace with a new fixture utilizing energy saver lamps and electronic ballast:

In order to fully understand the extent of the project, contractors should visit the job site and estimate the time to perform the renovation work in accordance with applicable national, state and local regulation and codes.

The contractor may need to add extra labor hours to his estimate if he discovers unknown concealed conditions such as: contaminated asbestos ceiling, broken acoustical ceiling tile and need to repair, patch and touch-up paint all the damaged or disturbed areas, tasks normally assigned to general contractors. In addition, the owner could request that the contractors salvage the materials removed and turn over the materials to the owner or dispose of the materials to a reclamation station. The normal removal item is 0.5 labor hour for a lighting fixture and 1.5 labor-hours for new installation time. Revise the estimate times from 2 labor-hours work up to a minimum 4 labor-hours work for just fluorescent lighting fixture.

For removal of large concentrations of lighting fixtures in the same area, apply an "economy of scale" to reduce estimating labor hours.

R16060-800 Grounding

Grounding – When taking off grounding systems, identify separately the type and size of wire.

Example:

Bare copper & size

Bare aluminum & size

Insulated copper & size

Insulated aluminum & size

Count the number of ground rods and their size.

Example:

1.	8' grounding rod – 5/8" dia.	20 Ea.
2.	10' grounding rod – 5/8" dia.	12 Ea.
3.	15' grounding rod – 3/4" dia.	4 Ea.

Count the number of connections; the size of the largest wire will determine the productivity.

Example:

Braze a #2 wire to a #4/0 cable

The 4/0 cable will determine the L.H. and cost to be used.

Include individual connections to:
1. Ground rods
2. Building steel
3. Equipment
4. Raceways

Price does not include:
1. Excavation
2. Backfill
3. Sleeves or raceways used to protect grounding wires
4. Wall penetrations
5. Floor cutting
6. Core drilling

Job Conditions: Productivity is based on a ground floor area, using cable reels in an unobstructed area. Material staging area assumed to be within 100' of work being performed.

R16120-120 Armored Cable

Armored Cable – Quantities are taken off in the same manner as wire.

Bx Type Cable – Productivities are based on an average run of 50' before terminating at a box fixture, etc. Each 50' section includes field preparation of (2) ends with hacksaw, identification and tagging of wire. Set up is open coil type without reels attaching cable to snake and pulling across a suspended ceiling or open face wood or steel studding, price does not include drilling of studs.

Cable in Tray – Productivities are based on an average run of 100 L.F. with set up of pulling equipment for (2) 90° bends, attaching cable to pull-in means, identification and tagging of wires, set up of reels. Wire termination to breakers equipment, etc., are not included.

Job Conditions – Productivities are based on new construction to a height of 15' using rolling staging in an unobstructed area. Material staging is assumed to be within 100' of work being performed.

16 ELECTRICAL

REFERENCE NOS.

R16120-400   Fiber Optics

Fiber optic systems use optical fiber such as plastic, glass, or fused silica, a transparent material, to transmit radiant power (i.e., light) for control, communication, and signaling applications. The types of fiber optic cables can be nonconductive, conductive, or composite. The composite cables contain fiber optics and current-carrying electrical conductors. The configuration for one of the fiber optic systems is as follows:

The transceiver module acts as transmitting and receiving equipment in a common house, which converts electrical energy to light energy or vice versa.

Pricing the fiber optic system is not an easy task. The performance of the whole system will affect the cost significantly. New specialized tools and techniques decrease the installing cost tremendously. In the fiber optic section of Means Electrical Cost Data, a benchmark for labor-hours and material costs is set up so that users can adjust their costs according to unique project conditions.

Units for Measure: Fiber optic cable is measured in hundred linear feet (C.L.F.) or industry units of measure - meter (m) or kilometer (km). The connectors are counted as units (EA.)

Material Units: Generally, the material costs include only the cable. All the accessories shall be priced separately.

Labor Units: The following procedures are generally included for the installation of fiber optic cables:

* Receiving
* Material handling
* Setting up pulling equipment
* Measuring and cutting cable
* Pulling cable

These additional items are listed and extended: Terminations

Takeoff Procedure: Cable should be taken off by type, size, number of fibers, and number of terminations required. List the lengths of each type of cable on the takeoff sheets. Total and add 10% for waste. Transfer the figures to a cost analysis sheet and extend.

R16120-750   High Performance Cable

There are several categories used to describe high performance cable. The following information includes a description of categories CAT 3, 5, 5e, 6, and 7, and details classifications of frequency and specific standards. The category standards have evolved under the sponsorship of organizations such as the Telecommunication Industry Association (TIA), the Electronic Industries Alliance (EIA), the American National Standards Institute (ANSI), the International Organization for Standardization (ISO), and the International Electrotechnical Commission (IEC), all of which have catered to the increasing complexities of modern network technology. For network cabling, users must comply with national or international standards. A breakdown of these categories is as follows:

Category 3: Designed to handle frequencies up to 16 MHz.

Category 5: (TIA/EIA 568A) Designed to handle frequencies up to 100 MHz.

Category 5e: Additional transmission performance to exceed Category 5.

Category 6: Development by TIA and other international groups to handle frequencies of 250 MHz.

Category 7 (draft): Under development to handle a frequency range from 1 to 600 MHz.

R16120-800 Undercarpet Systems

Takeoff Procedure for Power Systems: List components for each fitting type, tap, splice, and bend on your quantity takeoff sheet. Each component must be priced separately. Start at the power supply transition fittings and survey each circuit for the components needed. List the quantities of each component under a specific circuit number. Use the floor plan layout scale to get cable footage.

Reading across the list, combine the totals of each component in each circuit and list the total quantity in the last column. Calculate approximately 5% for scrap for items such as cable, top shield, tape, and spray adhesive. Also provide for final variations that may occur on-site.

Suggested guidelines are:
1. Equal amounts of cable and top shield should be priced.
2. For each roll of cable, price a set of cable splices.
3. For every 1 ft. of cable, price 2-1/2 ft. of hold-down tape.
4. For every 3 rolls of hold-down tape, price 1 can of spray adhesive.

Adjust final figures wherever possible to accommodate standard packaging of the product. This information is available from the distributor.

Each transition fitting requires:
1. 1 base
2. 1 cover
3. 1 transition block

Each floor fitting requires:
1. 1 frame/base kit
2. 1 transition block
3. 2 covers (duplex/blank)

Each tap requires:
1. 1 tap connector for each conductor
2. 1 pair insulating patches
3. 2 top shield connectors

Each splice requires:
1. 1 splice connector for each conductor
2. 1 pair insulating patches
3. 3 top shield connectors

Each cable bend requires:
1. 2 top shield connectors

Each cable dead end (outside of transition block) requires:
1. 1 pair insulating patches

Labor does not include:
1. Patching or leveling uneven floors.
2. Filling in holes or removing projections from concrete slabs.
3. Sealing porous floors.
4. Sweeping and vacuuming floors.
5. Removal of existing carpeting.
6. Carpet square cut-outs.
7. Installation of carpet squares.

Takeoff Procedures for Telephone Systems: After reviewing floor plans identify each transition. Number or letter each cable run from that fitting.

Start at the transition fitting and survey each circuit for the components needed. List the cable type, terminations, cable length, and floor fitting type under the specific circuit number. Use the floor plan layout scale to get the cable footage. Add some extra length (next higher increment of 5 feet) to preconnectorized cable.

Transition fittings require:
1. 1 base plate
2. 1 cover
3. 1 transition block

Floor fittings require:
1. 1 frame/base kit
2. 2 covers
3. Modular jacks

Reading across the list, combine the list of components in each circuit and list the total quantity in the last column. Calculate the necessary scrap factors for such items as tape, bottom shield and spray adhesive. Also provide for final variations that may occur on-site.

Adjust final figures whenever possible to accommodate standard packaging. Check that items such as transition fittings, floor boxes, and floor fittings that are to utilize both power and telephone have been priced as combination fittings, so as to avoid duplication.

Make sure to include marking of floors and drilling of fasteners if fittings specified are not the adhesive type.

Labor does not include:
1. Conduit or raceways before transition of floor boxes
2. Telephone cable before transition boxes
3. Terminations before transition boxes
4. Floor preparation as described in power section

Be sure to include all cable folds when pricing labor.

Takeoff Procedure for Data Systems: Start at the transition fittings and take off quantities in the same manner as the telephone system, keeping in mind that data cable does not require top or bottom shields.

The data cable is simply cross-taped on the cable run to the floor fitting.

Data cable can be purchased in either bulk form in which case coaxial connector material and labor must be priced, or in preconnectorized cut lengths.

Data cable cannot be folded and must be notched at 1 inch intervals. A count of all turns must be added to the labor portion of the estimate. (Note: Some manufacturers have prenotched cable.)

Notching required:
1. 90 degree turn requires 8 notches per side.
2. 180 degree turn requires 16 notches per side.

Floor boxes, transition boxes, and fittings are the same as described in the power and telephone procedures.

Since undercarpet systems require special hand tools, be sure to include this cost in proportion to number of crews involved in the installation.

Job Conditions: Productivity is based on new construction in an unobstructed area. Staging area is assumed to be within 200' of work being performed.

R16120-900 Wire

Wire quantities are taken off by either measuring each cable run or by extending the conduit and raceway quantities times the number of conductors in the raceway. Ten percent should be added for waste and tie-ins. Keep in mind that the unit of measure of wire is C.L.F. not L.F. as in raceways so the formula would read:

$$\frac{\text{(L.F. Raceway x No. of Conductors) x 1.10}}{100} = \text{C.L.F.}$$

Price per C.L.F. of wire includes:
1. Set up wire coils or spools on racks
2. Attaching wire to pull in means
3. Measuring and cutting wire
4. Pulling wire into a raceway
5. Identifying and tagging

Price does not include:
1. Connections to breakers, panelboards or equipment
2. Splices

Job Conditions: Productivity is based on new construction to a height of 15′ using rolling staging in an unobstructed area. Material staging is assumed to be within 100′ of work being performed.

Economy of Scale: If more than three wires at a time are being pulled, deduct the following percentages from the labor of that grouping:

4-5 wires	25%
6-10 wires	30%
11-15 wires	35%
over 15	40%

If a wire pull is less than 100′ in length and is interrupted several times by boxes, lighting outlets, etc., it may be necessary to add the following lengths to each wire being pulled:

Junction box to junction box	2 L.F.
Lighting panel to junction box	6 L.F.
Distribution panel to sub panel	8 L.F.
Switchboard to distribution panel	12 L.F.
Switchboard to motor control center	20 L.F.
Switchboard to cable tray	40 L.F.

Measure of Drops and Riser: It is important when taking off wire quantities to include the wire for drops to electrical equipment. If heights of electrical equipment are not clearly stated, use the following guide:

	Bottom A.F.F.	Top A.F.F.	Inside Cabinet
Safety switch to 100A	5′	6′	2′
Safety switch 400 to 600A	4′	6′	3′
100A panel 12 to 30 circuit	4′	6′	3′
42 circuit panel	3′	6′	4′
Switch box	3′	3′6″	1′
Switchgear	0′	8′	8′
Motor control centers	0′	8′	8′
Transformers - wall mount	4′	8′	2′
Transformers - floor mount	0′	12′	4′

ELECTRICAL 16

REFERENCE NOS.

R16120-905 Maximum Circuit Length (approximate) for Various Power Requirements Assuming THW, Copper Wire @ 75° C, Based Upon a 4% Voltage Drop

Maximum Circuit Length: Table R16120-905 indicates typical maximum installed length a circuit can have and still maintain an adequate voltage level at the point of use. The circuit length is similar to the conduit length.

If the circuit length for an ampere load and a copper wire size exceeds the length obtained from Table R16120-905, use the next largest wire size to compensate voltage drop.

Example: A 130 ampere load at 480 volts, 3 phase, 3 wire with No. 1 wire can be run a maximum of 555 L.F. and provide satisfactory operation. If the same load is to be wired at the end of a 625 L.F. circuit, then a larger wire must be used.

Amperes	Wire Size	Maximum Circuit Length in Feet				
		2 Wire, 1 Phase		3 Wire, 3 Phase		
		120V	240V	240V	480V	600V
15	14*	50	105	120	240	300
	14	50	100	120	235	295
20	12*	60	125	145	290	360
	12	60	120	140	280	350
30	10*	65	130	155	305	380
	10	65	130	150	300	375
50	8	60	125	145	285	355
65	6	75	150	175	345	435
85	4	90	185	210	425	530
115	2	110	215	250	500	620
130	1	120	240	275	555	690
150	1/0	130	260	305	605	760
175	2/0	140	285	330	655	820
200	3/0	155	315	360	725	904
230	4/0	170	345	395	795	990
255	250	185	365	420	845	1055
285	300	195	395	455	910	1140
310	350	210	420	485	975	1220
380	500	245	490	565	1130	1415

*Solid Conductor

Note: The circuit length is the one-way distance between the origin and the load.

R16120-910   Minimum Copper and Aluminum Wire Size Allowed for Various Types of Insulation

Amperes	Copper		Aluminum		Amperes	Copper		Aluminum	
	THW THWN or XHHW	THHN XHHW *	THW XHHW	THHN XHHW *		THW THWN or XHHW	THHN XHHW *	THW XHHW	THHN XHHW *
15A	#14	#14	#12	#12	195	3/0	2/0	250kcmil	4/0
20	#12	#12	#10	#10	200	3/0	3/0	250kcmil	4/0
25	#10	#10	#10	#10	205	4/0	3/0	250kcmil	4/0
30	#10	#10	#8	#8	225	4/0	3/0	300kcmil	250kcmil
40	#8	#8	#8	#8	230	4/0	4/0	300kcmil	250kcmil
45	#8	#8	#6	#8	250	250kcmil	4/0	350kcmil	300kcmil
50	#8	#8	#6	#6	255	250kcmil	4/0	400kcmil	300kcmil
55	#6	#8	#4	#6	260	300kcmil	4/0	400kcmil	350kcmil
60	#6	#6	#4	#6	270	300kcmil	250kcmil	400kcmil	350kcmil
65	#6	#6	#4	#4	280	300kcmil	250kcmil	500kcmil	350kcmil
75	#4	#6	#3	#4	285	300kcmil	250kcmil	500kcmil	400kcmil
85	#4	#4	#2	#3	290	350kcmil	250kcmil	500kcmil	400kcmil
90	#3	#4	#2	#2	305	350kcmil	300kcmil	500kcmil	400kcmil
95	#3	#4	#1	#2	310	350kcmil	300kcmil	500kcmil	500kcmil
100	#3	#3	#1	#2	320	400kcmil	300kcmil	600kcmil	500kcmil
110	#2	#3	1/0	#1	335	400kcmil	350kcmil	600kcmil	500kcmil
115	#2	#2	1/0	#1	340	500kcmil	350kcmil	600kcmil	500kcmil
120	#1	#2	1/0	1/0	350	500kcmil	350kcmil	700kcmil	500kcmil
130	#1	#2	2/0	1/0	375	500kcmil	400kcmil	700kcmil	600kcmil
135	1/0	#1	2/0	1/0	380	500kcmil	400kcmil	750kcmil	600kcmil
150	1/0	#1	3/0	2/0	385	600kcmil	500kcmil	750kcmil	600kcmil
155	2/0	1/0	3/0	3/0	420	600kcmil	500kcmil		700kcmil
170	2/0	1/0	4/0	3/0	430		500kcmil		750kcmil
175	2/0	2/0	4/0	3/0	435		600kcmil		750kcmil
180	3/0	2/0	4/0	4/0	475		600kcmil		

*Dry Locations Only

Notes:

1. Size #14 to 4/0 is in AWG units (American Wire Gauge).
2. Size 250 to 750 is in kcmil units (Thousand Circular Mils).
3. Use next higher ampere value if exact value is not listed in table.
4. For loads that operate continuously increase ampere value by 25% to obtain proper wire size.
5. Refer to Table R16120-905 for the maximum circuit length for the various size wires.
6. Table R16120-910 has been written for estimating purpose only, based on ambient temperature of 30° C (86° F); for ambient temperature other than 30° C (86° F), ampacity correction factors will be applied.

ELECTRICAL 16

REFERENCE NOS.

R16120-915 Metric Equivalent, Wire

United States		European	
Size AWG or kcmil	Area Cir. Mils.(cmil) mm²	Size mm²	Area Cir. Mils.
18	1620/.82	.75	1480
16	2580/1.30	1.0	1974
14	4110/2.08	1.5	2961
12	6530/3.30	2.5	4935
10	10,380/5.25	4	7896
8	16,510/8.36	6	11,844
6	26,240/13.29	10	19,740
4	41,740/21.14	16	31,584
3	52,620/26.65	25	49,350
2	66,360/33.61	–	–
1	83,690/42.39	35	69,090
1/0	105,600/53.49	50	98,700
2/0	133,100/67.42	–	–
3/0	167,800/85.00	70	138,180
4/0	211,600/107.19	95	187,530
250	250,000/126.64	120	236,880
300	300,000/151.97	150	296,100
350	350,000/177.30	–	–
400	400,000/202.63	185	365,190
500	500,000/253.29	240	473,760
600	600,000/303.95	300	592,200
700	700,000/354.60	–	–
750	750,000/379.93	–	–

U.S. vs. European Wire – Approximate Equivalents

R16120-920 Size Required and Weight (Lbs./1000 L.F.) of Aluminum and Copper THW Wire by Ampere Load

Amperes	Copper Size	Aluminum Size	Copper Weight	Aluminum Weight
15	14	12	24	11
20	12	10	33	17
30	10	8	48	39
45	8	6	77	52
65	6	4	112	72
85	4	2	167	101
100	3	1	205	136
115	2	1/0	252	162
130	1	2/0	324	194
150	1/0	3/0	397	233
175	2/0	4/0	491	282
200	3/0	250	608	347
230	4/0	300	753	403
255	250	400	899	512
285	300	500	1068	620
310	350	500	1233	620
335	400	600	1396	772
380	500	750	1732	951

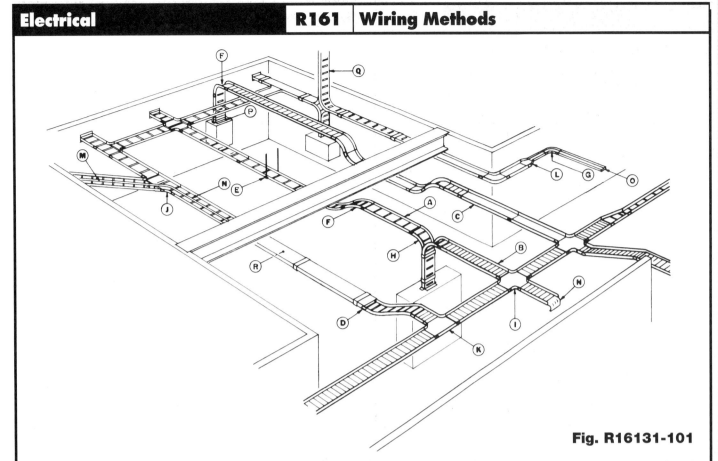

Fig. R16131-101

R16131-100 Cable Tray

Cable Tray - When taking off cable tray it is important to identify separately the different types and sizes involved in the system being estimated. (Fig. R16131-101)

A. – Ladder Type, galvanized or aluminum
B. – Trough Type, galvanized or aluminum
C. – Solid Bottom, galvanized or aluminum

The unit of measure is calculated in linear feet; do not deduct from this footage any length occupied by fittings, this will be the only allowance for scrap. Be sure to include all vertical drops to panels, switch gear, etc.

Hangers – Included in the linear footage of cable tray is

D. – 1 – Pair of connector plates per 12 L.F.
E. – 1 – Pair clamp type hangers and 4′ of 3/8″ threaded rod per 12 L.F.

Not included are structural supports, which must be priced in addition to the hangers. (Fig. R16131-102)

Fittings – Identify separately the different types of fittings

1.) Ladder Type, galvanized or aluminum
2.) Trough Type, galvanized or aluminum
3.) Solid Bottom Type, galvanized or aluminum

The configuration, radius and rung spacing must also be listed.
The unit of measure is "Ea." (Fig. R16131-102)

F. – Elbow, vertical Ea.
G. – Elbow, horizontal Ea.
H. – Tee, vertical Ea.
I. – Cross, horizontal Ea.
J. – Wye, horizontal Ea.
K. – Tee, horizontal Ea.
L. – Reducing fitting Ea.

Depending on the use of the system other examples of units which must be included are:

M. – Divider strip L.F.
N. – Drop-outs Ea.
O. – End caps Ea.
P. – Panel connectors Ea.

Wire and cable are not included and should be taken off separately, see Division 161.

Job Conditions – Unit prices are based on a new installation to a work plane of 15′ using rolling staging.

Add to labor for elevated installations (Fig. R16131-102)

15′ to 20′ High	10%
20′ to 25′ High	20%
25′ to 30′ High	25%
30′ to 35′ High	30%
35′ to 40′ High	35%
Over 40′ High	40%

Add these percentages for L.F. totals but not to fittings. Add percentages to only those quantities that fall in the different elevations, in other words, if the total quantity of cable tray is 200′ but only 75′ is above 15′ then the 10% is added to the 75′ only.

Linear foot costs do not include penetrations through walls and floors which must be added to the estimate. This section appears in Section 16131.

Cable Tray Covers

Covers – Cable tray covers are taken off in the same manner as the tray itself, making distinctions as to the type of cover. (Fig. R16131-101)

Q. – Vented, galvanized or aluminum
R. – Solid, galvanized or aluminum

Cover configurations are taken off separately noting type, specific radius and widths.

Note: Care should be taken to identify from plans and specifications exactly what is being covered. In many systems only vertical fittings are covered to retain wire and cable.

COST ANALYSIS

PROJECT: Cable Tray (Sample)

ARCHITECT: S. Proulx

TAKE OFF BY: JC QUANTITIES BY: JC PRICES BY: RSM EXTENSIONS BY: JM CHECKED BY: SP

SHEET NO. 1 of 1

ESTIMATE NO. 005

DATE 7/31/04

DESCRIPTION	SOURCE/LINE NUMBER			QUANTITY	UNIT	MATERIAL		LABOR-HOURS		LABOR	
						UNIT COST	TOTAL	UNIT	TOTAL	UNIT COST	TOTAL
Cable Tray:											
Ladder Type, galv. steel 4"D × 12"W w/9" rungs	16131	105	0930	240	LF	8.55	2052	.17	40.80	6.70	1608
@ 15'- 20' high, add 10%	16131	130	9910	110	LF	—	—	.017	1.87	.67	74
Solid Bottom Type galv. steel, 3"D × 18"W	16131	110	0260	120	LF	10.65	1278	.229	27.48	9.00	1080
@ 20'- 25' high, add 20%	16131	130	9920	55	LF	—	—	.046	2.53	1.80	99
Fittings:											
Vert. Elbow - Ladder 90°, 9" rung 24" rad., 12"W	16131	105	1590	8	Ea	68	544	2.222	17.78	87.50	700
Horiz. Elbow - Ladder 90°, 9" rung 24" rad., 12"W	16131	105	1140	4	Ea	72	288	2.222	8.89	87.50	350
Dropout 12"W	16131	105	2540	4	Ea	6.90	28	.615	2.46	24.50	98
Vert. Elbow - Solid Bottom 24" rad., 18"W	16131	110	0700	4	Ea	106	424	3.2	12.80	126	504
Horiz. Elbow - Solid	16131	110	0450	2	Ea	100	200	3.2	6.40	126	252
Dropout 18"W	16131	110	1610	2	Ea	14.15	28	.727	1.45	28.50	57
Subtotals (before O&P)							$4842		122.46		$4822
Total Bare Cost (before O&P)											$9664

R16132-205　Conduit To 15' High

List conduit by quantity, size, and type. Do not deduct for lengths occupied by fittings, since this will be allowance for scrap. Example:

- A. Aluminum — size
- B. Rigid Galvanized — size
- C. Steel Intermediate (IMC) — size
- D. Rigid Steel, plastic coated 20 Mil — size
- E. Rigid Steel, plastic coated 40 Mil — size
- F. Electric Metallic Tubing (EMT) — size
- G. PVC Schedule 40 — size

Types (A) thru (E) listed above contain the following per 100 L.F.:

1. (11) Threaded couplings
2. (11) Beam type hangers
3. (2) Factory sweeps
4. (2) Fiber bushings
5. (4) Locknuts
6. (2) Field threaded pipe terminations
7. (2) Removal of concentric knockouts

Type (F) contains per 100 L.F.

1. (11) Set screw couplings
2. (11) Beam clamps
3. (2) Field bends on 1/2" and 3/4" diameter
4. (2) Factory sweeps for 1" and above
5. (2) Set screw steel connectors
6. (2) Removal of concentric knockouts

Type (G) contains per 100 L.F.

1. (11) Field cemented couplings
2. (34) Beam clamps
3. (2) Factory sweeps

4. (2) Adapters
5. (2) Locknuts
6. (2) Removal of concentric knockouts

Labor-hours for all conduit to 15' include:

1. Unloading by hand
2. Hauling by hand to an area up to 200' from loading dock
3. Set up of rolling staging
4. Installation of conduit and fittings, as described in Conduit models (A) thru (G)

Not included in the material and labor are:

1. Staging rental or purchase
2. Structural Modifications
3. Wire
4. Junction boxes
5. Fittings in excess of those described in conduit models (A) thru (G)
6. Painting of conduit

Fittings

Only those fittings listed above are included in the linear foot totals, although they should be listed separately from conduit lengths, without prices, to insure proper quantities for material procurement.

If the fittings required exceed the quantities included in the model conduit runs, then material and labor costs must be added to the difference. If actual needs per 100 L.F. of conduit are: (2) sweeps, (4) LB's and (1) field bend.

Then, in this case (4) LB's and (1) field bend must be priced additionally.

R16132-206 Hangers

It is sometimes desirable to substitute an alternate style of hanger if the support being used is not the type described in the conduit models.

Example: The 3/4" RGS conduit assumes that supports are beam type as defined on line 16070-320-7960. The bare costs are:

16070-		M +	L =	T	
320-7960		$4.12	$9.85	$13.97	Jay Clamp

To substitute an alternate support, such as hang-on style with a 12" threaded rod and a concrete expansion shield, first add that assembly's component costs:

16070-		M	+	L	=	T
320-7830	Hanger & Rod	2.09		2.17		4.26
05090-						
380-0200	1/4" Expansion Shield	.96		2.93		3.89
340-0200	3/8" Drilled Hole	.07		4.19		4.26
340-1000	For Ceiling Inst.	—		1.68		1.68
	Totals	3.12		10.97		14.09

The difference in cost between the two types of supports is:

Hang-On	Jay Clamp	Increase/Each
$14.09	$13.97	$.12

Since the model assumes 11 supports per 100 L.F., the total increase would be:

$$11 \times \$.12 = \$1.32 \text{ per CLF; or } \$.01 \text{ per L.F.}$$

Another approach to hanger configurations would be to start with the conduit only as per section 16132-210 and add all the supports and any other items as separate lines. This procedure is most useful if the project involves racking many runs of conduit on a single hanger, for instance a trapeze type hanger.

Example: Five (5) 2" RGS conduits, 50 L.F. each, are to be run on trapeze hangers from one pull box to another. The run includes one 90° bend.

First, list the hanger's components to create an assembly cost for each 2' wide trapeze.

16070-		Q		M	+	L	=	T
320-3900	Channel	2	L.F.	8.64		9.00		17.64
320-4250	3/8" Spr. Nuts	2	Ea.	3.70		6.30		10.00
320-3300	3/8" Washers	2	Ea.	.20		—		.20
320-3050	3/8" Nuts	2	Ea.	.26		—		.26
320-2600	3/8" Rod	2	L.F.	3.32		3.16		6.48
05090-								
380-0400	3/8" Shields	2	Ea.	3.16		6.22		9.38
340-0300	1/2" Hole	4	In.	.28		21.20		21.48
	Trapeze Hgr. Totals			19.56		45.88		65.44

Next, list the components for the 50 ft. run, noting that 6 supports will be required.

16132-		Q		M	+	L	=	T
210-0600	2" RGS	250	L.F.	1,140.00		1,212.50		2,352.50
205-2130	2" Elbows	5	Ea.	97.50		132.50		230.00
250-1000	2" Locknuts	20	Ea.	34.60		—		34.60
250-1250	2" Bushings	10	Ea.	19.50		210.00		229.50
Assembly	2" Trap. HGR	6	Ea.	117.36		275.28		392.64
Total Installation				1,408.96		1,830.28		$3,239.24

Job Conditions: Productivities are based on new construction to 15' high, using scaffolding in an unobstructed area. Material storage is assumed to be within 100' of work being performed.

Add to labor for elevated installations:

15' to 20' High 10%	30' to 35 High 30%
20' to 25' High 20%	35' to 40 High 35%
25' to 30' High 25%	Over 40' High 40%

Add these percentages to the L.F. labor cost, but not to fittings. Add these percentages only to quantities exceeding the different height levels, not the total conduit quantities.

Linear foot price for labor does not include penetrations in walls or floors, and must be added to the estimate. This section appears in Section 16132-260.

R16132-210   Conductors in Conduit

Table below lists maximum number of conductors for various sized conduit using THW, TW or THWN insulations.

Copper Wire Size	1/2"			3/4"			1"			1-1/4"			1-1/2"			2"			2-1/2"			3"		3-1/2"		4"	
	TW	THW	THWN	TW	THW	THWN	TW	THW	THWN	TW	THW	THWN	TW	THW	THWN	TW	THW	THWN	TW	THW	THWN	THW	THWN	THW	THWN	THW	THWN
#14	9	6	13	15	10	24	25	16	39	44	29	69	60	40	94	99	65	154	142	93		143		192			
#12	7	4	10	12	8	18	19	13	29	35	24	51	47	32	70	78	53	114	111	76	164	117		157			
#10	5	4	6	9	6	11	15	11	18	26	19	32	36	26	44	60	43	73	85	61	104	95	160	127		163	
#8	2	1	3	4	3	5	7	5	9	12	10	16	17	13	22	28	22	36	40	32	51	49	79	66	106	85	136
#6		1	1		2	4		4	6		7	11		10	15		16	26		23	37	36	57	48	76	62	98
#4		1	1		1	2		3	4		5	7		7	9		12	16		17	22	27	35	36	47	47	60
#3		1	1		1	1		2	3		4	6		6	8		10	13		15	19	23	29	31	39	40	51
#2		1	1		1	1		2	3		4	5		5	7		9	11		13	16	20	25	27	33	34	43
#1					1	1		1	1		3	3		4	5		6	8		9	12	14	18	19	25	25	32
1/0					1	1		1	1		2	3		3	4		5	7		8	10	12	15	16	21	21	27
2/0					1	1		1	1		1	2		3	3		5	6		7	8	10	13	14	17	18	22
3/0					1	1		1	1		1	1		2	3		4	5		6	7	9	11	12	14	15	18
4/0						1		1	1		1	1		1	2		3	4		5	6	7	9	10	12	13	15
250 kcmil								1	1		1	1		1	1		2	3		4	4	6	7	8	10	10	12
300								1	1		1	1		1	1		2	3		3	4	5	6	7	8	9	11
350									1		1	1		1	1		1	2		3	3	4	5	6	7	8	9
400											1	1		1	1		1	1		2	3	4	5	5	6	7	8
500											1	1		1	1		1	1		1	2	3	4	4	5	6	7
600												1		1	1		1	1		1	1	3	3	4	4	5	5
700														1	1		1	1		1	1	2	3	3	3	4	5
750														1	1		1	1		1	1	2	2	3	3	4	4

R16132-215   Metric Equivalent, Conduit

U.S. vs. European Conduit – Approximate Equivalents			
United States		European	
Trade Size	Inside Diameter Inch/mm	Trade Size	Inside Diameter mm
½	.622/15.8	11	16.4
¾	.824/20.9	16	19.9
1	1.049/26.6	21	25.5
1¼	1.380/35.0	29	34.2
1½	1.610/40.9	36	44.0
2	2.067/52.5	42	51.0
2½	2.469/62.7		
3	3.068/77.9		
3½	3.548/90.12		
4	4.026/102.3		
5	5.047/128.2		
6	6.065/154.1		

16 ELECTRICAL

REFERENCE NOS.

R16132-221 Conduit Weight Comparisons (Lbs. per 100 ft.) Empty

Type	1/2"	3/4"	1"	1-1/4"	1-1/2"	2"	2-1/2"	3"	3-1/2"	4"	5"	6"
Rigid Aluminum	28	37	55	72	89	119	188	246	296	350	479	630
Rigid Steel	79	105	153	201	249	332	527	683	831	972	1314	1745
Intermediate Steel (IMC)	60	82	116	150	182	242	401	493	573	638		
Electrical Metallic Tubing (EMT)	29	45	65	96	111	141	215	260	365	390		
Polyvinyl Chloride, Schedule 40	16	22	32	43	52	69	109	142	170	202	271	350
Polyvinyl Chloride Encased Burial						38		67	88	105	149	202
Fibre Duct Encased Burial						127		164	180	206	400	511
Fibre Duct Direct Burial						150		251	300	354		
Transite Encased Burial						160		240	290	330	450	550
Transite Direct Burial						220		310		400	540	640

R16132-222 Conduit Weight Comparisons (Lbs. per 100 ft.) with Maximum Cable Fill*

Type	1/2"	3/4"	1"	1-1/4"	1-1/2"	2"	2-1/2"	3"	3-1/2"	4"	5"	6"
Rigid Galvanized Steel (RGS)	104	140	235	358	455	721	1022	1451	1749	2148	3083	4343
Intermediate Steel (IMC)	84	113	186	293	379	611	883	1263	1501	1830		
Electrical Metallic Tubing (EMT)	54	116	183	296	368	445	641	930	1215	1540		

*Conduit & Heaviest Conductor Combination

R16132-230 Conduit In Concrete Slab

List conduit by quantity, size and type.

Example:
   A. Rigid galvanized steel + size
   B. P.V.C. + size

Rigid galvanized steel (A) contains per 100 L.F.:
   1. (20) Ties to slab reinforcing
   2. (11) Threaded steel couplings
   3. (2) Factory sweeps
   4. (2) Field threaded conduit terminations

   5. (2) Fiber bushings + locknuts
   6. (2) Removal of concentric knockouts

P.V.C. (B) contains per 100 L.F.:
   1. (20) Ties to slab reinforcing
   2. (11) Field cemented couplings
   3. (2) Factory sweeps
   4. (2) Adapters
   5. (2) Removal of concentric knockouts

R16132-240 Conduit In Trench

Conduit in trench is galvanized steel and contains per 100 L.F.:
   1. (11) Threaded couplings
   2. (2) Factory sweeps
   3. (2) Fiber bushings + (4) locknuts
   4. (2) Field threaded conduit terminations
   5. (2) Removal of concentric knockouts

Note:

Conduit in sections 16132-230 and 16132-240 do not include:
   1. Floor cutting
   2. Excavation or backfill
   3. Grouting or patching

Conduit fittings in excess of those listed in the above Conduit models must be added. (Refer to R16132-205 and R16131-100 for Procedure example.)

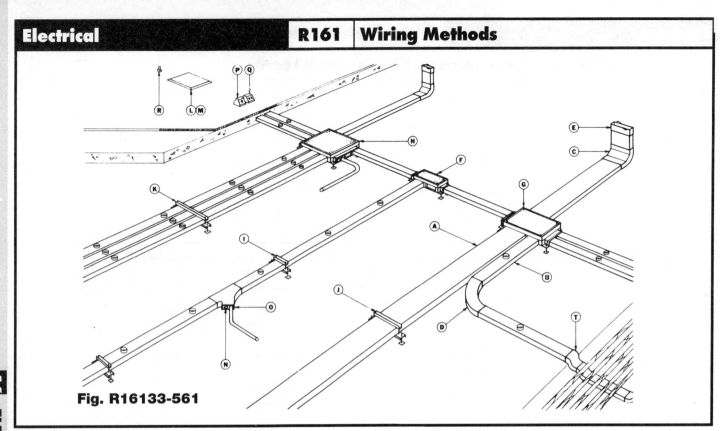

Fig. R16133-561

R16133-560 Underfloor Duct

When pricing Underfloor Duct it is important to identify and list each component, since costs vary significantly from one type of fitting to another. Do not deduct boxes or fittings from linear foot totals; this will be your allowance for scrap.

The first step is to identify the system as either:

      FIG. R16133-561 Single Level

      FIG. R16133-562 Dual Level

Single Level System

Include on your "takeoff sheet" the following unit price items, making sure to distinquish between Standard and Super duct:

  A. Feeder duct (blank) in L.F.
  B. Distribution duct (Inserts 2′ on center) in L.F.
  C. Elbows (Vertical) Ea.
  D. Elbows (Horizontal) Ea.
  E. Cabinet connector Ea.
  F. Single duct junction box Ea.
  G. Double duct junction box Ea.
  H. Triple duct junction box Ea.
  I. Support, single cell Ea.
  J. Support, double cell Ea.
  K. Support, triple cell Ea.
  L. Carpet pan Ea.
  M. Terrazzo pan Ea.
  N. Insert to conduit adapter Ea.
  O. Conduit adapter Ea.
  P. Low tension outlet Ea.
  Q. High tension outlet Ea.
  R. Galvanized nipple Ea.
  S. Wire per C.L.F.
  T. Offset (Duct type) Ea.

Dual Level System + Labor
see next page

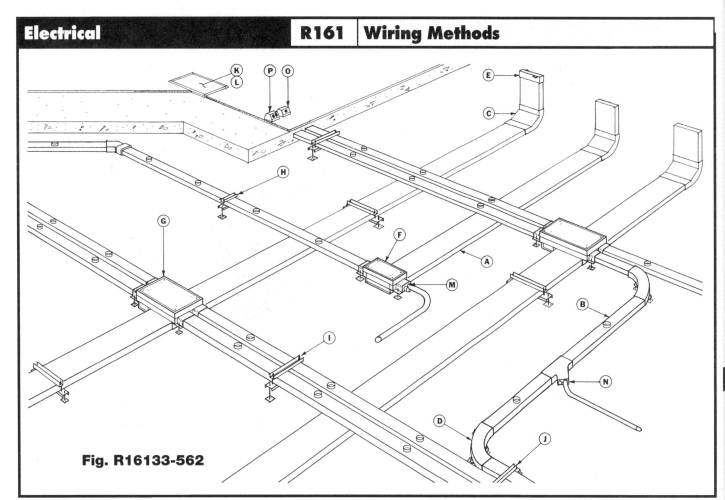

Fig. R16133-562

R16133-560 Underfloor Duct (cont.)

Dual Level

Include the following when "taking off" Dual Level systems:

Distinguish between Standard and Super duct.

A. Feeder duct (blank) in L.F.
B. Distribution duct (Inserts 2' on center) in L.F.
C. Elbows (Vertical) Ea.
D. Elbows (Horizontal) Ea.
E. Cabinet connector Ea.
F. Single duct, 2 level, junction box Ea.
G. Double duct, 2 level, junction box Ea.
H. Support, single cell Ea.
I. Support, double cell Ea.
J. Support, triple cell Ea.
K. Carpet pan Ea.
L. Terrazzo pan Ea.
M. Insert to conduit adapter Ea.
N. Conduit adapter Ea.
O. Low tension outlet Ea.
P. High tension outlet Ea.
Q. Wire per C.L.F.

Note: Make sure to include risers in linear foot totals. High tension outlets include box, receptacle, covers and related mounting hardware.

Labor-hours for both Single and Dual Level systems include:

1. Unloading and uncrating
2. Hauling up to 200' from loading dock
3. Measuring and marking
4. Setting raceway and fittings in slab or on grade
5. Leveling raceway and fittings

Labor-hours do not include:

1. Floor cutting
2. Excavation or backfill
3. Concrete pour
4. Grouting or patching
5. Wire or wire pulls
6. Additional outlets after concrete is poured
7. Piping to or from Underfloor Duct

Note: Installation is based on installing up to 150' of duct. If quantities exceed this, deduct the following percentages:

1. 150' to 250' - 10%
2. 250' to 350' - 15%
3. 350' to 500' - 20%
4. over 500' - 25%

Deduct these percentages from labor only.

Deduct these percentages from straight sections only.

Do not deduct from fittings or junction boxes.

Job Conditions: Productivity is based on new construction.

Underfloor duct to be installed on first three floors.

Material staging area within 100' of work being performed.

Area unobstructed and duct not subject to physical damage.

R16134-150 Wireway

When "taking off" Wireway, list by size and type.

Example:

1. Screw cover, unflanged + size
2. Screw cover, flanged + size
3. Hinged cover, flanged + size
4. Hinged cover, unflanged + size

Each 10' length on Wireway contains:

1. 10' of cover either screw or hinged type
2. (1) Coupling or flange gasket
3. (1) Wall type mount

All fittings must be priced separately.

Substitution of hanger types is done the same as described in R16132-205, "HANGERS", keeping in mind that the wireway model is based on 10' sections instead of a 100' conduit run.

Labor-hours for wireway include:

1. Unloading by hand
2. Hauling by hand up to 100' from loading dock
3. Measuring and marking
4. Mounting wall bracket using (2) anchor type lead fasteners
5. Installing wireway on brackets, to 15' high (For higher elevations use factors in R16132-205)

Job Conditions: Productivity is based on new construction, to a height of 15' using rolling staging in an unobstructed area.

Material staging area is assumed to be within 100' of work being performed. For other heights see R16132-210.

R16136-600   Outlet Boxes

Outlet boxes should be included on the same takeoff sheet as branch piping or devices to better explain what is included in each circuit.

Each unit price in this section is a stand alone item and contains no other component unless specified. For example, to estimate a duplex outlet, components that must be added are:

1. 4″ square box
2. 4″ plaster ring
3. Duplex receptacle from Div. 16140-910
4. Device cover

The method of mounting outlet boxes is (2) plastic shield fasteners.

Outlet boxes plastic, labor-hours include:
1. Marking box location on wood studding
2. Mounting box

Economy of Scale – For large concentrations of plastic boxes in the same area deduct the following percentages from labor-hour totals:

1	to	10	0%
11	to	25	20%
26	to	50	25%
51	to	100	30%
	over	100	35%

Note: It is important to understand that these percentages are not used on the total job quantities, but only areas where concentrations exceed the levels specified.

R16136-700   Pull Boxes and Cabinets

List cabinets and pull boxes by NEMA type and size.

Example:	TYPE	SIZE
	NEMA 1	6″W x 6″H x 4″D
	NEMA 3R	6″W x 6″H x 4″D

Labor-hours for wall mount (indoor or outdoor) installations include:
1. Unloading and uncrating
2. Handling of enclosures up to 200′ from loading dock using a dolly or pipe rollers
3. Measuring and marking
4. Drilling (4) anchor type lead fasteners using a hammer drill
5. Mounting and leveling boxes

Note: A plywood backboard is not included.

Labor-hours for ceiling mounting include:
1. Unloading and uncrating
2. Handling boxes up to 100′ from loading dock

3. Measuring and marking
4. Drilling (4) anchor type lead fasteners using a hammer drill
5. Installing and leveling boxes to a height of 15′ using rolling staging

Labor-hours for free standing cabinets include:
1. Unloading and uncrating
2. Handling of cabinets up to 200′ from loading dock using a dolly or pipe rollers
3. Marking of floor
4. Drilling (4) anchor type lead fasteners using a hammer drill
5. Leveling and shimming

Labor-hours for telephone cabinets include:
1. Unloading and uncrating
2. Handling cabinets up to 200′ using a dolly or pipe rollers
3. Measuring and marking
4. Mounting and leveling, using (4) lead anchor type fasteners

R16136-705   Weight Comparisons of Common Size Cast Boxes in Lbs.

Size NEMA 4 or 9	Cast Iron	Cast Aluminum	Size NEMA 7	Cast Iron	Cast Aluminum
6″ x 6″ x 6″	17	7	6″ x 6″ x 6″	40	15
8″ x 6″ x 6″	21	8	8″ x 6″ x 6″	50	19
10″ x 6″ x 6″	23	9	10″ x 6″ x 6″	55	21
12″ x 12″ x 6″	52	20	12″ x 6″ x 6″	100	37
16″ x 16″ x 6″	97	36	16″ x 16″ x 6″	140	52
20″ x 20″ x 6″	133	50	20″ x 20″ x 6″	180	67
24″ x 18″ x 8″	149	56	24″ x 18″ x 8″	250	93
24″ x 24″ x 10″	238	88	24″ x 24″ x 10″	358	133
30″ x 24″ x 12″	324	120	30″ x 24″ x 10″	475	176
36″ x 36″ x 12″	500	185	30″ x 24″ x 12″	510	189

R16140-910   Wiring Devices

Wiring devices should be priced on a separate takeoff form which includes boxes, covers, conduit and wire.

Labor-hours for devices include:
1. Stripping of wire
2. Attaching wire to device using terminators on the device itself, lugs, set screws etc.
3. Mounting of device in box

Labor-hours do not include:
1. Conduit
2. Wire
3. Boxes
4. Plates

Economy of Scale – for large concentrations of devices in the same area deduct the following percentages from labor-hours:

1	to	10	0%
11	to	25	20%
26	to	50	25%
51	to	100	30%
	over	100	35%

ELECTRICAL 16

REFERENCE NOS.

R16140-910 Wiring Devices (cont.)

NEMA No.	15 R	20 R	30 R	50 R	60 R
1 125V 2 Pole, 2 Wire					
2 250V 2 Pole, 2 Wire					
5 125V 2 Pole, 3 Wire					
6 250V 2 Pole, 3 Wire					
7 277V, AC 2 Pole, 3 Wire					
10 125/250V 3 Pole, 3 Wire					
11 3 Phase 250V 3 Pole, 3 Wire					
14 125/250V 3 Pole, 4 Wire					
15 3 Phase 250V 3 Pole, 4 Wire					
18 3 Phase 208Y/120V 4 Pole, 4 Wire					

R16140-910 Wiring Devices (cont.)

NEMA No.	15 R	20 R	30 R	NEMA No.	15 R	20 R	30 R
L 1 125V 2 Pole, 2 Wire	◯			**L 13** 3 Phase 600V 3 Pole, 3 Wire			◯
L 2 250V 2 Pole, 2 Wire		◯ 15A		**L 14** 125/250V 3 Pole, 4 Wire		◯	◯
L 5 125 V 2 Pole, 3 Wire	◯	◯	◯	**L 15** 3 Phase 250 V 3 Pole, 4 Wire		◯	◯
L 6 250 V 2 Pole, 3 Wire	◯	◯	◯	**L 16** 3 Phase 480V 3 Pole, 4 Wire		◯	◯
L 7 227 V, AC 2 Pole, 3 Wire	◯	◯	◯	**L 17** 3 Phase 600 V 3 Pole, 4 Wire			◯
L 8 480 V 2 Pole, 3 Wire		◯	◯	**L 18** 3 Phase 208Y/120V 4 Pole, 4 Wire		◯	◯
L 9 600 V 2 Pole, 3 Wire		◯	◯	**L 19** 3 Phase 480Y/277V 4 Pole, 4 Wire		◯	◯
L 10 125 /250V 3 Pole, 3 Wire		◯	◯	**L 20** 3 Phase 600Y/347V 4 Pole, 4 Wire		◯	◯
L 11 3 Phase 250 V 3 Pole, 3 Wire		◯	◯	**L 21** 3 Phase 208Y/120V 4 Pole, 5 Wire		◯	◯
L 12 3 Phase 480 V 3 Pole, 3 Wire		◯	◯	**L 22** 3 Phase 480Y/277V 4 Pole, 5 Wire		◯	◯
				L 23 3 Phase 600Y/347V 4 Pole, 5 Wire		◯	◯

R16150-275 Motor Connections

Motor connections should be listed by size and type of motor.
Included in the material and labor cost is:

1. (2) Flex connectors
2. 18″ of flexible metallic wireway
3. Wire identification and termination
4. Test for rotation
5. (2) or (3) Conductors

. Price does not include:

1. Mounting of motor
2. Disconnect Switch
3. Motor Starter
4. Controls
5. Conduit or wire ahead of flex

Note: When "Taking off" Motor connections, it is advisable to list connections on the same quantity sheet as Motors, Motor Starters and controls.

R16230-450 Generator Weight (Lbs.) by kW

3 Phase 4 Wire /480 Volt			
Gas		Diesel	
kW	Lbs.	kW	Lbs.
7.5	600	30	1800
10	630	50	2230
15	960	75	2250
30	1500	100	3840
65	2350	125	4030
85	2570	150	5500
115	4310	175	5650
170	6530	200	5930
		250	6320
		300	7840
		350	8220
		400	10750
		500	11900

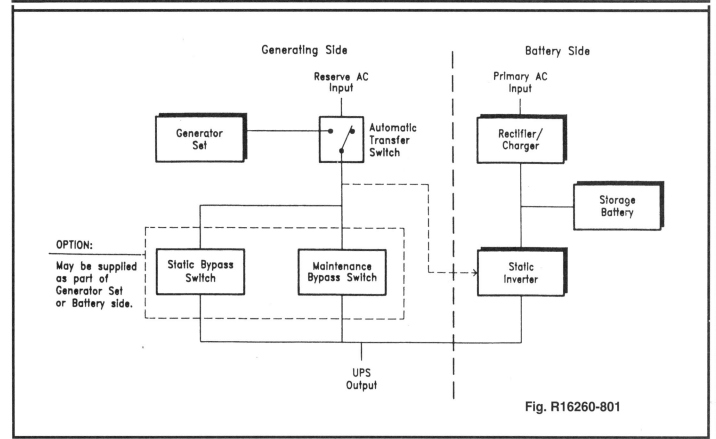

Fig. R16260-801

R16260-800 Uninterruptible Power Supply Systems

General: Uninterruptible Power Supply (UPS) Systems are used to provide power for legally required standby systems. They are also designed to protect computers and provide optional additional coverage for any or all loads in a facility. Figure R16260-801 shows a typical configuration for a UPS System with generator set.

Cost Determination:

It is recommended that UPS System material costs be obtained from equipment manufacturers as a package. Installation costs should be obtained from a vendor or contractor.

The recommended procedure for pricing UPS Systems would be to identify:
1. Frequency – by Hertz (Hz.)
   For example: 50, 60, and 400/415 Hz.
2. Apparent power and real power
   For example: 200 kVA/170 kW
3. Input/Output voltage   For example: 120V, 208V or 240V

4. Phase – either single or three-phase
5. Options and accessories – For extended run time more batteries would be required. Other accessories include battery cabinets, battery racks, remote control panel, power distribution unit (PDU), and warranty enhancement plans.

Note:
1. Larger systems can be configured by paralleling two or more standard small size single modules.
   For example: two 15 kVA modules, combined together and configured to 30 kVA.
2. Maximum input current during battery recharge is typically 15% higher than normal current.
3. UPS Systems weights vary depending on options purchased and power rating in kVA.

ELECTRICAL 16

REFERENCE NOS.

R16270-600 Oil Filled Transformers

Transformers in this section include:
1. Rigging (as required)
2. Rental of crane and operator
3. Setting of oil filled transformer
4. (4) Anchor bolts, nuts and washers in concrete pad

Price does not include:
1. Primary and secondary terminations
2. Transformer pad
3. Equipment grounding
4. Cable
5. Conduit locknuts or bushings

DRY TYPE TRANSFORMERS Section 16270-200
BUCK-BOOST TRANSFORMERS Section 16270-100
ISOLATING TRANSFORMERS Section 16280-340

Transformers in these sections include:
1. Unloading and uncrating
2. Hauling transformer to within 200' of loading dock
3. Setting in place
4. Wall mounting hardware
5. Testing

Price does not include:
1. Structural supports
2. Suspension systems
3. Welding or fabrication
4. Primary & secondary terminations

Add the following percentages to the labor for ceiling mounted transformers:

10' to 15'	= + 15%
15' to 25'	= + 30%
Over 25'	= + 35%

Job Conditions: Productivities are based on new construction. Installation is assumed to be on the first floor, in an obstructed area to a height of 10'. Material staging area is within 100' of final transformer location.

R16270-605 Transformer Weight (Lbs.) by kVA

Oil Filled 3 Phase 5/15 KV To 480/277			
kVA	Lbs.	kVA	Lbs.
150	1800	1000	6200
300	2900	1500	8400
500	4700	2000	9700
750	5300	3000	15000

Dry 240/480 To 120/240 Volt			
1 Phase		3 Phase	
kVA	Lbs.	kVA	Lbs.
1	23	3	90
2	36	6	135
3	59	9	170
5	73	15	220
7.5	131	30	310
10	149	45	400
15	205	75	600
25	255	112.5	950
37.5	295	150	1140
50	340	225	1575
75	550	300	1870
100	670	500	2850
167	900	750	4300

R16290-800 Switchboard Instruments

Switchboard instruments are added to the price of switchboards according to job specifications. This equipment is usually included when ordering "Gear" from the manufacturer and will arrive factory installed.

Included in the labor cost is:
1. Internal wiring connections
2. Wire identification
3. Wire tagging

Transition sections include:
1. Uncrating
2. Hauling sections up to 100' from loading dock
3. Positioning sections
4. Leveling sections
5. Bolting enclosures
6. Bolting vertical bus bars

Price does not include:
1. Equipment pads
2. Steel channels embedded or grouted in concrete
3. Special knockouts
4. Rigging

Job Conditions: Productivity is based on new construction, equipment to be installed on the first floor within 200' of the loading dock.

R16310-600 Average Transmission Line Material Requirements (Per Mile)

Conductor: 397,500 – Cir. Mil., 26/7–ACSR

Terrain:		Flat				Rolling				Mountain			
Item		69kV	161kV	161kV	500kV	69kV	161kV	161kV	500kV	69kV	161kV	161kV	500kV
Pole Type	Unit	Wood	Wood	Steel	Steel	Wood	Wood	Steel	Steel	Wood	Wood	Steel	Steel
Structures	Ea.	12[1]				9[2]				7[2]			
Poles	Ea.	12				18				14			
Crossarms	Ea.	24[3]				9[4]				7[4]			
Conductor	Ft.	15,990				15,990				15,990			
Insulators[5]	Ea.	180				135				105			
Ground Wire	Ft.	5,330				10,660				10,660			

Conductor: 636,000 – Cir. Mil., 26/7–ACSR

Item	Unit	69kV Wood	161kV Wood	161kV Steel	500kV Steel	69kV Wood	161kV Wood	161kV Steel	500kV Steel	69kV Wood	161kV Wood	161kV Steel	500kV Steel
Structures	Ea.	13[1]	11[6]			10[2]	9[2]	6[9]		8[2]	8[2]	6[9]	
Excavation	C.Y.	–	–			–	–	120		–	–	120	
Concrete	C.Y.	–	–			–	–	10		–	–	10	
Steel Towers	Tons	–	–			–	–	32		–	–	32	
Poles	Ea.	13	11			20	18	–		16	16	–	
Crossarms	Ea.	26[3]	33[7]			10[4]	9[8]	–		8[4]	8[8]	–	
Conductor	Ft.	15,990	15,990			15,990	15,990	15,990		15,990	15,990	15,990	
Insulators[5]	Ea.	195	165			150	297	297		120	264	297	
Ground Wire	Ft.	5330	5330			10,660	10,660	10,660		10,660	10,660	10,660	

Conductor: 954,000 – Cir. Mil., 45/7–ACSR

Item	Unit	69kV Wood	161kV Wood	161kV Steel	500kV Steel	69kV Wood	161kV Wood	161kV Steel	500kV Steel	69kV Wood	161kV Wood	161kV Steel	500kV Steel
Structures	Ea.	14[1]	12[6]		4[11]	10[2]	9[2]	6[9]	4[11]	8[2]	8[2]	6[9]	4[13]
Excavation	C.Y.	–	–		200	–	–	125	214	–	–	125	233
Concrete	C.Y.	–	–		20	–	–	10	21	–	–	10	21
Steel Towers	Tons	–	–		57	–	–	33	57	–	–	33	63
Poles	Ea.	14	12		–	20	18	–	–	16	16	–	–
Crossarms	Ea.	28[10]	36[7]		–	10[4]	9[8]	–	–	8[4]	8[8]	–	–
Conductor	Ft.	15,990	15,990		47,970[12]	15,990	15,990	15,990	47,970[12]	15,990	15,990	15,990	47,970[12]
Insulators[5]	Ea.	210	180		288	150	297	297	288	120	264	297	576
Ground Wire	Ft.	5330	5330		10,660	10,660	10,660	10,660	10,660	10,660	10,660	10,660	10,660

Conductor: 1,351,500 – Cir. Mil., 45/7–ACSR

Item	Unit	Flat	Rolling
Structures	Ea.	8[14]	8[14]
Excavation	C.Y.	220	220
Concrete	C.Y.	28	28
Steel Towers	Ton	46	46
Conductor	Ft.	31,680[15]	31,680[15]
Insulators	Ea.	528	528
Ground Wire	Ft.	10,660	10,660

1. Single pole two-arm suspension type construction
2. Two-pole wood H-frame construction
3. 4¾" x 5¾" x 8' and 4¾" x 5¾" x 10' wood crossarm
4. 6" x 8" x 26'-0" wood crossarm
5. 5¾" x 10" disc insulator
6. Single pole construction with 3 fiberglass crossarms (5 fog- type insulators per phase)
7. 7'-0" fiberglass crossarms
8. 6" x 10" x 35'-0" wood crossarm
9. Laced steel tower, single circuit construction
10. 5" x 7" x 8' and 5" x 7" x 10' wood crossarm
11. Laced steel tower, single circuit 500-kV construction
12. Bundled conductor (3 sub-conductors per phase)
13. Laced steel tower, single circuit restrained phases (500-kV)
14. Laced steel tower, double circuit construction
15. Both sides of double circuit strung

Note: To allow for sagging, a mile (5280 Ft.) of transmission line uses 5330 Ft. of conductor per wire (called a wire mile).

ELECTRICAL 16

REFERENCE NOS.

R16410-800   Safety Switches

List each Safety Switch by type, ampere rating, voltage, single or three phase, fused or nonfused.

Example:

  A. General duty, 240V, 3-pole, fused
  B. Heavy Duty, 600V, 3-pole, nonfused
  C. Heavy Duty, 240V, 2-pole, fused
  D. Heavy Duty, 600V, 3-pole, fused

Also include NEMA enclosure type and identify as:

1. Indoor
2. Weatherproof
3. Explosionproof

Installation of Safety Switches includes:

1. Unloading and uncrating
2. Handling disconnects up to 200' from loading dock
3. Measuring and marking location
4. Drilling (4) anchor type lead fasteners, using a hammer drill
5. Mounting and leveling Safety Switch
6. Installing (3) fuses
7. Phasing and tagging line and load wires

Price does not include:

1. Modifications to enclosure
2. Plywood backboard
3. Conduit or wire
4. Fuses
5. Termination of wires

Job Conditions: Productivities are based on new construction to an installed height of 6' above finished floor.

Material staging area is assumed to be within 100' of work in progress.

R16415-600   Automatic Transfer Switches

When taking off Automatic transfer switches identify by voltage, amperage and number of poles.

Example: Automatic transfer switch, 480V, 3 phase, 30A, NEMA 1 enclosure

Labor-hours for transfer switches include:

1. Unloading, uncrating and handling switches up to 200' from loading dock
2. Measuring and marking
3. Drilling (4) lead type anchors, using a hammer drill
4. Mounting and leveling
5. Circuit identification
6. Testing and load balancing

Labor-hours do not include:

1. Modifications in enclosure
2. Structural supports
3. Additional lugs
4. Plywood backboard
5. Painting or lettering
6. Conduit runs to or from transfer switch
7. Wire
8. Termination of wires

R16440-500   Load Centers and Panelboards

When pricing Load Centers list panels by size and type. List Breakers in a separate column of the "Quantity Sheet," and define by phase and ampere rating.

Material and Labor prices include breakers; for example: A 100 Amp. 3-wire, 120/240V, 18 circuit panel w/main breaker as described in 16440-500-3900 in the unit cost file contains 18 single pole 20A breakers.

If you do not choose to include a full panel of single pole breakers, use the following method to adjust material and labor costs.

Example: In an 18 circuit panel only 16 single pole breakers are required.

    18 breakers included
    16 breakers needed
     2 breakers need to be adjusted

In order to adjust the (2) single pole breakers not required, go to the unit price section of this book which in this example is 16440-700, and price the unit cost of a 20A single pole breaker.

Example:

16440-700-2000 1 pole breaker = $ 8.20   for material
                              $26.50   for labor
                              $34.70   total material and labor

Take the material price of the breaker and multiply by .50.

        $8.20   x .50 = $4.10

Take the labor price and multiply by .60.

        $26.50   x .60 = $15.90

The adjustment per breaker will be $4.10 for material and $15.90 for labor.

Take the adjusted material and labor costs and multiply by the number of breakers not required, in this case (2).

Example:

    $ 4.10 x 2 = $ 8.20 Material
    $15.90 x 2 = $31.80 Labor
                         $40.00 To be deducted from cost of panel

The adjusted material and labor cost of the panel will be:

Material	Labor	Total
$370.00	$395.00	$765.00
-8.20	-31.80	-40.00
$361.80	$363.20	$725.00 Adjusted Load Center

Labor-hours for Load Center installation includes:
1. Unloading, uncrating, and handling enclosures 200' from unloading area.
2. Measuring and marking
3. Drilling (4) lead anchor type fasteners using a hammer drill
4. Mounting and leveling panel to a height of 6'
5. Preparation and termination of feeder cable to lugs or main breaker
6. Branch circuit identification
7. Lacing using tie wraps
8. Testing and load balancing
9. Marking panel directory

Not included in the material and labor are:
1. Modifications to enclosure
2. Structural supports
3. Additional lugs
4. Plywood backboards
5. Painting or lettering

Note: Knockouts are included in the price of terminating pipe runs and need not be added to the Load Center costs.

Job Conditions: Productivity is based on new construction to a height of 6', in a unobstructed area. Material staging area is assumed to be within 100' of work being performed.

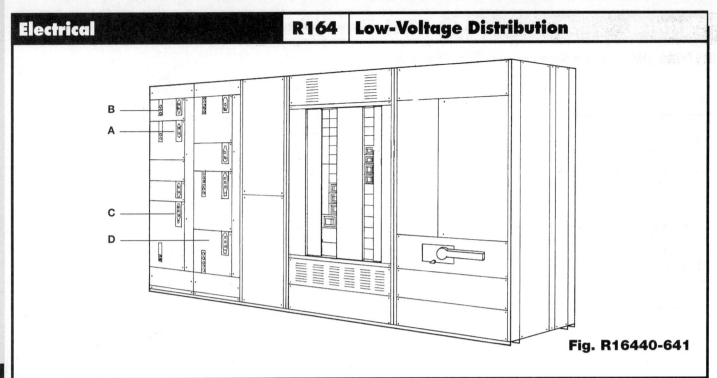

Fig. R16440-641

R16440-640 Motor Control Centers

When taking off Motor Control Centers, list the size, type and height of structures.

Example:
1. 600A, 22,000 RMS, 72" high
2. 600A, back to back, 72" high

Next take off individual starters; the number of structures can also be determined by adding the height in inches of starters divided by the height of the structure, and list on the same quantity sheet as the structures. Identify starters by Type, Horsepower rating, Size and Height in inches.

Example:
A. Class l, Type B, FVNR starter, 25 H.P., 18" high.
   Add to the list with Starters, factory installed controls.

Example:

B. Pilot lights
C. Push buttons
D. Auxiliary contacts

Identify starters and structures as either copper or aluminum and by the NEMA type of enclosure.

When pricing starters and structures, be sure to add or deduct adjustments, using lines 16440-640-1100 thru 16440-640-1900 in the unit price section.

Included in the cost of Motor Control Structures are:
1. Uncrating
2. Hauling to location within 100' of loading dock
3. Setting structures
4. Leveling
5. Aligning
6. Bolting together structure frames
7. Bolting horizontal bus bars

Labor-hours do not include:
1. Equipment pad
2. Steel channels embedded or grouted in concrete
3. Pull boxes
4. Special knockouts
5. Main switchboard section
6. Transition section
7. Instrumentation
8. External control wiring
9. Conduit or wire

Material for Starters includes:
1. Circuit breaker or fused disconnect
2. Magnetic motor starter
3. Control transformer
4. Control fuse and fuse block

Labor-hours for Starters include:
1. Handling
2. Installing starter within structure
3. Internal wiring connections
4. Lacing within enclosure
5. Testing
6. Phasing

Job Conditions: Productivity is based on new construction. Motor control location assumed to be on first floor in an unobstructed area. Material staging area within 100' of final location.

Note: Additional labor-hours must be added if M.C.C. is to be installed on other than the first floor, or if rigging is required.

R16440-660 Motor Starters and Controls

Motor starters should be listed on the same "Quantity Sheet" as Motors and Motor Connections. Identify each starter by:

1. Size
2. Voltage
3. Type

Example:

A. FVNR, 480V, 2HP, Size 00
B. FVNR, 480V, 5HP, Size 0, Combination type
C. FVR, 480V, Size 2

The NEMA type of enclosure should also be identified.

Included in the labor-hours are:

1. Unloading, uncrating and handling of starters up to 200' from loading area
2. Measuring and marking
3. Drilling (4) anchor type lead fasteners, using a hammer drill
4. Mounting and leveling starter
5. Connecting wire or cable to line and load sides of starter (when already lugged)
6. Installation of (3) thermal type heaters
7. Testing

The following is not included unless specified in the unit price description.

1. Control transformer
2. Controls, either factory or field installed
3. Conduit and wire to or from starter
4. Plywood backboard
5. Cable terminations

The following material and labor has been included for Combination type starters:

1. Unloading, uncrating and handling of starter up to 200' of loading dock
2. Measuring and marking
3. Drilling (4) anchor type lead fasteners using a hammer drill
4. Mounting and leveling
5. Connecting prepared cable conductors
6. Installation of (3) dual element cartridge type fuses
7. Installation of (3) thermal type heaters
8. Test for rotation

MOTOR CONTROLS

When pricing motor controls make sure you consider the type of control system being utilized. If the controls are factory installed and located in the enclosure itself, then you would add to your starter price the items 16440-660-5200 thru 16440-660-5800 in the unit cost file. If control voltage is different from line voltage, add the material and labor cost of a control transformer.

For external control of starters include the following items:

1. Raceways
2. Wire & terminations
3. Control enclosures
4. Fittings
5. Push button stations
6. Indicators

Job Conditions: Productivity is based on new construction to a height of 10'.

Material staging area is assumed to be within 100' of work being performed.

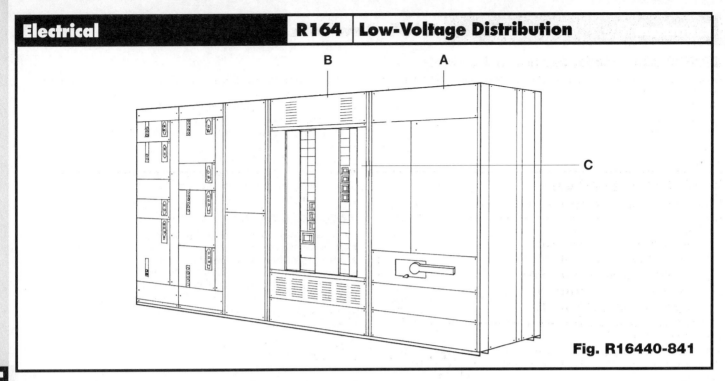

Fig. R16440-841

R16440-800   Distribution Section

After "Taking off" the Switchboard section, include on the same "Quantity sheet" the Distribution section; identify by:

1. Voltage
2. Ampere rating
3. Type

Example:

(Fig. R16440-841) (B) Distribution Section

Included in the labor costs of the Distribution section is:

1. Uncrating
2. Hauling to 200' of loading dock
3. Setting of distribution panel
4. Leveling & shimming
5. Anchoring of equipment to pad or floor
6. Bolting of horizontal bus bars between
7. Testing of equipment

Not included in the Distribution section is:

1. Breakers (C)
2. Equipment pads
3. Steel channels embedded or grouted in concrete
4. Pull boxes
5. Special knockouts
6. Transition section
7. Conduit or wire

R16440-820 Feeder Section

List quantities on the same sheet as the Distribution section. Identify breakers by: (Fig. R16440-841)
1. Frame type
2. Number of poles
3. Ampere rating

Installation includes:
1. Handling

2. Placing breakers in distribution panel
3. Preparing wire or cable
4. Lacing wire
5. Marking each phase with colored tape
6. Marking panel legend
7. Testing
8. Balancing

R16440-840 Switchgear

It is recommended that "Switchgear" or those items contained in the following sections be quoted from equipment manufacturers as a package price.

16029-800 Switchboard instruments

16440-800 Switchboard distribution sections

16440-820 Switchboard feeder sections

16440-840 Switchboard Service Disc.

16440-860 Switchboard In-plant Dist.

Included in these sections are the most common types and sizes of factory assembled equipment.

The recommended procedure for low voltage switchgear would be to price (Fig. R16440-841) (A) Main Switchboard Div. 16440-840 or 16440-860.

Identify by:
1. Voltage
2. Amperage
3. Type

Example:
1. 120/208V, 4-wire, 600A, nonfused
2. 277/480V, 4-wire, 600A, nonfused
3. 120/208V, 4-wire, 400A w/fused switch & CT compartment
4. 277/480V, 4-wire, 400A, w/fused switch & CT compartment

5. 120/208V, 4-wire, 800A, w/pressure switch & CT compartment
6. 277/480V, 4-wire, 800A, w/molded CB & CT compartment

Included in the labor costs for Switchboards are:
1. Uncrating
2. Hauling to 200' of loading dock
3. Setting equipment
4. Leveling and shimming
5. Anchoring
6. Cable identification
7. Testing of equipment

Not included in the Switchboard price is:
1. Rigging
2. Equipment pads
3. Steel channels embedded or grouted in concrete
4. Special knockouts
5. Transition or Auxiliary sections
6. Instrumentation
7. External control
8. Conduit and wire
9. Conductor terminations

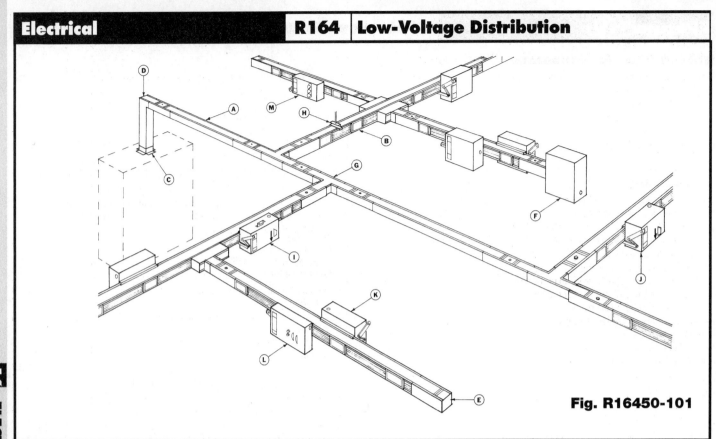

Fig. R16450-101

R16450-100   Aluminum Bus Duct

When taking off bus duct identify the system as either:

1. Aluminum (16450-100)
2. Copper (16450-320 and 16450-360)

List straight lengths by type and size (Fig. R16450-101)

  A. Plug-in — 800 A
  B. Feeder — 800 A

Do not measure thru fittings as you would on conduit, since there is no such thing as allowance for scrap in bus duct systems.

If upon taking off linear foot quantities of bus duct you find your quantities are not divisible by 10 ft., then the remainder must be priced as a special item and quoted from the manufactuer. Do not use the linear foot cost of material for these items, but you can use the total cost for L.F. labor.

Example:

	Quantity
800 A Feeder Bus (Alum.)	52 ft.
52 ÷ 10 =	5 — 10 ft. sections
	1 — 2 ft. section

From line 16450-100-0400

Material = 50 x $133.00 = $6,650.00 + price of 2 ft. section
Labor   = 52 x $ 22.50 = $1,170.00

Identify fittings by type and ampere rating

Example: (Fig. R16450-101)

C. Switchboard stub 800 A

D. Elbows 800 A

E. End box 800 A

F. Cable tap box 800 A

G. Tee Fittings 800 A

H. Hangers

Plug-in Units - List separately plug-in units and identify by type and ampere rating (Fig. R16450-101)

I. Plug-in switches 600 Volt 3 phase 60 A

J. Plug-in molded case C.B. 60 A

K. Combination starter FVNR NEMA 1

L. Combination contactor & fused switch NEMA 1

M. Combination fusible switch & lighting control 60 A

Labor-hours for feeder and plug-in sections include:

1. Unloading and uncrating
2. Hauling up to 200 ft. from loading dock
3. Measuring and marking
4. Set up of rolling staging
5. Installing hangers
6. Hanging and bolting sections
7. Aligning and leveling
8. Testing

Labor-hours do not include:

1. Modifications to existing structure for hanger supports
2. Threaded rod in excess of 2 ft.
3. Welding
4. Penetrations thru walls
5. Staging rental

Deduct the following percentages from labor only:

        150 ft. to 250 ft. — 10%
        251 ft. to 350 ft. — 15%
        351 ft. to 500 ft. — 20%
           Over 500 ft. — 25%

Deduct percentage only if runs are contained in the same area.

Example — If the job entails running 100 ft. in 5 different locations do not deduct 20%, but if the duct is being run in 1 area and the quantity is 500 ft. then you would deduct 20%.

Deduct only from straight lengths, not fittings or plug-in units.

R16450-100 Aluminum Bus Duct (cont.)

Add to labor for elevated installations:

15 ft. to 20 ft. high	10%
21 ft. to 25 ft. high	20%
26 ft. to 30 ft. high	30%
31 ft. to 35 ft. high	40%
36 ft. to 40 ft. high	50%
Over 40 ft. high	60%

Bus Duct Fittings:

Labor-hours for fittings include:
1. Unloading and uncrating
2. Hauling up to 200 ft. from loading dock
3. Installing, fitting, and bolting all ends to in place sections

Plug-in units include:
1. Unloading and uncrating
2. Hauling up to 200 ft. from loading dock
3. Installing plug-in into in place duct
4. Set up of rolling staging
5. Connection load wire to lugs
6. Marking wire
7. Checking phase rotation

Labor-hours for plug-ins do not include:
1. Conduit runs from plug-in
2. Wire from plug-in
3. Conduit termination

Economy of Scale - For large concentrations of plug-in units in the same area deduct the following percentages:

11	to	25	15%
26	to	50	20%
51	to	75	25%
76	to	100	30%
100	and	over	35%

Job Conditions: Productivities are based on new construction, in an unobstructed first floor area to a height of 15 ft.; using rolling staging.

Material staging area is within 100 ft. of work being performed.
Add to the duct fittings and hangers:
  Plug-in switches (Fused)
  Plug-in breakers
  Combination starters
  Combination contactors
  Combination fusible switch and lighting control

Labor-hours for duct and fittings include:
1. Unloading and uncrating
2. Hauling up to 200 ft. from loading dock
3. Measuring and marking
4. Installing duct runs
5. Leveling
6. Sound testing

Labor-hours for plug-ins include:
1. Unloading and uncrating
2. Hauling up to 200 ft. from loading dock
3. Installing plug-ins
4. Preparing wire
5. Wire connections and marking

R16450-105 Weight (Lbs./L.F.) of 4 Pole Aluminum and Copper Bus Duct by Ampere Load

Amperes	Aluminum Feeder	Copper Feeder	Aluminum Plug–In	Copper Plug–In
225			7	7
400			8	13
600	10	10	11	14
800	10	19	13	18
1000	11	19	16	22
1350	14	24	20	30
1600	17	26	25	39
2000	19	30	29	46
2500	27	43	36	56
3000	30	48	42	73
4000	39	67		
5000		78		

ELECTRICAL 16

REFERENCE NOS.

R16510-105 Comparison - Cost of Operation of High Intensity Discharge Lamps

Lamp Type	Wattage	Life (Hours)	1 Circuit Wattage	Average Initial Lumens	2 L.L.D.	3 Mean Lumens	4 One Year (4000 Hr.) Cost of Operation
M.V.	100 DX	24,000	125	4,000	61%	2,440	$ 40.00
L.P.S.	SOX-35	18,000	65	4,800	100%	4,800	20.80
H.P.S.	LU-70	12,000	84	5,800	90%	5,220	26.90
L.H.	No Equivalent						
M.V.	175 DX	24,000	210	8,500	66%	5,676	67.20
M.V.	250 DX	24,000	295	13,000	66%	7,986	94.40
L.P.S.	SOX-55	18,000	82	8,000	100%	8,000	26.25
H.P.S.	LU-100	12,000	120	9,500	90%	8,550	38.40
L.H.	No Equivalent						
M.V.	400 DX	24,000	465	24,000	64%	14,400	148.80
L.P.S.	SOX-90	18,000	141	13,500	100%	13,500	45.10
H.P.S.	LU-150	16,000	188	16,000	90%	14,400	60.15
L.H.	LH-175	7,500	210	14,000	73%	10,200	67.20
M.V.	No Equivalent						
L.P.S.	SOX-135	18,000	147	22,500	100%	22,500	47.05
H.P.S.	LU-250	20,000	310	25,500	92%	23,205	99.20
L.H.	LH-250	7,500	295	20,500	78%	16,000	94.40
M.V.	1000 DX	24,000	1,085	63,000	61%	37,820	347.20
L.P.S.	SOX-180	18,000	248	33,000	100%	33,000	79.35
H.P.S.	LU-400	20,000	480	50,000	90%	45,000	153.60
L.H.	LH400	15,000	465	34,000	72%	24,600	148.80
H.P.S.	LU-1000	15,000	1,100	140,000	91%	127,400	352.00

1. Includes ballast losses and average lamp watts
2. Lamp lumen depreciation (% of initial light output at 70% rated life)
3. Lamp lumen output at 70% rated life (L.L.D. x initial)
4. Based on average cost of $.080 per k. W. Hr.

M.V.   = Mercury Vapor
L.P.S. = Low pressure sodium
H.P.S. = High pressure sodium
L.H.   = Metal halide

R16510-110 For Other than Regular Cool White (CW) Lamps

		Multiply Material Costs as Follows:				
Regular Lamps	Cool white deluxe (CWX)	x 1.35	Energy Saving Lamps	Cool white (CW/ES)	x 1.35
	Warm white deluxe (WWX)	x 1.35		Cool white deluxe (CWX/ES)	x 1.65
	Warm white (WW)	x 1.30		Warm white (WW/ES)	x 1.55
	Natural (N)	x 2.05		Warm white deluxe (WWX/ES)	x 1.65

R16510-120 Lamp Comparison Chart with Enclosed Floodlight, Ballast, & Lamp for Pole Mounting

Type	Watts	Initial Lumens	Lumens per Watt	Lumens @ 40% Life	Life (Hours)
Incandescent	150	2,880	19	85	750
	300	6,360	21	84	750
	500	10,850	22	80	1,000
	1,000	23,740	24	80	1,000
	1,500	34,400	23	80	1,000
Tungsten	500	10,950	22	97	2,000
Halogen	1,500	35,800	24	97	2,000
Fluorescent	40	3,150	79	88	20,000
Cool	110	9,200	84	87	12,000
White	215	16,000	74	81	12,000
Deluxe	250	12,100	48	86	24,000
Mercury	400	22,500	56	85	24,000
	1,000	63,000	63	75	24,000
Metal	175	14,000	80	77	7,500
Halide	400	34,000	85	75	15,000
	1,000	100,000	100	83	10,000
	1,500	155,000	103	92	1,500
High	70	5,800	83	90	20,000
Pressure	100	9,500	95	90	20,000
Sodium	150	16,000	107	90	24,000
	400	50,000	125	90	24,000
	1,000	140,000	140	90	24,000
Low	55	4,600	131	98	18,000
Pressure	90	12,750	142	98	18,000
Sodium	180	33,000	183	98	18,000

Color: High Pressure Sodium — Slightly Yellow
       Low Pressure Sodium — Yellow
       Mercury Vapor — Green-Blue
       Metal Halide — Blue White

Note: Pole not included.

R16510-200 Energy Efficiency Rating for Luminaires

The energy efficiency program for luminaires recommends the use of a
metric called the Luminaire Efficacy Rating (LER). The LER value expresses
the total luminaires generated by a lamp, to the watts consumed by the lamp.

Lamp Type	Lumens per Watt
Incandescent	17
Tungsten Halogen	14 - 20
Fluorescent	50 - 104
Metal Halide	64 - 96
High Pressure Sodium	76 - 116

ELECTRICAL 16

REFERENCE NOS.

R16510-440 Interior Lighting Fixtures

When taking off interior lighting fixtures, it is advisable to set up your quantity work sheet to conform to the lighting schedule as it appears on the print. Include the alpha-numeric code plus the symbol on your work sheet.

Take off a particular section or floor of the building and count each type of fixture before going on to another type. It would also be advantageous to include on the same work sheet the pipe, wire, fittings and circuit number associated with each type of lighting fixture. This will help you identify the costs associated with any particular lighting system and in turn make material purchases more specific as to when and how much to order under the classification of lighting.

By taking off lighting first you can get a complete "WALK THRU" of the job. This will become helpful when doing other phases of the project.

Materials for a recessed fixture include:

1. Fixture
2. Lamps
3. 6' of jack chain
4. (2) S hooks
5. (2) Wire nuts

Labor for interior recessed fixtures include:

1. Unloading by hand
2. Hauling by hand to an area up to 200' from loading dock
3. Uncrating
4. Layout
5. Installing fixture
6. Attaching jack chain & S hooks
7. Connecting circuit power
8. Reassembling fixture
9. Installing lamps
10. Testing

Material for surface mounted fixtures includes:

1. Fixture
2. Lamps
3. Either (4) lead type anchors, (4) toggle bolts, or (4) ceiling grid clips
4. (2) Wire nuts

Material for pendent mounted fixtures includes:

1. Fixture
2. Lamps
3. (2) Wire nuts

4. Rigid pendents as required by type of fixtures
5. Canopies as required by type of fixture

Labor hours include the following for both surface and pendent fixtures:

1. Unloading by hand
2. Hauling by hand to an area up to 200' from loading dock
3. Uncrating
4. Layout and marking
5. Drilling (4) holes for either lead anchors or toggle bolts using a hammer drill
6. Installing fixture
7. Leveling fixture
8. Connecting circuit power
9. Installing lamps
10. Testing

Labor for surface or pendent fixtures does not include:

1. Conduit
2. Boxes or covers
3. Connectors
4. Fixture whips
5. Special support
6. Switching
7. Wire

Economy of Scale: For large concentrations of lighting fixtures in the same area deduct the following percentages from labor:

25	to	50	fixtures	15%
51	to	75	fixtures	20%
76	to	100	fixtures	25%
101	and over		30%

Job Conditions: Productivity is based on new construction in a unobstructed first floor location, using rolling staging to 15' high.

Material staging is assumed to be within 100' of work being performed.

Add the following percentages to labor for elevated installations:

15'	to	20'	high	10%
21'	to	25'	high	20%
26'	to	30'	high	30%
31'	to	35'	high	40%
36'	to	40'	high	50%
41'	and over		60%

R17100-100   Square Foot Project Size Modifier

One factor that affects the S.F. cost of a particular building is the size. In general, for buildings built to the same specifications in the same locality, the larger building will have the lower S.F. cost. This is due mainly to the decreasing contribution of the exterior walls plus the economy of scale usually achievable in larger buildings. The Area Conversion Scale shown below will give a factor to convert costs for the typical size building to an adjusted cost for the particular project.

The Square Foot Base Size lists the median costs, most typical project size in our accumulated data and the range in size of the projects.

The Size Factor for your project is determined by dividing your project area in S.F. by the typical project size for the particular Building Type. With this factor, enter the Area Conversion Scale at the appropriate Size Factor and determine the appropriate cost multiplier for your building size.

Example: Determine the cost per S.F. for a 100,000 S.F. Mid-rise apartment building.

$$\frac{\text{Proposed building area} = 100,000 \text{ S.F.}}{\text{Typical size from below} = 50,000 \text{ S.F.}} = 2.00$$

Enter Area Conversion scale at 2.0, intersect curve, read horizontally the appropriate cost multiplier of .94. Size adjusted cost becomes .94 x $77.00 = $72.40 based on national average costs.

Note: For Size Factors less than .50, the Cost Multiplier is 1.1
      For Size Factors greater than 3.5, the Cost Multiplier is .90

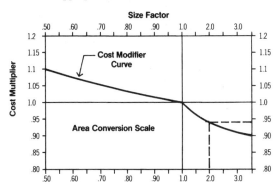

Square Foot Base Size						
Building Type	Median Cost per S.F.	Typical Size Gross S.F.	Typical Range Gross S.F.	Building Type	Median Cost per S.F.	Typical Size Gross S.F.	Typical Range Gross S.F.
Apartments, Low Rise	$ 60.50	21,000	9,700 - 37,200	Jails	$184.00	40,000	5,500 - 145,000
Apartments, Mid Rise	77.00	50,000	32,000 - 100,000	Libraries	114.00	12,000	7,000 - 31,000
Apartments, High Rise	87.00	145,000	95,000 - 600,000	Living, Assisted	99.00	32,300	23,500 - 50,300
Auditoriums	101.00	25,000	7,600 - 39,000	Medical Clinics	106.00	7,200	4,200 - 15,700
Auto Sales	75.50	20,000	10,800 - 28,600	Medical Offices	97.50	6,000	4,000 - 15,000
Banks	135.00	4,200	2,500 - 7,500	Motels	74.00	40,000	15,800 - 120,000
Churches	92.50	17,000	2,000 - 42,000	Nursing Homes	103.00	23,000	15,000 - 37,000
Clubs, Country	95.00	6,500	4,500 - 15,000	Offices, Low Rise	86.00	20,000	5,000 - 80,000
Clubs, Social	90.00	10,000	6,000 - 13,500	Offices, Mid Rise	85.00	120,000	20,000 - 300,000
Clubs, YMCA	106.00	28,300	12,800 - 39,400	Offices, High Rise	109.00	260,000	120,000 - 800,000
Colleges (Class)	118.00	50,000	15,000 - 150,000	Police Stations	136.00	10,500	4,000 - 19,000
Colleges (Science Lab)	173.00	45,600	16,600 - 80,000	Post Offices	101.00	12,400	6,800 - 30,000
College (Student Union)	132.00	33,400	16,000 - 85,000	Power Plants	750.00	7,500	1,000 - 20,000
Community Center	96.00	9,400	5,300 - 16,700	Religious Education	84.00	9,000	6,000 - 12,000
Court Houses	129.00	32,400	17,800 - 106,000	Research	142.00	19,000	6,300 - 45,000
Dept. Stores	56.00	90,000	44,000 - 122,000	Restaurants	122.00	4,400	2,800 - 6,000
Dormitories, Low Rise	100.00	25,000	10,000 - 95,000	Retail Stores	60.00	7,200	4,000 - 17,600
Dormitories, Mid Rise	126.00	85,000	20,000 - 200,000	Schools, Elementary	88.50	41,000	24,500 - 55,000
Factories	54.50	26,400	12,900 - 50,000	Schools, Jr. High	91.50	92,000	52,000 - 119,000
Fire Stations	96.00	5,800	4,000 - 8,700	Schools, Sr. High	98.00	101,000	50,500 - 175,000
Fraternity Houses	93.50	12,500	8,200 - 14,800	Schools, Vocational	89.50	37,000	20,500 - 82,000
Funeral Homes	105.00	10,000	4,000 - 20,000	Sports Arenas	74.00	15,000	5,000 - 40,000
Garages, Commercial	68.00	9,300	5,000 - 13,600	Supermarkets	60.00	44,000	12,000 - 60,000
Garages, Municipal	86.00	8,300	4,500 - 12,600	Swimming Pools	139.00	20,000	10,000 - 32,000
Garages, Parking	36.00	163,000	76,400 - 225,300	Telephone Exchange	163.00	4,500	1,200 - 10,600
Gymnasiums	91.00	19,200	11,600 - 41,000	Theaters	88.50	10,500	8,800 - 17,500
Hospitals	160.00	55,000	27,200 - 125,000	Town Halls	104.00	10,800	4,800 - 23,400
House (Elderly)	82.50	37,000	21,000 - 66,000	Warehouses	44.50	25,000	8,000 - 72,000
Housing (Public)	76.00	36,000	14,400 - 74,400	Warehouse & Office	47.00	25,000	8,000 - 72,000
Ice Rinks	109.00	29,000	27,200 - 33,600				

SQUARE FOOT 17

REFERENCE NOS.

Crews

Crew No.	Bare Costs		Incl. Subs O & P		Cost Per Labor-Hour	

Crew A-1	Hr.	Daily	Hr.	Daily	Bare Costs	Incl. O&P
1 Building Laborer	$26.00	$208.00	$40.50	$324.00	$26.00	$40.50
1 Concrete saw, gas manual		42.80		47.10	5.35	5.89
8 L.H., Daily Totals		$250.80		$371.10	$31.35	$46.39

Crew A-1A	Hr.	Daily	Hr.	Daily	Bare Costs	Incl. O&P
1 Skilled Worker	$33.65	$269.20	$52.35	$418.80	$33.65	$52.35
1 Shot Blaster, 20"		274.30		301.75	34.29	37.72
8 L.H., Daily Totals		$543.50		$720.55	$67.94	$90.07

Crew A-1B	Hr.	Daily	Hr.	Daily	Bare Costs	Incl. O&P
1 Laborers, (Semi-Skilled)	$26.00	$208.00	$40.50	$324.00	$26.00	$40.50
1 Concr. saw, gas, self-prop.		92.20		101.40	11.53	12.68
8 L.H., Daily Totals		$300.20		$425.40	$37.53	$53.18

Crew A-1C	Hr.	Daily	Hr.	Daily	Bare Costs	Incl. O&P
1 Building Laborer	$26.00	$208.00	$40.50	$324.00	$26.00	$40.50
1 Brush saw		18.80		20.70	2.35	2.59
8 L.H., Daily Totals		$226.80		$344.70	$28.35	$43.09

Crew A-1D	Hr.	Daily	Hr.	Daily	Bare Costs	Incl. O&P
1 Building Laborer	$26.00	$208.00	$40.50	$324.00	$26.00	$40.50
1 Vibrating plate, gas, 18"		28.60		31.45	3.58	3.93
8 L.H., Daily Totals		$236.60		$355.45	$29.58	$44.43

Crew A-1E	Hr.	Daily	Hr.	Daily	Bare Costs	Incl. O&P
1 Building Laborer	$26.00	$208.00	$40.50	$324.00	$26.00	$40.50
1 Vibrating plate, gas, 21"		41.40		45.55	5.18	5.69
8 L.H., Daily Totals		$249.40		$369.55	$31.18	$46.19

Crew A-1F	Hr.	Daily	Hr.	Daily	Bare Costs	Incl. O&P
1 Building Laborer	$26.00	$208.00	$40.50	$324.00	$26.00	$40.50
1 Rammer/tamper, gas, 6" - 11"		35.40		38.95	4.43	4.87
8 L.H., Daily Totals		$243.40		$362.95	$30.43	$45.37

Crew A-1G	Hr.	Daily	Hr.	Daily	Bare Costs	Incl. O&P
1 Building Laborer	$26.00	$208.00	$40.50	$324.00	$26.00	$40.50
1 Rammer/tamper, gas, 13" - 18"		35.40		38.95	4.43	4.87
8 L.H., Daily Totals		$243.40		$362.95	$30.43	$45.37

Crew A-1H	Hr.	Daily	Hr.	Daily	Bare Costs	Incl. O&P
1 Building Laborer	$26.00	$208.00	$40.50	$324.00	$26.00	$40.50
1 Pressure washer		55.60		61.15	6.95	7.65
8 L.H., Daily Totals		$263.60		$385.15	$32.95	$48.15

Crew A-1J	Hr.	Daily	Hr.	Daily	Bare Costs	Incl. O&P
1 Building Laborer	$26.00	$208.00	$40.50	$324.00	$26.00	$40.50
1 Rototiller		70.90		78.00	8.86	9.75
8 L.H., Daily Totals		$278.90		$402.00	$34.86	$50.25

Crew A-1K	Hr.	Daily	Hr.	Daily	Bare Costs	Incl. O&P
1 Building Laborer	$26.00	$208.00	$40.50	$324.00	$26.00	$40.50
1 Lawn aerator		24.45		26.90	3.06	3.36
8 L.H., Daily Totals		$232.45		$350.90	$29.06	$43.86

Crew A-1L	Hr.	Daily	Hr.	Daily	Bare Costs	Incl. O&P
1 Building Laborer	$26.00	$208.00	$40.50	$324.00	$26.00	$40.50
1 Powwr blower/vacuum		24.45		26.90	3.06	3.36
8 L.H., Daily Totals		$232.45		$350.90	$29.06	$43.86

Crew A-1M	Hr.	Daily	Hr.	Daily	Bare Costs	Incl. O&P
1 Building Laborer	$26.00	$208.00	$40.50	$324.00	$26.00	$40.50
1 Snow blower		70.90		78.00	8.86	9.75
8 L.H., Daily Totals		$278.90		$402.00	$34.86	$50.25

Crew A-2	Hr.	Daily	Hr.	Daily	Bare Costs	Incl. O&P
2 Laborers	$26.00	$416.00	$40.50	$648.00	$25.90	$40.05
1 Truck Driver (light)	25.70	205.60	39.15	313.20		
1 Light Truck, 1.5 Ton		122.00		134.20	5.08	5.59
24 L.H., Daily Totals		$743.60		$1095.40	$30.98	$45.64

Crew A-2A	Hr.	Daily	Hr.	Daily	Bare Costs	Incl. O&P
2 Laborers	$26.00	$416.00	$40.50	$648.00	$25.90	$40.05
1 Truck Driver (light)	25.70	205.60	39.15	313.20		
1 Light Truck, 1.5 Ton		122.00		134.20		
1 Concrete Saw		92.20		101.40	8.93	9.82
24 L.H., Daily Totals		$835.80		$1196.80	$34.83	$49.87

Crew A-3	Hr.	Daily	Hr.	Daily	Bare Costs	Incl. O&P
1 Truck Driver (heavy)	$26.45	$211.60	$40.30	$322.40	$26.45	$40.30
1 Dump Truck, 12 Ton		325.00		357.50	40.63	44.69
8 L.H., Daily Totals		$536.60		$679.90	$67.08	$84.99

Crew A-3A	Hr.	Daily	Hr.	Daily	Bare Costs	Incl. O&P
1 Truck Driver (light)	$25.70	$205.60	$39.15	$313.20	$25.70	$39.15
1 Pickup Truck (4x4)		82.00		90.20	10.25	11.28
8 L.H., Daily Totals		$287.60		$403.40	$35.95	$50.43

Crew A-3B	Hr.	Daily	Hr.	Daily	Bare Costs	Incl. O&P
1 Equip. Oper. (medium)	$33.65	$269.20	$50.70	$405.60	$30.05	$45.50
1 Truck Driver (heavy)	26.45	211.60	40.30	322.40		
1 Dump Truck, 16 Ton		476.60		524.25		
1 F.E. Loader, 3 C.Y.		295.60		325.15	48.26	53.09
16 L.H., Daily Totals		$1253.00		$1577.40	$78.31	$98.59

Crew A-3C	Hr.	Daily	Hr.	Daily	Bare Costs	Incl. O&P
1 Equip. Oper. (light)	$32.15	$257.20	$48.40	$387.20	$32.15	$48.40
1 Wheeled Skid Steer Loader		183.80		202.20	22.98	25.27
8 L.H., Daily Totals		$441.00		$589.40	$55.13	$73.67

Crew A-3D	Hr.	Daily	Hr.	Daily	Bare Costs	Incl. O&P
1 Truck Driver, Light	$25.70	$205.60	$39.15	$313.20	$25.70	$39.15
1 Pickup Truck (4x4)		82.00		90.20		
1 Flatbed Trailer, 25 Ton		88.60		97.45	21.33	23.46
8 L.H., Daily Totals		$376.20		$500.85	$47.03	$62.61

Crew A-3E	Hr.	Daily	Hr.	Daily	Bare Costs	Incl. O&P
1 Equip. Oper. (crane)	$34.80	$278.40	$52.40	$419.20	$30.63	$46.35
1 Truck Driver (heavy)	26.45	211.60	40.30	322.40		
1 Pickup Truck (4x4)		82.00		90.20	5.13	5.64
16 L.H., Daily Totals		$572.00		$831.80	$35.76	$51.99

Crew A-3F	Hr.	Daily	Hr.	Daily	Bare Costs	Incl. O&P
1 Equip. Oper. (crane)	$34.80	$278.40	$52.40	$419.20	$30.63	$46.35
1 Truck Driver (heavy)	26.45	211.60	40.30	322.40		
1 Pickup Truck (4x4)		82.00		90.20		
1 Tractor, 6x2, 40 Ton Cap.		303.00		333.30		
1 Lowbed Trailer, 75 Ton		171.80		189.00	34.80	38.28
16 L.H., Daily Totals		$1046.80		$1354.10	$65.43	$84.63

Crew No.	Bare Costs		Incl. Subs O & P		Cost Per Labor-Hour	
Crew A-3G	**Hr.**	**Daily**	**Hr.**	**Daily**	**Bare Costs**	**Incl. O&P**
1 Equip. Oper. (crane)	$34.80	$278.40	$52.40	$419.20	$30.63	$46.35
1 Truck Driver (heavy)	26.45	211.60	40.30	322.40		
1 Pickup Truck (4x4)		82.00		90.20		
1 Tractor, 6x4, 45 Ton Cap.		331.60		364.75		
1 Lowbed Trailer, 75 Ton		171.80		189.00	36.59	40.25
16 L.H., Daily Totals		$1075.40		$1385.55	$67.22	$86.60
Crew A-4	**Hr.**	**Daily**	**Hr.**	**Daily**	**Bare Costs**	**Incl. O&P**
2 Carpenters	$33.00	$528.00	$51.40	$822.40	$31.87	$49.08
1 Painter, Ordinary	29.60	236.80	44.45	355.60		
24 L.H., Daily Totals		$764.80		$1178.00	$31.87	$49.08
Crew A-5	**Hr.**	**Daily**	**Hr.**	**Daily**	**Bare Costs**	**Incl. O&P**
2 Laborers	$26.00	$416.00	$40.50	$648.00	$25.97	$40.35
.25 Truck Driver (light)	25.70	51.40	39.15	78.30		
.25 Light Truck, 1.5 Ton		30.50		33.55	1.69	1.86
18 L.H., Daily Totals		$497.90		$759.85	$27.66	$42.21
Crew A-6	**Hr.**	**Daily**	**Hr.**	**Daily**	**Bare Costs**	**Incl. O&P**
1 Instrument Man	$33.65	$269.20	$52.35	$418.80	$32.85	$50.40
1 Rodman/Chainman	32.05	256.40	48.45	387.60		
1 Laser Transit/Level		59.50		65.45	3.72	4.09
16 L.H., Daily Totals		$585.10		$871.85	$36.57	$54.49
Crew A-7	**Hr.**	**Daily**	**Hr.**	**Daily**	**Bare Costs**	**Incl. O&P**
1 Chief Of Party	$41.25	$330.00	$64.10	$512.80	$35.65	$54.97
1 Instrument Man	33.65	269.20	52.35	418.80		
1 Rodman/Chainman	32.05	256.40	48.45	387.60		
1 Laser Transit/Level		59.50		65.45	2.48	2.73
24 L.H., Daily Totals		$915.10		$1384.65	$38.13	$57.70
Crew A-8	**Hr.**	**Daily**	**Hr.**	**Daily**	**Bare Costs**	**Incl. O&P**
1 Chief Of Party	$41.25	$330.00	$64.10	$512.80	$34.75	$53.34
1 Instrument Man	33.65	269.20	52.35	418.80		
2 Rodmen/Chainmen	32.05	512.80	48.45	775.20		
1 Laser Transit/Level		59.50		65.45	1.86	2.05
32 L.H., Daily Totals		$1171.50		$1772.25	$36.61	$55.39
Crew A-9	**Hr.**	**Daily**	**Hr.**	**Daily**	**Bare Costs**	**Incl. O&P**
1 Asbestos Foreman	$36.50	$292.00	$57.50	$460.00	$36.06	$56.80
7 Asbestos Workers	36.00	2016.00	56.70	3175.20		
64 L.H., Daily Totals		$2308.00		$3635.20	$36.06	$56.80
Crew A-10	**Hr.**	**Daily**	**Hr.**	**Daily**	**Bare Costs**	**Incl. O&P**
1 Asbestos Foreman	$36.50	$292.00	$57.50	$460.00	$36.06	$56.80
7 Asbestos Workers	36.00	2016.00	56.70	3175.20		
64 L.H., Daily Totals		$2308.00		$3635.20	$36.06	$56.80
Crew A-10A	**Hr.**	**Daily**	**Hr.**	**Daily**	**Bare Costs**	**Incl. O&P**
1 Asbestos Foreman	$36.50	$292.00	$57.50	$460.00	$36.17	$56.97
2 Asbestos Workers	36.00	576.00	56.70	907.20		
24 L.H., Daily Totals		$868.00		$1367.20	$36.17	$56.97
Crew A-10B	**Hr.**	**Daily**	**Hr.**	**Daily**	**Bare Costs**	**Incl. O&P**
1 Asbestos Foreman	$36.50	$292.00	$57.50	$460.00	$36.13	$56.90
3 Asbestos Workers	36.00	864.00	56.70	1360.80		
32 L.H., Daily Totals		$1156.00		$1820.80	$36.13	$56.90

Crew No.	Bare Costs		Incl. Subs O & P		Cost Per Labor-Hour	
Crew A-10C	**Hr.**	**Daily**	**Hr.**	**Daily**	**Bare Costs**	**Incl. O&P**
3 Asbestos Workers	$36.00	$864.00	$56.70	$1360.80	$36.00	$56.70
1 Flatbed Truck		122.00		134.20	5.08	5.59
24 L.H., Daily Totals		$986.00		$1495.00	$41.08	$62.29
Crew A-10D	**Hr.**	**Daily**	**Hr.**	**Daily**	**Bare Costs**	**Incl. O&P**
2 Asbestos Workers	$36.00	$576.00	$56.70	$907.20	$34.00	$52.45
1 Equip. Oper. (crane)	34.80	278.40	52.40	419.20		
1 Equip. Oper. Oiler	29.20	233.60	44.00	352.00		
1 Hydraulic Crane, 33 Ton		649.00		713.90	20.28	22.31
32 L.H., Daily Totals		$1737.00		$2392.30	$54.28	$74.76
Crew A-11	**Hr.**	**Daily**	**Hr.**	**Daily**	**Bare Costs**	**Incl. O&P**
1 Asbestos Foreman	$36.50	$292.00	$57.50	$460.00	$36.06	$56.80
7 Asbestos Workers	36.00	2016.00	56.70	3175.20		
2 Chipping Hammers		32.00		35.20	.50	.55
64 L.H., Daily Totals		$2340.00		$3670.40	$36.56	$57.35
Crew A-12	**Hr.**	**Daily**	**Hr.**	**Daily**	**Bare Costs**	**Incl. O&P**
1 Asbestos Foreman	$36.50	$292.00	$57.50	$460.00	$36.06	$56.80
7 Asbestos Workers	36.00	2016.00	56.70	3175.20		
1 Large Prod. Vac. Loader		503.05		553.35	7.86	8.65
64 L.H., Daily Totals		$2811.05		$4188.55	$43.92	$65.45
Crew A-13	**Hr.**	**Daily**	**Hr.**	**Daily**	**Bare Costs**	**Incl. O&P**
1 Equip. Oper. (light)	$32.15	$257.20	$48.40	$387.20	$32.15	$48.40
1 Large Prod. Vac. Loader		503.05		553.35	62.88	69.17
8 L.H., Daily Totals		$760.25		$940.55	$95.03	$117.57
Crew B-1	**Hr.**	**Daily**	**Hr.**	**Daily**	**Bare Costs**	**Incl. O&P**
1 Labor Foreman (outside)	$28.00	$224.00	$43.60	$348.80	$26.67	$41.53
2 Laborers	26.00	416.00	40.50	648.00		
24 L.H., Daily Totals		$640.00		$996.80	$26.67	$41.53
Crew B-2	**Hr.**	**Daily**	**Hr.**	**Daily**	**Bare Costs**	**Incl. O&P**
1 Labor Foreman (outside)	$28.00	$224.00	$43.60	$348.80	$26.40	$41.12
4 Laborers	26.00	832.00	40.50	1296.00		
40 L.H., Daily Totals		$1056.00		$1644.80	$26.40	$41.12
Crew B-3	**Hr.**	**Daily**	**Hr.**	**Daily**	**Bare Costs**	**Incl. O&P**
1 Labor Foreman (outside)	$28.00	$224.00	$43.60	$348.80	$27.76	$42.65
2 Laborers	26.00	416.00	40.50	648.00		
1 Equip. Oper. (med.)	33.65	269.20	50.70	405.60		
2 Truck Drivers (heavy)	26.45	423.20	40.30	644.80		
1 F.E. Loader, T.M., 2.5 C.Y.		761.00		837.10		
2 Dump Trucks, 16 Ton		953.20		1048.50	35.71	39.28
48 L.H., Daily Totals		$3046.60		$3932.80	$63.47	$81.93
Crew B-3A	**Hr.**	**Daily**	**Hr.**	**Daily**	**Bare Costs**	**Incl. O&P**
4 Laborers	$26.00	$832.00	$40.50	$1296.00	$27.53	$42.54
1 Equip. Oper. (med.)	33.65	269.20	50.70	405.60		
1 Hyd. Excavator, 1.5 C.Y.		683.40		751.75	17.09	18.79
40 L.H., Daily Totals		$1784.60		$2453.35	$44.62	$61.33
Crew B-3B	**Hr.**	**Daily**	**Hr.**	**Daily**	**Bare Costs**	**Incl. O&P**
2 Laborers	$26.00	$416.00	$40.50	$648.00	$28.03	$43.00
1 Equip. Oper. (med.)	33.65	269.20	50.70	405.60		
1 Truck Driver (heavy)	26.45	211.60	40.30	322.40		
1 Backhoe Loader, 80 H.P.		230.80		253.90		
1 Dump Truck, 16 Ton		476.60		524.25	22.11	24.32
32 L.H., Daily Totals		$1604.20		$2154.15	$50.14	$67.32

Crew No.	Bare Costs		Incl. Subs O & P		Cost Per Labor-Hour	
Crew B-3C	Hr.	Daily	Hr.	Daily	Bare Costs	Incl. O&P
3 Laborers	$26.00	$624.00	$40.50	$972.00	$27.91	$43.05
1 Equip. Oper. (med.)	33.65	269.20	50.70	405.60		
1 F.E. Crawler Ldr, 4 C.Y.		1061.00		1167.10	33.16	36.47
32 L.H., Daily Totals		$1954.20		$2544.70	$61.07	$79.52

Crew B-4	Hr.	Daily	Hr.	Daily	Bare Costs	Incl. O&P
1 Labor Foreman (outside)	$28.00	$224.00	$43.60	$348.80	$26.41	$40.98
4 Laborers	26.00	832.00	40.50	1296.00		
1 Truck Driver (heavy)	26.45	211.60	40.30	322.40		
1 Tractor, 4 x 2, 195 H.P.		215.20		236.70		
1 Platform Trailer		120.60		132.65	7.00	7.70
48 L.H., Daily Totals		$1603.40		$2336.55	$33.41	$48.68

Crew B-5	Hr.	Daily	Hr.	Daily	Bare Costs	Incl. O&P
1 Labor Foreman (outside)	$28.00	$224.00	$43.60	$348.80	$28.47	$43.86
4 Laborers	26.00	832.00	40.50	1296.00		
2 Equip. Oper. (med.)	33.65	538.40	50.70	811.20		
1 Air Compr., 250 C.F.M.		127.60		140.35		
2 Air Tools & Accessories		22.40		24.65		
2-50 Ft. Air Hoses, 1.5" Dia.		10.00		11.00		
1 F.E. Loader, T.M., 2.5 C.Y.		761.00		837.10	16.45	18.09
56 L.H., Daily Totals		$2515.40		$3469.10	$44.92	$61.95

Crew B-5A	Hr.	Daily	Hr.	Daily	Bare Costs	Incl. O&P
1 Foreman	$28.00	$224.00	$43.60	$348.80	$28.03	$43.08
6 Laborers	26.00	1248.00	40.50	1944.00		
2 Equip. Oper. (med.)	33.65	538.40	50.70	811.20		
1 Equip. Oper. (light)	32.15	257.20	48.40	387.20		
2 Truck Drivers (heavy)	26.45	423.20	40.30	644.80		
1 Air Compr. 365 C.F.M.		157.60		173.35		
2 Pavement Breakers		22.40		24.65		
8 Air Hoses w/Coup.,1"		27.60		30.35		
2 Dump Trucks, 12 Ton		650.00		715.00	8.93	9.83
96 L.H., Daily Totals		$3548.40		$5079.35	$36.96	$52.91

Crew B-5B	Hr.	Daily	Hr.	Daily	Bare Costs	Incl. O&P
1 Powderman	$33.65	$269.20	$52.35	$418.80	$30.05	$45.77
2 Equip. Oper. (med.)	33.65	538.40	50.70	811.20		
3 Truck Drivers (heavy)	26.45	634.80	40.30	967.20		
1 F.E. Ldr. 2-1/2 CY		295.60		325.15		
3 Dump Trucks, 16 Ton		1429.80		1572.80		
1 Air Compr. 365 C.F.M.		157.60		173.35	39.23	43.15
48 L.H., Daily Totals		$3325.40		$4268.50	$69.28	$88.92

Crew B-5C	Hr.	Daily	Hr.	Daily	Bare Costs	Incl. O&P
3 Laborers	$26.00	$624.00	$40.50	$972.00	$28.57	$43.65
1 Equip. Oper. (medium)	33.65	269.20	50.70	405.60		
2 Truck Drivers (heavy)	26.45	423.20	40.30	644.80		
1 Equip. Oper. (crane)	34.80	278.40	52.40	419.20		
1 Equip. Oper. Oiler	29.20	233.60	44.00	352.00		
2 Dump Trucks, 16 Ton		953.20		1048.50		
1 F.E. Crawler Ldr, 4 C.Y.		1061.00		1167.10		
1 Hyd. Crane, 25 Ton		575.20		632.70	40.46	44.51
64 L.H., Daily Totals		$4417.80		$5641.90	$69.03	$88.16

Crew B-6	Hr.	Daily	Hr.	Daily	Bare Costs	Incl. O&P
2 Laborers	$26.00	$416.00	$40.50	$648.00	$28.05	$43.13
1 Equip. Oper. (light)	32.15	257.20	48.40	387.20		
1 Backhoe Loader, 48 H.P.		207.80		228.60	8.66	9.52
24 L.H., Daily Totals		$881.00		$1263.80	$36.71	$52.65

Crew B-6A	Hr.	Daily	Hr.	Daily	Bare Costs	Incl. O&P
.5 Labor Foreman (outside)	$28.00	$112.00	$43.60	$174.40	$29.46	$45.20
1 Laborer	26.00	208.00	40.50	324.00		
1 Equip. Oper. (med.)	33.65	269.20	50.70	405.60		
1 Vacuum Trk.,5000 Gal.		324.10		356.50	16.21	17.83
20 L.H., Daily Totals		$913.30		$1260.50	$45.67	$63.03

Crew B-6B	Hr.	Daily	Hr.	Daily	Bare Costs	Incl. O&P
2 Labor Foremen (out)	$28.00	$448.00	$43.60	$697.60	$26.67	$41.53
4 Laborers	26.00	832.00	40.50	1296.00		
1 Winch Truck		317.80		349.60		
1 Flatbed Truck		122.00		134.20		
1 Butt Fusion Machine		433.60		476.95	18.20	20.02
48 L.H., Daily Totals		$2153.40		$2954.35	$44.87	$61.55

Crew B-7	Hr.	Daily	Hr.	Daily	Bare Costs	Incl. O&P
1 Labor Foreman (outside)	$28.00	$224.00	$43.60	$348.80	$27.61	$42.72
4 Laborers	26.00	832.00	40.50	1296.00		
1 Equip. Oper. (med.)	33.65	269.20	50.70	405.60		
1 Chipping Machine		164.80		181.30		
1 F.E. Loader, T.M., 2.5 C.Y.		761.00		837.10		
2 Chain Saws, 36"		66.80		73.50	20.68	22.75
48 L.H., Daily Totals		$2317.80		$3142.30	$48.29	$65.47

Crew B-7A	Hr.	Daily	Hr.	Daily	Bare Costs	Incl. O&P
2 Laborers	$26.00	$416.00	$40.50	$648.00	$28.05	$43.13
1 Equip. Oper. (light)	32.15	257.20	48.40	387.20		
1 Rake w/Tractor		184.90		203.40		
2 Chain Saws, 18"		37.60		41.35	9.27	10.20
24 L.H., Daily Totals		$895.70		$1279.95	$37.32	$53.33

Crew B-8	Hr.	Daily	Hr.	Daily	Bare Costs	Incl. O&P
1 Labor Foreman (outside)	$28.00	$224.00	$43.60	$348.80	$28.68	$43.83
2 Laborers	26.00	416.00	40.50	648.00		
2 Equip. Oper. (med.)	33.65	538.40	50.70	811.20		
1 Equip. Oper. Oiler	29.20	233.60	44.00	352.00		
2 Truck Drivers (heavy)	26.45	423.20	40.30	644.80		
1 Hyd. Crane, 25 Ton		624.20		686.60		
1 F.E. Loader, T.M., 2.5 C.Y.		761.00		837.10		
2 Dump Trucks, 16 Ton		953.20		1048.50	36.54	40.19
64 L.H., Daily Totals		$4173.60		$5377.00	$65.22	$84.02

Crew B-9	Hr.	Daily	Hr.	Daily	Bare Costs	Incl. O&P
1 Labor Foreman (outside)	$28.00	$224.00	$43.60	$348.80	$26.40	$41.12
4 Laborers	26.00	832.00	40.50	1296.00		
1 Air Compr., 250 C.F.M.		127.60		140.35		
2 Air Tools & Accessories		22.40		24.65		
2-50 Ft. Air Hoses, 1.5" Dia.		10.00		11.00	4.00	4.40
40 L.H., Daily Totals		$1216.00		$1820.80	$30.40	$45.52

Crew B-9A	Hr.	Daily	Hr.	Daily	Bare Costs	Incl. O&P
2 Laborers	$26.00	$416.00	$40.50	$648.00	$26.15	$40.43
1 Truck Driver (heavy)	26.45	211.60	40.30	322.40		
1 Water Tanker		118.00		129.80		
1 Tractor		215.20		236.70		
2-50 Ft. Disch. Hoses		7.50		8.25	14.20	15.62
24 L.H., Daily Totals		$968.30		$1345.15	$40.35	$56.05

Crew No.	Bare Costs Hr.	Daily	Incl. Subs O&P Hr.	Daily	Bare Costs	Incl. O&P
Crew B-9B	Hr.	Daily	Hr.	Daily	Bare Costs	Incl. O&P
2 Laborers	$26.00	$416.00	$40.50	$648.00	$26.15	$40.43
1 Truck Driver (heavy)	26.45	211.60	40.30	322.40		
2-50 Ft. Disch. Hoses		7.50		8.25		
1 Water Tanker		118.00		129.80		
1 Tractor		215.20		236.70		
1 Pressure Washer		46.40		51.05	16.13	17.74
24 L.H., Daily Totals		$1014.70		$1396.20	$42.28	$58.17
Crew B-9C	Hr.	Daily	Hr.	Daily	Bare Costs	Incl. O&P
1 Labor Foreman (outside)	$28.00	$224.00	$43.60	$348.80	$26.40	$41.12
4 Laborers	26.00	832.00	40.50	1296.00		
1 Air Compr., 250 C.F.M.		127.60		140.35		
2-50 Ft. Air Hoses, 1.5" Dia.		10.00		11.00		
2 Breaker, Pavement, 60 lb.		22.40		24.65	4.00	4.40
40 L.H., Daily Totals		$1216.00		$1820.80	$30.40	$45.52
Crew B-9D	Hr.	Daily	Hr.	Daily	Bare Costs	Incl. O&P
1 Labor Foreman (Outside)	$28.00	$224.00	$43.60	$348.80	$26.40	$41.12
4 Common Laborers	26.00	832.00	40.50	1296.00		
1 Air Compressor, 250 CFM		127.60		140.35		
2 Air hoses, 1.5" x 50'		10.00		11.00		
2 Air tamper		47.00		51.70	4.62	5.08
40 L.H., Daily Totals		$1240.60		$1847.85	$31.02	$46.20
Crew B-10	Hr.	Daily	Hr.	Daily	Bare Costs	Incl. O&P
1 Equip. Oper. (med.)	$33.65	$269.20	$50.70	$405.60	$31.10	$47.30
.5 Laborer	26.00	104.00	40.50	162.00		
12 L.H., Daily Totals		$373.20		$567.60	$31.10	$47.30
Crew B-10A	Hr.	Daily	Hr.	Daily	Bare Costs	Incl. O&P
1 Equip. Oper. (med.)	$33.65	$269.20	$50.70	$405.60	$31.10	$47.30
.5 Laborer	26.00	104.00	40.50	162.00		
1 Roll. Compact., 2K Lbs.		130.80		143.90	10.90	11.99
12 L.H., Daily Totals		$504.00		$711.50	$42.00	$59.29
Crew B-10B	Hr.	Daily	Hr.	Daily	Bare Costs	Incl. O&P
1 Equip. Oper. (med.)	$33.65	$269.20	$50.70	$405.60	$31.10	$47.30
.5 Laborer	26.00	104.00	40.50	162.00		
1 Dozer, 200 H.P.		863.40		949.75	71.95	79.15
12 L.H., Daily Totals		$1236.60		$1517.35	$103.05	$126.45
Crew B-10C	Hr.	Daily	Hr.	Daily	Bare Costs	Incl. O&P
1 Equip. Oper. (med.)	$33.65	$269.20	$50.70	$405.60	$31.10	$47.30
.5 Laborer	26.00	104.00	40.50	162.00		
1 Dozer, 200 H.P.		863.40		949.75		
1 Vibratory Roller, Towed		610.20		671.20	122.80	135.08
12 L.H., Daily Totals		$1846.80		$2188.55	$153.90	$182.38
Crew B-10D	Hr.	Daily	Hr.	Daily	Bare Costs	Incl. O&P
1 Equip. Oper. (med.)	$33.65	$269.20	$50.70	$405.60	$31.10	$47.30
.5 Laborer	26.00	104.00	40.50	162.00		
1 Dozer, 200 H.P.		863.40		949.75		
1 Sheepsft. Roller, Towed		77.80		85.60	78.43	86.28
12 L.H., Daily Totals		$1314.40		$1602.95	$109.53	$133.58
Crew B-10E	Hr.	Daily	Hr.	Daily	Bare Costs	Incl. O&P
1 Equip. Oper. (med.)	$33.65	$269.20	$50.70	$405.60	$31.10	$47.30
.5 Laborer	26.00	104.00	40.50	162.00		
1 Tandem Roller, 5 Ton		108.20		119.00	9.02	9.92
12 L.H., Daily Totals		$481.40		$686.60	$40.12	$57.22

Crew No.	Bare Costs Hr.	Daily	Incl. Subs O&P Hr.	Daily	Bare Costs	Incl. O&P
Crew B-10F	Hr.	Daily	Hr.	Daily	Bare Costs	Incl. O&P
1 Equip. Oper. (med.)	$33.65	$269.20	$50.70	$405.60	$31.10	$47.30
.5 Laborer	26.00	104.00	40.50	162.00		
1 Tandem Roller, 10 Ton		183.60		201.95	15.30	16.83
12 L.H., Daily Totals		$556.80		$769.55	$46.40	$64.13
Crew B-10G	Hr.	Daily	Hr.	Daily	Bare Costs	Incl. O&P
1 Equip. Oper. (med.)	$33.65	$269.20	$50.70	$405.60	$31.10	$47.30
.5 Laborer	26.00	104.00	40.50	162.00		
1 Sheepsft. Roll., 130 H.P.		762.60		838.85	63.55	69.91
12 L.H., Daily Totals		$1135.80		$1406.45	$94.65	$117.21
Crew B-10H	Hr.	Daily	Hr.	Daily	Bare Costs	Incl. O&P
1 Equip. Oper. (med.)	$33.65	$269.20	$50.70	$405.60	$31.10	$47.30
.5 Laborer	26.00	104.00	40.50	162.00		
1 Diaphr. Water Pump, 2"		46.20		50.80		
1-20 Ft. Suction Hose, 2"		3.55		3.90		
2-50 Ft. Disch. Hoses, 2"		6.30		6.95	4.67	5.14
12 L.H., Daily Totals		$429.25		$629.25	$35.77	$52.44
Crew B-10I	Hr.	Daily	Hr.	Daily	Bare Costs	Incl. O&P
1 Equip. Oper. (med.)	$33.65	$269.20	$50.70	$405.60	$31.10	$47.30
.5 Laborer	26.00	104.00	40.50	162.00		
1 Diaphr. Water Pump, 4"		87.40		96.15		
1-20 Ft. Suction Hose, 4"		7.30		8.05		
2-50 Ft. Disch. Hoses, 4"		10.50		11.55	8.77	9.64
12 L.H., Daily Totals		$478.40		$683.35	$39.87	$56.94
Crew B-10J	Hr.	Daily	Hr.	Daily	Bare Costs	Incl. O&P
1 Equip. Oper. (med.)	$33.65	$269.20	$50.70	$405.60	$31.10	$47.30
.5 Laborer	26.00	104.00	40.50	162.00		
1 Centr. Water Pump, 3"		53.80		59.20		
1-20 Ft. Suction Hose, 3"		5.45		6.00		
2-50 Ft. Disch. Hoses, 3"		7.50		8.25	5.56	6.12
12 L.H., Daily Totals		$439.95		$641.05	$36.66	$53.42
Crew B-10K	Hr.	Daily	Hr.	Daily	Bare Costs	Incl. O&P
1 Equip. Oper. (med.)	$33.65	$269.20	$50.70	$405.60	$31.10	$47.30
.5 Laborer	26.00	104.00	40.50	162.00		
1 Centr. Water Pump, 6"		229.80		252.80		
1-20 Ft. Suction Hose, 6"		14.50		15.95		
2-50 Ft. Disch. Hoses, 6"		23.80		26.20	22.34	24.58
12 L.H., Daily Totals		$641.30		$862.55	$53.44	$71.88
Crew B-10L	Hr.	Daily	Hr.	Daily	Bare Costs	Incl. O&P
1 Equip. Oper. (med.)	$33.65	$269.20	$50.70	$405.60	$31.10	$47.30
.5 Laborer	26.00	104.00	40.50	162.00		
1 Dozer, 80 H.P.		300.20		330.20	25.02	27.52
12 L.H., Daily Totals		$673.40		$897.80	$56.12	$74.82
Crew B-10M	Hr.	Daily	Hr.	Daily	Bare Costs	Incl. O&P
1 Equip. Oper. (med.)	$33.65	$269.20	$50.70	$405.60	$31.10	$47.30
.5 Laborer	26.00	104.00	40.50	162.00		
1 Dozer, 300 H.P.		1099.00		1208.90	91.58	100.74
12 L.H., Daily Totals		$1472.20		$1776.50	$122.68	$148.04
Crew B-10N	Hr.	Daily	Hr.	Daily	Bare Costs	Incl. O&P
1 Equip. Oper. (med.)	$33.65	$269.20	$50.70	$405.60	$31.10	$47.30
.5 Laborer	26.00	104.00	40.50	162.00		
1 F.E. Loader, T.M., 1.5 C.Y		300.60		330.65	25.05	27.56
12 L.H., Daily Totals		$673.80		$898.25	$56.15	$74.86

CREWS

Crew No.	Bare Costs Hr.	Daily	Incl. Subs O & P Hr.	Daily	Cost Per Labor-Hour Bare Costs	Incl. O&P
Crew B-100	Hr.	Daily	Hr.	Daily	Bare Costs	Incl. O&P
1 Equip. Oper. (med.)	$33.65	$269.20	$50.70	$405.60	$31.10	$47.30
.5 Laborer	26.00	104.00	40.50	162.00		
1 F.E. Loader, T.M., 2.25 C.Y.		538.00		591.80	44.83	49.32
12 L.H., Daily Totals		$911.20		$1159.40	$75.93	$96.62

Crew No.	Bare Costs Hr.	Daily	Incl. Subs O & P Hr.	Daily	Cost Per Labor-Hour Bare Costs	Incl. O&P
Crew B-10P	Hr.	Daily	Hr.	Daily	Bare Costs	Incl. O&P
1 Equip. Oper. (med.)	$33.65	$269.20	$50.70	$405.60	$31.10	$47.30
.5 Laborer	26.00	104.00	40.50	162.00		
1 F.E. Loader, T.M., 2.5 C.Y.		761.00		837.10	63.42	69.76
12 L.H., Daily Totals		$1134.20		$1404.70	$94.52	$117.06

Crew No.	Bare Costs Hr.	Daily	Incl. Subs O & P Hr.	Daily	Cost Per Labor-Hour Bare Costs	Incl. O&P
Crew B-10Q	Hr.	Daily	Hr.	Daily	Bare Costs	Incl. O&P
1 Equip. Oper. (med.)	$33.65	$269.20	$50.70	$405.60	$31.10	$47.30
.5 Laborer	26.00	104.00	40.50	162.00		
1 F.E. Loader, T.M., 5 C.Y.		1061.00		1167.10	88.42	97.26
12 L.H., Daily Totals		$1434.20		$1734.70	$119.52	$144.56

Crew No.	Bare Costs Hr.	Daily	Incl. Subs O & P Hr.	Daily	Cost Per Labor-Hour Bare Costs	Incl. O&P
Crew B-10R	Hr.	Daily	Hr.	Daily	Bare Costs	Incl. O&P
1 Equip. Oper. (med.)	$33.65	$269.20	$50.70	$405.60	$31.10	$47.30
.5 Laborer	26.00	104.00	40.50	162.00		
1 F.E. Loader, W.M., 1 C.Y.		192.00		211.20	16.00	17.60
12 L.H., Daily Totals		$565.20		$778.80	$47.10	$64.90

Crew No.	Bare Costs Hr.	Daily	Incl. Subs O & P Hr.	Daily	Cost Per Labor-Hour Bare Costs	Incl. O&P
Crew B-10S	Hr.	Daily	Hr.	Daily	Bare Costs	Incl. O&P
1 Equip. Oper. (med.)	$33.65	$269.20	$50.70	$405.60	$31.10	$47.30
.5 Laborer	26.00	104.00	40.50	162.00		
1 F.E. Loader, W.M., 1.5 C.Y.		236.20		259.80	19.68	21.65
12 L.H., Daily Totals		$609.40		$827.40	$50.78	$68.95

Crew No.	Bare Costs Hr.	Daily	Incl. Subs O & P Hr.	Daily	Cost Per Labor-Hour Bare Costs	Incl. O&P
Crew B-10T	Hr.	Daily	Hr.	Daily	Bare Costs	Incl. O&P
1 Equip. Oper. (med.)	$33.65	$269.20	$50.70	$405.60	$31.10	$47.30
.5 Laborer	26.00	104.00	40.50	162.00		
1 F.E. Loader, W.M., 2.5 C.Y.		295.60		325.15	24.63	27.10
12 L.H., Daily Totals		$668.80		$892.75	$55.73	$74.40

Crew No.	Bare Costs Hr.	Daily	Incl. Subs O & P Hr.	Daily	Cost Per Labor-Hour Bare Costs	Incl. O&P
Crew B-10U	Hr.	Daily	Hr.	Daily	Bare Costs	Incl. O&P
1 Equip. Oper. (med.)	$33.65	$269.20	$50.70	$405.60	$31.10	$47.30
.5 Laborer	26.00	104.00	40.50	162.00		
1 F.E. Loader, W.M., 5.5 C.Y.		674.80		742.30	56.23	61.86
12 L.H., Daily Totals		$1048.00		$1309.90	$87.33	$109.16

Crew No.	Bare Costs Hr.	Daily	Incl. Subs O & P Hr.	Daily	Cost Per Labor-Hour Bare Costs	Incl. O&P
Crew B-10V	Hr.	Daily	Hr.	Daily	Bare Costs	Incl. O&P
1 Equip. Oper. (med.)	$33.65	$269.20	$50.70	$405.60	$31.10	$47.30
.5 Laborer	26.00	104.00	40.50	162.00		
1 Dozer, 700 H.P.		3047.00		3351.70	253.92	279.31
12 L.H., Daily Totals		$3420.20		$3919.30	$285.02	$326.61

Crew No.	Bare Costs Hr.	Daily	Incl. Subs O & P Hr.	Daily	Cost Per Labor-Hour Bare Costs	Incl. O&P
Crew B-10W	Hr.	Daily	Hr.	Daily	Bare Costs	Incl. O&P
1 Equip. Oper. (med.)	$33.65	$269.20	$50.70	$405.60	$31.10	$47.30
.5 Laborer	26.00	104.00	40.50	162.00		
1 Dozer, 105 H.P.		434.00		477.40	36.17	39.78
12 L.H., Daily Totals		$807.20		$1045.00	$67.27	$87.08

Crew No.	Bare Costs Hr.	Daily	Incl. Subs O & P Hr.	Daily	Cost Per Labor-Hour Bare Costs	Incl. O&P
Crew B-10X	Hr.	Daily	Hr.	Daily	Bare Costs	Incl. O&P
1 Equip. Oper. (med.)	$33.65	$269.20	$50.70	$405.60	$31.10	$47.30
.5 Laborer	26.00	104.00	40.50	162.00		
1 Dozer, 410 H.P.		1473.00		1620.30	122.75	135.03
12 L.H., Daily Totals		$1846.20		$2187.90	$153.85	$182.33

Crew No.	Bare Costs Hr.	Daily	Incl. Subs O & P Hr.	Daily	Cost Per Labor-Hour Bare Costs	Incl. O&P
Crew B-10Y	Hr.	Daily	Hr.	Daily	Bare Costs	Incl. O&P
1 Equip. Oper. (med.)	$33.65	$269.20	$50.70	$405.60	$31.10	$47.30
.5 Laborer	26.00	104.00	40.50	162.00		
1 Vibratory Drum Roller		323.60		355.95	26.97	29.66
12 L.H., Daily Totals		$696.80		$923.55	$58.07	$76.96

Crew No.	Bare Costs Hr.	Daily	Incl. Subs O & P Hr.	Daily	Cost Per Labor-Hour Bare Costs	Incl. O&P
Crew B-11A	Hr.	Daily	Hr.	Daily	Bare Costs	Incl. O&P
1 Equipment Oper. (med.)	$33.65	$269.20	$50.70	$405.60	$29.83	$45.60
1 Laborer	26.00	208.00	40.50	324.00		
1 Dozer, 200 H.P.		863.40		949.75	53.96	59.36
16 L.H., Daily Totals		$1340.60		$1679.35	$83.79	$104.96

Crew No.	Bare Costs Hr.	Daily	Incl. Subs O & P Hr.	Daily	Cost Per Labor-Hour Bare Costs	Incl. O&P
Crew B-11B	Hr.	Daily	Hr.	Daily	Bare Costs	Incl. O&P
1 Equipment Oper. (light)	$32.15	$257.20	$48.40	$387.20	$29.08	$44.45
1 Laborer	26.00	208.00	40.50	324.00		
1 Air Powered Tamper		23.50		25.85		
1 Air Compr. 365 C.F.M.		157.60		173.35		
2-50 Ft. Air Hoses, 1.5" Dia.		10.00		11.00	11.94	13.14
16 L.H., Daily Totals		$656.30		$921.40	$41.02	$57.59

Crew No.	Bare Costs Hr.	Daily	Incl. Subs O & P Hr.	Daily	Cost Per Labor-Hour Bare Costs	Incl. O&P
Crew B-11C	Hr.	Daily	Hr.	Daily	Bare Costs	Incl. O&P
1 Equipment Oper. (med.)	$33.65	$269.20	$50.70	$405.60	$29.83	$45.60
1 Laborer	26.00	208.00	40.50	324.00		
1 Backhoe Loader, 48 H.P.		207.80		228.60	12.99	14.29
16 L.H., Daily Totals		$685.00		$958.20	$42.82	$59.89

Crew No.	Bare Costs Hr.	Daily	Incl. Subs O & P Hr.	Daily	Cost Per Labor-Hour Bare Costs	Incl. O&P
Crew B-11K	Hr.	Daily	Hr.	Daily	Bare Costs	Incl. O&P
1 Equipment Oper. (med.)	$33.65	$269.20	$50.70	$405.60	$29.83	$45.60
1 Laborer	26.00	208.00	40.50	324.00		
1 Trencher, 8' D., 16" W.		1310.00		1441.00	81.88	90.06
16 L.H., Daily Totals		$1787.20		$2170.60	$111.71	$135.66

Crew No.	Bare Costs Hr.	Daily	Incl. Subs O & P Hr.	Daily	Cost Per Labor-Hour Bare Costs	Incl. O&P
Crew B-11L	Hr.	Daily	Hr.	Daily	Bare Costs	Incl. O&P
1 Equipment Oper. (med.)	$33.65	$269.20	$50.70	$405.60	$29.83	$45.60
1 Laborer	26.00	208.00	40.50	324.00		
1 Grader, 30,000 Lbs.		432.00		475.20	27.00	29.70
16 L.H., Daily Totals		$909.20		$1204.80	$56.83	$75.30

Crew No.	Bare Costs Hr.	Daily	Incl. Subs O & P Hr.	Daily	Cost Per Labor-Hour Bare Costs	Incl. O&P
Crew B-11M	Hr.	Daily	Hr.	Daily	Bare Costs	Incl. O&P
1 Equipment Oper. (med.)	$33.65	$269.20	$50.70	$405.60	$29.83	$45.60
1 Laborer	26.00	208.00	40.50	324.00		
1 Backhoe Loader, 80 H.P.		230.80		253.90	14.43	15.87
16 L.H., Daily Totals		$708.00		$983.50	$44.26	$61.47

Crew No.	Bare Costs Hr.	Daily	Incl. Subs O & P Hr.	Daily	Cost Per Labor-Hour Bare Costs	Incl. O&P
Crew B-11N	Hr.	Daily	Hr.	Daily	Bare Costs	Incl. O&P
1 Labor Foreman	$28.00	$224.00	$43.60	$348.80	$28.22	$42.98
2 Equipment Operators (med.)	33.65	538.40	50.70	811.20		
6 Truck Drivers (hvy.)	26.45	1269.60	40.30	1934.40		
1 F.E. Loader, 5.5 C.Y.		674.80		742.30		
1 Dozer, 400 H.P.		1473.00		1620.30		
6 Off Hwy. Tks. 50 Ton		7356.00		8091.60	132.00	145.20
72 L.H., Daily Totals		$11535.80		$13548.60	$160.22	$188.18

Crew No.	Bare Costs Hr.	Daily	Incl. Subs O & P Hr.	Daily	Cost Per Labor-Hour Bare Costs	Incl. O&P
Crew B-11Q	Hr.	Daily	Hr.	Daily	Bare Costs	Incl. O&P
1 Equipment Operator (med.)	$33.65	$269.20	$50.70	$405.60	$31.10	$47.30
.5 Laborer	26.00	104.00	40.50	162.00		
1 Dozer, 140 H.P.		542.20		596.40	45.18	49.70
12 L.H., Daily Totals		$915.40		$1164.00	$76.28	$97.00

Crew No.	Bare Costs		Incl. Subs O & P		Cost Per Labor-Hour	
Crew B-11R	Hr.	Daily	Hr.	Daily	Bare Costs	Incl. O&P
1 Equipment Operator (med.)	$33.65	$269.20	$50.70	$405.60	$31.10	$47.30
.5 Laborer	26.00	104.00	40.50	162.00		
1 Dozer, 215 H.P.		863.40		949.75	71.95	79.15
12 L.H., Daily Totals		$1236.60		$1517.35	$103.05	$126.45

Crew No.	Bare Costs		Incl. Subs O & P		Cost Per Labor-Hour	
Crew B-11S	Hr.	Daily	Hr.	Daily	Bare Costs	Incl. O&P
1 Equipment Operator (med.)	$33.65	$269.20	$50.70	$405.60	$31.10	$47.30
.5 Laborer	26.00	104.00	40.50	162.00		
1 Dozer, 300 H.P.		1099.00		1208.90		
1 Ripper, Beam & 1 Shank		70.60		77.65	97.47	107.21
12 L.H., Daily Totals		$1542.80		$1854.15	$128.57	$154.51

Crew No.	Bare Costs		Incl. Subs O & P		Cost Per Labor-Hour	
Crew B-11T	Hr.	Daily	Hr.	Daily	Bare Costs	Incl. O&P
1 Equipment Operator (med.)	$33.65	$269.20	$50.70	$405.60	$31.10	$47.30
.5 Laborer	26.00	104.00	40.50	162.00		
1 Dozer, 410 H.P.		1473.00		1620.30		
1 Ripper, Beam & 2 Shanks		80.40		88.45	129.45	142.40
12 L.H., Daily Totals		$1926.60		$2276.35	$160.55	$189.70

Crew No.	Bare Costs		Incl. Subs O & P		Cost Per Labor-Hour	
Crew B-11U	Hr.	Daily	Hr.	Daily	Bare Costs	Incl. O&P
1 Equipment Operator (med.)	$33.65	$269.20	$50.70	$405.60	$31.10	$47.30
.5 Laborer	26.00	104.00	40.50	162.00		
1 Dozer, 520 H.P.		1909.00		2099.90	159.08	174.99
12 L.H., Daily Totals		$2282.20		$2667.50	$190.18	$222.29

Crew No.	Bare Costs		Incl. Subs O & P		Cost Per Labor-Hour	
Crew B-11V	Hr.	Daily	Hr.	Daily	Bare Costs	Incl. O&P
3 Laborers	$26.00	$624.00	$40.50	$972.00	$26.00	$40.50
1 Roll. Compact., 2K Lbs.		130.80		143.90	5.45	6.00
24 L.H., Daily Totals		$754.80		$1115.90	$31.45	$46.50

Crew No.	Bare Costs		Incl. Subs O & P		Cost Per Labor-Hour	
Crew B-11W	Hr.	Daily	Hr.	Daily	Bare Costs	Incl. O&P
1 Equipment Operator (med.)	$33.65	$269.20	$50.70	$405.60	$27.01	$41.18
1 Common Laborer	26.00	208.00	40.50	324.00		
10 Truck Drivers, Heavy	26.45	2116.00	40.30	3224.00		
1 Dozer, 200 H.P.		863.40		949.75		
1 Vib. roller, smth, towed, 23 Ton		610.20		671.20		
10 Dump Truck, 10 Ton		3250.00		3575.00	49.20	54.12
96 L.H., Daily Totals		$7316.80		$9149.55	$76.21	$95.30

Crew No.	Bare Costs		Incl. Subs O & P		Cost Per Labor-Hour	
Crew B-11Y	Hr.	Daily	Hr.	Daily	Bare Costs	Incl. O&P
1 Labor Foreman (Outside)	$28.00	$224.00	$43.60	$348.80	$28.77	$44.24
5 Common Laborers	26.00	1040.00	40.50	1620.00		
3 Equipment Operator (med.)	33.65	807.60	50.70	1216.80		
1 Dozer, 80 H.P.		300.20		330.20		
2 Wlk-Beh. Comp., 2-Drum, 1 Ton		261.60		287.75		
4 Vibratory plate, gas, 21"		165.60		182.15	10.10	11.11
72 L.H., Daily Totals		$2799.00		$3985.70	$38.87	$55.35

Crew No.	Bare Costs		Incl. Subs O & P		Cost Per Labor-Hour	
Crew B-12A	Hr.	Daily	Hr.	Daily	Bare Costs	Incl. O&P
1 Equip. Oper. (crane)	$34.80	$278.40	$52.40	$419.20	$32.00	$48.20
1 Equip. Oper. Oiler	29.20	233.60	44.00	352.00		
1 Hyd. Excavator, 1 C.Y.		487.40		536.15	30.46	33.51
16 L.H., Daily Totals		$999.40		$1307.35	$62.46	$81.71

Crew No.	Bare Costs		Incl. Subs O & P		Cost Per Labor-Hour	
Crew B-12B	Hr.	Daily	Hr.	Daily	Bare Costs	Incl. O&P
1 Equip. Oper. (crane)	$34.80	$278.40	$52.40	$419.20	$32.00	$48.20
1 Equip. Oper. Oiler	29.20	233.60	44.00	352.00		
1 Hyd. Excavator, 1.5 C.Y.		683.40		751.75	42.71	46.98
16 L.H., Daily Totals		$1195.40		$1522.95	$74.71	$95.18

Crew No.	Bare Costs		Incl. Subs O & P		Cost Per Labor-Hour	
Crew B-12C	Hr.	Daily	Hr.	Daily	Bare Costs	Incl. O&P
1 Equip. Oper. (crane)	$34.80	$278.40	$52.40	$419.20	$32.00	$48.20
1 Equip. Oper. Oiler	29.20	233.60	44.00	352.00		
1 Hyd. Excavator, 2 C.Y.		887.20		975.90	55.45	61.00
16 L.H., Daily Totals		$1399.20		$1747.10	$87.45	$109.20

Crew No.	Bare Costs		Incl. Subs O & P		Cost Per Labor-Hour	
Crew B-12D	Hr.	Daily	Hr.	Daily	Bare Costs	Incl. O&P
1 Equip. Oper. (crane)	$34.80	$278.40	$52.40	$419.20	$32.00	$48.20
1 Equip. Oper. Oiler	29.20	233.60	44.00	352.00		
1 Hyd. Excavator, 3.5 C.Y.		1965.00		2161.50	122.81	135.09
16 L.H., Daily Totals		$2477.00		$2932.70	$154.81	$183.29

Crew No.	Bare Costs		Incl. Subs O & P		Cost Per Labor-Hour	
Crew B-12E	Hr.	Daily	Hr.	Daily	Bare Costs	Incl. O&P
1 Equip. Oper. (crane)	$34.80	$278.40	$52.40	$419.20	$32.00	$48.20
1 Equip. Oper. Oiler	29.20	233.60	44.00	352.00		
1 Hyd. Excavator, .5 C.Y.		323.00		355.30	20.19	22.21
16 L.H., Daily Totals		$835.00		$1126.50	$52.19	$70.41

Crew No.	Bare Costs		Incl. Subs O & P		Cost Per Labor-Hour	
Crew B-12F	Hr.	Daily	Hr.	Daily	Bare Costs	Incl. O&P
1 Equip. Oper. (crane)	$34.80	$278.40	$52.40	$419.20	$32.00	$48.20
1 Equip. Oper. Oiler	29.20	233.60	44.00	352.00		
1 Hyd. Excavator, .75 C.Y.		458.80		504.70	28.68	31.54
16 L.H., Daily Totals		$970.80		$1275.90	$60.68	$79.74

Crew No.	Bare Costs		Incl. Subs O & P		Cost Per Labor-Hour	
Crew B-12G	Hr.	Daily	Hr.	Daily	Bare Costs	Incl. O&P
1 Equip. Oper. (crane)	$34.80	$278.40	$52.40	$419.20	$32.00	$48.20
1 Equip. Oper. Oiler	29.20	233.60	44.00	352.00		
1 Power Shovel, .5 C.Y.		450.90		496.00		
1 Clamshell Bucket, .5 C.Y.		33.40		36.75	30.27	33.30
16 L.H., Daily Totals		$996.30		$1303.95	$62.27	$81.50

Crew No.	Bare Costs		Incl. Subs O & P		Cost Per Labor-Hour	
Crew B-12H	Hr.	Daily	Hr.	Daily	Bare Costs	Incl. O&P
1 Equip. Oper. (crane)	$34.80	$278.40	$52.40	$419.20	$32.00	$48.20
1 Equip. Oper. Oiler	29.20	233.60	44.00	352.00		
1 Power Shovel, 1 C.Y.		792.00		871.20		
1 Clamshell Bucket, 1 C.Y.		44.00		48.40	52.25	57.48
16 L.H., Daily Totals		$1348.00		$1690.80	$84.25	$105.68

Crew No.	Bare Costs		Incl. Subs O & P		Cost Per Labor-Hour	
Crew B-12I	Hr.	Daily	Hr.	Daily	Bare Costs	Incl. O&P
1 Equip. Oper. (crane)	$34.80	$278.40	$52.40	$419.20	$32.00	$48.20
1 Equip. Oper. Oiler	29.20	233.60	44.00	352.00		
1 Power Shovel, .75 C.Y.		604.50		664.95		
1 Dragline Bucket, .75 C.Y.		18.60		20.45	38.94	42.84
16 L.H., Daily Totals		$1135.10		$1456.60	$70.94	$91.04

Crew No.	Bare Costs		Incl. Subs O & P		Cost Per Labor-Hour	
Crew B-12J	Hr.	Daily	Hr.	Daily	Bare Costs	Incl. O&P
1 Equip. Oper. (crane)	$34.80	$278.40	$52.40	$419.20	$32.00	$48.20
1 Equip. Oper. Oiler	29.20	233.60	44.00	352.00		
1 Gradall, 3 Ton, .5 C.Y.		823.40		905.75	51.46	56.61
16 L.H., Daily Totals		$1335.40		$1676.95	$83.46	$104.81

Crew No.	Bare Costs		Incl. Subs O & P		Cost Per Labor-Hour	
Crew B-12K	Hr.	Daily	Hr.	Daily	Bare Costs	Incl. O&P
1 Equip. Oper. (crane)	$34.80	$278.40	$52.40	$419.20	$32.00	$48.20
1 Equip. Oper. Oiler	29.20	233.60	44.00	352.00		
1 Gradall, 3 Ton, 1 C.Y.		958.80		1054.70	59.93	65.92
16 L.H., Daily Totals		$1470.80		$1825.90	$91.93	$114.12

Crews

CREWS

Crew No.	Bare Costs Hr.	Daily	Incl. Subs O & P Hr.	Daily	Cost Per Labor-Hour Bare Costs	Incl. O&P
Crew B-12L	Hr.	Daily	Hr.	Daily	Bare Costs	Incl. O&P
1 Equip. Oper. (crane)	$34.80	$278.40	$52.40	$419.20	$32.00	$48.20
1 Equip. Oper. Oiler	29.20	233.60	44.00	352.00		
1 Power Shovel, .5 C.Y.		450.90		496.00		
1 F.E. Attachment, .5 C.Y.		45.20		49.70	31.01	34.11
16 L.H., Daily Totals		$1008.10		$1316.90	$63.01	$82.31
Crew B-12M	Hr.	Daily	Hr.	Daily	Bare Costs	Incl. O&P
1 Equip. Oper. (crane)	$34.80	$278.40	$52.40	$419.20	$32.00	$48.20
1 Equip. Oper. Oiler	29.20	233.60	44.00	352.00		
1 Power Shovel, .75 C.Y.		604.50		664.95		
1 F.E. Attachment, .75 C.Y.		50.00		55.00	40.91	45.00
16 L.H., Daily Totals		$1166.50		$1491.15	$72.91	$93.20
Crew B-12N	Hr.	Daily	Hr.	Daily	Bare Costs	Incl. O&P
1 Equip. Oper. (crane)	$34.80	$278.40	$52.40	$419.20	$32.00	$48.20
1 Equip. Oper. Oiler	29.20	233.60	44.00	352.00		
1 Power Shovel, 1 C.Y.		792.00		871.20		
1 F.E. Attachment, 1 C.Y.		56.80		62.50	53.05	58.36
16 L.H., Daily Totals		$1360.80		$1704.90	$85.05	$106.56
Crew B-12O	Hr.	Daily	Hr.	Daily	Bare Costs	Incl. O&P
1 Equip. Oper. (crane)	$34.80	$278.40	$52.40	$419.20	$32.00	$48.20
1 Equip. Oper. Oiler	29.20	233.60	44.00	352.00		
1 Power Shovel, 1.5 C.Y.		915.20		1006.70		
1 F.E. Attachment, 1.5 C.Y.		65.60		72.15	61.30	67.43
16 L.H., Daily Totals		$1492.80		$1850.05	$93.30	$115.63
Crew B-12P	Hr.	Daily	Hr.	Daily	Bare Costs	Incl. O&P
1 Equip. Oper. (crane)	$34.80	$278.40	$52.40	$419.20	$32.00	$48.20
1 Equip. Oper. Oiler	29.20	233.60	44.00	352.00		
1 Crawler Crane, 40 Ton		915.20		1006.70		
1 Dragline Bucket, 1.5 C.Y.		30.20		33.20	59.09	65.00
16 L.H., Daily Totals		$1457.40		$1811.10	$91.09	$113.20
Crew B-12Q	Hr.	Daily	Hr.	Daily	Bare Costs	Incl. O&P
1 Equip. Oper. (crane)	$34.80	$278.40	$52.40	$419.20	$32.00	$48.20
1 Equip. Oper. Oiler	29.20	233.60	44.00	352.00		
1 Hyd. Excavator, 5/8 C.Y.		418.00		459.80	26.13	28.74
16 L.H., Daily Totals		$930.00		$1231.00	$58.13	$76.94
Crew B-12R	Hr.	Daily	Hr.	Daily	Bare Costs	Incl. O&P
1 Equip. Oper. (crane)	$34.80	$278.40	$52.40	$419.20	$32.00	$48.20
1 Equip. Oper. Oiler	29.20	233.60	44.00	352.00		
1 Hyd. Excavator, 1.5 C.Y.		683.40		751.75	42.71	46.98
16 L.H., Daily Totals		$1195.40		$1522.95	$74.71	$95.18
Crew B-12S	Hr.	Daily	Hr.	Daily	Bare Costs	Incl. O&P
1 Equip. Oper. (crane)	$34.80	$278.40	$52.40	$419.20	$32.00	$48.20
1 Equip. Oper. Oiler	29.20	233.60	44.00	352.00		
1 Hyd. Excavator, 2.5 C.Y.		1181.00		1299.10	73.81	81.19
16 L.H., Daily Totals		$1693.00		$2070.30	$105.81	$129.39
Crew B-12T	Hr.	Daily	Hr.	Daily	Bare Costs	Incl. O&P
1 Equip. Oper. (crane)	$34.80	$278.40	$52.40	$419.20	$32.00	$48.20
1 Equip. Oper. Oiler	29.20	233.60	44.00	352.00		
1 Crawler Crane, 75 Ton		1217.00		1338.70		
1 F.E. Attachment, 3 C.Y.		89.40		98.35	81.65	89.82
16 L.H., Daily Totals		$1818.40		$2208.25	$113.65	$138.02

Crew No.	Bare Costs Hr.	Daily	Incl. Subs O & P Hr.	Daily	Cost Per Labor-Hour Bare Costs	Incl. O&P
Crew B-12V	Hr.	Daily	Hr.	Daily	Bare Costs	Incl. O&P
1 Equip. Oper. (crane)	$34.80	$278.40	$52.40	$419.20	$32.00	$48.20
1 Equip. Oper. Oiler	29.20	233.60	44.00	352.00		
1 Crawler Crane, 75 Ton		1217.00		1338.70		
1 Dragline Bucket, 3 C.Y.		47.60		52.35	79.04	86.94
16 L.H., Daily Totals		$1776.60		$2162.25	$111.04	$135.14
Crew B-13	Hr.	Daily	Hr.	Daily	Bare Costs	Incl. O&P
1 Labor Foreman (outside)	$28.00	$224.00	$43.60	$348.80	$28.00	$43.14
4 Laborers	26.00	832.00	40.50	1296.00		
1 Equip. Oper. (crane)	34.80	278.40	52.40	419.20		
1 Equip. Oper. Oiler	29.20	233.60	44.00	352.00		
1 Hyd. Crane, 25 Ton		624.20		686.60	11.15	12.26
56 L.H., Daily Totals		$2192.20		$3102.60	$39.15	$55.40
Crew B-13A	Hr.	Daily	Hr.	Daily	Bare Costs	Incl. O&P
1 Foreman	$28.00	$224.00	$43.60	$348.80	$28.60	$43.80
2 Laborers	26.00	416.00	40.50	648.00		
2 Equipment Operators	33.65	538.40	50.70	811.20		
2 Truck Drivers (heavy)	26.45	423.20	40.30	644.80		
1 Crane, 75 Ton		1217.00		1338.70		
1 F.E. Lder, 3.75 C.Y.		1061.00		1167.10		
2 Dump Trucks, 12 Ton		650.00		715.00	52.29	57.51
56 L.H., Daily Totals		$4529.60		$5673.60	$80.89	$101.31
Crew B-13B	Hr.	Daily	Hr.	Daily	Bare Costs	Incl. O&P
1 Labor Foreman (outside)	$28.00	$224.00	$43.60	$348.80	$28.00	$43.14
4 Laborers	26.00	832.00	40.50	1296.00		
1 Equip. Oper. (crane)	34.80	278.40	52.40	419.20		
1 Equip. Oper. Oiler	29.20	233.60	44.00	352.00		
1 Hyd. Crane, 55 Ton		911.00		1002.10	16.27	17.89
56 L.H., Daily Totals		$2479.00		$3418.10	$44.27	$61.03
Crew B-13C	Hr.	Daily	Hr.	Daily	Bare Costs	Incl. O&P
1 Labor Foreman (outside)	$28.00	$224.00	$43.60	$348.80	$28.00	$43.14
4 Laborers	26.00	832.00	40.50	1296.00		
1 Equip. Oper. (crane)	34.80	278.40	52.40	419.20		
1 Equip. Oper. Oiler	29.20	233.60	44.00	352.00		
1 Crawler Crane, 100 Ton		1602.00		1762.20	28.61	31.47
56 L.H., Daily Totals		$3170.00		$4178.20	$56.61	$74.61
Crew B-14	Hr.	Daily	Hr.	Daily	Bare Costs	Incl. O&P
1 Labor Foreman (outside)	$28.00	$224.00	$43.60	$348.80	$27.36	$42.33
4 Laborers	26.00	832.00	40.50	1296.00		
1 Equip. Oper. (light)	32.15	257.20	48.40	387.20		
1 Backhoe Loader, 48 H.P.		207.80		228.60	4.33	4.76
48 L.H., Daily Totals		$1521.00		$2260.60	$31.69	$47.09
Crew B-15	Hr.	Daily	Hr.	Daily	Bare Costs	Incl. O&P
1 Equipment Oper. (med)	$33.65	$269.20	$50.70	$405.60	$28.44	$43.30
.5 Laborer	26.00	104.00	40.50	162.00		
2 Truck Drivers (heavy)	26.45	423.20	40.30	644.80		
2 Dump Trucks, 16 Ton		953.20		1048.50		
1 Dozer, 200 H.P.		863.40		949.75	64.88	71.37
28 L.H., Daily Totals		$2613.00		$3210.65	$93.32	$114.67
Crew B-16	Hr.	Daily	Hr.	Daily	Bare Costs	Incl. O&P
1 Labor Foreman (outside)	$28.00	$224.00	$43.60	$348.80	$26.61	$41.22
2 Laborers	26.00	416.00	40.50	648.00		
1 Truck Driver (heavy)	26.45	211.60	40.30	322.40		
1 Dump Truck, 16 Ton		476.60		524.25	14.89	16.38
32 L.H., Daily Totals		$1328.20		$1843.45	$41.50	$57.60

Crew No.	Bare Costs		Incl. Subs O & P		Cost Per Labor-Hour	
Crew B-17	Hr.	Daily	Hr.	Daily	Bare Costs	Incl. O&P
2 Laborers	$26.00	$416.00	$40.50	$648.00	$27.65	$42.42
1 Equip. Oper. (light)	32.15	257.20	48.40	387.20		
1 Truck Driver (heavy)	26.45	211.60	40.30	322.40		
1 Backhoe Loader, 48 H.P.		207.80		228.60		
1 Dump Truck, 12 Ton		325.00		357.50	16.65	18.32
32 L.H., Daily Totals		$1417.60		$1943.70	$44.30	$60.74
Crew B-18	Hr.	Daily	Hr.	Daily	Bare Costs	Incl. O&P
1 Labor Foreman (outside)	$28.00	$224.00	$43.60	$348.80	$26.67	$41.53
2 Laborers	26.00	416.00	40.50	648.00		
1 Vibrating Compactor		41.40		45.55	1.73	1.90
24 L.H., Daily Totals		$681.40		$1042.35	$28.40	$43.43
Crew B-19	Hr.	Daily	Hr.	Daily	Bare Costs	Incl. O&P
1 Pile Driver Foreman	$34.05	$272.40	$56.25	$450.00	$32.63	$52.11
4 Pile Drivers	32.05	1025.60	52.95	1694.40		
2 Equip. Oper. (crane)	34.80	556.80	52.40	838.40		
1 Equip. Oper. Oiler	29.20	233.60	44.00	352.00		
1 Crane, 40 Ton & Access.		915.20		1006.70		
60 L.F. Pile Leads		90.00		99.00		
1 Hammer, Diesel, 22k Ft-Lb		564.00		620.40	24.52	26.97
64 L.H., Daily Totals		$3657.60		$5060.90	$57.15	$79.08
Crew B-19A	Hr.	Daily	Hr.	Daily	Bare Costs	Incl. O&P
1 Pile Driver Foreman	$34.05	$272.40	$56.25	$450.00	$32.63	$52.11
4 Pile Drivers	32.05	1025.60	52.95	1694.40		
2 Equip. Oper. (crane)	34.80	556.80	52.40	838.40		
1 Equip. Oper. Oiler	29.20	233.60	44.00	352.00		
1 Crawler Crane, 75 Ton		1217.00		1338.70		
60 L.F. Leads, 25K Ft. Lbs.		132.00		145.20		
1 Hammer, Diesel, 41k Ft-Lb		662.60		728.85	31.43	34.57
64 L.H., Daily Totals		$4100.00		$5547.55	$64.06	$86.68
Crew B-20	Hr.	Daily	Hr.	Daily	Bare Costs	Incl. O&P
1 Labor Foreman (out)	$28.00	$224.00	$43.60	$348.80	$29.22	$45.48
1 Skilled Worker	33.65	269.20	52.35	418.80		
1 Laborer	26.00	208.00	40.50	324.00		
24 L.H., Daily Totals		$701.20		$1091.60	$29.22	$45.48
Crew B-20A	Hr.	Daily	Hr.	Daily	Bare Costs	Incl. O&P
1 Labor Foreman	$28.00	$224.00	$43.60	$348.80	$31.33	$47.79
1 Laborer	26.00	208.00	40.50	324.00		
1 Plumber	39.60	316.80	59.45	475.60		
1 Plumber Apprentice	31.70	253.60	47.60	380.80		
32 L.H., Daily Totals		$1002.40		$1529.20	$31.33	$47.79
Crew B-21	Hr.	Daily	Hr.	Daily	Bare Costs	Incl. O&P
1 Labor Foreman (out)	$28.00	$224.00	$43.60	$348.80	$30.01	$46.47
1 Skilled Worker	33.65	269.20	52.35	418.80		
1 Laborer	26.00	208.00	40.50	324.00		
.5 Equip. Oper. (crane)	34.80	139.20	52.40	209.60		
.5 S.P. Crane, 5 Ton		158.90		174.80	5.68	6.24
28 L.H., Daily Totals		$999.30		$1476.00	$35.69	$52.71

Crew No.	Bare Costs		Incl. Subs O & P		Cost Per Labor-Hour	
Crew B-21A	Hr.	Daily	Hr.	Daily	Bare Costs	Incl. O&P
1 Labor Foreman	$28.00	$224.00	$43.60	$348.80	$32.02	$48.71
1 Laborer	26.00	208.00	40.50	324.00		
1 Plumber	39.60	316.80	59.45	475.60		
1 Plumber Apprentice	31.70	253.60	47.60	380.80		
1 Equip. Oper. (crane)	34.80	278.40	52.40	419.20		
1 S.P. Crane, 12 Ton		482.20		530.40	12.06	13.26
40 L.H., Daily Totals		$1763.00		$2478.80	$44.08	$61.97
Crew B-22	Hr.	Daily	Hr.	Daily	Bare Costs	Incl. O&P
1 Labor Foreman (out)	$28.00	$224.00	$43.60	$348.80	$30.33	$46.87
1 Skilled Worker	33.65	269.20	52.35	418.80		
1 Laborer	26.00	208.00	40.50	324.00		
.75 Equip. Oper. (crane)	34.80	208.80	52.40	314.40		
.75 S.P. Crane, 5 Ton		238.35		262.20	7.95	8.74
30 L.H., Daily Totals		$1148.35		$1668.20	$38.28	$55.61
Crew B-22A	Hr.	Daily	Hr.	Daily	Bare Costs	Incl. O&P
1 Labor Foreman (out)	$28.00	$224.00	$43.60	$348.80	$29.42	$45.53
1 Skilled Worker	33.65	269.20	52.35	418.80		
2 Laborers	26.00	416.00	40.50	648.00		
.75 Equipment Oper. (crane)	34.80	208.80	52.40	314.40		
.75 Crane, 5 Ton		238.35		262.20		
1 Generator, 5 KW		37.00		40.70		
1 Butt Fusion Machine		433.60		476.95	18.66	20.52
38 L.H., Daily Totals		$1826.95		$2509.85	$48.08	$66.05
Crew B-22B	Hr.	Daily	Hr.	Daily	Bare Costs	Incl. O&P
1 Skilled Worker	$33.65	$269.20	$52.35	$418.80	$29.83	$46.42
1 Laborer	26.00	208.00	40.50	324.00		
1 Electro Fusion Machine		172.80		190.10	10.80	11.88
16 L.H., Daily Totals		$650.00		$932.90	$40.63	$58.30
Crew B-23	Hr.	Daily	Hr.	Daily	Bare Costs	Incl. O&P
1 Labor Foreman (outside)	$28.00	$224.00	$43.60	$348.80	$26.40	$41.12
4 Laborers	26.00	832.00	40.50	1296.00		
1 Drill Rig, Wells		2877.00		3164.70		
1 Light Truck, 3 Ton		160.40		176.45	75.94	83.53
40 L.H., Daily Totals		$4093.40		$4985.95	$102.34	$124.65
Crew B-23A	Hr.	Daily	Hr.	Daily	Bare Costs	Incl. O&P
1 Labor Foreman (outside)	$28.00	$224.00	$43.60	$348.80	$29.22	$44.93
1 Laborer	26.00	208.00	40.50	324.00		
1 Equip. Operator (medium)	33.65	269.20	50.70	405.60		
1 Drill Rig, Wells		2877.00		3164.70		
1 Pickup Truck, 3/4 Ton		75.80		83.40	123.03	135.34
24 L.H., Daily Totals		$3654.00		$4326.50	$152.25	$180.27
Crew B-23B	Hr.	Daily	Hr.	Daily	Bare Costs	Incl. O&P
1 Labor Foreman (outside)	$28.00	$224.00	$43.60	$348.80	$29.22	$44.93
1 Laborer	26.00	208.00	40.50	324.00		
1 Equip. Operator (medium)	33.65	269.20	50.70	405.60		
1 Drill Rig, Wells		2877.00		3164.70		
1 Pickup Truck, 3/4 Ton		75.80		83.40		
1 Pump, Cntfgl, 6"		229.80		252.80	132.61	145.87
24 L.H., Daily Totals		$3883.80		$4579.30	$161.83	$190.80
Crew B-24	Hr.	Daily	Hr.	Daily	Bare Costs	Incl. O&P
1 Cement Finisher	$31.55	$252.40	$46.45	$371.60	$30.18	$46.12
1 Laborer	26.00	208.00	40.50	324.00		
1 Carpenter	33.00	264.00	51.40	411.20		
24 L.H., Daily Totals		$724.40		$1106.80	$30.18	$46.12

CREWS

Crew No.	Bare Costs Hr.	Daily	Incl. Subs O & P Hr.	Daily	Cost Per Labor-Hour Bare Costs	Incl. O&P
Crew B-25	Hr.	Daily	Hr.	Daily	Bare Costs	Incl. O&P
1 Labor Foreman	$28.00	$224.00	$43.60	$348.80	$28.27	$43.56
7 Laborers	26.00	1456.00	40.50	2268.00		
3 Equip. Oper. (med.)	33.65	807.60	50.70	1216.80		
1 Asphalt Paver, 130 H.P		1457.00		1602.70		
1 Tandem Roller, 10 Ton		183.60		201.95		
1 Roller, Pneumatic Wheel		244.60		269.05	21.42	23.57
88 L.H., Daily Totals		$4372.80		$5907.30	$49.69	$67.13

Crew B-25B	Hr.	Daily	Hr.	Daily	Bare Costs	Incl. O&P
1 Labor Foreman	$28.00	$224.00	$43.60	$348.80	$28.72	$44.16
7 Laborers	26.00	1456.00	40.50	2268.00		
4 Equip. Oper. (medium)	33.65	1076.80	50.70	1622.40		
1 Asphalt Paver, 130 H.P		1457.00		1602.70		
2 Rollers, Steel Wheel		367.20		403.90		
1 Roller, Pneumatic Wheel		244.60		269.05	21.55	23.71
96 L.H., Daily Totals		$4825.60		$6514.85	$50.27	$67.87

Crew B-25C	Hr.	Daily	Hr.	Daily	Bare Costs	Incl. O&P
1 Labor Foreman	$28.00	$224.00	$43.60	$348.80	$28.88	$44.42
3 Laborers	26.00	624.00	40.50	972.00		
2 Equip. Oper. (medium)	33.65	538.40	50.70	811.20		
1 Asphalt Paver, 130 H.P		1457.00		1602.70		
1 Rollers, Steel Wheel		183.60		201.95	34.18	37.60
48 L.H., Daily Totals		$3027.00		$3936.65	$63.06	$82.02

Crew B-26	Hr.	Daily	Hr.	Daily	Bare Costs	Incl. O&P
1 Labor Foreman (outside)	$28.00	$224.00	$43.60	$348.80	$29.09	$45.06
6 Laborers	26.00	1248.00	40.50	1944.00		
2 Equip. Oper. (med.)	33.65	538.40	50.70	811.20		
1 Rodman (reinf.)	37.10	296.80	61.25	490.00		
1 Cement Finisher	31.55	252.40	46.45	371.60		
1 Grader, 30,000 Lbs.		432.00		475.20		
1 Paving Mach. & Equip.		1580.00		1738.00	22.86	25.15
88 L.H., Daily Totals		$4571.60		$6178.80	$51.95	$70.21

Crew B-27	Hr.	Daily	Hr.	Daily	Bare Costs	Incl. O&P
1 Labor Foreman (outside)	$28.00	$224.00	$43.60	$348.80	$26.50	$41.28
3 Laborers	26.00	624.00	40.50	972.00		
1 Berm Machine		188.40		207.25	5.89	6.48
32 L.H., Daily Totals		$1036.40		$1528.05	$32.39	$47.76

Crew B-28	Hr.	Daily	Hr.	Daily	Bare Costs	Incl. O&P
2 Carpenters	$33.00	$528.00	$51.40	$822.40	$30.67	$47.77
1 Laborer	26.00	208.00	40.50	324.00		
24 L.H., Daily Totals		$736.00		$1146.40	$30.67	$47.77

Crew B-29	Hr.	Daily	Hr.	Daily	Bare Costs	Incl. O&P
1 Labor Foreman (outside)	$28.00	$224.00	$43.60	$348.80	$28.00	$43.14
4 Laborers	26.00	832.00	40.50	1296.00		
1 Equip. Oper. (crane)	34.80	278.40	52.40	419.20		
1 Equip. Oper. Oiler	29.20	233.60	44.00	352.00		
1 Gradall, 3 Ton, 1/2 C.Y.		823.40		905.75	14.70	16.17
56 L.H., Daily Totals		$2391.40		$3321.75	$42.70	$59.31

Crew B-30	Hr.	Daily	Hr.	Daily	Bare Costs	Incl. O&P
1 Equip. Oper. (med.)	$33.65	$269.20	$50.70	$405.60	$28.85	$43.77
2 Truck Drivers (heavy)	26.45	423.20	40.30	644.80		
1 Hyd. Excavator, 1.5 C.Y.		683.40		751.75		
2 Dump Trucks, 16 Ton		953.20		1048.50	68.19	75.01
24 L.H., Daily Totals		$2329.00		$2850.65	$97.04	$118.78

Crew B-31	Hr.	Daily	Hr.	Daily	Bare Costs	Incl. O&P
1 Labor Foreman (outside)	$28.00	$224.00	$43.60	$348.80	$27.80	$43.30
3 Laborers	26.00	624.00	40.50	972.00		
1 Carpenter	33.00	264.00	51.40	411.20		
1 Air Compr., 250 C.F.M.		127.60		140.35		
1 Sheeting Driver		7.20		7.90		
2-50 Ft. Air Hoses, 1.5" Dia.		10.00		11.00	3.62	3.98
40 L.H., Daily Totals		$1256.80		$1891.25	$31.42	$47.28

Crew B-32	Hr.	Daily	Hr.	Daily	Bare Costs	Incl. O&P
1 Laborer	$26.00	$208.00	$40.50	$324.00	$31.74	$48.15
3 Equip. Oper. (med.)	33.65	807.60	50.70	1216.80		
1 Grader, 30,000 Lbs.		432.00		475.20		
1 Tandem Roller, 10 Ton		183.60		201.95		
1 Dozer, 200 H.P.		863.40		949.75	46.22	50.84
32 L.H., Daily Totals		$2494.60		$3167.70	$77.96	$98.99

Crew B-32A	Hr.	Daily	Hr.	Daily	Bare Costs	Incl. O&P
1 Laborer	$26.00	$208.00	$40.50	$324.00	$31.10	$47.30
2 Equip. Oper. (medium)	33.65	538.40	50.70	811.20		
1 Grader, 30,000 Lbs.		432.00		475.20		
1 Roller, Vibratory, 29,000 Lbs.		436.80		480.50	36.20	39.82
24 L.H., Daily Totals		$1615.20		$2090.90	$67.30	$87.12

Crew B-32B	Hr.	Daily	Hr.	Daily	Bare Costs	Incl. O&P
1 Laborer	$26.00	$208.00	$40.50	$324.00	$31.10	$47.30
2 Equip. Oper. (medium)	33.65	538.40	50.70	811.20		
1 Dozer, 200 H.P.		863.40		949.75		
1 Roller, Vibratory, 29,000 Lbs.		436.80		480.50	54.18	59.59
24 L.H., Daily Totals		$2046.60		$2565.45	$85.28	$106.89

Crew B-32C	Hr.	Daily	Hr.	Daily	Bare Costs	Incl. O&P
1 Labor Foreman	$28.00	$224.00	$43.60	$348.80	$30.16	$46.12
2 Laborers	26.00	416.00	40.50	648.00		
3 Equip. Oper. (medium)	33.65	807.60	50.70	1216.80		
1 Grader, 30,000 Lbs.		432.00		475.20		
1 Roller, Steel Wheel		183.60		201.95		
1 Dozer, 200 H.P.		863.40		949.75	30.81	33.89
48 L.H., Daily Totals		$2926.60		$3840.50	$60.97	$80.01

Crew B-33A	Hr.	Daily	Hr.	Daily	Bare Costs	Incl. O&P
1 Equip. Oper. (med.)	$33.65	$269.20	$50.70	$405.60	$31.46	$47.79
.5 Laborer	26.00	104.00	40.50	162.00		
.25 Equip. Oper. (med.)	33.65	67.30	50.70	101.40		
1 Scraper, Towed, 7 C.Y.		170.00		187.00		
1.25 Dozer, 300 H.P.		1373.75		1511.15	110.27	121.29
14 L.H., Daily Totals		$1984.25		$2367.15	$141.73	$169.08

Crew B-33B	Hr.	Daily	Hr.	Daily	Bare Costs	Incl. O&P
1 Equip. Oper. (med.)	$33.65	$269.20	$50.70	$405.60	$31.46	$47.79
.5 Laborer	26.00	104.00	40.50	162.00		
.25 Equip. Oper. (med.)	33.65	67.30	50.70	101.40		
1 Scraper, Towed, 10 C.Y.		189.10		208.00		
1.25 Dozer, 300 H.P.		1373.75		1511.15	111.63	122.80
14 L.H., Daily Totals		$2003.35		$2388.15	$143.09	$170.59

Crews

Crew No.	Bare Costs Hr.	Daily	Incl. Subs O & P Hr.	Daily	Cost Per Labor-Hour Bare Costs	Incl. O&P
Crew B-33C	Hr.	Daily	Hr.	Daily	Bare Costs	Incl. O&P
1 Equip. Oper. (med.)	$33.65	$269.20	$50.70	$405.60	$31.46	$47.79
.5 Laborer	26.00	104.00	40.50	162.00		
.25 Equip. Oper. (med.)	33.65	67.30	50.70	101.40		
1 Scraper, Towed, 12 C.Y.		189.10		208.00		
1.25 Dozer, 300 H.P.		1373.75		1511.15	111.63	122.80
14 L.H., Daily Totals		$2003.35		$2388.15	$143.09	$170.59

Crew No.	Hr.	Daily	Hr.	Daily	Bare Costs	Incl. O&P
Crew B-33D	Hr.	Daily	Hr.	Daily	Bare Costs	Incl. O&P
1 Equip. Oper. (med.)	$33.65	$269.20	$50.70	$405.60	$31.46	$47.79
.5 Laborer	26.00	104.00	40.50	162.00		
.25 Equip. Oper. (med.)	33.65	67.30	50.70	101.40		
1 S.P. Scraper, 14 C.Y.		1468.00		1614.80		
.25 Dozer, 300 H.P.		274.75		302.25	124.48	136.93
14 L.H., Daily Totals		$2183.25		$2586.05	$155.94	$184.72

Crew No.	Hr.	Daily	Hr.	Daily	Bare Costs	Incl. O&P
Crew B-33E	Hr.	Daily	Hr.	Daily	Bare Costs	Incl. O&P
1 Equip. Oper. (med.)	$33.65	$269.20	$50.70	$405.60	$31.46	$47.79
.5 Laborer	26.00	104.00	40.50	162.00		
.25 Equip. Oper. (med.)	33.65	67.30	50.70	101.40		
1 S.P. Scraper, 24 C.Y.		2297.00		2526.70		
.25 Dozer, 300 H.P.		274.75		302.25	183.70	202.07
14 L.H., Daily Totals		$3012.25		$3497.95	$215.16	$249.86

Crew No.	Hr.	Daily	Hr.	Daily	Bare Costs	Incl. O&P
Crew B-33F	Hr.	Daily	Hr.	Daily	Bare Costs	Incl. O&P
1 Equip. Oper. (med.)	$33.65	$269.20	$50.70	$405.60	$31.46	$47.79
.5 Laborer	26.00	104.00	40.50	162.00		
.25 Equip. Oper. (med.)	33.65	67.30	50.70	101.40		
1 Elev. Scraper, 11 C.Y.		805.40		885.95		
.25 Dozer, 300 H.P.		274.75		302.25	77.15	84.87
14 L.H., Daily Totals		$1520.65		$1857.20	$108.61	$132.66

Crew No.	Hr.	Daily	Hr.	Daily	Bare Costs	Incl. O&P
Crew B-33G	Hr.	Daily	Hr.	Daily	Bare Costs	Incl. O&P
1 Equip. Oper. (med.)	$33.65	$269.20	$50.70	$405.60	$31.46	$47.79
.5 Laborer	26.00	104.00	40.50	162.00		
.25 Equip. Oper. (med.)	33.65	67.30	50.70	101.40		
1 Elev. Scraper, 20 C.Y.		1597.00		1756.70		
.25 Dozer, 300 H.P.		274.75		302.25	133.70	147.07
14 L.H., Daily Totals		$2312.25		$2727.95	$165.16	$194.86

Crew No.	Hr.	Daily	Hr.	Daily	Bare Costs	Incl. O&P
Crew B-33H	Hr.	Daily	Hr.	Daily	Bare Costs	Incl. O&P
.25 Laborer	$26.00	$52.00	$40.50	$81.00	$32.33	$48.94
1 Equipment Operator (med.)	33.65	269.20	50.70	405.60		
.2 Equipment Operator (med.)	33.65	53.84	50.70	81.12		
1 Scraper, 32-44 C.Y.		2700.00		2970.00		
.2 Dozer, 298 kw		294.60		324.05	258.16	283.97
11. L.H., Daily Totals		$3369.64		$3861.77	$290.49	$332.91

Crew No.	Hr.	Daily	Hr.	Daily	Bare Costs	Incl. O&P
Crew B-33J	Hr.	Daily	Hr.	Daily	Bare Costs	Incl. O&P
1 Equipment Operator (med.)	$33.65	$269.20	$50.70	$405.60	$33.65	$50.70
1 Scraper 17 C.Y.		1468.00		1614.80	183.50	201.85
8 L.H., Daily Totals		$1737.20		$2020.40	$217.15	$252.55

Crew No.	Hr.	Daily	Hr.	Daily	Bare Costs	Incl. O&P
Crew B-34A	Hr.	Daily	Hr.	Daily	Bare Costs	Incl. O&P
1 Truck Driver (heavy)	$26.45	$211.60	$40.30	$322.40	$26.45	$40.30
1 Dump Truck, 12 Ton		325.00		357.50	40.63	44.69
8 L.H., Daily Totals		$536.60		$679.90	$67.08	$84.99

Crew No.	Hr.	Daily	Hr.	Daily	Bare Costs	Incl. O&P
Crew B-34B	Hr.	Daily	Hr.	Daily	Bare Costs	Incl. O&P
1 Truck Driver (heavy)	$26.45	$211.60	$40.30	$322.40	$26.45	$40.30
1 Dump Truck, 16 Ton		476.60		524.25	59.58	65.53
8 L.H., Daily Totals		$688.20		$846.65	$86.03	$105.83

Crew No.	Hr.	Daily	Hr.	Daily	Bare Costs	Incl. O&P
Crew B-34C	Hr.	Daily	Hr.	Daily	Bare Costs	Incl. O&P
1 Truck Driver (heavy)	$26.45	$211.60	$40.30	$322.40	$26.45	$40.30
1 Truck Tractor, 40 Ton		303.00		333.30		
1 Dump Trailer, 16.5 C.Y.		103.20		113.50	50.78	55.85
8 L.H., Daily Totals		$617.80		$769.20	$77.23	$96.15

Crew No.	Hr.	Daily	Hr.	Daily	Bare Costs	Incl. O&P
Crew B-34D	Hr.	Daily	Hr.	Daily	Bare Costs	Incl. O&P
1 Truck Driver (heavy)	$26.45	$211.60	$40.30	$322.40	$26.45	$40.30
1 Truck Tractor, 40 Ton		303.00		333.30		
1 Dump Trailer, 20 C.Y.		116.40		128.05	52.43	57.67
8 L.H., Daily Totals		$631.00		$783.75	$78.88	$97.97

Crew No.	Hr.	Daily	Hr.	Daily	Bare Costs	Incl. O&P
Crew B-34E	Hr.	Daily	Hr.	Daily	Bare Costs	Incl. O&P
1 Truck Driver (heavy)	$26.45	$211.60	$40.30	$322.40	$26.45	$40.30
1 Truck, Off Hwy., 25 Ton		920.40		1012.45	115.05	126.56
8 L.H., Daily Totals		$1132.00		$1334.85	$141.50	$166.86

Crew No.	Hr.	Daily	Hr.	Daily	Bare Costs	Incl. O&P
Crew B-34F	Hr.	Daily	Hr.	Daily	Bare Costs	Incl. O&P
1 Truck Driver (heavy)	$26.45	$211.60	$40.30	$322.40	$26.45	$40.30
1 Truck, Off Hwy., 22 C.Y.		946.60		1041.25	118.33	130.16
8 L.H., Daily Totals		$1158.20		$1363.65	$144.78	$170.46

Crew No.	Hr.	Daily	Hr.	Daily	Bare Costs	Incl. O&P
Crew B-34G	Hr.	Daily	Hr.	Daily	Bare Costs	Incl. O&P
1 Truck Driver (heavy)	$26.45	$211.60	$40.30	$322.40	$26.45	$40.30
1 Truck, Off Hwy., 34 C.Y.		1226.00		1348.60	153.25	168.58
8 L.H., Daily Totals		$1437.60		$1671.00	$179.70	$208.88

Crew No.	Hr.	Daily	Hr.	Daily	Bare Costs	Incl. O&P
Crew B-34H	Hr.	Daily	Hr.	Daily	Bare Costs	Incl. O&P
1 Truck Driver (heavy)	$26.45	$211.60	$40.30	$322.40	$26.45	$40.30
1 Truck, Off Hwy., 42 C.Y.		1310.00		1441.00	163.75	180.13
8 L.H., Daily Totals		$1521.60		$1763.40	$190.20	$220.43

Crew No.	Hr.	Daily	Hr.	Daily	Bare Costs	Incl. O&P
Crew B-34J	Hr.	Daily	Hr.	Daily	Bare Costs	Incl. O&P
1 Truck Driver (heavy)	$26.45	$211.60	$40.30	$322.40	$26.45	$40.30
1 Truck, Off Hwy., 60 C.Y.		1663.00		1829.30	207.88	228.66
8 L.H., Daily Totals		$1874.60		$2151.70	$234.33	$268.96

Crew No.	Hr.	Daily	Hr.	Daily	Bare Costs	Incl. O&P
Crew B-34K	Hr.	Daily	Hr.	Daily	Bare Costs	Incl. O&P
1 Truck Driver (heavy)	$26.45	$211.60	$40.30	$322.40	$26.45	$40.30
1 Truck Tractor, 240 H.P.		331.60		364.75		
1 Low Bed Trailer		171.80		189.00	62.93	69.22
8 L.H., Daily Totals		$715.00		$876.15	$89.38	$109.52

Crew No.	Hr.	Daily	Hr.	Daily	Bare Costs	Incl. O&P
Crew B-34N	Hr.	Daily	Hr.	Daily	Bare Costs	Incl. O&P
1 Truck Driver (heavy)	$26.45	$211.60	$40.30	$322.40	$26.45	$40.30
1 Dump Truck, 12 Ton		325.00		357.50		
1 Flatbed Trailer, 40 Ton		120.60		132.65	55.70	61.27
8 L.H., Daily Totals		$657.20		$812.55	$82.15	$101.57

Crew B-35

	Bare Costs		Incl. Subs O & P		Cost Per Labor-Hour	
Crew B-35	Hr.	Daily	Hr.	Daily	Bare Costs	Incl. O&P
1 Laborer Foreman (out)	$28.00	$224.00	$43.60	$348.80	$31.88	$48.72
1 Skilled Worker	33.65	269.20	52.35	418.80		
1 Welder (plumber)	39.60	316.80	59.45	475.60		
1 Laborer	26.00	208.00	40.50	324.00		
1 Equip. Oper. (crane)	34.80	278.40	52.40	419.20		
1 Equip. Oper. Oiler	29.20	233.60	44.00	352.00		
1 Electric Welding Mach.		77.75		85.55		
1 Hyd. Excavator, .75 C.Y.		458.80		504.70	11.18	12.30
48 L.H., Daily Totals		$2066.55		$2928.65	$43.06	$61.02

Crew B-35A

	Bare Costs		Incl. Subs O & P		Cost Per Labor-Hour	
Crew B-35A	Hr.	Daily	Hr.	Daily	Bare Costs	Incl. O&P
1 Laborer Foreman (out)	$28.00	$224.00	$43.60	$348.80	$31.04	$47.54
2 Laborers	26.00	416.00	40.50	648.00		
1 Skilled Worker	33.65	269.20	52.35	418.80		
1 Welder (plumber)	39.60	316.80	59.45	475.60		
1 Equip. Oper. (crane)	34.80	278.40	52.40	419.20		
1 Equip. Oper. Oiler	29.20	233.60	44.00	352.00		
1 Welder, 300 amp		75.20		82.70		
1 Crane, 75 Ton		1217.00		1338.70	23.08	25.38
56 L.H., Daily Totals		$3030.20		$4083.80	$54.12	$72.92

Crew B-36

	Bare Costs		Incl. Subs O & P		Cost Per Labor-Hour	
Crew B-36	Hr.	Daily	Hr.	Daily	Bare Costs	Incl. O&P
1 Labor Foreman (outside)	$28.00	$224.00	$43.60	$348.80	$29.46	$45.20
2 Laborers	26.00	416.00	40.50	648.00		
2 Equip. Oper. (med.)	33.65	538.40	50.70	811.20		
1 Dozer, 200 H.P.		863.40		949.75		
1 Aggregate Spreader		41.40		45.55		
1 Tandem Roller, 10 Ton		183.60		201.95	27.21	29.93
40 L.H., Daily Totals		$2266.80		$3005.25	$56.67	$75.13

Crew B-36A

	Bare Costs		Incl. Subs O & P		Cost Per Labor-Hour	
Crew B-36A	Hr.	Daily	Hr.	Daily	Bare Costs	Incl. O&P
1 Labor Foreman (outside)	$28.00	$224.00	$43.60	$348.80	$30.66	$46.77
2 Laborers	26.00	416.00	40.50	648.00		
4 Equip. Oper. (med.)	33.65	1076.80	50.70	1622.40		
1 Dozer, 200 H.P.		863.40		949.75		
1 Aggregate Spreader		41.40		45.55		
1 Roller, Steel Wheel		183.60		201.95		
1 Roller, Pneumatic Wheel		244.60		269.05	23.80	26.18
56 L.H., Daily Totals		$3049.80		$4085.50	$54.46	$72.95

Crew B-36B

	Bare Costs		Incl. Subs O & P		Cost Per Labor-Hour	
Crew B-36B	Hr.	Daily	Hr.	Daily	Bare Costs	Incl. O&P
1 Labor Foreman (outside)	$28.00	$224.00	$43.60	$348.80	$30.13	$45.96
2 Laborers	26.00	416.00	40.50	648.00		
4 Equip. Oper. (medium)	33.65	1076.80	50.70	1622.40		
1 Truck Driver, Heavy	26.45	211.60	40.30	322.40		
1 Grader, 30,000 Lbs.		432.00		475.20		
1 F.E. Loader, crl, 1.5 C.Y.		354.40		389.85		
1 Dozer, 300 H.P.		1099.00		1208.90		
1 Roller, Vibratory		436.80		480.50		
1 Truck, Tractor, 240 H.P.		331.60		364.75		
1 Water Tanker, 5000 Gal.		118.00		129.80	43.31	47.64
64 L.H., Daily Totals		$4700.20		$5990.60	$73.44	$93.60

Crew B-36C

	Bare Costs		Incl. Subs O & P		Cost Per Labor-Hour	
Crew B-36C	Hr.	Daily	Hr.	Daily	Bare Costs	Incl. O&P
1 Labor Foreman (outside)	$28.00	$224.00	$43.60	$348.80	$31.08	$47.20
3 Equip. Oper. (medium)	33.65	807.60	50.70	1216.80		
1 Truck Driver, Heavy	26.45	211.60	40.30	322.40		
1 Grader, 30,000 Lbs.		432.00		475.20		
1 Dozer, 300 H.P.		1099.00		1208.90		
1 Roller, Vibratory		436.80		480.50		
1 Truck, Tractor, 240 H.P.		331.60		364.75		
1 Water Tanker, 5000 Gal.		118.00		129.80	60.44	66.48
40 L.H., Daily Totals		$3660.60		$4547.15	$91.52	$113.68

Crew B-37

	Bare Costs		Incl. Subs O & P		Cost Per Labor-Hour	
Crew B-37	Hr.	Daily	Hr.	Daily	Bare Costs	Incl. O&P
1 Labor Foreman (outside)	$28.00	$224.00	$43.60	$348.80	$27.36	$42.33
4 Laborers	26.00	832.00	40.50	1296.00		
1 Equip. Oper. (light)	32.15	257.20	48.40	387.20		
1 Tandem Roller, 5 Ton		108.20		119.00	2.25	2.48
48 L.H., Daily Totals		$1421.40		$2151.00	$29.61	$44.81

Crew B-38

	Bare Costs		Incl. Subs O & P		Cost Per Labor-Hour	
Crew B-38	Hr.	Daily	Hr.	Daily	Bare Costs	Incl. O&P
1 Labor Foreman (outside)	$28.00	$224.00	$43.60	$348.80	$29.16	$44.74
2 Laborers	26.00	416.00	40.50	648.00		
1 Equip. Oper. (light)	32.15	257.20	48.40	387.20		
1 Equip. Oper. (medium)	33.65	269.20	50.70	405.60		
1 Backhoe Loader, 48 H.P.		207.80		228.60		
1 Hyd.Hammer, (1200 lb)		112.00		123.20		
1 F.E. Loader (170 H.P.)		438.40		482.25		
1 Pavt. Rem. Bucket		50.00		55.00	20.21	22.23
40 L.H., Daily Totals		$1974.60		$2678.65	$49.37	$66.97

Crew B-39

	Bare Costs		Incl. Subs O & P		Cost Per Labor-Hour	
Crew B-39	Hr.	Daily	Hr.	Daily	Bare Costs	Incl. O&P
1 Labor Foreman (outside)	$28.00	$224.00	$43.60	$348.80	$27.36	$42.33
4 Laborers	26.00	832.00	40.50	1296.00		
1 Equip. Oper. (light)	32.15	257.20	48.40	387.20		
1 Air Compr., 250 C.F.M.		127.60		140.35		
2 Air Tools & Accessories		22.40		24.65		
2-50 Ft. Air Hoses, 1.5" Dia.		10.00		11.00	3.33	3.67
48 L.H., Daily Totals		$1473.20		$2208.00	$30.69	$46.00

Crew B-40

	Bare Costs		Incl. Subs O & P		Cost Per Labor-Hour	
Crew B-40	Hr.	Daily	Hr.	Daily	Bare Costs	Incl. O&P
1 Pile Driver Foreman (out)	$34.05	$272.40	$56.25	$450.00	$32.63	$52.11
4 Pile Drivers	32.05	1025.60	52.95	1694.40		
2 Equip. Oper. (crane)	34.80	556.80	52.40	838.40		
1 Equip. Oper. Oiler	29.20	233.60	44.00	352.00		
1 Crane, 40 Ton		915.20		1006.70		
1 Vibratory Hammer & Gen.		1403.00		1543.30	36.22	39.84
64 L.H., Daily Totals		$4406.60		$5884.80	$68.85	$91.95

Crew B-41

	Bare Costs		Incl. Subs O & P		Cost Per Labor-Hour	
Crew B-41	Hr.	Daily	Hr.	Daily	Bare Costs	Incl. O&P
1 Labor Foreman (outside)	$28.00	$224.00	$43.60	$348.80	$26.91	$41.76
4 Laborers	26.00	832.00	40.50	1296.00		
.25 Equip. Oper. (crane)	34.80	69.60	52.40	104.80		
.25 Equip. Oper. Oiler	29.20	58.40	44.00	88.00		
.25 Crawler Crane, 40 Ton		228.80		251.70	5.20	5.72
44 L.H., Daily Totals		$1412.80		$2089.30	$32.11	$47.48

Crew B-42

Crew No.	Bare Costs		Incl. Subs O & P		Cost Per Labor-Hour	
Crew B-42	Hr.	Daily	Hr.	Daily	Bare Costs	Incl. O&P
1 Labor Foreman (outside)	$28.00	$224.00	$43.60	$348.80	$29.14	$46.07
4 Laborers	26.00	832.00	40.50	1296.00		
1 Equip. Oper. (crane)	34.80	278.40	52.40	419.20		
1 Equip. Oper. Oiler	29.20	233.60	44.00	352.00		
1 Welder	37.15	297.20	66.55	532.40		
1 Hyd. Crane, 25 Ton		624.20		686.60		
1 Gas Welding Machine		75.20		82.70		
1 Horz. Boring Csg. Mch.		381.60		419.75	16.89	18.58
64 L.H., Daily Totals		$2946.20		$4137.45	$46.03	$64.65

Crew B-43

Crew B-43	Hr.	Daily	Hr.	Daily	Bare Costs	Incl. O&P
1 Labor Foreman (outside)	$28.00	$224.00	$43.60	$348.80	$28.33	$43.58
3 Laborers	26.00	624.00	40.50	972.00		
1 Equip. Oper. (crane)	34.80	278.40	52.40	419.20		
1 Equip. Oper. Oiler	29.20	233.60	44.00	352.00		
1 Drill Rig & Augers		2877.00		3164.70	59.94	65.93
48 L.H., Daily Totals		$4237.00		$5256.70	$88.27	$109.51

Crew B-44

Crew B-44	Hr.	Daily	Hr.	Daily	Bare Costs	Incl. O&P
1 Pile Driver Foreman	$34.05	$272.40	$56.25	$450.00	$32.23	$51.67
4 Pile Drivers	32.05	1025.60	52.95	1694.40		
2 Equip. Oper. (crane)	34.80	556.80	52.40	838.40		
1 Laborer	26.00	208.00	40.50	324.00		
1 Crane, 40 Ton, & Access.		915.20		1006.70		
45 L.F. Leads, 15K Ft. Lbs.		67.50		74.25	15.35	16.89
64 L.H., Daily Totals		$3045.50		$4387.75	$47.58	$68.56

Crew B-45

Crew B-45	Hr.	Daily	Hr.	Daily	Bare Costs	Incl. O&P
1 Equip. Oper. (med.)	$33.65	$269.20	$50.70	$405.60	$30.05	$45.50
1 Truck Driver (heavy)	26.45	211.60	40.30	322.40		
1 Dist. Tank Truck, 3K Gal.		240.40		264.45		
1 Tractor, 4 x 2, 250 H.P.		314.20		345.60	34.66	38.13
16 L.H., Daily Totals		$1035.40		$1338.05	$64.71	$83.63

Crew B-46

Crew B-46	Hr.	Daily	Hr.	Daily	Bare Costs	Incl. O&P
1 Pile Driver Foreman	$34.05	$272.40	$56.25	$450.00	$29.36	$47.27
2 Pile Drivers	32.05	512.80	52.95	847.20		
3 Laborers	26.00	624.00	40.50	972.00		
1 Chain Saw, 36" Long		33.40		36.75	.70	.77
48 L.H., Daily Totals		$1442.60		$2305.95	$30.06	$48.04

Crew B-47

Crew B-47	Hr.	Daily	Hr.	Daily	Bare Costs	Incl. O&P
1 Blast Foreman	$28.00	$224.00	$43.60	$348.80	$28.72	$44.17
1 Driller	26.00	208.00	40.50	324.00		
1 Equip. Oper. (light)	32.15	257.20	48.40	387.20		
1 Crawler Type Drill, 4"		634.20		697.60		
1 Air Compr., 600 C.F.M.		286.00		314.60		
2-50 Ft. Air Hoses, 3" Dia.		35.40		38.95	39.82	43.80
24 L.H., Daily Totals		$1644.80		$2111.15	$68.54	$87.97

Crew B-47A

Crew B-47A	Hr.	Daily	Hr.	Daily	Bare Costs	Incl. O&P
1 Drilling Foreman	$28.00	$224.00	$43.60	$348.80	$30.67	$46.67
1 Equip. Oper. (heavy)	34.80	278.40	52.40	419.20		
1 Oiler	29.20	233.60	44.00	352.00		
1 Quarry Drill		827.40		910.15	34.48	37.92
24 L.H., Daily Totals		$1563.40		$2030.15	$65.15	$84.59

Crew B-47C

Crew No.	Bare Costs		Incl. Subs O & P		Cost Per Labor-Hour	
Crew B-47C	Hr.	Daily	Hr.	Daily	Bare Costs	Incl. O&P
1 Laborer	$26.00	$208.00	$40.50	$324.00	$29.08	$44.45
1 Equip. Oper. (light)	32.15	257.20	48.40	387.20		
1 Air Compressor, 750 CFM		305.00		335.50		
2-50' Air Hoses, 3"		35.40		38.95		
1 Air Track Drill, 4"		634.20		697.60	60.91	67.00
16 L.H., Daily Totals		$1439.80		$1783.25	$89.99	$111.45

Crew B-47E

Crew B-47E	Hr.	Daily	Hr.	Daily	Bare Costs	Incl. O&P
1 Laborer Foreman	$28.00	$224.00	$43.60	$348.80	$26.50	$41.28
3 Laborers	26.00	624.00	40.50	972.00		
1 Truck, Flatbed, 3 Ton		160.40		176.45	5.01	5.51
32 L.H., Daily Totals		$1008.40		$1497.25	$31.51	$46.79

Crew B-48

Crew B-48	Hr.	Daily	Hr.	Daily	Bare Costs	Incl. O&P
1 Labor Foreman (outside)	$28.00	$224.00	$43.60	$348.80	$28.88	$44.27
3 Laborers	26.00	624.00	40.50	972.00		
1 Equip. Oper. (crane)	34.80	278.40	52.40	419.20		
1 Equip. Oper. Oiler	29.20	233.60	44.00	352.00		
1 Equip. Oper. (light)	32.15	257.20	48.40	387.20		
1 Centr. Water Pump, 6"		229.80		252.80		
1-20 Ft. Suction Hose, 6"		14.50		15.95		
1-50 Ft. Disch. Hose, 6"		11.90		13.10		
1 Drill Rig & Augers		2877.00		3164.70	55.95	61.55
56 L.H., Daily Totals		$4750.40		$5925.75	$84.83	$105.82

Crew B-49

Crew B-49	Hr.	Daily	Hr.	Daily	Bare Costs	Incl. O&P
1 Labor Foreman (outside)	$28.00	$224.00	$43.60	$348.80	$30.02	$46.56
3 Laborers	26.00	624.00	40.50	972.00		
2 Equip. Oper. (crane)	34.80	556.80	52.40	838.40		
2 Equip. Oper. Oilers	29.20	467.20	44.00	704.00		
1 Equip. Oper. (light)	32.15	257.20	48.40	387.20		
2 Pile Drivers	32.05	512.80	52.95	847.20		
1 Hyd. Crane, 25 Ton		624.20		686.60		
1 Centr. Water Pump, 6"		229.80		252.80		
1-20 Ft. Suction Hose, 6"		14.50		15.95		
1-50 Ft. Disch. Hose, 6"		11.90		13.10		
1 Drill Rig & Augers		2877.00		3164.70	42.70	46.97
88 L.H., Daily Totals		$6399.40		$8230.75	$72.72	$93.53

Crew B-50

Crew B-50	Hr.	Daily	Hr.	Daily	Bare Costs	Incl. O&P
2 Pile Driver Foremen	$34.05	$544.80	$56.25	$900.00	$31.23	$50.04
6 Pile Drivers	32.05	1538.40	52.95	2541.60		
2 Equip. Oper. (crane)	34.80	556.80	52.40	838.40		
1 Equip. Oper. Oiler	29.20	233.60	44.00	352.00		
3 Laborers	26.00	624.00	40.50	972.00		
1 Crane, 40 Ton		915.20		1006.70		
60 L.F. Leads, 15K Ft. Lbs.		90.00		99.00		
1 Hammer, 15K Ft. Lbs.		359.80		395.80		
1 Air Compr., 600 C.F.M.		286.00		314.60		
2-50 Ft. Air Hoses, 3" Dia.		35.40		38.95		
1 Chain Saw, 36" Long		33.40		36.75	15.36	16.89
112 L.H., Daily Totals		$5217.40		$7495.80	$46.59	$66.93

Crew B-51

Crew B-51	Hr.	Daily	Hr.	Daily	Bare Costs	Incl. O&P
1 Labor Foreman (outside)	$28.00	$224.00	$43.60	$348.80	$26.28	$40.79
4 Laborers	26.00	832.00	40.50	1296.00		
1 Truck Driver (light)	25.70	205.60	39.15	313.20		
1 Light Truck, 1.5 Ton		122.00		134.20	2.54	2.80
48 L.H., Daily Totals		$1383.60		$2092.20	$28.82	$43.59

CREWS

Crew No.	Bare Costs Hr.	Bare Costs Daily	Incl. Subs O & P Hr.	Incl. Subs O & P Daily	Cost Per Labor-Hour Bare Costs	Cost Per Labor-Hour Incl. O&P
Crew B-52	Hr.	Daily	Hr.	Daily	Bare Costs	Incl. O&P
1 Carpenter Foreman	$35.00	$280.00	$54.55	$436.40	$30.42	$47.13
1 Carpenter	33.00	264.00	51.40	411.20		
3 Laborers	26.00	624.00	40.50	972.00		
1 Cement Finisher	31.55	252.40	46.45	371.60		
.5 Rodman (reinf.)	37.10	148.40	61.25	245.00		
.5 Equip. Oper. (med.)	33.65	134.60	50.70	202.80		
.5 F.E. Ldr., T.M., 2.5 C.Y.		380.50		418.55	6.79	7.47
56 L.H., Daily Totals		$2083.90		$3057.55	$37.21	$54.60
Crew B-53	Hr.	Daily	Hr.	Daily	Bare Costs	Incl. O&P
1 Equip. Oper. (light)	$32.15	$257.20	$48.40	$387.20	$32.15	$48.40
1 Trencher, Chain, 12 H.P.		59.40		65.35	7.43	8.17
8 L.H., Daily Totals		$316.60		$452.55	$39.58	$56.57
Crew B-54	Hr.	Daily	Hr.	Daily	Bare Costs	Incl. O&P
1 Equip. Oper. (light)	$32.15	$257.20	$48.40	$387.20	$32.15	$48.40
1 Trencher, Chain, 40 H.P.		208.60		229.45	26.08	28.68
8 L.H., Daily Totals		$465.80		$616.65	$58.23	$77.08
Crew B-54A	Hr.	Daily	Hr.	Daily	Bare Costs	Incl. O&P
.17 Labor Foreman (outside)	$28.00	$38.08	$43.60	$59.30	$32.83	$49.67
1 Equipment Operator (med.)	33.65	269.20	50.70	405.60		
1 Wheel Trencher, 67 H.P.		767.40		844.15	81.99	90.19
9.36 L.H., Daily Totals		$1074.68		$1309.05	$114.82	$139.86
Crew B-54B	Hr.	Daily	Hr.	Daily	Bare Costs	Incl. O&P
.25 Labor Foreman (outside)	$28.00	$56.00	$43.60	$87.20	$32.52	$49.28
1 Equipment Operator (med.)	33.65	269.20	50.70	405.60		
1 Wheel Trencher, 150 H.P.		1326.00		1458.60	132.60	145.86
10 L.H., Daily Totals		$1651.20		$1951.40	$165.12	$195.14
Crew B-55	Hr.	Daily	Hr.	Daily	Bare Costs	Incl. O&P
2 Laborers	$26.00	$416.00	$40.50	$648.00	$25.90	$40.05
1 Truck Driver (light)	25.70	205.60	39.15	313.20		
1 Auger, 4" to 36" Dia		614.80		676.30		
1 Flatbed 3 Ton Truck		160.40		176.45	32.30	35.53
24 L.H., Daily Totals		$1396.80		$1813.95	$58.20	$75.58
Crew B-56	Hr.	Daily	Hr.	Daily	Bare Costs	Incl. O&P
1 Laborer	$26.00	$208.00	$40.50	$324.00	$29.08	$44.45
1 Equip. Oper. (light)	32.15	257.20	48.40	387.20		
1 Crawler Type Drill, 4"		634.20		697.60		
1 Air Compr., 600 C.F.M.		286.00		314.60		
1-50 Ft. Air Hose, 3" Dia.		17.70		19.45	58.62	64.48
16 L.H., Daily Totals		$1403.10		$1742.85	$87.70	$108.93
Crew B-57	Hr.	Daily	Hr.	Daily	Bare Costs	Incl. O&P
1 Labor Foreman (outside)	$28.00	$224.00	$43.60	$348.80	$29.36	$44.90
2 Laborers	26.00	416.00	40.50	648.00		
1 Equip. Oper. (crane)	34.80	278.40	52.40	419.20		
1 Equip. Oper. (light)	32.15	257.20	48.40	387.20		
1 Equip. Oper. Oiler	29.20	233.60	44.00	352.00		
1 Power Shovel, 1 C.Y.		792.00		871.20		
1 Clamshell Bucket, 1 C.Y.		44.00		48.40		
1 Centr. Water Pump, 6"		229.80		252.80		
1-20 Ft. Suction Hose, 6"		14.50		15.95		
20-50 Ft. Disch. Hoses, 6"		238.00		261.80	27.46	30.21
48 L.H., Daily Totals		$2727.50		$3605.35	$56.82	$75.11

Crew No.	Bare Costs Hr.	Bare Costs Daily	Incl. Subs O & P Hr.	Incl. Subs O & P Daily	Cost Per Labor-Hour Bare Costs	Cost Per Labor-Hour Incl. O&P
Crew B-58	Hr.	Daily	Hr.	Daily	Bare Costs	Incl. O&P
2 Laborers	$26.00	$416.00	$40.50	$648.00	$28.05	$43.13
1 Equip. Oper. (light)	32.15	257.20	48.40	387.20		
1 Backhoe Loader, 48 H.P.		207.80		228.60		
1 Small Helicopter, w/pilot		2108.00		2318.80	96.49	106.14
24 L.H., Daily Totals		$2989.00		$3582.60	$124.54	$149.27
Crew B-59	Hr.	Daily	Hr.	Daily	Bare Costs	Incl. O&P
1 Truck Driver (heavy)	$26.45	$211.60	$40.30	$322.40	$26.45	$40.30
1 Truck, 30 Ton		215.20		236.70		
1 Water tank, 6000 Gal.		118.00		129.80	41.65	45.82
8 L.H., Daily Totals		$544.80		$688.90	$68.10	$86.12
Crew B-59A	Hr.	Daily	Hr.	Daily	Bare Costs	Incl. O&P
2 Laborers	$26.00	$416.00	$40.50	$648.00	$26.15	$40.43
1 Truck Driver (heavy)	26.45	211.60	40.30	322.40		
1 Water tank, 5K w/pum		118.00		129.80		
1 Truck, 30 Ton		215.20		236.70	13.88	15.27
24 L.H., Daily Totals		$960.80		$1336.90	$40.03	$55.70
Crew B-60	Hr.	Daily	Hr.	Daily	Bare Costs	Incl. O&P
1 Labor Foreman (outside)	$28.00	$224.00	$43.60	$348.80	$29.76	$45.40
2 Laborers	26.00	416.00	40.50	648.00		
1 Equip. Oper. (crane)	34.80	278.40	52.40	419.20		
2 Equip. Oper. (light)	32.15	514.40	48.40	774.40		
1 Equip. Oper. Oiler	29.20	233.60	44.00	352.00		
1 Crawler Crane, 40 Ton		915.20		1006.70		
45 L.F. Leads, 15K Ft. Lbs.		67.50		74.25		
1 Backhoe Loader, 48 H.P.		207.80		228.60	21.26	23.38
56 L.H., Daily Totals		$2856.90		$3851.95	$51.02	$68.78
Crew B-61	Hr.	Daily	Hr.	Daily	Bare Costs	Incl. O&P
1 Labor Foreman (outside)	$28.00	$224.00	$43.60	$348.80	$27.63	$42.70
3 Laborers	26.00	624.00	40.50	972.00		
1 Equip. Oper. (light)	32.15	257.20	48.40	387.20		
1 Cement Mixer, 2 C.Y.		161.00		177.10		
1 Air Compr., 160 C.F.M.		82.00		90.20	6.08	6.68
40 L.H., Daily Totals		$1348.20		$1975.30	$33.71	$49.38
Crew B-62	Hr.	Daily	Hr.	Daily	Bare Costs	Incl. O&P
2 Laborers	$26.00	$416.00	$40.50	$648.00	$28.05	$43.13
1 Equip. Oper. (light)	32.15	257.20	48.40	387.20		
1 Loader, Skid Steer		143.20		157.50	5.97	6.56
24 L.H., Daily Totals		$816.40		$1192.70	$34.02	$49.69
Crew B-63	Hr.	Daily	Hr.	Daily	Bare Costs	Incl. O&P
4 Laborers	$26.00	$832.00	$40.50	$1296.00	$27.23	$42.08
1 Equip. Oper. (light)	32.15	257.20	48.40	387.20		
1 Loader, Skid Steer		143.20		157.50	3.58	3.94
40 L.H., Daily Totals		$1232.40		$1840.70	$30.81	$46.02
Crew B-64	Hr.	Daily	Hr.	Daily	Bare Costs	Incl. O&P
1 Laborer	$26.00	$208.00	$40.50	$324.00	$25.85	$39.83
1 Truck Driver (light)	25.70	205.60	39.15	313.20		
1 Power Mulcher (small)		103.20		113.50		
1 Light Truck, 1.5 Ton		122.00		134.20	14.08	15.48
16 L.H., Daily Totals		$638.80		$884.90	$39.93	$55.31

Crew B-65

Crew No.	Bare Costs Hr.	Daily	Incl. Subs O & P Hr.	Daily	Cost Per Labor-Hour Bare Costs	Incl. O&P
1 Laborer	$26.00	$208.00	$40.50	$324.00	$25.85	$39.83
1 Truck Driver (light)	25.70	205.60	39.15	313.20		
1 Power Mulcher (large)		202.20		222.40		
1 Light Truck, 1.5 Ton		122.00		134.20	20.26	22.29
16 L.H., Daily Totals		$737.80		$993.80	$46.11	$62.12

Crew B-66

Crew No.	Bare Costs Hr.	Daily	Incl. Subs O & P Hr.	Daily	Bare Costs	Incl. O&P
1 Equip. Oper. (light)	$32.15	$257.20	$48.40	$387.20	$32.15	$48.40
1 Backhoe Ldr. w/Attchmt.		168.40		185.25	21.05	23.16
8 L.H., Daily Totals		$425.60		$572.45	$53.20	$71.56

Crew B-67

Crew No.	Bare Costs Hr.	Daily	Incl. Subs O & P Hr.	Daily	Bare Costs	Incl. O&P
1 Millwright	$34.35	$274.80	$50.70	$405.60	$33.25	$49.55
1 Equip. Oper. (light)	32.15	257.20	48.40	387.20		
1 Forklift		239.80		263.80	14.99	16.49
16 L.H., Daily Totals		$771.80		$1056.60	$48.24	$66.04

Crew B-68

Crew No.	Bare Costs Hr.	Daily	Incl. Subs O & P Hr.	Daily	Bare Costs	Incl. O&P
2 Millwrights	$34.35	$549.60	$50.70	$811.20	$33.62	$49.93
1 Equip. Oper. (light)	32.15	257.20	48.40	387.20		
1 Forklift		239.80		263.80	9.99	10.99
24 L.H., Daily Totals		$1046.60		$1462.20	$43.61	$60.92

Crew B-69

Crew No.	Bare Costs Hr.	Daily	Incl. Subs O & P Hr.	Daily	Bare Costs	Incl. O&P
1 Labor Foreman (outside)	$28.00	$224.00	$43.60	$348.80	$28.33	$43.58
3 Laborers	26.00	624.00	40.50	972.00		
1 Equip Oper. (crane)	34.80	278.40	52.40	419.20		
1 Equip Oper. Oiler	29.20	233.60	44.00	352.00		
1 Truck Crane, 80 Ton		1020.00		1122.00	21.25	23.38
48 L.H., Daily Totals		$2380.00		$3214.00	$49.58	$66.96

Crew B-69A

Crew No.	Bare Costs Hr.	Daily	Incl. Subs O & P Hr.	Daily	Bare Costs	Incl. O&P
1 Labor Foreman	$28.00	$224.00	$43.60	$348.80	$28.53	$43.71
3 Laborers	26.00	624.00	40.50	972.00		
1 Equip. Oper. (medium)	33.65	269.20	50.70	405.60		
1 Concrete Finisher	31.55	252.40	46.45	371.60		
1 Curb Paver		573.60		630.95	11.95	13.15
48 L.H., Daily Totals		$1943.20		$2728.95	$40.48	$56.86

Crew B-69B

Crew No.	Bare Costs Hr.	Daily	Incl. Subs O & P Hr.	Daily	Bare Costs	Incl. O&P
1 Labor Foreman	$28.00	$224.00	$43.60	$348.80	$28.53	$43.71
3 Laborers	26.00	624.00	40.50	972.00		
1 Equip. Oper. (medium)	33.65	269.20	50.70	405.60		
1 Cement Finisher	31.55	252.40	46.45	371.60		
1 Curb/Gutter Paver		687.40		756.15	14.32	15.75
48 L.H., Daily Totals		$2057.00		$2854.15	$42.85	$59.46

Crew B-70

Crew No.	Bare Costs Hr.	Daily	Incl. Subs O & P Hr.	Daily	Bare Costs	Incl. O&P
1 Labor Foreman (outside)	$28.00	$224.00	$43.60	$348.80	$29.56	$45.31
3 Laborers	26.00	624.00	40.50	972.00		
3 Equip. Oper. (med.)	33.65	807.60	50.70	1216.80		
1 Motor Grader, 30,000 Lb.		432.00		475.20		
1 Grader Attach., Ripper		70.60		77.65		
1 Road Sweeper, S.P.		420.00		462.00		
1 F.E. Loader, 1-3/4 C.Y.		236.20		259.80	20.69	22.76
56 L.H., Daily Totals		$2814.40		$3812.25	$50.25	$68.07

Crew B-71

Crew No.	Bare Costs Hr.	Daily	Incl. Subs O & P Hr.	Daily	Bare Costs	Incl. O&P
1 Labor Foreman (outside)	$28.00	$224.00	$43.60	$348.80	$29.56	$45.31
3 Laborers	26.00	624.00	40.50	972.00		
3 Equip. Oper. (med.)	33.65	807.60	50.70	1216.80		
1 Pvmt. Profiler, 750 H.P.		4351.00		4786.10		
1 Road Sweeper, S.P.		420.00		462.00		
1 F.E. Loader, 1-3/4 C.Y.		236.20		259.80	89.41	98.36
56 L.H., Daily Totals		$6662.80		$8045.50	$118.97	$143.67

Crew B-72

Crew No.	Bare Costs Hr.	Daily	Incl. Subs O & P Hr.	Daily	Bare Costs	Incl. O&P
1 Labor Foreman (outside)	$28.00	$224.00	$43.60	$348.80	$30.08	$45.99
3 Laborers	26.00	624.00	40.50	972.00		
4 Equip. Oper. (med.)	33.65	1076.80	50.70	1622.40		
1 Pvmt. Profiler, 750 H.P.		4351.00		4786.10		
1 Hammermill, 250 H.P.		1315.00		1446.50		
1 Windrow Loader		805.60		886.15		
1 Mix Paver 165 H.P.		1674.00		1841.40		
1 Roller, Pneu. Tire, 12 T.		244.60		269.05	131.10	144.21
64 L.H., Daily Totals		$10315.00		$12172.40	$161.18	$190.20

Crew B-73

Crew No.	Bare Costs Hr.	Daily	Incl. Subs O & P Hr.	Daily	Bare Costs	Incl. O&P
1 Labor Foreman (outside)	$28.00	$224.00	$43.60	$348.80	$31.03	$47.26
2 Laborers	26.00	416.00	40.50	648.00		
5 Equip. Oper. (med.)	33.65	1346.00	50.70	2028.00		
1 Road Mixer, 310 H.P.		1610.00		1771.00		
1 Roller, Tandem, 12 Ton		183.60		201.95		
1 Hammermill, 250 H.P.		1315.00		1446.50		
1 Motor Grader, 30,000 Lb.		432.00		475.20		
.5 F.E. Loader, 1-3/4 C.Y.		118.10		129.90		
.5 Truck, 30 Ton		107.60		118.35		
.5 Water Tank 5000 Gal.		59.00		64.90	59.77	65.75
64 L.H., Daily Totals		$5811.30		$7232.60	$90.80	$113.01

Crew B-74

Crew No.	Bare Costs Hr.	Daily	Incl. Subs O & P Hr.	Daily	Bare Costs	Incl. O&P
1 Labor Foreman (outside)	$28.00	$224.00	$43.60	$348.80	$30.19	$45.94
1 Laborer	26.00	208.00	40.50	324.00		
4 Equip. Oper. (med.)	33.65	1076.80	50.70	1622.40		
2 Truck Drivers (heavy)	26.45	423.20	40.30	644.80		
1 Motor Grader, 30,000 Lb.		432.00		475.20		
1 Grader Attach., Ripper		70.60		77.65		
2 Stabilizers, 310 H.P.		2122.00		2334.20		
1 Flatbed Truck, 3 Ton		160.40		176.45		
1 Chem. Spreader, Towed		66.00		72.60		
1 Vibr. Roller, 29,000 Lb.		436.80		480.50		
1 Water Tank 5000 Gal.		118.00		129.80		
1 Truck, 30 Ton		215.20		236.70	56.58	62.24
64 L.H., Daily Totals		$5553.00		$6923.10	$86.77	$108.18

Crew B-75

Crew No.	Bare Costs Hr.	Daily	Incl. Subs O & P Hr.	Daily	Bare Costs	Incl. O&P
1 Labor Foreman (outside)	$28.00	$224.00	$43.60	$348.80	$30.72	$46.74
1 Laborer	26.00	208.00	40.50	324.00		
4 Equip. Oper. (med.)	33.65	1076.80	50.70	1622.40		
1 Truck Driver (heavy)	26.45	211.60	40.30	322.40		
1 Motor Grader, 30,000 Lb.		432.00		475.20		
1 Grader Attach., Ripper		70.60		77.65		
2 Stabilizers, 310 H.P.		2122.00		2334.20		
1 Dist. Truck, 3000 Gal.		240.40		264.45		
1 Vibr. Roller, 29,000 Lb.		436.80		480.50	58.96	64.86
56 L.H., Daily Totals		$5022.20		$6249.60	$89.68	$111.60

Crew No.	Bare Costs		Incl. Subs O & P		Cost Per Labor-Hour	

Crew B-76	Hr.	Daily	Hr.	Daily	Bare Costs	Incl. O&P
1 Dock Builder Foreman	$34.05	$272.40	$56.25	$450.00	$32.57	$52.20
5 Dock Builders	32.05	1282.00	52.95	2118.00		
2 Equip. Oper. (crane)	34.80	556.80	52.40	838.40		
1 Equip. Oper. Oiler	29.20	233.60	44.00	352.00		
1 Crawler Crane, 50 Ton		1260.00		1386.00		
1 Barge, 400 Ton		273.20		300.50		
1 Hammer, 15K Ft. Lbs.		359.80		395.80		
60 L.F. Leads, 15K Ft. Lbs.		90.00		99.00		
1 Air Compr., 600 C.F.M.		286.00		314.60		
2-50 Ft. Air Hoses, 3" Dia.		35.40		38.95	32.01	35.21
72 L.H., Daily Totals		$4649.20		$6293.25	$64.58	$87.41

Crew B-77	Hr.	Daily	Hr.	Daily	Bare Costs	Incl. O&P
1 Labor Foreman	$28.00	$224.00	$43.60	$348.80	$26.34	$40.85
3 Laborers	26.00	624.00	40.50	972.00		
1 Truck Driver (light)	25.70	205.60	39.15	313.20		
1 Crack Cleaner, 25 H.P.		44.40		48.85		
1 Crack Filler, Trailer Mtd.		157.00		172.70		
1 Flatbed Truck, 3 Ton		160.40		176.45	9.05	9.95
40 L.H., Daily Totals		$1415.40		$2032.00	$35.39	$50.80

Crew B-78	Hr.	Daily	Hr.	Daily	Bare Costs	Incl. O&P
1 Labor Foreman	$28.00	$224.00	$43.60	$348.80	$26.28	$40.79
4 Laborers	26.00	832.00	40.50	1296.00		
1 Truck Driver (light)	25.70	205.60	39.15	313.20		
1 Paint Striper, S.P.		134.20		147.60		
1 Flatbed Truck, 3 Ton		160.40		176.45		
1 Pickup Truck, 3/4 Ton		75.80		83.40	7.72	8.49
48 L.H., Daily Totals		$1632.00		$2365.45	$34.00	$49.28

Crew B-79	Hr.	Daily	Hr.	Daily	Bare Costs	Incl. O&P
1 Labor Foreman	$28.00	$224.00	$43.60	$348.80	$26.34	$40.85
3 Laborers	26.00	624.00	40.50	972.00		
1 Truck Driver (light)	25.70	205.60	39.15	313.20		
1 Thermo. Striper, T.M.		549.20		604.10		
1 Flatbed Truck, 3 Ton		160.40		176.45		
2 Pickup Trucks, 3/4 Ton		151.60		166.75	21.53	23.68
40 L.H., Daily Totals		$1914.80		$2581.30	$47.87	$64.53

Crew B-80	Hr.	Daily	Hr.	Daily	Bare Costs	Incl. O&P
1 Labor Foreman	$28.00	$224.00	$43.60	$348.80	$27.96	$42.91
1 Laborer	26.00	208.00	40.50	324.00		
1 Truck Driver (light)	25.70	205.60	39.15	313.20		
1 Equip. Oper. (light)	32.15	257.20	48.40	387.20		
1 Flatbed Truck, 3 Ton		160.40		176.45		
1 Fence Post Auger, T.M.		357.40		393.15	16.18	17.80
32 L.H., Daily Totals		$1412.60		$1942.80	$44.14	$60.71

Crew B-80A	Hr.	Daily	Hr.	Daily	Bare Costs	Incl. O&P
3 Laborers	$26.00	$624.00	$40.50	$972.00	$26.00	$40.50
1 Flatbed Truck, 3 Ton		160.40		176.45	6.68	7.35
24 L.H., Daily Totals		$784.40		$1148.45	$32.68	$47.85

Crew B-80B	Hr.	Daily	Hr.	Daily	Bare Costs	Incl. O&P
3 Laborers	$26.00	$624.00	$40.50	$972.00	$27.54	$42.48
1 Equip. Oper. (light)	32.15	257.20	48.40	387.20		
1 Crane, Flatbed Mnt.		201.80		222.00	6.31	6.94
32 L.H., Daily Totals		$1083.00		$1581.20	$33.85	$49.42

Crew B-81	Hr.	Daily	Hr.	Daily	Bare Costs	Incl. O&P
1 Laborer	$26.00	$208.00	$40.50	$324.00	$28.70	$43.83
1 Equip. Oper. (med.)	33.65	269.20	50.70	405.60		
1 Truck Driver (heavy)	26.45	211.60	40.30	322.40		
1 Hydromulcher, T.M.		196.00		215.60		
1 Tractor Truck, 4x2		215.20		236.70	17.13	18.85
24 L.H., Daily Totals		$1100.00		$1504.30	$45.83	$62.68

Crew B-82	Hr.	Daily	Hr.	Daily	Bare Costs	Incl. O&P
1 Laborer	$26.00	$208.00	$40.50	$324.00	$29.08	$44.45
1 Equip. Oper. (light)	32.15	257.20	48.40	387.20		
1 Horiz. Borer, 6 H.P.		62.80		69.10	3.93	4.32
16 L.H., Daily Totals		$528.00		$780.30	$33.01	$48.77

Crew B-83	Hr.	Daily	Hr.	Daily	Bare Costs	Incl. O&P
1 Tugboat Captain	$33.65	$269.20	$50.70	$405.60	$29.83	$45.60
1 Tugboat Hand	26.00	208.00	40.50	324.00		
1 Tugboat, 250 H.P.		440.60		484.65	27.54	30.29
16 L.H., Daily Totals		$917.80		$1214.25	$57.37	$75.89

Crew B-84	Hr.	Daily	Hr.	Daily	Bare Costs	Incl. O&P
1 Equip. Oper. (med.)	$33.65	$269.20	$50.70	$405.60	$33.65	$50.70
1 Rotary Mower/Tractor		222.20		244.40	27.78	30.55
8 L.H., Daily Totals		$491.40		$650.00	$61.43	$81.25

Crew B-85	Hr.	Daily	Hr.	Daily	Bare Costs	Incl. O&P
3 Laborers	$26.00	$624.00	$40.50	$972.00	$27.62	$42.50
1 Equip. Oper. (med.)	33.65	269.20	50.70	405.60		
1 Truck Driver (heavy)	26.45	211.60	40.30	322.40		
1 Aerial Lift Truck, 80'		501.80		552.00		
1 Brush Chipper, 130 H.P.		164.80		181.30		
1 Pruning Saw, Rotary		10.55		11.60	16.93	18.62
40 L.H., Daily Totals		$1781.95		$2444.90	$44.55	$61.12

Crew B-86	Hr.	Daily	Hr.	Daily	Bare Costs	Incl. O&P
1 Equip. Oper. (med.)	$33.65	$269.20	$50.70	$405.60	$33.65	$50.70
1 Stump Chipper, S.P.		87.20		95.90	10.90	11.99
8 L.H., Daily Totals		$356.40		$501.50	$44.55	$62.69

Crew B-86A	Hr.	Daily	Hr.	Daily	Bare Costs	Incl. O&P
1 Equip. Oper. (medium)	$33.65	$269.20	$50.70	$405.60	$33.65	$50.70
1 Grader, 30,000 Lbs.		432.00		475.20	54.00	59.40
8 L.H., Daily Totals		$701.20		$880.80	$87.65	$110.10

Crew B-86B	Hr.	Daily	Hr.	Daily	Bare Costs	Incl. O&P
1 Equip. Oper. (medium)	$33.65	$269.20	$50.70	$405.60	$33.65	$50.70
1 Dozer, 200 H.P.		863.40		949.75	107.93	118.72
8 L.H., Daily Totals		$1132.60		$1355.35	$141.58	$169.42

Crew B-87	Hr.	Daily	Hr.	Daily	Bare Costs	Incl. O&P
1 Laborer	$26.00	$208.00	$40.50	$324.00	$32.12	$48.66
4 Equip. Oper. (med.)	33.65	1076.80	50.70	1622.40		
2 Feller Bunchers, 50 H.P.		864.80		951.30		
1 Log Chipper, 22" Tree		1207.00		1327.70		
1 Dozer, 105 H.P.		434.00		477.40		
1 Chainsaw, Gas, 36" Long		33.40		36.75	63.48	69.83
40 L.H., Daily Totals		$3824.00		$4739.55	$95.60	$118.49

Crew No.	Bare Costs		Incl. Subs O & P		Cost Per Labor-Hour	
Crew B-88	**Hr.**	**Daily**	**Hr.**	**Daily**	**Bare Costs**	**Incl. O&P**
1 Laborer	$26.00	$208.00	$40.50	$324.00	$32.56	$49.24
6 Equip. Oper. (med.)	33.65	1615.20	50.70	2433.60		
2 Feller Bunchers, 50 H.P.		864.80		951.30		
1 Log Chipper, 22″ Tree		1207.00		1327.70		
2 Log Skidders, 50 H.P.		1490.80		1639.90		
1 Dozer, 105 H.P.		434.00		477.40		
1 Chainsaw, Gas, 36″ Long		33.40		36.75	71.96	79.16
56 L.H., Daily Totals		$5853.20		$7190.65	$104.52	$128.40

Crew B-89	Hr.	Daily	Hr.	Daily	Bare Costs	Incl. O&P
1 Equip. Oper. (light)	$32.15	$257.20	$48.40	$387.20	$28.93	$43.78
1 Truck Driver (light)	25.70	205.60	39.15	313.20		
1 Truck, Stake Body, 3 Ton		160.40		176.45		
1 Concrete Saw		92.20		101.40		
1 Water Tank, 65 Gal.		13.60		14.95	16.64	18.30
16 L.H., Daily Totals		$729.00		$993.20	$45.57	$62.08

Crew B-89A	Hr.	Daily	Hr.	Daily	Bare Costs	Incl. O&P
1 Skilled Worker	$33.65	$269.20	$52.35	$418.80	$29.83	$46.42
1 Laborer	26.00	208.00	40.50	324.00		
1 Core Drill (large)		106.00		116.60	6.63	7.29
16 L.H., Daily Totals		$583.20		$859.40	$36.46	$53.71

Crew B-89B	Hr.	Daily	Hr.	Daily	Bare Costs	Incl. O&P
1 Equip. Oper. (light)	$32.15	$257.20	$48.40	$387.20	$28.93	$43.78
1 Truck Driver, Light	25.70	205.60	39.15	313.20		
1 Wall Saw, Hydraulic, 10 H.P.		75.10		82.60		
1 Generator, Diesel, 100 KW		179.60		197.55		
1 Water Tank, 65 Gal.		13.60		14.95		
1 Flatbed Truck, 3 Ton		160.40		176.45	26.79	29.47
16 L.H., Daily Totals		$891.50		$1171.95	$55.72	$73.25

Crew B-90	Hr.	Daily	Hr.	Daily	Bare Costs	Incl. O&P
1 Labor Foreman (outside)	$28.00	$224.00	$43.60	$348.80	$27.90	$42.81
3 Laborers	26.00	624.00	40.50	972.00		
2 Equip. Oper. (light)	32.15	514.40	48.40	774.40		
2 Truck Drivers (heavy)	26.45	423.20	40.30	644.80		
1 Road Mixer, 310 H.P.		1610.00		1771.00		
1 Dist. Truck, 2000 Gal.		208.00		228.80	28.41	31.25
64 L.H., Daily Totals		$3603.60		$4739.80	$56.31	$74.06

Crew B-90A	Hr.	Daily	Hr.	Daily	Bare Costs	Incl. O&P
1 Labor Foreman	$28.00	$224.00	$43.60	$348.80	$30.66	$46.77
2 Laborers	26.00	416.00	40.50	648.00		
4 Equip. Oper. (medium)	33.65	1076.80	50.70	1622.40		
2 Graders, 30,000 Lbs.		864.00		950.40		
1 Roller, Steel Wheel		183.60		201.95		
1 Roller, Pneumatic Wheel		244.60		269.05	23.08	25.38
56 L.H., Daily Totals		$3009.00		$4040.60	$53.74	$72.15

Crew B-90B	Hr.	Daily	Hr.	Daily	Bare Costs	Incl. O&P
1 Labor Foreman	$28.00	$224.00	$43.60	$348.80	$30.16	$46.12
2 Laborers	26.00	416.00	40.50	648.00		
3 Equip. Oper. (medium)	33.65	807.60	50.70	1216.80		
1 Roller, Steel Wheel		183.60		201.95		
1 Roller, Pneumatic Wheel		244.60		269.05		
1 Road Mixer, 310 H.P.		1610.00		1771.00	42.46	46.71
48 L.H., Daily Totals		$3485.80		$4455.60	$72.62	$92.83

Crew No.	Bare Costs		Incl. Subs O & P		Cost Per Labor-Hour	
Crew B-91	**Hr.**	**Daily**	**Hr.**	**Daily**	**Bare Costs**	**Incl. O&P**
1 Labor Foreman (outside)	$28.00	$224.00	$43.60	$348.80	$30.13	$45.96
2 Laborers	26.00	416.00	40.50	648.00		
4 Equip. Oper. (med.)	33.65	1076.80	50.70	1622.40		
1 Truck Driver (heavy)	26.45	211.60	40.30	322.40		
1 Dist. Truck, 3000 Gal.		240.40		264.45		
1 Aggreg. Spreader, S.P.		771.20		848.30		
1 Roller, Pneu. Tire, 12 Ton		244.60		269.05		
1 Roller, Steel, 10 Ton		183.60		201.95	22.50	24.75
64 L.H., Daily Totals		$3368.20		$4525.35	$52.63	$70.71

Crew B-92	Hr.	Daily	Hr.	Daily	Bare Costs	Incl. O&P
1 Labor Foreman (outside)	$28.00	$224.00	$43.60	$348.80	$26.50	$41.28
3 Laborers	26.00	624.00	40.50	972.00		
1 Crack Cleaner, 25 H.P.		44.40		48.85		
1 Air Compressor		52.40		57.65		
1 Tar Kettle, T.M.		40.95		45.05		
1 Flatbed Truck, 3 Ton		160.40		176.45	9.32	10.25
32 L.H., Daily Totals		$1146.15		$1648.80	$35.82	$51.53

Crew B-93	Hr.	Daily	Hr.	Daily	Bare Costs	Incl. O&P
1 Equip. Oper. (med.)	$33.65	$269.20	$50.70	$405.60	$33.65	$50.70
1 Feller Buncher, 50 H.P.		432.40		475.65	54.05	59.46
8 L.H., Daily Totals		$701.60		$881.25	$87.70	$110.16

Crew B-94A	Hr.	Daily	Hr.	Daily	Bare Costs	Incl. O&P
1 Laborer	$26.00	$208.00	$40.50	$324.00	$26.00	$40.50
1 Diaph. Water Pump, 2″		46.20		50.80		
1-20 Ft. Suction Hose, 2″		3.55		3.90		
2-50 Ft. Disch. Hoses, 2″		6.30		6.95	7.01	7.71
8 L.H., Daily Totals		$264.05		$385.65	$33.01	$48.21

Crew B-94B	Hr.	Daily	Hr.	Daily	Bare Costs	Incl. O&P
1 Laborer	$26.00	$208.00	$40.50	$324.00	$26.00	$40.50
1 Diaph. Water Pump, 4″		87.40		96.15		
1-20 Ft. Suction Hose, 4″		7.30		8.05		
2-50 Ft. Disch. Hoses, 4″		10.50		11.55	13.15	14.47
8 L.H., Daily Totals		$313.20		$439.75	$39.15	$54.97

Crew B-94C	Hr.	Daily	Hr.	Daily	Bare Costs	Incl. O&P
1 Laborer	$26.00	$208.00	$40.50	$324.00	$26.00	$40.50
1 Centr. Water Pump, 3″		53.80		59.20		
1-20 Ft. Suction Hose, 3″		5.45		6.00		
2-50 Ft. Disch. Hoses, 3″		7.50		8.25	8.34	9.18
8 L.H., Daily Totals		$274.75		$397.45	$34.34	$49.68

Crew B-94D	Hr.	Daily	Hr.	Daily	Bare Costs	Incl. O&P
1 Laborer	$26.00	$208.00	$40.50	$324.00	$26.00	$40.50
1 Centr. Water Pump, 6″		229.80		252.80		
1-20 Ft. Suction Hose, 6″		14.50		15.95		
2-50 Ft. Disch. Hoses, 6″		23.80		26.20	33.51	36.86
8 L.H., Daily Totals		$476.10		$618.95	$59.51	$77.36

Crew B-95A	Hr.	Daily	Hr.	Daily	Bare Costs	Incl. O&P
1 Equip. Oper. (crane)	$34.80	$278.40	$52.40	$419.20	$30.40	$46.45
1 Laborer	26.00	208.00	40.50	324.00		
1 Hyd. Excavator, 5/8 C.Y.		418.00		459.80	26.13	28.74
16 L.H., Daily Totals		$904.40		$1203.00	$56.53	$75.19

Crew No.	Bare Costs Hr.	Daily	Incl. Subs O & P Hr.	Daily	Cost Per Labor-Hour Bare Costs	Incl. O&P
Crew B-95B	Hr.	Daily	Hr.	Daily	Bare Costs	Incl. O&P
1 Equip. Oper. (crane)	$34.80	$278.40	$52.40	$419.20	$30.40	$46.45
1 Laborer	26.00	208.00	40.50	324.00		
1 Hyd. Excavator, 1.5 C.Y.		683.40		751.75	42.71	46.98
16 L.H., Daily Totals		$1169.80		$1494.95	$73.11	$93.43
Crew B-95C	Hr.	Daily	Hr.	Daily	Bare Costs	Incl. O&P
1 Equip. Oper. (crane)	$34.80	$278.40	$52.40	$419.20	$30.40	$46.45
1 Laborer	26.00	208.00	40.50	324.00		
1 Hyd. Excavator, 2.5 C.Y.		1181.00		1299.10	73.81	81.19
16 L.H., Daily Totals		$1667.40		$2042.30	$104.21	$127.64
Crew C-1	Hr.	Daily	Hr.	Daily	Bare Costs	Incl. O&P
3 Carpenters	$33.00	$792.00	$51.40	$1233.60	$31.25	$48.68
1 Laborer	26.00	208.00	40.50	324.00		
32 L.H., Daily Totals		$1000.00		$1557.60	$31.25	$48.68
Crew C-2	Hr.	Daily	Hr.	Daily	Bare Costs	Incl. O&P
1 Carpenter Foreman (out)	$35.00	$280.00	$54.55	$436.40	$32.17	$50.11
4 Carpenters	33.00	1056.00	51.40	1644.80		
1 Laborer	26.00	208.00	40.50	324.00		
48 L.H., Daily Totals		$1544.00		$2405.20	$32.17	$50.11
Crew C-2A	Hr.	Daily	Hr.	Daily	Bare Costs	Incl. O&P
1 Carpenter Foreman (out)	$35.00	$280.00	$54.55	$436.40	$31.93	$49.28
3 Carpenters	33.00	792.00	51.40	1233.60		
1 Cement Finisher	31.55	252.40	46.45	371.60		
1 Laborer	26.00	208.00	40.50	324.00		
48 L.H., Daily Totals		$1532.40		$2365.60	$31.93	$49.28
Crew C-3	Hr.	Daily	Hr.	Daily	Bare Costs	Incl. O&P
1 Rodman Foreman	$39.10	$312.80	$64.55	$516.40	$33.96	$54.87
4 Rodmen (reinf.)	37.10	1187.20	61.25	1960.00		
1 Equip. Oper. (light)	32.15	257.20	48.40	387.20		
2 Laborers	26.00	416.00	40.50	648.00		
3 Stressing Equipment		48.60		53.45		
.5 Grouting Equipment		78.47		86.30	1.99	2.18
64 L.H., Daily Totals		$2300.27		$3651.35	$35.95	$57.05
Crew C-4	Hr.	Daily	Hr.	Daily	Bare Costs	Incl. O&P
1 Rodman Foreman	$39.10	$312.80	$64.55	$516.40	$37.60	$62.08
3 Rodmen (reinf.)	37.10	890.40	61.25	1470.00		
3 Stressing Equipment		48.60		53.45	1.52	1.67
32 L.H., Daily Totals		$1251.80		$2039.85	$39.12	$63.75
Crew C-5	Hr.	Daily	Hr.	Daily	Bare Costs	Incl. O&P
1 Rodman Foreman	$39.10	$312.80	$64.55	$516.40	$35.93	$57.99
4 Rodmen (reinf.)	37.10	1187.20	61.25	1960.00		
1 Equip. Oper. (crane)	34.80	278.40	52.40	419.20		
1 Equip. Oper. Oiler	29.20	233.60	44.00	352.00		
1 Hyd. Crane, 25 Ton		624.20		686.60	11.15	12.26
56 L.H., Daily Totals		$2636.20		$3934.20	$47.08	$70.25
Crew C-6	Hr.	Daily	Hr.	Daily	Bare Costs	Incl. O&P
1 Labor Foreman (outside)	$28.00	$224.00	$43.60	$348.80	$27.26	$42.01
4 Laborers	26.00	832.00	40.50	1296.00		
1 Cement Finisher	31.55	252.40	46.45	371.60		
2 Gas Engine Vibrators		52.00		57.20	1.08	1.19
48 L.H., Daily Totals		$1360.40		$2073.60	$28.34	$43.20

Crew No.	Bare Costs Hr.	Daily	Incl. Subs O & P Hr.	Daily	Cost Per Labor-Hour Bare Costs	Incl. O&P
Crew C-7	Hr.	Daily	Hr.	Daily	Bare Costs	Incl. O&P
1 Labor Foreman (outside)	$28.00	$224.00	$43.60	$348.80	$28.04	$43.03
5 Laborers	26.00	1040.00	40.50	1620.00		
1 Cement Finisher	31.55	252.40	46.45	371.60		
1 Equip. Oper. (med.)	33.65	269.20	50.70	405.60		
1 Equip. Oper. (oiler)	29.20	233.60	44.00	352.00		
2 Gas Engine Vibrators		52.00		57.20		
1 Concrete Bucket, 1 C.Y.		15.60		17.15		
1 Hyd. Crane, 55 Ton		911.00		1002.10	13.59	14.95
72 L.H., Daily Totals		$2997.80		$4174.45	$41.63	$57.98
Crew C-7A	Hr.	Daily	Hr.	Daily	Bare Costs	Incl. O&P
1 Labor Foreman (outside)	$28.00	$224.00	$43.60	$348.80	$26.36	$40.84
5 Laborers	26.00	1040.00	40.50	1620.00		
2 Truck Drivers(Heavy)	26.45	423.20	40.30	644.80		
2 Conc. Transit Mixers		1418.40		1560.25	22.16	24.38
64 L.H., Daily Totals		$3105.60		$4173.85	$48.52	$65.22
Crew C-7B	Hr.	Daily	Hr.	Daily	Bare Costs	Incl. O&P
1 Labor Foreman (outside)	$28.00	$224.00	$43.60	$348.80	$27.75	$42.81
5 Laborers	26.00	1040.00	40.50	1620.00		
1 Equipment Operator (heavy)	34.80	278.40	52.40	419.20		
1 Equipment Oiler	29.20	233.60	44.00	352.00		
1 Conc. Bucket, 2 C.Y.		24.80		27.30		
1 Truck Crane, 165 Ton		2034.00		2237.40	32.17	35.39
64 L.H., Daily Totals		$3834.80		$5004.70	$59.92	$78.20
Crew C-7C	Hr.	Daily	Hr.	Daily	Bare Costs	Incl. O&P
1 Labor Foreman (outside)	$28.00	$224.00	$43.60	$348.80	$28.16	$43.44
5 Laborers	26.00	1040.00	40.50	1620.00		
2 Equipment Operator (medium)	33.65	538.40	50.70	811.20		
2 Wheel Loader, 4 C.Y.		876.80		964.50	13.70	15.07
64 L.H., Daily Totals		$2679.20		$3744.50	$41.86	$58.51
Crew C-7D	Hr.	Daily	Hr.	Daily	Bare Costs	Incl. O&P
1 Labor Foreman (outside)	$28.00	$224.00	$43.60	$348.80	$27.38	$42.40
5 Laborers	26.00	1040.00	40.50	1620.00		
1 Equipment Operator (med.)	33.65	269.20	50.70	405.60		
1 Concrete Conveyer		145.20		159.70	2.59	2.85
56 L.H., Daily Totals		$1678.40		$2534.10	$29.97	$45.25
Crew C-8	Hr.	Daily	Hr.	Daily	Bare Costs	Incl. O&P
1 Labor Foreman (outside)	$28.00	$224.00	$43.60	$348.80	$28.96	$44.10
3 Laborers	26.00	624.00	40.50	972.00		
2 Cement Finishers	31.55	504.80	46.45	743.20		
1 Equip. Oper. (med.)	33.65	269.20	50.70	405.60		
1 Concrete Pump (small)		719.20		791.10	12.84	14.13
56 L.H., Daily Totals		$2341.20		$3260.70	$41.80	$58.23
Crew C-8A	Hr.	Daily	Hr.	Daily	Bare Costs	Incl. O&P
1 Labor Foreman (outside)	$28.00	$224.00	$43.60	$348.80	$28.18	$43.00
3 Laborers	26.00	624.00	40.50	972.00		
2 Cement Finishers	31.55	504.80	46.45	743.20		
48 L.H., Daily Totals		$1352.80		$2064.00	$28.18	$43.00

Crew No.	Bare Costs		Incl. Subs O & P		Cost Per Labor-Hour	

Crew C-8B

	Hr.	Daily	Hr.	Daily	Bare Costs	Incl. O&P
1 Labor Foreman (outside)	$28.00	$224.00	$43.60	$348.80	$27.93	$43.16
3 Laborers	26.00	624.00	40.50	972.00		
1 Equipment Operator	33.65	269.20	50.70	405.60		
1 Vibrating Screed		37.65		41.40		
1 Vibratory Roller		436.80		480.50		
1 Dozer, 200 H.P.		863.40		949.75	33.45	36.79
40 L.H., Daily Totals		$2455.05		$3198.05	$61.38	$79.95

Crew C-8C

	Hr.	Daily	Hr.	Daily	Bare Costs	Incl. O&P
1 Labor Foreman (outside)	$28.00	$224.00	$43.60	$348.80	$28.53	$43.71
3 Laborers	26.00	624.00	40.50	972.00		
1 Cement Finisher	31.55	252.40	46.45	371.60		
1 Equipment Operator (med.)	33.65	269.20	50.70	405.60		
1 Shotcrete Rig, 12 CY/hr		227.40		250.15	4.74	5.21
48 L.H., Daily Totals		$1597.00		$2348.15	$33.27	$48.92

Crew C-8D

	Hr.	Daily	Hr.	Daily	Bare Costs	Incl. O&P
1 Labor Foreman (outside)	$28.00	$224.00	$43.60	$348.80	$29.43	$44.74
1 Laborer	26.00	208.00	40.50	324.00		
1 Cement Finisher	31.55	252.40	46.45	371.60		
1 Equipment Operator (light)	32.15	257.20	48.40	387.20		
1 Compressor, 250 CFM		127.60		140.35		
2 Hoses, 1", 50'		6.90		7.60	4.20	4.62
32 L.H., Daily Totals		$1076.10		$1579.55	$33.63	$49.36

Crew C-8E

	Hr.	Daily	Hr.	Daily	Bare Costs	Incl. O&P
1 Labor Foreman (outside)	$28.00	$224.00	$43.60	$348.80	$29.43	$44.74
1 Laborer	26.00	208.00	40.50	324.00		
1 Cement Finisher	31.55	252.40	46.45	371.60		
1 Equipment Operator (light)	32.15	257.20	48.40	387.20		
1 Compressor, 250 CFM		127.60		140.35		
2 Hoses, 1", 50'		6.90		7.60		
1 Concrete Pump (small)		719.20		791.10	26.68	29.35
32 L.H., Daily Totals		$1795.30		$2370.65	$56.11	$74.09

Crew C-10

	Hr.	Daily	Hr.	Daily	Bare Costs	Incl. O&P
1 Laborer	$26.00	$208.00	$40.50	$324.00	$29.70	$44.47
2 Cement Finishers	31.55	504.80	46.45	743.20		
24 L.H., Daily Totals		$712.80		$1067.20	$29.70	$44.47

Crew C-10B

	Hr.	Daily	Hr.	Daily	Bare Costs	Incl. O&P
3 Laborers	$26.00	$624.00	$40.50	$972.00	$28.22	$42.88
2 Cement Finishers	31.55	504.80	46.45	743.20		
1 Concrete mixer, 10 CF		105.20		115.70		
2 Concrete finisher, 48" dia		54.40		59.85	3.99	4.39
40 L.H., Daily Totals		$1288.40		$1890.75	$32.21	$47.27

Crew C-11

	Hr.	Daily	Hr.	Daily	Bare Costs	Incl. O&P
1 Struc. Steel Foreman	$39.15	$313.20	$70.15	$561.20	$36.23	$62.87
6 Struc. Steel Workers	37.15	1783.20	66.55	3194.40		
1 Equip. Oper. (crane)	34.80	278.40	52.40	419.20		
1 Equip. Oper. Oiler	29.20	233.60	44.00	352.00		
1 Truck Crane, 150 Ton		1610.00		1771.00	22.36	24.60
72 L.H., Daily Totals		$4218.40		$6297.80	$58.59	$87.47

Crew C-12

	Hr.	Daily	Hr.	Daily	Bare Costs	Incl. O&P
1 Carpenter Foreman (out)	$35.00	$280.00	$54.55	$436.40	$32.47	$50.28
3 Carpenters	33.00	792.00	51.40	1233.60		
1 Laborer	26.00	208.00	40.50	324.00		
1 Equip. Oper. (crane)	34.80	278.40	52.40	419.20		
1 Hyd. Crane, 12 Ton		612.00		673.20	12.75	14.03
48 L.H., Daily Totals		$2170.40		$3086.40	$45.22	$64.31

Crew C-13

	Hr.	Daily	Hr.	Daily	Bare Costs	Incl. O&P
1 Struc. Steel Worker	$37.15	$297.20	$66.55	$532.40	$35.77	$61.50
1 Welder	37.15	297.20	66.55	532.40		
1 Carpenter	33.00	264.00	51.40	411.20		
1 Gas Welding Machine		75.20		82.70	3.13	3.45
24 L.H., Daily Totals		$933.60		$1558.70	$38.90	$64.95

Crew C-14

	Hr.	Daily	Hr.	Daily	Bare Costs	Incl. O&P
1 Carpenter Foreman (out)	$35.00	$280.00	$54.55	$436.40	$32.19	$50.44
5 Carpenters	33.00	1320.00	51.40	2056.00		
4 Laborers	26.00	832.00	40.50	1296.00		
4 Rodmen (reinf.)	37.10	1187.20	61.25	1960.00		
2 Cement Finishers	31.55	504.80	46.45	743.20		
1 Equip. Oper. (crane)	34.80	278.40	52.40	419.20		
1 Equip. Oper. Oiler	29.20	233.60	44.00	352.00		
1 Crane, 80 Ton, & Tools		1020.00		1122.00	7.08	7.79
144 L.H., Daily Totals		$5656.00		$8384.80	$39.27	$58.23

Crew C-14A

	Hr.	Daily	Hr.	Daily	Bare Costs	Incl. O&P
1 Carpenter Foreman (out)	$35.00	$280.00	$54.55	$436.40	$33.14	$52.00
16 Carpenters	33.00	4224.00	51.40	6579.20		
4 Rodmen (reinf.)	37.10	1187.20	61.25	1960.00		
2 Laborers	26.00	416.00	40.50	648.00		
1 Cement Finisher	31.55	252.40	46.45	371.60		
1 Equip. Oper. (med.)	33.65	269.20	50.70	405.60		
1 Gas Engine Vibrator		26.00		28.60		
1 Concrete Pump (small)		719.20		791.10	3.73	4.10
200 L.H., Daily Totals		$7374.00		$11220.50	$36.87	$56.10

Crew C-14B

	Hr.	Daily	Hr.	Daily	Bare Costs	Incl. O&P
1 Carpenter Foreman (out)	$35.00	$280.00	$54.55	$436.40	$33.08	$51.79
16 Carpenters	33.00	4224.00	51.40	6579.20		
4 Rodmen (reinf.)	37.10	1187.20	61.25	1960.00		
2 Laborers	26.00	416.00	40.50	648.00		
2 Cement Finishers	31.55	504.80	46.45	743.20		
1 Equip. Oper. (med.)	33.65	269.20	50.70	405.60		
1 Gas Engine Vibrator		26.00		28.60		
1 Concrete Pump (small)		719.20		791.10	3.58	3.94
208 L.H., Daily Totals		$7626.40		$11592.10	$36.66	$55.73

Crew C-14C

	Hr.	Daily	Hr.	Daily	Bare Costs	Incl. O&P
1 Carpenter Foreman (out)	$35.00	$280.00	$54.55	$436.40	$31.63	$49.56
6 Carpenters	33.00	1584.00	51.40	2467.20		
2 Rodmen (reinf.)	37.10	593.60	61.25	980.00		
4 Laborers	26.00	832.00	40.50	1296.00		
1 Cement Finisher	31.55	252.40	46.45	371.60		
1 Gas Engine Vibrator		26.00		28.60	.23	.26
112 L.H., Daily Totals		$3568.00		$5579.80	$31.86	$49.82

CREWS

Crew No.	Bare Costs		Incl. Subs O & P		Cost Per Labor-Hour	

Crew C-14D	Hr.	Daily	Hr.	Daily	Bare Costs	Incl. O&P
1 Carpenter Foreman (out)	$35.00	$280.00	$54.55	$436.40	$32.82	$51.22
18 Carpenters	33.00	4752.00	51.40	7401.60		
2 Rodmen (reinf.)	37.10	593.60	61.25	980.00		
2 Laborers	26.00	416.00	40.50	648.00		
1 Cement Finisher	31.55	252.40	46.45	371.60		
1 Equip. Oper. (med.)	33.65	269.20	50.70	405.60		
1 Gas Engine Vibrator		26.00		28.60		
1 Concrete Pump (small)		719.20		791.10	3.73	4.10
200 L.H., Daily Totals		$7308.40		$11062.90	$36.55	$55.32

Crew C-14E	Hr.	Daily	Hr.	Daily	Bare Costs	Incl. O&P
1 Carpenter Foreman (out)	$35.00	$280.00	$54.55	$436.40	$32.63	$51.85
2 Carpenters	33.00	528.00	51.40	822.40		
4 Rodmen (reinf.)	37.10	1187.20	61.25	1960.00		
3 Laborers	26.00	624.00	40.50	972.00		
1 Cement Finisher	31.55	252.40	46.45	371.60		
1 Gas Engine Vibrator		26.00		28.60	.30	.33
88 L.H., Daily Totals		$2897.60		$4591.00	$32.93	$52.18

Crew C-14F	Hr.	Daily	Hr.	Daily	Bare Costs	Incl. O&P
1 Laborer Foreman (out)	$28.00	$224.00	$43.60	$348.80	$29.92	$44.81
2 Laborers	26.00	416.00	40.50	648.00		
6 Cement Finishers	31.55	1514.40	46.45	2229.60		
1 Gas Engine Vibrator		26.00		28.60	.36	.40
72 L.H., Daily Totals		$2180.40		$3255.00	$30.28	$45.21

Crew C-14G	Hr.	Daily	Hr.	Daily	Bare Costs	Incl. O&P
1 Laborer Foreman (out)	$28.00	$224.00	$43.60	$348.80	$29.46	$44.34
2 Laborers	26.00	416.00	40.50	648.00		
4 Cement Finishers	31.55	1009.60	46.45	1486.40		
1 Gas Engine Vibrator		26.00		28.60	.46	.51
56 L.H., Daily Totals		$1675.60		$2511.80	$29.92	$44.85

Crew C-14H	Hr.	Daily	Hr.	Daily	Bare Costs	Incl. O&P
1 Carpenter Foreman (out)	$35.00	$280.00	$54.55	$436.40	$32.61	$50.93
2 Carpenters	33.00	528.00	51.40	822.40		
1 Rodman (reinf.)	37.10	296.80	61.25	490.00		
1 Laborer	26.00	208.00	40.50	324.00		
1 Cement Finisher	31.55	252.40	46.45	371.60		
1 Gas Engine Vibrator		26.00		28.60	.54	.60
48 L.H., Daily Totals		$1591.20		$2473.00	$33.15	$51.53

Crew C-15	Hr.	Daily	Hr.	Daily	Bare Costs	Incl. O&P
1 Carpenter Foreman (out)	$35.00	$280.00	$54.55	$436.40	$31.02	$48.11
2 Carpenters	33.00	528.00	51.40	822.40		
3 Laborers	26.00	624.00	40.50	972.00		
2 Cement Finishers	31.55	504.80	46.45	743.20		
1 Rodman (reinf.)	37.10	296.80	61.25	490.00		
72 L.H., Daily Totals		$2233.60		$3464.00	$31.02	$48.11

Crew C-16	Hr.	Daily	Hr.	Daily	Bare Costs	Incl. O&P
1 Labor Foreman (outside)	$28.00	$224.00	$43.60	$348.80	$30.77	$47.91
3 Laborers	26.00	624.00	40.50	972.00		
2 Cement Finishers	31.55	504.80	46.45	743.20		
1 Equip. Oper. (med.)	33.65	269.20	50.70	405.60		
2 Rodmen (reinf.)	37.10	593.60	61.25	980.00		
1 Concrete Pump (small)		719.20		791.10	9.99	10.99
72 L.H., Daily Totals		$2934.80		$4240.70	$40.76	$58.90

Crew C-17	Hr.	Daily	Hr.	Daily	Bare Costs	Incl. O&P
2 Skilled Worker Foremen	$35.65	$570.40	$55.45	$887.20	$34.05	$52.97
8 Skilled Workers	33.65	2153.60	52.35	3350.40		
80 L.H., Daily Totals		$2724.00		$4237.60	$34.05	$52.97

Crew C-17A	Hr.	Daily	Hr.	Daily	Bare Costs	Incl. O&P
2 Skilled Worker Foremen	$35.65	-$570.40	$55.45	$887.20	$34.06	$52.96
8 Skilled Workers	33.65	2153.60	52.35	3350.40		
.125 Equip. Oper. (crane)	34.80	34.80	52.40	52.40		
.125 Crane, 80 Ton, & Tools		127.50		140.25	1.57	1.73
81 L.H., Daily Totals		$2886.30		$4430.25	$35.63	$54.69

Crew C-17B	Hr.	Daily	Hr.	Daily	Bare Costs	Incl. O&P
2 Skilled Worker Foremen	$35.65	$570.40	$55.45	$887.20	$34.07	$52.96
8 Skilled Workers	33.65	2153.60	52.35	3350.40		
.25 Equip. Oper. (crane)	34.80	69.60	52.40	104.80		
.25 Crane, 80 Ton, & Tools		255.00		280.50		
.25 Walk Behind Power Tools		6.80		7.50	3.19	3.51
82 L.H., Daily Totals		$3055.40		$4630.40	$37.26	$56.47

Crew C-17C	Hr.	Daily	Hr.	Daily	Bare Costs	Incl. O&P
2 Skilled Worker Foremen	$35.65	$570.40	$55.45	$887.20	$34.08	$52.95
8 Skilled Workers	33.65	2153.60	52.35	3350.40		
.375 Equip. Oper. (crane)	34.80	104.40	52.40	157.20		
.375 Crane, 80 Ton & Tools		382.50		420.75	4.61	5.07
83 L.H., Daily Totals		$3210.90		$4815.55	$38.69	$58.02

Crew C-17D	Hr.	Daily	Hr.	Daily	Bare Costs	Incl. O&P
2 Skilled Worker Foremen	$35.65	$570.40	$55.45	$887.20	$34.09	$52.94
8 Skilled Workers	33.65	2153.60	52.35	3350.40		
.5 Equip. Oper. (crane)	34.80	139.20	52.40	209.60		
.5 Crane, 80 Ton & Tools		510.00		561.00	6.07	6.68
84 L.H., Daily Totals		$3373.20		$5008.20	$40.16	$59.62

Crew C-17E	Hr.	Daily	Hr.	Daily	Bare Costs	Incl. O&P
2 Skilled Worker Foremen	$35.65	$570.40	$55.45	$887.20	$34.05	$52.97
8 Skilled Workers	33.65	2153.60	52.35	3350.40		
1 Hyd. Jack with Rods		77.55		85.30	.97	1.07
80 L.H., Daily Totals		$2801.55		$4322.90	$35.02	$54.04

Crew C-18	Hr.	Daily	Hr.	Daily	Bare Costs	Incl. O&P
.125 Labor Foreman (out)	$28.00	$28.00	$43.60	$43.60	$26.22	$40.84
1 Laborer	26.00	208.00	40.50	324.00		
1 Concrete Cart, 10 C.F.		48.80		53.70	5.42	5.96
9 L.H., Daily Totals		$284.80		$421.30	$31.64	$46.80

Crew C-19	Hr.	Daily	Hr.	Daily	Bare Costs	Incl. O&P
.125 Labor Foreman (out)	$28.00	$28.00	$43.60	$43.60	$26.22	$40.84
1 Laborer	26.00	208.00	40.50	324.00		
1 Concrete Cart, 18 C.F.		74.80		82.30	8.31	9.14
9 L.H., Daily Totals		$310.80		$449.90	$34.53	$49.98

Crew C-20	Hr.	Daily	Hr.	Daily	Bare Costs	Incl. O&P
1 Labor Foreman (outside)	$28.00	$224.00	$43.60	$348.80	$27.90	$42.91
5 Laborers	26.00	1040.00	40.50	1620.00		
1 Cement Finisher	31.55	252.40	46.45	371.60		
1 Equip. Oper. (med.)	33.65	269.20	50.70	405.60		
2 Gas Engine Vibrators		52.00		57.20		
1 Concrete Pump (small)		719.20		791.10	12.05	13.26
64 L.H., Daily Totals		$2556.80		$3594.30	$39.95	$56.17

Crew C-21

Crew No.	Bare Costs Hr.	Bare Costs Daily	Incl. Subs O&P Hr.	Incl. Subs O&P Daily	Cost Per Labor-Hour Bare Costs	Cost Per Labor-Hour Incl. O&P
1 Labor Foreman (outside)	$28.00	$224.00	$43.60	$348.80	$27.90	$42.91
5 Laborers	26.00	1040.00	40.50	1620.00		
1 Cement Finisher	31.55	252.40	46.45	371.60		
1 Equip. Oper. (med.)	33.65	269.20	50.70	405.60		
2 Gas Engine Vibrators		52.00		57.20		
1 Concrete Conveyer		145.20		159.70	3.08	3.39
64 L.H., Daily Totals		$1982.80		$2962.90	$30.98	$46.30

Crew C-22

Crew No.	Bare Costs Hr.	Bare Costs Daily	Incl. Subs O&P Hr.	Incl. Subs O&P Daily	Cost Per Labor-Hour Bare Costs	Cost Per Labor-Hour Incl. O&P
1 Rodman Foreman	$39.10	$312.80	$64.55	$516.40	$37.24	$61.26
4 Rodmen (reinf.)	37.10	1187.20	61.25	1960.00		
.125 Equip. Oper. (crane)	34.80	34.80	52.40	52.40		
.125 Equip. Oper. Oiler	29.20	29.20	44.00	44.00		
.125 Hyd. Crane, 25 Ton		78.03		85.85	1.86	2.04
42 L.H., Daily Totals		$1642.03		$2658.65	$39.10	$63.30

Crew C-23

Crew No.	Bare Costs Hr.	Bare Costs Daily	Incl. Subs O&P Hr.	Incl. Subs O&P Daily	Cost Per Labor-Hour Bare Costs	Cost Per Labor-Hour Incl. O&P
2 Skilled Worker Foremen	$35.65	$570.40	$55.45	$887.20	$33.72	$52.14
6 Skilled Workers	33.65	1615.20	52.35	2512.80		
1 Equip. Oper. (crane)	34.80	278.40	52.40	419.20		
1 Equip. Oper. Oiler	29.20	233.60	44.00	352.00		
1 Crane, 90 Ton		1416.00		1557.60	17.70	19.47
80 L.H., Daily Totals		$4113.60		$5728.80	$51.42	$71.61

Crew C-24

Crew No.	Bare Costs Hr.	Bare Costs Daily	Incl. Subs O&P Hr.	Incl. Subs O&P Daily	Cost Per Labor-Hour Bare Costs	Cost Per Labor-Hour Incl. O&P
2 Skilled Worker Foremen	$35.65	$570.40	$55.45	$887.20	$33.72	$52.14
6 Skilled Workers	33.65	1615.20	52.35	2512.80		
1 Equip. Oper. (crane)	34.80	278.40	52.40	419.20		
1 Equip. Oper. Oiler	29.20	233.60	44.00	352.00		
1 Truck Crane, 150 Ton		1610.00		1771.00	20.13	22.14
80 L.H., Daily Totals		$4307.60		$5942.20	$53.85	$74.28

Crew C-25

Crew No.	Bare Costs Hr.	Bare Costs Daily	Incl. Subs O&P Hr.	Incl. Subs O&P Daily	Cost Per Labor-Hour Bare Costs	Cost Per Labor-Hour Incl. O&P
2 Rodmen (reinf.)	$37.10	$593.60	$61.25	$980.00	$29.02	$48.35
2 Rodmen Helpers	20.95	335.20	35.45	567.20		
32 L.H., Daily Totals		$928.80		$1547.20	$29.02	$48.35

Crew C-27

Crew No.	Bare Costs Hr.	Bare Costs Daily	Incl. Subs O&P Hr.	Incl. Subs O&P Daily	Cost Per Labor-Hour Bare Costs	Cost Per Labor-Hour Incl. O&P
2 Cement Finishers	$31.55	$504.80	$46.45	$743.20	$31.55	$46.45
1 Concrete Saw		92.20		101.40	5.76	6.34
16 L.H., Daily Totals		$597.00		$844.60	$37.31	$52.79

Crew C-28

Crew No.	Bare Costs Hr.	Bare Costs Daily	Incl. Subs O&P Hr.	Incl. Subs O&P Daily	Cost Per Labor-Hour Bare Costs	Cost Per Labor-Hour Incl. O&P
1 Cement Finisher	$31.55	$252.40	$46.45	$371.60	$31.55	$46.45
1 Portable Air Compressor		18.60		20.45	2.33	2.56
8 L.H., Daily Totals		$271.00		$392.05	$33.88	$49.01

Crew D-1

Crew No.	Bare Costs Hr.	Bare Costs Daily	Incl. Subs O&P Hr.	Incl. Subs O&P Daily	Cost Per Labor-Hour Bare Costs	Cost Per Labor-Hour Incl. O&P
1 Bricklayer	$34.25	$274.00	$52.15	$417.20	$29.93	$45.58
1 Bricklayer Helper	25.60	204.80	39.00	312.00		
16 L.H., Daily Totals		$478.80		$729.20	$29.93	$45.58

Crew D-2

Crew No.	Bare Costs Hr.	Bare Costs Daily	Incl. Subs O&P Hr.	Incl. Subs O&P Daily	Cost Per Labor-Hour Bare Costs	Cost Per Labor-Hour Incl. O&P
3 Bricklayers	$34.25	$822.00	$52.15	$1251.60	$30.99	$47.30
2 Bricklayer Helpers	25.60	409.60	39.00	624.00		
.5 Carpenter	33.00	132.00	51.40	205.60		
44 L.H., Daily Totals		$1363.60		$2081.20	$30.99	$47.30

Crew D-3

Crew No.	Bare Costs Hr.	Bare Costs Daily	Incl. Subs O&P Hr.	Incl. Subs O&P Daily	Cost Per Labor-Hour Bare Costs	Cost Per Labor-Hour Incl. O&P
3 Bricklayers	$34.25	$822.00	$52.15	$1251.60	$30.90	$47.10
2 Bricklayer Helpers	25.60	409.60	39.00	624.00		
.25 Carpenter	33.00	66.00	51.40	102.80		
42 L.H., Daily Totals		$1297.60		$1978.40	$30.90	$47.10

Crew D-4

Crew No.	Bare Costs Hr.	Bare Costs Daily	Incl. Subs O&P Hr.	Incl. Subs O&P Daily	Cost Per Labor-Hour Bare Costs	Cost Per Labor-Hour Incl. O&P
1 Bricklayer	$34.25	$274.00	$52.15	$417.20	$29.40	$44.64
2 Bricklayer Helpers	25.60	409.60	39.00	624.00		
1 Equip. Oper. (light)	32.15	257.20	48.40	387.20		
1 Grout Pump, 50 C.F./hr		110.90		122.00		
1 Hoses & Hopper		15.60		17.15		
1 Accessories		11.70		12.85	4.32	4.75
32 L.H., Daily Totals		$1079.00		$1580.40	$33.72	$49.39

Crew D-5

Crew No.	Bare Costs Hr.	Bare Costs Daily	Incl. Subs O&P Hr.	Incl. Subs O&P Daily	Cost Per Labor-Hour Bare Costs	Cost Per Labor-Hour Incl. O&P
1 Bricklayer	$34.25	$274.00	$52.15	$417.20	$34.25	$52.15
8 L.H., Daily Totals		$274.00		$417.20	$34.25	$52.15

Crew D-6

Crew No.	Bare Costs Hr.	Bare Costs Daily	Incl. Subs O&P Hr.	Incl. Subs O&P Daily	Cost Per Labor-Hour Bare Costs	Cost Per Labor-Hour Incl. O&P
3 Bricklayers	$34.25	$822.00	$52.15	$1251.60	$30.05	$45.81
3 Bricklayer Helpers	25.60	614.40	39.00	936.00		
.25 Carpenter	33.00	66.00	51.40	102.80		
50 L.H., Daily Totals		$1502.40		$2290.40	$30.05	$45.81

Crew D-7

Crew No.	Bare Costs Hr.	Bare Costs Daily	Incl. Subs O&P Hr.	Incl. Subs O&P Daily	Cost Per Labor-Hour Bare Costs	Cost Per Labor-Hour Incl. O&P
1 Tile Layer	$31.50	$252.00	$46.25	$370.00	$27.98	$41.08
1 Tile Layer Helper	24.45	195.60	35.90	287.20		
16 L.H., Daily Totals		$447.60		$657.20	$27.98	$41.08

Crew D-8

Crew No.	Bare Costs Hr.	Bare Costs Daily	Incl. Subs O&P Hr.	Incl. Subs O&P Daily	Cost Per Labor-Hour Bare Costs	Cost Per Labor-Hour Incl. O&P
3 Bricklayers	$34.25	$822.00	$52.15	$1251.60	$30.79	$46.89
2 Bricklayer Helpers	25.60	409.60	39.00	624.00		
40 L.H., Daily Totals		$1231.60		$1875.60	$30.79	$46.89

Crew D-9

Crew No.	Bare Costs Hr.	Bare Costs Daily	Incl. Subs O&P Hr.	Incl. Subs O&P Daily	Cost Per Labor-Hour Bare Costs	Cost Per Labor-Hour Incl. O&P
3 Bricklayers	$34.25	$822.00	$52.15	$1251.60	$29.93	$45.58
3 Bricklayer Helpers	25.60	614.40	39.00	936.00		
48 L.H., Daily Totals		$1436.40		$2187.60	$29.93	$45.58

Crew D-10

Crew No.	Bare Costs Hr.	Bare Costs Daily	Incl. Subs O&P Hr.	Incl. Subs O&P Daily	Cost Per Labor-Hour Bare Costs	Cost Per Labor-Hour Incl. O&P
1 Bricklayer Foreman	$36.25	$290.00	$55.20	$441.60	$31.30	$47.55
1 Bricklayer	34.25	274.00	52.15	417.20		
2 Bricklayer Helpers	25.60	409.60	39.00	624.00		
1 Equip. Oper. (crane)	34.80	278.40	52.40	419.20		
1 Truck Crane, 12.5 Ton		482.20		530.40	12.06	13.26
40 L.H., Daily Totals		$1734.20		$2432.40	$43.36	$60.81

Crew D-11

Crew No.	Bare Costs Hr.	Bare Costs Daily	Incl. Subs O&P Hr.	Incl. Subs O&P Daily	Cost Per Labor-Hour Bare Costs	Cost Per Labor-Hour Incl. O&P
1 Bricklayer Foreman	$36.25	$290.00	$55.20	$441.60	$32.03	$48.78
1 Bricklayer	34.25	274.00	52.15	417.20		
1 Bricklayer Helper	25.60	204.80	39.00	312.00		
24 L.H., Daily Totals		$768.80		$1170.80	$32.03	$48.78

Crew D-12

Crew No.	Bare Costs Hr.	Bare Costs Daily	Incl. Subs O&P Hr.	Incl. Subs O&P Daily	Cost Per Labor-Hour Bare Costs	Cost Per Labor-Hour Incl. O&P
1 Bricklayer Foreman	$36.25	$290.00	$55.20	$441.60	$30.43	$46.34
1 Bricklayer	34.25	274.00	52.15	417.20		
2 Bricklayer Helpers	25.60	409.60	39.00	624.00		
32 L.H., Daily Totals		$973.60		$1482.80	$30.43	$46.34

CREWS

Crew No.	Bare Costs		Incl. Subs O & P		Cost Per Labor-Hour	
	Hr.	Daily	Hr.	Daily	Bare Costs	Incl. O&P
Crew D-13					$31.58	$48.19
1 Bricklayer Foreman	$36.25	$290.00	$55.20	$441.60		
1 Bricklayer	34.25	274.00	52.15	417.20		
2 Bricklayer Helpers	25.60	409.60	39.00	624.00		
1 Carpenter	33.00	264.00	51.40	411.20		
1 Equip. Oper. (crane)	34.80	278.40	52.40	419.20		
1 Truck Crane, 12.5 Ton		482.20		530.40	10.05	11.05
48 L.H., Daily Totals		$1998.20		$2843.60	$41.63	$59.24

Crew No.	Hr.	Daily	Hr.	Daily	Bare Costs	Incl. O&P
Crew E-1					$36.15	$61.70
1 Welder Foreman	$39.15	$313.20	$70.15	$561.20		
1 Welder	37.15	297.20	66.55	532.40		
1 Equip. Oper. (light)	32.15	257.20	48.40	387.20		
1 Gas Welding Machine		75.20		82.70	3.13	3.45
24 L.H., Daily Totals		$942.80		$1563.50	$39.28	$65.15

Crew No.	Hr.	Daily	Hr.	Daily	Bare Costs	Incl. O&P
Crew E-2					$35.96	$61.82
1 Struc. Steel Foreman	$39.15	$313.20	$70.15	$561.20		
4 Struc. Steel Workers	37.15	1188.80	66.55	2129.60		
1 Equip. Oper. (crane)	34.80	278.40	52.40	419.20		
1 Equip. Oper. Oiler	29.20	233.60	44.00	352.00		
1 Crane, 90 Ton		1416.00		1557.60	25.29	27.81
56 L.H., Daily Totals		$3430.00		$5019.60	$61.25	$89.63

Crew No.	Hr.	Daily	Hr.	Daily	Bare Costs	Incl. O&P
Crew E-3					$37.82	$67.75
1 Struc. Steel Foreman	$39.15	$313.20	$70.15	$561.20		
1 Struc. Steel Worker	37.15	297.20	66.55	532.40		
1 Welder	37.15	297.20	66.55	532.40		
1 Gas Welding Machine		75.20		82.70	3.13	3.45
24 L.H., Daily Totals		$982.80		$1708.70	$40.95	$71.20

Crew No.	Hr.	Daily	Hr.	Daily	Bare Costs	Incl. O&P
Crew E-4					$37.65	$67.45
1 Struc. Steel Foreman	$39.15	$313.20	$70.15	$561.20		
3 Struc. Steel Workers	37.15	891.60	66.55	1597.20		
1 Gas Welding Machine		75.20		82.70	2.35	2.59
32 L.H., Daily Totals		$1280.00		$2241.10	$40.00	$70.04

Crew No.	Hr.	Daily	Hr.	Daily	Bare Costs	Incl. O&P
Crew E-5					$36.52	$63.60
2 Struc. Steel Foremen	$39.15	$626.40	$70.15	$1122.40		
5 Struc. Steel Workers	37.15	1486.00	66.55	2662.00		
1 Equip. Oper. (crane)	34.80	278.40	52.40	419.20		
1 Welder	37.15	297.20	66.55	532.40		
1 Equip. Oper. Oiler	29.20	233.60	44.00	352.00		
1 Crane, 90 Ton		1416.00		1557.60		
1 Gas Welding Machine		75.20		82.70	18.64	20.50
80 L.H., Daily Totals		$4412.80		$6728.30	$55.16	$84.10

Crew No.	Hr.	Daily	Hr.	Daily	Bare Costs	Incl. O&P
Crew E-6					$36.57	$63.80
3 Struc. Steel Foremen	$39.15	$939.60	$70.15	$1683.60		
9 Struc. Steel Workers	37.15	2674.80	66.55	4791.60		
1 Equip. Oper. (crane)	34.80	278.40	52.40	419.20		
1 Welder	37.15	297.20	66.55	532.40		
1 Equip. Oper. Oiler	29.20	233.60	44.00	352.00		
1 Equip. Oper. (light)	32.15	257.20	48.40	387.20		
1 Crane, 90 Ton		1416.00		1557.60		
1 Gas Welding Machine		75.20		82.70		
1 Air Compr., 160 C.F.M.		82.00		90.20		
2 Impact Wrenches		28.80		31.70	12.52	13.77
128 L.H., Daily Totals		$6282.80		$9928.20	$49.09	$77.57

Crew No.	Bare Costs		Incl. Subs O & P		Cost Per Labor-Hour	
	Hr.	Daily	Hr.	Daily	Bare Costs	Incl. O&P
Crew E-7					$36.52	$63.60
1 Struc. Steel Foreman	$39.15	$313.20	$70.15	$561.20		
4 Struc. Steel Workers	37.15	1188.80	66.55	2129.60		
1 Equip. Oper. (crane)	34.80	278.40	52.40	419.20		
1 Equip. Oper. Oiler	29.20	233.60	44.00	352.00		
1 Welder Foreman	39.15	313.20	70.15	561.20		
2 Welders	37.15	594.40	66.55	1064.80		
1 Crane, 90 Ton		1416.00		1557.60		
2 Gas Welding Machines		150.40		165.45	19.58	21.54
80 L.H., Daily Totals		$4488.00		$6811.05	$56.10	$85.14

Crew No.	Hr.	Daily	Hr.	Daily	Bare Costs	Incl. O&P
Crew E-8					$36.28	$62.88
1 Struc. Steel Foreman	$39.15	$313.20	$70.15	$561.20		
4 Struc. Steel Workers	37.15	1188.80	66.55	2129.60		
1 Welder Foreman	39.15	313.20	70.15	561.20		
4 Welders	37.15	1188.80	66.55	2129.60		
1 Equip. Oper. (crane)	34.80	278.40	52.40	419.20		
1 Equip. Oper. Oiler	29.20	233.60	44.00	352.00		
1 Equip. Oper. (light)	32.15	257.20	48.40	387.20		
1 Crane, 90 Ton		1416.00		1557.60		
4 Gas Welding Machines		300.80		330.90	16.51	18.16
104 L.H., Daily Totals		$5490.00		$8428.50	$52.79	$81.04

Crew No.	Hr.	Daily	Hr.	Daily	Bare Costs	Incl. O&P
Crew E-9					$36.57	$63.80
2 Struc. Steel Foremen	$39.15	$626.40	$70.15	$1122.40		
5 Struc. Steel Workers	37.15	1486.00	66.55	2662.00		
1 Welder Foreman	39.15	313.20	70.15	561.20		
5 Welders	37.15	1486.00	66.55	2662.00		
1 Equip. Oper. (crane)	34.80	278.40	52.40	419.20		
1 Equip. Oper. Oiler	29.20	233.60	44.00	352.00		
1 Equip. Oper. (light)	32.15	257.20	48.40	387.20		
1 Crane, 90 Ton		1416.00		1557.60		
5 Gas Welding Machines		376.00		413.60	14.00	15.40
128 L.H., Daily Totals		$6472.80		$10137.20	$50.57	$79.20

Crew No.	Hr.	Daily	Hr.	Daily	Bare Costs	Incl. O&P
Crew E-10					$38.15	$68.35
1 Welder Foreman	$39.15	$313.20	$70.15	$561.20		
1 Welder	37.15	297.20	66.55	532.40		
1 Gas Welding Machines		75.20		82.70		
1 Truck, 3 Ton		160.40		176.45	14.73	16.20
16 L.H., Daily Totals		$846.00		$1352.75	$52.88	$84.55

Crew No.	Hr.	Daily	Hr.	Daily	Bare Costs	Incl. O&P
Crew E-11					$29.56	$50.38
2 Painters, Struc. Steel	$30.05	$480.80	$56.30	$900.80		
1 Building Laborer	26.00	208.00	40.50	324.00		
1 Equip. Oper. (light)	32.15	257.20	48.40	387.20		
1 Air Compressor 250 C.F.M.		127.60		140.35		
1 Sand Blaster		15.60		17.15		
1 Sand Blasting Accessories		11.70		12.85	4.84	5.32
32 L.H., Daily Totals		$1100.90		$1782.35	$34.40	$55.70

Crew No.	Hr.	Daily	Hr.	Daily	Bare Costs	Incl. O&P
Crew E-12					$35.65	$59.28
1 Welder Foreman	$39.15	$313.20	$70.15	$561.20		
1 Equip. Oper. (light)	32.15	257.20	48.40	387.20		
1 Gas Welding Machine		75.20		82.70	4.70	5.17
16 L.H., Daily Totals		$645.60		$1031.10	$40.35	$64.45

Crew No.	Hr.	Daily	Hr.	Daily	Bare Costs	Incl. O&P
Crew E-13					$36.82	$62.90
1 Welder Foreman	$39.15	$313.20	$70.15	$561.20		
.5 Equip. Oper. (light)	32.15	128.60	48.40	193.60		
1 Gas Welding Machine		75.20		82.70	6.27	6.89
12 L.H., Daily Totals		$517.00		$837.50	$43.09	$69.79

CREWS

Crew No.	Bare Costs		Incl. Subs O & P		Cost Per Labor-Hour	
Crew E-14	Hr.	Daily	Hr.	Daily	Bare Costs	Incl. O&P
1 Welder Foreman	$39.15	$313.20	$70.15	$561.20	$39.15	$70.15
1 Gas Welding Machine		75.20		82.70	9.40	10.34
8 L.H., Daily Totals		$388.40		$643.90	$48.55	$80.49
Crew E-16	Hr.	Daily	Hr.	Daily	Bare Costs	Incl. O&P
1 Welder Foreman	$39.15	$313.20	$70.15	$561.20	$38.15	$68.35
1 Welder	37.15	297.20	66.55	532.40		
1 Gas Welding Machine		75.20		82.70	4.70	5.17
16 L.H., Daily Totals		$685.60		$1176.30	$42.85	$73.52
Crew E-17	Hr.	Daily	Hr.	Daily	Bare Costs	Incl. O&P
1 Structural Steel Foreman	$39.15	$313.20	$70.15	$561.20	$38.15	$68.35
1 Structural Steel Worker	37.15	297.20	66.55	532.40		
1 Power Tool		4.40		4.85	.28	.30
16 L.H., Daily Totals		$614.80		$1098.45	$38.43	$68.65
Crew E-18	Hr.	Daily	Hr.	Daily	Bare Costs	Incl. O&P
1 Structural Steel Foreman	$39.15	$313.20	$70.15	$561.20	$36.85	$64.10
3 Structural Steel Workers	37.15	891.60	66.55	1597.20		
1 Equipment Operator (med.)	33.65	269.20	50.70	405.60		
1 Crane, 20 Ton		647.45		712.20	16.19	17.80
40 L.H., Daily Totals		$2121.45		$3276.20	$53.04	$81.90
Crew E-19	Hr.	Daily	Hr.	Daily	Bare Costs	Incl. O&P
1 Structural Steel Worker	$37.15	$297.20	$66.55	$532.40	$36.15	$61.70
1 Structural Steel Foreman	39.15	313.20	70.15	561.20		
1 Equip. Oper. (light)	32.15	257.20	48.40	387.20		
1 Power Tool		4.40		4.85		
1 Crane, 20 Ton		647.45		712.20	27.16	29.88
24 L.H., Daily Totals		$1519.45		$2197.85	$63.31	$91.58
Crew E-20	Hr.	Daily	Hr.	Daily	Bare Costs	Incl. O&P
1 Structural Steel Foreman	$39.15	$313.20	$70.15	$561.20	$36.11	$62.41
5 Structural Steel Workers	37.15	1486.00	66.55	2662.00		
1 Equip. Oper. (crane)	34.80	278.40	52.40	419.20		
1 Oiler	29.20	233.60	44.00	352.00		
1 Power Tool		4.40		4.85		
1 Crane, 40 Ton		772.50		849.75	12.14	13.35
64 L.H., Daily Totals		$3088.10		$4849.00	$48.25	$75.76
Crew E-22	Hr.	Daily	Hr.	Daily	Bare Costs	Incl. O&P
1 Skilled Worker Foreman	$35.65	$285.20	$55.45	$443.60	$34.32	$53.38
2 Skilled Workers	33.65	538.40	52.35	837.60		
24 L.H., Daily Totals		$823.60		$1281.20	$34.32	$53.38
Crew E-24	Hr.	Daily	Hr.	Daily	Bare Costs	Incl. O&P
3 Structural Steel Workers	$37.15	$891.60	$66.55	$1597.20	$36.28	$62.59
1 Equipment Operator (medium)	33.65	269.20	50.70	405.60		
1-25 Ton Crane		624.20		686.60	19.51	21.46
32 L.H., Daily Totals		$1785.00		$2689.40	$55.79	$84.05
Crew E-25	Hr.	Daily	Hr.	Daily	Bare Costs	Incl. O&P
1 Welder Foreman	$39.15	$313.20	$70.15	$561.20	$39.15	$70.15
1 Cutting Torch		20.00		22.00		
1 Gases		64.80		71.30	10.60	11.66
8 L.H., Daily Totals		$398.00		$654.50	$49.75	$81.81

Crew No.	Bare Costs		Incl. Subs O & P		Cost Per Labor-Hour	
Crew F-3	Hr.	Daily	Hr.	Daily	Bare Costs	Incl. O&P
4 Carpenters	$33.00	$1056.00	$51.40	$1644.80	$33.36	$51.60
1 Equip. Oper. (crane)	34.80	278.40	52.40	419.20		
1 Hyd. Crane, 12 Ton		612.00		673.20	15.30	16.83
40 L.H., Daily Totals		$1946.40		$2737.20	$48.66	$68.43
Crew F-4	Hr.	Daily	Hr.	Daily	Bare Costs	Incl. O&P
4 Carpenters	$33.00	$1056.00	$51.40	$1644.80	$32.67	$50.33
1 Equip. Oper. (crane)	34.80	278.40	52.40	419.20		
1 Equip. Oper. Oiler	29.20	233.60	44.00	352.00		
1 Hyd. Crane, 55 Ton		911.00		1002.10	18.98	20.88
48 L.H., Daily Totals		$2479.00		$3418.10	$51.65	$71.21
Crew F-5	Hr.	Daily	Hr.	Daily	Bare Costs	Incl. O&P
1 Carpenter Foreman	$35.00	$280.00	$54.55	$436.40	$33.50	$52.19
3 Carpenters	33.00	792.00	51.40	1233.60		
32 L.H., Daily Totals		$1072.00		$1670.00	$33.50	$52.19
Crew F-6	Hr.	Daily	Hr.	Daily	Bare Costs	Incl. O&P
2 Carpenters	$33.00	$528.00	$51.40	$822.40	$30.56	$47.24
2 Building Laborers	26.00	416.00	40.50	648.00		
1 Equip. Oper. (crane)	34.80	278.40	52.40	419.20		
1 Hyd. Crane, 12 Ton		612.00		673.20	15.30	16.83
40 L.H., Daily Totals		$1834.40		$2562.80	$45.86	$64.07
Crew F-7	Hr.	Daily	Hr.	Daily	Bare Costs	Incl. O&P
2 Carpenters	$33.00	$528.00	$51.40	$822.40	$29.50	$45.95
2 Building Laborers	26.00	416.00	40.50	648.00		
32 L.H., Daily Totals		$944.00		$1470.40	$29.50	$45.95
Crew G-1	Hr.	Daily	Hr.	Daily	Bare Costs	Incl. O&P
1 Roofer Foreman	$30.45	$243.60	$51.50	$412.00	$26.59	$44.97
4 Roofers, Composition	28.45	910.40	48.10	1539.20		
2 Roofer Helpers	20.95	335.20	35.45	567.20		
1 Application Equipment		138.40		152.25		
1 Tar Kettle/Pot		52.50		57.75		
1 Crew Truck		109.10		120.00	5.36	5.89
56 L.H., Daily Totals		$1789.20		$2848.40	$31.95	$50.86
Crew G-2	Hr.	Daily	Hr.	Daily	Bare Costs	Incl. O&P
1 Plasterer	$29.95	$239.60	$45.50	$364.00	$27.25	$41.73
1 Plasterer Helper	25.80	206.40	39.20	313.60		
1 Building Laborer	26.00	208.00	40.50	324.00		
1 Grouting Equipment		110.90		122.00	4.62	5.08
24 L.H., Daily Totals		$764.90		$1123.60	$31.87	$46.81
Crew G-3	Hr.	Daily	Hr.	Daily	Bare Costs	Incl. O&P
2 Sheet Metal Workers	$38.80	$620.80	$59.50	$952.00	$32.40	$50.00
2 Building Laborers	26.00	416.00	40.50	648.00		
32 L.H., Daily Totals		$1036.80		$1600.00	$32.40	$50.00
Crew G-4	Hr.	Daily	Hr.	Daily	Bare Costs	Incl. O&P
1 Labor Foreman (outside)	$28.00	$224.00	$43.60	$348.80	$26.67	$41.53
2 Building Laborers	26.00	416.00	40.50	648.00		
1 Light Truck, 1.5 Ton		122.00		134.20		
1 Air Compr., 160 C.F.M.		82.00		90.20	8.50	9.35
24 L.H., Daily Totals		$844.00		$1221.20	$35.17	$50.88

Crew G-5

Crew No.	Bare Costs Hr.	Daily	Incl. Subs O&P Hr.	Daily	Cost Per Labor-Hour Bare Costs	Incl. O&P
1 Roofer Foreman	$30.45	$243.60	$51.50	$412.00	$25.85	$43.72
2 Roofers, Composition	28.45	455.20	48.10	769.60		
2 Roofer Helpers	20.95	335.20	35.45	567.20		
1 Application Equipment		138.40		152.25	3.46	3.81
40 L.H., Daily Totals		$1172.40		$1901.05	$29.31	$47.53

Crew G-6A

Crew No.	Bare Costs Hr.	Daily	Incl. Subs O&P Hr.	Daily	Cost Per Labor-Hour Bare Costs	Incl. O&P
2 Roofers Composition	$28.45	$455.20	$48.10	$769.60	$28.45	$48.10
1 Small Compressor		13.10		14.40		
2 Pneumatic Nailers		39.00		42.90	3.26	3.58
16 L.H., Daily Totals		$507.30		$826.90	$31.71	$51.68

Crew G-7

Crew No.	Bare Costs Hr.	Daily	Incl. Subs O&P Hr.	Daily	Cost Per Labor-Hour Bare Costs	Incl. O&P
1 Carpenter	$33.00	$264.00	$51.40	$411.20	$33.00	$51.40
1 Small Compressor		13.10		14.40		
1 Pneumatic Nailer		19.50		21.45	4.08	4.48
8 L.H., Daily Totals		$296.60		$447.05	$37.08	$55.88

Crew H-1

Crew No.	Bare Costs Hr.	Daily	Incl. Subs O&P Hr.	Daily	Cost Per Labor-Hour Bare Costs	Incl. O&P
2 Glaziers	$32.05	$512.80	$48.45	$775.20	$34.60	$57.50
2 Struc. Steel Workers	37.15	594.40	66.55	1064.80		
32 L.H., Daily Totals		$1107.20		$1840.00	$34.60	$57.50

Crew H-2

Crew No.	Bare Costs Hr.	Daily	Incl. Subs O&P Hr.	Daily	Cost Per Labor-Hour Bare Costs	Incl. O&P
2 Glaziers	$32.05	$512.80	$48.45	$775.20	$30.03	$45.80
1 Building Laborer	26.00	208.00	40.50	324.00		
24 L.H., Daily Totals		$720.80		$1099.20	$30.03	$45.80

Crew H-3

Crew No.	Bare Costs Hr.	Daily	Incl. Subs O&P Hr.	Daily	Cost Per Labor-Hour Bare Costs	Incl. O&P
1 Glazier	$32.05	$256.40	$48.45	$387.60	$28.30	$43.28
1 Helper	24.55	196.40	38.10	304.80		
16 L.H., Daily Totals		$452.80		$692.40	$28.30	$43.28

Crew J-1

Crew No.	Bare Costs Hr.	Daily	Incl. Subs O&P Hr.	Daily	Cost Per Labor-Hour Bare Costs	Incl. O&P
3 Plasterers	$29.95	$718.80	$45.50	$1092.00	$28.29	$42.98
2 Plasterer Helpers	25.80	412.80	39.20	627.20		
1 Mixing Machine, 6 C.F.		90.00		99.00	2.25	2.48
40 L.H., Daily Totals		$1221.60		$1818.20	$30.54	$45.46

Crew J-2

Crew No.	Bare Costs Hr.	Daily	Incl. Subs O&P Hr.	Daily	Cost Per Labor-Hour Bare Costs	Incl. O&P
3 Plasterers	$29.95	$718.80	$45.50	$1092.00	$28.67	$43.35
2 Plasterer Helpers	25.80	412.80	39.20	627.20		
1 Lather	30.55	244.40	45.20	361.60		
1 Mixing Machine, 6 C.F.		90.00		99.00	1.88	2.06
48 L.H., Daily Totals		$1466.00		$2179.80	$30.55	$45.41

Crew J-3

Crew No.	Bare Costs Hr.	Daily	Incl. Subs O&P Hr.	Daily	Cost Per Labor-Hour Bare Costs	Incl. O&P
1 Terrazzo Worker	$31.60	$252.80	$46.40	$371.20	$28.45	$41.78
1 Terrazzo Helper	25.30	202.40	37.15	297.20		
1 Terrazzo Grinder, Electric		59.55		65.50		
1 Terrazzo Mixer		121.60		133.75	11.32	12.45
16 L.H., Daily Totals		$636.35		$867.65	$39.77	$54.23

Crew J-4

Crew No.	Bare Costs Hr.	Daily	Incl. Subs O&P Hr.	Daily	Cost Per Labor-Hour Bare Costs	Incl. O&P
1 Tile Layer	$31.50	$252.00	$46.25	$370.00	$27.98	$41.08
1 Tile Layer Helper	24.45	195.60	35.90	287.20		
16 L.H., Daily Totals		$447.60		$657.20	$27.98	$41.08

Crew K-1

Crew No.	Bare Costs Hr.	Daily	Incl. Subs O&P Hr.	Daily	Cost Per Labor-Hour Bare Costs	Incl. O&P
1 Carpenter	$33.00	$264.00	$51.40	$411.20	$29.35	$45.28
1 Truck Driver (light)	25.70	205.60	39.15	313.20		
1 Truck w/Power Equip.		160.40		176.45	10.03	11.03
16 L.H., Daily Totals		$630.00		$900.85	$39.38	$56.31

Crew K-2

Crew No.	Bare Costs Hr.	Daily	Incl. Subs O&P Hr.	Daily	Cost Per Labor-Hour Bare Costs	Incl. O&P
1 Struc. Steel Foreman	$39.15	$313.20	$70.15	$561.20	$34.00	$58.62
1 Struc. Steel Worker	37.15	297.20	66.55	532.40		
1 Truck Driver (light)	25.70	205.60	39.15	313.20		
1 Truck w/Power Equip.		160.40		176.45	6.68	7.35
24 L.H., Daily Totals		$976.40		$1583.25	$40.68	$65.97

Crew L-1

Crew No.	Bare Costs Hr.	Daily	Incl. Subs O&P Hr.	Daily	Cost Per Labor-Hour Bare Costs	Incl. O&P
1 Electrician	$39.40	$315.20	$58.60	$468.80	$39.50	$59.03
1 Plumber	39.60	316.80	59.45	475.60		
16 L.H., Daily Totals		$632.00		$944.40	$39.50	$59.03

Crew L-2

Crew No.	Bare Costs Hr.	Daily	Incl. Subs O&P Hr.	Daily	Cost Per Labor-Hour Bare Costs	Incl. O&P
1 Carpenter	$33.00	$264.00	$51.40	$411.20	$28.77	$44.75
1 Carpenter Helper	24.55	196.40	38.10	304.80		
16 L.H., Daily Totals		$460.40		$716.00	$28.77	$44.75

Crew L-3

Crew No.	Bare Costs Hr.	Daily	Incl. Subs O&P Hr.	Daily	Cost Per Labor-Hour Bare Costs	Incl. O&P
1 Carpenter	$33.00	$264.00	$51.40	$411.20	$36.05	$55.23
.5 Electrician	39.40	157.60	58.60	234.40		
.5 Sheet Metal Worker	38.80	155.20	59.50	238.00		
16 L.H., Daily Totals		$576.80		$883.60	$36.05	$55.23

Crew L-3A

Crew No.	Bare Costs Hr.	Daily	Incl. Subs O&P Hr.	Daily	Cost Per Labor-Hour Bare Costs	Incl. O&P
1 Carpenter Foreman (outside)	$35.00	$280.00	$54.55	$436.40	$36.27	$56.20
.5 Sheet Metal Worker	38.80	155.20	59.50	238.00		
12 L.H., Daily Totals		$435.20		$674.40	$36.27	$56.20

Crew L-4

Crew No.	Bare Costs Hr.	Daily	Incl. Subs O&P Hr.	Daily	Cost Per Labor-Hour Bare Costs	Incl. O&P
2 Skilled Workers	$33.65	$538.40	$52.35	$837.60	$30.62	$47.60
1 Helper	24.55	196.40	38.10	304.80		
24 L.H., Daily Totals		$734.80		$1142.40	$30.62	$47.60

Crew L-5

Crew No.	Bare Costs Hr.	Daily	Incl. Subs O&P Hr.	Daily	Cost Per Labor-Hour Bare Costs	Incl. O&P
1 Struc. Steel Foreman	$39.15	$313.20	$70.15	$561.20	$37.10	$65.04
5 Struc. Steel Workers	37.15	1486.00	66.55	2662.00		
1 Equip. Oper. (crane)	34.80	278.40	52.40	419.20		
1 Hyd. Crane, 25 Ton		624.20		686.60	11.15	12.26
56 L.H., Daily Totals		$2701.80		$4329.00	$48.25	$77.30

Crew L-5A

Crew No.	Bare Costs Hr.	Daily	Incl. Subs O&P Hr.	Daily	Cost Per Labor-Hour Bare Costs	Incl. O&P
1 Structural Steel Foreman	$39.15	$313.20	$70.15	$561.20	$37.06	$63.91
2 Structural Steel Workers	37.15	594.40	66.55	1064.80		
1 Equip. Oper. (crane)	34.80	278.40	52.40	419.20		
1 Crane,SP, 25 Ton		575.20		632.70	17.98	19.77
32 L.H., Daily Totals		$1761.20		$2677.90	$55.04	$83.68

Crew L-6

Crew No.	Bare Costs Hr.	Daily	Incl. Subs O&P Hr.	Daily	Cost Per Labor-Hour Bare Costs	Incl. O&P
1 Plumber	$39.60	$316.80	$59.45	$475.60	$39.53	$59.17
.5 Electrician	39.40	157.60	58.60	234.40		
12 L.H., Daily Totals		$474.40		$710.00	$39.53	$59.17

Crew No.	Bare Costs Hr.	Daily	Incl. Subs O & P Hr.	Daily	Cost Per Labor-Hour Bare Costs	Incl. O&P
Crew L-7	Hr.	Daily	Hr.	Daily	Bare Costs	Incl. O&P
2 Carpenters	$33.00	$528.00	$51.40	$822.40	$31.91	$49.31
1 Building Laborer	26.00	208.00	40.50	324.00		
.5 Electrician	39.40	157.60	58.60	234.40		
28 L.H., Daily Totals		$893.60		$1380.80	$31.91	$49.31
Crew L-8	Hr.	Daily	Hr.	Daily	Bare Costs	Incl. O&P
2 Carpenters	$33.00	$528.00	$51.40	$822.40	$34.32	$53.01
.5 Plumber	39.60	158.40	59.45	237.80		
20 L.H., Daily Totals		$686.40		$1060.20	$34.32	$53.01
Crew L-9	Hr.	Daily	Hr.	Daily	Bare Costs	Incl. O&P
1 Labor Foreman (inside)	$26.50	$212.00	$41.30	$330.40	$30.08	$48.48
2 Building Laborers	26.00	416.00	40.50	648.00		
1 Struc. Steel Worker	37.15	297.20	66.55	532.40		
.5 Electrician	39.40	157.60	58.60	234.40		
36 L.H., Daily Totals		$1082.80		$1745.20	$30.08	$48.48
Crew L-10	Hr.	Daily	Hr.	Daily	Bare Costs	Incl. O&P
1 Structural Steel Foreman	$39.15	$313.20	$70.15	$561.20	$37.03	$63.03
1 Structural Steel Worker	37.15	297.20	66.55	532.40		
1 Equip. Oper. (crane)	34.80	278.40	52.40	419.20		
1 Hyd. Crane, 12 Ton		612.00		673.20	25.50	28.05
24 L.H., Daily Totals		$1500.80		$2186.00	$62.53	$91.08
Crew M-1	Hr.	Daily	Hr.	Daily	Bare Costs	Incl. O&P
3 Elevator Constructors	$41.60	$998.40	$62.20	$1492.80	$39.52	$59.10
1 Elevator Apprentice	33.30	266.40	49.80	398.40		
5 Hand Tools		70.00		77.00	2.19	2.41
32 L.H., Daily Totals		$1334.80		$1968.20	$41.71	$61.51
Crew M-3	Hr.	Daily	Hr.	Daily	Bare Costs	Incl. O&P
1 Electrician Foreman (out)	$41.40	$331.20	$61.55	$492.40	$35.46	$53.35
1 Common Laborer	26.00	208.00	40.50	324.00		
.25 Equipment Operator, Medium	33.65	67.30	50.70	101.40		
1 Elevator Constructor	41.60	332.80	62.20	497.60		
1 Elevator Apprentice	33.30	266.40	49.80	398.40		
.25 Crane, SP, 4 x 4, 20 ton		143.60		157.95	4.22	4.65
34 L.H., Daily Totals		$1349.30		$1971.75	$39.68	$58.00
Crew M-4	Hr.	Daily	Hr.	Daily	Bare Costs	Incl. O&P
1 Electrician Foreman (out)	$41.40	$331.20	$61.55	$492.40	$35.18	$52.92
1 Common Laborer	26.00	208.00	40.50	324.00		
.25 Equipment Operator, Crane	34.80	69.60	52.40	104.80		
.25 Equipment Operator, Oiler	29.20	58.40	44.00	88.00		
1 Elevator Constructor	41.60	332.80	62.20	497.60		
1 Elevator Apprentice	33.30	266.40	49.80	398.40		
.25 Crane, Hyd, SP, 4WD, 40 Ton		215.30		236.85	5.98	6.58
36 L.H., Daily Totals		$1481.70		$2142.05	$41.16	$59.50
Crew Q-1	Hr.	Daily	Hr.	Daily	Bare Costs	Incl. O&P
1 Plumber	$39.60	$316.80	$59.45	$475.60	$35.65	$53.53
1 Plumber Apprentice	31.70	253.60	47.60	380.80		
16 L.H., Daily Totals		$570.40		$856.40	$35.65	$53.53
Crew Q-1C	Hr.	Daily	Hr.	Daily	Bare Costs	Incl. O&P
1 Plumber	$39.60	$316.80	$59.45	$475.60	$34.98	$52.58
1 Plumber Apprentice	31.70	253.60	47.60	380.80		
1 Equip. Oper. (medium)	33.65	269.20	50.70	405.60		
1 Trencher, Chain		1310.00		1441.00	54.58	60.04
24 L.H., Daily Totals		$2149.60		$2703.00	$89.56	$112.62

Crew No.	Bare Costs Hr.	Daily	Incl. Subs O & P Hr.	Daily	Cost Per Labor-Hour Bare Costs	Incl. O&P
Crew Q-2	Hr.	Daily	Hr.	Daily	Bare Costs	Incl. O&P
2 Plumbers	$39.60	$633.60	$59.45	$951.20	$36.97	$55.50
1 Plumber Apprentice	31.70	253.60	47.60	380.80		
24 L.H., Daily Totals		$887.20		$1332.00	$36.97	$55.50
Crew Q-3	Hr.	Daily	Hr.	Daily	Bare Costs	Incl. O&P
1 Plumber Foreman (inside)	$40.10	$320.80	$60.20	$481.60	$37.75	$56.67
2 Plumbers	39.60	633.60	59.45	951.20		
1 Plumber Apprentice	31.70	253.60	47.60	380.80		
32 L.H., Daily Totals		$1208.00		$1813.60	$37.75	$56.67
Crew Q-4	Hr.	Daily	Hr.	Daily	Bare Costs	Incl. O&P
1 Plumber Foreman (inside)	$40.10	$320.80	$60.20	$481.60	$37.75	$56.67
1 Plumber	39.60	316.80	59.45	475.60		
1 Welder (plumber)	39.60	316.80	59.45	475.60		
1 Plumber Apprentice	31.70	253.60	47.60	380.80		
1 Electric Welding Mach.		77.75		85.55	2.43	2.67
32 L.H., Daily Totals		$1285.75		$1899.15	$40.18	$59.34
Crew Q-5	Hr.	Daily	Hr.	Daily	Bare Costs	Incl. O&P
1 Steamfitter	$39.75	$318.00	$59.65	$477.20	$35.78	$53.70
1 Steamfitter Apprentice	31.80	254.40	47.75	382.00		
16 L.H., Daily Totals		$572.40		$859.20	$35.78	$53.70
Crew Q-6	Hr.	Daily	Hr.	Daily	Bare Costs	Incl. O&P
2 Steamfitters	$39.75	$636.00	$59.65	$954.40	$37.10	$55.68
1 Steamfitter Apprentice	31.80	254.40	47.75	382.00		
24 L.H., Daily Totals		$890.40		$1336.40	$37.10	$55.68
Crew Q-7	Hr.	Daily	Hr.	Daily	Bare Costs	Incl. O&P
1 Steamfitter Foreman (inside)	$40.25	$322.00	$60.40	$483.20	$37.89	$56.86
2 Steamfitters	39.75	636.00	59.65	954.40		
1 Steamfitter Apprentice	31.80	254.40	47.75	382.00		
32 L.H., Daily Totals		$1212.40		$1819.60	$37.89	$56.86
Crew Q-8	Hr.	Daily	Hr.	Daily	Bare Costs	Incl. O&P
1 Steamfitter Foreman (inside)	$40.25	$322.00	$60.40	$483.20	$37.89	$56.86
1 Steamfitter	39.75	318.00	59.65	477.20		
1 Welder (steamfitter)	39.75	318.00	59.65	477.20		
1 Steamfitter Apprentice	31.80	254.40	47.75	382.00		
1 Electric Welding Mach.		77.75		85.55	2.43	2.67
32 L.H., Daily Totals		$1290.15		$1905.15	$40.32	$59.53
Crew Q-9	Hr.	Daily	Hr.	Daily	Bare Costs	Incl. O&P
1 Sheet Metal Worker	$38.80	$310.40	$59.50	$476.00	$34.92	$53.58
1 Sheet Metal Apprentice	31.05	248.40	47.65	381.20		
16 L.H., Daily Totals		$558.80		$857.20	$34.92	$53.58
Crew Q-10	Hr.	Daily	Hr.	Daily	Bare Costs	Incl. O&P
2 Sheet Metal Workers	$38.80	$620.80	$59.50	$952.00	$36.22	$55.55
1 Sheet Metal Apprentice	31.05	248.40	47.65	381.20		
24 L.H., Daily Totals		$869.20		$1333.20	$36.22	$55.55
Crew Q-11	Hr.	Daily	Hr.	Daily	Bare Costs	Incl. O&P
1 Sheet Metal Foreman (inside)	$39.30	$314.40	$60.30	$482.40	$36.99	$56.74
2 Sheet Metal Workers	38.80	620.80	59.50	952.00		
1 Sheet Metal Apprentice	31.05	248.40	47.65	381.20		
32 L.H., Daily Totals		$1183.60		$1815.60	$36.99	$56.74

Crew No.	Bare Costs		Incl. Subs O & P		Cost Per Labor-Hour	
Crew Q-12	Hr.	Daily	Hr.	Daily	Bare Costs	Incl. O&P
1 Sprinkler Installer	$39.25	$314.00	$59.40	$475.20	$35.33	$53.45
1 Sprinkler Apprentice	31.40	251.20	47.50	380.00		
16 L.H., Daily Totals		$565.20		$855.20	$35.33	$53.45
Crew Q-13	Hr.	Daily	Hr.	Daily	Bare Costs	Incl. O&P
1 Sprinkler Foreman (inside)	$39.75	$318.00	$60.15	$481.20	$37.41	$56.61
2 Sprinkler Installers	39.25	628.00	59.40	950.40		
1 Sprinkler Apprentice	31.40	251.20	47.50	380.00		
32 L.H., Daily Totals		$1197.20		$1811.60	$37.41	$56.61
Crew Q-14	Hr.	Daily	Hr.	Daily	Bare Costs	Incl. O&P
1 Asbestos Worker	$36.00	$288.00	$56.70	$453.60	$32.40	$51.03
1 Asbestos Apprentice	28.80	230.40	45.35	362.80		
16 L.H., Daily Totals		$518.40		$816.40	$32.40	$51.03
Crew Q-15	Hr.	Daily	Hr.	Daily	Bare Costs	Incl. O&P
1 Plumber	$39.60	$316.80	$59.45	$475.60	$35.65	$53.53
1 Plumber Apprentice	31.70	253.60	47.60	380.80		
1 Electric Welding Mach.		77.75		85.55	4.86	5.35
16 L.H., Daily Totals		$648.15		$941.95	$40.51	$58.88
Crew Q-16	Hr.	Daily	Hr.	Daily	Bare Costs	Incl. O&P
2 Plumbers	$39.60	$633.60	$59.45	$951.20	$36.97	$55.50
1 Plumber Apprentice	31.70	253.60	47.60	380.80		
1 Electric Welding Mach.		77.75		85.55	3.24	3.56
24 L.H., Daily Totals		$964.95		$1417.55	$40.21	$59.06
Crew Q-17	Hr.	Daily	Hr.	Daily	Bare Costs	Incl. O&P
1 Steamfitter	$39.75	$318.00	$59.65	$477.20	$35.78	$53.70
1 Steamfitter Apprentice	31.80	254.40	47.75	382.00		
1 Electric Welding Mach.		77.75		85.55	4.86	5.35
16 L.H., Daily Totals		$650.15		$944.75	$40.64	$59.05
Crew Q-17A	Hr.	Daily	Hr.	Daily	Bare Costs	Incl. O&P
1 Steamfitter	$39.75	$318.00	$59.65	$477.20	$35.45	$53.27
1 Steamfitter Apprentice	31.80	254.40	47.75	382.00		
1 Equip. Oper. (crane)	34.80	278.40	52.40	419.20		
1 Truck Crane, 12 Ton		612.00		673.20		
1 Electric Welding Mach.		77.75		85.55	28.74	31.61
24 L.H., Daily Totals		$1540.55		$2037.15	$64.19	$84.88
Crew Q-18	Hr.	Daily	Hr.	Daily	Bare Costs	Incl. O&P
2 Steamfitters	$39.75	$636.00	$59.65	$954.40	$37.10	$55.68
1 Steamfitter Apprentice	31.80	254.40	47.75	382.00		
1 Electric Welding Mach.		77.75		85.55	3.24	3.56
24 L.H., Daily Totals		$968.15		$1421.95	$40.34	$59.24
Crew Q-19	Hr.	Daily	Hr.	Daily	Bare Costs	Incl. O&P
1 Steamfitter	$39.75	$318.00	$59.65	$477.20	$36.98	$55.33
1 Steamfitter Apprentice	31.80	254.40	47.75	382.00		
1 Electrician	39.40	315.20	58.60	468.80		
24 L.H., Daily Totals		$887.60		$1328.00	$36.98	$55.33
Crew Q-20	Hr.	Daily	Hr.	Daily	Bare Costs	Incl. O&P
1 Sheet Metal Worker	$38.80	$310.40	$59.50	$476.00	$35.82	$54.58
1 Sheet Metal Apprentice	31.05	248.40	47.65	381.20		
.5 Electrician	39.40	157.60	58.60	234.40		
20 L.H., Daily Totals		$716.40		$1091.60	$35.82	$54.58

Crew No.	Bare Costs		Incl. Subs O & P		Cost Per Labor-Hour	
Crew Q-21	Hr.	Daily	Hr.	Daily	Bare Costs	Incl. O&P
2 Steamfitters	$39.75	$636.00	$59.65	$954.40	$37.67	$56.41
1 Steamfitter Apprentice	31.80	254.40	47.75	382.00		
1 Electrician	39.40	315.20	58.60	468.80		
32 L.H., Daily Totals		$1205.60		$1805.20	$37.67	$56.41
Crew Q-22	Hr.	Daily	Hr.	Daily	Bare Costs	Incl. O&P
1 Plumber	$39.60	$316.80	$59.45	$475.60	$35.65	$53.53
1 Plumber Apprentice	31.70	253.60	47.60	380.80		
1 Truck Crane, 12 Ton		612.00		673.20	38.25	42.08
16 L.H., Daily Totals		$1182.40		$1529.60	$73.90	$95.61
Crew Q-22A	Hr.	Daily	Hr.	Daily	Bare Costs	Incl. O&P
1 Plumber	$39.60	$316.80	$59.45	$475.60	$33.03	$49.99
1 Plumber Apprentice	31.70	253.60	47.60	380.80		
1 Laborer	26.00	208.00	40.50	324.00		
1 Equip. Oper. (crane)	34.80	278.40	52.40	419.20		
1 Truck Crane, 12 Ton		612.00		673.20	19.13	21.04
32 L.H., Daily Totals		$1668.80		$2272.80	$52.16	$71.03
Crew Q-23	Hr.	Daily	Hr.	Daily	Bare Costs	Incl. O&P
1 Plumber Foreman	$41.60	$332.80	$62.45	$499.60	$38.28	$57.53
1 Plumber	39.60	316.80	59.45	475.60		
1 Equip. Oper. (medium)	33.65	269.20	50.70	405.60		
1 Power Tool		4.40		4.85		
1 Crane, 20 Ton		647.45		712.20	27.16	29.88
24 L.H., Daily Totals		$1570.65		$2097.85	$65.44	$87.41
Crew R-1	Hr.	Daily	Hr.	Daily	Bare Costs	Incl. O&P
1 Electrician Foreman	$39.90	$319.20	$59.35	$474.80	$34.53	$51.89
3 Electricians	39.40	945.60	58.60	1406.40		
2 Helpers	24.55	392.80	38.10	609.60		
48 L.H., Daily Totals		$1657.60		$2490.80	$34.53	$51.89
Crew R-1A	Hr.	Daily	Hr.	Daily	Bare Costs	Incl. O&P
1 Electrician	$39.40	$315.20	$58.60	$468.80	$31.98	$48.35
1 Helper	24.55	196.40	38.10	304.80		
16 L.H., Daily Totals		$511.60		$773.60	$31.98	$48.35
Crew R-2	Hr.	Daily	Hr.	Daily	Bare Costs	Incl. O&P
1 Electrician Foreman	$39.90	$319.20	$59.35	$474.80	$34.57	$51.96
3 Electricians	39.40	945.60	58.60	1406.40		
2 Helpers	24.55	392.80	38.10	609.60		
1 Equip. Oper. (crane)	34.80	278.40	52.40	419.20		
1 S.P. Crane, 5 Ton		317.80		349.60	5.68	6.24
56 L.H., Daily Totals		$2253.80		$3259.60	$40.25	$58.20
Crew R-3	Hr.	Daily	Hr.	Daily	Bare Costs	Incl. O&P
1 Electrician Foreman	$39.90	$319.20	$59.35	$474.80	$38.68	$57.66
1 Electrician	39.40	315.20	58.60	468.80		
.5 Equip. Oper. (crane)	34.80	139.20	52.40	209.60		
.5 S.P. Crane, 5 Ton		158.90		174.80	7.95	8.74
20 L.H., Daily Totals		$932.50		$1328.00	$46.63	$66.40
Crew R-4	Hr.	Daily	Hr.	Daily	Bare Costs	Incl. O&P
1 Struc. Steel Foreman	$39.15	$313.20	$70.15	$561.20	$38.00	$65.68
3 Struc. Steel Workers	37.15	891.60	66.55	1597.20		
1 Electrician	39.40	315.20	58.60	468.80		
1 Gas Welding Machine		75.20		82.70	1.88	2.07
40 L.H., Daily Totals		$1595.20		$2709.90	$39.88	$67.75

Crew No.	Bare Costs		Incl. Subs O & P		Cost Per Labor-Hour	

Crew R-5	Hr.	Daily	Hr.	Daily	Bare Costs	Incl. O&P
1 Electrician Foreman	$39.90	$319.20	$59.35	$474.80	$34.05	$51.21
4 Electrician Linemen	39.40	1260.80	58.60	1875.20		
2 Electrician Operators	39.40	630.40	58.60	937.60		
4 Electrician Groundmen	24.55	785.60	38.10	1219.20		
1 Crew Truck		109.10		120.00		
1 Tool Van		129.35		142.30		
1 Pickup Truck, 3/4 Ton		75.80		83.40		
.2 Crane, 55 Ton		182.20		200.40		
.2 Crane, 12 Ton		122.40		134.65		
.2 Auger, Truck Mtd.		575.40		632.95		
1 Tractor w/Winch		254.40		279.85	16.46	18.11
88 L.H., Daily Totals		$4444.65		$6100.35	$50.51	$69.32

Crew R-6	Hr.	Daily	Hr.	Daily	Bare Costs	Incl. O&P
1 Electrician Foreman	$39.90	$319.20	$59.35	$474.80	$34.05	$51.21
4 Electrician Linemen	39.40	1260.80	58.60	1875.20		
2 Electrician Operators	39.40	630.40	58.60	937.60		
4 Electrician Groundmen	24.55	785.60	38.10	1219.20		
1 Crew Truck		109.10		120.00		
1 Tool Van		129.35		142.30		
1 Pickup Truck, 3/4 Ton		75.80		83.40		
.2 Crane, 55 Ton		182.20		200.40		
.2 Crane, 12 Ton		122.40		134.65		
.2 Auger, Truck Mtd.		575.40		632.95		
1 Tractor w/Winch		254.40		279.85		
3 Cable Trailers		480.75		528.85		
.5 Tensioning Rig		157.45		173.20		
.5 Cable Pulling Rig		916.00		1007.60	34.12	37.54
88 L.H., Daily Totals		$5998.85		$7810.00	$68.17	$88.75

Crew R-7	Hr.	Daily	Hr.	Daily	Bare Costs	Incl. O&P
1 Electrician Foreman	$39.90	$319.20	$59.35	$474.80	$27.11	$41.64
5 Electrician Groundmen	24.55	982.00	38.10	1524.00		
1 Crew Truck		109.10		120.00	2.27	2.50
48 L.H., Daily Totals		$1410.30		$2118.80	$29.38	$44.14

Crew R-8	Hr.	Daily	Hr.	Daily	Bare Costs	Incl. O&P
1 Electrician Foreman	$39.90	$319.20	$59.35	$474.80	$34.53	$51.89
3 Electrician Linemen	39.40	945.60	58.60	1406.40		
2 Electrician Groundmen	24.55	392.80	38.10	609.60		
1 Pickup Truck, 3/4 Ton		75.80		83.40		
1 Crew Truck		109.10		120.00	3.85	4.24
48 L.H., Daily Totals		$1842.50		$2694.20	$38.38	$56.13

Crew R-9	Hr.	Daily	Hr.	Daily	Bare Costs	Incl. O&P
1 Electrician Foreman	$39.90	$319.20	$59.35	$474.80	$32.04	$48.44
1 Electrician Lineman	39.40	315.20	58.60	468.80		
2 Electrician Operators	39.40	630.40	58.60	937.60		
4 Electrician Groundmen	24.55	785.60	38.10	1219.20		
1 Pickup Truck, 3/4 Ton		75.80		83.40		
1 Crew Truck		109.10		120.00	2.89	3.18
64 L.H., Daily Totals		$2235.30		$3303.80	$34.93	$51.62

Crew R-10	Hr.	Daily	Hr.	Daily	Bare Costs	Incl. O&P
1 Electrician Foreman	$39.90	$319.20	$59.35	$474.80	$37.01	$55.31
4 Electrician Linemen	39.40	1260.80	58.60	1875.20		
1 Electrician Groundman	24.55	196.40	38.10	304.80		
1 Crew Truck		109.10		120.00		
3 Tram Cars		340.35		374.40	9.36	10.30
48 L.H., Daily Totals		$2225.85		$3149.20	$46.37	$65.61

Crew R-11	Hr.	Daily	Hr.	Daily	Bare Costs	Incl. O&P
1 Electrician Foreman	$39.90	$319.20	$59.35	$474.80	$35.36	$53.09
4 Electricians	39.40	1260.80	58.60	1875.20		
1 Equip. Oper. (crane)	34.80	278.40	52.40	419.20		
1 Helper	24.55	196.40	38.10	304.80		
1 Common Laborer	26.00	208.00	40.50	324.00		
1 Crew Truck		109.10		120.00		
1 Crane, 12 Ton		612.00		673.20	11.27	12.39
64 L.H., Daily Totals		$2983.90		$4191.20	$46.63	$65.48

Crew R-12	Hr.	Daily	Hr.	Daily	Bare Costs	Incl. O&P
1 Carpenter Foreman	$33.50	$268.00	$52.20	$417.60	$30.94	$48.82
4 Carpenters	33.00	1056.00	51.40	1644.80		
4 Common Laborers	26.00	832.00	40.50	1296.00		
1 Equip. Oper. (med.)	33.65	269.20	50.70	405.60		
1 Steel Worker	37.15	297.20	66.55	532.40		
1 Dozer, 200 H.P.		863.40		949.75		
1 Pickup Truck, 3/4 Ton		75.80		83.40	10.67	11.74
88 L.H., Daily Totals		$3661.60		$5329.55	$41.61	$60.56

Crew R-13	Hr.	Daily	Hr.	Daily	Bare Costs	Incl. O&P
1 Electrician Foreman	$39.90	$319.20	$59.35	$474.80	$37.33	$55.67
3 Electricians	39.40	945.60	58.60	1406.40		
.25 Equip. Oper. (crane)	34.80	69.60	52.40	104.80		
1 Equipment Oiler	29.20	233.60	44.00	352.00		
.25-1 Hyd. Crane, 33 Ton		162.25		178.50	3.86	4.25
42 L.H., Daily Totals		$1730.25		$2516.50	$41.19	$59.92

Crew R-15	Hr.	Daily	Hr.	Daily	Bare Costs	Incl. O&P
1 Electrician Foreman	$39.90	$319.20	$59.35	$474.80	$38.28	$57.02
4 Electricians	39.40	1260.80	58.60	1875.20		
1 Equipment Operator	32.15	257.20	48.40	387.20		
1 Aerial Lift Truck		250.20		275.20	5.21	5.73
48 L.H., Daily Totals		$2087.40		$3012.40	$43.49	$62.75

Crew R-18	Hr.	Daily	Hr.	Daily	Bare Costs	Incl. O&P
.25 Electrician Foreman	$39.90	$79.80	$59.35	$118.70	$30.30	$46.04
1 Electrician	39.40	315.20	58.60	468.80		
2 Helpers	24.55	392.80	38.10	609.60		
26 L.H., Daily Totals		$787.80		$1197.10	$30.30	$46.04

Crew R-19	Hr.	Daily	Hr.	Daily	Bare Costs	Incl. O&P
.5 Electrician Foreman	$39.90	$159.60	$59.35	$237.40	$39.50	$58.75
2 Electricians	39.40	630.40	58.60	937.60		
20 L.H., Daily Totals		$790.00		$1175.00	$39.50	$58.75

Crew R-21	Hr.	Daily	Hr.	Daily	Bare Costs	Incl. O&P
1 Electrician Foreman	$39.90	$319.20	$59.35	$474.80	$39.38	$58.59
3 Electricians	39.40	945.60	58.60	1406.40		
.1 Equip. Oper. (med.)	33.65	26.92	50.70	40.56		
.1 Hyd. Crane 25 Ton		57.52		63.25	1.75	1.93
32. L.H., Daily Totals		$1349.24		$1985.01	$41.13	$60.52

Crew R-22	Hr.	Daily	Hr.	Daily	Bare Costs	Incl. O&P
.66 Electrician Foreman	$39.90	$210.67	$59.35	$313.37	$33.10	$49.91
2 Helpers	24.55	392.80	38.10	609.60		
2 Electricians	39.40	630.40	58.60	937.60		
37.28 L.H., Daily Totals		$1233.87		$1860.57	$33.10	$49.91

Crew No.	Bare Costs		Incl. Sub O & P		Cost Per Labor-Hour	
Crew R-30	Hr.	Daily	Hr.	Daily	Bare Costs	Incl. O&P
.25 Electrician	$41.40	$82.80	$61.55	$123.10	$31.31	$47.69
1 Electrician	39.40	315.20	58.60	468.80		
2 Laborers, (Semi-Skilled)	26.00	416.00	40.50	648.00		
26 L.H., Daily Totals		$814.00		$1239.90	$31.31	$47.69
Crew R-31	Hr.	Daily	Hr.	Daily	Bare Costs	Incl. O&P
1 Electrician	$39.40	$315.20	$58.60	$468.80	$39.40	$58.60
1 Core Drill, Elec, 2.5 HP		47.45		52.20	5.93	6.52
8 L.H., Daily Totals		$362.65		$521.00	$45.33	$65.12
Crew W-41E	Hr.	Daily	Hr.	Daily	Bare Costs	Incl. O&P
1 Laborers, (Semi-Skilled)	$26.00	$208.00	$40.50	$324.00	$34.56	$52.47
1 Plumber	39.60	316.80	59.45	475.60		
.5 Plumber	41.60	166.40	62.45	249.80		
20 L.H., Daily Totals		$691.20		$1049.40	$34.56	$52.47

Historical Cost Indexes

The table below lists both the Means Historical Cost Index based on Jan. 1, 1993 = 100 as well as the computed value of an index based on Jan. 1, 2004 costs. Since the Jan. 1, 2004 figure is estimated, space is left to write in the actual index figures as they become available through either the quarterly "Means Construction Cost Indexes" or as printed in the "Engineering News-Record." To compute the actual index based on Jan. 1, 2004 = 100, divide the Historical Cost Index for a particular year by the actual Jan. 1, 2004 Construction Cost Index. Space has been left to advance the index figures as the year progresses.

Year	Historical Cost Index Jan. 1, 1993 = 100		Current Index Based on Jan. 1, 2004 = 100		Year	Historical Cost Index Jan. 1, 1993 = 100	Current Index Based on Jan. 1, 2004 = 100		Year	Historical Cost Index Jan. 1, 1993 = 100	Current Index Based on Jan. 1, 2004 = 100	
	Est.	Actual	Est.	Actual		Actual	Est.	Actual		Actual	Est.	Actual
Oct 2004					July 1989	92.1	69.3		July 1971	32.1	24.1	
July 2004					1988	89.9	67.6		1970	28.7	21.6	
April 2004					1987	87.7	65.9		1969	26.9	20.2	
Jan 2004	133.0		100.0	100.0	1986	84.2	63.3		1968	24.9	18.7	
July 2003		132.0	99.2		1985	82.6	62.1		1967	23.5	17.7	
2002		128.7	96.8		1984	82.0	61.6		1966	22.7	17.1	
2001		125.1	94.1		1983	80.2	60.3		1965	21.7	16.3	
2000		120.9	90.9		1982	76.1	57.3		1964	21.2	15.9	
1999		117.6	88.4		1981	70.0	52.6		1963	20.7	15.6	
1998		115.1	86.5		1980	62.9	47.3		1962	20.2	15.2	
1997		112.8	84.8		1979	57.8	43.5		1961	19.8	14.9	
1996		110.2	82.9		1978	53.5	40.2		1960	19.7	14.8	
1995		107.6	80.9		1977	49.5	37.2		1959	19.3	14.5	
1994		104.4	78.5		1976	46.9	35.3		1958	18.8	14.1	
1993		101.7	76.5		1975	44.8	33.7		1957	18.4	13.8	
1992		99.4	74.8		1974	41.4	31.1		1956	17.6	13.2	
1991		96.8	72.8		1973	37.7	28.3		1955	16.6	12.5	
1990		94.3	70.9		1972	34.8	26.2		1954	16.0	12.0	

Adjustments to Costs

The Historical Cost Index can be used to convert National Average building costs at a particular time to the approximate building costs for some other time.

Example:

Estimate and compare construction costs for different years in the same city.

To estimate the National Average construction cost of a building in 1970, knowing that it cost $900,000 in 2004:

INDEX in 1970 = 28.7

INDEX in 2004 = 133.0

Note: The City Cost Indexes for Canada can be used to convert U.S. National averages to local costs in Canadian dollars.

Time Adjustment using the Historical Cost Indexes:

$$\frac{\text{Index for Year A}}{\text{Index for Year B}} \times \text{Cost in Year B} = \text{Cost in Year A}$$

$$\frac{\text{INDEX 1970}}{\text{INDEX 2004}} \times \text{Cost 2004} = \text{Cost 1970}$$

$$\frac{28.7}{133.0} \times \$900,000 = .216 \times \$900,000 = \$194,400$$

The construction cost of the building in 1970 is $194,400.

How to Use the City Cost Indexes

What you should know before you begin

Means City Cost Indexes (CCI) are an extremely useful tool to use when you want to compare costs from city to city and region to region.

This publication contains average construction cost indexes for 316 major U.S. and Canadian cities and Location Factors covering over 930 three-digit zip code locations.

Keep in mind that a City Cost Index number is a *percentage ratio* of a specific city's cost to the national average cost of the same item at a stated time period.

In other words, these index figures represent relative construction *factors* (or, if you prefer, multipliers) for Material and Installation costs, as well as the weighted average for Total In Place costs for each CSI MasterFormat division. Installation costs include both labor and equipment rental costs. When estimating equipment rental rates only, for a specific location, use 01590 EQUIPMENT RENTAL index.

The 30 City Average Index is the average of 30 major U.S. cities and serves as a National Average.

Index figures for both material and installation are based on the 30 major city average of 100 and represent the cost relationship as of July 1, 2003. The index for each division is computed from representative material and labor quantities for that division. The weighted average for each city is a weighted total of the components listed above it, but does not include relative productivity between trades or cities.

As changes occur in local material prices, labor rates and equipment rental rates, the impact of these changes should be accurately measured by the change in the City Cost Index for each particular city (as compared to the 30 City Average).

> *Therefore, if you know (or have estimated) building costs in one city today, you can easily convert those costs to expected building costs in another city.*

> *In addition, by using the Historical Cost Index, you can easily convert National Average building costs at a particular time to the approximate building costs for some other time. The City Cost Indexes can then be applied to calculate the costs for a particular city.*

Quick Calculations

Location Adjustment Using the City Cost Indexes:

$$\frac{\text{Index for City A}}{\text{Index for City B}} \times \text{Cost in City B} = \text{Cost in City A}$$

Time Adjustment for the National Average Using the Historical Cost Index:

$$\frac{\text{Index for Year A}}{\text{Index for Year B}} \times \text{Cost in Year B} = \text{Cost in Year A}$$

Adjustment from the National Average:

$$\frac{\text{Index for City A}}{100} \times \text{National Average Cost} = \text{Cost in City A}$$

Since each of the other RSMeans publications contains many different items, any *one* item multiplied by the particular city index may give incorrect results. However, the larger the number of items compiled, the closer the results should be to actual costs for that particular city.

The City Cost Indexes for Canadian cities are calculated using Canadian material and equipment prices and labor rates, in Canadian dollars. Therefore, indexes for Canadian cities can be used to convert U.S. National Average prices to local costs in Canadian dollars.

How to use this section

1. Compare costs from city to city.

In using the Means Indexes, remember that an index number is not a fixed number but a *ratio:* It's a percentage ratio of a building component's cost at any stated time to the National Average cost of that same component at the same time period. Put in the form of an equation:

$$\frac{\text{Specific City Cost}}{\text{National Average Cost}} \times 100 = \text{City Index Number}$$

Therefore, when making cost comparisons between cities, do not subtract one city's index number from the index number of another city and read the result as a percentage difference. Instead, divide one city's index number by that of the other city. The resulting number may then be used as a multiplier to calculate cost differences from city to city.

The formula used to find cost differences between cities for the purpose of comparison is as follows:

$$\frac{\text{City A Index}}{\text{City B Index}} \times \text{City B Cost (Known)} = \text{City A Cost (Unknown)}$$

In addition, you can use *Means CCI* to calculate and compare costs division by division between cities using the same basic formula. (Just be sure that you're comparing similar divisions.)

2. Compare a specific city's construction costs with the National Average.

When you're studying construction location feasibility, it's advisable to compare a prospective project's cost index with an index of the National Average cost.

For example, divide the weighted average index of construction costs of a specific city by that of the 30 City Average, which = 100.

$$\frac{\text{City Index}}{100} = \% \text{ of National Average}$$

As a result, you get a ratio that indicates the relative cost of construction in that city in comparison with the National Average.

3. Convert U.S. National Average to actual costs in Canadian City.

$$\frac{\text{Index for Canadian City}}{100} \times \text{National Average Cost} = \text{Cost in Canadian City in \$ CAN}$$

4. Adjust construction cost data based on a National Average.

When you use a source of construction cost data which is based on a National Average (such as *Means cost data publications*), it is necessary to adjust those costs to a specific location.

$$\frac{\text{City Index}}{100} \times \begin{array}{c}\text{"Book" Cost Based on}\\\text{National Average Costs}\end{array} = \begin{array}{c}\text{City Cost}\\\text{(Unknown)}\end{array}$$

5. When applying the City Cost Indexes to demolition projects, use the appropriate division index. For example, for removal of existing doors and windows, use the Division 8 index.

What you might like to know about how we developed the Indexes

To create a reliable index, RSMeans researched the building type most often constructed in the United States and Canada. Because it was concluded that no one type of building completely represented the building construction industry, nine different types of buildings were combined to create a composite model.

The exact material, labor and equipment quantities are based on detailed analysis of these nine building types, then each quantity is weighted in proportion to expected usage. These various material items, labor hours, and equipment rental rates are thus combined to form a composite building representing as closely as possible the actual usage of materials, labor and equipment used in the North American Building Construction Industry.

The following structures were chosen to make up that composite model:

1. Factory, 1 story
2. Office, 2–4 story
3. Store, Retail
4. Town Hall, 2–3 story
5. High School, 2–3 story
6. Hospital, 4–8 story
7. Garage, Parking
8. Apartment, 1–3 story
9. Hotel/Motel, 2–3 story

For the purposes of ensuring the timeliness of the data, the components of the index for the composite model have been streamlined. They currently consist of:

• specific quantities of 66 commonly used construction materials;
• specific labor-hours for 21 building construction trades; and
• specific days of equipment rental for 6 types of construction equipment (normally used to install the 66 material items by the 21 trades.)

A sophisticated computer program handles the updating of all costs for each city on a quarterly basis. Material and equipment price quotations are gathered quarterly from over 316 cities in the United States and Canada. These prices and the latest negotiated labor wage rates for 21 different building trades are used to compile the quarterly update of the City Cost Index.

The 30 major U.S. cities used to calculate the National Average are:

Atlanta, GA	Memphis, TN
Baltimore, MD	Milwaukee, WI
Boston, MA	Minneapolis, MN
Buffalo, NY	Nashville, TN
Chicago, IL	New Orleans, LA
Cincinnati, OH	New York, NY
Cleveland, OH	Philadelphia, PA
Columbus, OH	Phoenix, AZ
Dallas, TX	Pittsburgh, PA
Denver, CO	St. Louis, MO
Detroit, MI	San Antonio, TX
Houston, TX	San Diego, CA
Indianapolis, IN	San Francisco, CA
Kansas City, MO	Seattle, WA
Los Angeles, CA	Washington, DC

F.Y.I.: The CSI MasterFormat Divisions

1. General Requirements
2. Site Construction
3. Concrete
4. Masonry
5. Metals
6. Wood & Plastics
7. Thermal & Moisture Protection
8. Doors & Windows
9. Finishes
10. Specialties
11. Equipment
12. Furnishings
13. Special Construction
14. Conveying Systems
15. Mechanical
16. Electrical

The information presented in the CCI is organized according to the Construction Specifications Institute (CSI) MasterFormat.

What the CCI does not indicate

The weighted average for each city is a total of the components listed above weighted to reflect typical usage, but it does *not* include the productivity variations between trades or cities.

In addition, the CCI does not take into consideration factors such as the following:

• managerial efficiency
• competitive conditions
• automation
• restrictive union practices
• unique local requirements
• regional variations due to specific building codes

City Cost Indexes

DIVISION		UNITED STATES 30 CITY AVERAGE			BIRMINGHAM			HUNTSVILLE			MOBILE			MONTGOMERY			TUSCALOOSA		
		MAT.	INST.	TOTAL	MAT.	INST.	TOTAL	MAT.	INST.	TOTAL	MAT.	INST.	TOTAL	MAT.	INST.	TOTAL	MAT.	INST.	TOTAL
01590	EQUIPMENT RENTAL	.0	100.0	100.0	.0	101.4	101.4	.0	101.3	101.3	.0	97.8	97.8	.0	97.8	97.8	.0	101.3	101.3
02	SITE CONSTRUCTION	100.0	100.0	100.0	85.9	93.3	91.5	83.9	92.8	90.6	95.3	86.3	88.5	95.7	86.8	89.0	84.4	92.1	90.2
03100	CONCRETE FORMS & ACCESSORIES	100.0	100.0	100.0	91.9	76.1	78.1	93.7	70.5	73.3	93.7	54.8	59.6	92.2	49.6	54.9	93.6	40.9	47.4
03200	CONCRETE REINFORCEMENT	100.0	100.0	100.0	92.4	87.1	89.2	92.4	79.2	84.4	95.4	55.2	71.1	95.4	85.6	89.5	92.4	85.9	88.5
03300	CAST-IN-PLACE CONCRETE	100.0	100.0	100.0	93.5	69.8	83.6	88.3	67.4	79.6	93.3	56.3	77.8	95.0	50.7	76.5	92.0	48.4	73.8
03	CONCRETE	100.0	100.0	100.0	91.0	77.1	83.9	88.6	72.3	80.2	91.4	57.0	73.8	92.1	58.6	74.9	90.4	54.0	71.7
04	MASONRY	100.0	100.0	100.0	85.7	78.1	80.9	85.6	66.6	73.7	86.2	54.0	66.0	86.8	37.5	55.9	85.9	39.8	57.1
05	METALS	100.0	100.0	100.0	97.7	95.8	97.0	97.3	91.4	95.1	95.9	80.9	90.3	95.4	92.5	94.3	96.4	93.1	95.1
06	WOOD & PLASTICS	100.0	100.0	100.0	93.1	76.5	84.2	92.4	71.2	81.0	92.4	54.9	72.2	90.7	50.3	68.9	92.4	39.9	64.1
07	THERMAL & MOISTURE PROTECTION	100.0	100.0	100.0	95.7	81.5	88.8	95.4	75.7	85.8	95.4	70.2	83.1	95.1	63.4	79.7	95.3	62.2	79.2
08	DOORS & WINDOWS	100.0	100.0	100.0	98.5	79.1	93.6	98.5	67.8	90.7	98.5	55.1	87.5	98.5	59.4	88.6	98.5	59.3	88.5
09200	PLASTER & GYPSUM BOARD	100.0	100.0	100.0	102.1	76.4	84.6	98.7	71.0	79.8	98.7	54.2	68.4	98.7	49.4	65.2	98.7	38.7	57.9
095,098	CEILINGS & ACOUSTICAL TREATMENT	100.0	100.0	100.0	95.0	76.4	82.4	95.0	71.0	78.7	95.0	54.2	67.4	95.0	49.4	64.2	95.0	38.7	56.9
09600	FLOORING	100.0	100.0	100.0	104.3	52.5	91.0	104.3	45.4	89.1	114.7	56.9	99.8	112.9	29.4	91.4	104.3	43.2	88.6
097,099	WALL FINISHES, PAINTS & COATINGS	100.0	100.0	100.0	95.3	72.0	81.4	95.3	66.2	77.9	100.2	58.7	75.3	95.3	57.7	72.8	95.3	49.7	68.0
09	FINISHES	100.0	100.0	100.0	99.0	71.1	83.9	98.4	65.1	80.4	103.5	55.2	77.4	102.6	45.6	71.8	98.4	40.7	67.2
10 - 14	TOTAL DIV. 10000 - 14000	100.0	100.0	100.0	100.0	80.8	95.9	100.0	78.9	95.5	100.0	73.3	94.3	100.0	71.2	93.8	100.0	68.5	93.2
15	MECHANICAL	100.0	100.0	100.0	99.8	67.8	85.1	99.8	64.9	83.8	99.8	63.2	83.0	99.8	43.1	73.7	99.8	37.9	71.3
16	ELECTRICAL	100.0	100.0	100.0	97.6	65.6	78.9	97.8	69.2	81.1	97.8	53.3	71.8	97.6	66.8	79.6	97.8	65.6	79.0
01 - 16	WEIGHTED AVERAGE	100.0	100.0	100.0	96.6	76.5	86.9	96.2	72.6	84.8	97.2	62.8	80.6	97.1	59.0	78.7	96.3	57.1	77.4

DIVISION		ALASKA ANCHORAGE			FAIRBANKS			JUNEAU			ARIZONA FLAGSTAFF			MESA/TEMPE			PHOENIX		
		MAT.	INST.	TOTAL	MAT.	INST.	TOTAL	MAT.	INST.	TOTAL	MAT.	INST.	TOTAL	MAT.	INST.	TOTAL	MAT.	INST.	TOTAL
01590	EQUIPMENT RENTAL	.0	118.4	118.4	.0	118.4	118.4	.0	118.4	118.4	.0	94.7	94.7	.0	97.5	97.5	.0	98.1	98.1
02	SITE CONSTRUCTION	143.9	134.0	136.5	127.3	134.0	132.3	139.5	134.0	135.4	82.2	101.2	96.4	85.5	104.0	99.3	85.9	104.9	100.1
03100	CONCRETE FORMS & ACCESSORIES	133.3	118.0	119.9	135.2	124.0	125.4	134.9	118.0	120.1	103.1	63.4	68.3	100.7	65.3	69.6	101.9	70.6	74.4
03200	CONCRETE REINFORCEMENT	137.9	108.1	119.9	115.9	108.1	111.2	102.6	108.1	105.9	101.6	75.2	85.7	100.0	73.4	83.9	101.5	79.9	92.5
03300	CAST-IN-PLACE CONCRETE	193.1	117.7	161.6	160.9	118.2	143.1	193.3	117.7	162.0	94.7	79.8	88.5	101.4	70.6	88.5	101.5	79.9	92.5
03	CONCRETE	151.3	115.4	132.9	127.9	118.3	123.0	147.7	115.4	131.2	119.0	71.2	94.5	100.2	68.5	84.0	99.8	74.6	86.9
04	MASONRY	220.7	124.7	160.6	214.1	124.7	158.2	223.4	124.7	161.6	102.5	57.0	74.0	107.2	54.4	74.1	95.1	68.1	78.2
05	METALS	129.4	103.2	119.5	129.5	103.5	119.7	129.7	103.2	119.7	97.8	69.1	86.9	98.6	69.1	87.5	100.0	71.8	89.4
06	WOOD & PLASTICS	115.0	116.0	115.5	115.2	123.8	119.8	115.0	116.0	115.5	108.6	62.5	83.8	103.7	71.4	86.3	104.8	71.5	86.9
07	THERMAL & MOISTURE PROTECTION	199.5	119.0	160.3	195.5	121.7	159.5	196.2	119.0	158.6	107.4	67.8	88.1	105.8	65.5	86.2	105.7	71.0	88.8
08	DOORS & WINDOWS	127.0	111.8	123.1	124.1	116.0	122.0	124.1	111.8	121.0	102.2	66.4	93.1	99.1	67.1	91.0	100.2	71.3	92.9
09200	PLASTER & GYPSUM BOARD	136.6	116.3	122.8	136.6	124.3	128.2	136.6	116.3	122.8	95.2	61.5	72.2	99.6	70.5	79.8	100.1	70.7	80.1
095,098	CEILINGS & ACOUSTICAL TREATMENT	126.4	116.3	119.5	126.4	124.3	124.9	126.4	116.3	119.5	107.2	61.5	76.2	108.0	70.5	82.6	108.0	70.7	82.7
09600	FLOORING	164.6	132.1	156.2	164.6	132.1	156.2	164.6	132.1	156.2	95.7	53.9	84.9	97.8	67.2	89.9	98.1	70.2	90.9
097,099	WALL FINISHES, PAINTS & COATINGS	167.8	110.4	133.4	167.8	124.7	141.9	167.8	110.4	133.4	94.8	50.4	68.2	105.2	54.8	75.0	105.2	62.3	79.5
09	FINISHES	157.0	119.9	137.0	154.5	126.1	139.1	155.4	119.9	136.2	98.1	59.5	77.2	100.4	64.3	80.9	100.6	69.2	83.7
10 - 14	TOTAL DIV. 10000 - 14000	100.0	119.7	104.2	100.0	120.8	104.5	100.0	119.7	104.2	100.0	76.5	95.0	100.0	73.0	94.2	100.0	78.1	95.3
15	MECHANICAL	100.4	108.8	104.3	100.4	114.7	107.0	100.4	105.9	102.9	100.0	77.2	89.6	100.2	69.1	85.9	100.2	77.3	89.6
16	ELECTRICAL	153.3	116.7	131.9	155.8	116.7	133.0	155.8	116.7	133.0	100.6	49.9	70.9	96.4	62.6	76.7	104.2	65.0	81.3
01 - 16	WEIGHTED AVERAGE	135.1	116.5	126.1	131.0	119.2	125.3	134.4	115.9	125.5	102.0	68.9	86.1	99.7	69.2	85.0	100.0	74.8	87.9

DIVISION		ARIZONA PRESCOTT			TUCSON			ARKANSAS FORT SMITH			JONESBORO			LITTLE ROCK			PINE BLUFF		
		MAT.	INST.	TOTAL	MAT.	INST.	TOTAL	MAT.	INST.	TOTAL	MAT.	INST.	TOTAL	MAT.	INST.	TOTAL	MAT.	INST.	TOTAL
01590	EQUIPMENT RENTAL	.0	94.7	94.7	.0	97.5	97.5	.0	85.9	85.9	.0	107.7	107.7	.0	85.9	85.9	.0	85.9	85.9
02	SITE CONSTRUCTION	70.0	100.6	92.9	82.3	104.6	99.0	77.4	84.0	82.3	101.0	99.5	99.9	77.2	84.0	82.3	79.5	84.0	82.9
03100	CONCRETE FORMS & ACCESSORIES	98.4	58.0	63.0	101.4	70.1	73.9	99.8	43.5	50.4	85.7	49.6	54.0	93.6	61.4	65.3	77.8	61.2	63.2
03200	CONCRETE REINFORCEMENT	101.6	73.0	84.3	97.1	75.2	83.9	97.6	76.4	84.8	93.3	50.8	67.6	97.8	73.9	83.4	97.7	73.9	83.3
03300	CAST-IN-PLACE CONCRETE	94.6	65.8	82.6	104.3	79.8	94.1	89.9	68.7	81.0	85.6	58.7	74.3	89.9	68.8	81.1	82.4	68.7	76.7
03	CONCRETE	103.7	63.6	83.2	101.1	74.2	87.3	87.0	59.2	72.7	83.7	54.7	68.8	86.6	66.7	76.4	84.6	66.5	75.3
04	MASONRY	103.0	60.6	76.5	97.0	56.9	71.9	96.8	55.0	70.6	91.9	48.3	64.6	95.1	55.0	70.0	117.3	55.0	78.3
05	METALS	97.8	67.8	86.5	98.6	69.8	87.7	97.2	72.6	87.9	91.4	77.3	86.1	96.8	72.1	87.5	95.9	71.9	86.8
06	WOOD & PLASTICS	104.2	56.7	78.6	104.0	71.5	86.5	104.0	42.2	70.7	90.0	50.8	68.2	100.9	65.8	82.0	82.7	65.8	73.6
07	THERMAL & MOISTURE PROTECTION	105.6	63.6	85.2	107.0	64.6	86.4	99.3	50.0	75.3	109.1	53.4	82.0	98.0	52.5	75.8	98.0	52.5	75.8
08	DOORS & WINDOWS	102.2	60.3	91.6	96.1	71.3	89.8	97.1	47.4	84.5	98.7	49.6	86.2	97.1	60.4	87.8	92.5	60.4	84.3
09200	PLASTER & GYPSUM BOARD	93.1	55.5	67.6	100.4	70.7	80.2	86.1	41.3	55.6	93.5	49.7	63.7	86.1	65.6	72.1	80.1	65.6	70.2
095,098	CEILINGS & ACOUSTICAL TREATMENT	107.2	55.5	72.2	109.4	70.7	83.1	98.1	41.3	59.6	98.7	49.7	65.5	98.1	65.6	76.1	94.2	65.6	74.8
09600	FLOORING	94.0	53.7	83.6	97.2	53.9	86.1	116.5	72.9	105.2	77.8	47.4	70.0	117.7	72.9	106.2	105.4	72.9	97.0
097,099	WALL FINISHES, PAINTS & COATINGS	94.8	50.4	68.2	102.8	50.4	71.3	96.6	60.2	74.7	85.0	54.1	66.5	96.6	61.7	75.7	96.6	61.7	75.7
09	FINISHES	96.1	55.9	74.3	100.3	64.8	81.1	99.0	50.2	72.6	89.7	49.7	68.1	99.3	64.2	80.4	94.0	64.2	77.9
10 - 14	TOTAL DIV. 10000 - 14000	100.0	75.4	94.7	100.0	78.1	95.3	100.0	67.3	93.0	100.0	58.2	91.0	100.0	70.4	93.6	100.0	70.4	93.6
15	MECHANICAL	100.2	75.0	88.6	100.1	71.1	86.8	100.1	44.7	74.6	100.3	44.6	74.6	100.1	58.6	81.0	100.1	47.2	75.8
16	ELECTRICAL	100.2	48.0	69.7	99.2	63.3	78.2	95.9	67.0	79.0	101.8	49.6	71.3	96.5	72.8	82.7	95.0	72.8	82.0
01 - 16	WEIGHTED AVERAGE	99.6	66.4	83.6	99.0	71.2	85.6	96.4	58.6	78.2	95.6	56.7	76.9	96.3	65.9	81.6	95.8	63.6	80.3

DIVISION		ARKANSAS			CALIFORNIA														
		TEXARKANA			ANAHEIM			BAKERSFIELD			FRESNO			LOS ANGELES			OAKLAND		
		MAT.	INST.	TOTAL	MAT.	INST.	TOTAL	MAT.	INST.	TOTAL	MAT.	INST.	TOTAL	MAT.	INST.	TOTAL	MAT.	INST.	TOTAL
01590	EQUIPMENT RENTAL	.0	86.6	86.6	.0	102.1	102.1	.0	99.5	99.5	.0	99.5	99.5	.0	97.6	97.6	.0	103.5	103.5
02	SITE CONSTRUCTION	96.1	84.8	87.7	102.6	109.4	107.7	107.6	106.5	106.7	109.1	106.1	106.9	96.4	108.0	105.1	154.4	104.7	117.2
03100	CONCRETE FORMS & ACCESSORIES	86.1	41.1	46.6	104.3	124.1	121.7	97.5	124.0	120.7	100.2	131.5	127.7	106.2	123.8	121.7	107.4	136.1	132.5
03200	CONCRETE REINFORCEMENT	97.2	49.2	68.2	109.0	113.7	111.8	104.0	113.7	109.8	107.6	113.7	111.3	106.0	113.9	110.8	97.6	114.6	107.9
03300	CAST-IN-PLACE CONCRETE	89.8	45.8	71.4	105.8	120.4	111.9	101.0	119.4	108.7	110.6	115.1	112.4	85.9	117.8	99.2	132.7	120.3	127.5
03	CONCRETE	83.6	45.2	64.0	109.4	119.9	114.8	106.0	119.4	112.9	111.4	121.3	116.5	101.2	118.8	110.2	123.8	125.4	124.6
04	MASONRY	96.8	33.9	57.4	89.9	119.9	108.7	108.8	119.7	115.6	112.5	109.7	110.8	99.7	123.9	114.9	149.0	124.9	133.9
05	METALS	88.3	60.5	77.7	110.0	100.5	106.4	104.4	100.0	102.7	106.3	100.3	104.0	111.3	98.8	106.6	101.5	105.9	103.2
06	WOOD & PLASTICS	91.7	44.0	66.0	91.8	121.4	107.8	84.1	121.5	104.3	97.2	135.1	117.6	89.8	121.0	106.6	100.9	138.0	120.9
07	THERMAL & MOISTURE PROTECTION	98.8	43.7	72.0	121.4	117.9	119.7	105.6	114.2	109.8	101.3	111.2	106.2	111.6	120.0	115.7	113.2	127.3	120.1
08	DOORS & WINDOWS	97.5	42.5	83.6	102.8	116.4	106.2	101.6	112.9	104.5	101.9	121.7	106.9	96.5	116.2	101.5	106.3	126.5	111.5
09200	PLASTER & GYPSUM BOARD	83.9	43.2	56.2	99.0	122.0	114.6	99.0	122.0	114.6	97.9	135.9	123.8	90.6	122.0	112.0	98.7	138.4	125.7
095,098	CEILINGS & ACOUSTICAL TREATMENT	99.4	43.2	61.3	117.3	122.0	120.4	117.3	122.0	120.4	117.3	135.9	129.9	111.4	122.0	118.6	113.8	138.4	130.5
09600	FLOORING	108.2	45.2	92.0	122.6	104.9	118.1	116.8	95.4	111.3	133.2	140.0	135.0	114.5	104.9	112.0	115.1	121.5	116.8
097,099	WALL FINISHES, PAINTS & COATINGS	96.6	37.9	61.4	111.1	108.3	109.4	110.6	95.4	101.5	134.3	98.2	112.6	99.8	109.8	105.8	115.8	140.6	130.7
09	FINISHES	96.9	41.6	67.0	113.9	118.9	116.6	114.2	116.0	115.2	121.5	130.9	126.6	107.6	118.7	113.6	116.7	134.8	126.5
10 - 14	TOTAL DIV. 10000 - 14000	100.0	41.6	87.4	100.0	116.2	103.5	100.0	128.0	106.0	100.0	128.4	106.1	100.0	115.0	103.2	100.0	131.5	106.8
15	MECHANICAL	100.1	39.4	72.2	100.1	112.1	105.7	100.1	105.7	102.7	100.1	111.2	105.2	100.0	112.0	105.6	100.2	128.0	113.0
16	ELECTRICAL	97.6	41.3	64.7	95.3	107.1	102.2	95.2	97.2	96.4	94.6	95.6	95.2	107.7	112.3	110.4	109.7	139.6	127.2
01 - 16	WEIGHTED AVERAGE	95.2	46.7	71.8	103.8	113.3	108.4	103.2	109.9	106.5	105.1	112.4	108.6	102.9	114.0	108.3	111.1	125.7	118.2

DIVISION		CALIFORNIA																	
		OXNARD			REDDING			RIVERSIDE			SACRAMENTO			SAN DIEGO			SAN FRANCISCO		
		MAT.	INST.	TOTAL	MAT.	INST.	TOTAL	MAT.	INST.	TOTAL	MAT.	INST.	TOTAL	MAT.	INST.	TOTAL	MAT.	INST.	TOTAL
01590	EQUIPMENT RENTAL	.0	98.1	98.1	.0	99.1	99.1	.0	100.6	100.6	.0	103.1	103.1	.0	97.4	97.4	.0	108.9	108.9
02	SITE CONSTRUCTION	109.1	104.2	105.4	114.2	105.4	107.6	100.2	107.1	105.4	117.4	111.1	112.7	102.9	100.8	101.3	156.7	111.3	122.7
03100	CONCRETE FORMS & ACCESSORIES	103.0	124.1	121.5	102.4	131.6	128.0	104.7	124.0	121.6	105.7	132.1	128.9	106.3	106.0	106.1	107.7	137.3	133.7
03200	CONCRETE REINFORCEMENT	104.0	113.4	109.7	100.8	113.8	108.7	103.1	113.5	109.4	98.7	113.9	107.9	106.2	113.4	110.6	111.2	115.0	113.5
03300	CAST-IN-PLACE CONCRETE	107.2	119.5	112.4	121.4	115.1	118.8	104.7	120.4	111.3	114.4	116.2	115.1	112.9	107.5	110.6	132.6	122.0	128.1
03	CONCRETE	109.5	119.4	114.6	119.9	121.3	120.6	108.3	119.8	114.2	114.6	121.9	118.3	114.8	107.3	111.0	125.3	126.6	126.0
04	MASONRY	113.7	114.2	114.0	116.4	111.6	113.4	86.4	117.4	105.8	119.4	111.6	114.5	100.5	106.2	104.1	149.2	128.1	136.0
05	METALS	103.9	99.6	102.3	108.3	100.6	105.4	109.9	100.2	106.2	97.6	100.0	98.5	109.8	99.4	105.9	107.0	107.6	107.2
06	WOOD & PLASTICS	91.6	121.5	107.7	92.2	135.1	115.3	91.8	121.4	107.8	95.1	135.4	116.8	97.8	102.1	100.1	100.9	138.2	121.0
07	THERMAL & MOISTURE PROTECTION	110.6	113.7	112.1	111.4	112.0	111.7	119.2	114.5	116.9	124.3	113.7	119.1	113.9	103.9	109.0	113.2	131.8	122.3
08	DOORS & WINDOWS	100.4	116.5	104.5	103.2	123.0	108.3	102.8	116.4	106.2	117.8	123.9	119.4	104.3	104.9	104.5	110.8	130.3	115.8
09200	PLASTER & GYPSUM BOARD	99.0	122.0	114.6	98.1	135.9	123.8	98.4	122.0	114.4	96.8	135.9	123.4	100.9	101.9	101.6	100.5	138.4	126.3
095,098	CEILINGS & ACOUSTICAL TREATMENT	117.3	122.0	120.4	122.5	135.9	131.6	114.7	122.0	119.6	120.3	135.9	130.9	112.6	101.9	105.3	118.7	138.4	132.1
09600	FLOORING	116.8	104.9	113.8	116.1	111.0	114.8	121.3	104.9	117.1	117.9	114.8	117.1	110.8	107.6	110.0	115.1	127.7	118.4
097,099	WALL FINISHES, PAINTS & COATINGS	110.0	101.6	104.9	110.0	115.8	113.5	107.4	108.3	108.0	113.1	115.8	114.7	109.6	108.3	108.8	115.8	153.4	138.3
09	FINISHES	114.0	118.2	116.2	115.1	127.8	122.0	112.3	118.9	115.8	116.6	128.8	123.2	110.7	106.1	108.2	118.0	137.9	128.8
10 - 14	TOTAL DIV. 10000 - 14000	100.0	116.4	103.5	100.0	128.5	106.1	100.0	116.1	103.5	100.0	129.4	106.3	100.0	110.5	102.3	100.0	132.2	106.9
15	MECHANICAL	100.1	112.2	105.7	100.1	109.4	104.4	100.0	112.1	105.6	100.1	113.0	106.0	100.2	107.3	103.5	100.2	156.1	125.9
16	ELECTRICAL	99.1	99.8	99.5	103.4	106.7	105.3	95.4	103.5	100.1	101.7	106.7	104.6	98.9	98.4	98.6	106.3	154.6	134.5
01 - 16	WEIGHTED AVERAGE	104.3	110.9	107.5	107.3	113.4	110.3	103.2	112.2	107.6	107.5	115.0	111.1	105.1	104.3	104.7	112.5	135.5	123.6

DIVISION		CALIFORNIA												COLORADO					
		SAN JOSE			SANTA BARBARA			STOCKTON			VALLEJO			COLORADO SPRINGS			DENVER		
		MAT.	INST.	TOTAL	MAT.	INST.	TOTAL	MAT.	INST.	TOTAL	MAT.	INST.	TOTAL	MAT.	INST.	TOTAL	MAT.	INST.	TOTAL
01590	EQUIPMENT RENTAL	.0	99.9	99.9	.0	99.5	99.5	.0	99.1	99.1	.0	103.7	103.7	.0	95.5	95.5	.0	100.6	100.6
02	SITE CONSTRUCTION	148.7	99.7	112.0	109.1	106.5	107.1	107.0	105.4	105.8	114.0	111.2	111.9	95.9	97.3	97.0	94.4	106.6	103.5
03100	CONCRETE FORMS & ACCESSORIES	105.2	136.2	132.4	103.8	124.1	121.6	103.2	131.4	128.0	106.8	134.5	131.1	90.4	85.4	86.0	99.5	85.5	87.2
03200	CONCRETE REINFORCEMENT	101.1	114.6	109.3	104.0	113.5	109.8	104.5	113.8	110.1	97.4	114.7	107.8	98.7	86.5	91.3	98.1	86.7	91.2
03300	CAST-IN-PLACE CONCRETE	126.8	119.6	123.8	106.8	119.4	112.1	106.7	115.0	110.2	119.0	117.3	118.3	101.1	89.4	96.2	94.6	88.9	92.2
03	CONCRETE	119.1	125.3	122.3	109.3	119.4	114.5	109.3	121.2	115.4	116.9	123.5	120.2	108.4	87.1	97.5	101.8	87.0	94.2
04	MASONRY	149.9	124.4	134.0	109.2	117.2	114.2	113.3	111.5	112.2	84.7	125.3	110.1	108.8	84.5	93.6	108.5	87.1	95.1
05	METALS	109.8	107.2	108.8	103.7	99.9	102.3	105.9	100.4	103.8	100.8	101.9	101.3	100.8	89.2	96.4	104.3	89.4	98.7
06	WOOD & PLASTICS	99.5	137.7	120.1	91.6	121.5	107.7	93.5	135.1	115.9	92.9	137.7	117.1	93.1	87.2	89.9	100.7	87.1	93.3
07	THERMAL & MOISTURE PROTECTION	107.2	130.0	118.3	106.9	111.7	109.2	110.9	110.3	110.6	126.6	125.5	126.1	103.8	86.5	95.4	103.1	84.5	94.0
08	DOORS & WINDOWS	92.7	128.4	101.8	101.6	116.5	105.4	100.9	123.7	106.7	119.2	130.0	122.0	98.4	90.6	96.4	99.8	90.5	97.4
09200	PLASTER & GYPSUM BOARD	99.2	138.4	125.9	99.0	122.0	114.6	100.0	135.9	124.4	97.3	138.4	125.3	92.5	86.9	88.7	102.3	86.9	91.8
095,098	CEILINGS & ACOUSTICAL TREATMENT	108.1	138.4	128.6	117.3	122.0	120.4	117.3	135.9	129.9	120.3	138.4	132.6	102.8	86.9	92.1	101.5	86.9	91.6
09600	FLOORING	113.8	127.7	117.4	116.8	104.5	113.6	116.8	110.7	115.2	122.5	127.7	123.9	108.3	92.7	104.3	109.3	97.9	106.3
097,099	WALL FINISHES, PAINTS & COATINGS	111.6	138.7	127.9	110.0	101.6	104.9	110.0	104.5	106.7	112.4	140.3	129.1	106.9	55.3	76.0	106.9	76.6	88.7
09	FINISHES	113.1	135.7	125.3	114.2	118.1	116.3	114.1	126.3	120.7	115.5	135.4	126.2	100.1	83.4	91.1	101.3	86.8	93.5
10 - 14	TOTAL DIV. 10000 - 14000	100.0	130.6	106.6	100.0	116.5	103.5	100.0	128.5	106.1	100.0	129.3	106.3	100.0	91.4	98.1	100.0	90.9	98.0
15	MECHANICAL	100.1	143.6	120.1	100.1	112.2	105.7	100.1	104.8	102.3	100.1	119.1	108.8	100.2	83.1	92.3	100.1	84.9	93.1
16	ELECTRICAL	107.9	146.6	130.5	92.3	108.6	101.8	102.7	113.9	109.2	98.1	102.7	100.8	98.4	91.5	94.4	100.0	95.2	97.2
01 - 16	WEIGHTED AVERAGE	109.3	129.7	119.2	103.4	112.6	107.9	104.9	113.3	108.9	106.0	118.9	112.2	101.3	87.6	94.7	101.3	89.9	95.8

COST INDEXES

449

COLORADO / CONNECTICUT

DIVISION		FORT COLLINS MAT.	INST.	TOTAL	GRAND JUNCTION MAT.	INST.	TOTAL	GREELEY MAT.	INST.	TOTAL	PUEBLO MAT.	INST.	TOTAL	BRIDGEPORT MAT.	INST.	TOTAL	BRISTOL MAT.	INST.	TOTAL
01590	EQUIPMENT RENTAL	.0	97.0	97.0	.0	99.7	99.7	.0	97.0	97.0	.0	96.6	96.6	.0	101.8	101.8	.0	101.8	101.8
02	SITE CONSTRUCTION	105.8	99.9	101.4	124.1	102.3	107.8	92.5	98.6	97.0	115.9	96.6	101.5	101.9	104.8	104.0	101.0	104.7	103.8
03100	CONCRETE FORMS & ACCESSORIES	98.9	79.9	82.2	109.5	79.9	83.5	96.4	51.1	56.6	104.3	85.5	87.8	104.0	115.0	113.6	104.0	114.6	113.3
03200	CONCRETE REINFORCEMENT	102.9	84.7	91.9	106.6	84.6	93.3	102.4	80.0	88.9	99.5	86.5	91.7	109.1	125.7	119.1	109.1	125.6	119.1
03300	CAST-IN-PLACE CONCRETE	109.8	77.8	96.4	116.8	83.4	102.8	91.8	62.7	79.6	104.7	90.1	98.6	107.1	121.4	113.1	100.4	121.3	109.1
03	CONCRETE	115.7	80.2	97.5	116.7	82.1	99.0	99.4	61.4	79.9	105.6	87.4	96.3	110.2	119.1	114.8	106.8	118.9	113.0
04	MASONRY	123.4	62.8	85.5	148.1	63.9	95.4	116.9	41.7	69.8	106.8	84.5	92.8	106.7	121.0	115.7	98.3	121.0	112.5
05	METALS	100.1	84.6	94.2	101.1	83.8	94.6	100.1	81.0	92.8	102.7	90.1	97.9	98.4	122.6	107.6	98.4	122.3	107.4
06	WOOD & PLASTICS	100.5	82.6	90.8	108.7	82.7	94.7	98.0	50.3	72.3	104.5	87.6	95.4	101.0	112.4	107.1	101.0	112.4	107.1
07	THERMAL & MOISTURE PROTECTION	103.7	73.5	89.0	105.1	68.6	87.3	102.8	61.2	82.5	104.5	85.9	95.4	102.1	121.5	111.6	102.3	117.4	109.7
08	DOORS & WINDOWS	95.8	87.9	93.8	100.8	87.3	97.3	95.8	69.2	89.0	95.2	90.8	94.1	107.2	125.3	111.8	107.2	117.0	109.7
09200	PLASTER & GYPSUM BOARD	101.1	82.3	88.3	112.8	82.3	92.0	100.0	49.1	65.4	89.7	86.9	87.8	108.5	111.8	110.7	108.5	111.8	110.7
095,098	CEILINGS & ACOUSTICAL TREATMENT	96.3	82.3	86.8	102.0	82.3	88.7	96.3	49.1	64.3	112.4	86.9	95.1	105.9	111.8	109.9	105.9	111.8	109.9
09600	FLOORING	109.1	74.4	100.2	116.7	74.4	105.8	107.7	74.4	99.1	113.0	97.9	109.1	100.3	119.8	105.3	100.3	119.8	105.3
097,099	WALL FINISHES, PAINTS & COATINGS	106.9	54.8	75.7	117.3	73.0	90.7	106.9	34.2	63.3	117.3	51.2	77.7	93.0	113.0	105.0	93.0	113.0	105.0
09	FINISHES	100.7	76.2	87.5	108.9	78.7	92.6	99.4	52.1	73.9	106.4	84.3	94.4	104.3	114.9	110.1	104.4	114.9	110.1
10 - 14	TOTAL DIV. 10000 - 14000	100.0	89.1	97.7	100.0	89.8	97.8	100.0	80.0	95.8	100.0	92.2	98.3	100.0	117.0	103.7	100.0	117.0	103.7
15	MECHANICAL	100.1	83.0	92.2	100.1	73.1	87.7	100.1	74.9	88.5	100.1	72.4	87.3	100.2	107.7	103.6	100.2	107.6	103.6
16	ELECTRICAL	93.7	87.2	89.9	94.8	68.0	79.1	93.7	87.2	89.9	95.9	78.1	85.5	102.8	109.9	107.0	102.8	112.7	108.6
01 - 16	WEIGHTED AVERAGE	102.5	82.1	92.7	106.3	78.1	92.7	99.6	70.5	85.6	101.8	83.7	93.1	103.1	114.3	108.5	102.2	114.2	108.0

CONNECTICUT

DIVISION		HARTFORD MAT.	INST.	TOTAL	NEW BRITAIN MAT.	INST.	TOTAL	NEW HAVEN MAT.	INST.	TOTAL	NORWALK MAT.	INST.	TOTAL	STAMFORD MAT.	INST.	TOTAL	WATERBURY MAT.	INST.	TOTAL
01590	EQUIPMENT RENTAL	.0	101.8	101.8	.0	101.8	101.8	.0	102.4	102.4	.0	101.8	101.8	.0	101.8	101.8	.0	101.8	101.8
02	SITE CONSTRUCTION	101.4	104.7	103.9	101.2	104.7	103.9	101.1	105.6	104.5	101.7	104.8	104.0	102.3	104.8	104.2	101.4	104.8	103.9
03100	CONCRETE FORMS & ACCESSORIES	102.7	114.6	113.1	104.3	114.7	113.4	103.8	114.9	113.5	104.0	115.0	113.7	103.9	115.2	113.9	104.0	114.9	113.5
03200	CONCRETE REINFORCEMENT	109.1	125.6	119.1	109.1	125.6	119.1	109.1	125.6	119.1	109.1	125.8	119.2	109.1	125.8	119.2	109.1	125.6	119.1
03300	CAST-IN-PLACE CONCRETE	101.5	121.3	109.8	102.0	121.3	110.1	103.7	121.4	111.1	105.4	123.0	112.8	107.1	123.1	113.8	107.1	121.4	113.1
03	CONCRETE	107.3	118.8	113.2	107.7	118.9	113.4	124.2	119.0	121.6	109.4	119.7	114.6	110.2	119.8	115.1	110.2	119.0	114.7
04	MASONRY	98.4	121.0	112.5	98.4	121.0	112.5	98.5	121.0	112.6	98.6	122.8	113.7	98.8	122.8	113.8	98.8	121.0	112.7
05	METALS	99.0	122.2	107.8	95.1	122.3	105.4	95.3	122.5	105.6	98.4	122.8	107.7	98.4	123.1	107.7	98.4	122.5	107.5
06	WOOD & PLASTICS	101.0	112.4	107.1	101.0	112.4	107.1	101.0	112.4	107.1	101.0	112.4	107.1	101.0	112.4	107.1	101.0	112.4	107.1
07	THERMAL & MOISTURE PROTECTION	100.9	117.4	108.9	102.3	119.1	110.5	102.4	119.1	110.6	102.3	122.2	112.0	102.3	122.2	112.0	102.3	119.1	110.5
08	DOORS & WINDOWS	107.2	117.0	109.7	107.2	117.0	109.7	107.2	125.3	111.8	107.2	125.3	111.8	107.2	125.3	111.8	107.2	125.3	111.8
09200	PLASTER & GYPSUM BOARD	108.5	111.8	110.7	108.5	111.8	110.7	108.5	111.8	110.7	108.5	111.8	110.7	108.5	111.8	110.7	108.5	111.8	110.7
095,098	CEILINGS & ACOUSTICAL TREATMENT	105.9	111.8	109.9	105.9	111.8	109.9	105.9	111.8	109.9	105.9	111.8	109.9	105.9	111.8	109.9	105.9	111.8	109.9
09600	FLOORING	100.3	119.8	105.3	100.3	119.8	105.3	100.3	119.8	105.3	100.3	119.8	105.3	100.3	119.8	105.3	100.3	119.8	105.3
097,099	WALL FINISHES, PAINTS & COATINGS	93.0	113.0	105.0	93.0	113.0	105.0	93.0	113.0	105.0	93.0	113.0	105.0	93.0	113.0	105.0	93.0	113.0	105.0
09	FINISHES	104.3	114.9	110.1	104.4	114.9	110.1	104.4	114.9	110.1	104.4	114.9	110.1	104.4	114.9	110.1	104.2	114.9	110.0
10 - 14	TOTAL DIV. 10000 - 14000	100.0	117.0	103.7	100.0	117.0	103.7	100.0	117.0	103.7	100.0	117.0	103.7	100.0	117.2	103.7	100.0	117.0	103.7
15	MECHANICAL	100.2	107.6	103.6	100.2	107.6	103.6	100.2	107.6	103.6	100.2	107.7	103.6	100.2	107.7	103.6	100.2	107.6	103.6
16	ELECTRICAL	102.2	107.1	105.1	102.9	112.7	108.6	102.7	112.7	108.5	102.8	112.7	108.6	102.8	143.1	126.3	102.1	109.9	106.7
01 - 16	WEIGHTED AVERAGE	102.2	113.4	107.6	101.8	114.2	107.8	103.8	114.7	109.1	102.5	115.0	108.5	102.6	119.3	110.7	102.5	114.2	108.2

D.C. / DELAWARE / FLORIDA

DIVISION		WASHINGTON MAT.	INST.	TOTAL	WILMINGTON MAT.	INST.	TOTAL	DAYTONA BEACH MAT.	INST.	TOTAL	FORT LAUDERDALE MAT.	INST.	TOTAL	JACKSONVILLE MAT.	INST.	TOTAL	MELBOURNE MAT.	INST.	TOTAL
01590	EQUIPMENT RENTAL	.0	102.4	102.4	.0	118.2	118.2	.0	97.8	97.8	.0	89.4	89.4	.0	97.8	97.8	.0	97.8	97.8
02	SITE CONSTRUCTION	101.5	89.4	92.4	85.7	111.5	105.0	118.1	86.8	94.7	103.4	74.0	81.4	118.2	87.4	95.2	126.2	87.3	97.1
03100	CONCRETE FORMS & ACCESSORIES	100.1	80.2	82.6	98.3	102.5	102.0	94.2	69.7	72.7	92.2	70.8	73.4	93.9	55.3	60.0	88.7	73.0	75.0
03200	CONCRETE REINFORCEMENT	99.9	88.5	93.0	96.9	98.8	98.1	95.4	83.3	88.1	95.4	74.6	82.8	95.4	52.4	69.4	96.4	83.4	88.6
03300	CAST-IN-PLACE CONCRETE	112.7	84.7	101.0	79.0	97.0	86.5	90.5	70.2	82.0	95.0	71.4	85.1	91.4	60.5	78.5	105.2	74.6	92.4
03	CONCRETE	110.2	84.7	97.1	97.5	100.9	99.2	90.2	73.7	81.7	92.3	73.0	82.4	90.6	58.2	74.0	100.0	76.8	88.1
04	MASONRY	91.4	81.5	85.2	108.7	86.5	94.8	86.5	64.1	72.5	86.8	69.7	76.1	86.1	53.0	65.4	84.0	71.6	76.2
05	METALS	97.0	107.8	101.1	98.2	115.8	104.9	98.2	95.5	97.2	97.6	91.4	95.2	97.2	81.5	91.3	106.1	96.1	102.3
06	WOOD & PLASTICS	97.4	80.6	88.3	96.5	104.0	100.5	92.9	72.1	81.7	89.0	68.6	78.0	92.9	54.2	72.1	87.7	72.1	79.3
07	THERMAL & MOISTURE PROTECTION	97.0	82.3	89.8	101.2	105.6	103.4	95.4	70.2	83.1	95.4	70.1	83.1	95.7	59.4	78.0	95.6	76.0	86.1
08	DOORS & WINDOWS	102.0	90.0	99.0	94.2	107.5	97.6	100.9	69.5	92.9	98.5	65.6	90.1	100.9	51.4	88.3	100.1	74.1	93.5
09200	PLASTER & GYPSUM BOARD	106.1	80.0	88.4	109.0	103.9	105.5	98.7	71.9	80.4	98.2	68.2	77.8	98.7	53.5	67.9	95.2	71.9	79.3
095,098	CEILINGS & ACOUSTICAL TREATMENT	101.9	80.0	87.1	101.9	103.9	103.2	95.0	71.9	79.4	95.0	68.2	76.9	95.0	53.5	66.9	91.1	71.9	78.1
09600	FLOORING	112.0	97.1	108.1	83.5	93.1	86.0	118.6	72.1	106.6	118.6	63.4	104.4	118.6	52.1	101.4	115.2	72.1	104.1
097,099	WALL FINISHES, PAINTS & COATINGS	113.2	89.7	99.1	90.3	100.0	96.1	111.3	74.4	89.1	107.5	51.5	73.9	111.3	49.6	74.3	111.3	95.6	101.9
09	FINISHES	102.1	84.1	92.4	100.9	100.2	100.5	107.5	70.6	87.6	105.3	66.9	84.6	107.5	53.5	78.4	105.7	75.0	89.1
10 - 14	TOTAL DIV. 10000 - 14000	100.0	91.6	98.2	100.0	105.8	101.2	100.0	82.0	96.1	100.0	89.9	97.8	100.0	77.2	95.1	100.0	85.1	96.8
15	MECHANICAL	100.0	90.6	95.6	100.4	120.7	109.7	99.8	75.1	88.4	99.8	72.2	87.1	99.8	53.2	78.3	99.8	79.5	90.5
16	ELECTRICAL	95.4	95.7	95.6	98.6	104.7	102.1	97.8	66.3	79.4	97.8	72.8	83.2	97.4	66.9	79.6	97.6	74.1	83.9
01 - 16	WEIGHTED AVERAGE	100.2	89.9	95.2	98.9	106.8	102.7	98.7	74.7	87.2	97.9	73.5	86.2	98.6	62.4	81.1	100.8	79.0	90.3

DIVISION		MIAMI MAT.	INST.	TOTAL	ORLANDO MAT.	INST.	TOTAL	PANAMA CITY MAT.	INST.	TOTAL	PENSACOLA MAT.	INST.	TOTAL	ST. PETERSBURG MAT.	INST.	TOTAL	TALLAHASSEE MAT.	INST.	TOTAL
								FLORIDA											
01590	EQUIPMENT RENTAL	.0	89.4	89.4	.0	97.8	97.8	.0	97.8	97.8	.0	97.8	97.8	.0	97.8	97.8	.0	97.8	97.8
02	SITE CONSTRUCTION	102.6	73.8	81.1	118.7	86.5	94.6	133.4	84.1	96.5	130.9	86.5	97.7	119.1	86.0	94.3	119.6	86.0	94.4
03100	CONCRETE FORMS & ACCESSORIES	92.5	70.8	73.5	93.9	55.3	60.0	93.0	28.8	36.6	83.9	53.0	56.8	91.6	49.1	54.3	93.8	40.5	47.0
03200	CONCRETE REINFORCEMENT	95.4	74.6	82.8	95.4	80.6	86.5	99.6	51.3	70.4	102.1	51.6	71.6	98.8	59.0	74.8	95.4	51.9	69.1
03300	CAST-IN-PLACE CONCRETE	92.5	71.6	83.8	98.4	70.0	86.5	96.1	35.6	70.8	96.1	56.5	79.5	102.4	57.0	83.4	94.7	49.8	76.0
03	CONCRETE	91.0	73.1	81.8	91.8	66.8	79.0	99.2	37.2	67.5	98.1	55.7	76.4	96.4	55.5	75.4	92.2	47.9	69.6
04	MASONRY	84.6	70.1	75.5	90.7	64.1	74.1	90.9	28.4	51.8	88.0	51.4	65.1	134.4	50.8	82.1	89.0	40.0	58.3
05	METALS	101.4	90.9	97.5	107.3	93.9	102.2	96.4	66.9	85.3	96.4	80.9	90.5	100.2	82.2	93.4	98.1	79.8	91.2
06	WOOD & PLASTICS	89.0	68.6	78.0	92.9	53.5	71.6	91.8	28.9	57.9	83.1	53.5	67.2	90.5	49.1	68.2	91.1	38.9	63.0
07	THERMAL & MOISTURE PROTECTION	99.4	71.5	85.8	95.7	68.3	82.4	96.0	32.4	65.0	95.7	54.5	75.6	95.3	51.2	73.8	95.7	48.0	72.5
08	DOORS & WINDOWS	98.5	66.2	90.3	100.9	58.0	90.0	98.5	27.6	80.5	98.4	51.9	86.6	99.5	48.1	86.4	99.5	42.5	85.0
09200	PLASTER & GYPSUM BOARD	98.2	68.2	77.8	102.1	52.7	68.5	96.9	27.5	49.7	92.8	52.8	65.6	96.4	48.2	63.6	98.7	37.7	57.2
095,098	CEILINGS & ACOUSTICAL TREATMENT	95.0	68.2	76.9	95.0	52.7	66.4	89.8	27.5	47.6	89.8	52.8	64.7	89.8	48.2	61.6	95.0	37.7	56.2
09600	FLOORING	126.0	64.2	110.0	118.6	72.1	106.6	118.0	20.0	92.7	112.2	55.3	97.5	116.9	55.2	101.0	118.6	40.5	98.5
097,099	WALL FINISHES, PAINTS & COATINGS	107.5	51.5	73.9	111.3	57.0	78.7	111.3	25.4	59.8	111.3	57.6	79.1	111.3	48.0	73.3	111.3	40.8	69.1
09	FINISHES	107.8	67.1	85.8	108.0	57.7	80.8	107.3	26.3	63.6	104.7	53.8	77.2	105.7	49.9	75.5	107.7	39.7	70.9
10 - 14	TOTAL DIV. 10000 - 14000	100.0	89.9	97.8	100.0	79.5	95.6	100.0	48.9	89.0	100.0	54.7	90.3	100.0	57.3	90.8	100.0	64.0	92.2
15	MECHANICAL	99.8	76.0	88.9	99.8	56.4	79.8	99.8	26.1	65.9	99.8	51.8	77.7	99.8	51.7	77.6	99.8	41.3	72.9
16	ELECTRICAL	98.2	76.2	85.4	98.4	46.3	67.9	96.2	34.7	60.2	100.8	56.0	74.6	98.2	50.0	70.0	98.4	42.9	65.9
01 - 16	WEIGHTED AVERAGE	98.6	74.8	87.1	100.6	64.6	83.2	99.8	39.1	70.6	99.5	59.2	80.1	102.2	57.6	80.7	99.1	50.6	75.7

DIVISION		FLORIDA TAMPA MAT.	INST.	TOTAL	ALBANY MAT.	INST.	TOTAL	ATLANTA MAT.	INST.	TOTAL	AUGUSTA MAT.	INST.	TOTAL	COLUMBUS MAT.	INST.	TOTAL	MACON MAT.	INST.	TOTAL
									GEORGIA										
01590	EQUIPMENT RENTAL	.0	97.8	97.8	.0	90.2	90.2	.0	92.9	92.9	.0	92.2	92.2	.0	90.2	90.2	.0	102.6	102.6
02	SITE CONSTRUCTION	119.3	86.6	94.8	103.6	75.9	82.9	104.6	95.9	98.1	100.9	92.9	94.9	103.5	76.3	83.2	104.3	94.1	96.6
03100	CONCRETE FORMS & ACCESSORIES	95.3	77.7	79.9	93.6	44.4	50.5	91.4	81.4	82.6	88.6	53.9	58.1	94.8	56.9	61.5	92.5	57.5	61.8
03200	CONCRETE REINFORCEMENT	95.4	82.6	93.8	94.8	90.8	92.4	99.1	92.6	95.2	104.8	79.2	89.3	95.4	91.0	92.7	97.8	91.0	93.7
03300	CAST-IN-PLACE CONCRETE	100.1	63.5	84.8	96.7	42.9	74.2	99.9	76.6	90.2	94.4	50.4	76.0	96.3	43.8	74.4	95.0	47.1	75.0
03	CONCRETE	95.0	76.7	85.6	93.0	54.3	73.2	98.5	81.7	89.9	94.4	57.9	75.7	93.0	60.2	76.2	92.5	61.7	76.7
04	MASONRY	87.6	78.1	81.6	88.7	32.9	53.8	89.8	72.1	78.8	90.0	41.5	59.7	88.8	36.3	55.9	102.6	39.6	63.2
05	METALS	101.1	97.4	99.7	95.5	89.4	93.2	91.4	80.7	87.4	89.8	71.0	82.7	95.9	91.1	94.1	90.8	91.5	91.1
06	WOOD & PLASTICS	94.2	80.6	86.9	92.4	44.3	66.4	90.5	84.2	87.1	88.1	55.5	70.5	93.8	61.0	76.1	101.1	59.9	78.9
07	THERMAL & MOISTURE PROTECTION	95.6	64.3	80.4	95.5	52.2	74.4	91.7	77.5	84.8	91.1	51.4	71.8	95.1	54.8	75.5	93.9	58.6	76.7
08	DOORS & WINDOWS	100.9	73.7	94.0	98.5	51.2	86.5	100.0	79.6	94.8	94.1	56.6	84.6	98.5	60.5	88.8	96.9	60.8	87.7
09200	PLASTER & GYPSUM BOARD	98.7	80.6	86.4	98.4	43.3	60.9	117.1	84.2	94.7	116.1	54.7	74.3	98.4	60.5	72.6	107.6	59.3	74.7
095,098	CEILINGS & ACOUSTICAL TREATMENT	95.0	80.6	85.3	93.7	43.3	59.6	96.0	84.2	88.0	97.5	54.7	68.5	93.7	60.5	71.2	87.8	59.3	68.5
09600	FLOORING	118.6	55.2	102.2	118.6	33.8	96.7	84.4	81.1	83.5	83.2	43.1	72.9	118.6	39.4	98.2	92.5	39.7	78.9
097,099	WALL FINISHES, PAINTS & COATINGS	111.3	48.0	73.3	107.5	42.7	68.6	90.4	80.6	84.5	90.4	40.6	60.5	107.5	40.9	67.6	109.2	50.0	73.7
09	FINISHES	107.6	71.0	87.8	105.1	41.1	70.5	93.1	81.5	86.8	92.6	50.1	69.6	105.0	51.8	76.2	91.7	53.0	70.8
10 - 14	TOTAL DIV. 10000 - 14000	100.0	80.8	95.9	100.0	74.0	94.4	100.0	84.6	96.7	100.0	71.8	93.9	100.0	76.1	94.9	100.0	77.4	95.1
15	MECHANICAL	99.8	76.5	89.1	99.8	46.9	75.5	99.9	80.1	90.8	99.9	68.6	85.5	99.8	38.4	71.6	99.8	43.2	73.8
16	ELECTRICAL	97.2	50.0	69.6	92.3	62.8	75.1	96.6	85.3	90.0	98.4	51.3	70.8	93.3	40.7	62.6	91.9	62.0	74.4
01 - 16	WEIGHTED AVERAGE	99.8	74.8	87.8	97.4	55.6	77.2	96.8	82.0	89.7	95.4	61.1	78.9	97.5	54.2	76.6	95.9	60.7	78.9

DIVISION		GEORGIA SAVANNAH MAT.	INST.	TOTAL	VALDOSTA MAT.	INST.	TOTAL	HAWAII HONOLULU MAT.	INST.	TOTAL	IDAHO BOISE MAT.	INST.	TOTAL	LEWISTON MAT.	INST.	TOTAL	POCATELLO MAT.	INST.	TOTAL
01590	EQUIPMENT RENTAL	.0	91.3	91.3	.0	90.2	90.2	.0	99.3	99.3	.0	101.4	101.4	.0	94.3	94.3	.0	101.4	101.4
02	SITE CONSTRUCTION	104.1	77.3	84.0	114.4	76.5	86.1	138.9	106.5	114.7	78.8	103.0	96.9	83.9	96.9	93.7	80.0	103.0	97.2
03100	CONCRETE FORMS & ACCESSORIES	93.3	54.6	59.3	81.0	47.2	51.4	106.3	145.7	140.8	98.4	82.2	84.2	113.7	69.3	74.8	98.3	81.7	83.7
03200	CONCRETE REINFORCEMENT	101.1	79.4	88.0	101.2	45.0	67.3	104.0	120.7	114.1	98.2	77.9	85.9	111.5	96.4	102.4	98.6	77.4	85.8
03300	CAST-IN-PLACE CONCRETE	93.3	52.8	76.4	94.6	53.4	77.4	197.0	130.3	169.1	99.0	93.4	96.6	107.3	88.1	99.3	98.2	93.2	96.1
03	CONCRETE	92.0	60.1	75.7	97.8	50.7	73.7	154.7	134.2	144.2	104.7	85.1	94.6	113.1	81.2	96.8	100.9	84.7	92.6
04	MASONRY	91.9	54.9	68.7	94.9	49.9	66.7	134.4	131.7	132.7	136.9	76.9	99.3	138.0	87.8	106.5	132.6	68.5	92.5
05	METALS	96.2	85.9	92.3	95.7	74.0	87.5	115.7	108.4	112.9	112.8	78.2	99.7	96.5	88.2	93.3	113.1	77.0	99.4
06	WOOD & PLASTICS	106.8	52.3	77.4	80.1	43.2	60.2	95.0	149.6	124.4	96.8	80.8	88.2	106.6	62.9	83.0	96.8	80.8	88.2
07	THERMAL & MOISTURE PROTECTION	95.5	54.5	75.5	95.3	59.0	77.6	113.2	131.8	122.2	97.1	82.1	89.8	167.6	81.7	125.7	97.3	72.3	85.1
08	DOORS & WINDOWS	99.6	52.7	87.7	93.8	40.0	80.1	105.9	137.7	113.9	94.7	77.6	90.4	115.0	68.9	103.3	94.7	70.7	88.6
09200	PLASTER & GYPSUM BOARD	98.7	51.5	66.6	91.3	42.1	57.8	119.3	150.9	140.8	88.2	80.1	82.7	155.2	61.8	91.6	88.2	80.1	82.7
095,098	CEILINGS & ACOUSTICAL TREATMENT	95.0	51.5	65.5	89.8	42.1	57.5	117.3	150.9	140.0	112.4	80.1	90.5	159.6	61.8	93.4	112.4	80.1	90.5
09600	FLOORING	118.6	50.8	101.1	110.4	40.6	92.4	167.8	136.4	159.7	100.3	58.5	89.5	138.4	97.3	127.8	100.6	58.5	89.8
097,099	WALL FINISHES, PAINTS & COATINGS	108.2	50.7	73.7	107.5	37.0	65.2	104.5	149.9	131.8	103.9	53.8	73.9	129.9	76.6	97.9	103.9	56.0	75.2
09	FINISHES	105.6	53.2	77.3	101.4	44.2	70.5	135.5	145.9	141.1	99.7	74.8	86.2	168.8	74.0	117.6	99.8	75.0	86.4
10 - 14	TOTAL DIV. 10000 - 14000	100.0	74.8	94.6	100.0	73.5	94.3	100.0	122.9	104.9	100.0	91.1	98.1	100.0	93.1	98.5	100.0	91.0	98.1
15	MECHANICAL	99.8	49.2	76.5	99.8	43.3	73.8	100.1	120.6	109.5	100.1	79.5	90.6	101.4	86.8	94.7	100.1	79.5	90.6
16	ELECTRICAL	94.7	54.6	71.3	90.8	31.4	56.1	113.2	122.9	118.9	84.0	79.7	81.5	85.3	87.7	86.7	86.6	76.8	80.9
01 - 16	WEIGHTED AVERAGE	98.1	59.5	79.5	97.5	50.4	74.8	117.6	126.1	121.7	101.6	81.8	92.1	112.2	84.7	99.0	101.3	79.8	90.9

DIVISION		IDAHO TWIN FALLS			ILLINOIS CHICAGO			DECATUR			EAST ST. LOUIS			JOLIET			PEORIA		
		MAT.	INST.	TOTAL	MAT.	INST.	TOTAL	MAT.	INST.	TOTAL	MAT.	INST.	TOTAL	MAT.	INST.	TOTAL	MAT.	INST.	TOTAL
01590	EQUIPMENT RENTAL	.0	101.4	101.4	.0	91.5	91.5	.0	102.2	102.2	.0	108.8	108.8	.0	89.3	89.3	.0	101.4	101.4
02	SITE CONSTRUCTION	86.8	100.9	97.3	94.7	90.5	91.6	85.9	95.9	93.4	103.0	96.4	98.1	94.7	86.6	90.9	94.8	95.0	95.0
03100	CONCRETE FORMS & ACCESSORIES	101.2	36.5	44.5	103.4	138.7	134.3	98.1	106.9	105.8	91.9	106.4	104.6	104.3	136.6	132.6	96.2	110.6	108.9
03200	CONCRETE REINFORCEMENT	107.1	50.3	72.8	100.0	142.6	125.7	98.4	98.4	98.4	104.1	105.0	104.6	100.0	131.1	118.8	98.4	101.1	100.0
03300	CAST-IN-PLACE CONCRETE	100.8	46.2	78.0	111.1	134.4	120.9	98.7	106.1	101.8	92.4	111.4	100.3	111.0	124.3	116.6	96.4	107.7	101.1
03	CONCRETE	109.9	43.6	75.9	109.6	137.0	123.6	97.4	105.3	101.5	88.1	108.8	98.7	109.6	130.3	120.2	96.2	108.0	102.3
04	MASONRY	136.0	40.2	76.0	92.4	135.5	119.4	68.1	94.6	84.7	72.7	111.2	96.8	93.9	125.1	113.5	111.4	108.1	109.4
05	METALS	113.0	63.3	94.2	96.2	122.9	106.3	95.6	105.9	99.5	93.1	120.9	103.6	94.3	115.1	102.2	95.6	108.9	100.7
06	WOOD & PLASTICS	99.5	36.5	65.5	104.5	137.7	122.4	100.3	106.3	103.5	97.0	104.0	100.8	106.4	137.7	123.3	100.3	108.6	104.7
07	THERMAL & MOISTURE PROTECTION	98.4	48.1	73.9	100.4	131.2	115.4	97.3	96.1	96.7	92.3	104.7	98.3	100.2	125.5	112.5	97.2	106.8	101.9
08	DOORS & WINDOWS	98.0	37.3	82.6	103.7	138.7	112.6	96.2	104.0	98.2	86.0	111.5	92.5	101.6	135.6	110.2	96.2	105.9	98.7
09200	PLASTER & GYPSUM BOARD	87.7	34.5	51.5	101.4	138.6	126.7	105.0	106.3	105.9	101.8	104.0	103.3	98.8	138.7	125.9	105.0	108.6	107.5
095,098	CEILINGS & ACOUSTICAL TREATMENT	105.9	34.5	57.6	100.6	138.6	126.3	96.2	106.3	103.0	91.0	104.0	99.8	100.6	138.7	126.4	96.2	108.6	104.6
09600	FLOORING	101.9	58.5	90.7	83.6	132.8	96.2	96.8	88.0	94.5	105.5	108.7	106.4	83.1	120.9	92.9	96.8	103.8	98.6
097,099	WALL FINISHES, PAINTS & COATINGS	103.9	33.0	61.4	80.6	133.6	112.4	91.3	100.6	96.9	99.4	93.2	95.7	78.4	119.1	102.8	91.3	99.4	96.2
09	FINISHES	99.5	39.9	67.3	90.6	137.5	115.9	95.4	103.2	99.6	95.2	100.4	99.9	90.0	132.8	113.1	95.4	108.4	102.4
10 - 14	TOTAL DIV. 10000 - 14000	100.0	51.4	89.5	100.0	127.6	105.9	100.0	99.2	99.8	100.0	100.7	100.1	100.0	125.8	105.5	100.0	99.5	99.9
15	MECHANICAL	100.1	37.6	71.4	100.0	123.5	110.8	100.0	101.4	100.7	100.0	99.2	99.6	100.2	121.8	110.1	100.0	103.6	101.7
16	ELECTRICAL	88.7	32.7	56.0	100.0	124.6	114.4	102.6	84.7	92.1	99.3	95.3	97.0	99.5	110.5	105.9	101.6	92.2	96.1
01 - 16	WEIGHTED AVERAGE	103.3	47.2	76.3	99.6	126.4	112.5	96.2	99.0	97.5	93.8	104.3	98.8	99.1	120.4	109.4	98.6	103.5	101.0

DIVISION		ILLINOIS ROCKFORD			SPRINGFIELD			INDIANA ANDERSON			BLOOMINGTON			EVANSVILLE			FORT WAYNE		
		MAT.	INST.	TOTAL	MAT.	INST.	TOTAL	MAT.	INST.	TOTAL	MAT.	INST.	TOTAL	MAT.	INST.	TOTAL	MAT.	INST.	TOTAL
01590	EQUIPMENT RENTAL	.0	101.4	101.4	.0	102.2	102.2	.0	96.9	96.9	.0	86.1	86.1	.0	121.5	121.5	.0	96.9	96.9
02	SITE CONSTRUCTION	93.9	95.3	94.9	92.2	95.9	95.0	87.5	96.8	94.5	77.4	95.9	91.2	82.5	130.4	118.4	88.4	96.8	94.7
03100	CONCRETE FORMS & ACCESSORIES	101.1	117.1	115.2	100.4	107.6	106.7	92.6	78.4	80.1	103.3	79.0	82.0	92.3	78.2	80.0	92.0	77.9	79.6
03200	CONCRETE REINFORCEMENT	98.4	133.3	119.5	98.4	98.4	98.4	96.4	83.1	88.3	88.4	83.0	85.1	96.9	83.2	88.6	96.4	79.9	86.4
03300	CAST-IN-PLACE CONCRETE	98.7	100.0	99.2	91.7	102.7	96.3	104.2	83.1	95.4	102.2	79.8	92.9	97.6	92.5	95.5	110.7	78.6	97.3
03	CONCRETE	97.6	114.4	106.2	94.1	104.5	99.4	97.8	81.5	89.4	105.5	79.7	92.3	105.2	84.4	94.5	101.0	79.1	89.8
04	MASONRY	82.7	112.2	101.2	69.3	95.6	85.8	91.7	79.9	84.3	90.3	78.3	82.8	86.5	82.4	83.9	91.8	80.2	84.5
05	METALS	95.6	124.1	106.4	95.6	106.2	99.6	102.6	90.2	97.9	97.7	76.6	89.7	91.1	88.0	89.9	102.6	88.6	97.3
06	WOOD & PLASTICS	100.3	116.5	109.0	103.3	106.3	104.9	111.2	77.1	92.8	123.2	77.8	98.7	97.8	75.2	85.6	112.1	76.6	93.0
07	THERMAL & MOISTURE PROTECTION	97.2	112.5	104.6	96.8	101.5	99.1	100.6	78.7	89.9	91.9	83.0	87.5	95.1	86.5	90.9	100.3	85.2	93.0
08	DOORS & WINDOWS	96.2	118.9	102.0	96.2	104.0	98.2	99.3	80.8	94.6	103.2	81.2	97.6	95.6	78.5	91.3	99.3	77.4	93.8
09200	PLASTER & GYPSUM BOARD	105.0	116.8	113.1	105.0	106.3	105.9	102.4	77.1	85.2	99.4	77.9	84.8	95.3	73.6	80.5	101.3	76.9	84.5
095,098	CEILINGS & ACOUSTICAL TREATMENT	96.2	116.8	110.2	96.2	106.3	103.0	97.8	77.1	83.8	84.2	77.9	79.9	95.0	73.6	80.5	97.8	76.6	83.4
09600	FLOORING	96.8	97.1	96.9	97.0	88.2	94.7	86.9	92.2	88.3	103.5	92.2	100.6	96.7	87.0	94.2	86.9	83.1	85.9
097,099	WALL FINISHES, PAINTS & COATINGS	91.3	112.2	103.8	91.3	93.0	92.4	91.1	69.7	78.3	89.4	89.0	89.1	96.1	90.9	93.0	91.1	77.1	82.7
09	FINISHES	95.5	113.0	104.9	95.5	102.6	99.3	90.7	79.7	84.8	96.2	82.3	88.7	95.4	80.6	87.4	90.5	78.5	84.0
10 - 14	TOTAL DIV. 10000 - 14000	100.0	109.6	102.1	100.0	99.7	99.9	100.0	86.8	97.2	100.0	86.4	97.1	100.0	89.1	97.7	100.0	88.9	97.6
15	MECHANICAL	100.0	108.4	103.9	100.0	104.0	101.9	99.8	83.7	92.4	99.7	85.9	93.3	99.9	87.3	94.1	99.8	81.4	91.3
16	ELECTRICAL	101.6	114.3	109.0	102.5	92.5	96.7	85.6	90.1	88.2	100.3	83.7	90.6	96.0	86.9	90.7	86.2	84.0	84.9
01 - 16	WEIGHTED AVERAGE	97.2	111.9	104.3	96.0	100.7	98.3	97.0	85.0	91.2	99.1	83.2	91.4	96.6	89.1	93.0	97.5	83.1	90.6

DIVISION		INDIANA GARY			INDIANAPOLIS			MUNCIE			SOUTH BEND			TERRE HAUTE			IOWA CEDAR RAPIDS		
		MAT.	INST.	TOTAL	MAT.	INST.	TOTAL	MAT.	INST.	TOTAL	MAT.	INST.	TOTAL	MAT.	INST.	TOTAL	MAT.	INST.	TOTAL
01590	EQUIPMENT RENTAL	.0	96.9	96.9	.0	91.7	91.7	.0	96.3	96.3	.0	107.0	107.0	.0	121.5	121.5	.0	95.7	95.7
02	SITE CONSTRUCTION	88.1	98.7	96.0	87.5	100.5	97.2	77.5	96.0	91.4	87.7	96.5	94.3	83.9	130.4	118.7	86.3	94.8	92.7
03100	CONCRETE FORMS & ACCESSORIES	93.0	104.5	103.1	93.2	87.3	88.0	90.1	78.1	79.6	97.8	80.6	82.7	93.7	80.6	82.2	103.8	81.2	84.0
03200	CONCRETE REINFORCEMENT	96.4	102.0	99.8	95.7	85.0	89.2	97.8	83.1	88.9	96.4	78.3	85.4	96.9	82.7	88.3	98.3	84.6	90.0
03300	CAST-IN-PLACE CONCRETE	108.8	104.8	107.1	100.2	87.8	95.0	107.4	82.7	97.1	101.3	85.1	94.5	94.5	93.3	94.0	106.2	82.6	96.3
03	CONCRETE	100.1	104.2	102.2	100.2	86.6	93.2	103.3	81.3	92.0	92.1	83.2	87.6	108.3	85.6	96.7	99.8	83.0	91.2
04	MASONRY	93.1	96.0	94.9	98.7	85.9	90.7	92.5	79.9	84.6	90.1	79.9	83.7	93.4	80.7	85.4	106.9	82.1	91.4
05	METALS	102.6	103.1	102.8	104.7	79.6	95.2	99.3	90.5	96.0	101.2	102.9	101.9	91.9	88.1	90.4	88.4	94.1	90.5
06	WOOD & PLASTICS	110.2	105.0	107.4	107.7	87.3	96.7	113.0	77.0	93.6	111.4	80.2	94.6	99.9	79.3	88.8	110.4	80.1	94.1
07	THERMAL & MOISTURE PROTECTION	99.6	100.5	100.1	99.4	87.0	93.4	93.5	79.0	86.4	99.4	85.1	92.4	95.2	84.3	89.9	99.8	83.0	91.6
08	DOORS & WINDOWS	99.3	106.2	101.1	107.1	86.8	101.9	98.2	80.7	93.7	92.7	79.7	89.4	96.2	81.8	92.5	99.4	83.1	95.3
09200	PLASTER & GYPSUM BOARD	98.3	105.8	103.4	98.2	87.4	90.8	94.1	77.1	82.5	101.9	80.2	87.1	95.3	77.9	83.4	103.3	79.8	87.3
095,098	CEILINGS & ACOUSTICAL TREATMENT	97.8	105.8	103.2	96.5	87.4	90.3	85.5	77.1	79.8	97.8	80.2	85.9	95.0	77.9	83.4	110.3	79.8	89.7
09600	FLOORING	86.9	103.4	91.1	86.7	92.2	88.1	95.8	92.2	94.9	86.9	80.5	85.2	96.7	90.8	95.2	123.6	60.5	107.3
097,099	WALL FINISHES, PAINTS & COATINGS	91.1	105.0	99.4	91.1	92.2	91.8	89.4	69.7	77.6	91.1	83.5	86.6	96.1	90.5	92.7	108.0	79.9	91.1
09	FINISHES	90.1	104.4	97.8	90.2	88.9	89.5	92.9	79.6	85.7	90.6	80.7	85.3	95.4	83.3	88.9	111.6	76.3	92.5
10 - 14	TOTAL DIV. 10000 - 14000	100.0	104.7	101.0	100.0	89.7	97.8	100.0	86.3	97.1	100.0	82.5	96.2	100.0	90.7	98.0	100.0	86.9	97.2
15	MECHANICAL	99.8	98.0	99.0	99.8	86.2	93.5	99.6	83.7	92.3	99.8	79.6	90.5	99.9	81.2	91.3	100.4	86.4	94.0
16	ELECTRICAL	91.6	106.6	100.4	101.3	90.0	94.7	90.7	79.4	84.1	96.9	86.7	90.9	93.8	83.7	87.9	97.0	85.3	90.2
01 - 16	WEIGHTED AVERAGE	97.9	101.9	99.8	100.2	88.0	94.3	97.3	83.4	90.6	96.3	85.2	91.0	97.4	87.9	92.8	99.3	85.3	92.6

City Cost Indexes

		IOWA																
	DIVISION	COUNCIL BLUFFS			DAVENPORT			DES MOINES			DUBUQUE			SIOUX CITY			WATERLOO		
		MAT.	INST.	TOTAL	MAT.	INST.	TOTAL	MAT.	INST.	TOTAL	MAT.	INST.	TOTAL	MAT.	INST.	TOTAL	MAT.	INST.	TOTAL
01590	EQUIPMENT RENTAL	.0	95.4	95.4	.0	99.8	99.8	.0	101.7	101.7	.0	94.3	94.3	.0	99.8	99.8	.0	99.8	99.8
02	SITE CONSTRUCTION	91.3	91.5	91.5	84.6	98.7	95.2	78.1	100.3	94.7	84.5	91.8	90.0	93.6	95.7	95.2	85.1	94.9	92.4
03100	CONCRETE FORMS & ACCESSORIES	80.6	60.6	63.1	103.4	89.9	91.5	105.5	81.1	84.1	82.5	69.9	71.5	103.8	64.6	69.4	104.4	51.2	57.7
03200	CONCRETE REINFORCEMENT	100.3	80.0	88.1	98.3	98.0	98.1	98.3	81.3	88.0	96.9	84.3	89.3	98.3	69.0	80.6	98.3	84.1	89.7
03300	CAST-IN-PLACE CONCRETE	110.5	76.6	96.3	102.3	103.9	103.0	106.5	81.3	96.0	104.0	90.6	98.4	105.3	59.0	85.9	106.2	50.0	82.7
03	CONCRETE	101.7	71.0	86.0	97.8	96.8	97.3	98.8	82.0	90.2	96.1	80.7	88.2	99.3	64.6	81.5	99.8	58.5	78.6
04	MASONRY	106.5	69.4	83.3	104.4	90.4	95.6	100.6	82.8	89.4	108.0	74.2	86.8	99.7	61.4	75.7	101.2	61.4	76.3
05	METALS	93.7	90.7	92.5	88.4	103.2	94.0	88.3	94.1	90.5	86.9	93.2	89.3	88.4	85.3	87.2	88.4	91.9	89.7
06	WOOD & PLASTICS	87.2	56.7	70.8	110.4	86.9	97.8	111.9	80.2	94.8	89.3	67.4	77.5	110.4	63.2	84.9	111.1	51.3	78.9
07	THERMAL & MOISTURE PROTECTION	99.0	67.3	83.5	99.2	90.4	94.9	100.1	81.2	90.9	99.3	71.6	85.8	99.2	59.4	79.8	98.8	55.5	77.7
08	DOORS & WINDOWS	98.5	63.8	89.7	99.4	92.0	97.5	99.4	83.2	95.3	98.5	78.1	93.3	99.4	64.5	90.6	95.0	63.7	87.1
09200	PLASTER & GYPSUM BOARD	94.1	55.8	68.0	103.3	86.6	91.9	99.6	79.6	86.0	94.6	66.8	75.7	103.3	62.2	75.3	103.3	50.0	67.0
095,098	CEILINGS & ACOUSTICAL TREATMENT	106.4	55.8	72.1	110.3	86.6	94.2	109.0	79.6	89.1	106.4	66.8	79.6	110.3	62.2	77.7	110.3	50.0	69.5
09600	FLOORING	98.5	49.4	85.9	109.9	82.8	102.9	109.7	45.1	93.0	112.6	42.1	94.4	110.4	63.8	98.4	111.6	64.1	99.3
097,099	WALL FINISHES, PAINTS & COATINGS	99.5	74.4	84.4	103.7	97.9	100.2	103.7	81.0	90.1	106.7	79.9	90.6	104.9	65.0	81.0	104.9	38.0	64.8
09	FINISHES	100.7	58.6	77.9	106.7	88.8	97.0	105.1	73.8	88.2	105.9	64.6	83.6	108.1	64.0	84.3	107.4	51.4	77.2
10 - 14	TOTAL DIV. 10000 - 14000	100.0	78.9	95.5	100.0	91.0	98.1	100.0	87.7	97.4	100.0	83.7	96.5	100.0	82.5	96.2	100.0	76.7	95.0
15	MECHANICAL	100.4	79.6	90.8	100.4	97.3	99.0	100.4	83.5	92.6	100.4	81.7	91.8	100.4	84.8	93.2	100.4	48.5	76.5
16	ELECTRICAL	103.9	84.5	92.5	93.4	92.6	92.9	97.7	86.4	91.1	102.0	73.1	85.1	97.0	80.7	87.5	97.0	66.3	79.1
01 - 16	WEIGHTED AVERAGE	99.6	76.3	88.4	98.1	94.7	96.5	98.1	85.0	91.7	98.3	78.9	88.9	98.7	75.7	87.6	98.0	64.0	81.6

		KANSAS															KENTUCKY		
	DIVISION	DODGE CITY			KANSAS CITY			SALINA			TOPEKA			WICHITA			BOWLING GREEN		
		MAT.	INST.	TOTAL	MAT.	INST.	TOTAL	MAT.	INST.	TOTAL	MAT.	INST.	TOTAL	MAT.	INST.	TOTAL	MAT.	INST.	TOTAL
01590	EQUIPMENT RENTAL	.0	103.3	103.3	.0	99.9	99.9	.0	103.3	103.3	.0	101.4	101.4	.0	103.3	103.3	.0	96.2	96.2
02	SITE CONSTRUCTION	109.3	92.8	96.9	89.8	91.0	90.7	99.0	93.2	94.7	91.7	90.6	90.9	93.0	93.5	93.4	69.4	100.4	92.6
03100	CONCRETE FORMS & ACCESSORIES	94.7	45.0	51.1	100.2	95.1	95.7	90.3	47.7	52.9	100.0	46.7	53.2	96.7	60.3	64.7	82.7	83.0	82.9
03200	CONCRETE REINFORCEMENT	106.6	59.9	78.4	101.3	84.1	90.9	106.0	81.9	91.4	98.4	92.0	94.5	98.4	83.1	89.2	87.8	70.2	77.2
03300	CAST-IN-PLACE CONCRETE	114.6	58.9	91.3	89.8	94.1	91.6	99.4	50.4	78.9	90.8	54.9	75.8	86.6	59.2	75.1	89.2	82.8	86.5
03	CONCRETE	115.1	54.2	83.9	95.7	93.1	94.4	101.6	56.7	78.6	93.6	59.7	76.3	91.3	65.5	78.1	96.6	80.6	88.4
04	MASONRY	103.1	43.5	65.8	102.4	84.6	91.2	117.6	41.9	70.2	97.2	52.4	69.2	90.9	67.8	76.5	94.0	75.5	82.4
05	METALS	92.9	79.6	87.9	97.8	95.6	97.0	92.7	89.8	91.6	95.7	95.9	95.8	95.7	92.0	94.3	95.1	78.3	88.7
06	WOOD & PLASTICS	93.4	45.1	67.4	99.8	96.0	97.8	89.7	48.6	67.5	97.2	43.0	67.9	95.2	59.1	75.7	94.1	82.5	87.8
07	THERMAL & MOISTURE PROTECTION	98.2	50.7	75.1	96.2	91.5	93.9	97.6	56.4	77.5	97.5	70.0	84.1	97.1	67.9	82.9	85.0	80.2	82.6
08	DOORS & WINDOWS	96.1	48.7	84.1	95.0	92.5	94.4	96.1	54.3	85.5	96.2	61.2	87.4	96.2	65.7	88.5	95.3	75.5	90.3
09200	PLASTER & GYPSUM BOARD	102.5	43.4	62.3	98.3	95.8	96.6	101.5	47.0	64.4	105.0	41.2	61.6	105.0	57.8	72.9	90.8	82.3	85.0
095,098	CEILINGS & ACOUSTICAL TREATMENT	89.7	43.4	58.4	96.2	95.8	95.9	89.7	47.0	60.8	96.2	41.2	58.9	96.2	57.8	70.2	91.1	82.3	85.2
09600	FLOORING	96.1	53.2	85.1	87.1	56.8	79.3	94.1	38.9	79.9	97.7	49.4	85.2	96.8	82.5	93.1	92.5	65.5	85.6
097,099	WALL FINISHES, PAINTS & COATINGS	91.3	51.5	67.4	99.4	92.3	95.2	91.3	40.6	60.9	91.3	69.3	78.1	91.3	65.4	75.8	96.1	77.1	84.7
09	FINISHES	94.7	46.8	68.8	93.9	87.8	90.6	93.3	44.5	66.9	95.7	47.5	69.6	95.5	64.6	78.8	92.8	79.2	85.4
10 - 14	TOTAL DIV. 10000 - 14000	100.0	53.3	89.9	100.0	89.5	97.7	100.0	70.1	93.6	100.0	74.3	94.5	100.0	75.0	94.6	100.0	67.3	93.0
15	MECHANICAL	100.0	49.6	76.8	99.9	91.2	95.9	100.0	37.5	71.3	100.0	72.9	87.6	100.0	68.8	85.7	99.9	84.4	92.8
16	ELECTRICAL	98.1	52.3	71.3	106.0	106.1	106.0	97.7	75.4	84.7	103.2	78.0	88.5	100.4	75.4	85.8	94.1	83.8	88.1
01 - 16	WEIGHTED AVERAGE	100.0	56.5	79.0	98.3	92.9	95.7	98.6	59.0	79.5	97.6	69.4	84.0	96.7	73.0	85.3	95.4	82.2	89.0

		KENTUCKY									LOUISIANA								
	DIVISION	LEXINGTON			LOUISVILLE			OWENSBORO			ALEXANDRIA			BATON ROUGE			LAKE CHARLES		
		MAT.	INST.	TOTAL	MAT.	INST.	TOTAL	MAT.	INST.	TOTAL	MAT.	INST.	TOTAL	MAT.	INST.	TOTAL	MAT.	INST.	TOTAL
01590	EQUIPMENT RENTAL	.0	103.7	103.7	.0	96.2	96.2	.0	121.5	121.5	.0	86.6	86.6	.0	86.4	86.4	.0	86.4	86.4
02	SITE CONSTRUCTION	77.4	102.5	96.2	66.7	100.3	91.8	82.0	130.0	117.9	103.2	85.6	90.0	111.7	85.6	92.2	113.3	85.3	92.4
03100	CONCRETE FORMS & ACCESSORIES	97.1	70.6	73.9	91.5	82.6	83.7	88.3	67.2	69.7	80.7	45.8	50.1	100.7	50.1	56.3	101.1	51.0	57.1
03200	CONCRETE REINFORCEMENT	96.9	92.8	94.4	96.9	93.8	95.0	87.9	96.2	92.9	99.1	64.0	77.9	98.3	52.5	70.6	98.3	52.4	70.6
03300	CAST-IN-PLACE CONCRETE	96.3	82.5	90.6	93.2	81.4	88.3	92.0	79.2	86.6	93.9	45.0	73.5	87.3	54.0	73.4	92.5	55.5	77.0
03	CONCRETE	98.5	79.3	88.7	96.6	84.4	90.3	106.7	77.2	91.6	88.6	50.1	68.9	95.1	52.7	73.4	97.7	53.6	75.1
04	MASONRY	90.5	49.6	64.9	92.1	79.3	84.1	90.9	53.6	67.6	114.7	55.6	77.7	95.5	57.3	71.6	93.3	58.4	71.5
05	METALS	96.6	88.1	93.4	96.6	88.0	87.5	86.5	89.0	87.5	87.1	73.8	82.1	100.6	66.7	87.7	100.1	66.9	87.5
06	WOOD & PLASTICS	103.9	73.3	87.4	102.6	82.5	91.7	94.1	66.2	79.1	86.4	45.6	64.4	110.0	50.3	77.8	108.0	51.0	77.3
07	THERMAL & MOISTURE PROTECTION	95.7	79.8	87.9	85.2	81.8	83.5	95.2	69.9	82.9	99.4	56.0	78.2	97.4	57.1	77.8	100.6	57.3	79.5
08	DOORS & WINDOWS	96.2	79.7	92.0	96.2	86.2	93.7	93.6	77.9	89.6	99.2	51.4	87.1	104.8	50.4	91.0	104.8	52.4	91.5
09200	PLASTER & GYPSUM BOARD	96.8	71.7	79.7	98.7	82.3	87.5	90.2	64.4	72.7	81.2	44.8	56.4	102.5	49.5	66.4	102.5	50.2	66.9
095,098	CEILINGS & ACOUSTICAL TREATMENT	95.0	71.7	79.2	95.0	82.3	86.4	82.0	64.4	70.1	96.8	44.8	61.6	101.9	49.5	66.4	101.9	50.2	66.9
09600	FLOORING	97.5	45.5	84.1	96.0	73.7	90.3	94.9	65.5	87.4	105.6	71.7	96.9	110.0	70.3	99.7	110.2	55.0	95.9
097,099	WALL FINISHES, PAINTS & COATINGS	96.1	61.6	75.4	96.1	80.7	86.8	96.1	102.8	100.1	96.6	43.5	64.7	98.9	49.8	69.5	98.9	45.9	67.1
09	FINISHES	96.0	64.4	78.9	95.4	80.5	87.3	91.8	70.0	80.0	95.7	49.9	70.9	104.3	53.7	77.0	104.4	50.9	75.5
10 - 14	TOTAL DIV. 10000 - 14000	100.0	86.7	97.1	100.0	89.8	97.8	100.0	89.5	97.7	100.0	75.9	94.8	100.0	69.8	93.5	100.0	70.1	93.6
15	MECHANICAL	99.9	50.4	77.1	99.9	83.1	92.1	99.9	55.3	79.4	100.1	39.8	72.3	100.1	45.2	74.9	100.1	58.8	81.1
16	ELECTRICAL	94.7	52.7	70.2	94.8	83.8	88.4	93.4	77.5	84.1	95.8	58.5	74.0	100.2	52.3	72.2	100.2	65.6	80.0
01 - 16	WEIGHTED AVERAGE	96.7	67.9	82.9	95.9	85.0	90.7	95.5	75.5	85.9	96.7	56.0	77.1	100.6	56.5	79.4	100.9	61.1	81.7

COST INDEXES

DIVISION		LOUISIANA									MAINE						
		MONROE			NEW ORLEANS			SHREVEPORT			AUGUSTA			BANGOR			LEWISTON	
		MAT.	INST.	TOTAL	MAT.	INST.	TOTAL	MAT.	INST.	TOTAL	MAT.	INST.	TOTAL	MAT.	INST.	TOTAL	MAT.	INST.	TOTAL
01590	EQUIPMENT RENTAL	.0	86.6	86.6	.0	87.6	87.6	.0	86.6	86.6	.0	101.8	101.8	.0	101.8	101.8	.0	101.8	101.8
02	SITE CONSTRUCTION	103.2	85.4	89.9	115.8	87.8	94.8	101.9	85.3	89.5	82.8	102.7	97.7	82.7	101.1	96.5	81.3	101.1	96.1
03100	CONCRETE FORMS & ACCESSORIES	80.0	46.5	50.6	100.6	70.0	73.8	100.9	49.4	55.7	102.5	67.5	71.7	96.4	83.3	84.9	102.6	83.3	85.7
03200	CONCRETE REINFORCEMENT	98.0	64.2	77.6	98.3	66.3	79.0	97.6	65.2	78.0	88.8	115.5	104.9	88.8	115.8	105.1	109.1	115.8	113.1
03300	CAST-IN-PLACE CONCRETE	93.9	49.4	75.3	91.6	61.4	78.9	92.6	53.2	76.1	83.0	65.2	75.6	83.0	67.2	76.4	94.5	67.2	83.1
03	CONCRETE	88.5	51.9	69.7	97.2	66.7	81.6	88.4	54.7	71.1	100.6	75.7	87.9	99.7	83.3	91.3	103.8	83.3	93.3
04	MASONRY	109.5	59.9	78.4	95.3	62.0	74.4	101.7	52.7	71.0	94.6	54.0	69.1	112.6	62.0	80.9	96.2	62.0	74.8
05	METALS	87.1	73.5	81.9	103.0	74.6	92.2	87.9	73.5	82.5	94.4	86.5	91.4	94.2	84.7	90.6	97.5	84.7	92.7
06	WOOD & PLASTICS	85.6	46.1	64.3	104.3	73.4	87.6	105.9	50.1	75.8	100.0	66.0	81.6	94.1	86.0	89.7	100.0	86.0	92.4
07	THERMAL & MOISTURE PROTECTION	99.4	58.9	79.7	101.4	65.7	84.0	98.3	57.5	78.4	102.7	54.4	79.1	102.5	61.6	82.6	102.3	61.6	82.5
08	DOORS & WINDOWS	99.2	55.9	88.2	105.4	70.6	96.6	97.1	54.3	86.2	104.0	64.3	93.9	103.9	77.0	97.1	107.2	77.0	99.5
09200	PLASTER & GYPSUM BOARD	80.7	45.3	56.6	101.4	73.2	82.2	86.1	49.4	61.2	107.2	64.0	77.8	104.3	84.6	90.9	109.1	84.6	92.4
095,098	CEILINGS & ACOUSTICAL TREATMENT	96.8	45.3	61.9	100.4	73.2	82.0	98.1	49.4	65.1	98.1	64.0	75.0	96.8	84.6	88.5	105.9	84.6	91.5
09600	FLOORING	105.2	45.4	89.8	110.6	53.9	96.2	116.3	67.2	103.6	100.3	45.6	86.2	97.8	56.9	87.3	100.3	56.9	89.1
097,099	WALL FINISHES, PAINTS & COATINGS	96.6	51.9	69.8	100.7	65.9	79.8	96.6	43.5	64.7	93.0	37.6	59.7	93.0	34.6	58.0	93.0	34.6	58.0
09	FINISHES	95.5	45.9	68.7	104.6	67.0	84.3	100.1	51.6	73.9	101.4	59.3	78.7	100.0	73.6	85.8	102.8	73.6	87.0
10-14	TOTAL DIV. 10000 - 14000	100.0	67.4	93.0	100.0	74.9	94.6	100.0	69.7	93.5	100.0	67.9	93.1	100.0	88.2	97.5	100.0	88.2	97.5
15	MECHANICAL	100.1	53.8	78.8	100.1	64.7	83.8	100.1	56.6	80.1	100.2	74.9	88.5	100.2	73.9	88.1	100.2	73.9	88.1
16	ELECTRICAL	98.3	58.0	74.8	100.5	69.6	82.5	96.6	69.6	80.8	102.6	84.1	91.8	100.1	84.0	90.7	102.7	84.0	91.8
01-16	WEIGHTED AVERAGE	96.6	59.0	78.5	101.5	69.4	86.0	96.5	61.6	79.6	99.4	74.6	87.4	99.9	79.2	89.9	100.8	79.2	90.4

DIVISION		MAINE			MARYLAND						MASSACHUSETTS							
		PORTLAND			BALTIMORE			HAGERSTOWN			BOSTON			BROCKTON			FALL RIVER	
		MAT.	INST.	TOTAL	MAT.	INST.	TOTAL	MAT.	INST.	TOTAL	MAT.	INST.	TOTAL	MAT.	INST.	TOTAL	MAT.	INST.	TOTAL
01590	EQUIPMENT RENTAL	.0	101.8	101.8	.0	102.9	102.9	.0	98.8	98.8	.0	108.7	108.7	.0	104.0	104.0	.0	105.2	105.2
02	SITE CONSTRUCTION	80.0	101.1	95.8	97.8	93.1	94.3	87.5	88.7	88.4	87.8	109.0	103.7	85.5	104.9	100.0	84.6	105.0	99.9
03100	CONCRETE FORMS & ACCESSORIES	101.6	83.3	85.6	103.3	75.0	78.4	91.2	76.2	78.1	108.4	132.5	129.5	108.1	120.4	118.9	108.1	120.5	119.0
03200	CONCRETE REINFORCEMENT	109.1	115.8	113.1	99.9	89.8	93.8	88.3	75.9	80.8	107.9	141.0	127.9	109.1	140.6	128.1	109.1	125.8	119.2
03300	CAST-IN-PLACE CONCRETE	86.0	67.2	78.2	95.8	78.9	88.7	81.5	62.5	73.5	107.1	140.1	120.9	102.0	132.4	114.7	98.6	135.9	114.2
03	CONCRETE	99.5	83.3	91.2	99.1	80.3	89.5	84.7	72.5	78.5	113.3	135.9	125.0	110.5	127.7	119.3	108.8	126.1	117.7
04	MASONRY	93.7	62.0	73.9	93.1	74.9	81.7	97.4	74.5	83.1	115.5	141.6	131.8	111.3	130.3	123.2	111.1	133.4	125.1
05	METALS	97.5	84.7	92.7	95.3	100.4	97.2	94.3	91.7	93.3	100.4	124.5	109.5	97.4	120.9	106.3	97.4	115.4	104.2
06	WOOD & PLASTICS	100.0	86.0	92.4	101.6	76.0	87.8	89.8	75.8	82.3	103.8	132.4	119.2	102.7	119.5	111.8	102.7	119.9	112.0
07	THERMAL & MOISTURE PROTECTION	102.1	61.6	82.4	93.7	83.4	88.7	92.9	72.4	82.9	102.9	138.0	120.0	102.7	127.2	114.6	102.6	122.0	112.1
08	DOORS & WINDOWS	107.2	77.0	99.5	93.8	84.7	91.5	91.0	77.4	87.6	103.1	133.5	110.9	101.1	122.9	106.6	101.1	119.7	105.8
09200	PLASTER & GYPSUM BOARD	109.1	84.6	92.4	110.6	75.7	86.8	105.5	75.7	85.2	112.5	132.4	126.0	105.7	119.1	114.8	105.7	119.1	114.8
095,098	CEILINGS & ACOUSTICAL TREATMENT	105.9	84.6	91.5	98.0	75.7	82.9	99.4	75.7	83.3	104.5	132.4	123.4	107.4	119.1	115.3	107.4	119.1	115.3
09600	FLOORING	100.3	56.9	89.1	93.2	83.2	90.6	88.3	77.5	85.5	100.9	145.3	112.3	101.4	145.3	112.7	101.1	145.3	112.5
097,099	WALL FINISHES, PAINTS & COATINGS	93.0	34.6	58.0	96.2	73.3	82.5	96.2	42.4	63.9	96.1	147.2	126.8	95.5	123.0	112.0	95.5	123.0	112.0
09	FINISHES	102.9	73.6	87.1	95.4	75.7	84.8	92.7	72.8	81.9	103.2	136.6	121.3	103.0	125.2	115.0	103.0	125.5	115.1
10-14	TOTAL DIV. 10000 - 14000	100.0	88.2	97.5	100.0	86.9	97.2	100.0	89.4	97.7	100.0	124.1	105.2	100.0	120.7	104.4	100.0	121.6	104.7
15	MECHANICAL	100.2	73.9	88.1	99.9	84.5	92.8	99.9	88.0	94.4	100.2	123.8	111.1	100.2	108.6	104.0	100.2	108.5	104.0
16	ELECTRICAL	103.0	84.0	91.9	100.3	90.5	94.6	97.1	81.8	88.2	98.5	121.3	111.9	98.2	100.1	99.3	97.7	100.0	99.1
01-16	WEIGHTED AVERAGE	100.1	79.2	90.0	97.5	84.9	91.4	94.5	81.1	88.1	102.9	128.1	115.0	101.5	116.6	108.8	101.2	116.0	108.3

DIVISION		MASSACHUSETTS																
		HYANNIS			LAWRENCE			LOWELL			NEW BEDFORD			PITTSFIELD			SPRINGFIELD	
		MAT.	INST.	TOTAL	MAT.	INST.	TOTAL	MAT.	INST.	TOTAL	MAT.	INST.	TOTAL	MAT.	INST.	TOTAL	MAT.	INST.	TOTAL
01590	EQUIPMENT RENTAL	.0	104.0	104.0	.0	104.0	104.0	.0	101.8	101.8	.0	105.2	105.2	.0	101.8	101.8	.0	101.8	101.8
02	SITE CONSTRUCTION	82.3	104.9	99.2	86.1	104.9	100.2	85.2	104.9	99.9	84.2	105.0	99.8	86.3	103.6	99.3	85.7	103.8	99.2
03100	CONCRETE FORMS & ACCESSORIES	98.1	120.1	117.4	107.9	120.6	119.0	104.3	120.7	118.7	108.1	120.5	119.0	104.2	94.5	95.7	104.4	112.6	111.6
03200	CONCRETE REINFORCEMENT	87.5	125.8	110.6	108.2	127.4	119.8	109.1	127.4	120.2	109.1	125.8	119.2	90.7	107.9	101.1	109.1	123.9	118.0
03300	CAST-IN-PLACE CONCRETE	93.0	135.2	110.7	102.8	136.1	116.7	93.5	136.2	111.3	95.7	135.9	112.5	102.0	109.9	105.3	97.3	111.4	103.2
03	CONCRETE	100.7	125.7	113.5	110.8	126.5	118.9	100.8	126.4	113.9	107.4	126.1	117.0	102.9	102.0	102.5	102.7	113.6	108.3
04	MASONRY	110.3	133.5	124.8	110.5	130.3	122.9	96.1	130.3	117.4	110.8	133.4	125.0	96.7	104.5	101.6	96.4	106.8	102.9
05	METALS	93.9	114.8	101.8	94.9	115.7	102.8	94.9	112.8	101.7	97.4	115.4	104.2	94.7	99.6	96.6	97.3	106.5	100.8
06	WOOD & PLASTICS	93.5	119.5	107.5	102.7	119.5	111.8	101.9	119.5	111.4	102.7	119.9	112.0	101.9	92.2	96.7	101.9	115.2	109.0
07	THERMAL & MOISTURE PROTECTION	102.1	123.0	112.3	102.6	127.7	114.9	102.4	127.4	114.5	102.6	122.0	112.1	102.4	105.4	103.9	102.4	108.7	105.5
08	DOORS & WINDOWS	97.2	119.5	102.9	101.1	119.4	105.7	107.2	119.4	110.3	101.1	119.7	105.8	107.2	100.2	105.4	107.2	116.9	109.7
09200	PLASTER & GYPSUM BOARD	99.9	119.1	113.0	109.1	119.1	115.9	109.1	119.1	115.9	105.7	119.1	114.8	109.1	91.0	96.8	109.1	114.6	112.8
095,098	CEILINGS & ACOUSTICAL TREATMENT	98.3	119.1	112.4	105.9	119.1	114.8	105.9	119.1	114.8	107.4	119.1	115.3	105.9	91.0	95.8	105.9	114.6	111.8
09600	FLOORING	97.3	145.3	109.7	100.3	145.3	111.9	100.3	145.3	111.9	101.1	145.3	112.5	100.5	116.4	104.6	100.2	116.4	104.4
097,099	WALL FINISHES, PAINTS & COATINGS	95.5	123.0	112.0	93.1	123.0	111.0	93.0	123.0	111.0	95.5	123.0	112.0	93.0	98.9	96.5	94.8	98.9	97.2
09	FINISHES	99.0	125.2	113.2	102.6	125.2	114.8	102.6	125.2	114.8	102.9	125.5	115.1	102.7	98.4	100.3	102.7	112.4	107.9
10-14	TOTAL DIV. 10000 - 14000	100.0	120.7	104.4	100.0	120.8	104.5	100.0	120.8	104.5	100.0	121.6	104.7	100.0	101.8	100.4	100.0	105.8	101.2
15	MECHANICAL	100.2	108.4	104.0	100.2	109.8	104.6	100.2	120.8	109.7	100.2	108.5	104.0	100.2	97.3	98.9	100.2	98.8	99.5
16	ELECTRICAL	94.4	100.0	97.7	102.1	124.4	115.1	102.7	124.4	115.4	98.5	100.0	99.4	102.7	87.4	93.8	102.7	91.4	96.1
01-16	WEIGHTED AVERAGE	98.4	115.8	106.8	101.5	119.5	110.2	100.2	121.4	110.4	101.1	116.0	108.3	100.5	98.6	99.6	100.8	104.9	102.8

COST INDEXES

454

City Cost Indexes

MASSACHUSETTS / MICHIGAN

DIVISION		WORCESTER			ANN ARBOR			DEARBORN			DETROIT			FLINT			GRAND RAPIDS		
		MAT.	INST.	TOTAL	MAT.	INST.	TOTAL	MAT.	INST.	TOTAL	MAT.	INST.	TOTAL	MAT.	INST.	TOTAL	MAT.	INST.	TOTAL
01590	EQUIPMENT RENTAL	.0	101.8	101.8	.0	112.4	112.4	.0	112.4	112.4	.0	97.8	97.8	.0	112.4	112.4	.0	105.3	105.3
02	SITE CONSTRUCTION	85.2	104.8	99.9	79.6	95.3	91.4	79.4	95.5	91.5	94.1	97.3	96.5	69.7	94.3	88.2	82.6	88.1	86.7
03100	CONCRETE FORMS & ACCESSORIES	104.9	120.1	118.2	97.5	114.5	112.4	97.4	123.7	120.5	98.8	123.8	120.7	101.9	96.9	97.5	97.3	82.1	84.0
03200	CONCRETE REINFORCEMENT	109.1	139.6	127.5	95.7	122.3	111.8	95.7	122.8	112.1	95.1	122.8	111.8	95.7	121.9	111.5	96.9	88.8	92.0
03300	CAST-IN-PLACE CONCRETE	93.5	131.6	109.4	94.6	114.7	103.0	92.5	120.5	104.2	101.5	120.5	109.5	95.3	96.5	95.8	96.5	103.1	99.2
03	CONCRETE	100.9	126.9	114.2	92.7	116.6	104.9	91.6	122.7	107.6	96.2	121.3	109.0	93.3	102.4	98.0	98.6	90.0	94.2
04	MASONRY	96.1	130.1	117.4	99.6	109.6	105.8	99.4	120.5	112.6	98.4	120.5	112.2	99.6	95.5	97.0	95.0	57.8	71.7
05	METALS	97.3	117.1	104.8	100.8	126.8	110.6	100.9	128.2	111.2	101.4	105.6	103.0	100.9	123.8	109.6	97.6	82.0	91.7
06	WOOD & PLASTICS	102.4	119.5	111.6	102.0	114.1	108.5	102.0	123.9	113.8	102.5	123.9	114.1	106.1	98.0	101.7	98.2	82.9	90.0
07	THERMAL & MOISTURE PROTECTION	102.4	121.9	111.9	97.9	112.0	104.7	96.8	124.1	110.1	95.3	124.1	109.3	96.1	95.6	95.9	92.0	64.6	78.7
08	DOORS & WINDOWS	107.2	126.3	112.0	96.4	114.1	100.9	96.4	119.4	102.2	98.0	120.4	103.7	96.4	101.6	97.7	94.2	73.6	89.0
09200	PLASTER & GYPSUM BOARD	109.1	119.1	115.9	104.7	113.2	110.5	104.7	123.2	117.3	104.7	123.2	117.3	106.3	96.5	99.6	95.3	77.3	83.1
095,098	CEILINGS & ACOUSTICAL TREATMENT	105.9	119.1	114.8	102.0	113.2	109.6	102.0	123.2	116.4	103.5	123.2	116.9	102.0	96.5	98.3	95.0	77.3	83.0
09600	FLOORING	100.3	142.3	111.1	87.6	114.2	94.5	87.2	125.0	96.9	87.3	125.0	97.0	87.4	77.9	85.0	96.2	46.0	83.3
097,099	WALL FINISHES, PAINTS & COATINGS	93.0	123.0	111.0	86.5	106.5	98.5	86.5	113.2	102.5	88.4	113.2	103.3	86.5	83.3	84.6	96.1	44.0	64.9
09	FINISHES	102.6	124.6	114.5	95.1	113.7	105.1	94.9	123.2	110.2	96.1	123.2	110.7	94.6	91.6	93.0	95.3	71.2	82.3
10-14	TOTAL DIV. 10000 - 14000	100.0	109.1	102.0	100.0	113.3	102.9	100.0	116.0	103.4	100.0	116.0	103.4	100.0	98.2	99.6	100.0	107.7	101.7
15	MECHANICAL	100.2	107.9	103.7	100.0	106.9	103.1	100.0	117.5	108.0	100.0	119.2	108.8	100.0	96.8	98.5	99.9	59.0	81.1
16	ELECTRICAL	102.7	102.5	102.6	94.2	91.3	92.5	94.2	118.1	108.2	95.3	118.0	108.6	94.2	92.4	93.1	94.3	63.7	76.4
01-16	WEIGHTED AVERAGE	100.5	116.0	108.0	97.1	108.6	102.7	97.0	118.5	107.4	98.3	116.8	107.2	96.9	98.6	97.7	96.8	72.4	85.0

MICHIGAN / MINNESOTA

DIVISION		KALAMAZOO			LANSING			MUSKEGON			SAGINAW			DULUTH			MINNEAPOLIS		
		MAT.	INST.	TOTAL	MAT.	INST.	TOTAL	MAT.	INST.	TOTAL	MAT.	INST.	TOTAL	MAT.	INST.	TOTAL	MAT.	INST.	TOTAL
01590	EQUIPMENT RENTAL	.0	105.3	105.3	.0	112.4	112.4	.0	105.3	105.3	.0	112.4	112.4	.0	102.0	102.0	.0	106.4	106.4
02	SITE CONSTRUCTION	83.2	88.3	87.0	86.1	94.1	92.1	81.2	88.2	86.4	72.0	94.1	88.5	82.6	102.9	97.8	82.4	109.2	102.5
03100	CONCRETE FORMS & ACCESSORIES	96.5	91.9	92.4	101.1	94.4	95.2	97.4	88.6	89.7	97.5	95.6	95.9	99.4	123.9	120.9	100.1	138.1	133.4
03200	CONCRETE REINFORCEMENT	96.9	90.4	93.0	95.7	121.7	111.4	97.5	90.2	93.1	95.7	121.3	111.2	98.1	107.2	103.6	98.3	133.1	119.3
03300	CAST-IN-PLACE CONCRETE	98.2	103.6	100.4	94.7	96.8	95.6	95.9	100.6	97.9	93.5	96.5	94.7	107.3	107.4	107.4	105.7	130.8	116.2
03	CONCRETE	102.0	94.8	98.3	93.0	101.4	97.3	96.5	92.3	94.3	92.1	101.7	97.0	99.3	115.6	107.6	100.3	135.0	118.0
04	MASONRY	97.7	86.9	91.0	92.9	95.9	94.8	94.8	81.3	86.4	101.1	88.8	93.4	108.5	117.3	114.0	108.5	140.8	128.7
05	METALS	97.9	86.0	93.4	99.6	123.6	108.7	95.8	85.6	92.0	100.9	122.5	109.1	95.2	125.4	106.6	97.1	142.3	114.2
06	WOOD & PLASTICS	99.9	91.9	95.6	104.7	93.2	98.5	96.5	88.7	92.3	97.6	96.2	96.9	111.9	126.3	119.7	112.3	136.1	125.2
07	THERMAL & MOISTURE PROTECTION	91.6	87.3	89.5	96.8	94.9	95.9	90.7	80.8	85.9	96.7	91.9	94.4	100.6	117.0	108.6	100.5	137.5	118.5
08	DOORS & WINDOWS	90.9	86.0	89.7	96.4	100.2	97.4	90.2	84.9	88.8	95.9	100.7	97.1	97.2	123.6	103.9	100.3	145.6	111.8
09200	PLASTER & GYPSUM BOARD	95.3	86.5	89.3	108.1	91.7	96.9	88.2	83.2	84.8	104.7	94.7	97.9	100.0	127.6	118.8	100.0	137.5	125.6
095,098	CEILINGS & ACOUSTICAL TREATMENT	95.0	86.5	89.3	102.0	91.7	95.0	100.2	83.2	88.7	102.0	94.7	97.1	85.5	127.6	114.0	85.5	137.5	120.7
09600	FLOORING	96.2	73.0	90.2	95.8	86.3	93.3	95.4	84.7	92.6	87.6	61.9	81.0	106.3	126.6	111.5	103.7	124.9	109.1
097,099	WALL FINISHES, PAINTS & COATINGS	96.1	79.9	86.4	100.5	97.4	98.6	95.3	63.8	76.4	86.5	83.6	84.8	93.5	106.2	101.1	99.8	131.1	118.6
09	FINISHES	95.3	86.9	90.8	99.9	92.7	96.0	94.6	84.9	89.4	94.8	87.6	90.9	99.6	123.2	112.6	99.1	135.3	118.7
10-14	TOTAL DIV. 10000 - 14000	100.0	110.3	102.2	100.0	102.6	100.6	100.0	109.0	101.9	100.0	98.7	99.7	100.0	104.7	101.0	100.0	112.0	102.6
15	MECHANICAL	99.9	89.3	95.0	100.0	95.6	98.0	99.7	92.2	96.3	100.0	94.2	97.3	100.1	98.1	99.2	100.0	121.9	110.1
16	ELECTRICAL	94.0	74.6	82.6	93.1	91.6	92.2	94.4	70.0	80.1	95.1	86.4	90.1	101.6	98.9	100.0	102.6	111.9	108.0
01-16	WEIGHTED AVERAGE	97.0	87.4	92.4	97.2	98.3	97.7	95.6	85.9	90.9	96.9	95.7	96.3	99.1	110.8	104.8	99.9	127.7	113.4

MINNESOTA / MISSISSIPPI

DIVISION		ROCHESTER			SAINT PAUL			ST. CLOUD			BILOXI			GREENVILLE			JACKSON		
		MAT.	INST.	TOTAL	MAT.	INST.	TOTAL	MAT.	INST.	TOTAL	MAT.	INST.	TOTAL	MAT.	INST.	TOTAL	MAT.	INST.	TOTAL
01590	EQUIPMENT RENTAL	.0	102.0	102.0	.0	102.0	102.0	.0	101.7	101.7	.0	98.3	98.3	.0	98.3	98.3	.0	98.3	98.3
02	SITE CONSTRUCTION	81.7	102.1	97.0	84.6	103.7	98.9	77.5	105.6	98.5	105.4	86.4	91.2	109.8	86.2	92.1	101.8	86.2	90.1
03100	CONCRETE FORMS & ACCESSORIES	100.0	108.4	107.3	90.7	135.7	130.2	83.8	131.6	125.7	96.0	44.0	50.4	77.3	36.4	41.4	91.3	41.6	47.7
03200	CONCRETE REINFORCEMENT	98.1	132.4	118.8	94.7	133.1	117.9	101.5	132.4	120.1	95.4	63.9	76.4	103.4	46.0	68.7	95.4	49.9	67.9
03300	CAST-IN-PLACE CONCRETE	103.6	101.2	102.6	106.6	129.9	116.4	96.4	127.8	109.5	105.2	47.7	81.1	105.6	41.8	78.9	103.2	44.3	78.6
03	CONCRETE	97.5	111.5	104.7	102.1	133.6	118.2	89.8	130.8	110.8	97.5	51.0	73.7	100.4	42.2	70.6	96.2	46.1	70.5
04	MASONRY	107.7	112.9	111.0	118.5	140.8	132.4	110.1	126.1	120.1	89.2	39.3	58.0	132.0	40.8	74.9	91.3	40.8	59.7
05	METALS	95.1	139.6	111.9	93.9	141.8	112.0	95.2	138.8	111.7	97.2	84.8	92.5	95.5	76.4	88.3	97.1	78.3	90.0
06	WOOD & PLASTICS	112.3	108.0	110.0	103.1	133.4	119.4	94.3	129.6	113.4	96.4	44.1	68.2	77.3	35.8	54.9	91.6	42.5	65.1
07	THERMAL & MOISTURE PROTECTION	100.4	107.2	103.7	100.4	135.6	117.5	98.3	121.3	109.5	95.3	48.8	72.6	95.4	43.7	70.2	94.9	45.0	70.6
08	DOORS & WINDOWS	97.2	130.3	105.6	94.4	144.1	107.0	91.7	142.0	104.5	98.5	50.2	86.2	97.9	40.6	83.3	98.9	45.3	85.3
09200	PLASTER & GYPSUM BOARD	100.0	108.7	106.0	95.3	134.8	122.2	88.2	131.1	117.4	100.7	43.1	61.5	91.3	34.5	52.7	100.7	41.4	60.4
095,098	CEILINGS & ACOUSTICAL TREATMENT	85.5	108.7	101.2	82.9	134.8	118.1	70.1	131.1	111.4	95.0	43.1	59.9	89.8	34.5	52.4	95.0	41.4	58.7
09600	FLOORING	106.1	85.4	100.8	99.1	124.9	105.7	97.9	124.9	104.8	118.6	43.0	99.1	109.0	36.8	90.4	118.6	42.6	99.0
097,099	WALL FINISHES, PAINTS & COATINGS	95.9	102.3	99.8	99.8	124.3	114.5	103.6	131.1	120.1	107.5	37.6	65.6	107.5	37.0	65.2	107.5	37.0	65.2
09	FINISHES	99.6	103.9	102.0	96.8	132.9	116.3	92.6	130.7	113.2	105.6	42.7	71.6	100.6	36.3	65.8	105.6	41.4	70.9
10-14	TOTAL DIV. 10000 - 14000	100.0	100.0	100.0	100.0	111.2	102.4	100.0	109.2	102.0	100.0	58.6	91.1	100.0	56.8	90.7	100.0	57.7	90.9
15	MECHANICAL	100.0	102.6	101.2	100.0	117.1	107.9	99.7	120.2	109.1	99.8	57.9	80.5	99.8	36.9	70.8	99.8	36.5	70.7
16	ELECTRICAL	101.5	87.7	93.5	100.8	110.3	106.4	99.4	110.3	105.8	97.8	47.5	68.4	96.9	39.5	63.4	98.3	39.5	63.9
01-16	WEIGHTED AVERAGE	98.8	107.6	103.0	99.2	125.5	111.8	96.1	123.3	109.2	98.8	56.1	78.2	100.6	47.4	75.0	98.7	49.0	74.7

City Cost Indexes

DIVISION		MISSISSIPPI MERIDIAN			MISSOURI CAPE GIRARDEAU			MISSOURI COLUMBIA			MISSOURI JOPLIN			MISSOURI KANSAS CITY			MISSOURI SPRINGFIELD		
		MAT.	INST.	TOTAL	MAT.	INST.	TOTAL	MAT.	INST.	TOTAL	MAT.	INST.	TOTAL	MAT.	INST.	TOTAL	MAT.	INST.	TOTAL
01590	EQUIPMENT RENTAL	.0	98.3	98.3	.0	108.6	108.6	.0	108.8	108.8	.0	106.0	106.0	.0	103.9	103.9	.0	102.5	102.5
02	SITE CONSTRUCTION	100.7	86.2	89.8	93.3	92.9	93.0	101.0	94.6	96.2	100.4	98.5	99.0	93.8	97.1	96.3	94.2	93.3	93.5
03100	CONCRETE FORMS & ACCESSORIES	75.8	35.4	40.4	84.1	77.8	78.6	85.6	75.0	76.3	103.5	67.8	72.1	102.8	103.9	103.8	101.5	68.0	72.1
03200	CONCRETE REINFORCEMENT	102.1	49.3	70.2	102.9	86.7	93.1	102.4	110.3	107.2	107.9	75.3	88.2	102.5	110.9	107.6	98.4	109.5	105.1
03300	CAST-IN-PLACE CONCRETE	99.7	44.3	76.5	85.9	87.2	86.4	88.3	70.7	81.0	101.4	74.5	90.2	94.2	105.3	98.9	101.2	65.3	86.2
03	CONCRETE	92.7	43.3	67.4	86.7	84.5	85.6	84.7	82.0	83.3	99.0	72.8	85.6	95.2	106.1	100.8	98.9	76.1	87.2
04	MASONRY	88.9	30.3	52.2	113.2	78.2	91.3	130.4	86.0	102.6	97.1	62.2	75.3	101.7	100.8	101.1	86.4	67.4	74.5
05	METALS	95.5	77.3	88.6	92.8	111.5	99.9	93.1	121.3	103.8	97.4	91.4	95.2	100.9	113.1	105.5	96.5	105.6	99.9
06	WOOD & PLASTICS	76.2	34.4	53.6	80.7	75.3	77.8	91.4	71.1	80.4	102.8	66.0	83.0	102.7	103.7	103.2	100.3	69.3	83.6
07	THERMAL & MOISTURE PROTECTION	94.8	41.5	68.8	95.1	82.5	89.0	92.4	83.9	88.3	99.0	65.5	82.7	98.2	104.4	101.2	96.5	72.3	84.7
08	DOORS & WINDOWS	97.8	39.3	83.0	93.8	77.7	89.7	90.8	93.6	91.5	91.6	73.5	87.0	97.8	106.7	100.0	96.2	79.6	92.0
09200	PLASTER & GYPSUM BOARD	91.3	33.0	51.6	92.1	74.4	80.0	100.2	70.1	79.7	109.5	64.9	79.1	103.9	103.6	103.7	105.0	68.2	80.0
095,098	CEILINGS & ACOUSTICAL TREATMENT	89.8	33.0	51.4	88.2	74.4	78.9	91.0	70.1	76.9	94.0	64.9	74.3	99.3	103.6	102.2	96.2	68.2	77.3
09600	FLOORING	107.9	29.9	87.8	92.9	81.3	89.9	102.4	94.1	100.3	112.2	54.7	97.4	92.1	94.1	92.6	105.4	54.7	92.5
097,099	WALL FINISHES, PAINTS & COATINGS	107.5	36.6	64.9	101.0	81.0	89.0	99.4	77.2	86.1	92.4	47.8	65.7	97.3	104.5	101.7	93.7	68.5	78.6
09	FINISHES	99.6	33.8	64.0	90.2	77.8	83.5	93.8	76.1	84.2	101.8	62.8	80.7	96.9	101.9	99.6	98.6	65.0	80.4
10-14	TOTAL DIV. 10000 - 14000	100.0	56.8	90.7	100.0	74.7	94.5	100.0	92.1	98.3	100.0	84.4	96.6	100.0	97.7	99.5	100.0	85.5	96.9
15	MECHANICAL	99.8	36.6	70.7	100.0	97.2	98.7	100.0	95.4	97.9	100.2	59.9	81.6	100.0	105.6	102.6	100.0	70.0	86.2
16	ELECTRICAL	96.5	62.6	76.7	100.5	113.1	107.9	96.8	85.6	90.2	93.6	66.1	77.5	105.5	108.7	107.4	102.3	71.4	84.2
01-16	WEIGHTED AVERAGE	96.8	49.3	73.9	96.0	92.2	94.1	96.7	90.4	93.6	98.0	71.1	85.0	99.4	104.8	102.0	98.0	76.4	87.6

DIVISION		MISSOURI ST. JOSEPH			MISSOURI ST. LOUIS			MONTANA BILLINGS			MONTANA BUTTE			MONTANA GREAT FALLS			MONTANA HELENA		
		MAT.	INST.	TOTAL	MAT.	INST.	TOTAL	MAT.	INST.	TOTAL	MAT.	INST.	TOTAL	MAT.	INST.	TOTAL	MAT.	INST.	TOTAL
01590	EQUIPMENT RENTAL	.0	102.4	102.4	.0	109.8	109.8	.0	98.7	98.7	.0	98.4	98.4	.0	98.4	98.4	.0	98.4	98.4
02	SITE CONSTRUCTION	95.3	93.5	93.9	92.9	96.7	95.8	84.2	96.4	93.3	90.9	95.1	94.0	94.4	95.7	95.4	96.3	95.6	95.8
03100	CONCRETE FORMS & ACCESSORIES	102.7	89.1	90.7	99.4	107.2	106.2	97.8	67.7	71.4	84.3	65.4	67.7	103.7	66.2	70.8	103.8	66.8	71.4
03200	CONCRETE REINFORCEMENT	101.3	99.6	100.3	94.1	111.6	104.7	98.3	77.7	85.8	106.6	77.8	89.2	98.3	77.7	85.9	101.7	67.4	81.0
03300	CAST-IN-PLACE CONCRETE	94.2	100.0	96.6	85.9	111.3	96.5	117.2	70.8	97.8	120.7	70.9	99.9	127.5	58.6	98.7	129.9	69.2	104.5
03	CONCRETE	95.0	95.6	95.3	86.7	110.4	98.9	104.9	71.6	87.9	104.4	70.6	87.1	110.4	66.7	88.1	112.1	68.7	89.9
04	MASONRY	101.2	83.5	90.1	94.6	110.9	104.8	121.1	68.5	88.1	121.0	69.5	88.8	124.7	70.4	90.7	120.1	67.7	87.3
05	METALS	99.5	105.7	101.8	95.2	125.0	106.5	97.1	88.4	93.8	96.4	88.2	93.3	97.5	88.2	94.0	96.7	82.6	91.4
06	WOOD & PLASTICS	103.4	89.3	95.8	95.0	105.0	100.4	102.7	67.8	83.9	90.9	66.2	77.6	110.3	66.0	86.4	110.4	66.0	86.5
07	THERMAL & MOISTURE PROTECTION	98.8	92.0	95.5	95.0	106.5	100.6	99.8	69.5	85.0	99.0	69.7	84.7	100.0	67.9	84.3	100.0	69.1	84.9
08	DOORS & WINDOWS	96.5	96.1	96.4	92.7	112.0	97.6	99.2	66.4	90.9	96.0	65.5	88.2	99.4	65.5	90.8	98.9	62.7	89.7
09200	PLASTER & GYPSUM BOARD	110.4	88.8	95.7	99.2	104.9	103.1	103.3	67.2	78.7	95.0	65.6	75.0	103.3	65.3	77.5	102.7	65.3	77.3
095,098	CEILINGS & ACOUSTICAL TREATMENT	98.0	88.8	91.7	92.1	104.9	100.8	110.3	67.2	81.1	107.7	65.6	79.2	110.3	65.3	79.9	107.7	65.3	79.0
09600	FLOORING	94.4	87.6	92.7	100.2	99.4	100.0	108.3	53.3	94.1	99.6	42.4	84.9	108.3	57.6	95.3	108.3	55.7	94.8
097,099	WALL FINISHES, PAINTS & COATINGS	93.0	82.7	86.8	101.0	108.3	105.4	99.5	63.8	78.1	99.5	43.5	65.9	99.5	48.9	69.1	99.5	54.3	72.4
09	FINISHES	98.0	87.7	92.4	94.3	105.4	100.3	105.8	64.3	83.4	101.2	58.2	77.9	105.9	62.3	82.4	105.4	63.2	82.6
10-14	TOTAL DIV. 10000 - 14000	100.0	92.4	98.4	100.0	102.5	100.5	100.0	76.5	94.9	100.0	75.3	94.7	100.0	76.3	94.9	100.0	62.7	92.0
15	MECHANICAL	100.2	93.2	97.0	100.0	104.8	102.3	100.4	78.3	90.2	100.4	73.0	87.8	100.0	76.5	89.4	100.4	75.6	89.0
16	ELECTRICAL	104.9	86.6	94.2	103.9	113.1	109.3	96.0	77.6	85.2	106.1	72.3	86.4	96.7	72.5	82.6	96.7	72.5	82.5
01-16	WEIGHTED AVERAGE	99.2	92.1	95.8	96.0	108.8	102.2	101.1	76.2	89.1	101.1	73.4	87.8	102.5	74.2	88.9	102.3	73.2	88.3

DIVISION		MONTANA MISSOULA			NEBRASKA GRAND ISLAND			NEBRASKA LINCOLN			NEBRASKA NORTH PLATTE			NEBRASKA OMAHA			NEVADA CARSON CITY		
		MAT.	INST.	TOTAL	MAT.	INST.	TOTAL	MAT.	INST.	TOTAL	MAT.	INST.	TOTAL	MAT.	INST.	TOTAL	MAT.	INST.	TOTAL
01590	EQUIPMENT RENTAL	.0	98.4	98.4	.0	101.4	101.4	.0	101.4	101.4	.0	101.4	101.4	.0	90.6	90.6	.0	101.4	101.4
02	SITE CONSTRUCTION	74.3	94.9	89.7	97.0	91.2	92.6	88.0	91.2	90.4	97.8	90.0	91.9	77.7	89.9	86.8	61.9	104.4	93.7
03100	CONCRETE FORMS & ACCESSORIES	88.5	61.2	64.6	97.8	52.7	58.2	102.8	48.5	55.2	97.7	49.0	55.0	96.7	75.1	77.8	98.2	98.6	98.6
03200	CONCRETE REINFORCEMENT	108.5	78.6	90.4	107.2	76.3	88.5	98.4	77.3	85.7	108.6	76.1	89.0	103.1	77.8	87.8	106.2	112.2	109.8
03300	CAST-IN-PLACE CONCRETE	88.9	67.2	79.8	115.3	60.2	92.3	104.1	62.8	86.8	115.3	53.9	89.6	108.1	78.0	95.5	115.2	87.9	103.8
03	CONCRETE	82.2	67.6	74.8	108.3	61.1	84.1	100.5	60.3	79.9	108.5	57.3	82.2	101.6	76.8	88.9	110.3	97.4	103.7
04	MASONRY	146.6	62.4	93.9	104.5	51.5	71.3	94.8	68.7	78.5	89.9	41.8	59.8	100.0	77.7	86.1	134.6	83.1	102.3
05	METALS	96.5	87.5	93.1	93.0	84.9	89.9	96.0	86.4	92.4	93.2	84.8	90.0	98.6	77.5	90.6	103.6	102.4	103.1
06	WOOD & PLASTICS	95.3	62.3	77.5	97.1	47.3	70.2	101.4	41.3	69.0	96.9	47.3	70.2	94.8	76.3	84.8	96.1	100.0	98.2
07	THERMAL & MOISTURE PROTECTION	98.1	72.0	85.4	96.9	59.3	78.6	97.4	62.2	80.2	96.9	48.9	73.5	92.5	72.6	82.8	103.0	90.2	96.8
08	DOORS & WINDOWS	96.0	64.3	87.9	90.0	51.8	80.3	95.4	51.4	84.2	89.3	55.0	80.6	98.7	69.4	91.3	94.4	108.3	97.9
09200	PLASTER & GYPSUM BOARD	97.0	61.5	72.9	101.5	45.6	63.5	105.0	39.5	60.4	101.8	45.6	63.6	112.4	76.1	87.7	86.6	99.8	95.6
095,098	CEILINGS & ACOUSTICAL TREATMENT	107.7	61.5	76.4	89.7	45.6	59.8	96.2	39.5	57.8	91.0	45.6	60.3	134.0	76.1	94.8	105.9	99.8	101.8
09600	FLOORING	101.7	66.3	92.6	95.0	37.8	80.3	96.8	45.5	83.6	94.9	39.3	80.6	123.3	50.3	104.5	101.7	69.4	93.4
097,099	WALL FINISHES, PAINTS & COATINGS	99.5	48.9	69.1	91.3	44.2	63.1	91.3	44.6	63.3	91.3	53.4	68.6	148.6	75.9	105.0	103.9	84.2	92.1
09	FINISHES	101.3	60.4	79.2	93.8	47.4	68.7	95.9	45.6	68.7	94.0	47.0	68.6	121.3	70.5	93.8	98.0	91.3	94.4
10-14	TOTAL DIV. 10000 - 14000	100.0	58.7	91.1	100.0	78.4	95.4	100.0	77.6	95.2	100.0	58.5	91.1	100.0	80.5	95.8	100.0	120.5	104.4
15	MECHANICAL	100.4	71.0	86.9	100.0	80.3	90.9	100.0	80.3	90.9	100.0	76.0	89.0	99.7	81.1	91.2	100.1	90.1	95.5
16	ELECTRICAL	103.4	75.2	86.9	94.6	69.2	79.7	102.4	69.2	83.0	97.3	45.8	67.2	91.8	85.3	88.0	93.8	91.4	92.4
01-16	WEIGHTED AVERAGE	99.2	72.0	86.1	97.8	68.3	83.6	98.1	69.8	84.5	97.2	61.8	80.1	100.2	79.1	90.0	101.2	94.8	98.2

COST INDEXES

456

DIVISION		NEVADA						NEW HAMPSHIRE									NEW JERSEY		
		LAS VEGAS			RENO			MANCHESTER			NASHUA			PORTSMOUTH			CAMDEN		
		MAT.	INST.	TOTAL	MAT.	INST.	TOTAL	MAT.	INST.	TOTAL	MAT.	INST.	TOTAL	MAT.	INST.	TOTAL	MAT.	INST.	TOTAL
01590	EQUIPMENT RENTAL	.0	101.4	101.4	.0	101.4	101.4	.0	101.8	101.8	.0	101.8	101.8	.0	101.8	101.8	.0	99.7	99.7
02	SITE CONSTRUCTION	61.3	106.1	94.8	61.6	104.4	93.6	86.1	100.6	96.9	87.4	100.6	97.3	81.8	99.6	95.1	90.7	105.2	101.6
03100	CONCRETE FORMS & ACCESSORIES	93.7	108.9	107.0	98.6	98.7	98.7	103.8	66.4	70.9	104.4	66.4	71.0	91.0	61.5	65.2	104.4	127.6	124.7
03200	CONCRETE REINFORCEMENT	98.1	119.9	111.2	98.1	119.8	111.2	109.1	94.2	100.1	109.1	94.2	100.1	87.2	94.0	91.3	109.1	107.1	107.9
03300	CAST-IN-PLACE CONCRETE	108.7	111.3	109.8	117.3	87.9	105.1	102.9	100.5	101.9	94.5	100.5	97.0	89.6	93.8	91.3	82.2	127.5	101.1
03	CONCRETE	105.8	111.4	108.7	110.5	98.9	104.5	108.1	83.6	95.6	103.9	83.6	93.5	95.4	79.1	87.0	97.8	122.4	110.4
04	MASONRY	125.4	92.5	104.8	134.0	83.1	102.1	96.2	95.7	95.9	96.4	95.7	96.0	91.9	84.8	87.4	89.7	123.7	111.0
05	METALS	104.3	106.8	105.2	104.2	105.6	104.7	97.3	88.8	94.1	97.3	88.8	94.1	93.9	86.6	91.1	97.2	98.5	97.7
06	WOOD & PLASTICS	92.1	106.9	100.1	96.4	100.0	98.4	101.9	58.2	78.4	101.9	58.2	78.3	88.6	58.2	72.2	101.9	127.9	115.9
07	THERMAL & MOISTURE PROTECTION	105.5	100.1	102.9	103.0	90.2	96.8	102.2	92.0	97.2	102.6	92.0	97.4	102.1	90.0	96.2	102.0	123.1	112.3
08	DOORS & WINDOWS	94.7	114.1	99.7	94.7	108.7	98.3	107.2	67.7	97.2	107.2	67.7	97.2	108.1	61.5	96.3	107.2	117.9	109.9
09200	PLASTER & GYPSUM BOARD	87.1	107.0	100.6	88.2	99.8	96.1	109.1	56.0	73.0	109.1	56.0	73.0	100.7	56.0	70.3	109.1	127.7	121.8
095,098	CEILINGS & ACOUSTICAL TREATMENT	112.4	107.0	108.7	112.4	99.8	103.9	105.9	56.0	72.1	105.9	56.0	72.1	96.8	56.0	69.2	105.9	127.7	120.7
09600	FLOORING	101.7	82.6	96.8	101.7	69.4	93.4	100.5	107.9	102.4	100.3	107.9	102.2	94.8	107.9	98.2	100.3	128.8	107.6
097,099	WALL FINISHES, PAINTS & COATINGS	103.9	108.4	106.6	103.9	84.2	92.1	93.0	93.4	93.2	93.0	93.4	93.2	93.0	41.6	62.2	93.0	123.4	111.2
09	FINISHES	99.2	103.5	101.5	99.4	91.3	95.0	102.8	75.3	88.0	103.0	75.3	88.1	98.0	66.7	81.1	103.3	128.0	116.6
10 - 14	TOTAL DIV. 10000 - 14000	100.0	110.4	102.2	100.0	120.5	104.4	100.0	85.0	96.8	100.0	85.0	96.8	100.0	80.6	95.8	100.0	116.3	103.5
15	MECHANICAL	100.1	111.9	105.5	100.1	90.2	95.5	100.2	89.0	95.0	100.2	89.0	95.0	100.2	83.1	92.3	100.2	113.4	106.3
16	ELECTRICAL	95.7	106.3	101.9	93.8	91.4	92.4	102.9	77.4	88.0	102.7	77.4	87.9	100.1	77.4	86.9	102.7	122.2	114.1
01 - 16	WEIGHTED AVERAGE	100.6	106.7	103.5	101.5	95.4	98.5	101.5	85.4	93.7	101.0	85.4	93.5	98.3	80.7	89.8	100.0	117.3	108.4

DIVISION		NEW JERSEY															NEW MEXICO		
		ELIZABETH			JERSEY CITY			NEWARK			PATERSON			TRENTON			ALBUQUERQUE		
		MAT.	INST.	TOTAL	MAT.	INST.	TOTAL	MAT.	INST.	TOTAL	MAT.	INST.	TOTAL	MAT.	INST.	TOTAL	MAT.	INST.	TOTAL
01590	EQUIPMENT RENTAL	.0	101.8	101.8	.0	99.7	99.7	.0	101.8	101.8	.0	101.8	101.8	.0	99.2	99.2	.0	116.0	116.0
02	SITE CONSTRUCTION	105.1	105.6	105.5	90.7	105.1	101.5	109.4	105.6	106.5	103.3	105.6	104.6	92.5	105.6	102.3	79.1	116.6	103.4
03100	CONCRETE FORMS & ACCESSORIES	115.0	126.4	125.0	104.4	127.3	124.5	102.5	126.5	123.5	103.1	126.4	123.5	103.2	127.7	124.7	96.4	70.1	73.3
03200	CONCRETE REINFORCEMENT	84.0	125.2	108.9	109.1	125.1	118.8	109.1	125.2	118.8	109.1	125.2	118.8	109.1	117.5	114.2	98.8	66.5	79.3
03300	CAST-IN-PLACE CONCRETE	90.5	125.5	105.1	82.2	127.0	100.9	95.9	125.6	108.3	104.7	125.5	113.4	97.3	123.4	108.2	107.8	75.6	94.3
03	CONCRETE	101.6	124.8	113.5	97.8	125.5	112.0	104.5	124.8	114.9	109.0	124.8	117.1	105.2	123.0	114.3	105.7	72.4	88.6
04	MASONRY	111.3	119.9	116.7	89.7	122.6	110.3	98.5	119.9	111.9	94.1	119.9	110.3	89.3	119.6	108.3	122.0	65.8	86.8
05	METALS	93.7	108.7	99.4	97.2	105.6	100.4	97.2	108.8	101.6	97.2	108.7	101.6	98.0	102.4	99.7	105.3	88.1	98.8
06	WOOD & PLASTICS	116.1	127.9	122.5	101.9	127.9	115.9	103.8	127.9	116.8	103.8	127.9	116.8	101.9	127.8	115.9	96.8	72.2	83.6
07	THERMAL & MOISTURE PROTECTION	102.9	117.3	110.0	102.0	118.5	110.0	102.2	117.3	109.6	102.8	120.9	111.6	100.6	119.9	110.0	101.8	72.9	87.7
08	DOORS & WINDOWS	108.2	122.4	111.8	107.2	122.4	111.0	113.3	122.4	115.6	113.3	122.4	115.6	107.2	120.5	110.6	94.7	73.9	89.4
09200	PLASTER & GYPSUM BOARD	112.1	127.7	122.7	109.1	127.7	121.8	109.1	127.7	121.8	109.1	127.7	121.8	109.1	127.7	121.8	91.5	70.9	77.5
095,098	CEILINGS & ACOUSTICAL TREATMENT	96.8	127.7	117.7	105.9	127.7	120.7	105.9	127.7	120.7	105.9	127.7	120.7	105.9	127.7	120.7	112.4	70.9	84.3
09600	FLOORING	105.0	128.8	111.1	100.3	128.8	107.6	100.5	128.8	107.8	100.3	128.8	107.6	100.5	121.7	105.9	101.7	66.4	92.6
097,099	WALL FINISHES, PAINTS & COATINGS	92.9	123.5	111.2	93.0	123.5	111.3	92.9	123.5	111.2	92.9	123.5	111.2	93.0	123.5	111.3	103.9	56.8	75.7
09	FINISHES	104.3	127.0	116.6	103.3	127.7	116.5	103.9	127.0	116.4	103.5	127.0	116.2	103.4	126.5	115.9	100.3	67.9	82.8
10 - 14	TOTAL DIV. 10000 - 14000	100.0	91.8	98.2	100.0	92.9	98.5	100.0	91.8	98.2	100.0	91.8	98.2	100.0	116.2	103.5	100.0	80.1	95.7
15	MECHANICAL	100.2	121.6	110.1	100.2	122.1	110.3	100.2	122.7	110.6	100.2	120.7	109.6	100.2	120.2	109.4	100.1	72.5	87.4
16	ELECTRICAL	98.7	124.1	113.6	104.2	128.4	118.3	104.0	128.4	118.2	104.2	124.1	115.8	102.9	134.9	121.6	89.0	76.7	81.8
01 - 16	WEIGHTED AVERAGE	101.6	119.4	110.2	100.1	120.3	109.9	102.8	120.3	111.2	102.9	119.3	110.8	101.0	120.4	110.3	100.5	77.2	89.3

DIVISION		NEW MEXICO												NEW YORK					
		FARMINGTON			LAS CRUCES			ROSWELL			SANTA FE			ALBANY			BINGHAMTON		
		MAT.	INST.	TOTAL	MAT.	INST.	TOTAL	MAT.	INST.	TOTAL	MAT.	INST.	TOTAL	MAT.	INST.	TOTAL	MAT.	INST.	TOTAL
01590	EQUIPMENT RENTAL	.0	116.0	116.0	.0	86.6	86.6	.0	116.0	116.0	.0	116.0	116.0	.0	115.8	115.8	.0	115.9	115.9
02	SITE CONSTRUCTION	84.9	111.6	104.9	90.8	86.8	87.8	89.5	111.6	106.0	78.9	111.6	103.4	70.3	106.8	97.6	92.6	90.0	90.6
03100	CONCRETE FORMS & ACCESSORIES	96.3	70.1	73.3	93.6	68.4	71.5	96.3	70.0	73.2	96.3	70.1	73.3	98.0	92.6	93.3	103.0	79.4	82.3
03200	CONCRETE REINFORCEMENT	107.9	66.5	82.9	101.5	60.2	76.5	107.1	60.6	79.0	105.9	66.5	82.1	102.6	92.0	96.2	101.6	89.9	94.5
03300	CAST-IN-PLACE CONCRETE	108.4	75.6	94.7	94.3	65.8	82.4	99.8	75.6	89.7	101.9	75.6	90.9	88.9	103.6	95.1	106.2	89.3	99.2
03	CONCRETE	109.1	72.4	90.3	85.8	66.7	76.0	109.5	71.2	89.9	103.5	72.4	87.6	101.5	97.2	99.3	101.0	86.9	93.8
04	MASONRY	126.8	65.8	88.6	110.7	60.6	79.4	127.1	65.8	88.7	119.1	65.8	85.7	89.6	100.0	96.1	106.4	81.5	90.8
05	METALS	104.4	88.1	98.2	98.6	76.2	90.2	104.4	85.2	97.1	104.4	88.1	98.2	108.0	107.0	103.1	95.8	116.3	103.6
06	WOOD & PLASTICS	96.8	72.2	83.6	87.7	70.8	78.6	96.8	72.2	83.6	96.8	72.2	83.6	97.8	90.9	94.1	106.7	78.1	91.3
07	THERMAL & MOISTURE PROTECTION	102.7	72.9	88.2	88.5	65.9	77.5	103.0	72.9	88.4	102.1	72.9	87.9	90.6	95.5	92.9	101.4	83.0	92.4
08	DOORS & WINDOWS	97.6	73.9	91.6	87.1	71.3	83.1	93.6	72.1	88.1	93.8	73.9	88.7	95.6	85.8	93.1	90.7	76.0	87.0
09200	PLASTER & GYPSUM BOARD	86.0	70.9	75.8	88.5	70.9	76.5	86.0	70.9	75.8	86.0	70.9	75.8	101.7	90.4	94.0	108.4	76.9	86.9
095,098	CEILINGS & ACOUSTICAL TREATMENT	103.3	70.9	81.4	102.1	70.9	81.0	103.3	70.9	81.4	103.3	70.9	81.4	98.9	90.4	93.1	98.9	76.9	84.0
09600	FLOORING	101.7	66.4	92.6	134.0	67.9	117.0	101.7	66.4	92.6	101.7	66.4	92.6	86.5	97.9	89.4	97.4	83.4	93.8
097,099	WALL FINISHES, PAINTS & COATINGS	103.9	56.8	75.7	97.3	56.8	73.0	103.9	56.8	75.7	103.9	56.8	75.7	85.1	80.5	82.3	90.3	81.5	85.0
09	FINISHES	98.2	67.9	81.8	115.0	67.2	89.1	98.6	67.9	82.0	98.0	67.9	81.7	95.4	92.1	93.6	96.8	80.0	87.7
10 - 14	TOTAL DIV. 10000 - 14000	100.0	80.1	95.7	100.0	75.9	94.8	100.0	80.1	95.7	100.0	80.1	95.7	100.0	92.3	98.3	100.0	89.9	97.8
15	MECHANICAL	100.1	72.5	87.4	100.3	72.0	87.3	100.1	72.4	87.4	100.1	72.5	87.4	100.4	92.0	96.5	100.6	85.3	93.6
16	ELECTRICAL	86.7	76.7	80.9	87.5	62.7	73.0	87.7	76.7	81.3	89.0	76.7	81.8	100.9	92.9	96.2	101.4	83.3	90.8
01 - 16	WEIGHTED AVERAGE	101.2	77.2	89.6	96.9	69.9	83.9	101.0	76.7	89.3	99.6	77.2	88.8	97.8	96.2	97.0	98.7	87.0	93.1

COST INDEXES

NEW YORK

DIVISION		BUFFALO MAT.	INST.	TOTAL	HICKSVILLE MAT.	INST.	TOTAL	NEW YORK MAT.	INST.	TOTAL	RIVERHEAD MAT.	INST.	TOTAL	ROCHESTER MAT.	INST.	TOTAL	SCHENECTADY MAT.	INST.	TOTAL
01590	EQUIPMENT RENTAL	.0	93.4	93.4	.0	117.4	117.4	.0	117.8	117.8	.0	117.4	117.4	.0	116.8	116.8	.0	115.8	115.8
02	SITE CONSTRUCTION	94.6	93.5	93.8	111.5	130.4	125.7	132.4	128.4	129.4	111.7	130.4	125.7	72.5	107.5	98.7	70.0	106.6	97.4
03100	CONCRETE FORMS & ACCESSORIES	102.6	113.9	112.5	89.6	155.5	147.4	111.2	180.8	172.2	93.8	155.5	148.0	102.6	97.3	98.0	102.6	91.5	92.9
03200	CONCRETE REINFORCEMENT	101.0	102.8	102.1	103.1	188.7	154.8	109.4	189.2	157.6	105.1	188.8	155.6	101.9	88.6	93.9	101.3	92.0	95.7
03300	CAST-IN-PLACE CONCRETE	119.6	119.1	119.4	99.4	156.2	123.1	124.0	161.1	139.5	97.5	156.2	122.0	113.8	103.0	109.3	97.5	102.0	99.4
03	CONCRETE	108.6	112.8	110.8	105.0	160.4	133.4	121.7	173.3	148.1	103.9	160.5	132.9	114.6	98.7	106.4	106.0	96.2	101.0
04	MASONRY	102.6	116.2	111.1	109.2	157.2	139.2	106.4	163.0	141.8	115.8	157.2	141.7	100.9	99.7	100.1	91.0	97.1	94.8
05	METALS	103.2	93.8	99.6	110.4	138.7	121.1	110.2	141.6	122.1	110.4	140.2	121.7	106.7	109.4	107.7	100.8	107.0	103.1
06	WOOD & PLASTICS	108.7	113.6	111.3	88.9	157.5	125.9	108.4	187.7	151.2	92.7	157.5	127.6	105.1	97.6	101.1	103.1	90.9	96.5
07	THERMAL & MOISTURE PROTECTION	98.9	105.2	102.0	103.9	149.8	126.3	104.9	158.4	131.0	103.9	149.8	126.3	95.5	99.6	97.5	90.7	94.2	92.4
08	DOORS & WINDOWS	93.1	102.5	95.5	88.3	158.1	106.0	97.1	175.4	117.0	88.3	159.0	106.3	96.4	89.3	94.6	95.6	85.8	93.1
09200	PLASTER & GYPSUM BOARD	96.5	113.7	108.2	96.2	159.1	139.0	111.4	189.7	164.7	97.3	159.1	139.3	98.2	97.5	97.7	101.7	90.4	94.0
095,098	CEILINGS & ACOUSTICAL TREATMENT	93.7	113.7	107.2	83.3	159.1	134.6	113.0	189.7	165.0	83.3	159.1	134.6	97.8	97.5	97.6	98.9	90.4	93.1
09600	FLOORING	95.8	115.1	100.8	97.2	93.7	96.3	97.9	165.3	115.3	98.5	93.7	97.3	85.1	89.3	86.2	86.5	97.9	89.4
097,099	WALL FINISHES, PAINTS & COATINGS	94.9	114.3	106.5	112.8	151.3	135.9	97.4	153.2	130.9	112.8	151.3	135.9	92.6	99.0	96.4	85.1	80.5	82.3
09	FINISHES	94.0	115.0	105.3	106.1	142.8	125.9	108.8	176.5	145.4	106.6	142.8	126.2	94.7	96.1	95.4	95.1	91.4	93.1
10-14	TOTAL DIV. 10000 - 14000	100.0	109.8	102.1	100.0	127.9	106.0	100.0	142.5	109.2	100.0	127.9	106.0	100.0	96.8	99.3	100.0	91.1	98.1
15	MECHANICAL	99.9	97.3	98.7	99.8	150.0	122.9	100.3	163.9	129.6	99.8	150.1	123.0	99.9	91.1	95.9	100.4	90.4	95.8
16	ELECTRICAL	99.8	97.3	98.4	102.1	150.5	130.4	107.6	176.8	148.1	103.0	150.5	130.8	104.0	92.5	97.3	101.2	92.9	96.3
01-16	WEIGHTED AVERAGE	100.1	103.8	101.9	102.3	148.3	124.5	106.9	163.2	134.0	102.8	148.5	124.8	101.3	97.4	99.4	98.4	95.3	96.9

NEW YORK / NORTH CAROLINA

DIVISION		SYRACUSE MAT.	INST.	TOTAL	UTICA MAT.	INST.	TOTAL	WATERTOWN MAT.	INST.	TOTAL	WHITE PLAINS MAT.	INST.	TOTAL	YONKERS MAT.	INST.	TOTAL	ASHEVILLE MAT.	INST.	TOTAL
01590	EQUIPMENT RENTAL	.0	115.8	115.8	.0	115.8	115.8	.0	115.8	115.8	.0	117.5	117.5	.0	117.5	117.5	.0	93.1	93.1
02	SITE CONSTRUCTION	91.3	106.8	102.9	68.0	105.7	96.2	76.1	107.8	99.8	122.4	125.4	124.7	131.8	125.2	126.9	102.6	72.4	80.0
03100	CONCRETE FORMS & ACCESSORIES	101.8	88.1	89.8	103.3	83.8	86.2	84.8	91.4	90.6	111.9	140.4	136.9	111.0	140.4	136.8	95.9	44.6	50.8
03200	CONCRETE REINFORCEMENT	102.6	90.5	95.3	102.6	82.6	90.5	103.2	90.6	95.6	102.1	188.0	154.0	106.1	188.0	155.5	94.3	40.4	61.8
03300	CAST-IN-PLACE CONCRETE	98.6	96.5	97.7	90.2	93.8	91.7	105.0	88.3	98.0	109.3	137.1	120.9	122.3	137.1	128.5	94.3	48.6	75.2
03	CONCRETE	104.6	92.5	98.4	102.5	88.2	95.2	116.5	91.2	103.5	109.1	147.0	128.5	120.6	147.0	134.1	98.9	47.2	72.4
04	MASONRY	96.5	95.2	95.7	89.6	91.9	91.0	90.7	96.9	94.6	101.2	136.1	123.1	106.0	136.1	124.9	78.9	38.2	53.4
05	METALS	100.6	105.9	102.6	98.8	102.6	100.2	98.9	105.6	101.4	99.7	136.9	113.8	106.9	137.0	118.3	93.6	75.8	86.9
06	WOOD & PLASTICS	103.1	86.8	94.3	103.1	82.6	92.1	85.1	92.1	88.9	110.5	141.3	127.1	109.2	141.3	126.5	96.6	46.4	69.5
07	THERMAL & MOISTURE PROTECTION	98.9	95.0	97.0	90.7	93.8	92.2	91.0	95.5	93.2	105.7	139.3	122.1	106.1	139.3	122.2	95.8	43.7	70.4
08	DOORS & WINDOWS	93.6	81.6	90.5	95.6	77.2	90.9	95.6	83.0	92.4	91.7	150.3	106.6	95.2	150.3	109.2	91.9	42.8	79.4
09200	PLASTER & GYPSUM BOARD	101.7	86.2	91.1	101.7	81.9	88.2	94.4	91.6	92.5	106.8	142.0	130.8	111.1	142.0	132.1	105.4	44.7	64.1
095,098	CEILINGS & ACOUSTICAL TREATMENT	98.9	86.2	90.3	98.9	81.9	87.4	98.9	91.6	94.0	90.9	142.0	125.5	111.7	142.0	132.2	89.5	44.7	59.1
09600	FLOORING	88.0	93.0	89.3	86.5	93.1	88.2	79.0	93.1	82.7	94.1	165.3	112.4	93.7	165.3	112.1	100.7	43.0	85.8
097,099	WALL FINISHES, PAINTS & COATINGS	90.4	87.6	88.7	85.1	87.3	86.4	85.1	75.7	79.5	95.0	151.3	128.7	95.0	151.3	128.7	108.7	40.3	67.7
09	FINISHES	96.6	88.7	92.3	95.2	85.3	89.8	92.3	89.6	90.9	101.9	146.5	126.0	106.9	146.5	128.3	96.5	43.9	68.1
10-14	TOTAL DIV. 10000 - 14000	100.0	102.6	100.6	100.0	88.9	97.6	100.0	88.5	97.5	100.0	124.0	105.2	100.0	132.0	106.9	100.0	66.2	92.7
15	MECHANICAL	100.4	87.3	94.4	100.4	83.6	92.7	100.4	79.7	90.9	100.6	130.2	114.2	100.6	130.3	114.2	100.0	44.6	74.5
16	ELECTRICAL	101.2	85.7	92.2	101.2	84.9	91.6	101.2	82.1	90.0	95.2	139.8	121.3	105.5	139.8	125.5	97.7	42.0	65.1
01-16	WEIGHTED AVERAGE	99.3	92.7	96.1	97.6	89.4	93.6	99.1	90.5	95.0	101.0	137.7	118.7	105.8	137.9	121.2	96.2	49.8	73.8

NORTH CAROLINA

DIVISION		CHARLOTTE MAT.	INST.	TOTAL	DURHAM MAT.	INST.	TOTAL	FAYETTEVILLE MAT.	INST.	TOTAL	GREENSBORO MAT.	INST.	TOTAL	RALEIGH MAT.	INST.	TOTAL	WILMINGTON MAT.	INST.	TOTAL
01590	EQUIPMENT RENTAL	.0	93.1	93.1	.0	99.5	99.5	.0	99.5	99.5	.0	99.5	99.5	.0	99.5	99.5	.0	93.1	93.1
02	SITE CONSTRUCTION	102.7	72.4	80.0	102.5	82.4	87.4	100.5	82.4	86.9	102.3	82.4	87.4	103.5	82.4	87.7	103.5	72.4	80.2
03100	CONCRETE FORMS & ACCESSORIES	103.7	44.6	51.9	98.0	44.9	51.4	93.4	44.9	50.8	98.1	44.9	51.4	100.8	44.9	51.8	97.5	44.9	51.3
03200	CONCRETE REINFORCEMENT	94.7	40.4	61.9	94.7	56.0	71.3	93.7	56.0	71.0	94.7	56.0	71.3	94.7	56.0	71.3	95.0	56.0	71.5
03300	CAST-IN-PLACE CONCRETE	96.7	48.6	76.6	94.5	48.7	75.4	91.2	48.7	73.5	93.7	48.7	74.9	100.0	48.7	78.6	93.8	48.7	75.0
03	CONCRETE	99.9	47.2	72.9	98.4	50.3	73.8	96.0	50.3	72.6	98.0	50.3	73.6	101.4	50.3	75.2	98.8	50.3	74.0
04	MASONRY	83.5	38.2	55.1	80.2	38.2	53.9	83.9	38.2	55.3	80.0	38.2	53.9	84.6	38.2	55.6	69.6	38.2	50.0
05	METALS	94.8	75.9	87.7	94.9	82.6	90.2	96.0	82.6	90.9	95.7	82.5	90.7	94.9	82.6	90.3	93.9	82.6	89.6
06	WOOD & PLASTICS	105.2	46.4	73.5	98.6	46.4	70.4	93.8	46.4	68.2	98.6	46.4	70.4	101.8	46.4	71.9	98.2	46.4	70.2
07	THERMAL & MOISTURE PROTECTION	95.8	44.7	70.9	96.5	44.3	71.0	96.1	44.3	70.8	96.5	44.3	71.0	96.3	44.3	70.9	95.8	44.3	70.7
08	DOORS & WINDOWS	96.0	42.8	82.5	96.0	48.8	84.0	92.0	48.8	81.0	96.0	48.8	84.0	92.8	48.8	81.6	92.1	48.8	81.1
09200	PLASTER & GYPSUM BOARD	111.2	44.7	65.9	111.2	44.7	65.9	105.7	44.7	64.2	111.2	44.7	65.9	111.2	44.7	65.9	106.2	44.7	64.3
095,098	CEILINGS & ACOUSTICAL TREATMENT	94.7	44.7	60.8	94.7	44.7	60.8	90.8	44.7	59.6	94.7	44.7	60.8	94.7	44.7	60.8	90.8	44.7	59.6
09600	FLOORING	104.5	43.0	88.7	104.7	43.0	88.8	100.8	43.0	85.9	104.7	43.0	88.8	104.7	43.0	88.8	101.7	43.0	86.6
097,099	WALL FINISHES, PAINTS & COATINGS	108.7	40.3	67.7	108.7	40.3	67.7	108.7	40.3	67.7	108.7	40.3	67.7	108.7	40.3	67.7	108.7	40.3	67.7
09	FINISHES	99.5	43.9	69.5	99.7	43.9	69.5	96.9	43.9	68.2	99.7	43.9	69.5	99.7	43.9	69.6	97.2	43.9	68.4
10-14	TOTAL DIV. 10000 - 14000	100.0	66.2	92.7	100.0	69.1	93.3	100.0	69.1	93.3	100.0	66.2	92.7	100.0	69.1	93.3	100.0	69.1	93.3
15	MECHANICAL	100.0	44.6	74.5	100.0	44.8	74.6	100.0	44.8	74.6	100.0	44.8	74.6	100.0	44.8	74.6	100.0	44.8	74.6
16	ELECTRICAL	99.8	37.7	63.5	96.0	40.2	63.4	92.3	40.2	61.8	96.9	40.2	63.7	96.7	40.2	63.6	97.5	40.2	64.0
01-16	WEIGHTED AVERAGE	97.8	49.2	74.4	97.1	51.9	75.3	96.0	51.9	74.7	97.2	51.8	75.3	97.4	51.9	75.5	95.8	51.0	74.2

COST INDEXES

City Cost Indexes

DIVISION		NORTH CAROLINA WINSTON-SALEM			NORTH DAKOTA BISMARCK			FARGO			GRAND FORKS			MINOT			OHIO AKRON		
		MAT.	INST.	TOTAL	MAT.	INST.	TOTAL	MAT.	INST.	TOTAL	MAT.	INST.	TOTAL	MAT.	INST.	TOTAL	MAT.	INST.	TOTAL
01590	EQUIPMENT RENTAL	.0	99.5	99.5	.0	98.4	98.4	.0	98.4	98.4	.0	98.4	98.4	.0	98.4	98.4	.0	97.6	97.6
02	SITE CONSTRUCTION	102.7	82.3	87.5	87.1	96.0	93.7	86.6	96.0	93.6	95.1	94.4	94.6	92.6	96.0	95.1	101.1	107.0	105.5
03100	CONCRETE FORMS & ACCESSORIES	99.6	44.7	51.4	96.4	52.8	58.1	97.2	53.1	58.5	93.0	46.2	51.9	87.7	59.9	63.3	100.0	99.5	99.5
03200	CONCRETE REINFORCEMENT	94.7	40.5	61.9	106.6	77.3	88.9	98.3	76.4	85.0	105.0	76.6	87.9	108.6	77.4	89.8	94.9	97.4	96.4
03300	CAST-IN-PLACE CONCRETE	96.7	48.6	76.6	103.3	57.9	84.3	108.3	60.2	88.2	105.8	56.0	85.0	105.8	56.4	85.1	99.4	106.2	102.2
03	CONCRETE	99.6	47.2	72.8	98.8	60.2	79.1	110.4	61.0	85.1	105.3	56.5	80.3	103.7	62.9	82.8	100.1	100.4	100.2
04	MASONRY	80.2	38.2	53.9	104.6	68.9	82.2	107.0	51.5	72.2	106.2	53.0	72.9	107.3	70.3	84.1	90.5	103.1	98.4
05	METALS	94.9	75.9	87.7	94.3	81.8	89.6	94.2	80.4	89.0	94.3	79.0	88.5	94.6	81.9	89.8	94.3	82.9	90.0
06	WOOD & PLASTICS	98.6	46.4	70.4	88.6	48.1	66.7	88.6	48.8	67.1	85.3	44.2	63.1	80.6	57.7	68.2	94.0	99.0	96.7
07	THERMAL & MOISTURE PROTECTION	96.5	44.3	71.0	100.3	58.4	79.9	101.0	55.1	78.7	101.1	53.2	77.8	100.8	59.6	80.7	104.5	100.4	102.5
08	DOORS & WINDOWS	96.0	44.5	82.9	99.6	52.3	87.6	99.6	52.7	87.7	99.6	47.3	86.3	99.7	57.6	89.0	104.2	99.4	103.0
09200	PLASTER & GYPSUM BOARD	111.2	44.7	65.9	120.3	46.9	70.3	120.3	47.6	70.8	118.7	42.9	67.1	117.2	56.6	76.1	100.3	98.5	99.1
095,098	CEILINGS & ACOUSTICAL TREATMENT	94.7	44.7	60.8	142.3	46.9	77.7	142.3	47.6	78.2	142.3	42.9	75.0	142.3	56.8	84.4	100.6	98.5	99.1
09600	FLOORING	104.7	43.0	88.8	108.9	81.6	101.8	108.7	48.7	93.2	107.1	48.7	92.0	104.3	84.3	99.2	103.4	95.4	101.4
097,099	WALL FINISHES, PAINTS & COATINGS	108.7	40.3	67.7	99.5	40.1	63.9	99.5	75.9	85.4	99.5	30.7	58.2	99.5	33.6	60.0	106.7	122.7	116.3
09	FINISHES	99.7	43.9	69.5	114.5	55.8	82.8	114.3	53.6	81.5	114.2	44.3	76.4	112.9	61.3	85.0	102.6	101.2	101.9
10 - 14	TOTAL DIV. 10000 - 14000	100.0	66.2	92.7	100.0	78.8	95.4	100.0	78.9	95.5	100.0	51.3	89.5	100.0	80.1	95.7	100.0	97.0	99.4
15	MECHANICAL	100.0	44.6	74.5	100.6	66.5	84.9	100.6	70.8	86.9	100.6	48.5	76.6	100.6	63.9	83.8	99.9	101.7	100.7
16	ELECTRICAL	96.9	40.2	63.7	97.7	72.8	83.2	97.5	72.5	82.9	101.6	74.2	85.6	105.7	70.6	85.2	98.0	94.0	95.7
01 - 16	WEIGHTED AVERAGE	97.3	50.6	74.8	100.2	68.9	85.1	101.6	67.6	85.2	101.6	60.3	81.7	101.6	69.6	86.2	99.3	99.0	99.2

DIVISION		OHIO CANTON			CINCINNATI			CLEVELAND			COLUMBUS			DAYTON			LORAIN		
		MAT.	INST.	TOTAL	MAT.	INST.	TOTAL	MAT.	INST.	TOTAL	MAT.	INST.	TOTAL	MAT.	INST.	TOTAL	MAT.	INST.	TOTAL
01590	EQUIPMENT RENTAL	.0	97.6	97.6	.0	102.6	102.6	.0	97.9	97.9	.0	96.6	96.6	.0	96.9	96.9	.0	97.6	97.6
02	SITE CONSTRUCTION	101.2	107.3	105.7	72.9	108.7	99.7	101.1	106.9	105.4	89.1	102.1	98.8	71.7	108.4	99.2	100.5	106.1	104.7
03100	CONCRETE FORMS & ACCESSORIES	100.0	88.8	90.2	99.7	91.6	92.6	100.2	106.4	105.7	94.9	87.9	88.8	99.6	83.2	85.2	100.1	107.8	106.9
03200	CONCRETE REINFORCEMENT	94.9	81.7	86.9	93.2	88.7	90.4	95.4	97.6	96.7	104.1	88.6	94.7	93.2	84.1	87.7	94.9	97.5	96.5
03300	CAST-IN-PLACE CONCRETE	100.3	104.2	101.9	84.7	93.8	88.5	97.5	114.6	104.6	90.4	96.0	92.7	77.2	87.6	81.5	94.6	109.0	100.6
03	CONCRETE	100.6	92.0	96.2	92.2	91.8	92.0	99.2	106.4	102.9	96.6	90.4	93.5	88.5	84.5	86.4	97.7	105.0	101.5
04	MASONRY	91.3	95.0	93.6	82.8	96.1	91.1	94.9	110.5	104.6	95.7	95.8	95.8	82.2	87.9	85.8	87.5	102.3	96.8
05	METALS	94.3	75.8	87.3	93.6	89.5	92.1	95.7	85.3	91.8	95.5	83.4	90.9	93.0	78.7	87.6	94.8	83.3	90.5
06	WOOD & PLASTICS	94.1	87.6	90.6	98.7	90.3	94.1	93.1	103.5	98.7	100.7	86.1	92.8	99.9	81.4	89.9	94.0	110.2	102.7
07	THERMAL & MOISTURE PROTECTION	105.2	97.1	101.2	89.6	97.6	93.5	103.3	114.5	108.8	99.8	97.2	98.5	95.3	89.6	92.5	105.1	109.2	107.1
08	DOORS & WINDOWS	98.6	79.6	93.8	95.9	87.2	93.7	95.0	101.8	96.8	99.5	85.1	95.9	96.2	80.1	92.1	98.6	105.5	100.4
09200	PLASTER & GYPSUM BOARD	100.8	86.8	91.3	97.6	90.3	92.6	99.5	103.1	102.0	101.3	85.6	90.6	97.6	81.2	86.4	100.3	110.1	106.9
095,098	CEILINGS & ACOUSTICAL TREATMENT	100.6	86.8	91.2	92.4	90.3	91.0	99.3	103.1	101.9	94.9	85.6	88.6	93.9	81.2	85.3	100.6	110.1	107.0
09600	FLOORING	103.7	86.9	99.3	105.1	104.3	104.9	103.2	108.5	104.6	98.1	94.7	97.2	107.8	87.4	102.5	103.7	113.9	106.3
097,099	WALL FINISHES, PAINTS & COATINGS	106.7	88.5	95.8	103.8	97.7	100.1	106.7	122.7	116.3	99.5	102.7	101.5	103.8	92.4	97.0	106.7	122.7	116.3
09	FINISHES	102.8	87.5	94.5	96.3	94.5	95.3	102.3	108.4	105.6	99.5	90.3	94.5	97.5	84.5	90.4	102.7	111.2	107.3
10 - 14	TOTAL DIV. 10000 - 14000	100.0	84.6	96.7	100.0	94.9	98.9	100.0	109.0	101.9	100.0	94.8	98.9	100.0	92.4	98.4	100.0	106.5	101.4
15	MECHANICAL	99.9	87.7	94.3	99.8	93.1	96.7	99.9	109.2	104.2	100.0	92.6	96.6	100.9	87.4	94.7	99.9	91.9	96.2
16	ELECTRICAL	96.9	93.7	95.0	97.2	83.1	88.9	97.4	109.4	104.4	100.3	90.5	94.6	95.9	87.2	90.8	97.0	109.4	104.2
01 - 16	WEIGHTED AVERAGE	98.7	90.4	94.7	95.0	93.0	94.0	98.5	106.2	102.2	98.3	91.9	95.2	94.8	87.6	91.3	98.2	101.8	99.9

DIVISION		OHIO SPRINGFIELD			TOLEDO			YOUNGSTOWN			OKLAHOMA ENID			LAWTON			MUSKOGEE		
		MAT.	INST.	TOTAL	MAT.	INST.	TOTAL	MAT.	INST.	TOTAL	MAT.	INST.	TOTAL	MAT.	INST.	TOTAL	MAT.	INST.	TOTAL
01590	EQUIPMENT RENTAL	.0	96.9	96.9	.0	99.7	99.7	.0	97.6	97.6	.0	77.0	77.0	.0	78.1	78.1	.0	86.6	86.6
02	SITE CONSTRUCTION	72.2	107.4	98.6	88.3	103.1	99.4	100.9	107.2	105.6	109.7	89.4	94.5	104.8	91.0	94.5	95.1	85.4	87.8
03100	CONCRETE FORMS & ACCESSORIES	99.6	86.3	87.9	94.9	101.2	100.4	100.0	93.0	93.8	95.6	41.9	48.5	99.7	57.4	62.6	101.3	37.9	45.7
03200	CONCRETE REINFORCEMENT	93.2	84.1	87.7	104.1	91.2	96.3	94.9	97.5	96.5	97.3	84.7	89.7	97.6	84.7	89.8	97.3	40.3	62.8
03300	CAST-IN-PLACE CONCRETE	80.9	87.6	83.7	90.4	107.7	97.6	98.4	105.3	101.3	92.6	53.0	76.1	89.6	53.1	74.3	83.3	40.8	65.6
03	CONCRETE	90.3	85.9	88.0	96.6	101.0	98.9	99.6	97.2	98.4	90.4	54.0	71.8	86.8	60.9	73.5	82.4	40.5	60.9
04	MASONRY	82.5	88.0	85.9	105.8	101.5	103.1	90.6	97.6	95.0	102.6	62.1	77.2	96.7	62.1	75.0	112.4	44.6	70.0
05	METALS	92.9	78.6	87.5	95.3	88.3	92.7	94.3	82.7	89.9	93.9	68.5	84.3	95.9	68.6	85.6	93.8	60.1	81.1
06	WOOD & PLASTICS	101.2	85.7	92.8	100.7	101.2	100.9	94.0	91.0	92.4	100.9	39.0	67.5	104.0	59.9	80.2	105.9	38.6	69.6
07	THERMAL & MOISTURE PROTECTION	95.2	90.1	92.7	101.5	106.8	104.1	105.3	99.7	102.6	99.7	62.3	81.5	99.4	64.5	82.4	99.2	47.1	73.8
08	DOORS & WINDOWS	94.2	83.1	91.3	97.5	95.3	96.9	98.6	97.6	98.4	95.6	53.3	84.8	97.1	64.7	88.9	95.5	37.9	80.9
09200	PLASTER & GYPSUM BOARD	97.6	85.6	89.4	101.3	101.1	101.2	100.3	90.3	93.5	84.3	38.2	52.9	86.1	59.7	68.1	85.3	37.6	52.8
095,098	CEILINGS & ACOUSTICAL TREATMENT	93.9	85.6	88.2	94.9	101.1	99.1	100.6	90.3	93.6	90.3	38.2	55.0	98.1	59.7	72.1	90.3	37.6	54.6
09600	FLOORING	107.8	87.4	102.5	97.2	99.0	97.7	103.7	101.9	103.2	113.5	55.9	98.7	116.5	55.9	100.8	117.7	45.0	99.0
097,099	WALL FINISHES, PAINTS & COATINGS	103.8	92.4	97.0	99.6	113.9	108.1	106.7	103.5	104.8	96.6	58.9	74.0	96.6	58.9	74.0	96.6	37.4	61.1
09	FINISHES	97.5	87.0	91.8	99.2	102.1	100.8	102.7	95.2	98.7	98.3	44.4	69.2	100.5	56.6	76.8	98.7	39.1	66.5
10 - 14	TOTAL DIV. 10000 - 14000	100.0	93.0	98.5	100.0	98.3	99.6	100.0	96.7	99.3	100.0	67.8	93.1	100.0	70.5	93.7	100.0	66.9	92.9
15	MECHANICAL	100.9	87.4	94.7	100.0	102.1	101.0	99.9	93.1	96.8	100.1	65.9	84.4	100.1	65.9	84.4	100.1	31.2	68.4
16	ELECTRICAL	95.9	87.3	90.9	101.0	103.3	102.3	97.0	92.4	94.3	95.3	71.2	81.2	97.6	66.5	79.4	94.8	36.5	60.7
01 - 16	WEIGHTED AVERAGE	94.8	88.3	91.7	98.7	100.6	99.6	98.6	95.1	96.9	97.3	63.5	81.0	97.3	66.3	82.4	96.5	45.0	71.7

COST INDEXES

459

DIVISION		OKLAHOMA						OREGON										
		OKLAHOMA CITY			TULSA			EUGENE			MEDFORD			PORTLAND			SALEM		
		MAT.	INST.	TOTAL	MAT.	INST.	TOTAL	MAT.	INST.	TOTAL	MAT.	INST.	TOTAL	MAT.	INST.	TOTAL	MAT.	INST.	TOTAL
01590	EQUIPMENT RENTAL	.0	78.3	78.3	.0	86.6	86.6	.0	99.5	99.5	.0	99.5	99.5	.0	99.5	99.5	.0	99.5	99.5
02	SITE CONSTRUCTION	106.1	91.6	95.3	101.6	86.4	90.2	108.3	104.8	105.7	118.2	104.8	108.1	109.4	104.8	106.0	108.4	104.8	105.7
03100	CONCRETE FORMS & ACCESSORIES	100.9	49.4	55.7	101.0	50.0	56.2	103.2	106.8	106.4	98.0	106.6	105.5	104.5	107.1	106.8	104.4	107.0	106.6
03200	CONCRETE REINFORCEMENT	97.6	84.7	89.8	97.6	84.6	89.8	106.8	98.3	101.7	98.6	98.2	98.4	102.1	98.5	99.9	107.9	98.4	102.2
03300	CAST-IN-PLACE CONCRETE	96.9	56.6	80.0	90.8	51.8	74.5	104.6	105.6	105.0	108.0	105.5	106.9	112.3	105.7	109.5	108.5	105.6	107.3
03	CONCRETE	90.6	58.6	74.2	87.5	58.2	72.5	108.5	104.4	106.4	115.4	104.2	109.7	111.9	104.6	108.2	110.7	104.5	107.5
04	MASONRY	98.6	64.3	77.1	96.0	64.4	76.2	113.5	103.6	107.3	110.3	103.6	106.1	114.7	106.3	109.4	118.3	106.3	110.8
05	METALS	96.1	68.6	85.7	96.6	83.4	91.6	94.8	96.8	95.6	94.3	96.5	95.2	93.5	97.3	94.9	95.4	97.1	96.1
06	WOOD & PLASTICS	105.9	48.0	74.7	104.7	49.5	74.9	91.6	106.8	99.8	86.4	106.8	97.4	93.1	106.8	100.5	93.1	106.8	100.5
07	THERMAL & MOISTURE PROTECTION	98.4	64.2	81.8	99.2	61.9	81.0	110.3	94.3	102.5	111.3	91.2	101.5	110.2	102.3	106.3	110.2	98.5	104.5
08	DOORS & WINDOWS	97.1	58.2	87.2	97.1	58.5	87.3	98.8	106.7	100.8	101.5	106.7	102.8	96.3	106.7	98.9	98.1	106.7	100.3
09200	PLASTER & GYPSUM BOARD	86.1	47.4	59.8	86.1	48.8	60.7	99.0	106.9	104.3	95.3	106.9	103.2	97.9	106.9	104.0	99.5	106.9	104.5
095,098	CEILINGS & ACOUSTICAL TREATMENT	98.1	47.4	63.8	98.1	48.8	64.7	117.3	106.9	110.2	114.7	106.9	109.4	117.3	106.9	110.2	123.8	106.9	112.3
09600	FLOORING	116.5	55.9	100.8	116.3	58.1	101.3	115.3	101.5	111.8	112.8	101.5	109.9	115.3	101.5	111.8	115.3	101.5	111.8
097,099	WALL FINISHES, PAINTS & COATINGS	96.6	58.9	74.0	96.6	55.3	71.8	115.4	74.6	91.0	115.4	64.7	85.0	115.4	74.6	91.0	115.4	74.6	91.0
09	FINISHES	100.6	50.2	73.4	99.9	51.1	73.6	113.9	102.5	107.7	112.8	101.3	106.6	113.8	102.5	107.7	115.1	102.5	108.3
10-14	TOTAL DIV. 10000 - 14000	100.0	69.9	93.5	100.0	70.6	93.7	100.0	109.8	102.1	100.0	109.7	102.1	100.0	109.8	102.1	100.0	109.8	102.1
15	MECHANICAL	100.1	67.1	84.9	100.1	64.8	83.8	100.1	107.0	103.3	100.1	106.9	103.3	100.1	107.0	103.3	100.1	107.0	103.3
16	ELECTRICAL	96.5	71.2	81.7	97.6	47.6	68.3	103.2	101.5	102.2	107.9	91.1	98.1	103.1	108.3	106.2	102.9	101.5	102.1
01-16	WEIGHTED AVERAGE	97.9	65.9	82.5	97.3	63.1	80.8	103.1	103.5	103.3	104.5	101.8	103.2	103.1	105.0	104.0	103.7	103.9	103.8

DIVISION		PENNSYLVANIA																
		ALLENTOWN			ALTOONA			ERIE			HARRISBURG			PHILADELPHIA			PITTSBURGH		
		MAT.	INST.	TOTAL	MAT.	INST.	TOTAL	MAT.	INST.	TOTAL	MAT.	INST.	TOTAL	MAT.	INST.	TOTAL	MAT.	INST.	TOTAL
01590	EQUIPMENT RENTAL	.0	115.8	115.8	.0	115.8	115.8	.0	115.8	115.8	.0	115.0	115.0	.0	94.5	94.5	.0	115.8	115.8
02	SITE CONSTRUCTION	90.2	106.6	102.4	94.2	106.8	103.6	90.8	107.4	103.2	81.0	105.0	99.0	100.0	94.2	95.6	98.7	109.6	106.9
03100	CONCRETE FORMS & ACCESSORIES	101.7	109.1	108.2	82.8	89.4	88.6	100.4	98.2	98.4	93.5	85.0	86.1	101.5	131.3	127.7	96.9	99.8	99.4
03200	CONCRETE REINFORCEMENT	102.6	104.5	103.7	99.6	96.9	98.0	101.6	96.5	98.5	102.6	96.8	99.1	98.6	138.7	122.8	108.2	106.6	107.2
03300	CAST-IN-PLACE CONCRETE	89.3	97.7	92.8	99.5	89.5	95.3	97.9	87.4	93.5	98.6	88.5	94.4	97.4	125.7	109.2	97.3	98.7	97.9
03	CONCRETE	98.8	105.3	102.1	93.7	92.3	93.0	92.5	95.4	94.0	101.0	90.0	95.4	107.1	130.1	118.9	95.1	102.0	98.6
04	MASONRY	94.4	93.5	93.8	96.9	83.2	88.3	86.8	93.2	90.8	94.9	84.3	88.3	97.2	129.2	117.3	89.9	99.5	95.9
05	METALS	98.7	122.1	107.6	92.3	117.9	102.0	92.5	118.6	102.4	100.5	117.8	107.1	100.7	124.0	109.5	97.2	124.9	107.6
06	WOOD & PLASTICS	103.1	112.9	108.4	81.0	91.8	86.8	98.5	99.1	98.8	96.4	84.5	90.0	99.3	131.6	116.7	94.0	99.0	96.7
07	THERMAL & MOISTURE PROTECTION	98.9	110.4	104.5	97.5	92.7	95.1	97.6	98.4	98.0	102.8	103.1	102.9	101.8	131.0	116.0	96.9	101.9	99.4
08	DOORS & WINDOWS	93.6	109.0	97.5	88.1	97.8	90.5	88.2	96.0	90.2	93.6	93.9	93.6	96.6	135.7	106.5	93.9	109.2	97.8
09200	PLASTER & GYPSUM BOARD	101.7	113.0	109.4	93.0	91.2	91.8	101.7	98.8	99.7	101.7	83.7	89.4	100.5	132.3	122.2	90.6	98.7	96.1
095,098	CEILINGS & ACOUSTICAL TREATMENT	98.9	113.0	108.5	95.0	91.2	92.5	98.9	98.8	98.8	98.9	83.7	88.6	100.7	132.3	122.1	89.4	98.7	95.7
09600	FLOORING	88.0	86.7	87.7	82.1	61.9	76.9	90.0	92.1	90.5	88.2	81.6	86.5	85.6	130.5	97.1	96.8	102.2	98.2
097,099	WALL FINISHES, PAINTS & COATINGS	90.4	96.2	93.9	85.6	115.3	103.4	97.1	91.3	93.6	90.4	86.9	88.3	92.1	136.2	118.6	96.5	117.4	109.0
09	FINISHES	96.6	103.9	100.5	93.0	86.6	89.5	98.2	96.3	97.2	95.5	84.0	89.3	100.2	131.4	117.1	95.4	101.6	98.8
10-14	TOTAL DIV. 10000 - 14000	100.0	102.8	100.6	100.0	100.6	100.1	100.0	104.8	101.0	100.0	92.8	98.4	100.0	122.3	104.8	100.0	105.2	101.1
15	MECHANICAL	100.4	103.2	101.7	99.8	82.5	91.8	99.8	93.5	96.9	100.4	88.6	95.0	100.2	127.1	112.6	100.0	99.7	99.9
16	ELECTRICAL	101.2	85.9	92.3	88.9	102.6	96.9	90.8	86.8	88.4	100.4	79.7	88.3	96.4	124.8	113.0	95.8	102.6	99.8
01-16	WEIGHTED AVERAGE	98.2	102.8	100.4	94.4	94.3	94.3	94.5	97.3	95.9	98.3	91.4	95.0	100.2	125.0	112.1	96.7	104.5	100.5

DIVISION		PENNSYLVANIA									PUERTO RICO			RHODE ISLAND			SOUTH CAROLINA		
		READING			SCRANTON			YORK			SAN JUAN			PROVIDENCE			CHARLESTON		
		MAT.	INST.	TOTAL	MAT.	INST.	TOTAL	MAT.	INST.	TOTAL	MAT.	INST.	TOTAL	MAT.	INST.	TOTAL	MAT.	INST.	TOTAL
01590	EQUIPMENT RENTAL	.0	118.0	118.0	.0	115.8	115.8	.0	115.0	115.0	.0	89.3	89.3	.0	103.7	103.7	.0	99.1	99.1
02	SITE CONSTRUCTION	97.6	110.7	107.4	90.7	106.6	102.6	80.3	105.0	98.8	116.2	92.4	98.4	80.2	104.1	98.1	95.7	81.4	85.0
03100	CONCRETE FORMS & ACCESSORIES	101.0	86.4	88.2	101.8	87.7	89.4	81.1	82.5	82.3	103.2	20.9	31.0	106.7	114.0	113.1	98.0	39.8	46.9
03200	CONCRETE REINFORCEMENT	96.9	96.7	96.8	102.6	99.4	100.7	101.6	96.8	98.7	196.0	12.8	85.4	109.1	117.7	114.3	94.7	67.2	78.1
03300	CAST-IN-PLACE CONCRETE	71.2	89.9	79.0	93.3	91.4	92.5	87.9	86.3	87.2	107.1	32.5	75.9	87.7	115.0	99.1	83.4	50.6	69.7
03	CONCRETE	93.7	91.1	92.4	100.8	92.7	96.7	99.1	88.2	93.5	111.1	24.1	66.5	103.3	114.6	109.1	92.9	50.6	71.2
04	MASONRY	98.6	84.7	89.9	94.7	95.6	95.2	94.2	84.2	87.9	212.0	17.9	90.5	106.3	123.7	117.2	88.2	37.3	56.3
05	METALS	97.5	117.8	105.2	100.6	119.5	107.7	97.8	117.7	105.4	129.0	31.9	92.2	97.3	110.2	102.2	94.9	82.7	90.3
06	WOOD & PLASTICS	99.3	84.6	91.4	103.1	85.3	93.5	87.9	81.2	84.3	92.8	21.2	54.2	102.6	112.8	108.1	98.6	39.1	66.5
07	THERMAL & MOISTURE PROTECTION	101.8	103.2	102.5	98.8	101.5	100.1	97.1	101.1	99.1	169.2	24.8	98.9	101.1	113.7	107.2	96.0	45.4	71.3
08	DOORS & WINDOWS	94.2	91.4	93.5	93.6	95.0	93.9	90.4	92.1	90.8	154.6	17.7	119.8	101.1	114.7	104.5	96.0	44.1	82.8
09200	PLASTER & GYPSUM BOARD	109.0	84.0	92.0	101.7	84.6	90.0	96.5	80.4	85.5	278.6	18.7	101.8	105.7	112.2	110.1	111.2	37.2	60.9
095,098	CEILINGS & ACOUSTICAL TREATMENT	101.9	84.0	89.8	98.9	84.6	89.2	92.4	80.4	84.3	383.3	18.7	136.4	107.4	112.2	110.6	94.7	37.2	55.8
09600	FLOORING	83.7	87.9	84.8	88.0	85.4	87.3	84.0	87.2	84.8	201.7	17.8	154.4	101.1	116.3	105.0	104.7	42.7	88.7
097,099	WALL FINISHES, PAINTS & COATINGS	90.3	98.3	95.1	90.4	98.3	95.1	90.4	78.8	83.4	202.2	16.3	90.7	95.5	117.0	108.4	108.7	40.8	68.0
09	FINISHES	100.8	87.1	93.4	96.6	87.7	91.8	92.4	82.1	86.8	255.2	20.1	128.1	102.9	114.9	109.4	99.9	39.9	67.5
10-14	TOTAL DIV. 10000 - 14000	100.0	97.4	99.4	100.0	98.9	99.8	100.0	92.3	98.3	100.0	22.4	83.3	100.0	107.5	101.6	100.0	67.6	93.0
15	MECHANICAL	100.4	101.3	100.8	100.4	90.4	95.8	100.4	88.5	94.9	105.0	16.4	64.2	100.2	104.2	102.1	100.0	46.1	75.2
16	ELECTRICAL	98.8	83.5	89.8	101.2	87.8	93.4	93.0	74.5	82.2	138.8	16.3	67.2	98.2	97.8	98.0	96.8	34.5	60.4
01-16	WEIGHTED AVERAGE	98.2	95.7	97.0	98.7	95.4	97.1	96.1	90.0	93.1	140.4	27.1	85.8	100.1	109.5	104.6	96.7	50.4	74.4

COST INDEXES

SOUTH CAROLINA / SOUTH DAKOTA

DIVISION		COLUMBIA			FLORENCE			GREENVILLE			SPARTANBURG			ABERDEEN			PIERRE	
		MAT.	INST.	TOTAL	MAT.	INST.	TOTAL	MAT.	INST.	TOTAL	MAT.	INST.	TOTAL	MAT.	INST.	TOTAL	MAT.	INST.	TOTAL
01590	EQUIPMENT RENTAL	.0	99.1	99.1	.0	99.1	99.1	.0	99.1	99.1	.0	99.1	99.1	.0	98.4	98.4	.0	98.4	98.4
02	SITE CONSTRUCTION	95.0	81.4	84.8	106.8	81.4	87.8	101.6	81.0	86.2	101.4	81.0	86.1	83.1	93.9	91.2	81.4	93.9	90.8
03100	CONCRETE FORMS & ACCESSORIES	102.8	42.2	49.6	82.8	42.1	47.1	97.8	41.8	48.7	101.7	41.8	49.2	97.0	41.6	48.4	95.2	43.3	49.6
03200	CONCRETE REINFORCEMENT	94.7	67.2	78.1	94.3	67.2	77.9	94.3	40.9	62.0	94.3	40.9	62.0	105.2	50.5	72.1	104.7	62.7	79.3
03300	CAST-IN-PLACE CONCRETE	78.7	50.3	66.8	71.0	50.5	62.5	71.0	50.4	62.4	71.0	50.4	62.4	108.0	52.9	80.8	97.9	48.5	77.2
03	CONCRETE	90.8	51.5	70.7	94.0	51.6	72.3	92.4	46.5	68.9	92.7	46.5	69.0	98.2	48.7	72.8	95.8	50.2	72.4
04	MASONRY	86.9	38.2	56.4	73.0	37.3	50.6	70.9	37.3	49.9	73.0	37.3	50.6	108.4	59.8	78.0	104.8	57.9	75.5
05	METALS	94.8	82.6	90.2	93.8	82.5	89.5	93.6	72.9	85.8	93.7	72.9	85.8	106.0	72.2	93.2	105.9	77.4	95.1
06	WOOD & PLASTICS	104.2	42.6	71.0	83.2	42.6	61.3	98.4	42.6	68.3	102.6	42.6	70.2	102.6	39.7	68.7	100.8	42.0	69.1
07	THERMAL & MOISTURE PROTECTION	95.9	45.2	71.2	96.3	45.7	71.7	96.3	45.7	71.6	96.3	45.7	71.7	98.8	54.0	77.0	99.0	51.7	75.9
08	DOORS & WINDOWS	96.0	45.9	83.3	92.0	45.9	80.3	92.0	39.8	78.7	92.0	39.8	78.7	95.2	42.6	81.8	98.4	47.3	85.5
09200	PLASTER & GYPSUM BOARD	111.2	40.7	63.3	100.5	40.7	59.8	106.4	40.7	61.7	108.5	40.7	62.4	104.6	38.3	59.5	103.3	40.6	60.7
095,098	CEILINGS & ACOUSTICAL TREATMENT	94.7	40.7	58.2	90.8	40.7	56.9	89.5	40.7	56.5	89.5	40.7	56.5	107.2	38.3	60.6	105.9	40.6	61.7
09600	FLOORING	104.5	42.7	88.6	94.7	42.7	81.3	102.1	43.5	87.0	103.8	43.5	88.3	109.2	61.0	96.8	108.3	43.5	91.6
097,099	WALL FINISHES, PAINTS & COATINGS	108.7	40.8	68.0	108.7	40.8	68.0	108.7	40.8	68.0	108.7	40.8	68.0	99.5	36.7	61.8	99.5	42.0	65.0
09	FINISHES	99.9	41.9	68.1	95.6	41.9	66.5	98.2	42.0	67.8	99.0	42.0	68.2	105.5	44.2	72.4	104.7	42.6	71.1
10 - 14	TOTAL DIV. 10000 - 14000	100.0	68.1	93.1	100.0	68.0	93.1	100.0	68.1	93.1	100.0	68.1	93.1	100.0	61.1	91.6	100.0	72.7	94.1
15	MECHANICAL	100.0	38.9	71.9	100.0	38.9	71.9	100.0	38.8	71.8	100.0	38.8	71.8	100.3	41.5	73.2	100.0	41.5	72.4
16	ELECTRICAL	97.8	36.4	61.9	94.4	20.8	51.4	97.2	33.1	59.7	97.2	33.1	59.7	102.2	63.1	79.4	97.0	53.9	71.8
01 - 16	WEIGHTED AVERAGE	96.5	49.8	74.0	94.9	47.5	72.1	95.1	47.4	72.1	95.4	47.4	72.3	100.8	56.3	79.4	100.1	55.4	78.5

SOUTH DAKOTA / TENNESSEE

DIVISION		RAPID CITY			SIOUX FALLS			CHATTANOOGA			JACKSON			JOHNSON CITY			KNOXVILLE		
		MAT.	INST.	TOTAL	MAT.	INST.	TOTAL	MAT.	INST.	TOTAL	MAT.	INST.	TOTAL	MAT.	INST.	TOTAL	MAT.	INST.	TOTAL
01590	EQUIPMENT RENTAL	.0	98.4	98.4	.0	99.6	99.6	.0	104.5	104.5	.0	104.9	104.9	.0	97.4	97.4	.0	97.4	97.4
02	SITE CONSTRUCTION	81.5	93.7	90.6	82.5	95.8	92.5	104.6	96.6	98.6	102.7	95.9	97.6	115.0	85.4	92.8	90.9	85.4	86.8
03100	CONCRETE FORMS & ACCESSORIES	107.8	38.5	46.9	95.3	44.2	50.5	95.1	48.9	54.5	85.3	39.9	45.4	80.4	49.7	53.5	93.9	49.8	55.2
03200	CONCRETE REINFORCEMENT	98.3	62.8	76.8	98.3	62.8	76.8	95.2	45.7	65.3	99.3	45.4	66.8	99.3	45.9	67.0	95.2	45.9	65.4
03300	CAST-IN-PLACE CONCRETE	97.2	45.8	75.7	101.3	51.5	80.4	101.6	50.9	80.4	101.2	44.0	77.3	81.8	55.7	70.9	95.4	51.1	76.9
03	CONCRETE	95.5	47.2	70.7	95.5	51.6	73.0	93.8	50.9	71.8	95.4	44.5	69.3	100.0	52.9	75.9	90.9	51.4	70.7
04	MASONRY	104.8	53.8	72.9	102.1	57.9	74.4	97.7	45.8	65.2	111.7	35.6	64.1	109.8	45.6	69.6	75.3	45.6	56.7
05	METALS	108.0	77.8	96.5	108.5	77.8	96.9	96.2	79.2	89.8	94.2	77.4	87.9	93.9	78.7	88.1	96.4	79.1	89.9
06	WOOD & PLASTICS	110.3	37.5	71.1	100.8	42.6	69.5	98.4	50.1	72.3	83.8	39.8	60.1	73.8	51.2	61.6	86.6	51.2	67.5
07	THERMAL & MOISTURE PROTECTION	99.4	49.8	75.3	98.9	51.3	75.7	100.6	53.3	77.6	99.4	41.9	71.4	96.4	50.9	74.2	93.9	51.1	73.0
08	DOORS & WINDOWS	99.4	44.9	85.6	99.4	47.7	86.3	101.4	49.9	88.3	102.1	42.0	86.9	97.1	51.1	85.4	93.7	51.1	82.9
09200	PLASTER & GYPSUM BOARD	104.5	36.1	57.9	104.5	41.3	61.5	84.6	49.2	60.5	88.1	38.6	54.4	88.9	50.4	62.7	96.1	50.4	65.0
095,098	CEILINGS & ACOUSTICAL TREATMENT	111.1	36.1	60.3	111.1	41.3	63.9	102.3	49.2	66.4	95.7	38.6	57.0	92.9	50.4	64.1	99.4	50.4	66.2
09600	FLOORING	108.3	76.6	100.1	108.3	75.0	99.8	103.9	53.7	90.9	92.8	26.0	75.6	97.8	55.2	86.8	103.4	55.2	90.9
097,099	WALL FINISHES, PAINTS & COATINGS	99.5	42.0	65.0	99.5	42.0	65.0	109.5	48.9	73.2	96.6	33.2	58.5	107.3	57.4	77.4	107.3	57.4	77.4
09	FINISHES	105.9	45.4	73.2	105.8	49.3	75.3	100.0	49.8	72.8	95.9	36.6	63.9	99.6	51.5	73.6	94.6	51.5	71.3
10 - 14	TOTAL DIV. 10000 - 14000	100.0	70.1	93.6	100.0	72.9	94.2	100.0	58.9	91.2	100.0	55.5	90.4	100.0	67.7	93.0	100.0	67.7	93.0
15	MECHANICAL	100.3	37.1	71.2	100.3	38.7	71.9	99.9	48.2	76.1	100.0	59.7	81.4	99.8	60.2	81.6	99.8	61.9	82.3
16	ELECTRICAL	97.6	53.9	72.1	96.7	76.6	84.9	103.9	56.2	76.0	101.5	54.4	74.0	92.3	54.4	70.2	100.0	63.6	78.7
01 - 16	WEIGHTED AVERAGE	100.7	54.1	78.3	100.5	59.7	80.8	99.2	57.5	79.1	99.2	55.0	77.9	98.6	59.3	79.7	95.1	60.8	78.6

TENNESSEE / TEXAS

DIVISION		MEMPHIS			NASHVILLE			ABILENE			AMARILLO			AUSTIN			BEAUMONT		
		MAT.	INST.	TOTAL	MAT.	INST.	TOTAL	MAT.	INST.	TOTAL	MAT.	INST.	TOTAL	MAT.	INST.	TOTAL	MAT.	INST.	TOTAL
01590	EQUIPMENT RENTAL	.0	102.3	102.3	.0	105.5	105.5	.0	86.6	86.6	.0	86.6	86.6	.0	86.1	86.1	.0	88.9	88.9
02	SITE CONSTRUCTION	96.4	91.8	93.0	96.5	99.3	98.6	102.0	85.6	89.7	102.4	86.5	90.5	88.3	84.6	85.5	96.3	87.7	89.8
03100	CONCRETE FORMS & ACCESSORIES	95.4	64.4	68.2	94.1	67.6	70.9	97.9	45.5	51.9	101.6	55.2	60.9	99.5	58.6	63.6	103.5	56.5	62.2
03200	CONCRETE REINFORCEMENT	93.6	67.6	77.9	96.2	66.0	78.0	97.6	54.9	71.8	97.6	57.5	73.3	96.0	55.3	71.4	92.8	46.8	65.1
03300	CAST-IN-PLACE CONCRETE	90.3	66.6	80.4	94.6	71.8	85.0	94.2	44.7	73.5	99.0	51.2	79.0	88.1	52.3	73.2	86.2	58.1	74.4
03	CONCRETE	88.3	67.5	77.6	91.0	70.3	80.4	89.0	48.1	68.0	91.6	55.1	72.9	79.6	56.4	67.8	85.1	56.0	70.2
04	MASONRY	84.7	77.5	80.2	86.2	67.9	74.8	99.6	56.1	72.3	101.6	54.5	72.1	95.0	58.1	71.9	104.2	60.9	77.1
05	METALS	95.7	95.3	95.6	98.9	92.4	96.4	97.2	68.9	86.5	97.2	70.1	87.0	96.1	67.9	85.4	99.1	66.9	86.9
06	WOOD & PLASTICS	93.1	65.4	78.2	93.9	68.9	80.4	101.1	45.1	70.9	103.9	57.9	79.1	99.8	62.2	79.5	113.1	57.4	83.1
07	THERMAL & MOISTURE PROTECTION	96.8	73.8	85.6	97.3	68.7	83.4	99.4	52.8	76.7	102.0	51.8	77.5	96.5	57.3	77.4	103.0	61.5	82.8
08	DOORS & WINDOWS	100.8	68.1	92.5	98.7	69.1	91.1	92.8	48.0	81.4	92.8	53.2	82.8	93.9	61.3	85.6	97.4	52.8	86.1
09200	PLASTER & GYPSUM BOARD	94.3	64.8	74.2	85.8	68.5	74.0	86.1	44.3	57.6	86.1	57.4	66.6	89.8	61.8	70.8	88.2	56.8	66.9
095,098	CEILINGS & ACOUSTICAL TREATMENT	95.0	64.8	74.6	91.0	68.5	75.7	98.1	44.3	61.6	98.1	57.4	70.6	93.4	61.8	72.0	99.6	56.8	70.6
09600	FLOORING	98.7	49.0	85.9	106.0	80.7	99.5	116.5	71.9	105.0	116.3	69.4	104.2	101.4	53.2	89.0	110.6	75.5	101.6
097,099	WALL FINISHES, PAINTS & COATINGS	99.1	61.2	76.4	112.8	72.4	88.6	95.4	58.4	73.2	95.4	40.5	62.5	92.3	45.4	64.2	91.6	54.5	69.4
09	FINISHES	94.3	60.6	76.1	102.9	70.7	85.5	100.0	51.4	73.7	100.0	56.3	76.4	93.5	56.2	73.4	93.1	59.7	75.0
10 - 14	TOTAL DIV. 10000 - 14000	100.0	74.4	94.5	100.0	75.3	94.7	100.0	73.4	94.3	100.0	70.0	93.5	100.0	70.4	93.6	100.0	78.1	95.3
15	MECHANICAL	99.9	75.9	88.9	99.8	75.9	88.8	100.1	48.4	76.3	100.1	54.8	79.2	99.9	58.9	81.0	100.0	66.7	84.7
16	ELECTRICAL	100.6	79.5	88.3	100.7	68.1	81.6	97.6	51.9	70.9	98.4	63.9	78.2	95.2	72.1	81.7	94.7	67.9	79.0
01 - 16	WEIGHTED AVERAGE	96.4	76.2	86.7	97.9	75.7	87.2	97.3	56.1	77.4	97.9	60.9	80.1	94.5	63.6	79.6	97.0	65.4	81.7

COST INDEXES

City Cost Indexes

TEXAS

DIVISION	CORPUS CHRISTI MAT.	INST.	TOTAL	DALLAS MAT.	INST.	TOTAL	EL PASO MAT.	INST.	TOTAL	FORT WORTH MAT.	INST.	TOTAL	HOUSTON MAT.	INST.	TOTAL	LAREDO MAT.	INST.	TOTAL
01590 EQUIPMENT RENTAL	.0	94.9	94.9	.0	96.9	96.9	.0	86.6	86.6	.0	86.6	86.6	.0	99.2	99.2	.0	86.1	86.1
02 SITE CONSTRUCTION	123.3	79.9	90.8	129.6	86.2	97.1	102.5	85.3	89.6	101.9	85.9	89.9	125.1	85.9	95.8	88.6	84.4	85.5
03100 CONCRETE FORMS & ACCESSORIES	103.5	41.3	48.9	97.8	59.5	64.2	98.4	50.4	56.2	100.2	59.1	64.1	91.2	68.5	71.3	91.9	41.3	47.5
03200 CONCRETE REINFORCEMENT	94.1	53.2	69.4	102.4	59.0	76.2	97.6	51.9	70.0	97.6	58.9	74.2	95.5	64.5	76.8	96.0	53.3	70.2
03300 CAST-IN-PLACE CONCRETE	104.5	48.5	81.1	97.4	57.9	80.9	96.5	39.2	72.5	92.6	53.5	76.3	88.8	70.2	81.0	80.4	55.8	70.1
03 CONCRETE	89.8	48.0	68.4	88.7	60.4	74.2	90.2	47.6	68.4	88.4	57.8	72.7	84.4	69.9	77.0	79.4	49.6	64.1
04 MASONRY	86.2	51.0	64.1	101.7	62.1	76.9	98.6	50.9	68.8	95.0	62.0	74.3	99.8	65.8	78.5	95.3	52.6	68.5
05 METALS	95.6	81.2	90.1	95.5	85.1	91.6	97.0	64.8	84.8	97.0	70.8	87.1	103.6	91.8	99.2	97.0	66.1	85.3
06 WOOD & PLASTICS	113.9	40.9	74.6	102.7	60.4	79.9	102.3	54.2	76.3	109.9	60.4	83.2	98.0	69.1	82.4	90.3	40.8	63.6
07 THERMAL & MOISTURE PROTECTION	99.4	49.5	75.1	94.8	61.3	78.5	98.9	53.9	76.9	99.8	56.8	78.8	94.5	69.0	82.1	95.3	53.3	74.8
08 DOORS & WINDOWS	101.0	42.8	86.2	105.7	57.3	93.4	92.8	49.1	81.7	87.3	57.3	79.7	107.1	67.7	97.1	94.5	43.5	81.6
09200 PLASTER & GYPSUM BOARD	91.1	39.8	56.1	90.9	60.0	69.9	86.1	53.6	64.0	86.1	60.0	68.4	86.2	68.8	74.3	87.3	39.8	54.9
095,098 CEILINGS & ACOUSTICAL TREATMENT	98.6	39.8	58.8	96.5	60.0	71.8	98.1	53.6	68.0	98.1	60.0	72.3	99.3	68.8	78.6	93.4	39.8	57.1
09600 FLOORING	114.6	52.6	98.6	106.5	64.5	95.7	116.5	69.1	104.3	150.9	51.6	125.3	98.7	70.9	91.5	97.9	52.6	86.2
097,099 WALL FINISHES, PAINTS & COATINGS	105.8	44.5	69.0	104.9	57.8	76.7	95.4	38.8	61.5	96.6	57.7	73.3	100.0	63.9	78.4	92.3	49.4	66.6
09 FINISHES	101.0	43.0	69.7	101.5	60.1	79.2	100.0	53.1	74.7	111.8	57.5	82.4	98.2	68.4	82.1	92.1	43.5	65.9
10-14 TOTAL DIV. 10000 - 14000	100.0	74.4	94.5	100.0	77.6	95.2	100.0	68.2	93.1	100.0	77.6	95.2	100.0	82.8	96.3	100.0	66.4	92.8
15 MECHANICAL	99.9	49.5	76.7	99.8	67.5	85.0	100.1	36.9	71.0	100.1	62.8	82.9	100.0	73.3	87.7	99.8	44.5	74.4
16 ELECTRICAL	93.0	55.7	71.2	98.7	66.5	79.9	96.5	59.1	74.7	96.4	64.8	77.9	96.0	70.0	80.8	95.2	65.1	77.6
01-16 WEIGHTED AVERAGE	97.7	55.5	77.4	99.5	67.9	84.3	97.3	54.0	76.4	97.5	64.5	81.6	99.4	73.7	87.0	94.4	55.2	75.5

TEXAS / UTAH

DIVISION	LUBBOCK MAT.	INST.	TOTAL	ODESSA MAT.	INST.	TOTAL	SAN ANTONIO MAT.	INST.	TOTAL	WACO MAT.	INST.	TOTAL	WICHITA FALLS MAT.	INST.	TOTAL	LOGAN MAT.	INST.	TOTAL
01590 EQUIPMENT RENTAL	.0	96.8	96.8	.0	86.6	86.6	.0	89.0	89.0	.0	86.6	86.6	.0	86.6	86.6	.0	100.4	100.4
02 SITE CONSTRUCTION	133.0	83.8	96.1	102.0	86.0	90.0	88.3	88.8	88.7	100.9	85.9	89.6	101.6	85.6	89.6	90.9	101.1	98.5
03100 CONCRETE FORMS & ACCESSORIES	98.7	44.7	51.3	97.9	42.6	49.4	91.9	57.0	61.2	99.8	43.3	50.2	99.8	45.8	52.4	101.9	63.7	68.4
03200 CONCRETE REINFORCEMENT	97.0	54.8	71.5	99.7	54.7	72.5	96.0	53.5	70.3	97.6	55.1	71.9	97.6	53.4	70.9	98.6	77.5	85.9
03300 CAST-IN-PLACE CONCRETE	94.4	51.3	76.4	94.2	47.0	74.5	78.9	61.1	71.5	85.0	55.2	72.5	90.8	50.8	74.1	91.1	74.6	84.2
03 CONCRETE	87.8	51.0	68.9	89.3	47.5	67.9	78.6	58.5	68.3	84.5	50.7	67.2	87.4	50.1	68.1	109.1	70.4	89.3
04 MASONRY	98.9	52.8	70.1	99.6	53.2	70.5	95.2	59.3	72.7	96.4	61.2	74.4	96.8	59.8	73.6	121.1	64.2	85.5
05 METALS	101.3	84.0	94.7	97.4	68.1	86.3	97.6	68.9	86.7	97.1	68.7	86.4	97.1	69.5	86.7	105.0	76.9	94.3
06 WOOD & PLASTICS	102.7	45.2	71.7	101.1	42.8	69.7	90.3	57.0	72.4	108.8	39.8	71.6	108.8	45.0	74.4	88.4	62.1	74.2
07 THERMAL & MOISTURE PROTECTION	90.9	54.1	73.0	99.4	49.3	74.9	95.3	65.1	80.6	100.1	52.9	77.1	100.1	56.8	79.0	102.5	70.8	87.0
08 DOORS & WINDOWS	104.2	46.1	89.5	92.8	44.4	80.5	96.5	56.6	86.4	87.3	41.6	75.6	87.3	47.9	77.3	89.3	62.7	82.6
09200 PLASTER & GYPSUM BOARD	86.4	44.2	57.7	86.1	41.9	56.1	87.3	56.4	66.3	86.1	38.8	53.9	86.1	44.2	57.6	88.2	60.8	69.6
095,098 CEILINGS & ACOUSTICAL TREATMENT	99.4	44.2	62.0	98.1	41.9	60.1	93.4	56.4	68.3	98.1	38.8	58.0	98.1	44.2	61.6	112.4	60.8	77.5
09600 FLOORING	107.8	43.7	91.3	116.5	43.2	97.6	97.9	56.6	87.3	151.2	40.1	122.5	151.7	78.4	132.8	101.7	64.8	92.2
097,099 WALL FINISHES, PAINTS & COATINGS	107.0	36.3	64.6	95.4	36.3	60.0	92.3	49.4	66.6	96.6	37.4	61.1	98.4	59.3	75.0	103.9	51.2	72.3
09 FINISHES	101.7	42.8	69.9	100.0	41.4	68.3	92.2	55.6	72.4	111.8	40.8	73.4	112.1	52.8	80.1	101.1	61.8	79.8
10-14 TOTAL DIV. 10000 - 14000	100.0	73.1	94.2	100.0	67.3	93.0	100.0	71.4	93.8	100.0	74.9	94.6	100.0	68.3	93.2	100.0	80.4	95.8
15 MECHANICAL	99.7	51.8	77.7	100.1	40.5	72.7	99.8	69.9	86.0	100.1	58.3	80.9	100.1	52.8	78.3	100.1	73.7	88.0
16 ELECTRICAL	96.1	48.9	68.5	97.6	45.5	67.1	95.3	65.1	77.7	97.6	65.7	78.9	101.5	59.7	77.1	93.8	79.4	85.4
01-16 WEIGHTED AVERAGE	99.6	56.5	78.8	97.3	51.5	75.2	94.6	65.7	80.7	97.2	59.3	78.9	99.9	58.9	79.1	101.0	74.0	88.1

UTAH / VERMONT / VIRGINIA

DIVISION	OGDEN MAT.	INST.	TOTAL	PROVO MAT.	INST.	TOTAL	SALT LAKE CITY MAT.	INST.	TOTAL	BURLINGTON MAT.	INST.	TOTAL	RUTLAND MAT.	INST.	TOTAL	ALEXANDRIA MAT.	INST.	TOTAL
01590 EQUIPMENT RENTAL	.0	100.4	100.4	.0	99.2	99.2	.0	100.4	100.4	.0	101.8	101.8	.0	101.8	101.8	.0	101.0	101.0
02 SITE CONSTRUCTION	79.0	101.1	95.6	87.3	99.1	96.2	78.7	101.0	95.4	78.9	99.5	94.3	78.9	99.5	94.3	113.7	85.8	92.9
03100 CONCRETE FORMS & ACCESSORIES	101.9	63.7	68.4	103.4	63.8	68.6	100.0	63.8	68.2	93.8	53.9	58.8	104.8	54.0	60.2	92.8	74.5	76.8
03200 CONCRETE REINFORCEMENT	102.8	77.5	87.5	106.6	77.5	89.0	100.3	77.5	86.6	109.1	57.3	77.8	109.1	57.4	77.9	84.1	83.6	83.8
03300 CAST-IN-PLACE CONCRETE	92.4	74.6	85.0	91.2	74.6	84.2	100.3	74.6	89.6	99.0	67.0	85.6	94.3	67.0	82.9	97.6	79.7	90.1
03 CONCRETE	98.8	70.4	84.3	108.5	70.5	89.0	118.2	70.5	93.7	105.5	59.6	82.0	103.9	59.7	81.2	102.0	79.2	90.3
04 MASONRY	114.3	64.2	82.9	127.0	64.2	87.7	132.4	64.2	89.7	97.1	64.0	76.4	85.0	64.0	71.8	85.7	74.1	78.4
05 METALS	105.4	76.9	94.6	102.3	77.0	92.7	104.2	77.0	93.9	98.0	68.9	87.0	97.3	69.0	86.6	94.9	97.1	95.7
06 WOOD & PLASTICS	88.4	62.1	74.2	89.9	62.1	74.9	88.5	62.1	74.3	89.9	52.2	69.6	102.4	52.2	75.3	97.6	75.1	85.4
07 THERMAL & MOISTURE PROTECTION	100.9	70.8	86.2	104.9	70.8	88.3	103.8	70.8	87.7	102.0	58.9	81.0	101.7	60.1	81.4	95.0	79.6	87.5
08 DOORS & WINDOWS	89.3	62.7	82.6	93.7	62.7	85.8	89.3	62.7	82.6	107.2	48.5	92.3	107.2	48.5	92.3	96.0	77.8	91.3
09200 PLASTER & GYPSUM BOARD	88.2	60.8	69.6	86.8	60.8	69.1	88.2	60.8	69.6	109.1	49.8	68.7	109.1	49.8	68.7	111.2	74.2	86.0
095,098 CEILINGS & ACOUSTICAL TREATMENT	112.4	60.8	77.5	104.6	60.8	75.0	112.4	60.8	77.5	105.9	49.8	67.9	105.9	49.8	67.9	94.7	74.2	80.8
09600 FLOORING	101.7	64.8	92.2	102.5	64.8	92.8	101.2	64.8	91.8	100.3	74.9	93.7	100.3	74.9	93.7	104.7	86.1	99.9
097,099 WALL FINISHES, PAINTS & COATINGS	103.9	51.2	72.3	103.9	65.3	80.8	103.9	65.3	80.8	93.0	42.4	62.6	93.0	42.4	62.6	119.8	87.4	100.3
09 FINISHES	99.9	61.8	79.3	100.1	63.4	80.3	100.3	63.4	80.4	102.0	55.9	77.1	101.9	55.9	77.0	100.4	77.9	88.3
10-14 TOTAL DIV. 10000 - 14000	100.0	80.4	95.8	100.0	80.4	95.8	100.0	80.4	95.8	100.0	90.1	97.9	100.0	90.1	97.9	100.0	82.8	96.3
15 MECHANICAL	100.1	73.7	88.0	100.1	73.7	88.0	100.2	73.7	88.0	100.2	70.9	86.7	100.2	70.9	86.7	100.0	85.5	93.3
16 ELECTRICAL	93.8	79.4	85.4	94.5	79.4	85.6	97.2	79.4	86.8	103.7	70.1	84.1	102.7	70.1	83.6	97.0	92.6	94.4
01-16 WEIGHTED AVERAGE	98.9	74.0	86.9	101.3	74.0	88.1	102.5	74.2	88.9	101.0	68.2	85.1	100.0	68.2	84.7	98.2	83.9	91.3

COST INDEXES

VIRGINIA

DIVISION		ARLINGTON			NEWPORT NEWS			NORFOLK			PORTSMOUTH			RICHMOND			ROANOKE		
		MAT.	INST.	TOTAL	MAT.	INST.	TOTAL	MAT.	INST.	TOTAL	MAT.	INST.	TOTAL	MAT.	INST.	TOTAL	MAT.	INST.	TOTAL
01590	EQUIPMENT RENTAL	.0	99.5	99.5	.0	104.6	104.6	.0	105.3	105.3	.0	104.5	104.5	.0	104.5	104.5	.0	99.5	99.5
02	SITE CONSTRUCTION	124.8	82.9	93.4	106.5	84.7	90.2	105.7	85.8	90.8	105.0	84.3	89.5	107.1	85.1	90.6	104.1	81.2	86.9
03100	CONCRETE FORMS & ACCESSORIES	91.6	71.9	74.3	98.0	49.3	55.2	102.5	49.5	56.0	85.3	49.3	53.7	98.8	59.3	64.2	97.8	38.3	45.6
03200	CONCRETE REINFORCEMENT	94.9	66.6	77.8	94.7	71.6	80.7	94.7	71.6	80.7	94.3	71.6	80.6	94.7	70.4	80.0	94.7	66.2	77.5
03300	CAST-IN-PLACE CONCRETE	94.9	76.3	87.1	95.5	57.6	79.7	98.3	57.6	81.3	94.6	57.3	79.0	101.6	57.4	83.1	107.3	49.0	82.9
03	CONCRETE	106.8	73.6	89.8	99.0	58.1	78.0	100.7	58.2	78.9	97.6	58.0	77.3	102.1	62.3	81.7	104.9	49.2	76.3
04	MASONRY	98.6	68.7	79.9	89.1	54.6	67.5	95.8	54.6	70.0	94.2	54.5	69.3	88.0	52.4	65.7	89.5	41.3	59.3
05	METALS	93.8	89.0	92.0	94.9	90.1	93.1	94.1	90.3	92.7	94.1	89.5	92.4	94.9	90.1	93.1	94.7	83.5	90.5
06	WOOD & PLASTICS	94.2	75.1	83.9	98.6	49.4	72.0	103.9	49.4	74.5	85.8	49.4	66.2	99.7	61.9	79.3	98.6	37.8	65.8
07	THERMAL & MOISTURE PROTECTION	97.0	74.2	85.9	96.0	52.0	74.5	95.7	52.0	74.4	96.0	52.0	74.5	95.5	56.1	76.3	95.9	47.5	72.3
08	DOORS & WINDOWS	93.9	72.8	88.6	96.0	52.4	84.9	96.0	55.2	85.6	96.0	55.2	85.7	96.0	57.9	86.3	96.0	45.2	83.1
09200	PLASTER & GYPSUM BOARD	107.7	74.2	84.9	111.2	46.9	67.5	111.2	46.9	67.5	103.0	46.9	64.8	111.2	59.8	76.2	111.2	35.9	60.0
095,098	CEILINGS & ACOUSTICAL TREATMENT	90.8	74.2	79.6	94.7	46.9	62.3	94.7	46.9	62.3	94.7	46.9	62.3	94.7	59.8	71.1	94.7	35.9	54.9
09600	FLOORING	102.9	73.4	95.3	104.7	42.8	88.8	104.5	42.8	88.6	95.9	42.8	82.2	104.5	68.0	95.1	104.7	38.4	87.6
097,099	WALL FINISHES, PAINTS & COATINGS	119.8	89.1	101.4	108.7	44.9	70.4	108.7	58.0	78.3	108.7	39.3	67.1	108.7	58.7	78.7	108.7	35.0	64.5
09	FINISHES	99.7	74.3	86.0	99.7	46.9	71.2	99.7	48.4	72.0	95.7	46.3	69.0	99.6	61.1	78.8	99.6	37.3	65.9
10 - 14	TOTAL DIV. 10000 - 14000	100.0	76.7	95.0	100.0	72.2	94.0	100.0	72.2	94.0	100.0	72.2	94.0	100.0	73.5	94.3	100.0	66.7	92.8
15	MECHANICAL	100.0	79.4	90.5	100.0	60.5	81.8	100.0	64.5	83.6	100.0	64.4	83.6	100.0	64.4	83.6	100.0	42.2	73.4
16	ELECTRICAL	94.2	89.6	91.5	96.9	58.8	74.6	97.0	61.0	75.9	94.7	61.0	75.0	97.8	72.4	82.9	96.9	36.4	61.5
01 - 16	WEIGHTED AVERAGE	99.2	79.0	89.4	97.8	62.2	80.6	98.3	63.7	81.7	97.1	63.2	80.8	98.2	67.6	83.5	98.4	49.9	75.0

WASHINGTON

DIVISION		EVERETT			RICHLAND			SEATTLE			SPOKANE			TACOMA			VANCOUVER		
		MAT.	INST.	TOTAL	MAT.	INST.	TOTAL	MAT.	INST.	TOTAL	MAT.	INST.	TOTAL	MAT.	INST.	TOTAL	MAT.	INST.	TOTAL
01590	EQUIPMENT RENTAL	.0	104.2	104.2	.0	89.9	89.9	.0	104.0	104.0	.0	89.9	89.9	.0	104.2	104.2	.0	97.4	97.4
02	SITE CONSTRUCTION	95.0	115.4	110.3	104.6	89.4	93.3	99.3	113.2	109.7	105.4	89.5	93.5	97.9	115.4	111.0	111.5	100.9	103.6
03100	CONCRETE FORMS & ACCESSORIES	109.8	100.2	101.4	116.7	82.0	86.3	94.1	100.8	100.0	123.8	82.1	87.2	94.1	100.4	99.6	98.5	98.1	98.1
03200	CONCRETE REINFORCEMENT	105.4	91.4	96.9	98.9	86.9	91.7	99.4	91.5	94.6	99.5	87.0	91.9	99.4	91.5	94.6	102.3	91.3	95.6
03300	CAST-IN-PLACE CONCRETE	98.2	106.6	101.7	115.9	85.3	103.1	103.1	108.2	105.2	120.0	85.4	105.5	101.0	108.0	104.0	112.8	100.5	107.7
03	CONCRETE	98.0	100.2	99.1	107.2	83.9	95.3	100.3	101.1	100.7	109.9	84.0	96.6	99.2	100.8	100.0	110.5	97.3	103.8
04	MASONRY	133.2	102.2	113.8	114.8	85.3	96.4	127.7	102.2	111.7	116.4	86.0	97.4	127.5	102.2	111.7	128.2	100.5	110.8
05	METALS	98.5	87.2	94.2	92.8	82.6	89.0	100.0	89.5	96.0	92.6	82.7	88.9	99.5	87.4	94.9	98.0	89.3	94.7
06	WOOD & PLASTICS	99.4	99.1	99.2	97.5	81.6	88.9	84.4	99.1	92.3	106.2	81.6	92.9	83.2	99.1	91.8	80.9	97.8	90.0
07	THERMAL & MOISTURE PROTECTION	107.4	99.1	103.3	173.4	81.6	128.7	107.2	100.3	103.8	170.6	83.3	128.1	107.0	100.3	103.7	109.8	92.8	101.5
08	DOORS & WINDOWS	98.4	97.1	98.1	112.1	76.6	103.1	100.6	97.1	99.7	111.9	77.0	103.0	99.0	97.1	98.5	95.9	95.3	95.7
09200	PLASTER & GYPSUM BOARD	103.0	98.9	100.2	129.6	80.9	96.4	97.2	98.9	98.4	133.2	80.9	97.6	100.3	98.9	99.4	101.8	98.0	99.3
095,098	CEILINGS & ACOUSTICAL TREATMENT	112.5	98.9	103.3	118.6	80.9	93.0	113.5	98.9	103.6	118.6	80.9	93.0	115.1	98.9	104.2	123.1	98.0	106.1
09600	FLOORING	118.8	105.7	115.5	110.3	47.0	94.0	110.8	105.7	109.5	114.0	76.7	104.4	111.4	105.7	109.9	117.1	86.5	109.2
097,099	WALL FINISHES, PAINTS & COATINGS	111.8	83.9	95.1	113.2	76.4	91.1	111.8	95.0	101.7	113.2	76.4	91.1	111.8	95.0	101.7	119.3	70.6	90.1
09	FINISHES	114.2	99.4	106.2	127.7	74.2	98.8	111.2	100.6	105.5	129.4	80.2	102.8	112.0	100.6	105.9	113.1	92.8	102.1
10 - 14	TOTAL DIV. 10000 - 14000	100.0	100.9	100.2	100.0	93.5	98.6	100.0	106.1	101.3	100.0	93.5	98.6	100.0	106.1	101.3	100.0	90.2	97.9
15	MECHANICAL	99.9	99.0	99.5	100.5	97.1	98.9	99.9	113.0	105.9	100.5	85.5	93.6	100.0	97.1	98.6	100.1	103.2	101.5
16	ELECTRICAL	103.2	99.1	100.8	102.5	90.1	95.3	102.9	112.8	108.7	94.8	85.8	89.5	102.9	100.9	101.7	112.7	105.9	108.7
01 - 16	WEIGHTED AVERAGE	103.0	100.0	101.5	107.4	86.6	97.4	103.0	105.2	104.1	107.2	84.6	96.3	102.7	100.3	101.5	105.0	98.7	102.0

DIVISION		WASHINGTON			WEST VIRGINIA												WISCONSIN		
		YAKIMA			CHARLESTON			HUNTINGTON			PARKERSBURG			WHEELING			EAU CLAIRE		
		MAT.	INST.	TOTAL	MAT.	INST.	TOTAL	MAT.	INST.	TOTAL	MAT.	INST.	TOTAL	MAT.	INST.	TOTAL	MAT.	INST.	TOTAL
01590	EQUIPMENT RENTAL	.0	104.2	104.2	.0	99.5	99.5	.0	99.5	99.5	.0	99.5	99.5	.0	99.5	99.5	.0	100.7	100.7
02	SITE CONSTRUCTION	101.4	114.2	111.0	101.8	85.9	89.9	103.2	86.7	90.8	108.2	85.9	91.5	108.8	85.6	91.4	83.5	102.6	97.8
03100	CONCRETE FORMS & ACCESSORIES	96.4	94.8	95.0	105.8	91.3	93.1	99.0	91.1	92.1	88.0	90.8	90.5	90.0	87.7	88.0	99.5	98.3	98.5
03200	CONCRETE REINFORCEMENT	101.9	90.3	94.9	94.7	86.3	89.6	94.7	89.1	91.3	93.3	92.1	92.6	92.7	92.4	92.5	101.8	102.6	102.3
03300	CAST-IN-PLACE CONCRETE	107.9	84.2	98.0	93.5	108.2	99.6	100.8	108.9	104.2	94.2	97.3	95.5	94.2	98.4	96.0	98.6	97.7	98.2
03	CONCRETE	104.7	89.8	97.0	98.4	96.9	97.6	101.7	97.6	99.6	102.1	94.0	98.0	102.1	93.1	97.5	97.1	99.2	98.2
04	MASONRY	119.3	75.5	91.9	86.7	88.9	88.1	89.1	96.3	93.6	76.6	82.4	80.2	99.8	87.9	92.3	92.7	99.2	96.8
05	METALS	98.4	82.5	92.4	94.9	99.9	96.8	95.0	101.4	97.4	93.8	101.3	96.6	93.9	102.3	97.1	92.8	104.3	97.2
06	WOOD & PLASTICS	85.6	95.1	90.2	106.9	90.3	97.9	98.6	89.1	93.5	88.2	90.2	89.3	90.0	87.0	88.4	114.9	97.8	105.6
07	THERMAL & MOISTURE PROTECTION	107.2	79.9	93.9	95.9	90.0	93.0	96.2	92.5	94.4	96.4	85.1	90.9	96.5	85.7	91.2	98.5	91.2	94.9
08	DOORS & WINDOWS	98.6	83.9	94.8	97.2	83.8	93.8	96.0	83.7	92.8	96.6	83.1	93.2	97.5	86.4	94.7	100.9	96.8	99.9
09200	PLASTER & GYPSUM BOARD	100.4	98.9	99.4	111.2	89.9	96.7	111.2	88.6	95.9	104.6	89.8	94.5	105.2	86.4	92.4	114.4	98.0	103.3
095,098	CEILINGS & ACOUSTICAL TREATMENT	111.2	98.9	102.9	94.7	89.9	91.4	94.7	88.6	90.6	90.8	89.8	90.1	90.8	86.4	87.9	106.4	98.0	100.7
09600	FLOORING	112.7	70.7	101.9	104.5	101.2	103.7	104.5	113.4	106.8	99.1	87.8	96.2	100.2	106.0	101.7	91.4	109.5	96.0
097,099	WALL FINISHES, PAINTS & COATINGS	111.8	76.4	90.6	108.7	99.6	103.3	108.7	93.0	99.3	108.7	97.7	102.1	108.7	88.6	96.6	89.7	72.0	79.1
09	FINISHES	111.9	88.8	99.5	99.7	94.1	96.7	99.5	95.4	97.3	96.6	91.0	93.6	97.1	91.0	93.8	99.9	97.7	98.7
10 - 14	TOTAL DIV. 10000 - 14000	100.0	101.9	100.4	100.0	96.8	99.3	100.0	97.1	99.4	100.0	96.7	99.3	100.0	102.6	100.6	100.0	98.7	99.7
15	MECHANICAL	100.0	96.7	98.5	100.0	85.2	93.2	100.0	84.3	92.8	100.0	90.4	95.6	100.0	89.9	95.3	100.3	87.4	94.4
16	ELECTRICAL	106.8	90.1	97.0	97.0	86.6	90.9	96.9	85.1	90.0	97.5	93.6	95.2	93.8	91.0	92.2	103.5	88.5	94.7
01 - 16	WEIGHTED AVERAGE	103.2	91.2	97.4	97.7	90.3	94.2	98.0	91.2	94.8	97.1	90.9	94.1	98.3	91.2	94.9	98.3	95.6	97.0

COST INDEXES

		WISCONSIN																
	DIVISION	GREEN BAY			KENOSHA			LA CROSSE			MADISON			MILWAUKEE			RACINE		
		MAT.	INST.	TOTAL	MAT.	INST.	TOTAL	MAT.	INST.	TOTAL	MAT.	INST.	TOTAL	MAT.	INST.	TOTAL	MAT.	INST.	TOTAL
01590	EQUIPMENT RENTAL	.0	98.4	98.4	.0	97.9	97.9	.0	100.7	100.7	.0	100.2	100.2	.0	86.4	86.4	.0	100.2	100.2
02	SITE CONSTRUCTION	85.9	98.5	95.3	90.9	101.6	98.9	77.7	102.6	96.3	85.6	105.0	100.1	86.9	93.3	91.7	85.9	105.5	100.6
03100	CONCRETE FORMS & ACCESSORIES	116.8	97.6	99.9	113.1	105.0	106.0	83.1	97.9	96.1	102.7	98.5	99.0	105.3	109.3	108.8	102.2	104.9	104.6
03200	CONCRETE REINFORCEMENT	98.3	91.6	94.3	104.8	96.6	99.9	100.4	90.4	94.3	98.3	90.6	93.7	96.6	96.8	96.7	98.3	96.6	97.3
03300	CAST-IN-PLACE CONCRETE	102.0	99.8	101.1	114.1	101.1	108.7	88.7	97.1	92.2	102.9	100.3	101.8	102.9	107.4	104.7	102.9	100.7	102.0
03	CONCRETE	98.5	97.5	98.0	105.5	102.0	103.7	88.0	96.5	92.4	98.5	97.8	98.1	98.7	105.5	102.2	98.4	101.9	100.2
04	MASONRY	124.3	98.2	108.0	99.0	109.7	105.7	91.9	99.2	96.5	103.3	102.2	102.6	102.8	115.1	110.5	102.8	109.6	107.1
05	METALS	95.3	99.4	96.8	95.7	100.4	97.4	92.7	98.7	95.0	96.0	97.0	96.4	97.6	91.0	95.1	96.1	100.3	97.7
06	WOOD & PLASTICS	125.2	97.8	110.4	120.5	104.0	111.6	98.1	97.8	97.9	115.6	97.7	105.9	118.3	108.2	112.8	115.8	104.0	109.4
07	THERMAL & MOISTURE PROTECTION	100.8	89.3	95.2	97.2	104.7	100.9	97.7	90.3	94.1	96.8	98.4	97.6	96.1	107.1	101.4	97.5	102.5	99.9
08	DOORS & WINDOWS	99.2	90.1	96.9	96.8	103.1	98.4	100.9	86.7	97.3	101.7	97.0	100.5	104.0	105.3	104.3	101.7	103.1	102.0
09200	PLASTER & GYPSUM BOARD	98.6	98.0	98.2	99.5	104.6	103.0	106.7	98.0	100.8	112.2	98.0	102.6	111.3	108.6	109.5	111.3	104.6	106.7
095,098	CEILINGS & ACOUSTICAL TREATMENT	101.5	98.0	99.2	88.9	104.6	99.5	99.9	98.0	98.6	101.4	98.0	99.1	97.5	108.6	105.0	97.5	104.6	102.3
09600	FLOORING	109.3	109.5	109.4	110.5	115.0	111.6	83.9	109.5	90.5	94.0	110.3	98.2	96.7	114.7	101.4	94.0	115.0	99.4
097,099	WALL FINISHES, PAINTS & COATINGS	99.5	78.2	86.7	103.9	94.8	98.4	89.7	72.0	79.1	93.5	94.6	94.2	96.0	105.9	101.9	93.5	90.7	91.8
09	FINISHES	104.1	98.1	100.9	101.4	106.1	104.0	94.9	97.7	96.4	99.1	100.4	99.8	99.4	110.4	105.4	98.3	105.7	102.3
10 - 14	TOTAL DIV. 10000 - 14000	100.0	95.0	98.9	100.0	97.5	99.5	100.0	98.7	99.7	100.0	98.4	99.7	100.0	99.7	99.9	100.0	97.5	99.5
15	MECHANICAL	100.6	86.1	93.9	100.2	97.5	99.0	100.3	89.3	95.2	99.9	92.9	96.7	99.9	99.0	99.5	99.9	97.5	98.8
16	ELECTRICAL	97.0	88.1	91.8	95.3	89.3	91.8	103.8	88.5	94.9	96.5	88.6	91.9	94.4	99.1	97.2	92.7	97.5	95.5
01 - 16	WEIGHTED AVERAGE	100.6	93.7	97.2	99.2	100.4	99.8	96.4	94.6	95.5	98.8	96.9	97.9	99.2	102.3	100.7	98.4	101.8	100.1

		WYOMING									CANADA								
	DIVISION	CASPER			CHEYENNE			ROCK SPRINGS			CALGARY, ALBERTA			EDMONTON, ALBERTA			HALIFAX, NOVA SCOTIA		
		MAT.	INST.	TOTAL	MAT.	INST.	TOTAL	MAT.	INST.	TOTAL	MAT.	INST.	TOTAL	MAT.	INST.	TOTAL	MAT.	INST.	TOTAL
01590	EQUIPMENT RENTAL	.0	101.4	101.4	.0	101.4	101.4	.0	101.4	101.4	.0	107.0	107.0	.0	107.0	107.0	.0	98.6	98.6
02	SITE CONSTRUCTION	79.9	100.6	95.3	79.7	100.5	95.3	83.0	100.1	95.8	113.1	105.4	107.3	116.9	105.4	108.3	96.1	96.0	96.1
03100	CONCRETE FORMS & ACCESSORIES	98.9	45.2	51.8	99.0	45.0	51.6	104.1	31.7	40.6	122.9	79.3	84.7	120.7	79.3	84.4	90.7	68.8	71.5
03200	CONCRETE REINFORCEMENT	105.1	53.3	73.8	98.6	53.4	71.3	107.1	53.0	74.5	161.6	68.4	105.3	161.6	68.4	105.3	148.5	50.9	89.5
03300	CAST-IN-PLACE CONCRETE	102.5	59.4	84.5	102.5	59.3	84.5	103.0	44.6	78.6	191.5	89.7	149.0	206.9	89.7	157.9	178.4	70.3	133.2
03	CONCRETE	103.9	52.4	77.5	103.2	52.3	77.1	104.7	41.3	72.3	160.8	81.5	120.1	168.3	81.5	123.8	152.1	66.7	108.4
04	MASONRY	109.4	40.5	66.2	106.6	42.7	66.6	174.5	33.0	86.0	180.3	76.4	115.3	181.4	76.4	115.7	160.2	71.8	104.9
05	METALS	104.7	64.2	89.4	106.0	64.2	90.2	104.7	62.7	88.8	124.2	87.3	110.2	125.9	87.3	111.3	107.9	73.9	95.0
06	WOOD & PLASTICS	97.4	43.4	68.3	96.7	43.4	68.0	103.4	30.8	64.2	113.3	79.1	94.9	110.1	79.1	93.4	85.1	68.2	76.0
07	THERMAL & MOISTURE PROTECTION	102.7	56.0	79.9	102.7	56.5	80.2	103.4	42.5	73.7	126.1	82.5	104.9	126.1	82.5	104.9	105.3	70.0	88.1
08	DOORS & WINDOWS	93.5	45.9	81.4	94.3	45.9	82.0	100.6	39.0	85.0	91.7	75.7	87.7	91.7	75.7	87.7	82.4	63.2	77.6
09200	PLASTER & GYPSUM BOARD	86.3	41.7	55.9	88.2	41.7	56.5	91.5	28.6	48.7	155.9	77.9	102.8	157.5	77.9	103.4	185.1	67.4	105.0
095,098	CEILINGS & ACOUSTICAL TREATMENT	104.6	41.7	62.0	112.4	41.7	64.5	104.0	28.6	53.2	121.4	77.9	92.0	122.7	77.9	92.4	99.3	67.4	77.7
09600	FLOORING	101.7	44.8	87.0	101.7	72.0	94.0	104.0	43.9	88.5	132.7	89.6	121.6	132.7	89.6	121.6	110.0	65.1	98.4
097,099	WALL FINISHES, PAINTS & COATINGS	103.9	58.7	76.8	103.8	58.7	76.8	103.9	35.7	63.0	109.3	78.9	91.1	109.5	78.9	91.1	109.4	62.3	81.1
09	FINISHES	98.5	45.5	69.9	100.1	51.0	73.6	99.9	33.5	64.0	124.5	81.0	101.0	125.1	81.0	101.3	115.4	67.5	89.5
10 - 14	TOTAL DIV. 10000 - 14000	100.0	80.3	95.8	100.0	80.3	95.8	100.0	53.6	90.0	140.0	90.9	129.4	140.0	90.9	129.4	140.0	66.1	124.1
15	MECHANICAL	100.1	70.8	86.6	100.1	46.3	75.4	100.1	32.7	69.1	101.5	80.4	91.8	101.6	80.4	91.8	101.3	71.4	87.6
16	ELECTRICAL	93.9	65.3	77.2	93.9	54.6	70.9	93.0	47.1	66.2	117.6	89.7	101.3	113.5	89.7	99.6	127.3	72.7	95.4
01 - 16	WEIGHTED AVERAGE	99.7	61.9	81.5	99.9	56.4	78.9	104.5	46.2	76.4	123.4	84.6	104.7	124.4	84.6	105.2	116.5	72.5	95.3

		CANADA																	
	DIVISION	HAMILTON, ONTARIO			KITCHENER, ONTARIO			LAVAL, QUEBEC			LONDON, ONTARIO			MONTREAL, QUEBEC			OSHAWA, ONTARIO		
		MAT.	INST.	TOTAL	MAT.	INST.	TOTAL	MAT.	INST.	TOTAL	MAT.	INST.	TOTAL	MAT.	INST.	TOTAL	MAT.	INST.	TOTAL
01590	EQUIPMENT RENTAL	.0	103.1	103.1	.0	103.0	103.0	.0	99.2	99.2	.0	103.2	103.2	.0	101.0	101.0	.0	103.0	103.0
02	SITE CONSTRUCTION	109.3	103.8	105.2	98.3	103.2	102.0	91.1	97.1	95.5	109.3	103.5	105.0	90.8	97.4	95.7	110.9	103.4	105.3
03100	CONCRETE FORMS & ACCESSORIES	122.3	90.1	94.1	116.4	82.6	86.7	131.0	84.4	90.1	128.4	85.5	90.8	131.2	84.7	90.4	123.6	86.6	91.0
03200	CONCRETE REINFORCEMENT	174.8	91.3	124.4	110.9	91.1	98.8	157.2	85.8	114.0	123.8	89.7	103.2	159.8	85.8	115.1	174.8	91.9	124.7
03300	CAST-IN-PLACE CONCRETE	157.2	97.7	132.3	146.7	78.7	118.3	140.1	92.2	120.1	157.2	95.3	131.3	137.5	93.8	119.2	169.7	84.3	134.0
03	CONCRETE	145.1	93.0	118.4	125.0	83.1	103.6	135.0	87.5	110.7	139.5	89.8	114.1	134.0	88.2	110.6	151.4	86.8	118.3
04	MASONRY	167.1	94.0	121.3	162.9	89.8	117.2	160.4	84.0	112.6	167.1	91.2	119.6	163.7	84.1	113.9	165.6	93.3	120.3
05	METALS	115.6	91.0	106.3	114.5	90.4	105.4	106.9	87.1	99.4	113.5	90.5	104.8	116.5	87.6	105.6	106.9	90.7	100.8
06	WOOD & PLASTICS	114.1	89.1	100.6	108.5	81.0	93.7	126.5	84.5	103.9	114.1	83.9	97.8	126.5	84.8	104.0	115.1	84.8	98.8
07	THERMAL & MOISTURE PROTECTION	110.6	91.2	101.2	108.7	86.6	98.0	105.2	88.6	97.1	113.4	89.4	101.7	106.1	89.2	97.9	110.1	87.2	98.9
08	DOORS & WINDOWS	91.7	88.7	91.0	83.1	82.7	83.0	91.7	74.8	87.4	92.6	84.9	90.7	91.7	75.0	87.5	90.8	86.0	89.6
09200	PLASTER & GYPSUM BOARD	210.5	88.8	127.7	162.7	80.6	106.8	158.1	84.1	107.7	210.2	83.5	124.0	163.6	84.1	109.5	166.3	84.4	110.6
095,098	CEILINGS & ACOUSTICAL TREATMENT	124.0	88.8	100.2	99.3	80.6	86.6	99.3	84.1	89.0	122.7	83.5	96.2	122.7	84.1	96.5	99.3	84.4	89.2
09600	FLOORING	132.7	95.6	123.2	128.4	95.6	119.9	132.7	96.3	123.4	134.9	95.6	124.7	132.7	96.3	123.4	132.7	98.1	123.8
097,099	WALL FINISHES, PAINTS & COATINGS	109.4	102.4	105.2	109.4	93.8	100.0	109.4	89.1	97.2	109.4	102.4	105.2	109.4	89.1	97.2	109.4	107.6	108.3
09	FINISHES	131.5	92.6	110.5	118.6	86.2	101.1	119.5	87.5	102.2	132.0	89.2	108.8	124.7	87.7	104.7	121.2	91.0	104.9
10 - 14	TOTAL DIV. 10000 - 14000	140.0	97.1	130.8	140.0	94.8	130.3	140.0	84.1	128.0	140.0	95.8	130.5	140.0	84.9	128.1	140.0	96.3	130.6
15	MECHANICAL	101.7	88.6	95.7	101.3	84.8	93.8	101.3	77.2	90.2	101.6	85.0	94.0	101.8	77.3	90.5	101.3	86.5	94.5
16	ELECTRICAL	128.8	97.6	110.6	125.5	94.2	107.2	125.3	76.2	96.6	128.7	94.4	108.7	123.4	76.2	95.8	125.7	95.0	107.7
01 - 16	WEIGHTED AVERAGE	120.8	93.4	107.6	114.9	89.0	102.4	115.9	83.7	100.4	120.1	90.8	105.9	117.7	84.0	101.5	118.8	91.2	105.5

City Cost Indexes

CANADA

DIVISION		OTTAWA, ONTARIO			QUEBEC, QUEBEC			REGINA, SASKATCHEWAN			SASKATOON, SASKATCHEWAN			ST CATHARINES, ONTARIO			ST JOHNS, NEWFOUNDLAND		
		MAT.	INST.	TOTAL	MAT.	INST.	TOTAL	MAT.	INST.	TOTAL	MAT.	INST.	TOTAL	MAT.	INST.	TOTAL	MAT.	INST.	TOTAL
01590	EQUIPMENT RENTAL	.0	103.0	103.0	.0	101.4	101.4	.0	98.6	98.6	.0	98.6	98.6	.0	100.4	100.4	.0	98.6	98.6
02	SITE CONSTRUCTION	109.6	103.4	104.9	92.5	97.4	96.2	109.4	95.0	98.6	104.5	95.1	97.5	98.8	99.1	99.0	111.9	94.7	99.1
03100	CONCRETE FORMS & ACCESSORIES	121.1	86.4	90.7	131.1	85.0	90.6	102.8	58.1	63.6	102.8	58.0	63.5	113.9	84.4	88.0	100.6	59.1	64.2
03200	CONCRETE REINFORCEMENT	174.8	89.6	123.3	148.5	85.9	110.7	114.1	67.4	85.9	119.2	67.4	87.9	111.4	91.2	99.2	148.5	52.8	90.7
03300	CAST-IN-PLACE CONCRETE	159.5	95.2	132.6	151.5	94.3	127.6	162.8	67.6	123.0	148.0	67.6	114.4	140.1	83.3	116.3	179.6	66.6	132.4
03	CONCRETE	146.2	90.2	117.5	139.7	88.5	113.5	132.5	63.9	97.3	125.6	63.8	94.0	121.7	85.5	103.1	166.1	61.4	112.4
04	MASONRY	167.2	90.8	119.3	164.7	84.1	114.2	166.8	62.6	101.6	167.1	62.6	101.7	162.4	87.2	115.3	160.3	61.9	98.7
05	METALS	115.6	90.5	106.1	115.5	87.9	105.1	106.0	74.2	94.0	106.1	74.1	94.0	105.9	90.5	100.1	107.8	71.6	94.1
06	WOOD & PLASTICS	113.5	85.6	98.4	127.0	84.9	104.3	94.7	56.8	74.2	93.1	56.8	73.5	106.2	85.4	95.0	94.1	58.2	74.8
07	THERMAL & MOISTURE PROTECTION	110.6	87.9	99.5	105.5	89.4	97.7	105.3	63.8	85.1	104.2	62.9	84.1	108.7	87.9	98.6	108.2	61.4	85.4
08	DOORS & WINDOWS	91.7	85.7	90.2	91.7	82.4	89.3	86.6	55.6	78.8	85.6	55.6	78.0	82.6	85.4	83.3	98.0	55.7	87.2
09200	PLASTER & GYPSUM BOARD	256.2	85.2	139.9	215.5	84.1	126.1	146.0	55.6	84.5	146.0	55.6	84.5	147.5	85.0	105.0	191.4	57.1	100.0
095,098	CEILINGS & ACOUSTICAL TREATMENT	109.7	85.2	93.1	105.8	84.1	91.1	96.7	55.6	68.8	96.7	55.6	68.8	96.7	85.0	88.8	101.9	57.1	71.5
09600	FLOORING	132.7	94.3	122.8	132.7	96.3	123.4	120.1	61.2	104.9	120.1	61.2	104.9	126.6	95.6	118.6	114.9	55.4	99.6
097,099	WALL FINISHES, PAINTS & COATINGS	109.4	95.8	101.2	109.4	89.1	97.2	109.4	64.9	82.7	109.4	55.3	77.0	109.4	95.8	101.3	109.4	60.0	79.8
09	FINISHES	134.6	89.0	110.0	128.1	87.8	106.3	114.5	59.1	84.6	114.3	58.0	83.9	115.6	88.0	100.7	119.4	58.5	86.5
10-14	TOTAL DIV. 10000-14000	140.0	93.7	130.0	140.0	85.1	128.2	140.0	63.8	123.6	140.0	63.8	123.6	140.0	72.2	125.4	140.0	64.5	123.8
15	MECHANICAL	101.7	85.3	94.1	101.6	77.4	90.5	101.3	64.2	84.2	101.3	64.2	84.2	101.3	83.1	92.9	101.3	60.7	82.7
16	ELECTRICAL	124.4	95.4	107.5	127.2	76.2	97.4	128.4	65.9	91.9	128.6	65.9	92.0	127.5	95.5	108.8	124.2	61.2	87.4
01-16	WEIGHTED AVERAGE	120.8	90.9	106.4	119.0	84.4	102.3	115.2	66.9	91.9	114.1	66.8	91.3	113.1	88.3	101.2	120.7	64.8	93.7

CANADA

DIVISION		THUNDER BAY, ONTARIO			TORONTO, ONTARIO			VANCOUVER, B C			WINDSOR, ONTARIO			WINNIPEG, MANITOBA					
		MAT.	INST.	TOTAL	MAT.	INST.	TOTAL	MAT.	INST.	TOTAL	MAT.	INST.	TOTAL	MAT.	INST.	TOTAL	MAT.	INST.	TOTAL
01590	EQUIPMENT RENTAL	.0	100.4	100.4	.0	103.2	103.2	.0	111.3	111.3	.0	100.4	100.4	.0	105.3	105.3	.0	.0	.0
02	SITE CONSTRUCTION	104.6	99.2	100.6	110.5	104.2	105.8	114.9	107.8	109.6	94.7	99.4	98.2	108.9	100.6	102.7	.0	.0	.0
03100	CONCRETE FORMS & ACCESSORIES	123.6	85.5	90.1	123.9	96.9	100.2	115.5	88.1	91.5	123.6	86.1	90.7	123.1	69.9	76.4	.0	.0	.0
03200	CONCRETE REINFORCEMENT	99.5	90.5	94.1	171.2	92.2	123.5	161.6	83.7	114.5	109.1	89.8	97.4	161.6	57.9	99.0	.0	.0	.0
03300	CAST-IN-PLACE CONCRETE	154.2	95.3	129.6	163.2	105.1	139.0	153.2	97.4	129.9	143.5	96.5	123.8	157.5	74.7	122.9	.0	.0	.0
03	CONCRETE	131.6	89.9	110.3	147.8	98.7	122.7	145.3	90.9	117.4	123.8	90.5	106.7	143.4	69.9	105.8	.0	.0	.0
04	MASONRY	163.3	88.0	116.2	186.2	102.2	133.6	175.1	91.5	122.8	162.7	94.6	120.1	169.0	65.4	104.1	.0	.0	.0
05	METALS	105.7	89.6	99.6	117.6	92.8	108.2	122.2	92.7	111.0	105.7	90.8	100.1	124.9	77.3	106.9	.0	.0	.0
06	WOOD & PLASTICS	115.1	85.8	99.3	115.1	95.5	104.5	113.3	86.1	98.6	115.1	84.3	98.5	113.2	70.8	90.4	.0	.0	.0
07	THERMAL & MOISTURE PROTECTION	109.0	87.1	98.3	110.6	98.2	104.6	113.9	92.5	103.5	108.8	91.0	100.1	105.2	72.5	89.3	.0	.0	.0
08	DOORS & WINDOWS	81.6	84.9	82.5	90.8	94.0	91.6	94.2	84.0	91.7	81.4	84.7	82.2	91.7	63.5	84.6	.0	.0	.0
09200	PLASTER & GYPSUM BOARD	175.6	85.4	114.2	166.9	95.4	118.2	152.4	85.2	106.6	168.8	83.9	111.0	157.6	69.4	97.6	.0	.0	.0
095,098	CEILINGS & ACOUSTICAL TREATMENT	95.4	85.4	88.6	101.9	95.4	97.5	101.9	85.2	90.5	95.4	83.9	87.6	101.9	69.4	79.9	.0	.0	.0
09600	FLOORING	132.7	55.1	112.7	132.7	101.4	124.7	132.7	92.8	122.5	132.7	96.4	123.4	132.7	68.0	116.1	.0	.0	.0
097,099	WALL FINISHES, PAINTS & COATINGS	109.4	97.9	102.5	112.4	107.6	109.5	109.3	101.8	104.8	109.4	97.5	102.2	109.5	57.0	78.0	.0	.0	.0
09	FINISHES	121.3	82.1	100.1	122.1	99.0	109.7	121.1	90.0	104.3	120.1	89.2	103.4	121.0	68.4	92.6	.0	.0	.0
10-14	TOTAL DIV. 10000-14000	140.0	73.3	125.6	140.0	99.6	131.3	140.0	94.8	130.3	140.0	74.6	125.9	140.0	67.8	124.5	.0	.0	.0
15	MECHANICAL	101.3	83.8	93.3	101.6	94.6	98.4	101.5	87.5	95.1	101.3	86.5	94.5	101.6	70.9	87.4	.0	.0	.0
16	ELECTRICAL	125.5	93.6	106.9	130.9	98.3	111.8	130.3	86.0	104.4	125.7	95.5	108.0	128.4	73.1	96.0	.0	.0	.0
01-16	WEIGHTED AVERAGE	114.9	88.1	102.0	121.7	98.0	110.2	121.8	91.0	107.0	113.5	90.8	102.5	120.7	73.2	97.8	.0	.0	.0

Location Factors

Costs shown in *Means cost data publications* are based on National Averages for materials and installation. To adjust these costs to a specific location, simply multiply the base cost by the factor and divide by 100 for that city. The data is arranged alphabetically by state and postal zip code numbers. For a city not listed, use the factor for a nearby city with similar economic characteristics.

STATE/ZIP	CITY	MAT.	INST.	TOTAL
ALABAMA				
350-352	Birmingham	96.6	76.5	86.9
354	Tuscaloosa	96.3	57.1	77.4
355	Jasper	97.4	54.2	76.6
356	Decatur	96.3	59.4	78.5
357-358	Huntsville	96.2	72.6	84.8
359	Gadsden	97.1	60.6	79.5
360-361	Montgomery	97.1	59.0	78.7
362	Anniston	96.1	48.9	73.4
363	Dothan	97.0	52.1	75.3
364	Evergreen	96.3	53.4	75.6
365-366	Mobile	97.2	62.8	80.6
367	Selma	96.6	54.3	76.2
368	Phenix City	97.4	57.8	78.3
369	Butler	96.8	52.1	75.2
ALASKA				
995-996	Anchorage	135.1	116.5	126.1
997	Fairbanks	131.0	119.2	125.3
998	Juneau	134.4	115.9	125.5
999	Ketchikan	145.5	115.9	131.2
ARIZONA				
850,853	Phoenix	100.0	74.8	87.9
852	Mesa/Tempe	99.7	69.2	85.0
855	Globe	100.6	64.5	83.2
856-857	Tucson	99.0	71.2	85.6
859	Show Low	100.7	65.2	83.5
860	Flagstaff	102.0	68.9	86.1
863	Prescott	99.6	66.4	83.6
864	Kingman	98.2	67.6	83.4
865	Chambers	98.2	64.4	81.9
ARKANSAS				
716	Pine Bluff	95.8	63.6	80.3
717	Camden	94.1	41.4	68.7
718	Texarkana	95.2	46.7	71.8
719	Hot Springs	93.2	40.8	68.0
720-722	Little Rock	96.3	65.9	81.6
723	West Memphis	95.6	56.7	76.9
724	Jonesboro	95.6	56.7	76.9
725	Batesville	94.2	51.4	73.6
726	Harrison	95.6	51.4	74.3
727	Fayetteville	92.6	36.1	65.4
728	Russellville	94.1	49.5	72.6
729	Fort Smith	96.4	58.6	78.2
CALIFORNIA				
900-902	Los Angeles	102.9	114.0	108.3
903-905	Inglewood	99.2	108.3	103.6
906-908	Long Beach	100.9	108.2	104.4
910-912	Pasadena	101.2	108.4	104.7
913-916	Van Nuys	105.0	108.5	106.7
917-918	Alhambra	103.8	108.5	106.1
919-921	San Diego	105.1	104.3	104.7
922	Palm Springs	101.2	105.1	103.1
923-924	San Bernardino	98.8	105.5	102.0
925	Riverside	103.2	112.2	107.6
926-927	Santa Ana	101.0	108.0	104.4
928	Anaheim	103.8	113.3	108.4
930	Oxnard	104.3	110.9	107.5
931	Santa Barbara	103.4	112.6	107.9
932-933	Bakersfield	103.2	109.9	106.5
934	San Luis Obispo	105.0	107.8	106.4
935	Mojave	101.6	105.1	103.3
936-938	Fresno	105.1	112.4	108.6
939	Salinas	105.9	117.5	111.5
940-941	San Francisco	112.5	135.5	123.6
942,956-958	Sacramento	107.5	115.0	111.1
943	Palo Alto	106.0	127.2	116.2
944	San Mateo	109.2	127.3	117.9
945	Vallejo	106.0	118.9	112.2
946	Oakland	111.1	125.7	118.2
947	Berkeley	110.7	124.6	117.4
948	Richmond	110.4	123.9	116.9
949	San Rafael	111.8	124.6	118.0
950	Santa Cruz	110.1	117.9	113.9

STATE/ZIP	CITY	MAT.	INST.	TOTAL
CALIFORNIA (CONT'D)				
951	San Jose	109.3	129.7	119.2
952	Stockton	104.9	113.3	108.9
953	Modesto	104.9	112.9	108.7
954	Santa Rosa	106.1	123.8	114.6
955	Eureka	107.3	106.6	107.0
959	Marysville	106.3	113.0	109.5
960	Redding	107.3	113.4	110.3
961	Susanville	107.3	112.9	110.0
COLORADO				
800-802	Denver	101.3	89.9	95.8
803	Boulder	99.2	85.8	92.8
804	Golden	101.8	85.4	93.9
805	Fort Collins	102.5	82.1	92.7
806	Greeley	99.6	70.5	85.6
807	Fort Morgan	100.1	85.5	93.1
808-809	Colorado Springs	101.3	87.6	94.7
810	Pueblo	101.8	83.7	93.1
811	Alamosa	104.3	80.7	92.9
812	Salida	104.3	80.6	92.9
813	Durango	104.8	79.7	92.7
814	Montrose	103.2	79.6	91.8
815	Grand Junction	106.3	78.1	92.7
816	Glenwood Springs	104.1	85.6	95.2
CONNECTICUT				
060	New Britain	101.8	114.2	107.8
061	Hartford	102.2	113.4	107.6
062	Willimantic	102.5	113.1	107.6
063	New London	99.2	114.2	106.5
064	Meriden	101.8	114.7	108.0
065	New Haven	103.8	114.7	109.1
066	Bridgeport	103.1	114.3	108.5
067	Waterbury	102.5	114.2	108.2
068	Norwalk	102.5	115.0	108.5
069	Stamford	102.6	119.3	110.7
D.C.				
200-205	Washington	100.2	89.9	95.2
DELAWARE				
197	Newark	99.5	106.8	103.0
198	Wilmington	98.9	106.8	102.7
199	Dover	99.6	106.8	103.1
FLORIDA				
320,322	Jacksonville	98.6	62.4	81.1
321	Daytona Beach	98.7	74.7	87.2
323	Tallahassee	99.1	50.6	75.7
324	Panama City	99.8	39.1	70.6
325	Pensacola	99.5	59.2	80.1
326,344	Gainesville	100.3	58.5	80.2
327-328,347	Orlando	100.6	64.6	83.2
329	Melbourne	100.8	79.0	90.3
330-332,340	Miami	98.6	74.8	87.1
333	Fort Lauderdale	97.9	73.5	86.2
334,349	West Palm Beach	96.8	70.6	84.2
335-336,346	Tampa	99.8	74.8	87.8
337	St. Petersburg	102.2	57.6	80.7
338	Lakeland	98.9	74.5	87.1
339,341	Fort Myers	98.2	66.2	82.8
342	Sarasota	100.0	68.7	84.9
GEORGIA				
300-303,399	Atlanta	96.8	82.0	89.7
304	Statesboro	97.0	46.4	72.6
305	Gainesville	95.8	59.1	78.1
306	Athens	95.0	64.0	80.0
307	Dalton	97.2	52.5	75.7
308-309	Augusta	95.4	61.1	78.9
310-312	Macon	95.9	60.7	78.9
313-314	Savannah	98.1	59.5	79.5
315	Waycross	97.9	53.2	76.4
316	Valdosta	97.5	50.4	74.8
317	Albany	97.4	55.6	77.2
318-319	Columbus	97.5	54.2	76.6

STATE/ZIP	CITY	MAT.	INST.	TOTAL
HAWAII				
967	Hilo	115.9	126.1	120.8
968	Honolulu	117.6	126.1	121.7
STATES & POSS.				
969	Guam	196.3	55.8	128.6
IDAHO				
832	Pocatello	101.3	79.8	90.9
833	Twin Falls	103.3	47.2	76.3
834	Idaho Falls	100.5	57.0	79.5
835	Lewiston	112.2	84.7	99.0
836-837	Boise	101.6	81.8	92.1
838	Coeur d'Alene	111.9	58.5	86.2
ILLINOIS				
600-603	North Suburban	99.1	119.6	109.0
604	Joliet	99.1	120.4	109.4
605	South Suburban	99.1	117.4	107.9
606	Chicago	99.6	126.4	112.5
609	Kankakee	95.6	108.6	101.9
610-611	Rockford	97.2	111.9	104.3
612	Rock Island	95.2	99.6	97.3
613	La Salle	96.7	102.0	99.2
614	Galesburg	96.4	102.2	99.2
615-616	Peoria	98.6	103.5	101.0
617	Bloomington	95.7	102.9	99.2
618-619	Champaign	99.6	101.3	100.4
620-622	East St. Louis	93.8	104.3	98.8
623	Quincy	95.7	97.1	96.4
624	Effingham	95.0	99.5	97.1
625	Decatur	96.2	99.0	97.5
626-627	Springfield	96.0	100.7	98.3
628	Centralia	93.1	100.4	96.6
629	Carbondale	92.7	98.1	95.3
INDIANA				
460	Anderson	97.0	85.0	91.2
461-462	Indianapolis	100.2	88.0	94.3
463-464	Gary	97.9	101.9	99.8
465-466	South Bend	96.3	85.2	91.0
467-468	Fort Wayne	97.5	83.1	90.6
469	Kokomo	95.6	84.8	90.4
470	Lawrenceburg	93.4	82.6	88.2
471	New Albany	95.0	77.1	86.3
472	Columbus	97.3	82.8	90.3
473	Muncie	97.3	83.4	90.6
474	Bloomington	99.1	83.2	91.4
475	Washington	95.6	87.9	91.9
476-477	Evansville	96.6	89.1	93.0
478	Terre Haute	97.4	87.9	92.8
479	Lafayette	97.0	81.8	89.7
IOWA				
500-503,509	Des Moines	98.1	85.0	91.7
504	Mason City	96.8	65.3	81.6
505	Fort Dodge	97.0	61.3	79.8
506-507	Waterloo	98.0	64.0	81.6
508	Creston	97.5	68.5	83.5
510-511	Sioux City	98.7	75.7	87.6
512	Sibley	97.9	54.6	77.0
513	Spencer	100.0	52.6	77.1
514	Carroll	96.6	58.0	78.0
515	Council Bluffs	99.6	76.3	88.4
516	Shenandoah	96.7	53.9	76.0
520	Dubuque	98.3	78.9	88.9
521	Decorah	97.7	56.2	77.7
522-524	Cedar Rapids	99.3	85.3	92.6
525	Ottumwa	97.8	72.7	85.7
526	Burlington	96.7	76.1	86.8
527-528	Davenport	98.1	94.7	96.5
KANSAS				
660-662	Kansas City	98.3	92.9	95.7
664-666	Topeka	97.6	69.4	84.0
667	Fort Scott	97.4	67.5	83.0
668	Emporia	97.2	61.3	79.9
669	Belleville	99.3	56.5	78.6
670-672	Wichita	96.7	73.0	85.3
673	Independence	98.9	53.5	77.0
674	Salina	98.6	59.0	79.5
675	Hutchinson	94.0	52.6	74.0
676	Hays	98.6	56.5	78.3
677	Colby	99.2	56.5	78.6

STATE/ZIP	CITY	MAT.	INST.	TOTAL
KANSAS (CONT'D)				
678	Dodge City	100.0	56.5	79.0
679	Liberal	98.4	47.0	73.6
KENTUCKY				
400-402	Louisville	95.9	85.0	90.7
403-405	Lexington	96.7	67.9	82.9
406	Frankfort	95.9	70.1	83.4
407-409	Corbin	94.6	46.0	71.2
410	Covington	95.6	94.8	95.2
411-412	Ashland	94.0	97.6	95.7
413-414	Campton	95.5	46.5	71.9
415-416	Pikeville	96.8	64.1	81.0
417-418	Hazard	94.8	46.2	71.4
420	Paducah	93.7	88.8	91.3
421-422	Bowling Green	95.4	82.2	89.0
423	Owensboro	95.5	75.5	85.9
424	Henderson	93.2	91.7	92.4
425-426	Somerset	92.5	48.0	71.0
427	Elizabethtown	92.0	85.0	88.6
LOUISIANA				
700-701	New Orleans	101.5	69.4	86.0
703	Thibodaux	101.7	66.1	84.5
704	Hammond	98.6	65.3	82.6
705	Lafayette	100.7	56.5	79.4
706	Lake Charles	100.9	61.1	81.7
707-708	Baton Rouge	100.6	56.5	79.4
710-711	Shreveport	96.5	61.6	79.6
712	Monroe	96.6	59.0	78.5
713-714	Alexandria	96.7	56.0	77.1
MAINE				
039	Kittery	97.6	74.8	86.6
040-041	Portland	100.1	79.2	90.0
042	Lewiston	100.8	79.2	90.4
043	Augusta	99.4	74.6	87.4
044	Bangor	99.9	79.2	89.9
045	Bath	98.9	74.8	87.3
046	Machias	98.4	74.1	86.7
047	Houlton	98.5	77.9	88.6
048	Rockland	97.4	72.8	85.6
049	Waterville	98.9	69.4	84.7
MARYLAND				
206	Waldorf	98.3	74.0	86.6
207-208	College Park	98.2	81.2	90.0
209	Silver Spring	97.5	78.6	88.4
210-212	Baltimore	97.5	84.9	91.4
214	Annapolis	97.0	80.9	89.2
215	Cumberland	94.3	79.2	87.0
216	Easton	95.9	46.8	72.2
217	Hagerstown	94.5	81.1	88.1
218	Salisbury	96.4	56.4	77.1
219	Elkton	93.4	69.2	81.7
MASSACHUSETTS				
010-011	Springfield	100.8	104.9	102.8
012	Pittsfield	100.5	98.6	99.6
013	Greenfield	98.8	101.2	99.9
014	Fitchburg	97.5	115.4	106.1
015-016	Worcester	100.5	116.0	108.0
017	Framingham	96.8	121.2	108.6
018	Lowell	100.2	121.4	110.4
019	Lawrence	101.5	119.5	110.2
020-022, 024	Boston	102.9	128.1	115.0
023	Brockton	101.5	116.6	108.8
025	Buzzards Bay	96.0	115.8	105.6
026	Hyannis	98.4	115.8	106.8
027	New Bedford	101.1	116.0	108.3
MICHIGAN				
480,483	Royal Oak	94.8	109.4	101.8
481	Ann Arbor	97.1	108.6	102.7
482	Detroit	98.3	116.8	107.2
484-485	Flint	96.9	98.6	97.7
486	Saginaw	96.9	95.7	96.3
487	Bay City	96.6	95.8	96.2
488-489	Lansing	97.2	98.3	97.7
490	Battle Creek	96.6	90.9	93.8
491	Kalamazoo	97.0	87.4	92.4
492	Jackson	95.3	94.2	94.8
493,495	Grand Rapids	96.8	72.4	85.0
494	Muskegon	95.6	85.9	90.9

Location Factors

STATE/ZIP	CITY	MAT.	INST.	TOTAL
MICHIGAN (CONT'D)				
496	Traverse City	94.7	76.3	85.8
497	Gaylord	96.1	80.4	88.5
498-499	Iron Mountain	98.2	89.5	94.0
MINNESOTA				
550-551	Saint Paul	99.2	125.5	111.8
553-555	Minneapolis	99.9	127.9	113.4
556-558	Duluth	99.1	110.8	104.8
559	Rochester	98.8	107.6	103.0
560	Mankato	97.3	104.6	100.8
561	Windom	96.0	83.2	89.8
562	Willmar	95.3	89.9	92.7
563	St. Cloud	96.1	123.3	109.2
564	Brainerd	97.1	103.5	100.2
565	Detroit Lakes	99.3	101.6	100.4
566	Bemidji	98.6	100.2	99.4
567	Thief River Falls	97.6	96.4	97.0
MISSISSIPPI				
386	Clarksdale	97.0	31.3	65.3
387	Greenville	100.6	47.4	75.0
388	Tupelo	98.6	39.6	70.2
389	Greenwood	98.4	33.8	67.3
390-392	Jackson	98.7	49.0	74.7
393	Meridian	96.8	49.3	73.9
394	Laurel	98.5	34.7	67.7
395	Biloxi	98.8	56.1	78.2
396	McComb	96.8	53.4	75.9
397	Columbus	98.5	39.0	69.8
MISSOURI				
630-631	St. Louis	96.0	108.8	102.2
633	Bowling Green	95.8	92.2	94.1
634	Hannibal	94.6	87.8	91.3
635	Kirksville	97.2	80.8	89.3
636	Flat River	96.9	96.6	96.8
637	Cape Girardeau	96.0	92.2	94.1
638	Sikeston	94.6	87.0	90.9
639	Poplar Bluff	94.1	87.7	91.0
640-641	Kansas City	99.4	104.8	102.0
644-645	St. Joseph	99.2	92.1	95.8
646	Chillicothe	96.7	72.8	85.2
647	Harrisonville	96.2	100.7	98.3
648	Joplin	98.0	71.1	85.0
650-651	Jefferson City	95.3	89.1	92.3
652	Columbia	96.7	90.4	93.6
653	Sedalia	96.3	86.9	91.8
654-655	Rolla	94.9	80.2	87.8
656-658	Springfield	98.0	76.4	87.6
MONTANA				
590-591	Billings	101.1	76.2	89.1
592	Wolf Point	102.1	71.8	87.5
593	Miles City	99.8	72.4	86.6
594	Great Falls	102.5	74.2	88.9
595	Havre	100.7	72.1	86.9
596	Helena	102.3	73.2	88.3
597	Butte	101.1	73.4	87.8
598	Missoula	99.2	72.0	86.1
599	Kalispell	98.6	71.3	85.4
NEBRASKA				
680-681	Omaha	100.2	79.1	90.0
683-685	Lincoln	98.1	69.8	84.5
686	Columbus	96.9	52.7	75.6
687	Norfolk	99.1	64.5	82.4
688	Grand Island	97.8	68.3	83.6
689	Hastings	97.5	60.3	79.6
690	Mccook	97.3	50.3	74.7
691	North Platte	97.2	61.8	80.1
692	Valentine	100.7	40.9	71.9
693	Alliance	100.4	38.1	70.4
NEVADA				
889-891	Las Vegas	100.6	106.7	103.5
893	Ely	101.7	83.7	93.0
894-895	Reno	101.5	95.4	98.5
897	Carson City	101.2	94.8	98.2
898	Elko	100.3	83.3	92.1
NEW HAMPSHIRE				
030	Nashua	101.0	85.4	93.5
031	Manchester	101.5	85.4	93.7

STATE/ZIP	CITY	MAT.	INST.	TOTAL
NEW HAMPSHIRE (CONT'D)				
032-033	Concord	99.0	85.4	92.4
034	Keene	98.3	54.2	77.0
035	Littleton	98.3	63.9	81.8
036	Charleston	97.8	50.9	75.2
037	Claremont	96.9	50.9	74.7
038	Portsmouth	98.3	80.7	89.8
NEW JERSEY				
070-071	Newark	102.8	120.3	111.2
072	Elizabeth	101.6	119.4	110.2
073	Jersey City	100.1	120.3	109.9
074-075	Paterson	102.9	119.3	110.8
076	Hackensack	100.1	119.9	109.6
077	Long Branch	99.7	119.8	109.4
078	Dover	100.4	119.2	109.5
079	Summit	100.5	119.4	109.6
080,083	Vineland	98.0	118.9	108.1
081	Camden	100.0	117.3	108.4
082,084	Atlantic City	98.9	117.7	108.0
085-086	Trenton	101.0	120.4	110.3
087	Point Pleasant	100.4	119.2	109.4
088-089	New Brunswick	100.8	118.4	109.3
NEW MEXICO				
870-872	Albuquerque	100.5	77.2	89.3
873	Gallup	101.1	77.2	89.6
874	Farmington	101.2	77.2	89.6
875	Santa Fe	99.6	77.2	88.8
877	Las Vegas	99.4	77.2	88.7
878	Socorro	98.9	77.2	88.4
879	Truth/Consequences	98.7	72.1	85.9
880	Las Cruces	96.9	69.9	83.9
881	Clovis	99.5	76.6	88.5
882	Roswell	101.0	76.7	89.3
883	Carrizozo	101.8	77.2	89.9
884	Tucumcari	100.5	76.6	89.0
NEW YORK				
100-102	New York	106.9	163.2	134.0
103	Staten Island	103.0	159.3	130.2
104	Bronx	100.4	159.3	128.8
105	Mount Vernon	101.3	138.4	119.2
106	White Plains	101.0	137.7	118.7
107	Yonkers	105.8	137.9	121.2
108	New Rochelle	101.9	138.4	119.5
109	Suffern	101.9	127.3	114.1
110	Queens	102.4	159.2	129.8
111	Long Island City	104.3	159.2	130.8
112	Brooklyn	104.5	160.3	131.4
113	Flushing	105.2	159.2	131.3
114	Jamaica	103.1	159.4	130.2
115,117,118	Hicksville	102.3	148.3	124.5
116	Far Rockaway	105.4	159.2	131.3
119	Riverhead	102.8	148.5	124.8
120-122	Albany	97.8	96.2	97.0
123	Schenectady	98.4	95.3	96.9
124	Kingston	102.8	116.5	109.4
125-126	Poughkeepsie	101.9	121.4	111.3
127	Monticello	101.1	116.0	108.3
128	Glens Falls	93.2	90.1	91.7
129	Plattsburgh	98.6	85.5	92.3
130-132	Syracuse	99.3	92.7	96.1
133-135	Utica	97.6	89.4	93.6
136	Watertown	99.1	90.5	95.0
137-139	Binghamton	98.7	87.0	93.1
140-142	Buffalo	100.1	103.8	101.9
143	Niagara Falls	98.4	106.0	102.1
144-146	Rochester	101.3	97.4	99.4
147	Jamestown	97.6	88.1	93.0
148-149	Elmira	97.0	87.4	92.4
NORTH CAROLINA				
270,272-274	Greensboro	97.2	51.8	75.3
271	Winston-Salem	97.3	50.6	74.8
275-276	Raleigh	97.4	51.9	75.5
277	Durham	97.1	51.9	75.3
278	Rocky Mount	95.8	37.3	67.6
279	Elizabeth City	96.6	39.8	69.2
280	Gastonia	97.3	48.9	74.0
281-282	Charlotte	97.8	49.2	74.4
283	Fayetteville	96.0	51.9	74.7
284	Wilmington	95.8	51.0	74.2
285	Kinston	94.4	36.8	66.7

Location Factors

STATE/ZIP	CITY	MAT.	INST.	TOTAL
NORTH CAROLINA (CONT'D)				
286	Hickory	94.6	36.0	66.3
287-288	Asheville	96.2	49.8	73.8
289	Murphy	95.9	35.1	66.6
NORTH DAKOTA				
580-581	Fargo	101.6	67.6	85.2
582	Grand Forks	101.6	60.3	81.7
583	Devils Lake	101.6	62.4	82.7
584	Jamestown	101.6	55.9	79.6
585	Bismarck	100.2	68.9	85.1
586	Dickinson	102.5	64.3	84.0
587	Minot	101.6	69.6	86.2
588	Williston	100.8	63.4	82.7
OHIO				
430-432	Columbus	98.3	91.9	95.2
433	Marion	95.5	90.8	93.2
434-436	Toledo	98.7	100.6	99.6
437-438	Zanesville	95.7	85.8	90.9
439	Steubenville	97.5	95.2	96.4
440	Lorain	98.2	101.8	99.9
441	Cleveland	98.5	106.2	102.2
442-443	Akron	99.3	99.0	99.2
444-445	Youngstown	98.6	95.1	96.9
446-447	Canton	98.7	90.4	94.7
448-449	Mansfield	96.3	94.2	95.3
450	Hamilton	94.7	91.6	93.2
451-452	Cincinnati	95.0	93.0	94.0
453-454	Dayton	94.8	87.6	91.3
455	Springfield	94.8	88.3	91.7
456	Chillicothe	94.3	94.8	94.5
457	Athens	97.8	79.5	88.9
458	Lima	98.4	88.2	93.5
OKLAHOMA				
730-731	Oklahoma City	97.9	65.9	82.5
734	Ardmore	95.7	65.3	81.0
735	Lawton	97.3	66.3	82.4
736	Clinton	97.3	63.5	81.0
737	Enid	97.3	63.5	81.0
738	Woodward	95.9	63.6	80.3
739	Guymon	97.1	34.9	67.1
740-741	Tulsa	97.3	63.1	80.8
743	Miami	94.6	67.4	81.5
744	Muskogee	96.5	45.0	71.7
745	Mcalester	94.2	56.1	75.8
746	Ponca City	94.7	63.4	79.6
747	Durant	94.8	63.7	79.8
748	Shawnee	96.5	62.0	79.9
749	Poteau	93.7	64.5	79.6
OREGON				
970-972	Portland	103.1	105.0	104.0
973	Salem	103.7	103.9	103.8
974	Eugene	103.1	103.5	103.3
975	Medford	104.5	101.8	103.2
976	Klamath Falls	105.5	101.8	103.7
977	Bend	104.3	103.6	103.9
978	Pendleton	98.3	102.1	100.1
979	Vale	95.8	95.0	95.4
PENNSYLVANIA				
150-152	Pittsburgh	96.7	104.5	100.5
153	Washington	94.2	103.5	98.7
154	Uniontown	94.4	100.5	97.4
155	Bedford	95.5	92.3	93.9
156	Greensburg	95.6	99.6	97.5
157	Indiana	94.3	98.6	96.4
158	Dubois	95.8	95.7	95.7
159	Johnstown	95.4	97.0	96.2
160	Butler	92.6	102.2	97.2
161	New Castle	92.6	102.2	97.3
162	Kittanning	93.2	103.8	98.3
163	Oil City	92.6	97.5	94.9
164-165	Erie	94.5	97.3	95.9
166	Altoona	94.4	94.3	94.3
167	Bradford	96.1	94.9	95.6
168	State College	95.8	94.7	95.3
169	Wellsboro	96.9	89.9	93.5
170-171	Harrisburg	98.3	91.4	95.0
172	Chambersburg	96.4	88.5	92.6
173-174	York	96.1	90.0	93.1
175-176	Lancaster	95.0	88.3	91.8

STATE/ZIP	CITY	MAT.	INST.	TOTAL
PENNSYLVANIA (CONT'D)				
177	Williamsport	93.4	83.9	88.8
178	Sunbury	95.8	90.6	93.3
179	Pottsville	94.8	92.1	93.5
180	Lehigh Valley	96.5	105.4	100.8
181	Allentown	98.2	102.8	100.4
182	Hazleton	95.8	94.4	95.1
183	Stroudsburg	95.7	98.8	97.2
184-185	Scranton	98.7	95.4	97.1
186-187	Wilkes-Barre	95.4	93.3	94.4
188	Montrose	95.0	94.6	94.8
189	Doylestown	95.3	116.1	105.3
190-191	Philadelphia	100.2	125.0	112.1
193	Westchester	97.2	114.5	105.6
194	Norristown	96.0	119.5	107.4
195-196	Reading	98.2	95.7	97.0
PUERTO RICO				
009	San Juan	140.4	27.1	85.8
RHODE ISLAND				
028	Newport	99.8	109.5	104.4
029	Providence	100.1	109.5	104.6
SOUTH CAROLINA				
290-292	Columbia	96.5	49.8	74.0
293	Spartanburg	95.4	47.4	72.3
294	Charleston	96.7	50.4	74.4
295	Florence	94.9	47.5	72.1
296	Greenville	95.1	47.4	72.1
297	Rock Hill	95.3	34.9	66.2
298	Aiken	96.4	72.5	84.9
299	Beaufort	97.2	38.5	68.9
SOUTH DAKOTA				
570-571	Sioux Falls	100.5	59.7	80.8
572	Watertown	99.7	53.6	77.5
573	Mitchell	98.6	53.2	76.7
574	Aberdeen	100.8	56.3	79.4
575	Pierre	100.1	55.4	78.5
576	Mobridge	99.3	53.6	77.2
577	Rapid City	100.7	54.1	78.3
TENNESSEE				
370-372	Nashville	97.9	75.7	87.2
373-374	Chattanooga	99.2	57.5	79.1
375,380-381	Memphis	96.4	76.2	86.7
376	Johnson City	98.6	59.3	79.7
377-379	Knoxville	95.1	60.8	78.6
382	Mckenzie	97.5	51.3	75.2
383	Jackson	99.2	55.0	77.9
384	Columbia	95.9	58.7	77.9
385	Cookeville	97.3	53.1	76.0
TEXAS				
750	Mckinney	99.6	58.7	79.9
751	Waxahackie	99.5	59.4	80.2
752-753	Dallas	99.5	67.9	84.3
754	Greenville	99.7	43.0	72.4
755	Texarkana	99.1	54.3	77.5
756	Longview	99.3	45.1	73.2
757	Tyler	100.0	57.6	79.5
758	Palestine	95.5	45.5	71.4
759	Lufkin	96.5	49.5	73.8
760-761	Fort Worth	97.5	64.5	81.6
762	Denton	97.9	54.7	77.1
763	Wichita Falls	97.9	58.9	79.1
764	Eastland	97.0	44.6	71.7
765	Temple	95.7	53.0	75.1
766-767	Waco	97.2	59.3	78.9
768	Brownwood	98.0	42.0	71.0
769	San Angelo	97.6	50.4	74.9
770-772	Houston	99.4	73.7	87.0
773	Huntsville	98.3	42.9	71.6
774	Wharton	99.8	48.0	74.8
775	Galveston	97.6	71.8	85.2
776-777	Beaumont	97.0	65.4	81.7
778	Bryan	94.9	66.2	81.1
779	Victoria	100.1	50.7	76.3
780	Laredo	94.4	55.2	75.5
781-782	San Antonio	94.6	65.7	80.7
783-784	Corpus Christi	97.7	55.5	77.4
785	Mc Allen	98.3	49.5	74.8
786-787	Austin	94.5	63.6	79.6

Location Factors

STATE/ZIP	CITY	MAT.	INST.	TOTAL
TEXAS (CONT'D)				
788	Del Rio	97.5	35.3	67.5
789	Giddings	94.6	44.8	70.6
790-791	Amarillo	97.9	60.9	80.1
792	Childress	97.6	55.1	77.1
793-794	Lubbock	99.6	56.5	78.8
795-796	Abilene	97.3	56.1	77.4
797	Midland	100.5	52.9	77.6
798-799,885	El Paso	97.3	54.0	76.4
UTAH				
840-841	Salt Lake City	102.5	74.2	88.9
842,844	Ogden	98.9	74.0	86.9
843	Logan	101.0	74.0	88.0
845	Price	101.7	52.6	78.1
846-847	Provo	101.3	74.0	88.1
VERMONT				
050	White River Jct.	99.9	49.4	75.6
051	Bellows Falls	98.2	51.9	75.9
052	Bennington	98.6	52.3	76.2
053	Brattleboro	99.1	52.4	76.6
054	Burlington	101.0	68.2	85.1
056	Montpelier	98.4	68.2	83.8
057	Rutland	100.0	68.2	84.7
058	St. Johnsbury	100.1	52.2	77.0
059	Guildhall	98.5	51.7	75.9
VIRGINIA				
220-221	Fairfax	98.5	80.3	89.7
222	Arlington	99.2	79.0	89.4
223	Alexandria	98.2	83.9	91.3
224-225	Fredericksburg	96.9	68.4	83.1
226	Winchester	97.7	56.9	78.0
227	Culpeper	97.5	57.8	78.3
228	Harrisonburg	97.9	49.6	74.6
229	Charlottesville	98.1	62.0	80.7
230-232	Richmond	98.2	67.6	83.5
233-235	Norfolk	98.3	63.7	81.7
236	Newport News	97.8	62.2	80.6
237	Portsmouth	97.1	63.2	80.8
238	Petersburg	97.9	67.6	83.3
239	Farmville	97.2	43.6	71.4
240-241	Roanoke	98.4	49.9	75.0
242	Bristol	96.9	50.2	74.4
243	Pulaski	96.6	44.4	71.5
244	Staunton	97.4	48.0	73.6
245	Lynchburg	97.5	52.6	75.9
246	Grundy	96.9	46.6	72.6
WASHINGTON				
980-981,987	Seattle	103.0	105.2	104.1
982	Everett	103.0	100.0	101.5
983-984	Tacoma	102.7	100.3	101.5
985	Olympia	102.5	100.2	101.4
986	Vancouver	105.0	98.7	102.0
988	Wenatchee	103.8	86.4	95.4
989	Yakima	103.2	91.2	97.4
990-992	Spokane	107.2	84.6	96.3
993	Richland	107.4	86.6	97.4
994	Clarkston	106.1	84.7	95.8
WEST VIRGINIA				
247-248	Bluefield	95.7	80.1	88.2
249	Lewisburg	97.3	83.3	90.5
250-253	Charleston	97.7	90.3	94.2
254	Martinsburg	97.2	78.0	87.9
255-257	Huntington	98.0	91.2	94.8
258-259	Beckley	95.4	88.9	92.3
260	Wheeling	98.3	91.2	94.9
261	Parkersburg	97.1	90.9	94.1
262	Buckhannon	97.1	90.7	94.0
263-264	Clarksburg	97.6	90.3	94.1
265	Morgantown	97.7	90.8	94.4
266	Gassaway	96.9	90.6	93.9
267	Romney	96.9	85.0	91.1
268	Petersburg	96.8	87.9	92.5
WISCONSIN				
530,532	Milwaukee	99.2	102.3	100.7
531	Kenosha	99.2	100.4	99.8
534	Racine	98.4	101.8	100.1
535	Beloit	98.3	98.8	98.6
537	Madison	98.8	96.9	97.9

STATE/ZIP	CITY	MAT.	INST.	TOTAL
WISCONSIN (CONT'D)				
538	Lancaster	96.7	94.2	95.5
539	Portage	95.1	95.7	95.4
540	New Richmond	96.8	97.2	97.0
541-543	Green Bay	100.6	93.7	97.2
544	Wausau	95.8	92.4	94.2
545	Rhinelander	99.4	92.7	96.1
546	La Crosse	96.4	94.6	95.5
547	Eau Claire	98.3	95.6	97.0
548	Superior	96.7	97.5	97.1
549	Oshkosh	96.4	91.7	94.1
WYOMING				
820	Cheyenne	99.9	56.4	78.9
821	Yellowstone Nat'l Park	99.3	50.6	75.8
822	Wheatland	101.0	49.5	76.2
823	Rawlins	102.4	45.5	75.0
824	Worland	100.1	46.2	74.1
825	Riverton	101.3	48.8	76.0
826	Casper	99.7	61.9	81.5
827	Newcastle	99.9	45.5	73.7
828	Sheridan	100.9	54.4	78.5
829-831	Rock Springs	104.5	46.2	76.4
CANADIAN FACTORS (reflect Canadian currency)				
ALBERTA				
	Calgary	123.4	84.6	104.7
	Edmonton	124.4	84.6	105.2
	Fort McMurray	117.1	84.1	101.2
	Lethbridge	118.0	83.6	101.4
	Lloydminster	117.1	84.1	101.2
	Medicine Hat	117.2	83.6	101.0
	Red Deer	117.6	83.6	101.2
BRITISH COLUMBIA				
	Kamloops	117.9	87.0	103.0
	Prince George	119.6	87.0	103.9
	Vancouver	121.8	91.0	107.0
	Victoria	119.8	87.5	104.2
MANITOBA				
	Brandon	117.1	72.9	95.8
	Portage la Prairie	117.1	72.9	95.8
	Winnipeg	120.7	73.2	97.8
NEW BRUNSWICK				
	Bathurst	115.2	64.8	90.9
	Dalhousie	115.2	64.8	90.9
	Fredericton	115.6	69.0	93.2
	Moncton	114.9	64.8	90.7
	Newcastle	115.3	64.8	90.9
	Saint John	117.0	69.0	93.9
NEWFOUNDLAND				
	Corner Brook	121.1	64.8	93.9
	St. John's	120.7	64.8	93.7
NORTHWEST TERRITORIES				
	Yellowknife	114.1	82.3	98.8
NOVA SCOTIA				
	Dartmouth	116.8	71.4	94.9
	Halifax	116.5	72.5	95.3
	New Glasgow	116.5	71.4	94.7
	Sydney	113.7	71.4	93.3
	Yarmouth	116.3	71.4	94.7
ONTARIO				
	Barrie	120.2	89.4	105.4
	Brantford	119.3	93.2	106.7
	Cornwall	119.3	89.7	105.0
	Hamilton	120.8	93.4	107.6
	Kingston	119.8	90.1	105.5
	Kitchener	114.9	89.0	102.4
	London	120.1	90.8	105.9
	North Bay	119.3	87.7	104.0
	Oshawa	118.8	91.2	105.5
	Ottawa	120.8	90.9	106.4
	Owen Sound	120.4	87.7	104.6
	Peterborough	119.3	89.5	104.9
	Sarnia	119.5	93.5	107.0
	St. Catharines	113.1	88.3	101.2
	Sudbury	113.4	87.9	101.1

Location Factors

STATE/ZIP	CITY	MAT.	INST.	TOTAL
ONTARIO (CONT'D)				
	Thunder Bay	114.9	88.1	102.0
	Toronto	121.7	98.0	110.2
	Windsor	113.5	90.8	102.5
PRINCE EDWARD ISLAND				
	Charlottetown	118.4	60.5	90.5
	Summerside	118.6	60.5	90.6
QUEBEC				
	Cap-de-la-Madeleine	116.6	83.9	100.8
	Charlesbourg	116.6	83.9	100.8
	Chicoutimi	115.7	83.6	100.2
	Gatineau	115.8	83.7	100.3
	Laval	115.9	83.7	100.4
	Montreal	117.7	84.0	101.5
	Quebec	119.0	84.4	102.3
	Sherbrooke	116.2	83.7	100.5
	Trois Rivieres	116.8	83.9	100.9
SASKATCHEWAN				
	Moose Jaw	114.5	66.9	91.5
	Prince Albert	113.8	66.7	91.1
	Regina	115.2	66.9	91.9
	Saskatoon	114.1	66.8	91.3
YUKON				
	Whitehorse	114.1	65.8	90.8

Abbreviation	Meaning
A	Area Square Feet; Ampere
ABS	Acrylonitrile Butadiene Stryrene; Asbestos Bonded Steel
A.C.	Alternating Current; Air-Conditioning; Asbestos Cement; Plywood Grade A & C
A.C.I.	American Concrete Institute
AD	Plywood, Grade A & D
Addit.	Additional
Adj.	Adjustable
af	Audio-frequency
A.G.A.	American Gas Association
Agg.	Aggregate
A.H.	Ampere Hours
A hr.	Ampere-hour
A.H.U.	Air Handling Unit
A.I.A.	American Institute of Architects
AIC	Ampere Interrupting Capacity
Allow.	Allowance
alt.	Altitude
Alum.	Aluminum
a.m.	Ante Meridiem
Amp.	Ampere
Anod.	Anodized
Approx.	Approximate
Apt.	Apartment
Asb.	Asbestos
A.S.B.C.	American Standard Building Code
Asbe.	Asbestos Worker
A.S.H.R.A.E.	American Society of Heating, Refrig. & AC Engineers
A.S.M.E.	American Society of Mechanical Engineers
A.S.T.M.	American Society for Testing and Materials
Attchmt.	Attachment
Avg.	Average
A.W.G.	American Wire Gauge
AWWA	American Water Works Assoc.
Bbl.	Barrel
B. & B.	Grade B and Better; Balled & Burlapped
B. & S.	Bell and Spigot
B. & W.	Black and White
b.c.c.	Body-centered Cubic
B.C.Y.	Bank Cubic Yards
BE	Bevel End
B.F.	Board Feet
Bg. cem.	Bag of Cement
BHP	Boiler Horsepower; Brake Horsepower
B.I.	Black Iron
Bit.; Bitum.	Bituminous
Bk.	Backed
Bkrs.	Breakers
Bldg.	Building
Blk.	Block
Bm.	Beam
Boil.	Boilermaker
B.P.M.	Blows per Minute
BR	Bedroom
Brg.	Bearing
Brhe.	Bricklayer Helper
Bric.	Bricklayer
Brk.	Brick
Brng.	Bearing
Brs.	Brass
Brz.	Bronze
Bsn.	Basin
Btr.	Better
BTU	British Thermal Unit
BTUH	BTU per Hour
B.U.R.	Built-up Roofing
BX	Interlocked Armored Cable
c	Conductivity, Copper Sweat
C	Hundred; Centigrade
C/C	Center to Center, Cedar on Cedar
Cab.	Cabinet
Cair.	Air Tool Laborer
Calc	Calculated
Cap.	Capacity
Carp.	Carpenter
C.B.	Circuit Breaker
C.C.A.	Chromate Copper Arsenate
C.C.F.	Hundred Cubic Feet
cd	Candela
cd/sf	Candela per Square Foot
CD	Grade of Plywood Face & Back
CDX	Plywood, Grade C & D, exterior glue
Cefi.	Cement Finisher
Cem.	Cement
CF	Hundred Feet
C.F.	Cubic Feet
CFM	Cubic Feet per Minute
c.g.	Center of Gravity
CHW	Chilled Water; Commercial Hot Water
C.I.	Cast Iron
C.I.P.	Cast in Place
Circ.	Circuit
C.L.	Carload Lot
Clab.	Common Laborer
C.L.F.	Hundred Linear Feet
CLF	Current Limiting Fuse
CLP	Cross Linked Polyethylene
cm	Centimeter
CMP	Corr. Metal Pipe
C.M.U.	Concrete Masonry Unit
CN	Change Notice
Col.	Column
CO₂	Carbon Dioxide
Comb.	Combination
Compr.	Compressor
Conc.	Concrete
Cont.	Continuous; Continued
Corr.	Corrugated
Cos	Cosine
Cot	Cotangent
Cov.	Cover
C/P	Cedar on Paneling
CPA	Control Point Adjustment
Cplg.	Coupling
C.P.M.	Critical Path Method
CPVC	Chlorinated Polyvinyl Chloride
C.Pr.	Hundred Pair
CRC	Cold Rolled Channel
Creos.	Creosote
Crpt.	Carpet & Linoleum Layer
CRT	Cathode-ray Tube
CS	Carbon Steel, Constant Shear Bar Joist
Csc	Cosecant
C.S.F.	Hundred Square Feet
CSI	Construction Specifications Institute
C.T.	Current Transformer
CTS	Copper Tube Size
Cu	Copper, Cubic
Cu. Ft.	Cubic Foot
cw	Continuous Wave
C.W.	Cool White; Cold Water
Cwt.	100 Pounds
C.W.X.	Cool White Deluxe
C.Y.	Cubic Yard (27 cubic feet)
C.Y./Hr.	Cubic Yard per Hour
Cyl.	Cylinder
d	Penny (nail size)
D	Deep; Depth; Discharge
Dis.;Disch.	Discharge
Db.	Decibel
Dbl.	Double
DC	Direct Current
DDC	Direct Digital Control
Demob.	Demobilization
d.f.u.	Drainage Fixture Units
D.H.	Double Hung
DHW	Domestic Hot Water
Diag.	Diagonal
Diam.	Diameter
Distrib.	Distribution
Dk.	Deck
D.L.	Dead Load; Diesel
DLH	Deep Long Span Bar Joist
Do.	Ditto
Dp.	Depth
D.P.S.T.	Double Pole, Single Throw
Dr.	Driver
Drink.	Drinking
D.S.	Double Strength
D.S.A.	Double Strength A Grade
D.S.B.	Double Strength B Grade
Dty.	Duty
DWV	Drain Waste Vent
DX	Deluxe White, Direct Expansion
dyn	Dyne
e	Eccentricity
E	Equipment Only; East
Ea.	Each
E.B.	Encased Burial
Econ.	Economy
E.C.Y	Embankment Cubic Yards
EDP	Electronic Data Processing
EIFS	Exterior Insulation Finish System
E.D.R.	Equiv. Direct Radiation
Eq.	Equation
Elec.	Electrician; Electrical
Elev.	Elevator; Elevating
EMT	Electrical Metallic Conduit; Thin Wall Conduit
Eng.	Engine, Engineered
EPDM	Ethylene Propylene Diene Monomer
EPS	Expanded Polystyrene
Eqhv.	Equip. Oper., Heavy
Eqlt.	Equip. Oper., Light
Eqmd.	Equip. Oper., Medium
Eqmm.	Equip. Oper., Master Mechanic
Eqol.	Equip. Oper., Oilers
Equip.	Equipment
ERW	Electric Resistance Welded
E.S.	Energy Saver
Est.	Estimated
esu	Electrostatic Units
E.W.	Each Way
EWT	Entering Water Temperature
Excav.	Excavation
Exp.	Expansion, Exposure
Ext.	Exterior
Extru.	Extrusion
f.	Fiber stress
F	Fahrenheit; Female; Fill
Fab.	Fabricated
FBGS	Fiberglass
F.C.	Footcandles
f.c.c.	Face-centered Cubic
f'c.	Compressive Stress in Concrete; Extreme Compressive Stress
F.E.	Front End
FEP	Fluorinated Ethylene Propylene (Teflon)
F.G.	Flat Grain
F.H.A.	Federal Housing Administration
Fig.	Figure
Fin.	Finished
Fixt.	Fixture
Fl. Oz.	Fluid Ounces
Flr.	Floor
F.M.	Frequency Modulation; Factory Mutual
Fmg.	Framing
Fndtn.	Foundation
Fori.	Foreman, Inside

Foro.	Foreman, Outside	J	Joule	M.C.F.M.	Thousand Cubic Feet per Minute
Fount.	Fountain	J.I.C.	Joint Industrial Council	M.C.M.	Thousand Circular Mils
FPM	Feet per Minute	K	Thousand; Thousand Pounds;	M.C.P.	Motor Circuit Protector
FPT	Female Pipe Thread		Heavy Wall Copper Tubing, Kelvin	MD	Medium Duty
Fr.	Frame	K.A.H.	Thousand Amp. Hours	M.D.O.	Medium Density Overlaid
F.R.	Fire Rating	KCMIL	Thousand Circular Mils	Med.	Medium
FRK	Foil Reinforced Kraft	KD	Knock Down	MF	Thousand Feet
FRP	Fiberglass Reinforced Plastic	K.D.A.T.	Kiln Dried After Treatment	M.F.B.M.	Thousand Feet Board Measure
FS	Forged Steel	kg	Kilogram	Mfg.	Manufacturing
FSC	Cast Body; Cast Switch Box	kG	Kilogauss	Mfrs.	Manufacturers
Ft.	Foot; Feet	kgf	Kilogram Force	mg	Milligram
Ftng.	Fitting	kHz	Kilohertz	MGD	Million Gallons per Day
Ftg.	Footing	Kip.	1000 Pounds	MGPH	Thousand Gallons per Hour
Ft. Lb.	Foot Pound	KJ	Kiljoule	MH, M.H.	Manhole; Metal Halide; Man-Hour
Furn.	Furniture	K.L.	Effective Length Factor	MHz	Megahertz
FVNR	Full Voltage Non-Reversing	K.L.F.	Kips per Linear Foot	Mi.	Mile
FXM	Female by Male	Km	Kilometer	MI	Malleable Iron; Mineral Insulated
Fy.	Minimum Yield Stress of Steel	K.S.F.	Kips per Square Foot	mm	Millimeter
g	Gram	K.S.I.	Kips per Square Inch	Mill.	Millwright
G	Gauss	kV	Kilovolt	Min., min.	Minimum, minute
Ga.	Gauge	kVA	Kilovolt Ampere	Misc.	Miscellaneous
Gal.	Gallon	K.V.A.R.	Kilovar (Reactance)	ml	Milliliter, Mainline
Gal./Min.	Gallon per Minute	KW	Kilowatt	M.L.F.	Thousand Linear Feet
Galv.	Galvanized	KWh	Kilowatt-hour	Mo.	Month
Gen.	General	L	Labor Only; Length; Long;	Mobil.	Mobilization
G.F.I.	Ground Fault Interrupter		Medium Wall Copper Tubing	Mog.	Mogul Base
Glaz.	Glazier	Lab.	Labor	MPH	Miles per Hour
GPD	Gallons per Day	lat	Latitude	MPT	Male Pipe Thread
GPH	Gallons per Hour	Lath.	Lather	MRT	Mile Round Trip
GPM	Gallons per Minute	Lav.	Lavatory	ms	Millisecond
GR	Grade	lb.; #	Pound	M.S.F.	Thousand Square Feet
Gran.	Granular	L.B.	Load Bearing; L Conduit Body	Mstz.	Mosaic & Terrazzo Worker
Grnd.	Ground	L. & E.	Labor & Equipment	M.S.Y.	Thousand Square Yards
H	High; High Strength Bar Joist;	lb./hr.	Pounds per Hour	Mtd.	Mounted
	Henry	lb./L.F.	Pounds per Linear Foot	Mthe.	Mosaic & Terrazzo Helper
H.C.	High Capacity	lbf/sq.in.	Pound-force per Square Inch	Mtng.	Mounting
H.D.	Heavy Duty; High Density	L.C.L.	Less than Carload Lot	Mult.	Multi; Multiply
H.D.O.	High Density Overlaid	L.C.Y.	Loose Cubic Yards	M.V.A.	Million Volt Amperes
Hdr.	Header	Ld.	Load	M.V.A.R.	Million Volt Amperes Reactance
Hdwe.	Hardware	LE	Lead Equivalent	MV	Megavolt
Help.	Helpers Average	LED	Light Emitting Diode	MW	Megawatt
HEPA	High Efficiency Particulate Air	L.F.	Linear Foot	MXM	Male by Male
	Filter	Lg.	Long; Length; Large	MYD	Thousand Yards
Hg	Mercury	L & H	Light and Heat	N	Natural; North
HIC	High Interrupting Capacity	LH	Long Span Bar Joist	nA	Nanoampere
HM	Hollow Metal	L.H.	Labor Hours	NA	Not Available; Not Applicable
H.O.	High Output	L.L.	Live Load	N.B.C.	National Building Code
Horiz.	Horizontal	L.L.D.	Lamp Lumen Depreciation	NC	Normally Closed
H.P.	Horsepower; High Pressure	L-O-L	Lateralolet	N.E.M.A.	National Electrical Manufacturers
H.P.F.	High Power Factor	lm	Lumen		Assoc.
Hr.	Hour	lm/sf	Lumen per Square Foot	NEHB	Bolted Circuit Breaker to 600V.
Hrs./Day	Hours per Day	lm/W	Lumen per Watt	N.L.B.	Non-Load-Bearing
HSC	High Short Circuit	L.O.A.	Length Over All	NM	Non-Metallic Cable
Ht.	Height	log	Logarithm	nm	Nanometer
Htg.	Heating	L.P.	Liquefied Petroleum; Low Pressure	No.	Number
Htrs.	Heaters	L.P.F.	Low Power Factor	NO	Normally Open
HVAC	Heating, Ventilation & Air-	LR	Long Radius	N.O.C.	Not Otherwise Classified
	Conditioning	L.S.	Lump Sum	Nose.	Nosing
Hvy.	Heavy	Lt.	Light	N.P.T.	National Pipe Thread
HW	Hot Water	Lt. Ga.	Light Gauge	NQOD	Combination Plug-on/Bolt on
Hyd.;Hydr.	Hydraulic	L.T.L.	Less than Truckload Lot		Circuit Breaker to 240V.
Hz.	Hertz (cycles)	Lt. Wt.	Lightweight	N.R.C.	Noise Reduction Coefficient
I.	Moment of Inertia	L.V.	Low Voltage	N.R.S.	Non Rising Stem
I.C.	Interrupting Capacity	M	Thousand; Material; Male;	ns	Nanosecond
ID	Inside Diameter		Light Wall Copper Tubing	nW	Nanowatt
I.D.	Inside Dimension; Identification	M²CA	Meters Squared Contact Area	OB	Opposing Blade
I.F.	Inside Frosted	m/hr; M.H.	Man-hour	OC	On Center
I.M.C.	Intermediate Metal Conduit	mA	Milliampere	OD	Outside Diameter
In.	Inch	Mach.	Machine	O.D.	Outside Dimension
Incan.	Incandescent	Mag. Str.	Magnetic Starter	ODS	Overhead Distribution System
Incl.	Included; Including	Maint.	Maintenance	O.G.	Ogee
Int.	Interior	Marb.	Marble Setter	O.H.	Overhead
Inst.	Installation	Mat; Mat'l.	Material	O & P	Overhead and Profit
Insul.	Insulation/Insulated	Max.	Maximum	Oper.	Operator
I.P.	Iron Pipe	MBF	Thousand Board Feet	Opng.	Opening
I.P.S.	Iron Pipe Size	MBH	Thousand BTU's per hr.	Orna.	Ornamental
I.P.T.	Iron Pipe Threaded	MC	Metal Clad Cable	OSB	Oriented Strand Board
I.W.	Indirect Waste	M.C.F.	Thousand Cubic Feet	O. S. & Y.	Outside Screw and Yoke

Abbreviation	Meaning
Ovhd.	Overhead
OWG	Oil, Water or Gas
Oz.	Ounce
P.	Pole; Applied Load; Projection
p.	Page
Pape.	Paperhanger
P.A.P.R.	Powered Air Purifying Respirator
PAR	Parabolic Reflector
Pc., Pcs.	Piece, Pieces
P.C.	Portland Cement; Power Connector
P.C.F.	Pounds per Cubic Foot
P.C.M.	Phase Contract Microscopy
P.E.	Professional Engineer; Porcelain Enamel; Polyethylene; Plain End
Perf.	Perforated
Ph.	Phase
P.I.	Pressure Injected
Pile.	Pile Driver
Pkg.	Package
Pl.	Plate
Plah.	Plasterer Helper
Plas.	Plasterer
Pluh.	Plumbers Helper
Plum.	Plumber
Ply.	Plywood
p.m.	Post Meridiem
Pntd.	Painted
Pord.	Painter, Ordinary
pp	Pages
PP; PPL	Polypropylene
P.P.M.	Parts per Million
Pr.	Pair
P.E.S.B.	Pre-engineered Steel Building
Prefab.	Prefabricated
Prefin.	Prefinished
Prop.	Propelled
PSF; psf	Pounds per Square Foot
PSI; psi	Pounds per Square Inch
PSIG	Pounds per Square Inch Gauge
PSP	Plastic Sewer Pipe
Pspr.	Painter, Spray
Psst.	Painter, Structural Steel
P.T.	Potential Transformer
P. & T.	Pressure & Temperature
Ptd.	Painted
Ptns.	Partitions
Pu	Ultimate Load
PVC	Polyvinyl Chloride
Pvmt.	Pavement
Pwr.	Power
Q	Quantity Heat Flow
Quan.; Qty.	Quantity
Q.C.	Quick Coupling
r	Radius of Gyration
R	Resistance
R.C.P.	Reinforced Concrete Pipe
Rect.	Rectangle
Reg.	Regular
Reinf.	Reinforced
Req'd.	Required
Res.	Resistant
Resi.	Residential
Rgh.	Rough
RGS	Rigid Galvanized Steel
R.H.W.	Rubber, Heat & Water Resistant; Residential Hot Water
rms	Root Mean Square
Rnd.	Round
Rodm.	Rodman
Rofc.	Roofer, Composition
Rofp.	Roofer, Precast
Rohe.	Roofer Helpers (Composition)
Rots.	Roofer, Tile & Slate
R.O.W.	Right of Way
RPM	Revolutions per Minute
R.S.	Rapid Start
Rsr	Riser
RT	Round Trip
S.	Suction; Single Entrance; South
SCFM	Standard Cubic Feet per Minute
Scaf.	Scaffold
Sch.; Sched.	Schedule
S.C.R.	Modular Brick
S.D.	Sound Deadening
S.D.R.	Standard Dimension Ratio
S.E.	Surfaced Edge
Sel.	Select
S.E.R.;	Service Entrance Cable
S.E.U.	Service Entrance Cable
S.F.	Square Foot
S.F.C.A.	Square Foot Contact Area
S.F.G.	Square Foot of Ground
S.F. Hor.	Square Foot Horizontal
S.F.R.	Square Feet of Radiation
S.F. Shlf.	Square Foot of Shelf
S4S	Surface 4 Sides
Shee.	Sheet Metal Worker
Sin.	Sine
Skwk.	Skilled Worker
SL	Saran Lined
S.L.	Slimline
Sldr.	Solder
SLH	Super Long Span Bar Joist
S.N.	Solid Neutral
S-O-L	Socketolet
sp	Standpipe
S.P.	Static Pressure; Single Pole; Self-Propelled
Spri.	Sprinkler Installer
spwg	Static Pressure Water Gauge
Sq.	Square; 100 Square Feet
S.P.D.T.	Single Pole, Double Throw
SPF	Spruce Pine Fir
S.P.S.T.	Single Pole, Single Throw
SPT	Standard Pipe Thread
Sq. Hd.	Square Head
Sq. In.	Square Inch
S.S.	Single Strength; Stainless Steel
S.S.B.	Single Strength B Grade
sst	Stainless Steel
Sswk.	Structural Steel Worker
Sswl.	Structural Steel Welder
St.; Stl.	Steel
S.T.C.	Sound Transmission Coefficient
Std.	Standard
STK	Select Tight Knot
STP	Standard Temperature & Pressure
Stpi.	Steamfitter, Pipefitter
Str.	Strength; Starter; Straight
Strd.	Stranded
Struct.	Structural
Sty.	Story
Subj.	Subject
Subs.	Subcontractors
Surf.	Surface
Sw.	Switch
Swbd.	Switchboard
S.Y.	Square Yard
Syn.	Synthetic
S.Y.P.	Southern Yellow Pine
Sys.	System
t.	Thickness
T	Temperature; Ton
Tan	Tangent
T.C.	Terra Cotta
T & C	Threaded and Coupled
T.D.	Temperature Difference
T.E.M.	Transmission Electron Microscopy
TFE	Tetrafluoroethylene (Teflon)
T. & G.	Tongue & Groove; Tar & Gravel
Th.; Thk.	Thick
Thn.	Thin
Thrded	Threaded
Tilf.	Tile Layer, Floor
Tilh.	Tile Layer, Helper
THHN	Nylon Jacketed Wire
THW.	Insulated Strand Wire
THWN;	Nylon Jacketed Wire
T.L.	Truckload
T.M.	Track Mounted
Tot.	Total
T-O-L	Threadolet
T.S.	Trigger Start
Tr.	Trade
Transf.	Transformer
Trhv.	Truck Driver, Heavy
Trlr	Trailer
Trlt.	Truck Driver, Light
TV	Television
T.W.	Thermoplastic Water Resistant Wire
UCI	Uniform Construction Index
UF	Underground Feeder
UGND	Underground Feeder
U.H.F.	Ultra High Frequency
U.L.	Underwriters Laboratory
Unfin.	Unfinished
URD	Underground Residential Distribution
US	United States
USP	United States Primed
UTP	Unshielded Twisted Pair
V	Volt
V.A.	Volt Amperes
V.C.T.	Vinyl Composition Tile
VAV	Variable Air Volume
VC	Veneer Core
Vent.	Ventilation
Vert.	Vertical
V.F.	Vinyl Faced
V.G.	Vertical Grain
V.H.F.	Very High Frequency
VHO	Very High Output
Vib.	Vibrating
V.L.F.	Vertical Linear Foot
Vol.	Volume
VRP	Vinyl Reinforced Polyester
W	Wire; Watt; Wide; West
w/	With
W.C.	Water Column; Water Closet
W.F.	Wide Flange
W.G.	Water Gauge
Wldg.	Welding
W. Mile	Wire Mile
W-O-L	Weldolet
W.R.	Water Resistant
Wrck.	Wrecker
W.S.P.	Water, Steam, Petroleum
WT., Wt.	Weight
WWF	Welded Wire Fabric
XFER	Transfer
XFMR	Transformer
XHD	Extra Heavy Duty
XHHW; XLPE	Cross-Linked Polyethylene Wire Insulation
XLP	Cross-linked Polyethylene
Y	Wye
yd	Yard
yr	Year
Δ	Delta
%	Percent
~	Approximately
Ø	Phase
@	At
#	Pound; Number
<	Less Than
>	Greater Than

INDEX

Reed Construction Data, a leading worldwide provider of total construction information solutions, is comprised of three main product groups designed specifically to help construction professionals advance their businesses with timely, accurate and actionable project, product, and cost data. Reed Construction Data is a division of Reed Business Information, a member of the Reed Elsevier plc group of companies.

The *Project, Product, and Cost & Estimating* divisions offer a variety of innovative products and services designed for the full spectrum of design, construction, and manufacturing professionals. Through it's *International* companies, Reed Construction Data's reputation for quality construction market data is growing worldwide.

Cost Information
RSMeans, the undisputed market leader and authority on construction costs, publishes current cost and estimating information in annual cost books and on the CostWorks CD-ROM. RSMeans furnishes the construction industry with a rich library of complementary reference books and a series of professional seminars that are designed to sharpen professional skills and maximize the effective use of cost estimating and management tools. RSMeans also provides construction cost consulting for Owners, Manufacturers, Designers, and Contractors.

Project Data
Reed Construction Data provides complete, accurate and relevant project information through all stages of construction. Customers are supplied industry data through leads, project reports, contact lists, plans and specifications surveys, market penetration analyses and sales evaluation reports. Any of these products can pinpoint a county, look at a state, or cover the country. Data is delivered via paper, e-mail, CD-ROM or the Internet.

Building Product Information
The First Source suite of products is the only integrated building product information system offered to the commercial construction industry for comparing and specifying building products. These print and online resources include *First Source*, CSI's SPEC-DATA™, CSI's MANU-SPEC™, First Source CAD, and Manufacturer Catalogs. Written by industry professionals and organized using CSI's MasterFormat™, construction professionals use this information to make better design decisions.

FirstSourceONL.com combines Reed Construction Data's project, product and cost data with news and information from Reed Business Information's *Building Design & Construction* and *Consulting-Specifying Engineer,* this industry-focused site offers easy and unlimited access to vital information for all construction professionals.

International
BIMSA/Mexico provides construction project news, product information, cost-data, seminars and consulting services to construction professionals in Mexico. Its subsidiary, PRISMA, provides job costing software.

Byggfakta Scandinavia AB, founded in 1936, is the parent company for the leaders of customized construction market data for Denmark, Estonia, Finland, Norway and Sweden. Each company fully covers the local construction market and provides information across several platforms including subscription, ad-hoc basis, electronically and on paper.

Reed Construction Data Canada serves the Canadian construction market with reliable and comprehensive project and product information services that cover all facets of construction. Core services include: *BuildSource, BuildSpec, BuildSelect,* product selection and specification tools available in print and on the Internet; Building Reports, a national construction project lead service; CanaData, statistical and forecasting information; *Daily Commercial News,* a construction newspaper reporting on news and projects in Ontario; and *Journal of Commerce,* reporting news in British Columbia and Alberta.

Cordell Building Information Services, with its complete range of project and cost and estimating services, is Australia's specialist in the construction information industry. Cordell provides in-depth and historical information on all aspects of construction projects and estimation, including several customized reports, construction and sales leads, and detailed cost information among others.

For more information, please visit our Web site at www.reedconstructiondata.com.

Reed Construction Data Corporate Office
30 Technology Parkway South
Norcross, GA 30092-2912
(800) 322-6996
(800) 895-8661 (fax)
info@reedbusiness.com
www.reedconstructiondata.com

Means Project Cost Report

By filling out and returning the Project Description, you can receive a discount of $20.00 off any one of the Means products advertised in the following pages. The cost information required includes all items marked (✔) except those where no costs occurred. The sum of all major items should equal the Total Project Cost.

$20.00 Discount per product for each report you submit.

DISCOUNT PRODUCTS AVAILABLE—FOR U.S. CUSTOMERS ONLY—STRICTLY CONFIDENTIAL

Project Description (No remodeling projects, please.)

✔ Type Building _____

✔ Location _____

Capacity _____

✔ Frame _____

✔ Exterior _____

✔ Basement: full ☐ partial ☐ none ☐ crawl ☐

✔ Height in Stories _____

✔ Total Floor Area _____

Ground Floor Area _____

✔ Volume in C.F. _____

% Air Conditioned _____ Tons _____

Comments _____

Owner _____

Architect _____

General Contractor _____

✔ Bid Date _____

Typical Bay Size _____

✔ Labor Force: _____ % Union _____ % Non-Union

✔ Project Description (Circle one number in each line)
   1. Economy 2. Average 3. Custom 4. Luxury
   1. Square 2. Rectangular 3. Irregular 4. Very Irregular

	✔ **Total Project Cost**		$	
A	✔ **General Conditions**		$	
B	✔ **Site Work**		$	
BS	Site Clearing & Improvement			
BE	Excavation	(	C.Y.)	
BF	Caissons & Piling	(	L.F.)	
BU	Site Utilities			
BP	Roads & Walks Exterior Paving	(	S.Y.)	
C	✔ **Concrete**		$	
C	Cast in Place	(	C.Y.)	
CP	Precast	(	S.F.)	
D	✔ **Masonry**		$	
DB	Brick	(	M)	
DC	Block	(	M)	
DT	Tile	(	S.F.)	
DS	Stone	(	S.F.)	
E	✔ **Metals**		$	
ES	Structural Steel	(	Tons)	
EM	Misc. & Ornamental Metals			
F	✔ **Wood & Plastics**		$	
FR	Rough Carpentry	(	MBF)	
FF	Finish Carpentry			
FM	Architectural Millwork			
G	✔ **Thermal & Moisture Protection**		$	
GW	Waterproofing-Dampproofing	(	S.F.)	
GN	Insulation	(	S.F.)	
GR	Roofing & Flashing	(	S.F.)	
GM	Metal Siding/Curtain Wall	(	S.F.)	
H	✔ **Doors and Windows**		$	
HD	Doors	(	Ea.)	
HW	Windows	(	S.F.)	
HH	Finish Hardware			
HG	Glass & Glazing	(	S.F.)	
HS	Storefronts	(	S.F.)	

J	✔ **Finishes**		$	
JL	Lath & Plaster	(	S.Y.)	
JD	Drywall	(	S.F.)	
JM	Tile & Marble	(	S.F.)	
JT	Terrazzo	(	S.F.)	
JA	Acoustical Treatment	(	S.F.)	
JC	Carpet	(	S.Y.)	
JF	Hard Surface Flooring	(	S.F.)	
JP	Painting & Wall Covering	(	S.F.)	
K	✔ **Specialties**		$	
KB	Bathroom Partitions & Access.	(	S.F.)	
KF	Other Partitions	(	S.F.)	
KL	Lockers	(	Ea.)	
L	✔ **Equipment**		$	
LK	Kitchen			
LS	School			
LO	Other			
M	✔ **Furnishings**		$	
MW	Window Treatment			
MS	Seating	(	Ea.)	
N	✔ **Special Construction**		$	
NA	Acoustical	(	S.F.)	
NB	Prefab. Bldgs.	(	S.F.)	
NO	Other			
P	✔ **Conveying Systems**		$	
PE	Elevators	(	Ea.)	
PS	Escalators	(	Ea.)	
PM	Material Handling			
Q	✔ **Mechanical**		$	
QP	Plumbing	(No. of fixtures	)	
QS	Fire Protection (Sprinklers)			
QF	Fire Protection (Hose Standpipes)			
QB	Heating, Ventilating & A.C.			
QH	Heating & Ventilating	(BTU Output	)	
QA	Air Conditioning	(	Tons)	
R	✔ **Electrical**		$	
RL	Lighting	(	S.F.)	
RP	Power Service			
RD	Power Distribution			
RA	Alarms			
RG	Special Systems			
S	✔ **Mech./Elec. Combined**		$	

Product Name _____

Product Number _____

Your Name _____

Title _____

Company _____

☐ Company

☐ Home   Street Address _____

City, State, Zip _____

☐ Please send _____ forms.

Please specify the Means product you wish to receive. Complete the address information as requested and return this form with your check (product cost less $20.00) to address below.

RSMeans Company, Inc.,
Square Foot Costs Department
P.O. Box 800
Kingston, MA 02364-9988

**For more information
visit Means Web Site
at www.rsmeans.com**

Reed Construction Data/RSMeans . . . a tradition of excellence in Construction Cost Information and Services since 1942.

Book Selection Guide

The following table provides definitive information on the content of each cost data publication. The number of lines of data provided in each unit price or assemblies division, as well as the number of reference tables and crews is listed for each book. The presence of other elements such as an historical cost index, city cost indexes, square foot models or cross-referenced index is also indicated. You can use the table to help select the Means' book that has the quantity and type of information you most need in your work.

Unit Cost Divisions	Building Construction Costs	Mechanical	Electrical	Repair & Remodel.	Square Foot	Site Work Landsc.	Assemblies	Interior	Concrete Masonry	Open Shop	Heavy Construc.	Light Commercial	Facil. Construc.	Plumbing	Western Construction Costs	Residential
1	1102	821	906	1012		1044		846	1013	1101	1050	751	1527	873	1101	709
2	2536	1320	460	1455		7430		545	1383	2495	4645	679	4194	1574	2518	758
3	1404	89	69	725		1239		204	1781	1400	1401	207	1315	47	1404	238
4	809	22	0	617		650		564	1017	785	589	373	1032	0	795	303
5	1788	212	167	916		723		909	647	1756	991	754	1770	287	1770	726
6	1232	82	78	1215		448		1216	312	1223	598	1371	1343	47	1571	1473
7	1260	155	71	1232		464		486	404	1259	346	954	1314	165	1260	746
8	1785	28	0	1862		294		1621	620	1771	49	1187	1980	0	1787	1166
9	1575	47	0	1417		234		1671	359	1532	156	1318	1792	49	1566	1206
10	858	47	25	493		194		697	179	860	0	391	879	232	858	215
11	1018	322	169	500		134		812	44	925	107	217	1077	291	925	103
12	307	0	0	46		210		1416	27	298	0	62	1420	0	298	63
13	1084	950	359	446		367		846	75	1068	258	445	1698	889	1050	193
14	342	55	0	255		36		291	0	342	30	12	341	33	340	6
15	1981	12970	617	1769		1562		1174	59	1992	1762	1198	10672	9474	2011	826
16	1282	470	9614	1003		739		1108	55	1300	760	1081	9381	415	1231	552
17	427	354	427	0		0		0	0	427	0	0	427	356	427	0
Totals	20790	17950	12967	14963		15768		14406	7975	20534	12642	11000	42162	14742	20912	9283

Assembly Divisions	Building Construction Costs	Mechanical	Electrical	Repair & Remodel.	Square Foot	Site Work Landsc.	Assemblies	Interior	Concrete Masonry	Open Shop	Heavy Construc.	Light Commercial	Facil. Construc.	Plumbing	Western Construction Costs	Asm Div	Residential
A		19	0	192	150	540	612	0	550		542	149	24	0		1	374
B		0	0	810	2470	0	5578	333	1914		0	2020	145	0		2	217
C		0	0	635	810	0	1204	1514	145		0	710	266	0		3	588
D		1031	782	685	1814	0	2430	752	0		0	1310	1031	881		4	871
E		0	0	85	255	0	292	5	0		0	255	5	0		5	393
F		0	0	0	123	0	126	0	0		0	123	3	0		6	350
G		465	172	332	111	1836	584	0	482		432	110	87	540		7	299
																8	760
																9	80
																10	0
																11	0
																12	0
Totals		1515	954	2739	5733	2376	10826	2604	3091		974	4677	1561	1421			3932

Reference Section	Building Construction Costs	Mechanical	Electrical	Repair & Remodel.	Square Foot	Site Work Landsc.	Assemblies	Interior	Concrete Masonry	Open Shop	Heavy Construc.	Light Commercial	Facil. Construc.	Plumbing	Western Construction Costs	Residential
Tables	129	45	85	69	4	80	219	46	71	130	61	57	104	51	131	42
Models					102							43				32
Crews	410	410	410	391		410		410	410	393	410	393	391	410	410	393
City Cost Indexes	yes	yes	yes	yes	yes	yes	yes	yes	yes	yes	yes	yes	yes	yes	yes	yes
Historical Cost Indexes	yes	yes	yes	yes	yes	yes	yes	yes	yes	yes	yes	yes	yes	yes	yes	no

Annual Cost Guides

For more information
visit Means Web Site
at www.rsmeans.com

Means Building Construction Cost Data 2004

Available in Both Softbound and Looseleaf Editions

The "Bible" of the industry comes in the standard softcover edition or the looseleaf edition.

Many customers enjoy the convenience and flexibility of the looseleaf binder, which increases the usefulness of *Means Building Construction Cost Data 2004* by making it easy to add and remove pages. You can insert your own cost information pages, so everything is in one place. Copying pages for faxing is easier also. Whichever edition you prefer, softbound or the convenient looseleaf edition, you'll be eligible to receive *The Change Notice* FREE. Current subscribers receive *The Change Notice* via e-mail.

$108.95 per copy, Softbound
Catalog No. 60014

$136.95 per copy, Looseleaf
Catalog No. 61014

Means Building Construction Cost Data 2004

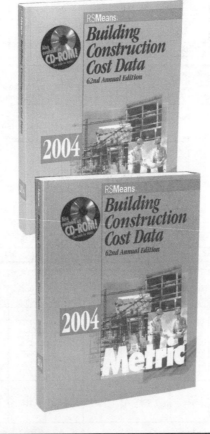

Offers you unchallenged unit price reliability in an easy-to-use arrangement. Whether used for complete, finished estimates or for periodic checks, it supplies more cost facts better and faster than any comparable source. Over 23,000 unit prices for 2004. The City Cost Indexes cover over 930 areas, for indexing to any project location in North America. Order and get *The Change Notice* FREE. You'll have year-long access to the Means Estimating **HOTLINE** FREE with your subscription. Expert assistance when using Means data is just a phone call away.

$108.95 per copy
Over 650 pages, illustrated, available Oct. 2003
Catalog No. 60014

Means Building Construction Cost Data 2004

Metric Version

The Federal Government has stated that all federal construction projects must now use metric documentation. The *Metric Version* of *Means Building Construction Cost Data 2004* is presented in metric measurements covering all construction areas. Don't miss out on these billion dollar opportunities. Make the switch to metric today.

$108.95 per copy
Over 650 pages, illus., available Dec. 2003
Catalog No. 63014

For more information
visit Means Web Site
at www.rsmeans.com

Annual Cost Guides

Means Mechanical Cost Data 2004

• HVAC • Controls

Total unit and systems price guidance for mechanical construction...materials, parts, fittings, and complete labor cost information. Includes prices for piping, heating, air conditioning, ventilation, and all related construction.

Plus new 2004 unit costs for:
• Over 2500 installed HVAC/controls assemblies
• "On Site" Location Factors for over 930 cities and towns in the U.S. and Canada
• Crews, labor and equipment

$108.95 per copy
Over 600 pages, illustrated, available Oct. 2003
Catalog No. 60024

Means Plumbing Cost Data 2004

Comprehensive unit prices and assemblies for plumbing, irrigation systems, commercial and residential fire protection, point-of-use water heaters, and the latest approved materials. This publication and its companion, *Means Mechanical Cost Data*, provide full-range cost estimating coverage for all the mechanical trades.

$108.95 per copy
Over 500 pages, illustrated, available Oct. 2003
Catalog No. 60214

Means Electrical Cost Data 2004

Pricing information for every part of electrical cost planning: More than 15,000 unit and systems costs with design tables; clear specifications and drawings; engineering guides and illustrated estimating procedures; complete labor-hour and materials costs for better scheduling and procurement; the latest electrical products and construction methods.
• A Variety of Special Electrical Systems including Cathodic Protection
• Costs for maintenance, demolition, HVAC/mechanical, specialties, equipment, and more

$108.95 per copy
Over 450 pages, illustrated, available Oct. 2003
Catalog No. 60034

Means Electrical Change Order Cost Data 2004

You are provided with electrical unit prices exclusively for pricing change orders based on the recent, direct experience of contractors and suppliers. Analyze and check your own change order estimates against the experience others have had doing the same work. It also covers productivity analysis and change order cost justifications. With useful information for calculating the effects of change orders and dealing with their administration.

$108.95 per copy
Over 450 pages, available Oct. 2003
Catalog No. 60234

Means Facilities Maintenance & Repair Cost Data 2004

Published in a looseleaf format, *Means Facilities Maintenance & Repair Cost Data* gives you a complete system to manage and plan your facility repair and maintenance costs and budget efficiently. Guidelines for auditing a facility and developing an annual maintenance plan. Budgeting is included, along with reference tables on cost and management and information on frequency and productivity of maintenance operations.

The only nationally recognized source of maintenance and repair costs. Developed in cooperation with the Army Corps of Engineers.

$238.95 per copy
Over 600 pages, illustrated, available Dec. 2003
Catalog No. 60304

Means Square Foot Costs 2004

It's Accurate and Easy To Use!

• **Updated 2004 price information,** based on nationwide figures from suppliers, estimators, labor experts and contractors.

• **"How-to-Use" Sections,** with **clear examples** of commercial, residential, industrial, and institutional structures.

• Realistic graphics, offering true-to-life illustrations of building projects.

• Extensive information on using square foot cost data, including **sample estimates** and **alternate pricing methods.**

$119.95 per copy
Over 450 pages, illustrated, available Nov. 2003
Catalog No. 60054

Annual Cost Guides

For more information
visit Means Web Site
at www.rsmeans.com

Means Repair & Remodeling Cost Data 2004

Commercial/Residential

You can use this valuable tool to estimate commercial and residential renovation and remodeling.

Includes: New costs for hundreds of unique methods, materials and conditions that only come up in repair and remodeling. PLUS:

• Unit costs for over 16,000 construction components
• Installed costs for over 90 assemblies
• Costs for 300+ construction crews
• Over 930 "On Site" localization factors for the U.S. and Canada.

$95.95 per copy
Over 600 pages, illustrated, available Nov. 2003
Catalog No. 60044

Means Residential Cost Data 2004

Contains square foot costs for 30 basic home models with the look of today, plus hundreds of custom additions and modifications you can quote right off the page. With costs for the 100 residential systems you're most likely to use in the year ahead. Complete with blank estimating forms, sample estimates and step-by-step instructions.

$95.95 per copy
Over 550 pages, illustrated, available Oct. 2003
Catalog No. 60174

Means Assemblies Cost Data 2004

Means Assemblies Cost Data 2004 takes the guesswork out of preliminary or conceptual estimates. Now you don't have to try to calculate the assembled cost by working up individual components costs. We've done all the work for you.

Presents detailed illustrations, descriptions, specifications and costs for every conceivable building assembly—240 types in all—arranged in the easy-to-use UNIFORMAT II system. Each illustrated "assembled" cost includes a complete grouping of materials and associated installation costs including the installing contractor's overhead and profit.

$179.95 per copy
Over 550 pages, illustrated, available Oct. 2003
Catalog No. 60064

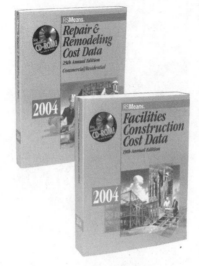

Means Facilities Construction Cost Data 2004

For the maintenance and construction of commercial, industrial, municipal, and institutional properties. Costs are shown for new and remodeling construction and are broken down into materials, labor, equipment, overhead, and profit. Special emphasis is given to sections on mechanical, electrical, furnishings, site work, building maintenance, finish work, and demolition. More than 45,000 unit costs, plus assemblies and reference sections are included.

$263.95 per copy
Over 1150 pages, illustrated, available Nov. 2003
Catalog No. 60204

Means Light Commercial Cost Data 2004

Specifically addresses the light commercial market, which is an increasingly specialized niche in the industry. Aids you, the owner/designer/contractor, in preparing all types of estimates, from budgets to detailed bids. Includes new advances in methods and materials. Assemblies section allows you to evaluate alternatives in the early stages of design/planning.

Over 13,000 unit costs for 2004 ensure you have the prices you need...when you need them.

$96.95 per copy
Over 600 pages, illustrated, available Nov. 2003
Catalog No. 60184

Means Site Work & Landscape Cost Data 2004

Means Site Work & Landscape Cost Data 2004 is organized to assist you in all your estimating needs. Hundreds of fact-filled pages help you make accurate cost estimates efficiently.

Updated for 2004!

• Demolition features—including ceilings, doors, electrical, flooring, HVAC, millwork, plumbing, roofing, walls and windows
• State-of-the-art segmental retaining walls
• Flywheel trenching costs and details
• Updated Wells section
• Landscape materials, flowers, shrubs and trees

$108.95 per copy
Over 600 pages, illustrated, available Nov. 2003
Catalog No. 60284

For more information
visit Means Web Site
at www.rsmeans.com

Annual Cost Guides

Means Open Shop Building Construction Cost Data 2004

The latest costs for accurate budgeting and estimating of new commercial and residential construction... renovation work... change orders... cost engineering. *Means Open Shop BCCD* will assist you to...
• Develop benchmark prices for change orders
• Plug gaps in preliminary estimates, budgets
• Estimate complex projects
• Substantiate invoices on contracts
• Price ADA-related renovations

$108.95 per copy
Over 650 pages, illustrated, available Dec. 2003
Catalog No. 60154

Means Heavy Construction Cost Data 2004

A comprehensive guide to heavy construction costs. Includes costs for highly specialized projects such as tunnels, dams, highways, airports, and waterways. Information on different labor rates, equipment, and material costs is included. Has unit price costs, systems costs, and numerous reference tables for costs and design. Valuable not only to contractors and civil engineers, but also to government agencies and city/town engineers.

$108.95 per copy
Over 450 pages, illustrated, available Nov. 2003
Catalog No. 60164

Means Building Construction Cost Data 2004
Western Edition

This regional edition provides more precise cost information for western North America. Labor rates are based on union rates from 13 western states and western Canada. Included are western practices and materials not found in our national edition: tilt-up concrete walls, glu-lam structural systems, specialized timber construction, seismic restraints, landscape and irrigation systems.

$108.95 per copy
Over 600 pages, illustrated, available Dec. 2003
Catalog No. 60224

Means Heavy Construction Cost Data 2004
Metric Version

Make sure you have the Means industry standard metric costs for the federal, state, municipal and private marketplace. With thousands of up-to-date metric unit prices in tables by CSI standard divisions. Supplies you with assemblies costs using the metric standard for reliable cost projections in the design stage of your project. Helps you determine sizes, material amounts, and has tips for handling metric estimates.

$108.95 per copy
Over 450 pages, illustrated, available Dec. 2003
Catalog No. 63164

Means Construction Cost Indexes 2004

Who knows what 2004 holds? What materials and labor costs will change unexpectedly? By how much?
• Breakdowns for 316 major cities.
• National averages for 30 key cities.
• Expanded five major city indexes.
• Historical construction cost indexes.

$237.95 per year
$59.50 individual quarters
Catalog No. 60144 A,B,C,D

Means Interior Cost Data 2004

Provides you with prices and guidance needed to make accurate interior work estimates. Contains costs on materials, equipment, hardware, custom installations, furnishings, labor costs ... every cost factor for new and remodel commercial and industrial interior construction, including updated information on office furnishings, plus more than 50 reference tables. For contractors, facility managers, owners.

$108.95 per copy
Over 550 pages, illustrated, available Oct. 2003
Catalog No. 60094

Means Concrete & Masonry Cost Data 2004

Provides you with cost facts for virtually all concrete/masonry estimating needs, from complicated formwork to various sizes and face finishes of brick and block, all in great detail. The comprehensive unit cost section contains more than 8,500 selected entries. Also contains an assemblies cost section, and a detailed reference section which supplements the cost data.

$99.95 per copy
Over 450 pages, illustrated, available Dec. 2003
Catalog No. 60114

Means Labor Rates for the Construction Industry 2004

Complete information for estimating labor costs, making comparisons and negotiating wage rates by trade for over 300 cities (United States and Canada). With 46 construction trades listed by local union number in each city, and historical wage rates included for comparison. No similar book is available through the trade.

Each city chart lists the county and is alphabetically arranged with handy visual flip tabs for quick reference.

$239.95 per copy
Over 300 pages, available Dec. 2003
Catalog No. 60124

Reference Books

For more information visit Means Web Site at www.rsmeans.com

Residential & Light Commercial Construction Standards, 2nd Edition
By RSMeans and Contributing Authors

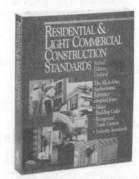

New, updated second edition of this unique collection of industry standards that define quality construction. For contractors, subcontractors, owners, developers, architects, engineers, attorneys, and insurance personnel, this book provides authoritative requirements and recommendations compiled from the nation's leading professional associations, industry publications, and building code organizations. This one-stop reference is enhanced by helpful commentary from respected practitioners, including identification of items most frequently targeted for construction defect claims. The new second edition provides the latest building code requirements.

$59.95 per copy
600 pages, illustrated, Softcover
Catalog No. 67322A

Designing & Building with the IBC, 2nd Edition
By Rolf Jensen & Associates, Inc.

This updated comprehensive guide helps building professionals make the transition to the 2003 *International Building Code*®. Includes a side-by-side code comparison of the IBC 2003 to the IBC 2000 and the three primary model codes, a quick-find index, and professional code commentary.

- The Quick-Find Index lets you instantly identify the IBC equivalent to the model code sections you are familiar with: the *BOCA*® *National Building Code, Uniform Building Code*™, or *Standard Building Code*©.
- The Code Comparison highlights the differences between the IBC 2003, IBC 2000, and the model codes — in an easy-to-read, side-by-side format.
- The Code Commentary points out issues to watch out for in making the transition to the new international code.

With illustrations, abbreviations key, and an extensive Resource section.

$99.95 per copy
Over 875 pages
Catalog No. 67328A

Means Estimating Handbook, 2nd Edition
By RSMeans

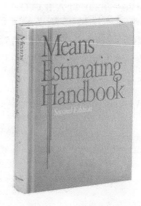

Updated new Second Edition answers virtually any estimating technical question - all organized by CSI MasterFormat. This comprehensive reference covers the full spectrum of technical data required to estimate construction costs. The book includes information on sizing, productivity, equipment requirements, code-mandated specifications, design standards and engineering factors.

This reference—widely used in the industry for tasks ranging from routine estimates to special cost analysis projects—has been completely updated with new and expanded technical information and an enhanced format for ease-of-use.

$99.95 per copy
Over 900 pages, Hardcover
Catalog No. 67276A

For more information
visit Means Web Site
at www.rsmeans.com

Reference Books

Value Engineering: Practical Applications

. . . For Design, Construction, Maintenance & Operations

By Alphonse Dell'Isola, PE

A tool for immediate application—for engineers, architects, facility managers, owners, and contractors. Includes: Making the Case for VE—The Management Briefing, Integrating VE into Planning and Budgeting, Conducting Life Cycle Costing, Integrating VE into the Design Process, Using VE Methodology in Design Review and Consultant Selection, Case Studies, VE Workbook, and a Life Cycle Costing program on disk.

$79.95 per copy
Over 450 pages, illustrated, Hardcover
Catalog No. 67319

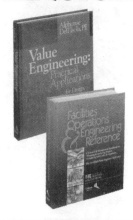

Facilities Operations & Engineering Reference

By the Association for Facilities Engineering and RSMeans

An all-in-one technical referance for planning and managing facility projects and solving day-to-day operations problems. Selected as the official Certified Plant Engineer reference, this handbook covers financial analysis, maintenance, HVAC and energy efficiency, and more.

$109.95 per copy
Over 700 pages, illustrated, Hardcover
Catalog No. 67318

Builder's Essentials: Advanced Framing Methods

By Scot Simpson

A highly illustrated, "framer-friendly" approach to advanced framing elements. Provides expert, but easy-to-interpret, instruction for laying out and framing complex walls, roofs, and stairs, and special requirements for earthquake and hurricane protection. Also helps bring framers up to date on the latest building code changes, and provides tips on the lead framer's role and responsibilities, how to prepare for a job, and how to get the crew started.

$24.95 per copy
250 pages, illustrated, Softcover
Catalog No. 67330

HVAC: Design Criteria, Options, Selection

Expanded 2nd Edition

By William H. Rowe III, AIA, PE

Includes Indoor Air Quality, CFC Removal, Energy Efficient Systems, and Special Systems by Building Type. Helps you solve a wide range of HVAC system design and selection problems effectively and economically. Gives you clear explanations of the latest ASHRAE standards.

$84.95 per copy
Over 600 pages, illustrated, Hardcover
Catalog No. 67306

Total Productive Facilities Management

By Richard W. Sievert, Jr.

This book provides a comprehensive program to:
• Achieve business goals by optimizing facility resources
• Implement best practices through benchmarking, evaluation & project management
• Increase your value to the organization

$29.98 per copy
Over 270 pages, illustrated, Hardcover
Catalog No. 67321

Facilities Maintenance Management

By Gregory H. Magee, PE

Now you can get successful management methods and techniques for all aspects of facilities maintenance. This comprehensive reference explains and demonstrates successful management techniques for all aspects of maintenance, repair, and improvements for buildings, machinery, equipment, and grounds. Plus, guidance for outsourcing and managing internal staffs.

$86.95 per copy
Over 280 pages with illustrations, Hardcover
Catalog No. 67249

Cost Planning & Estimating for Facilities Maintenance

In this unique book, a team of facilities management authorities shares their expertise at:
• Evaluating and budgeting maintenance operations
• Maintaining & repairing key building components
• Applying *Means Facilities Maintenance & Repair Cost Data* to your estimating

Covers special maintenance requirements of the 10 major building types.

$89.95 per copy
Over 475 pages, Hardcover
Catalog No. 67314

The Building Professional's Guide to Contract Documents

New 3rd Edition

By Waller S. Poage, AIA, CSI, CVS

This comprehensive treatment of Contract Documents is an important reference for owners, design professionals, contractors, and students.

• Structure your Documents for Maximum Efficiency
• Effectively communicate construction requirements to all concerned
• Understand the Roles and Responsibilities of Construction Professionals
• Improve Methods of Project Delivery

$64.95 per copy, 400 pages
Diagrams and construction forms, Hardcover
Catalog No. 67261A

For more information
visit Means Web Site
at www.rsmeans.com

Reference Books

Builder's Essentials: Plan Reading & Material Takeoff
By Wayne J. DelPico
For Residential and Light Commercial Construction

A valuable tool for understanding plans and specs, and accurately calculating material quantities.
Step-by-step instructions and takeoff procedures based on a full set of working drawings.

$35.95 per copy
Over 420 pages, Softcover
Catalog No. 67307

Planning & Managing Interior Projects
2nd Edition
By Carol E. Farren, CFM

Addresses changes in technology and business, guiding you through commercial design and construction from initial client meetings to post-project administration. Includes: evaluating space requirements, alternative work models, telecommunications and data management, and environmental issues.

$69.95 per copy
Over 400 pages, illustrated, Hardcover
Catalog No. 67245A

Builder's Essentials: Best Business Practices for Builders & Remodelers:
An Easy-to-Use Checklist System
By Thomas N. Frisby

A comprehensive guide covering all aspects of running a construction business, with more than 40 user–friendly checklists. This book provides expert guidance on: increasing your revenue and keeping more of your profit; planning for long-term growth; keeping good employees and managing subcontractors.

$29.95 per copy
Over 220 pages, Softcover
Catalog No. 67329

Builder's Essentials: Framing & Rough Carpentry 2nd Edition
By Scot Simpson

A complete training manual for apprentice and experienced carpenters. Develop and improve your skills with "framer-friendly," easy-to-follow instructions, and step-by-step illustrations. Learn proven techniques for framing walls, floors, roofs, stairs, doors, and windows. Updated guidance on standards, building codes, safety requirements, and more. Also available in Spanish!

$24.95 per copy
Over 150 pages, Softcover
Catalog No. 67298A   Spanish Catalog No. 67298AS

How to Estimate with Means Data & CostWorks
By RSMeans and Saleh Mubarak
New 2nd Edition!

Learn estimating techniques using Means cost data. Includes an instructional version of Means CostWorks CD–ROM with Sample Building Plans.
The step-by-step guide takes you through all the major construction items. Over 300 sample estimating problems are included.

$59.95 per copy
Over 190 pages, Softcover
Catalog No. 67324A

Unit Price Estimating Methods
New 3rd Edition

This new edition includes up-to-date cost data and estimating examples, updated to reflect changes to the CSI numbering system and new features of Means cost data. It describes the most productive, universally accepted ways to estimate, and uses checklists and forms to illustrate shortcuts and timesavers. A model estimate demonstrates procedures. A new chapter explores computer estimating alternatives.

$59.95 per copy
Over 350 pages, illustrated, Hardcover
Catalog No. 67303A

Interior Home Improvement Costs, 8th Edition

Provides 66 updated estimates for the most popular projects including estimates for home offices, in-law apartments, and remodeling for disabled residents. Includes guidance for remodeling older homes as well as attic and basement conversions, kitchen and bath remodeling, lighting and security; fireplaces; storage; stairs, new floors, and walls and ceilings.

$19.95 per copy
250 pages, illustrated, Softcover
Catalog No. 67308D

Exterior Home Improvement Costs, 8th Edition

Provides quick estimates for 64 projects including room additions and garages; windows and doors; dormers, roofs, and skylights; decks and pergolas; paving and patios; painting and siding; walls, fences, and porches; landscaping and driveways. It also contains price comparison sheets for material and equipment.

$19.95 per copy
270 pages, illustrated, Softcover
Catalog No. 67309D

Means Landscape Estimating Methods
4th Edition
By Sylvia H. Fee

This revised edition offers expert guidance for preparing accurate estimates for new landscape construction and grounds maintenance. Includes a complete project estimate featuring the latest equipment and methods, and **two chapters on Life Cycle Costing, and Landscape Maintenance Estimating.**

$62.95 per copy
Over 300 pages, illustrated, Hardcover
Catalog No. 67295B

Means Environmental Remediation Estimating Methods, 2nd Edition
By Richard R. Rast

Guidelines for estimating 50 standard remediation technologies. Use it to prepare preliminary budgets, develop estimates, compare costs and solutions, estimate liability, review quotes, negotiate settlements.

A valuable support tool for *Means Environmental Remediation Unit Price* and *Assemblies* books.

$99.95 per copy
Over 750 pages, illustrated, Hardcover
Catalog No. 64777A

For more information
visit Means Web Site
at www.rsmeans.com

Reference Books

Means Illustrated Construction Dictionary, Condensed Edition, 2nd Edition
By RSMeans

Recognized in the industry as the best resource of its kind, this has been further enhanced with updates to existing terms and the addition of hundreds of new terms and illustrations... in keeping with the most recent developments in the industry.
The best portable dictionary for office or field use—an essential tool for contractors, architects, insurance and real estate personnel, homeowners, and anyone who needs quick, clear definitions for construction terms.

$59.95 per copy
Over 500 pages
Catalog No. 67282A

Means Repair and Remodeling Estimating
New 4th Edition
By Edward B. Wetherill & RSMeans

Focuses on the unique problems of estimating renovations of existing structures. It helps you determine the true costs of remodeling through careful evaluation of architectural details and a site visit.
New section on disaster restoration costs.

$69.95 per copy
Over 450 pages, illustrated, Hardcover
Catalog No. 67265B

Facilities Planning & Relocation
New, lower price and user-friendly format.
By David D. Owen

A complete system for planning space needs and managing relocations. Includes step-by-step manual, over 50 forms, and extensive reference section on materials and furnishings.

$89.95 per copy
Over 450 pages, Softcover
Catalog No. 67301

Means Square Foot & Assemblies Estimating Methods
New 3rd Edition!

Develop realistic Square Foot and Assemblies Costs for budgeting and construction funding. The new edition features updated guidance on square foot and assemblies estimating using UNIFORMAT II. An essential reference for anyone who performs conceptual estimates.

$69.95 per copy
Over 300 pages, illustrated, Hardcover
Catalog No. 67145B

Means Electrical Estimating Methods
3rd Edition

Expanded new edition includes sample estimates and cost information in keeping with the latest version of the CSI MasterFormat and UNIFORMAT II. Complete coverage of Fiber Optic and Uninterruptible Power Supply electrical systems, broken down by components and explained in detail. Includes a new chapter on computerized estimating methods. A practical companion to *Means Electrical Cost Data*.

$64.95 per copy
Over 325 pages, Hardcover
Catalog No. 67230A

Means Mechanical Estimating Methods
New 3rd Edition

This guide assists you in making a review of plans, specs, and bid packages with suggestions for takeoff procedures, listings, substitutions and pre-bid scheduling. Includes suggestions for budgeting labor and equipment usage. Compares materials and construction methods to allow you to select the best option.

$64.95 per copy
Over 350 pages, illustrated, Hardcover
Catalog No. 67294A

Successful Estimating Methods:
From Concept to Bid
By John D. Bledsoe, PhD, PE

A highly practical, all-in-one guide to the tips and practices of today's successful estimator. Presents techniques for all types of estimates, and advanced topics such as life cycle cost analysis, value engineering, and automated estimating. Estimate spreadsheets available at Means Web site.

$32.48 per copy
Over 300 pages, illustrated, Hardcover
Catalog No. 67287

Means Spanish/English Construction Dictionary
By RSMeans, The International Conference of Building Officials (ICBO), and Rolf Jensen & Associates (RJA)

Designed to facilitate communication among Spanish- and English-speaking construction personnel—improving performance and job-site safety. Features the most common words and phrases used in the construction industry, with easy-to-follow pronunciations. Includes extensive building systems and tools illustrations.

$22.95 per copy
250 pages, illustrated, Softcover
Catalog No. 67327

Means Productivity Standards for Construction
Expanded Edition (*Formerly Man-Hour Standards*)

Here is the working encyclopedia of labor productivity information for construction professionals, with labor requirements for thousands of construction functions in CSI MasterFormat.
Completely updated, with over 3,000 new work items.

$49.98 per copy
Over 800 pages, Hardcover
Catalog No. 67236A

Project Scheduling & Management for Construction
New 2nd Edition
By David R. Pierce, Jr.

A comprehensive yet easy-to-follow guide to construction project scheduling and control—from vital project management principles through the latest scheduling, tracking, and controlling techniques. The author is a leading authority on scheduling with years of field and teaching experience at leading academic institutions. Spend a few hours with this book and come away with a solid understanding of this essential management topic.

$64.95 per copy
Over 250 pages, illustrated, Hardcover
Catalog No. 67247A

Reference Books

For more information visit Means Web Site at www.rsmeans.com

Cyberplaces: The Internet Guide for AECs & Facility Managers
By Paul Doherty, AIA, CSI
2nd Edition
$59.95 per copy
Catalog No. 67317A

Concrete Repair and Maintenance Illustrated
By Peter H. Emmons
$69.95 per copy
Catalog No. 67146

Superintending for Contractors:
How to Bring Jobs in On-Time, On-Budget
By Paul J. Cook
$35.95 per copy
Catalog No. 67233

HVAC Systems Evaluation
By Harold R. Colen, PE
$84.95 per copy
Catalog No. 67281

Basics for Builders: How to Survive and Prosper in Construction
By Thomas N. Frisby
$34.95 per copy
Catalog No. 67273

Successful Interior Projects Through Effective Contract Documents
By Joel Downey & Patricia K. Gilbert
Now $24.98 per copy, limited quantity
Catalog No. 67313

Building Spec Homes Profitably
By Kenneth V. Johnson
$29.95 per copy
Catalog No. 67312

Estimating for Contractors
How to Make Estimates that Win Jobs
By Paul J. Cook
$35.95 per copy
Catalog No. 67160

Means Forms for Building Construction Professionals
$47.48 per copy
Catalog No. 67231

Understanding Building Automation Systems
By Reinhold A. Carlson, PE & Robert Di Giandomenico
$29.98 per copy, limited quantity
Catalog No. 67284

Managing Construction Purchasing
By John G. McConville, CCC, CPE
Now $19.98 per copy, limited quantity
Catalog No. 67302

Fundamentals of the Construction Process
By Kweku K. Bentil, AIC
Now $34.98 per copy, limited quantity
Catalog No. 67260

Means ADA Compliance Pricing Guide
$29.98 per copy
Catalog No. 67310

Means Facilities Maintenance Standards
By Roger W. Liska, PE, AIC
Now $79.95 per copy
Catalog No. 67246

How to Estimate with Metric Units
Now $9.98 per copy, limited quantity
Catalog No. 67304

Means Plumbing Estimating Methods, 2nd Edition
Now $59.95 per copy, limited quantity
Catalog No. 67283A

For more information
visit Means Web Site
at www.rsmeans.com

Reference Books

Preventive Maintenance for Higher Education Facilities

By Applied Management Engineering, Inc.

An easy-to-use system to help facilities professionals establish the value of PM, and to develop and budget for an appropriate PM program for their college or university. Features interactive campus building models typical of those found in different-sized higher education facilities, such as dormitories, classroom buildings, laboratories, athletic facilities, and more. Includes PM checklists linked to each piece of equipment or system in hard copy and electronic format, along with required labor hours to complete the PM tasks. Helps users select and develop the best possible PM plan within budget.

$149.95 per copy
150 pages, Hardcover
Catalog No. 67337

Preventive Maintenance Guidelines for School Facilities

By John C. Maciha

This new publication is a complete PM program for K-12 schools that ensures sustained security, safety, property integrity, user satisfaction, and reasonable ongoing expenditures.

Includes schedules for weekly, monthly, semiannual, and annual maintenance. Comes as a 3-ring binder with hard copy and electronic forms available at Means website, and a laminated wall chart.

$149.95 per copy
Over 225 pages, Hardcover
Catalog No. 67326

Historic Preservation: Project Planning & Estimating
By Swanke Hayden Connell Architects

- *Managing Historic Restoration, Rehabilitation, and Preservation Building Projects and*
- *Determining and Controlling Their Costs*

The authors explain:

- How to determine whether a structure qualifies as historic
- Where to obtain funding and other assistance
- How to evaluate and repair more than 75 historic building materials
- How to properly research, document, and manage the project to meet code, agency, and other special requirements

$99.95 per copy
Over 675 pages, Hardcover
Catalog No. 67323

Means Illustrated Construction Dictionary, 3rd Edition

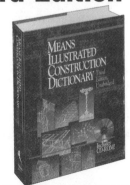

Long regarded as the Industry's finest, the Means *Illustrated Construction Dictionary* is now even better. With the addition of over 1,000 new terms and hundreds of new illustrations, it is the clear choice for the most comprehensive and current information.

The companion CD-ROM that comes with this new edition adds many extra features: larger graphics, expanded definitions, and links to both CSI MasterFormat numbers and product information.

- 19,000 construction words, terms, phrases, symbols, weights, measures, and equivalents
- 1,000 new entries
- 1,200 helpful illustrations
- Easy-to-use format, with thumbtabs

$99.95 per copy
Over 790 pages, illustrated Hardcover
Catalog No. 67292A

For more information
visit Means Web Site
at www.rsmeans.com

Seminars

Means CostWorks Training

This one-day seminar course has been designed with the intention of assisting both new and existing users to become more familiar with CostWorks program. The class is broken into two unique sections: (1) A one-half day presentation on the function of each icon; and each student will be shown how to use the software to develop a cost estimate. (2) Hands-on estimating exercises that will ensure that each student thoroughly understands how to use CostWorks. You must bring your own laptop computer to this course.

CostWorks Benefits/Features:
- Estimate in your own spreadsheet format
- Power of Means National Database
- Database automatically regionalized
- Save time with keyword searches
- Save time by establishing common estimate items in "Bookmark" files
- Customize your spreadsheet template
- Hot key to Product Manufacturers' listings and specs
- Merge capability for networking environments
- View crews and assembly components
- AutoSave capability
- Enhanced sorting capability

Unit Price Estimating

This interactive two-day seminar teaches attendees how to interpret project information and process it into final, detailed estimates with the greatest accuracy level.

The single most important credential an estimator can take to the job is the ability to visualize construction in the mind's eye, and thereby estimate accurately.

Some Of What You'll Learn:
- Interpreting the design in terms of cost
- The most detailed, time tested methodology for accurate "pricing"
- Key cost drivers—material, labor, equipment, staging and subcontracts
- Understanding direct and indirect costs for accurate job cost accounting and change order management

Who Should Attend: Corporate and government estimators and purchasers, architects, engineers...and others needing to produce accurate project estimates.

Square Foot and Assemblies Cost Estimating

This two-day course teaches attendees how to quickly deliver accurate square foot estimates using limited budget and design information.

Some Of What You'll Learn:
- How square foot costing gets the estimate done faster
- Taking advantage of a "systems" or "assemblies" format
- The Means "building assemblies/square foot cost approach"
- How to create a very reliable preliminary and systems estimate using bare-bones design information

Who Should Attend: Facilities managers, facilities engineers, estimators, planners, developers, construction finance professionals...and others needing to make quick, accurate construction cost estimates at commercial, government, educational and medical facilities.

Repair and Remodeling Estimating

This two-day seminar emphasizes all the underlying considerations unique to repair/remodeling estimating and presents the correct methods for generating accurate, reliable R&R project costs using the unit price and assemblies methods.

Some Of What You'll Learn:
- Estimating considerations—like labor-hours, building code compliance, working within existing structures, purchasing materials in smaller quantities, unforeseen deficiencies
- Identifies problems and provides solutions to estimating building alterations
- Rules for factoring in minimum labor costs, accurate productivity estimates and allowances for project contingencies
- R&R estimating examples are calculated using unite prices and assemblies data

Who Should Attend: Facilities managers, plant engineers, architects, contractors, estimators, builders...and others who are concerned with the proper preparation and/or evaluation of repair and remodeling estimates.

Mechanical and Electrical Estimating

This two-day course teaches attendees how to prepare more accurate and complete mechanical/electrical estimates, avoiding the pitfalls of omission and double-counting, while understanding the composition and rationale within the Means Mechanical/Electrical database.

Some Of What You'll Learn:
- The unique way mechanical and electrical systems are interrelated
- M&E estimates, conceptual, planning, budgeting and bidding stages
- Order of magnitude, square foot, assemblies and unit price estimating
- Comparative cost analysis of equipment and design alternatives

Who Should Attend: Architects, engineers, facilities managers, mechanical and electrical contractors...and others needing a highly reliable method for developing, understanding and evaluating mechanical and electrical contracts.

Plan Reading and Material Takeoff

This two-day program teaches attendees to read and understand construction documents and to use them in the preparation of material takeoffs.

Some of What You'll Learn:
- Skills necessary to read and understand typical contract documents—blueprints and specifications
- Details and symbols used by architects and engineers
- Construction specifications' importance in conjunction with blueprints
- Accurate takeoff of construction materials and industry-accepted takeoff methods

Who Should Attend: Facilities managers, construction supervisors, office managers...and other responsible for the execution and administration of a construction project including government, medical, commercial, educational or retail facilities.

Facilities Maintenance and Repair Estimating

This two-day course teaches attendees how to plan, budget, and estimate the cost of ongoing and preventive maintenance and repair for existing buildings and grounds.

Some Of What You'll Learn:
- The most financially favorable maintenance, repair and replacement scheduling and estimating
- Auditing and value engineering facilities
- Preventive planning and facilities upgrading
- Determining both in-house and contract-out service costs; annual, asset-protecting M&R plan

Who Should Attend: Facility managers, maintenance supervisors, buildings and grounds superintendents, plant managers, planners, estimators...and others involved in facilities planning and budgeting.

Scheduling and Project Management

This two-day course teaches attendees the most current and proven scheduling and management techniques needed to bring projects in on time and on budget.

Some Of What You'll Learn:
- Crucial phases of planning and scheduling
- How to establish project priorities, develop realistic schedules and management techniques
- Critical Path and Precedence Methods
- Special emphasis on cost control

Who Should Attend: Construction project managers, supervisors, engineers, estimators, contractors...and others who want to improve their project planning, scheduling and management skills.

Advanced Project Management

This two-day seminar will teach you how to effectively manage and control the entire design-build process and allow you to take home tangible skills that will be immediately applicable on existing projects.

Some Of What You'll Learn:
- Value engineering, bonding, fast-tracking and bid package creation
- How estimates and schedules can be integrated to provide advanced project management tools
- Cost engineering, quality control, productivity measurement and improvement
- Front loading a project and predicting its cash flow

Who Should Attend: Owners, project managers, architectural and engineering managers, construction managers, contractors...and anyone else who is responsible for the timely design and completion of construction projects.

Seminars

For more information
visit Means Web Site
at www.rsmeans.com

2004 Means Seminar Schedule

Location	Dates
Las Vegas, NV	March TBD
Washington, DC	April TBD
Denver, CO	May TBD
San Francisco, CA	June 7-10
Cape Cod, MA	September TBD
Washington, DC	September TBD
Las Vegas, NV	October TBD
Orlando, FL	November TBD
Atlantic City, NJ	November 15-18
San Diego, CA	December 13-16

Note: Call for exact dates and details.

Registration Information

Register Early... Save up to $125! Register 30 days before the start date of a seminar and save $125 off your total fee. *Note: This discount can be applied only once per order. It cannot be applied to team discount registrations or any other special offer.*

How to Register Register by phone today! Means toll-free number for making reservations is: **1-800-334-3509, ext. 5115.**

Individual Seminar Registration Fee $895. Individual CostWorks Training Registration Fee $349. To register by mail, complete the registration form and return with your full fee to: Seminar Division, Reed Construction Data, RSMeans Seminars, 63 Smiths Lane, Kingston, MA 02364.

Federal Government Pricing All Federal Government employees save 25% off regular seminar price. Other promotional discounts cannot be combined with Federal Government discount.

Team Discount Program Two to four seminar registrations: $760 per person–Five or more seminar registrations: $710 per person–Ten or more seminar registrations: Call for pricing.

Consecutive Seminar Offer One individual signing up for two separate courses at the same location during the designated time period pays only $1,435. You get the second course for only $540 (**a 40% discount**). Payment must be received at least ten days prior to seminar dates to confirm attendance.

Refund Policy Cancellations will be accepted up to ten days prior to the seminar start. There are no refunds for cancellations received later than ten working days prior to the first day of the seminar. A $150 processing fee will be applied for all cancellations. Written notice of cancellation is required . Substitutions can be made at anytime before the session starts. **No-shows are subject to the full seminar fee.**

AACE Approved Courses The RSMeans Construction Estimating and Management Seminars described and offered to you here have each been approved for 14 hours (1.4 recertification credits) of credit by the AACE International Certification Board toward meeting the continuing education requirements for re-certification as a Certified Cost Engineer/Certified Cost Consultant.

AIA Continuing Education We are registered with the AIA Continuing Education System (AIA/CES) and is committed to developing quality learning activities in accordance with the CES criteria. RSMeans seminars meet the AIA/CES criteria for Quality Level 2. AIA members will receive (28) learning units (LUs) for each two-day RSMeans Course.

Daily Course Schedule The first day of each seminar session begins at 8:30 A.M. and ends at 4:30 P.M. The second day is 8:00 A.M.–4:00 P.M. Participants are urged to bring a hand-held calculator since many actual problems will be worked out in each session.

Continental Breakfast Your registration includes the cost of a continental breakfast, a morning coffee break, and an afternoon break. These informal segments will allow you to discuss topics of mutual interest with other members of the seminar. (You are free to make your own lunch and dinner arrangements.)

Hotel/Transportation Arrangements RSMeans has arranged to hold a block of rooms at each hotel hosting a seminar. To take advantage of special group rates when making your reservation, be sure to mention that you are attending the Means Seminar. You are, of course, free to stay at the lodging place of your choice. (**Hotel reservations and transportation arrangements should be made directly by seminar attendees.**)

Important Class sizes are limited, so please register as soon as possible.

Note: Pricing subject to change.

Registration Form Call 1-800-334-3509 x5115 to register or FAX 1-800-632-6732. Visit our Web site www.rsmeans.com

Please register the following people for the Means Construction Seminars as shown here. Full payment or deposit is enclosed, and we understand that we must make our own hotel reservations if overnight stays are necessary.

☐ Full payment of $ _____ enclosed.

☐ Bill me

Name of Registrant(s)
(To appear on certificate of completion)

_____

_____

_____

P.O. #: _____
GOVERNMENT AGENCIES MUST SUPPLY PURCHASE ORDER NUMBER

Firm Name _____

Address _____

City/State/Zip _____

Telephone No. fax No. _____

E-mail Address _____

Charge our registration(s) to: ☐ MasterCard ☐ VISA ☐ American Express ☐ Discover

Account No. _____ Exp. Date _____

Cardholder's Signature _____

Seminar Name	City	Dates

Please mail check to: Seminar Division, Reed Construction Data, RSMeans Seminars, 63 Smiths Lane, P.O.Box 800, Kingston, MA 02364 USA

MeansData™

CONSTRUCTION COSTS FOR SOFTWARE APPLICATIONS
Your construction estimating software is only as good as your cost data.

A proven construction cost database is a mandatory part of any estimating package. The following list of software providers can offer you MeansData™ as an added feature for their estimating systems. See the table below for what types of products and services they offer (match their numbers). Visit online at **www.rsmeans.com/demosource/** for more information and free demos. Or call their numbers listed below.

1. **3D International**
   713-871-7000
   venegas@3di.com

2. **4Clicks-Solutions, LLC**
   719-574-7721
   mbrown@4clicks-solutions.com

3. **Aepco, Inc.**
   301-670-4642
   blueworks@aepco.com

4. **Applied Flow Technology**
   800-589-4943
   info@aft.com

5. **ArenaSoft Estimating**
   888-370-8806
   info@arenasoft.com

6. **ARES Corporation**
   925-299-6700
   sales@arescorporation.com

7. **BSD - Building Systems Design, Inc.**
   888-273-7638
   bsd@bsdsoftlink.com

8. **CMS - Computerized Micro Solutions**
   800-255-7407
   cms@proest.com

9. **Corecon Technologies, Inc.**
   714-895-7222
   sales@corecon.com

10. **Estimating Systems, Inc.**
    800-967-8527
    esipulsar@adelphia.net

11. **G/C Emuni**
    514-633-8339
    rpa@gcei.com

12. **LUQS International**
    888-682-5573
    info@luqs.com

13. **MAESTRO Estimator**
    Schwaab Technology Solutions, Inc.
    281-578-3039
    Stefan@schwaabtech.com

14. **Magellan K-12**
    936-441-1744
    sam.wilson@magellan-K12.com

15. **MC² - Management Computer**
    800-225-5622
    vkeys@mc2-ice.com

16. **Maximus Asset Solutions**
    800-659-9001
    assetsolutions@maximus.com

17. **Prism Computer Corp.**
    800-774-7622
    famis@prismcc.com

18. **Quest Solutions, Inc.**
    800-452-2342
    info@questsolutions.com

19. **Shaw Beneco Enterprises, Inc.**
    877-719-4748
    inquire@beneco.com

20. **Timberline Software Corp.**
    800-628-6583
    product.info@timberline.com

21. **TMA Systems, LLC**
    800-862-1130
    sales@tmasys.com

22. **US Cost, Inc.**
    800-372-4003
    sales@uscost.com

23. **Vanderweil Facility Advisors**
    671-451-5100
    info@VFA.com

24. **Vertigraph, Inc.**
    800-989-4243
    info-request@vertigraph.com

25. **WinEstimator, Inc.**
    800-950-2374
    sales@winest.com

TYPE	1	2	3	4	5	6	7	8	9	10	11	12	13	14	15	16	17	18	19	20	21	22	23	24	25
BID					●		●				●				●			●		●				●	●
Estimating		●			●	●	●	●	●	●	●	●	●		●			●	●	●		●		●	●
DOC/JOC/SABER		●			●		●			●						●		●	●						●
IDIQ		●								●						●		●							●
Asset Mgmt.														●		●	●						●		●
Facility Mgmt.	●		●										●	●		●	●				●		●		
Project Mgmt.	●	●					●		●				●					●	●						
TAKE-OFF					●				●			●			●			●		●		●		●	●
EARTHWORK					●				●			●			●			●						●	
Pipe Flow				●																					
HVAC/Plumbing					●							●													
Roofing					●																				
Design	●				●													●	●						●
Other Offers/Links:																									
Accounting/HR		●			●													●	●						
Scheduling					●								●							●		●		●	●
CAD														●						●					●
PDA																				●					●
Lt. Versions		●																		●		●			
Consulting	●	●			●											●	●			●		●	●		
Training		●			●		●		●	●						●		●	●	●	●	●	●	●	●

Reseller applications now being accepted. Call Carol Polio Ext. 5107.

FOR MORE INFORMATION
CALL 1-800-448-8182, EXT. 5107 OR FAX 1-800-632-6732

For more information
visit Means Web Site
at www.rsmeans.com

New Titles

Building Security: Strategies & Costs
By David Owen

Seeking ways to identify your facility's risk, balance security needs with safe access and a building that functions efficiently? Theft, assault, corporate espionage, unauthorized systems access, terrorism, hurricanes and tornadoes, sabotage, vandalism, fire, explosions, and other threats... All are considerations as building owners, facility managers, and design and construction professionals seek ways to meet security needs.

This comprehensive resource will help you evaluate your facility's security needs — and design and budget for the materials and devices needed to fulfill them. The text and cost data will help you to:

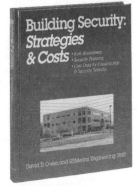

- Identify threats, probability of occurrence, and the potential losses— to determine and address your real vulnerabilities.
- Perform a detailed risk assessment of an existing facility, and prioritize and budget for security enhancement.
- Evaluate and price security systems and construction solutions, so you can make cost-effective choices.

Includes over 130 pages of Means Cost Data for installation of security systems and materials, plus a review of more than 50 security devices and construction solutions—how they work, and how they compare.

$89.95 per copy
Over 400 pages, Hardcover
Catalog No. 37339

Green Building: Project Planning & Cost Estimating
By RSMeans and Contributing Authors

Written by a team of leading experts in sustainable design, this new book is a complete guide to planning and estimating green building projects, a growing trend in building design and construction – commercial, industrial, institutional and residential. It explains:

- All the different criteria for "green-ness"
- What criteria your building needs to meet to get a LEED, Energy Star, or other recognized rating for green buildings

- How the project team works differently on a green versus a traditional building project
- How to select and specify green products
- How to evaluate the cost and value of green products versus conventional ones — not only for their first (installation) cost, but their cost over time (in maintenance and operation).

Features an extensive Green Building Cost Data section, which details the available products, how they are specified, and how much they cost.

$89.95 per copy
350 pages, illustrated, Hardcover
Catalog No. 67338

Life Cycle Costing for Facilities
By Alphonse Dell'Isola and Dr. Steven Kirk

Guidance for achieving higher quality design and construction projects at lower costs!

Facility designers and owners are frustrated with cost-cutting efforts that yield the cheapest product, but sacrifice quality. Life Cycle Costing, properly done, enables them to achieve both—high quality, incorporating innovative design, and costs that meet their budgets.

The authors show how LCC can work for a broad variety of projects – from several types of buildings, to roads and bridges, to HVAC and electrical upgrades, to materials and equipment procurement. Case studies include:

- Health care and nursing facilities
- College campus and high schools
- Office buildings, courthouses, and banks
- Exterior walls, elevators, lighting, HVAC, and more

The book's extensive cost section provides maintenance and replacement costs for facility elements.

$99.95 per copy
450 pages, Hardcover
Catalog No. 67341

2004 Order Form

Qty.	Book No.	COST ESTIMATING BOOKS	Unit Price	Total
	60064	Assemblies Cost Data 2004	$179.95	
	60014	Building Construction Cost Data 2004	108.95	
	61014	Building Const. Cost Data–Looseleaf Ed. 2004	136.95	
	63014	Building Const. Cost Data–Metric Version 2004	108.95	
	60224	Building Const. Cost Data–Western Ed. 2004	108.95	
	60114	Concrete & Masonry Cost Data 2004	99.95	
	60144	Construction Cost Indexes 2004	237.95	
	60144A	Construction Cost Index–January 2004	59.50	
	60144B	Construction Cost Index–April 2004	59.50	
	60144C	Construction Cost Index–July 2004	59.50	
	60144D	Construction Cost Index–October 2004	59.50	
	60344	Contr. Pricing Guide: Resid. R & R Costs 2004	39.95	
	60334	Contr. Pricing Guide: Resid. Detailed 2004	39.95	
	60324	Contr. Pricing Guide: Resid. Sq. Ft. 2004	39.95	
	64024	ECHOS Assemblies Cost Book 2004	179.95	
	64014	ECHOS Unit Cost Book 2004	119.95	
	54004	ECHOS (Combo set of both books)	251.95	
	60234	Electrical Change Order Cost Data 2004	108.95	
	60034	Electrical Cost Data 2004	108.95	
	60204	Facilities Construction Cost Data 2004	263.95	
	60304	Facilities Maintenance & Repair Cost Data 2004	238.95	
	60164	Heavy Construction Cost Data 2004	108.95	
	63164	Heavy Const. Cost Data–Metric Version 2004	108.95	
	60094	Interior Cost Data 2004	108.95	
	60124	Labor Rates for the Const. Industry 2004	239.95	
	60184	Light Commercial Cost Data 2004	96.95	
	60024	Mechanical Cost Data 2004	108.95	
	60154	Open Shop Building Const. Cost Data 2004	108.95	
	60214	Plumbing Cost Data 2004	108.95	
	60044	Repair and Remodeling Cost Data 2004	95.95	
	60174	Residential Cost Data 2004	95.95	
	60284	Site Work & Landscape Cost Data 2004	108.95	
	60054	Square Foot Costs 2004	119.95	
		REFERENCE BOOKS		
	67147A	ADA in Practice	59.98	
	67310	ADA Pricing Guide	29.98	
	67273	Basics for Builders: How to Survive and Prosper	34.95	
	67330	Bldrs Essentials: Adv. Framing Methods	24.95	
	67329	Bldrs Essentials: Best Bus. Practices for Bldrs	29.95	
	67298A	Bldrs Essentials: Framing/Carpentry 2nd Ed.	24.95	
	67298AS	Bldrs Essentials: Framing/Carpentry Spanish	24.95	
	67307	Bldrs Essentials: Plan Reading & Takeoff	35.95	
	67261A	Bldg. Prof. Guide to Contract Documents 3rd Ed.	64.95	
	67339	Building Security: Strategies & Costs	89.95	
	67312	Building Spec Homes Profitably	29.95	
	67146	Concrete Repair & Maintenance Illustrated	69.95	
	67314	Cost Planning & Est. for Facil. Maint.	89.95	
	67317A	Cyberplaces: The Internet Guide 2nd Ed.	59.95	
	67328A	Designing & Building with the IBC, 2nd Ed.	99.95	
	7230B	Electrical Estimating Methods 3rd Ed.	64.95	

Qty.	Book No.	REFERENCE BOOKS (Cont.)	Unit Price	Total
	64777A	Environmental Remediation Est. Methods 2nd Ed.	$ 99.95	
	67160	Estimating for Contractors	35.95	
	67276A	Estimating Handbook 2nd Ed.	99.95	
	67249	Facilities Maintenance Management	86.95	
	67246	Facilities Maintenance Standards	79.95	
	67318	Facilities Operations & Engineering Reference	109.95	
	67301	Facilities Planning & Relocation	89.95	
	67231	Forms for Building Const. Professional	47.48	
	67260	Fundamentals of the Construction Process	34.98	
	67323	Historic Preservation: Proj. Planning & Est.	99.95	
	67308D	Home Improvement Costs–Int. Projects 8th Ed.	19.95	
	67309D	Home Improvement Costs–Ext. Projects 8th Ed.	19.95	
	67324A	How to Est. w/Means Data & CostWorks 2nd Ed.	59.95	
	67304	How to Estimate with Metric Units	9.98	
	67306	HVAC: Design Criteria, Options, Select. 2nd Ed.	84.95	
	67281	HVAC Systems Evaluation	84.95	
	67282A	Illustrated Construction Dictionary, Condensed	59.95	
	67292A	Illustrated Construction Dictionary, w/CD-ROM	99.95	
	67295B	Landscape Estimating 4th Ed.	62.95	
	67341	Life Cycle Costing	99.95	
	67302	Managing Construction Purchasing	19.98	
	67294A	Mechanical Estimating 3rd Ed.	64.95	
	67245A	Planning and Managing Interior Projects 2nd Ed.	69.95	
	67283A	Plumbing Estimating Methods 2nd Ed.	59.95	
	67326	Preventive Maint. Guidelines for School Facil.	149.95	
	67337	Preventive Maint. for Higher Education Facilities	149.95	
	67236A	Productivity Standards for Constr.–3rd Ed.	49.98	
	67247A	Project Scheduling & Management for Constr.	64.95	
	67265B	Repair & Remodeling Estimating 4th Ed.	69.95	
	67322A	Resi. & Light Commercial Const. Stds. 2nd Ed.	59.95	
	67327	Spanish/English Construction Dictionary	22.95	
	67145A	Sq. Ft. & Assem. Estimating Methods 3rd Ed.	69.95	
	67287	Successful Estimating Methods	32.48	
	67313	Successful Interior Projects	24.98	
	67233	Superintending for Contractors	35.95	
	67321	Total Productive Facilities Management	29.98	
	67284	Understanding Building Automation Systems	29.98	
	67303A	Unit Price Estimating Methods 3rd Ed.	59.95	
	67319	Value Engineering: Practical Applications	79.95	

MA residents add 5% state sales tax		
Shipping & Handling**		
Total (U.S. Funds)*		

Prices are subject to change and are for U.S. delivery only. *Canadian customers may call for current prices. **Shipping & handling charges: Add 7% of total order for check and credit card payments. Add 9% of total order for invoiced orders.

Send Order To: **ADDV-1001**

Name (Please Print) _____

Company _____

☐ **Company**

☐ **Home**   Address _____

_____

City/State/Zip _____

Phone # _____ P.O. # _____

(Must accompany all orders being billed)